ADVANCES IN TRANSPORTATION GEOTECHNICS

PROCEEDINGS OF THE 1ST INTERNATIONAL CONFERENCE ON TRANSPORTATION GEOTECHNICS, NOTTINGHAM, UK, 25–27 AUGUST 2008

Advances in Transportation Geotechnics

Edited by

Ed Ellis, Hai-Sui Yu & Glenn McDowell
Nottingham Centre for Geomechanics, University of Nottingham, UK

Andrew Dawson & Nick Thom
Nottingham Transportation Engineering Centre, University of Nottingham, UK

CRC Press is an imprint of the
Taylor & Francis Group, an **informa** business
A BALKEMA BOOK

CRC Press/Balkema is an imprint of the Taylor & Francis Group, an informa business

Typeset by Charon Tec Ltd (A Macmillan Company), Chennai, India
Printed and bound in Great Britain by Antony Rowe (A CPI Group Company), Chippenham, Wiltshire

Published by: CRC Press/Balkema
P.O. Box 447, 2300 AK Leiden, The Netherlands
e-mail: Pub.NL@taylorandfrancis.com
www.crcpress.com – www.taylorandfrancis.co.uk – www.balkema.nl

ISBN: 978-0-415-47590-7 (Hardback)
ISBN: 978-0-203-88594-9 (eBook)

Advances in Transportation Geotechnics – Ellis, Yu, McDowell, Dawson & Thom (eds)
© 2008 Taylor & Francis Group, London, ISBN 978-0-415-47590-7

Table of Contents

3 Slope instability, stabilisation, and asset management

4 Construction on soft ground

5 Interaction with structures and geogrid reinforced soil

6 Effect of climate change and vegetation

7 *Highways, pavements and subgrade*

8 *Rail*

9 *Soil improvement*

Advances in Transportation Geotechnics – Ellis, Yu, McDowell, Dawson & Thom (eds)

Preface

The first International Conference on Transportation Geotechnics (ICTG-1) was held in Nottingham, England from 25–27 August 2008 to bring together researchers and engineers who are interested in geotechnics and particularly its application to transportation infrastructure. The concept of ICTG was initiated by ISSMGE Technical Committee TC3 on Geotechnics of Pavements. The 7th International Symposium of Unbound Aggregates and Roads (UNBAR 7) has been included as a theme of this conference.

This book contains 5 invited keynote lectures by outstanding experts in the field and 101 contributed papers from 22 countries. The papers are divided into 9 themes:

1. UNBAR
2. Slope instability, stabilisation and asset management
3. Construction on soft ground
4. Interaction with structures and geogrid reinforced soil
5. Effect of climate change and vegetation
6. Highways, pavements and subgrade
7. Rail
8. Soil improvement
9. Characterisation and recycling of geomaterials

We would like to thank all contributors for their high quality papers. The conference would not have been a success without the dedicated work of the organising committee and many members of the international advisory board. Special thanks go to Caroline Dolby, Merilyn Kay, Karen Medd and Ting Ting Gan for providing the administrative support for the organisation of the conference.

It is hoped that the ICTG-1 will be a solid 'foundation' for subsequent international conferences on this subject.

Professor Hai-Sui Yu
Conference Chairman, University of Nottingham

Dr Ed Ellis
Conference Secretary General, University of Nottingham

Advances in Transportation Geotechnics – Ellis, Yu, McDowell, Dawson & Thom (eds)
© 2008 Taylor & Francis Group, London, ISBN 978-0-415-47590-7

The Organising and Advisory Committees

Local Organising Committee:
Professor Hai-Sui Yu (Chair)
Dr Ed Ellis (Secretary General)
Mr Andrew Dawson
Professor Glenn McDowell
Dr Nick Thom

ISSMGE TC3 Core Members:
Professor António Gomes Correia (Portugal, Chair)
Dr Dietmar Adam (Austria, Secretary)
Professor Tuncer E. Edil (USA)
Professor Andreas Loizos (Greece)
Dr Yoshitsugu Momoya (Japan)
Professor Soheil Nazarian (USA)
Dr Ing M. Raithel (Germany)
Mr Alain Quibel (France)
Professor Hai-Sui Yu (UK)

1 Keynotes

Advances in Transportation Geotechnics – Ellis, Yu, McDowell, Dawson & Thom (eds)
© 2008 Taylor & Francis Group, London, ISBN 978-0-415-47590-7

Innovations in design and construction of granular pavements and railways

A. Gomes Correia
Department of Civil Engineering, University of Minho, Guimarães, Portugal

ABSTRACT: This paper describes some of the work related with the International Technical Committee TC3 – Geotechnics of Pavements of ISSMGE. For brevity, some topics are selected to be described in some detail, while others are acknowledged for reference purposes. These topics cover: (1) Data Mining tools in transporttation geotechnics showing the capabilities to predict real-value from several attributes and also the possibility to develop a formal updating framework to reduce uncertainty and increase reliability of deformability modulus updating in pavement and rail track structures; (2) Design aspects related with the mechanist approach in the framework of soil mechanics; (3) construction and quality control aspects covering compactor technologies and advanced tools for continuous compaction control and bearing capacity surveys during and after construction. These contributions aim to promote the use of some powerful available tools in the design and construction processes with impacts in the reduction of maintenance costs.

1 INTRODUCTION

Improvements in a number of areas related to both routine and sophisticated design and construction affect the economic success of transportation earthworks. These areas include modelling of the behaviour of embankments and earth foundations, design of earthworks, preparation and quality control of pavement subgrade and earthwork quality control methods and procedures. Furthermore, modelling, design and construction of granular layers can also have an important effect in pavement and rail track performances.

Design and construction of pavements and railways for high speed trains is moving from merely empirical procedures towards mechanistic approaches based on a sounder theoretical basis. This will facilitate the use of new materials in transportation infrastructure under various actions and climatic and traffic (load and speed) conditions. In addition, this will lead to a future challenge to implement a common framework between road, railway and geotechnical engineers.

In the framework of the International Technical Committee – TC3 – Geotechnics of pavements of the International Society for Soil Mechanics and Geotechnical Engineering (ISSMGE) some contributions were done to cover some of the gaps identified in these areas. Improvements on it could have a tremendous impact on pavement maintenance costs.

In this context the European Technical Committee ETC11 (1997–2001) and further on the International Committee TC3 (2002–2005 and 2005–2009) of ISSMGE promote several workshops and seminar reported in the European and International Conferences proceedings (XII ECSMGE 1999, XVI ICSMGE, Osaka 2005; XIV ECSMGE, Madrid 2007) and in several publications (Gomes Correia and Quibel, 2000; Gomes Correia and Brandl, 2001; Gomes Correia and Loizos, 2004; Gomes Correia et al., 2007). Some of the inputs are summarised and updated hereafter.

In this context some innovative technologies and methods are presented covering the design and construction phases of granular pavements and railways.

Data Mining techniques are presented as a transversal tool that can successfully be used in a broad spectrum of transportation geotechnics. An example application to compaction is presented as well as a proposal for a formal updating framework to reduce uncertainty and increase reliability of deformability modulus updating in pavement and rail track structures.

Design aspects are acknowledged briefly in the framework of the mechanistic approach with emphases in the peculiarities of soil and granular materials behaviours and shakedown concept.

Concerning construction aspects, innovations of compaction technologies and quality control techniques, as well as correlations between parameters used in control compaction, are presented, what would be highly beneficial to many projects.

2 DATA MINING

Data Mining(DM), also known as Knowledge Discovery in Databases (KDD), arose due to the advances of Information Technology, leading to an exponential growth of business, scientific and engineering databases (Flood and Kartam, 1994; Fayyad et al., 1996; Cortez, 2008).

The DM goal is to extract valuable information, such as trends and patterns, which can be used to improve decision making and optimize success. DM tools have the potential to analyze the raw data and extract high-level information for the decision-maker, mainly when vast or/and complex data is present. DM uses several methods, each one with his own purposes and capabilities. In particular, Neural Networks (NNs) and Support Vector Machines (SVMs) are useful for supervised nonlinear learning, where the intention is to model an unknown function that maps several input features with one output variable (Cortez, 2008). NNs and more recently SVMs can be used for two important DM goals: classification (labeling a data item into one of several predefined classes) and regression (estimate a real-value from several attributes.

The use of NNs in Transportation geotechnics (i.e., geomechanical and pavement systems) has already been successfully applied during the last decade, despite a lack of understanding and scepticism. The contributions of A2K05(3) TRB Subcommittee on Neural Nets and Other Computational Intelligence–Based Modeling Systems, sponsored by A2K05 Committee on Modeling techniques in geomechanics are relevant in this field (TR, 1999). This Committee reports that NNs have been successfully applied in a full spectrum of transportation geotechnic tasks such as: site characterization; foundation engineering; soil liquefaction; constitutive modelling for soils, granular materials and asphalts; pavement structural models; distress models; backcalculating pavement layer moduli; pavement design and pavement performance prediction.

A recent application of these methods has been applied to the database available in the Technical guide (GTR) SETRA/LCPC Construction of embankments and capping layers (Marques et al., 2008; SETRA/LCPC, 2000). GTR contains a soil classification system (based in intrinsic properties and moisture conditions) and provides the thickness of the compacted layer, the speed of compaction, and the Q/S ratio, depending, for each type of soil and type of compaction equipment, on the required use of soils (embankment or capping layer) and the energy of compaction. Q/S is the ratio, expressed in m3/m2, between the volume Q of soil compacted over a certain period of time and the area S swept by the compactor.

Figure 1 represents the matrix of relation between these variables of GTR database, showing the non linearity of the problem. Consequently, SVMs and NNs are appropriate DM methods for this domain. Indeed, the NN and SVM methods were applied successfully to the prediction of field compaction required parameters. Figure 2 is an example of the good prediction of Q/S parameters by NNs with the independent parameters: soil type, type of compaction equipment and energy (Marques et al., 2008).

Despite the good performance of these methods in many situations, a further issue that needs to be

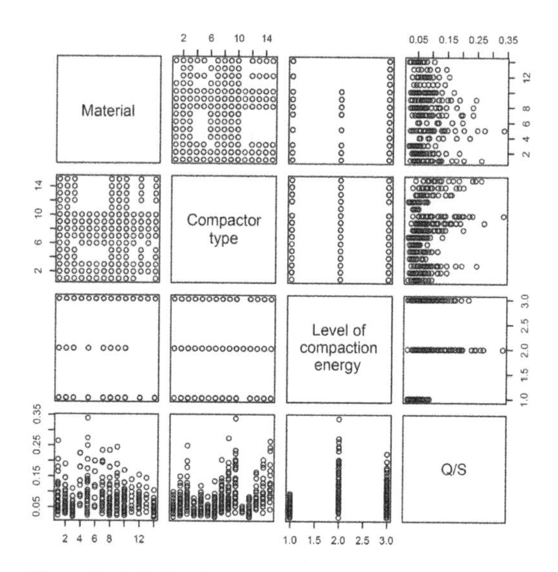

Figure 1. Matrix of relation between some GTR database parameters.

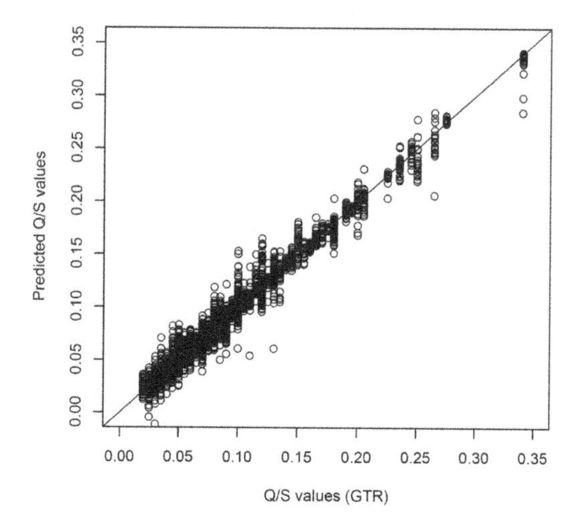

Figure 2. GTR database Q/S values for soil embankments versus predicted values by NNs with parameters: soil type, compactor type and level of compaction energy (Marques et al., 2008).

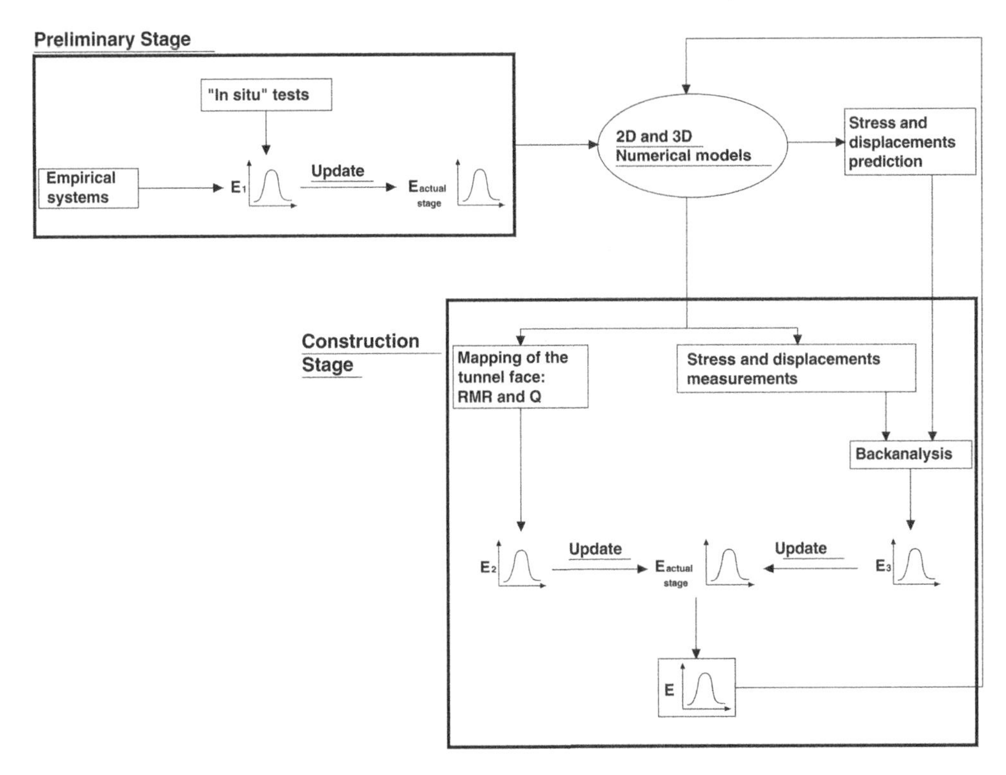

Figure 3. Several stages of updating of deformability modulus (E) during the preliminary and construction stages of an underground structure (Miranda, 2007).

given some attention in the future development of DM methods is to include treatment of uncertainties associated with geotechnical engineering parameters. In fact, NNs models that have been developed to date in the field of geotechnical engineering are essentially deterministic. Consequently, procedures that incorporate such uncertainties into DM methods are essential, as they will provide more realistic solutions. Miranda, 2007 applied Bayesian methodology for updating geomecahnical parameters and uncertainty quantification in the scope of a huge underground construction in Portugal (Powerhouse – Venda Nova II). It is believed that the same framework can be applied in more geotechnical works in the Transportation field. In this framework formal assignment of uncertainties and the updating procedure in order to reduce them and improve future predictions are two critical aspects. The first has been already performed in many areas of geotechnical engineering and decision aids for tunnelling. It is a procedure and a computer code which allows formalizing uncertainties related with geological and construction aspects that can be adapted for other geotechnical transportation works (Einstein et al., 1999, Einstein, 2004, Min et al., 2005). The second is much less addressed and only a few works in updating have been developed in geotechnics, mainly based on Bayes' theorem. Einstein (1988) used the observation of cracks in pavements to refine uncertainties concerning surface creep in slopes. Karam (2005) also used a Bayesian approach in order to update costs in the construction of tunnels. Figures 3 and 4 exemplified the schema process that can be implemented in a formal updating framework (Miranda, 2007; Faber, 2005).

For pavement or rail track structures this schema can be applied as follow: (1) in the initial stages, based on preliminary geotechnical surveys and studies, the value of moduli (E) can be evaluated based on the empirical rules (as material classifications and CBR correlations); (2) as the project advances, more geotechnical information is gathered from in situ and laboratory tests (Resilient modulus test protocols) which can be used to update the initial predictions; (3) the geomechanical parameters are used in the numerical models for design purposes allowing to calculate, among other things, stresses and displacements in the structures; (4) during construction, new information concerning E can be obtained from several sources (for instance field measurements used in back analysis calculations, like FWD – Falling Weight

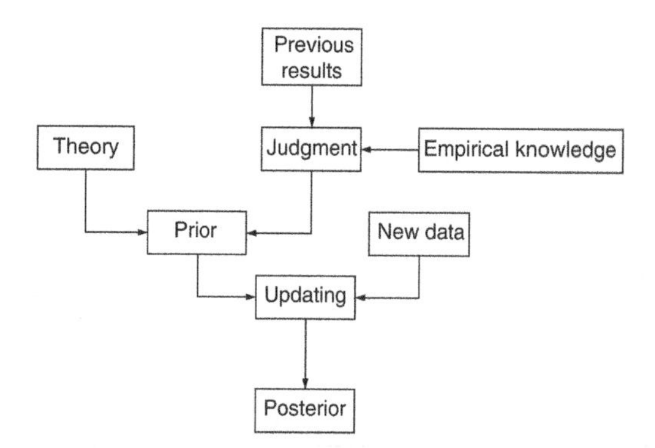

Figure 4. Bayesian framework for the updating (Miranda, 2007, adapted from Faber, 2005).

Deflectometer – Tests). This information can be used to update the value of E in a dynamic process that improves the prediction about the parameter as the quantity of data increases reducing uncertainties. Using Bayesian framework, different levels of prior information and types of probability density function for the data can be considered and results compared to evaluate the sensitivity of the posterior updated distribution to prior assumptions. As demonstrated in Miranda (2007) work, this framework allowed a significant decrease of the uncertainty and increased the reliability of the interval range of deformability modulus updating in an underground structure.

It can be concluded that DM has a number of significant benefits that make it a powerful and practical tool for solving many problems in the field of transportation geotechnics, such as site investigation, laboratory characterisation, material modelling, design, construction, maintenance and management.

3 DESIGN

In the process of design, material models are an important component in the response model. Associated with the material models, it is necessary to take into account the tests needed to obtain their parameters. This chain of operations must be always borne in mind at the design stage in order to ensure a good planning of the tests needed for the models. This process will be completed by choosing the performance models and relevant design criteria (Gomes Correia, 2001). The complexity of all the process was reported in COST 337 action and summarized by Gomes Correia (2001), and Gomes Correia and Lacasse (2006). The detail of this is not described here for the sake of brevity. In this section, only the main challenges from a geotechnical point of view are evocated (Brown, 1996, 2004, Gomes Correia, 2004, Gomes Correia and Lacasse, 2006, Yu et al., 2005, Wermeister et al., 2001).

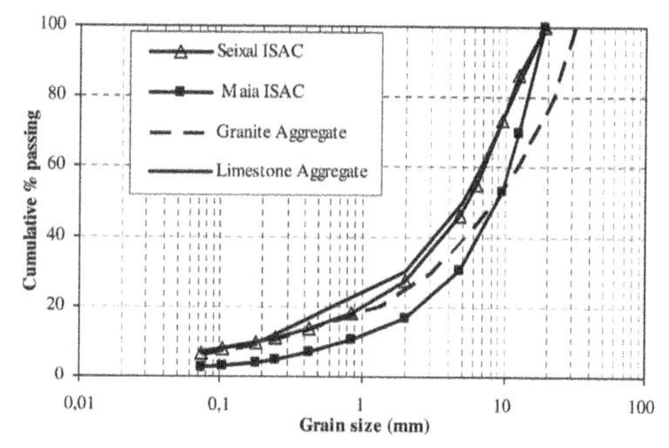

a) Grain size distributions of natural crushed aggregates (granite 0/31.5 and limestone 0/19) and two Portuguese ISACs.

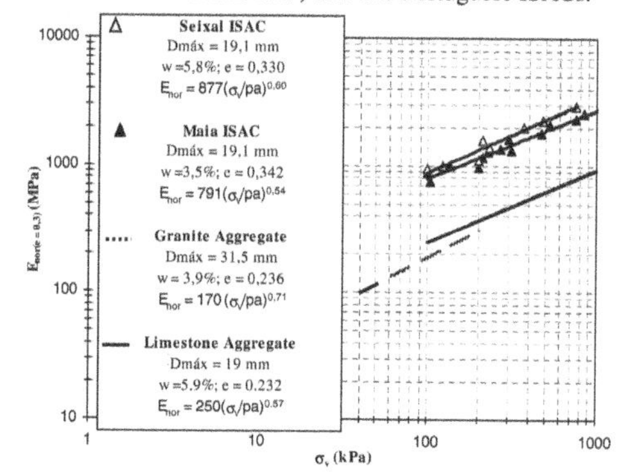

b) Small strain Young's moduli, normalised for a constant void ratio (e=0.3) in function of vertical stress.

Figure 5. Comparison of the Young's moduli of natural crushed aggregates and steel slags aggregates (ISACs).

In which concerns characterization and specifications of subgrade soils and unbound granular materials for pavements and railways an urgent mutation of conventional procedures is necessary to pursue the changes in advance design and new construction technologies. Moreover, it is also necessary to optimize the use of traditional materials and allow the utilisation of new materials. In this context the use of performance based tests should be more used in practical applications.

At laboratory level the precision cyclic triaxial test seems to be a good compromise, facilitated by the existing CEN standard (EN 13286-7, 2004).

Figure 5 show the comparison of the mechanical performance, in terms of the small strain Young modulus, of natural crushed aggregates used in pavement base courses, with two Portuguese steel slags (so called Inert Aggregate for Construction – ISAC). This shows much better mechanical properties (stiffness) of ISACs which, with acceptable environmental properties, leads to be a novel Portuguese material to be used in geotechnical works, and particularly in transportation infrastructures (Gomes Correia, et al., 2008).

Test protocol of EN 13286-7 (2004), allows to determine parameters for modelling as well for ranking materials based in the resistance to permanent deformation.

Physical model tests are also very useful mainly those simulating traffic loading by a moving wheel (Momoya, Sekine 2005); these tests are more representative of loading conditions inducing the rotation of planes of principal stresses, not possible by the cyclic triaxial test. An alternative laboratory tests is the cyclic hollow cylinder (Brown, 2007), but limited in scale size for coarse aggregates, as ballasts. Permanent deformations under traffic loading will be more representative by this type of physical tests and consequently more appropriated for calibration of numerical models and validation of design methods. An interesting research field is to use Data Mining to train with the results of the physical tests, for instance, permanent deformations, and then use it in the model prediction of real pavement performance.

The knowledge of the mechanical behaviour of materials can be supported by the soil mechanics framework, non saturated soil mechanics and soil dynamics to move from the empirical rules used in routine design of pavements and rail tracks to a mechanistic approach. The use of new materials, different loading conditions, different environmental conditions and new technologies in construction require design methods based in a mechanistic-performance basis. Pavements mechanistic models should be developed considering the peculiarities of soils and unbound granular materials: non linearity, pre-straining, state conditions, effective stress concept; number of cycles. For rail tracks dynamic aspects should be considered (integrating ground, embankment and track), as well as residual settlements very restrictive for high speed trains (Gomes Correia and Lacasse, 2006). In this context the new developments using the shakedown concept for pavement design are very promising (Yu et al., 2005). In fact, for design purposes, there is a maximum load level associated with a quasi elastic behaviour which must be determined and then not exceeded, if the onset of permanent deformation is to be prevented. Such behaviour can be explained by consideration of "shakedown" concept.

Figure 6 shows four types of response of elastic/plastic structure to repeated loading (after Johnson, 1992; Collins and Boulbibance, 2000, presented by Yu et al., 2005). It can be observed that the elastic shakedown response is developed if after some cycles plastic flow ceases to develop further and the accumulated dissipated energy in the whole structure remains bounded such that the structure responds purely elastically to the applied variable loads, one says that the structure "shakes down". The maximum load for which elastic shakedown can be achieved is known as shakedown limit.

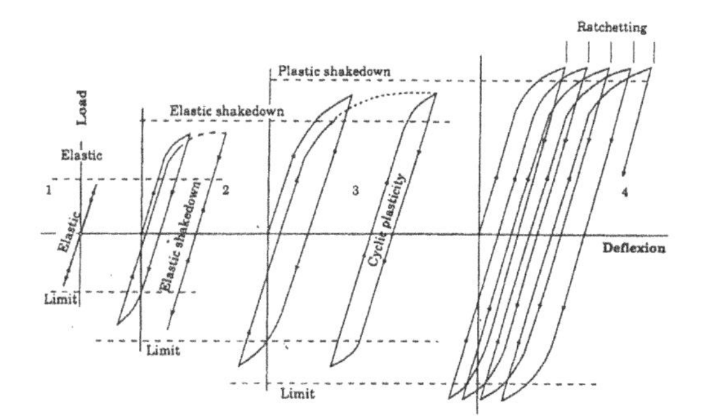

Figure 6. Four types of response of elastic/plastic structure to repeated loading (after Johnson, 1992; Collins and Boulbibance, 2000, presented by Yu et al., 2005.

Anyway, material models and structural models should be calibrated and validated for design by full scale tests with specific instrumentation. Vibration measurements in ballast are of relevance to understand its performance and conclude about countermeasures applications.

4 CONSTRUCTION

4.1 *Compactors technologies*

Recent developments that have taken place in compaction technology use high energy impact compactors and intelligent compaction rollers. Some of them integrate instrumentation allowing direct measurement of the engineering properties during compaction in real time, which can be used for improving compaction uniformity and effective compaction effort. These equipments appear to provide a much greater depth of compaction, allowing for placement of thicker fills and consequently a significantly increased of rate of embankment construction. Furthermore, the use of instrumentation on the compaction equipment provides 100 percent quality control coverage leading to the effective implementation of warrantees and guarantees for both earthworks and pavements.

These developments pretend accelerate construction of embankments and granular layers respecting compaction requirements of modern pavements and rail tracks in a cost-effective manner.

Pinard (2001) shows that the particularity features of all impact compactors is their non-circular compacting masses which have a series of points alternating with flat compacting faces. Figure 7 illustrates how the faces vary in relation to their shape (3, 4 or 5-sided), mass and configuration.

This author presents several advantages of these impact compactors as a result of their much higher

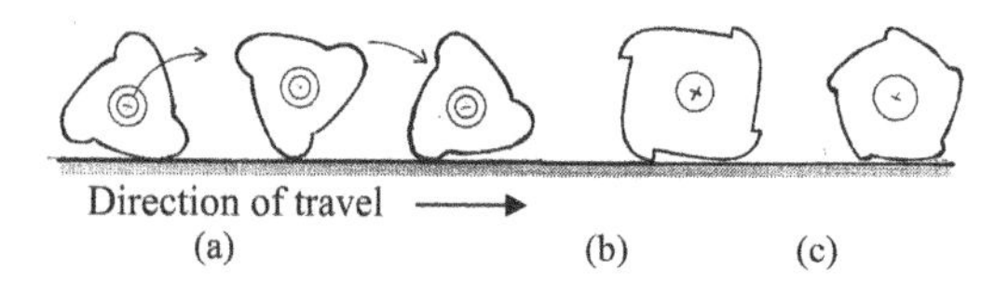

Figure 7. Schema of: (a) 3-sided impact compactor in operation, (b) 4-sided and (c) 5-sided impact compactor profiles (Pinard, 2001).

Table 1. Established CCC-systems, CCC-values and producers.

CCC-systems	CCC-values	Producers
Compactometer	CMV	Geodynamik
Terrameter	OMEGA E_{vib}	Bomag
ACE	k_B	Ammann

"energy density" and speed of operation over conventional vibratory rollers, such as:

- greater depth of compaction;
- greater speed (productivity) of compaction;
- less use of compaction water;
- higher compacted layer stiffness;
- more effective compaction of rock fill;
- more effective "proof compaction";
- more effective compaction of "collapsible" soils.

Pinard (2001) referred that the above advantages have been quantified from a number of comparative trials and at the southern runway at Hong Kong's international airport at Chek Lap Kok. Adam and Markiewicz (2000) and Brandl (2001a, c) also corroborate some of these findings with rollers equipped with dynamic excited polygonal shaped drums.

However, conventional vibratory rollers have also been object of several developments. One of the main technological developments concerns energy variability and efficiency by the use of two counter-rotating weights in the drum rather than the conventional single, one-directional eccentric weight. The weights rotate in opposite directions and only come together in a common direction in the downward vertical inclination. This eliminates unwanted and wasteful movements in the lateral and upward directions that occur with conventional compaction drums. Internally, the entire counterweight assembly is rotated to adjust the direction of the point where the two weights act together.

Adam and Kopf (2000) describe that the "Vario" roller developed by Bomag company can be used both for dynamic compression compaction (like a vibratory roller), for dynamic shear compaction (like an oscillatory roller), and a combination of these two possibilities, depending only on the adjustable force direction.

With these technologies onboard monitoring system were applied to vibratory rollers that automatically adjust their energy output to a defined control criteria to achieve compaction optimisation (so called intelligent compaction – IC). Several companies, for example, Ammann, Bomag and Dynapac all have IC machines ready for work in soils and aggregates, and Ammann and Bomag have asphalt models as well.

The intelligent compaction supposes neither to under-compact nor over-compact materials, allowing an optimised and a highly uniform compaction (Brandl, 2001c).

The development of the automatic control of the operation modes of different dynamic rollers was done in connection with a roller-integrated continuous compaction control system (CCC) measuring the dynamic interaction between dynamic roller and material. This allows the definition of dynamic compaction values by different producers which have proven suitable for checking on real time the compaction state of the material. Adam (2007) summarises the definitions of different CCC values associated with different systems of different producers (Table 1).

4.2 *Advanced continuous surveys and spot tests for quality control during and after construction*

Increase in life-time and serviceability of transport infrastructures (roads, airports and railways) is essential to reduce maintenance and consequently save costs and contribute for the quality of life (Gomes Correia and Lacasse, 2006). A major concern affecting these aspects is the assessment of quality of materials used in the structure and substructure (subgrade and embankment), as well as the quality control during and after construction. This of course assuming a proper design is done. These topics are developed hereafter.

4.2.1 *Spot tests, continuous compaction control equipments and continuous bearing capacity meters*

An inventory of the in-situ tests for compaction control during and after construction and its relative performances were reported by Gomes Correia and Lacasse (2006) and referred hereafter.

The assessment of subgrade materials was carried out and reported by the PIARC TC C12 on earthworks, drainage and subgrade. Specifically for unbound granular materials, an inventory of the in-situ mechanical tests used at the European level was also realized during the COST 337 action www.cordis.lu/cost-transport/src/cost337.htm. From these two surveys it can be concluded the following (Gomes Correia, 2004; Gomes Correia and Lacasse, 2007):

(1) To understand the role of in situ tests, it is necessary to look at the role of design methods; in this context index tests (e.g. density, water content, CBR) are oriented towards empirical design methods, while sound mechanical tests are more suitable to mechanistic design methods.
(2) The in situ CBR test is still used in several countries, despite its empirical character and its questionable relationship with material performance. Index tests, like CBR and dynamic penetration tests (DPT) or dynamic cone penetrometers (DCP), are generally used as control or ranking tools, being their results related to empirical knowledge.
(3) The likely performance of the subgrade soils and UGM should be assessed by those tests that simulate the traffic loading and from which load deformation criteria can be obtained (stiffness).
(4) The two mechanical performance tests used almost everywhere is the static plate bearing test (SPLT) and the Falling Weight Deflectometer (FWD).
(5) Other common mechanical in situ methods are: Ménard Pressuremeter, dynamic plate loading tests, French dynaplaque and Lacroix deflectograph.
(6) The use of dynamic plate loading tests is increasing, because they are simpler and faster than static plate bearing tests. Various types of equipment are available for these tests, such as, for example: Light falling weight deflectometer – LFWD (low frequency impulse) and stiffness gauge (low-strain, high-frequency vibratory). It may be stressed that these tests differ by their size, measurement principle and consequently by the parameters determined. Figure 8 show both these two equipments.
(7) More advanced performance-related in situ tests are continuous measurement equipments like the instrumented compaction rollers (continuous compaction control) and continuous bearing capacity meters.

With respect to the mechanical performance tests for construction quality control, it must be encouraged to move from the spot (local) measurements to the continuous measurements where theoretical investigations and field observations are now available and put in practical applications (Thurner and Sandstrom, 2000; Adam, 1999; Quibel, 1999; Adam and Kopf, 2000; Krober et al., 2001; Brandl, 2001a,b).

As described in Brandl (2001c), "the hitherto used conventional methods of soil mechanics have proved suitable for decades but are based only on spot checking by more or less random or subjective selection or along grids". Thurner and Sandstrom (2000) referred that according to national compaction standards in different countries one sample (spot test) is taken on 2 000 m^3 of compacted soil, which means a relation between sample volume and compacted volume of 1:1.000.000.

Figure 8. Light portable falling weight deflectometer – LFWD- (left) and stiffness gauge (right) equipments.

Continuous measurements of thickness, stiffness or deflections become important where high uniformity and homogeneity is required, as it should be the case for embankments of roads, airports and particularly railways, as well as for the structural layers of their structures. In this context, several methods are now available:

- Roller-integrated continuous compaction control (CCC) that could has also the advantage that continuous compaction optimisation can be done during the compaction process, since the roller is the measuring tool (IC), as mentioned previously. This technology represents an improvement, since control data are already available during the compaction process. These data have to be calibrated by means of representative conventional tests. Comprehensive field tests have disclosed that the correlation between dynamic compaction values (CMV, OMEGA, Evib, (CC-values in Table 1)) depends strongly on the contact between roller drum and surface of the layer being compacted. Consequently, no generally valid correlation to conventional properties (modulus of load plate tests, Proctor compaction degree e.g.) can be obtained. But a clear correlation does exist, if the specific material and roller operation are taken into account. This was one of the important outputs deliveries of TC3 (ISSMGE) that has been accomplished with the technical contractual provisions and recommendations presented by Adam (2007), in order to promote the use of CCC. During the formal presentation of these recommendations in the Osaka, TC3 workshop (Gomes Correia et al., 2006), it was stressed that the conventional methods of compaction control for earthworks, roads and railways are not sufficient any more for high quality projects.

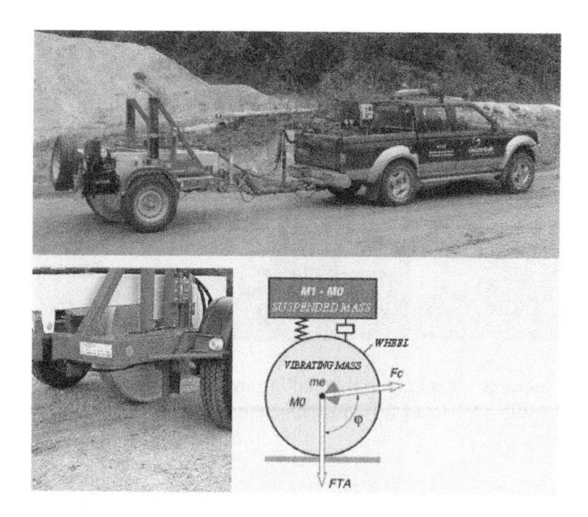

Figure 9. "Portancemètre" equipment: a) general view, b) instrumented rolling wheel, c) and principles.

- The "Portancemètre" (Quibel, 1999), in contrast, requires separate external test equipment, where the roller itself is the measurement tool, and therefore is post-compaction control (fig. 9). It consists of a wheel (1 m diameter, 200 mm wide), equipped with a vibratory loading system (like a roller) and instrumented with accelerometers. The system allows continuous determination of the stiffness of the evaluated layer, at a speed of 3 km/h (depth of action about 600 mm). The wheel is mounted on a trailer that can be towed by a vehicle (fig. 9). This equipment has the advantage, in relation to the roller-integrated continuous compaction control, of applying the load actions under normalized conditions (same boundary conditions) and consequently is more appropriated for standardization.
- The rolling dynamic deflectometer (RDD) presented by Lee and Stokoe (2005) is a truck-mounted, electro-mechanical system, measuring continuous deflections while moving at approximately 1.6 km/h on pavements (roads and airports). This device has been developed to be used at project level as an alternative to the rolling wheel deflectometer (RWD) moving at 88 km/h that is more adapted for network level. This last device uses laser technology to measure deflections, similar to the high speed deflectograph (HSD) of the Danish Road Directorate, reported by Rasmussen et al. (2002).
- RSMVT, a moving rail car to measure track modulus – rolling stiffness measurement vehicle – using last technologies (Berggren, 2005). The RSMV is a two-axle freight wagon that can excite the track dynamically through two oscillating masses above one of the wheelsets. The measuring axle has a static axle load of 180 kN and the dynamic axle load amplitude can be varied up to 60 kN in the frequency range of 3–50 Hz. The stiffness is given as a complex quantity and thus presents both the magnitude and the phase.
- GPR, Ground penetrating radar, is a non-destructive, based on the propagation of electromagnetic radiation, also designated by electromagnetic waves or radio waves, through the ground or other dielectric media. GPR is being actively used and studied for several decades by diverse researchers (Saarenketo, 1992, 1999, Clark et al. 2004). This technique has become nowadays an accepted method for in-situ monitoring and with the development of air-launched horn antennas (high speed GPR) also for traffic infrastructure surveys (roads and railways). The georadar allows to obtain a significant number of parameters that are very relevant to analyse and assess the conditions of construction layers and embankments. It includes the assessment of the layer thicknesses, dielectric and magnetic properties, voids and moisture conditions of materials (Narayanan et al. 2004, Clark et al. 2001). An application for railways is presented by Fernandes et al. (2008).
- SASW, Spectral Analysis of Surface Waves, is based on the measurement of the speed of propagation of surface waves in the material. It is a seismic method that can nondestructively determine modulus profiles of pavement sections (Nazarian and Desai, 1993). SASW allows determination of the thickness of the layer and its stiffness in terms of shear modulus, G0, under small strains. Its main advantage, in comparison with the other techniques, is the greater depth of investigation that can be achieved and the capability to determine multi-layer stiffness (Nazarian et al., 1987; Stokoe et al., 1999). Moreover, its results can be compared with values of small shear strain modulus obtained in the laboratory. In fact SASW is the only test that will provide continuity among design, laboratory and construction. In addition, giving a practically intrinsic material property, small strain shear modulus (G_0), it can be used to correlate with other test results, mainly with index properties, or to extrapolate for other stress/strain levels (Gomes Correia et al., 2004).

4.2.2 *Discussion*

From this brief inventory it can be concluded that a lot of in situ tests are being used at the design, construction and after construction stages. Unfortunately it must be pointed out that the drawback of most routine in-situ tests lies in the fact that the stress and strain distribution necessary for the identification of constitutive laws is unknown. Consequently only the tests with well-known boundary conditions can be expected to be used universally, while index tests will apply only for the cases for which they were established and different correlations may be required for different materials (Atkinson and Sallfors, 1991).

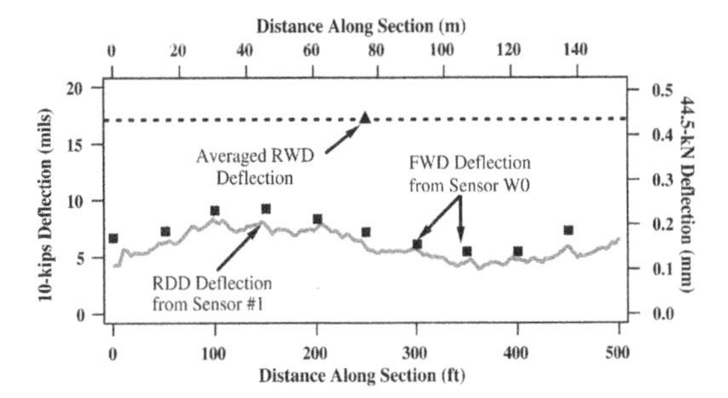

Figure 10. Comparison of results from different field loading tests (Lee, Stokoe and Bay, 2005).

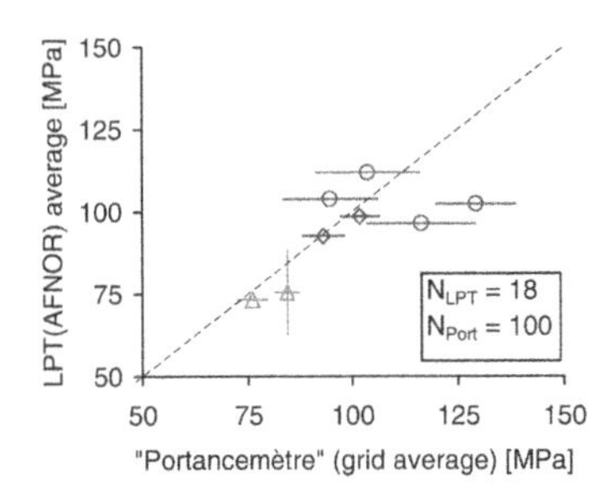

Figure 11. Comparison of deformability modulus Ev obtained by "Portancemètre" and static load plate test according NF P 94-117-1 (2000).

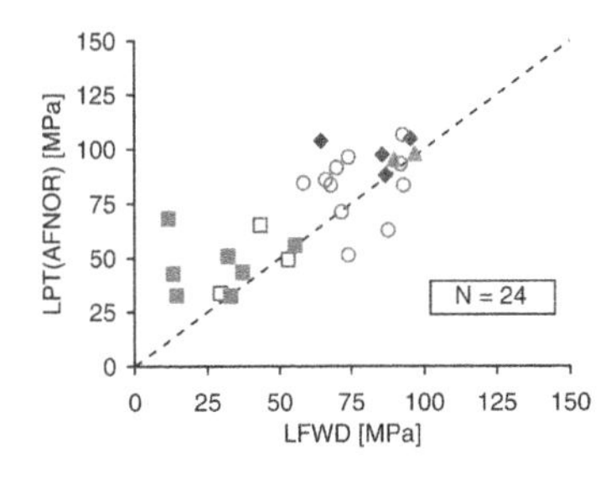

Figure 12. Comparison of deformability modulus Ev obtained by "Light falling weight deflectometer – LFW" and static load plate test according NF P 94-117-1 (2000).

In practice it is expected that different results can be obtained by these different tests, using different types of loadings (strain rates), different induced stresses and different boundary conditions. This have been reported by several authors (Flemming and Rogers, 1995; Gomes Correia et al, 2004).

Figure 10 from Lee and Stokoe (2005) illustrate these realities.

However, if proper calibration between test equipments is done, comparable results can be expected for a particular site and equipment as referred by Gomes Correia et al. 2008 (figs. 11 and 12).

Therefore, any prediction of pavement performance based in direct values obtained by simplified analysis of field test results would be different. A more sound analysis of tests results must be developed, taking into account the non-linear behaviour of soils and unbound materials, as well as strain rate effects (Gomes Correia et al., 2004).

Furthermore, the use of field results in design should be corrected to take into account the design moisture content or suction of material for the service life of the structure, as also mentioned by Edil and Sawangsurija (2005). Roller-integrated continuous compaction control (CCC) referred before facilitates significantly a sophisticated compaction control of homogeneity and should increasingly substitute the hitherto way of conventional quality assessment and acceptance testing (Brandl, 2001b, c). It is recommended that the use of these methods should be obligatory to guaranty a uniform stiffness for soils, as well for all the pavement layers and for sub-ballast layers in rail track. Brandl (2001b) reported that with the use of CCC (IC) on high-quality fill materials and intensively compacted embankment settlements may be of the order of 0.1% to 0.3% of its height (H). This is a very important improvement in relation to the local control methods where embankment-settlements obtained are typically of order of 1%.

5 CONCLUSIONS

Design and construction of road pavements and rail tracks have an important influence on their performance and consequently on their maintenance costs. The optimisation of these costs requires the use of best technologies and methods available. It should be referred that the large number of cycles of variable loads associated with variables environmental conditions and the complex behaviour of soil and pavement structural materials makes pavements and rail tracks among the most difficult geotechnical structures to analyze, design and construct. The advanced available tools in the analysis, design and construction processes presented in this paper are contributions aiming to impact on theories and in practice to be used and developed.

ACKNOWLEDGEMENTS

The author would like to acknowledge the partial support from the Portuguese Foundation for the Science and Technology by the project number POCI/ECM/6114/2004. Profs. P. Cortez and T. Miranda and Mr. R. Marques are also acknowledging by their contributions to the Data Mining session, as well as Mr. J. P. Martins and Mrs. S. Ferreira for the help in preparing some Figures.

REFERENCES

Adam, D. 2007. Roller-integrated continuous compaction control (CCC). Technical contractual provisions & recomendations. *Design and Construction of Pavements and Rail Tracks*. (Gomes Correia, Momoya, Tatsuoka, eds). Balkema, Leiden, The Netherlands, 111–138.

Adam, D., Kopf, F. 2000. Sophisticated compaction technologies and continuous compaction control. *Compaction of Soils and Granular Materials* (Gomes Correia & Quibel, eds). Presses de l' Ecole Nationale des Ponts et Chaussées, Paris, 207–220.

Adam, D., Markiewicz, R. 2001. Compaction behaviour and depth effect of the Polygon-Drum. *Geotechnics for Roads, Rail Tracks and Earth Structures* (Gomes Correia & Brandl, eds), Balkema, The Netherlands, 27–36.

Atkinson, J.H., Sallfors, G. 1991. Experimental determination of stress-strain-time characteristics in laboratory and in situ tests. *Proc. 10th ECSMFE, Deformation of Soils and Displacements of Structures*, Associazione Geotecnica Italiana, Vol. III, 915–956.

Berggren, E. 2005. *Dynamic track stiffness measurement. A new tool for condition monitoring of track substructure*. Royal Institute of Technology. Thesis.

Brandl, H. 2001a. Compaction of soil and other granular material – interactions. *Geotechnics for Roads, Rail Tracks and Earth Structures* (Gomes Correia & Brandl, Eds), Balkema, The Netherlands, 3–11.

Brandl, H. 2001b. High embankments instead of bridges and bridges foundations in embankments. *Geotechnics for Roads, Rail Tracks and Earth Structures* (Gomes Correia & Brandl, Eds), Balkema, The Netherlands, 13–26.

Brandl, H. 2001c. The importance of optimum compaction of soil and other granular materials. *Geotechnics for Roads, Rail Tracks and Earth Structures* (Gomes Correia & Brandl, Eds). Balkema, The Netherlands, 47–66.

Brown, S.F. 1996. *36th Rankine lecture: Soil Mechanics in Pavement Engineering*, Géotechnique, 46: 3, 383–426.

Brown, S.F. 2004. Design considerations for pavement and rail track foundations. *Geotechnics in Pavement and Railway Design and Construction* (Gomes Correia & Loizos, Eds), MillPress, Rotterdam, Netherlands.

Brown, S.F. 2007. The effect of shear reversal on the accumulation of plastic strain in granular materials under cyclic loading. *Design and Construction of Pavements and Rail Tracks*. (Gomes Correia, Momoya, Tatsuoka, eds.). Balkema, Leiden, The Netherlands, 89–108.

Clark, M.R., Gordon, M., Forde, M.C. 2004. *Issues over high-speed non-invasive monitoring of railway trackbed*. NDT&E International, 37: 131–139.

Cortez, P. 2008. RMiner: Data Mining with Neural Networks and Support Vector Machines using R, In R. Rajesh (Ed.), *Introduction to Advanced Scientific Softwares and Toolboxes*, International Association for Engineering (IAEng), In press.

Edil, T.B., Sawangsuriya, A. 2005. Earthwork quality control using soil stiffness. Proc. XVI ICSMGE, Millpress, Rotterdam, The Netherlands.

Einstein, H. 1988. Landslide risk assessment procedure. In Proc. 5th Int. Symp. on Landslides. Keynote paper. Lausanne, Switzerland.

Einstein, H. 2004. *The Decision Aids for Tunnelling (DAT) – an update*. Transportation Research Record, 1892: 199–207.

Einstein, H., C. Indermitte, J. Sinfield, F. Descoeudres, and J. Dudt. 1999. *The Decision Aids for Tunnelling*. Transportation Research Record, 1656.

EN 13286-7 - 2004. Unbound and hydraulically bound mixtures – Part 7: Cyclic load triaxial test for unbound mixtures.

Faber, M. 2005. *Risk and safety in civil, surveying and environmental engineering. Swiss Federal Institute of Technology*. Lecture notes. Zurich, 394 p.

Fayyad, U., Piatetsky-Shapiro, G. and Smyth, P. 1996. *Advances in Knowledge Discovery and Data Mining*. MIT Press.

Fernandes, F., Oliveira, M. Gomes Correia, A., Lourenço, P.B., Caldeira, L. 2008. Assessment of layer thickness and uniformity in railway embankments with GPR. 1st Conference on Transportation Geotechnics.

Flemming, P. and Rogers, C. 1995. *Assessment of pavement foundations during construction*. Paper 10625, Proc. Inst. Civil Engineers Transportation, 111, pp. 105–115.

Flood, I., and Kartam, N. 1994. *Neural networks in civil engineering I: Principles and understanding*. J. Computing in Civil Engrg, ASCE, 8(2), 131–148.

Goh, A. T. C. 1995. *Empirical design in geotechnics using neural networks*. Geotechnique, 45(4), 709–714.

Gomes Correia A., Ferreira S. M. R., Roque A. J., Castro F., Cavalheiro A. 2008. Environmental and engineering properties for processed Portuguese steel slags, Green5 Conference, Vilnius, Lithuania (submitted).

Gomes Correia, & Brandl, H. (Editors). *Geotechnics for Roads, Rail Tracks and Earth Structures*. Balkema, Lisse, The Netherlands, 2001.

Gomes Correia, A & Quibel, Alain (Editors). *Compaction of Soils and Granular Materials*. Presses de l' Ecole Nationale des Ponts et Chaussées, Paris, 2000.

Gomes Correia, A., Momoya, Y., F. Tatsuoka (Editors). *Design and Construction of Pavements and Rail Tracks. Geotechnical aspects and processed materials*. Balkema, Leiden, The Netherlands, 2007.

Gomes Correia, A. 2001. Soil Mechanics in routine and advanced pavement and rail track rational design. *Geotechnics for Roads, Rail tracks and Earth Structures* (Gomes Correia & Brandl, eds.). Balkema, pp. 165–187.

Gomes Correia, A. 2004. Evaluation of mechanical properties of unbound granular materials for pavements and rail tracks, keynote lecture. *Pavement and railway Design and Construction* (Gomes Correia & Loizos, eds.), Millpress, 35–60.

Gomes Correia, A., Lacasse, S. 2006. Marine and transportation geotechnical engineering – Technical Session 2e. Proc. XVI ICSMGE, Millpress, Rotterdam, The Netherlands, vol. 5, 3045–3069.

Gomes Correia, A., Lacasse, S., Ohtsuka, S. 2006. Marine and transportation geotechnical engineering – session 2e, General report. Proc. XVI ICSMGE, Millpress, Rotterdam, The Netherlands, vol. 5, 3185–3188.

Gomes Correia, A., Momoya, Y., Tatsuoka, F. 2006. TC 3 Workshop – Geotechnical aspects related to foundation layers of pavements and rail track. Proc. XVI ICSMGE, Millpress, Rotterdam, The Netherlands, vol. 5, 3259–3262.

Gomes Correia, A., Viana da Fonseca, A. and Gambin, M. 2004. Routine and advanced analysis of mechanical in situ tests. Results on saprolitic soils from granites more or less

mixed in Portugal. Keynote lecture, ISC'2, Porto. (Viana da Fonseca and Mayne, eds.), Millpress, Rotterdam.
Gomes Correia, A., Loizos, A. (Editors). *Geotechnics in Pavement and Railway Design and Construction*, Mill-Press, Rotterdam, Netherlands, Athens, 2004.
Karam, J. 2005. *Decision Aids for Tunnel Exploration*. Master's thesis, Massachusetts Institute of Technology, Massachusetts, USA.
Kröber, W., Floss, E.H.R., Wallrath, W. 2001. Dynamic soil stiffness as quality criterion for soil compaction. *Geotechnics for Roads, Rail tracks and Earth Structures* (Gomes Correia and Brandl, eds.). Balkema, pp. 189–199.
Lee, J.L., Stokoe, II, K.H., Bay, J.A. 2005. The rolling dynamic deflectometer: A tool for continuous deflection profiling of pavements. Proc. XVI ICSMGE, Millpress, Rotterdam, The Netherlands, vol. 3, 1745–1748.
Marques, R., Gomes Correia, A., Cortez, P. 2008. Data Mining in the "Technical Guide for Road Earthworks" - GTR (in Portuguese). 11th Portuguese Geotechnical Congress, Coimbra, Portugal, vol. III, 285–291.
Min, S., H. Einstein, Lee, J. 2005. *Application of the Decision Aids for Tunnelling (DAT) to update excavation cost/time information*. KSCE J. of Civil Engineering, 9(4).
Miranda, T. 2007. *Geomechanical parameters evaluation in underground structures. Artificial intelligence, Bayesian probabilities and inverse methods*. PhD thesis. University of Minho, Guimarães, Portugal, 291p.
Mohamed A. Shahin, Mark B. Jaksa and Holger R. Maier 2001. *Artificial Neural Network Applications in Geotechnical Engineering*. Australian Geomechanics – March 2001.
Momoya, Y., Sekine, E. 2005. Deformation characteristics of railway asphalt roadbed under a moving wheel load. XVI ICSMGE, Millpress, Rotterdam, The Netherlands.
Narayanan, R.M., Jakub, J.W., Li, D., Elias S.E.G. 2004. Railroad track modulus estimation using ground penetrating radar measurements. NDT&E International, 37: 141–151.
Nazarian, S. and Desai, M.R. *1993 Automated Surface Wave Method: Field Testing*, Journal of the Geotechnical Engineering Division, ASCE, Vol. 119, No. GT7, ASCE.
NF P 94-117-1 2000. *«Sols: reconnaissance et essais. Portance des plates-formes. Partie 1: Module sous chargement statique à la plaque (EV2)»*. Association Française de Normalisation (AFNOR).
Pinard, M.I. 2001. Developments in compaction technology. *Geotechnics for Roads, Rail Tracks and Earth Structures* (Gomes Correia & Brandl, eds.). Balkema, Lisse, The Netherlands, 37–46.
Quibel, A. 1999. New in situ devices to evaluate bearing capacity and compaction of unbound granular materials. *Unbound Granular Materials. Laboratory Testing, In-situ Testing and Modelling* (Gomes Correia, ed.), A.A. Balkema, Rotterdam, The Netherlands, 141–151.
Rasmussen, S., Krarup, J.A. and Hildebrand, G. 2002. Non-contact deflection measurement at high speed. *Bearing Capacity of Roads, Railways and Airfields* (Gomes Correia & Branco, eds.), Balkema, 1, pp. 53–60.
Saarenketo, T. 1992. Ground penetrating radar. Applications in road design and construction in Finish Lapland. Geological Survey of Finland, Special paper 15, 161–167.
Saarenketo, T. 1999. Road analysis, an advanced integrated survey method for road condition evaluation. *Unbound Granular Materials. Laboratory Testing, In-situ Testing and Modelling*. Gomes Correia (ed.), A.A. Balkema, Rotterdam, The Netherlands, 125–133.
SETRA/LCPC 2000. *Construction of embankments and capping layers*, Fascicules 1 and 2, Technical guide SETRA/LCPC.
Stokoe, II K.H., Allen, J.J., Bueno, J.L., Kalinski, M.E. and Myers, M.L. 1999. In-situ stiffness and density measurements of thik-lift unbound aggregate bases. *Unbound Granular Materials, Laboratory Testing, In-situ Testing and Modelling* (Gomes Correia ed.), Balkema, pp. 85–96.
Thurner H.F. and Sandstrom, A.J. 2000. Continuous compaction control, CCC. *Compaction of Soils and Granular Materials* (Gomes Correia & Quibel, Editors). Presses de l'Ecole Nationale des Ponts et Chaussées, Paris, 237–246.
TR 1999. *Use of Artificial Neural Networks in Geomechanical and Pavement Systems*. Transportation Research, Circular N. E-C012, TRB / National Research Council.
Wermeister S., Dawson A.R., Frohmut W. 2001. *Pavement deformation behavior of granular materials and the shakedown concept*. Transportation Research Board. Washington D.C. York, NY.
Yu, Hai-Sui, Li, H., Juspi, S. 2005. *Shakedown approach to pavement analysis and design*. Personal presentation at the TC3 workshop, XVI ICSMGE, Osaka.

Advances in Transportation Geotechnics – Ellis, Yu, McDowell, Dawson & Thom (eds)
© 2008 Taylor & Francis Group, London, ISBN 978-0-415-47590-7

Unified constitutive modeling for pavement materials with emphasis on creep, rate and interface behavior

Chandrakant S. Desai
The University of Arizona, Tucson, Arizona, USA

ABSTRACT: The importance of unified constitutive models for defining the behavior of materials in rigid and flexible pavements has been recognized. This paper presents such a unified modeling approach called the Disturbed State Concept (DSC) that can account for simultaneous occurrence of elastic, plastic and creep deformations, microcracking leading to fracture, failure and softening, and healing or strengthening. Emphasis is given on creep and rate dependence behavior and specialized DSC model for interfaces and joints. Typical validation of the DSC model at specimen level for a soil is presented, and validations for boundary value problems are available in other publications. It is believed that the DSC model can provide a unique and unified approach for constitutive modeling of pavement materials.

1 INTRODUCTION

Appropriate characterization of the mechanical behavior of pavement materials has been recognized as vital for realistic analysis, design and maintenance of pavement structures. Many specific models such as those based on the theories of elasticity (e.g., resilient modulus), plasticity, viscoplasticity, fracture and damage have been proposed in order to define respective behavioral features. However, the elastic, plastic and creep strains, microcracking leading to facture, softening or healing, occur simultaneously in the pavement under mechanical and environmental (moisture, temperature, chemical) loadings. Hence, although the need for unified model(s) for such integrated behavior has been recognized, there are hardly any such unified models available. A unified model based on the disturbed state concept (DSC) is described in this paper. It is capable of addressing various behavioral features in a simultaneous and holistic manner, both for flexible and rigid pavements.

The behavior of a pavement under such variety of features is rather complex, and it is recognized that the modern computer methods such as the finite element, should be invoked. Hence, the unified DSC model has been implemented in two- and three-dimensional finite element procedures (codes) with the aim to solve the pavements problems in the integrated manner.

The objective of this paper is to present brief details of the DSC approach, with emphasis on the multicomponent-DSC (MDSC) model for various versions of the creep behavior, which is important for pavement materials. The MDSC, which includes elastic (e), viscoelastic (ve) elastoviscoplastic (evp) and viscoelastic viscoplastic (vevp) versions, provides a compact and unified approach compared to other models intended for specific ve, evp or vevp behavior (Desai, 2001, 2002, 2007). The behavior of bound (and unbound) materials can be significantly affected by the rate of loading. Hence, a novel procedure to include the strain rate dependence is also presented in the context of a version of the MDSC creep model.

The behavior of pavement is influenced by the response of interfaces, e.g., between the bound material and the base material. The mathematical framework of the proposed DSC model can be specialized for interfaces and joints. A brief description of the DSC model for interfaces in pavements is given.

Statements of the implementation of the DSC model in 2-D and 3-D finite element procedures are followed by reference citations for applications of analysis of 2-D and 3-D pavement systems for permanent deformation (rutting), microcracking and fracture, reflection and thermal cracking and healing.

2 REVIEW

It is not intended to present comprehensive reviews of available constitutive models for unbound and bound pavement materials because of the length limitation. However, a brief review is presented below.

The resilient modulus (RM) has been used commonly to characterize the behavior of asphalt under repeated loading (Uzan 1985; Witczak & Uzan 1988; Huang 1993). It results essentially in a piecewise linear (elastic) approach to simulate the nonlinear behavior. Although it may be appropriate for computing displacements in pavements, it is not suitable for other important features such as plastic and creep deformations, volume change (dilation), stress path effects, microcracking leading to fracture and softening. It is often claimed that the RM approach is suitable for unbound materials such as soils. However, it is believed that the RM model cannot be suitable for the realistic behavior of soils because it cannot account for plastic and creep strains, volume change, stress path and softening aspects of the soil behavior. Plasticity, viscoelasticity, viscoplasticity, fractures and damage models have been proposed for specific behavior of materials (Lytton et al. 1993; Schapery 1990, 1999; Chelab et al. 2002, 2003; Tashman et al. 2004; Gibson et al. 2003; Gibson & Schwartz 2006). However, they may not be appropriate for the combined behavior involving elastic, irreversible (plastic) and creep, strains, microcracking leading to fracture, softening (damage), and healing, because each addresses only specific behavior, e.g., elastoplastic, viscoelastic, viscoplastic, sometimes with fracture and damage, which may not account for realistic discontinuities and combined behavior. The empirical, mechanistic-empirical and mechanics approaches including DSC model and applications are presented by Desai (2002, 2007).

3 THE DISTURBED STATE CONCEPT (DSC)

The DSC is considered to be a unified approach that is capable of accounting for the combined behavior involving elastic, plastic and creep strains, microcracking leading to softening and disturbance (damage) and healing. It is capable of providing a unified framework for a general mechanistic procedure for constitutive modeling of materials and interfaces in pavements (Desai 2002, 2007). Recently, various investigators have used and/or combined other models with the HISS (Bonaquist & Witczak 1997, Scarpas et al. 1997, Gibson & Schwartz 2006) plasticity model, which is a subset of the DSC addressed in this paper. However, such combinations may not allow for the behavior of pavement materials in a unified way.

The DSC is based on the idea that the behavior of a deforming material can be expressed in terms of the behavior of its components: the relative intact (RI) or continuum part and the microcracked (or healing) part called the fully adjusted (FA). In the absence of initial cracks or defects, the entire material element is in the RI or continuum state initially. As deformation progresses, the RI part transforms continuously into the FA part, and at the limiting ultimate condition, the entire material approaches the FA state. Figure 1(a) shows a schematic of the RI and FA states. The disturbance acts as the coupling and interpolation mechanism, which allows definition of the actual or observed behavior in terms of that of the material in the RI and FA states. The transformation in the material from the RI to the FA state takes place due to the microstructural changes caused by the relative motions such as translation, rotation and interpenetration, and softening or healing at the microlevel. The disturbance expresses the effect of the microstructural changes through the weighted behavior of the RI and FA parts. Hence, it is not necessary to define material response at the particle level as it is required in the approaches based on micromechanics (Mühlhaus 1995). In fact, the DSC allows for the interaction and coupling between the RI and FA parts and avoids the need for the definition of particle level response (as in the micromechanics approach), which is difficult or impossible to measure at this time.

3.1 *Governing equations*

Based on the equilibrium of forces on a material element, the incremental constitutive equations are derived as (Desai 2001):

$$d\underset{\sim}{\sigma}^{a} = (1-D)\,d\underset{\sim}{\sigma}^{i} + D\,d\underset{\sim}{\sigma}^{c} + dD\left(\underset{\sim}{\sigma}^{c} - \underset{\sim}{\sigma}^{i}\right) \quad (1a)$$

or

$$d\underset{\sim}{\sigma}^{a} = (1-D)\underset{\sim}{C}^{i}\,d\underset{\sim}{\varepsilon}^{i} + D\underset{\sim}{C}^{c}\,d\underset{\sim}{\varepsilon}^{c} + dD\left(\underset{\sim}{\sigma}^{i} - \underset{\sim}{\sigma}^{i}\right) \quad (1b)$$

or

$$d\underset{\sim}{\sigma}^{a} = \underset{\sim}{C}^{DSC}\,d\underset{\sim}{\varepsilon} \quad (1c)$$

where a, i and c denote the observed, RI and FA responses, respectively, $\underset{\sim}{\sigma}$ and $\underset{\sim}{\varepsilon}$ denote the stress and strain vectors, respectively, $\underset{\sim}{C}$ denotes the constitutive or stress-strain matrix, D denotes the disturbance and dD denotes increment or rate. As a simplification, D is assumed to be a scalar in a weighted sense. It can, however, be expressed as a tensor, D_{ij}, if the test data to define the directional values of D is available (Desai 2001).

3.2 *Hierarchical options*

In the *single mathematical framework*, the DSC approach is capable of allowing for elastic, plastic and

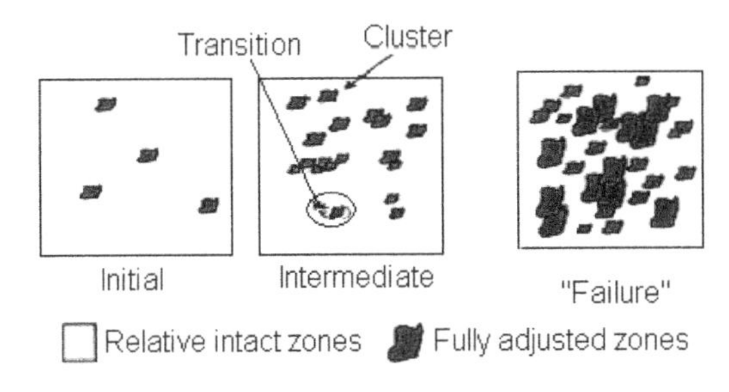

(a) RI and FA

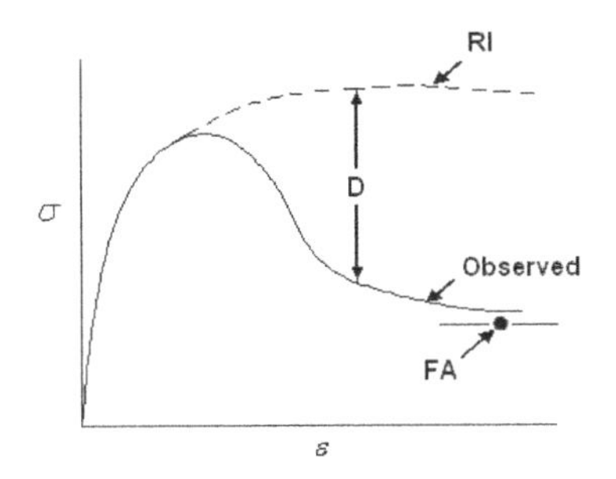

(b) Stress-strain behavior with disturbance

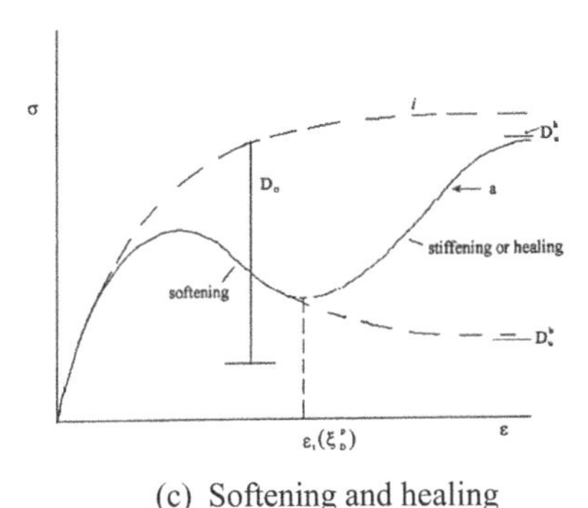

(c) Softening and healing

Figure 1. Schematic of disturbed state concept.

creep strains, microcracking, fracture and disturbance (damage) and stiffening or healing under mechanical and environmental loading. This is considered to be a unique advantage compared to other available models. A major advantage of the DSC is that various specialized versions such as elasticity, plasticity, creep, microcracking, degradation or softening and healing or stiffening can be specialized from Eq. (1). If there is no disturbance (damage) due to microcracking and

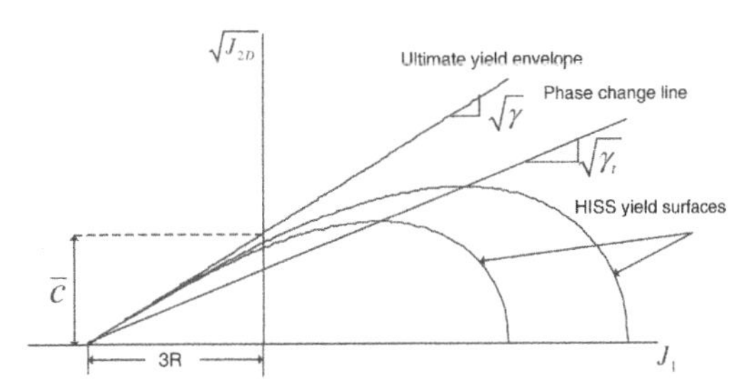

Figure 2. HISS yield surface.

fracture, D = 0 and Eq. (1) reduces to the classical continuum incremental equations as

$$d\underset{\sim}{\sigma}^i = \underset{\sim}{C}^i \, d\underset{\sim}{\varepsilon}^i \tag{1d}$$

where $\underset{\sim}{C}^i$ can represent elastic, elastoplastic or elastoviscoplastic response. If $D \neq 0$, the model can include damage and softening, and healing, Fig. 1. The user can choose an appropriate option for a given pavement material (or interface) and needs to input only parameters relevant to that option. For instance, the bound asphalt material can be characterized by using the DSC with elastoviscoplastic model, whereas the unbound materials can be treated as elastoplastic.

If it is assumed that the "damaged" part carries no stress at all, Eq. (1) reduces to the conventional continuum damage model (Kachanor 1986). However, such a model does not allow for the interaction between the damaged and undamaged parts and can suffer from deficiencies like spurious mesh dependence. Many workers have introduced external "enhancements" in the damage model (e.g., Bazant 1994) to allow for the interaction. The DSC model allows the interaction implicitly, and possesses a number of advantages compared with the other (enhanced) models (Desai 2001).

3.3 *Relative intact (RI) response*

In the general DSC equations, Eq. (1), $\underset{\sim}{C}^i$ represents the behavior of the RI material. It can be characterized as elastic, nonlinear elastic or elastoplastic hardening; for the latter, the unified hierarchical single surface (HISS) plasticity model can be used. The yield function, F, Fig. 2, in the HISS model for isotropic hardening is given by (Desai et al. 1986, Desai 2001).

$$F = \bar{J}_{2D} - \left(-\alpha \bar{J}_1^n + \gamma \bar{J}_1^2\right)\left(1 - \beta S_r\right)^{-0.5} = 0 \tag{2a}$$

$$= \bar{J}_{2D} - F_1 \cdot F_2 = 0 \tag{2b}$$

where $\bar{J}_{2D} = J_{2D}/p_a^2$ is the nondimensionalized second invariant (J_{2D}) of the deviatoric stress tensor, S_{ij}, p_a is the atmospheric pressure constant, $\overline{J}_1(= J_1 + \frac{\overline{c}}{\sqrt{\gamma}})/p_a$ is the nondimensionalized first invariant of the total stress tensor, $\sigma_{ij}, \overline{c}$ = proportional to cohesive strength, Fig. 2, in which 3R is the extension of F so as to include $\overline{c}$, S_r is proportional to the stress ratio $J_{3D}/J_{2D}^{3/2}$, J_{3D} is the third invariant of S_{ij}, n is the parameter associated with the phase change from contractive to dilative response, γ and β are associated with the ultimate yield surface, Fig. 2, and α is the hardening or growth function. One of the forms for the latter is given by

$$\alpha = \frac{a_1}{\xi^{\eta_1}} \tag{3a}$$

or

$$\alpha = \frac{a_1}{\left(\xi^* + \xi\right)^{\eta_1}} \tag{3b}$$

where $\xi = \int (d\varepsilon_{ij}^p \cdot d\varepsilon_{ij}^p)^{1/2}$ is the trajectory of accumulated total plastic strains and a_1 and η_1 are the hardening parameters. ξ^* = initial ξ introduced due to the extension of the yield surface by 3R, Fig. 2 (Li 2003); for cohesionless materials $\xi^* = 0$. ξ can be decomposed as

$$\xi = \xi_v + \xi_D$$
$$\xi_v = \frac{1}{\sqrt{3}}\varepsilon_{ij}^D \tag{4}$$
$$\xi_D = \int \left(dE_{ij}^p \cdot dE_{ij}^p\right)^{1/2}$$

ε_{ij}^p, E_{ij}^p and ε_{ii}^p are the total, deviatoric and volumetric plastic strain tensors, respectively. It may be noted that the HISS function is defined (Fig. 2) for compressive behavior in the positive J_1-space; thus it is not valid in the negative J_1-space, i.e., for tensile yielding. In fact, as compressive and tensile yield surfaces are usually different, and they may not plot continuous in the stress space, Fig. 2 (Desai 2007).

Various classical and other plasticity models such as von Mises, Drucker-Prager, Mohr-Coulomb, critical state, Veemer and Cap models are included as special cases of F in Eq. (2) (Desai 2001). Thus, the general DSC model can provide a selection of the foregoing plasticity models as specialized versions.

3.4 *Fully adjusted (FA) state*

The behavior of the material in the FA state can be defined by using various models. It can be assumed to possess no strength like in the classical damage model (Kachanov 1986), which is not realistic because it ignores the interaction between materials in the RI and FA states. It can be assumed to possess only hydrostatic strength, or it can be assumed to be at the critical state at which it deforms under constant shear without change in the volume, for given mean pressure; then the critical state equations can be used to define the FA response (Schofield and Wroth 1968; Desai 2001).

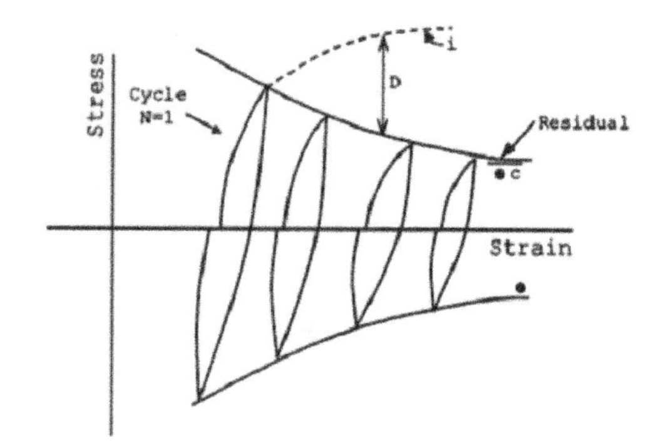

Figure 3. Cyclic stress-strain response with disturbance.

3.5 *Disturbance*

The disturbance, (D), which can be expressed in terms of ξ or (plastic) work, w, can characterize microcracking leading to fracture and degradation or healing. A simple expression for D is given by

$$D = D_u\left(1 - e^{-A\xi_D^Z}\right) \tag{5}$$

where D_u, A and Z are the disturbance parameters. Figures 1 and 3 show schematics of quasistatic and cyclic tests with the disturbance, respectively. Such test data can be obtained from uniaxial, shear or triaxial tests.

3.6 *Thermal effects*

The thermal effects involve responses due to the temperature change (ΔT) and the dependence of material parameters on the temperature. The incremental strain vector as can be expressed as (Desai 2001):

$$d\underset{\sim}{\varepsilon}^t(T) = d\underset{\sim}{\varepsilon}^e(T) + d\underset{\sim}{\varepsilon}^p(T) + d\underset{\sim}{\varepsilon}(T) \tag{6}$$

where $\underset{\sim}{\varepsilon}$ is the strain vector, t, e, and p denote total, elastic and plastic strains, T denotes temperature dependence and $d\underset{\sim}{\varepsilon}(T)$ is the strain vector due to the temperature change, ΔT.

The parameters in F, Eqs. (2, 3, 5), are expressed as the function of T. The temperature dependence is

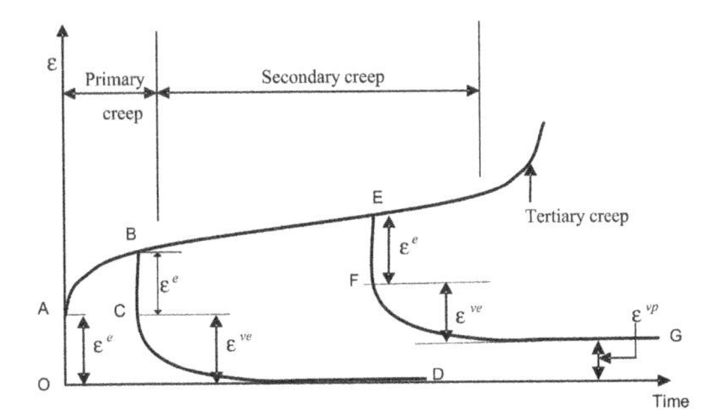

Figure 4. Schematic of creep behavior.

expressed by using a single function (Desai et al. 1997, Desai 2001).

$$P(T) = p(T_r)\left(\frac{T}{T_r}\right)^{\lambda} \tag{7}$$

where P is *any* parameter (elastic, plastic, creep, disturbance), T_r is the reference temperature, e.g., room temperature (= 27°C or 300 K), and λ is a parameter. The values of $P(T_r)$ and λ are found from laboratory test data at different temperatures.

4 CREEP BEHAVIOR

Both viscoelastic and elastoviscoplastic creep can be important for pavement materials, particularly for asphalt concrete. A schematic of the creep behavior is depicted in Fig. 4. Instantaneous elastic deformation (O–A) takes place upon the application of a (constant) stress. The viscoelastic (ve) (primary) creep (A–B–C–D) occurs after the application of the stress. The secondary (evp) creep (B–E–F–G) is typified by an approximate constant strain rate (BE). The tertiary creep occurs after the secondary creep, where failure can take place. Usually, asphalt concrete exhibits both primary and secondary creep, which is termed as vevp. For some materials (e.g., some metals), the primary creep may be negligible and the material can be defined only with the evp model.

4.1 *Multicomponent DSC (MDSC) model*

The idea of the MDSC model has been proposed to evolve a unified and compact model for the creep behavior including primary and secondary creep (Desai 2001) in the context of the DSC. The general MDSC model can include the Overlay model proposed by Zienkiewicz et al. (1972) and Pande et al. (1977) as a special case.

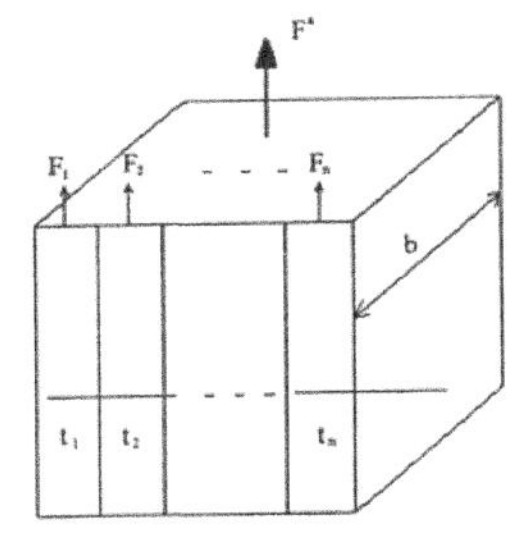

Figure 5. Multicomponent disturbed state concept (MDSC) for creep behavior.

In the MDSC, the material element can be divided into n number of components with different thicknesses (t_i, i = 1, 2, ... n), Fig. 5. The general model is derived as (Desai 2001):

$$d\underset{\sim}{\sigma}^a = t_1 \underset{\sim}{C}^1 d\underset{\sim}{\varepsilon}_1 + t_2 \underset{\sim}{C}^2 d\underset{\sim}{\varepsilon}_2 + \ldots . t_n \underset{\sim}{C}^n d\underset{\sim}{\varepsilon}_n \tag{8}$$

where C^i (i = 1, 2, ... n) are constitutive matrices for components. If it is assumed that the strains in all components are equal, Eq. (8) reduces to the overlay model (Pande et al. 1977).

Figure 6(a) shows a schematic of the MDSC-Overlay model in which each unit is called an overlay. Each component of the overlay can simulate different aspects of the material response; the spring, slider and dashpot represent the elastic, plastic and viscous responses, respectively. The viscoelastic (ve), elastoviscoplastic (evp) and viscoelasticviscoplastic (vevp) models are depicted in Fig. 6(b), (c) and (d), respectively. The well-known Perzyna model (1966) is essentially the same as the evp model, which is described below:

According to the evp (Perzyna) model, the viscoplastic strain rate is expressed as

$$\begin{aligned} \dot{\varepsilon}_{ij}^{vp} &= \Gamma\left(\frac{F}{F_o}\right)^N \frac{\partial F}{\partial \sigma_{ij}}, \quad \text{if } F > O \\ &= O, \qquad\qquad\quad \text{if } F \le O \end{aligned} \tag{9}$$

where $\dot{\varepsilon}_{ij}^{vp}$ is the viscoplastic strain rate tensor, Γ and N are material parameters, σ_{ij} is the stress tensor, F is the plastic yield function and F_o is a nondimensionalizing factor.

4.2 *Mechanics of viscoplastic response*

Figure 7 shows schematics of the time variation of applied (constant) stress, σ, total stress (ε), plastic strain trajectory or accumulated plastic strain (ξ), and the yield function (F) during the creep response, which has a nonzero positive value at t = 0 implying that the

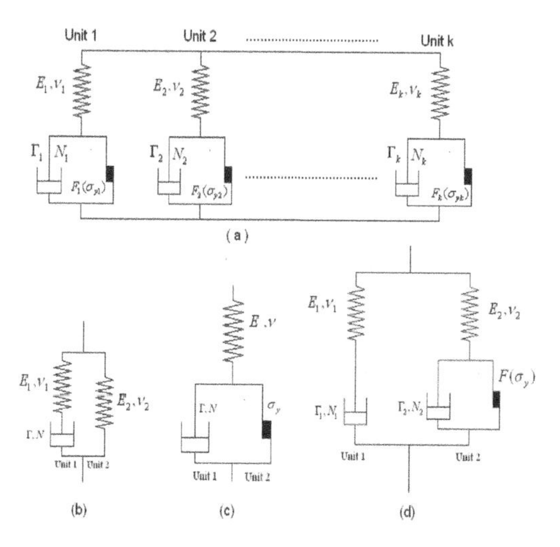

Figure 6. MDSC and ve, evp and vevp models.

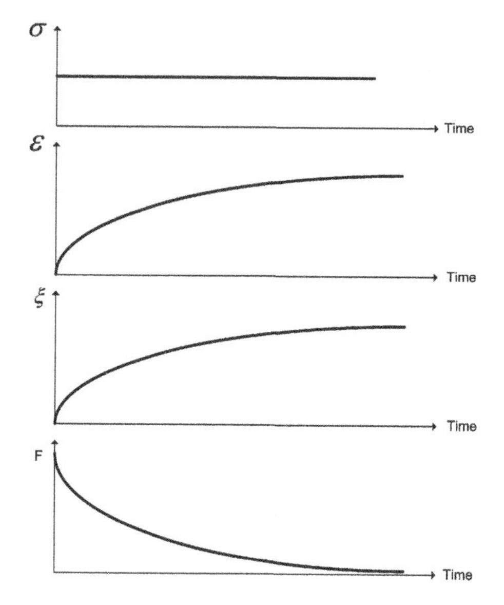

Figure 7. Mechanism of behavior for evp model.

material possesses a potential to deform under applied stress (Samtani and Desai 1991). As time passes, the (accumulated) viscoplastic strain (ξ) increases, but at greater times, its rate approaches zero. The yield function, F, approaches zero at greater times, as the strain rate tends to zero. Thus, in the evp model, it is necessary that the yield function, F, should be defined such that $F = 0$ corresponds to the *steady state* according to the steady strain state under the applied stress increment. In practice, such state can be obtained by performing the stress-strain tests at a very low or "static" strain rate. This aspect assumes much importance when the strain rate effect is considered.

Stress-strain tests at the static strain rate are usually not available, and often, test data at higher strain rate are used to find the parameters for F, which may not be consistent with the (viscoplastic) model. A procedure to develop static stress-strain data based on the test data from creep tests has been proposed, Fig. 8; details are given by Sane & Desai (2007). Brief description is given below:

The strain rate in a creep test under a constant stress usually becomes very small, and the asymptotic value, denoted by heavy dot, can be assumed to represent the steady state strain, Fig. 8(a). Thus, the applied constant stress and the asymptotic strain can be considered to be a point on the steady state stress-strain curve. Then, other constant stresses and the corresponding values of the asymptotic strains can provide points on the steady state stress-strain curve (Fig. 8b), under very low strain rate. Then the parameters for the "static" yield surface, F_S, can be obtained from such static curve, Fig. 8b, which is required for this viscoplastic model.

4.3 *Rate dependence*

The yield surface for "dynamic" loading, F_d, Fig. 9, that is, strain rate greater than the static rate, can be expressed as (Baladi & Rohani 1984);

$$F_d\left(\sigma,\ \alpha,\ \dot{\varepsilon}\right) = \frac{F_s\left(\sigma,\ \alpha\right) - F_R\left(\dot{\varepsilon}\right)}{F_o} = 0 \qquad (10)$$

where F_S is the yield surface at static condition and F_o is the term to render F_d nondimensional. The details of derivation of the rate dependence factor F_R are given in Sane & Desai (2007). Hence, the material behavior at any strain rate can be predicted by knowing F_s and F_R.

4.4 *Viscoelastic viscoplastic (vevp) model*

This model can be general and allows for both viscoelastic (ve) and elastoviscoplastic (evp) responses of materials like asphalt concrete.

Mechanism: Assume that in early times, the viscoplastic strains are negligible; this situation is feasible if the plastic strains do not take place until the yield stress, σ_y, is not exceeded. For the continuous yielding plasticity, models such as HISS (Eq. 2), the plastic strains may occur almost from the start, but they can be assumed to be small. The parameters for the vevp model, Fig. 6(d), can be adopted as a combination of the ve and evp models, Fig. 6(b, c, d). Details for determination of the parameters for the MDSC models are given in Desai (2001).

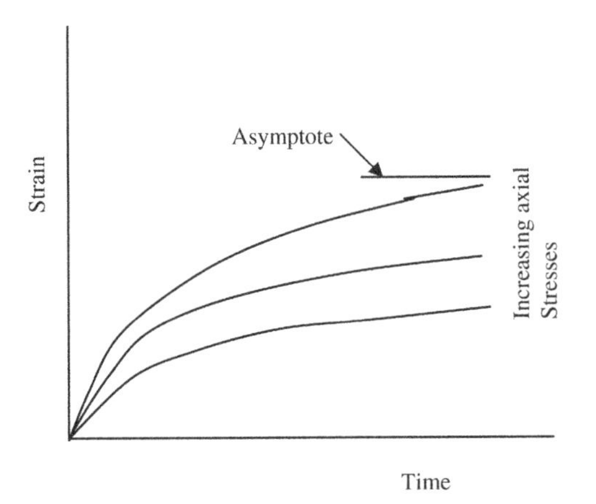

(a) Schematic of creep tests under different axial stresses

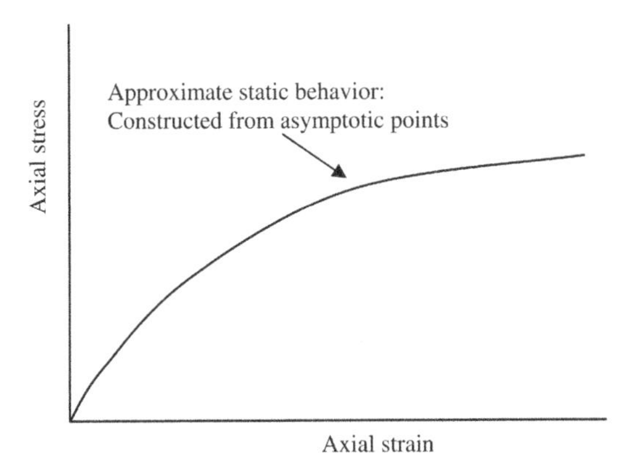

(b) Approximate static behavior

Figure 8. Approximate static behavior from creep tests.

5 VALIDATIONS FOR MDSC MODEL AT SPECIMEN LEVEL

The validations for the creep and rate dependent models are presented here for a soil (Sky Pilot glacial till), which is used here mainly to show the validations of the proposed models. If appropriate test data is available, the models can be used for bound and unbound materials in pavements. Comprehensive shear (triaxial) and creep tests under various confining pressures and constant stresses, respectively, were conducted on the soil (Sane & Desai 2007, Sane et al. 2008). These tests are used to find the parameters for the DSC model, which are shown in Table 1(a) for the DSC/HISS model, and in Table 1(b) for the vevp model.

Figure 10(a) shows the comparisons between predicted and observed shear stress-strain behavior at the specimen level for a typical confining pressure, $\sigma_3 = 400$ kPa, which was used to find the parameters: validation for test data used to find parameters is called Level 1. Validation (Level 2) for the independent

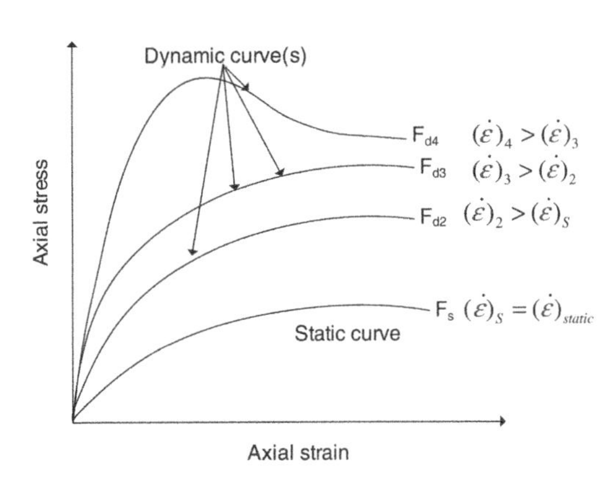

Figure 9. Schematic of stress-strain behavior under different loading rates.

Table 1(a). DSC/HISS parameters for Sky Pilot till.

Category	Symbol	Sky-Pilot Till
Elastic	E	50470
	ν	0.45
Plastic – HISS	γ	0.0092
	β	0.52
	n	6.85
	$\bar{c}$	16.67
Plastic – hardening	α_1	1.35e−7
	η_1	0.14
FA (critical) state	$\bar{m}$	0.09
	λ	0.16
	e_o	0.52
Disturbance	A	5.5
	Z	1.0

Table 1(b). Parameters for vevp model.

vevp creep parameters	Confining pressure ($\sigma_2 = \sigma_3$)		
	100 kPa	200 kPa	400 kPa
E_1 (kPa)	6666	24244	30000
E_2 (kPa)	42500	83333	63333
ν_1	0.45	0.45	0.45
ν_2	0.45	0.45	0.45
Γ_1(kPa^{-1}min^{-1})	0.001446	0.001841	4.24E−06
Γ_2(kPa^{-1}min^{-1})	0.006	0.006	0.006
N_1	2.28	4.00	1.40
N_2	1.63	2.20	2.0

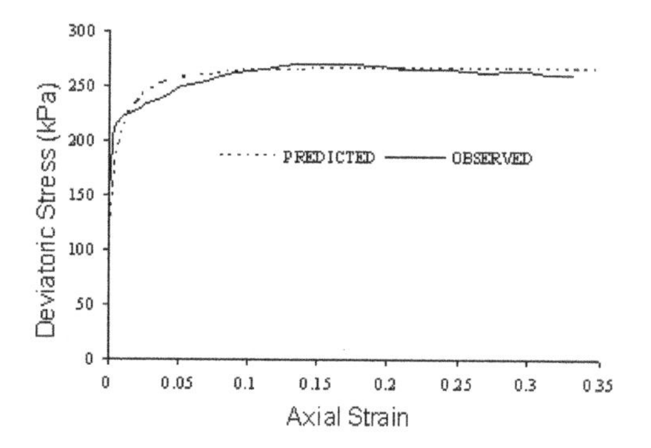

(a) Level 1 validation: confining pressure = 400 kPa

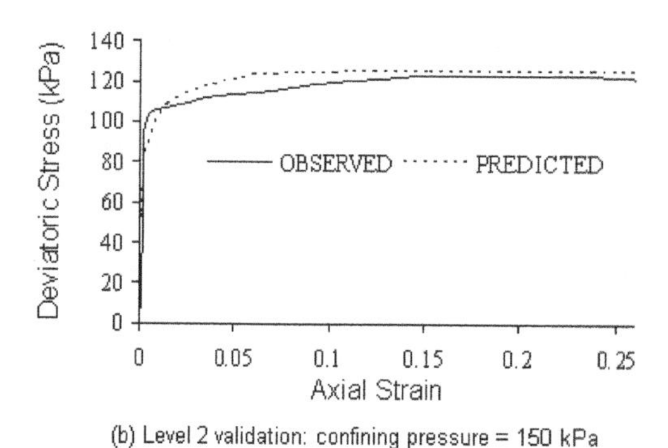

(b) Level 2 validation: confining pressure = 150 kPa

Figure 10. Comparison between predictions and test data.

behavior not used to find the parameters is shown in Fig. 10(b), for typical $\sigma_3 = 150$ kPa.

Finite element analysis for the evp and vevp models were performed using the DSC-SST2D (Desai, 1999) code, for which the mesh for the test specimen is shown in Fig. 11(a). Figure 11(b) shows Level 1 predictions in comparison with the creep test for typical applied stress = 182 kPa and $\sigma_3 = 400$ kPa. An independent prediction (Level 2) for applied stress = 45 kPa and $\sigma_3 = 300$ kPa is shown in Fig. 11(c). The vevp model involves greater strains compared to those for the evp model, and shows improved correlation with the test data. It is believed that asphalt concrete would exhibit both viscoelastic and viscoplastic strains, and can be modeled using the vevp model.

Figure 12(a) shows the static curve constructed for the creep test at $\sigma_3 = 300$ kPa in comparison with tests performed at three strain rates. It can be seen that the constructed static curve is near to that for strain rate $\dot{\varepsilon} = 0.005$ in/min (0.123 mm/min), thereby indicating that $\dot{\varepsilon} = 0.005$ in/min can be considered as the approximate static strain rate. Figure 12(b) shows the comparison between predicted and observed rate dependent triaxial data on the basis of the model, Eq. (10).

6 DSC FOR INTERFACES AND JOINTS

Interfaces in composite material systems such as pavements play a significant role in the overall behavior of pavement structures. Very often, in soil-structure interaction problems, the behavior of interfaces have been characterized by different models than those for the adjacent soil and structural material. For example, a plasticity model is used for soils, whereas for interfaces a linear or bilinear model is used. Use of such different models can be inconsistent with the mechanisms of deformation of the composite system.

The same unified framework for the DSC for solids (asphalt concrete, reinforced concrete, subbase and subgrade materials) can be specialized for interfaces and joints, thereby maintaining the consistency in deformation of the composite systems. DSC model for interfaces and joints is presented in various publications (Desai et al. 1984, Desai & Ma 1992, Samtani et al. 1996, Desai 2001, Desai et al. 2005). Here, a brief description is presented.

Figure 13(a) shows two-dimensional interface between the bound and unbound materials. A material element of the interface is shown in Fig. 13(b) with RI and FA components distributed over the "thin" interface zone.

The same mathematical framework for solids, Eq. (1), can be specialized for interfaces or joints. Accordingly, for the two-dimensional elements, the DSC equations are given by (Desai, 2001).

$$
\begin{aligned}
d\tau^a &= (1-D)d\tau^i + D\tau^c + dD(\tau^c - \tau^i) \\
d\sigma_n^a &= (1-D)\,d\sigma_n^i + D\,d\sigma_n^c + dD\,(\sigma_n^c - \sigma_n^i)
\end{aligned}
\tag{11}
$$

where τ and σ_n are shear and normal stresses.

Details of the DSC model for interfaces and joints, validations and applications are given in Desai & Ma (1992) and Desai (2001).

7 VALIDATIONS: BOUNDARY VALUE PROBLEMS

The DSC models for solids, i.e., bound and unbound materials, and interfaces (joints) have been implemented in two- and three-dimensional finite element procedures including static, cyclic and dynamic behavior, and dry and saturated materials. The computer procedures have been employed to predict behavior of field and simulated (in the laboratory) boundary value problems involving practical situations. They include problems from civil and mechanical engineering, electronic packaging and glaciology. Details are given in a number of publications and included in Desai (2001). Applications relevant to pavements are given in Desai (2002, 2007).

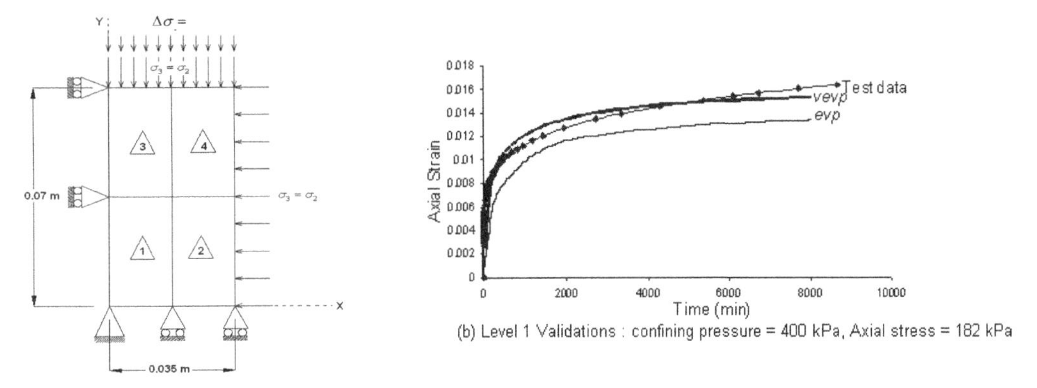

(a) Finite element mesh and loading for creep predictions

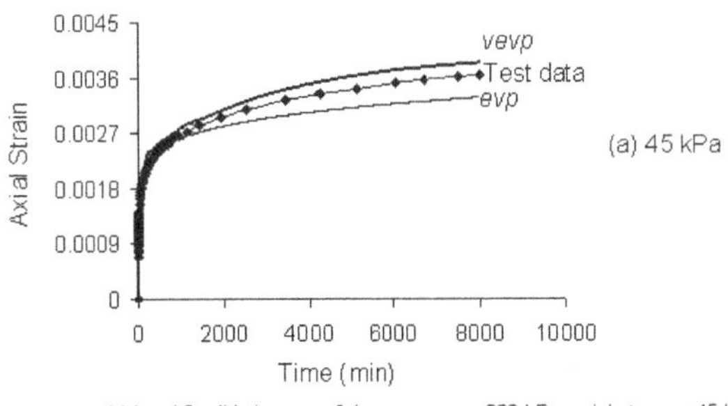

Figure 11. Mesh and predictions for creep behavior using evp and vevp models.

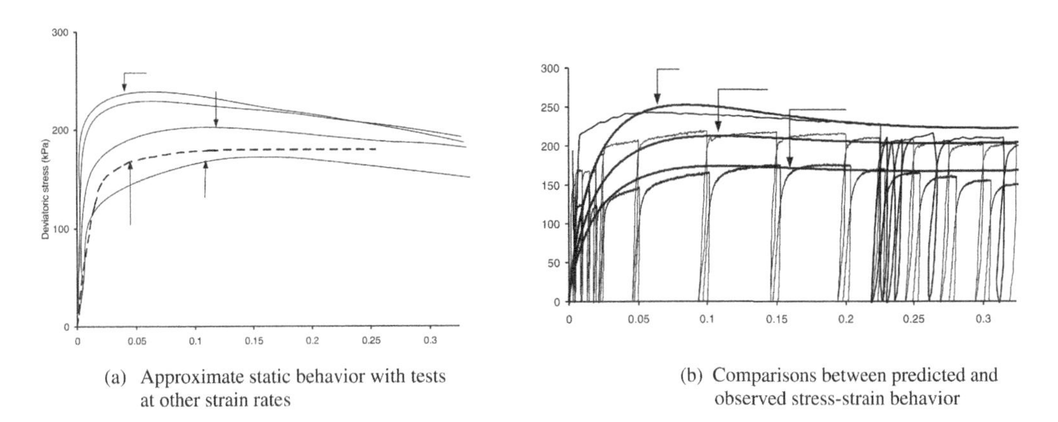

(a) Approximate static behavior with tests at other strain rates

(b) Comparisons between predicted and observed stress-strain behavior

Figure 12. Static curve and predictions for soil.

8 CONCLUSIONS

The DSC is a general procedure for defining behavior of engineering materials including those in pavement structures. It is hierarchical, and can account for elastic, plastic, and creep deformations, microcracking leading to fracture, softening and healing. It is extensively validated for specimen and boundary value problem levels for a wide range of engineering problems including pavements. Here, emphasis is given to creep, rate dependent and interface behavior, which are important for pavement problems. It is believed that

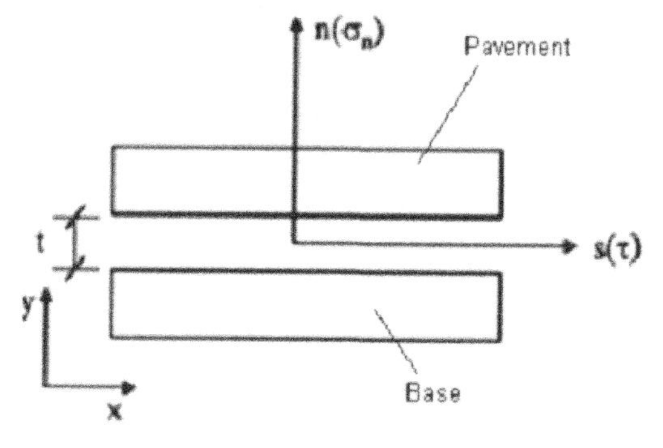

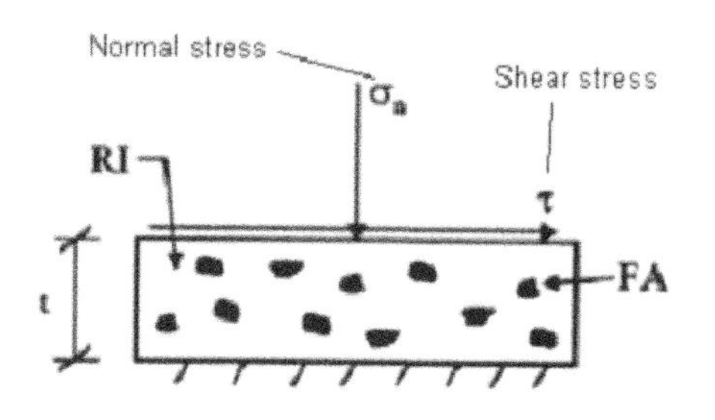

Figure 13. Interface zone with RI and FA states in DSC.

the DSC is perhaps the only unified procedure available that can handle the simultaneous and integrated occurrence of the important behavioral features of both bound and unbound material in pavement engineering.

ACKNOWLEDGMENTS

The research results presented here have been supported by a number of agencies such as the National Science Foundation, Department of Transportation, and Air Force Office of Scientific Research. Dr. S. Sane contributed toward the results presented here on creep and rate dependence.

REFERENCES

Baladi, G.Y. & Rohani, B. 1984. Development of an elastic viscoplastic constitutive relationship for earth material. In Chap. 2 Desai, C.S. & Gallagher, R.H. (eds), *Mechanics of Engineering Materials*, John Wiley, UK: 23–43.

Bazant, Z.P. 1994. Nonlocal damage theory based on micromechanics of crack interactions. *Journal of Engineering Mechanics, ASCE* 120: 593–617.

Bonaquist, R.J. & Witczak, M.W. 1997. A comprehensive consititutive model for granular materials in flexible pavement structures. *Proc. 8th int. conf. on asphalt pavements*. Seattle, WA, USA: 783–802.

Chehab, G.R. et al. 2002. Time temperature superposition principle for asphalt mixtures with growing damage in tension state. *Asphalt Paving Technology*. Association of Asphalt Paving Technologies 71: 559–593.

Chehab, G.R. et al. 2003. Characterization of asphalt concrete in uniaxial tension using viscoelastoplastic continuum damage model. *Asphalt Paving Technology*. Association of Asphalt Paving Technologies 72: 315–355.

Desai, C.S. 1999. DSC-SST2D code for two-dimensional static repetitive and dynamic analysis. *User's manual I to III*. Tucson, AZ, USA.

Desai, C.S. 2001. *Mechanics of materials and interfaces: the disturbed state concept*. CRC Press: Boca Raton, FL, USA.

Desai, C.S. 2002. Mechanistic pavement analysis and design using unified material and computer model. Keynote paper, *3rd int. symp. on 3-D finite element for pavement analysis, design and research, Amsterdam*. Netherlands.

Desai, C.S. 2007. Unified DSC constitutive model for pavement materials with numerical implementation. *Int. J. Geomechanics, ASCE* 7(2): 83–101.

Desai, C.S. et al. 1984. Thin-layer element for interfaces and joints. *Int. J. Num. Analyt. Meth. in Geomech.* 8(1): 19–43.

Desai, C.S. et al. 1986. A hierarchical approach for constitutive modeling of geologic materials. *Int. J. Num. Analyt. Meth. in Geomech.* 10(3): 225–257.

Desai, C.S. et al. 1997. Thermomechanical response of materials and interfaces in electronic packaging: parts I and II. *J. of Elect. Packaging, ASME* 119(4): 294–300, 301–309.

Desai, C.S. & Ma, Y. 1992. Modelling of joints and interfaces using the disturbed state concept. *Int. J. Num. & Analyt. Meth. in Geomech.* 16(9): 623–653.

Desai, C.S. et al. 2005. Cyclic testing and constitutive modeling of saturated sand-concrete interfaces using the Disturbed State Concept. *Int. J. of Geomech.* 5(4): 286–294.

Gibson, N.H. et al. 2003. Viscoelastic, viscoplastic, and damage modeling of asphalt concrete in unconfined compression. *Transportation Research Record* No. 1860: 3–15.

Gibson, N. & Schwartz, C.W. 2006. Three-dimensional viscoplastic characterization of asphalt concrete utilizing Perzyna and HISS methodologies. 10th *int. conf. of asphalt pavements*. Quebec, 1:205–214.

Huang, Y.H. 1993. *Pavement analysis and design*. Prentice Hall: Englewood Cliffs, NJ, USA.

Kachanov, L.M. 1986. *Introduction to continuum damage mechanics*. Martinus Nijhoft Publishers: Dordrecht, Netherlands.

Li, H.B. 2003. FEM analysis with DSC modeling for materials in chip-substrate systems. *Ph.D. dissertation*. Dept. of Civil Eng. and Eng. Mechs., Univ. of Arizona, Tucson, AZ, USA.

Lytton, R. L. et al. 1993. Asphalt concrete pavement distress prediction: laboratory testing analysis, calibration and validation. *Report no. A357*. Project SHRP RF.7157-2, Texas A&M Univ., College Station, TX, USA.

Mühlhaus, H.B. (ed.) 1995. *Continuum models for materials with microstructure*. John Wiley & Sons, UK.

Pande, G.N. et al. 1977. Overlay models in time dependent nonlinear material analysis. *Computer and Structures* 7: 435–443.

Perzyna, P. 1966. Funddamental problems in viscoplasticity. *Adv. in Appl. Mech.* 9: 243–277.

Samtani, N.C. & Desai, C.S. 1991. Constitutive modeling and finite element analysis of slow moving landslides using hierarchical viscoplastic material model. *Report to National Science Foundation*. Dept. of Civil Eng. and Eng. Mechs., Univ. of Arizona, Tucson, AZ, USA.

Samtani, N.C. et al. 1996. An interface model to describe viscoplastic behavior. *Int. J. Num. & Analyt. Meth. in Geomech.* 20(4): 231–252.

Sane, S.M. & Desai, C.S. 2007. Disturbed state concept based constitutive modeling for reliability analysis of lead free solders in electronic packaging and for prediction of glacial motion. *Report to NSF*. Dept. of Civil Eng. and Eng. Mechs., Univ. of Arizona, Tucson, AZ, USA.

Sane, S.M. et al. 2008. Disturbed state constitutive modeling of two Pleistocene tills. *Quaternary Science Reviews*. 27: 267–283.

Scarpas, A. et al. 1997. Finite element simulation of damage development in asphalt concrete pavements. *Proc. 8th int. conf. on asphalt pavements*. Univ. of Washington, Seattle, WA, USA: 673–692.

Schapery, R.A. 1990. A theory of mechanical behavior of elastic media with growing damages and other changes in structure. *J. Mech. Phys. Solids* 28: 215–253.

Schapery, R.A. 1999. Nonlinear viscoelastic and viscoplastic constitutive equations with growing damage. *Int. J. of Fracture* 97: 33–66.

Schofield, A.N. & Wroth, C.P. 1968. *Critical state soil mechanics*. McGraw-Hill: London, UK.

Schwartz, C.W. & Gibson, N.H. 2006. A viscoelastic-viscoplastic-continuum damage constitutive model for asphalt concrete. *Final Report, NCHRP Project 9–19*. National Cooperative Highway Research Program, Transportation Research Board, National Research Council, Washington, DC: February.

Tashman, L. et al. 2004. Damage evolution in triaxial compression tests in HMA at high temperatures. *Asphalt Paving Tech.*, Assoc. of Asphalt Paving Tech. 73: 53–87.

Uzan, J. 1985. Characterization of granular materials. *NRC TRB 1022 Transp. Res. Board*. Washington, DC: 52–59.

Witczak, M.W. & Uzan, J. 1988. The universal airport pavement design system. Granulat material characterization. *Reports I to IV*. Univ. of Maryland, College Park, MD, USA.

Zienkiewicz, O.C. et al. 1972. Composites and Overlay models in numerical analysis of elasto-plastic continua. *Proc. int. symp. on foundations of plasticity*. Warsaw, Poland.

Advances in Transportation Geotechnics – Ellis, Yu, McDowell, Dawson & Thom (eds)
© 2008 Taylor & Francis Group, London, ISBN 978-0-415-47590-7

Applications of reinforced soil for transport infrastructure

A. McGown
University of Strathclyde and University of Nottingham, UK

S.F. Brown
University of Nottingham, UK

ABSTRACT: Reinforced soil has been used for many years in geotechnical engineering and a number of its applications have a role in the construction of transportation infrastructure, including applications for foundations, embankments, slopes, walls, pavements, railway track-beds and other geotechnical structures. The reinforcing elements available are manufactured from a range of basic materials and by a number of very different manufacturing processes, so they exhibit a variety of properties. The various applications have significantly different operational requirements and there are many different design methods available. The variety of reinforcing products and their critical engineering properties are described, together with the test methods employed to assess them. Operational requirements are considered for each application type and the importance of construction effects is highlighted. The design methods presently available for the various applications are critically assessed against a background of experience and improved understanding from research.

1 INTRODUCTION

Polymeric fabrics, nets and meshes were first introduced into civil engineering in the late 1960's and, since then, there has been continuous development of the range of products, their applications, the design methods available and the construction techniques employed. *Geosynthetics* is the general term applied to the wide range of products which are now available. They are manufactured by many specialized processes using a number of different polymers (Koerner, 1998). This allows Geosynthetics to perform one or more functions, including *Containment, Separation, Filtration, Erosion Control, Drainage and Reinforcement.* The result is that Geosynthetics assume very different operational roles in different applications and specific properties are required in order for them to fulfill their various functions. Thus, there is a need to carefully match the operational properties of Geosynthetics to the operational requirements of the application in which they are to be used. In this paper, only reinforcement applications for Geosynthetics are considered.

In order to assess the properties of Geosynthetics which are important for their soil reinforcement function, many test methods have been developed. Some of these are suitable only for quality control purposes (*Index Tests*) but others can be usefully employed to determine the operational behaviour of Geosynthetics (*Performance Tests*), (Murray and McGown, 1982). Only properties obtained from performance testing should be used for design purposes or to compare the operational behaviour of different Geosynthetic products.

Design methods are available to allow the operational requirements of the various soil reinforcement applications to be matched to the performance properties of Geosynthetics. The first methods employed, some of which are still in use, were purely empirical, often based on field trials involving specific products functioning in particular conditions. However, analytical design methods quickly emerged based on established, classical, geotechnical and pavement *Limit Equilibrium Methods.* More recently, design methods based on *Limit State Analysis* and *Numerical Methods* have been developed. Some current design codes are based on a combination of Limit Equilibrium and Limit State principles and these are termed *Hybrid Methods.*

Few of the design methods available to date, fully take into account the construction techniques employed. However, construction processes can greatly influence short or long term stability and construction or post-construction deformation behaviour. Thus, detailed consideration of construction techniques is essential and should form part of the overall design approach. Specific construction aspects to be considered include: *Material Transport Handling and*

Laying, Compaction Methods and Levels, Construction Sequence and Period.

2 GESYNTHETIC REINFORCING MATERIALS AND THEIR PROPERTIES

2.1 *Basic types*

Geosynthetic reinforcing products are manufactured in the form of *Geotextiles, Straps/Strips, Geonets, Geomeshes* and *Geogrids*, some of which are illustrated in Figure 1.

Geogrids may be formed by structural members comprising fibres, filaments or straps/strips connected at junctions formed by entanglement, heat or chemical bonding or welding. Another process involves the stretching of punched sheets in one or two directions to form uniaxial or biaxial geogrids with integral junctions.

The most commonly used polymers are polyethylene, polypropylene and polyester, which are all thermoplastic materials. These polymers can be repeatedly heated to their softening point, shaped or worked as required, and then cooled to preserve their shape or form. For reinforcing products, stretching or drawing is involved at some stage, in order that the amorphous structure with random molecular orientation of the basic polymer is modified to produce a certain degree of crystallinity through aligned molecular orientation. However, the degree of crystallinity in any product must be controlled. Although increasing crystallinity increases tensile stiffness and strength and chemical resistance, it is also associated with a decrease in strain at failure, bending stiffness, impact resistance and stress crack resistance. Thus, the improvements in load carrying capacity and tensile stiffness have to be balanced against the development of product brittleness.

An important attribute of Geosynthetic reinforcing products is that they can be manufactured to exhibit their tensile load carrying capacity and tensile stiffness in certain combinations of direction as follows:

- Only one direction, (*Uniaxial Products*)
- In one principal direction and one orthogonal secondary direction, (*Anisotropic Biaxial Products*)
- In two equal and orthoganol directions, (*Isotropic Biaxial Products*).

A recent development has been the production of *Isotropic Triaxial Geogrids,* which possess three directions of equal tensile load carrying capacity and stiffness but which are not 'triaxial' in the normal three dimensional sense.

2.2 *Properties and tests*

For the purposes of product identification and quality control, Geosynthetic reinforcing products are

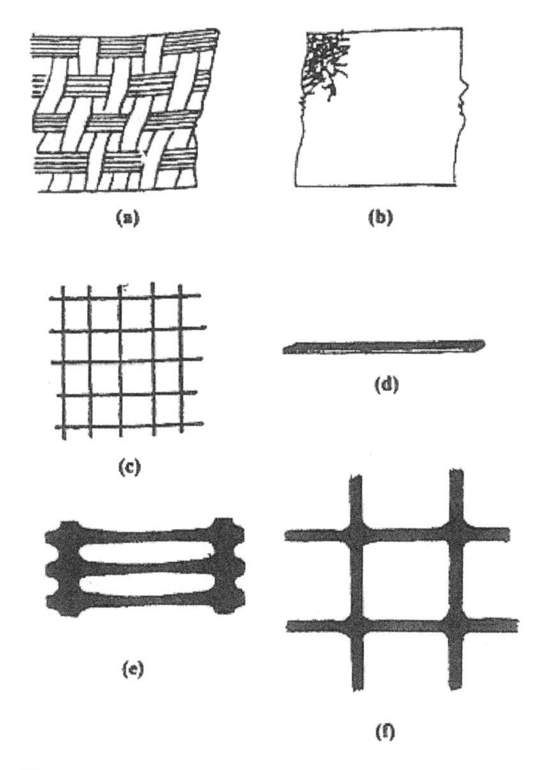

Figure 1. Forms of Geosynthetic Reinforcing Products: a) Woven Geotextile, b) Non-woven Geotextile, c) Geonet, d) Strap, e) Uniaxial Geogrid and f) Biaxial Geogrid.

characterized in their *As Manufactured* or *As Delivered* condition, using Index Test data. National and international standards exist which specify such tests in detail (e.g. BS/EN, ISO and ASTM) and include procedures to measure the following:

- Specific Gravity
- Mass per Unit Area
- Physical Dimensions
- Bending Stiffness / Flexibility
- Tensile Stiffness and Strength

It is important to realize that Index Test data do not provide a basis for the prediction or assessment of the load-strain-time behaviour of the products in soil under operational conditions. In particular, it should be noted that the Index Tests to determine tensile stiffness and strength are carried out in air at a standard temperature and relative humidity. Uniaxial loads are applied with the test specimen extended at a constant rate of strain, (*CRS Testing*), and the rate of strain employed may differ between different test standards. Nevertheless, correlations between Index test data and the operational load-strain behaviour of some Geosynthetic reinforcing products have been suggested. However, these correlations should not be relied upon for design purposes or to compare the operational behaviour of different products, as they can be greatly influenced by several site specific factors, including:

- Damage during installation
- Loading regime
- Confinement in soil
- Temperature

Thus for design purposes, it is necessary to use data obtained from Performance Tests which have been carried out in a manner and under conditions appropriate to the operational situation. The principal properties of Geosynthetic reinforcing products which should be assessed by Performance Tests are:

- Tensile stiffness and strength
- Surface friction and interlock

Such Performance Tests should be carried out at the operational temperature and the influences of physical damage and chemical degradation over time should be assessed. This is normally achieved by testing materials in their as-manufactured condition at different temperatures when in air, or in soil if the operational behaviour of the Geosynthetic is likely to be significantly influenced by in-soil confinement. Additionally, some limited amount of separate specialist testing is undertaken to allow *Risk Factors/Partial Factors* to be established, which will allow the measured properties of the materials in their as-manufactured condition to be modified to take account of possible physical damage or chemical degradation.

Given that the load-strain behaviour of polymeric materials is time dependent, it is necessary to test the Geosynthetics using an appropriate loading regime. This means either applying a series of uniaxial or biaxial loads over a long period of time, (*Creep Testing*), or applying a series of repeated loading cycles, (*Cyclic Testing*), depending on the application. The data obtained from these tests are then plotted either as load against strain for different specific periods of time or for different numbers of load cycles, McGown et al (1984). Thus, for any period of time or number of load cycles the appropriate tensile stiffness and/or strength of the product can be determined.

The surface friction characteristics of sheets, straps and strips are determined by *Direct Shear Testing* with soil or granular materials in direct contact with the surface of the product. This allows determination of the *Coefficient of Direct Sliding*. For nets, meshes and grids, the mode of interaction is quite different, (Jewell et al 1984), the interaction with the soil being partly developed through bearing stresses against the transverse structural members (*Ribs*) and partly by direct shear along the surfaces of the material. Thus, for these materials, the *Coefficient of Bond/Interlock* should be measured when the product is embedded in the soil in which it will operate. This is often achieved by *Pull-out Testing* (Jewell, 1996) but the data so obtained needs to be very carefully interpreted, (Ziegler and Timmers, 2004 and McGown and Kupec, 2008).

As a consequence of the complexities of performance testing, many large-scale model and full-scale trials of Geosynthetic Reinforced Soil Structures have been carried out and back analysed to provide operational properties for Geosynthetic reinforcing materials. However, great care must be taken when using these data, to ensure that the operational conditions within the models or field trials are known and are similar to the site specific conditions in which the Geosynthetics will be embedded in practice.

3 SOIL AND GRANULAR MATERIALS REINFORCEMENT APPLICATIONS IN TRANSPORTATION INFRASTRUCTURE

Soil or granular material reinforcement applications may be split into two main categories, (Bonaparte et al, 1985). *Geosynthetic Reinforced Earth Structures,* including walls, steep slopes and embankments and *Geosynthetic Reinforced Load Supporting Structures,* which include unpaved roads and permanent road foundations, load supporting pads and railway trackbeds. The applications commonly employed within transportation infrastructure, together with their loading regimes and some of their more important operational conditions, are as follows:

- Soil walls and steep slopes:

Typical applications include retaining structures, bridge abutments, interchanges, noise barriers, boulder barriers, landslide repairs, excavations, flood dikes.

Loading regime involves construction equipment plus essentially dead loads with some live, impact or seismic loading.

Operational conditions are low strain levels over long-term design lives when in engineered fills.

- Embankments on weak foundations:

Typical applications include embankments over soft soils, potential voids and cavities.

Loading regimes involve construction equipment plus essentially dead loads with some live, impact or seismic loading.

Operational conditions are moderate strain levels over medium-term design lives when in engineered fills.

- Load supporting pads:

Typical applications include construction equipment areas, heavy vehicle site access, fabrication yards and staging areas.

Loading regimes include construction equipment and heavy vehicles plus dead and live loads.

Operational conditions are moderate to high strain levels over short-term design lives at the interface between natural soils and engineered fills.

– Unpaved roads on weak foundations:
Typical applications include access and haul roads.
Loading regimes involve construction equipment plus essentially live (traffic) loads.
Operational conditions are moderate to high strain levels over short-term design lifetimes at the interface between natural soil and engineered fills.
– Permanent pavement foundations:
Typical applications include paved road foundations, industrial storage yards, parking areas and low-volume high-performance roads.
Loading regimes involve construction equipment and heavy vehicles plus dead and live (traffic) loads.
Operational conditions are very low strain levels over long-term design lives at the interface between natural soil and road materials or within road materials.
– Railways:
Typical applications include railway track-beds in ballast or at the interface between natural soil and ballast.
Loading regimes involve construction equipment plus essentially live (traffic) loads.
Operational conditions are low strain levels over long-term design lives at the interface between natural soil and ballast or within ballast.

This demonstrates that the operational requirements for Geosynthetics embedded within reinforced earth structures and reinforced load supporting structures are very different. Additionally, quite different design methods and construction techniques are used for these two groups of geosynthetic reinforcing applications. Consequently, they are dealt with separately in the following two sections.

4 THE DESIGN AND CONSTRUCTION OF GEOSYNTHETIC REINFORCED EARTH STRUCTURES

4.1 *Basic types*

Geosynthetic Reinforced Earth Structures may be defined as those for which the primary design consideration is stability of the structure under its own weight, (Bonaparte et al, 1985). Further, it can be suggested that soil walls, steep slopes and embankments over weak foundations all fall into this category. However, it should be noted that these structures may be subject to additional live, impact or seismic loading.

Typically, these structures are built with alternating horizontal layers of compacted granular soils/materials and Geosynthetic reinforcement, for a distance back from the lateral boundaries. In some situations the reinforcement may extend across the whole width of the structure.

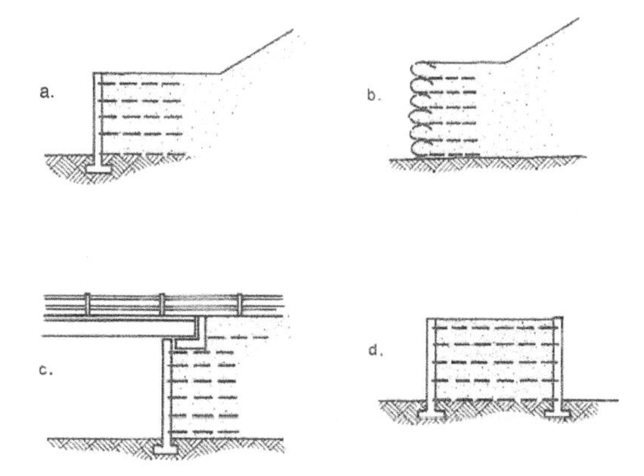

Figure 2. Reinforced Soil Walls: (a) Concrete Faced Retaining Wall; (b) "Wrap-around" Faced Retaining Wall; (c) Bridge Abutment and (d) Dike. (after Bonaparte et al, 1985).

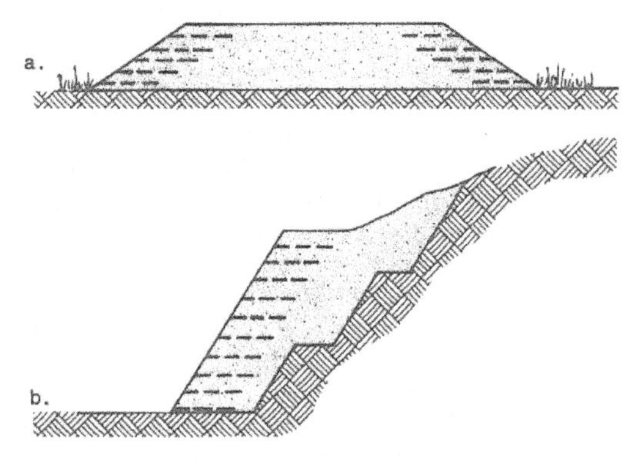

Figure 3. Reinforced Steep Slopes: (a) Steep Sided Embankment and (b) Landslide Repair. (after Bonaparte et al, 1985).

With reinforced soil walls and steep slopes, a facing may be necessary for aesthetic reasons and to prevent localized soil erosion or perhaps to resist lateral pressures. The reinforcement strengthens the soil by adding tensile stiffness and strength, thus increasing local and overall stability, and modifying the local and/or overall deformations of the structure. Various application types are shown in Figures 2 and 3.

Recently soil walls and steep slopes have been constructed with many different facing types and with variable angle boundaries, including stepped facings and combined sections of soil walls and steep slopes. Thus, the distinction made in many design standards and codes between soil walls, steep slopes and shallow slopes is becoming more difficult to justify, (Kupec et al, 2008).

The design of embankments is generally split into two groupings as shown in Figure 4. The first grouping relates to embankments initially built over weak and compressible foundation soils. Due to the process of

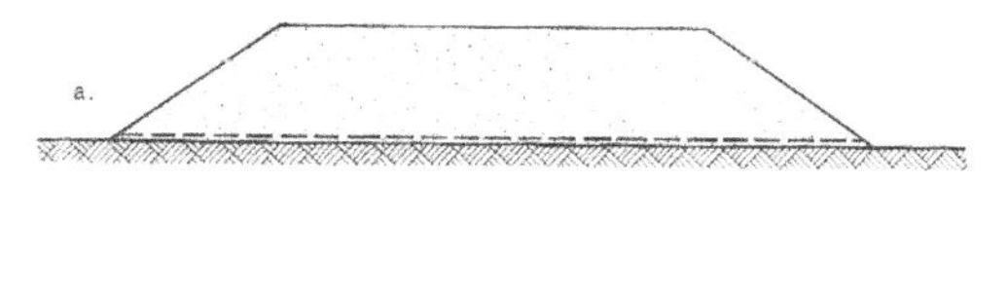

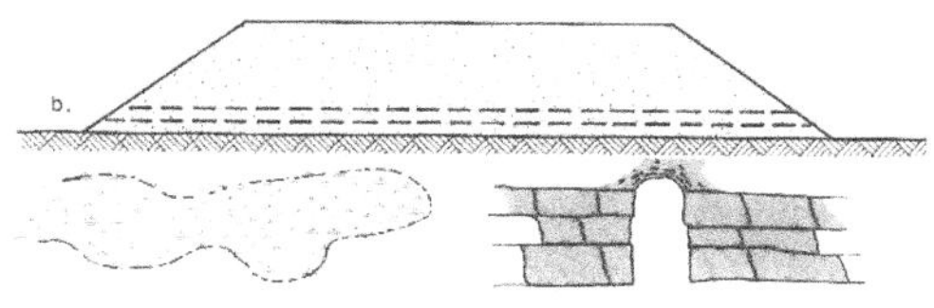

Figure 4. Embankments on Weak Soils: (a) Embankment on a Uniformly Weak and Compressible Soil and (b) Embankment over Locally Weak Soils or Rocks and Cavities (after Bonaparte et al, 1985).

consolidation, the sub-soil will gradually increase in strength and reduce in compressibility with time. Thus, the reinforcement requirements will be greatest at the end of construction and gradually decrease with time.

The second grouping relates to embankments built over initially strong foundation soils or rocks, which may suffer post construction localized collapse or sudden large deformations.

The load-strain-time behaviour of these two types of embankment is entirely different and the reinforcement requirements are therefore different. Further, it has been shown that the deformation mechanisms, hence the reinforcement requirements of embankments over potential voids and cavities, vary with the stiffness of the reinforced layer of the embankment, (Saathoff et al, 2002). The use of Geogrids to form a three dimensional mattress beneath an embankment is an example of the use of a very stiff Geosynthetic reinforced layer and the influence stiffness has on the deformation mechanism of the embankment, (Edgar, 1985 and Jenner et al, 1988).

A further application of Geosynthetic reinforcement is within embankments supported on piles. The functions of the Geosynthetic reinforcement layer, (or layers), are to span across the gaps between the pile caps and to reinforce the lateral boundaries. Once again the deformation mechanisms are different and so require separate design approaches, (Love and Milligan, 2003).

4.2 *Design methods*

Since the introduction of Geosynthetic Reinforced Earth Structures in the early 1970's, the design methods have gradually evolved but the resulting designs have not varied greatly. Indeed, if anything they have become more conservative, (Berg et al, 1998). The reason for this is that as each design method was developed it was "*calibrated*" to ensure that the results were similar to those from design methods previously employed.

The principal steps in the development of design methods are set out below.

4.2.1 *Empirical methods*

The modern use of reinforced soil structures started in the late 1960's with the introduction of *Terra Armee,* (Reinforced Earth), (Schlosser and Vidal, 1969). Essentially, this was a system of retaining wall construction using closely specified granular soils reinforced by strips of steel attached to hexagonal concrete facing panels. In the early to mid 1970's some Geosynthetic products, (Geotextiles and Straps), were employed as substitutes for the metal strips and various types of alternative facing units and connections were introduced. The soils were represented by their peak shear strength and the Geosynthetics were modeled as pseudo metallic materials by heavily factoring their short term, (CRS test), properties. The interaction of the soil and Geosynthetic was modeled using a coefficient of direct sliding. The specification of the type of granular soils used was relaxed and the applications were extended to include steep slopes and embankments, the latter often not requiring a facing panel.

The first trial Geosynthetic Reinforced Earth Structures were dimensioned using the previously developed design approach employed for Terra Armee and then monitored. Based on the outcomes of these trials, the dimensioning of subsequent structures was modified to ensure their long term stability and to limit their construction and post construction deformability. The design procedures so developed were therefore empirical and often directly linked to the use of specific Geosynthetic products.

4.2.2 *Limit equilibrium methods*

The next step in the evolution of design methods was the introduction of more general approaches based on Limit Equilibrium principles. In these methods the granular soils/materials to be reinforced were represented by their peak shear strengths. The Geosynthetic reinforcements were initially represented by factored short term (CRS test) strength properties and, later, by factored sustained loading (creep test) strength properties. The interaction of the soil and Geosynthetic was initially modeled using the coefficient of direct sliding but as Geogrids and other types of Geosynthetics were introduced in the 1980's which were capable of developing interlock, so their interaction was sometimes modeled using a coefficient of bond/interlock (Jewell et al, 1984). Little or no account was taken of the deformation characteristics of the soil, the Geosynthetic or their interactive behaviour. Thus, large *Global Factors of Safety* and *Material Factors* were employed to ensure explicitly that collapse did not occur and to ensure implicitly that deformation levels under working conditions were acceptable.

As implemented in many national and international design standards or codes, these Limit Equilibrium Methods were semi-empirical and produced very conservative designs, (Berg et al, 1998 and Greenway et al, 1999). Loads and strain levels developed in the Geosynthetic reinforcements under working conditions were generally very small, (Yogarajah, 1993) and, as a result, failures have been very few, (Giroud, 1999). In fact, any failures are normally associated with overall external stability or deformability rather than with the internal stability of the reinforced soil block.

4.2.3 *Limit State Approach*

In the early 1990's, the Limit State Approach was introduced into geotechnical engineering and its application to Geosynthetic Reinforced Earth Structures has been given detailed consideration, (McGown et al, 1993 and McGown et al, 1998). This indicated that the following factors could be taken into account:

- The nature and size of the structure, including any special requirements for the structure itself or its surroundings.
- The nature of the existing ground and groundwater conditions.
- The nature of the operational environment, including any seismic effects.

To set the design requirements, various *Geotechnical Categories* were introduced, (Eurocode 7 1995), viz. Category 1 for small, simple structures, Category 2 for conventional structures and Category 3 for structures with abnormal risks or on difficult ground.

For each design problem, different performance criteria are required both for the construction period and for the design lifetime of the structure. Therefore, whenever a structure or part of a structure operates at a level equal to any of the preset performance criteria, it is said to reach a *Limit State*. Two levels of Limit States are identified, the *Ultimate Limit States* and the *Serviceability Limit States*. The former are concerned with loss of equilibrium of the structure or rupture of components and the latter with loss of functional or aesthetic utility of the structure or of a component. No global Factors of Safety are applied but *Partial Factors* are used for both material properties and actions.

Designs based on the Limit State Approach require the development of the following:

- Calculation Models
- Operational Material Properties
- Operational Actions
- Accurate Geometrical Data
- Limiting Values of Deformation of the Structure and Components.

In addition a detailed assessment has to be made of the Partial Factors to be used. All this requires a detailed understanding of the operational mechanisms within and around the structure and the operational behaviour of the materials forming and surrounding the structure. If the Limit State Approach is fully complied with, then it has been shown to produce outcome designs which are far less conservative than those derived from current international and national standards and codes, (McGown et al, 1998).

4.2.4 *Hybrid methods*

Rather than move completely away from Empirical or Limit Equilibrium design methods to design methods based on the "true" Limit State Approach, many national and international design standards and codes are being developed, or have been developed, based on design methods which incorporate both Limit Equilibrium and Limit State principles. These Hybrid, (*Transitional*), methods incorporate many of the principles of the Limit State Approach but include both Global and Partial Factors of Safety. The choice of values for these factors is generally prescribed and, to some extent, the values are specified in order to ensure that outcome designs are similar to those from pre-existing Limit Equilibrium methods. Thus these design methods are also "calibrated" to ensure little change in outcome design. Therefore, Hybrid design methods are semi-empirical.

4.2.5 *Numerical Modeling*

A considerable number of researchers are applying various Numerical Modeling techniques to the analysis of both the stability and deformability of Geosynthetic Reinforced Earth Structures. To date, no design standards or codes have emerged based on this approach but they are frequently employed to back-analyse the behaviour of existing structures. An important aspect of many of these numerical methods is that they do not model the construction stage, yet for many Reinforced Earth Structures, this stage can be critical and significantly influence the post-construction behaviour of the structure or some of its components.

The above review of the various approaches to design developed since the introduction of Geosynthetic Reinforced Earth Structures indicates that the outcomes have not significantly changed, because of the "calibration" process undertaken at each stage of the development. Furthermore, since all the design methods are at least semi-empirical, changing or omitting prescribed input parameters, Global Factors of Safety or Partial Factors is not recommended.

In all the various design methods, consideration is given to internal, overall and external equilibrium. These determine the type of reinforcing material to be used, reinforcement spacing and reinforcement length. The detailed analysis to quantify each of these is detailed in each design standard or code. Since the underlying principles used for these standards are not all the same, it is always necessary to clearly identify which standard or code is being used and not to combine parts of one with another.

4.3 *Construction methods*

Generally in Civil Engineering, designers do not specify the method of construction. However, in Geotechnical Engineering, particularly in relation to structures built over weak or compressible ground, construction procedures are often viewed as part of the design process and are specified. With Geosynthetic Reinforced Earth Structures, it is very often important for designers, if not to fully specify the method of construction, then at least to provide guidance to the contractor on the delivery, storage and placement of Geosynthetics and on the compaction and other construction techniques and rates of construction to be employed.

Full or partial specification of construction procedures is necessary to ensure that the Geosynthetic reinforcement and facings, if used, will:

- Be in good condition when they are used.
- Survive the construction process without significant damage.
- Be located within tolerances and not be displaced.
- Not be loaded in excess of their design loads.

Overlaps and joints should be specified with regard to location and size and guidance should be given to ensure that they are not disrupted by the construction process. Any connections between the Geosynthetics and facings must be specified in a similar manner.

Once in position, no secondary works should be undertaken which may damage or cut through the Geosynthetic reinforcements.

5 DESIGN PRINCIPLES FOR GEOSYNTHETIC REINFORCED LOAD SUPPORTING STRUCTURES

5.1 *Background*

Geosynthetic Reinforced Load Supporting Structures are those usually stable under their own weight and for which the primary design consideration is the support of applied loads, (usually traffic loads), with limited deformations, (Bonaparte et al, 1985).

The use of Geosynthetics in roads started through the rather unscientific application of various Geotextiles, both woven and unwoven. While these materials did have some merits through their separation and filtration functions, they were not effective for reinforcement, (Brown et al, 1982). This arose because of the low stiffness of the fabrics and their poor interaction with the surrounding materials resulting in an inability to reduce the tensile plastic strains that accumulate during repeated loading. Later work, (Chan et al, 1989), demonstrated that a medium stiffness geogrid is more effective at reducing rutting in a granular pavement than a much stiffer woven Geotextile.

These research findings suggested that the mechanisms involved in the reinforcement of granular materials in pavement and similar constructions need to be much better understood and some progress has been made in this area but it has been rather slow. Consequently, the design procedures used in practice for haul roads, permanent road foundations and railway track-beds have remained largely empirical. Such methods have involved the use of a design procedure for an unreinforced pavement, (eg. Powell et al, 1984), with thickness reductions introduced for the reinforced case based on data from field trials. Unreinforced design theory has also been used with a theoretical adjustment to the stress distribution in the pavement to allow for the effect of reinforcement, (Giroud and Noiray, 1981). Another approach has been based on bearing capacity theory, adjusted to accommodate the reinforcement through its ability to reduce tensile strain at the bottom of the granular layer, (Houlsby and Jewell 1990).

In all these methods, a key factor that has not been accounted for, other than empirically is that of repeated loading, which is significant for roads and runways but may be less important for load pads unless vibratory loading is involved. Another significant shortcoming is that the methods appear to have focused on reducing the stress level in the subgrade on the assumption that this is the major source of plastic strain development that causes surface rutting. It has however been shown that the plastic strains developing in the granular layer are likely to be the more important, (Dawson et al, 1994). In railways, the loss of riding quality resulting from the development of differential settlement is largely caused by strains in the ballast, (Selig and Waters, 1994).

A key difference between road or rail reinforcement and that for walls, slopes and embankments, is that the strains involved are much smaller and the design criteria are almost wholly deformation related. It is not a Limit Equilibrium problem and, consequently, the reinforcement operates under conditions well below failure loads and strains. The key properties are stiffness rather than strength and an ability to provide strain continuity with the surrounding granular material or soil.

5.2 *Design concepts for reinforced pavements and rail track-beds*

5.2.1 *Introduction*

In the absence of a well-founded design method for pavements and railway track-beds, this section is devoted to a presentation of design concepts that are considered important for future developments. They are based on recent research dealing with the use of biaxial geogrids for the reinforcement of railway ballast, (Brown et al, 2007a and 2007b).

5.2.2 *Biaxial Geogrid characteristics*

This research has revealed the key parameters which influence the effectiveness of the reinforcement. They were somewhat easier to determine when applying Geogrids to ballast, which is an essentially single sized aggregate, than they would be for a graded material. Repeated-load experiments at full-scale under conditions that simulated the situation in a ballast layer identified the following Geogrid properties as having an influence on the effectiveness of the reinforcing action:

- The ratio of Geogrid aperture size to aggregate nominal size. This was defined as the Aspect Ratio and a value of 1.4 was identified and supported by theoretical analysis using the Discrete Element Method, (McDowell et al, 2006).
- The resilient in-plane stiffness of the Geogrid.
- The cross-sectional profile of the Geogrid ribs.
- The interaction of cross-plane rib stiffness and overburden pressure, since a rib that is too stiff appears to inhibit compaction and hence reduce the interlock potential.
- The nature of the Geogrid junctions.

Two further issues were also apparent from this research. The use of a Geotextile in conjunction with the Geogrid, whether bonded or not, appeared to inhibit good interlock. It was also shown that ythatplacement of the Geogrid within, rather than at the base of the layer being reinforced, improved the reinforcing action.

Geogrids are supplied commercially with a quoted Index Test tensile strength, but their performance in pavements and railways is dictated more by their operational resilient stiffness. Some low strain cyclic load tests on biaxial Geogrids have shown that the relationship between these two parameters is non-linear (Brown et al, 2007b). Thus future specification requirements may need to include resilient stiffness, and a suitable test will need to be developed.

5.2.3 *Design concepts*

The load spreading ability of a granular layer depends on the stress levels to which it is subjected as the material is markedly non-linear. Live load induced stresses will depend on the stiffness of the subgrade as well as the layer thickness. The introduction of a Geogrid has been shown to have no influence on the resilient response of the structure, as illustrated by Figure 5, (Brown et al, 2007a). This shows similar levels of subgrade stress and resilient deflection for rail track-beds with and without a Geogrid and emphasizes that the role of the Geogrid is to limit the development of tensile plastic strains that accumulate under repeated loading rather than to stiffen the structure. This is an area that merits further research. The concepts are important as they simplify the approach that can be taken to design based on the use of theoretical analysis.

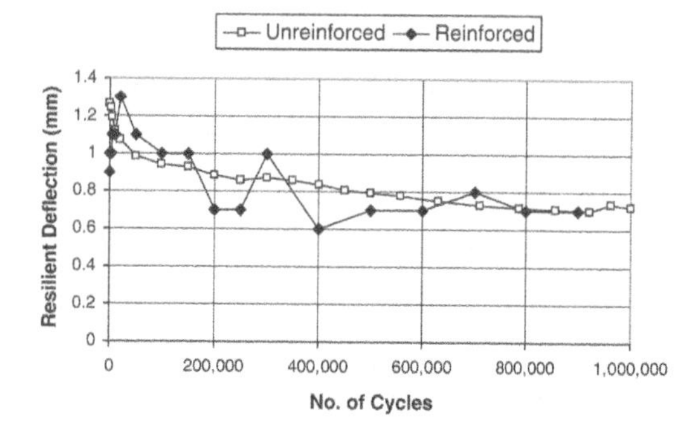

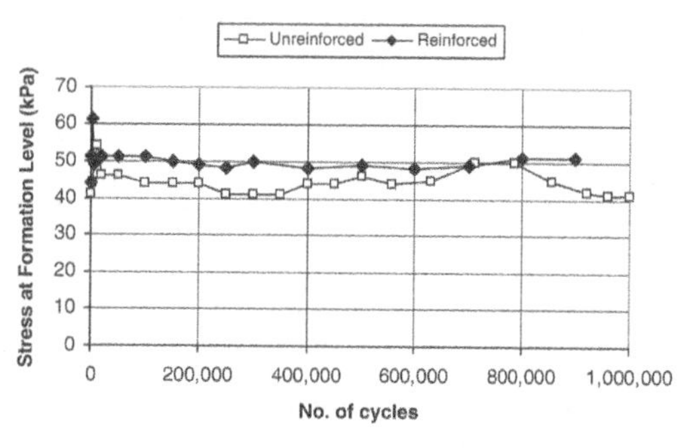

Figure 5. Resilient deflection and subgrade vertical stress in reinforced and unreinforced rail track-beds (after Brown et al, 2007a).

A well-established approach to the computation of rut depths in pavements is to conduct an elastic stress analysis and then to separately determine the vertical plastic strain distribution through the structure. Summation of these strains with depth yields the surface deformation, which is the principal design criterion. This approach has recently been incorporated into the new Mechanistic-Empirical Pavement Design Guide in the USA, (ARA, 2006).

Another concept is to consider a reinforced granular layer as a composite material having the same resilient characteristics as when unreinforced but with enhanced resistance to plastic strains. This approach requires research in order to quantify the plastic properties for particular composites and to correctly design them. From the point of view of theoretical analysis, it may be possible to adopt an elasto-plastic model and conduct the computations in one stage rather than the two used for the separative method outlined above.

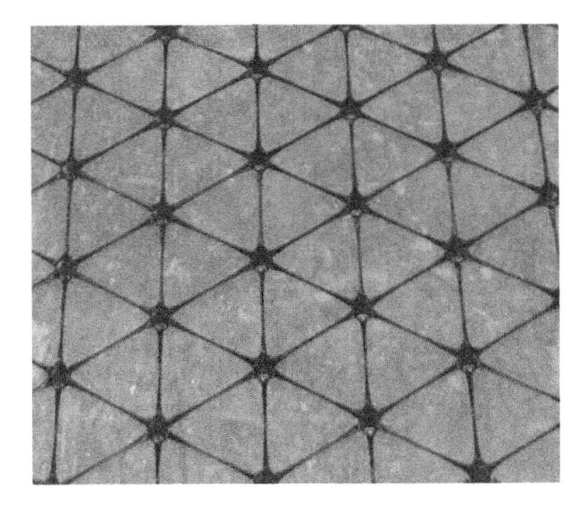

Figure 6. 'Triaxial' Geogrid.

5.2.4 *New developments*

For axisymetrical loading situations, such as a road pavements, the introduction of the 'triaxial' Geogrids (Figure 6) with their triangular apertures, presents interesting possibilities which are the subject of current investigations at Nottingham.

The concept of an Aspect Ratio needs further study, particularly for applications involving graded aggregates to determine which particle size is the dominant one with respect to effective interlock with the Geogrid. Equally, realistic tests such as the Composite Element Test, (Brown et al, 2007b) used for railway ballast applications will need to be developed for the axisymetric case. This will allow the parameters that influence reinforcement to be studied under realistic stress conditions and for the composite material idea to be taken further. Once such element tests have been completed, the performance of the most promising solutions can be investigated using full-scale wheel loading experiments. These can also serve to validate evolving design methods.

6 CONCLUDING DISCUSSION

This review of reinforced soil applications for transport infrastructure has demonstrated that although the earth structure applications (slopes and walls) have been very successfully used for many years, their evolution has been rather conservative with each new development tied to previous ones via a "calibration", which has, in effect, resulted in little real change to the design outcomes. Hence, design remains empirical and conservative. This situation has arisen because of the lack of fundamental research studies on geosynthetic-soil interactions involving effective experiments to improve understanding of the basic mechanisms involved. Such work has recently been restarted for the load bearing structure applications (roads and railways) and has proved useful in establishing an improved basis for design and for Geogrid selection for particular applications and the development of new products. Use of the Discrete Element Method, though time consuming at present, can greatly assist in exploring different application scenarios once the computational inputs have been validated. This is also a subject of ongoing research at Nottingham.

As the mechanics become clearer, the application of numerical analysis can assist in the development of improved design methods. These will require validation through field or other full-scale testing incorporating careful monitoring of performance.

In summary, it can be concluded that although reinforced soil may appear to be a mature subject with well-established materials and design methods, in reality it requires a new injection of fundamental research to develop more economic and safer design methods for the future.

REFERENCES

ARA Inc ERES Consultants Division. 2004. Guide for mechanistic-empirical design of new and rehabilitated pavement structures. *Report to NCHRP, Project 1-37A.*

Berg, R.R., Allen T.M. and Bell, J.R. 1998. Design procedures for reinforced soil walls – a historical perspective. *Proceedings 6th International Conference on Geosynthetics. Atlanta.* 2. 491–496.

Bonaparte, R., Holtz, R.D. and Giroud, J-P. 1985. Soil reinforcement design using geotextiles and geogrids. *Proceedings of the ASTM Symposium on Geotextile Testing and the Design Engineer.* ASTM Special Technical Publication; 952: 1987. 69–116.

Brown, S.F, Jones, C.P.D. and Broderick, B.V. 1982 Use of non-woven fabrics in permanent road pavements. *Proceedings of the Institution of Civil Engineers*, Pt. 2, 73: 541–563.

Brown, S.F., Brodrick, B.V., Thom, N.H and McDowell, G.R. 2007a. The Nottingham Railway Test Facility. *Proceedings of the Institution of Civil Engineers – Transport.* 160(TR2): 59–65.

Brown, S.F., Kwan, J. and Thom, N.H. 2007b. Identifying the key parameters that influence geogrid reinforcement of railway ballast. *Geotextiles and Geomembranes* 25(6): 326–335.

Chan, F.W.K, Barksdale, R.D. and Brown, S.F. 1989. Aggregate base reinforcement of surfaced pavements. *Geotextiles and Geomembranes* 8(3):165–189.

Dawson, A.R, Little, P.H. and Brown, S.F. 1994. Rutting behaviour in geosynthetic-reinforced unsurfaced pavements. *Proceedings 5th International Conference on Geotextiles, Geomembranes and related products.* 1: 143–146.

Edgar, S. 1984. The use of high tensile polymer grid mattress on the Musselburgh and Portobello by-pass. *Proceedings Conference on Polymer Grid Reinforcement. London.* Polymer Grid Reinforcement. Thomas Telford: 1985. 103–111.

Eurocode 7. 1995. Geotechnical design. *DD ENV 1997-1. 1995. General Rules.*

Greenway, D., Bell, R. and Vandre, B. 1999. Snailback wall – first fabric wall revisited at 25 year milestone. *Proceedings International Conference on Geosynthetics. Boston.* 2. 905–919.

Giroud, J-P. and Noiray, L. 1981. Geotextile reinforced unpaved road design. *Proceedings ASCE* 107(GT9): 1233–1254.

Giroud, J-P. 1999. Lessons learned from failures associated with geotextiles. *Proceedings International Conference on Geosynthetics: Geosynthetics '99. Boston.* 2. 905–919.

Houlsby, G.T., and Jewell, R.A. 1990. Design of reinforced unpaved roads for small rut depths. *Geotextiles, Geomembranes and related products.* Den Hoedt (ed), Balkema, Rotterdam: 171–176.

Jenner, C.G., Bush, D.I. and Bassett, R.H. 1988. The use of slip line fields to assess the improvement in bearing capacity given by a cellular foundation mattress installed at the base of an embankment. *Proceedings International Symposium on Theory and Practice of Earth Reinforcement. Fukuoka.* 209–214.

Jewell, R.A., Milligan, G.W.E., Sarsby, R.W. and Dubois, D. 1984. Interaction between soil and geogrids. *Proceedings Conference on Polymer Grid Reinforcement. London.* Polymer Grid Reinforcement. Thomas Telford: 1985. 18–30.

Jewell, R.A. 1996. Soil reinforcement with geotextiles. *Special Publication 123.* CIRIA. London. 23–29.

Koerner. R.M. 1998. *Designing with Geosynthetics – 4th Edition.* Prentice Hall. New Jersey.USA.

Kupec, J., McGown, A, and Jenner, C.G. 2008. The design of geosynthetic soil retaining structures with variable angle lateral boundaries. *Proceedings 4th European Geosynthetics Conference, Edinburgh.* (In press).

Love, J. and Milligan, G.W.E. 2003. Design methods for basally reinforced pile-supported embankments over soft ground. *Ground Engineering.* 36. 3. 39–43.

McDowell, G.R., Harireche, O., Konietsky, H., Brown, S.F., and Thom, N.H. 2006. Discrete element modeling of geogrid-reinforced aggregates, *Proceedings of the Institution of Civil Engineers – Geotechnical Engineering.* 159(GE1): 35–48.

McGown, A., Andrawes, K.Z., Yeo, K.C. and Dubois, D. 1984. The load-strain-time behaviour of Tensar geogrids. *Proceedings Conference on Polymer Grid Reinforcement. London.* Polymer Grid Reinforcement. Thomas Telford: 1985. 11–17.

McGown, A., Andrawes, K.Z., Paul, J. and Austin, R.A. 1993. Limit state design of reinforced walls, slopes and embankments. *Danish Geotechnical Society. ISLSD93.* 1/3. 275–284.

McGown, A., Andrawes, K.Z., Pradhan, S. and Khan, A.J. 1998. Limit state design of geosynthetic reinforced soil structures. *Proceedings 6th International Conference on Geosynthetics. Atlanta.* 1. 143–179.

McGown, A. and Kupec, J. 2008. Modeling geogrid behaviour: the influence of junctions on the behaviour of various types of geogrids. *Proceedings 4th European Geosynthetics Conference, Edinburgh.* (In press).

Murray, R.T. and McGown, A. 1982. The selection of testing procedures for the specification of geotextiles. *Proceedings 2nd International Conference on Geotextiles. Las Vegas.* 2. 291–296.

Powell, W.D, Potter, J.F, Mayhew, H.C. and Nunn, M.E. 1984. The structural design of bituminous roads. *TRRL Report 1132.*

Saathoff, F., Vollmert, L., Wittemoller, J., Stelljes, K and Klompmaker, J. 2002. *Proceedings 7th International Conference on Geosynthetics. Nice.* 1. 355–358.

Schlosser, F. and Vidal, H. 1969. La Terra Armee. *Bulletin de Liaison des Laboratoirie. Routiers Ponts et Chaussees.* 41. 101–144.

Selig, E.T and Waters, J. 1994, *Track geotechnology and substructure management,* Thomas Telford.

Yogarajah, I. 1993. Effects of construction procedures on the behaviour of geogrid reinforced walls. *PhD. Thesis. University of Strathclyde, Glasgow, UK.*

Ziegler, M. and Timmers, V. 2004. A new approach to design of geogrid reinforcement. *Proceedings 3rd European Conference on Geosynthetics. Munich.* 2. 661–666.

Advances in Transportation Geotechnics – Ellis, Yu, McDowell, Dawson & Thom (eds)
© 2008 Taylor & Francis Group, London, ISBN 978-0-415-47590-7

Recent research on railway track sub-base behaviour

W. Powrie, J.A. Priest & C.R.I. Clayton
University of Southampton, UK

ABSTRACT: This paper describes some recent and current research on the behaviour of railway track. The main focus is on the deflections of the track as trains pass, and how these may be linked to the stress paths experienced by the ballast and the underlying sub-base. Three questions are addressed: (1) what are the stress paths that soil elements below a railway track are subjected to during train passage?; (2) what is the influence of principal stress rotation on the deformation and failure behaviour of these materials?; and (3) what are the effects of train speed? It is shown that (1) the relevant stress paths may involve principal stress rotation; (2) principal stress rotation may have a significant effect on the behaviour of some soils; and (3) track loading and deflections must be expected to increase with train speed, which may increase load cycling effects. While these conclusions confirm what has been known or suspected for some time, the contribution of the work described in this paper is in beginning to quantify and define boundaries to the significance of these effects.

1 INTRODUCTION

The last ten years have seen significant growth in passenger and freight traffic on railways across the world. In addition to an increase in overall traffic volume, there have been moves to higher line speeds on passenger routes and heavier axle loads for freight vehicles. The increase in demand has led to the expansion of existing networks, construction of dedicated high speed lines and extensive rehabilitation of existing track to meet the demands of modern railway traffic. However, traditional ballasted track remains the mainstay of most new, classic and upgraded railways.

Traditional ballasted railway track systems consist of the superstructure (rails, fastening system and sleepers) and the substructure (ballast, sub-ballast and subgrade). Although the components of the superstructure have changed over the years, with the introduction of new grades of rail steel and new types of sleeper and rail fastenings, the ballast and sub-base have remained substantially the same.

The role of the ballast and sub-ballast is to ensure that the stresses due to train loading transmitted to the subgrade are reduced, through spreading, sufficiently to prevent failure of the subgrade by excessive deformation (plastic strain) or progressive shearing of the soil (Selig and Waters, 1994). Historically, the depth of the granular layer(s) needed to prevent failure was estimated on the basis of practical experience (Heath *et al.*, 1972), with standards generally using equations developed empirically – for example, the American Railway Engineering Association manual (AREA, 1996) and the Union Internationale des Chemins de fer Code 719R (UIC, 1994).

Modifications to the AREA design method were proposed by Raymond (1985) to take some account of the subgrade type. However, stresses in the subgrade were calculated on the basis of a homogenous halfspace subject to static loading, with no allowance for the effects of repeated loading cycles.

Heath et al. (1972) suggested specifying the ballast thickness so as to limit the stress in the subgrade to the 'threshold stress', defined as the deviator stress above which plastic strains began to increase rapidly, determined in triaxial element tests. In analysis, the subsoil was still assumed to be homogeneous, with no facility to assess the effect of layers of different stiffness on track and sub-base performance.

The Geotrack approach proposed by Li & Selig (1998a; 1998b) used an analysis based on a multilayered linear elastic finite element model to calculate stresses in the subgrade taking account of the potentially different stiffnesses of the ballast, sub-ballast and natural ground. Li & Selig (1998a) also carried out extensive laboratory testing using a cyclic triaxial apparatus to investigate the influence of factors such as the number of loading cycles and the ratio of deviatoric stress to soil strength on the cumulative plastic strain for various soil types. The results from the numerical model and the laboratory tests were then used to determine the ballast layer thickness required to limit the cumulative plastic strain to a specified value for a given number of cycles of a particular design load.

Despite these advances, problems with railway tracks requiring excessive continuous maintenance remain. A study comparing current railway track design methods was carried out by Burrow et al. (2007), with reference to various sites in the UK. At one site, on a mixed traffic (~50% freight) railway near Leominster, Herefordshire with a design speed of 128 km/h, the actual depth of the ballast layer ranged from 0.9 m to 1.3 m. This was greater than indicated by all of the design methods considered, yet frequent maintenance is required at this site to maintain satisfactory line and level.

Gräbe (2002) presents data relating to the Broodsnyerplass to Richards Bay COALlink line in South Africa, which was originally intended to carry 21 million tonnes of coal per year in vehicles with an axle load of 20 tonnes. The line has been continually upgraded and now carries over 60 million tonnes per year with axle loads of 26 tonnes. In 1999, 58% of capital expenditure was associated with the construction of new track foundations (sub-ballast and subgrade). Up to 68% of total maintenance costs over the period 1999–2004 were associated with failure of the track foundations, even though the line was built using the most up to date standards.

The examples cited by Burrow et al. (2007) and Gräbe (2002) tend to confirm the contention of (Shahu et al., 2000) that current track system design methods may only be suitable for low axle loads, line speeds and train frequencies. There are several factors that could account for this. Soil parameters are often obtained from cyclic triaxial tests, in which only the major (axial) principal stress is cycled (Heath et al., 1972; Li and Selig, 1998a); in reality, a soil element within the subgrade is subjected to complex loadings that involve a rotation of the principle stress as a train passes (Brown, 1996). Cyclic loading with principal stress rotation can affect both the soil stiffness and the rate accumulation of plastic strain (Chan and Brown, 1994; Lekarp et al., 2000, 2000b; Gräbe, 2002), with the rate of accumulation of plastic strain being greater for some soils when subjected to cyclic loading with principal stress rotation than cycling the axial stress alone (Gräbe, 2002). Also, the models used to estimate subgrade stresses are generally based on static analyses, and may not fully represent the effects of a moving load.

This Paper describes current research being carried out under the auspices of the Universities' Centre for Rail Systems Research, Rail Research UK, into the behaviour of railway track sub-base systems. The research addresses the following four questions. (1) What are the stress paths that soil elements below a railway track are subjected to during train passage? (2) What is the influence of principal stress rotation on the deformation and failure behaviour of these materials? (3) What are the effects of train speed? and (4) What happens in reality?

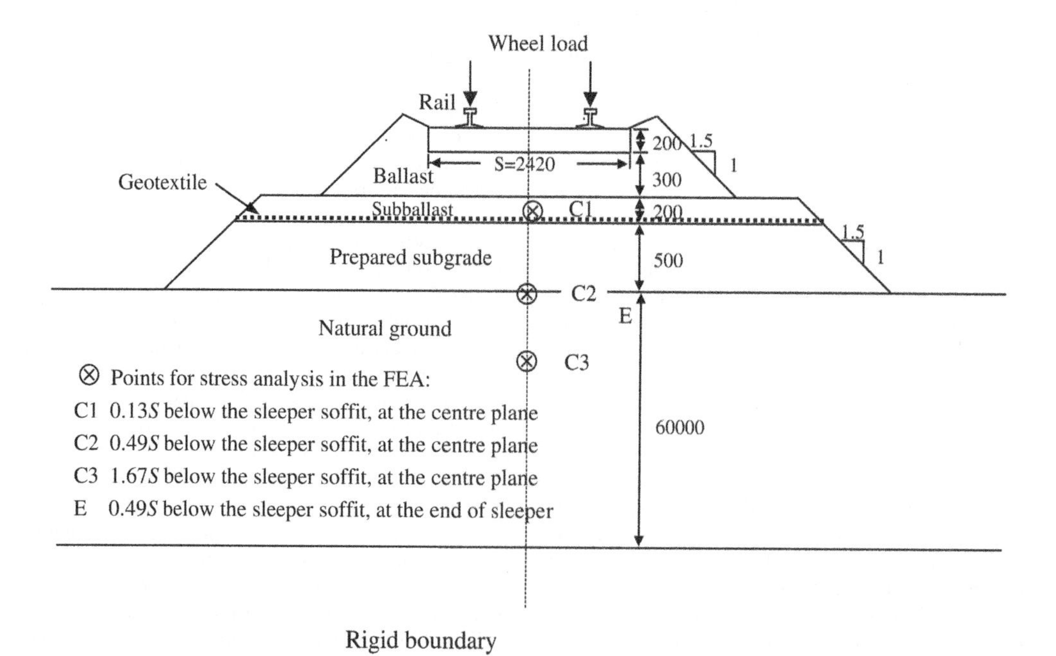

Figure 1. Schematic cross section of a typical track structure (dimensions in mm); from Powrie et al. (2007).

2 STRESS PATHS IN THE BALLAST AND SUB-BASE

2.1 *Method*

A series of static, 3D finite element analyses was carried out using the program ABAQUS (Hibbitt et al., 2005), to investigate the stress paths followed by soil and ballast elements below the typical track structure (with the geotextile omitted) shown in Figure 1.

Materials were generally assumed to behave elastically with deformation at constant volume, modelling an undrained, pre-failure response interpreted in terms of total stresses. The loads modelled represented (in terms of both magnitude and wheelbase) the MBA freight wagons used by English, Welsh and Scottish Railways (EWS) to carry heavy bulk materials such as coal and aggregates.

The wheel load was taken as 125 kN acting vertically. No enhancement factors to account, empirically, for dynamic magnification were applied; the effect of a lateral load due to braking or acceleration was investigated but is no reported here. The 3D finite element mesh is shown in Figure 2: the width of 16.9 m modelled the distance between the centrelines of adjacent wagons in an infinitely long train. The analyses are described in full by Powrie et al. (2007).

2.2 *Results*

Typical contours of vertical stress increase at the centre plane of the mesh are indicated in Figure 3, which shows clearly the transition with depth from axle loading to bogie loading to wagon loading. It is perhaps easiest to visualize the effects of principal stress rotation by plotting the direction of the major principal stress increment due to the wheel load as a function of distance from the centreline of a single box car within a train, for soil or ballast elements at different depths below the sleeper soffit. These graphs are shown in Figure 4 for ballast/soil element at depths of $0.13S$, $0.49S$ and $1.67S$ below the sleeper soffit, where S is the sleeper length (=2.42 m). Figure 4 also shows clearly the transition from axle loading (apparent at a depth of $0.13S$) to bogie loading (at $0.49S$) to wagon loading (at $1.67S$); and that at shallower depth, both the number and the magnitude of load cycles is increased.

2.3 *Application of results to laboratory testing*

In a conventional triaxial test, only the vertical stress σ_{zz} and the horizontal stress $\sigma_{xx} = \sigma_{yy}$ can be controlled, and all of these are constrained to be principal stresses. Thus there is no way in which the complex pattern of stress changes experienced by railway track sub-bases can be approximated in this apparatus.

Powrie et al. (2007) show how the calculated changes in vertical and horizontal normal and shear stresses may be approximated in the hollow cylinder apparatus, by superimposing variations in the axial load, torque and internal and external cell pressures onto a suitable initial stress state in a cyclic hollow

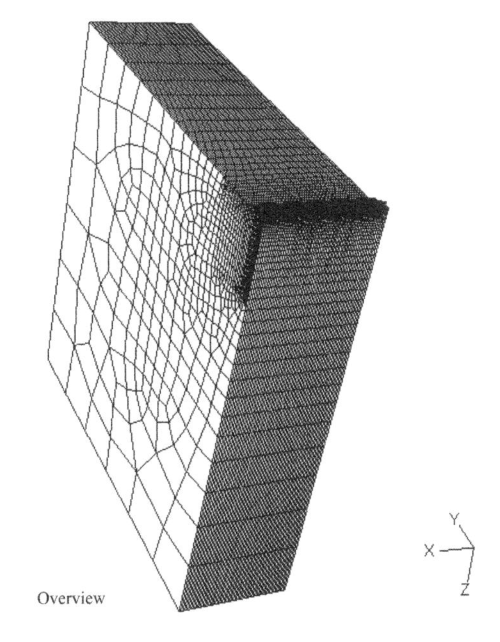

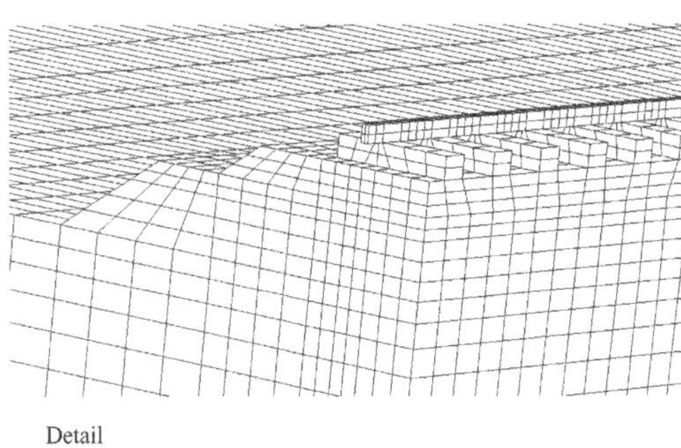

Figure 2. 3D finite element mesh (overview and detail); from Powrie et al. (2007).

Figure 3. Typical contours of vertical stress increment at the centre plane; from Powrie et al. (2007).

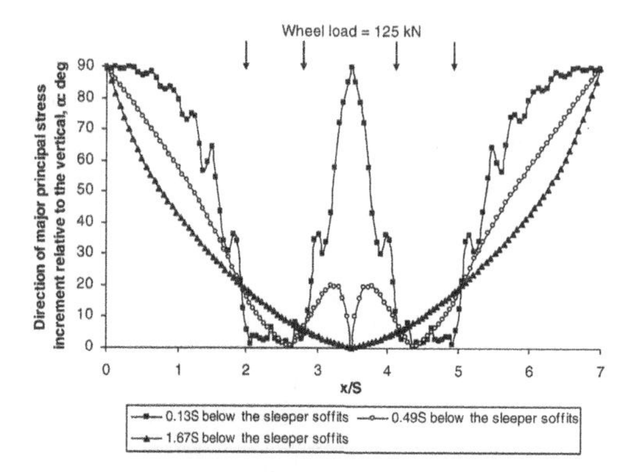

Figure 4. Variation of major principal stress increment direction with longitudinal distance at different depths on the centre plane (3D finite element analysis, with Young's modulus $E = 30 + 10.89.(z/S)$ and $\upsilon = 0.49$ for the natural ground, without initial stresses); from Powrie et al. (2007).

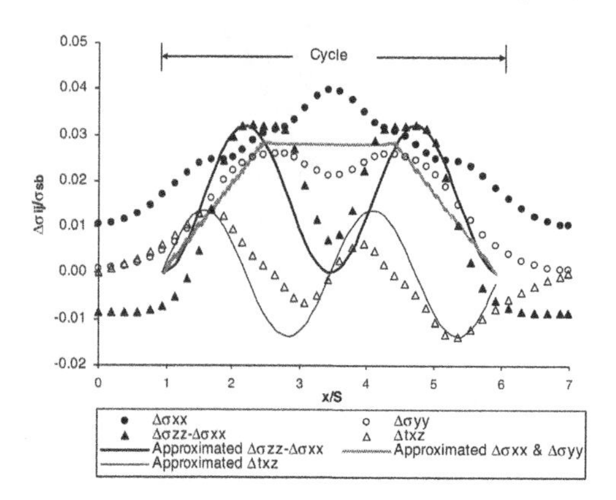

Figure 5. Approximation in hollow cylinder apparatus of the stress path followed by a soil element at the surface of the natural ground (depth 0.49S), during train passage; from Powrie et al. (2007).

cylinder apparatus. The calculated and approximate experimental stress paths are shown in Figure 5, for an element at a depth of 0.49S, corresponding to the top of the natural soil layer in Figure 1. The element at this depth is probably the most onerously loaded in comparison with its likely strength properties.

The behaviour of a soil element subjected to simulated train loading in the hollow cylinder apparatus will be considered in Section 3.

3 THE EFFECT OF CYCLIC PRINCIPAL STRESS ROTATION ON SOIL BEHAVIOUR

3.1 *Experiments by Gräbe (2002)*

Gräbe (2002) carried out cyclic tests on samples of a material representative of an engineered sub-base used

Table 1. Material properties.

Specimen	Sand vol.	Silt vol.	Clay vol.	P.L %	L.L %	P.I %	Activity A
A	78	14	7	14	25	11	1.4
B	68	18	14	14	31	17	1.2
C	54	22	24	14	37	21	0.9
D	73	16	11	16	28	14	1.3

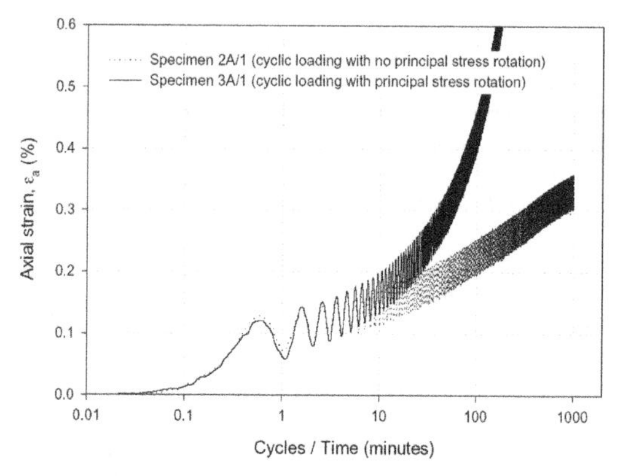

Figure 6. Data from (a) cyclic triaxial and (b) cyclic hollow cylinder tests on samples of the same soil, showing the potentially damaging effects of the latter; from Gräbe (2002).

on the Broodsnyerplass to Richards Bay COALlink line in South Africa, both with and without principal stress rotation.

The material properties are summarised in Table 1. The soil was consolidated from a slurry to a preconsolidation pressure of up to 300 kPa, before being subjected to up to 1000 cycles of undrained loading. The axial deviator stress was cycled between 0 and 30 kPa; in the tests with principal stress rotation, the shear stress on the horizontal plane was cycled between +7 kPa and −7 kPa. The results of some of these tests are indicated in Figure 6, and show that for certain types of soil at least the impact of principal stress rotation can be very significant. This particular material appeared stable under cyclic vertical loading in the absence of principal stress rotation, whereas with principal stress rotation plastic strains began to accumulate. Identification of the types of soil that are susceptible to the effects of principal stress rotation in cyclic loading is a subject of continuing research at Southampton.

4 EFFECTS OF TRAIN SPEED

4.1 *Method*

Yang et al. (in review) carried out 2D dynamic finite element analyses using ABAQUS (Hibbitt et al., 1995)

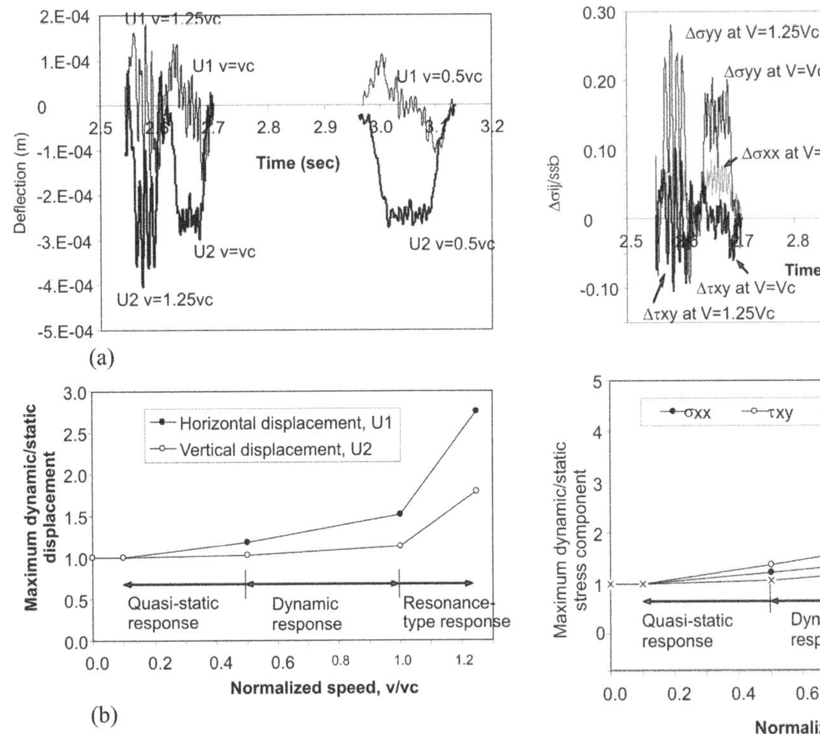

Figure 7. (a) Displacement of the natural ground surface plotted against time during the passage of a bogie pair at the ends of adjacent wagons at three different train speeds; (b) maximum displacement against train speed; from Yang et al. (in review).

Figure 8. (a) Stress increments at ground surface plotted against time during the passage of a bogie pair at the ends of adjacent wagons at three different train speeds; (b) maximum stress increment against train speed; from Yang et al. (in review).

to investigate the amplification effect of train speed on stresses in and deflections of the sub-base. They validated their finite element model with reference to measurements of ground movements made 60 km south of Vryheid on the Broodsnyerplass to Richards Bay COALlink line in South Africa, as discussed briefly in Section 5 of this Paper and in full by Priest et al. (in review).

4.2 *Results and implications*

Some of the results of the analyses of Yang et al. (in review) are summarised in Figures 7 and 8. Figure 7a shows the vertical deflection at the surface of the natural ground during passage of a pair of adjoining bogies at the ends of adjacent wagons as a function of time, for three different train speeds.

The train speeds are normalised with respect to the Rayleigh wave velocity, v_c, in the natural ground, which is generally equal to about 90–95% of the shear wave velocity, $v_s = \sqrt{(G/\rho)}$ where G is the shear modulus and ρ is the bulk density of the soil. Figure 7b shows the variation in maximum vertical displacement with normalised train speed. Figure 8 shows the corresponding data for the vertical normal stress increment $\Delta\sigma_{yy}$, horizontal normal stress increment $\Delta\sigma_{xx}$, and horizontal shear stress increment, $\Delta\tau_{xy}$.

The results presented in Figures 7 and 8 confirm previous research (e.g. Woldringh and New, 1999; Madshus and Kaynia, 2000) which has indicated that a static analysis will usually suffice for train speeds less than about $0.5v_c$. However, the static analysis will at a train speed of $0.5v_c$ underestimate the shear stress by about 30%. Further results presented by Yang et al. (in review) indicate that superimposed acceleration or braking loads may bring the stress state of the soil element closer to the failure envelope, and suggest that a static analysis might underestimate the adverse effects of dynamic loading particularly where a sub-soil susceptible to the effects of principal stress rotation is subjected to acceleration or braking loads over a large number of cycles.

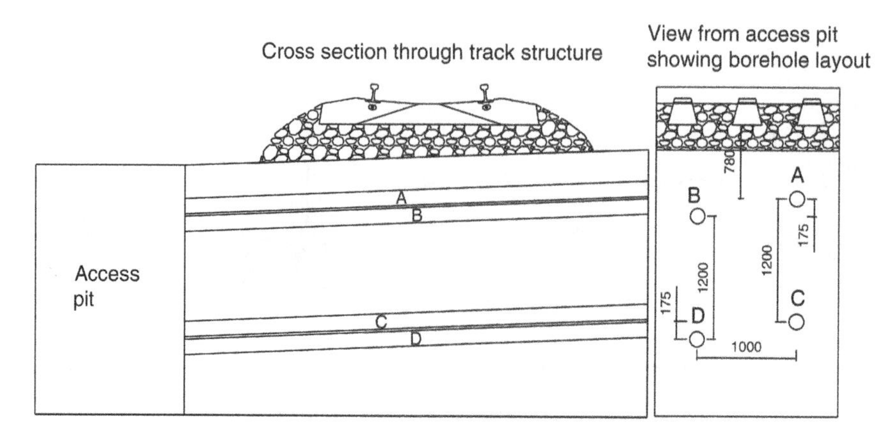

Figure 9. Schematic cross section and side elevation of the borehole layout for geophone measurements of vertical, lateral and longitudinal ground velocities on the COALlink line, South Africa; from Priest et al. (in review).

5 FIELD DATA

5.1 *Method and apparatus*

Bowness et al. (2007) describe the development of two independent methods of measuring the dynamic displacements of railway track as trains pass. One system combines relatively simple remote video monitoring and particle image velocimetry (PIV; Adrian, 1991) to assess displacements directly. The second uses low frequency geophones to measure ground velocities, which are then integrated (following appropriate filtering) to determine displacements. Remote video monitoring measures the absolute displacement from a remote datum, while the geophones give displacement relative to the unloaded ("at rest") position. Video monitoring requires site of a target and can therefore only be used to measure track or ground surface displacements. Geophones can by installed in the ground to measure sub-surface velocities, in any direction depending on the orientation of the geophone. The methods are described in full and validated by Bowness et al. (2007).

5.2 *Results from the COALlink line, South Africa*

The field investigation was carried out on a section of track 60 km south of Vryheid on the COALlink line, South Africa. This line runs from the coal fields in the Transvaal region, east of Johannesburg, to the port of Richards Bay. The section of track chosen was rehabilitated in 2004, during which extensive instrumentation was built into the track formation by Transnet Freight Rail (formerly Spoornet Engineering) as part of an investigation into formation design life (Gräbe et al., 2005). The field investigation described in this paper was undertaken on a section of track close to the Transnet Freight Rail instrumented section, so that comparisons could be made with their measurements.

The site is located in a cutting though a weathered hard mud-rock with closely spaced discontinuities, forming part of the Vreyheid formation. Rehabilitation of the track foundation involved excavating 800 mm of the old formation, compacting of the underlying subgrade to a defined specification and then placing of 800 mm of engineered fill in four equal layers (with a cross fall of 1:25) before laying the ballast and superstructure. A detailed description of the formation design and the material properties of the engineered fill are given by Gräbe *et al.* (2005). The minimum depth of ballast beneath the sleeper along the centreline of the track is about 300 mm.

To measure displacements of the ground below the track, four horizontal boreholes were installed within the ground at different depths. The boreholes ran transversely across the line of the track and were accessed via a pit at one side of the track, at the depths indicated in Figure 9. Each borehole was installed at an angle of 1:25 to the horizontal to correspond with the fall of the engineered fill layers. Flexible plastic liners were grouted into the boreholes to allow insertion and retrieval of a specially designed carrier, which could be pushed into any desired position within the borehole, on which three geophones were mounted orthogonally. During measurements the carrier was fixed within the borehole using a contact arm extended on to the wall of the borehole by means of pneumatic pistons attached to the carrier. Full details of the instrumentation are given by Priest et al. (in review). Figure 10 shows the vertical and longitudinal ground movements measured using the geophones at different depths below the track, during the passage of four bogies at successive ends of three adjacent coal wagons having a 20-tonne axle load.

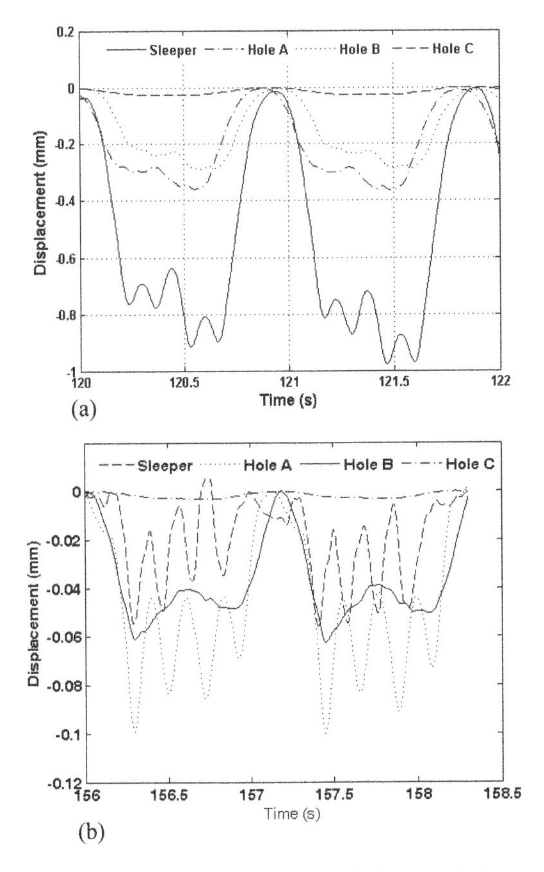

Figure 10. (a) Vertical and (b) longitudinal ground displacements measured using geophones at different depths on the COALlink line, South Africa, during the passage of four bogies at the ends of three successive 20-tonne axle load coal wagons; from Priest et al. (in review).

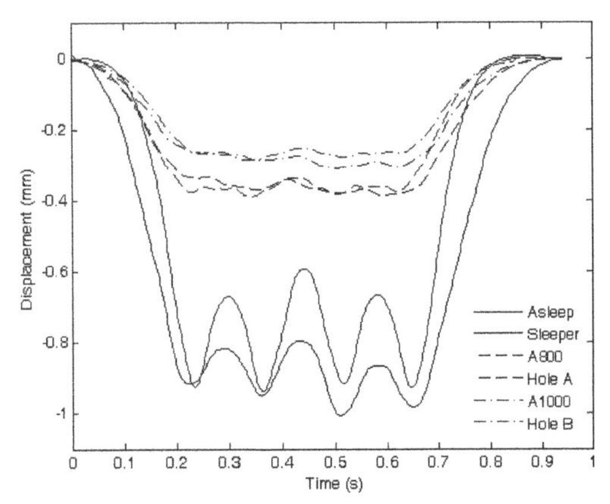

Figure 11. Comparison of measured vertical ground movements with those calculated using a 2D dynamic finite element analysis; from Priest et al. (in review).

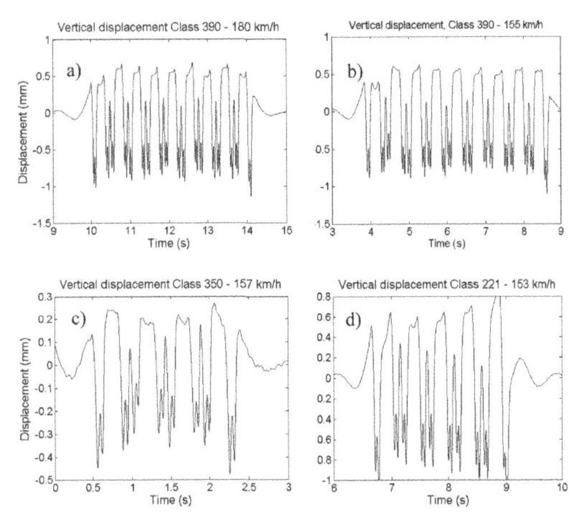

Figure 12. Vertical displacement of sleeper at the high rail end for a variety of train classes and speed; (a) Class 390 at 155 km/h, (b) Class 390 at 180 km/h, (c) Class 350 at 157 km/h, (d) Class 221 at 153 km/h; from Priest et al. (2008).

The vertical displacements measured at the sleeper were similar to those determined using video imaging and PIV, while the vertical displacements measured at depth were broadly consistent with those made using multi-depth deflectometers (a form of lvdt) installed by Transnet (Priest et al., in review). Figure 10a shows very clearly the attenuation of vertical displacement in terms of its frequency as well as magnitude, with the effects of individual axles being apparent at the sleeper; bogies at holes A and B (up to about 1 m below the sleeper soffit); but only bogie pairs at hole C (about 2 m below the sleeper soffit). Figure 10b suggests that the longitudinal displacements are dominated by the frequency of passing axles to a greater depth.

Figure 11 shows that the measured vertical displacements are in reasonably close agreement with those calculated by the 2D dynamic finite element analysis described by Yang et al. (in review), using appropriate soil parameters and loadings as discussed by Priest et al.

5.3 *Results from the West Coast Main Line, UK*

Data of vertical and lateral sleeper movements during the passage of trains of different axle loading at different speeds on canted, curved track on the UK West Coast Mail Line (WCML) at Weedon, Northamptonshire, are presented in a paper to this conference by Priest et al. (2008). These data are reproduced here in Figures 12 and 13. Figure 12 shows that the Class 350 train, which has the lowest axle load

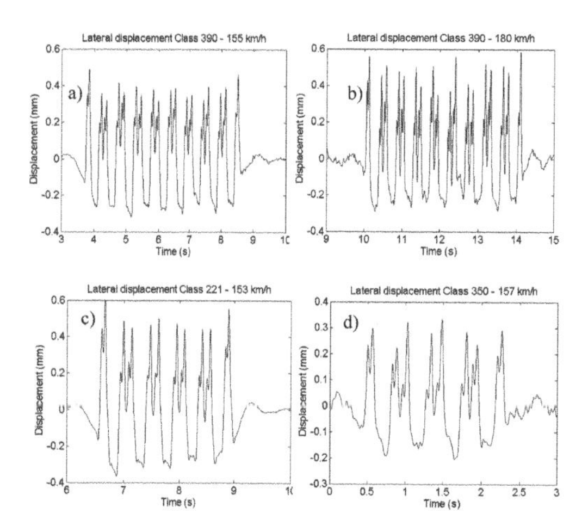

Figure 13. Lateral displacement of sleeper at the high rail end for a variety of train classes and speed; (a) Class 390 at 155 km/h, (b) Class 390 at 180 km/h, (c) Class 350 at 157 km/h, (d) Class 221 at 153 km/h; from Priest et al. (2008).

(12 tonne) and one of the lowest speeds, produces the least vertical displacement.

The Class 221 was travelling at the lowest speed, and gave similar vertical displacements to the Class 390 (a tilting Pendolino). However, the Class 221 also has the largest axle load (14 tonne), which offsets the reduction in speed. The difference in lateral displacements between the Class 350 and the Class 390 shown in Figure 13 is much greater, owing to the much more significant effect of speed (than mass) on the lateral (centrifugal) force. The Class 390 train travelling at180 km/h produces the largest lateral displacement of the track.

These data indicate quite graphically the effects of axle load and speed on vertical and lateral sleeper movements as trains pass. What is as yet not clear, however, is how the cumulative effect of many thousands of load cycles will vary with speed and axle load. This point is discussed by Priest et al. (2008) in relation to the introduction of Class 390 tilting Pendolino trains on the WCML, where they replaced trains formed of Class 87 electric locomotives hauling Mk3 trailers. The Pendolino train has a smaller axle load than a Class 87 locomotive, as a result of which the centrifugal force exerted by each when curving at maximum permitted speed is similar. However, in the case of a conventional train, this force is exerted only by the four axles of the Class 87 locomotive; the lateral force associated with the lighter trailing vehicles is much smaller. In the case of a Pendolino, the maximum lateral force is exerted by all 36 axles in the train, giving ten times the rate of cycling for high lateral load events. The impact of this, if any, on the long-term maintenance requirements of the track is as yet unknown.

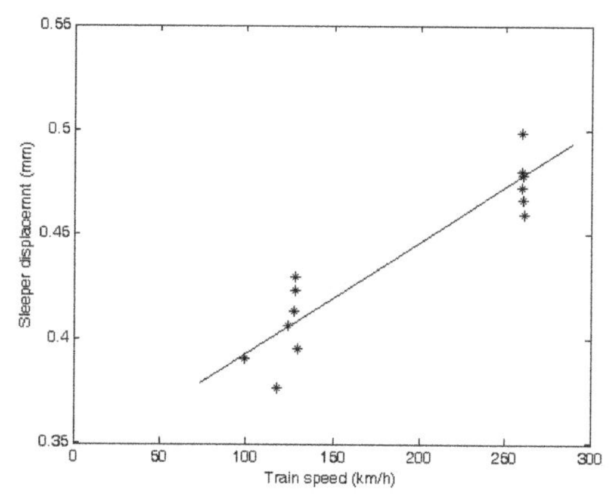

Figure 14. Vertical displacement vs train speed for a single sleeper on UK HS1; from Priest and Powrie (in preparation).

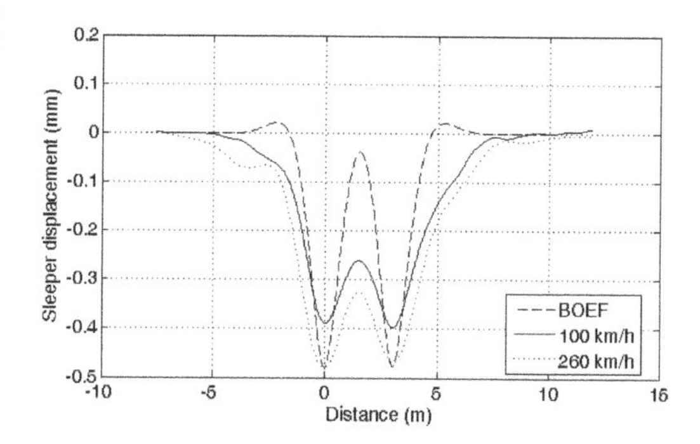

Figure 15. Measured and calculated sleeper displacements for a single bogie with a 15.36 tonne axle load (deflection is plotted against equivalent distance); from Priest and Powrie (in preparation).

5.4 *Results from High Speed 1 (formerly the Channel Tunnel Rail Link), UK*

Measurements of sleeper deflections at two different train speeds, about 120 km/hr and 260 km/hr, were obtained on a section of the UK High Speed 1 rail line (HS1, formerly the Channel Tunnel Rail Link) and reported by Priest and Powrie (in preparation). These data are shown in Figure 14, and show a ~17% increase in vertical displacement at the higher train speed. This is in contrast to current codes of practice: for example, UIC 719 makes no allowance for dynamic magnification effects, while the AREA design code would suggest an increase in deflection of 57% for this difference in speed.

A further indication of the effect of train speed is given in Figure 15, which compares the measured sleeper displacements at speeds of 100 km/hr and 260 km/hr with those calculated using a simple beam on an elastic foundation (BOEF) approach, plotted as a function of equivalent distance for a single bogie with a 15.36 tonne axle load.

These measurements clearly show that train speed, even if sub-critical, has an effect. The lack of agreement between the observations and current design guidance as to the nature of the effect, however, suggests that it is not well understood and that further investigation is required.

6 CONCLUSIONS

This Paper has set out to describe some recent and current research into the dynamic behaviour of railway track, with a particular emphasis on the stresses and displacements experienced by the underlying ground. It has described a finite element investigation into the stress paths that soil elements below a railway track are subjected to during train passage; considered the influence of principal stress rotation on the deformation and failure behaviour of these materials and the effects of train speed; and finally illustrated some of the aspects of behaviour under investigation with reference to field data from South Africa and the UK.

It has been shown, by means of a combination of analysis, laboratory element testing and field observations, that

(1) the relevant stress paths may involve principal stress rotation;
(2) principal stress rotation may have a significant effect on the behaviour of some soils; and
(3) track loading and deflections must be expected to increase with train speed, which may increase load cycling effects.

While these conclusions confirm what has been known or suspected for some time, the contribution of the work described in this paper is in beginning to quantify and define boundaries to the significance of these effects. Clearly, further work is required not least in

(1) identifying the range of soil types, and loading conditions, for which the effects of principal stress rotation may be especially significant;
(2) developing a better understanding of the effects of train speed on stress paths and ground movements, so that dynamic magnification effects may be more reliably allowed for.

REFERENCES

Adrian, R.J. 1991. Particle imaging techniques for experimental fluid mechanics. *Ann. Rev. Fluid Mech.* 23: 261–304

American Railway Engineering Association (AREA). 1996. *Manual for railway engineering, Vol. 1.* Washington D.C., U.S.A.: AREA.

Bowness, D., Lock, A.C., Powrie, W., Priest, J.A. and Richards, D.J 2007. Monitoring the dynamic displacements of railway track. *Proc. IMechE Part F: J. of Rail and Rapid Transport* 221 (F1): 13–22.

Brown, S. F. 1996. Soil mechanics in pavement engineering. *Geotechnique*, 46(3), 384–426.

Burrows, M.P.N., Bowness, D. and Ghataora, G.S. 2007. A comparison of railway track foundation design methods. *Proc. IMechE Part F: J. of Rail and Rapid Transport* 221 (F1): 1–12.

Chan, F.W.K. and Brown, S.F. 1994. Significance of principal stress rotation in pavements. In, *Proc. of the 13th Int. Conference on Soil Mechanics and Foundation Engineering, New Delhi, India,*: 1823–1826. Rotterdam: Balkema.

Gräbe, P.J. 2002. *Resilient and permanent deformation of railway foundations under principal stress rotation.* PhD dissertation, University of Southampton.

Gräbe, P.J., Clayton, C.R.I. and Shaw, F.J. 2005. Deformation measurement on a heavy haul track formation. *Proc. of the 8th Int Heavy Haul Conf.*, Rio de Janeiro, Brazil: 287–295.

Heath, D.L., Shenton, M.J., Sparrow, R.W. and Waters, J.M. 1972. Design of conventional rail track foundations. *Proc. Instn Civ. Engrs.* 51(February): 251–267.

Hibbitt, Karlsson and Sorensen Inc. 2005. *ABAQUS/Explicit user's manuals, version 6.6.* Pawtucket, RI: Hibbitt, Karlsson and Sorensen, Inc.

Lekarp, F., Isacsson, U. and Dawson, A.R. 2000a. State of the art, I: Resilient response of unbound aggregates. *J. Transportation Engineering* 126(1): 66–75.

Lekarp, F., Isacsson, U. and Dawson, A.R. 2000b. State of the art. II: Permanent strain response of unbound material. *J. Transportation Engineering* 126(1): 76–83.

Li, D. and Selig, E.T. 1998a. Method for railroad track foundation design. I: Development. *ASCE Journal of Geotechnical and Geoenvironmental Engineering* 124(4): 316–322.

Li, D. and Selig, E.T. 1998b. Method for railway track foundation design. II: Applications. *ASCE Journal of Geotechnical and Geoenvironmental Engineering* 124(4): 323–329.

Madshus, C. and Kaynia, A. M. 2000. High-speed railway lines on soft ground: dynamic behaviour at critical train speed. *Journal of Sound and Vibration*, 231(3), 689–701.

Powrie, W., Yang, L.A. and Clayton, C.R.I. 2007. Stress changes in the ground below ballasted railway track during train passage. *Proc. IMechE Part F: J. of Rail and Rapid Transport* 221 (F2): 247–261.

Priest, J.A., Powrie, W., Yang, L.A., Gräbe, P.J. and Clayton, C.R.I. in review. Dynamic ground deformations below a ballasted railway track. Submitted for possible publication to *Géotechnique*, 2007.

Priest, J.A., Le Pen, L., Powrie, W., Mak, P. and Burstow, M. 2008. Performance of canted ballasted track during curving of high speed trains. *Proc. Int Conf on Transportation geotechnics*, Nottingham. Rotterdam: Balkema.

Priest, J.A. and Powrie, W. Determination of dynamic track modulus from measurement of track velocity during train passage. In preparation.

Selig, E.T. and Waters, J.M. 1994. *Track geotechnology and substructure management.* London: Thomas Telford.

Shahu, J.T., Yudhir and Kameswara Rao, N.S.V. 2000. A rational approach for design of railroad track foundation. *Soils and Foundations* 40(6): 1–10.

UIC (1994). *UIC Code 719R. Earthworks and track-bed layers for railway lines*. Paris, France: International Union of Railways.

Woldringh, R. F. and New, B. M. (1999). "Embankment design for high speed trains on soft soils." *Proceedings of 12th European Soil Mechanics and Geotechnical Engineering Conference on Geotechnical Engineering for Transportation Infrastructure,* Amsterdam, The Netherlands, 1703–1712.

Yang, L.A., Powrie, W. and Priest, J.A. (in review). Dynamic stress analysis of a ballasted railway track bed during train passage. Submitted for possible publication to the *ASCE Journal of Geotechnical and Geoenvironmental Engineering*, 2007.

Advances in Transportation Geotechnics – Ellis, Yu, McDowell, Dawson & Thom (eds)
© 2008 Taylor & Francis Group, London, ISBN 978-0-415-47590-7

The effect of soil viscosity on the behaviour of reinforced embankments

R.K. Rowe & C. Taechakumthorn
GeoEngineering Centre at Queen's-RMC, Department of Civil Engineering, Queen's University, Canada

ABSTRACT: For conventional soils, a slower construction rate leads to higher embankment stability. In contrast, rate-sensitive soils show higher short-term embankment stability with a faster construction rate; however this may also give rise to long-term problems. For example, rate-sensitive soils exhibit significant creep induced deformations and this can contribute to the development of excessive long-term reinforcement strains. Delayed creep induced excess pore water pressures may also give rise to minimum stability some time after the end of construction and post-construction unreinforced embankment failures have been reported in the literature as a result of soil viscosity. Prefabricated vertical drains (PVDs) have been shown to substantially reduce the effect of creep induced excess pore pressures and increase the rate of strength gain due to consolidation of the soil. It is shown herein that PVDs may also enhance the effectiveness of geosynthetic reinforcement at minimizing differential settlement and lateral deformations of the foundation soils. This paper highlights the findings from earlier work by the authors by presenting a numerical study of the behaviour of geosynthetic reinforced embankments constructed over rate-sensitive soils. The effect of soil viscosity is addressed using a elasto-viscoplastic constitutive model. The effect of various factors such as construction rate, reinforcement stiffness, and PVD spacing is investigated, both during and following construction. The benefit of using PVDs combined with the use of geosynthetic reinforcement to improve the performance of reinforced embankments is highlighted.

1 INTRODUCTION

Stability and the time required for consolidation are two key considerations with respect to the construction of embankments over soft clay deposits. Several techniques have been developed to improve embankment stability and accelerate consolidation. These techniques include the use of geosynthetic basal reinforcement to reduce the outward shear force on the foundation soil (e.g. Rowe, 1984; Fowler and Koerner, 1987; Jewell, 1987; Rowe and Soderman, 1987; Rowe and Li, 1999; Bergado et al., 2002; Shen et al., 2005; Rowe and Taechakumthorn, 2007a; Kelln et al., 2007 and many others) and the use of prefabricated vertical drains (PVDs) to accelerate pore pressure dissipation in the clay (e.g. Crawford et al., 1992; Bergado et al., 1997; Chai and Miura, 1999; Li and Rowe, 1999; Indraratna and Redana, 2000; Bo, 2004; Chai et al., 2004; Rujikiatkamjorn et al., 2007 amongst many). For the case of a conventional soft clay deposit, the combined use of geosynthetic reinforcement with PVDs has been studied by several researchers and been shown to allow the rapid construction of higher embankments than would have been possible with the use of either method alone (Li and Rowe, 2001).

There have been many studies of the time-dependent behaviour of rate-sensitive soils (e.g. Lo and Morin, 1972; Vaid and Campanella, 1997; Vaid et al., 1979; Graham et al., 1983; Kabbaj et al., 1988; Leroueil, 1988). The performance of reinforced and unreinforced embankments constructed over rate-sensitive soil has also been investigated by both field studies (Rowe et al., 1995, 1996) and numerical analysis (Hinchberger and Rowe, 1998; Rowe and Hinchberger, 1998; Rowe and Li, 2002; Rowe and Taechakumthorn, 2007a). A study of a test embankment on a rate-sensitive soil at Sackville, New Brunswick (Rowe et al., 1996) showed that in order to accurately capture the time-dependent behaviour of the test embankment, the constitutive model need to consider the effect of soil viscosity. Rowe and Hinchberger (1998) proposed an elasto-viscoplastic constitutive model and demonstrated that the model can adequately describe the field behaviour of the test embankment.

Rowe and Li (2002) showed that the long-term stability of a reinforced embankment on rate-sensitive soil decreases after the end of construction due to the generation of delayed creep induced pore pressures. Rowe and Taechakumthorn (2007b) demonstrated the significance of PVDs in reducing the effect of delayed creep induced pore pressures and improving the stability of reinforced embankments over rate-sensitive soil. However, there has been very little research into the effects of the combined use of geosynthetic reinforcement and PVDs on the performance of embankments on rate-sensitive soil.

This paper uses the results of finite element analyses to illustrate a number of important issues related to the behaviour of the reinforced embankments on rate-sensitive soil deposits, both during and following the construction. In doing so it builds on a number of recent papers by the authors (Rowe and Taechakumthorn 2007a,b, 2008 and Taechakumthorn and Rowe 2007). The effect of the presence of PVDs combined with the use of geosynthetic reinforcement will be discussed. The interacting effects of soil viscosity, reinforcement stiffness, construction rate and PVD spacing will also be examined with respect to the time-dependent responses of the excess pore pressures and reinforcement strains. Finally, the effect of the construction rate on the surface settlement, heave and lateral deformation of the foundation will be highlighted.

2 FINITE ELEMENT MODELING AND MODEL PARAMETERS

2.1 *Mesh discretisation*

To simulate the construction of the reinforced embankment, a version of finite element program AFENA (Carter and Balaam, 1990), as adapted by Rowe and Hinchberger (1998), was adopted. A series of small strain finite element analyses were conducted. The effect of PVDs was modeled using drainage elements (Russell, 1990) as implemented by Li and Rowe (2001).

The results presented herein were obtained for embankments with the 2H:1V side slopes overlaying a 15 m soft, rate-sensitive, clay deposit above a rigid and permeable layer. A typical finite element mesh (Fig. 1) comprised 1815 six-noded triangle elements discretizing the embankment and the subsoil as well as a series of bar elements to discretize the geosynthetic reinforcement and two-noded rigid-plastic interface joint elements (Rowe and Soderman, 1985) to model fill/reinforcement and fill/foundation interfaces. PVDs were modeled using two-noded drainage elements (Li and Rowe, 2001).

The embankment centerline and far field boundary, located 100 m away from the centerline, were taken to be smooth-rigid boundaries. The bottom boundary was assumed to be rough-rigid with a free drainage

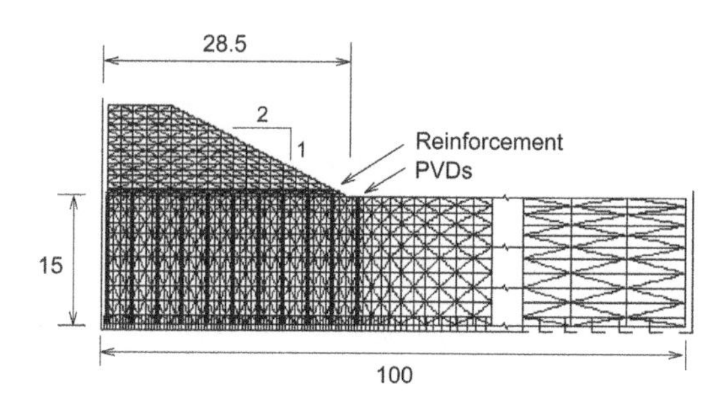

Figure 1. Finite element mesh discretisation.

boundary. The construction of reinforced embankments was simulated by gradually turning on the gravity in the fill material in 0.75 thick lifts at a rate corresponding to the construction rate of the embankment. The PVDs were fully penetrating in a square pattern. The effect of the smear zone was addressed by assuming that the ratio of equivalent radian of smear zone and vertical drains, s, was equal to 4. Details regarding modelling of PVDs and smear zone are provided in a subsequent sub-section.

2.2 *Constitutive model for rate-sensitive soils*

The elasto-viscoplastic constitutive model utilized herein (Rowe and Hinchberger, 1998), is fully coupled with Biot consolidation theory (Biot, 1941). This model also incorporates the Perzyna's theory of overstress viscoplasticity (Perzyan, 1963). The model is based on an elliptical yield cap model (Chen and Mizuno, 1990), a Drucker-Prager failure envelope and concepts drawn from critical state soil mechanics (Roscoe and Schofield, 1963). The constitutive model and computer program adopted in this study were successfully verified with the results from test embankments constructed over soft rate-sensitive soil in Sackville (Rowe and Hinchberger, 1998) and Gloucester (Hinchberger and Rowe, 1998). The main features of the model are summarized below. Additional details regarding the model were provided by Hinchberger (1996) and Rowe and Hinchberger (1998).

According to Perzyna's overstress theory of viscoplasticity, the governing equation was expressed in terms of strain-rate tensor:

$$\dot{\varepsilon}_{ij} = \frac{\dot{S}_{ij}}{2G} + \frac{1}{3K}\dot{\sigma}_{ii} + \gamma^{vp}\left\langle \phi(F) \right\rangle \frac{\partial f}{\partial \sigma_{ij}} \quad (1)$$

where: S_{ij} is deviatoric stress; G is shear modulus; σ_{ij} is sum of principle stress stresses; K is bulk modulus and γ^{vp} is viscoplastic fluidity parameter. The terms $\phi(F)$ and f are flow and an elliptical plastic potential functions in $\sigma'_m - \sqrt{2J_2}$ stress space (where σ'_m is mean effective stress and J_2 is second invariant of deviatoric stress tensor), respectively which can be expressed as:

$$\phi(F) = \left(\frac{\sigma'^{(s)}_{my} + \sigma'^{(d)}_{os}}{\sigma'^{(s)}_{my}}\right)^n - 1 \quad (2)$$

$$f = (\sigma'_m - l)^2 + 2J_2R^2 - (\sigma'_{my} - l)^2 = 0 \quad (3)$$

where: $\sigma'^{(s)}_{my}$ or σ'_{my} is the intercept of the elliptical cap yield surface with the σ'_m axis ; $\sigma'^{(d)}_{os}$ is overstress; n is strain-rate exponent; l is mean effective stress corresponding to the center of the ellipse and R is ratio between major and minor of the ellipse.

The elastic bulk modulus, K, and shear modulus, G, are function of the mean effective stress given by:

$$K = \frac{1+e}{\kappa}\sigma'_m \quad (4)$$

$$G = \frac{3(1-2\nu')K}{2(1+\nu')} \quad (5)$$

where: e is void ratio; κ is recompression index and ν is Poisson's ratio.

2.3 *Drainage elements*

Previous research has shown that the 3-D system of a soil foundation with PVDs can be reasonable modeled using a 2-D plane strain approximation. Hird et al. (1992) demonstrated that the average degree of consolidation on the horizontal plane, U_h, at depth z and time t under instantaneous loading obtained from an axisymmetric analysis could be matched to that obtained in a plane strain analysis by geometric matching, permeability matching or a mix of these two techniques. Li and Rowe (2001) adopted the permeability matching scheme by modifying the horizontal hydraulic conductivity of the soil and the equivalent discharge capacity of vertical drains under plane strain conditions. The same matching scheme was also adopted herein, viz:

$$k_{pl} = \frac{2k_{ax}}{3\left[\ln\left(\frac{n}{s}\right) + \left(\frac{k_{ax}}{k_s}\right)\ln(s) - \frac{3}{4}\right]} \quad (6)$$

$$Q_w = \left(\frac{2}{\pi R}\right)q_w \quad (7)$$

$$n = \frac{R}{r_w}, \quad s = \frac{r_s}{r_w} \quad and \quad q_w = \pi k_w r_w^2 \quad (8)$$

where: k_{pl}, k_{ax}, k_s and k_w are the hydraulic conductivities of: the bulk soil in the horizontal direction for plane strain conditions, the bulk soil under axial symmetric conditions, the soil in the smear zone (assumed to be isotropic and the same as the vertical hydraulic conductivity of soil), and the vertical drain, respectively. r_w, r_s and R are the equivalent radius of the vertical drain, smear zone and influence zone, respectively. Q_w and q_w are the equivalent discharge capacities for the plane strain and axisymmetric unit cells, respectively. Details of the drainage elements have been provided by Li and Rowe (2001).

Table 1. Details of foundation soil properties.

Soil Parameter	Soil RS1	Soil RS2
Failure envelope, $M_{N/C}$ (ϕ')	0.96 (29°)	0.96 (29°)
Cohesion intercept, c_k (kPa)	0	0
Failure envelope, $M_{O/C}$	0.75	0.75
Aspect ratio, R	1.25	1.25
Compression index, λ	0.16	0.16
Recompression index, κ	0.034	0.034
Coefficient of at rest earth pressure, K'_o	0.75	0.75
Poisson's ratio, υ	0.3	0.3
Reference hydraulic conductivity, k_{vo} (m/s)	2×10^{-9}	2×10^{-9}
Hydraulic conductivity ratio, k_h/k_v	4	4
Unit weight, γ (kN/m^3)	17	17
Initial void ratio, e_o	1.50	1.50
Viscoplastic fluidity, γ^{vp} (1/hour)	2.0×10^{-5}	1.0×10^{-7}
Strain rate exponent, n	20	30

3 MODEL PARAMETERS

3.1 *Foundation soils properties*

As in a number of previous papers by the authors (Rowe and Li, 2002; Rowe and Taechakumthorn, 2007a,b), the two soft rate-sensitive soils examined in this study, denoted as RS1 and RS2, have parameters as listed in Table 1 together with preconsolidation and initial vertical stress profiles as shown in Figure 2. The constitutive parameters used for soil RS1 are similar to those for the subsoil below the Sackville test embankment (Rowe and Hinchberger, 1998). Soil RS2 was assumed to have the same properties as Soil RS1 except for the rate-sensitive parameters (the viscoplasticity and the strain-rate exponent).

The hydraulic conductivity of soft rate-sensitive clay was taken to be a function of void ratio as given by:

$$k_v = k_{vo}\exp\left(\frac{e-e_o}{C_k}\right) \quad (9)$$

where: C_k is hydraulic conductivity change index ($C_k = 0.2$). The hydraulic conductivity was considered

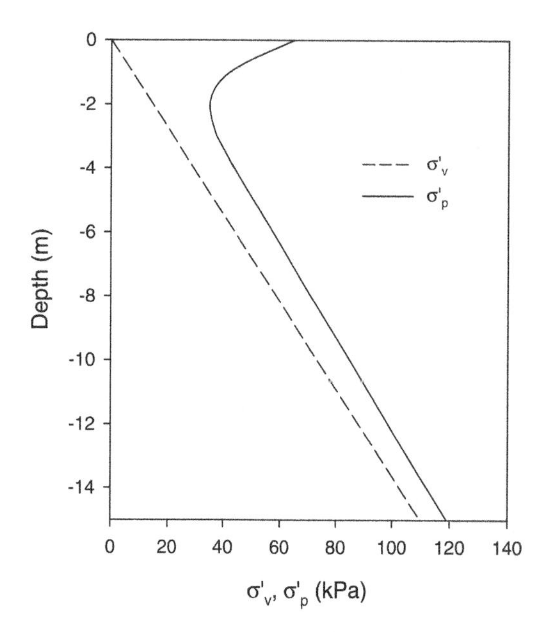

Figure 2. Preconsolidation pressure and initial vertical stress profiles (based on Rowe and Li, 2002).

to be anisotropic with the ratio of horizontal to vertical hydraulic conductivity (k_h/k_v) of 4.

3.2 *Embankment fill properties and rate of construction*

The embankment fill was assumed to be a purely frictional granular soil with a friction angle of 37°, a dilation angle of 6°, and a unit weight of 20 kN/m^3. The nonlinear elastic behaviour of the fill was model using Janbu's equation (1963) as:

$$\frac{E}{P_a} = K\left(\frac{\sigma_3}{P_a}\right)^m \qquad (10)$$

where: E is Young's modulus; P_a is the atmospheric pressure; σ_3 is the minor principle stress and K and m are material constants selected to be 300 and 0.5, respectively.

The construction rates of the cases examined in this study were 2, 4, 6, 8 and 10 m/month.

3.3 *Interface parameters and stiffness of reinforcements*

The interactions between fill/reinforcement and fill/foundation were assumed to have frictional angle of 37°. The axial tensile stiffness of the geosynthetic reinforcement was varied with values of 0 (no reinforcement), 500, 750, 1000, 2000, 4000 and 8000 kN/m being considered.

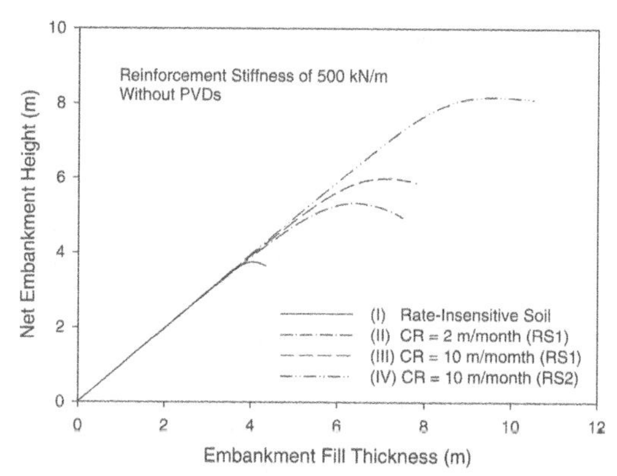

Figure 3. The effect of soil viscosity and construction rate on the short-term stability of reinforced embankments.

3.4 *PVDs properties*

PVDs with a typical rectangular cross section of 100 mm × 4 mm (Holtz, 1987) were considered. The equivalent diameter (d_w) of a circular drain is equal to 52 mm based on Rixner et al. (1986): $d_w = (b+t)/2$, where: b and t are width and thickness of PVDs, respectively. The PVD spacings (S) of 1, 2 and 3 m examined in this paper are in the typical range used in practice (Holtz, 1987). The discharge capacity of the PVDs was taken to be a 100 m^3/year (Rixner et al., 1986).

4 RESULTS AND DISCUSSIONS

4.1 *Effects on short-term stability of the reinforced embankment – No PVDs*

A series of finite element analyses were conducted to simulate the construction of reinforced embankments on different rate-sensitive soils. The effects of different reinforcement stiffness, construction rate, and PVD spacing were investigated.

The short-term failure height of a reinforced embankment can be assessed in terms of the failure height which is defined as the height of the embankment fill at which any further attempt to add fill material will not result in an increase in the net embankment height (i.e. the height above the original ground surface). The failure height can be obtained by plotting the relationship between the net embankment height (fill thickness minus settlement) against the fill thickness as shown in Figure 3.

The results in Figure 3 show the effect of soil viscosity and construction rate on the stability of the reinforced embankment examined. For Case I which represents a rate-insensitive (non-viscous) soil, a conventional elasto-plastic constitutive model was employed having the same soil properties as the other

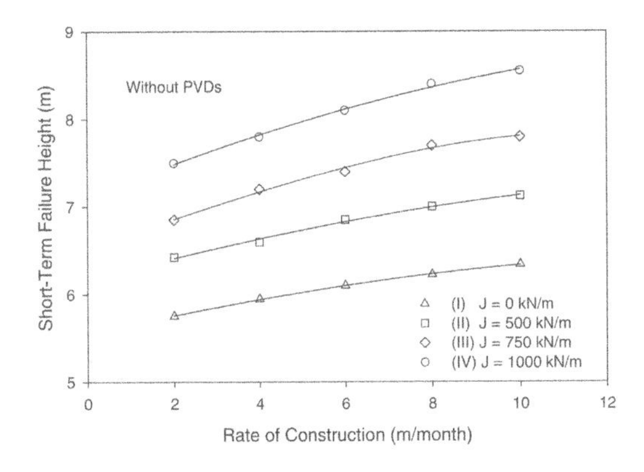

Figure 4. The effect of reinforcement stiffness and construction rate on the short-term embankment stability (Soil RS1).

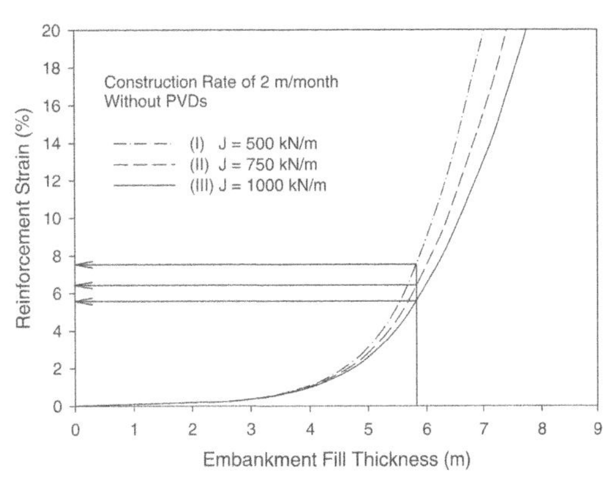

Figure 5. Short-term reinforcement strain versus fill thickness (Soil RS1).

soils except for soil viscosity. In all cases there was a linear initial response followed by non-linear response until the calculated failure heights of 4.05, 6.45, 7.20, and 9.45 m were reached for Cases I, II, III and IV, respectively. The differences in short-term failure height between Case I, III and IV illustrate the effect of soil viscosity. The short-term strength of the viscous soil allowed the soil to carry extra load with the more viscous soil (RS2) exhibiting the higher short-term undrained shear strength and hence a greater short-term failure height than the less viscous soil (RS1, Case III) or inviscous soil (Case I). Figure 3 also demonstrate the effect of construction rate with the faster construction rate of 10 m/month (Case III) resulting in a 1.75 m higher failure height (7.2 m) than obtained at the slower rate of 2 m/month (6.45 m-Case II). However, the shear strength decreased with time as will be discussed later.

The main function of geosynthetic reinforcement is to reduce outward shear stress on the foundation soil, and if stiff enough, induce inward shear stress on the foundation soil. In so doing, the reinforcement increases the stability of the embankment. Figure 4 demonstrates the role of reinforcement in improving the short-term stability of the embankment. A series of unreinforced and reinforced embankments were numerically constructed at different construction rates until failure. The stiffer reinforcement resulted in higher failure height. The effect of construction rate on the rate-sensitive soil was also confirmed as the faster construction rate led to higher short-term embankment stability.

Case histories (Rowe et al., 1995, Rowe and Hinchberger, 1998 and Bergado et al., 2002) indicate that a geosynthetic reinforced embankment can be constructed to a height well beyond the point where there is significant plastic failure in the foundation soil and hence the point where an unreinforced embankment would fail.

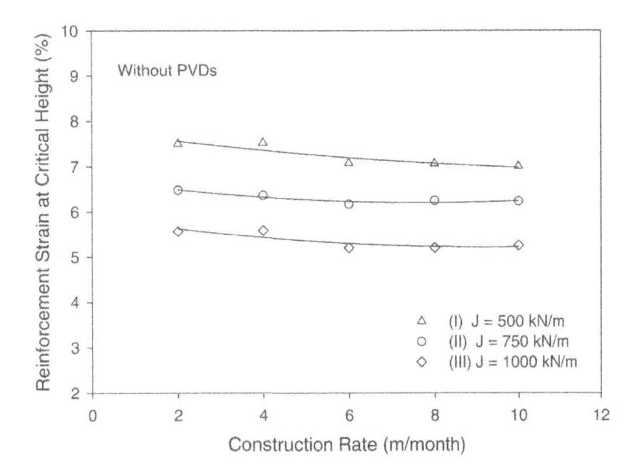

Figure 6. The effect of reinforcement stiffness and construction rate on the short-term reinforcement strains (Soil RS1).

Bergado et al. (2002) defined the term critical height as the failure height of the unreinforced embankment on the same soil foundation. Figure 5 shows the relationship between embankment fill thickness and corresponding reinforcement strain for the case of 2 m/month construction rate. The critical height of 5.75 m for the case in Figure 5 was obtained from the short-term failure height of the unreinforced embankment in Figure 4. The compatible reinforcement strains at the critical height were 7.5, 6.5 and 5.5% for reinforcement stiffness of 500, 750 and 1000 kN/m, respectively. The reinforcement strains calculated for the different reinforcement stiffnesses at the critical height were plotted against the corresponding construction rate in Figure 6. In these cases the effect of construction rate is small as the reinforcement strains are about 7.5, 6.5, and 5.5% for reinforcement stiffness of 500, 750 and 1000 kN/m, respectively regardless of applied construction rate.

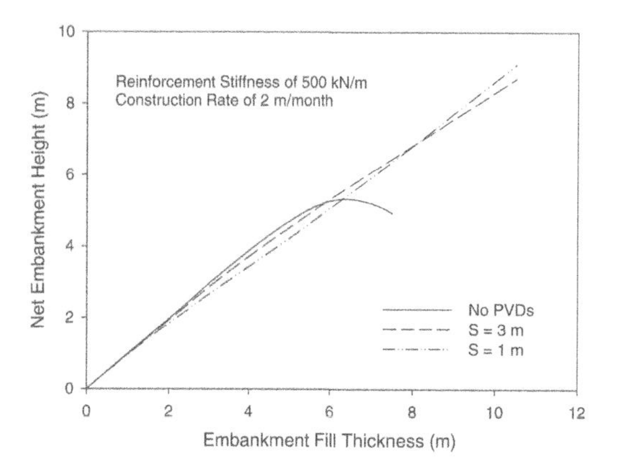

Figure 7. The effect of PVDs on the short-term stability of the reinforced embankments (Soil RS1).

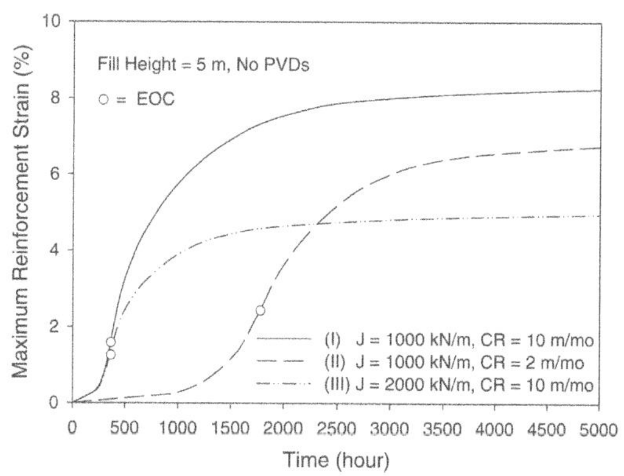

Figure 8. Effect of reinforcement stiffness and construction rate on the long-term reinforcement strains (Soil RS1).

4.2 *Effects of PVDs on short-term stability*

PVDs have been being used to accelerate the construction of high embankments on soft ground by accelerating pore pressure dissipation and hence the gain in shear strength of the foundation with time. As shown in Figure 7, the short-term stability of the reinforced embankments was significantly improved when PVDs were installed. The results showed no failure even when the embankment fill thickness was increased up to 10.5 m. During the initial stage of construction, the case of smaller PVD spacing (S = 1 m) showed larger settlement because the subsoil had a higher degree of partial consolidation compared to the case of larger PVD spacing (S = 3 m). However, the higher degree of partial consolidation resulted in less overstress remaining in the soil and accordingly, less viscoplastic deformation was developed. As a result, the case of smaller PVD spacing showed smaller settlement after the fill thickness exceeded 8.5 m.

4.3 *Effects on the long-term reinforcement strains*

As noted above, even though the viscoplastic characteristics of a rate-sensitive soil allowed the construction of higher embankments in the short-term, the overstress and creep deformation associated with the construction can be expected to cause long-term deformation and, potentially, stability problems. To investigate this problem, a conventional embankment design was performed using limit equilibrium analysis. The undrained shear strength profile of the soft clay deposit used in the design calculations corresponded to the plane strain shear strength established form numerical tests at a typical recommended strain rate of 0.5–1.0%/hour (Germaine and Ladd, 1988). The analysis shows that using the reinforcement force of 100 kN/m (i.e. $J = 2000$ kN/m at 5% strain) a 5.25 m height reinforced embankment should be stable with safety factor of 1.3. However, from the finite element modeling, the 5.25 m height reinforced embankment resulted in 6% long-term reinforcement strain, which implied that the limit equilibrium analysis overestimated the stability of the reinforced embankment. For this particular soil and a reinforcement stiffness of 2000 kN/m, only a 5 m high reinforced embankment could be constructed if the long-term reinforcement strain was to be limited to an allowable design strain of 5%.

Figure 8 shows the effect construction rate and reinforcement stiffness on the long-term mobilized reinforcement strain. For a rate-sensitive soil, a faster construction rate led to higher short-term shear strength of the soil and hence smaller end of construction (EOC) strains in the reinforcement. In Case I (faster construction rate, 10 m/month), the foundation soil carried most of the load at the end of construction and so the EOC reinforcement strain was only 1.6%. In contrast, at a lower construction rate (Case II, 2 m/month) the maximum EOC reinforcement strain was 2.6%. However the amount of overstress in the soil was reduced at this slower rate and this reduced the subsequent creep deformation as well as the delayed excess pore water pressure (as discussed later) and ultimately resulted in smaller long-term reinforcement stains (6.9% for Case II) than were obtained for the higher construction rate (8.3 % for Case I).

The results from Case I and III demonstrated the effect of reinforcement stiffness for a given construction rate (10 m/month) and as expected, the stiffer reinforcement (Case III) gave smaller strain both at the end of construction and long-term condition (1.3% and 4.9%, respectively) compare to those from Case I (1.6% and 8.3%).

PVDs greatly enhanced the rate of excess pore pressure dissipation and reduced the amount of overstress in the soil. Consequently, the effects of viscoplastic response such as long-term creep deformation, stress

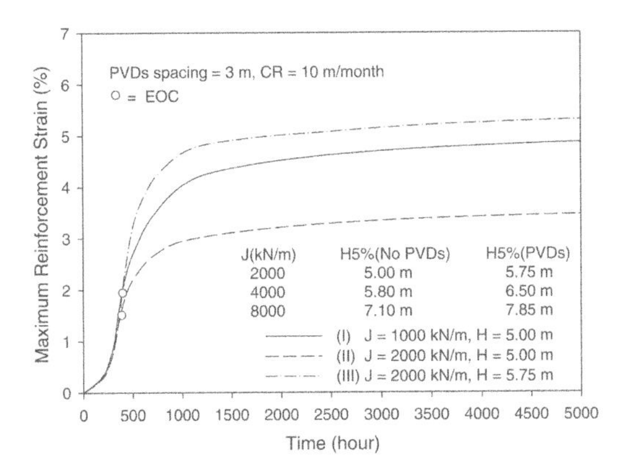

Figure 9. The effect of PVDs on long-term reinforcement strains (Soil RS1).

relaxation and delayed creep induced pore pressure are minimized. For the case of 5 m a high embankment with the reinforcement stiffness of 1000 kN/m constructed as slow as 2 m/month, the reinforcement strain still reached 6.9% for long-term conditions and this exceeds the typical design limit of 5% strain (Case II in Figure 8). In contrast when PVDs were installed with 3 m spacing (Case I in Figure 9), the faster construction rate of 10 m/month could be employed and the long-term reinforcement was limited to 4.8%. With stiffer reinforcement ($J = 2000$ kN/m), the PVDs reduced the long-term reinforcement strain from 5.0% to only 3.8% (Case III in Figure 8 versus Case II in Figure 9). If the reinforcement stiffness of 2000 kN/m was combined with PVDs, the reinforced embankment height could be increased up to 5.75 m without the long-term reinforcement strain exceeding 5.0% (Case III in Figure 9). For a 5% long-term strain and a PVD spacing of 3 m, reinforced embankments could be constructed up to 6.5 m and 7.85 m for reinforcement stiffness of 4000 and 8000 kN/m, respectively (see insert to Figure 9).

4.4 *Effects on the excess pore water pressure*

During and following embankment construction on a rate-sensitive soil, two processes occur simultaneously: (1) the generation of creep induced pore water pressures, and (2) the dissipation of excess pore pressures due to consolidation. Figures 10a and 10b show contours of the increase in excess pore pressure between the end of construction and 1 month after the construction of a 5 m high reinforced embankment with reinforcement stiffness of 1000 and 2000 kN/m, respectively. The shear induced generation of pore pressure was evident since the contours of increase in pore pressure formed along the potential failure zone even though there was pore pressure dissipation. The effective stress and shear strength was reduced after the end of construction thus the critical period with

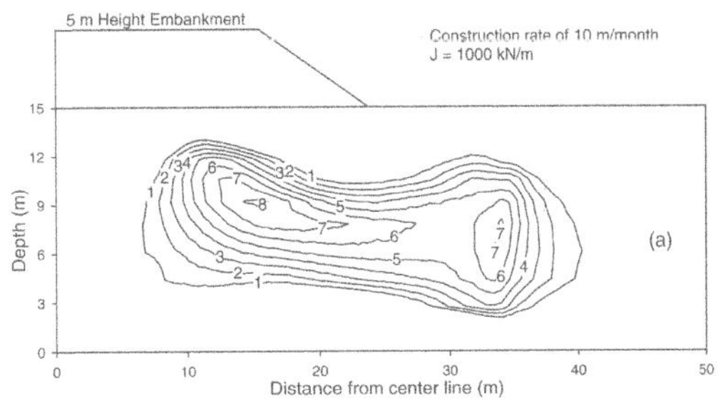

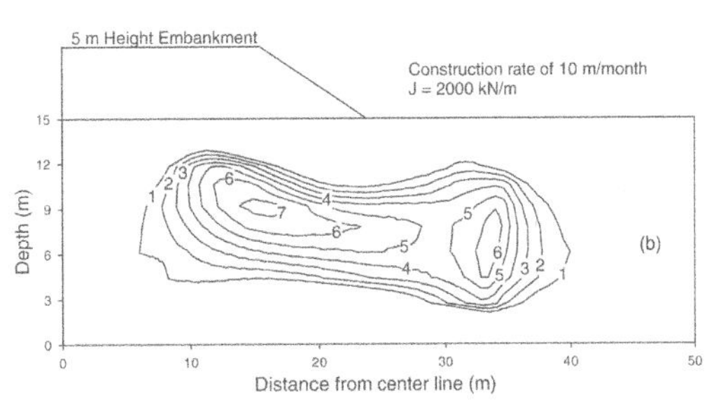

Figure 10. Increase in excess pore water pressure between end of construction and 1 month after the end of construction (Soil RS1).

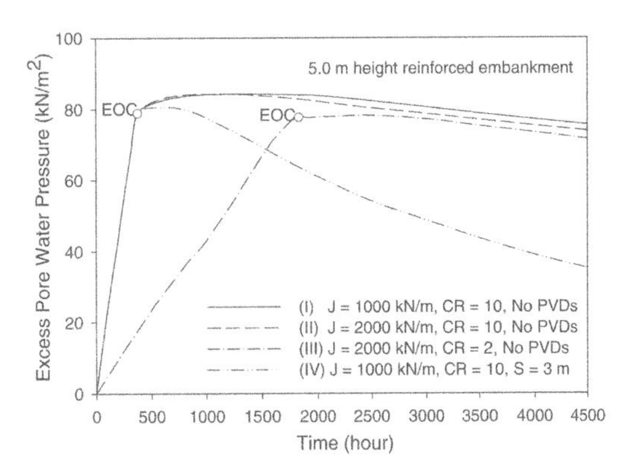

Figure 11. The effect of reinforcement stiffness and PVDs on the excess pore pressure dissipation (Soil RS1).

respect to the embankment stability (minimum factor of safety) occurred following the end of construction. Reinforcement provided a stabilizing force, as discussed earlier, and the stiffer reinforcement gave more restraint and hence a slight decrease in the creep induced pore pressure (compare Figures 10a and 10b).

The excess pore pressures at 6 m beneath the embankment crest, where the maximum increase in pore pressure occurred, are shown in Figure 11 for four cases. The excess pore pressures at the end of construction were approximately 80 kPa for all cases examined and kept increasing even when no more

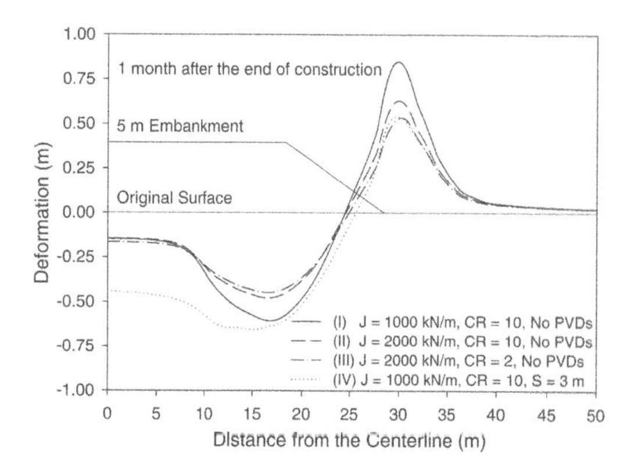

Figure 12. Effect of reinforcement stiffness and PVDs on the differential settlement and heave of the foundation (Soil RS1).

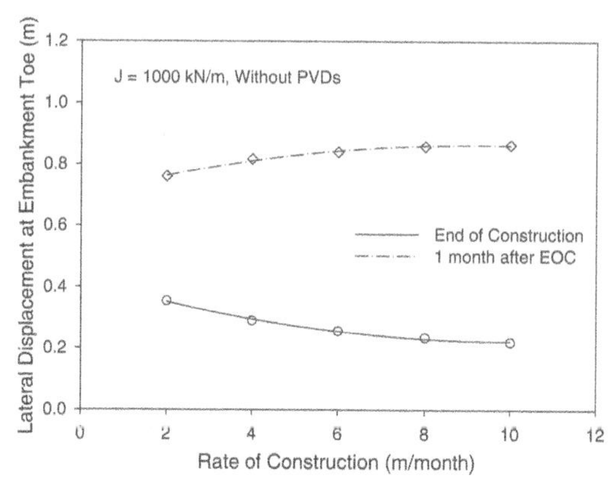

Figure 13. Effect of construction rate on the lateral displacement at the embankment toe (Soil RS1).

fill was added. The stiffer the reinforcement, resulted in slightly smaller creep induced pore pressures with time and hence slightly faster pore pressure dissipation (compare Cases I and II in Figure 11). Construction rate also affected the rate of excess pore pressure dissipation. A slower construction rate allowed higher degree of partial consolidation during the construction and this reduced the amount of overstress and long-term creep in the foundation.

The installation of PVDs significantly reduced the effect of delayed excess pore pressure build up in this rate-sensitive soil. Figure 11 also demonstrates that with the use of PVDs, the excess pore pressure promptly reduced following the end of construction (Cases IV).

4.5 *Effects on differential settlement and lateral deformation of the foundation*

Reinforcement can significantly reduce the differential settlement and toe heave for embankments constructed on rate-sensitive soil. For a fill thickness of 5 m on soil RS1, Figure 12 shows the ground surface profile at 1 month following the construction for several reinforced embankment scenarios.

The results in Case I and II illustrated the effect of reinforcement stiffness. The stiffer (J = 2000 kN/m) reinforcement (Case II) limited heave to 0.63 m and reduced the differential settlement between centerline and embankment crest to 0.34 m compared to 0.85 m and 0.47 m, respectively, for the less stiff reinforcement (J = 1000 kN/m, Case I). The effect of the construction rate on the differential settlement and heave of the foundation was insignificant.

The presence of PVDs reduced the differential settlement of the foundation for reinforcement stiffens of 1000 kN/m from 0.47 m (Case I) to 0.15 m (Case IV). The PVDs reduced the heave from 0.85 m (Case I) to 0.54 m (Case IV) and hence gave a very similar toe heave to that obtained with much stiffer (J = 2000 kN/m) reinforcement (Case II).

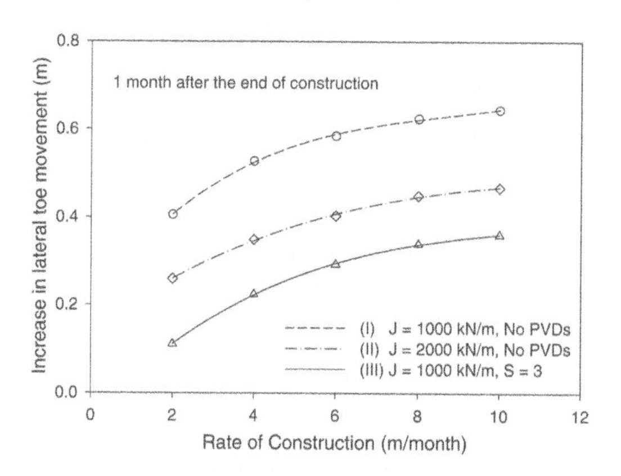

Figure 14. Effect of rate of construction on the increase in lateral displacement of the embankment toe between the end of construction and 1 month after the end of construction (Soil RS1).

Figure 13 shows the effect of construction rate on the lateral movement at the embankment toe. As previously discussed, faster construction rate led to higher short-term shear strength of the foundation and that resulted in smaller end of construction lateral movement at the toe (solid line in Figure 13. However, faster construction also led to higher overstress in the soil and this generated larger creep deformation that resulted in higher lateral movements 1 month after construction than was observed for slower construction rates (dashed line in Figure 13).

The increase in lateral toe movement, between the end of construction and 1 month later is shown in Figure 14. The higher construction rate resulted in a substantially greater post construction increase in lateral movement for the reasons described above. Stiffer reinforcement had a beneficial effect on reducing the lateral deformation of the embankment (compare Cases I and II). The combined use of geosynthetic reinforcement (J = 1000 kN/m) and PVDs (Case III) gave the lowest increase in post construction lateral movement of the three cases examined.

5 CONCLUSIONS

This paper has examined the time-dependent behaviour of reinforced embankments constructed over rate-sensitive soil. The results indicated that the viscoplastic characteristic of rate-sensitive soil had a significant effect on the performances of reinforced embankments, especially after the end of construction. The behaviour of reinforced embankments on rate-sensitive soil, within the range of cases and parameters considered in this paper, can be summarized as follow.

For the rate-sensitive soil, the faster the rate of construction, the higher is the short-term stability of the embankment. However, the consequent large overstress generated in the soil led to larger viscoplastic deformations and could result in failure after the end of construction. Geosynthetics reinforcement reduced the outward shear force on the foundation and, in so doing, minimized the long-term creep deformation and improved the short-term stability of the embankment. The short-term reinforcement strain at the critical height decreased with increasing reinforcement stiffness. PVDs greatly improved the short-term stability of the reinforced embankments by accelerating the consolidation process and hence increased the rate of strength gain in the foundation.

Faster construction rates resulted in higher short-term strength therefore less force was transferred to the reinforcement and only small strains were developed at the end of construction. However, the larger overstress associated with faster construction resulted in larger long-term viscoplastic deformation and larger long-term reinforcement strain. PVDs substantially improved the beneficial effects of reinforcement. With the combined used of reinforcement and PVDs, less stiff reinforcement could be used with the faster construction rate while limiting the long-term reinforcement strain to acceptable values.

Delayed creep induced pore pressures were generated in the rate-sensitive soil. Reinforcement could reduce creep deformation and the associated pore pressures, thus the rate of pore pressure dissipation was accelerated. Without PVDs, the excess pore pressures kept increasing until they reached a peak sometime after the end of construction; therefore, the critical period with respected to the stability of the embankment could occur after the construction was completed. PVDs resulted in rapidly reduced excess pore pressures after the end of construction.

The post-construction deformations (differential settlement, heave and lateral toe movement) of the reinforced embankments were minimized by using geosynthetic reinforcement. The stiffer the reinforcement, the smaller the post-construction deformations. Other things being equal, a higher construction rate led to higher overstress in the foundation and hence larger long-term creep and lateral toe movements. The effect of reinforcement on the long-term deformation of the embankments could be enhanced with the use of PVDs.

ACKNOWLEDGEMENT

The research reported in this paper was supported by the Natural Sciences and Engineering Research Council of Canada (NSERC).

REFERENCES

Bergado D.T., Balasubramaniam, A.S., Fannin, R.J., Anderson, L.R., and Holtz, R.D. 1997. Full scale field test of prefabricated vertical drains (PVD) on soft Bangkok clay and subsiding environmental. Ground improvements 1987-1997 Edited by V.R. Schaefer. ASCE, Geotechnical Special Publication 69: 372–393.

Bergado, D.T., Long, P.V., and Murthy, B.R. 2002. A case study of geotextile-reinforced embankment on soft ground. Geotextile and Geomembranes, 20 : 343–365.

Biot, M.A. 1941. General theory of three-dimensional consolidation. Journal of Applied Physics. 12: 155–164.

Bo, M. W. 2004. Discharge capacity of prefabricated vertical drain and their field measurements, Geotextiles and Geomembranes, 22 (1–2). Special Issue on Prefabricated Vertical Drains: 37–48.

Carter, J.P., and Balaam, N.P. 1990. AFENA– A general finite element algorithm: Users manual. School of Civil Engineering and Mining Engineering, University of Sydney, N.S.W. 2006, Australia.

Chai, J. C., Miura, N., and Nomura, T. 2004. Effect of hydraulic radius on long-term drainage capacity of geosynthetics drains, Geotextiles and Geomembranes, 22 (1–2). Special Issue on Prefabricated Vertical Drains: 3–16.

Chai, J.C., and Miura, N. 1999. Investigation of factors affecting vertical drain behaviour. Journal of Geotechnical and Geoenvironmental Engineering, 125(3): 216–226.

Chen, W.F., and Mizuno, E. 1990. Nonlinear Analysis in Soil Mechanics: Theory and Implementation. Elsevier Science Publishing Company Inc., New York, U.S.A.

Crawford, C.B., Fannin, R.J., deBoer, L.J. and Kern C.B. 1992. Experiences with prefabricated vertical (wick) drains at Vernon, B.C. Canadian Geotechnical Journal, 29: 67–79.

Fowler, J., and Koerner, R.M. 1987. Stabilization of very soft soils using geosynthetics. In Proceedings of Geosynthetics '87 Conference, New Orleans, Vol.1: 289–299.

Germaine, J.T., and Ladd, C.C. 1988. Triaxial testing of saturated cohesive soils. In Advanced triaxial testing of soil and Rock. ASTM STP 977: 421–459.

Graham, J., Crooks, H.A., and Bell, A.L. 1983. Time effects on the stress-strain behaviour of natural soft clays. Geotechnique, 33: 327–340.
Hinchberger, S.D. 1996. The behaviour of reinforced and unreinforced embankments on soft rate-sensitive foundation soils. Ph.D. thesis, Department of Civil Engineering, The University of Western Ontario, London, Ont.
Hinchberger, S.D., and Rowe, R.K. 1998. Modelling the rate-sensitive characteristics of the Gloucester foundation soil. Canadian Geotechnical Journal, 35: 769–789.
Hird, C.C., Pyrah, I.C., and Russell, D. 1992. Finite element method modelling of vertical drains beneath embankment on soft ground. Geotechnique, 42(3): 499–511.
Holtz, R.D. 1987. Preloading with prefabricated vertical strip drains. Geotextiles and Geomembranes, 6(1 3): 109 131.
Indraratna, B., and Redana, I.W. 2000. Numerical modeling of vertical drains with smear and well resistance installed in soft clay. Canadian Geotechnical Journal, 37: 132–145.
Jewell, R.A. 1987. Reinforced soil walls analysis and behaviour. The Application of Polymeric Reinforcement in Soil Retaining Structures, Jarret P.M., and McGown, A. (Edited). Kluwer Academic Publishers: 365–408.
Kabbaj, M., Tavenas, F. and Leroueil, S. 1988. In situ and laboratory stress-strain relationships. Geotechnique, 38(1): 83–100.
Kelln, C., Sharma, J., Hughes, D., and Gallagher, G. 2007. Deformation of a soft estuarine deposit under a geotextile reinforced embankment. Canadian Geotechnical Journal, 44: 603–617.
Leroueil, S. 1988. Tenth Canadian Geotechnical Colloquium: Recent developments in consolidation of natural clays. Canadian Geotechnical Journal, 25(1): 85–107.
Li, A.L., and Rowe, R.K. 1999. Reinforced embankments constructed on foundations with prefabricated vertical drains. In Proceedings of the 52nd Canadian Geotechnical Conference, Regina, Sask.: 411–418.
Li, A.L., and Rowe, R.K. 2001. Combined effects of reinforcement and prefabricated vertical drains on embankment performance. Canadian Geotechnical Journal, 38: 1266–1282.
Lo, K.Y., and Morin, J.P. 1972. Strength anisotropy and time effects of two sensitive clays. Canadian Geotechnical Journal, 9(3): 261–277.
Perzyna, P. 1963. The constitutive equations for rate-sensitive plastic materials" Quarterly of Applied Mathematics, 20(4): 321–332.
Rixner, J.J., Kraemer, S.R., and Smith, A.D. 1986. Prefabricated vertical drains. Vol.1: Engineering guideline. U.S. Federal Highway Administration, Report FHWA-RD-86/186.
Roscoe, K.H., and Schofield, A.N. 1963. Mechanical behaviour of an idealised "wet clay". Proceedings of the Second European Conference on Soil Mechanics: 47–54.
Rowe, R.K. 1984. Reinforced embankments: analysis and design. Journal of Geotechnical Engineering, ASCE, 110(GT2): 231–246.
Rowe, R.K., and Hinchberger, S.D. 1998. The significance of rate effects in modelling the Sackville test embankment. Canadian Geotechnical Journal, 33: 500–516.
Rowe, R.K., and Li, A.L. 1999. Reinforced embankments over soft foundations under undrained and partially drained conditions. Geotextiles and Geomembranes, 17(3): 129–146.
Rowe, R.K., and Li, A.L. 2002. Behaviour of reinforced embankments on soft rate sensitive soils. Geotechnique, 52(1): 29–40.
Rowe, R.K., and Soderman, K.L. 1985. An approximate method for estimating the stability for geotextiles reinforced embankments. Canadian Geotechnical Journal. 22(3): 392–398.
Rowe, R.K., and Soderman, K.L. 1987. Reinforcement of the embankments on soils whose strength increase with depth. In Proceedings of Geosynthetics '87 Conference, New Orleans, Vol.1: 266–277.
Rowe, R.K., and Taechakumthorn, C. 2007a. Behaviour of reinforced embankments on soft rate-sensitive foundation. Proceeding of Geosynthetics Conference 2007, Washington D.C., USA: 86–98
Rowe, R.K., and Taechakumthorn, C. 2007b. The counteracting effects of rate of construction on reinforced embankments on rate-sensitive clay. Proc. of 5th International Conference on Earth Reinforcement IS Kyushu 2007, Kyushu, Japan: 437–440.
Rowe, R.K. and Taechakumthorn, C. (2008) "Combined effect of PVDs and reinforcement on embankments over rate-sensitive soils", *Geotextiles and Geomembranes,* (in press).
Rowe, R.K., Gnanendran, C.T., Landva, A.O. and Valsangkar, A.J. 1995. Construction and performance of a full-scale geotextile reinforced test embankment, Sackville, New Brunswick. Canadian Geotechnical Journal. 32(3): 512–534.
Rowe, R.K., Gnanendran, C.T., Landva, A.O., and Valsangkar, A.J. 1996. Calculated and observed behaviour of a reinforced embankment over soft compressible soil. Canadian Geotechnical Journal, 33: 324–338.
Rujikiatkamjorn, C., Indraratna, B., and Chu, J. 2007. Numerical modelling of soft soil stabilized by vertical drains, combining surcharge and vacuum preloading for a storage yard. Canadian Geotechnical Journal, 44: 326–342.
Russell, D. 1990. An element to model thin, highly permeable materials in two dimensional finite element consolidation analyses. In Proceeding of the 2nd European Specialty Conference on Numerical Method in Geotechnical Engineering, Santander, Spain: 303–310.
Shen, S.L., Chai, J.C., Hong, Z.S., and F.X. 2005. Analysis of field performance of embankments on soft clay deposit with and without PVD-improvement, Geotextiles and Geomembranes, 23(6): 463–485.
Taechakumthorn, C. and Rowe, R.K. (2007). "The effect of PVDs and reinforcement on the behaviour of embankments on soft rate-sensitive soils," 60th Canadian Geotechnical Conference, Ottawa, October 1171–1176.
Vaid, Y.P., and Campanella, G. 1977. Time-dependent behaviour of undisturbed clay. Journal of the Geotechnical Engineering, 103(GT7): 693–709.
Vaid, Y.P., Robertson, P.K., and Campanella, R.G. 1979. Strain rate behaviour of Saint-Jean-Vianney clay. Canadian Geotechnical Journal, 16: 34–42.

2 UNBAR

Advances in Transportation Geotechnics – Ellis, Yu, McDowell, Dawson & Thom (eds)
© 2008 Taylor & Francis Group, London, ISBN 978-0-415-47590-7

Characterizing the resilient behaviour of treated municipal solid waste bottom ash blends for use in foundations

A.T. Ahmed & H.A. Khalid
Dept of Engineering, Brodie Tower, University of Liverpool, UK

ABSTRACT: Bottom ash waste is a residual material produced by incinerating Municipal Solid Waste. In this study, an experimental programme was undertaken to investigate the influence of enzyme treatment on the behaviour of bottom ash and limestone blends for use as foundation layers in flexible pavement structures. The research focused on evaluating the blends' resilient modulus, which is the most sought after property in analytical pavement design. Cyclic triaxial compression tests were adopted to determine the materials' resilient moduli. Emphasis was on examining the effect of various parameters, such as bottom ash content, enzyme content, moisture content and curing time on the resilient characteristics of the investigated blends and the parameters' impact on the resulting design of a typical pavement structure. Results showed that the bottom ash blends gave favourable performance as foundation layers in comparison with the control limestone blend.

1 INTRODUCTION

When Municipal Solid Waste is incinerated in waste plants, approximately 20–30% by weight and 10% by volume of the burnt waste remains as an inert gravely like ash known as Incinerator Bottom Ash Aggregate (IBAA). Traditionally, this ash has been landfilled but in recent years it has been increasingly used as a secondary aggregate. The main property which should be monitored when using IBAA blends as foundation layers is the resilient modulus. It is generally considered as an essential parameter for mechanistically-based pavement design procedures. Also, it is an appropriate measure of stiffness for unbound granular pavement materials (UGM).

In this study, it is aimed to evaluate the influence of a plant-based enzyme on the resilient modulus of IBAA-limestone blends, and the impact of this additive on the resulting pavement design.

2 MATERIALS

Three materials were used in this study: limestone, IBAA and enzyme solution. Limestone was chosen as the control aggregate in the mixtures. It was supplied in six sizes: 20, 14, 10, 6, 4 mm – dust and filler. IBAA was supplied in two sizes: 20–10 mm and 10 mm.

The used enzyme solution contained a mixture of plant-based proteins with 0.5% w/v Potassium sorbets as preservative. It had three different contents: 0.1, 0.3 and 0.5 g/l, protein/water. The enzyme was mixed at one unit enzyme solution to 500 units of water.

The samples were divided into four groups, coded as A, B, C and D. Group A was the control blend of limestone only. Group B had 30% bottom ash and 70% limestone. Group C had 50% of each of IBAA and limestone and group D had 80% IBAA. The replacement of limestone with IBAA in the various blends followed a simple approach based on matching particle size distributions and is explained elsewhere (Hassan & Khalid, 2007).

3 CYCLIC TRIAXIAL TEST (CTT)

The CTT was performed on cylindrical specimens, placed in a cell, under a confining pressure, $\sigma 3$, and a vertical stress, $\sigma 1$. The type of cyclic loading applied was constant confining pressure, where the axial stress was cycled. In this study, the CTT was used to examine the resilient modulus of unbound mixtures under procedures that simulate the physical conditions and stress states of these materials in pavement layers subjected to moving loads.

3.1 *Equipment*

A servo-pneumatic machine which uses air as pressure medium was adopted. The pressure chamber was a typical conventional triaxial cell of 135 mm diameter. The axial load applied to the specimen was monitored by a load cell with a sensitivity range which yields measurements of axial stress to an accuracy of ± 2 kPa. The load cell is placed outside the triaxial cell, according to TP46 procedure (FHWA, 1996), as shown in figure 1.

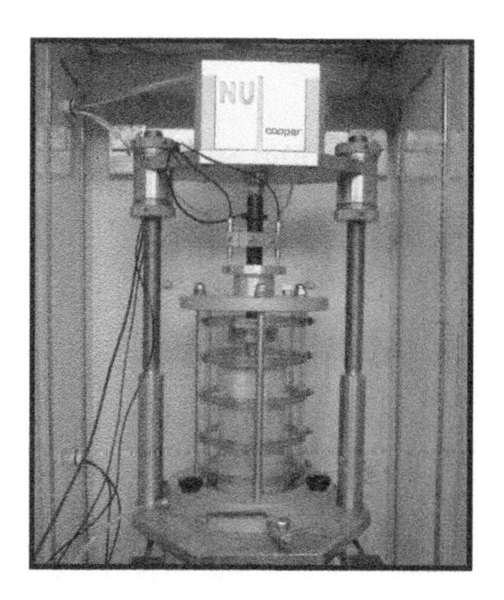

Figure 1. The triaxial test assembly.

3.2 *Sample preparation*

Cylindrical specimens of 100 mm diameter and 200 mm height were prepared in a split steel mould. Aggregates were placed in the mould in three layers and compacted using a vibrating hammer with a tamping foot attachment that had a diameter equal to the internal diameter of the compaction mould. The surface of each layer was manually roughened before adding the next layer on top; in this way a good layer interlock and a homogeneous sample was obtained. Specimens were kept in the steel mould for 24 hours within a plastic sheet in order to make them airtight. Meanwhile a membrane was placed in a membrane stretcher to which vacuum was applied. The membrane was then carefully placed on the specimen. The stretcher was removed from the membrane by switching off the vacuum. After placing the rubber membrane around the specimen, it was kept in a humid environment for seven days, to allow for uniform distribution of water within the specimen and pozzolanic reaction as well. After seven days, the specimen was attached to the top and bottom platens with rubber O-rings and then was installed in the triaxial cell.

3.3 *Resilient modulus test procedure*

In order to study the resilient behaviour, cured specimens were tested according to the AASHTO TP46 protocol (FHWA, 1996) for base/subbase materials in which a constant stress ratio was maintained by increasing both the principal stresses simultaneously. The procedure consisted of applying cyclic conditioning to the sample followed by a series of cyclic loadings along different stress paths. The objective of the conditioning was to eliminate the permanent deformations occurring during the initial load cycles of the test, and to obtain stable resilient behaviour independent of the number of cycles. Test results represent the average of three determinations.

4 RESILIENT MODULUS TEST RESULTS

The test programme was planned with the aim of emphasis on examining the effect of parameters such as stress level, bottom ash content, enzyme content, moisture content and curing time on the resilient modulus of the investigated blends. The expression "average resilient modulus" was used to refer to the average of the fifteen values of resilient modulus at different stress levels during the test each of which is an average of three determinations.

4.1 *K-θ Model*

The effect of different blend properties was studied using a well known curve-fitting tool named the K-θ model. It is a non-linear, stress-dependent power function model described by Seed et al (1962). The model is given as follows.

$$M_R = K_1\, \theta^{K_2} \tag{1}$$

where: M_R = resilient modulus; K_1, K_2 = Regression constant and θ = bulk stress = $\sigma 1 + 2\,\sigma 3$.

Although the model is out of date, it has been found very useful for design of new pavements (Thompson, 1992). Furthermore, the relationship between the resilient modulus and bulk stress provided by the K-θ model has been incorporated into none-linear pavement analysis procedures using finite elements and layered systems (Brown, 1978).

Here, the model has been adopted to exhibit the influence of various parameters on the material's resilient modulus, as will be shown in the following sections.

4.2 *Stress levels*

The deviator stress, confining pressure and bulk stress significantly affect the resilient modulus of UGM. To analyse the effect of stress state on resilient modulus values, test results were plotted against confining pressure at optimum moisture content (OMC) in figure 2. It can be seen that the resilient modulus increases with increase in confining pressure for all blends. This is in agreement with results on granular materials reported by Rahim & George (2005). Moreover, figure 3 shows the relation between resilient modulus and deviator stress. It is observed that, at constant confining pressure, the resilient modulus is constant or slightly increases with increase in deviator stress. Lekarp et al. (2000) reported that various researchers seem to agree that the resilient response is

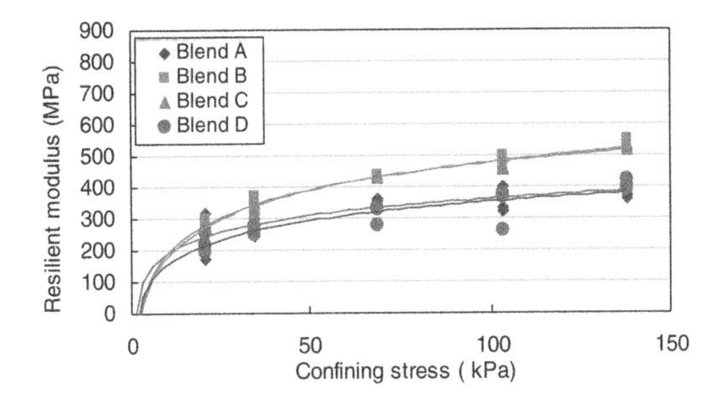

Figure 2. Resilient modulus versus confining stress at OMC.

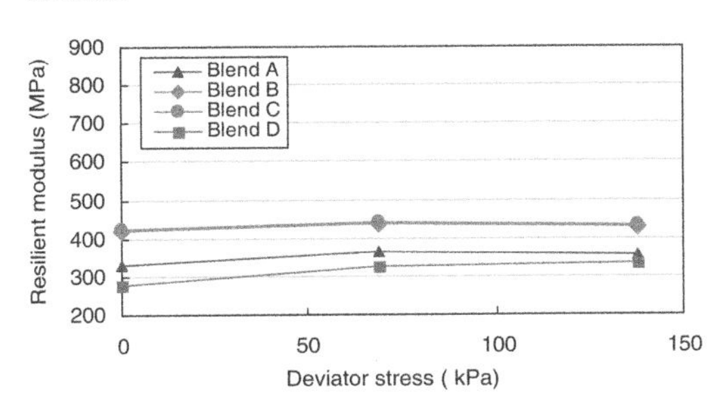

Figure 3. Resilient modulus versus deviator stress at constant 68 kPa confining stress.

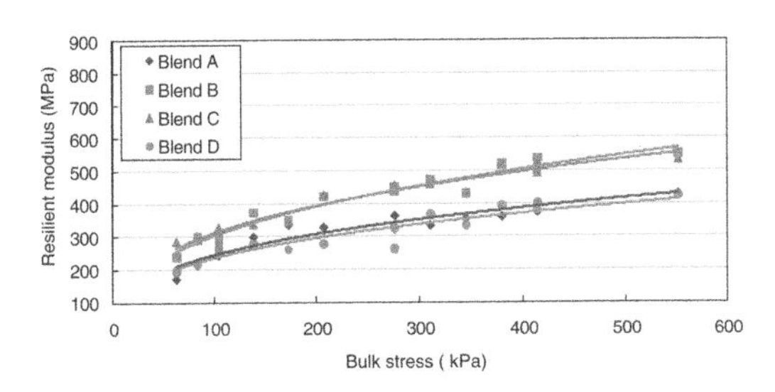

Figure 4. Resilient modulus versus bulk stress at OMC.

influenced most by the level of applied stresses. The resilient modulus increases markedly with confining pressure and sum of principal stresses, and slightly with deviator stress.

4.3 *Incinerator bottom ash content*

Figure 4 shows resilient modulus results against principal stress for blends with different IBAA contents at OMC. It can be clearly seen that blends with 30 and 50% IBAA have higher resilient modulus than the limestone blend. However, the 80% IBAA blend exhibits nearly the same resilient behaviour as the control blend. This means that adding IBAA up to 50% improves the blend's deformation characteristics, probably because IBAA provides better interlock between particles and pozzolanic reaction as well. On the other hand, the high water absorption of IBAA increases with increase in the IBAA content in the blend and this leads to a reduced amount of water that surrounds the particles and, as a result, the pozzolanic reaction decreases.

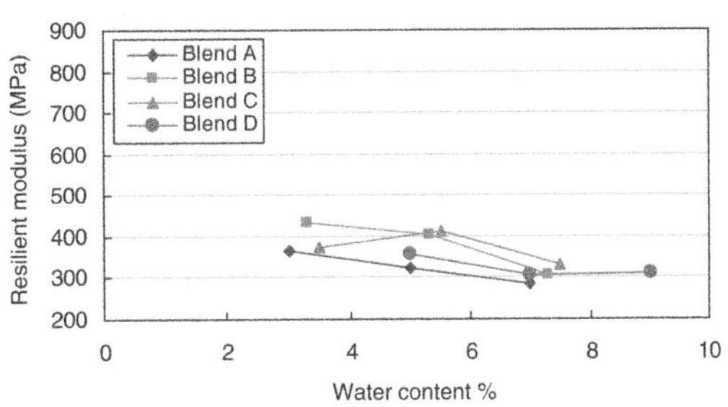

Figure 5. Average resilient modulus versus water content.

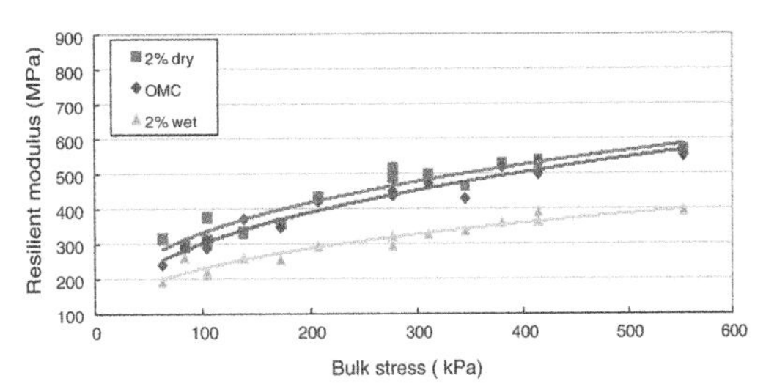

Figure 6. Resilient modulus versus bulk stress for blend B.

4.4 *Water content*

To investigate the water content effect, three contents were used, namely OMC, 2% less than OMC and 2% higher. From figure 5, it is observed that the average resilient modulus of blend A, i.e. limestone only, and 30% IBAA, blend B, increases with the decrease in water content. However, this trend is not clear with 50% and 80% IBAA, i.e. blends C and D. Probably this is due to the fact that the IBAA has high water absorption characteristics. Figure 6 shows that resilient modulus of blend B, at different stress levels, increases with decrease in water content. The available literature revealed that increase in water content above the optimum in UGM in the laboratory and in the field led to a decrease in resilient modulus (Hicks & Monismith, 1971). The combination of a high degree of saturation and low permeability, due to poor drainage, leads to high pore pressure and low effective stress and, consequently, low stiffness, low resistance to permanent deformation and low resilient deformation (Dawson et al. 1996).

4.5 *Curing time*

It was aimed to examine the pozzolanic reaction potential of IBAA when mixed with water, especially in the

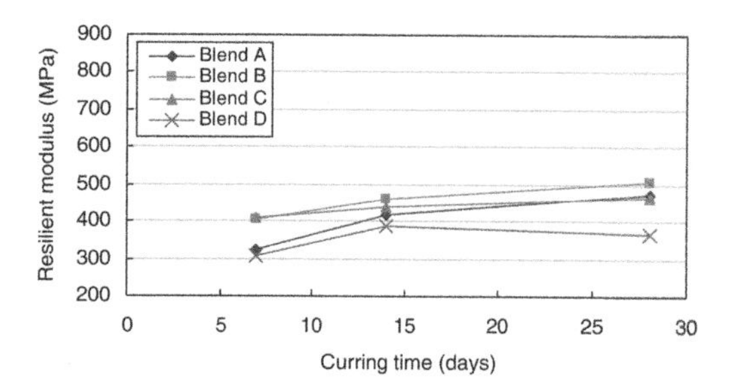

Figure 7. Average resilient modulus versus curing time at OMC.

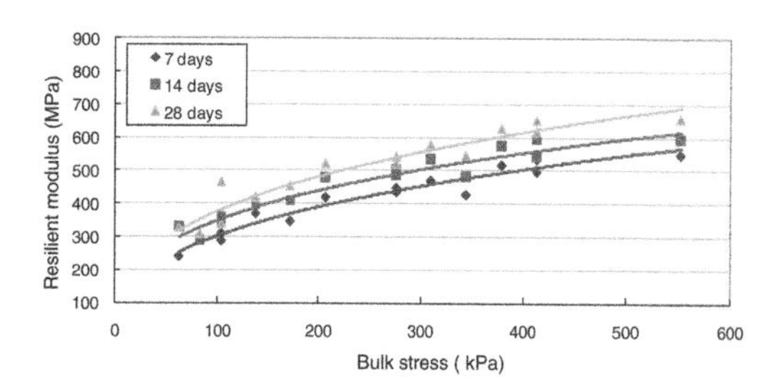

Figure 8. Resilient modulus versus bulk stress for blend B.

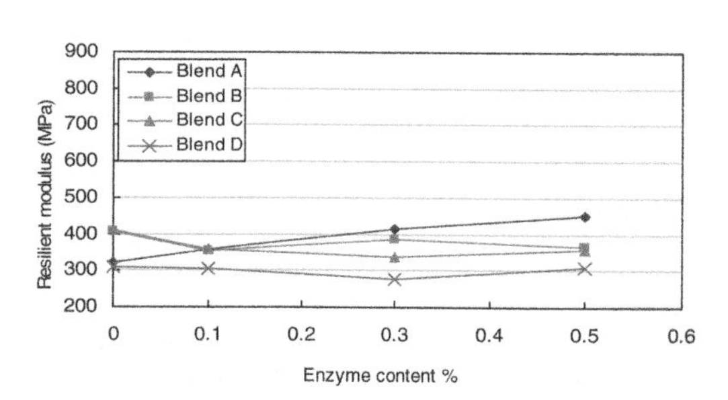

Figure 9. Average resilient modulus versus enzyme content at OMC and seven days curing.

presence of limestone. The specimens were tested at curing durations of 7, 14 and 28 days. Typical plots are presented in figure 7. It can be seen that blends A, B and C underwent a noticeable increase in the average resilient modulus values with time. However, the rate of increase decreased with time. Blend D, on the other hand, showed an improvement in the resilient modulus up to 14 days but decreased in the period from 14 to 28 days. Moreover, figure 8 shows that resilient modulus of blend B, at different stress levels, increased with time. This behaviour indicates that the pozzolanic reaction is improved with increase in IBAA content up to 50%. From these results and the results in section 4.3, there appears to be an optimum IBAA content to achieve highest performance at around 50%.

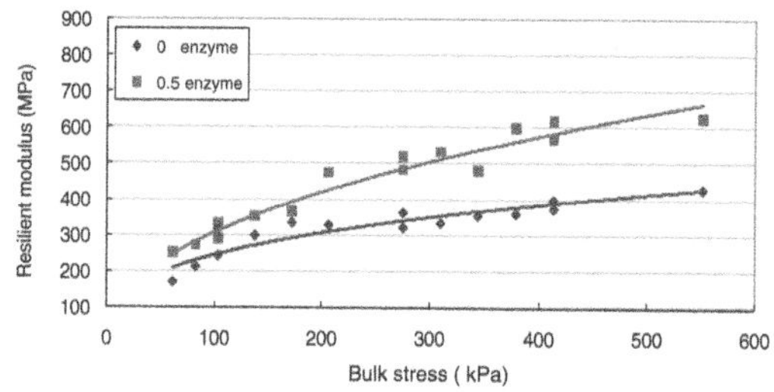

Figure 10. Resilient modulus versus bulk stress at seven days for blend A.

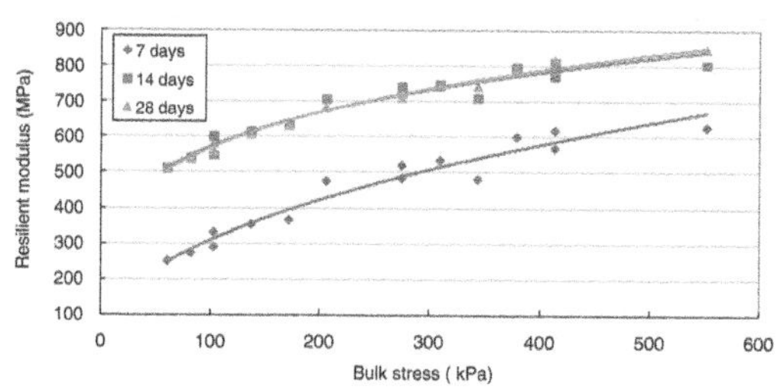

Figure 11. Resilient modulus versus bulk stress at 0.5% enzyme content for blend A.

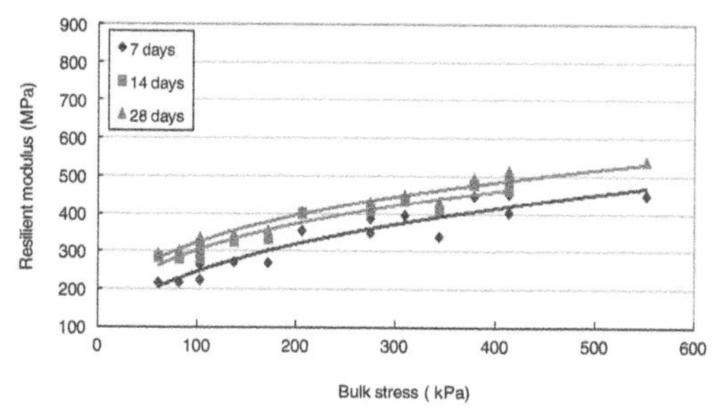

Figure 12. Resilient modulus versus bulk stress for enzyme treated blend C.

4.6 *Effect of enzyme addition*

Figure 9 shows that, at seven days, the average resilient modulus of blends B, C and D is not sensitive to the addition of the enzyme. However, figure 10 shows that the enzyme increased the average resilient modulus of blend A by 40%. Furthermore, from figure 11, it can be seen that the addition of the enzyme had a pronounced effect on the resilient modulus of blend A during the first 14 days of curing, where the average resilient modulus increased significantly from 7 to 14 days by 51%. But this increase was very small from 14 to 28 days. With regard to blends B, C and D, enzyme addition led to a small increase in resilient modulus after 14 days. Figure 12 shows the effect of curing time on resilient modulus for blend C. Blends B and D, although not shown here, exhibited similar trends. From the results, it can be clearly stated that the

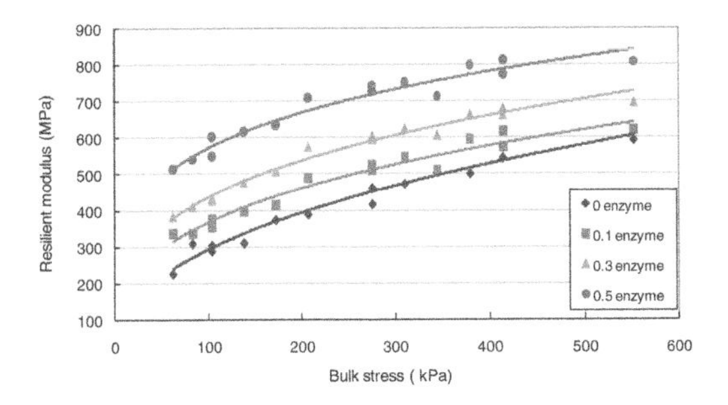

Figure 13. Resilient modulus versus bulk stress at 14 days for blend A.

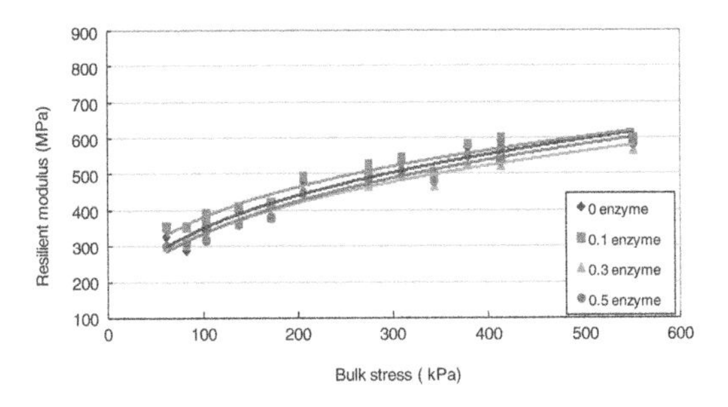

Figure 14. Resilient modulus versus bulk stress at 14 days for blend B.

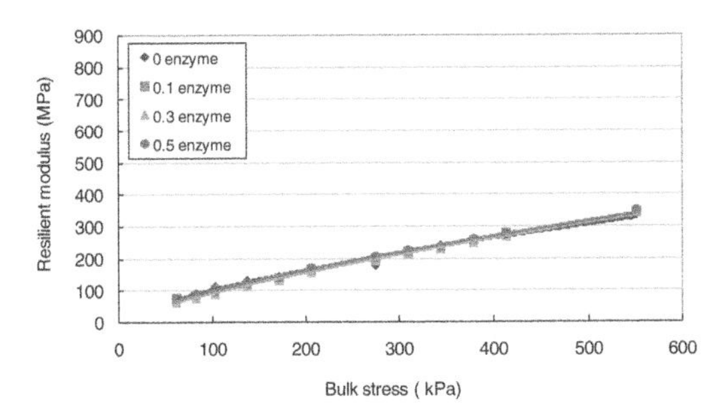

Figure 15. Resilient modulus versus bulk stress at 14 days for blend D.

blend type and IBBA content significantly affected the impact of the treatments. To examine the enzyme content effect, three levels were used, namely 0.1, 0.3 and 0.5. Figure 13 shows that, for blend A, the enzyme effect increases with increase in the enzyme content. However, figure 14 shows that the resilient modulus of blend B increased only slightly with enzyme addition. For blends C and D, the resilient modulus was either unaffected, as figure 15 shows, or decreased with increase in enzyme content. It was also noticed that the application rate suggested by Velasquez et al. (2006) and adopted here, e.g. up to 2 cc of enzyme solution with 1000 cc water, improves the effectiveness of the stabilization process of the enzyme, but only with blend A. However, this is not the case for blends

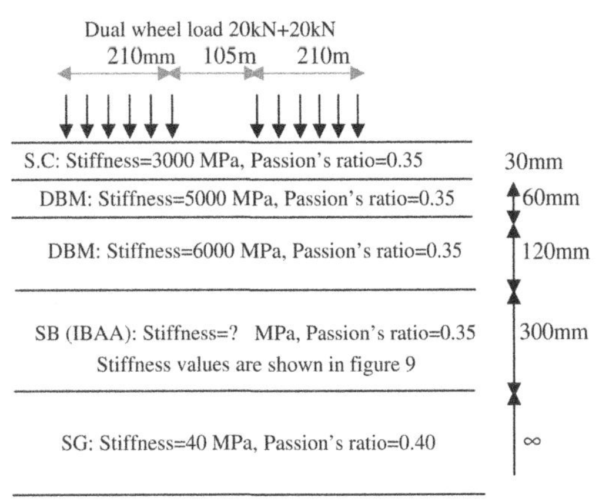

Figure 16. Layer properties and loading configuration.

B, C and D. The effectiveness of the enzyme will be further investigated in forthcoming work by addressing the reasons for lack of chemical and/or physical interaction with IBAA.

5 PAVEMENT MODEL AND ANALYTICAL EVALUATION

5.1 *Model concept*

The Bitumen Stress Analysis in Roads (BISAR) developed by Shell was used in the analysis. A typical pavement was adopted consisting of five layers as shown in figure 16. BISAR considers the pavement as an elastic multilayered system and the subgrade is assumed to extend to infinity, and all pavement layers are assumed to be infinite in the horizontal direction.

In this work, the evaluation of pavement life in terms of critical deformations and mixture characteristics was determined using the following equation (Brown & Brunton, 1985):

$$N_{cr}=f_r\,[3*10^9/\varepsilon^{3.57}] \qquad (2)$$

where N_{cr} = the number of load applications to failure, measured in millions of standard axles (msa); ε = vertical compressive strain at top of subgrade, and f_r = rut factor, f_r = 1.56 for Dense Bitumen Macadam.

5.2 *Analytical results*

Figure 17 shows pavement life for the different blends. The main objective of the analysis was to present the blend's effect in a more practitioner-oriented context by exhibiting the impact on the pavement's design life in msa. The results show the positive influence of enzyme treatment on the predicted design life of the pavement with blend A as subbase in contrast to

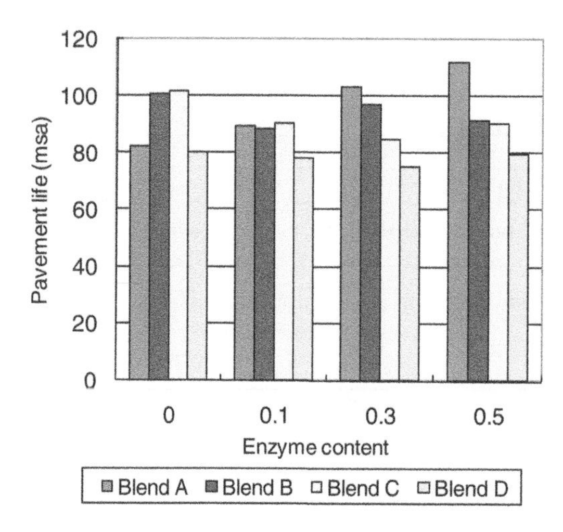

Figure 17. Effect of blend type and treatment on design life.

that for blends B, C and D, where little or no effect is apparent.

6 CONCULSIONS

The resilient modulus for all blends increased with increase in confining pressure and deviator stress. The resilient modulus of the limestone and 30% IBAA blends increased with decrease in water content. Blends with high limestone content underwent significant increase in the resilient modulus values with time. However, the rate of increase decreased. The used enzyme did not improve the resilient modulus of bottom ash blends B, C and D. However, it led to an increase in the resilient modulus of limestone blend A after 28 days on average by 46%.

Based on the resilient modulus results, the bottom ash blends showed good behaviour as a foundation layer in comparison with the control blend of limestone only.

It is recommended that further study is undertaken to examine the factors affecting the enzyme's physicochemical interaction potential with aggregates and, hence, its propensity to stabilise blends containing IBAA.

ACKNOWLEDGEMENT

The authors extend their gratitude to Aggregate Industries UK for the financial and technical support of this project.

REFERENCES

Brown, S. F. 1978. Material characteristic for analytical pavement design, in Developments in Highways Pavement Engineering (ed. P.S. Pell): 41–92. Applied Science Publishers.

Brown, S. F. & Brunton, J. M. 1985. An introduction to the analytical design of bituminous pavements. 3rd Ed. University Nottingham, Nottingham, UK.

Dawson, A., Thom, N. H. & Paute J. L. 1996. Mechanical characteristics of unbound granular materials as a function of condition. Flexible Pavements, Proc., Eur. Symp. Euroflex 1993, A G Correia, ed. 35–44. Netherlands, Rotterdam: Balkema.

Federal Highway Administration (FHWA), TP46. 1996. Resilient modulus of unbound granular base/subbase materials and subgrade soils. U.S. Department of Transportation, Federal Highway Administration, Washington, D.C.

Hassan, M. & Khalid, H. 2007. Incinerator bottom ash aggregate in bituminous mixtures. 4th Intern. Symp. Bituminous Mixtures and Pavement: 489–498. Thessaloniki, Greece.

Hicks, R G. & Monismith, C L. 1971. Factors influencing the resilient response of Granular materials, Highway Research Record. 345:15–31.

Hoff, I., Baklokk, L. & Aurstad, J. 2004. Influence of laboratory compaction method on unbound granular materials. 6th Intern. Symp. on Pavements Unbound.

Lekarp, F., Isacsson, U. & Dawson, A. 2000. State of the Art. I: Resilient response of unbound aggregate. Jl. of Transportation Engineering: 66–75.

Manual of Contract Documents for Highways Works (MCHW). 2005. Unbound cement and Other Hydraulically Bound Mixtures. Volume 1 Series 800, Road pavements.

National Cooperative Highway Research Program (NCHRP). 1998. Laboratory determination of resilient modulus for flexible pavement design, Web Doc. 14, Final Report.

Rahim, A. M. & George, K. P. 2005. Models to estimate subgrade resilient modulus for pavement design Intern. Jl. of Pavement Engineering. 6(2): 89–96.

Seed, H.B. Chan, C.K. & Lee, C.E. 1962. Resilience characteristics of subgrade soils and their relations to fatigue in asphalt pavements. Proc. Intern. Symp. on Structural Design of Asphalt Pavements. Ann Arbor, USA. 1:611–636.

Thompson, M. R. 1992. Report of the discussion group on backcalculation limitations and future improvements. No. 1377.

Velasquez, R., Marasteanu, M. & Hozalski, R. 2006. Investigation of the effectiveness and mechanisms of enzyme products for subgrade stabilization. Intern. Jl. of Pavement Engineering 7 (3): 213–220.

Advances in Transportation Geotechnics – Ellis, Yu, McDowell, Dawson & Thom (eds)
© 2008 Taylor & Francis Group, London, ISBN 978-0-415-47590-7

Rut depth prediction of granular pavements using the repeated load triaxial apparatus and application in New Zealand specifications for granular materials

G. Arnold & D. Arnold
Pavespec Limited, Auckland, New Zealand

A.R. Dawson
Nottingham Transportation Engineering Centre, (NTEC), University of Nottingham, UK

D.A.B. Hughes & D. Robinson
School of Planning, Architecture and Civil Engineering, The Queen's University of Belfast

S. Werkmeister
Technical University of Dresden, Germany

D. Alabaster
Engineering Policy Section, Transit New Zealand, Wellington, New Zealand

J. Ellis & R. Ashby
Stevenson Group Limited, Auckland, New Zealand

J. Lowe
Winstone Aggregates, Auckland, New Zealand

ABSTRACT: The rutting of granular pavements was studied by examining the permanent deformation behaviour of granular and subgrade materials used in a Northern Ireland (UK) pavement field trial and accelerated pavement tests at CAPTIF (Transit New Zealand's test track) located in Christchurch New Zealand. Repeated Load Triaxial (RLT) tests at many different combinations of confining stress and vertical cyclic stress for 50,000 loading cycles were conducted on the granular and subgrade materials. The aim of the RLT tests was to derive relationships between permanent strain and stress level. These relationships were later used in finite element models to predict rutting behaviour and magnitude for the pavements tested in Northern Ireland and the CAPTIF test track. Predicted rutting behaviour and magnitude were compared to actual rut depth measurements during full scale pavement tests to validate the methods used. It was found that the trend (long term tangential rate of rutting) in rut depth progression was accurately predicted for 11 out of the 17 full scale pavement tests, while the magnitude of rut depth predicted for these 11 tests was within 3mm of the measured values. Three of the poorly predicted test sections were still reasonably predicted while the other three poor predictions were considered outliers due to differences in moisture content between RLT tests and in the field. As a result of the success of this predictive method of assessing rutting in granular materials Transit New Zealand undertook a three year study to implement a RLT test into specifications for selecting road base aggregate. Eight New Zealand aggregates of known performance in the road were selected for RLT testing and associated rut depth prediction to confirm the RLT test can predict performance for a range of aggregate types. Results were positive and the RLT test was implemented into Transit New Zealand's specification for selecting basecourse aggregate. It is expected the benefits of this new test will reduce the number of early pavement rutting failures and allow the use of alternative materials such as local quarry overburden material modified with cement which show equivalent performance in the RLT test to traditional high quality granular materials.

1 INTRODUCTION

1.1 *Granular Pavement Layers*

Granular pavement layers play an important role in the pavement. They are required to provide a working platform for the construction of the asphalt base layers and reduce compressive stresses on the subgrade and tensile stresses in the asphalt base. For thin-surfaced pavements the unbound granular material (UGM) contributes to the full structural strength of the pavement. It is therefore important that the granular materials have adequate stiffness and do not deform. Material specifications usually ensure this is the case.

1.2 *Repeated Load Triaxial (RLT)*

The repeated load triaxial (RLT), hollow cylinder [Chan 1990] and k-mould [Semmelink et al, 1997] apparatuses can in various degrees simulate pavement loading on soils and granular materials. Permanent strain tests in the Repeated Load Triaxial (RLT) apparatus commonly show a wide range of performances for UGMs even though all comply with the same specification (Thom and Brown, 1989). Accelerated pavement tests on thinly sealed pavements show the same results and also report that 30% to 70% of the surface rutting is attributed to the granular layers (Arnold et al., 2001; Little, 1993; Pidwerbesky, 1996; Korkiala-Tanttu et al, 2003).

Furthermore, recycled aggregates and other materials considered suitable for use as unbound sub-base pavement layers can often fail the highway agency material specifications and thus restrict their use. There is potential of the permanent strain test in the RLT (or similar) apparatus to assess the suitability of these alternative materials for use at various depths within the pavement (e.g. sub-base and lower sub-base). Thus, current pavement design methods and material specifications should consider the repeated load deformation performance of the UGM layers.

2 RLT TESTS

Using the Repeat Load Triaxial (RLT) apparatus, Arnold (2004) studied the effect of different combinations of cyclic vertical and horizontal stress levels on a range of granular materials. The granular materials chosen were those used in full scale pavement tests in Northern Ireland, UK and at Transit New Zealand's indoor accelerated pavement testing facility CAPTIF. The subgrade silty clay soil used at CAPTIF was also tested in the RLT apparatus.

The aim of RLT tests was to determine the effect of stress condition on permanent strain. RLT permanent strain tests are time consuming and many tests are needed to cover the full spectra of stresses expected within the pavement. To cover the full spectra of stresses expected in a pavement (Jouve and Guezouli, 1993) a series of permanent strain tests was conducted at different combinations of cell pressure and cyclic vertical load. Most of the stresses calculated by Jouve and Guezouli (1993) show the mean principal stress ($p = (\sigma_1 + 2\sigma_3)/3$) to vary from 50 to 300 kPa and the deviatoric stress ($q = (\sigma_1 - \sigma_3)$) from 50 to 700 kPa. These ranges of stresses were confirmed by the authors with some pavement analysis using the CIRCLY linear elastic program (Wardle, 1980). Although at the base of the granular layers negative values of p were calculated, as a granular material has limited tensile strength negative values of mean principal stress were discounted. The results of the static shear failure tests conducted on the materials plotted in $p - q$ stress space were used as an approximate upper limit for testing stresses.

It is common that a new specimen is used for each stress level. However, to reduce testing time multi-stage tests were devised and conducted. These tests involved applying a range of stress conditions on one sample. After application of 50,000 load cycles (if the sample had not failed) new stress conditions were applied for another 50,000 cycles. These new stress conditions were always slightly more severe (i.e. closer to the yield line) than the previous stress conditions.

Test stresses were chosen by keeping the maximum value of p constant while increasing q for each new increasing stress level closer and occasionally above the static yield line. Three samples for each material were tested at three different values of maximum mean principal stress p (75, 150 and 250 kPa). This covered the full spectra of stresses in $p - q$ stress space so as to allow later interpolation of permanent strain behaviour in relation to stress level. A typical output and stress paths from a multi-stage RLT test is shown in Figure 1.

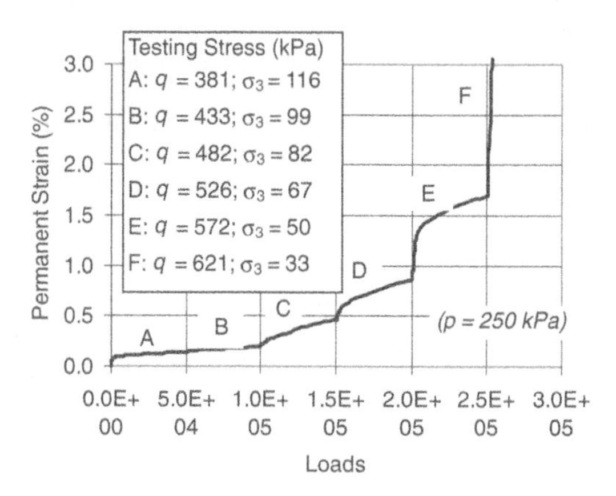

Figure 1. Typical RLT permanent strain test result for NI Good UGM.

3 MODELLING PERMANENT STRAIN

Results from Repeated Load Triaxial tests showed a high dependence on stress level. Plots of maximum deviator stress, q versus permanent strain rate for each multi-stage test analysed separately with mean principal stress, p constant do result in exponential relationships that fit well to the measured data (Figure 2). This prompted an investigation of an exponential model that could determine the secant permanent strain rate from the two maximum values of q, p stress.

Utilising the rate of deformation seems more appropriate for RLT multi-stage tests than using the accumulated sum of the permanent strain for each part of the multi-stage test. Adding the sum of all the permanent strains of all the previous stages will likely over-estimate the amount of deformation, for example by increasing the number of test stages conducted prior to that of the present stress level in question, and will lead to a higher magnitude of permanent strain. Also reviewing the RLT permanent strain results during the first 20,000 load cycles there is a "bedding in" phase until a more stable/equilibrium type state is achieved. It is argued that this equilibrium state is unaffected by differences in sample preparation and previous tests in the multi-stage tests. Therefore, relationships considering permanent strain rate were explored.

It was found that the scatter in results was reduced by using the secant rate of permanent strain between 25,000 and 50,000 load cycles in place of permanent strain magnitude when plotted against stress ratio (q/p). This rate of permanent strain was calculated in percent per 1 million load cycles. These units have some practical interpretation as a value of 5% per million load cycles can be approximately related to 5mm of deformation occurring for a 100mm layer after 1 million load applications.

Regression analysis was undertaken on the RLT test data with stress invariants p and q against the associated natural logarithm of the strain rate. The result was the first part (on the left of the subtraction sign) of

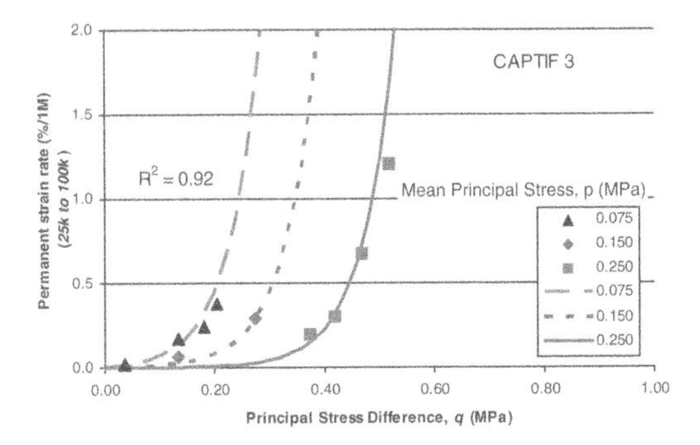

Figure 2. Example plot showing how exponential functions fit measured data for individual multi-stage tests.

Equation 1, while the part of Equation 1 being subtracted (i.e. on the right of the subtraction sign) was included to ensure that deviatoric stress, q, is zero if the resulting permanent strain rate is also zero. Permanent strain rate is thus defined by Equation 1:

$$\varepsilon_{p(\text{rate or magn})} = e^{(a)}\, e^{(bp)}\, e^{(cq)} - e^{(a)}\, e^{(bp)} = e^{(a)}\, e^{(bp)}\, (e^{(cq)} - 1) \qquad (1)$$

where,
e = 2.718282;
$\varepsilon_{p(\text{rate or magn})}$ = secant permanent strain rate or can be just permanent strain magnitude;
a, b & c = constants obtained by regression analysis fitted to the measured RLT data;
p = mean principal stress (MPa); and
q = mean principal stress difference (MPa).

In order to make the relationship dimensionally stable, stress invariants p and q in Equation 1 can be substituted by p/p_0 and q/q_0 respectively, where p_0 and q_0 are reference stresses. In this case, the reference stresses were taken as equal to 1 MPa.

To determine the total permanent strain for any given number of load cycles and stress condition it is proposed to describe the permanent strain data into four zones. It was observed in the New Zealand Accelerated Pavement Tests (Arnold, 2004) that the permanent strain rates changed during the life of the pavements, different values being associated with the early, mid, late and long term periods of trafficking. Similarly RLT permanent strain tests showed changing permanent strain rates during their loading. After studying RLT and accelerated pavement test results a power law equation of the form, $y = ax^c$ is fitted to each 50,000 RLT load cycle stage (Figure 1) to extend the permanent strain data to 1 Million load cycles. To limit the number of times Equation 1 is used to fit the RLT data, it was decided to break the RLT permanent strain data into the following four zones of different behaviour for use in calculating permanent strain at any given number of load applications (N):

1. Early behaviour (compaction important) – 0 – 25,000 load applications. The magnitude of permanent strain at 25,000 load applications, being the incremental amount $\varepsilon_p(25\,\text{k})$, is used for the reasons outlined above. Keeping the magnitude of permanent strain separate at 25,000 was useful when predicting rut depth in terms of identifying where the errors occurred.
2. Mid term behaviour – 25 k – 100 k load applications. The secant permanent strain rate between 25 k and 100 k load applications is used.
3. Late behaviour – 100 k – 1 M load applications. The secant permanent strain rate between 100k and 1M load applications is used.

4. Long term behaviour – > 1 M load applications. The secant permanent strain rate between 1 M and 2 M load applications is used as the permanent strain rate for all load applications greater than 1 M.

This assumes that the permanent strain rate remains constant after 1M load applications. The approach is appropriate as the aim is to calculate rut depth for pavements with thin surfacings where Arnold (2004) found the rate of rutting is linear. Further, assuming the permanent strain rate does not decrease after 1M load applications is a conservative estimate compared with the assumption of permanent strain rate continually decreasing with increasing load cycles (i.e. a power model, $y = ax^b$).

Hence, four versions (four sets of constants) of Equation 1 are determined and used for describing the permanent strain behaviour at any given number of loadings.

The RLT multi-stage permanent strain test results for 6 granular and 1 subgrade material were analysed by relating permanent strain rate with stress level for each stage. Microsoft Excel© Solver was used to determine the equation constants a and b by minimising the mean error being the difference in measured and calculated strain rates. For the CAPTIF 3 granular material the model fits the data well with a R^2 value of 0.92 (Figure 2). Similar goodness of fit to the model (Equation 1) were found for the other 5 granular and 1 subgrade materials tested. Mean errors were less than 1%/1M which, when applied to rut prediction, give an equivalent error of 1 mm per 100 mm thickness for every 1 million wheel passes. Overall the model (Equation 1) shows the correct trends in material behaviour as there is an increasing permanent strain rate with increasing deviatoric stress (q) while a higher load can be sustained with higher confining stress.

4 CALCULATING PAVEMENT RUT DEPTH

4.1 *Background*

To predict the surface rut depth of a granular pavement from the Equation 1 requires a series of steps. There are assumptions required in each step which significantly affects the magnitude of calculated rut depth. Steps and associated errors and assumptions are summarised below. The first five steps relate to the interpretation of RLT permanent strain tests as already described in Section 3 above. The final three steps involve pavement stress analysis, calculations and validation required to predict surface rut depth of a pavement.

4.2 *Pavement Stress Analysis*

Equation 1 is a model where for any number of load applications and stress condition the permanent strain can be calculated. Stress is therefore computed within the pavement under a wheel load for use in Equation 1 to calculate permanent strain. It is recognised from literature and RLT tests that the stiffness of granular and subgrade materials are highly non-linear. A non-linear finite element (FE) model, DEFPAV (Snaith et al., 1980) was used to compute stresses within the pavement. DEFPAV only approximated the non-linear characteristics of the granular and subgrade materials. Further, the small residual confining stresses considered to occur during compaction of the pavement layers were assumed to be nil in the analysis.

From the pavement stress analysis, the mean principal stress (p) and deviatoric stress (q) under the centre of the load are calculated for input into a spreadsheet along with depth for the calculation of rut depth. The calculated stresses have a direct influence on the magnitude of permanent strain calculated and resulting rut depth. Thus any errors in the calculation of stress will result in errors in the prediction of rut depth. Some errors in the calculation of stress from DEFPAV are a result of not considering the tensile stress limits of granular materials and the assumption of a single circular load of uniform stress approximating dual tyres, which do not have a uniform contact stress (de Beer et al, 2002).

4.3 *Surface rut depth calculation.*

The relationships derived from the RLT permanent strain tests are applied to the computed stresses in the FE analysis. Permanent strain calculated at each point under the centre of the load is multiplied by the associated depth increment and summed to obtain the surface rut depth.

4.4 *Validation*

The calculated surface rut depth with number of wheel load applications is compared with actual rut depth measurements from accelerated pavement tests in New Zealand (CAPTIF) and the Northern Ireland (NI) field trial. This comparison determines the amount of rut depth adjustment required at 25,000 cycles while the long term rate of rut depth progression and, in part, the initial rut depth at 25,000 cycles is governed by the magnitude of horizontal residual stress added. An iterative process is required to determine the initial rut depth adjustment and the amount of horizontal residual stress to add, in order that the calculated surface rut depth matches the measured values.

The result of this validation process is detailed fully in Arnold (2004). Overall the predictions of rut depth are good, particularly the trends in rut depth progression with increasing loading cycles (i.e. this relationship is sensibly the same for actual and predicted measurements in 11 out 17 analyses). Adjustment of up to a few millimetres to the predicted rut depth at 25,000 cycles is generally all that is needed to obtain an accurate prediction of rut depth. For the six tests with poor predictions 3 of these could be accurately predicted by applying a small residual stress of 17kPa to the stresses calculated in the pavement analysis. The other three poor results are due to an asphalt layer of 100mm and a moisture sensitive aggregate where the RLT test was at a higher moisture content than what actually occurred in the test pavement. The predictions were as expected in terms of accurately predicting the rate or rutting but inaccurately predicting the initial rutting due to secondary consolidation as the actual magnitude for the first 25,000 load applications from the RLT test is difficult to estimate, it being unknown whether the value is the cumulative or incremental value of permanent strain from the multi-stage RLT test (Figure 1). The method adopted was to take the incremental value of permanent strain at 25,000 cycles, which is considered to be a low estimate. This is the case for pavement Test Sections 3, 3a and 3b which all used the CAPTIF 3 granular material and for Test Sections 4, 5 and 6 where an additional amount of rutting was added to coincide with the measured values in the pavement test. The opposite occurs for the Test Sections 1, 1a, and 1b which all used the CAPTIF 1 material.

5 APPLICATION IN NEW ZEALAND SPECIFICATIONS

5.1 *Background*

Premature surface rutting occurs frequently on newly constructed thin-surfaced unbound granular pavements, common in New Zealand. It is now recognised that a key parameter that governs pavement longevity is the granular materials resistance to rutting within. However, in existing specifications, there is more focus on limits determined by a range of empirical tests (such as grading, broken faces, crushing resistance, amount of fines etc) than a direct measure of deformation resistance. The Repeated Load Triaxial test was thus investigated as a method to assess an aggregates resistance to rutting and thus suitability for use in the pavement for use in specifications. As the RLT test is performance-based another benefit and reason for it's use is the ability to assess alternative pavement materials as being suitable or otherwise as pavement base materials.

5.2 *RLT Test Development*

Based on the RLT test validated by Arnold (2004) research was undertaken to develop a more simplified RLT test and associated analysis that could determine a traffic loading limit for a granular base (Arnold et al. 2007). Ten different aggregates were tested using three different testing methodologies and analysis methods. The first RLT test method trialled was based on the method develop by Arnold (2004) as presented in this paper. A 6 stage RLT test was developed with the aim of covering most of the spectra of stresses expected to occur within the pavement with 50 thousand load cycles applied for each stage. Six stage RLT tests from subsequent research projects and commercial tests were analysed to predict rutting within a pavement profile tested at Transit New Zealand's accelerated pavement testing facility, CAPTIF, using the method developed by Arnold (2004). The results showed a reasonable ranking of materials performance along with traffic loading limits that were validated from CAPTIF results. It was found that a relationship between average slope in the RLT test from 25k to 50k of all the 6 stages could be related to a traffic loading limit, which was recommended for use in specifications.

The second method trialled was the Austroads method which applied 3 test stress stages at 10 thousand load cycles per stage (Vuong and Brimble, 2000) combined with a simplified analysis procedure. However, the Austroads method did not rank the materials performance as expected based on CAPTIF results and anecdotal pavement field performance and was disregarded for use in New Zealand (Arnold et al. 2007).

5.3 *Specification for RLT Testing*

A specification detailing the 6 stage RLT test, sample preparation and analysis was developed named TNZ T/15 (2007) *Specification for repeated load triaxial (rlt) testing of unbound and modified road base aggregates*. This specification is still currently under development to improve on the prediction of traffic loading limit from a simple RLT test parameter. In the meantime the specification is being used by Quarry owners to assess the performance of their existing granular products and to develop new granular materials sometimes modified with cement or similar additive. Analysis of the RLT results is conducted using the full rut depth prediction method developed by Arnold (2004) for a standard pavement cross-section used at the CAPTIF test track.

5.4 *Observations found from RLT Testing*

To gain a better understanding of granular material performance, RLT tests by Quarry owners are repeated at different moisture contents, densities and drained or un-drained conditions. A significant number of RLT tests have now been conducted by the larger quarry owners namely Stevenson and Winstone Aggregates. Typical results for a traditionally accepted high quality granular material meeting the requirements of

Table 1. Rut Depth Modelling Results for RLT Tests shown in Figure 3.

	Rut Depth Prediction for CAPTIF Pavement – 300 mm Aggregate over Subgrade CBR = 10			
	Total Pavement	In Granular Layers	In Granular Layers	In Granular Layers
Test No. (see Fig. 2)	N, ESAs to get 25 mm rut Million ESAs	N, ESAs to get 10 mm rut Million ESAs	Long term rate of rutting within aggregate mm per 1 M ESAs	Vertical Resilient Modulus in top layer MPa
1	3.04	9.4 (*Note 1*)	0.95	497
2	1.31	0.71	7.5	413
3	3.30	33 (*Note 1*)	0.26	749
4	2.75	5.5	1.5	485

ESAs = Equivalent Standard Axles (Heavy Axle Passes as defined by Austroads 2004 Pavement Design).
Note 1: This value is currently the recommended Traffic Loading Limit for this material as per TNZ T/15 (Transit, 2007).

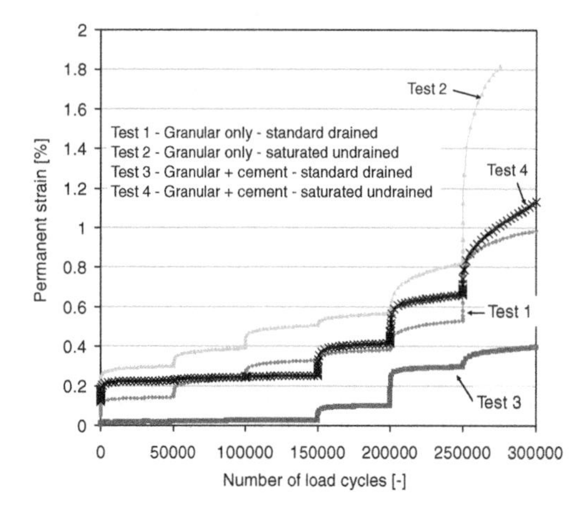

Figure 3. Typical RLT results for a New Zealand granular material.

TNZ M/4 specification are detailed in Table 1 and Figure 3. Results show that performance is poor if the granular material becomes saturated while this performance is improved with cement modification. Other observations found from RLT tests is that for some compliant granular materials poor performance is obtained which often supports the poor performance seen in the field.

6 DISCUSSION

This paper presented a constitutive relationship between permanent strain rate and stress level from multi-stage permanent strain tests of granular and subgrade materials in the Repeated Load Triaxial apparatus. From the constitutive relationship and stresses calculated within the pavement, a rut depth can be estimated. This resultant rut depth does not consider rutting in the asphalt base layer and its magnitude is only relevant to the moisture and compaction levels used in the RLT test. In New Zealand most pavements do not have a asphalt base layer but simply a 25mm or less asphalt or chipseal surfacing over granular materials. Moisture conditions of the granular and subgrade materials can change regularly which effects the amount of rutting. Therefore, the application of this rut depth method is recommended for ranking and classifying the predicted performance of granular material layers used in pavements. Transit New Zealand plan to revise their specification for granular materials to include the practical RLT test summarised in this paper to enable granular materials to be categorised as being suitable as base materials for either high, medium or low traffic situations in either dry/typical or saturated conditions. Classifying granular materials in this way will allow a range of materials to be used in appropriate locations (i.e. low or high traffic and wet or dry environments), including marginal materials (previously discarded); industrial by-products (e.g. Melter Slag) and waste materials (e.g. glass, recycled crushed concrete).

REFERENCES

Arnold, G. Werkmeister & S. Alabaster, D. 2007. *Performance tests for road aggregates and alternative materials*. Land Transport New Zealand, Wellington, New Zealand.

Arnold G. 2004. *Rutting of granular pavements*. PhD Thesis, University of Nottingham, Nottingham, UK.

Arnold, G. Alabaster & D. Steven, B. 2001. *Prediction of pavement performance from repeat load triaxial tests on granular materials*. Transfund New Zealand Research Report, Wellington, New Zealand.

de Beer, M. Fisher & C. Jooste, F.J. 2002. Evaluation of non-uniform tyre contact stresses on thin asphalt pavements.

Ninth International Conference on Asphalt Pavements, August 17–22, 2002, Copenhagen, Denmark.

Chan, F.W.K. 1990. *Permanent deformation resistance of granular layers in pavements*, Ph.D. Thesis, University of Nottingham, Nottingham, UK.

Jouve, P. & Guezouli, S. 1993. Comparison and harmonized development of some finite element programs. *Flexible Pavements*. Edited by A. Gomes Correia, Technical University of Lisbon. *Proceedings of the European Symposium Euroflex* 1993, Lisbon, Portugal 20–22 September 1993.

Korkiala-Tanttu, L. Laaksonen, & R. Törnqvist, J. 2003. *Effect of the spring and overload to the rutting of a low-volume road. HVS-Nordic-research*. Helsinki 2003. Finnish Road Administration. Finnra Reports 22/2003. 39 p. + app..

Little, P.H. 1993. *The design of unsurfaced roads using geosynthetics*, PhD thesis, Dept. of Civil Engineering, University of Nottingham.

Pidwerbesky, B. 1996. *Fundamental behaviour of unbound granular pavements subjected to various loading conditions and accelerated trafficking*. PhD Thesis, University of Canterbury, Christchurch, New Zealand, 1996.

Semmelink, C.J., Jooste, F.J. & de Beer, M. 1997. Use of the K-mould in determination of the elastic and shear properties of road materials for flexible pavements, *8th Int. Conf. on Asphalt Pavements*, August, Seattle, Washington, USA.

Snaith, M.S., McMullen, D., Freer-Hewish, R.J. & Shein, A. 1980. *Flexible pavement analysis*. Contracted Report to Sponsors, European Research Office of the U.S. Army.

Thom, N. & Brown, S. 1989. The mechanical properties of unbound aggregates from various sources. *Proceedings of the Third International Symposium on Unbound Aggregates in Roads, UNBAR 3*, Nottingham, United Kingdom, 11–13 April 1989.

Transit. 2007. TNZ T/15 *Specification for repeated load triaxial (RLT) testing of unbound and modified road base aggregates*. Transit New Zealand, Wellington, New Zealand.

Wardle L.J. 1980. Program *CIRCLY*, a computer program for the analysis of multiple complex loads on layered anisotropic media.

Vuong. B.T. & Brimble, R. 2000. *Austroads repeated load triaxial test method – determination of permanent deformation and resilient modulus characteristics of unbound granular materials under drained conditions*. APRG 00/33 (MA) June.

Advances in Transportation Geotechnics – Ellis, Yu, McDowell, Dawson & Thom (eds)
© 2008 Taylor & Francis Group, London, ISBN 978-0-415-47590-7

Assessment of sustainable highway geotechnics

J.M. Belton & R.P. Thompson
Coffey Geotechnics Ltd, Manchester, UK

A. Jukes
Highways Agency, UK

ABSTRACT: Sustainable development is a national and international government objective to balance the need for economic development against the protection of the environment and the conservation of resources for future generations. As a consequence sustainable development is included in Highways Agency mission statements. A contribution to these objectives has been provided in a study carried out jointly by EDGE Consultants UK Ltd (now Coffey Geotechnics Ltd) and the Norwegian Geotechnical Institute. An innovative sustainability assessment procedure was proposed for relative comparison of alternative highway geotechnics options, this being just one of a number of disciplines requiring consideration for a new or improved highway scheme. The study identifies five themes under which sustainability comparisons of geotechnical aspects of alternative highway schemes could be meaningfully assessed. A pilot study was undertaken to trial the sustainability assessment procedure and check its validity. This paper discussed the effectiveness and value of the proposed sustainability indicators.

1 INTRODUCTION

1.1 *Background*

Sustainable development is a national and international government objective to balance the need for economic development against the protection of the environment and the conservation of resources for future generations. In line with these objectives the UK Government Policy is to promote sustainable development and as a consequence sustainable development is included in Highways Agency mission statements. Sustainable development is a nebulous task and it is recognised that overriding principles and technical specifics must be balanced.

Key requirements in achieving the goals of sustainable construction include the development of meaningful indicators to enable appropriate and objective assessment of alternative options and to provide methods of measurement by which improvements can be gauged. Prior to this study no assessment system existed for the geotechnical aspects of highway design. It is recognized that highway geotechnics is just one of a number of disciplines requiring consideration for a new or improved highway scheme.

In response to Highway Agency recognition that indicator systems are required to enable sustainability decisions to be taken objectively an innovative sustainability assessment procedure for relative comparison of alternative highway geotechnics options is proposed as a useful tool to be incorporated into a wider suite of assessments.

For the purposes of the assessment Highway Geotechnics are deemed to include:

- Earthworks at grade, including ground improvements (e.g. alternative ground improvement techniques such as lime stabilisation against stone columns).
- Cuttings, including soft to intermediate support (e.g. comparison of unsupported shallow batters against steepened nailed/shotcreted slopes).
- Embankments (e.g. comparison of unsupported shallow side batters against steepened reinforced slopes).

The work was divided into two phases, Phase and Phase II.

1.2 *Contributors*

The sustainability assessment procedure was proposed in Phase I of the study carried out jointly by EDGE Consultants UK Ltd (now Coffey Geotechnics Ltd) and the Norwegian Geotechnical Institute. A pilot study of the sustainability assessment procedure was undertaken by EDGE in Phase II.

2 SUSTAINABILITY ASSESSMENT PROCEDURE

2.1 *Themes*

Amongst existing construction sustainability processes, the major themes consistently covered include initial land take (its effect on the environment, including improvements and mitigation provisions), construction, maintenance, facility operation, and ultimately decommissioning/reuse.

In keeping with these principals the following five themes were developed for the sustainability assessment procedure:

1. Land Take
2. Geotechnical construction
3. Geotechnical maintenance issues during the life of a highway
4. Highway usage, as affected by geotechnical earthworks design
5. Adaptation and Decommissioning.

A series of indicators were developed for each theme, which identified meaningful parameters for assessment. The input parameters for the themes are shown in Table 1 and discussed further below.

In order to provide an overall rational comparison of alternative schemes the result from the evaluation process for each theme, being a score between 0 and 1, is given a weighting and then aggregated to yield a single comparative score.

2.2 *Theme 1 – land take*

The areas between the hard shoulder and the toe of the supporting embankment or crest of cuttings are of primary interest in Theme 1. The development of the carriageway and hard shoulder areas will be largely irrespective of other land take considerations. The land take theme is of primary use in comparing alternative schemes when there is scope to optimise on some aspect e.g. by the avoidance of existing features and/or reuse of brownfield areas as opposed to greenfield areas. Many of the sustainability elements of land take are already considered in the planning process and the Environmental Impact Assessments. Theme 1 is intended to complement these issues with respect to geotechnics.

2.3 *Theme 2 – geotechnical construction*

Materials used for geotechnical construction can be split into two broad categories: earthworks materials and geo-structural components (such as polymers for soil reinforcement, steel for nails and reinforcement, lime/cement for stabilisation and stone for vibro columns etc). Theme 2 primarily considers the environmental impact of the production, handling and placement of these materials. The source/destination, volume and haul distance for each material are recorded and the theme score calculated from the volume/mass, construction and transport scores. A second calculation provides absolute measures of energy (excavation, transportation, placement/processing), volumes and percentages of re-cycled/secondary to primary aggregates. These outputs could be used for direct comparison and benchmarking of different road projects when normalised, say per km of highway.

2.4 *Theme 3 – geotechnical maintenance*

Whole life costs of earthwork slopes (embankments or cuttings) outwith the carriageway are considered in Theme 3. The calculation is based on work by

Table 1. Summary of proposed highway geotechnics sustainability assessment.

Theme	Input Parameters	Recording Procedure	Output	Proposed Weight
1	Biodiversity management Creation/protection of habitats Construction management	Qualitative subjective and efficiency	Points 0–1	5%
2	Volumes per source type Haul distances In-situ processing	Quantitative (m^3, & km) Ratios Recycled: natural Reused: landfill	Points 0–1 & Energy Volumes & Ratios	15%
3	Slope length for different slope angles and heights, per geology type	Quantitative (km/slope height and gradient)	Points 0–1	30%
4	Length of carriageway at different gradients	Quantitative (km/gradient)	Points 0–1 & Energy	45%
5	Percentage volumes, lengths and areas	Quantitative (%) & qualitative	Points 0–1	5%
Total				100%

Perry (1989) that demonstrated: Shallow failures are a significant maintenance problem on motorway earthworks; Failure rates are influenced by geology and slope geometry.

The theme score is calculated based on the lengths of slopes of different heights in specific geologies.

2.5 *Theme 4 – highway usage*

The effects of vehicle usage of a highway can encompass a wide range of 'sustainability' themes (economic, social and environmental). Most of the primary considerations are outwith geotechnical influence but will have been dealt with at an early stage of the scheme development and prior to detailed design. The principal consideration for which geotechnics can contribute to improved sustainability in highway usage is via the provision of appropriate road gradients by means of cuttings and embankments (and tunnels/bridges). The theme score is calculated according to the fuel consumption component of the gradient factors. This theme is of particular interest when alternative routes for a scheme are being compared.

2.6 *Theme 5 – adaptation and decommissioning*

Two topics are considered in respect of geotechnical earthworks: decommissioning the existing road and widening the existing road. The primary consideration for decommissioning is the suitability of the material used in the geotechnical construction for re-use.

There are issues that may be considered in the original design that have an impact on the sustainability of widening schemes. These include the ability to reuse materials and to widen the carriageway within the existing corridor. The theme score is based on these issues.

3 PILOT STUDY

3.1 *Contributors*

Phase II was undertaken to evaluate the effectiveness of the proposed sustainability assessment procedure for highway geotechnics. The sustainability indicators for each theme were assessed in terms of: Information available; Ease of interpretation; A way of comparing highway schemes with respect to sustainable geotechnics.

Projects at various stages of design and construction were consulted (see Table 1).

3.2 *Methodology*

A questionnaire was developed to enable the information required for the sustainability assessment procedure prescribed by the Phase I report to be collected easily. The questionnaire was intended to allow project teams of participating projects to provide the details needed without having studied the Phase I report. This approach was consistent with expected practice if sustainability assessments become commonplace. The questionnaire also asked for comments about the availability of information and ease of interpretation for each theme.

Calculations and qualitative assessments for indicator scores and calculation of theme and overall scores were undertaken by EDGE. Similar assumptions were therefore used in the interpretation of data and indicator requirements for all projects.

Theme scores for each project were not compared against each other as such a comparison would not be appropriate due to scheme specific influences. However, some analysis of scores achieved across a wide variety of projects was undertaken. This was necessary to assess the appropriateness of the indicators and weightings as a sustainability assessment tool.

4 MAIN POINTS OUT OF PILOT STUDY

The Phase II study yielded observations on overall concepts of implementing sustainability assessment as well as specific themes, indicators and definitions.

The concepts of implementing sustainability assessment may be primarily related to a single theme, or else affect more than one theme. They offer a philosophy to be adopted to encourage the assessment and application of sustainability as something more than a tick-box paper exercise.

4.1 *Acquisition*

Any new highway scheme requires land for the carriageway and any cuttings and embankments. A 'corridor' of land will be created which includes the carriageway itself. Typically the HA will own the land in this corridor – either before the scheme proceeds

Table 2. Range of participating projects.

Project	Stage	Type
1a	Preliminary Design	Upgrade–widening
1b	Preliminary Design	Upgrade–widening & new build
2	Detailed Design	New build
3	Detailed Design	Upgrade-widening
4	Construction	Upgrade–widening & new build
5	Preliminary Design	New Build
6	Preliminary Design	Upgrade-widening
7	Construction	Upgrade-widening
8	Preliminary Design	Upgrade-widening

or once acquired via a Compulsory Purchase Order (CPO).

The pilot study highlighted a trend amongst HA designers to consider only land acquired by CPO as 'land take' and so to not automatically recognise any loss of habitat within the HA's existing corridor. While this could potentially be an issue for new build projects, it is most evident on widening projects – especially where cuttings are steepened to allow increased pavement area.

To ensure account is taken of loss of habitat, regardless of land ownership, the area of 'land modified' should be considered as 'land take'.

4.2 *Forward thinking*

Potential conflict between economic/politically acceptable highway design and the sustainable benefits of designing to be adaptable to future demands was identified. Whole life cost of a highway may be a more cost-effective solution in the long term but the relatively high initial cost that invariably accompanies such design philosophy is often disconcerting or prohibitive to the client. Further, systematic attitudes amongst highway designers against adaptability of highway schemes were evident. It was noted that there is currently no requirement by the HA or Planning Authorities to allow for further introduction of services and/or widening and that in the absence of such a requirement it was unlikely that designers would consider doing so. It was further noted that projects subject to CPOs and Public Enquiries may find it difficult to successfully justify land take on such a basis.

Sustainability arguments in themselves may not yet be able to dominate during a Public Enquiry, but it should be encouraged that sustainability issues are considered alongside engineering and political factors. Leaving room to add extra lanes or additional services and designing structures that span the highway to accommodate a future widening of the carriageway may reduce the whole life costs of a project, but may also be considered to be encouraging increased highway use.

Present attitudes and, to an extent, policies may need to be addressed to ensure appropriate introduction of sustainability indicators so that they may be a driver for sustainable highway geotechnics design (rather than simply an assessment of the relative sustainability of highway schemes).

4.3 *New build*

The pilot study demonstrated a counter-intuitive point that in some instances a new build scheme may score more highly than an improvement scheme. It is proposed that the ability to choose the route for a new build highway allows benefit to be made of the topography of a site to best meet sustainability aims. Across all the projects included in the study, new build projects scored favorably for Themes 1 and 3 – Land Take and Geotechnical Maintenance, showing an apparent better appreciation of sustainability in geotechnics. Two factors are likely to contribute to this. The first is that widening schemes may be constrained by the existing embankment and cutting heights, which may not have been designed with sustainability in mind. The second is that new build highways may be subject to greater planning and public scrutiny such that a proportionate amount of additional effort may be given to fulfilling sustainability requirements. These findings also highlight the fact that geotechnical indictors should not be used in isolation but to enhance sustainability assessments of the whole highway. For example, intuitively construction of a new road seems less sustainable than adapting an existing highway therefore the social effects of relocating the road must also be included in the overall assessment.

4.4 *Good practice*

The pilot study highlighted the strength of existing guidance and good practice in encouraging design of sustainable highways. The Design Manual for Roads and Bridges (DMRB) advice note gives guidance of how natural conservation and biodiversity issues should be treated and is intended to assist in meeting the project specific commitments and targets as laid down in Environmental Action Plans and Biodiversity Actions Plans. The strength of such existing guidance potentially has a two-fold effect on the effectiveness of sustainability indictors for highways geotechnics.

Indictors that reflect statutory requirements or other established best practice that serves as a requirement for planning would be expected to be achieved by all projects. While enforcing consideration of sustainability, such consistency in scoring could potentially mask differentiation between alternative schemes. This study has not identified this as a significant problem.

It was also apparent from the pilot study that many project managers incorporate such levels of sustainability consideration as are required to meet statutory and local planning requirements but are not inclined to go over and above such guidance for fear of inadvertently falling foul of other regulations. For example many project managers would not consider use of long shallow slopes as they fear that planners would reject the proposal if it required larger areas of CPOs than others.

These challenges present an opportunity for sustainable geotechnics indicators to be adopted as part of a holistic suite of assessments.

4.5 *Project sequencing*

Theme 2 scores the sustainability aspects of geotechnical construction. Projects which have a balance of cut and fill material may be considered more sustainable than those which have a large excess of material or require much material to be imported. Two of the projects within the study were let as a single package under an ECI (early Contractor Involvement) Contract so that the contractor could programme the works to suit. One project required much imported material whereas the other had a surplus of excavated materials. For maximum sustainability and efficiency the second would have commenced first so that the excess excavated material could be moved directly to the other, adjacent, site. Due to other factors the programming of the works was not in the event conducive to material balance.

It is encouraging that the ECI Contract aims to allow contractors flexibility in programming the works although more may need to be done to allow sustainability to be a driver in programming decisions.

4.6 *Sustainability throughout design*

The assumption that more information will be available at later stages of a project allowing more accurate calculation of highway sustainable geotechnics indicator scores was not demonstrated by the pilot study. Two projects were at the construction stage and provided information for Themes 1 and 5 only. In both cases it was stated that volumes of materials used on site and final slope profiles would be available from the contractor at the end of the works. If project managers and designers were familiar with the indicator assessment from an early stage of design then it is more likely that an improved approach to this aspect would be taken.

It is important that sustainability is considered at the start of the project and reviewed throughout the detailed design process and ultimately to construction. It is intended that use of the indicators for sustainable geotechnics becomes a driver for sustainable design and forms an input factor in the decision process enabling scores to increase as the design process progresses.

4.7 *Summary of concept recommendations*

1. Land take should be considered as the area of modified land so that account of loss of habitat etc is taken regardless of land ownership.
2. Consideration of highway adaptation and sustainability drivers for land use should be encouraged.
3. Sustainable thinking above and beyond current requirements should be addressed.
4. The information required for the sustainability assessments should be identified early in a project to aid collation of required data and efficient completion of the questionnaire.

5 IMPLEMENTATION

5.1 *Theme scores and weightings*

Table 3 shows the un-weighted theme and overall indicator scores for each project included in the pilot study. The score for each theme is value between 0 and 1. The overall score is the sum of the theme scores, shown as a percentage. Where not all the information is available, the overall score has been calculated as a percentage of the available score.

The proposed theme weightings are shown in Table 1. Table 4 shows the scores with the proposed weightings applied to each theme. The overall score is a sum of the weighted theme scores, shown as a percentage of the available score.

Score comparisons between different projects should be treated with care. During a highways

Table 3. Summary of Sustainability Indicator Scores (unweighted).

Project	Theme 1	Theme 2	Theme 3	Theme 4	Theme 5	Overall
1a	0.554	0.882	0.819	–	0.799	76.35*
1b	0.554	0.961	0.843	–	0.799	78.92*
2	0.878	0.772	0.813	0.895	0.872	84.60
3	0.502	0.824	0.752	0.940	0.938	78.46
4	0.500	–	–	–	–	50.00*
5	0.697	0.623	0.818	0.896	0.860	77.88
6	0.719	0.551**	0.793	0.632	0.971	73.32
7	0.585	–	–	–	0.942	76.30*
8	0.313	–	–	–	0.943	62.80*

*Overall results calculated from available data only
**Haulage distance of 15 km assumed

Table 4. Summary of Sustainability Indicator Scores (weighted).

Project	Theme 1	Theme 2	Theme 3	Theme 4	Theme 5	Overall
1a	2.77	13.24	24.56	–	3.99	81.02*
1b	2.77	14.41	25.29	–	3.99	84.47*
2	4.39	11.58	24.40	40.29	4.36	85.02
3	2.51	12.36	22.55	42.31	4.69	84.00
4	2.50	–	–	–	–	50.00*
5	3.49	9.35	24.53	40.31	4.30	82.00
6	3.59	8.26**	23.78	28.42	4.86	69.00
7	2.92	–	–	–	4.71	76.3*
8	1.56	–	–	–	4.71	62.7*

*Overall results calculated from available data only
**Haulage distance of 15 km assumed

geotechnical sustainability assessment similar assumptions should be made for each option to ensure that comparable scores are achieved.

The study showed that for this wide range of projects, with their different site-specific factors, the relative ranking of the overall scores was stable under the proposed theme weightings. This gives confidence that the weightings are valid for assessments of highway geotechnics sustainability. The weightings were found to affect the dominant theme in some cases and to widen the range of overall scores thus enabling a clearer comparison of projects.

5.2 *Widening schemes*

The majority of projects included in the pilot study were widening schemes. The applicability of the sustainability indictors to widening schemes was not explicitly addressed in the Phase I report. The following guidelines for widening schemes were born out of the Phase 2 study:

- Land take may be considered as 'land modified'. In this way alternative widening schemes proposing different methods of embankment steepening could be compared in terms of the plan area of such works and account could be made of any ecologically valuable habitat on any existing cutting/embankment now lost to a steeper slope.
- Road gradients are likely to be fixed for widening schemes. If a widening option and a new build option were alternatives, all theme scores should be considered.
- For comparison of two or more alternative widening schemes that do not change the road gradients, Theme 4 (Highway Usage) need not be included, but inclusion would allow projects to be more generally comparable.

5.3 *Practicalities*

The Phase II study has shown that the majority of the questionnaire could be completed by a project manager with general overview knowledge of the project. Answers to the more specific questions can be obtained from specialists or contactors as appropriate. Engineers would benefit from increased knowledge of the advantages of embedding sustainability and environmental issues in the design process. If indicator assessments are to be routine the information required should be collated throughout the project, as it is more difficult to find the information retrospectively.

Independent assessment of the information provided would be appropriate to ensure that for a specific project similar assumptions are made for each alternative scheme. This is important, as some of the indicators are subjective.

An alternative approach may be for contractors to calculate their own indicator scores and apply the weightings for a particular scheme. Some auditing of this process would be required to ensure fair assessments are made.

5.4 *Summary of implementation recommendations*

1. The highway geotechnical sustainability assessment procedure is proposed for use in assessing alternative schemes, as part of the tender decision making process for alternative bids and to drive sustainability throughout the development of a single scheme.
2. A questionnaire is proposed for use by engineers and contractors.
3. Completed questionnaires for alternative schemes should be subject to an independent assessment to ensure that the indicator tables and scores are calculated according to similar assumptions.
4. The proposed theme weightings should be applied.
5. Auditing of indicator score calculations may be necessary to ensure fair evaluation.

5.5 *Areas for further study*

The following recommendations for further study are made:

1. Assess the effectiveness of the indicators to evaluate sustainability with respect to contaminated land.
2. Increase the list of Case Reference Soil Types for Theme 3 by interrogating the GDMS database.
3. Study the use of indicators as a tool for real tender decisions.
4. Extend to include highway geotechnical structures.
5. Integration into global indicators/HA Sustainable Development Action Plan.

6 CONCLUSION

The sustainability assessment procedure is outlined above and the initial testing that has been undertaken is discussed. It is proposed as a useful tool to be taken forward by the highways sector.

As it stands the assessment procedure is ready for use to allow comparison of alternative schemes or the tracking of sustainability of a single scheme though design and construction. It is recognised that it could become a more powerful tool if incorporated into a more holistic sustainability assessment. The implementation of any assessment procedure must therefore be critically reviewed to ensure the outcome correctly reflects and satisfies society and scheme specific needs.

ACKNOWLEDGEMENTS

The authors acknowledge with thanks the approval of the Highways Agency to publish this paper and also gratefully recognise the contributions given by the various project organisations. The authors also thank R. P. Hiller, formally of EDGE Consultants.

REFERENCES

CEEQUAL Ltd. 1993. The Civil Engineering Environmental Quality and Assessment Scheme.Scheme manual version 3.

EDGE Consultants UK Ltd & The Norwegian Geotechnical Institute. 2005. Sustainable Geotechnics. *Highways Agency.*

EDGE Consultants UK Ltd. 2007. Sustainable Geotechnics Phase II. *Highways Agency.*

Environment. 2004. Key Performance Indicators. Handbook 2004. Constructing Excellence, London.

Highways Agency. 2003. Building Better Roads: Towards Sustainable Construction. *Compiled by the Centre for Sustainability and TRL Ltd.*

Highways Agency. 2001. Design Manual for Roads and Bridges (DMRB). Volume 10 Environmental Design and Management.

Highways Agency. 2005. Design Manual for Roads and Bridges (DMRB). Volume 11 Environmental Assessment.

ICE. 2007. Sustainable Development Strategy and Action Plan for Civil Engineering. *ICE, ACE, CECA, CIRIA and Construction Products Association.*

Leiper, Q. 2006. Presidential Address 2006. *Proceedings of the ICE – Civil Engineering, 2007, 160, No. CEI.*

Parry, A.R. & Potter, J.F. 1995. Energy Consumption in Road Construction and Use. *Unpublished project Report PR/CE/48/95 E106A/HE.* TRL: Crowthorne.

Reid, J.M. & Clark, G.T. 2000. A Whole Life Cost Model for Earthwork Slopes. TRL Report 430. TRL: Crowthorne.

Advances in Transportation Geotechnics – Ellis, Yu, McDowell, Dawson & Thom (eds)
© 2008 Taylor & Francis Group, London, ISBN 978-0-415-47590-7

Using pavement trials: Evaluating rutting in forest roads in southern Scotland

L.A.T. Brito & A.R. Dawson
Nottingham Transportation Engineering Centre, University of Nottingham, England, UK

R.W.W. Tyrrell
Forestry Civil Engineering, Castle Douglas, Scotland, UK

ABSTRACT: Full scale pavement trials were carried out in South West Scotland as part of the research activities of the Roads Under Timber Transport Project (R.U.T.T). The main goals of these pavement trials were to evaluate the performance of aggregate under saturated conditions and relative pavement deformation caused by different timber haulage vehicles. A road segment simulating a standard forest road section was constructed on purpose-built facility located at the Ringour Quarry facility. Ten different trials were carried out combining three different aggregate materials and five types of vehicles. Tyre fitment, axle configuration and tyre pressure were assessed and demonstrated to play an important role on the study of rutting development. Preliminary conclusion drawn from the results suggested that management of tyre's inflation pressure and axle overload may be one of the most economic means of managing pavement deterioration in the forest road network.

1 INTRODUCTION

Permanent deformation is unquestionably one of the main distress mechanisms in low volume roads. Numerous road segments collapse after a small number of load cycles due to excessive rutting. This is especially true for those roads on which budget is limited, and therefore, construction standards are far from desirable, material quality is low, vehicle loads are high and weather unfavorable. Forest roads in Scotland are certainly a paradigm case.

In order to improve the forest roads' condition and to help to find an optimal balance between financial investment in the road and cost/effective timber haulage, the Forestry Commission (FC) through the Forestry Civil Engineering and the University of Nottingham through the Nottingham Transportation Engineering Centre (NTEC), sponsored by the Strategic Timber Transport Scheme, have developed the R.U.T.T – Roads Under Timber Transport – research project.

The R.U.T.T project, conceived as a two-year long study, aims to evaluate the damage in forest roads located in southern Scotland, by monitoring 21 forest road entrances and also in performing specific trafficking trials on purpose-built road constructions, the latter at the Ringour Quarry Facility.

The main objective of these trials was to evaluate the performance of aggregate under saturated conditions and the relative pavement deformation caused by different timber haulage vehicles. In particular different traffic types (axle arrangements, tyre arrangements, tyre inflation pressure) were investigated. The aim of the programme is to gather sufficient data to determine the principal controls affecting rutting and the sensitivity to changes in value of the road material type and condition.

2 TRIALS DESCRIPTION

Ringour, located off the A762 road, four miles from New Galloway, was originally a quarry for aggregate used in forest road construction. It was later transformed into a purpose built test facility by the Forestry Civil Engineering/FC for full scale pavement tests.

The site was used to construct specific road sections of approximately 30m length in a more controlled fashion. A pair of concrete walls was provided to act as a confinement of the pavement edges and to provide a datum from which to monitor rutting under controlled trafficking (Fig. 1). A watering system was assembled to promote the soaking of the road surface, in order to accelerate the rutting process (Fig. 2).

The trial section layout consisted of an unsurfaced pavement which comprised a base course of 300 mm, on which three different aggregate materials were tested, and a 300 mm layer of sub-base.

2.1 *Vehicles*

A variety of conventional and unconventional vehicle types were used to traffic the trial sections. The most common vehicles used were an articulated truck and trailer, the trailer being fitted with super-single tyres, and the other a FC rigid body vehicle with twin tyres. A third employed vehicle was a rigid body lorry with a trailer, a "Lorry & Drag", with twin tyres fitted on most axles.

Use has also been made of a bespoke vehicle currently used in Scotland, known as a Low Ground Pressure (LGP) vehicle, as a tool to reduced damage to forest roads by means of an evenly spaced configuration of tyres. Although the LGP vehicle has the same number of wheels as a twin-tyred lorry, the effect of load spreading caused by two separate sets of suspensions per axle and evenly spaced tyres across the chassis aims to allow a more evenly distribution load on the pavement, even if the road is cambered.

In addition, an articulated truck and trailer with a Tyre Pressure Control System (TPCS) was employed. This system allows the driver to operate on low pressure tyres when so controlled from the driver's cab. The TPCS system is of growing use in some countries in the forest business (Douglas et al, 2003). The variable pressure in the tyres has a direct effect on the area in contact with the pavement, and, therefore, the tyre-pavement contact pressure. At lower pressures the same vehicle may cross a softer pavement without causing rutting or wheel spin. The technique also seems to result in much less tyre wear (Munro & MacCulloch, 2007) and, perhaps, fuel use. The drawback is that the vehicle cannot travel at speed on conventional pavements without safety concerns and extra fuel use, so the pressure must then be increased to conventional levels. This system is credited in reducing the rutting damage in unsealed roads and also in improving ride quality for drivers.

Figure 1. Ringour Testing Facility before construction of the test section with major dimensions indicated and sections chainage.

Figure 2. Intermittent watering system during trial.

Figure 3 illustrates a summary of the vehicles employed during the trials. Note, especially, the use of vehicles with "super-single" tyres (single tyres nominally 385 mm wide as opposed to a pair of tyres each nominally 295 mm wide).

Figure 4a pictures a typical vehicle (FC Multi-lift) running on a trial; Figure 4b, a rear view of the LGP vehicle.

2.2 *Materials*

The granular materials used in this study were:

a. Type 1 – standard material for base and sub-base in road works in the United Kingdom – (Highways Agency, 2007) from Morrington Quarry,
b. a material from FC's Risk Quarry – a lower quality material of proven lower performance in regard to rutting observed in previous trials (Tyrrell, 2004),
c. a Type 1 material from the FC's Craignell Quarry,
d. a standard Type 3 (Highways Agency, 2007) material is also used.

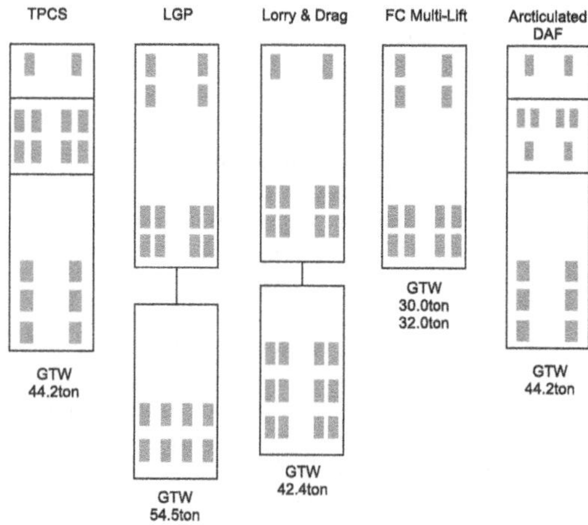

Figure 3. Tyre and axle arrangements for the vehicles used in the trial.

The first three of these were used in several of the trials as the surfacing aggregate whereas the last one was only used beneath the other aggregate layers to provide a cushion and mitigate the effects of the hard quarry floor on the surfacing. More details of the materials can be found elsewhere (Brito & Dawson 2007).

2.3 *Trial procedure*

A standard compaction protocol was followed (Highways Agency 2005) in order to reduce variability in compactions levels among the various trials. Between each trial, the test road's surface was prepared by loosening, re-grading and compacting the upper layer.

In summary, the trials consisted of 100 passes of one vehicle at a time in the test section, following the same wheel track. Measurements were taken of the changing surface profile at various number of passes, as described in Section 3 of this paper.

Figure 4. (a) FC "Multi-lift" truck trafficking a Ringour trial (b) "Low ground pressure" vehicle trailer with tyres equally spaced across pavement on stub-axles.

In some trials the vehicle re-trafficked the same wheel track, either at different moisture content, first dry and then in wet condition (Trials 1 & 2, 9 & 10), or at different tyre pressure, lower pressures first (Trial 7). Occasionally, two vehicles were run simultaneously in different wheel tracks during the same trial (Trials 4, 6, 9 & 10).

Throughout the test, the watering system was turned on to soak the surface for a defined period of time after each 10 vehicle passes. As weather varied amongst trials, and therefore evaporation and precipitation levels, moisture contents were determined in the aggregate material at 150 mm of depth at the end of each trial. The purpose of the water is to weaken the material perhaps promoting higher levels of permanent deformation. The combinations of materials, material condition, trafficking arrangements and abbreviation are shown in Table 1.

A Dynamic Cone Penetrometer (DCP) and a Mini FWD device were used in some of the trials for the assessment of compaction level and materials strength. They did not yield results with much spread in value, and are currently under further analysis.

On the basis of the axle loadings of each vehicle (obtained from weighbridges or by manufacturer's specifications and known loading), the equivalent damaging potential of each vehicle was estimated according to the familiar "4th Power Law" and the number of ESALs (Equivalent Standard Axle Loads) per vehicle pass was computed (Table 2). This "law" is known not to apply to the rutting of pavements without a bound course, but the alternative methods are no more reliable, so the 4th power computation is used on the basis that it is, at least, familiar.

One deficiency of the technique is immediately apparent. The TPCS (low tyre pressure) vehicle is computed to have the same damaging capacity as the same vehicle at higher tyre pressures because damage is assumed to be based solely on axle loading. Clearly this is erroneous as lowering tyre pressures unquestionably reduces pavement damage. However the same set of data also shows a relevant point – the so-called "Low ground pressure" vehicle causes more damage

Table 1. Loading and moisture matrix of Ringour Trials.

MATERIAL		Morrington Type 1		Risk Quarry Material		Craignell FC Type 1	
VEHICLE TYPE	Key	Dry	Wet	Dry	Wet	Dry	Wet
FC Multi-Lift with twin tyres	ML	#1	#2 #8*		#6	#9	#10
Artic DAF with super singles	Artic		#3		#6	#9	#10
Lorry & drag with super singles	L&D		#4				
LGP Vehicle	LGP		#5				
TPCS Vehicle	TPCS				#7		

represents the chronological order of the trial
*Trial with tyre pressures of 70 psi and after 110 psi

per pass than conventional vehicles. This apparent anomaly is not necessarily erroneous. Although the special trailer spreads the load better than conventional vehicles, the truck pulling the trailer almost certainly imposes significant damage which is greater than the damage saved by using the novel trailer.

3 PERMANENT DEFORMATION MONITORING

Permanent deformation monitoring was performed with a laser based device equipped with built-in Bluetooth transmitter, allowing connection with a handheld computer for data storage.

Two cross section monitoring stations T2 and T4 (see Figure 1) were set 6 m apart. On each station a portable 5.5 m long aluminum beam allowed height readings relative to a datum to be taken at 0.1 m spacings across the road.

At each cross section two rows of readings, 0.3 m apart, were taken for consistency. All four measured cross section profiles were plotted and the permanent deformation recorded for each wheel path at 0, 10, 20, 40, 70 and 100 vehicles passes. Figure 5 show a typical result of the permanent deformation monitoring.

Two methods were used to calculate the permanent deformation developed in the trials: rut depth and vertical surface deformation (VSD). The rut depth considers the upward shoving at the edges of the wheel path, and is determined by placing a straight edge across the surface. The VSD is the difference between the current cross section level and the start reference level of the pavement.

Although arguably the VSD may be considered best for describing surface deformation (Arnold 2004), the rut depth calculation yield better results for the analysis of the Ringour trials. This could be attributed to the high level of near surface deformation – considerable amount of heaving in the wheel paths edges. This is likely to be due to the high levels of moisture in the surface caused by watering mechanism used. Furthermore, as the surfacing materials used in the trials were somewhat dense graded, the water is expected to have permeated at low speed. Also, due to the rock floor in the quarry, subgrade rutting is not possible in these trials.

4 RESULTS

The observed rutting in the all trials is summarized in Figure 6. Those trials on which the vehicle run at different configurations in the same wheel track were plotted consecutively, whereas when two vehicles ran parallel to each other, the results are plotted from the origin.

The results of the analysis of permanent deformation as a function of number of ESALs are presented in Table 3. In this table the rutting rates are calculated considering only the last 60 vehicle passes in each trial – after initial settlement has occurred and rutting rates are more stable. The table also summarizes the various variables monitored in the trials and presents a ranking according to the rutting rates (per ESAL) which is illustrated by Figure 7.

5 DISCUSSION

For the pavements with low total rutting there is normally an initial high rate of rutting followed by a stabilizing response (e.g. Trials 1, 9 & 10). This type of response is very common (Dawson, 2008), and is

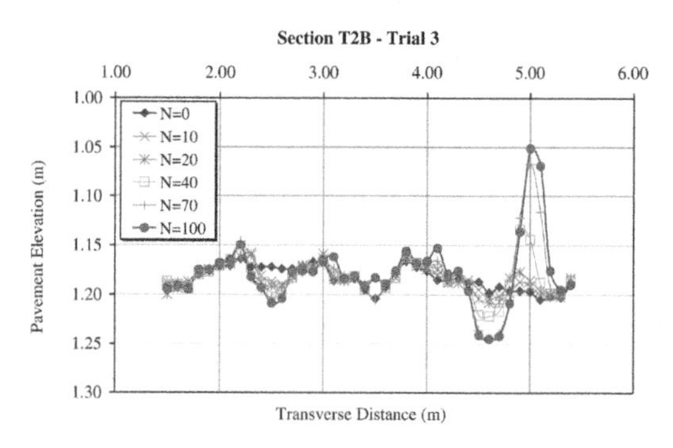

Figure 5. Example of a cross section profile measurement.

Table 2. Vehicles' ESAL (Equivalent Standard Axle Load) calculation according to 4th Power Law.

Vehicle	Trial	GTW (kN)	Front	Axles load (kN)				Rear	ESAL
ML	# 1,2,4,6,8	300400	62500	62500	87700	87700			3.63
ML	# 9	319900	62500	62500	97450	97450			5.15
Artic	# 3,6,9,10	442000	60000	95000	65000	74000	74000	74000	4.94
L&D	# 4	424000	75000	95000	95000	53000	53000	53000	5.33
LGP	# 5	545000	80000	80000	105000	105000	87500	87500	10.80
TPCS	# 7	442000	60000	95000	65000	74000	74000	74000	4.94

the response desired of the best pavements where traffic levels are to be highest and long-term performance must be assured.

Other pavements show an on-going development of rutting. In some cases this development is very large and rapid (e.g. Trial 3) but, for the most part, it is moderately fast. Such a response is appropriate for a less frequently trafficked pavement or one that only needs to provide service for a limited time.

Regarding the vehicle used, the much greater rutting that occurs under the FC's multi-lift vehicle is apparent when compared with that occurring due to the articulated vehicle used on the same pavements (Trials 6, 9 & 10). This is despite the fact that the articulated lorry has "super-single" tyres which, it is known from other studies, are more damaging than the twin tyres fitted to the multi-lift vehicle. Why is there this disparity in the present study? For Trial 6 this might be explained due to tyre inflation problems. Only by the time of Trial 8, it was found that the FC Multi-Lift vehicle had run from Trials 1 to 6 with the inner tyres of the rear-most axle with very low pressure (69 & 276 kPa). This was due to the difficult in reaching those tyres for pressure checking. Hence, as the rear axle didn't have tyres at the same pressures, this almost certainly resulted in local overloading and accelerated damage. For the other trials the reasoning is not so clear. It may relate to the more limited opportunity for load distribution along the length of a rigid-bodied lorry than along an articulated one. Table 2 does not indicate that the individual axles are very differently loaded, so a "rogue" axle loading does not seem to be the explanation in this case.

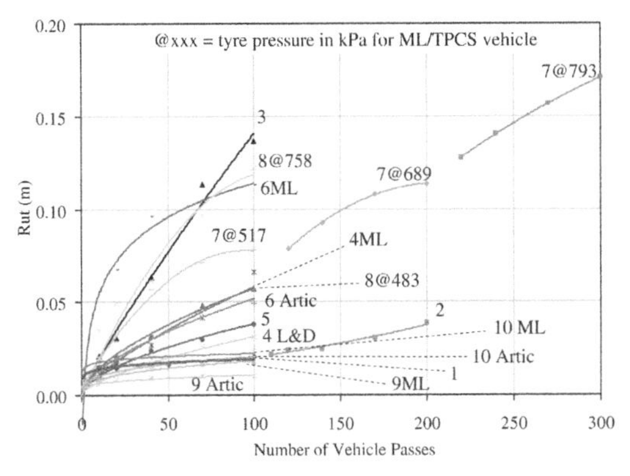

Figure 6. Rutting observed in trials at Ringour.

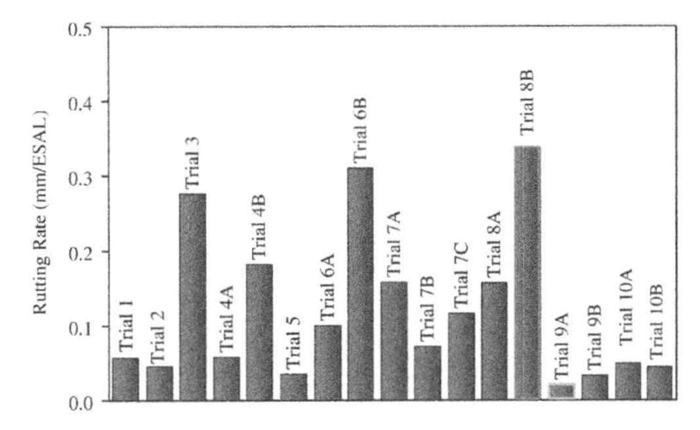

Figure 7. Rutting rates observed according to number of ESALs.

Table 3. Ringour Trials summary chart.

	Vehicle	Vehicle ESAL	Tyre Fitment	Mean Tyre Pressure (kPa)	Material tested	Moisture Condition	Moisture Content (%)	Rutting (mm)	No. of ESALs	Rutting Rate (mm/ESAL)	Ranking*
Trial 1	ML	3.63	Twin Tyres		Morring. T1	Dry		21	363	0.056	11
Trial 2	ML	3.63	Twin Tyres		Morring. T1	Wet	3.67	17	363	0.046	13
Trial 3	Artic	4.94	Sup. Singles	786	Morring. T1	Wet	4.80	137	494	0.276	3
Trial 4A	L&D	5.33	Twin Tyres	834	Morring. T1	Wet	4.14	31	533	0.058	10
Trial 4B	ML	3.63	Twin Tyres		Morring. T1	Wet		66	363	0.182	4
Trial 5	LGP	10.80	Twin Tyres	510	Morring. T1	Wet	5.17	38	1080	0.035	15
Trial 6A	Artic	4.94	Sup. Singles	724	Risk NS	Wet	6.20	49	494	0.099	8
Trial 6B	ML	3.63	Twin Tyres	779	Risk NS	Wet		113	363	0.311	2
Trial 7A	TPCS	4.94	Sup. Singles	517	Risk NS	Wet	5.53	78	494	0.158	5
Trial 7B	TPCS	4.94	Sup. Singles	689	Risk NS	Wet		35	494	0.072	9
Trial 7C	TPCS	4.94	Sup. Singles	793	Risk NS	Wet		58	494	0.116	7
Trial 8A	ML	3.63	Twin Tyres	483	Morring. T1	Wet	4.76	57	363	0.157	6
Trial 8B	ML	3.63	Twin Tyres	758	Morring. T1	Wet		122	363	0.336	1
Trial 9A	Artic	4.94	Sup. Singles	689	Craignell FC T1	Dry	3.48	10	494	0.020	17
Trial 9B	ML	5.15	Twin Tyres	689	Craignell FC T1	Dry		17	515	0.033	16
Trial 10A	Artic	4.94	Sup. Singles	689	Craignell FC T1	Wet	5.00	24	494	0.049	12
Trial 10B	ML	5.15	Twin Tyres	689	Craignell FC T1	Wet		23	515	0.045	14

*1 being the worst result – more rutting

The same explanation is almost certainly valid for the comparison of Trials 4A and 4B. Trial 4B caused approximately twice as much rutting as 4A, and a three times higher rutting rate, when related to number of ESALs. Nevertheless, a close study shows that the VSD levels were very similar for both vehicles; the heaving effect on the right hand side wheel path – trafficked by the FC Multi-Lift – was responsible for increasing the rutting measurement, suggesting higher shear stresses very near the surface. This may be explained by the later finding that the rear-most axle of the multi-lift was running virtually on two wheels – a factor that is likely to be the responsible for a more localised stress level between tyre and pavement, provoking higher shear stresses near the surface.

With the so-called "low ground pressure" vehicle, the novelty of spreading the load across the pavement through multiple wheels is largely negated by the heavily loaded axles of the rear of the tractor unit. Visibly, these appear to have been the chief cause of damage and the results in Figure 6 appear to confirm this although the lack of a direct comparison to the trafficking of the same pavement by another vehicle hinders interpretation.

Looking only at the rate of rutting during the last 60 passes of each trial (Table 3, last column), the (twin-tyred) Multi-Lift vehicle shows either a similar or greater rate than for the articulated vehicle (with "super-single" tyres on its trailer) – Trials 6, 9 & 10. Regarding the "low-ground pressure" vehicle, however, the story is somewhat different. Despite applying a very high number of equivalent standard axles compared to other vehicles, the rate of rutting is very low. This suggests that the vehicle doesn't prevent rut initiation but limits ongoing damage due to an inherent "kneading" action occasioned by the multiple wheels over the road width. Nevertheless this does lead to some looser aggregate on the pavement surface which can collect between the wheel tracks allowing higher than expected rutting, even when the VSD reading is low.

The story told by Table 3 concerning the effect of moisture and materials is interesting. There is a clear ranking of rutting from Risk (greatest) to Craignell (least). The relative moisture content has a lesser effect on rutting than the material chosen though some effect can be seen (e.g. compare Trials 9 & 10 where other parameters are constant). Trial 2 (ranked 13th in Table 3) behaved somewhat better than Trial 1 (ranked 11th), had the same lorry and the same construction, but was wet instead of dry. One possible reason is that, as the lorry trafficked the same wheel path during both trials, the initial movement that occurred in Trial 1 compacted the aggregate, limiting the opportunity for further deformation in Trial 2. Furthermore, following the assessment of Figure 6, it is possible to notice that the majority of the rutting from Trial 1 came from the early passes, resulting in a rather stable behaviour after 20 passes of the vehicle. Trial 2, however, presents an increasing rate of rutting throughout the test, resembling incremental collapse behaviour of the material. The relatively small permanent deformation registered is a valid response to the very low moisture content for a wet trial (3.7% – lowest of all trials – indicating that surface wetting did not measurably percolate into the aggregate). Considered is this way, the wetter condition is, after all, associated with poorer performance. Because the behaviour in the first trial probably affected the performance in the second, more disturbance and reconstruction was included between subsequent trials.

6 CONCLUSIONS

Some conclusions from the results and analysis presented in this study case can be drawn:

i. The means of loading and the type of aggregate used have a bigger influence on rutting performance than the relative moisture level of the aggregate (over the range of moisture conditions considered).
ii. Reduced axle overloading and controlling tyre pressures may be a more economic means of managing pavement deterioration than by using higher quality aggregates (the other meaningful option). This suggests that some kind of policing or QA system for truck operators might be worth considering.
iii. As far as rutting is concerned the type of vehicle appears to make some difference, but it has not been clearly established which factors of the different types of vehicles are most important in controlling propensity to rut. This aspect could usefully be investigated further, especially given the variety of options available and the need to have vehicles that can easily manoeuvre in the forest environment.

ACKNOWLEDGEMENTS

The work presented in this paper is part of the RUTT Project, funded by the Strategic Timber Transport Forum/Scotland. The authors wish to express their sincere thanks for all collaborants involved.

REFERENCES

Arnold G. 2004. Rutting of Granular Pavements. PhD Thesis, University of Nottingham, UK.

Brito, L.A.T & Dawson, A.R. 2007. Roads Under Timber Traffic. Interim Research Project Report. NTEC Report No. 07016. University of Nottingham. December 2007.

Dawson, A.R. (2008 – in press). Rut Accumulation in Low-Volume Pavements due to Mixed Traffic, paper accepted by *Transportation Research Record.*

Douglas R.A., Woodward W.D.H., Rogers R.J. 2003. Contact Pressures and Energies Beneath Soft Tires. Modelling Effects of Central Tire Inflation-Equipped Heavy-Truck Traffic on Road Surfaces. *Transportation Research Record* 1819. Paper No. LVR8-1015. p. 221–227.

Highways Agency 2005. Manual Of Contract Documents For Highway Works – Volume 1 Specification For Highway Works. Series 800 Road Pavements – (11/04) Unbound, Cement And Other Hydraulically Bound Mixtures. Amendment May 2005.

Highways Agency 2007. Manual of Contract Documents For Highway Works – Volume 2. Notes for Guidance On The Specification For Highway Works. Series Ng 800 Road Pavements – (11/04) Unbound, Cement And Other Hydraulically Bound Mixtures. Amendment November 2007.

Munro, R. & MacCulloch F. 2007. Tyre Pressure Control on Timber Haulage Vehicles. Roadex III Report on Taks B-2. December 2007.

Tyrrell, W. 2004. Pavement trials at Risk Quarry, Kirroughtree. *6th International Symposium on Unbound Materials – Unbar 6.* Proceedings Intern. Symp., Nottingham, 6–8 July 2004. England: Balkema.

Advances in Transportation Geotechnics – Ellis, Yu, McDowell, Dawson & Thom (eds)
© 2008 Taylor & Francis Group, London, ISBN 978-0-415-47590-7

Calcined clay aggregate: A feasible alternative for Brazilian road construction

Gustavo da Luz Lima Cabral & Laura Maria Goretti da Motta
COPPE/UFRJ, Rio de Janeiro, RJ, Brazil

Luiz Antônio Silveira Lopes & Álvaro Vieira
Institute of Military Engineering, Rio de Janeiro, RJ, Brazil

ABSTRACT: The shortage of natural aggregates in some regions of Brazil– particularly the Amazon region– constitutes a serious obstacle to highway engineering and, consequently, to the regional development. In recent years, research has been directed to obtain lower cost aggregates by way of the calcination of selected clay soils and via the analysis of the performance of these aggregates as paving materials. Said technology has attained results whose mechanical behaviour characteristics are similar to rock aggregates, which has resulted in a financially vantageous alternative for highway infrastructure in the Amazon region, and the North of Brazil in general. To put into practice the conceived methodology, samples of calcined clay aggregates originating from experimental production in a ceramic industry in the Amazon region were prepared. This paper presents the principal results obtained, as well as the quantification of the involved costs in this pioneering industrial production.

1 INTRODUCTION

The lack of natural aggregates in various regions of the country, particularly in the Amazon (Fig. 1), constitutes a serious obstacle to highway construction. Summing together the geologic scarcity of mineral deposits, the severe (and correct) environmental restrictions regarding their exploration, serve to aggravate this need, thus making the search for artificial aggregates an attractive alternative. In Brazil, the principal studies about technologies related to the discovery of artificial aggregates had continued to advance due to the efforts of scientific and strategic research towards this end.

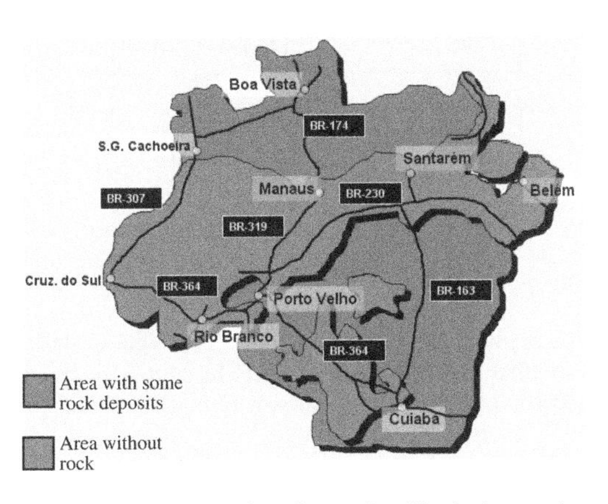

Figure 1. Shortage of rocks at Brazilian's Amazonian area (north area of the country). Principal cities (including Santarém) and federal roads.

Within this context, the Military Engineers of Brazilian Army have had to deal with this problem for some time, since several Engineering Battalions are located in this area, that, frequently are responsible for the construction and maintenance of the diverse highway corridors which are indispensable and aggregate value to regional development, national integration and help to maintain the strategic policies of the Amazon region.

Research about the production and use of ceramic aggregates in Brazil, had its beginning with studies being done at the Highway Research Institute (Instituto de Pesquisas Rodoviárias – IPR/DNER[1]), regarding the production viability of expanded clay aggregates in the northern region (DNER 1981, Fabrício 1986). The preliminary results indicated elevated production

[1] In Brazil, the DNER – Departamento Nacional de Estradas de Rodagem (currently the DNIT – Departamento Nacional de Infra-estrutura de Transportes) was the political organ responsible to execute transport policies as determined by the Federal Government.

costs for this type of aggregate, which made further studies not viable.

In the last 10 years, research done at the Institute of Military Engineering (Costa et al. 2000, Batista 2004, Cabral 2005) and at the Federal University of Rio de Janeiro – COPPE/UFRJ (Nascimento 2005) has been directed towards, with success, obtaining aggregates at lower cost, by way of calcination of selected clay soils and the analysis of their behaviour in asphalt and soils mixtures.

2 PROPOSED METHODOLOGY

With the intuition of searching for an industrial process for the calcined clay production, naturally, a great similarity was observed between the conventional ceramic industries. With this as a basis, the state of the art was researched, with emphasis placed on technical norms, laboratory tests, industrial equipment employed, fuel for furnaces, temperature and firing time of ceramic products.

After the literature search, it was possible to plan a methodology which could best represent and evaluate the characteristics of the raw material, the ceramic produced and at the same time, obtain the responses of the aggregate to be used in pavement. The proposed methodology is illustrated in Figure 2, which shows all the steps that involves the process: from the choice of raw materials, production, and employment of the calcined clay aggregate.

2.1 *The ceramic industry alternative*

Parallel to the beginning of the experimental phase of the study, which contemplated a complete application of the presented methodology, the principal national ceramic production poles were visited in the States of Rio de Janeiro and Pará.

Such knowledge was useful to conceive a plant design to manufacture the calcined clay aggregate, being proposed jointly with the methodology, similar in form to that elaborated by DNER (1981), whose research was focused on the expanded clay aggregate and that, due to motives of costs and low mechanical resistance, was deemed not viable for use in paving.

During these visits, another possibility for production was envisioned, and, consequently, a novel alternative was to be considered: to produce the calcined clay aggregate in a conventional industrial unit used to manufacture ceramic pieces, such as bricks and roof tiles.

This alternative was also included, due to peculiarities of the diverse municipalities located in the northern region of the country, affected by the scarcity of rock formations, and which do not possess the elevated financial resources required for the acquisition of a ready made plant factory. It was in light of these facts that the solution encountered was implemented and composed a line of action within the research.

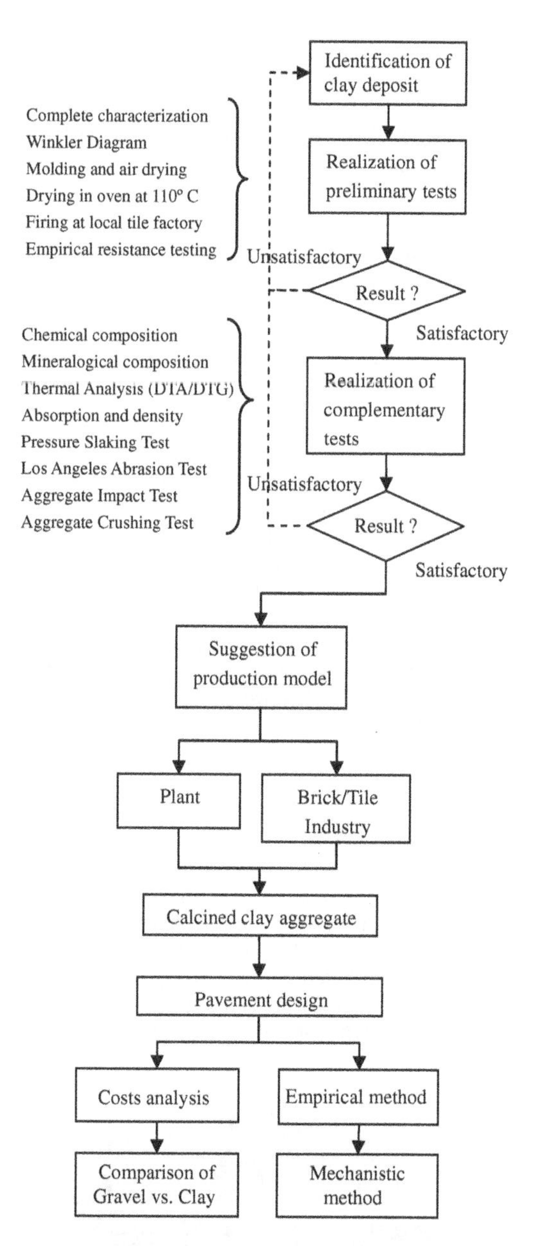

Figure 2. Flow diagram of methodology: steps of analysis of the raw material; choice of manufacturing model; aggregate employment.

2.2 *Experimental production*

As related, the possibility of evaluating an experimental production in a ceramic industry, seemed highly desirable. At this point, were then procured industries in the municipality of Santarém (located in the Amazon region) which had at their disposal a kiln with adequate control of firing temperature, to serve as a model within the framework of the proposed methodology.

During the visit to ceramic industries at Santarém, the technology was presented to the respective businessmen, and initially, was almost unanimously accepted, once it was demonstrated that the technology was not only a solution to a local engineering problem, but also represented a new alternative to the existing market of construction materials. The technical merits of the aggregates, their quality having been tested and established through subsequent research, combined with a lower cost to the final consumer in comparison to the acquisition of gravel based aggregates, proved to be an attractive alternative.

Believing that a pilot production could only be viable in the mid-term, nonetheless in less than three weeks contact was made by the owner of a ceramic factory, informing that they had already manufactured approximately 400 kg of aggregate, already crushed, and ready to be sent for evaluation in laboratory. Thus, in this manner was registered the first industrial production of the calcined clay aggregate, under technological control, in a production unit of ceramic pieces.

3 THE AGGREGATES PRODUCTION

To follow, are related the principal characteristics of the manufacture realized in the ceramic industry (in this work being denominated as the Production Unit or simply, PU) which was made available for the experimental production.

The clay is extracted and transported from the source (at the Amazon River) to the deposit under a covered patio at the PU during certain time, which a short curing or seasoning occurs. The mass is then prepared with conventional equipment common to ceramic factories (laminators, disintegrators, crushers and humidifiers) and then is extruded into bars 25 cm long, with the thickness and height approximately equal to a solid brick. Drying of the bars was satisfactory being performed at ambient temperature although a drying unit does exist at the PU.

The PU is equipped with tunnel type kilns for firing conventional ceramics. These pieces are piled in "wagons" which are moved over tracks into the kilns, which in turn, are closed and sealed for firing. The firing of the cylindrical bars was effected up to a threshold temperature of 950° C. These bars were piled in the upper part of the wagons, the same position as roof tiles, and were fired together with the pieces (bricks and tiles) of the conventional production at the PU.

Table 1. Steps of the experimental stage.

Item	Action taken
1st	Preliminary tests on raw material.
2nd	Complementary tests on raw material: chemical and mineralogical analysis, differential thermal analysis and thermogravimetric analysis.
3rd	Experimental production of 400 kg of calcined clay aggregates in a ceramic industry.
4th	Mechanical resistance evaluation of the calcined clay aggregates experimental lot.
5th	Costs analysis involved in the experimental production.
6th	Employment: soil-aggregate mixture; asphalt mixture; performance evaluation (laboratory).

The PU effects temperature control by way of sensors placed at diverse points within the tunnel furnace. The firing time lasted, on the average, 36 hours, but this was due to the necessity to take advantage of the total firing time used for the production of bricks and tiles, which, at this PU, generally requires this firing time. The fuel used was a mixture of firewood and leftovers from logging and wood manufacturing operations. Both are acquired at low cost due to the large number of wood manufacturers in the region near the PU. On the occasion of the visit, new burners were being installed to use sawdust and natural gas as fuels. A burning of one mobile wagon produces between 13,000 to 20,000 pieces, consuming on the average, 20 m^3 of firewood.

By convention, this sample of aggregates produced in the PU at Santarém, received the denomination of "Industrial Calcined Clay".

4 RESULTS OBTAINED

In the sequence of the research, there were realized diverse tests to verify the mechanical resistance of the "industrial calcined clay" aggregate, as well as its performance in soil aggregate and asphalt mixtures. To better understand the results obtained, Table 1 shows the operational sequence followed.

4.1 *Preliminary and complementary tests*

Besides the two samples of clay soils provenient from Santarém (light and dark), there was a third sample which, in truth is a mixture of 50 weight % of each. This was performed to simulate the practice of Santarém ceramic industries. These raw material were called "without addition", because other materials were in fact added in the test sequence to verify the absorption of the artificial aggregate.

Table 2. Characterization of the raw material (clays) from Santarém.

Sample	Fg %	Cs %	Ms %	Fs %	Si %	Cl %	LL %	PL %	PI %
Dark sample	0.0	0.0	0.35	13.7	12.8	73.1	59.8	30.6	29.2
Light sample	0.1	0.1	1.55	31.6	4.86	61.9	45.6	25.2	20.4
Mixture w/out addition	0.0	0.1	1.39	27.4	8.30	62.8	44.7	22.7	22.0

Fg = Fine crushed gravel; Cs = Coarse sand; Ms = Medium sand; Fs = Fine sand; Si = Silt; Cl = Clay; LL = Liquid Limit; PL = Plastic Limit; PI = Plasticity Index.

Table 3. Chemical composition by EDS and Fluorescence.

Sample (method)	(a) %	(b) %	(c) %	(d) %	(e) %	(f) %	(g) %	(h) %	(i) %
Methodology lower limit	0.10	15.0	11.9	0.08	0.01	0.01	0.10	0.01	0.01
Mixture w/out addition (EDS)	–	65.0	26.2	8.79	–	–	–	–	–
Mixture w/out addition (Fluorescence)	9.92	57.3	25.9	3.60	0.19	1.08	–	–	0.92
Methodology upper limit	27.0	77.8	56.0	9.62	20.1	3.50	16.3	11.8	16.9

Dash (–) = It was not identified a significant value of the chemical substance for the accomplished method

(a) = Ignition loss; (b) = SiO_2; (c) = Al_2O_3; (d) = Fe_2O_3; (e) = CaO; (f) = TiO_2; (g) = MgO; (h) = Na_2O; (i) = K_2O.

Table 4. Clay minerals present in raw material provenient from Santarém.

Sample	Present clay minerals	Predominant clay mineral
Light sample	Kaolinite, Illite, or Quartz	Kaolinite
Dark sample	Kaolinite and Illite	Illite
Mixture w/out addition	Kaolinite and Illite	Kaolinite

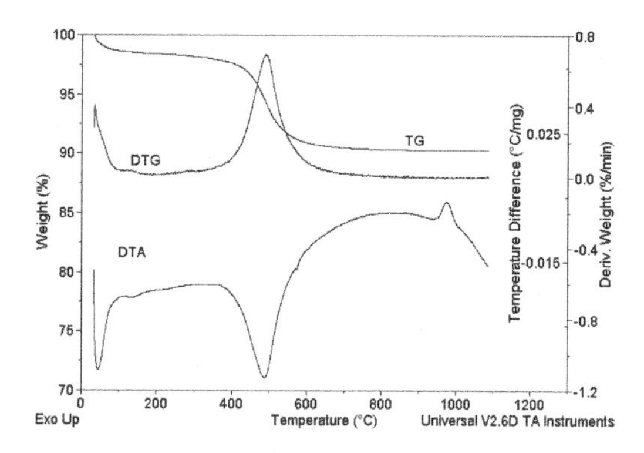

Figure 3. Curves of differential thermal analysis, thermogravimetric and derived thermogravimetric.

Being in accord with the methodology, whose limits are taken from the literature, the mixture of the two types of clays revealed chemical substances and clay minerals (Table 3 and Table 4) propense to the formation of a ceramic body in which the role of flux elements could diminish the porosity and absorption, and increase the mechanical resistance of the calcined clay aggregate.

Figure 3 shows the thermogravimetric (TG), derived thermogravimetric (DTG) and the differential thermal analysis (DTA) curves. They represent the thermal analysis of the raw material (clay) where its properties are studied as they change with temperature.

The DTA curve provides data on the transformations that have occurred, such as glass transitions, crystallization, melting and sublimation. Changes in the sample, either exothermic or endothermic, can be detected.

Table 5. Results of mechanical resistance tests with laboratory prepared aggregates of calcined clay (900°C), industrial calcined clay and commercial gravel provenient from Rio de Janeiro.

Test (Reference Method)	Sample 1	Sample 2	Sample 3	Limits*
AASHTO T 96	30%	28%	47%	Less than 50%
Tex-431-A	2.1%	1.2%	2.4%	Less than 6%
BS 812-110	23.5%	21.4%	27%	Less than 40%
AASHTO T 85	16%	11.3%	0.90%	Less than 15%
BS 812-111	114 kN	106 kN	89 kN	More than 60 kN
BS 812-112	19%	18.6%	20%	Less than 60%

*All the limits are suggested by the present methodology
Sample 1 = Calcined clay from laboratory at 900°C
Sample 2 = Industrial calcined clay
Sample 3 = Commercial gravel from Rio de Janeiro
AASHTO T 96 = Los Angeles abrasion test
Tex-431-A = Pressure slaking test of synthetic coarse aggregate
BS 812-110 = Aggregate crushing value – ACV
AASHTO T 85 = Specific gravity and absorption of coarse aggregate
BS 812-111 = Ten percent fines value – TFV
BS 812-112 = Aggregate impact value – AIV

The TG, and consequently, DTG curves are based on continuous recording of mass changes of a sample, as a function of a combination of temperature with time. The commonly investigated processes are: thermal stability and decomposition, dehydration, oxidation, determination of volatile content and other compositional analysis.

It should be pointed out what occurs on the TG and DTG after 500° C: the loss of mass is practically insignificant, and this fact may allow the firing process to be accelerated after passage through this critical temperature, once that would not be formed fissures, cracks or internal tension in the aggregate.

4.2 *Mechanical resistance of the industrial calcined clay (experimental lot)*

The following tests were realized: Los Angeles abrasion test (AASHTO T 96); Pressure slaking test of synthetic coarse aggregate (Texas DOT Method Tex-431-A); Aggregate crushing test (British Standards – BS 812–110); Specific gravity and absorption of coarse aggregate (AASHTO T 85); 10% Fines Aggregate Crushing Test (British Standards – BS 812–111); Aggregate impact test (British Standards – BS 812–112).

Comparatively, the same tests (with five samples) were performed with rock gravel aggregates and calcined clay aggregates made in laboratory ovens.

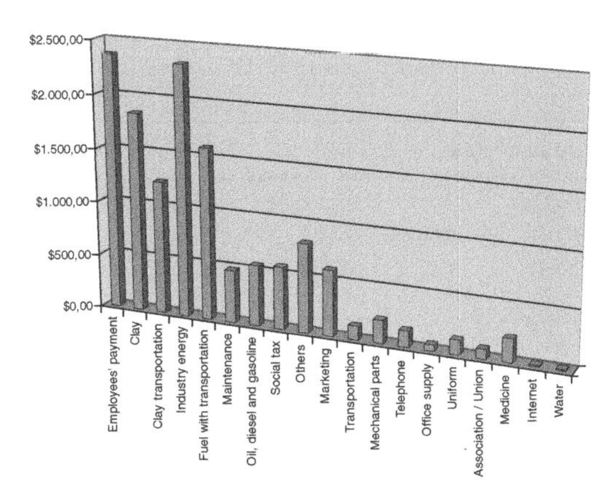

Figure 4. Costs (monthly) for solid brick production at the Production Unit (PU).

4.3 *Cost analysis of experimental production*

In order to estimate a manufacturing cost of the industrial calcined clay aggregate, some data relative to the productive process were necessary, as well as a monthly planning schedule (Fig. 4) relative to the PU costs, with information regarding overhead and material costs related to the production of solid brick manufactured in this ceramic industry.

Knowing that these annotations refers (in terms of weight) to a production of 486 metric tons of solid brick during the period of analysis, and parting from some basic considerations with respect to the transformation of the volume produced to crushed aggregates, calculations are presented in Table 6.

4.4 *Use as pavement material: soil-aggregate and asphalt mixture*

Tests were realized according to the following sequence:

a) Soil-aggregate mixture:
a.1 – Compaction Test;
a.2 – CBR Test;
a.3 – Resilient Modulus Test (M_R).
b) Asphalt mixture:
b.1 – Marshall method design;
b.2 – Indirect Tension Resistance test (R_T) and Resilient Modulus (M_R);
b.3 – Degradation test after Marshall compaction.

Table 6. Estimated cost of the industrial calcined clay aggregate in Brazilian reais (R$), US dollars (US$) and Euros (€).

Item description	BRL	USD	EUR
Cost/ton	R$ 60.78	US$ 30.65	€23.05
Specific apparent mass (ton/m^3)	1.05		
Conversion factor (lower limit)	0.25		
Conversion factor (upper limit)	0.60		
Cost/m^3 (solid)	R$ 63.82	US$ 32.19	€ 24.20
Cost/m^3 crushed (lower limit)	R$ 51.06	US$ 25.75	€ 19.37
Cost/m^3 crushed (upper limit)	R$ 39.89	US$ 20.12	€ 15.13

Table 7. Results of CBR and Resilient Modulus (M_R) of soil-aggregate mixtures.

Mixtures*	CBR %	M_R MPa
Mixture 1	86	153.6
Mixture 2	64	144.9
Mixture 3	84	192.3
Mixture 4	105	228.9

*All mixtures were designed with 50% of sub-grade soil and 50% of aggregates from the following sources:
Mixture 1: Aggregate of industrial calcined clay
Mixture 2: Aggregate of laboratory calcined clay at 900°C
Mixture 3: Aggregate of laboratory calcined clay at 1000°C
Mixture 4: Commercial gravel from Rio de Janeiro.

A soil-aggregate mixture was planned (Table 7) according to a composition of 50 wt. % of sub-grade soil (CBR = 15%) from the study region (Federal Highway BR-163 in the State of Pará), and 50 wt. % of clay aggregates. Comparatively, the same mixture (with an identical granulometric curve) was tested using the Rio de Janeiro gravel, which in turn, was being used as the basis for tests of mechanical resistance.

In addition to the gravel and the industrial calcined clay aggregate, two additional samples of aggregates were utilized, thus forming 4 distinct mixtures.

These last two samples were aggregates of calcined clay prepared in laboratory ovens at temperatures of 900°C and 1.000°C.

The asphalt reached an optimum content with 8.7% of bitumen (by weight of total mix) and a void content of 4.4%, according to Marshall specifications. The performance of the asphalt concrete designed only with bitumen (Penetration Grade 50/70) and 100% of industrial calcined clay aggregates is presented in Table 8.

Table 8. Comparative performance of asphalt concretes. Results of indirect tension resistance test (R_T) and Resilient Modulus (M_R).

Parameter	Reference 1* Asphalt concrete with gravel calcined clay	Reference 2** Asphalt concrete with laboratory clay aggregates	Asphalt concrete with 100% of industrial calcined
R_T – 25°C	0.81 MPa	0.65 MPa	0.76 MPa
R_T – 30°C	0.63 MPa	0.33 MPa	0.47 MPa
R_T – 35°C	0.42 MPa	0.22 MPa	0.33 MPa
M_R – 25°C	3520 MPa	2086 MPa	3225 MPa

*Reference 1 = Pinto (1991)
**Reference 2 = Batista (2004)

Table 9. Degradation Index of industrial calcined clay aggregates after Marshall compaction.

Sieve	Medium of specimens tested (passing %)	Designed asphalt mixture (passing %)	Difference
3/4″	100%	100%	0%
1/2″	85%	85%	0%
3/8″	78%	75%	3%
#4	58%	52%	6%
#10	34%	25%	9%
#40	17%	16%	1%
#80	13%	11%	2%
#200	11%	8%	3%
		Degradation Index (all sieves) = 4.0%	

Table 9 presents the last result regarding the asphalt mixture: the Degradation Index verified after compaction in Marshall equipment. Some authors (Carneiro & Silva 1979, Khan et al. 1998) registered that the Marshall compactor results in a more severe impact than that experienced in field. As such, limits of 6% were already cited as the most appropriate for acceptance of aggregate degradation, after compaction in the Marshall apparatus with bitumen.

5 DISCUSSION AND CONCLUSIONS

- The knowledge obtained with the ceramic manufacturers ratified the need to evaluate: the chemical composition, the mineralogy and the thermal analysis of the raw material;
- The raw material which possesses a significant content of flux elements (or still alkaline oxides), as well

as the clay minerals of the illite and kaolinite groups confers a satisfactory resistance upon the aggregate;

- Thermal analysis (differential and thermogravimetric) has proved to be a useful tool as how to thermally treat the aggregate, in other words, its "burning plan";
- The hypothesis of the production of calcined clay in a ceramic industry, consolidated during the visits, proved to be technically viable and financially advantageous. This alternative was mainly planned, in order to attend the small demands of municipalities from Brazilian northern region, which encounters logistical difficulties peculiar to this region, and does not have conditions to dispose elevated financial resources to acquire pre-fabricated plants;
- The produced aggregate in ceramic industry presented mechanical resistance similar to those observed made with natural aggregates. In addition to this fact, the degradation test confirmed the capacity of the aggregate to support severe mechanical compaction even superior than what is observed in field;
- An estimate and simple accounting of the production costs performed in Santarém demonstrated that one m^3 of the industrial calcined clay aggregate was approximately US$ 23.00/m^3 (€ 17.00/m^3). It must also be said that commercial gravel at the studied region costs more than US$ 50.00 (€ 35.50) per m^3;
- The bitumen content at the calcined clay aggregate asphalt (8% to 9% by weight of total mix) may appear elevated when compared to the amount used in conventional asphalt concretes with natural aggregates (4% a 7%). It should be remembered that the proportion of mass and volume differs between these two mixtures in particular, and it should be understood that the thickness of the layer observed in the pavement, in principle, should be the same;
- Taking as an example a conventional Marshall specimen, which for gravel mixtures, a mass of 1200 g (thickness approx. = 6 cm) is observed, for an asphalt concrete prepared with calcined clay, the mass corresponding to the same volume is approximately 900 g, including the bitumen. In other words, a bitumen content of 8% in an asphalt concrete with calcined clay, corresponds by weight to a bitumen content of 6.7% in a specimen with 1200 g and same thickness;
- Supposing now a 6.7% bitumen content, it would be increased by 1% compared to a conventional asphalt concrete with gravel (5.7%). It should be remembered at this point, that the relation of costs between acquisition of gravel in region of study, and the cost attained with experimental production, is greater than 100%. Considering also that the aggregates correspond, by weight, to more than 90% of both asphalt mixtures in question, it is trivial to conclude the financial advantage to the adoption of asphalt concrete with calcined clay, remembering yet again that similar performances were observed at both asphalt mixtures (indirect tension resistance and resilient modulus);
- Considerations made in the estimative calculation resulted in a median cost of US$ 23.00/m^3 (€ 17.00/m^3) for the calcined clay. According to data supplied by the owner of the industry responsible for the experimental production, there are various factors which are capable of causing an accentuated decrease in this value if a specific, dedicated production line were designed and constructed;
- The current objective to construct and evaluate two experimental tracks with this alternative material is being faced forcefully in order to propose this infrastructure transport solution to the diverse municipalities of the northern region of Brazil and, this way, qualify the construction of local roads which were previously not viable due to the "Amazon cost".

REFERENCES

Albuquerque, A. P. F. de. 2005. *Influence of the compacting energy and temperature in asphalt concretes mechanical properties(In Portuguese)*. M.Sc. Thesis. Rio de Janeiro: Institute of Military Engineering.

Asphalt Institute 1963. *Mix design methods for asphalt concrete and other hot-mix types.* 2nd Edition. 3rd Printing. Manual Series N° 2 (MS-2). College Park, Maryland.

Batista, F. G. da S. 2004. *Physical and mechanistic characterization of calcined clay aggregates produced with fine soils from BR-163(In Portuguese)*. M.Sc. Thesis. Rio de Janeiro: Institute of Military Engineering.

Cabral, G. da L. L. 2005. *Methodology for production and use of calcined clay aggregates in paving (In Portuguese)*. M.Sc. Thesis. Rio de Janeiro: Institute of Military Engineering.

Carneiro, F. B. L. & Silva, H. C. M. 1979. Aggregate degradation at granular base layers and at asphalt concrete surface layers (In Portuguese). 4th *Asphalt Meeting.* Rio de Janeiro: Instituto Brasileiro de Petróleo.

Costa, F.W.A, Silva M.A.V. da, Mello, M.A. 2000. *Amazon calcined clay aggregate: analysis of fatigue and resilience parameters in asphalt mixtures – Final Report (In Portuguese)*. Rio de Janeiro: Institute of Military Engineering.

DNER 1981. *Implantation Viability Research of an Expanded Clay Industry at the Amazon Region – Final Report (In Portuguese)*. Rio de Janeiro: Departamento Nacional de Estradas de Rodagem.

Fabrício, J. M. 1986. Development of a clay aggregate mobile plant (In Portuguese). 21st *Annual Pavement Congress, v. 1, pp. 150–188*. Salvador: Brazilian Pavement Association.

Grim, R. E. 1962. *Applied clay mineralogy*. New York: McGraw-Hill Book Company.

Khan, Z. A., Wahab, H. I. A., Asi, I.; Ramadhan, R. 1998. Comparative study of asphalt concrete laboratory compaction methods to simulate field compaction. *Construction and Building Materials 12, p. 373–384.*

Lee, Dah-Yinn & Kandhal, P. S. 1990. Absorption of asphalt into porous aggregates. *Strategic Highway Research Program.* SHRP-A/UIR-90-009.

Macedo, J. A. G., Lucena, F. B., Ferreira, H. C. 1987. Degradation of aggregates in hot asphalt mixtures. 22th *Annual Pavement Congress, v. 1, pp.132–144*, Maceió: Brazilian Pavement Association.

Nascimento, R. R. 2005. *Solutions for Rio Branco pavements with the calcined clay aggregate (In Portuguese).* M.Sc. Thesis. COPPE/UFRJ – Federal University of Rio de Janeiro.

Pinto, S 1991. *Study of the fatigue behavior of asphalt mixtures and application in structural evaluation of pavements (In Portuguese).* Ph.D. Thesis. COPPE/UFRJ – Federal University of Rio de Janeiro.

Advances in Transportation Geotechnics – Ellis, Yu, McDowell, Dawson & Thom (eds)
© 2008 Taylor & Francis Group, London, ISBN 978-0-415-47590-7

Modulus-based quality management of unbound pavement layers with seismic methods

M. Celaya, I. Abdallah & S. Nazarian
Center for Transportation Infrastructure Systems, The University of Texas at El Paso, El Paso, USA

ABSTRACT: The objective of any transportation project is to construct durable materials that perform well during the design life. The stiffness of an unbound layer and its variation with moisture are amongst the parameters that affect the performance of flexible pavements the most. The most common acceptance criterion in the field is to obtain the appropriate density; but this does not guarantee achieving the appropriate stiffness. A method to evaluate the stiffness of unbound layers is highly desirable. In this paper, a systematic quality management program based on seismic methods is presented. The protocol, which focuses on obtaining the elastic modulus of the material, consists of interrelated laboratory and field tests. A case study is presented to demonstrate the benefits of the proposed method.

1 INTRODUCTION

In the state of practice, most state highway agencies (SHAs) perform the quality management of the unbound materials on the basis of adequate gradation, plasticity, and nature of coarse aggregates. Even though these parameters have been valuable in ensuring durability, they do not provide any indication of the stiffness of the material, the primary parameter used in the structural design of pavements. The acceptance of the compacted layer is based on adequate density and moisture content, regardless of whether the desired modulus is achieved. To implement modulus-based quality management of aggregate layers, the gap among the design process, and laboratory and field testing should be addressed.

To provide continuity between the design and construction, appropriate laboratory tests need to be performed to obtain field target moduli and field methods need to be developed to evaluate moduli of the in-place materials. Nondestructive testing (NDT) methods allow for obtaining modulus in a rapid way for evaluating pavement layers. Seismic methods are particularly attractive since they are the only method where the lab and field tests yield the same parameters (low-strain elastic moduli) without a need for a transfer function or backcalulation. The Portable Seismic Property Analyzer (PSPA) is discussed here as a field device capable of performing almost real-time seismic tests on conventional compacted aggregate materials. The Free-free Resonant Column (FFRC) is the laboratory device used in this protocol to obtain target moduli on specimens prepared under laboratory conditions. With these seismic methods a systematic quality management program (QMP) can be implemented, where the minimum expected target values are obtained and potential problems during construction are identified.

Based on the previous work by Nazarian et al. (2002), some steps need to be taken in order to implement a comprehensive QMP. First, the suitability of a material is evaluated when design parameters are selected and pre-construction laboratory tests are carried out. Then, the acceptable design modulus is determined, based on the same tests. Finally, field modulus is measured on constructed materials, ensuring that acceptable level of target modulus is obtained.

2 TEST METHODOLOGY

2.1 *Laboratory testing*

In the proposed protocol, the FFRC (Figure 1) is used to obtain the modulus in the laboratory. In the FFRC (Nazarian et al. 2003), the specimen is prepared using the Proctor or other similar procedure. An accelerometer is securely placed on one end of the specimen, while the other end is impacted with a hammer connected to a load cell. The signals collected from the accelerometer and load cell are used to determine the resonant frequency, f. Then, the Young's modulus (E_{lab}) can be obtained:

$$E_{lab} = \rho(2fL)^2 \quad (1)$$

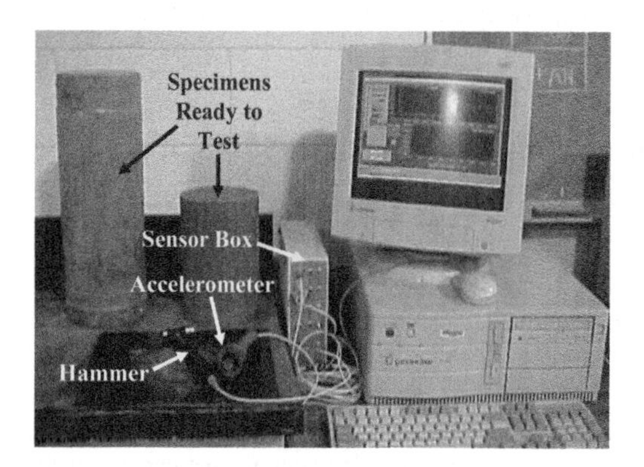

Figure 1. Free-free Resonant Column Test Setup.

where ρ = mass density and L = the length of the specimen. Since seismic tests are nondestructive, the specimen can also be tested later for strength (triaxial test), stiffness (resilient modulus test), or the variation in modulus with moisture under capillary suction.

2.2 *Field testing*

The PSPA, or its modified version for unbound materials, known as Dirt SPA (Figure 2), is the device used to measure modulus in the field. DSPA consists of two transducers (accelerometers) and a source packaged into a hand-portable system that performs high frequency seismic tests. The source package includes a transducer used for triggering and for advanced analysis purposes and the device is controlled from a computer that handles all the data acquisition system. The average modulus of the exposed surface layers can be estimated within a few seconds in the field using the Ultrasonic Surface Wave (USW) method (Nazarian et al. 1993).

In the USW method, the modulus of the top pavement layer is determined with the equation:

$$E = 2\rho(1+\upsilon)\left[V_R(1.13-0.16\upsilon)\right]^2 \quad (2)$$

where V_R = the surface wave velocity, ρ = mass density, and υ = Poisson's ratio. In this manner, the modulus of the top pavement layer is obtained without an inversion algorithm, since at wavelengths less than or equal to the thickness of the uppermost layer, the velocity of propagation is independent of wavelength.

Typical voltage outputs or time records of the three accelerometers for a base material are shown in Figure 3. These time records are converted to dispersion curve (variation in velocity with wavelength), which is then converted to modulus with wavelength (relabeled as depth for practical reasons in Figure 4).

Therefore, the operator can get a qualitative feel for the variation in modulus with depth. The displayed

Figure 2. Dirt Seismic Property Analyzer (DSPA).

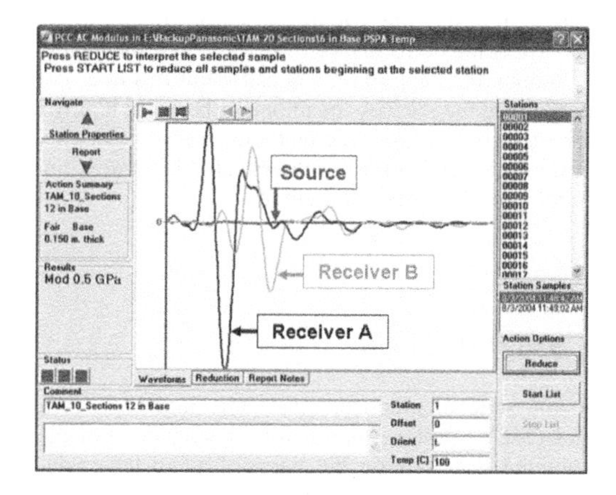

Figure 3. Typical Time Records (Base Material).

modulus is obtained by averaging the dispersion curve down to approximately the nominal thickness of the layer (150 mm in this case). To obtain good quality voltage outputs or time records, the surface of the material should be smooth and clean of debris, to allow good contact between the sensor and the material surface. This is not usually a problem since the DSPA is employed when the material is finished and accepted. When the accelerometers do not achieve a good contact with the material surface, the voltage outputs will be erratic with multiple peaks; or in the case of no contact with the surface, no apparent signal will appear in the time records. In those cases, the test is rejected, the DSPA is relocated and a new test is performed.

The phase spectrum, which can be considered as an intermediate step between the time records and the dispersion curve (bottom of Figure 4), is determined by conducting spectral analysis on the time records from the two receivers. Two phase spectra are shown, one measured from the time records, and the best estimation of the phase when the effect of the body waves are removed (interpolated). The latter is then used to compute the dispersion curve as described in Desai & Nazarian (1993).

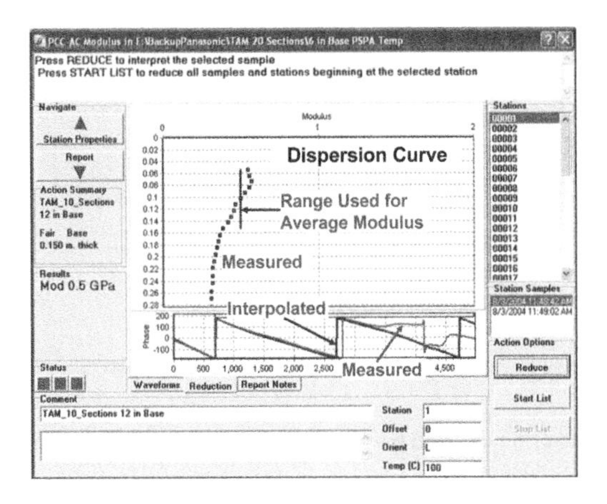

Figure 4. Reduced Data from Time Records from Figure 3.

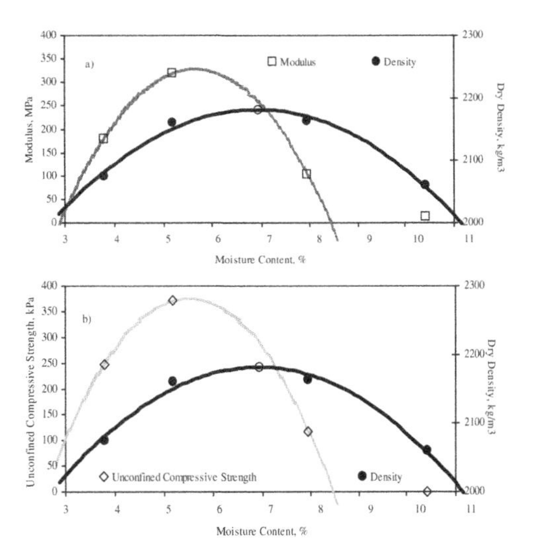

Figure 5. Moisture-Modulus and Moisture-Strength Curves.

3 QA/QC PROTOCOL

To accomplish a modulus-based quality management of unbound layers several steps need to be completed. These steps are briefly discussed next and the complete process can be found in Nazarian et al. (2002).

3.1 *Establish moisture-modulus curve*

Several specimens at different moisture contents need to be prepared with the same compaction energy to develop a moisture-density relationship. Specimens of 150 mm (diameter) by 300 mm (length) for materials that contain gravel and 100 mm by 200 mm for finer materials are recommended. In a same way, a moisture-modulus curve and the water content at which the maximum seismic modulus is obtained can be determined. Specimens compacted to develop the moisture-density relationship can be used to obtain the moisture-modulus curve.

As an example, a moisture-modulus curve is superimposed on the moisture-density curve for an unbound base in Figure 5a. The optimum moisture content (OMC) is about 6.9% and the maximum dry density (MDD) is 2182 kg/m^3. The maximum modulus of 330 MPa occurs at 5.7% moisture content, while the modulus at OMC is 214 MPa. One important observation from this figure is that by changing the moisture content by ±2.5% around the OMC, the dry density changes by less than 6%. In the same moisture range, the variation in modulus is more than 300%. The specimen at a moisture content of 10.5% was not stable outside the mold and yielded a modulus of about 14 MPa. By way of comparison, the variation in unconfined compressive strength of the same specimens with moisture is shown in Figure 5b. The trend in the variation in strength is quite similar to the modulus.

3.2 *Seasonal variation of modulus with moisture*

The variation in modulus with moisture and water retention potential are some of the major concerns with the unbound aggregate bases. Khoury & Zaman (2004) revealed that the variation in modulus with moisture obtained under a constant compactive effort is quite different from the moisture variation when the specimen is compacted at the optimum moisture content and then the moisture is allowed to change.

Nazarian et al. (2002) adapted a test to quantify the moisture sensitivity of unbound materials, by monitoring the moisture content and the seismic modulus. In the tests, a specimen compacted at optimum moisture content is dried in an oven set at 40°C for several days (usually two days), and then placed in a water bath for capillary wetting (typically six days). During this process, the specimen is weighed every day to determine the moisture change, and then tested with the FFRC device.

The variations in modulus and moisture content with time for drying and wetting cycles are then plotted, as shown in Figure 6. In general, as the specimen dries in the oven the modulus increases and the moisture content decreases. However, during the wetting stage, the moisture content increases and the modulus decreases. The magnitude of this drop is dependent on the mineralogy (especially fine content) of the material. The initial moisture content is about 6.5% for the case presented in Figure 6, while after two days of drying the moisture content is about 1.5%. After eight days of capillary saturation, the moisture content is about 5%, which is less than the initial moisture content and almost 2% less than the OMC. The seismic

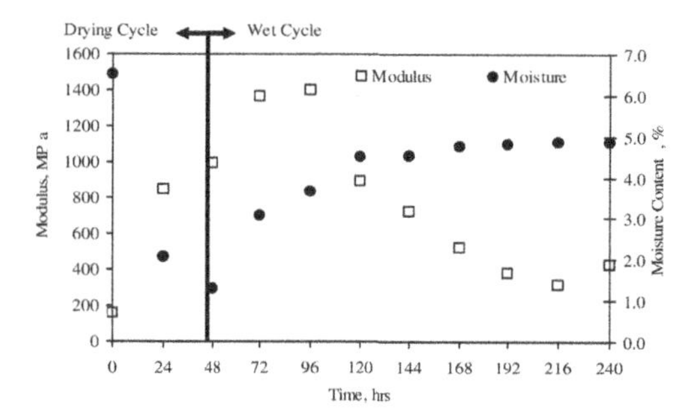

Figure 6. Moisture Susceptibility of Base Material.

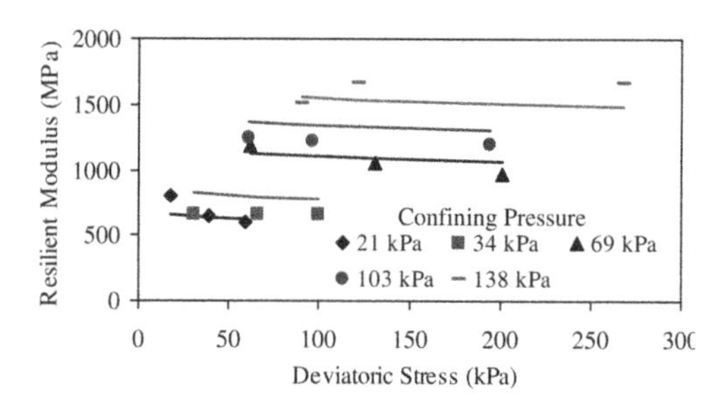

Figure 7. Resilient Modulus Results of Base Material.

modulus after 24 hours is 850 MPa, the peak modulus is about 1400 MPa, while the residual modulus, or modulus at the end of the wetting cycle, is about 400 MPa. The maximum modulus was obtained two days after the capillary saturation was initiated, showing that the absorbed moisture was localized to the bottom portion of the specimen, supported by the fact that the moisture content only increased from 1.3 to 3.6% from day 2 to day 4.

The modulus obtained at the end of the wetting cycle is considered as the residual modulus. Soils that do not exhibit low residual modulus compared to the modulus under the dry condition are assumed to be well-performing soils. Moreover, the seasonal variation of modulus can be estimated from the plot of modulus with moisture.

3.3 *Determine design modulus*

Seismic moduli need to be transformed to design moduli because seismic moduli are low-strain moduli; whereas the design moduli close to the applied load correspond to the high-strain moduli. Design modulus also depends on the thickness of the structure and on the state of stress under representative loads, and it can be related to seismic modulus through a nonlinear structural model (Abdallah et al. 2002). The material model adopted for bases and subgrades is in the form of:

$$E_{design} = E_{seis}\left(\frac{\sigma_{c\_ult}}{\sigma_{c\_init}}\right)^{k_2}\left(\frac{\sigma_{d\_ult}}{\sigma_{d\_init}}\right)^{k_3} \tag{3}$$

where E_{design} and E_{seis} (or k_1) are the design modulus and seismic modulus, respectively. Parameters σ_c and σ_d are the confining pressure and deviatoric stress at the representative depth, respectively. Subscripts "*ult*" and "*init*" correspond to the condition when the maximum truckload is applied to the pavement and the free-field condition, respectively. Parameters $k2$ and $k3$ are regression parameters that are preferably determined from resilient modulus laboratory tests on the specimen.

For the case presented in this paper, resilient modulus tests were carried out to obtain the regression parameters k_1, $k2$and $k3$. Laboratory results are presented in Figure 7 for several deviatoric stresses and confining pressures that varied from 21 to 138 kPa in this case. The regression parameters k_1, $k2$ and $k3$ are 386 MPa, 0.50 and −0.04, respectively.

3.4 *Field quality control*

The last step in the quality management protocol is to perform field tests using the DSPA. Tests are usually conducted at regular intervals or at any point subject to segregation, lack or excess moisture, or any other constructed related anomalies. Field seismic moduli should be greater than the representative seismic modulus obtained in the previous section. The field moduli obtained with the DSPA can be used to complete a statistical-based acceptance criterion, similar to moisture-density measurements.

4 CASE STUDY

Typical results from a base material under construction in Central Texas are introduced next. For the case presented, field tests were conducted at two different base sections. Section 1 corresponded to the outside lane of previously constructed base material and exposed to the environment for several days and Section 2 was located on two new lanes on the base accepted and constructed in the previous 24 hours.

Sieve analysis of the base material was performed in the lab. The gradation of the material is shown in Figure 8. Based on the gradation results the material constituents are distributed as follows: gravel 52%, coarse sand 29%, fine sand 17% and fines 2%. The Plasticity Index of the base was 5 based on the bar linear shrinkage method. As such, the base was classified as GW in the USCS and A-1-a in AASHTO classification system.

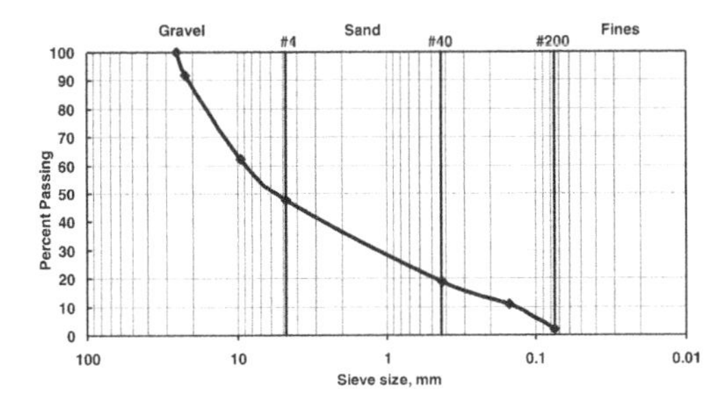

Figure 8. Gradation of Base Material from Site.

Table 1. Results of NDG Tests on the New Base.

	Section 1			Section 2		
	Average		COV*	Average		COV
	%	kg/m³	%	%	kg/m³	%
Moisture Content,	4.9		8.7	6.3		14.9
Dry Density,		2230	2.5		2172	1.6
Wet Density,		2337	2.6		2310	1.5

*Coefficient of Variation, %.

The information associated with Figures 5 and 6 is related to this site. The modulus after 24 hours of oven-drying, 850 MPa, is considered as the target modulus. To complete the quality control process, field tests were conducted with a DSPA on Section 1 at every 60 m on top of the base on a total of 22 locations and on the left wheel path (LWP), center and right wheel path (RWP) of the outside lane. In addition to the DSPA tests, a nuclear density gauge (NDG) was used to estimate the dry densities and moisture contents of the first 10 test locations of Section 1. For Section 2, outside and inside lanes were investigated with the DSPA at 6 locations with the nominal spacing between stations of 30 m. The NDG was used at every location of Section 2 on an alternating pattern from outside to inside lanes. The base layer was 300 mm thick in both sections, divided in two lifts of 150 mm. Only the top lift was the focus of this study.

The in-situ dry densities and moisture contents with the NDG are listed in Table 1. The averages dry densities and the average moisture contents were 2230 kg/m^3, and 4.9% for Section 1 and 2172 kg/m^3 and 6.3% for Section 2, respectively. On the average the dry density of Section 1 was about 48 kg/m^3 greater than the MDD, whereas the moisture content was about 2% below the OMC. For Section 2, the average density was about 10 kg/m^3 below the MDD and the average moisture content was about 0.6% below the OMC.

The average moduli at each station for both sections are shown in Figure 9. The average modulus of

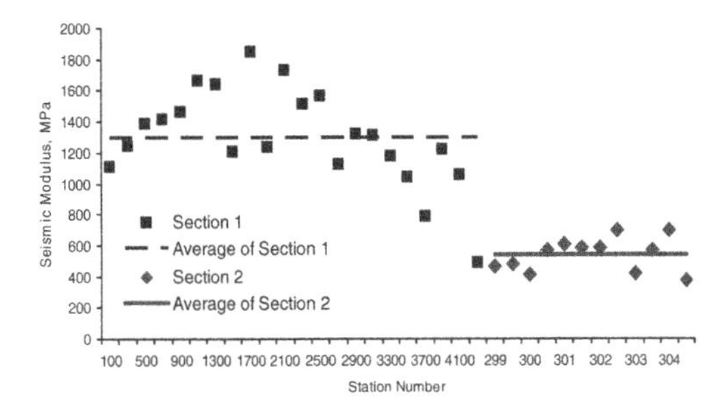

Figure 9. Variations of Measured Moduli along Sections.

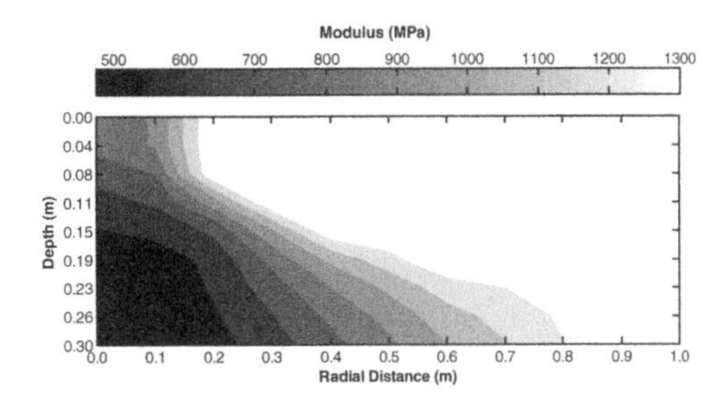

Figure 10. Variation in Nonlinear Modulus within the Base Layer for Section 1.

Section 1 was about 1298 MPa, with a coefficient of variation (COV) of 24.1%. For section 2 the average modulus was 539 MPa with a COV of 20.1%. The differences in the average moduli can be attributed to the difference in curing age as reflected in the different moisture contents measured for each section with the nuclear density gauge.

A software package called Seismic Modulus Analysis and Reduction Tool (SMART, Abdallah et al. 2002) was used to estimate the variation in the nonlinear base modulus with depth and horizontal distance due to a standard tandem axle. An asphalt layer of 50 mm thick on top of the base with a design modulus of 3.5 GPa, and a subgrade underneath with a modulus of 70 MPa were assumed in the design. The average seismic modulus from the DSPA for Section 1 along the regression parameters $k2$ and $k3$ measured from resilient modulus tests were used to estimate the nonlinear moduli (see Equation 3). The contour map of modulus from SMART is presented in Figure 10. At a given depth, the modulus increases (becomes closer to the linear elastic modulus) as the distance from the load increases. The minimum modulus occurs under the load near the interface of base and subgrade. For this section the minimum modulus directly under the load is equal to 475 MPa and the maximum 840 MPa. The weighted average modulus under the load, presumed to be similar to the modulus measured with a Falling Weight Deflectometer, is 600 MPa.

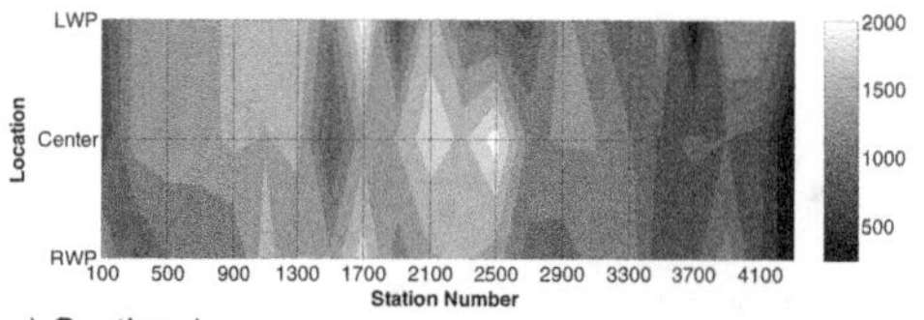

a) Section 1

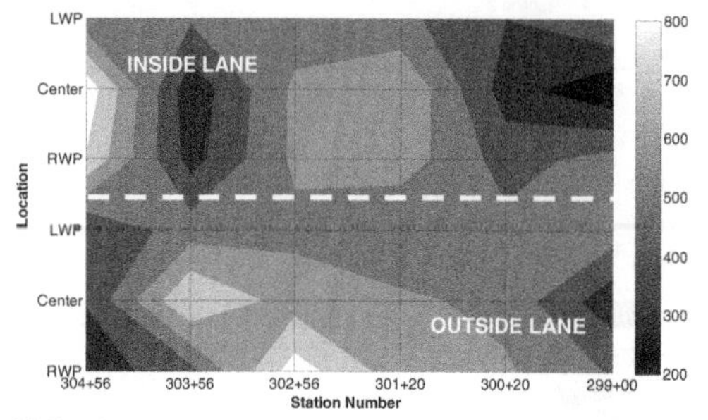

b) Section 2

Figure 11. Contour Plot of Measured Moduli along Sections.

Contour maps of the seismic moduli measured on both sections for all points tested with DSPA along the project are shown in Figure 11. These maps allow identifying stiffer and weaker areas in the longitudinal and transverse direction for the sections investigated.

5 CONCLUSIONS

In this paper, the foundation for a comprehensive quality management of unbound materials based on seismic technology is introduced. Seismic-based laboratory and field tests are briefly explained and the process of combining them to evaluate the appropriateness of a compacted layer from the stiffness point of view is discussed. Since seismic moduli are low-strain moduli, an algorithm is proposed to relate them to design moduli used in pavement analysis and design. A case study for a newly-constructed base is also presented to demonstrate the process.

ACKNOWLEDGEMENTS

The authors wish to gratefully acknowledge the support of the Texas Department of Transportation.

REFERENCES

Abdallah, I., Meshkani A., Yuan, D., and Nazarian S. (2002), "Design Program Using Seismic Moduli," Research Report 1780-4, Center for Highway Materials Research, UTEP, El Paso, TX.

Desai, M. and Nazarian, S. (1993), Automated Surface Wave Method: Field Testing, Journal of Geotechnical Engineering, ASCE, Vol. 119, No. 7, pp. 1094–1111.

Khoury, N. N., and Zaman, M. M. (2004), "Correlation between Resilient Modulus, Moisture Variation, and Soil Suction for Subgrade Soils," Transportation Research Record 1874, Washington, DC, pp. 99–107.

Nazarian, S., Baker, M. R., and Crain, K. (1993), "Fabrication and Testing of a Seismic Pavement Analyzer," SHRP Report H-375. SHRP, National Research Council, Washington, D.C.

Nazarian, S., Yuan, D., and Arellano, M. (2002), "Quality Management of Base and Subgrade Materials with Seismic Methods." Transportation Research Record No. 1786, Washington, D.C., pp. 3–10.

Nazarian, S., Williams, R., and Yuan, D. (2003), "A Simple Method for Determining Modulus of Base and Subgrade Materials," ASTM STP 1437, ASTM, West Conshohocken, PA, pp. 152–164.

Advances in Transportation Geotechnics – Ellis, Yu, McDowell, Dawson & Thom (eds)
© 2008 TRL Limited, ISBN 978-0-415-47590-7

A design method and performance based specification for UK road foundations

B.C.J. Chaddock & C. Roberts
Transport Research Laboratory, Crowthorne, UK

ABSTRACT: Research has been conducted since the 1980s by the Transport Research Laboratory on behalf of the Highways Agency to develop, for a wide range of materials, designs for road foundations and procedures for their in-situ assessment. In 2006, guidance was issued in Highways Design document HD 26 on the design of pavements for four classes of foundations that provide different degrees of support for the overlying pavement. Also issued as Interim Advice Note IAN 73 was design guidance for the foundation and a performance based specification, which aims to provide some assurance to the client of both the short and long term performance of the foundation. The main features of the design guidance and the performance based specification are outlined in this paper. A practical application of IAN 73 in the form of a two stage design method is then proposed for foundations whose subbases are slow setting, hydraulically bound materials that can act substantially as unbound granular materials in their early life but later cure to form hardened materials. This approach is illustrated by the retrospective application of aspects of the performance based specification to a foundation built recently on a UK trunk road. In addition, the potential range in structural performance of unbound granular foundations is illustrated by several case histories. The potential benefits for client and contractor of a performance based specification for road foundations are highlighted.

1 INTRODUCTION

The primary functional requirements of road foundations are the provision of a construction platform and long-term support to the pavement. The structural layers overlying the subgrade are designed to protect the soil from adverse environmental conditions as well as excessive stresses caused by traffic that can lead to foundation deformation. Subgrade soil is loaded by construction traffic on top of the foundation when the road is being built as well as by vehicles on the completed pavement during its in-service life. It is the purpose of foundation and pavement designs, material specifications and associated procedures to provide a framework within which roads can be economically constructed on soils of various strengths for different estimated amounts of construction and in-service traffic.

Prior to 2006, UK practice for the construction of pavement foundations employed a method specification, which prescribed the materials to be used for the constituent layers of the foundation, their thicknesses and how they should be compacted. This approach generally restricted the contractor's choice of materials to conventional materials with known behavior. For the majority of pavement types, there was no reduction in the thickness of the foundation or the pavement when superior materials were used in the foundation. But Nunn et al (1997) reported that, in other European countries, minimum elastic stiffness values at various levels in the foundation are specified as a performance measure. In addition, the design of the upper pavement layers was selected according to the magnitude of the foundation stiffness that is taken to be a measure of the support of the foundation for the pavement.

Within the UK, research has been carried out on behalf of the Highways Agency by the Transport Research Laboratory (TRL) and other organizations to develop a performance based specification for road foundations. Chaddock & Brown (1995) described research on the in-situ assessment by various tests of composite foundation structures that incorporated unbound granular sub-base and, if used, capping on the underlying soil. This research led to outline proposals for a performance based specification for road foundations. Fleming & Rogers (1999) described similar work for the underlying earthworks to develop a draft performance based specification for capping and subgrade. Chaddock and Merrill (2004) reported further developments of the individual test techniques and procedures to ensure compliance of the road foundations at construction. They suggested the establishment of various classes of foundations that are defined by their stiffness and offer different degrees of structural support to the overlying pavement. It was asserted that lower class foundations with the lower pavement support or stiffness could be built with unbound granular materials, whereas higher class foundations of higher pavement support or stiffness would require construction with hydraulically bound materials in their upper layers. Thinner pavements were proposed for the higher quality foundations. The main features of a proposed performance based specification for standard as well as superior foundations were outlined.

Based on these and other studies, guidance was issued in the document Highways Design (HD) 26 (2006) on the design of pavements for four foundation classes that provide different degrees of support for the pavement. Also, design guidance for the foundation and a performance based specification to provide the client some assurance of both short and long term performance of the foundation was issued as Interim Advice Note (IAN) 73 (2006).

The main features of the design guidance for road foundations and the associated performance based specification, which are currently under revision, are outlined in this paper. Illustrative studies by TRL on the structural performance of foundations constructed with unbound granular materials and a slow setting, hydraulically bound material, which can act substantially as an unbound granular material in its early life, are described in the paper in the context of the performance based specification.

2 DESIGN AND PERFORMANCE BASED SPECIFICATION

The upper layers of a road pavement are currently designed to criteria given in HD 26 (2006) that include the classification of road pavement foundations into one of four classes 1 to 4, where the higher the class number, the better is the foundation's support to the overlying pavement. IAN 73 (2006) provides design guidance for road pavement foundations in the form of two design approaches.

The first approach allows a limited number of conservative 'Restricted Designs' to be applied for foundation classes 2 and 3 and is particularly intended for use on small scale works.

The second approach allows 'Performance Designs' that cover all four foundation classes and provides more flexibility to the designer. The designs for a designated class of foundation are produced using a theoretically loaded, linear elastic, multi-layer model of the foundation to calculate the thicknesses of the component materials from estimates of their elastic properties. Criteria are adopted to limit subgrade deformation and foundation deflection. The method is described by Chaddock and Roberts (2006) and summarized in Section 5 of IAN 73 (2006). The acceptability of a particular Performance Design must be proven by making measurements on the constructed foundation. The procedures for this assessment are described in the performance based specification that is also given in Section 5 of IAN 73 (2006). The main elements of these procedures are given in the following summary:

Design:

- Design the foundation for the anticipated structural properties of soil subgrade and the required foundation class by selection of subbase and, if required, capping materials and their thicknesses.

Demonstration area:

- Measure the structural properties of subgrade and select the location of the demonstration area where the subgrade complies with design assumptions.
- Construct the demonstration area.
- Adjust the short-term foundation stiffness targets given in Table 1 when the structural properties of the soil in the demonstration area are better than the design values.
- Prove compliance of the foundation layers to the design by measuring their thicknesses, specified material properties and densities.
- Determine foundation stiffness by a dynamic plate test, which is described in Annex B of IAN 73 (2006), at pre-determined ages along the site before, and after, trafficking by construction vehicles. Calculate the running mean of 6 measurements; that is the average of test results 1 to 6, then results 2 to 7 etc., and show compliance with the adjusted short-term foundation stiffness targets.
- Demonstrate that foundation deformation complies with the requirements of the specification.

The foundation design should only be used in the main works when it has been proven in the demonstration area.

Main Works:

- Measure the structural properties of the subgrade and, where necessary, improve the soil to meet the design requirements.
- Construct the main works.
- Prove compliance of the foundation layers to the design by measuring their thicknesses, specified material properties and densities.
- Determine the foundation stiffness just prior to pavement construction and show compliance with the short-term foundation stiffness targets given in Table 1. Also, demonstrate that the foundation deformation complies with the specification.

The foundation stiffness targets for each foundation class at the time of construction are given in Table 1 for the four foundation classes. Average target values

Table 1. Foundation stiffness requirements.

Foundation stiffness, MPa, for Class:					
Long-term: design values:					
	1	2	3	4	
	50	100	200	400	
Short-term average targets:					
Unbound:	40	80		–	–
Bound:	50	100	Fast setting:	300	600
			Slow setting:	150	300
Short-term individual targets:					
Unbound:	25	50		–	–
Bound:	25	50	Fast setting:	150	300
			Slow setting:	75	150

for running means of 6 measurements of foundation stiffness and individual target values for single measurements of foundation stiffness are given for both unbound and bound materials, with bound materials for foundation classes 3 and 4 further subdivided into fast setting materials, such as those incorporating Portland cement, and slow setting materials. Foundations built with unbound granular materials are given lower average foundation stiffness targets than those foundations of the same class constructed with bound materials because only unbound granular materials are expected to increase in stiffness when confined by the pavement. Also, foundations built with slow setting bound materials are given lower foundation stiffness targets than those foundations of the same class constructed with fast setting bound materials because they would have developed a lower proportion of their ultimate stiffness at the time of test.

There are a variety of solutions for non-compliant foundations. For example, where the demonstration area is non-compliant but has a subgrade CBR strength at least as high as the design CBR, then either the foundation layers could be improved or the foundation redesigned, a new demonstration area constructed and, once again, tested for compliance. For a proven design, but a non-compliant main works, either the foundation layers could be improved, weak soil locally strengthened or, if uneconomic, the design structural properties of the subgrade could be reset so that the foundation could be redesigned, proven and reconstructed prior to retesting for compliance.

3 SLOW SETTING HYDRAULICALLY BOUND MATERIALS

The performance based specification was applied retrospectively to a foundation comprising a slow setting, slag bound material (SBM) subbase on a clay subgrade built for a trunk road in England and opened to traffic in 2006. The foundation design approach given in Appendix A of Nunn (2004) was followed to produce a foundation class 3, with the exception that a material of strength category C3/4 instead of C9/12 (MCHW 1 Series 800) was adopted. The foundation class 3 is for the long term support of the pavement. The foundation nominally comprised 270 mm of SBM subbase on a subgrade of CBR value 4%. This thickness of subbase was prescribed in HD 25 (1994), which is the predecessor of IAN 73 (2006), and is considered adequate to carry construction traffic to build a kilometre of road as long as the SBM behaves as well as unbound granular Type 1 subbase (MCHW 1 Series 800).

Table 2. Foundation stiffness in demonstration area.

Age, days	Foundation stiffness, MPa			
	Average		Individual	
	Lane 1	Lane 2	Lane 1	Lane 2
6	285	259	196–407	182–357
33	321	807	190–548	526–1101
Adjusted targets:	196		98	

A dual lane demonstration area of 60 m length was constructed and tested in late summer 2005. With regard to the primary measure of performance, the foundation stiffness, the average and individual values measured 6 and 33 days after construction are given in Table 2. These values are compared to the slow setting, target values for a foundation class 3 given in Table 1 after their correction to account for the in-situ CBR of the subgrade soil differing from its design value. The median CBR value of the subgrade was 12.1%, which resulted in a correction multiplication factor of 1.31.

For both series of tests, the average and individual values of foundation stiffness exceeded the adjusted targets. At a subbase age of 6 days, the lane 2 value was, on average, only 9% less than the lane 1 value; whereas, at an older age of 33 days, the lane 2 value was, on average, 150% greater than the lane 1 value. This difference at the older age was assumed to be caused by the degrading effect of construction traffic on subbase layer stiffness that was greater for lane 1 than lane 2 as the former lane was considered to have carried more traffic. When constructed under the warm environmental conditions of late summer, the foundation design was shown to have the potential of producing a foundation class 3.

The foundation stiffness for 1350 m of the main works in the eastbound carriageway of the road was tested according to IAN 73 (2006) and the individual and running mean of 6 measurements are shown in Figure 1. Except for a short 40 m length section, the carriageway, on average, complied with the unadjusted

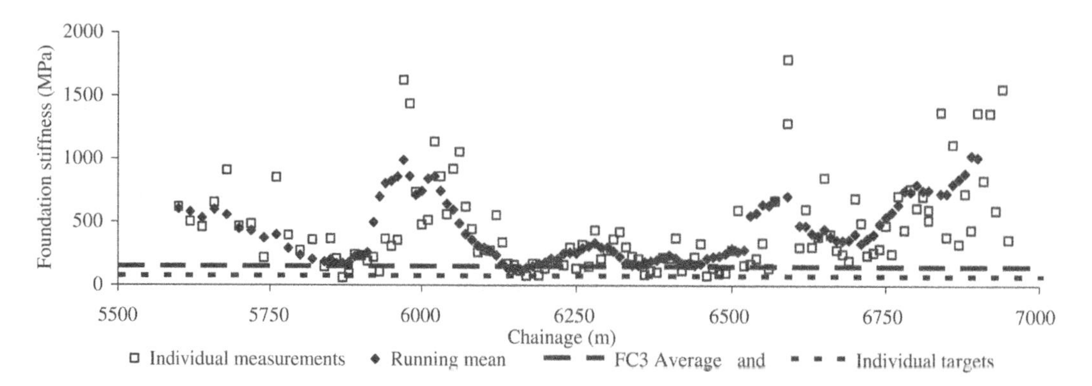

Figure 1. Eastbound main works foundation stiffness compliance at 33–64 days.

Table 3. Calculated eastbound layer stiffnesses.

Location	Age, days	Lane	Mean stiffnesses, MPa	
			Subgrade	Subbase
Demonstration Area	6	1	225	390
		2	233	326
	33	1	217	453
		2	268	2408
	65	1	236	879
		2	278	5338
Eastbound main works	33–64	1	250	952
		2	241	1599

targets. Further increase in foundation stiffness with time during the early life of the pavement is envisaged.

To estimate the performance of the SBM subbase, the deflection profiles measured by the falling weight deflectometer (FWD) were back-analysed using the programme HA MODULUS (2002) to calculate the layer stiffnesses of the components of the foundation. A two-layer model of the foundation, which comprised a SBM subbase on a soil subgrade, was adopted. Only subbase and subgrade stiffness values with acceptable goodness of fit values between the predicted and measured deflection profiles were accepted. The computed values of these layer stiffnesses are given in Table 3 for the demonstration area at various ages and for the eastbound carriageway main works.

The mean values of subbase stiffness at an age of 33 to 64 days in lanes 1 and 2 of the eastbound carriageway main works exceeded the mean values of subbase stiffness of the more heavily trafficked lane 1 of the demonstration area at both 33 and 64 days.

Less satisfactory was the performance of a 940 m section of the westbound carriageway built later in the year during the mid-autumn season in cooler conditions. Running mean values of foundation stiffness shown in Figure 2 had more extensive non-compliances of length 290 m than the eastbound carriageway. The subbase age at the time of testing varied from 46 to 59 days.

The subbase and subgrade stiffnesses of the westbound carriageway main works were also computed by back analysis of FWD deflection profiles and are given in Table 4.

The mean values of subbase stiffness at an age of 46 to 59 days in lanes 1 and 2 of the westbound carriageway were much lower than values derived from the eastbound carriageway tests at a similar age.

The temperatures measured at a depth of 300 mm in soil at nearby weather stations and presented in Figure 3 show, however, that the westbound carriageway was subjected to lower temperatures than the eastbound carriageway prior to paving the asphalt base. This temperature difference would explain the differences in the subbase stiffness between the east and westbound carriageways as the hardening of SBM is temperature and time dependent.

The running mean values of foundation stiffness of the westbound carriageway at an age of 46 to 59 days, however, did comply with the requirements of IAN 73 (2006) for a foundation of class 2 rather than class 3. This level of performance also occurred for all of the trial length except for about 70 m at a younger age of 8 to 21 days as shown in Figure 4.

The subbase and subgrade stiffnesses of the westbound carriageway main works at this earlier age were also computed by back analysis of FWD deflection profiles and are given in Table 5.

The analysis shows the mean stiffness of the subbase shortly after its construction to be approximately 150 MPa, which is the value assumed for unbound granular Type 1 subbase (MCHW 1 Series 800) in the design process described by Chaddock and Roberts (2006) that formed the basis of IAN 73 (2006) foundation designs.

Also obvious from comparison of Tables 4 and 5 is that the subbase in the westbound carriageway stiffened with time; that is approximately doubling

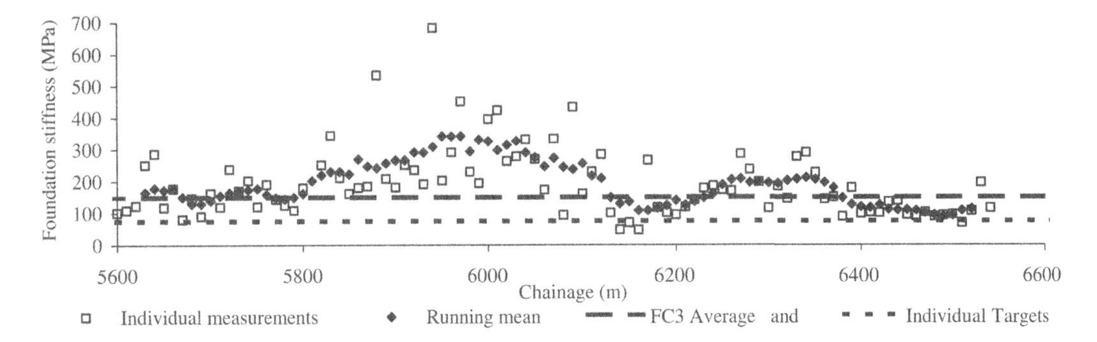

Figure 2. Westbound main works foundation stiffness compliance at 46–59 days.

Table 4. Calculated westbound layer stiffnesses.

Location	Age, days	Lane	Mean stiffnesses, MPa Subgrade	Subbase
Westbound main works	46–59	1	243	304
		2	216	352

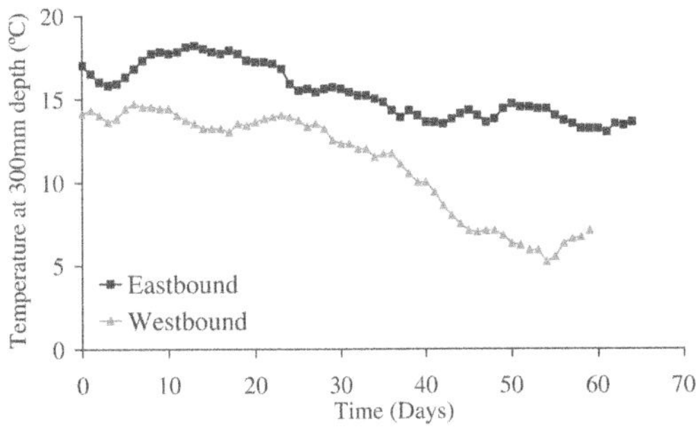

Figure 3. Temperatures since the start of subbase construction.

in stiffness in 38 days, whilst the subgrade stiffness remained almost unchanged. Hence, the increase in foundation stiffness with time is shown to be due to the hardening of the SBM subbase with time. Further increase in foundation stiffness with time during the early life of the pavement is envisaged.

The foundations in the trial areas provided a very good working platform for the transportation and compaction of road materials with only a very small area requiring remedial work.

4 TWO STAGE FOUNDATION DESIGN

Foundations comprised of a slow setting, hydraulically bound material (HBM) as subbase can therefore be seen to potentially pose foundation design difficulties. For instance, if the thickness of HBM is chosen for the long-term condition when the subbase is constrained in the completed pavement and is fully cured, then it may be too thin for the foundation to support direct trafficking by construction vehicles without deformation in the situation when the HBM hasn't hardened sufficiently prior to being trafficked. The choice of the subbase thickness for the construction traffic is difficult to judge because the hardening of the subbase is time and, as noted above, temperature dependent. This behavior of slow setting HBMs means that different thicknesses of subbase are required during different seasons of the year and for different construction practices, such as immediate and delayed paving of the pavement base. The prudent approach is to design the foundation to carry construction traffic as if the subbase was unbound during pavement construction. A foundation class 2 is normally suitable. Later, for correctly chosen HBMs, the foundation should qualify as a foundation class 3, or higher, in the completed pavement with consequential reductions in pavement thickness over a foundation class 2.

Although the foundation design would assume an unbound subbase, the HBM would often harden to some degree during pavement construction. Such behavior produces a higher quality foundation than that expected from the design and thereby reduces the risk of delays to pavement construction by temporary bans on foundation trafficking due, say, to a saturated subbase or by remedial works.

When the thickness of unbound subbase for an adequate construction platform exceeds the thickness of bound subbase required for the designated foundation class in the long term, then adoption of the prudent design will result in a foundation of higher long-term stiffness than specified. This additional stiffness of the foundation can either be taken by the client as a longer life pavement, or the overlying pavement layers can be thinned still further. This two stage approach therefore has several major advantages.

The case history described in this paper replicates, in part, from a design perspective, the two-stage design

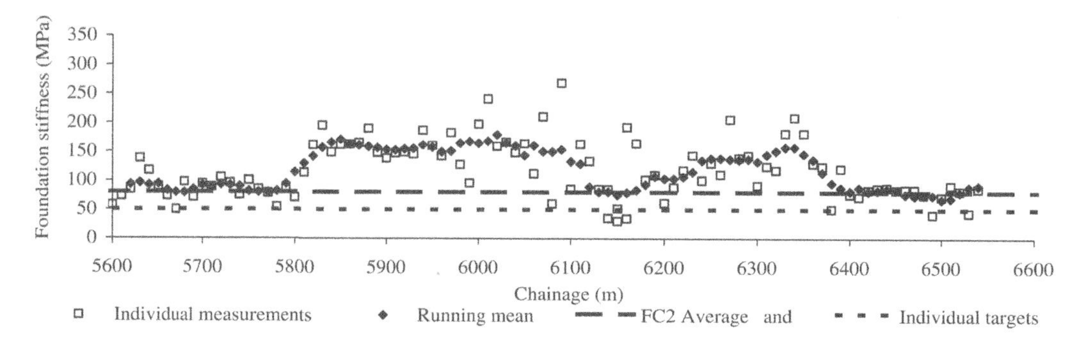

Figure 4. Westbound main works foundation stiffness at 8–21 days.

Table 5. Calculated westbound layer stiffnesses.

Location	Age, days	Lane	Mean stiffnesses, MPa	
			Subgrade	Subbase
Westbound main works	8–21	1	267	160
		2	224	139

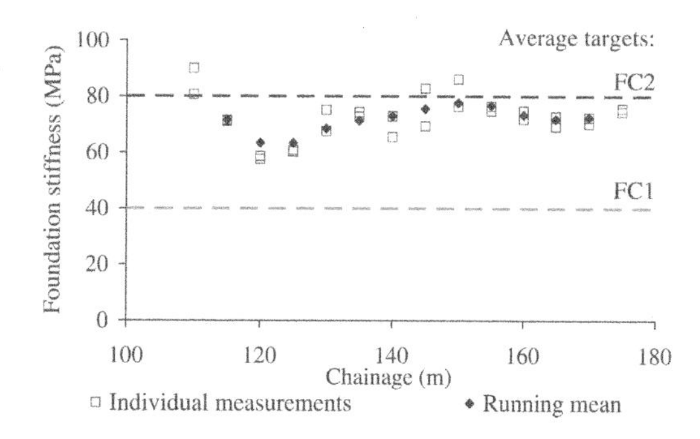

Figure 5. Foundation stiffness of main works – Case study 1.

process. The foundation can be considered to have been designed as a foundation of class FC2 for the construction phase. The pavement of design life 80 msa was thinned by 40 mm of HMB 35 asphalt to 270 mm on the reasonable expectation that the foundation will provide a foundation of class FC3 in the longer term.

5 TYPICAL BEHAVIOUR OF UNBOUND GRANULAR FOUNDATIONS

The importance and benefits of testing the foundation during construction is demonstrated in the following two case histories.

Prior to the publication of IAN 73 (2006), an unbound granular, crushed rock Type 1 subbase (MCHW Series 800) on a prepared "rock" subgrade was proposed as a foundation of class 3. The subbase thickness in a short 70 m section of the main works was estimated as about 200 mm from dynamic cone penetrometer (DCP) tests that are described in HD 29 (2008). FWD tests were carried out on the trial section. The individual and running mean of 6 readings are plotted in Figure 5. Retrospectively, average foundation stiffness targets of IAN 73 (2006) for unbound granular foundations of classes 1 and 2 have been chosen for comparison with the foundation stiffness measurements and are shown as FC1 and FC2 respectively in Figure 5.

The foundation only qualified as a foundation of class 1 and was substantially less stiff than the foundation stiffness that would have been required for an unbound foundation of class 3, if permitted on a departure from standards.

To determine the cause of the unexpectedly low value of foundation stiffness, in-situ and laboratory tests would have been required. Thereafter, modifications to the foundation materials and, or, design would have been needed. Alternative arrangements, however, were made instead of pursuing the production of a foundation of class 3.

This case history, however, demonstrates the benefits for the client of implementing a performance based specification because this approach can identify those sections of a road foundation that may not provide the long-term support to the pavement assumed during pavement design.

A performance based specification could also help to make best use of site conditions and available materials at specific sites. For example, a foundation consisting of an unbound granular Type 1 subbase on a chalk fill on a clay subgrade with flints was proposed as a foundation equivalent to conventional unbound granular Type 1 subbase on a capping foundation design to HD 25 (1994). The foundation design to this standard is normally expected to produce a foundation of class 2. Type 1 subbase of thickness 150 mm was laid on chalk fill of thickness 1 m. FWD tests were carried out on a 500 m length of the main works and the

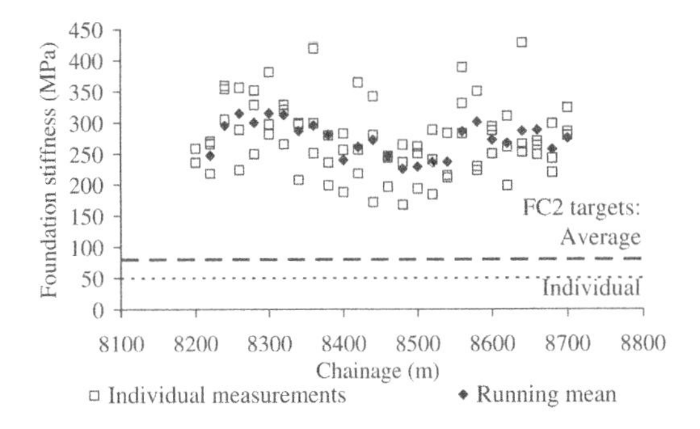

Figure 6. Foundation stiffness of main works – Case study 2.

individual and running mean of 6 readings are plotted in Figure 6.

Retrospectively, foundation stiffness targets of IAN 73 (2006) for the main works of an unbound granular foundation of class 2 have been selected for comparison with the foundation stiffness measurements and are shown as FC2 in Figure 6. The foundation complies with these requirements with a large margin of safety. Today, for a similar foundation constructed under IAN 73, a case could be made out by the contractor for the foundation to be classified a class higher, although, for an envisaged traffic of 80 msa, or greater, a departure from HD26 (2006) would be required as the subbase is unbound, not bound, as currently required.

A comparison of these two case histories, when taken together with earlier work by Chaddock and Brown (1995), demonstrates the potential variability of unbound granular foundations. Also obvious are the benefits of classifying foundations and applying a performance based specification. That is, best use of site conditions are made and economical foundation and pavement designs produced, whilst sub-standard foundations are avoided during road construction and the client is given more confidence in the long term behavior of the road.

6 SUMMARY AND CONCLUSIONS

- The design method and the performance based specification for road foundations of Highway Agency roads have been outlined. The use of the performance based specification on the four permitted quality classes of foundation should enable the adoption of a wider range of materials and more economical designs.
- The use of a proposed two-stage foundation design method for foundations built with slow setting hydraulically bound materials has been illustrated by its retrospective application to a major road construction. A foundation initially designed as a foundation of class 2 performed adequately during the pavement construction phase. The same foundation then showed its potential to achieve a higher quality foundation of class 3 in the longer term that permitted a thinner overlying asphalt pavement layer.
- The wide range in structural performance of unbound granular foundations has been illustrated.
- The value of the performance based specification in identifying sub-standard foundations during construction and in helping make best use of site conditions and available materials has been demonstrated.

ACKNOWLEDGEMENTS

Copyright Transport Research Laboratory 2008. This paper is presented with the permission of the Highways Agency and the Chief Executive of the Transport Research Laboratory. A grateful acknowledgement is extended to the staff of organizations, too numerous to mention individually, for their help in arranging the trials. The views expressed in this paper are not necessarily those of the Highways Agency.

REFERENCES

Chaddock, B.C.J. & Brown, A.J. 1995. In situ tests for road foundation assessment. In R. Jones and A. Dawson (eds.), *Unbound Aggregates in Roads (UNBAR 4); Proc. of Symp., Nottingham, 17–19 July 1995*. Rotterdam: Balkema.

Chaddock, B.C.J. & Merrill, D.M. 2004. An end product specification for road foundations. In A. Dawson (eds), *Pavements Unbound; Proc. of Symp., Nottingham, 6–8 July 2004*. London: Balkema.

Chaddock, B.C.J. & Roberts, C. 2006. Road foundation design for major UK highways. *PPR 127*. Crowthorne: Transport Research Laboratory.

Fleming, P.R. & Rogers, C.D.F. 1999. A performance based specification for subgrade and capping, *Contract Report No. 1*. Nottingham: Scott Wilson Pavement Engineering.

HD 25 1994. Foundations. *Design Manual for Roads and Bridges (7.2.2.)*. Norwich, TSO.

HD 26 2006. Pavement Design. *Design Manual for Roads and Bridges (7.2.3.)*. Norwich, TSO.

HD 29 2008. Data for pavement assessment. *Design Manual for Roads and Bridges (7.3.2.)*. Norwich, TSO.

IAN 73 2006. Design guidance for road pavement foundations (Draft HD25). Highways Agency. London.

Nunn, M.E., Brown, A.J., Weston, D & Nicholls, J.C. 1997. Design of long-life flexible pavements for heavy traffic, *TRL Report 250*. Crowthorne: Transport Research Laboratory.

Nunn, M.E 2004. Development of a versatile approach to flexible and flexible composite pavement design. *TRL Report 615*. Crowthorne: TRL Limited.

MCHW 1 Series 800. Specification for Highways Works. *Manual of Contract Documents for Highway Works*, Norwich, TSO.

HA MODULUS 2002. *Computer Program version 5.1*, Texas Transport Institute, USA.

Advances in Transportation Geotechnics – Ellis, Yu, McDowell, Dawson & Thom (eds)
© 2008 Taylor & Francis Group, London, ISBN 978-0-415-47590-7

The anti-shakedown effect

P.R. Donovan & E. Tutumluer
University of Illinois Department of Civil and Environmental Engineering

ABSTRACT: Repetitive loading along a channelized path causes unbound aggregate layers to densify, shakedown, and stabilize with little additional permanent or residual deformation per each wheel pass. Full-scale pavement tests conducted at the Federal Aviation Administration's National Airport Pavement Testing Facility (NAPTF) in the US indicated that a sequential aircraft wheel/gear wander pattern causes residual deformations to be recovered in the unbound aggregate layers. The downward residual deformation caused by a pass of heavily loaded landing gear is canceled by the upward residual deformation resulting from the pass of the same gear offset by wander. This interaction indicates shuffling or rearrangement of the particles in the unbound layer, which in turn reduces the strength of the layer causing future load applications to cause more residual deformation, the "anti-shakedown" effect. This paper analyzes the anti-shakedown effect of the contractive/dilative response of the unbound layers using multi-depth deflectometer data from NAPTF testing.

1 INTRODUCTION

1.1 *Observed anti-shakedown response*

Shakedown in unbound granular layers is well documented; application of additional loads along a channelized path causes unbound layers to densify, gain strength with time, and stabilize with little additional residual deformation. Full-scale airport pavement test data from the Federal Aviation Administration's (FAA's) National Airport Pavement Testing Facility (NAPTF) in the US used new generation aircraft loads on asphalt pavements and indicated that a sequential wander pattern caused residual deformations to be recovered, potentially reducing or even negating the shakedown effect. What has been seen is that the downward residual deformation (rutting) caused by a pass of heavily loaded landing gear is canceled by the upward residual deformation resulting from the pass of the same gear offset by wander. This interaction indicates a shuffling or rearrangement of the particles in the unbound aggregate base/subbase layers of the pavement system. The particle rearrangement in turn reduces the strength of the unbound layer causing future load applications to cause more residual deformation, the "anti-shakedown" effect.

This paper analyzes the anti-shakedown effect that was realized from the FAA's NAPTF testing when the pavements were subjected to loading with a wander pattern but no channelized traffic. Both the contractive and dilative responses of the unbound aggregate base/subbase layers collected using multi-depth deflectometers (MDDs) are analyzed to quantify the anti-shakedown effect. The focus will be on the data from the (M)edium strength subgrade section of the (F)lexible pavement test section built with a (C)onventional base course (MFC).

2 NAPTF TESTING

2.1 *Test facility*

The National Airport Test Facility (NAPTF) located at the William J. Hughes Technical Center close to the Atlantic City International Airport was built to analyze the affects of New Generation Aircraft (NGA) on pavements. NGA affect airfield pavements differently than older aircraft due to increased loads and changes to landing gear configurations. These differences require advanced airport pavement design procedures. The NAPTF was constructed to generate full-scale tests in support of the investigation of airport pavements subjected to complex NGA gear loading configurations such as the 6-wheel Boeing 777 (B777) dual-tridem landing gear. There were three main goals for the NAPTF to provide: (1) additional traffic data for incorporation in new thickness design procedures for airfield pavements, (2) full scale testing capabilities to examine response and failure information for use in airplane landing gear design and configuration studies, and (3) technical data for reexamining the CBR method of design for flexible airfield pavements. All three objectives were established to compare the damage done by the 6-wheel Boeing 777 (B777) type dual-tridem landing gear to dual and dual-tandem gear of older aircraft (Hayhoe, 2004; Hayhoe et al. 2004).

Tests are conducted using a specially designed 1.2-million-pound test vehicle which can apply loads of up to 75,000 pounds (34,020 kg) per wheel on two landing gears with up to six wheels per gear (total of 12 wheels for a load capacity of 900,000 lbs). The test vehicle is supported by rails on either side which allow the load to be varied according to the testing protocols. The vehicle can be configured to handle single, dual, dual-tandem, and dual-tridem loading configurations. The wheel and gear spacing can be varied. The maximum tire diameter is 56 in. (142 cm) and maximum tire width is 24 in. (61 cm). Vehicle control can be automatic or manual. Traffic tests were run in a fully automatic control mode at a travel speed of 5 mph (8 km/h). This speed represents aircraft taxiing from the gate to the takeoff position. It is during this maneuver that maximum damage occurs to the pavement because the aircraft is fully loaded and speed is low. Wheel loads are programmable along the travel lanes and the lateral positions of the landing gears are variable up to plus or minus 60 in (1524 mm) from the nominal travel lanes to simulate aircraft wander.

The first series of tests conducted are referred to as Construction Cycle 1 (CC1) tests and the first NGA aircraft to be analyzed was the Boeing 777 (B777). The B777 landing gear was tested in the north testing lane. The B777 has a six wheel dual-tridem configuration with dual wheel spacing of 54 in. (1372 mm) and tridem spacing of 57 in. (1448 mm). The wheel loads were set to 45,000 lbs (200.2 kN) and the tire pressure was 189 psi (1,303 kPa). Traffic was applied at 5 mph (8 km/h).

2.2 *Multi-depth deflectometer data*

Multi-depth deflectometers (MDDs) were used to record the deflection of the pavement system layers. The MDDs were located at various levels within the unbound aggregate layers and the subgrade. This paper will focus on the data from the medium strength subgrade and the B777 traffic lane of the CC1 testing. The layer details and general MDD locations are shown in Figure 1. In addition, several researchers have already reported about the analysis results and performances of the CC1 flexible pavement test sections using the MDD data (Gomez-Ramirez & Thompson 2001, Gopalakrishnan & Thompson 2004, Gopalakrishnan et al. 2006).

The MDDs work by recording the deflection of the individual sensors in relation to an "anchor" sensor that is buried below the zone of influence of the anticipated loads. The surface sensor is actually the only sensor to be directly connected to the anchor; the other sensors measure their deflection in relation to the surface sensor. The movement of the individual sensors is then calculated by subtracting the individual sensor reading from the surface sensor reading.

2.3 Wander pattern

To account for aircraft wander at NAPTF, the test passes or load applications were divided into nine wander positions spaced at intervals of 9.843 in. (250 mm). Each position was traveled a different number of times based on a normal distribution with a standard deviation that is typical of multiple gear passes in airport taxiways, 30.5 in. (775 mm). The nine wander positions cover 87% of all traffic (approximately 1.5 standard deviations). One complete wander pattern consists of 66 vehicle passes (33 east and 33 west). Figure 2 shows the location of the applied trafficking wander positions and gear wheels. Table 1 lists the percent of passes per wander position for each 66 pass pattern and the distance from the MDD to the center of each gear wheel for each wander position. During the 66 pass wander pattern there are five wander sequences based on the normal distribution of traffic as shown in Figure 3.

2.4 *Separation of response values*

Initially, data from the MDDs seemed erratic and random, however when the data were separated by wander

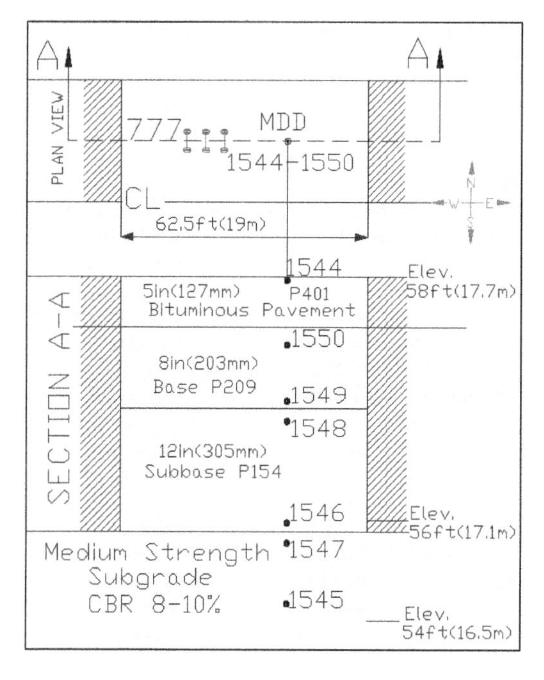

Figure 1. MFC section, B777 lane, MDD location and section details (not to scale).

position, travel direction, and wander sequence distinct patterns emerged. Figures 4 and 5 show the separated residual deflection data for wander position 0 in the P209 base and P154 subbase layers, respectively. As expected there is more residual deflection in the P154 layer (it is thicker and lower quality than the P209 layer). What is interesting to see in Figures 4 and 5 (and also seen with other wander position data) is that the first pass on each wander position in the West to East direction causes the most response and the return pass along the same wander position shows significantly less residual deflection down to almost zero. This is an indication that shakedown was occurring in the unbound aggregate layer, but because the wander position shifted every other pass, the layer did not stabilize.

It is also interesting to see in Figure 5 that if the wander pattern is kept narrow enough, shakedown does occur. As the 66 pass wander pattern goes from sequence 4 to sequence 5, the residual deflection caused by the West to East (W-E) pass on wander position 0 decreases 50%, yet, all other sequences have similar responses under wander position 0. The gear loading of sequence 5 has the three closest wander positions −1, 0, and 1 (see Figures 2 and 3 and Table 1).

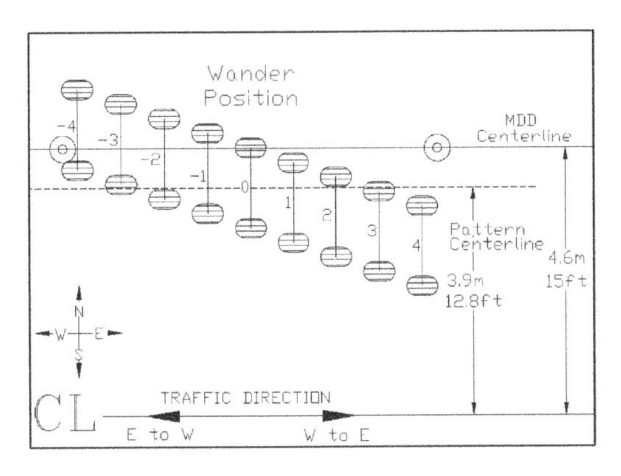

Figure 2. Wander positions and wheel centerlines.

3 ANALYSIS

3.1 *Residual Response of the subgrade layer*

As shown above, the response of the individual layers can be separated by wander position, traffic direction, and wander pattern sequence. To fully understand the influence wander sequence and wander position have on the residual responses, it was necessary to combine the data into 66 pass wander patterns. Figure 6 shows the 66 pass wander pattern for the subgrade layer.

The MDDs did not measure significant residual deformation in the subgrade layer with the residual

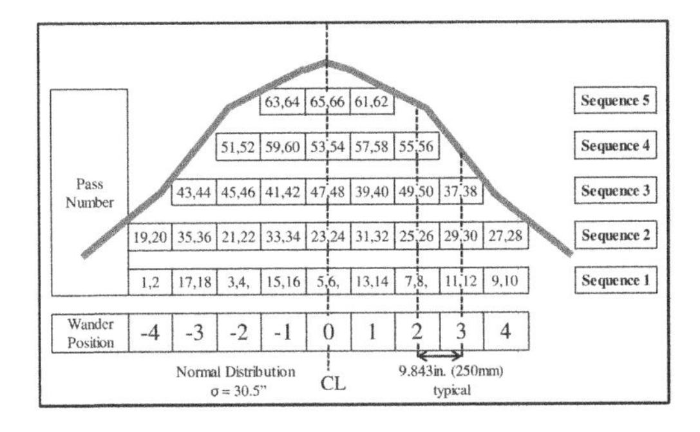

Figure 3. 66 pass wander pattern and 5 wander sequences.

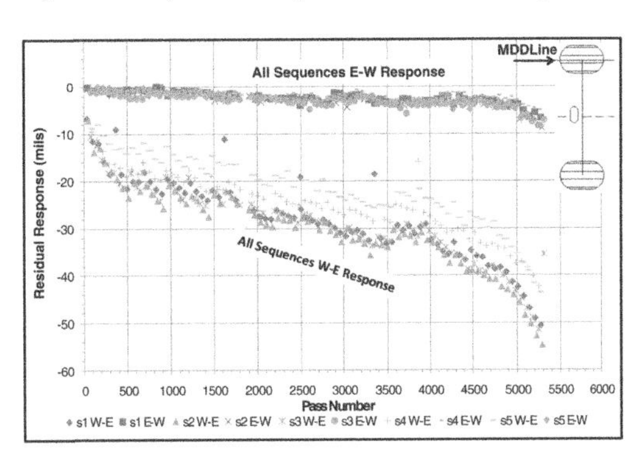

Figure 4. Residual response P209 layer, MFC section, B777 lane (1 mil = 0.001 in. = 0.0254 mm).

Table 1. B777 Gear wheel location in relation to MDD.

Wander Position	−4	−3	−2	−1	0	1	2	3	4
% of passes	6.1%	9.1%	12.1%	15.2%	15.2%	15.2%	12.1%	9.1%	6.1%
North Wheel	−39.97	−30.13	−20.29	−10.44	−0.6	9.24	19.09	28.93	38.77
inches (mm)	(−1015)	(−765)	(−515)	(−265)	(−15)	(235)	(485)	(735)	(985)
South Wheel	14.03	23.87	33.71	43.56	53.4	63.24	73.09	82.93	92.77
inches (mm)	(356)	(606)	(856)	(1106)	(1356)	(1606)	(1856)	(2106)	(2356)

*North MDD offset −26.4in(−671 mm) fromcenterlineof 777 "0" Pattern
*Wander Positions areseparated by 9.843 in (250 mm)

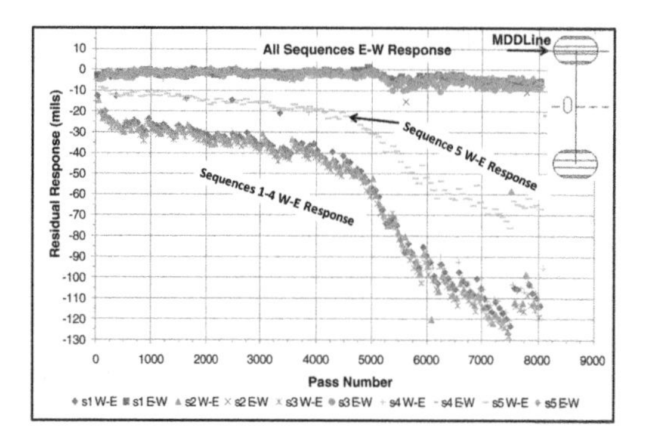

Figure 5. Residual response P154 layer, MFC section, B777 lane (1 mil = 0.001 in. = 0.0254 mm).

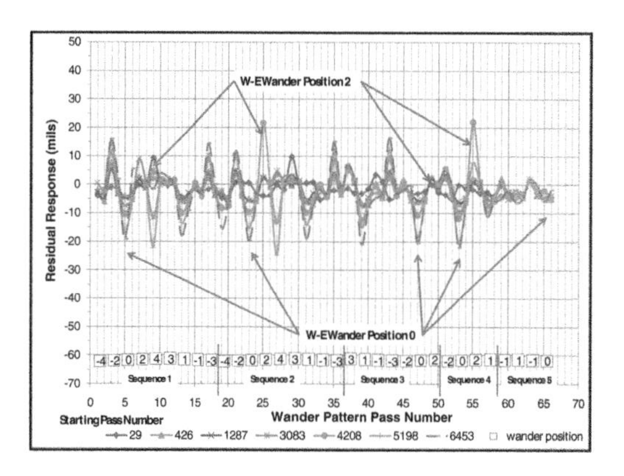

Figure 6. Subgrade residual response – 66 pass wander pattern and wander sequence, MFC section, B777 lane (1 mil = 0.001 in. = 0.0254 mm).

response values rarely exceeded ±20 mils (0.51 mm). This is considerably less than the responses measured in the unbound aggregate layers indicating that the unbound aggregate layers were responsible for the majority of the upward and downward residual response of the pavement system. As seen in Figure 6, the maximum contractive as well as dilative residual responses are caused by multiple wander positions indicating that the gear wheel interaction is higher for the subgrade layer (as expected).

3.2 *Residual responses of P209 and P154 layers*

Figures 7 and 8 show the 66 pass wander patterns for the P209 and the P154 layers, respectively. Both figures indicate that traffic in the W-E direction on wander row 0 produced the maximum downward residual deformation. This corresponded to the maximum load position with a wheel centerline located a mere 0.6 in. (1.5 cm) from the MDD centerline. Wander row 1 was the only other wander position to produce a consistent

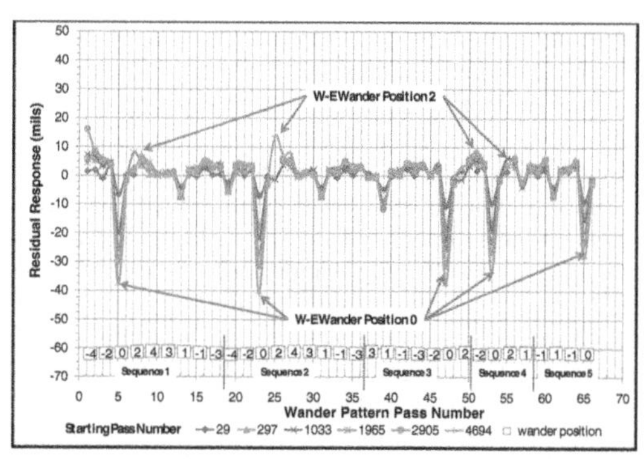

Figure 7. P209 residual response – 66 pass wander pattern and wander sequence, MFC section, B777 lane (1 mil = 0.001 in. = 0.0254 mm).

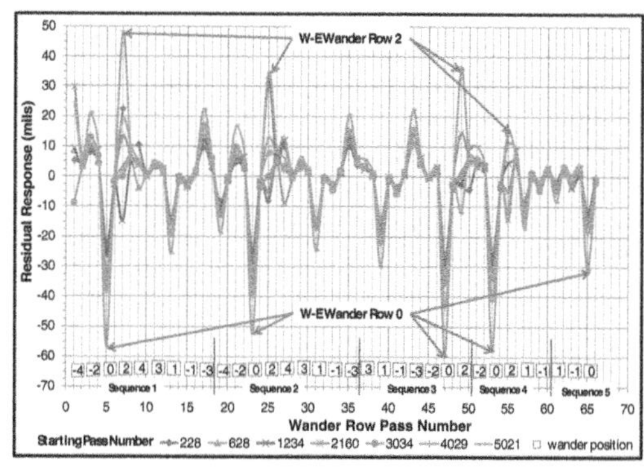

Figure 8. P154 residual response – 66 pass wander pattern and wander sequence, MFC section, B777 lane (1 mil = 0.001 in. = 0.0254 mm).

downward residual deformation of both the P154 and P209 layers.

All of the other wander positions contribute various amounts of upward residual deformation. Wander position 2 seems to provide the most upward deformation for both layers, but due to limited data points cannot be conclusively regarded as the wander position causing the most dilative effect.

Figures 7 and 8 show that as the number of passes increases both the upward and downward residual deformation values, in general, increase (note that the scale on the residual response axis is slightly different for Figures 7 and 8 so that the details are not lost due to the small graph size). This is in contrast to the shakedown concept where residual deformation per pass decreases with increasing repetitions for stable behavior.

It is the combination of the upward and downward residual deformation that proves the particles in the unbound aggregate layers are being rearranged.

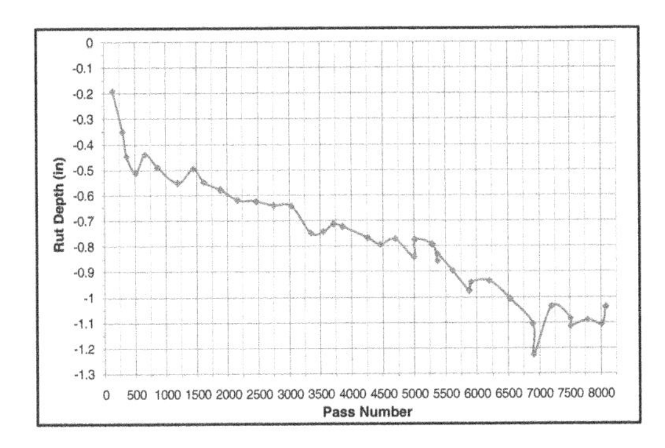

Figure 9. Rut depth measurements at the MDD location (1 in. = 25.4 mm).

Surface rutting was monitored manually throughout the traffic test program using a transverse surface profile (TSP) device, a rolling inclinometer, and straightedge rut depth measurements. The profile data were not measured after a specific wander position or wander sequence.

Figure 9 shows the rut depth at the MDD location at various passes. As shown, the rut depth does generally increase as expected; however there are numerous measurements where the rut depth decreased after an increase in the number of passes which should not have happened. There is no way to adequately compare the effect of wander position on rut depth using the manual rut depth measurements because the manual measurements do not cover a complete 66 pass wander pattern.

In future tests, the surface rut depth should be measured for each pass of a 66 pass wander pattern at some logical interval, for example every 10th complete pattern. This will provide another measure of the influence of wander on the surface rut depth.

4 UNBOUND AGGREGATE PARTICLE REARRANGEMENT

The NAPTF MDD data illustrate that wheel loads cause a dilative effect in unbound aggregate elements some distance away from the load centerline. Many factors combine to establish the dilative effect; the combination of wheel load, tire pressure, dual wheel spacing, tridem (or tandem) wheel spacing, trafficking speed, trafficking direction, and pavement system characteristics dictate where the dilative element or elements are. They also dictate the magnitude of the dilation. A strong stabilized layer may have little dilation while an unbound aggregate layer will dilate much more readily.

If there were no wander, the location of the unbound aggregate materials dilated or compressed by the loading would not change and the area of weakness or zone of stiffening would not differ. When locations of the dilating granular materials vary because of wander, more areas experience reduced strength. Thus, when loading returns to an area that has been weakened due to dilation and particle rearrangement more downward residual deformation occurs.

5 CONCLUSIONS

According to shakedown theory, the residual deformation per pass should decrease with increasing repetitions; however, this was not the case when load wander was introduced at the Federal Aviation Administration's (FAA's) National Airport Pavement Testing Facility (NAPTF) in the US on asphalt airport pavement test sections. The gear wheel loads typically caused dilative behavior in the unbound aggregate layers some distance away from the wheel centerline. When combined with wander, the dilative effect can cause an increase in rutting because wander induces weakness across the wander pattern and when the wheel load returns to an area weakened by dilation, more rutting occurs.

Individual aggregate particles in an unbound aggregate layer are designed to interlock to support loads. The dilative effect and particle rearrangement negates this interlock and reduces the load capacity of the layer. Future design procedures should incorporate this knowledge by ensuring any layer expected to experience dilatation is constructed or properly treated to prevent it.

ACKNOWLEDGEMENTS

This paper was prepared from a study conducted in the Center of Excellence for Airport Technology. Funding for the Center of Excellence is provided in part by the Federal Aviation Administration. The Center of Excellence is maintained at the University of Illinois at Urbana-Champaign in partnership with Northwestern University and the Federal Aviation Administration. Ms. Patricia Watts is the FAA Program Manager for Air Transportation Centers of Excellence and Dr. Satish Agrawal is the FAA Airport Technology Branch Manager.

The contents of this paper reflect the views of the authors who are responsible for the facts and accuracy of the data presented within. The contents do not necessarily reflect the official views and policies of the Federal Aviation Administration. This paper does not constitute a standard, specification, or regulation.

REFERENCES

Hayhoe, G.H. 2004. "Traffic testing results from the FAA's National Airport Pavement Test Facility," *Proceedings of the 2nd International Conference on Accelerated Pavement Testing*, University of Minnesota, Minneapolis, MN, October.

Hayhoe, G.F., Garg, N., & M. Dong. 2004. "Permanent deformations during traffic tests on flexible pavements at the National Airport Pavement Test Facility," *Proceedings of the 2003 Airfield Pavement Specialty Conference (APSC)*, Las Vegas, Nevada, September 21-24, 2003, Airfield Pavements: Challenges and New Technologies, ASCE Publication Edited by Moses Karakouzian: 147-169.

Gomez Ramirez, F.M., & M.R. Thompson. 2001. "Aircraft multiple wheel gear load interaction effects on airport flexible pavement responses," *Proceedings of the ASCE 2001 Airfield Pavement Specialty Conference*, Chicago, IL.

Gopalakrishnan, K., & M.R. Thompson. 2004. "Performance analysis of airport flexible pavements subjected to new generation aircraft," *FAA Center of Excellence (COE) Report No. 27*, Department of Civil Engineering, University of Illinois at Urbana-Champaign, Urbana, Illinois.

Gopalakrishnan, K, Thompson, M.R., & A. Manik. 2006. "Multi-depth deflectometer dynamic response measurements under simulated new generation aircraft gear loading," *Journal of Testing and Evaluation*, Volume 34, Issue 6, November.

Advances in Transportation Geotechnics – Ellis, Yu, McDowell, Dawson & Thom (eds)
© 2008 Taylor & Francis Group, London, ISBN 978-0-415-47590-7

Water in coarse granular materials: Resilient and retentive properties

J. Ekblad
Royal Institute of Technology, Division of Highway Engineering
Present affiliation: NCC Roads AB, Sweden

U. Isacsson
Royal Institute of Technology, Division of Highway Engineering

ABSTRACT: Granular material is, perhaps the most common construction material used in civil engineering, being an important constituent in road constructions, railways, embankments, foundations, buildings etc. This paper presents results from triaxial testing, at various water contents using constant confining pressure, of two different continuously graded granular materials with maximum particle size 90 mm and 63 mm, respectively. Furthermore, water retention properties of the unbound materials are presented and examples of water distributions in a common construction are shown. From the results presented, it can be concluded that increased water contents cause a reduction in resilient modulus and an increase in strain ratio. The distribution of water content in the vertical direction is highly nonlinear and the degree of saturation in the unbound layers of a road construction depends to a large degree on the level of the water table.

1 INTRODUCTION

In Highway engineering, the influence of environmental factors on the performance of roads is a complex matter. This is a concern both at the design stage when, due to the complexity, simplifications must be made and under service, causing problems for maintenance engineers. The behavior of unbound granular materials depends on water content. In road constructions granular materials are important constituents and the structural response of the road is partly determined by the unbound layers (base and subbase). Elevated water content is believed to adversely affect the long-term performance. Relationships between water content, properties of unbound granular materials and structural degradation of the road is indeed complex. This paper presents a study on properties, under repeated loading, of coarse granular materials related to water content. Furthermore, water retention properties, in terms of soil-water characteristic curves and distribution in a road structure are presented.

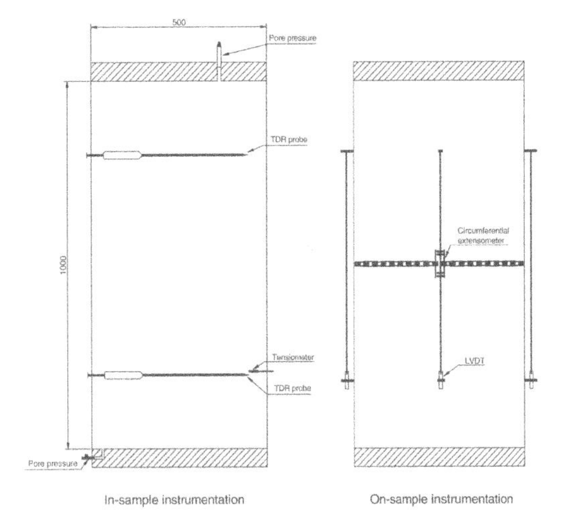

Figure 1. Specimen instrumentation (dimensions in mm).

2 EXPERIMENTAL

2.1 *Equipment*

A schematic view of the triaxial set-up is shown in Figure 1. The large-sample triaxial test facility has been developed at the Royal Institute of Technology (Lekarp & Isacsson 1998, Ekblad & Isacsson 2006). The sample outlined in Figure 1 is confined in an outer cell that is filled with either air or silicon oil. If silicon oil is used, the testing can be performed using repeated axial and confining stresses. However, the tests reported in this paper were performed using

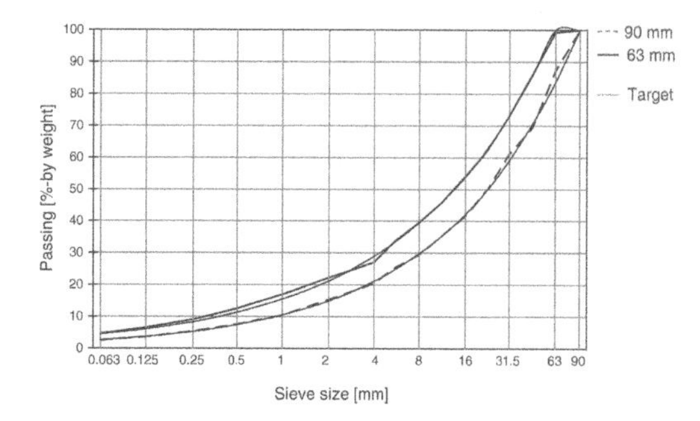

Figure 2. Particle size distributions denoted by maximum particle size.

Table 1. Sample properties.

D_{max} [mm]	Water state	Water content, mass fraction	Saturation
63	Low	0.014	0.14
	Retention limit	0.031	0.29
	Soaked/saturated	0.102	0.97
	Drained	0.032	0.31
90	Low	0.010	0.11
	Retention limit	0.020	0.21
	Soaked/saturated	0.099	1.00*
	Drained	0.038	0.40

* The determined value exceeded 1.

air as confining medium and conducted using static confining pressure of varying magnitude. To measure in-sample water contents, time-domain reflectometry probes (TDR) are buried during compaction and a ceramic cup connected to a tensiometer is inserted. Tubing, connected to the top and bottom plates, allows for water increase and decrease.

2.2 *Materials and samples*

The main difference between the two different gradings tested was maximum particle size: 90 mm and 63 mm, respectively. The crushed rock was quarried in Skärlunda, Sweden, and is characterized as a foliated medium grained granite.

It was considered important to use well-defined grading curves, hence the nominal particle size distribution was derived using the equation:

$$P=\left(\frac{d}{D_{Max}}\right)^{n} \tag{1}$$

where P is the percent smaller than d, D_{max} the maximum particle size and n the grading coefficient, describing the shape of the curve (Fuller 1905).

The two different gradings were part of two different test series: one studying the influence of grading and the other one influence of mica content on resilient properties (Ekblad & Isacsson 2006, Ekblad & Isacsson 2007b). There is a small difference in grading coefficient (n) between the two gradings: $D_{max} = 90$ mm the n equals 0.5 and for $D_{max} = 63$ mm n equals 0.45. However, in Figure 2 it can be seen that the main difference between the two gradings is a horizontal shift. The close concurrence to target grading, given by Equation 1, is also illustrated.

The samples were compacted in 10 layers using a 3-section split mold. Inside the mold, a rubber membrane was held in place by vacuum. The samples were compacted to 95% of maximum density as determined by ASTM D4253 (1996) using a vibratory table. After compaction and subsequent instrumentation, and prior to the resilient test schedule, each sample was conditioned using 20000 cycles and 280 kPa repeated deviator stress at 40 kPa constant confining pressure.

The basic test schedule comprised four stages of resilient testing. Firstly each sample is tested at a low water content state, corresponding to an approximate matric suction of 15 kPa, after which water was added to maximum retentive capacity and further on to saturated (or soaked) conditions. The samples were tested after each addition. Maximum retentive capacity corresponds to the maximum amount of water the sample can retain without draining, given the soil specific soil-water characteristic curve and sample geometry. After the soaked state, the samples were allowed to drain freely and tested for the fourth and final stage. In Table 1 water contents for the various stages are summarized.

To reach maximum retentive capacity when wetting, estimated amounts of water were added through the filters in the top plate and allowed to percolate the sample. The progression of water percolation was continuously monitored using the buried TDR-probes until equilibrium conditions were reached. Soaking, or saturation, was achieved by adding water through the bottom plate until water flowed through the top plate valve. After testing at maximum water content, the specimens were allowed to drain freely, after which they were tested again.

2.3 *Triaxial testing*

The samples were cylindrically confined using constant air pressure, while the deviator stress was repeated as a haversine stress pulse at 1 Hz. The stress state is further described by:

$$p=\frac{\sigma_1+2\sigma_3}{3} \tag{2}$$

$$q=\sigma_1-\sigma_3 \tag{3}$$

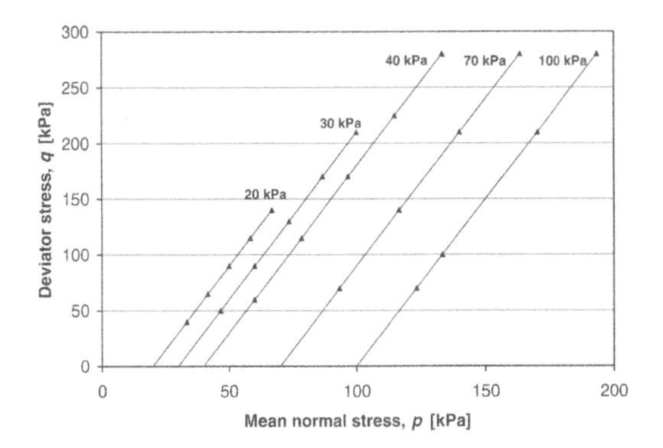

Figure 3. Stress conditions during constant confining stress triaxial tests. For the grading with maximum particle size 90 mm the highest 280 kPa deviator stress path was not used.

where σ_1 and σ_3 are principal stresses, p is mean normal stress and q deviatoric stress. Figure 3 summarizes the stress paths used in this investigation. When the samples became weaker, as a consequence of increasing water content, some of the more severe stress states were avoided; the stress paths with the highest ratio of deviator stress to confining pressure were excluded.

The sample response was described by two resilient parameters: resilient modulus, M_r, and strain ratio, R_ε, which are defined as:

$$M_r = \frac{\sigma_1 - \sigma_3}{\varepsilon_1} = \frac{q}{\varepsilon_1} \tag{4}$$

and

$$R_\varepsilon = -\frac{\varepsilon_3}{\varepsilon_1} \tag{5}$$

where ε_3 and ε_1 are the resilient radial and axial strains, respectively. Resilient modulus and strain ratio are determined as secant values in terms of cycled deviator stress.

The triaxial test mode was consolidated and undrained, i.e. induced pore pressure by increased confining pressure was allowed to dissipate while during repeated loading the water valves were closed. Only very small pore pressures, air or water, were observed, even for soaked/saturated conditions. Measured pore pressures were typically 1–3 kPa in the soaked state. In the unsaturated states pore air pressures were around 0.5 kPa.

3 RESULTS

Results are expressed in terms of total stresses, i.e. internally acting stresses, matric suction and pore pressures are disregarded. Measured values of these

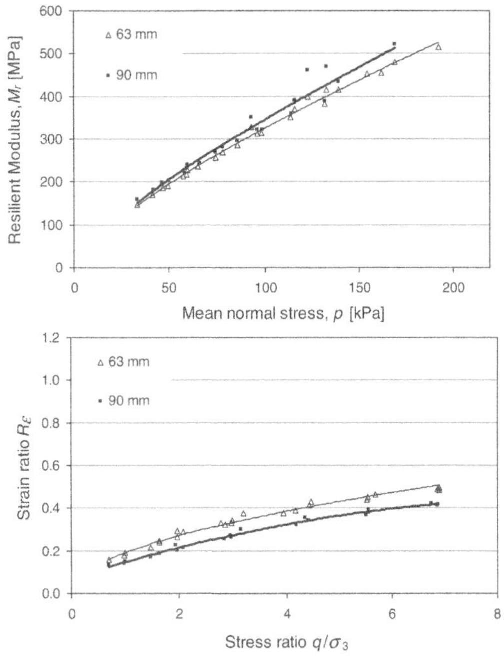

Figure 4. Resilient modulus and strain ratio at low water content state.

internal stresses were low (far below applied stresses). Using the extended effective stress framework would cause only small changes compared to using total stresses.

Figure 4 shows results from testing at the initial low water content state (cf. Table 1). To facilitate visual assessment of the results fitted lines are also shown in Figures 4–6; resilient modulus to a power law and strain ratio to a second degree polynomial.

At the initial low water content stage, the difference in resilient modulus between the gradings is fairly small (the 90 mm material being slightly stiffer). Concerning strain ratio, a difference is discernible. At each stress ratio the strain ratio is higher for the material with smaller maximum particle size.

The other extreme state was fully soaked or saturated samples. No direct measures to ensure full saturation could be undertaken, why it is believed that absolute saturation was not reached and hence, the term soaked was used. In Figure 5, the resilient response at soaked/saturated conditions for the two different gradings is visualized.

At soaked conditions, differences in resilient behavior appear. The 63 mm material shows lower resilient modulus and considerably higher strain ratio. Both gradings showed dilative behavior at higher stress ratios (deviator stress to confining pressure) i.e. increased shearing.

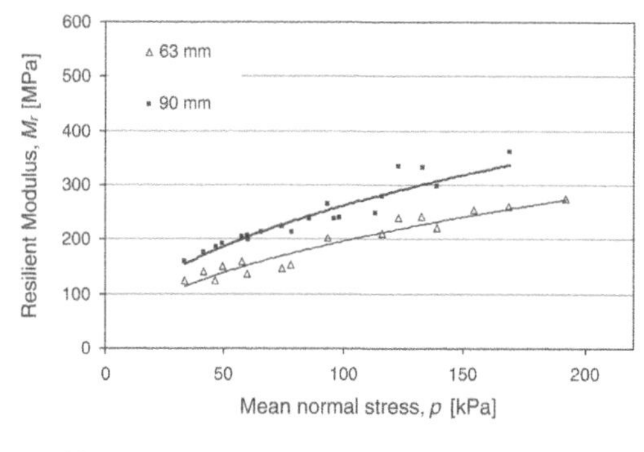

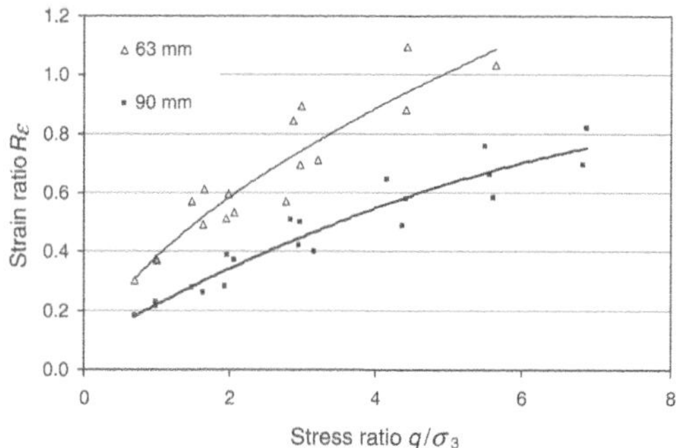

Figure 5. Resilient modulus and strain ratio at soaked/saturated conditions.

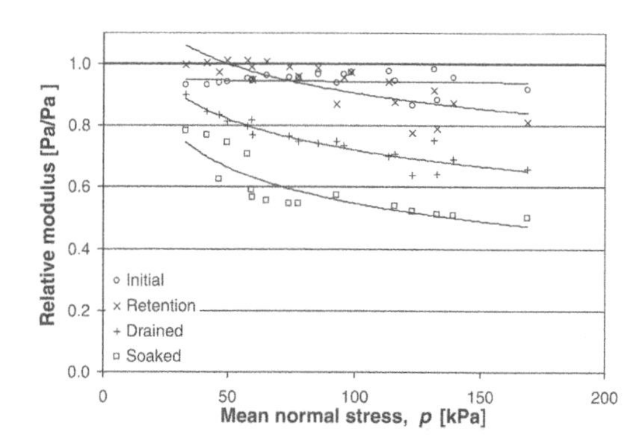

Figure 6. Relative modulus as a function of mean normal stress. 63 mm material at various water contents relative the 90 mm material at low water content.

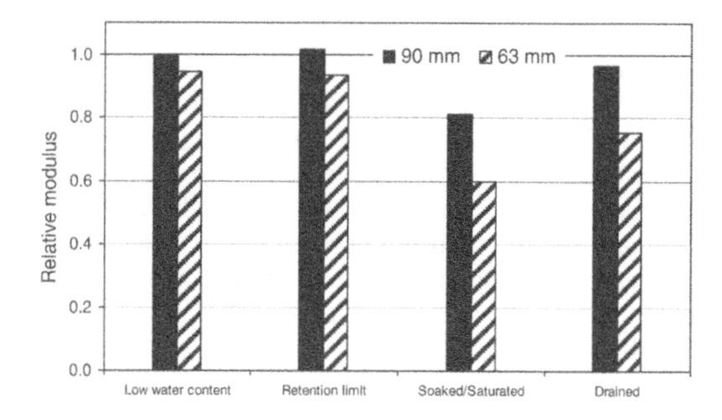

Figure 7. Summary of relative modulus for the two different gradings and various water contents (relative the 90 mm material at low water content.

Figures 4 and 5 show the two extreme states in terms of water: low water content and soaked state, respectively. The other two stages tested were maximum water content in wetting (retention limit) and drying (drained). Figures 6 and 7 summarize and compare all stages for the two gradings. To look at the relative change of mechanical response at varying water contents, normalized relative modulus as a function of mean normal stress is calculated and shown in Figure 6. The relative normalized modulus is calculated by comparing measured values for the two gradings at various water states with measured values for the 90 mm grading at initial low water content. In other words, all measurements are compared with the stiffest response measured for both gradings and water contents. This calculation is performed for all levels of mean normal stress. Figure 6 shows the 63 mm grading at various water states, related to the 90 mm grading at low water content.

In Figure 6 it can be seen that, compared to the stiffest response of the 90 mm grading, the 63 mm grading at initial water state shows approximately 10% lower modulus. At retention limit, the relative modulus varies between slightly stiffer at low levels of mean normal stress to approximately 15% softer at the high end of the stress range.

In Figure 7, the relative moduli for the different mean normal stresses are averaged to a single value i.e. the mean stiffness value for the whole stress range for each water state (shown in Figure 6) is calculated. In this way, the variability between gradings and water states, as a function of mean normal stress, is somewhat concealed, but it makes a general comparison easier.

Figure 7 indicates that the 90 mm grading is stiffer at all comparable water states. Compared to the initial state, the 90 mm grading suffers no decrease in modulus, when the water content is increased to the wetting retention limit. At soaked/saturated conditions, the average relative modulus drops to about 80% of the reference level and after draining, the original characteristic is almost regained, i.e. the average modulus is almost back to 1. A slightly different pattern can be observed for the 63 mm grading. The difference between the low water content state and the wetting retention limit is almost negligible and about 90% of the coarser grading (90 mm). However, at soaked conditions, the average relative modulus is about 60%

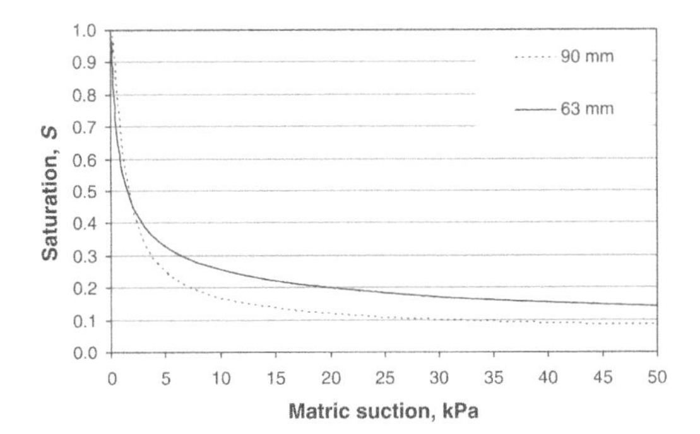

Figure 8. Soil-water characteristic curves for the two different gradings.

of the reference level, which corresponds to a drop of about 30%. Furthermore, after draining, the original modulus level is not fully recovered. The average relative modulus remains at slightly below 80% of the reference level. From Table 1, it can be noted that the water content after draining is higher than the corresponding retention limit when wetting. This is most probably a consequence of the hysteresis in the water retention characteristics between wetting and drying, or absorption and desorption. However, it can be noted that the difference between the retention limit and drained state is larger for the 90 mm grading than for the 63 mm grading. This is somewhat unexpected, given the soil-water characteristic curves of the two materials. This apparent anomaly might be partly caused by errors in estimating the water content for the large triaxial samples. It can also be noted that, despite the increased water content after draining, the 90 mm material almost fully recovers the initial stiffness, while the 63 mm material, showing a small increase in water content, loose stiffness at the drained state.

Concerning strain ratio, the pattern of differences between the gradings and water states are similar, but reversed, compared to resilient modulus, i.e. the measured strain ratios are in general higher for the 63 mm grading compared to the 90 mm grading and the strain ratio increases with increased water content for both materials. However, for the 90 mm grading the strain ratio does not fully recover after draining.

The actual water content for these materials, when used in road construction, depends on the water retention expressed by the soil-water characteristic curve, the geometry and the boundary conditions in terms of water table and water flux from various sources. For the materials described in this paper, the soil-water characteristic curves are shown in Figure 8. The curves shown represent the model fit (van Genuchten 1980) of measured results previously presented by Ekblad & Isacsson (2007a, c). Figure 8 expresses degree of saturation as a function of matric suction for a drying (desorption) path. As previously mentioned the soil-water characteristic curve is commonly hysteretic between wetting and drying, i.e. the level depends on the direction of change. Commonly, the wetting curve is below the drying curve. In this case, the drying curve is given since this curve represents a worst case scenario. In practice, actual behavior is a complex mix of drying and wetting.

From Figure 8, it can be observed that both curves are fairly steep, i.e. the saturation decreases comparably rapid with increased matric suction. Furthermore, for a given level of matric suction, the finer grading (63 mm) shows a higher degree of saturation compared to the coarser grading.

In a road construction, the materials described in Figure 2, can be used as base and subbase layers, respectively. According to Swedish guidelines (ATB Väg 2005), a common construction would comprise bituminous layers and underneath unbound layers: 80 mm base (63 mm grading) and 420 mm subbase (90 mm grading). Of course the actual thickness of these layers varies but the given ones are taken as typical values for use in the example of water distribution visualized in Figure 9. The values given in this figure represent steady-state degree of saturation during drying, given a specific level of matric suction determined by the distance above the water table; it does not represent a more thorough simulation of water content and flow. In Figure 9, three different levels of the water table are assumed. The water table is defined as the level where the matric suction equals zero (or the water pressure equals atmospheric pressure). No-flux conditions are supposed to exist at the top of the base layer. This is probably not an unrealistic constraint, since this surface is covered by a practically impermeable asphalt layer, at least as long as the asphalt surface remains uncracked. If the soil water is in a no-flux state, the matric suction (or potential) is balanced by the gravitational potential represented by the height above the water table. Figure 9 only describes unsaturated states of the unbound layers. In some cases e.g. during thawing in spring-time, it is conceivable that at least part of these layers reach saturation.

The uppermost part of Figure 9 shows the state where the water table is just below the unbound layers, i.e. the subgrade is saturated. It can be seen that in the subbase, the degree of saturation decreases fairly rapidly above the water table. The base layer is thinner and consequently, the saturation gradient is smaller. In the mid-part of Figure 9, the water table is lowered to 30 cm below the bottom of the subbase. According to Swedish design code, drainage is required to be located at least 30 cm below the subbase layer. Compared to the upper diagram, it can be seen that the degree of saturation in the subbase has decreased drastically. The corresponding decrease in the base layer

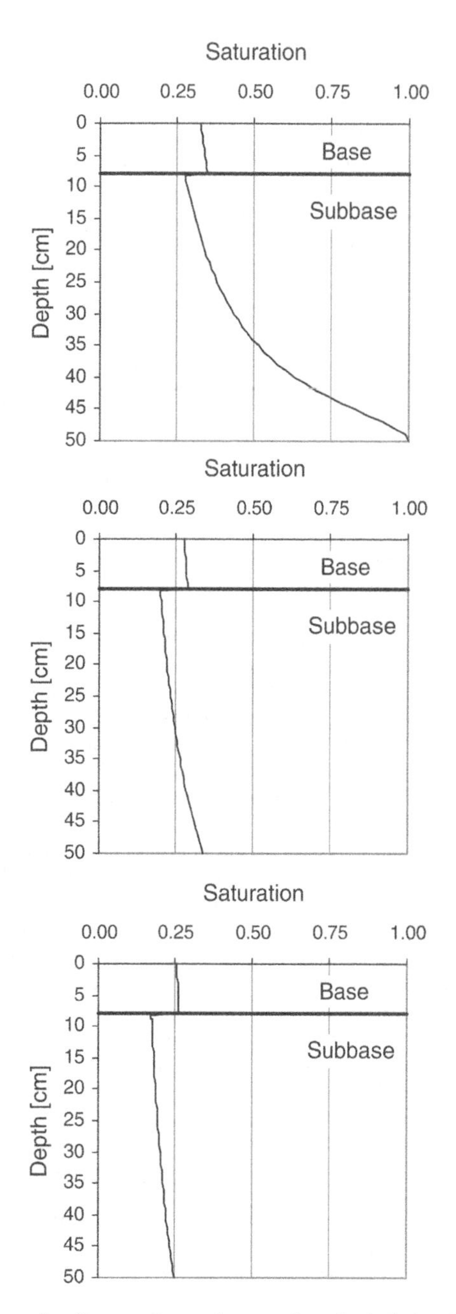

Figure 9. Degree of saturation at various depths below the bituminous layers for three water table levels: upper-just below subbase, middle-30 cm below subbase, lower-50 cm below subbase.

is smaller. Finally, the lowest part of Figure 9 shows the case when drainage is required in the subgrade; the Swedish design code requires that drainage is at least 50 cm below the subbase layer. In this configuration, a further decrease in saturation can be observed for both the layers.

4 DISCUSSION

This paper presents and summarizes findings concerning influence of water on resilient behavior of two different gradings, which differ in maximum particle size, while the shape of the grading was virtually unchanged. When comparing saturation levels in the triaxial samples with estimated levels in road construction layers, there exists a complicating circumstance. For the triaxial samples, the determined saturation represents sample averages. Given the height (100 cm) of the samples, there exists a considerable nonlinear distribution of the sample water. The sample deformation is measured in the mid 60 cm (cf. Figure 1) and in this part of the sample the degree of saturation is commonly lower than the sample average. This means that the saturation levels given in Table 1 at the retention limits (wetting and drying) are higher than corresponding levels in the actual part of deformation measurements. Furthermore, because of the height of the sample and the water retention properties of the materials, the range of unsaturated conditions is limited; at a certain point the sample cannot retain more water, which corresponds to the highest degree of saturation that is possible (except for full saturation) given the sample geometry. Therefore some caution is necessary when comparing the saturation levels shown in Figure 9 with the levels in the triaxial samples since the levels in these samples are probably overestimated.

A further circumstance worth notice is that, depending on the entire construction (including asphalt layers), the stresses experienced by the unbound layers are probably in the lower range of the stresses utilized in the stress path test schedule (the stresses decrease with increased depth). The loss of resilient modulus by increased water content is more pronounced in the higher stress range (cf. Figure 6). However, measuring or estimating the stresses in a road construction is indeed a complex matter and the outcome of a numerical analysis depends on models chosen and boundary conditions.

Based on the results and discussion given the following comments could be made:

- As long as the drainage keeps the unbound layers unsaturated, the main part of these layers remains at a fairly low degree of saturation.
- Given the prevailing stress and water retention conditions in the unbound layers, assuming drained conditions, the change in resilient modulus is rather moderate.

Of course, these statements can be invalidated by change in material properties, construction, stress and drainage conditions, and, possibly by a more elaborate analysis of the road construction and the prevailing conditions. It should also be noted that these comments are made in regard to resilient behavior. Accumulation of permanent deformation, which is also influenced by water content, is not treated.

REFERENCES

ASTM D4253. 1996. *Standard test methods for maximum index density and unit weight of soils using a vibratory table.*

ATB Väg. 2005. General Technical Construction Specifications for Roads, in Swedish. Swedish National Road Administration. http://www.vv.se/templates/page3____14328.aspx

Ekblad, J. & Isacsson, U. 2006. Influence of water on resilient properties of coarse granular materials. *Road Materials and Pavement Design* 7 (3): 369–404.

Ekblad, J. & Isacsson, U. 2007a. Time domain reflectometry measurements and soil-water characteristic curves of coarse granular materials used in road pavements. *Canadian Geotechnical Journal.* 44 (7): 858–872.

Ekblad, J. & Isacsson, U. 2007b. Influence of water and mica content on resilient properties of coarse granular materials. *International Journal of Pavement Engineering.* On-line publication: DOI: 10.1080/10298430701551193.

Ekblad, J. & Isacsson, U. 2007c. Influence of mica content on time domain reflectometry and soil-water characteristic curve of coarse granular materials. *Geotechnical Testing Journal.* Accepted for publication.

Fuller, W.B. 1905. Proportioning concrete. In *A treatise on concrete plain and reinforced*: 183–215, eds. Taylor, F.W. and Thompson, S.E.

van Genuchten, M.Th. 1980. A closed-form equation for predicting the hydraulic conductivity of unsaturated soils. *Soil Sci. Soc. Am. J.* 44: 892–898.

Lekarp, F. & Isacsson, U. Development of a large-scale triaxial apparatus for characterization of granular materials. *International Journal of Road Materials and Pavement Design.* 1 (2): 165–196.

Advances in Transportation Geotechnics – Ellis, Yu, McDowell, Dawson & Thom (eds)
© 2008 Taylor & Francis Group, London, ISBN 978-0-415-47590-7

Measurements and modelling of anisotropic elastic behaviour of unbound granular materials

A. Ezaoui & H. Di Benedetto
Département Génie Civil et Bâtiment – ENTPE – Université de Lyon, Vaulx en Velin, France

ABSTRACT: The elastic properties of granular materials, and more particularly those of sand, are discussed in this paper. The "quasi" elastic properties are observed experimentally in the small strain domain ($\varepsilon_{sa} < 10^{-5}$ m/m) for peculiar stress-strain conditions. The "quasi" elastic properties are determined using an accurate triaxial device allowing static and dynamic loadings. The experimental investigations consist in applying at different stress-strain states along isotropic and deviatoric triaxial stress paths, small axial cyclic loadings (strain single amplitude cycle $\varepsilon_{sa} \cong 10^{-5}$ m/m) and two types of dynamic waves, generated by piezoelectric sensors (compressive and shear waves in axial direction). First, the experimental technique and interpretation of results are described. The results from shear and compressive wave velocities are compared with those of small quasi static loading cycles (performed at small strain rate) during consolidation period. A good correlation between dynamic and static results can be noted and validates both testing procedure and existence of a quasi-elastic linear domain for granular materials. An anisotropic hypoelastic law, "DBGSP" model, is then presented. This model is used to simulate the evolution of elastic (or "quasi" elastic) tensor for different stress paths including classical triaxial tests and constant p/q loadings. Comparisons with experimental data obtain from our device and from the literature, confirm the ability of the model to catch the anisotropic linear elastic behaviour of sands and more generally of unbound granular materials used in road constructions.

1 INTRODUCTION

The purpose of this study is the experimental determination and modelling of the small strain characteristic properties of granular materials. Many researches, Hardin and Blandford (1989), Di Benedetto (1997), Tatsuoka et al. (1997), Tatsuoka and Shibuya (1991) among others, confirmed the existence of a "quasi" elastic domain for small strain amplitude lower than some 10^{-5} m/m ("very small" strain domain). The investigations presented in the first part of this paper are performed in this small strain domain using an accurate triaxial device (triaxial StaDy) coupling quasi cyclic static loadings and wave propagation tests. The tested material is dry Toyoura sand. This paper focuses on experimental campaign, which took part within an international round robin test on the measurement of shear modulus, G_{max} using bender elements. This "Round Robin Test" was organised by the Technical Committee 29 (TC29) of the International Society of Soil Mechanics and Geotechnical Engineering (ISSMGE) (Yamashita et al. (2007)). Two types of loading paths have been performed at ENTPE laboratory: isotropic compression test ($K = \sigma 3/\sigma 1 = 1$) and anisotropic compression test ($K = 0.5$).

These investigations consist in measuring secant stiffness from stress-strain curve (q, ε_z) at small strains ($\leq 0.001\%$) by applying small cyclic axial loadings and in measuring the shear and compressive waves velocities using piezoelectric elements. The testing device is briefly described in the next section. Then, the results obtained during the experimental campaign are exposed. The hypoelastic DBGSP model is introduced and the experimental data are used to calibrate and validate the abilities of the model.

Finally, the simulations with the anisotropic model DBGSP of different stress paths are compared with experimental data from literature on unbound granular materials (Missilac gravel) used in road engineering.

2 TESTING APPARATUS

2.1 *Loading system*

The axial loading system consists in an electromechanical test machine (ref. MTS DY36) which is controlled by closed loop feedback schemes. The analog to digital converter used for the control was realized by a 100 kHz, 16 bit/8 channels converter. The system imposes rate of loading by either stress or strain control

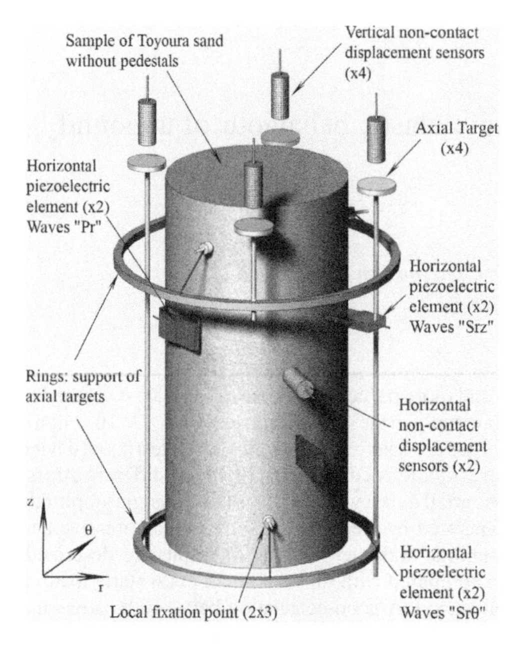

Figure 1. Schematic view of the sample with static displacement measuring devices.

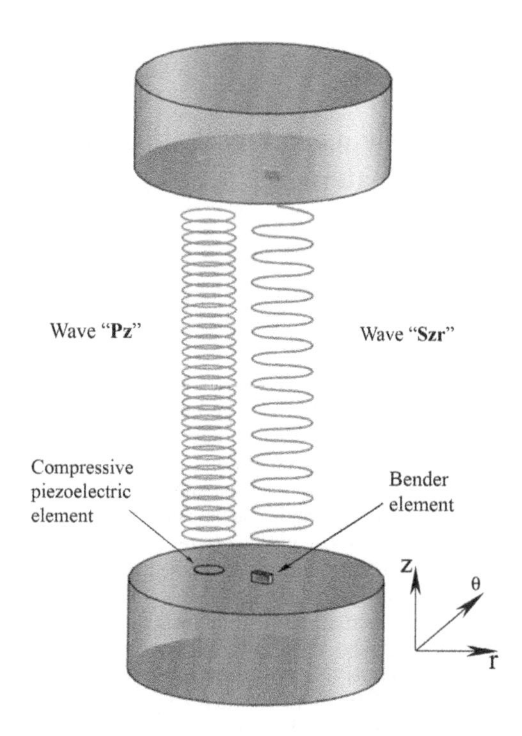

Figure 2. Schematic views of vertical dynamic loading system: shear and compressive vertical waves.

modes. The radial loading is applied by air pressure (Pham Van Bang (2004)).

2.2 *Static measuring device*

The triaxial cell has internal tie bar. A 10 kN load cell is placed inside the cell. Two displacement measuring devices were designed in order to obtain locally axial and radial displacements in the central part of the sample (Figure 1). Four axial displacement sensors (non contact type, 1 mm range) are fixed on mobile support and aim at four aluminum targets. These targets are fixed on two hung rings that are glued at 3 points of the specimen situated at the same horizontal level. The sensors can be moved from outside the cell by micrometric screws crossing the top plate. This allows the axial sensors to always remain inside their measuring range (1 mm). Two radial displacement sensors (non contact type, 1 mm range) are fixed on movable supports and aim at sheets of aluminum paper placed on the inner side of the neoprene membrane (thickness is 0.5 mm). Mobility of radial sensors is assured by micro-motor piloted from outside the cell.

2.3 *Dynamic testing system*

Four piezoelectric sensors (1 bender element and 1 compressive transducer situated in top and bottom platen) are employed for emission and reception of waves. The first bender element generates a wave, horizontally polarized, propagating along vertical direction, and the second one, records the arriving signal (figure 2). This dynamic shear mode (S_{zr} figure 2) gives the wave propagation velocity, noted Vs. The recording is carried out by an oscilloscope whose sampling rate is 2.10^5 per second.

The second mode (P_z figure 2) uses two compressive piezoelectric elements. The emitted compression wave, vertically polarized, propagates along the vertical direction and its velocity, Vp, is measured.

The excitation signal, produced by a function generator (ref. HP 33120A), is amplified (ref. power amplifier Bruel & kjaer 2713). This amplification is necessary because of the small transmission of the wave through the epoxy resin. Lastly, a calibration of bender element was carried out, in order to determine the time delay introduced in the measurements by the measurement chain (Brignoli et al. (1996)). The value of 8 μs was found for bender element and 2.5 μs for compressive transducer.

3 ISOTROPIC AND ANISOTROPIC CONSOLIDATION TESTS

3.1 *Testing Program*

The material used for this campaign is a Toyoura sand, provided by the Kitami Institute of Technology. The general characteristics of this sand are summarized in the table 1. The samples are prepared by

Table 1. Results of Toyoura sand grading, data from DGCB.

Diameter*:	D_{10} 0.15	D_{30} 0.175	D_{50} 0.19	D_{60} 0.2
Coefficient**:	C_u 1.33	C_c 1.02		
Void ratios :	e_{min} 0.605	e_{max} 0.977		

* Defined by x% passing particle size in weight.
** $Cu = D_{60}/D_{10}$ $Cc = (D_{30})^2/(D_{10}.D_{60})$.

air-pluviation method. Air dried sand is poured from a copper nozzle with a rectangular inner cross section of 1.5 mm × 15.0 mm (provided by the Japanese domestic committee of TC-29), while maintaining a constant drop height throughout the preparation. During this operation, small vibrations are applied around the specimen to reach the different void ratios e.

Before the mould is disassembled, a partial vacuum of 10 kPa is applied in the specimen. After the mould is dismantled, the vacuum is raised to 25 kPa (the specimen can deform freely). The partial vacuum is replaced with a cell pressure of 25 kPa while keeping the effective stress constant throughout the procedure. Two types of granular packing were tested. They correspond to denser (looser) samples at initial (after fabrication $\sigma_r = \sigma_z = 25$ kPa) void ratio $e_0 = 0.7$ ($e_0 = 0.8$). The experimental program consists in drained isotropic and anisotropic consolidation tests on dry Toyoura sand. Each of the four tests contains investigation points at: $\sigma_z = 50, 100, 200, 400$ kPa. For each investigation point the following steps are carried out:

1. Creep 10 minutes at constant stress state, including measurements of change in height and volume of the specimen.
2. Measurements of shear wave velocity using bender element.
3. Measurements of compressive wave velocity using compressive piezoelectric element.
4. Measurements of elastic parameters from stress-strain curves ($\varepsilon_{sa} = 0.001\%$) by applying axial cycling loading of small amplitude in drained conditions.

3.2 *Loading paths*

Figure 3 presents the followed stress paths for isotropic and anisotropic consolidation in the axes (σ_r, σ_z). It consists of i) Isotropic consolidation ($\sigma_z = \sigma_r$), ii) Anisotropic consolidation "close to" the consolidation stress line $K = \sigma_z/\sigma_r = 0.5$. The chosen non linear stress path for anisotropic consolidation is described equations (1) and (2) with $q = \sigma_z - \sigma_r$:

$$\Delta q = 0 \text{ and } \Delta\sigma_r = 25kPa \quad (1)$$

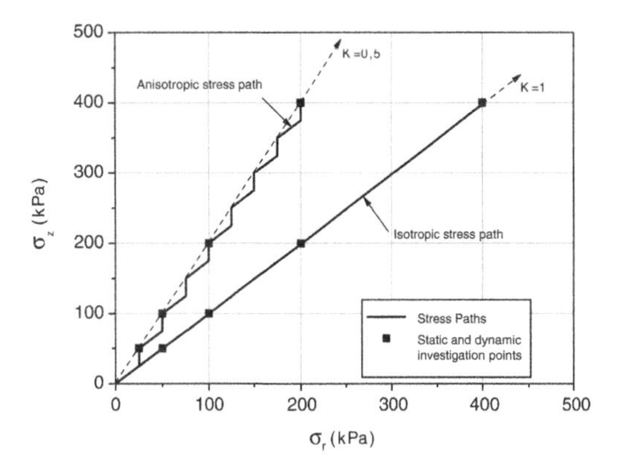

Figure 3. Stress paths for isotropic and anisotropic consolidation on dry Toyoura sand.

Table 2. Investigations points reached (×), for each test.

σ_z(kPa) Tests:	50	100	200	400
rrt.iso.78	x	x	x	x
rrt.iso.67	x	x	x	x
rrt.aniso.78	x	x	x	
rrt.aniso.66	x	x		

$$\Delta\sigma_r = 0 \text{ and } \Delta\sigma_z = 25kPa \quad (2)$$

The tests and the investigation points reached are listed in the table 2. Test nomenclature obeys to the following convention: rrt.tt.yy where characters "rrt" stands for Round Robin Test, "tt" for the type of test (isotropic or anisotropic consolidation), and "yy" for the initial void ratio e_0.

3.3 *Global stress-strain curves*

Figures 4 and 5 present the global curves of test rrt.iso.67. The investigations points (small static and dynamic loadings) are also indicated. Procedure for isotropic consolidation consists in maintaining a deviator stress q constant, equal to zero. This is verified in figure 4, where q is between +/− 2 kPa (except during axial cyclic loadings at investigation points). The evolution of axial strain is presented figure 5 (small cyclic loading are indicated).

3.4 *Investigation points: small static cyclic loadings*

The small axial static cyclic loadings performed at investigation points give directly the equivalent or static Young modulus E_z and the equivalent (or static)

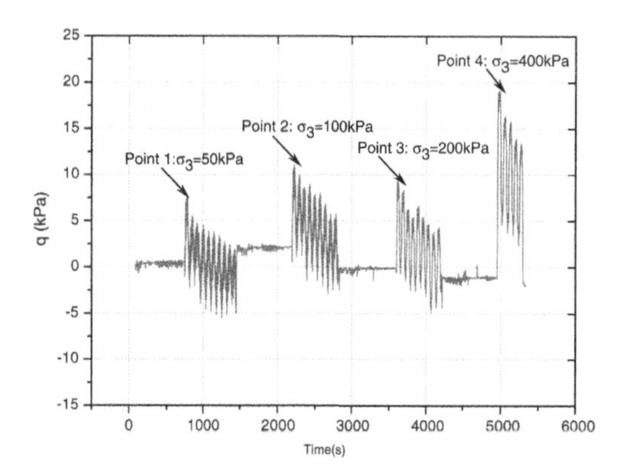

Figure 4. Evolution of deviator stress q with time, test rrt.iso.67.

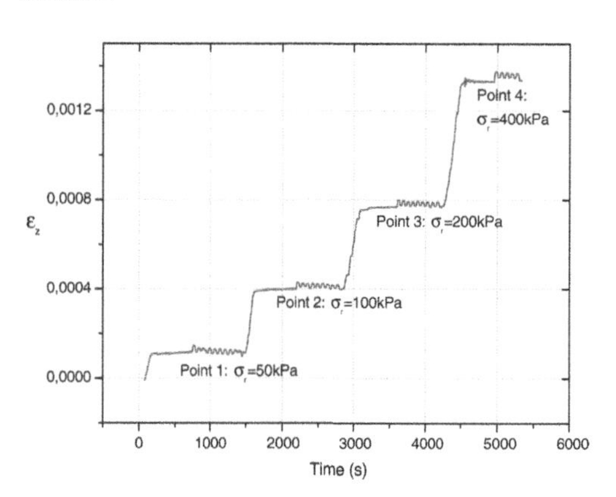

Figure 5. Evolution of axial strain ε_z with time, test rrt.iso.67.

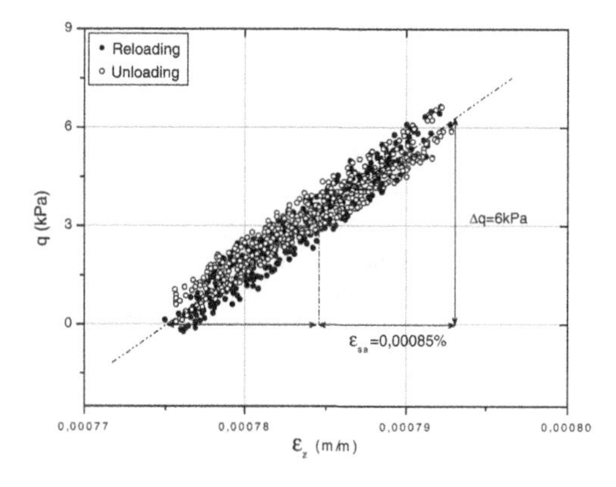

Figure 6. Six small cycles in the axes (ε_z,q) for rrt.iso.67. Test at $\sigma_r = 200$ kPa (Point 3 of figure 5).

Poisson ratio ν_{rz}, respectively by using linear fitting in the plans (ε_z,q) and (ε_z,ε_r). Examples of small cyclic "quasi" static results are presented figures 6 and 7.

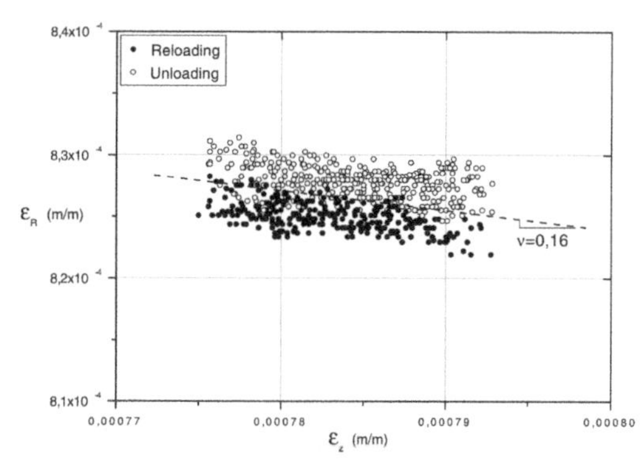

Figure 7. Six small cycles in the axes (ε_z,ε_r) for rrt.iso.67. Test at $\sigma_r = 200$ kPa (Point 3 of figure 5).

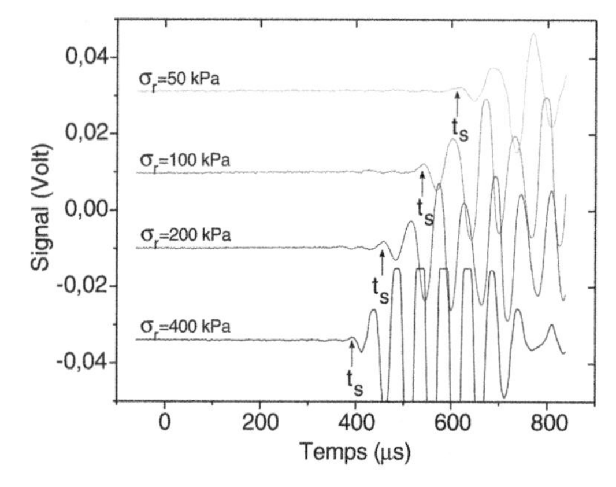

Figure 8. Output signals of S waves, on dry Toyoura sand. Tests rrt.iso.78, at 50, 100, 200, 400 kPa (isotropic stress state). The excitation frequencies are respectively 10, 10, 10 and 15 kHz.

3.5 *Dynamic investigations*

To determine the travel time "ts" of the shear waves, the invariant deflection of signals obtained at 10, 15 and 20 kHz is considered. Using this procedure we expect that near field effects are well considered.

The times "ts" determined with this method, are indicated figure 8 for different isotropic stress state (test rrt.iso.78).

The distance between the top of the two piezoelectric transducers, L', and the time ts' required by the S waves to cover the distance are used to calculate the propagation velocity, Vs, defined by:

$$V_S = \frac{L'}{t_S'} \tag{3}$$

where $L' = L - 2.lb$ with L length of the specimen, lb length of bender element, t's the flying travel time.

Same procedure has been realized for compressive waves (P):

$$V_P = \frac{L}{t_P'} \tag{4}$$

where tp′ is the flying time of P waves.

4 COMPARISON BETWEEN STATIC AND DYNAMIC MODULI OBTAINED DURING CONSOLIDATION TESTS

The dynamic moduli are given by inverse analysis from the measurement of the wave velocities through the sample. It is necessary to make assumptions on the soil behaviour to determine the dynamic moduli (Inverse analysis). Two cases are generally considered.

First, assuming the sample as a semi-infinite isotropic elastic medium; then the Young modulus E_z and the shear modulus G are related to the compression wave velocity V_p and to the shear wave velocity V_s as follows:

$$E = \frac{1-\nu-2\nu^2}{1-\nu}\rho.V_P^2 \tag{5}$$

$$G = \rho.V_S^2 \tag{6}$$

where ν is the Poisson ratio ($G = E/2(1+\nu)$). V_p and V_s are known experimentally.

Secondly, assuming the sample as semi-infinite transverse isotropic elastic medium, this is the similar assumption than for the quasi-static case. The relations linking moduli and wave velocities become:

$$\rho.V_P^2 = \frac{E_z^2(\nu_{rr}-1)}{(\nu_{rr}-1)E_z+2\nu_{rz}^2E_r} \tag{7}$$

$$\rho.V_S^2 = G \tag{8}$$

However, three more data are required to obtain the values of five parameters needed for this assumption. These data cannot be obtained with the presented apparatus. An improvement of this testing device consisting in introducing three pairs of bender elements on the lateral surface of the specimen (figure 1) and allowing to obtain the five parameters of the general transverse elastic tensor is presented in Ezaoui et al. (2006).

Figure 10 shows the comparison of quasi-static and dynamic moduli considering the first hypothesis mentioned above (isotropic case), for triaxial compression tests performed on Toyoura sand.

The slopes of the linear fits are closed to 1. A difference of 5% is obtained for the isotropic results and 11%

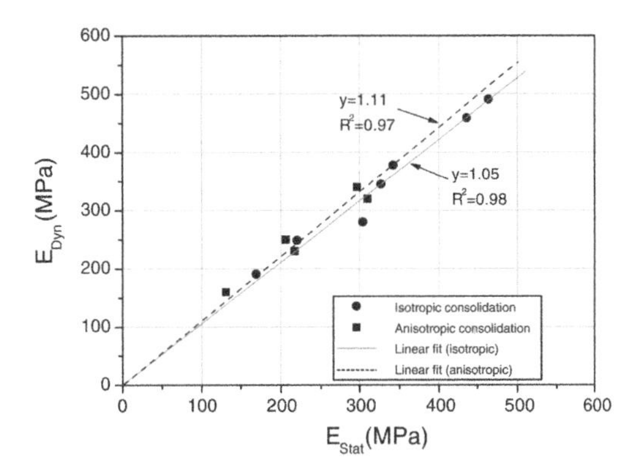

Figure 9. Comparison between static Young moduli and dynamic Young moduli, considering isotropic elastic hypothesis, for dry Toyoura sand, rrt.iso.78, rrt.iso.67, rrt.aniso.78, rrt.aniso.66.

for anisotropic results. These fits are based on the first hypothesis (eq. 5 and 6). The good accordance between static and dynamic axial Young modulus (figure 9) for isotropic stress states validate both procedures (wave propagation and small cyclic loadings) to obtain elastic parameters, and confirm the existence of an elastic domain.

In that case of presented loading paths, isotropic hypothesis gives good results. This is not the case for more general lading paths. For more details about inverse analysis in the anisotropic case and comparison between static and dynamic results on granular material, see Duttine et al. (2008).

5 ANISOTROPIC HYPOELASTIC MODEL FOR UNBOUND GRANULAR MATERIALS: "DBGSP"

5.1 *Hypoelastic model: DBGSP*

The hypoelastic model DBGSP, Di Benedetto (2006), Ezaoui et al. (2006) , is an improvement of DBGS model (Di Benedetto, Geoffroy, Sauzeat), developed at ENTPE (Di Benedetto et al. (2001), Sauzéat (2003)). The general formalism, linking the objective stress increment and the objective elastic strain increment, is:

$$\underline{d\varepsilon} = \underline{\underline{M}}(h).\underline{d\sigma} \tag{9}$$

where h is a set of history parameters. The hypoelastic models DBGS and DBGSP suppose the symmetry of the tensor $\underline{\underline{M}}$. This symmetry has been confirmed by experiments from T4cStaDy (Torsional apparatus on hollow cylinder specimen) on Toyoura sand, Cazacliu

(1996), Duttine (2005). The general formalism is written in equations (10) to (13):

$$\underline{\underline{M}}^{DBGSP} = \frac{1}{f(e)} \frac{\underline{\underline{S}}_v . \underline{\underline{\Gamma}} . \underline{\underline{\Sigma}}_p + {}^t(\underline{\underline{S}}_v . \underline{\underline{\Gamma}} . \underline{\underline{\Sigma}}_p)}{2} \tag{10}$$

In the principal axes of stress:

$$S_v = \begin{bmatrix} 1 & -n_0 & -n_0 & 0 & 0 & 0 \\ -n_0 & 1 & -n_0 & 0 & 0 & 0 \\ -n_0 & -n_0 & 1 & 0 & 0 & 0 \\ 0 & 0 & 0 & 1+n_0 & 0 & 0 \\ 0 & 0 & 0 & 0 & 1+n_0 & 0 \\ 0 & 0 & 0 & 0 & 0 & 1+n_0 \end{bmatrix} \tag{11}$$

$$S_p = P_{ref}^n \cdot \begin{bmatrix} \frac{1}{s_1^n} & 0 & 0 & 0 & 0 & 0 \\ 0 & \frac{1}{s_2^n} & 0 & 0 & 0 & 0 \\ 0 & 0 & \frac{1}{s_3^n} & 0 & 0 & 0 \\ 0 & 0 & 0 & \frac{1}{(s_2 \cdot s_3)^{n/2}} & 0 & 0 \\ 0 & 0 & 0 & 0 & \frac{1}{(s_1 \cdot s_3)^{n/2}} & 0 \\ 0 & 0 & 0 & 0 & 0 & \frac{1}{(s_1 \cdot s_2)^{n/2}} \end{bmatrix} \tag{12}$$

Where P_{ref} is a constant equal to 1 MPa.

In the principal axes of irreversible strain:

$$\underline{\underline{G}} = \begin{bmatrix} \frac{1}{\chi(\bar{e}_{ir,1}^{Tot})} & 0 & 0 & 0 & 0 & 0 \\ 0 & \frac{1}{\chi(\bar{e}_{ir,2}^{Tot})} & 0 & 0 & 0 & 0 \\ 0 & 0 & \frac{1}{\chi(\bar{e}_{ir,3}^{Tot})} & 0 & 0 & 0 \\ 0 & 0 & 0 & \frac{1}{\tilde{\chi}(\bar{e}_{ir,2}^{Tot}, \bar{e}_{ir,3}^{Tot})} & 0 & 0 \\ 0 & 0 & 0 & 0 & \frac{1}{\tilde{\chi}(\bar{e}_{ir,1}^{Tot}, \bar{e}_{ir,3}^{Tot})} & 0 \\ 0 & 0 & 0 & 0 & 0 & \frac{1}{\tilde{\chi}(\bar{e}_{ir,1}^{Tot}, \bar{e}_{ir,2}^{Tot})} \end{bmatrix} \tag{13}$$

With f(e) is a function of void ratio, ν_0 is a constant equal to the Poisson ratio at isotropic stress state, n a constant of the model and $\chi()$ a function of "total" irreversible deviatoric strain tensor (eq. 14). $\underline{\underline{\Gamma}}$ is identity for DBGS model.

$$\underline{\underline{\bar{\varepsilon}}}_{ir}^{Tot} = \underline{\underline{\bar{\varepsilon}}}_{ir}^{Mes} + \underline{\underline{\bar{\varepsilon}}}_{ir}^{Init} \tag{14}$$

Equation 14 presents the total irreversible deviatoric strain tensor, used in equation 13, as the sum of a "Measured" deviatoric strain tensor (corresponding to the deviatoric strain tensor measured during the test performed and linked to the material anisotropy involved during the test) and an "Initial" deviatoric strain tensor (corresponding to a deviatoric strain tensor applied from isotropic state, which would involve the same initial material anisotropy) (Ezaoui et al. (2006)).

Due to the specificity of the triaxial test, the symmetry of testing conditions is transverse isotropic. As a consequence, the hypoelastic law is transverse isotropic in the principal axes of stress and strain. For triaxial compression tests, the expression of the five parameters of tensor $\underline{\underline{M}}$ (equations 9 and 10) in the cylindrical (r,θ,z) axes are given in equations (15) to (19).

$$E_z = \frac{f(e)}{P_{ref}^n} . \chi(\bar{\varepsilon}_{ir,z}^{Tot}) . \sigma_z^n \tag{15}$$

$$E_r = \frac{f(e)}{P_{ref}^n} . \chi(\bar{\varepsilon}_{ir,r}^{Tot}) . \sigma_r^n \tag{16}$$

$$\nu_{rz} = \frac{\nu_0}{2} . (1 + \frac{\chi(\bar{\varepsilon}_{ir,z}^{Tot}) . \sigma_z^n}{\chi(\bar{\varepsilon}_{ir,r}^{Tot}) . \sigma_r^n}) \tag{17}$$

$$G = \frac{f(e)}{2(1+\nu_0) P_{ref}^n} . \tilde{\chi}(\bar{\varepsilon}_{ir,r}^{Tot}, \bar{\varepsilon}_{ir,z}^{Tot}) . (\sigma_r \sigma_z)^{n/2} \tag{18}$$

$$\nu_{rz} = \nu_0 \tag{19}$$

5.2 *Modelling of elastic parameters obtained during consolidation tests performed on Toyoura sand with DBGSP model*

Considering the test procedure and preparation of the specimen at initial stress state (before isotropic or anisotropic consolidation tests), it can be postulated as a first approximation that the sample do not exhibit initial material anisotropy (no preloading or tamping in a specific direction). Thus, the tensor $\underline{\underline{\Gamma}}$ becomes the unit tensor, and DBGSP model gives the same results as DBGS model.

$$\underline{\underline{\Gamma}} = \underline{\underline{1}} \tag{20}$$

Moreover, the volumetric strains are very small during theses consolidation tests, so the evolution of void ratio can be neglected and function f(e) is assumed to be constant. The simulations can be done with 3 constant only:

- ν_0
- $f(e) = f_0 =$ constant at a given void ratio
- n

For the tested Toyoura sand $\nu_0 = 0.21$ and $n = 0.42$. The equations 15, 17 and 18 become:

$$E_z = \frac{f_0}{P_{ref}^n} . \sigma_z^n \tag{21}$$

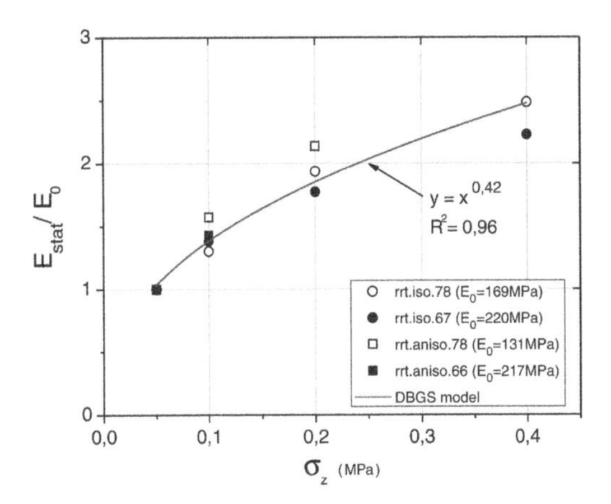

Figure 10. Evolution of static Young modulus with stress level σz, data from small axial cycles and simulation with DBGS model. $n = 0.42$ and E_0 is the modulus at $\sigma_z = 50$ kPa.

$$\nu_{rz} = \frac{\nu_0}{2}.(1 + \frac{\sigma_z^n}{\sigma_r^n}) \tag{22}$$

$$G = \frac{f_0}{2(1+\nu_0)P_{ref}^n}.(\sigma_r \sigma_z)^{n/2} \tag{23}$$

Figure 10 presents vertical Young modulus E^{stat} (obtained during "static" axial cyclic loadings) normalized by Young modulus at stress state $\sigma_z = 50$ kPa) E_0, versus axial stress σ_z. The static Young modulus is determined in the small axial strain domain (single strain amplitude of the cycles is less than $\varepsilon_{sa} = 0.001\%$, cf. figure 6) for each investigation points (table 2) as explained in previous section.

Figure 11 presents static Poisson ratios "ν_{rz}" versus stress σ_z. The static Poisson ratios are determined in the small strain domain (strain amplitude of the cycles is less than $\varepsilon_{sa} = 0.001\%$, figure 7) for each investigation points (table 2) as explained in previous section.

Figure 12 presents dynamic shear modulus normalized by G_0 (shear modulus at stress state $\sigma_z = 100$ kPa) versus stress product $\sigma_r \cdot \sigma_z$, determined using wave propagations with bender elements at different isotropic stress states.

Comparisons between experimental data and modelling (figures 10 to 12) confirm the ability of the model (DBGS) to simulate evolution of elastic parameters: Young modulus, Poisson ratio, and shear modulus.

The hypoelastic model DBGS seems to be relevant to describe the quasi elastic behaviour of slightly anisotropic sand and more generally of unbound granular material with few rheological constants.

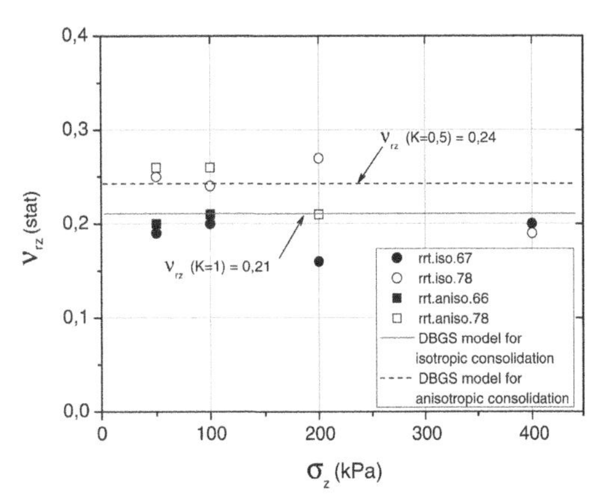

Figure 11. Evolution of static Poisson ratio ν_{rz} with stress product σ_z, experimental data from small axial cycles and simulation with DBGS model ($\nu_0 = 0.21$).

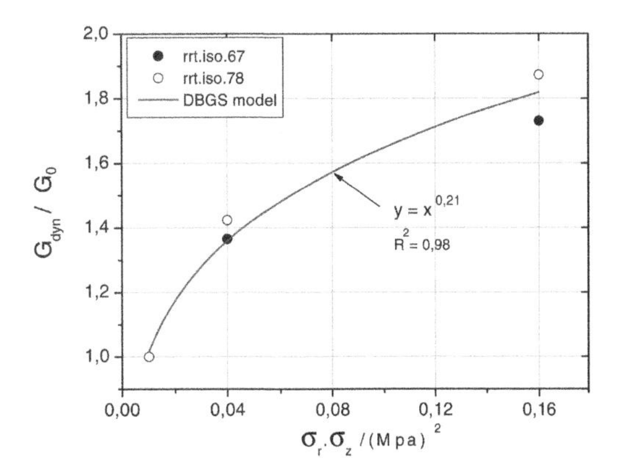

Figure 12. Evolution of dynamic shear modulus G with stress level $\sigma_r \cdot \sigma_z$, experimental data from wave propagations and simulation with DBGS model. $n = 0.42$ and G_0 is the shear modulus at $\sigma_z = 100$ kPa.

6 APPLICATION TO THE RESILIENT BEHAVIOUR OF ROAD GRANULAR MATERIALS

The purpose of this last section is to simulate the resilient behaviour (quasi-elastic) of road unbound granular materials. As the unbound granular materials used in road construction are strongly compacted in the vertical direction, the hypothesis of isotropy can no longer be postulated and the general DBGSP form of the hypoelastic model is used for the simulation.

6.1 *Material and test program*

The experimental study used hereafter has been realized on a 0/10 mm granular material called Missillac

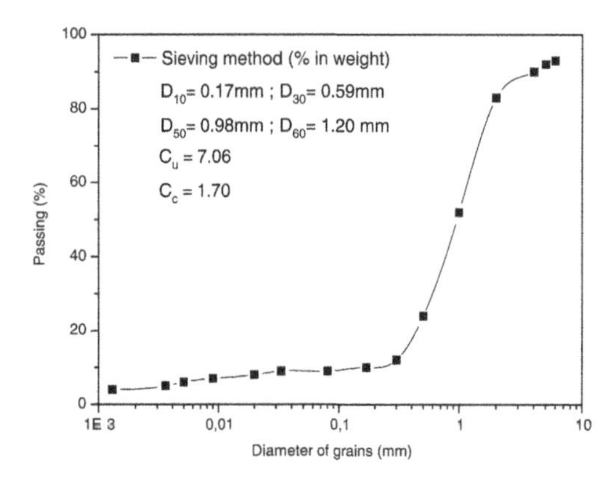

Figure 13. Grading curve of Missilac gravel, El Abd (2006).

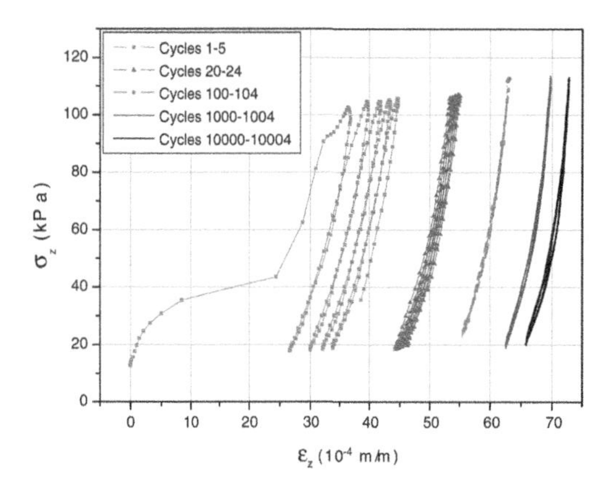

Figure 14. Conditioning phase in the axes (ε_z,σ_z): 10000 cycles (frequency = 1 Hz) applied with q/p = 2, El Abd (2006).

gravel. The grading curve is presented figure 13. All the experimental data are coming from El Abd (2006).

A standard procedure has been carried out, with cyclic loadings at q/p = 2. This conditioning phase consists in applying 10000 deviatoric cyclic loadings at q/p = 2 from σ_z = 20 kPa to 110 kPa (loading frequency is 1 Hz), in order to reach the stress-state domain where deformations are stable and "quasi" reversible. Figure 14 presents these conditioning cyclic loadings in the axes (ε_z,σ_z).

After this conditioning phase (10000 cycles), when resilient elastic domain is reached, different types of loading paths are followed. For each test, the ratio q/p is constant. It varies from 0 to 3. The experimental cyclic tests performed at q/p = constant, are given figure 15 in the axes (p,q) (mean pressure versus deviator stress).

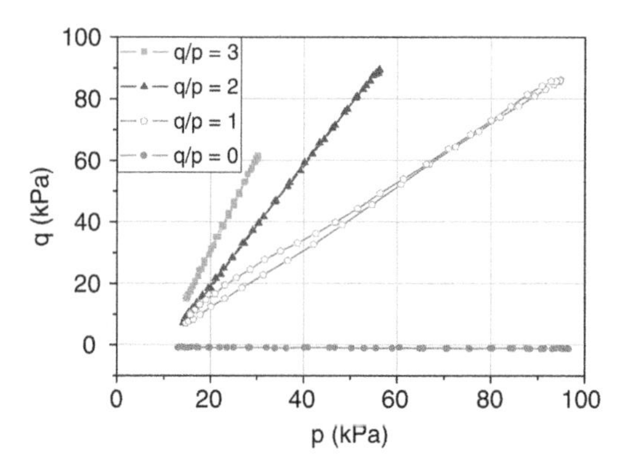

Figure 15. Cyclic stress paths in the axes (p,q), El Abd (2006).

6.2 *Simulation with DBGSP model*

The testing conditions (resilient state) considered in this section imply a strong material anisotropy as described in Hornych et al. (1998) and El Abd (2006). The high level of anisotropy is created by, i) the procedure to set the sample (proctor compaction) and ii) the high numbers of cyclic loadings in the vertical direction "z" during conditioning phase. After 10000 conditioning cycles, the void ratio and irreversible strain do not change anymore (cf. figure 14). Then, the function f(e) and the tensor $\underline{\underline{\Gamma}}$ (equations 15 to 19) can be considered as constant during the following cyclic stress paths given in figure 15. It comes from equations 15 to 19, that four constants only are necessary to obtain the different tangent elastic moduli and Poisson's ratios from DBGSP model:

- ν_0
- n
- $f(e).\chi(\overline{\varepsilon}_{ir,z}^{\mathrm{Tot}}) = \alpha =$ constant at fixed void ratio
- $\frac{\chi(\overline{\varepsilon}_{ir,z}^{\mathrm{Tot}})}{\chi(\overline{\varepsilon}_{ir,r}^{\mathrm{Tot}})} = \beta =$ constant

Equations 15, 16 and 17 become:

$$E_z = \frac{\alpha}{P_{ref}^n}.\sigma_z^n \tag{24}$$

$$E_r = \frac{1}{P_{ref}^n}.\frac{\alpha}{\beta}\sigma_r^n \tag{25}$$

$$\nu_{rz} = \frac{\nu_0}{2}.(1+\beta.\frac{\sigma_z^n}{\sigma_r^n}) \tag{26}$$

It should be underlined that these values are tangent properties which are integrated along the stress paths.

The average values, for "ν_0"and "n", obtained in case of sands (Silica, Hostun, Toyoura, Fontainebleau) or gravels are 0.2 and 0.5 respectively (Anhdan and

Table 3. Constants of DGBSP model (eq. 24 to 26).

ν_0	n	α(MPa)	β
0.2	0.5	530	1.84

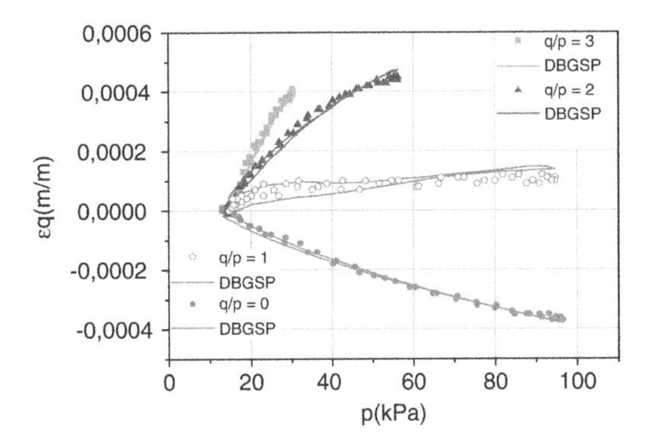

Figure 16. Cyclic loading in the axes (p,ε_q), for four constants q/p stress paths (figure 15), experimental data from El Abd (2006) and simulation with DBGSP model.

Koseki (2005), Belloti et al. (1996), Hoque (1996), Tatsuoka and Kohata (1995), Di Benedetto (2006)). These two values are chosen for Missillac gravel. They probably could be better optimized, considering complementary experimental results.

The two last constants, α and β, are determined from tests at q/p = 3 and q/p = 0 (figure 15). The constant α is evaluated from test q/p = 3 considering experimental global curve $\sigma_z - \varepsilon_z$ and the fact that constant β do not affect this curve (σ_r is constant for this stress path). The stress path σ_z is considered as input data for the model and the strain considered as output data of the model. Thus, the maximum strains (experimental and modelling) are compared for different constant α. Using this procedure, the maximum axial strains are identical for $\alpha = 530$ MPa. The second constant β is determined from test q/p = 0, considering the experimental global curves ε_q-p (deviator strain versus mean pressure) and same procedure than for constant α. The best fit gives the value 1.84 for β.

Table 3 summarizes the 4 constants used for our anisotropic elastic simulations on Missillac gravel.

Figure 16 (resp. 17) presents the results in terms of deviatoric strain (respectively volumetric strains) versus mean pressure "p".

It can be seen figures 16 and 17, that the anisotropic model DBGSP describes both ε_q and ε_{vol}quite well for all the tests at q/p = constant. We can quote that some cyclic modelling are not totally superimposed for loading and unloading. In fact, modelling strains (deviator or volumetric) are calculated from experimental

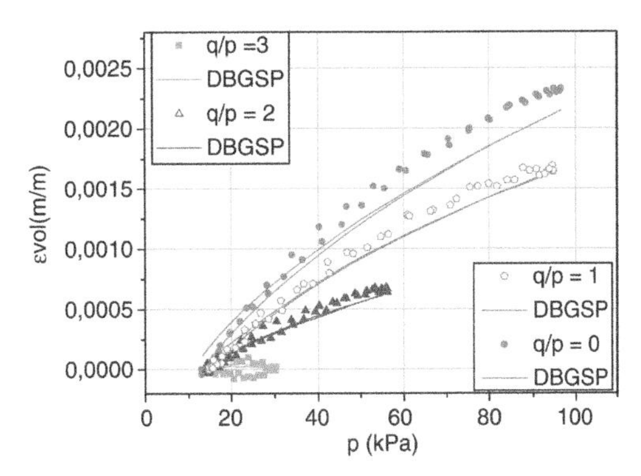

Figure 17. Cyclic loading in the axes (p,ε_{vol}), for four constant q/p stress paths (figure 15), experimental data from El Abd (2006) and simulation with DBGSP model.

followed stress paths which are not exactly the same during loading and unloading phases (cf. figure 15), which explain the non closed simulated stress strain loops.

The constant β allows evaluating material anisotropy. The value for $\beta = 1.84$ highlights the strong level of anisotropy ($\beta = 1$ for isotropic material) involved by the fabrication method and the conditioning phase (figure 14). It means that for an isotropic stress state, axial stiffness of the material is expected to be 1.84 higher than radial stiffness.

Moreover, the model allows to obtain directly all the elastic tensor terms (tangent terms) for general transverse isotropic assumption.

7 CONCLUSIONS

This paper deals with anisotropic elastic behaviour of unbound granular material. An experimental campaign performed on Toyoura sand (fine sand) is presented. During isotropic and anisotropic stress consolidation, investigations have been performed in the small strain domain using precise triaxial apparatus (triaxial StaDy) with dynamic (wave propagations) and local static measurement devices.

First, in the case of slightly anisotropic samples (Toyoura sand), the simulations with hypoelastic model "DBGS" show well accordance with experimental results in terms of axial Young modulus and Poisson's ratio. For this model, only stress tensor is taking into account for anisotropy effects.

Secondly the ability of the hypoelastic model to predict resilient behaviour of unbound granular materials is verified in case of Missillac gravel, for samples with a strong initial material anisotropy. The DBGSP model, an improved version of DBGS model, is able to predict deviatoric and volumetric strain during experimental stress paths performed at q/p = constant.

ACKNOWLEDGMENTS

The authors wish to thank P. Hornych from Laboratoire Centrale des Ponts et Chaussées (LCPC) (Nantes – France), for providing the experimental data on Missillac gravel.

REFERENCES

Anhdan, L. and J. Koseki "Small strain behaviour of dense granular soils by true triaxial tests." *Soils and Foundations* **45**(3), 2005, 21–38.

Belloti, R., M. Jamiolkowski, D. Lo Presti and D. A. O'neill "Anisotropy of small strain stiffness in Ticino sand." *Geotechnique* **46**(1), 1996, 115–131.

Brignoli, E., M. Gotti and K. H. Stokoe "Measurements of shear waves in laboratory specimen by means of piezoelectric transducers." *Geotechnical Testing Journal* **19**(4), 1996, 384–397.

Cazacliu, B. Comportement des sables en petites et moyennes déformations prototype d'essai de torsion compression sur cylindre creux, in french. *Thèse de doctorat*. Paris, ECP / ENTPE 241, 1996.

Di Benedetto, H. " Viscous effect and anisotropy for sand. (Panel discussion)". *Proc. of the 15th Int. Conf. of Soils Mechanics and Foundation Engineering*.1997, Hamburg. **4**.

Di Benedetto, H. "Small strain behaviour and viscous effects on sands and sand-clay mixtures". *Proc. Geotechnical Symposium in Roma*.2006, Roma, Springer, 159–190.

Di Benedetto, H., C. Sauzéat and H. Geoffroy. "Hollow cylinder test and modelling of prefailure behaviour of sand." *Proc. of the 2nd Int. Conference Albert Caquot*.2001, Paris, 8–16.

Duttine, A. Comportement des sables et des mélanges sable/argile sous sollicitations statiques et dynamiques avec et sans rotations d'axes.*Thèse de doctorat*. Lyon, Ecole doctorale MEGA, INSA / ENTPE 317, 2005.

Duttine, A., H. Di Benedetto, D. Pham Van Bang and A. Ezaoui "Anisotropic small strain elastic properties of sands and mixture of sand/clay measured by dynamic and static methods." *Soils and Foundations*, 2008, (To be published).

El Abd, A. Développement d'une méthode de prédiction des déformations de surface des chaussées à assises non traitées.*PhD Thesis*. Bordeaux, Bordeaux 1 277, 2006.

Ezaoui, A., H. Di Benedetto and D. Pham Van Bang. "Anisotropic behaviour of sand in the small strain domain. Experimental measurements and modelling." *Proc. Geotechnical Symposium in Roma*.2006, Roma, Springer, pp. 727–741.

Hardin, B. O. and G. E. Blandford "Elasticity of particulate materials." *Journal of Geotechnical Engineering* **115**(6), 1989, 788–805.

Hoque, E. Elastic deformation of sands in triaxial tests. *Phd Thesis*, University of Tokyo1996.

Hornych, P., A. Kazai and J. M. Piau. "Study of the resilient behaviour of unbound granular materials". *BCRA Conference*.1998, Trondheim.

Pham Van Bang, D. Comportement instantané et différé des sables des petites aux moyennes déformations: expérimentation et modélisation, in french. *Thèse de doctorat*. Lyon, Ecole doctorale MEGA, INSA / ENTPE238, 2004.

Sauzéat, C. Comportement des sables dans le domaine des petites et moyennes déformations, in french. *Thèse de doctorat*. Lyon, Ecole doctorale MEGA, INSA / ENTPE 331, 2003.

Tatsuoka, F., R. J. Jardine, D. Lo Presti, H. Di Benedetto and T. Kodata. "Characteristing the Pre-Failure Deformation Properties of Geomaterials." *XIV ICSMFE*.1997, Hamburg. **1**.

Tatsuoka, F. and Y. Kohata. "Stiffness of hard soils and soft rocks in engineering applications". *Proc. of Int. Symposium Pre-Failure Deformation of Geomaterials*.1995, Hamburg, Balkema. **2**, 947–1063.

Tatsuoka, F. and S. Shibuya. "Deformation characteristics of soils and rocks from field and laboratory test." *Proc. of the 9th Asian Regional Conf. on SMFE*.1991, Bangkok. **2**, 101–170.

Yamashita, S., T. Fujiwara, T. Kawaguchi, T. Mikami, Y. Nakata and S. Shibuya (2007). Report on international parallel test on the measurement of the shear modulus Gmax using bender elements. ISSMGE – TC29: 76.

Advances in Transportation Geotechnics – Ellis, Yu, McDowell, Dawson & Thom (eds)
© 2008 Taylor & Francis Group, London, ISBN 978-0-415-47590-7

Performance related design and construction of road foundations – Review of the recent changes to UK practice

P.R. Fleming & M.W. Frost
Department of Civil and building engineering, Loughborough University, England

P.J. Gilbert & P. Coney
Atkins Geotechnics, UK

ABSTRACT: Recently there has been a radical change to incorporate performance related design and compliance testing for UK highway foundations. New guidance has been introduced in the Highway Agency's Interim Advice Note 73/06. The potential rewards of this approach include the wider use of more sustainable marginal materials and savings due to the thinning of the upper structural pavement layers, as well as obtaining useful information relating to expected pavement life. This new framework relies heavily on performance-related testing during construction to assess if set performance targets have been met. This in turn requires very clear specifications for construction, to ensure the risks of non-compliance are managed and the potential for dispute is limited. Therefore the use of pre-construction site trials are proposed, which in some cases may be costly or impractical (due to accessibility of the location, or relevance of the subgrade conditions). A more holistic approach to pavement design is provided which whilst providing some benefits has resulted in little guidance on the prediction and management of sub-grade conditions which strongly influence the overall pavement performance. Combining the above with an 'observational' method would enhance the performance approach by utilising a greater requirement to understand the site conditions and permit simple and appropriate changes to be made during construction to overcome any variability encountered, and ensure small areas of low performance are adequately managed and remediated. This paper describes the development and key elements of the current performance related guidance, and describes case studies for the observational approach used in road construction schemes. It discusses the merits and limitations of both approaches, and proposes an appropriate step that could be made to better combine and integrate these procedures into a robust practical method for designing and specifying road foundations for the future.

1 INTRODUCTION

1.1 *Background*

The introduction of the new guidance in the Highways Agency's Interim Advice Note 73/06 (IAN 73, 2006) comprises a radical change in design and specification for UK highway foundations. The foundations permissible are separated into four classes, as defined by their stiffness. Two design processes, 'performance' and 'restricted' are described in the IAN. The potential rewards of the performance approach include: the efficient and wider use of more sustainable marginal materials; provide some assurance that the material performance assumptions made in the design are likely to be achieved; and recognise the structural contribution of improved foundations (with savings due to the thinning of the upper structural pavement layers, now permitted in HD26/06 (DMRB, 2006).

1.2 *Paper content*

In the following sections the developments and key elements of the performance related guidance are explained, and case studies for the observational approach used in recent construction schemes are outlined. The merits and limitations of these approaches to design are discussed, and proposals for an appropriate step are made to combine and integrate these procedures into an enhanced and robust practical method for design and specification.

This paper is intended as a catalyst for debate on the long-term future of road foundation design and specification in the UK.

2 PERFORMANCE RELATED SPECIFICATION

2.1 *Development and Key Elements*

The performance related specification and design guidance embodied in IAN 73 embodies a radical change from what was previously contained in the UK guidance (HD25/94, withdrawn February 2006) for road foundations.

In essence, three aspects of design and specification research and development have been implemented in a short space of time during 2006, and the latter two are embodied within IAN 73. These are:

1. new upper road pavement layer design guidance given in HD26/06 (thickness design based on forecast traffic but which permits a thinning of the layers for the higher stiffness class of foundations);
2. new design guidance for the four foundation classes (superseding HD25/94);
3. updated specification clauses for the MCHW (series 800) for material selection criteria, and laboratory and field testing requirements.

The IAN published in 2006 is currently undergoing a review by the HA and is expected to be re-released in 2008 for further discussion.

The four foundation classes permissible are defined by their long-term minimum stiffness "at top of foundation level":

- Class 1–50 MPa
- Class 2–100 MPa
- Class 3–200 MPa
- Class 4-400 MPa

It is evident that Class 2 represents the equivalent traditional granular sub-base on subgrade (with capping if required), whilst classes 3 and 4 represent superior stiffer foundations requiring the use of stabilised mixtures. Class 1 is only acceptable for minor roads.

A fundamental assumption within the new guidance is that the short-term stiffness (during construction) is an appropriate indicator of long-term performance – utilising some modification factors to adjust the short-term values to allow for confinement (of granular materials) or cracking (of stabilised materials) etc. The foundation 'Performance Design' method allows the designer to predict the likely foundation surface modulus (i.e. assuming all the foundation layers act as an homogeneous elastic half space under a dynamic plate test) that will be achieved by specific combinations of foundation layers over different types of natural ground (subgrade). The key difference here to previous advice for analytical design in LR1132 (Powell et al 1984) is the flexibility to choose an appropriate layer stiffness value for the subgrade (based on expected CBR), capping (granular or stabilised) and sub-base (granular or stabilised) to determine the likely foundation surface modulus. Thus, there are many more theoretical permutations and combinations of design thickness than previously permitted, notwithstanding practical aspects of tolerance and compactibility.

In addition to the introduction of routine insitu stiffness testing during construction, CBR testing of the subgrade is (still) required during construction (using a dynamic cone preferably) to check it is equal to or greater than the design value for each section of the scheme. Adequate material density is required for compacted granular layers, to avoid poor workmanship or problems of temporary high stiffness. A range of material compliance tests is still specified, including size range.

The 'Performance Design' guidance requires a demonstration area to be constructed and carefully evaluated, to confirm the materials and methods will meet the proposed foundation class. Lightweight insitu stiffness measuring devices may be utilised in the main works if they have been properly correlated with the full-scale (trailer mounted) Falling Weight Deflectometer, which is designated as the primary compliance test method. The selection of an appropriate section and subgrade area on which to construct the demonstration area is important, the IAN specifies how to adjust the measured foundation surface modulus to suit the specific demonstration subgrade conditions at the time of the trial. Failure requires re-design, construction and re-testing of the trial.

In addition to the stiffness testing, related to analytical design fundamentals (i.e. limiting stress criterion etc), there is also designated a trafficking trial in the demonstration area, with specific pass/fail criteria for cumulative rut depth. This is to demonstrate the design is also suitable to protect the subgrade from excessive deformation, and the foundation layers from excessive internal shearing, likely to be caused by construction traffic.

There will be significant cost increases to deliver this new approach, including:- the large amount of design and testing of the subgrade (at the state anticipated during construction), proposed foundation materials; site demonstration trial(s), construction compliance testing, and time programmed for approval of layers.

3 MERITS AND LIMITATIONS OF IAN 73

There are many potential benefits accrued from the introduction of performance related specifications for road design and construction. IAN 73 embodies the design philosophy set out in HD26/06 aimed at improved 'whole life value' and sustainability. Furthermore, the processes described in IAN 73 suggest the appropriate (re)use of materials within sound scientific designs and measured by new test technology.

Better assurance of performance provides better risk control, potential for innovation and long-term developments in materials and technology, and ultimately better user satisfaction. It could be argued that the realisation of these ideals has been a long time coming. However, the issue for debate is has this goal been delivered in the amalgam of documents, and might the changes bring other problems.

Industry has identified what it considers to be a number of potential limitations of the IAN 73. These include conservative thickness requirements for the 'restricted design' method, and costly extensive testing is required within the 'performance design method. For Class 2 foundations this is potentially inappropriate in some circumstances considering the design thickness of capping and sub-base is similar to that previously required by HD25/94.

In addition, there is limited advice on flexibility for the designer and constructor once on site, it appears assumed that the subgrade behaviour, and all the placed foundation materials and methods will be well defined, understood, and approved at the site demonstration trial stage. However, the proposed site demonstration trials are subject to many influencing variables, including physical access to appropriate subgrade conditions, suitable weather, (or allowance for wetting/drying effects), and the programming of complex trials with extensive monitoring prior to the main contract works.

Problems, disputes and uncertainty during the trials may have a large influence on the ongoing programme and cost. The assessment of suitable performance is largely limited to numerical and statistical analysis of the subgrade, capping and sub-base material field data. Potentially little scope has been allowed for detailed geotechnical assessment, nor allowance for uncertainty of the overall ground conditions (such as site specific factors relating to the insitu soils and groundwater) although this is perhaps a natural consequence of a specification based on formulaic end product testing.

However, it is clear that the problem facing the design engineer is the need to propose (suitable) pavement foundations, in circumstances of variable ground and groundwater conditions, allowing for the range of influencing factors that often occur in UK soils, as described further below.

3.1 *Sub-grade factors*

To a practicing geotechnical engineer it appears that the main reason to determine the four foundation classes described in IAN 73 is to enable the statistical approach of design of the upper bound layers that is enabled by the new HD26/06. This is a reasonable and valid aim as it should achieve the objective of delivering a holistic approach to pavement design. However, IAN 73 appears to assume that the sub-grade will be consistent, manageable and relatively problem free. This is often not the case as the sub-grade is influenced by very different factors when compared to the other engineered pavement layers. Therefore, a potential geotechnical criticism of the IAN is that it says too little about prediction and management of sub-grade conditions, reducing the art of pavement foundation design to the science of design and measurement of the placed foundation materials.

There appears to be a change in emphasis on what previously constituted the 'earthworks' (which included capping, and the pavement above formation), to now the pavement construction as a whole. Yet the earthworks are still required to deliver a consistent platform for the pavement construction. The result is the importance of the subgrade appears to have been reduced to be considered in a similar manner to any of the other foundation layers defined by a stiffness value. However unlike the other foundation layers the subgrade is variable and its performance is influenced by many factors which are not considered by IAN 73. The factors that apply are site specific but are likely to include some of the following:

- Soil type, grading, (especially for borderline cohesive/granular soils), variability,
- Permeability and stiffness,
- Horizontal and vertical geological variability
- Soil fabric (e.g. lamination)
- Likely presence of hard/soft spots
- Groundwater conditions and drainage
- Topography – transition zones for cut/fill
- Construction procedures adopted and skill of the site foreman to implement these
- Construction season, timing of drainage installation, exposure time of sub-grade, and quality control.

Yet from the above soil stiffness is just one variable. In addition, the Plasticity Index method of CBR prediction (retained in IAN) does not work well on glacial tills or mixed soils.

It would appear that much of the advice in the HD25/94 regarding subgrade issues has been omitted, including aiming for consistent formation stiffness over lengths of 500 m. In addition, little reference is made to the supporting information in HA44/91. In its present form IAN 73 has only 3 of its 62 pages dedicated to advice regarding sub-grade.

The 'Performance Designs' depend largely on demonstration areas to prove the performance of the proposed pavement, yet the subgrade conditions highlighted above will greatly influence the results. This raises some important questions regarding the adequacy of the trials, which include how many test areas should be constructed, and should there be one trial for each combination of ground and/or groundwater conditions?

Experience from a number of projects also shows that in practice the form of contract used can have the greatest influence on the approach taken to capping design. The move to design and build construction and partnering has resulted in very different pressures on the designer to those seen under more traditional re-measure forms of contract (Jarvis and Gilbert 2003). It is considered the geotechnical engineer must be given enough freedom to assess these various subgrade related factors, to help decide what is the appropriate design approach to deliver the foundation platform required, to then enable the successful optimum construction of the proposed bound layers. This appears to be a weakness of the current interim advice note, IAN 73.

3.2 *Design Benefits*

Within the IAN "Restricted Designs" are included which can be used when the detailed compliance testing for the "Performance Designs" is not appropriate. These designs have been deliberately set to be conservative which is understandable otherwise the performance specification might never be implemented. However, brief observation of the restricted design options suggests that they are possibly more conservative than the designs historically implemented in the UK, e.g. for a granular foundation (combined sub-base and capping):-

1. 300 mm required on good granular sub-grade (compared to traditional 150 mm),
2. 650 mm required on a good clay sub-grade (when around 400 mm may traditionally be expected).

The document does not account for this apparent caution. More significantly research undertaken on behalf of the Highways Agency to review the performance of existing roads, designed using the traditionally approaches set out in LR1132 & HD25/94, indicates that when the foundation design reflects the sub-grade conditions the performance is as expected and adequate (Gilbert et al 2004, and Gilbert et al, 2007).

The design charts given for the Performance Designs do not result in any significant reduction in foundation thickness compared to designs based on HD 25. Furthermore, the designer is encouraged away from the use of capping towards a single foundation layer, as it is only then that relatively thin foundations can be achieved. Yet in practice for many sites the use of capping is advisable to manage the sub-grade conditions. Therefore theoretical benefits of achieving a single foundation layer can only be delivered by construction during "good" construction condition. Designers may have a difficulty justifying the cost of extra testing if they can not offer a reduction in foundation materials and can only rely on savings being achieved in the bound layers design to HD26/06.

3.3 *Soil Mechanics and Psychology*

In recent years there have been major changes to how the industry is regulated, with the gradual move from the traditional contract arrangement of a Resident Engineer supervising a Contractor, to various forms of Design and Build contract where a Contractor is required to self certify their work. Anecdotally some engineers complain that this has resulted in a lower quality product (although this may be based on a subjective assessment of the past), but few consider why this change has come about. A potentially significant reason for the change is in response to the changing expectations of society, including the personal expectations of engineers, and the lack of experienced professionals within the industry. It is important to recognise these issues, and create engineering methods that can be implemented within the reality of the industry. For pavement engineering this tends to mean that many of those constructing the pavement foundation have limited experience in this field. A good aspect of the IAN is that it provides a useful description of many of the important factors that will influence performance. However, the language used in the document could be considered sufficiently complicated that it is likely to only be read by specialists in the field. This possible perception of lack of accessibility could prove a problem to the industry as it may lead to the construction team not understanding the intentions of the capping design and thus taking no ownership for its successful implementation.

Furthermore, it is apparent that a number of those experienced in pavement foundation design are finding the new 'performance design' approach complicated. This may result in a new form of specialist coming into existence who may give too much emphasis to the theory of the IAN, and due to lack of experience or support from other fields may neglect the many other influential aspects of a road construction project that lead to its success. This issue could be resolved by appropriate training, including widely disseminated feedback from projects utilising the performance based design approach.

It would therefore seem clear that as highway engineers we should not consider just the theoretical soil mechanics of the proposed new design method but also the psychological response of the industry that will receive it. If the method is too complicated for the majority of those in the industry then it may fail to deliver fit for purpose solutions as intended.

3.4 *Stiffness/test methods*

It is fair to say geotechnical and pavement engineers recognise that if a simple method is developed to model stiffness of the sub-grade and the pavement foundation layers then this would lead to a significant improvement in the theory behind foundation design. However, identifying a low cost test method that can be implemented both in the laboratory (at investigation stage) and in-situ (during construction) has proven difficult. The IAN has been released on the assumption that the Falling Weight Deflectometer (FWD) and/or Light Weight Deflectometer (LWD) can fulfil this role, but has retained subgrade CBR as a design/assessment measure through necessity.

The use of the (large and relatively expensive) FWD to control earthworks construction is not currently considered a realistic proposition for routine work; therefore the (portable) LWD approach has been developed. However, a review of a wide body of data from various LWD development trials on capping materials has shown that the stiffness values determined can vary significantly and this is complicated further by the large number of factors that can influence this variability, from the material properties and material states and this is a concern (Lambert, 2007).

Whilst the IAN focuses on stiffness, the designer will continue to have to divide the site into lengths of pavement of one expected sub-grade foundation requirement based on an assessment of the wide range of site issues described above. These lengths are ideally 500 m or more in length, along which the in-situ stiffness will vary (Fleming et al, 2000), the designer will therefore consider the stiffness data set as just one of the many variables.

During construction simple methods are required that enable the works to progress without interruptions (delays can lead to deterioration of the exposed sub-grade). This can be achieved by methods such as visual inspection backed up by a simple in-situ test (e.g. LWD or the dynamic cone penetrometer) to identify areas of softer than expected subgrade. Those on site could continue to use the empirical CBR as a subgrade condition descriptor which is widely understood and may still be adequate as a quick empirical control measure for the site works in partnership with other tests. This can be managed as part of the design process via an 'Observational Approach' that many in the industry are currently implementing (described further in Section 4).

One major benefit of the LWD testing is that it can be used for complementary laboratory trials of relatively unusual capping materials (e.g. recycled aggregates) prior to going to site. However, the cost of undertaking such laboratory trials is high due to the requirement to prepare large samples to obtain meaningful results (Lambert, 2007); this could restrict the materials considered and trialled to only those projects where there is scope for adequate savings to be made by investing in extra laboratory testing. However these tests are considered to have a role in material approval tests for suppliers. This leads to a major benefit of the IAN which is to use it for undertaking source approval paid for by the proposed supplier and then made available to all contractors who wish to consider using that material. This approach is widespread for other industry materials, and thus reduces the contractors costs to those for the in-situ testing during construction.

However the ongoing absence of advice on the methods to assess durability of unusual and recycled aggregates may continue to restrict the range of materials considered.

4 OBSERVATIONAL APPROACH

4.1 *Key Elements*

The flexibility required in design where variable site conditions are encountered during construction (or limited pre-construction site data is available) is embodied in the 'observational approach' (Nicholson et al, 1999). A similar methodology has been developed for a number of Design and Build schemes (Gilbert, 2004) and has proved appropriate for large highway foundation areas and allows swift construction.

This was achieved by including in contract documents full details of expected ground and groundwater conditions, predicted pavement foundation requirements, a defined regime of site inspection coupled with rapid insitu testing methods to obtain the data to identify locations where the conditions differ from those expected, and clear procedures developed to enable changes to be made on site based on the actual conditions encountered. This allows the designer to make realistic predictions of a range of pavement foundation requirements.

However for this approach to work trust and cooperation is required between all parties (Client, Contractor and designer), to allow the optimum foundation for the scheme to be built. Key construction issues for the project are identified and the design/specification prepared accordingly, to include/allow, for example, for variation in the amount of construction traffic using the capping (less than the normally assumed 1000 standard axles), exact timing of subgrade drainage installation (and hence the likely long-term equilibrium CBR predictions), the influence of compaction of the sub-formation to overcome disturbance during excavations, accommodation of stripping of subgrade after adverse weather (or if softening has occurred – incorporating associated thickening of capping). This needs effective site inspection regimes and quick short-term testing regimes to be utilised, with transition zones being given special attention particularly cut/fill intersections. The approach of the designer is to make maximum use of the materials available on site to form a suitable foundation for the pavement, this can extend past the limited classification of capping materials given in HD25/94 and is one of the aims of the IAN.

4.2 *Project example*

As previously described a simple Observational Approach to pavement foundation design has been used on various Design and Build road schemes over the past 12 years (Gilbert, 2004). The approach has been gradually developed, and was used as part of the strategy for the M6 Toll Road. It enabled a broadening of the acceptable range of capping materials used on the Glasgow Southern Orbital/M77 . Performance of these roads to date has been positive (Gilbert at al, 2007). Recently it has been developed to suit requirements in the Republic of Ireland by providing a flexible tool to enable a Contractor to keep construction options open and is further discussed below.

The N6 PPP scheme at Galway is a two lane dual carriageway currently under construction. The geology includes soft alluvium, Glacial Till and solid geology of limestone. The Glacial Till is of variable grading from cohesive material with high sand and gravel content, to a granular material with a significant fines content, changing dramatically over very short distances. One common factor is that the soil is highly moisture susceptible. Undulating topography along the route gives regular transitions from cut to fill and long sections of at-grade or on low embankment. The simplest way to manage these variable conditions has been to define a number of "Capping and Sub-base scenarios" based on:-

1. Alignment/topography condition (Cutting, at-grade/low embankment, embankment > 1 m, transition zone).
2. Sub-grade condition (principally soil type as groundwater is generally high).
3. Design CBR.
4. Foundation design options for each soil type: sub-base only or sub-base & capping.
5. Additional measures for soft ground.

Each earthwork drawing shows the Capping and Sub-base scenarios that are applicable to each length of main highway (leaving the foundation choice open to the Contractor), and identifies areas of continued uncertainty where special attention is required.

The scheme's Specification Appendix 6/7 then defines a set of requirements for the sub-grade to be implemented as part of the Contractors inspection and test plan. These are underpinned by:-

1. A visual inspection of the subgrade, including a review of the fines content and whether the design CBR is appropriate.
2. In-situ CBR testing by hand held dynamic probe (set of 3 tests at 20 m centres).
3. A defined procedure for re-assessment of design CBR values.

The approach described enables the observed ground conditions to be accommodated within the design, so that the construction is appropriate for the ground conditions. Any reduction in pavement foundation thickness is only permitted if there is an improvement in the ground conditions encountered; changes, based on short term in-situ CBR test results alone are not permitted as the design is based on predicted equilibrium water content CBR values.

Importantly the in-situ stiffness is used as a simple check of the expected ground conditions by comparison to predicted CBR; the LWD could be used for this purpose, but initially would still only be a tool to help identify variability (not a measure of absolute stiffness).

5 DISCUSSION – INTEGRATION OF THE TWO METHODS

The newly published advice regarding 'performance designs' for road foundations has incorporated many important facets including 'fit for purpose' materials and their fundamental properties in the design of adequate foundation thickness. The recognition of superior performance, and site compliance testing is well received and perhaps long overdue. However, it is also clear that the new guidance represents a sea change in the use of 'stiffness' both as a design and compliance parameter. It is also apparent that to avoid potential problems (as far as is practicable) the new guidance is very formulaic in its design steps, although the use of appropriate site trials should engender greater confidence (notwithstanding the comments in section 3.1). It undoubtedly requires the designer of a scheme to gather a relatively large volume of data prior to any detailed design and to make many assumptions, arguably the most significant is the uniformity and expected subgrade state during construction. The guidance appears to lack some flexibility in decision making *during* construction that could help deal with variable ground conditions or disputes if the measured values fall slightly below a foundation class requirements (for reasons other than poor materials or workmanship).

The observational approach, however, embraces much of the philosophy of the performance approach but with the added enhancement of anticipation of site variability and a pragmatic action plan to deal with the arising problems this may bring. The observational approach requires the designer to think through soil behaviour, review ground conditions, and develop a strategy to suit the project, rather than follow a formulaic design code.

It is clear that effective integration of the observational approach into a performance based specification is likely to be appropriate only where there is an appropriate form of contract to help share risks between the designer and constructor. However, where this is the case the maximum benefits can be accrued by both parties, and hence any site related problems overcome.

For the performance design method to effectively allow for variable subgrades, it implies that a demonstration trial is required for each possible eventuality. A potentially more practical scenario is to set out clearer guidelines within the advice regarding a framework for the interpretation of the subgrade (drawing more from the valuable insights given in previous HA design guidance), utilising the latest test techniques being developed within the observational approach and with the use of GIS databases for asset management. This would be a significant undertaking with regards to capturing the essential aspects of how experienced geotechnical engineers carry out their work with regard to highway design, but from an industry view this would add significantly to the advice.

Current research into earthworks performance specifications, a new draft British Standard (BS6031), and the use of intelligent compaction monitoring equipment; could usefully feed into any future review process. In addition, the move towards effective asset management has led to the demand for effective capture of site data (during construction) for future maintenance management and intervention strategies. This will become increasingly important for longer-term climate change impact research and mitigation strategies. Such data would make widening and reconstruction projects more effective, and could include the site measured performance related data on the foundation as well as the soils. Recent work has shown the benefit of such a geotechnical database for asset management of pavement foundations (Gilbert et al, 2007).

It is also interesting to note that research in Germany has highlighted a link between subgrade stiffness variability and the development of spatial variability in ride quality (Grabe et al, 2005). This points clearly to a possible long-term ride quality benefits of controlling foundation stiffness *consistency* and is similar to the effects observed in railway trackbed performance. This is where part of the original Department of Transport research regarding compliance testing of road foundations began in the 1980s (Cobbe 1986), the observational approach described herein also may help to address this.

The requirements of the IAN are perceived by some in the industry to be relatively complex and expensive. It will be interesting to see how widely and effectively it is implemented.

6 CONCLUSIONS & RECOMMENDATIONS

This paper considers the change in approach and philosophy to pavement foundation design that is proposed for the UK via IAN 73. The new code has two principal benefits:-

1. It provides a stiffness based philosophy that will enable the pavement foundation to be designed as part of an overall pavement solution. This should enable greater optimisation of design to be delivered for the more expensive bound layers (designed to HD26/06) using the true foundation support as an upper layer design input. This benefit is most achievable when stabilised materials are used for the foundation layers.
2. It also provides a framework that will enable utilisation of alternative foundation materials such as re-cycled and more marginal aggregates.

However, the drive to improve the theoretical aspects of pavement foundation design could potentially be to the detriment of the practical consideration of the geotechnical issues associated with the capping. It must be remembered that on many sites the primary objective of the capping is to provide a stable foundation upon which the other layers can be constructed, as the capping is used to manage sub-grade problems and variability.

The aim of using more re-cycled aggregates may potentially be more successful if the approach allowed the test approach listed as "performance designs" to form a material type/source approval.

From an industry perspective, the restricted designs allowed in IAN 73 are considered conservative, and seem to make little allowance for the experience gained with such standard designs or materials over many years. However the additional thickness from the new designs may be required to ensure stiffness targets are achieved, allowing for the observed site stiffness data variability.

Consequently on many sites there is the potential risk that the benefits of the IAN will not be achieved as thick standard foundations may be called for , and the extra costs involved will deliver no significant benefit over traditional designs.

This paper considers the possibility of modifying the approach within the IAN to incorporate sub-grade management to increase flexibility on site. This can be achieved on appropriate sites and projects with a move from a "performance" based approach to encompass an "observational" approach.

Due to the challenges of accurate long-term performance prediction and design, true performance specifications will only be achieved when these aspects can be addressed, which includes the requirement of a suitable form of contract and appropriate test methods.

It is a concern from some in industry that if the IAN were fully implemented designers would struggle with its requirements and contractors would not have the experience or background skills to implement it as intended.

To take the IAN forward it is proposed that it's status should be modified to possibly work in partnership with aspects of the old HD25/94, and that an industry review of IAN should be undertaken to look at it's impact, understanding and implementation. The feedback should then be used to update the advice and guidance, and training, in perhaps two years time.

REFERENCES

Brown, S.F., *Soil Mechanics in Pavement Engineering.* 36th Rankine Lecture of the British Geotechnical Society. Geotechnique, Vol. 46, No. 3, pp. 383-426. 1996.

Cobbe M. I., 1986, *Development of Acceptance Criteria for Subgrade Improvement Layers*, Report to Engineering Intelligence Division, Department of Transport, London.

Fleming, P.R., Rogers C.D.F., Thom, N., Frost, *A performance Specification for Pavement Foundations*, Transportation Geotechnics 2003, Thomas Telford, pp 161-176, ISBN 0 7277 3249 8.

Fleming, P.R., Frost, M.W. and Rogers, C.D., *A comparison of devices for measuring stiffness in situ*, Proceedings of Fifth International Conference on Unbound Aggregates in Roads, Nottingham, July 2000, pp 193–200.

Gilbert, P. J. 2004. *Practical Developments for Pavement Foundation Specification.* In proceedings of the international seminar on Geotechnics in Pavement and Railway Design and Construction, Eds. Gomes Correia & Loizos, Millpress, RotterDam.

Gilbert, P. J., Smith, V., and Gwede, D. 2007. *Geotechnical Asset Management of Pavement Foundations.* 2007 PIARC.

Grabe J., Mahutka K., 2005, *Long-term evenness of pavements with respect to soil deformations*, BCRA5, CD-ROM, Norway.

Jarvis, S., Gilbert, P. J., *Private Finance Initiative Infrastructure Projects – Implications on Geotechnical Design*, Transportation Geotechnics 2003, Thomas Telford, pp 117-129, ISBN 0 7277 3249 8.

Lambert, J.P., 2007, Novel *Assessment Test for Granular road Foundation Materials*, Engineering Doctorate thesis, CICE Loughborough University.

Nicholson, D., Tse, C-M. & Penny, C. 1999. *The Observational method in Ground Engineering.* CIRIA.

Powell, W.D., Potter, J.F., Mayhew, H.C., and Nunn, M.E. (1984): *The Structural Design of Bituminous Roads* (LR1132). Transport Research Laboratory, Crowthorne Berks. UK.

The Highways Agency. Design Manual for Roads and Bridges: Vol 4 Geotechnics and Drainage: Section *1 Earthworks*, Part 1, HA44/91. The Stationery Office, London, 2006.

The Highways Agency. Design Manual for Roads and Bridges: Vol 7 *Pavement Design and Maintenance*, Part 3, HD 26/06. The Stationery Office, London, 2006.

The Highways Agency. Interim advice note 73/06. *Design Guidance for Road Pavement Foundations* (Draft HD25). The Stationery Office, London, 2006.

Advances in Transportation Geotechnics – Ellis, Yu, McDowell, Dawson & Thom (eds)
 ISBN 978-0-415-47590-7

Utilization of waste foundry sand as pavement sub-base and fill material

A.G. Gedik
General Directorate of Highways, Istanbul, Turkey

M.A. Lav & A.H. Lav
Istanbul Technical University, Istanbul, Turkey

ABSTRACT: Foundry sand is consumed large amounts by manufacturing industries such as car manufacturing, iron-steel industries, alloy production and various branches of metallurgy industry. After the molding process, the material is considered as waste and stockpiled. Ever increasing stockpiling costs is becoming a concern in molding industry; hence, manufacturers are looking for a suitable way of utilization.

The aim of this study is to use waste foundry sand in highway constructions where high volumes of the material may be utilized. The steps of the study are as follows; firstly, each group of material is tested to determine index properties. Then, each group of the material compacted using various compaction efforts. Stabilization is considered to increase the strength of material and the sand is stabilized with cement and lime by 2%, 4%, 8%, and 10% by weight. Stabilized samples are later cured for 7-day, 14-day, and 28-day to observe the effect of stabilization by ultrasonic pulse velocity test (UT), unconfined compressive strength (UCS) test, and California bearing ratio (CBR) test.

1 INTRODUCTION

Foundry industry produces metal castings (Beeley, 2001). All over the world, foundries produce approximately 80 million tones castings, mainly for the automotive industry. Also, they have been used in other industries such as cement, agriculture, health, naval construction etc. (Yaylali, 2007). Casting is achieved by pouring molten metal into molds. In the process, sand is commonly utilized to resist pressure and heat of the molten metal. After a number of reuse, the sand is no longer suitable for molding because of physical and chemical breakdown. Foundries replace used sand by new one since it contains unspoiled binder and called waste foundry sand (WFS) (Clegg, 1991). Amount of produced WFS is approximately 9 million and 12 million tones in Europe and the USA, respectively (Guney et al., 2005). Majority of WFS are stockpiled, however stockpiles are generally exhausted in metropolitan areas and causing a problem to the foundry industry. For example, in Istanbul authorities are reluctant to provide foundries a dumping ground and charging transportation and dumping cost as high as 12\$/m^3. Under this circumstance, the foundry industries are desperate to find a way of utilization of WFS rather than stockpiling.

The main objective of this study is to investigate performance of stabilized WFS in highway construction as sub-base and fill material where there is a potential for high volume of utilization. As a result of this, a laboratory testing program was carried out on WFS samples stabilized with cement and lime separately to asses their suitability. The mixtures were prepared in the laboratory at optimum moisture content under specific compactive effort and subjected to mechanical tests. It should be noted that the study described in this article gives initial results of a project on the utilization of WFS. Based on the results, advanced geotechnical testing program including cyclic testing will be considered to understand the resilient behavior of the material under wheel loading.

2 CHARACTERISTICS OF WFS

At present there are two basic types of sand is used by the industry. Green sand contains clay (mainly bentonite) as the binder, and the chemically bounded sand that uses polymer for the same purpose (Patridge, 1998). Green sand consists of 85–95% silica, 0–12% clay, 2–10% carbonaceous additives such as sea coal, and 2–5% water (Fox et al, 1998). In this study, green sand containing 5.43% bentonite was investigated and the material was obtained from Atacelik Corporation located in Istanbul.

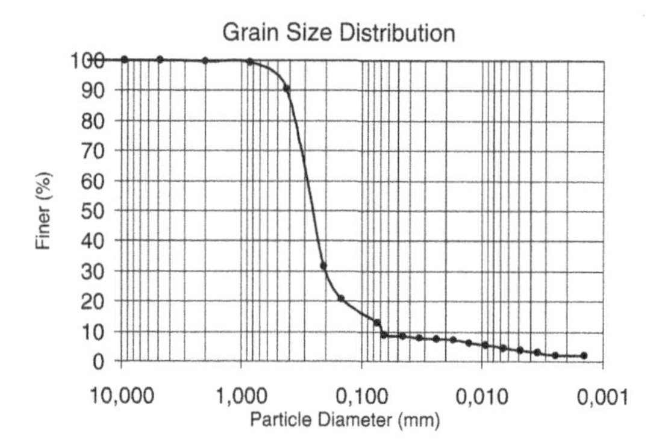

Figure 1. Grain size distribution of WFS.

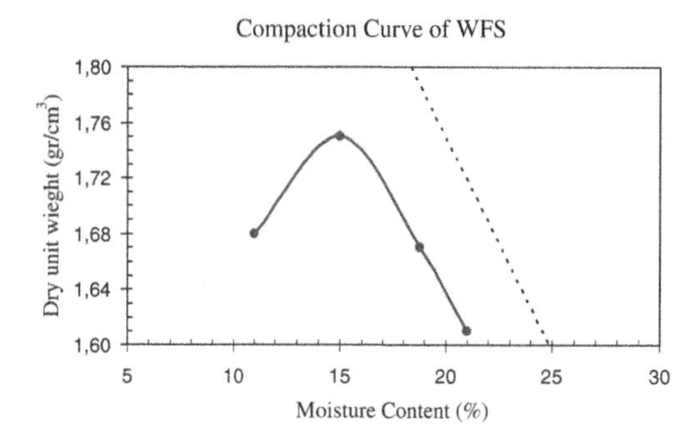

Figure 2. Showing the relationship between ρd_{max} and OMC.

Table 1.

Engineering Properties	Value
Classification (USCS)	SM
Classification (AASHTO)	A-2-4
G_s	2.56
OMC (%)	15
ρd_{max}	1.75
Fines Content (%)	13
Organic Content (%)	6.79

The grain size distribution of WFS is shown in Figure 1.

The specific gravity (G_s) of WFS was measured as 2.56 according to ASTM D 854. WFS used in this study was determined as nonplastic according to ASTM D 4318. The maximum dry density (ρd_{max}) and the optimum moisture content (OMC) are 1.75 and 15% respectively according to ASTM D 1557 as shown in Figure 2.

Briefly, basic properties of WFS are shown in Table 1.

In this study ordinary Portland cement and hydrated lime complying with ASTM C 150 and ASTM C 207 were used as stabilizing agent, respectively.

3 TEST PROCEDURES

Stabilization is a method used to improve the performance of highway materials such as bearing capacity, plasticity, permeability, and weathering (Kézdi, 1979). By treating waste materials in this fashion, such as WFS, they might satisfactorily perform in service as good as expensive highway materials such as gravel and crushed rock. There are several types of stabilization; however, in this study the material was stabilized with cement and lime separately as mentioned previously. Therefore, prepared samples are considered in two groups. The first group samples were stabilized with cement and the second group samples were stabilized with lime. For simplicity, the samples were labeled as C-WFS are referring cement stabilized samples (Group I) and L-WFS referring lime stabilized samples (Group II). All samples were wrapped in plastic bags and cured 7-day, 14-day, and 28-day in a humid room. (100% relative humidity and 21 ± 2°C temperature). Detail of sample preparation is given in Table 2.

In total, 96 samples were prepared for laboratory testing. Samples were prepared by the means of two different compactive efforts namely, standard Proctor (ASTM D 698) and modified Proctor (ASTM D 1557). Samples prepared by modified Proctor method were utilized only in CBR test. Other tests were performed on samples prepared by standard Proctor method. Details of testing are given as follows.

3.1 *Ultrasonic pulse velocity test (UT)*

The ultrasonic pulse velocity test is a measurement of the transit time of a longitudinal vibration pulse through a sample which has a known path length. The test is carried out by applying two transducers (transmitting and receiving) to the opposite end surfaces of the samples. A sufficient acoustical contact between the transducers and the surface of the sample is maintained by a couplant such as silicon grease. The pulse delay between the transducers is measured as the transmit time and the ultrasonic pulse velocity is calculated using the length of the sample (the distance between the transducers). Utilizing this approach and the following equation (Equation 1) dynamic Elasticity modulus of the samples were obtained and given in Figure 3, Figure 4 and, Figure 5.

$$E_d = 10^5 * V^2 * \frac{\Delta}{g} \tag{1}$$

Table 2. Labeling of samples.

Group I (C-WFS)			Group II (L-WFS)		
Cure duration of samples	Amount of cement	Duration of samples	Cure amount of samples	Range of lime	Samples
7-day, 14-day, 28-day	2%	C-WFS-I	7-day, 14-day, 28-day	2%	L-WFS-I
		C-WFS-II			L-WFS-II
		C-WFS-III			L-WFS-III
		C-WFS-IV			L-WFS-IV
	4%	C-WFS-I		4%	L-WFS-II
		C-WFS-II			L-WFS-II
		C-WFS-III			L-WFS-III
		C-WFS-IV			L-WFS-IV
	8%	C-WFS-I		8%	L-WFS-II
		C-WFS-II			L-WFS-II
		C-WFS-III			L-WFS-III
		C-WFS-IV			L-WFS-IV
	10%	C-WFS-I		10%	L-WFS-II
		C-WFS-II			L-WFS-II
		C-WFS-III			L-WFS-III
		C-WFS-IV			L-WFS-IV

Note. C: cement; L: lime; C-WFS: cement stabilization; and L-WFS: lime stabilization.

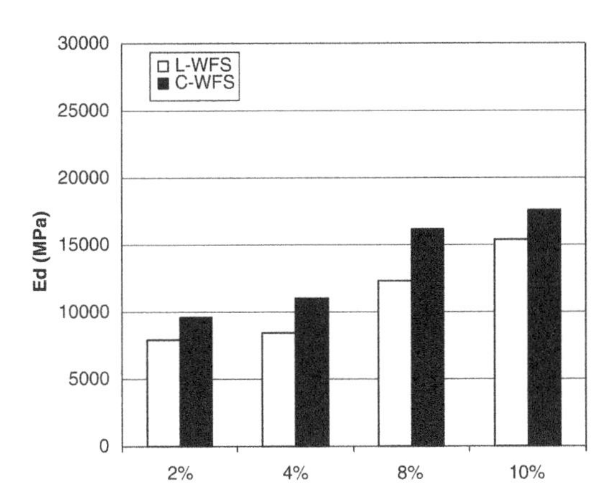

Figure 3. Showing the E_d values of samples cured 7-day.

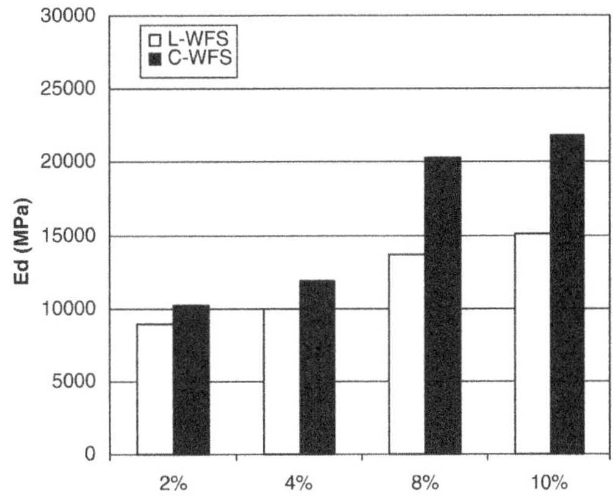

Figure 4. Showing the E_d values of samples cured 14-day.

where E_d = Dynamic Elasticity Modulus (kg/cm^2); V = Velocity (km/s); Δ = Density (kg/dm^3); and g = gravity (m/s^2).

3.2 *Unconfined compressive strength (UCS) test*

The strength gain over time is one of the most important parameter associated with performance of a stabilized road sub-base. The unconfined compressive strength test is, a widespread measure used to determine the relative response of the stabilized materials to the parameter mentioned above. The test involves the loading cylindrical samples to failure in simple compression, without lateral confinement. WFS stabilized with cement or lime with an agent content were 2%, 4%, 8%, and 10% by weight. The samples were tested after 7-day, 14-day, and 28-day curing as detailed above. Before being tested, all samples were capped to maintain a uniform stress distribution. The results of UCS test are given in Figure 6, Figure 7, and Figure 8.

3.3 *California bearing ratio (CBR) test*

California bearing ratio (CBR) test is an empirical method for determining strength of soils. Although the test is empirical, it is widely used for estimating the bearing capacity of pavement layers as well as

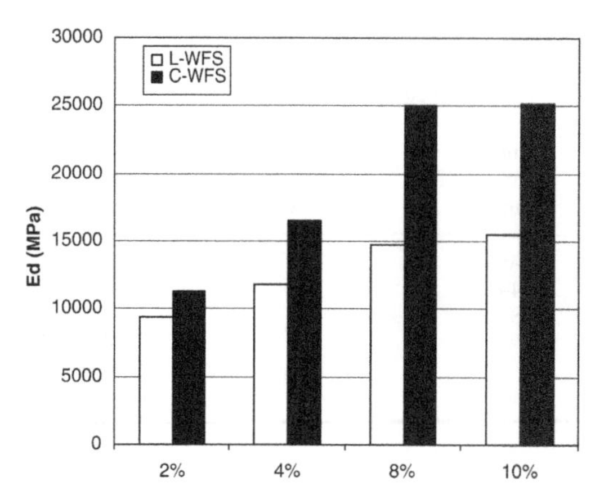

Figure 5. Showing the E_d values of samples cured 28-day.

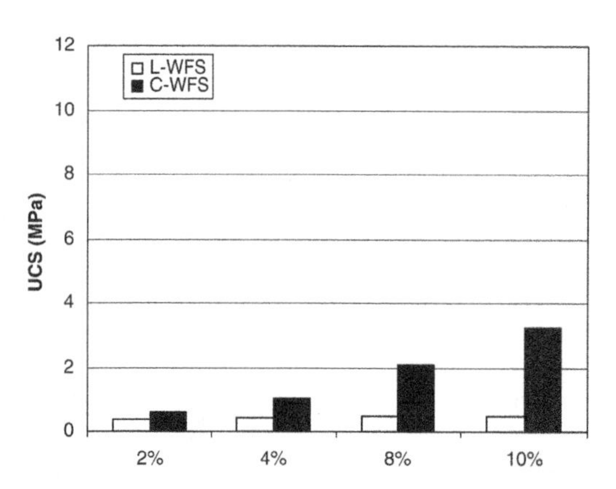

Figure 6. Showing the UCS values of samples cured 7-day.

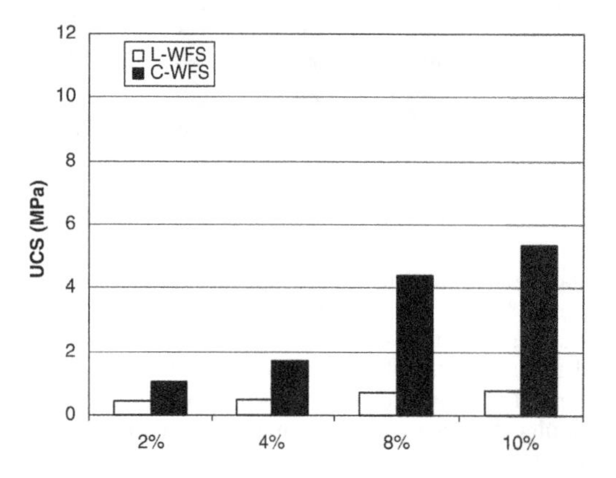

Figure 7. Showing the UCS values of samples cured 14-day.

subgrade. The test is achieved by applying a standard plunger into the sample at a fixed rate of penetration. During the test procedure, the force required to maintain this rate is measured and compared with the force

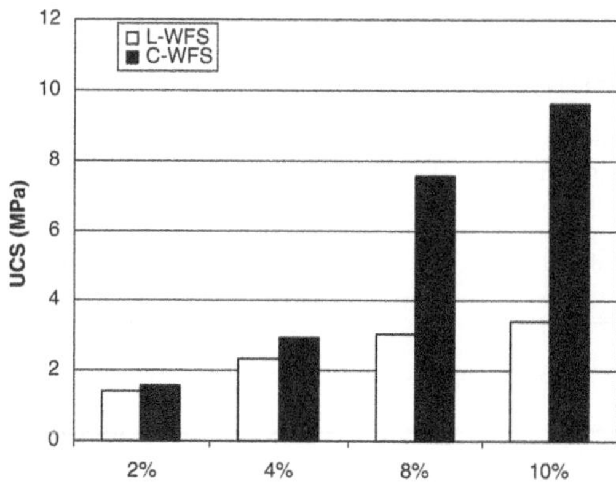

Figure 8. Showing the UCS values of samples cured 28-day.

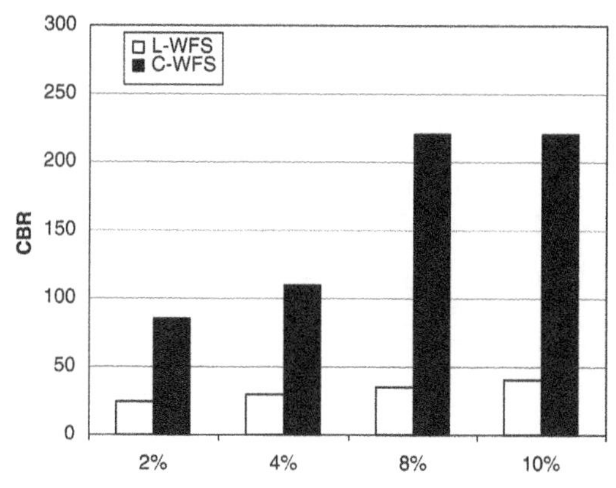

Figure 9. Showing the CBR values of samples cured 7-day.

maintaining the same rate on a standard rock. A CBR value of 100 is generally considered as an excellent pavement layer. In this study, the samples were tested in the mould as prepared and the results are given in Figure 9, Figure 10, and Figure 11.

4 DISCUSSIONS AND CONCLUSION

Both cement and lime favourable agents in soil stabilization. According to the results, it is apparent that cement is superior to lime as stabilization agent. However, cost of cement stabilization is quite expensive than lime stabilization in Turkey. Therefore, lime stabilization is considered as favourable application. However, according to the results, the unconfined compressive strength of lime stabilized samples is relatively similar and low at 7-day and 14-day cured samples. For 28-day cured samples the discrepancies between lime ratios become more apparent, especially 8% and 10% lime stabilized samples increased their strength, yet still relatively low. As expected, cement

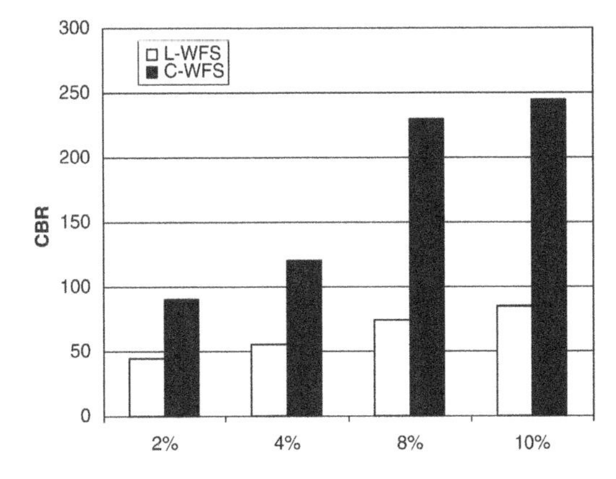

Figure 10. Showing the CBR values of samples cured 14-day.

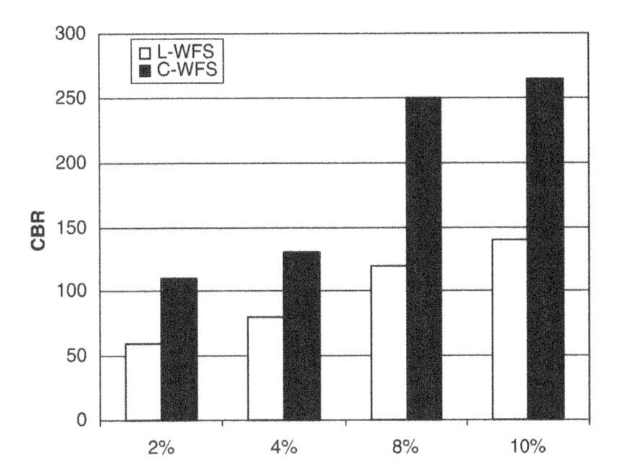

Figure 11. Showing the CBR values of samples cured 28-day.

stabilized samples gained higher strength than lime stabilized samples in the first 7 days and continued to increase its trend until 28 days. Besides, the ratio of strength gain between 8% and 10% cement and lime stabilized samples are significant especially at 14 and 28 days.

E_d and CBR values are in accord with the UCS values explained above. Particularly, regardless of agent type, amount and cure duration stabilized WFS showed excellent CBR results. 8% and 10% cement stabilized samples having more than doubled the strength that reference crushed rock. Based on the CBR results, it may be said that stabilized WFS could be considered as a quality pavement material.

Although the test results of cement stabilization are superior to lime stabilization, cement stabilized samples; particularly at higher cement ratios might be fragile. This would cause premature cracks in pavement layer and may be reflected to upper layers. Therefore, this drawback of cement stabilization should carefully be investigated and it is beyond the scope of this study.

In conclusion, this study aimed to asses the suitability of WFS as a road material and basic mechanical tests were performed. From the results obtained, WFS is a promising road material provided that further investigations will be carried out utilizing advanced geotechnical testing to establish the most appropriate cement and lime contents.

REFERENCES

ASSHTO Guide, 1993. *Guide for Design of Pavement Structures*. American Association of State Highway and Transportation Officials, Washington, DC.

ASTM, 2005. *Annual Book of ASTM Standards*. American Society for Testing and Materials.

Beeley P., 2001. *Foundry Technology*. University of Leeds, England.

Clegg A. J., 1991.*Precision casting processes*. New York: Pergamon Press.

Fox P. J. and Mast D. G., 1998. *Geotechnical Performance of A Highway Embankment Constructed Using Waste Foundry Sand*. Joint Highway Research Project, Final Report FHWA/IN/JTRP-98/18, Purdue University.

Guney Y., Aydilek A.H., and Demirkan M.M, 2005. *Geoenviromental Behavior of Foundry Sand Amended Mixtures for Highway Subbases*. Journal of Waste Management 26.

Kézdi Á., 1979. *Stabilized Earth Roads*. Technical University of Budapest, Budapest: Elsevier Scientific Publishing.

Partridge B. K., 1998. *Field Demonstration of Highway Embankment Constructed Using Waste Foundry Sand*. In partial Fulfillment of the Requirements for the Degree of Doctor of Philosophy, Purdue University.

Yaylali G., 2007. *Turkish Foundry Industry in 2007*. Journal of Turkish Metallurgical and Materials 13, Istanbul.

Advances in Transportation Geotechnics – Ellis, Yu, McDowell, Dawson & Thom (eds)
© 2008 Taylor & Francis Group, London, ISBN 978-0-415-47590-7

A study on permanent deformation of lateritic soils including the shakedown concept

A.C.R. Guimarães & L.M.G. da Motta
Universidade Federal do Rio de Janeiro, Brazil

ABSTRACT: This paper focuses a study on permanent and resilient deformation of three lateritic soils – a yellow clay from Rio de Janeiro, a laterite gravel from Amazon and a laterite gravel from Brasília – submitted to repeated load triaxial tests at several levels of stresses and number of cycles greater than 100,000. Thirty eight tests were made at different stress levels and compaction water contents near the optimum value. The occurrence of plastic shakedown or material shakedown was investigated. The evolution of permanent deformations according to different factors – number of load applications, moisture content, and state of applied stresses was observed. The variation of elastic deformation and resilient modulus with the number of load applications was also observed. Regression analyses were made to obtain a correlationship of permanent deformation with states of stresses and number of cycles.

1 INTRODUCTION

1.1 *Permanent deformation on tropical soils*

The accumulation of permanent deformation in asphalt pavement soils during its service life is associated to the structural defect known as dip of the wheel path.

In Brazil, thorough field researches such as Research on Interrelationship of Highway costs (Pesquisa de Inter-Relacionamento de Custos Rodoviários PICR), made by the GEIPOT (Brazilian Company of Planning in Transportation) between 1979 and 1981 and mentioned by Queiroz (1984), showed that values of the wheel path dip measured in 45 distances, hit a maximum value of 7,4 mm and an average value of 2,53 mm. Therefore, below the allowed maximum value of 1,27 cm, adopted by the FAA (Federal Aviation Administration).

It should be pointed that almost all the pavements assessed in the PICR research were dimensioned by the CBR method and this method tends to over dimension the pavement exactly on wheel path dip, because the method is based on the construction of layers over the subgrade to protect it from traffic load and the mechanical properties of the subgrade are assessed by its penetration resistance (CBR test), which does not represent the real field condition to which the soil is submitted. Besides, the soaking in water of the samples for four days is not compatible with the tropical environment.

Thus, studies on permanent deformation of Brazilian tropical soils remained for some years in a secondary plane of the fatigue studies of asphalt pavements. Studies by Motta (1991), Svenson (1980), Marangon (2004) and Santos (1998), among others should be enhanced.

However recent advances in Pavement Mechanics, allow a proper modeling of the behavior of permanent deformation in tropical soils and the present study is a small contribution to this purpose.

1.2 *The shakedown concept*

When a specimen is submitted to cyclic loading and after certain number of loadings is applied, the plastic deformation ends, it is said it does *shakedown*, and the non-existence of plastic deformation is justified by the equilibrium that appears in the stress field formed by the elastic deformation strains and the residual stresses. Therefore, the appearance of residual stresses is a primary condition for shakedown establishment.The purpose of the development of the shakedown theory is to determine the conditions and limits to which an engineering material is submitted for a certain load where the shakedown state occurs. In the specific situation of pavement layers, the purpose would be to determine the stress state (σ_d, σ_3) where a pavement layer, properly characterized, presents strictly elastic behavior, from a certain number N of applied loads.

The shakedown load of a pavement structure is the function of several factors such as cohesion "c" and the friction angle "ϕ" of the soils that compound the pavement under analysis. This is important mainly in the study of tropical soils because they have some special features such as the presence of Fe and Al oxides and hydroxides which confer aggregate properties and the subsequent increase of c and ϕ parameters. Thus, pavements constructed with tropical soils, would have a higher shakedown load and consequently a smaller value of permanent deformation as observed in the field.

The shakedown theory was initially developed for metals under sliding and rolling loads, and the first application for asphalt pavements was done by Sharp and Booker (1984). Raad et al (1988), (1989a) and (1989b) in an article of the Transportation Research Board (TRB) presented studies on pavement structural shakedown researches from numerical modeling. The principles of the shakedown theory, including its two basic theorems may be observed in Faria (1999) or in Collins and Boulbibane (2000).

The work conducted by Werkmeister et al (2001), involving the Dresden Technical University in Germany and the Nottinghan University in England enhanced the research of shakedown occurrence in the material through repeated load triaxial tests.

Werkmeister et al (2001) fulfilled several tests of permanent deformation of granular soils, at different stress levels, which showed three different behaviors or domains, called A, B and C, described below.

Level A – Shakedown or Plastic Accommodation. In this dominium the response is elastic for a finite number of load applications and after a post-compaction period, it becomes entirely elastic and there are no more plastic deformations. In the Dawson and Wellner model, this kind of behavior produces curves parallel to the axle of the accumulated permanent deformation rates. The pavement is called "in *shakedown*" and, consequently, the total accumulated permanent deformation is small.

Level B – Corresponds to an intermediate response level, that is, it is not possible to conclude if the material will collapse or if it is in the shakedown state.

Level C – Collapse. In this dominium, a successive increment of permanent deformation occurs for each loading cycle and the answer of the material is always plastic.

2 MATERIALS

In this work, three kinds of soils were selected: a yellow clay from Rio de Janeiro and a Laterite gravel from Brasília, both already tested by Guimarães (2001), and a Laterite gravel from Acre.

The Yellow Clay used in the present study was originated in the talus cut at the BR-040/RJ, km 111, and was used as final layer of the embankment of the experimental circular track of the Highway Research Institute (IPR/DNIT). It is a residual gnaisse soil previously characterized by Silva (2001), whose properties are described in table 1.

Table 1. Geotechnical characteristics of the Yellow Clay from Rio de Janeiro. (Silva, 2001).

Test	Unit	Value
Liquidness Limit (LL)	%	51,9
Plasticity Limit (PL)	%	23,1
Plasticity Index (PI)	%	28,8
Specific weight (γ)	g/dm^3	1.610
TRB Classification	A-7-6	–
Unified Classification	CH	–
MCT Classification	LA	–
Optimum moisture	%	20,7

The soil was classified as class LA by the MCT methodology presented by Nogami and Villibor (1995), and the equation 1 shows the resilient modulus value of the material.

$$RM = 68,6.(\sigma_d)^{-0,257} \quad h = 22,0\ \% \qquad (1)$$

Where RM = resilient modulus; σ_d = desviatoric tension.

The Laterite from Brasília is a granular material widely used in pavements works at the Federal District (DF). Its optimum compaction moisture is 17,5%, and presents non plastic behavior. Its resilient modulus is given by the equation 2:

$$RM = 769,5\ (\sigma_d)^{0,14}, \quad h = 17,5\% \qquad (2)$$

Where RM = resilient modulus; σ_d = desviatoric tension.

3 METHOD AND TESTING PROGRAM

A 5,0 kg sample was selected for each test, previously homogenized at the optimum compaction moisture, Normal Proctor energy was used for the Yellow Clay from Rio de Janeiro and Intermediate Proctor energy for Laterites from Brasília and Acre. The samples were kept in a moisture cabinet for at least 12 hours, from where they were removed according to the essay.

Then, the material was compacted in a cylindrical tripart mold with 10 cm radium and 20 cm height, in similar procedure as described by Medina and Motta (2005), to obtain the design density.

Repeated load triaxial tests were fulfilled with several values of deviatoric and confining stresses and for a number of applied loads (N) usually greater than 100.000 cycles, as shown in tables 4, 5 and 6.

Table 2. Stresses of the tests with the Yellow Clay from Rio de Janeiro.

Test	σ_d (kPa)	σ_3 (kPa)	w(%)	N
1	70	70	21,3	51.500
2	70	70	20,6	500.000
3	35	70	19,4	506.000
4	105	70	21,3	190.000
5	25	50	20,0	470.500
6	120	120	19,9	319.000
7	75	50	20,6	340.000
8	50	50	20,6	310.000
10	180	120	19,2	186.000
11	50	50	18,4	303.000
12	105	70	20,7	338.000
13	120	120	21,7	340.000
14	60	120	20,4	330.000

Table 3. Stresses of the tests fulfilled with the Laterite from Brasília.

Test	σ_d (kPa)	σ_3 (kPa)	N
1	113	75	1.000.000
2	105	105	201.700
3	75	75	392.100
4	70	70	530.000
5	75	150	508.500
6	150	100	319.500
7	150	150	200.200
8	100	200	647.200
9	200	150	472.000
10	200	100	532.200

3.1 *Rio de Janeiro Yellow Clay*

The test specimens of the Yellow Clay from Rio de Janeiro were intended to be molded at the optimum compaction moisture of 20,7%, with an energy equivalent to Normal Proctor test, however, it was observed as shown in table 2 that the moisture presented certain variations from test to test.

The tests were conducted beyond 100.000 cycles of applied load to characterize the eventual shakedown appearance or the plastic deformation accommodation.

3.2 *Brazilia Laterite Gravel*

Tests with the Laterite from Brasília were conducted in a similar way as those with the Yellow Clay from Rio de Janeiro, and their stresses and numbers of applied loads are presented in table 3.

An important aspect to be considered is the material moisture because this value may vary up to 1.5 percentage points with the soil fraction that is inserted in the capsule sent to the furnace. This is due to the fact that it is a granular material, whose gravel fraction is formed by aggregated of vesicular shape, of varying sizes, densities and porosities (that may reach up to 7,5%).

Table 4. Stresses of the tests with the Acre Laterite.

Test	σ_d(kPa)	σ_3 (kPa)	N
1	105	105	161.312
2	210		245.252
3	315		257.200
4	158		257.200
5	300	150	245.500
7	200		111.500
9	100		243.418
10	50	50	56.295
11	100		150.300
12	400	150	72.658
14	157,5	105	231.453

3.3 *Acre Laterite Gravel*

The tests conducted with the Laterite from Acre are similar to the others, except that a 1 Hz frequency was adopted, and the stresses and number of applied loads are shown in table 4.

4 TESTS RESULTS

4.1 *Rio de Janeiro Yellow Clay*

The occurrence of shakedown was researched analyzing the response levels of the test specimens to several stresses, in a similar way as proposed by Werkmeister (2001), already mentioned. It should be enhanced that those authors study was on granular materials, not including fine soils, silt or clay. The obtained results are shown figure 1.

The data in figure 1 show that all the tests with the Rio de Janeiro Yellow Clay presented a type "B" behavior, that is, it is not possible to conclude if the material will collapse or if it is in shakedown state.

The exception occurred in test 3, conducted with deviatoric stress σ_d of 35 kPa. Between 331.600 and 506.000 cycles, thus during more than 170.000 cycles, the same record of permanent deformation was obtained characterizing the accommodation of permanent deformation or shakedown.

Test 5 was realized with deviatoric stress, σ_d, 25 kPa thus lower stress than for test 3, and the same stress rate σ_d/σ_3 was maintained. However an accommodation level similar to test 3 was not observed, discarding the possibility of a lower shakedown limit, for that kind of soil.

Tests 4 to 12 were fulfilled with the same stress state however at different compaction moisture levels. Test

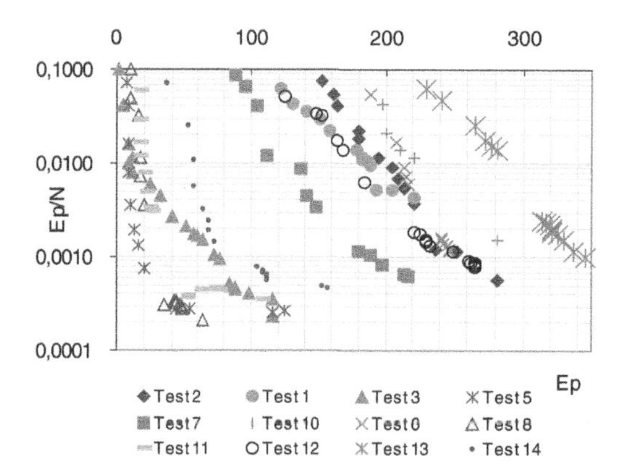

Figure 1. Dawson and Wellner Model Framework of some tests with the Rio de Janeiro Yellow Clay.

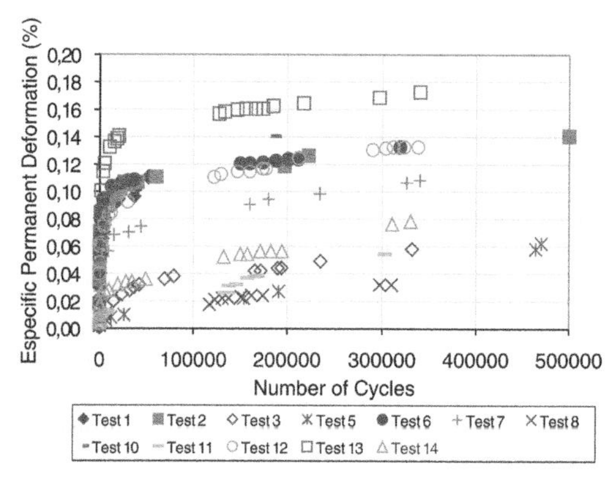

Figure 2. Variation of the Permanent Specific Deformation against the number of Cycles of Load Application for the Yellow Clay under study.

4 is a typical example of behavior type "B", because a constant incremental rate of permanent deformation was observed.

In test 12, the same reading of permanent deformation was obtained between cycles 312.000 and 338.000 (0,266 mm), however it is not possible to assure a shakedown situation because of the shape of the curve of Dawson and Wellner model, shown in figure 1. Then, in test 4 with material slightly more humid (h = 21,3%) a constant increment of small magnitude of permanent deformation was observed when compared with the test with a drier material (12).

When a situation of apparent accommodation is observed, such as in test 12, which does not correspond with the above mentioned model, it is called apparent shakedown.

Apparent shakedown was also verified in tests 8 and 10, and in the remaining tests: 1, 2, 6, 7, 11, 13 and 14 accommodation of permanent deformation was not observed, being typical examples of type "B" behavior.

4.1.1 *Permanent deformation assessment*

In Brazil, there is not still an agreement on the number of cycles of applied load necessary for permanent deformation tests. Usually 100.000 cycles are used when accommodation of deformation is not observed with successive blows.

Figure 2 presents the accumulated specific permanent deformation values for the test specimens of Rio de Janeiro Yellow Clay. All the curves present similar shape and show a high increment in the initial cycles, up to 50.000 cycles or less, and then a trend to asymptotic behavior.

Very reduced values of specific permanent deformation are observed, even when submitted to very high stresses for Brazilian highway subgrade standards.

For example, in test 10, the deviatoric stress applied, σ_d, was 180 kPa, while a typical work stress at the top of the subgrade is hardly higher than 50 kPa. The total specific permanent deformation in this case was only 0,15%. Tests 2 and 3 were conducted with the same confining stress, σ_3, equal to 70 kPa and different deviatoric stresses σ_d, of 70 and 50 kPa respectively. Test 2 presented a specific permanent deformation almost 43% higher than test 3, which corroborates the high influence of the deviatoric stress in the specific deformation of lateritic soils as observed by Cardoso (1987) among others.

The effect of the confining stress was observed through tests 2 and 7, conducted at the same deviatoric stress of 100 kPa. A 17% reduction was observed when the confining stress was reduced from 70 to 50 kPa. Comparing test 6, where test specimen with 19,9% moisture was used, with test 13 with test specimen with 21,7% moisture, it was possible to have an idea of the influence of this factor in the specific permanent deformation. The curves presented very similar shapes, but the test with the higher moisture test specimen presented total specific deformation near 33% higher than the lower moisture test specimen.

4.2 *Brasília laterite gravel*

Figure 3 shows the test results obtained with the Brasília Laterite Gravel according the Dawson and Wellner model for shakedown occurrence research.

Figure 3 analysis shows a significant difference between results obtained with Brasília Laterite and Rio de Janeiro Yellow Clay, because almost all the test specimens presented a type "A" behavior, according the classification proposed by Werkmeister (2001). Maybe this difference is due to the fact that Laterite is a predominantly granular material.

Test 1 was conducted with up to 1.000.000 cycles of applied loads and it was possible to make a good characterization of the shakedown situation as shown in figure 4. Total time to fulfill this test reached almost

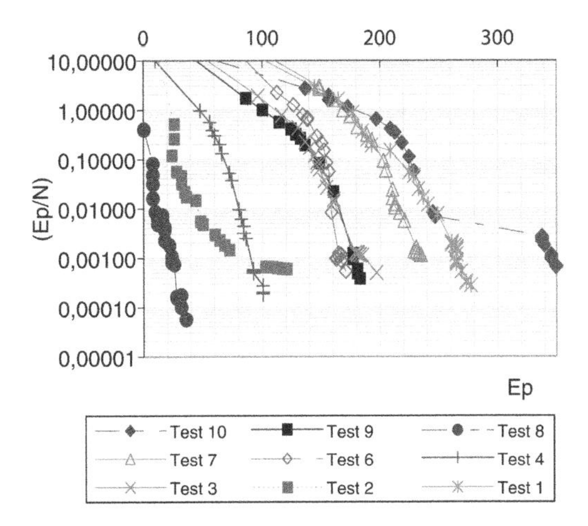

Figure 3. Dawson and Wellner Model Framework of some tests fulfilled with a Brasília Laterite Gravel.

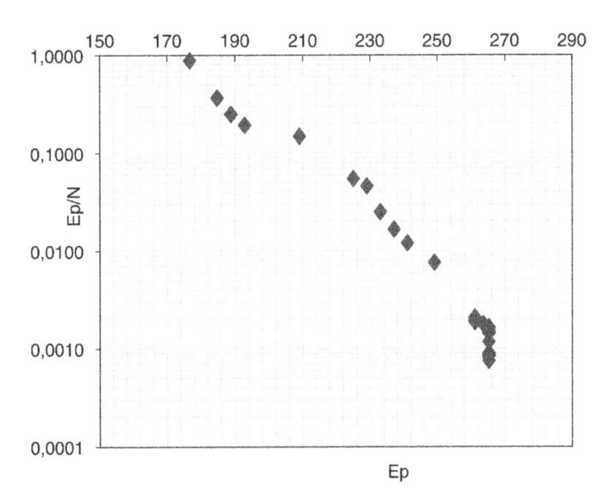

Figure 4. Dawson and Wellner Model Framework Test 1. Laterita Brasília.

six days of continuous use of repeated load triaxial equipment.

Test 2 presented a type "B" behavior, as shown quite clearly in figure 4. However, in test 3, exclusively the graphical analysis does not allow characterizing clearly the behavior, and it is necessary to use the recorded values. The last two records for 329.160 and 162.500 cycles were respectively 0,197 mm and 0,185 mm, then, there was no plastic accommodation. The same condition of increasing permanent deformation may be observed in tests 6, 7 and 8. In tests 4, 9 and 10 shakedown was verified.

4.2.1 *Permanent deformation assessment*

Figure 5 presents the specific permanent deformation values for Brazilia Laterite test specimens. All

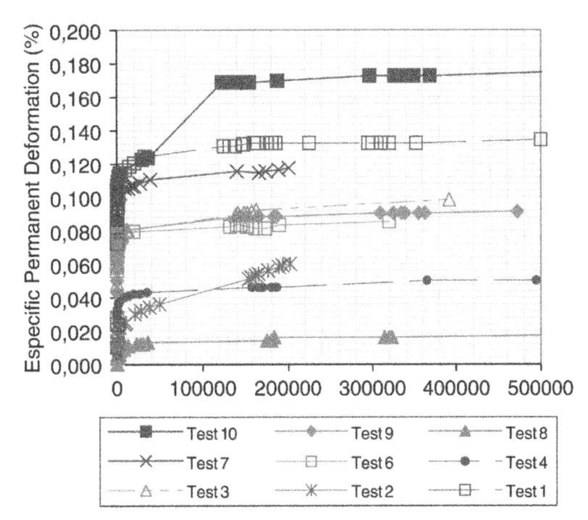

Figure 5. Variation of Specific Permanent Deformation against N for Brasília Laterite.

curves present similar shape with clear trend to permanent deformation accommodation, except for test number 2.

In test 10, fulfilled with σ_d equal to 200 kPa and σ_3 equal to 100 kPa, the highest total specific permanent deformation of 0,175% was observed. The highest deviatoric stress used in this work was applied in this test as well as in test 9.

In test 9, fulfilled with σ_d equal to 200 kPa and σ_3 equal to 150 kPa, the total specific permanent deformation was reduced to half. This fact proves the importance of the confining stress on the permanent deformation of Brazilia Laterite and can be explained, as suggested by Cardoso (1987), by the ratio σ_d/σ_3 that falls from 2,0 in test 10 to 1,33 in test 9.

Test 3 fulfilled with σ_d equal to 105 kPa, σ_3 equal to 105 kPa, was the only test whose variation curve of specific permanent deformation against N was always increasing. In this test, conducted at 201.700 load cycles, it is not possible to assure that this deformation will tend to accommodation, due to the relative small number of load applications, but all the other tests presented that trend to accommodation.

4.3 *Acre laterite gravel – shakedown research*

Figure 6 shows the results of the tests with the Acre Laterite, according with the Dawson and Wellner model for shakedown occurrence research.

Figure 6 shows that all the tests conducted with Acre Laterite, with the exception of test 12, presented a level A typical behavior, that is they did shakedown.

Test 12 presented an intermediate behavior, or level B, due to the number of applied load cycles which was 72.658 cycles, significantly smaller than the others. There was no time to present significantly reduced values of permanent deformation.

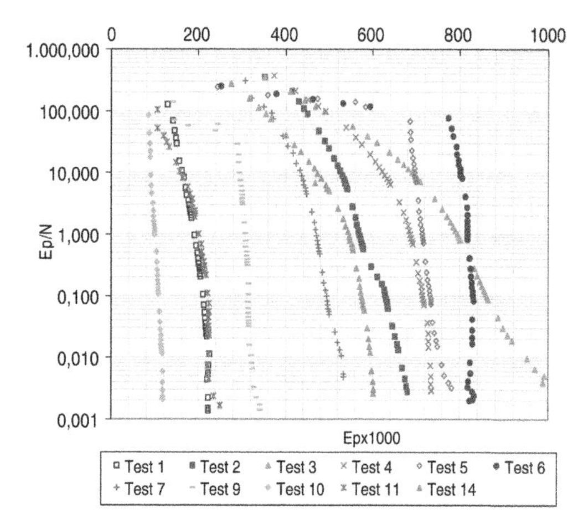

Figure 6. Research of Material Shakedown occurrence for Acre Laterite.

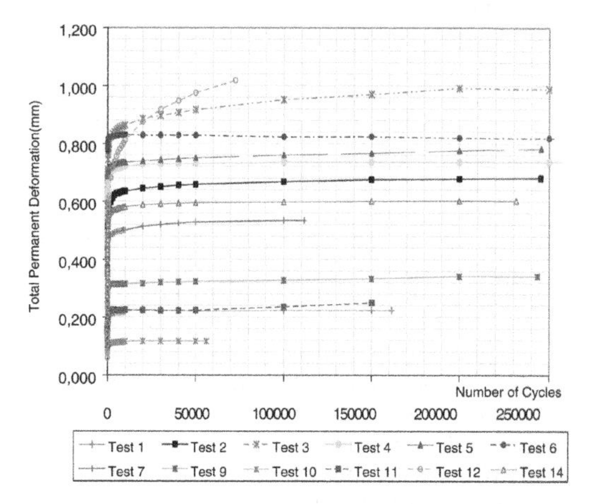

Figure 7. Variation of Total Permanent Deformation against the Number of Cycles of Applied Loads. Acre Laterite.

4.3.1 *Permanent deformation assessment*

Figure 7 presents the results of the permanent deformation tests fulfilled with the laterite from Acre.

Figure 7 shows that the highest obtained value of permanent deformation was 1,018 mm, for test 12 where a deviatoric stress of 400 kPa and confining stress of 150 kPa were used.

This stress state may be considered high for a field situation where an Acre laterite is used as base, related to the standard loading axle. In other words, even when a high stress state was used, the Acre laterite presented a low permanent deformation value, which characterizes it as a material with high resistance to permanent deformation.

From all the tests presented in figure 8, test 12 was the only one where it was not possible to identify a

Table 5. Results of Permanent Deformation Tests for Constant Confining Stress (105 kPa).

Test	σ_d (kPa)	$\varepsilon_p^{10.000}$	$\varepsilon_p^{100.000}$
Test 1	105	0,220	0,223
Test 14	157,5	0,581	0,598
Test 2	210	0,635	0,668
Test 3	315	0,864	0,952

clear situation of permanent deformation accommodation associated exclusively with a reduced number of applied load cycles. (72.658).

All the other tests showed a clear trend of total or accumulated permanent deformation accommodation with the increasing number of load application cycles. The tests presented in table 5 correspond to the same level of confining stress of 105 kPa, and deviatoric stress varying in such a way that the ratio σ_d/σ_3 took the following values: 1, 1.5, 2 e 3.

For the tests 1, 14, 2 and 3, presented in table 5, the initial permanent deformation represent respectively: 57,4%, 46%, 52,7% e 36,9% of the total permanent deformation. Then, the initial permanent deformation was about half the value of the permanent deformation for 100.000 cycles of applied load, except for deviatoric stress of 315 kPa where this percent was de 36,9%.

Another analysis of table 4 data, shows that the total permanent deformation value, $\varepsilon_p^{100.000}$, changed from 0,223 to 0,952, that is, the permanent deformation for σ_d equal to 3,15 was 4,26 times higher than the deformation obtained for σ_d equal to 105 kPa, enhancing the influence of the deviatoric stress on the total permanent deformation.

Besides, considering the other tests for ratios σ_d/σ_3 of 1.5, 2 e 3, the value of $\varepsilon_p^{100.000}$ increases in the following proportion: 2,68, 2,99 and 4,26, pointing a non-linear relation more sensible to the higher deviatoric stress value σ_d.

5 CONCLUSION

Plastic or permanent deformation accommodation or even shakedown were observed from certain number of applied load cycles for several moisture contents and different stress states in some test specimens of Rio de Janeiro Yellow Clay and mainly in test specimens of Laterite from Brasília and Acre.

The tests fulfilled with the lateritic tropical soils used in this work proved that the total permanent deformation, obtained through repeated load triaxial tests was highly influenced by the stress state and the moisture content of the test specimens.

On the other hand, the obtained values were very small, characteristic of soils with high resistance to permanent deformation and hence, recommended for road engineering.

Although the much higher values than gotten by Werkmeister (2003) for unbound granular materials, the models of Dawson and Wellner and Werkmeister and Dawson had revealed adequate for the study of permanent deformation of lateritic soils. The research continues using a larger set of stress level and another kind of lateritic soils.

ACKNOWLEDGMENTS

The authors thank some researches who have helped with shakedown occurrence researches and soil permanent deformation assessment sending their papers. They are: Niclas Odermatt, Erick Lekarp, Sabine Werkmeister, I. F. Collins, Greg Arnold and Inge Hoff.

REFERENCES

CARDOSO, H.S. (1987) "Procedure for Flexible Airfield Pavement Design Based on Permanent Deformation". Tese PhD. University of Maryland. EUA.

COLLINS, I. F e BOULBIBANE, M. (2000). Shakedown Under Moving Loads With Applications to Pavement Design and Wear. Developments in Theoretical Geomechanics (The J. Booker Memorial Volume). Ed by D. Smith e J. Carter, Balkema, pag 655–674.

FARIA, P.D.O (1999). Shakedown Analysis in Structural and Geotechnical Engineering. PhD Thesis. University of Wales, Swansea.

GUIMARÃES, A. C. R (2001). Estudo de deformação Permanente em Solos e a teoria do Shakedown Aplicada a Pavimentos Flexíveis. Tese de Mestrado da Universidade Federal do Rio de Janeiro (Coppe/UFRJ).

MARANGON, M. (2004). Proposição de Estruturas Típicas de Pavimentos para Região de Minas Gerais Utilizando Solos Lateríticos Locais a Partir da Pedologia, Classificação MCT e Resiliência. Tese de Doutorado. Programa de Engenharia Civil da COPPE/UFRJ. Rio de Janeiro,

MEDINA. J e MOTTA, L.M.G da (2005). Mecânica dos Pavimentos. 2ª Edição revisada e Ampliada.

MONISMITH, C. L. , OGAWA, N. , FREEME, C. R (1975). "Permanent Deformation Characteristics of Subgrade Soils Due to Repeated Loading". 54° Annual Meeting of TRB. Washington.

MOTTA, L. M. (1991). "Método e Dimensionamento de Pavimentos Flexíveis; Critério de Confiabilidade e Ensaios de Cargas Repetidas". Tese de Doutorado. Programa de Engenharia Civil – COPPE/UFRJ. Rio de Janeiro.

NOGAMI, J. S. e VILLIBOR, D. F. (1995). "Pavimentação de Baixo Custo com Solos Lateríticos". Editora Villibor. São Paulo.

QUEIROZ, C. A. V. (1984). Modelos de Previsão do Desempenho para a Gerência de Pavimentos no Brazil. Brasília, GEIPOT, 366P.

RAAD, L., WEICHERT, D., NAJM, W. (1988). Stability of Multilayer Systems Under Repeated Loads. Transportation Research Record, 1207: 181–186.

RAAD, L., WEICHERT, D., HAIDAR, A. (1989a). Analysis of Full-depth Asphalt Concrete Pavements Using Shakedown Theory. Transportation Research Record, 1227: 53–65.

RAAD, L., WEICHERT, D., HAIDAR, A. (1989b). Shakedown and fatigue of Pavements With Granular Bases, Transportation Research Record, 1227: 159–172.

SANTOS, J. D. G. (1998). "Contribuição ao Estudo dos Solos Lateríticos Granulares Como Camada de Pavimento". Tese de Doutorado. Programa de Engenharia Civil – COPPE/UFRJ. Rio de Janeiro.

SHARP, R. W e BOOKER, J. R. (1984). Shakedown of Pavements Under Moving Surface Loads. Journal of Transportation Engineering, ASCE, vol 110, pag 1–14.

SILVA. D. E. R. da (2001). Estudo do Reforço do Concreto de Cimento Portland (Whitetopping) na Pista Circular Experimental do Instituto de Pesquisas Rodoviárias. Tese de Doutorado. Programa de Engenharia Civil – COPPE/UFRJ.

SVENSON, M (1980). "Ensaios Triaxiais Dinâmicos de Solos Argilosos". Tese de Mestrado. Programa de Engenharia Civil - COPPE/UFRJ. Rio de Janeiro

WERKMEISTER, S., DAWSON, A. R., WELLNER, F. (2001). Permanent Deformation Behavior of Granular Materials and the Shakedown Concept. Transportation Research Record n° 01-01852, Washington, DC.

WERKMEISTER, S (2003). Permanent Deformation Behavior of Unbound Granular Materials in Pavements Construction. Ph.D Thesys Technischen Universitat Dresden.

Advances in Transportation Geotechnics – Ellis, Yu, McDowell, Dawson & Thom (eds)
© 2008 Taylor & Francis Group, London, ISBN 978-0-415-47590-7

New Danish test method for the light weight deflectometer (LWD)

C. Hejlesen & S. Baltzer
Danish Road Directorate, Danish Road Institute, Denmark

ABSTRACT: In the spring of 2007 The Danish Road Institute in cooperation with the Danish LWD-user group finished a national test method for determination of the E-modulus of unbound materials with the LWD. A LWD is handy equipment on site, and can be used in the early exploration and construction phases to obtain more knowledge at an early point in time which makes it possible to optimise the design of the pavement. The LWD is also an effective tool for describing the non-linearity of the subgrade.

1 INTRODUCTION

For design of new roads it is necessary to know the E-modulus of the different materials to be used in the pavement. In Denmark CBR-test or vane test are normally used to give a rating of the E-modulus of the natural subgrade. The Light Weight Deflectometer (LWD) offers a direct measure of the in situ E-modulus. A picture of a LWD is shown in Figure 1.

A precise knowledge of the E-modulus of the subgrade makes it possible to optimise the design of the pavement. An actual E-modulus measurement and therefore a final adjustment of the design today only takes place as a static Plata Load Test after the construction of the granular base layer. The LWD is handy equipment but also a quick and effective tool for testing at different levels of stress, which gives useful information on the non-linearity of the subgrade. With tests at different stress levels, it becomes important to compare results obtained by a LWD to results obtained using traditional equipment. (Baltzer et al. 2007).

The purpose of the test method is to determine the E-modulus, of unbound road materials in situ. The test method can be used for examinations to evaluate most types of unbound materials, which are used for construction works in Denmark. The method can also be used for testing of stabilized materials. (Hildebrand et al. 2007).

Figure 1. An example of a Light Weight Deflectometer used on a soft subgrade with a E-modulus of less then 10 MPa. (Baltzer et al. 2007).

2 RELATION TO OTHER MEASURING EQUIPMENT

LWD measurements can be used to great advantage to measure the E-modulus of soil. The traditional CBR and Standard Proctor tests will normally also be necessary to determine the sensitivity of the soil in regard to water content.

When compaction of unbound layers is checked, the LWD measurements cannot replace the compaction control with the Moisture-Density Gauges. Compaction control should ensure that the material is incorporated with optimum compaction, so that future settlement can be avoided. Only when the compaction has been completed is it relevant to measure the E-modulus. Thus the Moisture-Density Gauges and the LWD should be used together, since it then should be possible to ensure optimum workmanship and a good picture of the properties of the material.

The Falling Weight Deflectometer (FWD) measured on the same principles and gives approximately the same results as the LWD at the same level of stress. Measurement with the one or other equipment depends on the equipment available and whether it is possible to drive on the layer. For examination of roads with wearing courses or base layer made of bitumen or concrete the FWD should be used, since a great level of stress is required. When making LWD tests, typically the surface modulus (E_0) is determined. The surface modulus expresses an average E-modulus for the volume of soil, which is below the measurement point. On the other hand, the FWD evaluated the E-moduli for the individual layers in the construction. (Hildebrand et al. 2007).

3 PRINCIPLE

The LWD is handy equipment, which on the subgrade of a known load impact from a weight dropped from a given height onto a stiff, circular plate expresses the bearing capacity of the subgrade. This type of equipment has the potential to give simple measurements of the bearing capacity of the soil and compacted sand and gravel materials. Alternative and more unconventional unbound materials can also be tested with the LWD.

4 CALIBRATION

In order to ensure usable and reliable measurements with the LWD, the accelerometer/geophone of the equipment and the load cell should be calibrated at least once a year. The calibration period presupposes that the equipment has been treated and used in accordance to the User Instructions.

A certified calibration can only be carried out by the institutions or producers who have been authorised to do this. The calibration must be made with traceable measuring equipment. The calibration should be able to be verified and documented. (Hildebrand et al. 2007).

5 PROCEDURE

In a standard test, the plate diameter is 150 mm and the drop weight is 10 kg. Depending on the type of material to be measured, it may be necessary to use additional equipment such as extra drop weights or other plate sizes. In Table 1 find a summary of which stresses

Table 1. Stress levels for different materials.

Material	Stresses kPa	Note
Granular base layer	200–300	when using a 20 kg drop weight or a 200 mm plate
Subbase	100–200	
Firm subgrade	50–100	
Soft subgrade	10–60	low drop height

should be obtained for various materials to get reliable measurements. The recommended plate pressure is also an illustration of the typical load to which the individual layers are exposed, when the construction is completed.

When carrying out measurements on friction materials, a plate pressure is used corresponding to granular base or subbase layers. When measuring cohesive materials, a plate pressure corresponding to the subgrade should be used.

The measured reversible deformation (strain) should be between 500 and 1500 μm, but max. 2200 μm and min. 300 μm. Measuring precision for equipment which measures with variable height is ±2% in the area from 0–2200 μm.

On equipment where the drop height can be varied, it should also be ensured that the pulse time is as required. In the standard test a pulse time of 15–25 ms is used. If measurements are carried out at varying heights, the pulse time will change. For standard measurements, these changes in pulse time will be accepted, when the height is increased, even if they exceed the 15–25 ms. (Hildebrand et al. 2007).

6 ASSEMBLY AND EXECUTION

In order to obtain a correct and useable measurement with the LWD it is especially important to ensure that the plate has good contact with the ground, so that the contact pressure is distributed evenly during the impacts. It should also be ensured that the foot of the geophone/accelerometer can move freely.

The E-modulus of unbound materials is dependent on the stress which is applied during the measurement. On clay the E-modulus typically falls with increased load, which is also known as negative non-linear connection; the opposite applies on friction materials, but to a lesser degree. Thus, the best evaluations of the bearing capacity of a material are obtained by undertaking measurements at varying stresses. Depending on the material, it is recommended to make 2–4 varying stresses at each measuring point, always starting with the lowest.

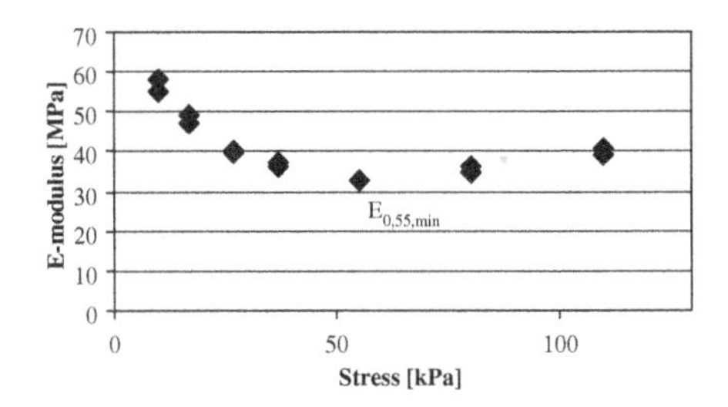

Figure 2. The E-modulus dependence of the stress for clay.

When the equipment is assembled correctly, repeated measurements are carried out at the same stress until three impacts with a mutual spread of less than 5% are made. (Hildebrand et al. 2007).

7 CALCULATIONS

The primary results from LWD measurements are E_0 which is calculated in the following way:

$$E_0 = \frac{f \cdot (1 - \upsilon^2) \sigma_0 \cdot a}{d_0} \tag{1}$$

where f = stress distribution factor; ν = Poisson's ratio; σ_0 = stress [MPa]; a = plate radius [m]; d_0 = deformation (strain) [m]. (Thagesen 1998).

To obtain a uniformly spread load on the soil, the stress distribution factor can be set at 2 and Poisson's ratio to 0.5. Thus the expression is reduced to:

$$E_0 = \frac{1{,}5 \cdot \sigma_0 \cdot a}{d_0} \tag{2}$$

The surface modulus is stated at the stress (in kPa) which is measured at i.e. $E_{o,100}$, if the stress was 100 kPa during the measurement.

Unbound road materials are only rarely linear elastic. The non-linearity of the material can be described by means of the parameters C_o and n.

Soil types can only be regarded as linear elastic to a point, which means that the E-modulus varies in relation to the load to which the soil type is exposed. It is therefore important to vary the stress levels which are used during the measurements, dependent on the material which is being measured and the future load to which it is exposed. A typical connection between the stress and the E-modulus for clay is shown in Figure 2, where the E-modulus is high at low stress levels, but decrease until the minimum point, after which it will have a constant value or slight increase until breaking point.

For granular base material, there is an increase of the E-modulus for increased plate load.

By means of a simple regression analysis, the surface modulus can be described as a power equation of the surface tension. This description makes it possible in later analyses (approximately) to calculate what the surface modulus will be, when other layers have been placed on top of the actual dividing layer. The regression analysis can be characterized as "good" or "doubtful" seen from an evaluation of the correlation coefficient. The border value for the correlation coefficient is, in this study, arbitrarily set at $R^2 = 0.7$.

$$E_0 = C_0 \cdot \left(\frac{\sigma_o}{100\,kPa} \right)^n \tag{3}$$

where C_o gives the surface modulus at a contact load of 100 kPa and the index n gives an indicative information about the general quality of the soil. If $n > 0$ the base behaves like a friction material, where the stiffness is increased by increasing dynamic contact load; if $n < 0$ is the reaction for a cohesive material, where the stiffness decreases by increasing dynamic contact load. (Hildebrand et al. 2007).

8 REPORTING

In connection with LWD measurements, comparability is important, in order to be able to compare with earlier tests and experiences and to be able to make useable before/after measurements. Furthermore, it is of importance that the recipient of the measurement results understands the conditions of the measurements.

The reports are therefore an important part of the measurement work and the following information should be included in the report:

1. Type of equipment, incl. physical equipment, additional equipment and possibly the data collection program
2. Company and employee carrying out the measurements
3. Date of measurement
4. Precise description of where measurements were made
5. Signature of the person who carried out the measurements and the controlling person
6. Description of the soil on which the experiments were carried out and relevant measurements were carried out.
7. Plan with indication of the measuring points, possibly with pictures of the area and the set-up.
8. Plate size (diameter)
9. Air temperature at the time of carrying out the measurements
10. Values for the stress distribution factor and Poisson's ratio which is used in the final reported measurements.

Table 2. Recommended stresses at reporting.

Material	Stresses	Note
Granular base layer	$E_{0,250}$	
Subbase	$E_{0,150}$	
Firm subgrade	$E_{0,100}$	
Soft subgrade	$E_{0,xx}$	xx is replaced by the stress at the minimum point, i.e. 60.

11. The results, stress, pulse time, deformation, and the surface E-modulus. The last is given with the stress used for the measurement, i.e. $E_{0,100} = 51$ MPa. It is recommended that measurements on the various road materials are reported at standard stress as in Table 2.
12. The non-linearity of the material can be reported by giving C_0 and n. If other functions are used to describe the non-linearity, this must be stated.

In general for all statements of figures or values, this should only be done with a suitable number of important figures. For example, stress, settlement and E-moduls should be stated without decimals in kPa, μm, and MPa. (Hildebrand et al. 2007).

9 CONCLUSION

The LWD can be used to determine the E-modulus for unbound and stabilized materials. Compared to other equipment – like the FWD and Static Plate Load test, the LWD is handy equipment, which makes it possible to use it in places where it is difficult to use heavier equipment, like soft subgrades.

When comparing the E-modulus determined by the LWD with E-modulus from other equipment the value is the approximately the same at the same stress level. At the same time it is easy to determine the non-linearity of different unbound materials.

REFERENCES

Baltzer, S. et al. 2007. *Praktisk anvendelse af minifaldlod til dimensionering af befæstelser*. DK-2640 Hedehusene, Danish Road Institute

Hildebrand, G. et al. 2007. *prVI 90-4: Målinger af overflade-modul med Minifaldlod*. DK-2640 Hedehusene, Danish Road Institute

Thagesen, B 1998. *Veje og stier*. DK-2800 Lyngby, Polyteknisk Forlag

Advances in Transportation Geotechnics – Ellis, Yu, McDowell, Dawson & Thom (eds)
© 2008 Taylor & Francis Group, London, ISBN 978-0-415-47590-7

Slotsgrus®: A new concept for footpaths with unbound surfaces

G. Hildebrand
The Danish Road Directorate, The Danish Road Institute, Hedehusene, Denmark

P. Kristoffersen & T. Dam
Forest & Landscape Denmark, University of Copenhagen, Frederiksberg, Denmark

ABSTRACT: Footpaths with surfaces made of unbound materials are mostly preferred to paths with surfacing of hot mix asphalt or surface treatment. The reason for this is both environmental and aesthetical. Traditionally, unbound paths are cheaper to construct but at the same they require more attention to maintenance processes such as keeping the required cross profile and surface structure and preventing the growth of weeds through the surface. For decades, unbound paths have been constructed with a traditional structure consisting of an unbound granular base under a thin surface layer made of clayey gravel. The latter material is most often an unspecified material which tends to become slippery when wet. Furthermore, re-profiling and mechanical weed removal tends to rip out the larger particles from the unbound base hence reducing the quality of the surface of the path.

In 2002 a new concept for unbound footpaths was developed by Forest & Landscape Denmark, University of Copenhagen for The Palaces and Properties Agency in Denmark. The principle is to use the same material both as base and surface layer. The concept, which is a registered trademark, is called Slotsgrus® (translates to 'Castle gravel' in English), and it has seen extensive use in Danish parks for paths and squares, etc. This includes one major Copenhagen park trafficked by horse carriages, service vehicles and women with high-heeled shoes.

In 2006 The Danish Road Institute conducted a major evaluation of the Slotsgrus® material and the principles for its use. Results from the project showed that the material has a bearing capacity similar to traditional Danish unbound base materials and that it is feasible to construct layers of up to 300 mm thickness in a single process using standard compaction equipment. As a result of the project, specifications and functional requirements for the material were established based on the new European standard for unbound aggregates.

1 INTRODUCTION

1.1 *A new type of unbound footpaths*

In 2001 and 2002 Forest & Landscape Denmark, University of Copenhagen conducted a development project for The Danish Palaces and Properties Agency with the objective of devising methods for construction and re-construction of footpaths in one of the parks in Copenhagen, Denmark. The project aimed at identifying and test materials suitable for construction of unbound park footpaths.

The unbound footpaths should be passable in all types of weather: rain, snow, freeze, thaw, and drought. This presented tough requirements to the materials, which one should be able to compact to a firm and hard surface. It would not be tolerated that the surface would be sticky in wet condition and neither would the surface be allowed to be dusty in dry condition.

The new footpaths should be maintainable without pesticides only using mechanical equipment that to some extent treats the surface and the upper parts of the structure of the path. From this follows that the unbound constructed material could be surface treated again and again and the surface could be compacted to a desired level after every treatment.

1.2 *Traditional footpath structures*

The project showed identified the prevailing structure for Danish footpaths to be one consisting of an unbound granular base (0–32 mm) and a surface course made of clayey gravel or stone powder. The very fine surface course materials did not allow the layer thickness of the surface layer to be more than 10–20 mm. Clayey gravel, especially, would tend to be unstable in wet condition leading to a soft path which would soil people's shoes. At the same time the granular base is often not suited as foundation for a thin, porous unbound surface layer as the larger aggregates in the base layer can be torn up and hence ends at the surface of the footpath. Mechanical surface treatments and re-profiling can also tear up the larger aggregates from the granular base layer.

Hence, arguments existed for developing a one-layer structure, where one material serves the purpose of both base and surface layer.

2 TESTING OF NEW MATERIALS

2.1 *Preliminary testing*

An introductory screening of candidate materials led to the selection of a total of 15 materials, which were constructed in Frederiksberg Have in Copenhagen, Denmark. Some of the materials were used as top layer with 50 mm thickness while others were constructed in one layer with a thickness of 100 mm. During 2002, the condition of all 15 sections was evaluated seven times based on registration of parameters like: ability to maintain the original profile, evenness, colour, surface hardness, smudging of e.g. shoes, dust emission, and growth of weeds. Furthermore, the amount of segregation of aggregates was registered and finally a user interview survey was also conducted.

Basically, a large part of the selected materials fulfilled the functional requirements set up for the unbound materials. One material performed marginally better than the others. Both when used as a top layer and when used as a one-layer structure. And the material could be used for both applications. Another advantage of the material is that is consists of three fractions: clayey gravel, stone powder, and small crushed aggregates. Furthermore, the material came from a pit without much chalk, which has influence on colour and durability of the material.

2.2 *Slotsgrus®*

The selected material was registered as a trademark by Forest & Landscape Denmark, which since 2003 has had a licensing agreement with one Danish gravel pit to produce the material. Slotsgrus® has seen major use in the last years, and it has now been applied at more than 600 cases of varying sizes.

The application of Slosgrus® in larger projects has led to the desire for a strict material specification including a design modulus for structural design. To establish this information Forest & Landscape Denmark contracted The Danish Road Institute under the Danish Road Directorate to perform the necessary tests.

Work conducted by the Danish Road Institute included laboratory testing of several batches of Slotsgrus®, field testing on in-service footpaths with Slotsgrus® of different ages as well as a field test to evaluate the practical constructability of the material.

Table 1. Results from laboratory testing of Slotsgrus®.

	Sample #1	Sample # 2	Sample # 3
Max. grain size (mm)	8	16	16
Fines (%)	9.7	10.7	9.8
Std. proctor density $\rho_{d,max}$ (Mg/m^3)	2.06	2.08	2.09
Std. proctor w_{opt} (%)	7.9	8.4	8.7
CBR at w_{opt} (%)	55*	75*	23
Density, vibrating table $\rho_{d,max}$ (Mg/m^3)	2.08	2.10	2.04
Water content, vibrating table (%)	9.0	8.9	8.8
Sand equivalent (%)	31	32	30
Natural water content w_{nat} (%)	4.2	4.7	4.6
Plasticity index (%)	NP	NP	NP
Los Angeles abrasion (63 μm–4 mm) (%)	31	31	31
Shape index (4–8 mm) (%)	13	16	19
Uncrushed aggregate (4–8 mm) C_r (%)	15	20	16

*Estimated value

3 LABORATORY TESTS

3.1 *Standard testing of Slotsgrus®*

Testing in the laboratory of three different samples of Slotsgrus® aimed at establishing core parameters for the material and at the same time evaluation of the uniformity of the material. Table 1 presents the results from the testing.

Table 1 shows that the material consists of material between 0 mm and 8 mm aggregate with some grains larger than 8 mm. There is a fines content (less than 63 μm) of about 10 percent, a sand equivalency value of 30 percent, and the material is non-plastic. The material can be compacted to a maximum dry density of about 2.080 Mg/m^3 at standard proctor as well as vibrating table compaction. The optimum water content just above 8 percent is also double the value of the natural water content. The material is durable with a Los Angeles abrasion value of 30 percent, a shape index of approximately 15 percent and an amount of round particles less than 20 percent.

Table 1 further shows that there is only very limited variation in the results from the three samples, which were sampled at the same time at the production site.

All tests were conducted according to common European standards.

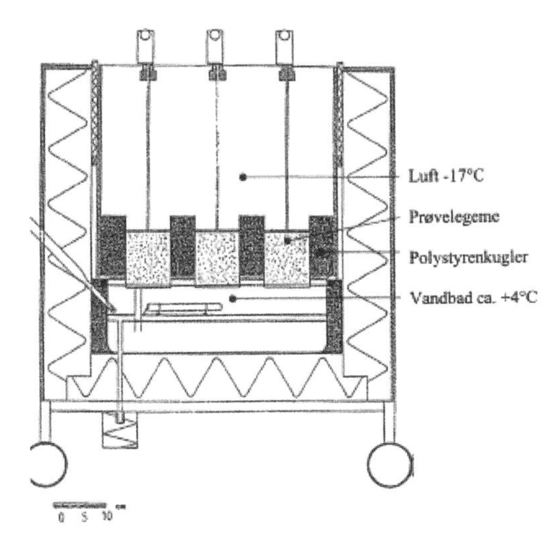

Figure 1. Testing apparatus for freeze-thaw test. The figure shows a cut through the test box (inner and outer box) with the location of the soil specimens. When the specimens are placed, all voids in the box are filled with polystyrene balls. Christensen & Palmquist (1976).

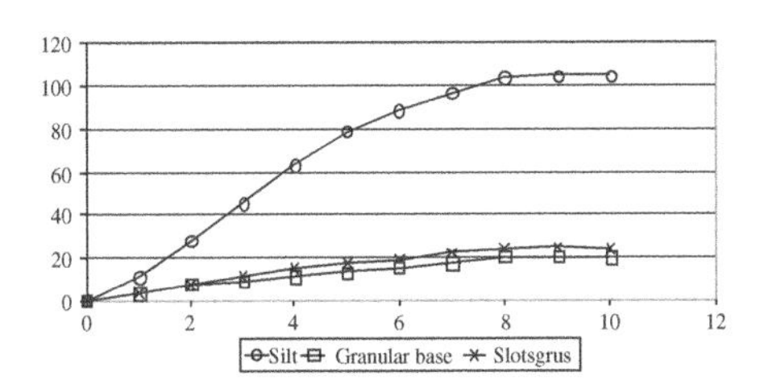

Figure 2. Freeze-thaw tests of Slotsgrus, granular base and silt.

3.2 *Freeze-thaw testing*

In addition to the standard tests mentioned above, a set of freeze-thaw tests were conducted at the Technical University of Denmark. The tests were conducted using the apparatus shown in Figure 1.

The principle of the freeze-thaw test is to apply frost to the top of a number of soil cores compacted in a standard proctor test and at the same time register the elongation (frost heave) of the cores. During the entire test, water suction is possible through the lower part of the cores. The entire test is conducted in a room with a constant temperature of −17 °C. Figure 2 shows the results of the freeze-thaw tests (averages for Slotsgrus® and granular base).

Figure 2 shows that the reference material (silt) has a very significant freeze-thaw reaction corresponding to a prolongation of approximately 90 percent of the original height of the core. There is only a limited freeze-thaw reaction for the traditional granular base (2 specimens) and for Slotsgrus® (6 specimens).

Figure 3. Light Weight Deflectometer testing in Frederiksberg Have, Copenhagen. Photo: P. Kristoffersen, Forest & Landscape Denmark.

The prolongation of the base was 17 percent of the original height, while the Slotsgrus® cores showed a prolongation of approximately 20 percent of the original height. Hence, it was concluded that Slotsgrus® is only frost susceptible to a very limited extent and to the same extent as the traditional granular base, which in practical terms is not seen as frost susceptible.

4 FUNCTIONAL TESTING ON AN IN-SERVICE PARK FOOTPATH

To evaluate the quality and especially the bearing capacity of Slotsgrus® pavements in service, 12 sections with Slotsgrus structures of different age (all in Frederiksberg Have) were selected. Within each section five testing points were selected at random. At all sixty testing points, the bearing capacity was determined using a Light Weight Deflectometer (see Figure 3):

Figure 4 illustrates the relation between elastic layer modulus of Slotsgrus® and pavement age. The average layer modulus was found to be 307 MPa with a

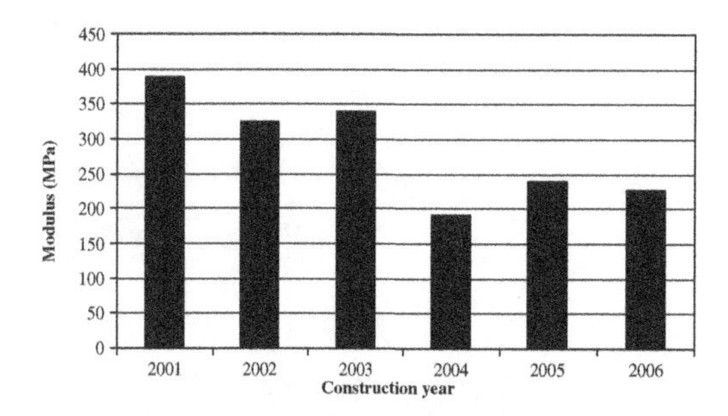

Figure 4. Elastic modulus of Slotsgrus® in relation to construction year. Based on Light Weight Deflectometer tests conducted in September 2006.

Figure 5. Roller compaction of Slotsgrus® at field test site. Photo: P. Kristoffersen, Forest & Landscape Denmark.

standard deviation of 83 MPa. The design modulus (85 percentile) is the determined as 252 MPa. The design modulus for traditional granular base in Denmark is 300 MPa, and using the Method of Equivalent thicknesses (Ullidtz (1987)), an equivalency factor of 1.06 is found, i.e. 10 mm of traditional base corresponds to 10.7 mm of Slotsgrus®. Hence, in practise the two materials are structurally equal.

The thickness of the structures tested ranged from 53 mm to 90 mm.

5 CONSTRUCTION OF A FIELD TEST SITE

To evaluate the practical constructability of the material a field tests was established. The test site had 15 3 m by 4 m cells, where material was constructed in five different layer thicknesses ranging from 50 mm to 300 mm after compaction. All cells were constructed as one-layer structures. Construction was made with a 410 kg plate vibrator and a 2740 kg roller, respectively (see Figure 5). The roller was used both with and without vibration.

Compaction control with a nuclear density gauge was conducted at five test points in all test cells after 0, 2, 4, 7 and 10 passes of either roller or plate vibrator. Table 2 shows one outcome of the work: guidelines to the relation between layer thickness and the needed number of passes with the roller or plate vibrator to obtain a degree of compaction of 95 percent standard proctor.

Table 2. Needed number of passes with the roller or plate vibrator to obtain a degree of compaction of 95 percent standard proctor.

Layer thickness (mm)	50	100	150	200	300
Roller with vibration	2	2	2	2	4
Roller without vibration	2	4	7	10	–
Plate vibrator	2	4	7	7	7

Table 3. Specifications for Slotsgrus®.

Maximum grain size	11.2 mm	
Fines	8–12% smaller than 0.063 mm	
Clay	The ratio between material passing the 0.002 mm and 0.063 mm sieves should be between 10% and 30%	
Grading	11.2 mm	100% (weight) passing
	8 mm	97–100%
	4 mm	72–82%
	2 mm	52–64%
	1 mm	40–47%
	0.5 mm	28–34%
	0.25 mm	18–23%
	0.125 mm	12–15%
	0.063 mm	8–12%
Sand equivalent	Not less than 30%	
Plasticity index	Non plastic	
Los Angeles abrasion (63 μm–4 mm)	Not higher than 35%	
Uncrushed aggregate (4–8 mm) C_r	Not higher than 20%	
Design E-value	250 MPa	
Compaction	Field test: nuclear density gauge Laboratory test: vibrating table Average degree of compaction: 95% No values less than 92%.	

The water content of the material is very important to be able to compact the material in the optimum way. The numbers for needed passes shown in Table 2 are given at water contents 2–3 percent below the optimum water content w_{opt} determined from the standard proctor test.

Compaction method has major influence on the evenness of the resulting surface of the Slotsgrus® structure. The plate vibrator gives a poorer evenness than the roller, especially when the roller is used with vibration and for structures thicker than 100 mm. The evenness of the surface shows no dependence on layer thickness up to thicknesses of 200 mm, but 300 mm thick layers seem to have a poorer evenness than the thinner structures.

6 CONCLUSIONS

The development of Slotsgrus® allows a simple and resource conscious way of designing footpaths. The material has been specified applying the concept of similar European specifications for unbound pavement materials (Table 3).

The same time the efficiency in the construction process can be improved through the use of the one-layer structure with a combined base and surface layer.

REFERENCES

Christensen, E. & K. Palmquist 1976. Frost i jord, teorier, kriterier, udstyr, resultater. IVTB Rapport nr. 4. The Technical University of Denmark. Lyngby, Denmark. (In Danish; English title: Frost in soils, theories, criteria, equipment, results).

Ullidtz, P. 1987. Pavement Analysis. Amsterdam, The Netherlands. Elsevier.

Advances in Transportation Geotechnics – Ellis, Yu, McDowell, Dawson & Thom (eds)
© 2008 Taylor & Francis Group, London, ISBN 978-0-415-47590-7

Contact stiffness affecting discrete element modelling of unbound aggregate granular assemblies

H. Huang, E. Tutumluer, Y.M.A. Hashash & J. Ghaboussi
Department of Civil and Environmental Engineering, University of Illinois, Urbana, Illinois, USA

ABSTRACT: This paper describes a preliminary study undertaken at the University of Illinois to investigate the impact of contact stiffness properties on aggregate assembly strength. An aggregate particle imaging based Discrete Element Modeling (DEM) methodology was employed to study the effects of normal and shear contact stiffnesses. Using the results of laboratory direct shear box tests on actual aggregate samples, direct shear box DEM simulations were calibrated to identify the effects of different contact stiffness combinations. It was found that as the normal to shear contact stiffness ratio increased, the shear strength of the aggregate assembly also increased. Further, the same stiffness ratio captured the shear behaviour at three different normal stress levels thus validating the DEM simulation results.

1 INTRODUCTION

Granular materials are widely used as unbound aggregate layers in the construction of various transportation infrastructure components such as highway and airport pavement base courses and railroad ballast layers. Strength, stability, and load transfer are primarily governed by inter-particle contact forces established from non-uniform spatial distributions, friction between particles governed by textural micro irregularities on the surfaces, and the shapes, sizes, and angularities of the individual particles in the granular assembly. A better insight into this load transfer through particle contacts is essential to properly model the stiffness and deformation behaviour of these constructed unbound aggregate layers.

In the past few decades, Discrete Element Modeling (DEM) has been widely utilized in many engineering practices including transportation and geotechnical engineering. It has advantages over traditional continuous modeling approaches due to its capability of dealing with discontinuities and contacts. Compared to finite element modeling of a continuum, DEM approach requires more complex calibration and validation steps which involve establishing normal and shear contact stiffness behaviour in addition to identifying strength, modulus, and Poison's ratio properties. Although contact mechanics is theoretically available (Hertz 1895; Kalker 1990; Johnson 1987) and contact stiffness for a few scenarios can be derived, contact stiffness in numerical solutions, such as a DEM simulation, has not been studied extensively because of its complexity. Recent research studies by Collop et al. (2004), Anthony et al. (2006), and Ng (2006) have all indicated that contact stiffness had a very important effect on the granular element assembly strength obtained from numerical solutions. The function of contact stiffness and its effect on strength, however, still remain to be fully investigated.

This paper describes an experimental study undertaken at the University of Illinois to explain the impact of contact stiffness on aggregate assembly strength. The objective is to investigate normal and shear contact stiffness properties by using a simplified mathematical model of the direct shear box tests. The methodology involves obtaining experimental results and then running DEM simulations of the granular assemblies tested in the shear box by utilizing a combined particle imaging and DEM approach. This image aided DEM approach, which utilized a DEM program BLOKS3D developed at the University of Illinois (Nezami et al. 2004), has been recently introduced to investigate effects of multi-scale aggregate morphological properties on performances of granular assemblies (Tutumluer et al., 2006a-b). Using results of laboratory direct shear box tests on actual aggregate samples of 3-dimensional particle sizes and shapes, direct shear box DEM simulations are calibrated to identify the effects of different contact normal and shear stiffness combinations. The respective roles played in the assembly strength and load transfer of each of the normal and shear contact stiffnesses are presented to establish guidelines for future DEM research on modeling granular assembly behaviour.

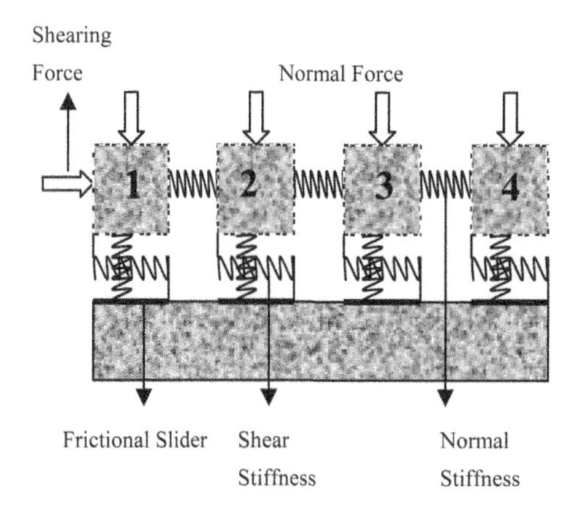

Figure 1. Simplified representation of a direct shear box test.

2 MATHEMATICAL MOTIVATION

Direct shear box was used in this study as the test procedure of choice since it is simple, reliable and widely used for particulate medium strength testing. In the shear box test, aggregate particles are pushed against each other to cause frictional resistance through aggregate interlock followed by aggregate sliding, rolling and even crushing. Therefore, interactions of individual particles and their hardness, shape, texture and angularity play significant roles in contributing to the granular assembly strength. The simplest case that one could model a shear box experiment would include few cubical blocks as quite large aggregate particles sliding on a rough surface as shown in Figure 1. The normal forces are applied on top of the four blocks indicated in Figure 1. The blocks are connected by three normal springs representing contacting blocks/aggregates in the horizontal direction. A shearing force is then applied on the first block to eventually move all four of them in the horizontal direction. It is worth noting that the contact between each block and the surface underneath consists of three parts: normal stiffness spring, shear stiffness spring, and a frictional slider. The frictional slider is a device that supplies frictional resistance between the block/aggregate and the surface. The block is allowed to only slide when the force transferred by the shear spring exceeds a certain value which usually depends on the roughness of the contact surface.

In this simplified shear model, if all four sliders start to slide at the same time, the shear reaction force will be the same no matter what the normal and shear stiffnesses are since the shear reaction force is simply the sum of the maximum friction of all four sliders. However, the slider underneath the first block in fact will be activated first which leads to the assumption made in this study as follows. As soon as the first slider allows the movement, the system of four blocks/aggregates will be in an unstable condition and the shearing force applied at that moment would be treated as the shear strength of the assembly. One can always argue that if the applied shearing force is more uniformly resisted by these four blocks (sliders), the higher strength the system will have. In other words, if the other springs deform in similar or larger amounts, overall the system will have a higher strength.

Using the simplified model and making the assumptions discussed above, the following equations can be derived to solve the behaviour of the system.

$$F_1 = K_s d_1 + (d_1 - d_2)K_n \tag{1}$$

$$F_2 = (d_1 - d_2)K_n \tag{2}$$

$$F_3 = (d_2 - d_3)K_n \tag{3}$$

$$F_4 = (d_3 - d_4)K_n \tag{4}$$

where F_1 is the horizontal shearing force applied on block 1, F_2 is the force of the normal spring between aggregate 1 and block 2, F_3 is the force of the normal spring between block 2 and block 3, F_4 is the force of the normal spring between block 3 and block 4, K_n is the normal spring stiffness, K_s is the shear spring stiffness, d_1 is the deformation of the shear spring under block 1, d_2 is the deformation of the shear spring under block 2, d_3 is the deformation of the shear spring under block 3, and finally, d_4 is the deformation of the shear spring under block 4. Also note that,

$$F_2 = K_s d_2 + (d_2 - d_3)K_n \tag{5}$$

$$F_3 = K_s d_3 + (d_3 - d_4)K_n \tag{6}$$

$$F_4 = K_s d_4 \tag{7}$$

From Equations 1 and 2, the following equation can be obtained

$$F_1 - F_2 = K_s d_1 \tag{8}$$

Also, from Equations 1 and 5,

$$F_1 - F_2 = (K_s + K_n)d_1 - (2K_n + K_s)d_2 + K_n d_3 \tag{9}$$

from Equations 2 and 3,

$$F_2 - F_3 = K_s d_2 \tag{10}$$

from Equations 2 and 6,

$$F_2 - F_3 = (K_s + K_n)d_2 - (2K_n + K_s)d_3 + K_n d_4 \tag{11}$$

and from Equations 4 and 7.

$$(d_3 - d_4)K_n = K_s d_4 \tag{12}$$

Using Equations 8 through 12, the relationship between the deformation of the first block and other three blocks are listed below:

$$d_2 = \frac{n(3n^2+4n+1)}{(2n+1)(2n^2+4n+1)} d_1 \quad (13)$$

$$d_3 = \frac{n^2}{2n^2+4n+1} d_1 \quad (14)$$

$$d_4 = \frac{n^3}{(2n+1)(2n^2+4n+1)} d_1 \quad (15)$$

where $n = K_n/K_s$.

From Equations 13, 14 and 15, one can observe that when the first shear spring's deformation is kept constant, the other shear spring deformations become larger when the ratio (n) of the contact stiffnesses becomes larger. This indicates that higher stiffness ratio yields more uniform deformation distribution among the blocks/aggregates. Hence, a higher stiffness ratio will increase the overall shear strength of granular assembly in direct shear testing. The concept and the assumptions described here of this simplified mathematical model will be further investigated for validity through DEM simulations of actual 3-dimensional aggregate particles characterized using imaging and tested in the laboratory through direct shear box testing.

3 MODELING OF SHEAR BOX TEST RESULTS

3.1 *Image aided DEM method*

Imaging technology provides detailed measurements of aggregate shape, texture and angularity properties and has been successfully used in the last two decades for quantifying aggregate morphology. A variety of imaging based aggregate morphological indices have been developed and linked to material strength and deformation properties in the recent years (Masad and Button 2000; Tutumluer et al. 2000). Among the various particle morphological indices, the flat and elongated (F&E) ratio, the angularity index (AI), and the surface texture (ST) index, all developed using University of Illinois Aggregate Image Analyzer (UIAIA), are key indices. They have been recently validated by measuring aggregate properties and successfully relating them to laboratory measured strength data and field rutting performances of both unbound aggregate and asphalt mixtures (Rao et al. 2001; Rao et al. 2002; Pan et al. 2004; Pan et al. 2005).

The UIAIA system features taking images of an individual aggregate particle from three orthogonal views to quantify imaging based F&E ratio, AI, and ST morphological indices. The image aided DEM approach then recreates the three-dimensional aggregate shapes as individual DEM elements based on the UIAIA processed top, front, and side views. This process can be easily performed using available computer aided design software and by changing the shapes of the top, front, and side aggregate 2-D images to establish representative elements with different shape properties, such as cubical, flat, flat and elongated, angular or rounded, in order to investigate effects of aggregate shape on the granular assembly strength. Figure 2 shows some representative elements generated in BLOKS3D which covers aggregates from angular to somewhat rounded shapes.

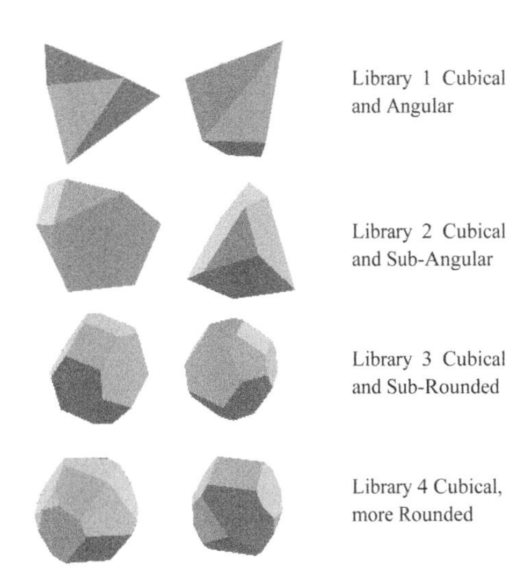

Figure 2. Some representative elements generated in BLOKS3D.

When dealing with uniform shaped aggregate particles in a granular assembly, such as by using one of the four library elements shown in Figure 2, the DEM simulations can be evaluated more effectively to investigate the effect of one input variable at a time, for example, the normal or shear contact stiffness, on the solution. No doubt different shaped DEM elements, especially textured ones, would also impact the behaviour by influencing these contact properties. In the next section, such analysis with the assessment of individual aggregate shape properties in a granular assembly will be shown to be quite useful in calibrating the DEM simulations of actual direct shear box tests.

3.2 *DEM simulation approach and its calibration*

The normal to shear spring stiffness ratio ($n = K_n/K_s$) was derived under the mathematical motivation section to control the shear strength of the granular assembly in a direct shear box test. Whether this simple concept

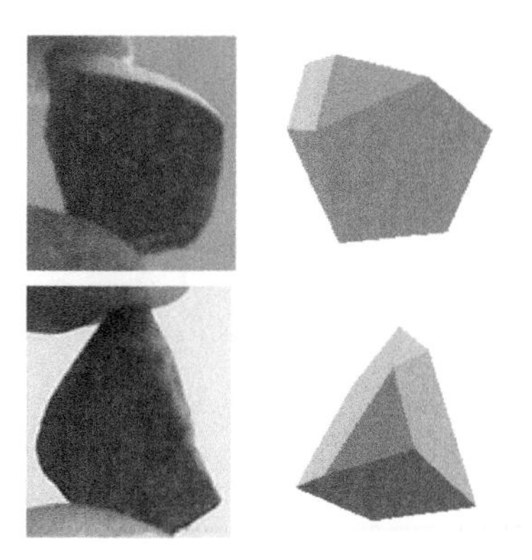

Figure 3. Comparisons of an actual tested aggregate particle with the element generated for BLOKS3D DEM program.

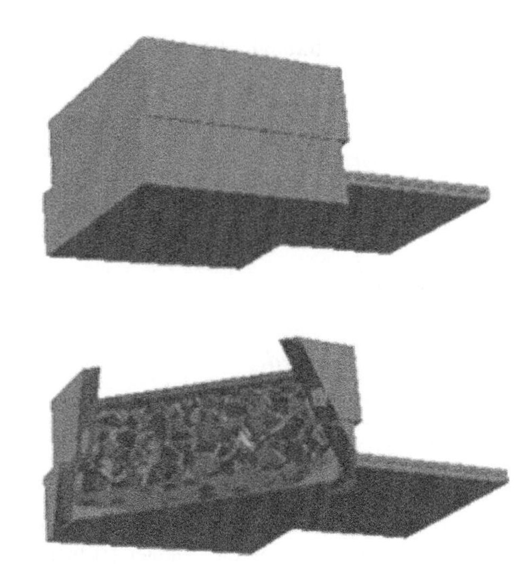

Figure 4. Direct shear box test set up using BLOKS3D.

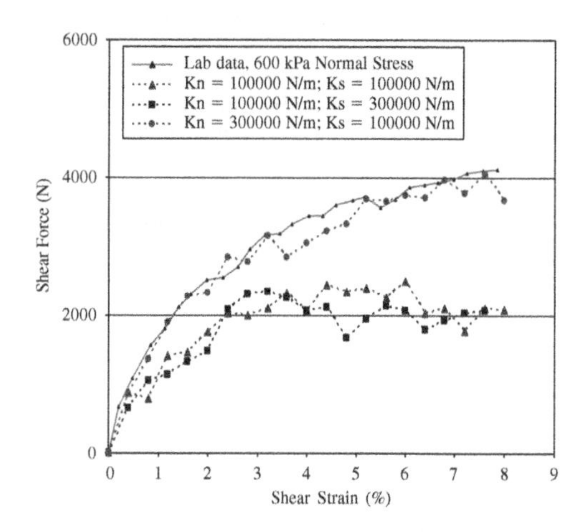

Figure 5. Calibrating contact stiffnesses in DEM simulations.

can be used to define the shear deformation behaviour is investigated in this section through BLOKS3D DEM simulations. The DEM calibration process aims to search for the best set of model input parameters that will match all the DEM simulations as the predicted deformations to the actual laboratory results.

For the calibration of the image aided DEM approach, nearly 300 pea gravel sized aggregate particles were first processed through the UIAIA and then representative DEM element shapes were created for an accurate modeling of the average shape properties in the granular assembly. Figure 3 shows the comparisons from two 2-D images between an actual aggregate particle tested in the direct shear test and the corresponding DEM representative element. In this particular case, all aggregate samples had uniform sizes ranging from 4.75 mm to 9.5 mm and were processed in UIAIA imaging equipment with average AI of 535 (sub-angular to angular) and F&E ratio of 1.4 to 1 obtained (Rao et al. 2002). This shape property is close to the particle shape in Library 2 which was chosen as the representative shape in DEM simulation.

A Humboldt HM-2560A type direct shear device, 100 mm by 100 mm square box approximately 30 mm deep, was used to perform strength tests on the UIAIA analyzed pea gravel sized aggregates. The aggregate sample was poured into the box and the normal pressure was applied with no stress conditioning. Similarly, in the DEM simulations particles were dropped in layers to completely fill the simulated shear box by using rigid blocks and the force contact equilibrium was established. Aggregate samples (at similar void ratios achieved in the tests around 0.56) were then sheared at a constant speed of 0.15 mm/sec. All tests were stopped when 9% strain was reached and shear stresses were graphed against measured shear strains. One low normal pressure of 30 kPa and two higher normal stresses of 400 kPa and 600 kPa were applied to cover typical road and railroad infrastructure normal load stress regimes. Figure 4 shows both the direct shear box set up and aggregate particles through a diagonal cut section as DEM elements inside the shear box.

The stiffness properties used in the initial DEM simulations were $K_n = 100000$ N/m and $K_s = 100000$ N/m with a stiffness ratio of $n = K_n/K_s = 1$. The simulation results from this first modeling attempt are compared with the laboratory results in Figure 5. The solid line shown in Figure 5 corresponds to the laboratory results for aggregates tested under an applied normal stress of 600 kPa. Relatively low shear forces were predicted for these stiffness properties. In the next step, only the shear stiffness was increased to 300000 N/m resulting in a stiffness ratio of $n = K_n/K_s = 1/3$. The DEM simulation results did not improve much and the

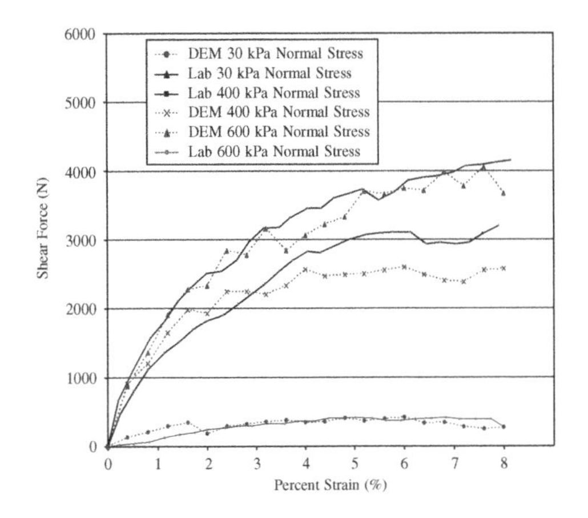

Figure 6. Validations of DEM simulation results at three different normal stress levels.

predicted shear forces were still lower than the laboratory obtained values. Finally, the normal stiffness was increased to 300000 N/m while keeping the shear stiffness value down to its low 100000 N/m, i.e., a stiffness ratio of $n = K_n/K_s = 3$. As a result of increasing the normal stiffness and thus the stiffness ratio, the predicted shear behaviour of the granular assembly was finally increased to match that of the laboratory obtained shear curve. Therefore, it is observed that normal contact stiffness is an important property in DEM simulations governing the shear behaviour and the ratio of normal to shear stiffnesses plays a critical role for determining the aggregate assembly strength. A higher stiffness ratio of $n = K_n/K_s$ clearly resulted in a definite increase in the shear strength from the DEM simulations.

Laboratory direct shear box tests were conducted for a low normal stress of 30 kPa and another high normal stress of 400 kPa under the same procedure as described previously. In addition to the 600 kPa normal stress validation, the DEM simulation results obtained by using the same stiffness ratio of 3 (Kn = 300000 N/m, Ks = 100000 N/m) are compared in Figure 6 with the laboratory results at all three normal stress levels. Although the DEM predictions for the intermediate 400 kPa normal stress did not perfectly match the laboratory results, in general, Figure 6 indicates a good match between the simulations and the laboratory results to validate the DEM approach and its appropriateness and applicability in solving contact stiffness problems in granular assemblies.

4 SUMMARY AND CONCLUSIONS

A simplified mathematical model introduced in this paper to describe normal and shear type contact stiffness behaviour in granular assemblies is validated by means of accurately modeling direct shear box aggregate test results using the Discrete Element Modeling (DEM) approach. The contact stiffness formulations derived from this simplified model reveals that the ratio between normal and shear contact stiffnesses affects the shear strength of the aggregate assembly. To test the mathematical approach for validity, DEM simulations were calibrated for the proper set of contact stiffness properties such that the DEM predictions had to match the laboratory shear box test results. It was found that as the normal to shear contact stiffness ratio also increased, the shear strength of the aggregate assembly increased. Further, the same stiffness ratio captured the shear behaviour at all three normal stress levels thus validating the DEM simulation results.

This study therefore demonstrated the important effect of contact stiffness properties, especially the normal to shear stiffness ratio, on the shearing behaviour of unbound aggregates. Either one of the normal or shear stiffness value is needed to calibrate the contact stiffness properties for DEM simulations. For this purpose, modulus tests should be conducted in the laboratory to determine modulus and deformation behaviour of aggregate materials.

REFERENCES

Antony, S.J., Moreno-Atanasio, R., & Hassanpour, A. 2006. "Influence of contact stiffnesses on the micromechanical characteristics of dense particulate systems subjected to shearing," *Applied Physics Letters*, Vol. 89.

Collop, A.C., McDowell, G. R., & Lee, Y. 2004. "Use of the Distinct Element Method to model the deformation behaviour of an idealized asphalt mixture," *The International Journal of Pavement Engineering*, Vol. 5(1), pp. 1–7.

Hertz, H. 1895. Ueber die Beruehrung elastischer Koerper (On Contact between Elastic Bodies), in *Gesammelte Werke (Collected Works)*, Vol. 1, Leipzig, Germany.

Johnson, K.L. 1987. Contact Mechanics, *Cambridge University Press*.

Kalker, J.J. 1990. Three-dimensional elastic bodies in rolling contact, *Kluwer Academic Publishers*, Dordrecht/Boston/London.

Masad, E. & Button, J.W. 2000. "Unified imaging approach for measuring aggregate angularity and texture," The *International Journal of Computer-Aided Civil and Infrastructure Engineering – Advanced Computer Technologies in Transportation Engineering*, Vol. 15, No. 4, pp. 273–280.

Ng,T.T. 2006. "Input parameters of Discrete Element Method," *ASCE Journal of Engineering Mechanics*, Vol. 132, pp. 723–729.

Pan, T., Tutumluer, E., & Carpenter, S.H. 2004. "Imaging based evaluation of coarse aggregate used in the NCAT pavement test track asphalt mixes," Proceedings of the *1st International Conference on Design and Construction of Long Lasting Asphalt Pavements*. Auburn, Alabama.

Pan, T., Tutumluer, E., & Carpenter, S.H. 2005. "Effect of coarse aggregate morphology on resilient modulus of hot-mix asphalt," In Transportation Research Record No. 1929, *Journal of the Transportation Research Board*, National Research Council, Washington, D.C.

Rao, C., Tutumluer, E., & Stefanski, J.A. 2001. "Flat and elongated ratios and gradation of coarse aggregates using a new image analyzer," ASTM *Journal of Testing and Standard*, Vol. 29, No. 5, pp. 79–89.

Rao, C., Tutumluer, E., & Kim, I.T. 2002. "Quantification of coarse aggregate angularity based on image analysis," Transportation Research Record No. 1787, *Journal of the Transportation Research Board*, National Research Council, Washington, D.C.

Tutumluer, E., Rao, C., & Stefanski, J.A. 2000. Video image analysis of aggregates, Final Project Report, FHWA-IL-UI-278, Civil Engineering Studies UILU-ENG-2000-2015, University of Illinois Urbana-Champaign, Urbana, Illinois.

Tutumluer, E., Huang, H., Hashash, Y., & Ghaboussi, J., 2006a. "Imaging based discrete element modeling of granular assemblies," In Proceedings of the Multiscale and Functionally Graded Materials Conference 2006 (FGM2006), Honolulu-Oahu, Hawaii, October 15–18.

Tutumluer, E., Huang, H., Hashash, Y., & Ghaboussi, J., 2006b. "Aggregate shape effects on ballast tamping and railroad track lateral stability," In Proceedings of the 2006 AREMA Annual Conference, Louisville, Kentucky September 17–20.

Advances in Transportation Geotechnics – Ellis, Yu, McDowell, Dawson & Thom (eds)
 ISBN 978-0-415-47590-7

NordFoU – Pavement performance models. Part 2: Project level, unbound material

A. Huvstig
SRA, Gothenburg, Sweden

S. Erlingsson
VTI, Linköping, Sweden

I. Hoff
SINTEF, Trondheim, Norway

R.G. Saba
NPRA, Trondheim, Norway

ABSTRACT: The NordFoU Pavement Performance Models project is a research project that is financed by the Nordic countries. This is an implementation project and the main objectives of this project are to:

Implement recent research results to test methods and models that should be used of normal engineers in the road design work to predict future performance.

The work consists of adapting existing performance prediction models, utilizing data from test and reference sections, improving the test methods and models and stimulates the development of expertise in the field of performance modelling (prediction of condition) for road structures.

The project aims to deliver a practical tool for calculating the future performance of pavements, which is much more easy to use, and probably gives better predictions, than the US "Design Guide",

In this work, some models for prediction of unbound materials have been evaluated in order to choose the most suitable models for unbound material in the Nordic roads.

1 BACKGROUND

A newly published investigation by Chalmers University of Technology in Sweden indicates that extra costs resulting from error and unnecessary works in the civil works industry amounts to more than 25% of the construction cost. There is every reason to believe that these costs are not any lower in the case of road construction. Developments in technology during the last 20 years have had a minimum impact on reducing costs in the civil works industry, while the mechanical industry has improved productivity and quality by almost 100% during this time.

In light of this, there is a great potential for savings, probably at least 30% of the production costs, over a five-year period. For the Nordic countries, this means possible savings of 100 to 150 million Euros every year.

- The most important means, for achieving these savings is access to contractors, consultants and clients with competent and motivated employees. In order to finance extension courses and to create motivation, what is needed are economic incentives related to lower Life Cycle Costs (LCC) that result from higher quality. The important key to all this is being able to predict the future performance of a road. Otherwise it is impossible to calculate LCC.
- It must also be possible to predict the future performance, if choosing to use performance based contracts, and not have to pay the contractor 10–20% extra for the risk involved. This means that test methods and prediction models for future performance are the most essential component in making these savings possible.
- As regards technical development, calculation instruments to predict future performance are necessary as it makes it possible to compare different innovative construction solutions and alternative or new materials.
- All these facts point out that the basic most important factor is that the prediction of future

performance is necessary if it is going to be possible to minimise the cost for new or rehabilitated road constructions.

If these models and test methods are going to be broadly used, an accepted implementation plan is necessary in each country. Experiences in Sweden indicate that the following measures provide good results:

- Contractors, road managers and consultants' staff must have knowledge, expertise, interest and motivation required for designing and building roads in the best way.
- Supplementary courses are necessary to improve knowledge and expertise.
- Economic incentives are necessary for motivation and also to finance extension courses.
- Co-operation, through the "win – win" concept, is of good help in developing motivation, interest and expertise.
- Work on "Active Design" at construction sites is a very good method for developing motivation, interest, expertise and economic incentives.
- Work involving the specified implementation of new technology in connection with real projects is very helpful in order to develop expertise and motivation.
- Work involving performance requirements promotes creativity and motivation.
- Work on minimising LCC is necessary if it is to be possible to choose the best quality materials and best means of execution the work. Minimising LCC is also the best way to obtain economic incentives. (It is not possible to minimise LCC and predict future performance, if there are no relevant test methods and models for this.)

2 THE PROJECT

The complete background to this project is described in the NordFoU Project: "Pavement Performance Models" and the paper entitled "Performance Prediction Models for Flexible Pavements: A State-of-the-art Report", Garba (2006).

An important initial phase of the work is to investigate existing prediction models and test methods for the prediction of future rutting, roughness and cracks on a road surface. This work does not include the development of new models and test methods.

The main objectives are to choose the best test methods and prediction models for future performance and make it possible to adapt these to the Nordic countries through user-friendly programs that can be combined in an easy and flexible way. It should be possible to combine the response model in a country with all the chosen and validated performance models. The ultimate goal is that these models and test methods should be broadly used in order to minimise LCC for new and rehabilitated roads through successful implementation.

When this project is finished, there should exist:

- One validated model and test methods for prediction of the elastic response in a road with consideration of nonlinearity in the unbound materials and the weight of the material.
- Validated and recommended models and test methods for prediction of future rutting.
- Validated and recommended models and test methods for prediction of the future roughness.
- Validated and recommended models and test methods for prediction of the future cracking in the asphalt layers.
- Recommendation of possible connections between the project model and a network model.
- Written documentation and Power Point material for education.
- Recommended connections from the predicted future functions to calculation of LCC for a certain project.
- Recommended economic incentives for a better quality, in order to improve the motivation.

When planning this project, it was decided that separate models for prediction of response and the different performance models should be used. One reason for this was that the individual countries should be able to use their own response model composed of suitable performance models. Another reason is based on the understanding and educational point of view. There are very few engineers working on road construction design in the Nordic countries. As regards the implementation, what is needed are extension courses for road design engineers, who must understand the different steps in the design process.

Over the years, response models for unbound and bituminous bound materials have been developed. It is possible today to choose highly sophisticated models, which can be used in the case of unbound materials to calculate response, taking into account the effect of nonlinearity, shear stress, shear and compression modulus and anisotropy. For bituminous bound materials, the models take into account the effect of viscosity, nonlinearity, temperature etc.

The response model should be easy to handle and entail a relatively short calculation time. One reason for this is past experiences when using "Design Guide" (which takes too long time to run). Another reason is that it must be possible to compare alternative design solutions.

3 PLATFORM FOR THE PROJECT

As a basis for the development of a system for prediction of future performance of a road project, research results from "Design Guide" (2004), SAMARIS (2006) or Gidel (2001) and other research projects were used, as presented in the "State-of-the-Art Report". Results from a Swedish ongoing project "Active Design", which is a joint research project run by SRA and SBUF (the Swedish construction industry's organisation for research and development), will be used in this project. Individual parts of the MMOPP (2007) model for prediction of roughness and rutting, etc developed in Denmark will also be validated through this project.

Knowledge and experience from the EN 13286-7:2004 Standard, "Cyclic load triaxial test for unbound mixtures", and the corresponding EN Standard for bituminous mixtures will also be used. The standard for unbound base materials describes how to evaluate triaxial tests results.

- The standard divides materials into different quality classes. These classes could be used as quality criteria for unbound material as regard how sensitive the material is to rutting.
- This standard also contains a method to calculate the "shake down" limit. This limit is not a distinct stress level where there is a change between decreasing permanent deformations for increasing load cycles (A) to increasing permanent deformations for increasing load cycles (C). The stress state between these two limits could be described as a transitional intermediate state, where the permanent deformations for increasing load cycles are more or less constant (B). These three zones (A, B and C) are described in the EN standard.

All research results from the Nordic countries should be used in this project, especially the recent developed of material models and methods for prediction of future performance.

Based on recent research results, a few programs with a particular structure and connections have already been developed and financed by SRA. These programs, and the knowledge related to them are going to be used as the basis for this project.

The structure of the programs consists of one finite element program to calculate of the elastic response, VagFEM, six Excel programs to calculate of permanent response in different layers of the road and two Excel programs to calculate of parameters from triaxial tests. Another platform for this program is MMOPP, which is described in the "State-of-the-art Report".

The Swedish "VagFEM" is a 3D FEM program designed to be simple and give a short user time, less than twenty minutes.

"VagFEM" may be briefly described as follows. The input data comprise road geometry, thickness of layers (see figure 1), position of loading, elasticity modulus (and viscosity) for the bituminous bound layers and linear elastic or nonlinear elastic resilient

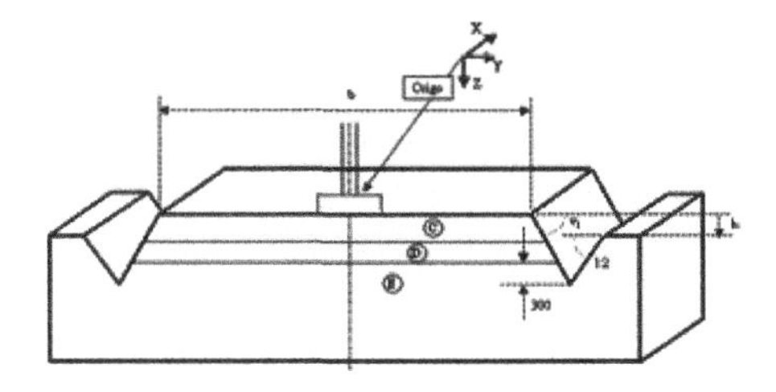

Figure 1. Road geometry in VagFEM.

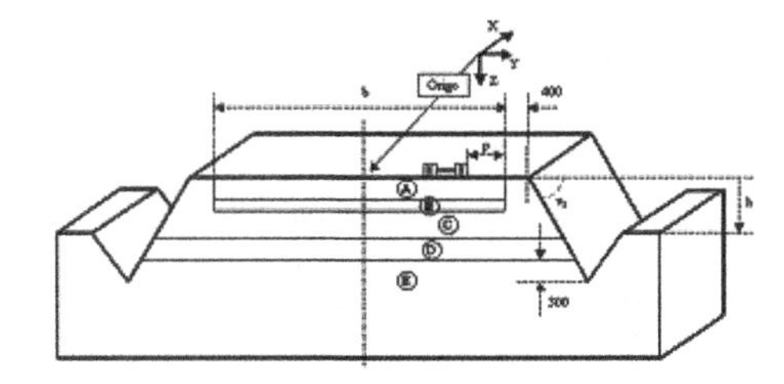

Figure 2. Geometry for the plate loading test in VagFEM.

modulus ($M_r = K_1 \Theta^{K2}$) for the unbound layers. The weight of the road material is included in the model.

VagFEM is built on modules from ABAQUS, which is also the program that carries out the calculations.

The output data comprises graphs of deformation, stresses and strains in different parts of the road structure.

"VagFEM" also includes a module for calculating step by step the deflection during plate loading tests, (see figure 2). These calculated deflections are compared with the actual deflections under the plate loading. In a kind of back calculation, the parameters K_1 and K_2 are changed in one or two steps, so the calculated deflections are similar to the measured plate loading deflections for all loads.

The purpose is to make it possible to calculate the resilient modulus for the unbound layers. To this end, the resilient modulus of the subgrade needs to be ascertained, which involves plate loading tests at the top of the subgrade and at the top of the base layer. The calculated values of the resilient modulus for these layers become input data for calculating the next layer.

The output data from the calculations in this module comprise the deflection beside the plate (see figure 3) and a graph for the various load steps under the plate (see figure 4). One possibility is to compare results from FWD with the curve in figure 4 in order to estimate the dynamic resilient modulus of the unbound layers.

4 SIMPLIFICATION OF MODELS

After an international workshop that took place on the 31st of May to 1st of June 2007, on Gullholmen

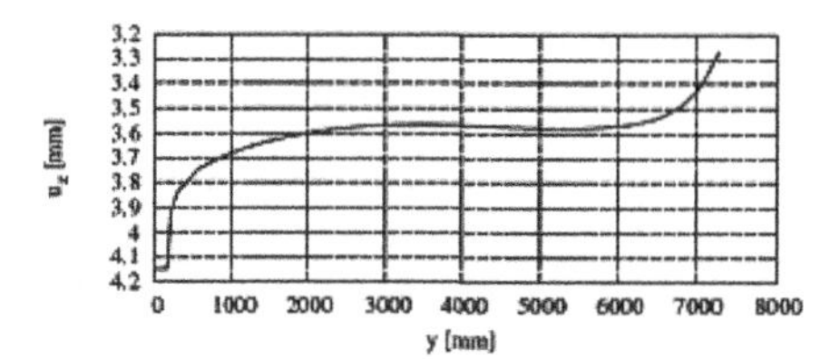

Figure 3. Deflection beside the plate at full load.

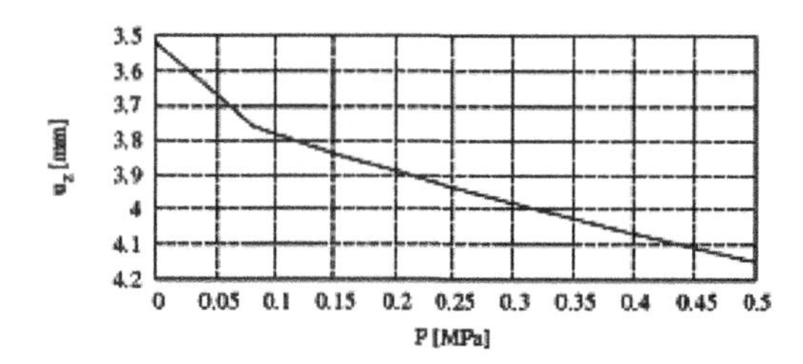

Figure 4. Deflection under the plate at the various load steps.

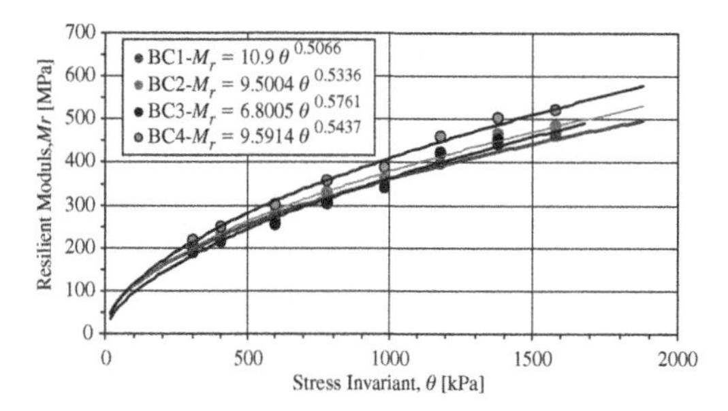

Figure 5. Calculated values on resilient modulus for unbound base material, HVS test north of Uddevalla. Erlingsson (2003).

in Sweden, the following was decided in order to simplify the calculations and still obtain a good prediction of the real stress and strain situation in a road construction.

- The nonlinearity of the unbound material and The weight of the pavement plus base materials should be included in the calculations of future performance. The reason for this is that these factors have a large influence on the stress and strain in the road structure, (see figure 5). Elastic stress and strain are used to calculate permanent deformations and roughness etc.
- The model for the nonlinearity of unbound material could be $M_r = K_1 \Theta^{K2}$. It is not necessary to include the influence of shear stress, $M_r = K_1 \Theta^{K2}(\tau + 1)^{K3}$, on a crushed base material in a simplified model. The effect of shear stress is more important on fine graded uncrushed material.
- In order to include the soil weight and nonlinearity factors, in a system for the prediction of future rutting, the calculation model should use a finite element program (FEM) to predict more realistic stress and strain.
- The model could be simplified to calculate rutting only in the vertical dimension, and in thin layers directly under the wheel load.
- For permanent deformations the influence from the lateral wandering of the wheel load could be estimated as a reduction factor. This reduction factor should be estimated from calculations and empirical research results.
- The simplest way to predict future performance is to use separate models for prediction of response and performance.
- It is not possible to use the internal angle of friction instead of the "shake down" limit. The internal angle of friction is higher because it is a strength property while the "shake down" load is a fatigue property.
- The influence of moisture on the unbound base layers could be estimated using the triaxial tests. The degree of moisture in the unbound materials could be measured in real roads on site, in order to obtain experience for choosing empirical values (as in USA).
- The influence of temperature on the elasticity modulus for bituminous bound layers should be estimated using "Master Curve" and estimated temperatures from site measurements.
- The reduction of the elasticity modulus for bituminous bound layers due to fatigue and the increase due to healing and ageing could be estimated from tests (HVS) and experiences from LTTP roads. A summation of these changes is difficult to make. One simplification could be to assume that the sum of these changes is zero.
- The best predictions are made using month-by-month incremental calculations of the different module for different layers in a road construction. This entails a great number of calculations, which could be reduced through using a summation of permanent deformations for the amount of heavy traffic on a road structure in a number of situations with different elasticity modules in the different layers.
- One possible simplification could be to classify unbound material according to their elastic properties and permanent deformation behavior (a first example is made by LCPC). One result from such a study was that certain materials have the same elastic properties but different permanent deformation behaviour. A conclusion drawn from this is that

it is not possible to predict the rutting only from the elastic properties of the material in a simplified model.

5 CHOSEN MODELS AND TEST METHODS

To meet these demands, VagFEM, an existing interface for the ABAQUS model, was proposed (see chapt 4).

The performance models could be divided into prediction models for rutting, cracks and roughness.

- For the prediction of rutting due to permanent deformations in bituminous bound and unbound materials, models developed through the Danish Road Directorate, MMOPP, Design Guide and the extensive research projects SAMARIS have been included as separate Excel programs in VagFEM. Also the Dresden model for unbound material, Werkmeister (2003) has been included as a separate Excel program.
- There already are models developed and used in the Nordic countries to predict rutting caused by wear from studded tires.
- For prediction of roughness, the MMOPP model could probably be used if it can be handled as a separate program.
- For prediction of cracking, a new model developed in Florida, should be included (a separate project)

The SAMARIS project has recommended a few laboratory test methods. Most of these are also recommended in the "Design Guide".

- It is recommended that tests for the prediction of rutting in unbound materials should be done in the laboratory using the triaxial test.
- For bituminous bound material the triaxial test is recommended for the prediction of rutting. "Wheel Track" could also possibly be used as a test method if it proves possible to back-calculate the results into some useful parameters.
- For the prediction of cracks in the bitumen, the tensile test and ageing test are recommended.

Some common test methods for site control are:

- Static plate loading.
- Dynamic plate loading.
- Troxler.
- Falling Weight Deflectometer (FWD).
- Tests on drilled out cores from the bituminous bound material.

The tests on sites that could provide useful input data for validation of the test methods and models will also be looked into. One test method that should be studied more thoroughly in this project is whether it would be possible to use the measurement equipment on compactors to calculate the resilient modulus and variations in different characteristics of the unbound layers.

Triaxial tests on bituminous bound material could be used to certify the asphalt recipes used by a specific contractor. This could result in a higher value for one particular asphalt due to improved quality and causing lower LCC. The improved quality should be verified through indirect tensile tests on cores drilled out on site.

All results from triaxial tests on unbound base materials, and perhaps subgrade materials, should be collected into a Nordic database. This would build up a knowledge database and make it unnecessary to test unbound materials in connection with smaller projects, where it should be possible to estimate values on the input data in the deterioration model.

6 INVESTIGATED MODELS FOR UNBOUND MATERIAL

The following material model has been investigated in connection to this project.

It is not possible to validate all models in this project, which is the main reason for only choosing three material models to be included in the project.

A statistical model from Design Guide (2004)

The model, used in Design Guide, NCHRP (2004) for the prediction of permanent deformations in the unbound layers is based on a model that is suggested by Tseng and Lytton, see below:

$$\delta_a(N) = \beta_{GB}\left(\frac{\varepsilon_0}{\varepsilon_r}\right)e^{-\left(\frac{\rho}{N}\right)^{\beta}}\varepsilon_v h$$

Where:
δ_a = permanent deformations for the layer (inch)
N = the amount of load cycles (passes)
β_{GB} = national calibration factor (=1,673)
$\varepsilon_0, \beta, \rho$ = material properties
ε_r = resilient strain from laboratory tests
h = layer thickness
ε_v = average value of the vertical resilient strain in the layer that is calculated in the primary response model.

The models in the Design Guide do not normally take the material characteristics from the triaxial tests into consideration. The values of the material properties are calculated as average values with help of statistics from rutting in real roads, water content and input data from response calculations like the actual stress and resilient modulus. The parameters are therefore calculated with help of the following equations:

Granular base

$$\log\left(\frac{\varepsilon_0}{\varepsilon_r}\right) = 0.80978 - 0.06626 \cdot W_c - 0.003077 \cdot \sigma_\theta + 0.000003 \cdot E_r$$

$$\log(\beta) = -0.9190 + 0.03105 \cdot W_c + 0.001806 \cdot \sigma_\theta - 0.0000015 \cdot E_r$$

$$\log(\rho) = -1.78667 + 1.45062 \cdot W_c + 0.0003784 \cdot \sigma_\theta^2 - 0.002074 \cdot W_c^2 \sigma_\theta - 0.0000105 \cdot E_r$$

Subgrade

$$\log\left(\frac{\varepsilon_0}{\varepsilon_r}\right) = -1.69867 + 0.09121 \cdot W_c - 0.11921 \cdot \sigma_d + 0.91219 \cdot \log E_r$$

$$\log(\beta) = -0.9730 - 0.0000278 \cdot W_c^2 \sigma_d + 0.017165 \cdot \sigma_d - 0.0000338 \cdot W_c^2 \sigma_\theta$$

$$\log(\rho) = 11.009 + 0.00068 \cdot W_c^2 \sigma_d - 0.40260 \cdot \sigma_d + 0.0000545 \cdot W_c^2 \sigma_\theta$$

This model should be validated in the project.

The MMOPP model

Some rather similar models have been developed in several countries. Those models are grounded on two important factors, which cause permanent deformation in unbound materials, the number of load repetitions and the stress level. These factors are included in the Danish model MMOPP:

$$\varepsilon_p = AN^B \left(\frac{\sigma_1}{\sigma'}\right)^C \quad (\text{for } \varepsilon_p < \varepsilon_0) \quad \text{primary creep,}$$

decreasing strain rate.

$$\varepsilon_p = \varepsilon_0 + (N - N_0) A^{\frac{1}{B}} B \varepsilon_0^{1-\frac{1}{B}} \left(\frac{\sigma_1}{\sigma'}\right)^{\frac{C}{B}} \quad (\text{for } \varepsilon_p > \varepsilon_0)$$

secondary creep, constant strain rate,
in which:

$$N_0 = \varepsilon_0^{\frac{1}{B}} A^{\frac{-1}{B}} \left(\frac{\sigma_1}{\sigma'}\right)^{\frac{-C}{B}}$$

Where:

ε_p = the plastic strain
N = the number of load repetitions
$\sigma 1$ = the major principal (vertical) stress
σ' = the reference stress (atmospheric pressure, 0,1 Mpa)
A,B,C = calibration constants

This model should be validated in the project.

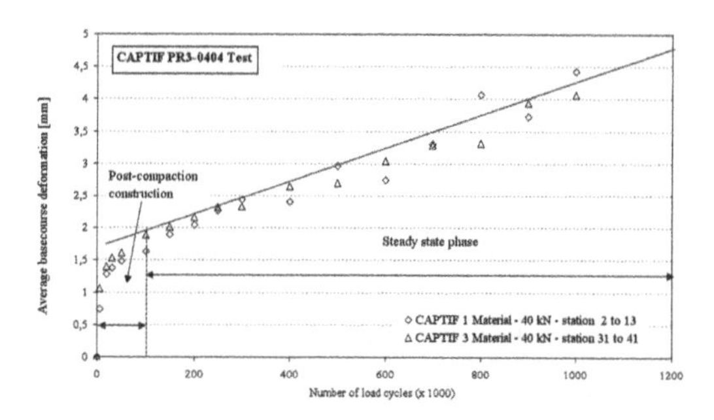

Figure 6. Number of average base course deformations versus number of load cycles. S. Werkmeister.

The Dresden/Huurman model

The plastic Dresden model Werkmeister (2003) is grounded on the so called "Shakedown" concept, which is based on that there exist a critical stress level, under which the material is stable. The model was first developed by Martin Huurman in Delft and further developed by S. Werkmeister in Dresden.

The Dresden model looks as follows:

$$\varepsilon_p(N) = A\left(\frac{N}{1000}\right)^B + C\left(e^{D\frac{N}{1000}} - 1\right)$$

Where:

ε_p = Permanent strain [-]
e = Base for the natural logarithm [-]
N = The number of load cycles
A, B, C, D = Stress dependant (σ_1, σ_3) model parameters

The plastic Dresden model describes the permanent strain as a function of the stress and the number of load repetitions. The material parameters (*A, B, C, D*) should be calculated through regression analysis from the results from the triaxial tests.

One limitation of the Dresden model is that today, it is impossible to calculate the parameters C and D as functions of the stress, which is the reason for not validating this model.

A simplified Dresden model

In connection to a large test program in New Zealand, S. Werkmeister has developed a simple model for the prediction of permanent deformations in unbound materials. This model has given a good prediction for material in both range A and B ("Shake Down" limits, which is described in code EN 13286-7:2004).

The plastic base course deformation of the pavements could be separated into two parts, an initial rapid post-construction compaction and the steady state phase (see figure 6). The steady state phase could be deemed to start when the change in plastic deformation accumulation becomes linear with respect to the

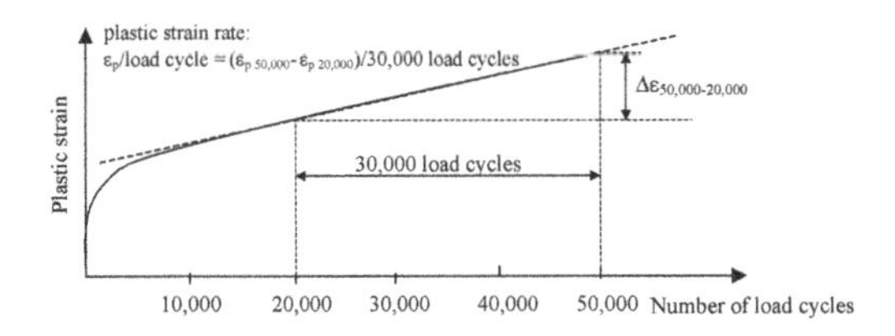

Figure 7. Determination of plastic strain rate value. S. Werkmeister.

number of load cycles and the elastic strains becomes constant. The post-construction compaction period was in this test completed after 25,000 to 100,000 load cycles when the elastic strains become constant (no shear movement).

This investigation is based on field results and triaxial test results. The model uses the axial elastic strain to predict the axial plastic strain rate per load cycle. The relationship has been developed using laboratory and field results, applied to the axial elastic strains calculated with a FEM program, and integrated over the depth of the base course layer and the number of load cycles in the tests to determine the plastic deformation (rut depth) occurring in the base course. The advantage of this simplified method to predict the rut depth of the base course is that elastic FEM calculation results (elastic strain distribution in the wheel path) are required only. The elastic FEM calculation process is less complicated and time consuming compared to plastic FE calculations.

The following exponential relationship between the elastic strain (ε_{el}) and plastic strain rate ($\dot{\varepsilon}_p$) can be determined as long as the shear stresses within the base course are sufficiently small and the material behaviour corresponds to either Range A or B. Excluded from the fit were the tests that failed prior to 50,000 load cycles, which was usually the highest stress level in each Multi-Stage (MS) test. Results where failure occurs do not follow the same trend as the other results due to significantly larger deformations/shear failure (Range C behaviour) and this mechanism of accumulation of plastic strain is different from the other test results.

$$\dot{\varepsilon}_p = E \cdot \varepsilon_e^{F}$$

Where:
$\dot{\varepsilon}_p$ [10^{-3}/cycle] Major principal plastic strain rate,
ε_{el} [10^{-3}] Major principal elastic strain,
E, F [-] Material parameters.
The parameters E and F are mainly dependant on the material, moisture content and degree of compaction and were determined for the test material in New Zealand.

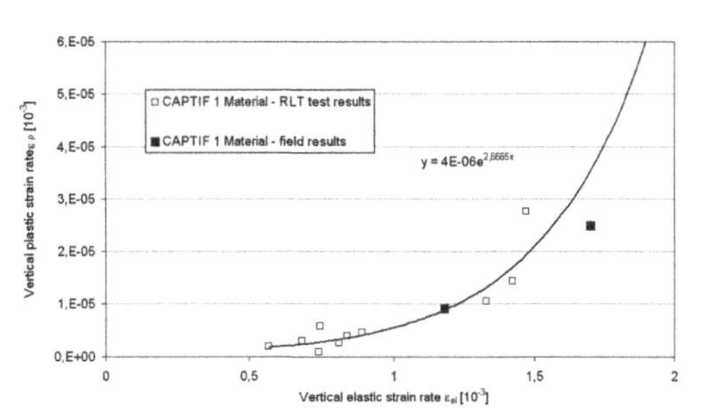

Figure 8. Axial elastic strain versus plastic strain rate for one test material, Triaxial test results and results from the test section in New Zealand. S. Werkmeister.

Figure 8 show the relationship between axial elastic strain and axial plastic strain rate per load cycle on a (ε_{el}) vs. ($\dot{\varepsilon}_p$) plot.

Only one type of material has been tested in this way. Experience from France indicates that two materials, with rather different elastic properties, could give the same permanent deformations in a road at the same loading, which is one reason for not validating this model.

Permanent deformations in the primary phase

In connection to the test program in New Zealand, S Werkmeister also developed a model in order to determine the permanent deformations in the primary phase (during the post compaction). Test data from this investigation were used to determine the post-construction compaction and to determine a relationship between the axial plastic strain rate during post-construction compaction and during the steady state (secondary) phase. The results showed that the number of load repetitions until completion of post-construction compaction is dependant on the elastic strains. In the investigation, a relationship between the axial elastic strains (ε_{el}) and the number of load cycles (N_{pc}) for completion of post-construction compaction could be determined.

$$N_{ep} = 211{,}71 \cdot \varepsilon_e^{0{,}8232}$$

Where:
N_{ep} [-] Number of load repetitions for completion of post-construction compaction.
ε_e [10^{-6}] Axial elastic strain.

The following exponential relationship between the plastic strain rate during the steady state phase ($\dot{\varepsilon}_p$) and the plastic strain rate during post-construction compaction ($\dot{\varepsilon}_{p\,pc}$) was found in this investigation for the unbound materials in this project:

$$R = 0.0042 \cdot \dot{\varepsilon}_p^{-0.6869}$$

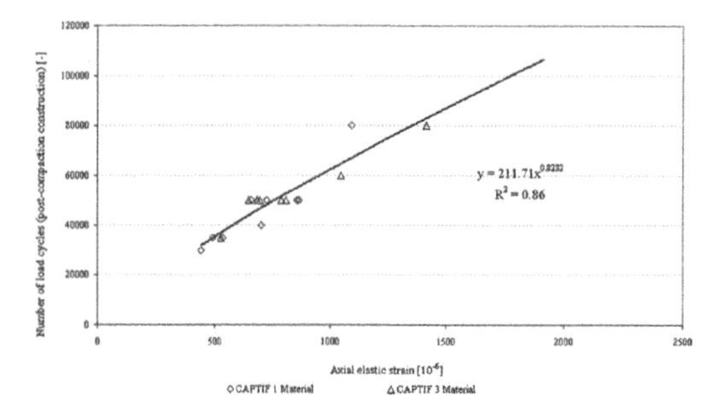

Figure 9. Axial elastic strain versus number of load cycles until post-construction compaction is completed. S. Werkmeister.

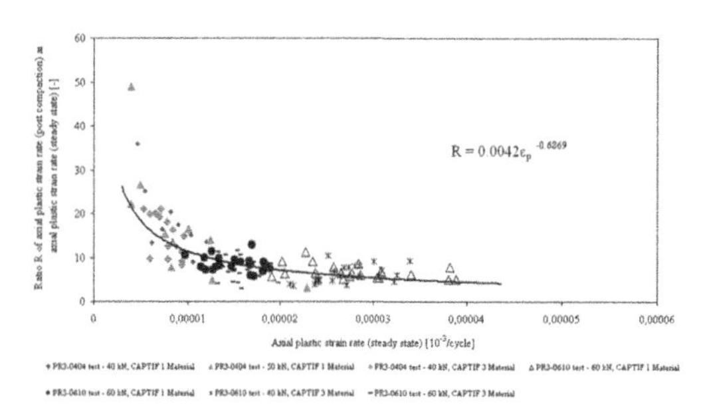

Figure 10. Axial plastic strain rate (steady state) versus the ratio of axial plastic strain rate (post-construction compaction) and axial plastic strain rate (steady state). S. Werkmeister.

Where:

$\dot{\varepsilon}_p$ [10^{-3}/cycle] Plastic strain rate per load cycle during steady state phase,

R [-] Ratio between plastic strain rate (post-construction compaction) and plastic strain rate (steady state).

This model should not be validated in this project. It is difficult to find validation data.

The LCPC (SAMARIS) model

LCPC has developed "A new approach for investigating the permanent deformation behaviour of unbound granular material using the repeated load triaxial apparatus", Gidel (2001).

The research is based on a lot of triaxial test on specimen from different unbound materials.

In the research, they found that stress, mineralogical nature of the material and moisture content has a large influence on the permanent strains, (see also figure 11–12). Further they found out that the final permanent deformation was not much influenced of if the tests were done in several loading steps or in one step (see figure 13 a and b).

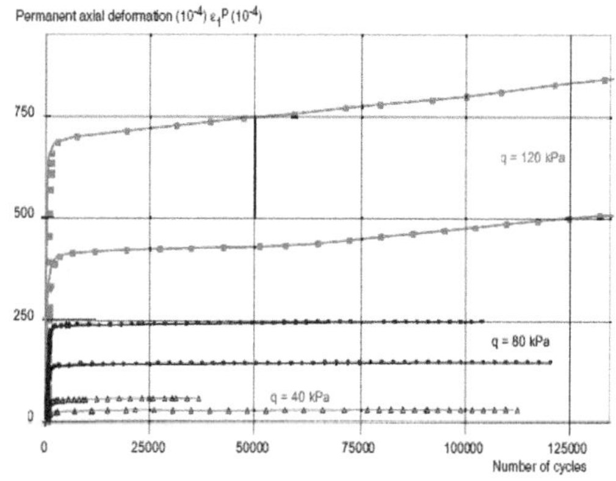

Figure 11. Influence of stress on permanent axial deformation.

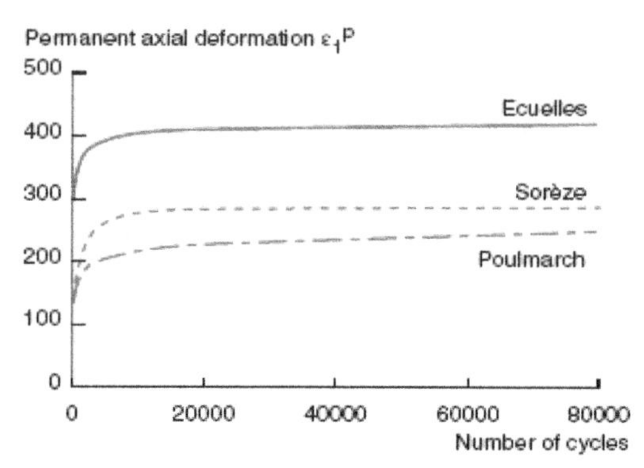

Figure 12. Permanent deformation for material of different quality.

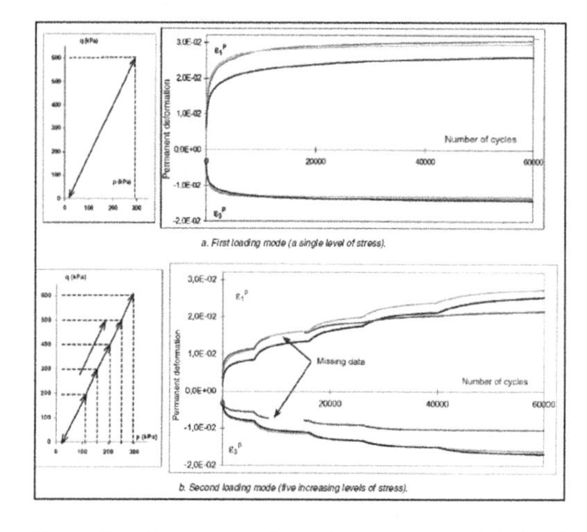

Figure 13. Permanent deformation with a. one single level of stress and b. five increasing levels of stress.

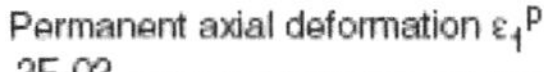

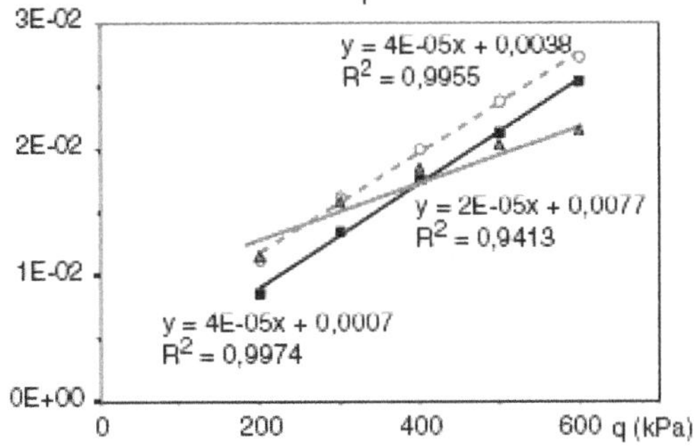

Figure 14. Permanent axial deformation at the end of a loading stage, plotted against q.

The final deformation of the specimen did not seem to be significantly affected by the preceding loading stages.

Finally it was found out that the permanent axial deformation at the end of a loading stage, varied almost linearly with q (or p if q/p was constant).

From these results an empirical deformation model was developed:

$$\varepsilon_1^p(N) = \varepsilon_1^{p0}[1-N]^{-B} \cdot \left(\frac{L_{max}}{p_a}\right)^n \cdot \frac{1}{m + \frac{s}{p_{max}} - \frac{q_{max}}{p_{max}}}$$

Where:

$\varepsilon_1{}^p$: permanent axial strain;
N: number of load cycles;
p_{max}, q_{max}: maximum values of the mean normal stress p and deviatoric stress q;

$$L_{max} = \sqrt{p_{max}{}^2 + q_{max}{}^2}$$

p_a: reference pressure equal to 100 kPa;
$\varepsilon_1{}^{p0}$, B, n model parameters;
m,s parameters of the failure line of the material, of equation q = m.p + s; (from experience, m = 2.5 to 2.6 and s = 20 kPa)

This model should be validated in the project.

7 MODELS FOR BITUMINOUS BOUND MATERIAL

The statistical model from Design Guide

$$\varepsilon_p / \varepsilon_r = a_1 \cdot N^{a2} \cdot T^{a3}$$

Where;
ε_p = Accumulated plastic strain at N repetitions of load
ε_r = Resilient strain of the asphalt material
N = Number of load repetitions
T = Temperature (10°C)
a_i = Non-linear regression coefficients (from NCHRP 1-37A)

The MMOPP model

This model is the same as for unbound material (see chapter 6).

8 VALIDATION OF MODELS AND TEST METHODS

The models and test methods will be validated to test road E6 bypass Falkenberg, HVS tests in Sweden, tests in LCPC (France) made in connection to SAMARIS project the MnROAD project and LTTP roads that have been under traffic for 5–15 years in Sweden, Norway and Denmark.

It should be of great interest to cooperate with other research programs, for example SAMARIS, and the partners in this project.

9 USEFUL NEW DEVELOPMENT IN CONNECTION TO THIS PROJECT

In addition to NordFoU, there are a lot of ongoing research projects in the Nordic countries. It is of great importance that all this research should be done in cooperation and with exchange of experience in order to achieve the same goals. There also could be cooperation with other countries. Such cooperation will also speed up the individual research projects and avoid unnecessary duplication of work.

A common plan for desirable research will also lead to ensuring that no important parts are missing, and that duplication of work is minimized.

Areas described in this chapter are important questions for a further development of deterioration models in order to predict future performance.

Special questions that ought to be solved concerning unbound material

a. Many response models produce the result that there are tensile stresses at the bottom of an unbound material layer, especially if the subgrade has a low resilient modulus (for example clay). Unbound materials cannot carry tensile stresses. Calculations using the VagFEM program with a material model from ABAQUS, which eliminates the tensile stresses in the calculation, becomes unstable, because the elements are moving away from each other. This is probably due to the geometry comprising slopes and ditches. The use of tensile stresses

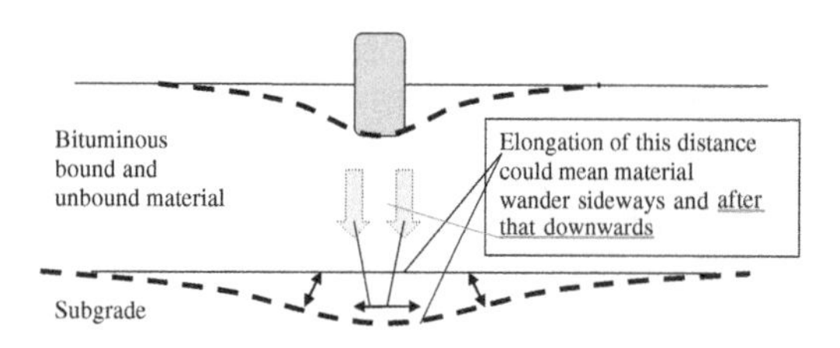

Figure 15. Elastic strain in the subgrade could be one reason for permanent deformation.

in a performance model also gives misleading values for future rutting. This could perhaps be solved by giving the tensile stresses the value of zero in a suitable way in the Excel programs that calculate future performance.

b. From previous experience it is known that unbound material with a high resilient modulus should not be used directly on a material (subgrade) with very low resilient modulus. The theories above also indicate this. In order to solve this problem in a practical way, a sub base layer can be inserted between the subgrade and the harder base layer (for example a layer of sand between a subgrade of clay and base layers of crushed rock). The use of stabilization of soft subgrade, which is common in many countries, points to the same problem.
c. a and b indicate that there could be a problem with permanent deformations in the bottom of the base layers on a soft subgrade. The reason for this could be that the base material could move sideways on the subgrade (see figure 15).
d. Compaction of unbound materials could give rise to a permanent horizontal stress in the material. The built-in horizontal stress from the compaction during construction probably relaxes to zero, but the heavy traffic on a road performs continuous compaction.
e. It is sometimes almost impossible to drain a dense subgrade, which means that the bearing capacity is very different and unpredictable during the year, depending on rainfall, frost/thaw and ground water level. Measurements in the field therefore do not give the true value.
f. On roads with a thin pavement, there will be large stress on the on the unbound layers, especially high up in the road. For this situation the nonlinear response from an unbound base material, has a large influence. Bituminous bound layers that are 4 cm thick has a different mechanistic function than thick layers (over 8 cm), where the thin pavement acts like a membrane, while the thicker pavement acts like a beam or slab. Layer thicknesses between 4 cm and 8 cm could therefore cause problems and ought to be avoided.
g. The resilient modulus is not the same for static and dynamic loads. In the "Active Design" project, there will be an investigation of possible connections that could be found through the triaxial tests. Dynamic plate loading could also be used in combination with the static plate loading test in order to investigate this question.
h. The resilient modulus that has been measured in triaxial tests is often different from the resilient modulus that can be measured in field, for example with help of back calculation from plate loading or FWD.
i. The first post-construction compaction of unbound base materials and subgrade is fairly rapid, but difficult to model (see chapter 6).
j. It is probably necessary to use different material models in order to describe a crushed sharp edged material (a typical base material) and a material with round and rather small grains (a typical subgrade).
k. It ought to be of great interest to determine whether it is possible to estimate the elastic and plastic (creep) properties of unbound material by means of aggregate curve, aggregate form aggregate size and geological properties.

10 SUMMATION

One very important factor in this project is the validation of test methods and models at real road projects. As the test method for prediction of future rutting, the triaxial tests for unbound material and bituminous bound materials, suggested in the Design Guide and the SAMARIS projects, should be used. The following measures should be carried out in order to conduct this NordFoU project.

- Conduct triaxial tests on a certain amount of real projects, which will also provide broader knowledge about material properties.
- Use the models that have already been developed, to predict future rutting. This will also promote the education and expertise that is necessary to conduct this NordFoU project.
- Monitor rutting development using vehicles equipped with laser technology (which is normal on many roads today).

When the project is completed in 2009, there should be a strong implementation phase, if the results from this project are to be of use for the Nordic countries.

The work includes the following implementation measures:

- Written course material that explains theories, facts and suitable ways to execute the works.
- Extension courses for consultant, contractor and road management staff.
- Development of possible economic incentives for contractors and, if possible also for the consultants,

in order to build roads of better quality or at a lower cost, something that could be measured and evaluated through using this new system.
- The use of performance requirements on some new road projects.

REFERENCES

Design Guide. NCHRP, Appendix GG 1: Calibration of permanent deformation models for flexible pavements. 2004.

Gidel G., Hornych P., Chauvin J-J., Breysse D. and Denis A. New approach for investigating the permanent deformation behaviour of unbound granular material using the repeated load triaxial apparatus. Bullentin des Laboratoires des Ponts et Chaussées 233, 2001.

Garba Saba R. Performance Prediction Models for Flexible Pavements: A State-of-the-art Report, 2006.

Huvstig, A. The development of a new road design method in Sweden. TRB. 2006.

MMOPP, Brugervejledning Danish Road Directorate. 2007.

SAMARIS EU project Work Package 5 Performance based specifications. Development and validation of a method of prediction of structural rutting of unbound pavement layers, 2006.

Werkmeister S. Permanent Deformation Behaviour of Unbound Granular materials in Pavement Constructions. Doctor Thesis Technischen Universität Dresden. 2003.

Advances in Transportation Geotechnics – Ellis, Yu, McDowell, Dawson & Thom (eds)
© 2008 Taylor & Francis Group, London, ISBN 978-0-415-47590-7

Influence of water content on mechanical behaviour of gravel under moving wheel loads

T. Ishikawa, M. Hosoda & S. Miura
Hokkaido University, Sapporo, Japan

E. Sekine
Railway Technical Research Institute, Tokyo, Japan

ABSTRACT: This paper presents an experimental study to evaluate the effects of water content on the mechanical behavior of gravel subjected to moving wheel loads. First, two types of small scale model tests which adopted a fixed-place loading method and a moving loading method were performed with a base course material under air-dried condition and saturated condition. Next, two types of multi-ring shear tests which could consider a change in direction of principal stresses like traffic loads were performed likewise. Based on test results, the relationships between the water content of test samples and the deformation-strength characteristics were examined. As the results, it was revealed that the shear strength of gravel decreased and the cumulative residual strain increased due to the saturation in both tests, regardless of loading methods. This indicates that the water content of gravel influences the mechanical behavior of granular roadbed strongly.

1 INTRODUCTION

In general, the granular roadbed at road and railway subjected to traffic loads gradually loses functions as a transportation facility, which should be normally maintained in service, with repeated vehicle passages. So, periodic maintenance activities of the roads and railways are required from the view-point of riding quality and safety. For performing more efficient maintenance, it is necessary to elucidate the cumulative irreversible (plastic) deformation characteristics of granular roadbed under moving wheel loads in detail.

In our recent studies (Ishikawa & Sekine 2002, Ishikawa & Sekine 2007), it was revealed that moving wheel loads has a strong influence on the mechanical behavior of coarse-grained base course materials under air-dried condition on the basis of cyclic loading tests simulating moving wheel loads. On the other hand, a number of studies have been made on the effect of water content on the mechanical behavior of roadbed as one of major assignments about highway engineering and railway engineering. For example, Coronado et al. (2005) and Ekblad & Isacsson (2006) report that the deformation modulus decreases with the increment of water content in triaxial tests of coarse granular road materials. However, little is known about the mechanical behavior of gravel subjected to moving wheel loads under both saturated and unsaturated conditions.

This paper presents an experimental study to evaluate the effects of water content on the mechanical behavior of a base course material subjected to moving wheel loads in case the degree of saturation varies due to rainfall and change of ground water level. In this paper, a series of laboratory element tests and model tests were performed with gravel under air-dried and saturated conditions. First, to understand the mechanical behavior of granular roadbed under traffic loads, two types of small scale model tests which adopt a fixed-place loading method and a moving loading method are performed. Here, the former is a method for applying pulsating compression vertical loads to a paved road, while the latter is a method for making a wheel with a constant vertical load travel on the paved road like actual traffic loading. Next, to examine the mechanical characteristics of the base course material subjected to moving wheel loads, two types of multi-ring shear tests which differ in loading method are performed. Here, the multi-ring shear test, which can apply torsional simple shear for gravel and then consider a change in direction of principal stresses like traffic loads, is newly developed as a laboratory element test (Ishikawa et al. 2007).

Table 1. Physical properties of soil samples.

Name	ρ_{dmax} g/cm^3	ρ_{dmin} g/cm^3	D_{max} mm	D_{50} mm	U_c
Crusher-run	2.22	1.72	37.5	10.8	27.3
Gravel	1.90	1.63	9.5	2.5	29.2

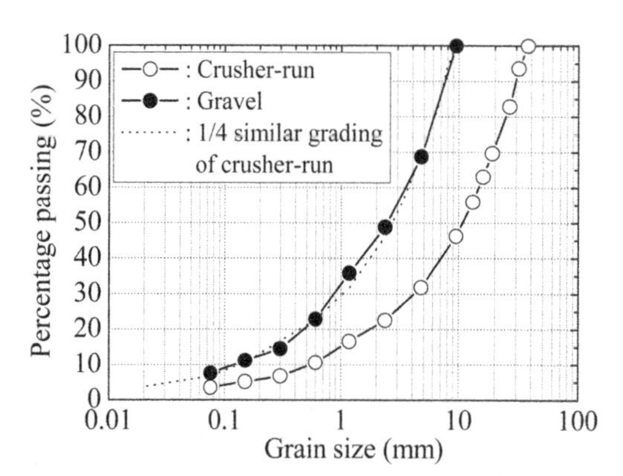

Figure 1. Grain size distribution of ballast.

2 TESTING METHODS

2.1 *Soil samples*

A test material is angular, hard andesite crushed stone, namely "crusher-run," used in Japanese roads as a base course material. Two types of test samples which had different mean grain size from each other were employed in this paper. Physical properties and gradation curves for the test samples are shown in Table 1 and Figure 1, respectively. The grading of actual crusher-run has a grain size distribution between approximately 40 mm and 0 mm. Accordingly, one test sample is actual crusher-run, and the other is similar grading gravel that has almost one-fourth mean grain size of original crusher-run by screening out particles larger than 9.5 mm in grain size. Here, the term "crusher-run" is used to refer to the former, and the term "gravel" is used to refer to the latter. The crusher-run was employed in small scale model tests, while in multi-ring shear tests, the gravel was employed for insuring experimental accuracy on the ratio of maximum particle size versus specimen size.

2.2 *Small scale model test*

2.2.1 *Test specimens*

A depth over 500 mm of crusher-run was placed within a rigid soil container, 1.2 m wide, 0.3 m deep and 0.6 m high, and compacted by tamping every layer of 50 mm in height with a wooden rammer so as to give some constant compaction energy until the degree of compaction for granular roadbed made of crusher-run reaches 90% (average dry density of granular roadbed $\rho_d = 1.86$ g/cm^3, relative density $D_r = 37.4\%$). Afterwards, for smoothing the edge face of granular roadbed, crusher-run smaller than 9.5 mm in grain size was scattered about 10 mm in thickness on the top surface of the compacted roadbed, and rolling compaction was conducted by a loading wheel (200 mm in diameter, 270 mm in deep) with constant vertical load equal to 1.0 kN. Here, a rigid steel plate is selected as for subgrade since stone particles may not penetrate it. To guarantee the granular roadbed to be in the plane strain state with the longitudinal section assumed to infinitely continue, a friction reduction layer composed of silicone grease and a transparent film was inserted between the acrylic side panels of soil container and granular roadbed. The granular roadbed was kept throughout the test under air-dried condition or field saturated condition, which was achieved by setting ground water level on 400 mm from the bottom of granular roadbed.

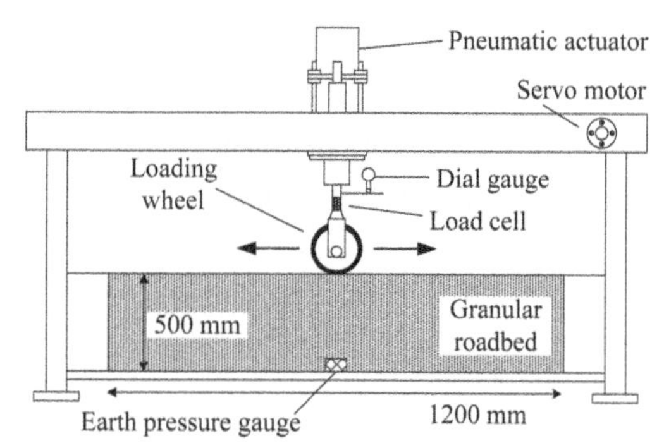

Figure 2. Moving constant loading test apparatus.

2.2.2 *Moving constant loading test*

Figure 2 shows the general test arrangement of the moving constant loading test. To roughly measure the pressure applied to granular roadbed, a two-component loadcell which could measure both the normal and shear components of the reaction force separately was installed to the loading rod as shown in Figure 2. Besides, a displacement transducer located on the loading wheel was employed to measure the settlement at the roadbed surface on which the loading wheel stands during cyclic loading. In addition, the roadbed pressure was measured with an earth pressure gauge located on the center of subgrade surface. At intervals of 0.1 seconds, all measuring data were recorded and stored in a computer through all test stages.

Cyclic loading was performed by a loading wheel as follows. First, the vertical pressure loaded on roadbed surface (point A in Figure 3) was estimated by performing numerical analysis of a Japanese paved road

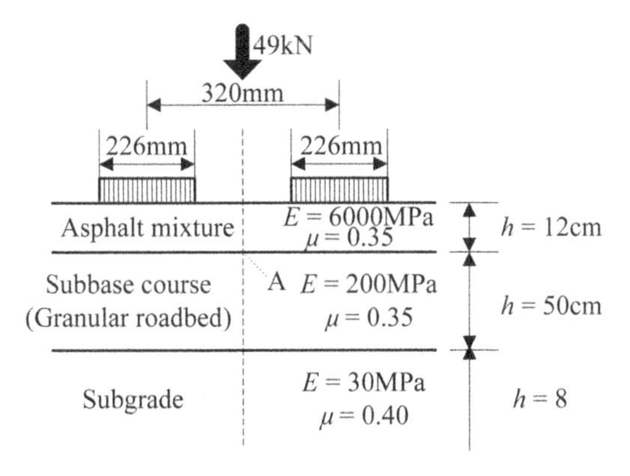

Figure 3. Numerical analysis of a Japanese paved road.

under design wheel loads of 49 kN as shown in Figure 3 with GAMES (General Analysis of Multi- layered Elastic Systems, Maina & Matsui 2004). Next, a vertical load of a loading wheel ($P_{max} = 2.23$ kN) was set so that applied pressure obtained from dividing the vertical load by the estimated ground contact area (about 70 mm × 270 mm) of the wheel was equal to the above-mentioned calculated pressure of 114.2 kPa. Vertical loads applied to the wheel through an electro-pneumatic actuator were increased step by step so that the uneven settlement of granular roadbed would not occur dramatically at the early stage of cyclic loading. Subsequently, a wheel with a given constant vertical load P_{max} travels along the rails at a constant speed of 384.0 mm/min and makes 100 round trips between the two ends of the soil container cyclically to simulate actual traffic loading. Here, the running speed of the loading wheel adopted for this test was much lower than the actual running speed of vehicles due to hardware constraints.

2.2.3 *Fixed-place cyclic loading test*

The general test arrangement of the fixed-place cyclic loading test is the same as that of the moving constant loading test, while the loading method is different as follows. Pulsating compression vertical loads P ranged from 0.19 to 2.23 kN were cyclically applied through a loading wheel to the center of granular roadbed so that the maximum applied vertical load was almost equivalent to the constant vertical load P_{max} in the moving constant loading test. A sinusoidal half wave was employed as a loading waveform, and vertical loads were increased step by step at the early stage of cyclic loading like moving constant loading tests. The number of loading cycles N_c was 200 cycles, and the loading frequency of 0.017 Hz was selected by considering the running speed of the loading wheel in the moving constant loading test.

In addition, bearing capacity tests of granular roadbed were performed as follows. Vertical load P was applied to the roadbed center through a footing, which has the area of the base equal to the estimated ground contact area, instead of the loading wheel, and it slowly increased from 0 kN to 5.58 kN at the constant loading speed of 2.23 kN/min, which was regarded as static loading. Here, stage loading was adopted pursuant to an actual plate loading test.

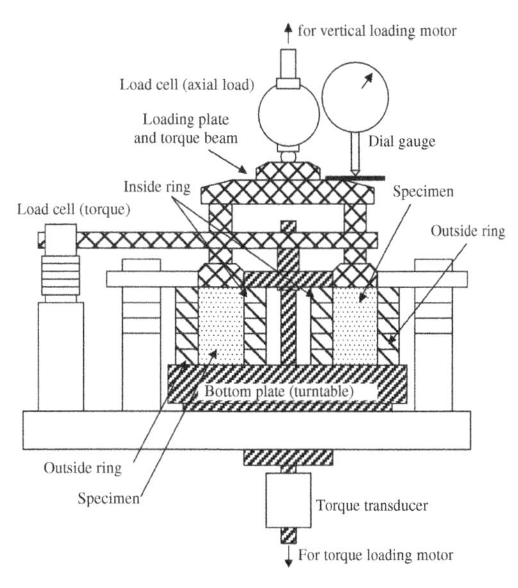

Figure 4. Multi-ring shear apparatus.

2.3 *Laboratory element test*

2.3.1 *Test specimens and test implements*

"Multi-ring shear apparatus" was developed experimentally as a kind of torsional simple shear test apparatus. A remarkable feature of the multi-ring shear apparatus is that it can evaluate the effect of rotating principal stress axes under shearing on the strength and deformation characteristics of granular materials. Figure 4 shows the schematic diagram of the multi-ring shear apparatus, and it is composed of a bottom plate, a loading plate, and rigid rings that support a specimen. The bottom plate supporting a specimen turns by a direct drive motor (DDM) for torque loading though the loading plate is fixed. Consequently, torsion (torque) can be loaded to a specimen confined by the bottom plate, the loading plate, inside rings and outside rings. In addition, vertical loads can be applied to a specimen by the DDM for vertical loading mounted on the loading plate. To decrease the friction between a specimen and rings as much as possible, the structure of inside and outside rings was designed as if each ring can move freely through loading. The width of a specimen was 60 mm (120 mm in inside diameter, 240 mm in outside diameter), and the height

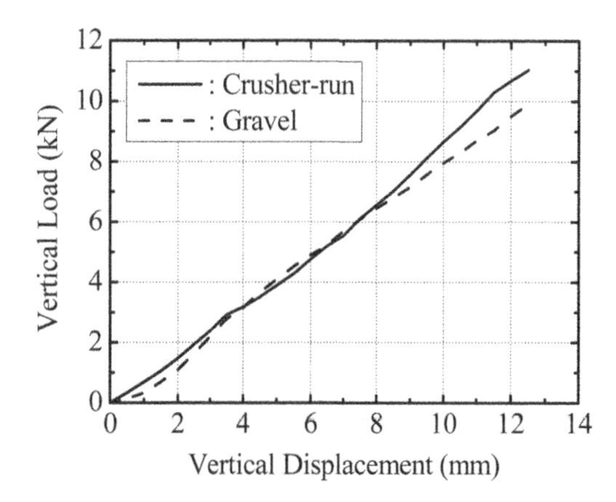

Figure 5. CBR test results of crusher-run and gravel.

is changeable within the range from 40 mm to 100 mm by changing the number of the rings which height is 20 mm. In this paper, the height of a specimen was set equal to 60 mm.

Specimens of gravel were prepared by tamping every layer of 20 mm in height with a wooden rammer so as to give some constant compaction energy until the relative density of specimens in multi-ring shear tests becomes almost equal to that of granular roadbed in the above-mentioned model tests, that is $D_r = 37.4\%$. The specimens were kept throughout the test under air-dried condition or field saturated condition. Here, a saturated specimen was achieved by soaking gravel, which had soaked in water for 96 hours or more before the test, with water in the multi-ring shear apparatus. For information, Figure 5 compares CBR test results of crusher-run and gravel, which have the same relative density, under air-dried condition. This figure indicates that the mechanical properties of gravel approximately coincide with those of crusher-run.

2.3.2 *Test procedures*

The loading process in cyclic loading tests under fully drained condition was performed as follows. In the multi-ring shear test adopting a moving loading mode (ML-multi-ring shear test), after consolidating a specimen of gravel one-dimensionally under the axial stress σ_a of 114.2 kPa, both shear stress $\tau_{a\theta}$ and axial stress σ_a in sinusoidal waveforms as shown in Figure 6 were cyclically applied to the specimen. Here, the waveforms of $\tau_{a\theta}$ and σ_a were estimated by performing FE analysis of the above-mentioned small scale model test with a two-dimensional FE model in the plane strain state as shown in Figure 7. In the multi-ring shear test adopting a fixed-place loading mode (FL-multi-ring shear test), after one-dimensional consolidation, only the axial stress σ_a in the sinusoidal waveform were cyclically applied to the specimen as shown in Figure 6. The ML-multi-ring shear test simulates the

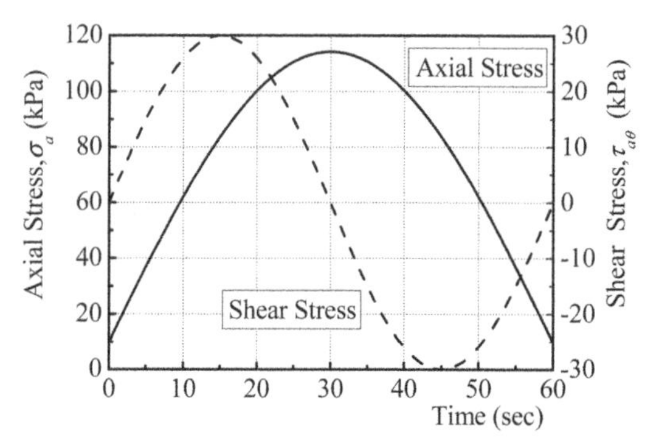

Figure 6. Loading conditions of multi-ring test.

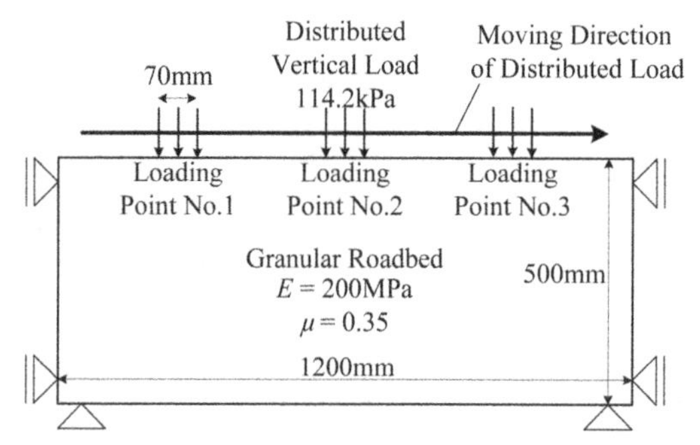

Figure 7. FE analysis of small scale model test.

moving constant loading test, and the FL-multi-ring shear test simulates the fixed-place cyclic loading test. In both cyclic loading shear tests, the number of loading cycles N_c was 200 cycles, and the loading frequency of 0.017 Hz was selected by referring the experimental conditions of the model tests. In the monotonic loading tests, after consolidation, the shear stress $\tau_{a\theta}$ was applied at the constant shear strain rate of 0.1%/min while keeping σ_a constant.

3 TEST RESULTS AND DISCUSSIONS

3.1 *Small scale model test*

3.1.1 *Effects of water content on bearing capacity*

The effect of the water content of base course materials on the load-displacement relations is discussed. Figure 8 shows the relations of crusher-run under air-dried and field saturated conditions between vertical load P and vertical displacement u at the center of the footing surface in bearing capacity tests. In this paper, the bearing capacity of granular roadbeds can be defined as the vertical displacement at the maximum vertical load of $P = 5$ kN. Accordingly, it is considered that a granular roadbed has stronger bearing capacity in case of obtaining smaller vertical displacement at $P = 5$ kN.

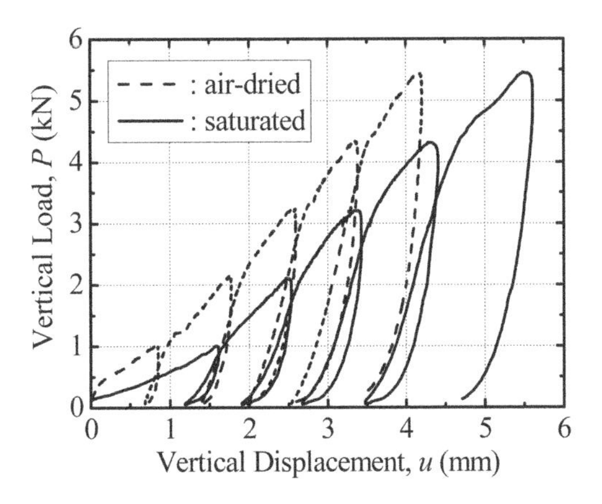

Figure 8. Comparison of bearing capacity of granular roadbed.

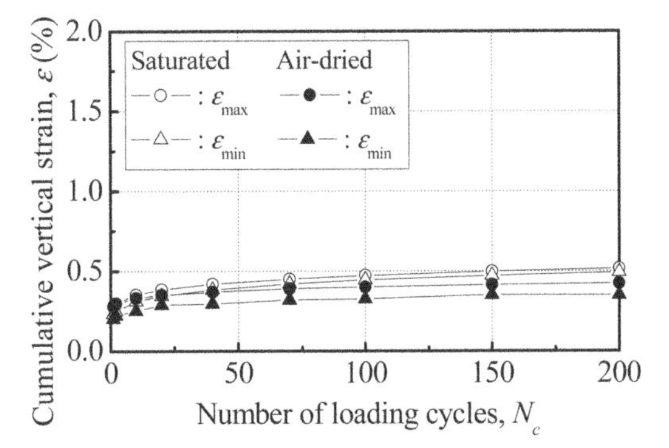

Figure 9. $\varepsilon - N_c$ relations in fixed-place cyclic loading tests.

From Figure 8, it is clearly shown that the bearing capacity of saturated granular roadbed is lower than that of air-dried granular roadbed, and that for plots with the same vertical load, the vertical displacement of saturated granular roadbed is much larger than that of air-dried granular roadbed. These results indicate that the water content of base course materials has a considerable influence on the decrease in the bearing capacity of granular roadbed.

3.1.2 *Effects of water content on cyclic deformation*
Figure 9 shows the relations in the fixed-place cyclic loading tests for air-dried and saturated granular roadbeds between the maximum vertical strain ε_{max} at loading P_{max}, the minimum vertical strain ε_{min} at unloading and the number of loading cycles N_c. Here, the vertical strain of small scale model tests is a kind of macro strain calculated by dividing the cumulative settlement at the roadbed surface on the loading point by the initial depth of granular roadbed (500 mm),

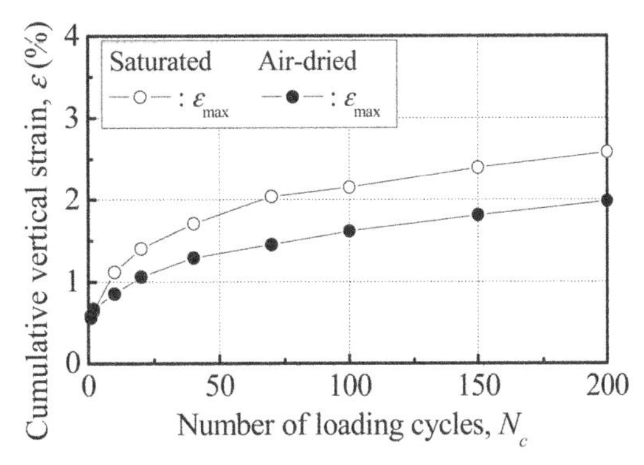

Figure 10. $\varepsilon_{max} - N_c$ relations in moving constant loading tests.

and ε_{min} may represent cumulative residual vertical strain. In Figure 9, both ε_{max} and ε_{min} increase slowly with the increment of N_c after the exponential increment at early stages of cyclic loading, and their rates of increase decrease with the increment of N_c. Moreover, the result that the strain amplitude ($\varepsilon_{max} - \varepsilon_{min}$) is constant even if the loading cycle increases indicates that the increase of ε_{max} mainly originates in the cumulative residual settlement of granular roadbed. According to these figures, the residual settlement of saturated granular roadbed is more likely to increase with the repetition of loads than that of air-dried granular roadbed.

Figure 10 shows the $\varepsilon_{max} - N_c$ relations in the moving constant loading tests for air-dried and saturated granular roadbeds. Here, the maximum vertical load applied to granular roadbed in the moving constant loading tests is generated when a loading wheel is just above the measuring point, and the minimum vertical load means unloading states as a loading wheel is far beyond the measuring point. When comparing Figure 10 with Figure 9, the $\varepsilon_{max} - N_c$ relation of moving constant loading tests resembles that of fixed-place cyclic loading tests in that the water content of base course materials noticeably influences the development of residual settlement for granular roadbed under cyclic loading. On the other hands, from a qualitative point of view, it is recognized that for plots with the same loading cycle, the amount of cumulative residual settlement caused by fixed-place cyclic loading is much smaller than the one caused by moving constant loading. This result indicates that the difference in loading methods has a considerable influence on the cyclic plastic deformation of granular roadbed. Moreover, Figure 10 indicates that the effect of the water content of base course materials on the development of residual settlement of granular roadbed appears more clearly in employing the moving load.

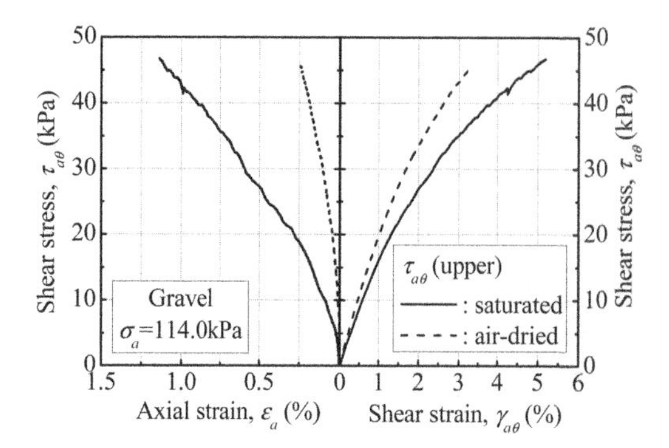

Figure 11. Comparison of bearing capacity of roadbed.

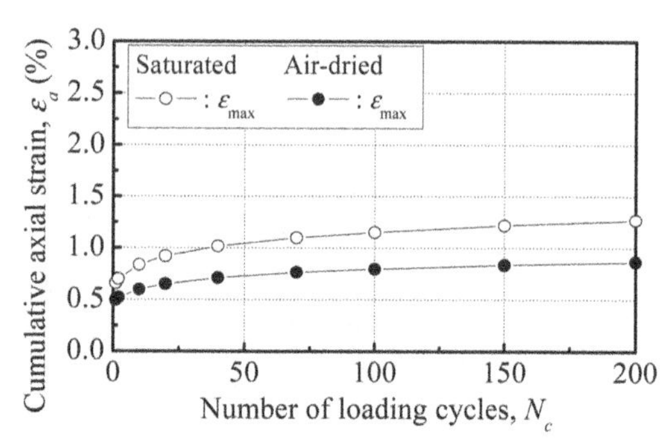

Figure 12. $(\varepsilon_a)_{\max}$ – N_c relations in FL-multi-ring shear tests.

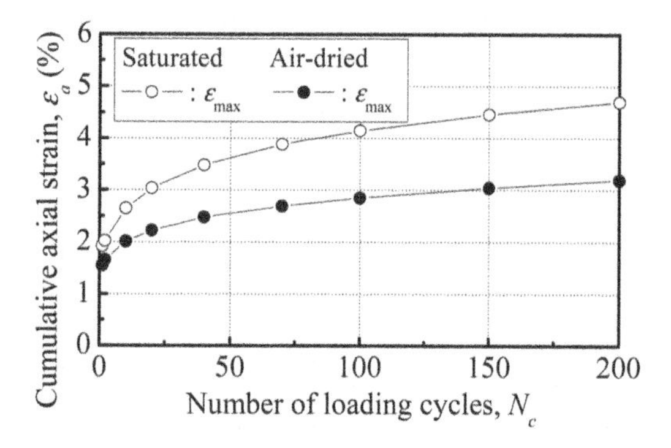

Figure 13. $(\varepsilon_a)_{\max}$ – N_c relations in ML-multi-ring shear tests.

3.2 *Laboratory element test*

3.2.1 *Effects of water content on shear strength*

The effect of the water content of base course materials on the deformation – strength characteristics is discussed. Figure 11 shows the relations of gravel under air-dried and field saturated conditions between shear stress $\tau_{a\theta}$ and shear strain $\gamma_{a\theta}$ in monotonic loading tests. It is observed that for plots with the same strain level, the shear strength of saturated gravel is a little smaller than that of air-dried gravel. Also, Figure 11 shows the relations between shear stress $\tau_{a\theta}$ and axial strain ε_a. Here, the change of ε_a means the volumetric change of a specimen because both radial strain ε_r and circumference strain ε_θ are zero at all times in the multi-ring shear test. In Figure 11, the volumetric change of gravel under shearing is altered from compression to dilation at early stages of monotonic loading regardless of the water content. However, the contraction for saturated gravel is much larger than air-dried gravel. These results indicate that the water content of gravel has a considerable influence on the deformation - strength characteristics of gravels in the multi-ring shear tests.

3.2.2 *Effects of water content on cyclic deformation*

First, the applicability of the multi-ring shear tests to an element test of granular roadbed subjected to moving wheel loads was examined. Figure 12 and Figure 13 shows the relations of air-dried and saturated gravels between cumulative axial strain $(\varepsilon_a)_{\max}$ at the maximum axial stress $(\sigma_a)_{\max}$ and number of loading cycles N_c in obtained from a FL-multi-ring shear tests and ML-multi-ring shear tests, respectively. Here, $(\sigma_a)_{\max}$ may represent the stress condition when the maximum vertical load is applied to granular roadbed in the moving constant loading test. When comparing Figure 12 and Figure 13 with Figure 9 and Figure 10, respectively, the cumulative strain obtained from multi-ring shear tests approximately coincides with the macro cumulative strain calculated from small scale model test results not only but qualitatively also quantitatively, irrespective of the loading mode. These demonstrate that the multi-ring shear test has high applicability to the estimation of deformation behavior of granular roadbed subjected to repeated moving wheel loads.

Next, the effect of the water content of base course materials on the cyclic plastic deformation characteristics is discussed. When comparing the results of air-dried gravel with those of saturated gravel in Figure 12 and Figure 13, it is observed that the cyclic plastic deformation for saturated gravels is much larger than the one for air-dried gravels, regardless of the loading mode. However, the difference between the results for air-dried and saturated gravels can hardly be distinguished in the FL-multi-ring shear test results, while the effect of the water content of gravel on the development of residual settlement remarkably comes to the surface in the ML-multi-ring shear test results. These indicate not only that the water content influences the mechanical behavior of gravel strongly, but also that the difference in loading methods, that is the difference whether the principal stress axes rotate or not, has a significant influence on an predictability of cyclic plastic deformation of base course materials subjected to moving wheel loads.

4 CONCLUSIONS

The following conclusions can be obtained;

- Cumulative residual strain obtained from multi-ring shear tests is almost equivalent to the macro strain calculated from load - displacement relationships in small scale model tests, regardless of the loading method.
- Shear strength and bearing capacity of base course materials decrease and cumulative residual strain increases due to saturation of test specimens in both small scale model tests and multi-ring shear tests, irrespective of the loading method.
- The water content of base course materials influences the mechanical behavior of granular roadbed strongly, and the effect on the development of residual settlement for granular roadbed under cyclic loading appears more clearly in employing the moving load.

REFERENCES

Coronado, O., Fleureau, J., Correia, A. & Caicedo, B. 2005. Influence of suction on the properties of two granular road materials. In I. Horvli (ed), *The 7th International Conference on the Bearing Capacity of Roads, Railways and Airfields; Proc. intern. conf., Trondheim, 27–29 June 2005.*: [1/1(CD-ROM)260].

Ekblad J. & Isacsson U. 2006. Influence of water on resilient properties of coarse granular materials. *Road Materials and Pavement Design* No.3: 369–404.

Ishikawa, T. & Sekine, E. 2002. Effects of moving wheel load on cyclic deformation of railroad ballast. *Railway Engineering-2002; Proc. intern. Symp., London, 3–4 July 2002.*: [1/1(CD-ROM)].

Ishikawa, T., Miura, S., & Sekine, E. 2007. Development and performance evaluation of multi-ring shear apparatus. In A. G. Correia et al. (eds), *Geotechnical Aspects and Processed Material in Design and Construction of Pavement and Rail Track; Proc. intern. wks., Osaka, 13 September 2005.*: 53–64. Rotterdam: Balkema.

Ishikawa, T. & Sekine, E. 2007. Effect evaluation of moving load on cyclic deformation of crushed stone. In N. Son (ed), *The 13th Asian Regional Conference on Soil Mechanics and Geotechnical Engineering; Proc. intern. conf., Kolkata, 10–14 December 2007.*: 121–126.

Maina, J.W. & Matsui, K. 2004. Developing software for elastic analysis of pavement structure responses to vertical and horizontal surface loadings. *Transportation Research Records* No.1896: 107–118.

Advances in Transportation Geotechnics – Ellis, Yu, McDowell, Dawson & Thom (eds)
 ISBN 978-0-415-47590-7

The evaluation of in-place recycled asphalt concrete as an unbound granular base material

EJ. Jeon, B. Steven & J. Harvey
University of California, Davis, USA

ABSTRACT: In-place recycling of asphalt concrete for use as an unbound granular base material was evaluated using comprehensive laboratory and field testing. Laboratory testing included particle size analysis, compaction, permeability, and static and repeated load triaxial tests. The triaxial tests were conducted with different sets of gradations, stress levels, and water contents to investigate the effect of the different gradations on the performance of pulverized material. Falling weight deflectometer test results were backcalculated to track the long-term performance of the pavement structure. Based on the laboratory and field test results, it was found that the performance of the pulverized material equaled or exceeded that of virgin aggregate base material.

1 INTRODUCTION

1.1 *Background*

The rehabilitation of roads in most developed countries is more common than new construction. However, virgin material sources are becoming scarce, resulting in increased construction costs due to increased hauling distances and other factors. The need to re-use existing in-situ road materials using rehabilitation techniques that will minimize traffic disruption is therefore increasing internationally.

A rehabilitation strategy that consists of in-situ reclaiming/pulverizing of existing failed asphalt concrete and a portion of the aggregate base to create a new granular layer, followed by asphalt concrete surfacing was used by the California Department of Transportation (Caltrans) at four pilot projects in northeastern California. For the pulverization process, a CMI RS-650 rubber tired reclaimer was used. This pulverized aggregate base material will be referred to as "pulverized material" or PAB. The thickness of the asphalt concrete surfacing is typically 120 to 165 mm. Since most highways in California were built up to 80 years ago and have thick layers of cracked asphalt concrete and many patches, this rehabilitation strategy is suitable for the state's highways. Advantages of this technique include reduced use of virgin aggregates, a reduction in construction traffic, and removal of the potential for reflective cracking from the underlying existing pavement. An initial life cycle cost analysis of this pavement rehabilitation technique when compared to the traditional approach of an asphalt concrete overlay over the existing pavement showed that the recycling option was cheaper in both the short term and over the life of the pavement (Bejarano & Harvey, 2004).

In this study, the performance of this rehabilitation strategy was evaluated based on comprehensive field and laboratory testing of materials from four pilot projects that were completed in the period from 2001 to 2006. The projects were in northeastern California at Alturas, Beckwourth, Poison Lake, and Cayton Creek.

1.2 *Scope*

The intent of this study is to evaluate the performance of pulverized material, especially the variability of gradation, and to compare it with a typical aggregate material in California. The range of gradations in the field was determined from laboratory sieve analyses from two of the projects. In order to investigate the effect of the different gradations on the performance of pulverized material, additional laboratory testing including compaction, permeability, static and repeated load triaxial tests with different sets of gradations, stress levels, and water contents were conducted. Repeated load triaxial tests were also undertaken on lime-and cement-stabilized samples to determine the potential benefits of lightly stabilizing the recycled material. Field performance was monitored regularly using a falling weight deflectometer and the results were backcalculated to track the long-term performance of pavement structure in terms of stiffness.

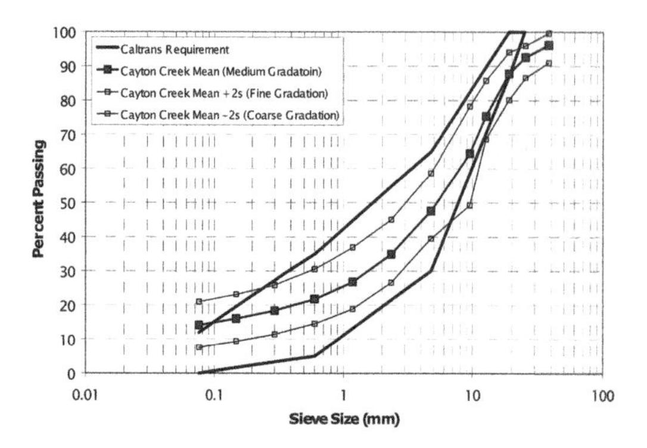

Figure 1. Three representative gradations of Cayton Creek pulverized material and Caltrans gradation requirement.

2 LABORATORY INDEX TESTING

2.1 *Particle size distribution*

The particle size distribution of the pulverized material obtained in the field was determined by wet sieving (ASTM C117) and dry sieving (ASTM C136). The gradation of Alturas and Beckwourth pulverized material met the requirements for Caltrans Aggregate Base (AB). In order to find the spatial variability of the gradation of the pulverized material, sieve analyses were conducted on material from different locations in the Poison Lake and Cayton Creek projects. Results indicate that the pulverized material generally met the requirements except for the large and small particle sizes. To determine the effect of variability of the gradation on the performance, three representative gradations were selected for additional testing. The three gradations were mean (Medium gradation), mean+2 standard deviations (Fine Gradation), mean-2 standard deviations (Coarse Gradation). Figure 1 shows the three representative gradations of the Cayton Creek pulverized material.

2.2 *Compaction and permeability tests*

Index tests were performed to obtain the basic properties of pulverized material. All the pulverized materials were classified as non-plastic, based on the results of Atterberg limit tests.

Compaction tests were conducted using the ASTM standard (ASTM D698), ASTM modified (ASTM D1577) and Caltrans (CTM216, (Caltrans 2000)) methods. The results of the compaction tests are summarized in Table 1. The test results show that the gradation has little effect on the density for the higher compactive effort (ASTM modified and Caltrans) while the effect is noticeable for the standard compactive effort. Since the compactive effort in the field is approximately equal to that used in the

Table 1. Summary compaction test results.

Material source	Gradation	Test method	γ_d^* (kg/m^3)	OMC** (%)
District 4 AB		Modified	2307	5.0
District 2 AB		Caltrans	2323	5.3
Alturas PAB		Caltrans	2100	6.5
Beckwourth PAB		Caltrans	2170	7.0
Poison Lake PAB		Caltrans	2017	6.0
Cayton Creek PAB	Coarse	Caltrans	2114	8.3
		Standard	2145	7.2
		Modified	2145	7.2
	Medium	Caltrans	2101	8.5
		Standard	2080	8.7
		Modified	2139	7.6
	Fine	Caltrans	2066	9.4
		Standard	2005	10.4
		Modified	2122	7.9

* γ_d = Maximum dry density.
** OMC = Optimum Moisture Content.

modified compaction test, the variability on the gradations should not affect the target densities used by the construction crews. However, the optimum moisture content of the pulverized material decreased as the coarseness increased. This result is expected as the amount of moisture required to provide lubrication of the particles during compaction increases as the percentage of finer particles increases.

The maximum dry density of pulverized material is less than that of a typical aggregate base material. This is due to the inclusion of approximately five percent by weight of asphalt binder in the original asphalt concrete prior to pulverization. Since the binder has a lower specific gravity than rock, the overall density will be lower than that for pure rock. The moisture contents and relative compactions of the triaxial test samples were determined from the results of the Caltrans test standard (CTM216).

The permeability of two pulverized materials, at two different compaction efforts, was measured using a constant head test (ASTM D2434). The test results are compared with a District 4 Class 2 aggregate base reported by Russo (2000). Figure 2 shows the summary of these test results for the pulverized materials and Class 2 aggregate base material.

As shown in Figure 2, permeability of the pulverized materials is generally higher than that of the typical aggregate material. Pulverized material with a coarse gradation has a higher permeability than the other gradations.

A higher permeability is better if the material could become saturated, as the water will then be able to quickly drain away. This is acceptable if the more permeable material can retain its strength under saturated

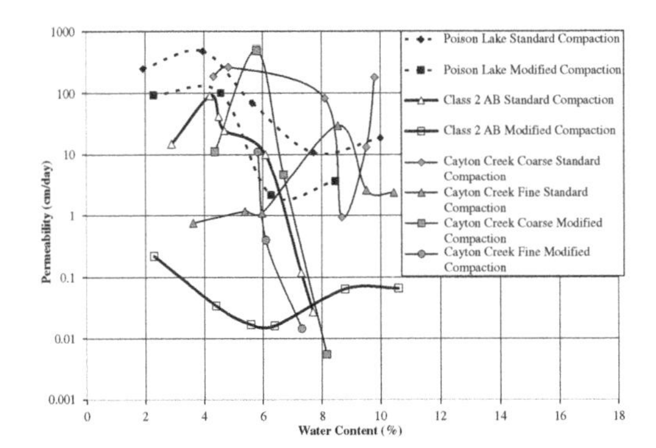

Figure 2. Permeability Test Results of pulverized and general aggregate base.

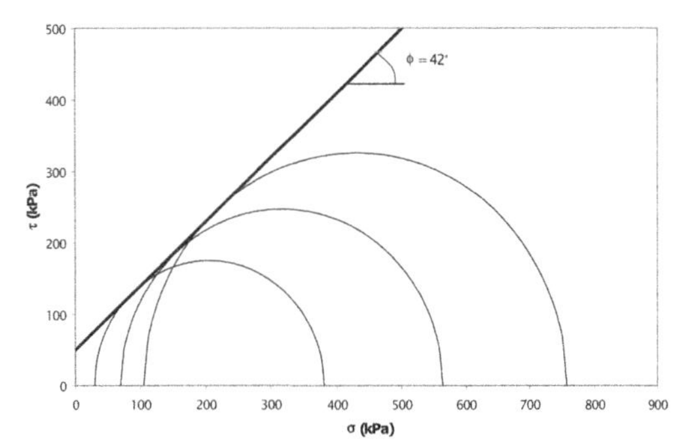

Figure 3. Static shear test results of Cayton Creek pulverized material (fine gradation, moisture content = 8% and relative compaction = 95%).

conditions. If the permeability of the material is lower, the amount of water entering the material will be lower.

3 TRIAXIAL TESTING

3.1 *Test Setup*

A servo-hydraulic triaxial test machine at the University of California Pavement Research Center (UCPRC) was used for all the triaxial testing (Heath, 2002). The samples had a diameter of 152.4 mm and a height of 300 mm. Samples were compacted using a vibratory hammer in five equal layers. The samples were tested under drained conditions. Samples were prepared with different combinations of material source, gradation, relative compaction, and moisture content. Test results of the pulverized material were compared with general aggregate materials reported by Heath (2002).

3.2 *Static shear triaxial testing*

In order to find the shear strength of material, static triaxial testing was conducted with different confining pressures. Samples were tested using displacement control at a rate of 0.5 mm/sec (10% axial strain per minute). Three confining pressures (35, 70, and 105 kPa), were used to cover the range of confinement to be encountered under field conditions. Mohr-Coulomb failure criteria were used to find the strength parameters. Test results of pulverized material from Cayton Creek are shown in Figure 3.

As shown in Figure 3, the failure envelope appears linear within the range of confining pressure as assumed by the Mohr-Coulomb failure criteria. The test results of pulverized materials were compared with Class 2 aggregate base materials and summarized in Table 2.

Dense-graded granular material, which is the general aggregate base material, is often non-plastic and

Table 2. Summary of static shear triaxial test results.

Material description	Gradation	RC* (%)	w** (%)	Friction angle (°)
District 4		95	3.0	44
Class 2 AB			4.9	46
			7.0	45
		97.5	5.6	49
		100	5.0	55
			5.5	53
Alturas		95	6.5	50
Pulverized AB		100	6.2	56
Beckwourth		95	4.5	47
Pulverized AB			6.3	41
			8.5	45
		92	6.7	39
Poison	Fine	95	5.6	48
Lake	Medium		5.9	50
Pulverized AB	Coarse		5.8	53
Cayton Creek	Fine	95	7.9	42
Pulverized AB	Medium		7.9	45
	Coarse		7.1	47

* Relative Compaction = Dry density/Maximum dry density.
** Moisture Content.

without true cohesion (Duncan et al, 1980). Although apparent cohesion is related to matric suction, which is an important characteristic of partially saturated soils, only friction angles were presented in Table 2 for the shear strength parameter. Based on the comparison of friction angles of the different materials, the shear strength of the pulverized material is similar to that of general aggregate base material.

The effects of moisture content, density, and gradation were also investigated. It was hard to find the effect of moisture content but it was clear that higher density results in higher friction angles. The effect of gradation of the pulverized material was also very clear in that the friction angle increased as the coarseness increased.

3.3 *Resilient modulus testing*

The resilient modulus of the granular base layer is important since it represents the traffic-induced load-carrying ability of the material. The resilient modulus is affected by numerous factors as summarized and reported by Lekarp et al. (2000a). According to Lekarp et al. (2000a), the stiffness of granular material is dependent on stress, density, grading, moisture content, and other factors. In order to find these effects, the resilient modulus tests of the pulverized and Class 2 aggregate base materials were performed with different combinations of moisture content and density. Resilient modulus tests were also undertaken on lime and cement stabilized samples in order to determine the potential benefits of adding lime to the pulverized material.

Resilient modulus testing of the pulverized material was conducted following the Strategic Highway Research Program (SHRP) test protocol P-46. The testing sequence for base materials in this test protocol consists of 15 blocks with different confining pressures and deviator stresses. Each block applies one hundred repetitions. These test results are then fitted to the generalized resilient modulus model proposed in NCHRP (2004) using linear regression techniques. The equation is given in Equation 1:

$$M_R = k_1 p_a \left(\frac{\theta}{p_a}\right)^{k_2} \left(\frac{\tau_{oct}}{p_a} + 1\right)^{k_3} \qquad (1)$$

where, M_R = resilient modulus; θ = bulk stress; p_a = atmospheric pressure to normalize stress; τ_{oct} = octahedral shear stress; k_1, k_2 and k_3 = regression constants obtained by fitting resilient modulus test data.

An average R^2 value of 0.93 was obtained. The resilient moduli at two reference stress state were calculated using Equation 1. One of the stress states (deviate stress = 103.4 kPa and confining pressure = 34.5 kPa) was recommended by Gudishala (2004) to calculate a general resilient modulus value for the base layer. He demonstrated the process of calculating stress states in the middle of granular base layer and considered the residual lateral stress developed in a granular base layer during the construction process. The reference resilient moduli and obtained regression model parameters for each material are summarized in Table 3.

Resilient moduli of four pulverized materials were compared with those of two Class 2 aggregate base materials in California, and crushed limestone and sand reported by Gudishala (2004).

Resilient modulus of the pulverized material was generally higher than general aggregate base material in California as well as crushed limestone and sand. This fact strongly supports the adequacy of the pulverized material as a granular base layer because the resilient modulus is an important material property in any mechanistic design or analysis for flexible pavements.

The effect of increasing the density was clear for all materials. Increasing the density of the samples resulted in higher values of resilient modulus.

Effects of moisture content and gradation were also investigated. Lekarp et al. (2000a) mentioned that the effect of moisture content of most granular materials is significant at saturation but the resilient moduli of dry and most partially saturated materials were similar. The test results agree with Lekarp, that is samples with low moisture contents have a higher resilient modulus than those close to or above optimum moisture content. Lekarp et al. (2000a) also reported that earlier literature was not clear on the impact of the fines content on the material stiffness. The test results also show that the effect of gradation was not clear for the pulverized material.

The test results of the pulverized material from Poison Lake show that lower levels of fines results in higher resilient moduli. However, the test results of Cayton Creek material show different results. The fine gradation samples have higher resilient moduli than the medium and coarse gradations except for the high moisture contents, which are greater than the optimum moisture content.

The tests were also conducted on the pulverized material samples stabilized by either lime or cement. The samples were stabilized with either three percent of lime or cement and cured for seven days. Both the lime and cement treatments resulted in an improvement of resilient moduli. The resilient modulus of the cement-treated sample was significantly higher than that of the other samples. However, it should be emphasized that the cost of stabilizing agents and possible problems such as cracking need be considered.

3.4 *Repeated load permanent deformation testing*

One of the important design philosophies for flexible pavements is the limitation of rut development in the pavement structure (Lekarp et al. 2000b). Rutting can occur either in the asphalt concrete layer or the unbound layer(s). Unbound base rutting is due to the permanent deformation of the unbound layer. Therefore, it is very important to understand the permanent deformation characteristics of the granular base layer. To understand permanent deformation characteristics of the pulverized material, repeated load permanent deformation testing was conducted at a number of stress states. Permanent deformation test

Table 3. Summary of resilient modulus triaxial test results.

Material description	w (%)	RC (%)	k_1	k_2	k_3	M_{R1}* (MPa)	M_{R2}** (MPa)
District 2	6.0	100	1069	0.644	−0.332	150	246
Class 2 AB	6.2	95	1072	0.674	−0.427	148	242
District 4	5.3	95	1905	0.560	−0.358	248	375
Class 2 AB	6.7	95	1197	0.633	−0.313	167	274
	7.2	95	1300	0.569	−0.286	176	273
	5.0	100	3475	0.716	−0.567	467	766
	4.2	100	4529	0.457	−0.612	495	633
Alturas PAB	4.7	100	3289	0.633	−0.762	385	554
	6.3	95	2259	0.584	−0.631	269	384
Beckwourth	4.6	95	2275	0.320	−0.198	265	336
PAB	8.4	95	931	0.614	−0.189	135	225
	6.5	95	2094	0.569	−0.517	258	375
Poison Lake	4.0	95	3589	0.464	−0.361	436	602
Coarse PAB	6.9	95	2559	0.630	−0.629	315	469
Poison Lake	5.9	95	2553	0.519	−0.565	297	408
Medium PAB	3.8	95	2754	0.414	−0.364	322	424
	7.3	95	2415	0.496	−0.567	277	371
	4.0	92	2612	0.303	−0.352	283	338
	5.9	100	2449	0.669	−0.539	322	509
Poison Lake PAB + 3% lime	7.1	100	6081	0.396	−0.540	655	807
Poison Lake	6.2	95	2553	0.499	−0.568	293	394
Fine PAB	6.7	95	2312	0.566	−0.691	265	366
Cayton Creek	6.8	95	1652	0.636	−0.248	237	395
Coarse PAB	5.3	95	2153	0.569	−0.249	295	463
	8.9	95	1528	0.625	−0.131	228	390
Cayton Creek	7.2	95	2163	0.530	−0.356	276	405
Medium PAB	8.0	92	2239	0.514	−0.510	266	368
Cayton Creek Medium	9.9	93	4355	0.444	−0.659	463	577
PAB + 3% lime	11	95	4385	0.429	−0.453	500	652
Cayton Creek Medium PAB + 3% cement	10	93	10,139	0.269	−0.209	1135	1368
Cayton Creek	8.2	95	2829	0.430	−0.441	325	425
Fine PAB	10.6	95	1858	0.519	−0.769	199	258
	6.0	95	4178	0.293	−0.441	433	501
Crushed limestone			1771	0.52	−0.326	227	333
Sand			1727	0.48	−0.47	203	276

* Modulus at $\sigma_d = 103.4$ kPa & $\sigma_c = 34.5$ kPa.
** Modulus at $\sigma_d = 206.8$ kPa & $\sigma_c = 103.4$ kPa.

results at a stress state of $p \approx 200$ kPa; $q \approx 340$ kPa; where p = mean stress; q = deviator stress; are shown in Figure 4. All samples were compacted by 95% relative compaction and close to optimum moisture content.

The results indicate that permanent strain of the pulverized material was similar to that of general aggregate material in California.

The effect of grading on permanent deformation was investigated by Thom and Brown (1988), who found that it varies with the compaction level. For the compacted samples, the effect of gradation is not clear and a preferred gradation to reduce permanent strain is not apparent.

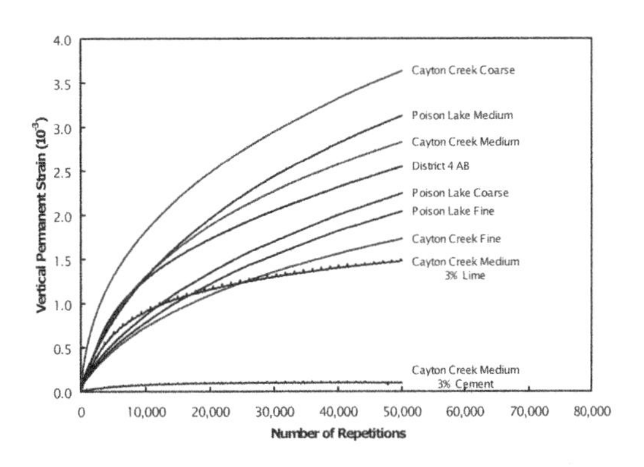

Figure 4. Permanent deformation test results.

The effect of gradation on permanent strain was not clear for the pulverized materials. For the pulverized material from Cayton Creek, the fine gradation sample had the smallest permanent strains among the results. On the other hands, coarse gradation has the maximum permanent strains. However, permanent strains of Poison Lake pulverized materials didn't have the same trends. It was found that fine gradation sample showed a more significant reduction of stiffness and permanent deformation resistance than the coarse gradation sample since more fines help to hold more water, resulting in higher pore pressures.

4 FIELD TESTING

To monitor the field performance of this rehabilitation strategy, field testing including Dynamic Cone Penetrometer (DCP) and Falling Weight Deflectometer (FWD) tests, and visual pavement condition survey were conducted on the pilot projects. DCP tests were conducted to obtain field measurements of thickness and strength of the unbound layers. In this study, the DCP test results were only used to obtain the thickness of each layer for the backcalculation process. FWD tests were conducted regularly to estimate pavement layer modulus using available mechanistic tools for pavement analysis. FWD tests were generally performed in the morning and afternoon to evaluate the stiffness with different temperatures since asphalt concrete stiffness changes with temperature.

The condition surveys showed that the performance of the pulverized sections was better than the control sections with little or no cracking or rutting.

Test results were backcalculated to obtain the stiffness of each layer. The backcalculation analyses were done using the CalBack computer program, which is currently being developed for Caltrans by the UCPRC. The backcalculated stiffness of the Alturas, Beckwourth, and Poison Lake rehabilitated roads were compared with a control section in Alturas project. This control section is a typical pavement structure in California. It consists of typical asphalt concrete, Class 2 aggregate base and subgrade. Backcalculation results of the granular layer are summarized in Table 4.

The results indicated that the backcalculated stiffnesses of the pulverized material were generally higher than that of the aggregate base material in Alturas. The average backcalculated stiffness of all three pulverized materials was higher than those of typical base material. Although the test results were only compared with one typical aggregate material with one specific pavement structure, this fact supports that the pavement structure with pulverized materials can generally reduce the elastic deformation and the stresses that cause rutting in the unbound layers.

Table 4. Backcalculation results for the FWD data.

		Alturas		Beckwourth	Poison Lake
		2001, 2005 & 2006		2005	2005 & 2006
Years		E^{*}_{AB} (MPa)	E^{**}_{PAB} (MPa)	E_{PAB} (MPa)	E_{PAB} (MPa)
Mean		164	240	259	279
Percentiles	25	95	151	136	231
	50	137	219	191	275
	75	223	316	310	324

* E_{AB} = Backcalculated stiffness of aggregate base in Alturas.
** E_{PAB} = Backcalculated stiffness of pulverized material.

5 SUMMARY AND CONCLUSIONS

This paper presents comprehensive laboratory and field testing results that evaluate the performance of a pavement rehabilitation strategy currently being trialed. The strategy includes pulverizing the failed asphalt concrete and a portion of the unbound base layer to create a granular base material. Test results were compared with different gradations of the pulverized materials and typical aggregate materials.

Basic index test results show that the pulverized material generally has lower density and higher permeability values than a typical aggregate material. It was found that both the static shear strength and the resilient modulus of the pulverized material were generally higher than a virgin aggregate material. The resilient moduli of the samples that were stabilized by either 3% lime or cement were higher than the unmodified materials as expected, however the higher cost of stabilization and potential issues with cracking should be considered prior to using lime or cement. In terms of permanent deformation properties, the pulverized material was not always superior to the virgin aggregate material.

Laboratory test results show that the performance of the pulverized material varied with changes in the gradation. However, this variation did not appear to significantly reduce the performance of the pulverized material when compared to a virgin material. Further research related to this variability is being performed.

Based on the backcalculation of FWD test results, it was observed that backcalculated stiffness of the pulverized material was generally higher than the typical Class 2 aggregate material although the test results were only compared with one typical pavement structure in California.

Based on laboratory and field test results, it is concluded that the pulverized material had equal or better performance than the typical aggregate material currently in use on California state highways. Since this rehabilitation strategy also costs less than the traditional approach of placing an asphalt concrete overlay on existing pavement, this rehabilitation strategy should be considered as one of the available options for flexible pavement rehabilitation.

ACKNOWLEGEMENTS

This paper describes research activities requested and sponsored by the California Department of Transportation (Caltrans), Division of Research and Innovation. Caltrans sponsorship is gratefully acknowledged. The contents of this paper reflect the views of the authors and do not reflect the official views or policies of the State of California or the Federal Highway Administration.

REFERENCES

Bejarano, M.O. & Harvey J.T. 2004. Deep in-situ recycling of asphalt concrete pavements as granular base – design, construction and performance evaluation experience in California. *Proceedings of 8th Conference on asphalt pavements for South Africa*, Sun City, North West Province, South Africa.

Caltrans 2000. Method of test for relative compaction of untreated and treated soils and aggregates. Test CTM-216, Caltrans Engineering Service Center, Transportation Laboratory, Sacramento, CA.

Gudishala, R. 2004. Development of resilient modulus prediction models for base and subgrade pavement layers from in situ devices test results. Master's thesis, Louisiana State University.

Heath, A.C. 2002. Modelling unsaturated granular pavement materials using bounding surface plasticity. PhD thesis, University of California at Berkeley.

Lekarp, F., Isacsson, U. & Dawson, A. 2000a. State of the art. I: resilient response of unbound aggregates. *ASCE Journal of Transportation Engineering* 126(1): 66–75.

Lekarp, F., Isacsson, U. & Dawson, A. 2000b. State of the art. II: permanent strain response of unbound aggreagates. *ASCE Journal of Transportation Engineering* 126(1): 76–83.

NCHRP 2004. Guide for Mechanistic-Empirical Design of New and Rehabilitated Pavement Structures. Report 1-37A, National Cooperative Highway Research Program.

Russo, M.A. 2000. Laboratory and field tests on aggregate base material for Caltrans accelerated pavement testing goal 5. Master's thesis, University of California at Berkeley

Thom, N. H., and Brown, S. F. 1988. The effect of grading and density on the mechanical properties of a crushed dolomitic limestone. *Proc., 14th ARRB Conf.*, Part 7, 94–100.

Advances in Transportation Geotechnics – Ellis, Yu, McDowell, Dawson & Thom (eds)

 ISBN 978-0-415-47590-7

Stress studies for permanent deformation calculations

L. Korkiala-Tanttu
VTT Technical Research Centre of Finland, Espoo, Finland

ABSTRACT: An analytical-mechanistic method for the calculation of permanent deformations in unbound pavement layers and subgrade has recently been developed in Technical Research Centre of Finland. The objective was to develop a relatively simple calculation method with a material model, which tie together permanent deformations to the most important effecting factors. The material model has been generated from the test results of accelerated pavement tests along with the complementary laboratory tests. This approach has created a new, important view to the research. The objective of this study is to compare the stress analysis done with the 3D and 2D modeling. The comparison of the 3D and 2D axisymmetric modeling showed that in the upper part of the pavement modeled with axisymmetric 2D overestimates stress responses. The stress analysis also proved that different non-linear elasto-plastic material models need separate material parameter C.

1 INTRODUCTION

Today need for evaluate of permanent deformations in pavement design is wide as well as global. The new procurement methods together with the functional requirements are underlining the demand for analytical and mechanical calculation methods for pavement rutting. The objective of the study was to compare stress responses of unbound pavement materials and subgrade analyzed with 2D and 3D models to give more confidence to the developed calculation method. Another objective was to study the material parameters of the calculation method The calculation method has been derived from accelerated pavement tests (APT) made in Finland with a Heavy Vehicle Simulator (HVS) and from the laboratory test results of Finnish deformation project. The studied unbound materials include crushed rock, sandy gravel, crushed gravel and sand. So far, the tested pavement materials have been the most common granular, unbound materials.

The previous stress response studies (Korkiala-Tanttu & Laaksonen 2004) have proven the benefits of the implementation of the elasto-plastic material models in the calculation of permanent deformations.

The new calculation method is based on the elasto-plastic stress responses and corresponding shear strength capacities of the material. The shear strength approach is used because the rutting is supposed to be dominated by the shear strength ratio.

The stress analysis has included three different calculation cases: the most common 2D axisymmetric, 2D plane strain (long continuous line loading) and true 3D cases. All the calculations have been conducted with the Plaxis code; 2D cases with Plaxis version 8.6 and 3D cases with Plaxis 3D version 2. The HVS test setup for Spring-Overload test (SO) was chosen as a test structure (Fig. 1). The wheel load is a dual wheel type.

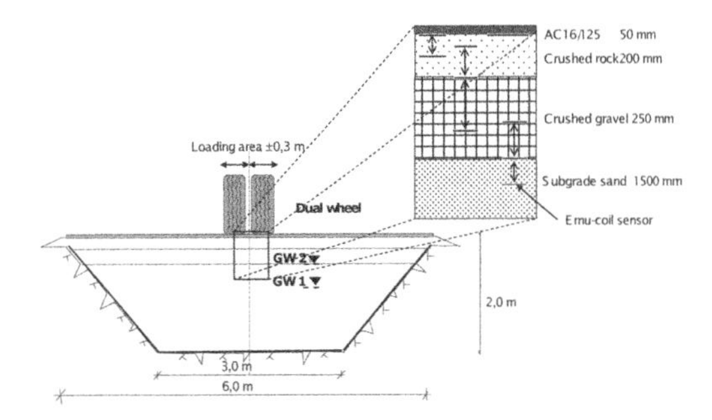

Figure 1. Cross section of the Spring-Overload test.

2 PREVIOUS STRESS RESPONSE COMPARISONS

The stress distribution studies of traffic load have shown that it is very important to calculate stresses in pavements with an elasto-plastic material model to avoid tensile stresses in unbound materials (Korkiala-Tanttu & Laaksonen 2004). The chosen material model drastically affects the stress distribution along with the permanent deformations, and also to some extent the resilient deformations.

Due to this, it is important to analyze stress responses for the permanent deformation calculations with a sophifisticated model than a conventional linear elastic material model. By using a linear elastic material model there is a high likelihood that the calculations will generate tensile stresses in the unbound pavement layers. The phenomenon is emphasized in pavement structures which are thinly paved or totally unpaved. These tensions will cause unrealistic stress concentrations with misleading information about permanent deformation sensitivity. Thus, the needed variables of the calculation method (shear stress ratio and material parameters) can only be defined from the elasto-plastic modeling.

The development of the calculation method for unbound granular materials and subgrade is presented in a previous paper by Korkiala-Tanttu (2005). The developed method is based on the number of loadings, shear stress ratio and material properties. In this method the permanent deformations will be calculated from the stress responses determined with a finite element program with a separated permanent deformation model. The basic equation (1) of the permanent deformation in unbound material is a relatively simple hyperbolic function.

$$\varepsilon_p = C \cdot N^b \cdot \frac{R}{1-R} \quad (1)$$

where ε_p = permanent vertical strain; C = permanent strain in the first loading cycle, a material state parameter; b = shear ratio parameter depending on the material; R = shear failure ratio = q/q_f, (q_f is defined in equation (2)); q = deviatoric stress, kPa.

3 MODELLING CASES

3.1 *Case 2D axisymmetric*

Axisymmetric geometry was generated from the test structure (Fig. 1) so that the axisymmetric geometry had the same area as the plain strain geometry. Likewise the dual wheel loading area was changed to the corresponding pressure area. The radius of the bottom of the test structure was 2.4 meter and the radius of the loading area was 0.2 meter.

3.2 *Case 2D plane strain*

For plain strain case the geometry corresponded to the actual cross section of the test structure (Fig. 1). The wheel load has been modeled as a continuous line loading. That is the reason why plain strain geometry is not widely applied to the pavement design. With infinite line loading, it is impossible to take into account the length (for heavy vehicles ~250 mm) of a wheel load. The plain strain modeling can only be used, when the shape of the geometry is studied.

Table 1. The modeling parameters for Mohr-Coulomb material of Spring-Overload test.

Material	Asphalt	Base course crushed rock	Subbase Crushed gravel	Subgrade Sand
Thickness, mm	50	200	250	1500
Modulus, MPa	5400	300-220-190	140-90	75
Poisson's ratio	0.3	0.35	0.35	0.35
Unit weight, kN/m^3	24	21.2	22.0	18.0
Cohesion, kPa	–	30	20	8
Friction angle (°)	–	43	45	36
Dilation angle (°)	–	13	15	6
K_0	1	0.32	0.30	0.42

3.3 *Case true 3D*

For the simplicity the sloped walls of the SO test structure were not taken into account in the 3D calculations. Thus the geometry had the average width of 4.5 meter. The length of the geometry was 10 meter. The dual wheel load was generated in compliance with the true pressure measurements of HVS tests. Otherwise the material layers and material parameters were the same as in 2D cases.

4 STRESS RESPONSE COMPARISONS

4.1 *Used material models for stress calculations*

Two different material models have been used for the axisymmetric and 3D calculations: linear elasto-plastic Mohr-Coulomb (MC) and non-linear elasto-plastic 'Hardening Soil' (HS). The hardening soil model (HS) is a non-linear elasto-plastic material model with Mohr-Coulomb failure criterion. A more detailed description of the material model is presented in Plaxis's manual (Brinkgreve 2002). The plain strain case has been studied only with MC material model. The analysis is based on the calculated deviatoric and vertical stress components.

4.2 *Material parameters*

The Table 1 presents the applied MC and Table 2 HS material parameters for in the modeling. It is important to note that the strength properties applied in the hardening soil model are higher than the normally applied static values. Several studies (Konrad & Juneau 2006, Hoff 1999, Courage 1999) have shown that the friction angle of well-compacted, partly saturated crushed rock in cyclic loading tests is typically between 50° to 60° and the apparent cohesion between 15 to 40 kPa.

Table 2. The modeling parameters for Hardening Soil material of Spring-Overload test (reference stress 100 kPa).

Material	Material model	DOC (%)/ w (%)	Friction angle°	Cohesion kN/m^2	Unloading/reloading modulus, MPa	Compression modulus, MPa	Deviatoric modulus, MPa
Asphalt concrete	LE*	–	–	–	–	–	5400*
Crushed rock	HS†	95.8/4.6	55	20	750	173	250
Sandy gravel	HS†	98.1/7.3	58	20	900	201	330
Sand (dry)	HS†	101.4/9.9	40	15	420	110	120
Sand (moist)	HS†	–	36	8	420	95	100

*linear elastic Young's modulus
†hardening soil

4.3 *Shear stress ratio*

According to Mohr-Coulomb's failure criteria, deviatoric stress in triaxial tests at failure q_f can be estimated with the help of equations 2–4. These equations are valid under the centre of the loading, where loading is axisymmetric and the angle of the major principal stress concurs with the vertical axis.

$$q_f = q_0 + M \cdot p' \quad (2)$$

$$M = \frac{6 \cdot \sin\phi}{3 - \sin\phi} \quad (3)$$

$$q_0 = \frac{c \cdot 6 \cdot \cos\phi}{3 - \sin\phi} \quad (4)$$

here q_f = deviatoric stress in failure; q_0 = deviatoric stress, when $p' = 0$; c = cohesion; M = slope of the failure line in p'–q space in triaxial test; p' = hyrostatic pressure; and φ = friction angle.

5 RESULTS

5.1 *Stress responses – deviatoric stress*

The calculated deviatoric stress responses with the 2D and 3D are compared with each other in Fig. 2. The stress components have been calculated under the centre line of the loading and the wheel load has been 50 kN.

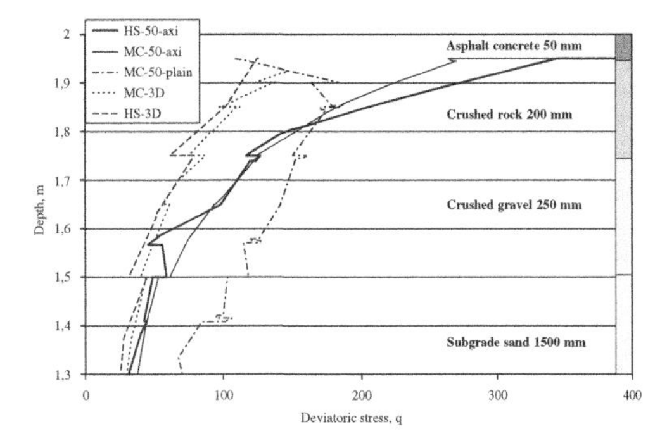

Figure 2. Comparison of the deviatoric stresses in the centre line of the loading (HS = Hardening soil, MC = Mohr-Coulomb) Spring-Overload test.

5.2 *Stress responses – vertical stress*

The calculated vertical stress responses with the 2D and 3D are compared with each other in Fig. 3. The stress components have been calculated under the centre line of the loading and the wheel load has been 50 kN. Due to the stress sign rules of Plaxis compression stresses have negative values. Also the measured earth pressure at the top of the subgrade sand is presented in the Fig. 3.

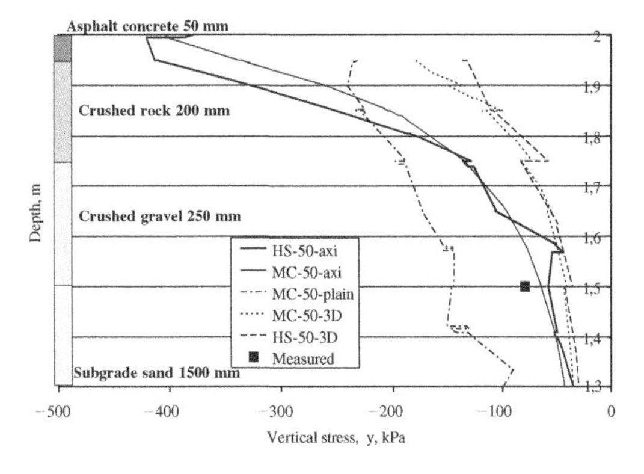

Figure 3. Comparison of the vertical stresses in the centre line of the loading Spring-Overload test.

5.3 *Stress responses – shear stress ratio*

The calculated shear stress ratios are compared with each other in Fig. 4. The shear stress ratio equation is only valid under the centre of the loading, where loading is axisymmetric and the angle of the major principal stress concurs with the vertical axis. Therefore, it can not be applied in the 3D results of two concurrent wheel loads, because the major principal

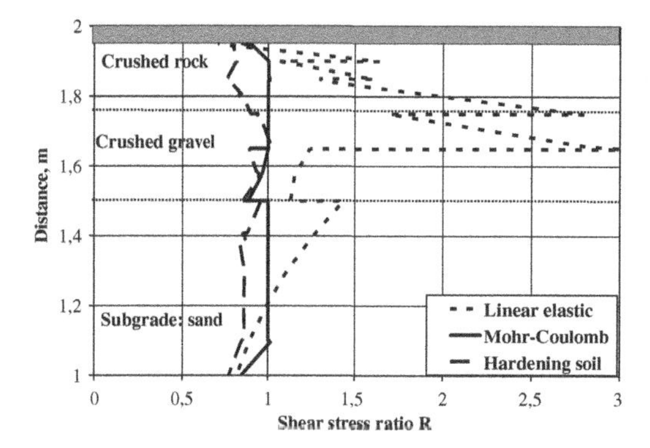

Figure 4. Comparison of the shear stress ratio in the centre line of the loading Spring-Overload test.

axis differs from the vertical axis even under the centre line of one wheel load.

6 DISCUSSION

6.1 *Deviatoric stresses*

The stress comparisons clearly show that the deviatoric stresses with HS material model give smaller deviatoric stresses in both 2D and 3D cases. This is quite natural, because the HS model has a non-linear, hyperbolic material model for the elastic deformations. The calculated deviatoric stresses in 3D for MC and HS models were close to each other. The average relative difference between 3D MC and HS calculated stresses was 11% and it varied between −1 to 21%. For 2D the difference was smallerer: in average MC defined stresses were about 4% bigger than HS defined.

At the depth of 500 mm the deviatoric stresses for 3D and 2D axisymmetric cases approached each other and the differences were less than 10 kPa. The differences between plain strain and axisymmetric modeling were the largest in the lower part of the structure. These results are all expected. In the plain strain case the deviatoric stresses decreased surprisingly slowly downwards. Thus, also deformations can easily be overestimated if plain strain modeling is used. After all, the plain strain modeling can only be recommended to be used when the shape of the pavement is analyzed (like shoulder and side slope steepness).

6.2 *Vertical stress*

The vertical stress component shows the same phenomenon as deviatoric stress comparison: stresses calculated with MC and HS models are relatively close to each other. The 2D and 3D stresses separate from each other in the upper part of the pavement (to the depth of about 500 mm). The 3D stress calculation is supposed to give more reliable results, because the load distribution can be modeled more correctly. It is also probable that with the single wheel load the difference between 3D and 2D is smaller in the upper part of the pavement.

Again the plain strain case's vertical stresses decrease very slowly downwards.

6.3 *Shear stress ratio*

If the HS and MC stress responses were reasonably close to each other, the shear stress ratios separated more clearly. The stress calculations for the permanent deformations should be done with HS model. If a suitable non-linear elasto-plastic model is not available, it is also possible to use MC type model. In that case the denominator in equation 1 should be of form (X-R), where X is about 1.05. This correction must be done, because otherwise the deformations become infinite.

6.4 *Material parameters*

The material parameter C for the permanent deformation method has been defined from the triaxial laboratory tests. Its values have also been fitted to match stress responses calculated with MC model. The problem with parameter definition was that the amount of full-scale tests was only two. Because in the axisymmetric 2D case the HS stress responses and especially shear strength ratio R are smaller than MC's, the parameter C needs redefining. Otherwise the method will give far too low deformations. Table 3 presents the material parameters used in the permanent deformation method for the MC model and Table 4 for HS model. The material parameter C has been estimated to have about 2...4 times larger values for the HS model than for the MC model. The value of parameter C has been defined from the Finnish accelerated pavement tests (Korkiala-Tanttu et al. 2003a & 2003b).

6.5 *Permanent deformations*

The permanent deformations have been calculated from the stress responses of 2D axisymmetric modeling case and they have been compared with the measured permanent strains for each layer. The permanent strains have been measured with Emu-Coils (Korkiala-Tanttu et al. 2003b). The 2D axisymmetric stress responses were chosen because shear stress ratio R can not be determined from the 3D results (see chapter 5.2). Both MC and HS models were applied to evaluate the differences between them. Fig. 5 illustrates the calculation results of the SO structure and the loading of 50 kN and Fig. 6 the loading of 70 kN.

The calculation approaches mostly underestimate the permanent strains especially for the high load of

Table 3. The parameters for permanent strain calculations for Mohr-Coulomb (MC) model.

Material	Parameter d	Parameter c	C (%)	DOC (%)	w (%)
HVS: Sand	0.16	0.21	0.0038 (±0.001)	95	8
HVS: Sandy gravel	0.18	0.15	0.0049 (±0.003)	97	5...7
HVS: Sandy gravel	0.18	0.15	0.0021 (±0.001)	100	5...7
HVS: Crushed rock	0.18	0.05	0.012 (±0.004)	97	4...5

Table 4. The parameters for permanent strain calculations for Hardening Soil (HS) model.

Material	Parameter d	Parameter c	C (%)	DOC (%)	w (%)
HVS: Sand	0.16	0.21	0.016 (±0.004)	95	8
HVS: Sand	0.16	0.21	0.035 (±0.01)	95	saturated
HVS: Sandy gravel	0.18	0.15	0.02 (±0.01)	97	5...7
HVS: Sandy gravel	0.18	0.15	0.008 (±0.003)	100	5...7
HVS: Crushed rock	0.18	0.05	0.048 (±0.016)	97	4...5

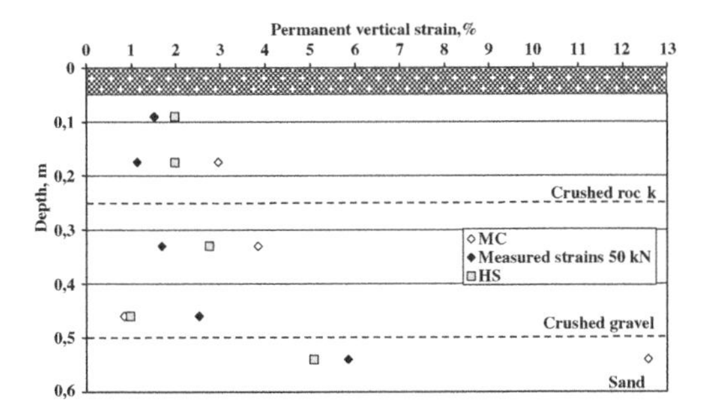

Figure 5. Comparison of the vertical strains in the centre line of SO structure with the 50 kN loading.

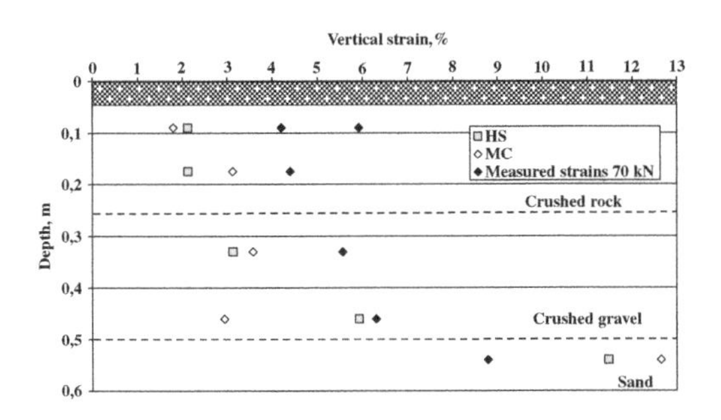

Figure 6. Comparison of the vertical strains in the centre line of SO structure with the 70 kN loading.

70 kN. Yet, for the subgrade sand the MC model gives the highest values for both load levels. In general the underestimation is much bigger for the MC model than for the HS model. The problem with the high load level is that when the maximum shear strength ratio 1 for MC model has been reached the permanent deformations will be the dependent on the load level. Another reason for the underestimation is that the method does not take into account the rotation of the principal axis. The studies of Kim & Tutumluer (2006) have been proven that the rotation of the principal axis has a significant effect on the permanent deformations.

The vertical strains are calculated from layer to layer, multiplied with the thickness of the layer and then they are summed up to get the total rut depth.

The measured error of Emu-Coil pairs according to Janoo et al. studies (1999) was within ±1 mm, which is also at the threshold limit of the ability to detect permanent deformations. This error corresponds to the %-unit error of ±0.5% to 1.25% depending on the distance of the coils.

7 CONCLUSIONS

From the stress analysis it can be concluded that:

- 2D axisymmetric modeling gives quite reasonable stress distributions in the lower part of the pavement structure,
- in the upper part 2D axisymmetric stresses overestimate the stress state especially for the dual wheel load,
- the differences between HS and MC analyzed stresses in 3D case is small,
- 2D plain strain modeling can be used for the scaling of different pavement geometries, but it is not recommend to be used in the deformation calculations because it overestimates greatly the stress state in the lower part of the pavement,

– 3D stress response can not be applied to the developed calculation method, because the maximum deviatoric stress calculation method is not valid in real 3D conditions.

The recommendation is that the stress responses for permanent deformation calculation should be done using 2D axisymmetric calculations. Preferably HS approach should be used. If MC is used, the denominator in equation 1 should be of form (X-R), where X is about 1.05. This correction has to be done, because otherwise the deformations will be infinite. Also the material parameter C should be chosen according to the used material model.

The permanent deformation method gives tolerable results for the normal load levels. For the high load levels it will probably underestimate the strains. The method suits also to the comparison of different pavement structures and their rutting sensitivity.

REFERENCES

Brinkgreve R.B.J. 2002. *Plaxis 2D Version 8 manual*. Material models, Delft, Balkema.

COURAGE. 1999. *Construction with Unbound Road Aggregates in Europe. Final report*, Contract No.:RO-97-SC.2056, 4th Framework Programme (1994–1998), Road Transport Research, http://www.civeng.nottingham.ac.uk/courage/, 123 p.

Hoff I. 1999. Material Properties of Unbound Aggregates for Pavement Structures, NTNU Norges teknisk-naturvetenskaplige universitet, Trondheim, PhD Thesis, 198 p.

Janoo V., Irwin L., Knuth K., Dawson A. R. and Eaton R. 1999. Use of inductive coils to measure dynamic and permanent pavement strains, *Proc. Accelerated Pavement Testing Conf.*, Reno, Nevada, USA. p.19.

Kim I. and Tutumluer E. 2006. Field Validation of Airport Pavement Granular Layer Rutting Predictions, *Annual Meeting of TRB*, Washington, DC, January, 26 p.

Korkiala-Tanttu L. and Laaksonen R. 2004. Modelling of the stress state and deformations of APT tests, *Proc. of the 2nd International Conference on Accelerated Pavement Testing, September 26.9.–29.9.2004*, Minneapolis, Minnesota, 22 p., CD-rom proceedings, http://mnroad.dot.state.mn.us/research/MnROAD_Project/index_files/pdfs/KorkialaTanthu_L.pdf.

Korkiala-Tanttu L. 2005. A new material model for permanent deformations in pavements, *Proc. of the Seventh Conference on Bearing Capacity of Roads and Airfields*, Trondheim 27.6.–29.6.2005, Editor Horvli I., 10 p.

Korkiala-Tanttu L., Jauhiainen P., Halonen P., Laaksonen R., Juvankoski M., Kangas H. and Sikiö J. 2003a. Effect of steepness of side-slope on rutting, Helsinki, *Finnish Road Administration, Finnra Reports 19/2003*, 40 p. + 17 app. http://alk.tiehallinto.fi/julkaisut/pdf/3200810e.pdf.

Korkiala-Tanttu L., Laaksonen, R., Törnqvist, J. 2003b. Effect of the spring and overload on the rutting of a low-volume road, HVS-Nordic-research, Helsinki, *Finnish Road Administration, Finnra Reports 22/2003*, 39 p. + app. http://alk.tiehallinto.fi/julkaisut/pdf/3200810e.pdf.

Konrad J.M., and Juneau O. 2006. Limit-State Curve of Base-Course Material and Its Relevance for Resilient Modulus Testing, *Journal of Geotechnical and Geoenvironmental Engineering*, Vol. 132, No 2 February 2006, 173–182.

Advances in Transportation Geotechnics – Ellis, Yu, McDowell, Dawson & Thom (eds)
© 2008 Taylor & Francis Group, London, ISBN 978-0-415-47590-7

Equivalence between dry bound macadam and other types of base layers for flexible pavements

A. Mateos
Transport Research Center of CEDEX, El Goloso, Madrid, Spain

C.M. Ribas Rotger
The Council of Majorca, Palma, Balearic Islands, Spain

ABSTRACT: This paper presents the results of the report made by CEDEX for the Government of the Balearic Islands in 2001. The purpose of the research was to allow for the design of pavements with dry bound macadam base layers in correspondence with those included in the Spanish Design Catalogue in force at that time and did not consider macadam any more. Macadam was used frequently in Spain until the 1970s, when its use began to decline, being replaced by the use of graded aggregates, whose production and laying were more easily mechanizable. The objective of the report was to establish an equivalence between the base layers made from graded aggregate and those made from macadam. This equivalence was established from two points of view: empirical and analytical. The conclusion was that the replacement of a graded aggregate base layer by a macadam layer not only is not harmful for the pavement performance but also improves it. Furthermore, this replacement allows for a reduction in the thickness of the macadam layer or bituminous layer.

1 INTRODUCTION

Dry bound macadam was used a lot in Spain until the 1970s. However, with the introduction of graded aggregates, whose production and laying were easier to mechanize, dry bound macadam was abandoned in most regions of the country. Despite macadam being no longer practically considered in the Spanish Design Catalogue of 1989, called "Instrucción 6.1 y 2 IC", and in the subsequent national design regulations, this material has been commonly used in the Balearic Islands. This is why the structural design of pavements with macadam base layer was made according to local experience.

In order to back up the use of macadam with a greater theoretical basis, in 2000 the Public Works, House and Transports Department of the Government of the Balearic Islands commissioned CEDEX (a research institution, attached to the Ministry for Public Works of Spain, that provides multidisciplinary support in the different areas of the Civil Engineering Sector) to produce the report "Equivalence between dry bound macadam and other types of base layers". This study was made by Mr. Ángel Mateos under the supervision of Mr. Aurelio Ruiz, and the coordination of Mr. Carlos Ribas on the part of the Balearic Government.

The content and the conclusions of that report are presented concisely hereinafter.

2 RESEARCH APPROACH

The main objective of the report was to settle an equivalence between base layers made from graded aggregate and those made from macadam. There was a need to establish a correspondence between the pavement sections of the Spanish Design Catalogue in force at that time and those having a macadam base layer.

Graded aggregate was the material present in the "Instrucción 6.1 y 2 IC" that was taken as reference to establish the correspondence given that the performance of a macadam layer is more similar to a quality granular material rather than a soil-cement or a bituminous base.

This equivalence was established from two points of view:

1. Empirical: the acquired experience with the performance of pavements built with macadam is compared to others with conventional base layers, such as graded aggregate or asphalt mixtures. The accumulated experience with the use of macadam as a pavement base layer was compiled: a bibliographical study as well as, especially, an analysis of

their own experience for the performance of these pavements in the Balearic Islands were carried out.
2. Analytical: it requires the mechanical modelling of the pavement structure and the application of several deterioration criteria, as common practice in analytical design. The correspondence between a pavement section with graded aggregates as base layer and another with macadam as base layer is established with the goal of satisfying the same deterioration criteria.

3 DRY BOUND MACADAM

This material came into being thanks to the Scottish engineer McAdam, at the end of the 1820s, to supplant the pavement system of the 18th century based on crushed stone mixed with earth. Since then its use was common, but in the 1960s its utilization began to decline.

The Spanish standard specifications for highway and bridge works (called "PG-3") considered the use of dry bound macadam in its article 502, which no longer exists. In the "PG-3", it is defined as a material made up of gap grading aggregates, prepared by laying and compacting the coarse aggregate whose voids are then filled up with a fine aggregate, called "recebo". The "PG-3" imposes a series of requirements related to the constituent materials (grading and resistance to wear for the coarse aggregate and grading and plasticity for the fine aggregate) as well as for the construction process and the finished surface.

The structural strength of a macadam layer is largely due to the assembly and friction of the coarse elements, in direct contact with one another, being fundamental to guarantee they fit appropriately. This is the reason for the requirements of compaction, face fracture and hardness demanded for the coarse aggregate.

The fine aggregate has to fill up the voids left by the coarse aggregate, giving stability to the whole. It does not increase perceptibly the internal friction of the macadam, although it reduces its deformability significantly. It also has an influence on the pavement permeability. This aggregate has to fulfill requirements related to grading and quality (plastic fine aggregates can produce deformations if water enters).

It is recommended to lay a granular subbase layer underneath macadam for several reasons:

- To reduce stresses on the subgrade.
- To avoid the collapse of the coarse aggregate or the punching of the subgrade.
- To avoid the contamination of the pavement with fines from the subgrade.
- To avoid the accumulation of water below the macadam layer.

This structural layout provides both an excellent support for the bituminous layer and an effective protection for the subgrade.

A peculiarity of these pavements is the possible circulation of water into the macadam layer. For this reason it is advisable to extend this layer under the shoulder to facilitate drainage.

4 USE OF MACADAM IN THE BALEARIC ISLANDS

This material has been traditionally used in the Balearic Islands, and there is an extensive accumulated experience with it. There are numerous quarries with wide experience in producing this material, as well as construction companies used to laying it. This report provides information about its characteristics and use in the Balearic Islands.

It was verified that the macadam used in the Balearic Islands meets sufficiently the specifications required by the "PG-3". It would only be necessary to make a few modifications so that the macadam produced fulfills completely the requirements of the "PG-3". The coarse aggregate used corresponds, in its practical entirety, to the grading envelope M1.

There are numerous road sections built with dry bound macadam in the Balearic Islands. The common section used consisted of a 200 mm thick macadam layer laid on a 150 mm thick subbase layer, and a bituminous layer of circa 120 mm thick. As a result of the analysis of eleven examples (including single carriageway roads and motorways, as well as different traffic and subgrade categories), the following tendencies can be deduced:

- The subbase layer used to be 150 mm thick when the subgrade category was E2 ($10 \leq CBR < 20$). For subgrades E1 ($5 \leq CBR < 10$) its thickness was increased to 200 or 250 mm. And for subgrades E3 ($CBR \geq 20$), sometimes its thickness was maintained and sometimes it was reduced or even the layer was eliminated.
- The macadam layer was generally 200 mm thick, although there were smaller thicknesses (170 mm) and larger (up to 300 mm).
- The bituminous layer generally had a mean thickness of 120 mm, which could range from 80 to 250 mm.

Aiming at studying the structural performance of pavements with macadam, sections considered as example and built before 1995 (five in all) were analyzed; the traffic volume carried ranging from the construction date to the date of the research, or date of its strengthening, was estimated, and a visual inspection was carried out in order to evaluate their state of deterioration. Sections built after

1995 were having an appropriate performance, without significant deteriorations detected.

5 STRUCTURAL CHARACTERIZATION

5.1 *Plate-bearing tests*

From this test, it is difficult to obtain a resilient modulus to use afterwards in the calculation. The stress level is very much higher than the one normally present in the material after construction. Moreover, the load effect itself produces a significant compaction, so the modulus measured in the second seating load (E_{v2}) is very much higher than that measured in the first one (E_{v1}).

Nevertheless, this test can be very useful to compare the structural capacity of macadam and graded aggregate. Nowadays, the "PG-3" requires for graded aggregate used as base layer to have a modulus E_{v2} between 100 (for heavy vehicles annual average daily traffic up to 199 heavy vehicles per day) and 180 MPa (for AADT greater than 799 heavy vehicles per day). The graded aggregates normally used as base layers usually comply with these requirements.

There were plate-bearing tests on macadam for a series of roads, which could be considered representative of the type of macadam used in the Balearic Islands. In most cases, only the first seating load was carried out (this result is conservative, since E_{v2} is always higher than E_{v1}). In any case, the values of E_{v1} were clearly higher than 100 MPa, and, when the second seating load was carried out, the modulus was very much higher than 180 MPa.

5.2 *Falling weight deflectometer tests*

Nowadays, this is one of the more widespread tests for structural characterization. The falling weight deflectometer applies a load on the pavement, measuring the deflection at several radial distances. This test allows to estimate the resilient modulus of several pavement layers by means of back-calculation.

Back-calculation is usually carried out through an iterative process which is conceptually simple:

1. It begins with an initial estimation of the resilient modulus of the layers.
2. With these moduli, the deflections are calculated by means of a structural calculation program.
3. The calculated deflections are compared to the measured ones, and, if they are sufficiently similar, they are accepted as valid.
4. In case the results are not satisfactory, the modulus is readjusted and the process is repeated again until the obtained results are satisfactory.

This back-calculation process implies a series of simplifications: materials are considered as linear elastic, dynamic effects are not taken into account, to presuppose a thickness for the layers and another simplifications. However, nowadays it is one of the most widely methods used for the structural characterization of the layers of flexible pavements.

This research was based on a condition survey carried out in 1995 on one of the analyzed examples of Balearic roads: the section C-717 (nowadays Ma19 A) – S'Arenal of the road PM-19 (nowadays Ma-19). This was, in turn, divided into two more sections due to different pavement structures, both with 12 cm of asphalt mixture over 20 cm of macadam, but with different subgrade and thickness of the subbase layer.

In the survey, points roughly 30 m distant from one another were tested, so that there were a large number of data. The average modulus obtained for the macadam was 720 MPa in one section and 910 MPa in the other one.

5.3 *References in the bibliography*

The bibliography related to the utilization of macadam layers is not abundant. Most of it comes from South Africa, where this type of material has been frequently used.

The South African design guide TRH4 considers equal macadam and graded aggregate layers. However, a study carried out in this country (Burrow 1975) verified that bases of macadam were having a higher structural performance than those of graded aggregate.

Also, a series of laboratory tests published in an article (Philips et al. 1993) proves that the resistant characteristics of macadam are superior to those of graded aggregate.

Some authors have pointed out that the substitution of a macadam layer for a graded aggregate one with the same thickness is too conservative, and that less thick macadam layers can be used (Philips et al. 1993).

Several articles with results of the falling weight deflectometer back-calculation on pavements with macadam layers were found of special interest:

- Visser et al. (1999): Average macadam moduli between 490 and 860 MPa.
- Evdorides et al. (1996): Macadam modulus = 700 MPa.

As the South African Catalogue, the Spanish Design Catalogue of 1975 for flexible pavements (called "Instrucción 6.1 IC 1975") also considered equal macadam and graded aggregate.

5.4 *Structural comparison: macadam/graded aggregate*

For graded aggregate, both several catalogues and the results of the falling weight deflectometer back-calculation infer a modulus which, depending on the conditions (material quality, moisture content, situation of the layer, etc.), can vary considerably, from the most pessimistic values (about 100–150 MPa) to the most optimistic ones (500–600 MPa).

For macadam, the range of moduli obtained from the falling weight deflectometer test is higher than that of graded aggregate. To summarize:

- Falling weight deflectometer tests in the Balearic Islands: 720–910 MPa.
- Visser et al. (1999): 490–860 MPa (depending on the quality of the support material).
- Evdorides et al. (1996): 700 MPa.

Thus, the mechanical features of macadam are significantly superior to those of graded aggregate, both due to its strength and its resilient modulus.

With a view to pavement modelling by means of a multi-layer linear elastic model, values for the macadam and graded aggregate moduli have to be established. From a conservative point of view, the graded aggregate modulus should be overestimated and the macadam modulus underestimated since the intention is to substitute the second one for the former. In view of the results obtained with the tests, as well as those compiled from the existing bibliography and normative, a value of 500 MPa for graded aggregate and 700 MPa for macadam can be considered reasonably conservative. These values are used for the analytical determination of the equivalence macadam/graded aggregate.

6 EMPIRICAL DETERMINATION OF THE EQUIVALENCE DRY BOUND MACADAM/GRADED AGGREGATE

As it has been mentioned, concerning the pavement sections of the oldest roads considered as example, the traffic carried ranging from the construction date to the date of the research or its strengthening was estimated, and a visual inspection was carried out in order to evaluate their state. With these data, the real pavement section was compared to the section that would have been necessary, using graded aggregate, according to the "Instrucción 6.1 y 2 IC". In practically all cases, the section with graded aggregate would have required thicker bituminous layers and similar or thicker base layers. By way of example, in the case of the motorway PM-19 (section C-717 – S'Arenal) the following sections were compared:

- Real section: subgrade E2/150 mm granular subbase/200 mm macadam/120 mm asphalt mixture.
- Required section according to the "Instrucción 6.1 y 2 IC": subgrade E2/250 mm natural graded aggregate/250 mm graded aggregate/200 mm asphalt mixture.

In short, pavement sections built with macadam were designed with bituminous surfacings much less thick than those considered in the "Instrucción 6.1 y 2 IC" for pavements with granular layers though having similar performance, with strengthenings in service periods between 8 and 13 years.

Therefore, after the analysis of the treatment of the macadam in several design catalogues, the performance of the sections built with macadam in the Balearic Islands, the results of several structural characterization tests carried out on macadam from the Balearic Islands and several bibliographic references about the structural capacity and performance of the sections with macadam, it can be concluded that the substitution of a macadam layer for a graded aggregate one with the same thickness not only is not harmful for the pavement performance but also improves it. This replacement may be even too conservative and could allow for a reduction of the thickness of the base and/or bituminous layer.

7 ANALYTICAL DETERMINATION OF THE EQUIVALENCE DRY BOUND MACADAM / GRADED AGGREGATE

7.1 *Analytical pavement design*

The analytical pavement design is usually carried out by means of obtaining the structural response of the pavement (stresses, strains, etc.) to the load from a vehicle, and applying several deterioration criteria afterwards. Consequently, it requires two main consecutive steps:

1. Structural response: obtaining the structural response to the load of a vehicle.
2. Deterioration: obtaining the pavement deterioration from the structural response.

The most used models in each one of the stages mentioned above are the multi-layer linear elastic as structural response model and the use of fatigue laws as deterioration model.

A fatigue law is a function that relates certain structural magnitudes to the number of allowable repetitions until the failure. Only two deterioration mechanisms are usually considered:

1. Fatigue cracking of the asphalt mixture. It is determined from the horizontal tensile strain at the bottom of the asphalt layer (ε_t).
2. Permanent deformations coming from the subgrade. It is determined from the vertical compressive strain at the top of the subgrade (ε_c).

For each one of the deterioration criteria or mechanisms, the number of allowable cycles until the exhaustion is obtained. The critical deterioration mechanism, that with the lowest number of cycles, will be the one that determines the serviceable life of the section. Depending on the section type, the critical deterioration mechanism will be the fatigue cracking of the mixture or the permanent deformations coming from the subgrade. For sections with a certain thickness of asphalt mixture (normally from 120–150 mm), the critical mechanism is the first one (cracking).

The mechanical equivalence between a section with a base layer made from graded aggregate and another with a base layer made from macadam is established with the goal of limiting the strains (ε_t and ε_c) to the same value for both sections. Nevertheless, if the critical mechanism is known a priori, the corresponding strain will determine the equivalence.

In sections with graded aggregate as base layer included in the "Instrucción 6.1 y 2 IC" for heavy vehicles annual average daily traffic greater than or equal to 50 heavy vehicles per day, the asphalt mixture is at least 150 mm thick. So the critical mechanism is cracking. Consequently, the substitution of a macadam base layer for a graded aggregate one will be established under the condition of limiting the horizontal strain at the bottom of the asphalt layer (ε_t) to the same value. For AADT lower than 50 heavy vehicles per day, the approach is different, as the asphalt mixture is 50 mm thick or there is simply a surface dressing, and the critical deterioration is the permanent deformation from the subgrade (ε_c).

7.2 *Structural model*

In the multi-layer linear elastic model, each layer has to be characterized by its thickness an its elastic properties (E and ν). The criteria for heir characterization were:

- Thicknesses (h): design thicknesses.
- Poisson ratios (ν). Common values for each material:
  - Asphalt mixture: $\nu = 0.35$
  - Graded aggregate: $\nu = 0.35$
  - Macadam: $\nu = 0.35$
  - Subgrade: $\nu = 0.40$
- Resilient moduli (E). Criteria validated by practice:
  - Asphalt mixture: 3000/6000/10000 MPa (according to the seasonal period)
  - Subgrade: E(MPa) = 10 × CBR
  - Granular layers (graded aggregates): according to the modulus of the underlying layer: $E_s = 0.2 \times E_i \times h^{0.45}$ (where E_s = modulus of the upper layer; E_i = modulus of the lower layer; and h = thickness of the upper layer in mm). This formula requires the establishment of maximum moduli for the materials, otherwise too high values could be obtained, especially for base layers. The maximum moduli assumed for each material were 500 MPa for graded aggregate and 300 MPa for natural graded aggregate.
  - Macadam: 700 MPa

Regarding the load, from the point of view of the establishment of the equivalence a single axle of 13 t with dual wheels and a tire pressure of 0.8 MPa was considered. The modelling of the load was done by means of two circular uniformly loaded areas.

7.3 *Justification of alternatives for the substitution of macadam for graded aggregate*

With these premises, the sections of the "Instrucción 6.1 y 2 IC" with graded aggregate base were analyzed (sections with a graded aggregate base on a natural graded aggregate subbase are the most similar to a typical macadam pavement), and alternative sections with macadam were calculated under the condition that the resulting critical strains were lower than those calculated for the original sections.

As the usual available material in the Balearic Islands comes from crushing, alternatives with macadam and a subbase made from graded aggregate, instead of natural graded aggregate, were proposed in those cases in which this improvement allowed for an additional reduction in the thickness of the asphalt mixture or any other layer. In any case, it is obvious that graded aggregate can be used instead of natural graded aggregate whichever it is the alternative proposed.

In general, the results were alternative sections with macadam in which the thickness of asphalt mixture of the original section was reduced in 20 or 30 mm.

In sections for heavy vehicles annual average daily traffic lower than 50 heavy vehicles per day, as the critical deterioration is the permanent deformation from the subgrade, the increment in the base layer modulus due to the substitution of macadam for graded aggregate does not have a significant effect. That is why in this case, from an analytical point of view, the substitution did not mean an improvement in the structural capacity of the section that could allow for a significant reduction of the thickness of some of the pavement layers.

As an example, the table prepared for the traffic category T2 (200–800 heavy vehicles per day) is shown below with the most appropriate alternatives from the point of view of construction:

8 CONCLUSIONS

From the performance of the sections built with macadam in the Balearic Islands, the results of several structural characterization tests carried out on macadam from the Balearic Islands, the consideration of the macadam in several design catalogues and the bibliographic references about the structural capacity

Table 1. Pavement sections for the traffic category T2.

	Section number							
Layers (thickness in cm)	212	212 m1*	212 m2*	222	222 m1*	222 m2*	232	232 m*
Asphalt mixture	25	23	22	20	18	17	20	18
Dry bound macadam		25	25		25	25		20
Graded aggregate	25		25	25		25	25	
Natural graded aggregate	25	25		25	25			15
Subgrade	E1			E2			E3	

*Alternative sections with macadam.

and performance of the sections with macadam base, the following conclusions can be drawn:

- The mechanical features of macadam are significantly superior to those of graded aggregate, both due to its strength and its resilient modulus.
- The replacement of a graded aggregate layer by a macadam layer with the same thickness not only is not harmful for the pavement performance but also improves it.
- This replacement is too conservative and less thick layers of macadam and/or asphalt mixture can be used.

Once the equivalence between dry bound macadam and graded aggregate has been quantified applying common criteria for the analytical pavement design, sections with macadam as base layer can be obtained as an alternative to the original sections of the "Instrucción 6.1 y 2 IC" with graded aggregate as base layer. The critical parameter of these macadam sections does not exceed the one from the original sections. Generally, the alternative sections allow for a reduction in the thickness of the bituminous layer.

REFERENCES

Burrow 1975. *Investigation of existing road pavements in the Transvaal*. South Africa.

CEDEX, Consejería de Obras Públicas, Vivienda y Transportes del Gobierno de las Islas Baleares 2001. *Equivalencias del macadam con los distintos tipos de bases*. Madrid.

Evdorides et al. 1996. A knowledge-based analysis process for road pavement condition assessment.

Philips et al. 1993. Comparative strength parameters of waterbound macadam and crushed stone basecourses.

Visser et al. 1999. Design Guidelines for Low-volume Macadam Pavements in South Africa.

Advances in Transportation Geotechnics – Ellis, Yu, McDowell, Dawson & Thom (eds)
© 2008 Taylor & Francis Group, London, ISBN 978-0-415-47590-7

Types and amounts of fines affecting aggregate behaviour

D. Mishra & E. Tutumluer
Department of Civil and Environmental Engineering, University of Illinois, Urbana, Illinois, USA

A.A. Butt
Engineering and Research International, Inc., Savoy, Illinois, USA

ABSTRACT: Construction of a working platform is often needed on soft, unstable soils to provide sufficient stability and adequate support for equipment mobility and paving operations without developing excessive rutting. Subgrade removal and aggregate replacement for cover is one of the most commonly used options for treating soft, unstable soils. Aggregate type and quality are important factors governing the required treatment/replacement thickness. This paper presents preliminary laboratory findings from an ongoing research project at the University of Illinois aimed to characterize behavior of three types of aggregates commonly used in Illinois for subgrade replacement and subbase. Samples were prepared with consistent engineered gradations, and their moisture-density and strength behavior were studied at different amount of fines, as well as with different types of fines. Plastic fines were found to reduce aggregate strength significantly, as compared to the same amount of non-plastic fines.

1 INTRODUCTION

1.1 *Importance of working platform*

Construction of a working platform is important on soft, unstable soils to facilitate paving operations without developing excessive rutting under the construction equipment. The subgrade must be sufficiently stable to: (i) prevent excessive rutting and shoving during construction, (ii) provide good support for placement and compaction of paving layers, (iii) limit pavement resilient deflections after construction, and (iv) restrict the development of excessive permanent deformation buildup during pavement service life, which can happen due to climatic effects, such as a significant moisture content increase during spring thaw. Construction of working platform plays an important role in pavement construction in the state of Illinois, USA, as the local soils are relatively soft (Tutumluer et al., 2005).

1.2 *Need and importance of project*

The Subgrade Stability Manual (SSM) currently used by the Illinois Department of Transportation (IDOT) recommends minimum levels of strength and stiffness that need to be achieved in the subgrade soil to a depth influenced by construction traffic to ensure adequate equipment mobility and prevent excessive rutting of vehicle tires (IDOT 2005). The subgrade treatment/replacement thickness is determined as a function of the soil's Immediate Bearing Value (IBV), which is a measure of soil strength obtained by conducting the standard bearing ratio test, commonly known as the California Bearing Ratio (CBR) according to ASTM D 1883, on molded soil samples immediately after compaction (without soaking). However, the SSM currently does not differentiate between different aggregate properties when recommending aggregate thickness.

Aggregate type and quality related to its fines content (smaller than, or minus, No. 200 sieve size or 0.075 mm) are important factors affecting strength and stiffness behavior and for determining aggregate cover or subbase placement thickness (Tutumluer & Seyhan 2000). Recent IDOT field experience has shown that aggregate properties have a significant effect on their performance in subgrade applications. Dense-graded aggregates with high fines contents and/or excessive Plasticity Index (PI) values may exhibit increased or high moisture sensitivity.

This paper presents laboratory findings from an ongoing research project at the University of Illinois aimed to characterize strength, stiffness, and deformation behavior of three types of aggregate materials, i.e., limestone, dolomite, and uncrushed gravel, commonly used in Illinois for subgrade replacement and subbase. Considering each aggregate type, the project aims at modifying or improving the current SSM cover thickness requirements by developing thickness correction factors for the amount of aggregate fines, PI of fines, moisture content in relation to the optimum moisture content, particle shape or angularity, and finally, number of wheel load applications during construction trafficking which will be studied later in the field validation. This paper mainly presents the test results from the initial laboratory moisture-density and unsoaked CBR or IBV testing phase of the research study.

2 SCOPE OF PROJECT

2.1 *Development of test matrix*

The initial phase of the research primarily focused on studying the effects of type and amount of fines on behavior of the three dense-graded aggregate materials, i.e., 100% crushed limestone and dolomite and the uncrushed gravel, engineered in the laboratory at four different target fines contents (4%, 8%, 12% and 16%). The fines used in the experimental matrix were primarily of two types. The first type had a plasticity index (PI) of zero, whereas the second type has a PI value of around 10.

2.2 *Engineering gradations*

One of the primary variables in any laboratory testing of aggregate materials is the grain size distribution. Differences in aggregate gradations can lead to significantly different behavior for the same aggregate type. Therefore, before conducting any parametric study on the aggregate properties affecting behavior, it was deemed important to keep the gradations consistent. This would enable the researchers to attribute the change in behavior to the induced changes in the variable parameters (e.g. fines percentage, and plasticity of fines).

Special attention was paid towards keeping the grain size distribution consistent while preparing samples for the test matrix. To engineer the gradations, sizing of the aggregate materials was undertaken first. Gradations were engineered in the laboratory, to prepare samples having 0%, 4%, 8%, 12%, and 16% fine material passing number 200 sieve size or 0.075 mm. This was needed to prepare laboratory aggregate specimens for all the moisture-density, IBV, rapid shear strength, and modulus and permanent deformation tests to be conducted following the experimental program. Moisture-density (Proctor) and IBV tests were then conducted as the first round of laboratory tests on the three aggregate materials.

Figure 1 shows the different engineered gradations used to prepare samples for the test matrix. It should be noted that the blending of aggregates for sample

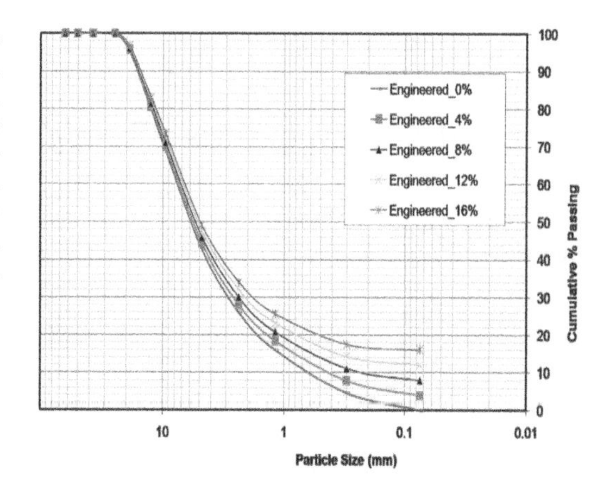

Figure 1. Engineered gradations.

preparation was done based on dry sieving results. However, wet sieving was conducted to check any differences between the results obtained from dry and wet sieving. It was noticed that a limestone sample blended with different fines to achieve 4% of material passing the sieve number 200, actually contained about 8.1% material passing the 200 sieve, based on wet sieving. This could be attributed to the significant amount of fines that remained stuck to the surfaces of larger particles during dry sieving and contributed towards changing the performance of the aggregate layer as a whole. Therefore, these fines should be considered while studying effects of fines on aggregate strength and deformation behavior. As a result, it was decided to report the results based on the actual fines content in the samples, rather than based on the target dry blended fines percentages. Accordingly, the reported fines contents of 4.4, 8.1, 11.8, 15.5 and 19.2%, were dry blended by adding 0, 4, 8, 12, and 16% of material passing the number 200 sieve, respectively.

3 LABORATORY TESTING

3.1 *Moisture-density test*

The objective of compaction is to improve the engineering properties of the soil mass; by compaction soil strength can be increased, bearing capacity of pavement subgrades can be improved, and undesirable volume changes, for example, caused by frost action, swelling and shrinkage, may be controlled (Holtz, 1990). Compaction characteristics for the three aggregate types were established using the method specified in ASTM D698. For each sample, a minimum of four moisture contents were used to conduct the Proctor tests and establish the maximum dry densities corresponding to the optimum moisture contents (OMCs). To better understand the moisture-density behavior of the materials; up to six moisture contents were used for some samples to establish the OMC and maximum dry density charts. It should be noted that at low fines contents, the samples can act as free draining material. Therefore, special care was taken to prevent the water from draining out of the mold during compaction.

3.2 *California Bearing Ratio (CBR) test*

The CBR test was performed on the same specimens used to obtain the density by following ASTM D1883 procedure with no soaking of the specimens. Conducting unsoaked CBR test gives expedited results. Moreover, as the aggregate behavior is being studied primarily on the wet side of OMC, there is no significant difference between the results obtained from soaked vs. unsoaked CBR tests. The change in unsoaked CBR or now IBV values with moisture content was studied at the previously listed fines contents to determine the strength variation trends of each material with moisture content.

4 TEST RESULTS

4.1 *Effect of non-plastic fines*

Standard Proctor tests and unsoaked CBR tests (IBV) were conducted on the samples prepared by adding different percentages of non-plastic fines, and the effect of percent fines content on the moisture-density as well as strength of aggregates was studied. The upper part of Figure 2 shows the moisture-density behavior of the limestone samples with non-plastic fines. It can be seen in Figure 2 that the maximum dry density values increase as the percentage of fines in the sample increases, with the highest maximum dry density value obtained at a fines content of 19.2%. As the addition of fines gradually fills the voids, the aggregate matrix continues to become denser thus making 19.2% sample the densest. The lower part of Figure 2 shows the change in unsoaked CBR with moisture content, for each of the Proctor samples. The dependence of unsoaked CBR with moisture is erratic at low fines contents (4.4%, and 8.1%). However, at higher fines contents, the CBR value decreases rapidly with the increase in moisture. It can also be seen that the rate of decrease in the CBR value with moisture is higher at higher fines contents.

This paper primarily compares the behavior of uncrushed gravel against that of the crushed limestone aggregate. The observed behavior of dolomite was similar to that of the limestone, and hence has not been presented in this paper due to space constraints. Figure 3 shows the moisture-density and unsoaked CBR-moisture content behavior of uncrushed gravel samples blended with different percentages of non-plastic fines. It can be seen that as the fines content

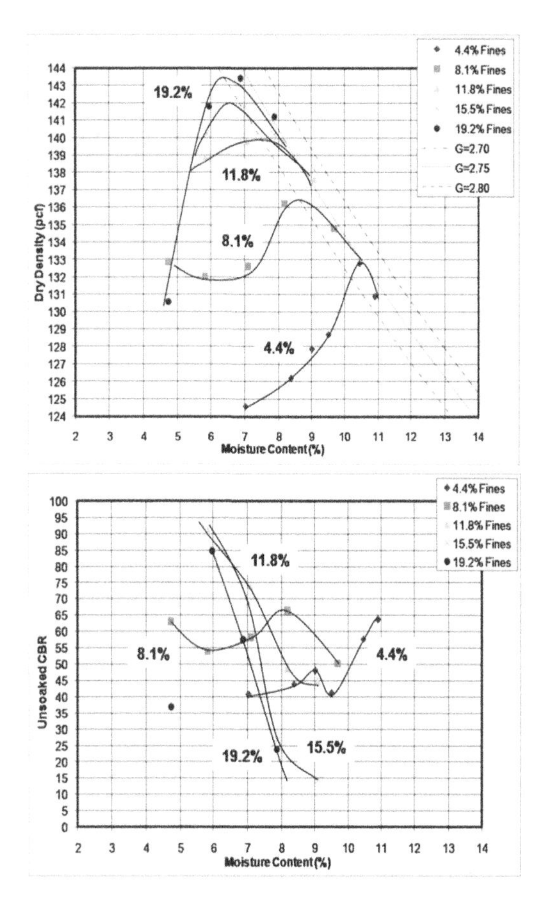

Figure 2. Effect of non-plastic fines on limestone.

increases from 2.9% to 14.5%, the maximum dry density value increases. However, as the fines content is further increased to 18.3%, the maximum dry density value decreases. This can be explained from the fact that the uncrushed gravel matrix, comprised of rounded aggregate particles, has a lower amount of voids when compared to that of the limestone (limestone particles are 100% crushed). Therefore, as the fines content increases beyond a certain point, all the voids in the uncrushed gravel matrix gets filled, and the coarse particles start getting displaced by the fines. This results in a reduction in the overall dry density, as the specific gravity of the fines is lower than that of the coarse particles.

A closer look at the unsoaked CBR relationship with moisture content suggests a behavior similar to that of the limestone, with the change in CBR with moisture being erratic at low fines contents (not shown here). However, as the fines content increases, there is a rapid decrease in CBR with increasing moisture content.

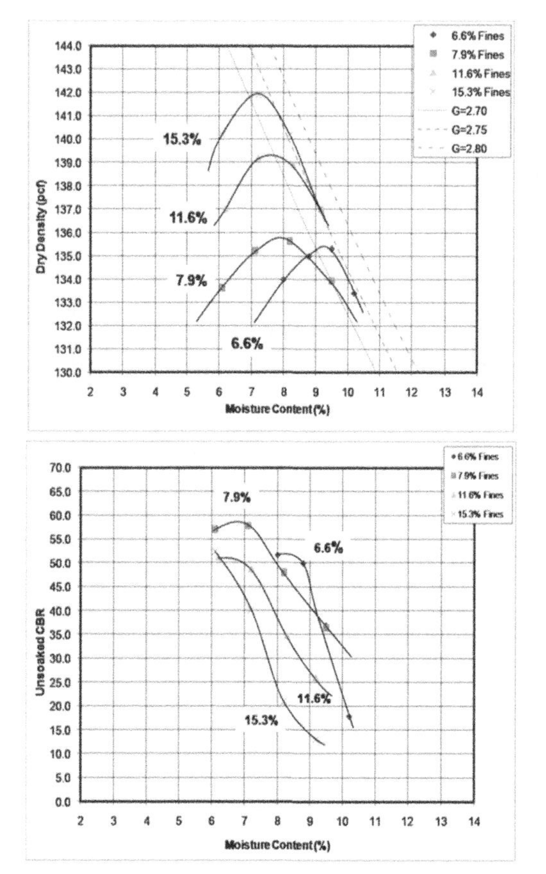

Figure 3. Effect of non-plastic fines on gravel.

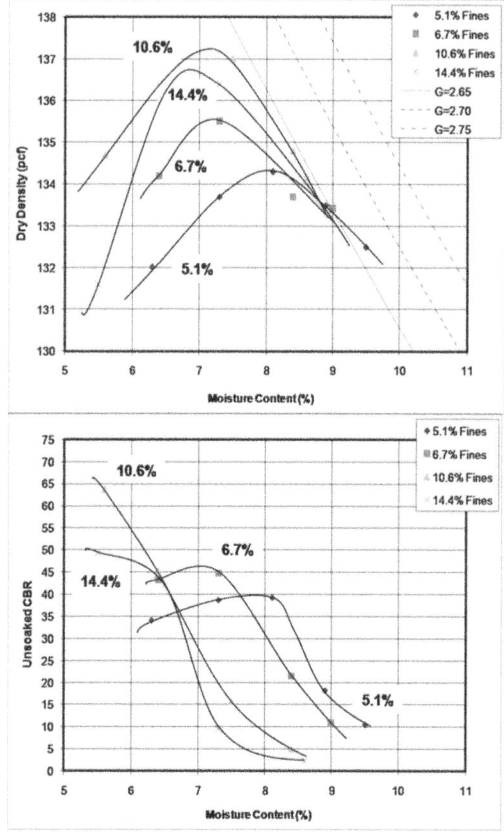

Figure 4. Effect of plastic fines on limestone.

4.2 *Effect of plastic fines*

Research studies in the past have shown that dense graded aggregate layers having highly plastic fines exhibited considerably higher moisture sensitivities when compared to the ones having only mineral filler type non-plastic fines. However, most of the highway agencies in the United States currently specify only the maximum amount of fines allowed in a pavement layer, and do not distinguish between the types of fines. Plastic fines from silty or clayey subgrade soils may often intrude into aggregate base during a pavement's service life. Therefore, it is important to quantify the effect of different types of fines on the behavior of an aggregate layer in order to modify the current specifications and better control the pavement behavior in the field.

To study the effect of plastic fines on aggregate behavior, test samples were prepared at different percentages of plastic fines. The plastic fines used in the study had PI values in the range of 8 to 14. Figure 4 shows the overall impact of plastic fines on the behavior of the limestone samples. The moisture-density behavior shows a pattern similar to that observed with the non-plastic fines. The attained maximum dry density values increase with increasing fines content from 6.6% to 15.3%. The lower part of Figure 4, i.e., graph relating unsoaked CBR with moisture content, clearly captures the different influences of plastic and non-plastic fines. Unlike in the case of non-plastic fines, the CBR value decreases rapidly with increasing moisture content, even at low fines contents. The preliminary findings outlined here certainly support the common belief that the type of fines affects aggregate behavior significantly. Therefore, limiting only the maximum amount of fines in an aggregate layer may not be the ideal method to ensure adequate pavement behavior.

Figure 5 shows the effect of plastic fines on uncrushed gravel samples. The moisture density curves exhibit a similar trend to that observed with non-plastic fines (maximum dry density falls at very high fines contents). However, the CBR-moisture content relationship exhibits a stronger effect of plastic fines on the strength properties of the aggregates, as already discussed for limestone.

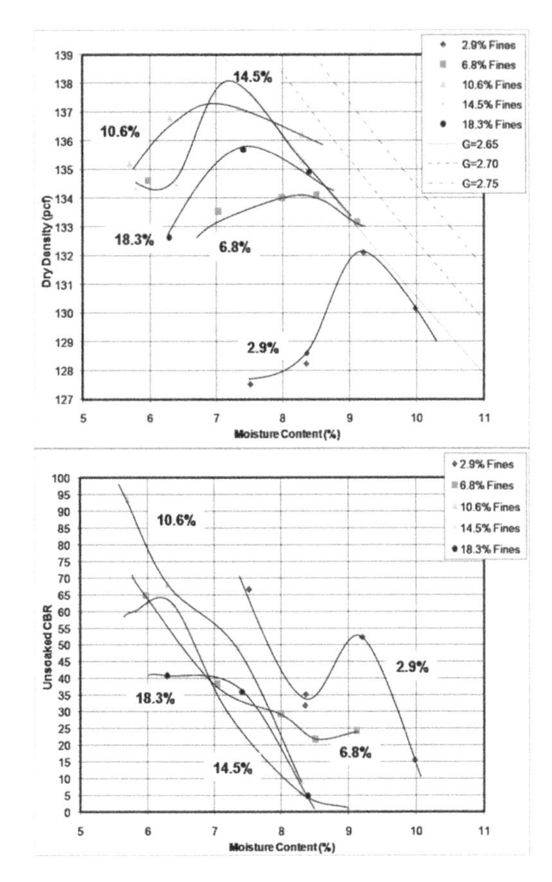

Figure 5. Effect of plastic fines on gravel.

4.3 *Dependence of CBR on moisture content*

The preliminary results of unsoaked CBR values were graphed next with percent fines at different moisture contents for all the three aggregate types. These relationships for the limestone and uncrushed gravel are shown in Figures 6 and 7, respectively. For both aggregate types, the CBR value at a given fines content is typically the lowest when the moisture content is at 110% of the OMC. Moreover, the figures also show that samples having plastic fines at 110% of OMC have the lowest CBR values. Thus, for any pavement layer, the combination of plastic fines and high moisture content will result in the lowest strength properties.

4.4 *Comparisons of plastic and non-plastic fines*

To compare and better evaluate the effect of fine type and amount of fines on aggregate strength, the CBR values of limestone and uncrushed gravel at the optimum moisture content were next graphed with the different percentages of fines (see Figure 8). It can be seen from Figure 8 that for limestone samples with

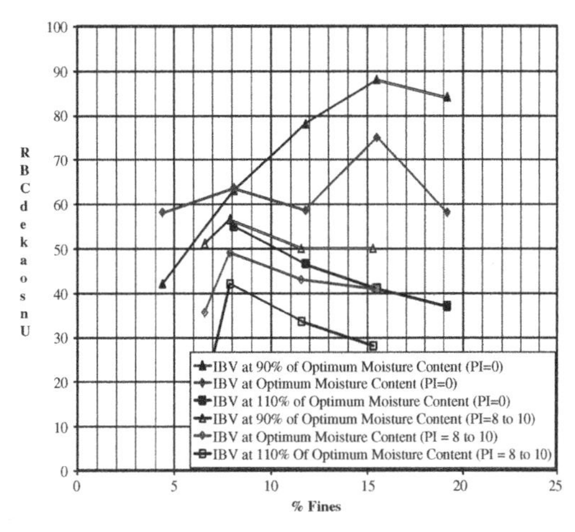

Figure 6. Relationship between unsoaked CBR and percent fines for limestone under different moisture conditions.

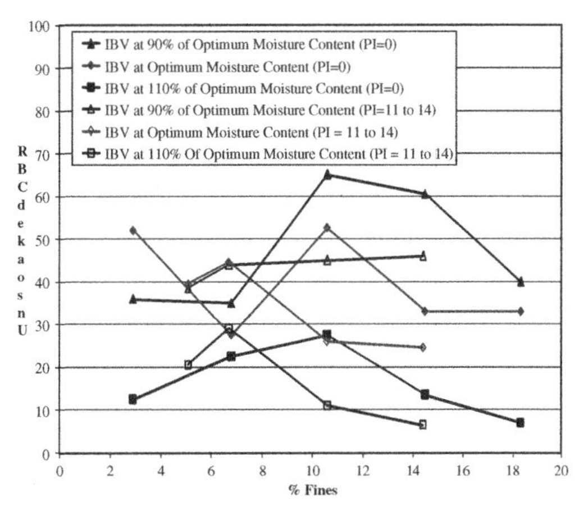

Figure 7. Relationship between unsoaked CBR and percent fines for gravel under different moisture conditions.

plastic fines, the CBR value is always less than the CBR value of samples at about the same percentage of non-plastic fines. A similar trend can be seen for uncrushed gravel samples at larger fines percentages. However, the effect of plastic fines on gravel is not as pronounced at low fines contents. Figure 8 also shows that for the same fines percentage, the crushed limestone yielded higher CBR values than the uncrushed gravel. This reinforces the common observation that crushed aggregate particles perform better in pavements when compared to uncrushed particles.

5 SUMMARY AND CONCLUSIONS

A current research project at the University of Illinois aims at studying the different types and amounts of

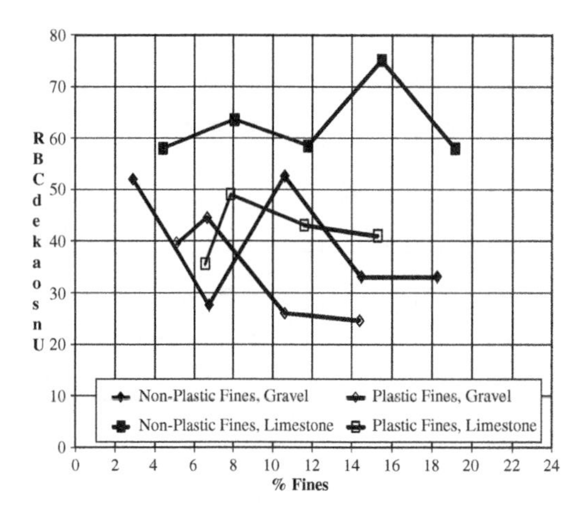

Figure 8. Relationship between unsoaked CBR and percent fines at optimum moisture content (OMC) conditions.

fines affecting the behavior of highway construction aggregates commonly used for subgrade replacement and subbase in the state of Illinois, USA. The initial laboratory phase of the study presented in this paper comprised of Proctor moisture-density and unsoaked CBR test results obtained from an experimental test matrix which considered both plastic and non-plastic fines (passing No. 200 sieve or 0.075 mm) blended in the engineered gradations of crushed and uncrushed aggregate materials at 4%, 8%, 12%, and 16% target fines content.

For non-plastic fines, the CBR behavior with moisture content was erratic at low fines contents. However, as the fines content was increased, the CBR decreased rapidly with increasing moisture content. This trend was observed for all the three types of materials tested (limestone, dolomite, and uncrushed gravel). From the test results, the presence of low amount of non-plastic fines in an aggregate layer may not adversely affect the pavement performance. However, for plastic fines, the CBR values showed rapid decreases with moisture contents even at low fines contents thus indicating increased moisture sensitivity. This means, the specifications currently used by most highway agencies in the US may need to be reconsidered for taking into consideration the effect of plasticity of fines.

Laboratory triaxial strength and modulus tests are currently underway to finally lead to the field validation phase of the study for developing improved subgrade stability aggregate cover requirements based on aggregate type and quality. After completion of the modulus and strength tests, the results from moisture-density tests can be related to the mechanical properties of the aggregates. This will help researchers as well as practitioners to better assess mechanical behavior and potentially predict the field behavior without conducting actual strength tests.

REFERENCES

Holtz, R.D. 1990. "Compaction concepts," Chapter 3, *Guide to Earthwork Compaction*, State of the Art Report 8, Transportation Research Board, Washington, D.C.

Illinois Department of Transportation – IDOT. 2005. "Subgrade Stability Manual," Bureau of Bridges and Structures, May 1, 2005, 27 pages.

Tutumluer, E. & Seyhan, U. 2000. "Effects of fines content on the anisotropic response and characterization of unbound aggregate bases," In *Unbound Aggregates in Road Construction*, Edited by A.R. Dawson, A.A. Balkema Publishers, Proceedings of the Unbound Aggregates in Roads (UNBAR5) Symposium, University of Nottingham, England, June 21–23, 2000, pp. 153–160.

Tutumluer, E., Thompson, M.R., Garcia, G. & Kwon, J. 2005, "Subgrade stability and pavement foundation requirements," In Proceedings of the 15th Colombian Symposium of Pavement Engineering, sponsored by Pontificia Universidad Javeriana in Bogota, Melgar, Colombia, March 9–12.

Advances in Transportation Geotechnics – Ellis, Yu, McDowell, Dawson & Thom (eds)
© 2008 Taylor & Francis Group, London, ISBN 978-0-415-47590-7

Using unbound aggregates resulting from amethyst mining in low volume roads

W. P. Núñez, J. A. Ceratti, R. Malysz & T. S. Retore
Federal University of Rio Grande do Sul, Porto Alegre, Brazil

ABSTRACT: Though presenting limitations that inhibit their use in heavily trafficked pavements, weathered rocks are generally suitable for low volume roads. This research was carried out to evaluate the mechanical behaviour of two kinds of weathered basalts with different alterability levels, produced from gallery opening in amethyst mines. The weaker material results from detonation of the layer with geodes; the stronger one comes from the layer above. The materials were crushed to obtain a uniform grain size distribution. Los Angeles abrasion, soundness, sand equivalent and resilient modulus tests were carried out; permanent deformation characteristics were evaluated. Tests results showed that both materials present durability, strength and strain characteristics suitable for use in low volume roads. Once drainage is provided, the weaker material may be used in pavements sub-base and the stronger one as base layer. A set of pavement structures for low volume roads is proposed and construction recommendations are presented.

1 INTRODUCTION

Brazil is one of the developing countries with longer unsurfaced roads network. More than 1.560.000 km (89.4% of existing roads) is used to transport agricultural production. Besides, nearly 36 million people still live in rural areas. In Rio Grande do Sul, Brazil southernmost state, a similar trend is observed: 18.4% of the population live in the countryside and 91.8% (more than 141,000 km) of the roads network is unpaved.

In the northern region of Rio Grande do Sul state; flanking the Uruguay river, amethyst mining is a very important economic activity. Ametista do Sul, a small county with no more than 8,000 inhabitants (including 2,200 miners), is the most important producer of that precious stone in South America. Conversely, the county's Human Development Index (HDI = 0.754), according to UNDP (2007) method, is one of the lowest in Rio Grande do Sul state (average HDI of 0.814). The complete lack of paved roads inhibits economic investments and turns more the difficult the access to education and health facilities.

As shown in Figure 1, geologically, Ametista do Sul county is in the Serra Geral Formation, the lowest portion of the Paraná Basin (a lava spill region covering parts of Brazil, Argentina, Uruguay and Paraguay). When miners open galleries in the basalt scarp looking for the amethyst geodes, a great volume of weathered rock is carelessly discarded, sometimes even obstructing water streams.

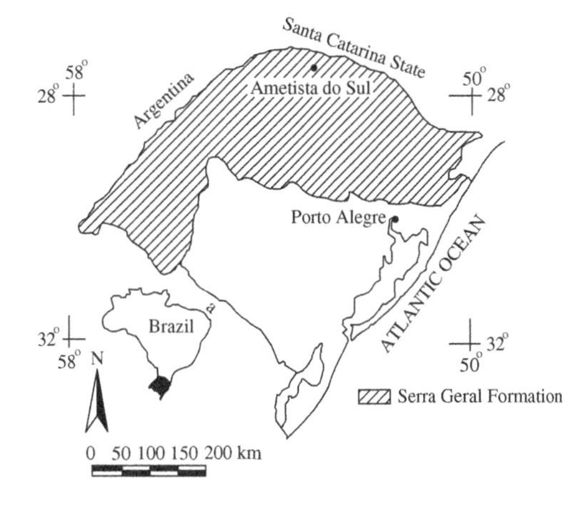

Figure 1. Ametista do Sul county in Rio Grande do Sul map.

Though presenting limitations that inhibit their use in heavily trafficked pavements, weathered basalts have proved to be suitable for low volume roads (Núñez et al, 2000).

2 BRIEF OVERVIEW

2.1 *Volcanism in Southern Brazil*

The volcanic activity that originated the lava spills of the Parana Basin covers an area of 1,200,000 km^2 and dates from the middle part of the lower Cretaceous age (120 to 130 million years ago). The rock composition varies from basic to acid, including several intermediates.

Spill package portions with rocks intensely fractured are frequently found with degradation processes predominating over decomposition ones. Many rock formations in Rio Grande do Sul State, particularly basalts, have undergone so much weathering that the interstitial basaltic glass has been altered to magnesium-saturated montmrillonite clay. Such occurrences are known as weathered basalts.

2.2 *Studies previously developed on volcanic materials in Southern Brazil*

In the Pavements Laboratory of the Federal University of Rio Grande do Sul (LAPAV) weathered volcanic rocks have been studied since 1992.

Núñez et al. (2000) presented the results of a comprehensive study on the use of weathered basalts for low-volume roads paving, including a qualifying criterion for weathered volcanic rocks proposed by Arnold (1993) and a design equation based on accelerated pavement testing (APT) previously developed (Núñez; 1999).

Soon afterwards, applying Arnold's acceptance criterion and Núñez's design equation and recommended procedures, the Rio Grande do Sul State Roads Department (DAER/RS) constructed a pavement 21-km long using acid weathered volcanic rocks in the base and sub-base layers. Núñez et al. (2003) evaluated this pavement performance a few years later and validated the reliability of the acceptance criterion and the design equation.

2.3 *Acceptance criterion for weathered volcanic rocks*

The degradation of volcanic rocks is related to their chemical-mineralogical composition and alteration state. According to Arnold (1993), the rock may be seen as a skeleton formed by rather sound and partially decomposed minerals.

Clays resulting from chemical decomposition of minerals with low resistance to weathering fill the cracks. Since some of those clays may be expansive, the rock degradation level is a function of the skeleton capacity to resist clay expansive forces.

Considering that in a pavement layer aggregates contact each other in angles and vertices, failure will be probably caused by tension forces inside the aggregate, as well as by clay expansion. Therefore, Arnold (1993) proposed the use of the point load test (ISRM, 1985) to qualify weathered volcanic rocks for paving purposes.

The point load strength index, Is(50), is computed for two sets of eight specimens each. One of the sets is 7-days soaked in water, while the other is air-dried. The ratio of soaked strength to dry strength is an indicative of the aggregate capability of resisting clays expansive forces.

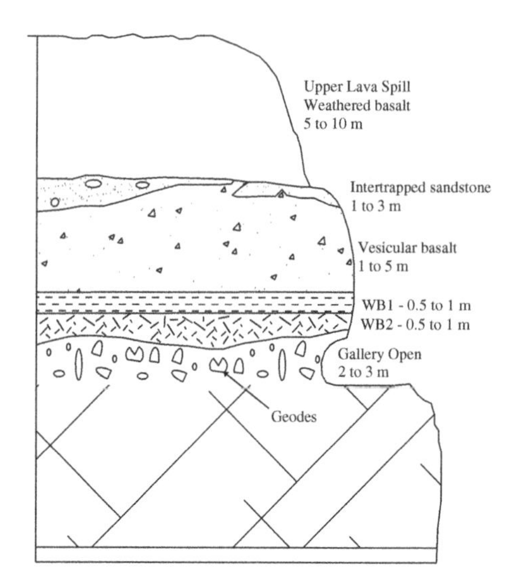

Figure 2. Lava spill statigraphy in the amethyst mining scarp.

The acceptance criterion defined 0.7 as the minimum value for the ratio of soaked to dry strength and 3.5 MPa for the point load strength index after soaking. Further details on the test itself and the application of the acceptance criterion to weathered volcanic rocks are presented elsewhere (Núñez et al., 2000; Wahys, 2003).

3 EXPERIMENTAL PROCEDURES

3.1 *Characterization of Volcanic weathered rocks resulting from amethyst mining*

In the area surrounding Ametista do Sul county, Scopel et al. (1986) found evidences that the rocks matrix had suffered hydrothermal alteration, confirmed by the presence of totally altered olivine and saponite type expansive clays.

Twelve basalt spills, 15.0 to 50.0 m-thick each, may be identified. In four of them amethysts geodes are found.

Figure 2 shows the typical spill statigraphy in the amethyst mining area. The following sequence of layers may be identified from the upper portion of the scarp:

a) Top weathered basalt, 10 to 15-m thick;
b) Intertrapped sandstone, 1 to 3-m thick;
c) Vesiculated basalt, 1 to 5-m thick;

Table 1. Air-dried Point Load Strength index of materials sampled in four sites in Ametista do Sul county.

Sampling site	Material	Air-dried Point load Strength Index (MPa)
Arceli	Fresh WB1	95
Arceli	WB1 stockpiled for 1 year	74
Ganzer	Fresh WB1	78
Ganzer	WB1 stockpiled for 2 year	81
Arceli	Fresh WB2	33
Arceli	WB1 stockpiled for 2 year	28
Ganzer	WB2	40
Ganzer	WB2 stockpiled for 2 year	31

d) Weathered basalt horizontally fractured, 0.5 to 1 m-thick;
e) Weathered basalt randomly fractured, 0.5 to 1 m-thick;
f) Weathered basalt with amethyst geodes in the upper portion, 6 to 8 m-thick.

The mining activity is developed in the last two layers. Galleries are excavated in the upper portion of layer "f" also involving part of layer "e". Local miners designate those materials as "cascalho" and "laje", respectively. From now on they will be referred as WB1 and WB2.

WB1 material is fractured in three directions with crack spacing of 10.0 cm. On the other hand, WB2 is a less fractured rock, with crack spacing of 30.0 cm.

In a previous study, Palma (2003) carried out point load tests in specimens of WB1 and WB2 materials, sampled in four different sites in Ametista do Sul county. Table 1 presents the materials air-dried point load strength indexes.

It may be seen that all samples of both materials present high values of point load strength. In spite of being more fractured, WB1 point-load strength index doubles those of WB2. The effect of weathering is also noticeable. In general (the exception is the Ganzer site WB1) the longer the period of open air stockpiling, the lower the sample point load strength.

3.2 *Samples preparation*

The materials were sampled in recently formed stockpiles, composed by fragments of different sizes resulting from recent detonation.

The fragments of both materials were crushed in laboratory in order to compose the grain size distribution shown in Figure 3. This open-graded grain size distribution was chosen due to the facility of being obtained in the field, with the use of a mobile stone crusher owned by the county administration.

In a parallel study on the resilient modulus and hydraulic conductivity coefficient of open-grade and dense-graded granular materials for road bases, Casagrande (2003) tested compacted specimens of sound basalt with several grains size distributions, including the one shown in Figure 3. In the following sections some results of laboratory tests carried out on specimens of sound and weathered basalts, in the same grain size distribution (Figure 3), are compared.

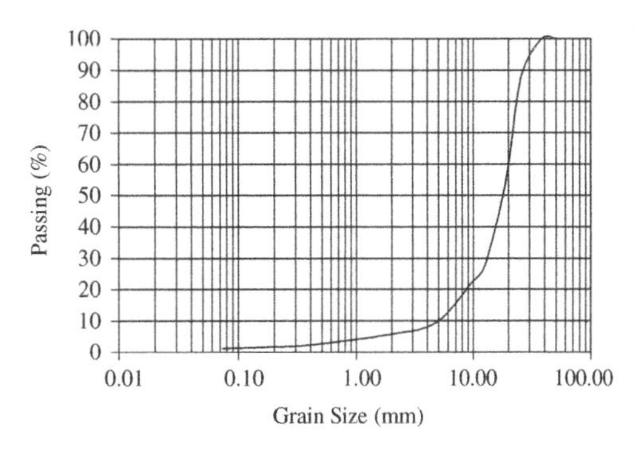

Figure 3. WB1 and WB2 grain size distributions obtained by laboratory crushing.

3.3 *Toughness, durability and particle geometric properties tests results*

When the altered rocks are used in pavement construction as base material, the secondary clay minerals, which have great affinity for water, may break down into plastic fines, that gradually permeates the entire rock mass. This degradation may be brought about by the action of water and weathering plus the abrasion of rock-against-rock as the entire mass is flexed under moving loads.

When the plastic fines permeate the base course, the latter loses stability rapidly and pavement failure soon follows. Failure is characterized by longitudinal cracks in the wheel paths followed by fatigue cracking and finally by breaking the overlying flexible pavement into separate and distinct pieces.

Due to the alterability of the studied materials it is most important to discuss the results of Los Angeles abrasion (toughness), sand equivalent (particle geometric properties), Washington state degradation index and soundness tests (durability), presented in Table 2.

It may be seen that WB1 results are in totally accordance to Brazilian standards. On the other hand, the sulfate soundness test result of WB2 is remarkably higher than the maximum acceptable. This could constitute a serious limitation to the use of that material, once that sulfate soundness test is considered as a fair predictor of aggregates performance. However, it must be kept in mind that weathered basalts have been proposed for low volume roads paving. Besides there is a consensus in Southern Brazil (Núñez, 1997) that the

Table 2. Laboratory quality tests results.

Test	Material WB1	WB2	Acceptable limits Sub-base	Base
Los Angeles	22.0%	18.4%	ND	≤40%
Sand Equivalent	43%	43%	≥25%	≥30%
Washington Degradation	67	64	≥15	≥25
Sulfate Soundness	4.7%	19.7%	≤12%	≤12%

ND = not defined

Table 3. Materials compaction parameters and CBR values.

Material	Property ω (%)	γ_d (kN/m^2)	CBR (%)
WB1	3.8	20.11	86
WB2	2.8	20.54	71

soundness is too severe for aggregates used in a region where freezing and thawing cycles never happen.

It is worthy of note that the results obtained by Casagrande (2003) on samples of fresh basalt rock were: Los Angeles abrasion of 16%; Sand Equivalent of 73.8% and Sulfate Soundness loss of 6.7%. As seen, excepting for sand Equivalent, the tests results on samples of WB1 and fresh basalt rock are quite similar, revealing the low alterability level of WB1. Conversely, WB2, where the amethyst geodes are found, is a more altered rock.

3.4 *Compaction parameters and Bearing Capacity*

Compaction and California Bearing Ratio (CBR) tests were carried out in both materials. Samples were compacted applying to the specimens AASHTO T 180 Modified Energy (2,693 kN.m/m^3) Table 3 presents compaction parameters and corresponding CBR values for both materials.

In view of constructions procedures adopting in the field, the material retained in the sieve 19.0 mm was not replaced by finer particles.

As shown in Table 3, the maximum dry unit weight (γ_D) achieved was somehow higher for WB2 than for WB1. On the other hand WB1 OMC is higher than that of WB2.

It must be observed that compaction curves were not typical (bell shaped), presenting irregular shapes either with two-peaks or without peak.

CBR values ranged from 71% to 86%. Following the tendency observed in Tables 1 and 2, WB1 presented higher CBR than WB2.

Brazilian standards determine that in order to be used in sub-bases the material CBR must be not less than 20%, while 80% is the minimum CBR for use in bases of heavily trafficked roads. For low volume roads bases, a CBR value of 60% meets the requirements. It is concluded that while WB1 material could be used in bases of both high and low volume roads, WB2 use is limited to bases of low volume roads. Obviously, both materials may be used in sub-bases of asphalt pavements, no matter the traffic volume.

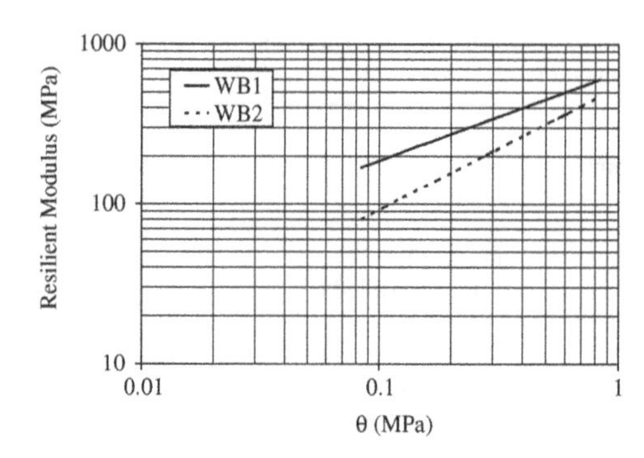

Figure 4. Materials resilient modulus as function of stress state.

Table 4. Parameters of resilient modulus models.

Material	Parameter K1	K2
WB1	663	0.55
WB2	540	0.77

3.5 *Resilient modulus tests results*

Resilient Modulus tests were carried out on following AASHTO TP46-94 standard, with loading frequency of 1 Hz and loading time of 0.1 s.

The specimens, 10.0 cm diameter and 20.0 high, were dynamically compacted in five layers, to achieve dry unit weights and moisture contents close to compaction parameters shown in Table 3.

Soil resilient modulus was modeled as a function of stress state.

$$RM = k_1\theta^{k_2}$$

where *RM* is the soil resilient modulus; θ is the sum of principal stresses; and k_1 and k_2 are the model parameters.

Figure 4 and Table 4 present tests results. High determination coefficient values (R^2) demonstrate that the model fits tests results rather well.

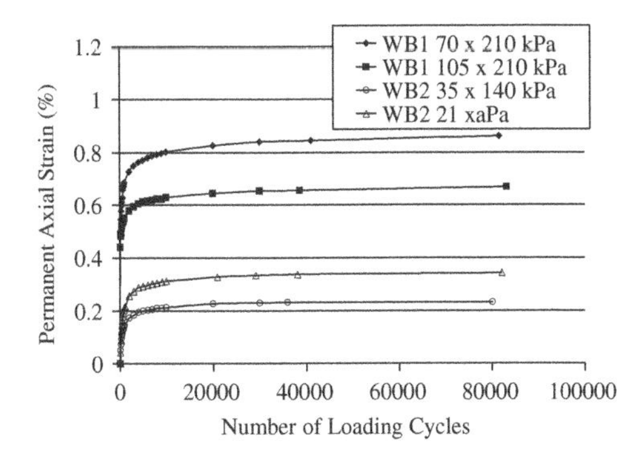

Figure 5. Effect of confining stress on the permanent axial strain evolutions in specimens of both materials.

Again, the results of WB1 specimens are better than those corresponding to WB2 specimens. For high stress levels, such as those possibly acting in bases of thinly surfaced or unsurfaced roads, WB1 resilient modulus may reach quite high values. Conversely, the resilient modulus of WB2 dependence on stress state (shown by a high k_2 value) would yield low modulus sub-bases. This fact somehow could limit the use of WB2 in pavements with thick asphalt layers, where fatigue is the main distress mechanism. However, as previously sated, the weathered basalts resulting from amethyst mining are meant to be used in unsurfaced or thinly surfaced low volume roads, where rutting and shear failure, not fatigue, are the most common distresses.

3.6 *Permanent deformation tests results*

Due to the lack of Brazilian standards, the repeated loading permanent deformation tests were carried out following procedures adopted in studies previously performed in the Pavement Laboratory of the Federal University of Rio Grande do Sul (Núñez et al., 2004).

The specimens, 10.0 cm diameter and 20.0 cm high, were dynamically compacted at optimum moisture content in order to achieve maximum dry unit weights. Approximately 80,000 loading cycles were applied to each specimen, at a frequency of 1 Hz.

Considering the high stress state acting in bases of unsurfaced or thinly surfaced pavements, WB1 specimens were tested under confining stresses (σ_3) of 70 and 105 kPa and deviator stress (σ_d) equal to 210 kPa.

Lower stresses were applied to WB2 specimens, since that material could be used in sub-bases: confining stresses (σ_3) of 21 and 35 kPa and deviator stress (σ_d) of 140 kPa.

As shown in Figure 5, the permanent deformation evolution in all specimens follows a pattern known as plastic shakedown. After a quite high initial buildup, deformation increases at a quite lower constant rate.

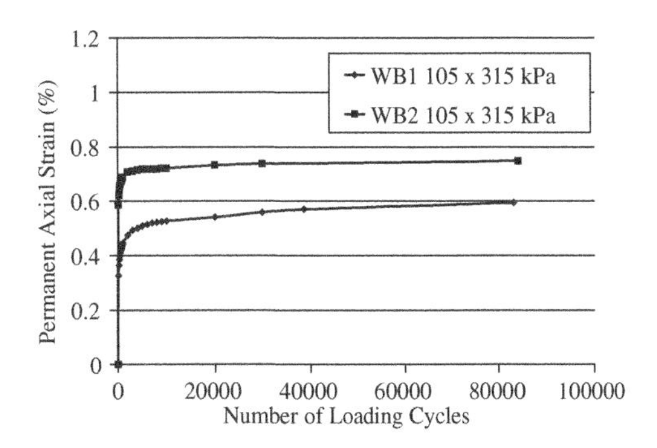

Figure 6. Comparison between permanent deformation evolution in both materials.

The effect of confining stress on permanent deformation evolution in specimens of both is clearly displayed. Figure 5 also shows that if properly compacted both materials will perform quite well in bases and sub-bases of unsurfaced or thinly surfaced pavements. Even when the highest deviator stress (210 kPa) was applied to WB1 specimen, at the end of the test the permanent deformation was less than 0.9% (1.8 mm in the 200 mm high specimen).

In order to compare the materials, Figure 6 shows the permanent deformation evolutions of specimens when the stress state $\sigma_3 = 105$ kPa and $\sigma_d = 140$ kPa was applied. It may be seen that WB1 and WB2 performed quite similarly. Concerning permanent deformation behavior, those weathered basalts are high-quality materials for road construction.

4 USE OF THE STUDIED MATERIALS IN LOW VOLUME ROADS

Based on tests results discussed in the previous section, three pavement structures, shown in Figure 7, were preliminarily designed, considering a traffic volume of 2.5×10^5 equivalent axle loads and subgrade CBR of 4%, 8% and 12%.

A 2.5 cm-thick surface treatment was adopted as wearing course, and the WB1 base layer thickness was fixed at 25.0 cm. As shown in Figure 7, the required WB2 sub-base thickness for subgrade CBR values of 4%, 8% and 12% are, respectively, 40.0 cm, 20.0 cm and 15.0 cm.

Using the resilient modulus models shown in Figure 4, a mechanistic analysis was carried out, using ELSYM5, a widely known software. Brazilian standard axle load (82 kN) and Poisson's ratios of 0.35 for weathered basalts and 0.45 for soil subgrades were considered.

Although it has been proved the inconsistency of relating soils and aggregates resilient modulus to CBR,

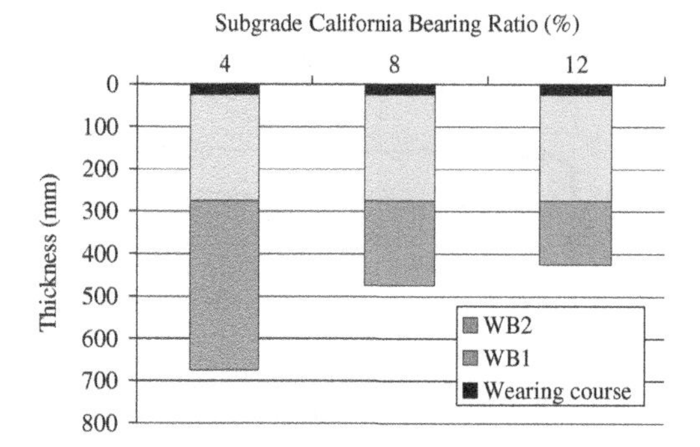

Figure 7. Pavement structures designed considering subgrade CBR.

Table 5. Structural answers computed using ELSYM 5.

$RM_{subgrade}$ (MPa)	Deflection (10^{-2} mm)	$\sigma_{v,base}$ (kPa)	$\sigma_{v,sub\text{-}base}$ (kPa)
50	59	247	24
100	46	183	49
150	38	197	82

moduli of 50 MPa, 100 MPa and 150 MPa were considered as corresponding to subgrades with CBR equal to 4%, 8% and 12%, respectively.

Table 5 presents the surface deflections and deviator stresses acting in the mid-depth of bases and sub-bases shown in Figure 7. It may be seen that surface deflections are rather low, in spite of the thin asphalt wearing course (2.5 cm-thick treatment) and rather low subgrades resilient modulus. Moreover, the vertical stresses acting in the base and sub-base layers are rather low. Because of that, the permanent deformation will not exceed a few millimeters, as shown in Figures 5 and 6.

Therefore, it may be concluded that the structures shown in Figure 7 would perform quite well.

5 CONSTRUCTION RECOMENDATIONS

This section presents a few recommendations aimed at ensuring that unsurfaced or lightly surfaced roads built with weathered basalts resulting of amethyst mining will perform in a good way.

a) It is mandatory to separate different materials stockpiles, thus avoiding contamination. It must be kept in mind that, due to its CBR value, the use of WB2 is limited to bases of low volume roads. Besides that material is more easily weathered and should not be open-air stocked for long periods.
b) In spite of the heterogeneity of the studied materials, the grain size distribution should be kept as close as possible to that shown in Figure 3.
c) Compaction must be severely controlled. With that purpose, it is highly recommended to measure deflections with Benkelman beam. Núñez (1997) verified that deflections lower than 80×10^{-2} mm on top of weathered basalts layers might be rather easily attained with proper compaction. Moreover, compaction should be controlled to target deflections not higher than 60×10^{-2} mm on the top of the base layer.
d) Topping the granular base with a surface treatment would not extraordinarily increase the pavement strength and resistance to permanent deformation. However, a thin asphalt layer will remarkably reduce water inflow, thus improving the subgrade and granular layers performances.
e) Finally, keep drainage properly working is mandatory. Although this may seem obvious, drainage, as well as compaction, are frequently neglected in low volume roads.

6 CONCLUSIONS

The results of a laboratory research on the use of weathered basalts resulting of amethyst mining in Southern Brazil were presented and discussed.

Two materials discarded during mining activities, identified as WB1 and the WB2, were studied.

Both materials performed quite well in permanent deformation laboratory test and the resilient modulus, corresponding to high stress levels that commonly happen in unsurfaced and lightly surfaced roads, were quite high.

Based on the results of toughness (Los Angeles abrasion), durability (sulfate soundness), strength (CBR) and deformability (resilient modulus and permanent deformation under repeated loading) tests, it is recommended the use of WB1 material in bases of any category of pavements.

The more altered material, WB2, may be also used in granular layers of lightly surfaced low volume road. However, it should not be used as base layer, since wetting and drying cycles would rapidly reduce its strength.

The adoption of a rather uniform grain size distribution, which was almost mandatory due to the need of using a rather simple mobile stone crusher, did not reduce the materials qualities and will facilitate their use at much lower cost.

The use of those materials in low volume roads paving might contribute to improve the economy and the HDI in Ametista do Sul county, Southern Brazil, also reducing the obstruction of water streams with discarded mining residues, thus helping to protect the environment.

REFERENCES

Arnold, G.P. 1993. Study of the mechanical behavior of weathered basalts in Rio Grande do Sul State with paving purposes. M.Sc. Thesis, Federal University of Rio Grande do Sul. (In Portuguese).

Casagrande, L.F. 2003. Study of fines content influence on hydraulic conductivity and elastic deformability of gravels. M.Sc. Thesis, Federal University of Rio Grande do Sul. (In Portuguese).

Núñez, W.P.; Ceratti, J.A.P.; Gehling, W.Y.Y. & Oliveira J.A., 1999. Full-scale load tests in Southern Brazil, Proc Int. Conf. Accelerated Pavement Testing, Reno, Nevada. (In CD-Rom).

Núñez, W.P.; Ceratti, J.A.P.; Arnold, G.P. & Oliveira, J. A., 2000. Weathered basalts, alternative aggregates for thin pavement bases, Proc. 5th Int. Symp. Unbound Aggregates in Road Construction (UNBAR5), University of Nottingham, Dawson (ed.), Balkema, Rotterdam, pp. 117–124.

Núñez, W.P.; Ceratti, J.A.P.; Arnold, G.P.; Oliveira, J.A.; Silveira, J. 2003. Paving low-volume roads weathered volcanic rocks; from APT to practice. Proc. 21st. ARRB Transport Research and 11st. Road Engineering Association of Asia and Australia Conference, Cains, Austalia.

Núñez, W.P.; Malysz, R.; Ceratti, J.A.P. & Gehling, W.Y.Y. 2004 Shear strength and permanent deformation of unbound aggregates used in Brazilian pavements. Proc. 6th Int. Symp. Unbound Aggregates in Roads (UNBAR6), University of Nottingham, Dawson (ed.), Balkema, Rotterdam, pp. 23–31.

Palma, P. S. 2003. Study of amethyst mining rejects in mineral district of Alto Uruguai, RS. Graduation Thesis, Geology Department, Federal University of Rio Grande do Sul. (In Portuguese).

Scopel, R.; Formoso, M.L. Dudoignon, P. & Meunier, A. 1986. Hydrothermal Alteration of Basalts, southern Paraná Basin – Brazil. In: Chemical Geology. University of Nottingham, n 38, pp. 21–34.

UNDP – United Nations Development Programme. 2007. Technical Report 1: Calculating the human development indices. *Human Development Report 2007/2008*: pp. 355–361. New York: Palgrave Macmillan.

Wahys, C.A.S.P. 2003. Study of alternative materials used in low costs paving at Rio Grande do Sul northeast region. M.Sc. Thesis, Federal University of Rio Grande do Sul. (In Portuguese).

Advances in Transportation Geotechnics – Ellis, Yu, McDowell, Dawson & Thom (eds)
© 2008 Taylor & Francis Group, London, ISBN 978-0-415-47590-7

Rampart roads in the peat lands of Ireland: Genesis, development and current performance

J.P. Osorio, E.R. Farrell & B.C. O'Kelly
Trinity College Dublin, Ireland

T. Casey
National Roads Authority, Dublin, Ireland

ABSTRACT: The construction and improvement of roads on peat lands has always been a challenging task in geotechnical engineering. Rampart roads, which are a particular feature of the bog roads in Ireland, are caused by the excavation of peat from the roadsides over many years for use as a domestic fuel. The historical development of rampart roads and current performance under the added traffic loading is discussed. Various road improvement methods have been applied to improve the road ability to carry traffic and the technical aspects and performance of the methods are discussed. In particular, the performance of the improvement methods is related to the geotechnical properties of the underlying bog foundation.

1 INTRODUCTION

According to Hobbs (1986), about 17.2% of the overall land surface of Ireland is covered with peat (Figure 1) although for some counties the peat coverage is significantly greater, for example, 36.1% in County Leitrim (Hammond, 1981). Hence, a significant part of the countries road network has been constructed across peat lands. In addition, peat harvesting for fuel in the vicinity of the bog roads has historically been common practice and has led to the creation of elevated roads, referred to as rampart roads.

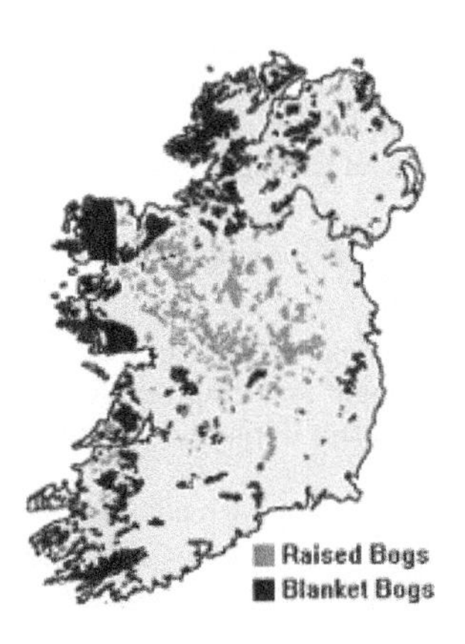

Figure 1. Peat land distribution in Ireland (Bord na Móna, 2001).

This paper presents an historical overview of the genesis and development of rampart roads in Ireland. The geotechnical properties of the bog foundation are discussed along with the different improvement methods that have been used, albeit with different levels of success, over the years.

2 RAMPART ROADS: HISTORICAL BACKGROUND

2.1 *Ancient bog roads*

Bog reclamation and peat harvesting in Ireland can be closely related to the development of the first bog roads. References in the Old Irish law texts (Feehan & O'Donovan, 1996) trace peat harvesting for use instead of wood as a fuel back to the seventh and eighth centuries although peat harvesting is know to have been common practice during the late Bronze Age (centuries leading up to 1000 BC).

The first roads constructed across the bogs in Ireland are called toghers and comprised oak planks that had been laid on sleepers and pegged into the bog foundation using vertical stakes (O'Keeffe, 1973). Some toghers incorporated a stone or gravel layer above a brushwood base. The toghers were mainly used as foot roads, and where wide enough, were sometimes used for wheeled vehicles and animal traffic.

During the seventeenth century, the role of the toghers was replaced with the construction of new permanent roads after ten Acts for the improvement of the road network were passed by the Parliament of James I. Of these, the Highway Act for the Amendment of Bridges and Toghers had the biggest impact (O'Keeffe, 1973).

2.2 *Bog roads network beginnings*

However, it was in the eighteenth century that the network of bog roads in Ireland began to be properly developed by landowners and the Grand Jurys in order to gain commercial and military access to the remote rural areas. The advances and developments in bog road construction by the middle of the nineteenth century have been reported by Parnell (1833) and Mullins & Mullins (1846). The usual construction procedure was as follows:

- Mark the centreline of the road and form the base on which the road materials are to be placed. After the centreline had been traced, a system of parallel and transversal open drains was excavated allowing sufficient drainage to achieve adequate bearing capacity and stability of the bog foundation.
- Peat material that had been excavated in forming the drains was placed in the form of a ridge above the road surface and allowed to naturally air dry. The drying period, which varied between two summer months and two years, depended on the water content of the *in situ* peat material.
- A sub-grade layer of stiff clay was placed and compacted above the dried peat material followed by a cover of stone/gravel as the road surface.

The procedure described differed slightly from one place to another depending on the site conditions and the construction requirements.

2.3 *Genesis of the rampart roads*

The improved roads across the peat lands provided better access and made peat harvesting easier. Peat was the main domestic fuel in Ireland during the eighteenth and nineteenth centuries with up to five million tonnes being harvested annually (Feehan & O'Donovan, 1996). The rapid cut away of the peat material in the vicinity of the roadsides resulted in an elevated road surface that in some cases was many meters above the surrounding ground (Figure 2). These elevated bog roads are referred to as rampart roads.

Rampart roads often undergo considerable distortion (Figure 3) due to the low shear strength and high compressibility of the bog foundation, which may pose a significant safety hazard. High maintenance budgets are necessary to keep bog roads in service. For example, Cuddy (1988) reported that the cost of maintaining a bog road at a similar performance level to that of a road constructed on a firm ground foundation was about ten times higher. According to Leebody (1911),

Figure 2. Rampart road on the N62 national secondary road, County Offaly, Ireland.

Figure 3. Extreme undulations on N62 rampart road, County Offaly, Ireland.

the authorities started to address the rampart road problem by the middle of the nineteenth century. In 1851, it was prohibited to sink any pit or hole within 9.0 m of the centreline of carriageways and public roads.

2.4 *Previous works on rampart roads*

2.4.1 *Leebody (1911)*

Leebody (1911) reported rampart road of between 1.5 and 9.0 m in height and proposed different ways of tackling the problem during the Second Irish Roads Congress.

In general, the rampart roads only become dangerously out of shape after a period of time has elapsed following peat harvesting (Leebody, 1911) since some time must elapse before aerobic decay of the remaining peat material can restart. Aerobic decay leads to deformations due to volumetric shrinkage and disintegration of the organic material (Hobbs, 1986). Moreover, peat harvesting operations required that the land area adjoining the rampart roads had been drained thereby allowing deeper harvesting. In many cases, the open drains along the roadsides were excavated deeper on one side causing the rampart to tilt towards the deeper drain. Additionally, rabbits were able to dig burrows close to the road surface as the peat decayed.

Leebody (1911) proposed three different approaches to deal with the rampart road problem. Planting a light fence with young trees was recommended where the rampart had not yet undergone large deformations. The tree root growth was expected to prevent the weakening of the rampart sides. Cutting a few meters from the top of the rampart and using the excavated material to reinforce the rampart sides was recommended in situations where the rampart had already experienced large deformations but rabbits had not started digging burrows. Planting a light fence with young trees was also recommended. The road surface was then rebuilt following the procedure described earlier in section 2.2.

Three different solutions were proposed in situations where rabbits had already started digging burrows. Firstly, exterminate the rabbits and reinforce the roadsides with a clay berm. Next, break and fill the burrows with a mixture of brushwood and clay. Restore the road surface only where needed, which allowed the road to remain open during the repair works. Secondly, and the most commonly applied solution, was to excavate the entire rampart such that the road surface was level with the surrounding ground, where possible. Otherwise, abandonment of the site and the construction of a new road on firm inorganic ground were recommended if the side slopes were too steep and heavy traffic was expected.

2.4.2 *Hanrahan (1953 & 1954)*

Hanrahan (1953 & 1954) presented research studies on bog roads, most of which were rampart roads of flexible construction (material ranging from clay to hand-placed pavement and with a varying layer thickness even within one cross-section). Nearly all of the sites had experienced some distortion, including: transverse or diagonal surface undulations; cracking or depressions due to poor drainage; lack of maintenance; thin road structure; large vegetation which increased the loads and rate of consolidation near the roadsides; humps and depressions at side road junctions and, in some cases, failure due to the low quality of the construction materials. An impervious road surface was cited as most importance since rapid degeneration appeared to occur following puncturing of the impermeable surface seal.

Hanrahan (1953) presented the following recommendations to improve the quality and safety of new and existing roads. Drainage should precede construction to increase the shear strength of the bog foundation (drainage after construction induces large consolidation settlements). Where possible, the construction or improvement of the bog roads should be carried out on pre-consolidated ground such as that produced on lowering the rampart road level. Trees and shrubs in the vicinity should be removed and replanted a safe distance away from the road. The road surface must be constructed using an impermeable material and should be regularly maintained in good condition. Uniform cross-sections along the length of the road are desirable and rapid changes in dimensions should be avoided. Cracks and depressions in the road surface may extend to such a depth that cavities may be found in the bog foundation. The cavities should be filled with an impermeable flexible material. Side drains should be piped, covered and located a safe distance from the roadside. Very steep vertical curves are undesirable due to the increase in applied loading caused by centrifugal force effects. The road material in contact with the bog foundation should be of uniform quality and impermeable, for example, a clay-sand mixture. Heavy static loads should be restricted on the road or verges.

2.4.3 *Hanrahan (1964, 1967 & 1976)*

The main road linking Edenderry and Rathangan, County Offaly, Ireland, was reconstructed between August 1953 and March 1954 following bearing and serviceability failures (road located on a bog foundation, 7.6 m in mean depth). Hanrahan (1964) reported that the failure occurred due to poor thickness control during the placement of the gravel base with excessive quantities placed at certain locations (anywhere between 0.30 and 0.76 m in thickness). Excessive loading caused overstressing of the bog foundation, slips, upheaval and lateral creep distortions along the road. The solution recommended was to replace part of the gravel layer with lightweight bales of compressed air-dried peat and thereby take advantage of the pre-consolidated bog foundation. Furthermore, Hanrahan (1967 & 1976) proposed a design method for road construction on bog foundations, considering three main requirements:

- The pavement must be adequately thick in order to reduce the stresses induced by traffic to a value that does not exceed the design shear strength of the underlying bog foundation.
- The bearing capacity of the bog foundation must not be exceeded by the combined weight of the pavement and traffic.
- Settlement or distortion of the road surface must not exceed values specified by local authorities.

The requirements listed above are very difficult to achieve on bog land due to its low shear strength and high compressibility. A gravel embankment that acts as a temporary surcharge was proposed to induce pre-consolidation in the bog foundation thereby increasing its shear strength and reducing its compressibility (Hanrahan 1967 & 1976). Sometimes the pre-consolidation period can be excessive and the permeability anisotropy may restrict the surcharge effect to a shallow depth. Hanrahan (1967 & 1976) suggested that the installation of vertical drains would accelerate the pre-consolidation process and produce a more uniform increase in the shear strength of the peat with depth. Part of the surcharge was removed to stop the pre-consolidation process when the required increase in shear strength had been achieved. A potential flooding problem may occur since the pre-consolidated area will most likely be depressed on removing the gravel surcharge. Hanrahan (1976) recommended placing the gravel layer over a lightweight fill to keep the road surface higher than the surrounding ground.

2.4.4 *Nerney (1985)*

A 2.0 km section of the R403 regional road (County Kildare, Ireland), which runs along a bog, had presented problems for decades requiring frequent and expensive repairs. Nerney (1985) conducted a study of the pavement layer and bog foundation. The work focused on three sections along the road that were representative of rampart, semi-rampart and level-profile cross-sections. A Dynaflect deflection survey using yielded Dynaflect maximum deflection (DMD) data – a measure of the overall strength of the road; surface curvature index (SCI) data – a measure of the strength of the upper road layers and Geophone 5 (G5) data – a measure of the strength of the sub-grade below a depth of about 1.0 m.

The high DMD values indicated a weak road structure overall. The SCI values indicated very strong and thick pavement structures, sometimes comparable with that of concrete slab values. However, the G5 values identified the bog foundation as the source of the structural weakness.

3 GEOTECHNICAL PROPERTIES

3.1 *Peat formation and drainage effect*

Peat deposits comprise partially decayed and fragmented plant remains that have accumulated under water (Mesri & Ajlouni, 2007). Two distinctive peat layers can be identified: (i) the uppermost layer known as the acrotelm which varies between 10 and 60 cm in thickness; (ii) the underlying catotelm layer permanently located below the groundwater table. In the acrotelm, the water contains sufficient oxygen from precipitation, flow and the atmosphere to support aerobic micro-flora, which maintains the aerobic decay process. As the oxygen concentration reduces with depth in the catotelm, the aerobic micro-flora population decreases and, conversely, the anaerobic micro-flora population increases. The anaerobic micro-flora have a slower metabolic activity, hence the decay process slows leading to the gradual accumulation of partially decayed plant material as peat (Hobbs, 1986).

Peat is a highly heterogeneous and anisotropic material and its geotechnical properties are generally extremely variable over small distances since peat is formed from different plant species and the decay process is not uniform throughout the bog mass. However, the geotechnical properties of peat (water content, unit weight, ignition loss, permeability, compressibility and shear strength, amongst others) are generally closely interrelated.

Land drainage for peat harvesting significantly alters the geotechnical properties inducing distortion and may potentially leading to instability in rampart roads. Further loading of the bog foundation by the road structure and traffic causes significant consolidation and may lead to shear failure of the ramparts. Drainage causes a lowering of the groundwater table increasing the thickness of the acrotelm layer (reduction in thickness of catotelm). Hence, zones that had been submerged (anaerobic state) regain oxygen and become repopulated with aerobic micro-flora, which significantly speeds up the decay process. Drainage has a significant impact on the geotechnical properties of both the acrotelm and catotelm layers. Several changes occur within the acrotelm, including: (i) the level of humification increases as the decay rate increases; (ii) volumetric shrinkage occurs due to air drying of the peat; (iii) the void ratio, water content and permeability values decrease while the unit weight, effective stress and shear strength values increase.

Meanwhile, the remaining peat in the catotelm layer is subjected to an increased state of effective stress that results in subsidence and may potentially lead to shear failure. Subsidence leads to the peat in the acrotelm re-submerging below the groundwater table and hence the loading situation changes on an ongoing basis (Cuddy, 1988). The dried out peat material undergoes a permanent material change due to oxidation and is unable to recover lost moisture on re-submergence (Hobbs, 1986).

3.2 *Water content*

Undrained peat from the Irish Midlands generally has water content values in the range of 650 to 1500% although values as low as 570% and as high as 4900% have been reported (Cuddy, 1988). The value of the water content can be reduced to about 1000% by shallow drainage or to about 700% by deep drainage (Hanrahan, 1954).

Clara bog in County Offaly, Ireland, is a raised bog and nature reserve that has no significant peat harvesting history. O'Loughlin (2001) reported water content values in the range of 1300 to 1500% with a mean water content value of 1400%. Hebib and Farrell (2003) studied the peat properties from Raheenmore and Ballydermot raised bogs in Ireland. The Raheenmore bog also had no significant peat harvesting history and had a mean water content value of 1200%. The Ballydermot bog had over 50 years of peat harvesting history and had a mean water content value of 850%.

3.3 *Compressibility*

The exceptionally high water content values and porous nature make peat extremely compressible. However, pore water pressure dissipation occurs simultaneously with secondary compression, the latter involving structural rearrangement, viscous processes and micro-pore water expulsion (Hobbs, 1986).

A temporary surcharge is applied to pre-consolidate the bog foundation and the surcharge is later removed to reduce the post-construction settlements to acceptable values. Surcharging produces a pre-consolidation pressure that is greater than the final vertical effective stress (Mesri & Ajlouni, 2007). According to Hanrahan (1976), pre-consolidation of the bog foundation has been used in Ireland as early as 1951 in the improvement of rampart roads.

O'Loughlin (2001) reported a mean pre-consolidation stress of 3 kPa and an initial void ratio in the range of 16 to 32 for the Clara bog peat. The mean pre-consolidation stress was 17 kPa and the void ratio was in the range of 10 to 16 for the Ballydermot bog peat. Hebib & Farrell (2003) also reported pre-consolidation stress and initial void ratio values of 15 kPa and 12, respectively, for the Ballydermot bog peat. The pre-consolidation stress and initial void ratio values were 5 kPa and 18, respectively, for the Raheenmore bog peat.

3.4 *Shear strength*

The shear strength of the bog foundation is increased by drainage (Hanrahan, 1954). Nerney (1985) reported that *insitu* vane measurements indicated higher shear strength values in the consolidated peat material beneath the road centreline and in the drier peat located above the groundwater table. Peat is a frictional material with high friction angle values. Mesri & Ajlouni (2007) reported effective friction angle values for peat in the range of 40° to 60°. Farrell & Hebib (1998) reported an effective friction angle of 55° for the Raheenmore bog peat.

3.5 *Permeability*

Permeability is an important engineering property that controls the consolidation rate. The coefficient of permeability (k) of peat for flow in the horizontal direction is generally greater than for the vertical direction. Hobbs (1986) reported horizontal-to-vertical coefficient of permeability ratios in the range of 1.7 to 7.5. However, after vertical loading, the horizontal permeability can be up to 300 times greater than the vertical permeability due to the general horizontal alignment of the constituent fibres (Cuddy, 1988).

The rapid decrease in the k value under loading is another important feature. Hanrahan (1954) reported initial k values of $4x\,10^{-6}$ m/s for a peat material. However, under an applied stress of 55 kPa, the k value had reduced to $2x\,10^{-8}$ m/s after a period of two days, and to 8×10^{-11} m/s after seven months.

4 CURRENT PERFORMANCE AND IMPROVEMENT METHODS

4.1 *Maintenance techniques in Ireland*

Over the last 30 years, the construction and improvement of bog roads and rampart roads has radically changed with the introduction of new materials such as geosynthetics, super-lightweight fills and very flexible mixed-bituminous materials.

Davitt et al. (2000) carried out a survey on the current performance and maintenance techniques used on the bog roads in Ireland. The survey was carried out on national, regional and local roads and, in total, responses were received from 12 local authorities. A summary of the preferred maintenance techniques used to improve and widen bog roads is presented in Table 1.

4.2 *Crushed stone and bituminous overlay*

Overlaying the existing pavement with crushed stone, hot-mixed or cold-mixed bituminous materials are the most popular maintenance techniques. The main disadvantage of overlaying the existing pavement is the increase in weight applied to the compressible bog foundation so that the technique really only provides

Table 1. Preferred maintenance techniques for bog roads used by County Councils in Ireland.

Maintenance technique	National roads	Regional roads	Local roads
Overlay existing pavement with hot-mix bituminous material or crushed stone.	First option	First option	First option
Reinforce the pavement incorporating geosynthetic with bituminous overlay.	Second option	Second option	Not used
Replace peat using granular fill.	Third Option	Not used	Not used

a temporary improvement. Most counties where hot-mixed bituminous materials are used reported higher unit costs and longer life spans than for crushed stone overlays. Cold-mixed bituminous materials were reported by four counties with the advantage of easier installation and similar unit costs than for hot-mixed bituminous materials. However, cold-mixed bituminous materials were relatively new to Ireland in 2000 (when survey was carried out) and hence no life span comparison was available.

Bituminous materials are stronger and structurally more efficient than crushed stone overlays and can be applied at about half the depth to achieve a similar structural contribution. Hence, the bituminous layer applies less weight to the bog foundation and most likely cause less settlement (Davitt et al., 2000).

4.3 *Geosynthetic combined with unbound or bituminous overlay materials*

In Ireland, geosynthetic reinforcement of the pavement over the last 15 years has proven to be successful and cost efficient giving longer life spans compared to crushed stone and bituminous overlays. All of the survey counties that had used geosynthetics reinforcement reported an increase in the pavement strength without adding any appreciable weight. The technique has succeeded in maintaining lightly trafficked roads over bog land for more than 10 years whereas crushed stone or bituminous overlays have had to be reapplied at intervals of between three and four years (Davitt et al., 2000).

4.4 *Lightweight and super-lightweight materials*

Lightweight materials (e.g. pulverized fuel ash, PFA) have been used to reduce the weight when improving bog roads that have become excessively deformed or critically unstable due to successive overlays of gravel/stone. The unit weight of PFA is about 16 kN/m^3 compared to that of crushed stone of about 22 kN/m^3 (Davitt & Killen, 1996). Nevertheless, research for new materials has increased due to the relatively high unit costs of PFA.

Super-lightweight materials including expanded polystyrene (EPS) have also been used due to its lower unit weight compared to traditional materials or lightweight fills. The unit weight of EPS is between 0.2 and 0.3 kN/m^3. Each EPS block (typically $3 \times 1.2 \times 0.6$ m) can be handled by a single person making the installation process much faster and easier. However, EPS is the most costly of the super-lightweight fills although it gives the maximum advantage in load reduction. A granular fill is usually placed on top to prevent floatation of the EPS blocks (Davitt & Killen, 1996).

5 SUMMARY AND CONCLUSIONS

Rampart roads, a particular feature of bog roads in Ireland, pose a significant safety hazard to road users. Drainage and peat harvesting have lead to changes in the geotechnical properties of the bog foundation inducing distortion and instability of the rampart roads. The authorities started to first address the problem in the mid 1800's. The existing rampart road network has to be improved and widened in accordance with the present traffic demand and economical growth.

An extensive research program on the current performance and geotechnical properties of rampart roads must be conducted since no large scale research on the subject has been undertaken since Hanrahan (1953 & 1954). In particular, the effectiveness of the maintenance, improvement and construction methods used for bog and rampart roads have to be assessed.

ACKNOWLEDGEMENTS

The research project 'Design approach for improving rampart roads' is funded by a Research Programme Fellowship (2007) from the National Roads Authority (Ireland). The first author would like to kindly acknowledge Universidad de Antioquia for the leave of absence granted to work on this project and research awards in the form of an Ussher Award from Trinity College Dublin and the Geotechnical Trust Fund award from Institution of Engineers of Ireland.

REFERENCES

Bord na Móna, 2001. *The peatlands of Ireland.* Dublin: Bord na Móna.

Cuddy, T. 1988. *The behaviour of bog road pavements.* MAI Thesis, Trinity College Dublin.

Davitt, S. & Killeen, R.C. 1996. *Maintenance techniques for bog roads.* Dublin: An Foras Forbartha.

Davitt, S., Lynch, J., Mullaney, D. & Wall, W. 2000. *Guidelines on the rehabilitation of roads over peat.* Dublin: Department of Environment and Local Government.

Farrell, E.R. & Hebib, S. 1998. The determination of the geotechnical parameters of organic soils. In: E. Yanagisawa, N. Moroto and T. Mitachi (ed.), *Problematic soils; Proc. Intern. Symp., Japan 28–30 October 1998.* Rotterdam: Balkema.

Feehan, J. and O'Donovan, G. 1996. *The bogs of Ireland: an introduction to the natural, cultural and industrial heritage of Irish peatlands.* Dublin: University College Dublin.

Hammond, R.F. 1981. *The peatlands of Ireland.* Dublin: An Foras Taluntais.

Hanrahan, E.T. 1953. The mechanical properties of peat with special reference to road construction. *Transactions of the Institution of Civil Engineers of Ireland* 78: 179–215.

Hanrahan, E.T. 1954. Investigation of some physical properties of peat. *Géotechnique* 4(3): 108–123.
Hanrahan, E.T. 1964. A road failure on peat. *Géotechnique* 14(3): 185–202.
Hanrahan, E.T. 1967. A design method for construction on peat foundations. *Transactions of the Institution of Civil Engineers of Ireland* 92: 32–55.
Hanrahan, E.T. 1976. Bog roads. *Irish Engineers, Institution of Engineers of Ireland* 29(10): 3–5.
Hebib, S. & Farrell, E.R. 2003. Some experiences on the stabilization of Irish peats. *Canadian Geotechnical Journal* 40(1): 107–120.
Hobbs, N.B. 1986. Mire morphology and the properties and behaviour of some British and foreign peats. *Quarterly Journal of Engineering Geology* 19(1): 7–80.
Leebody, J.W. 1911. "Bog" or "peat" borne roads. *Second Irish Road Congress*. Dublin: Cahill & Co.
Mesri, G. & Ajlouni, M. 2007. Engineering properties of fibrous peats. *ASCE Geotechnical and Geoenvironmental Engineering* 133(7): 850–866.
Mullins, B. & Mullins, M.B. 1846. On the origin and reclamation of peat bog, with some observations on the construction of roads, railways, and canals in bog. *Transactions of the Institution of Civil Engineers of Ireland* 2: 1–48.
Nerney, J. 1985. *A subsurface investigation of a road constructed on peat*. Dublin: An Foras Forbartha.
O'Keeffe, P.J. 1973. The development of Ireland's road network. *Transactions Institution of Engineers of Ireland* 98: 33–112.
O'Loughlin, C. 2001. *The one-dimensional compression of fibrous peat and other organic soils*. PhD Thesis, Trinity College Dublin.
Parnell, H. 1833. *A treatise on roads*. London: Longman, Rees, Orme, Brown, Green & Longman.

Advances in Transportation Geotechnics – Ellis, Yu, McDowell, Dawson & Thom (eds)
© 2008 Taylor & Francis Group, London, ISBN 978-0-415-47590-7

The effect of compaction level on the performance of thin surfaced granular pavements

J. Patrick
Opus Central Laboratories, Lower Hutt, New Zealand

S. Werkmeister
Dresden University, Dresden, Germany

M. Gribble
Opus Central Laboratories, Lower Hutt, New Zealand

D. Alabaster
Transit New Zealand, Christchurch, New Zealand

ABSTRACT: There have been a number of high profile early rutting failures in New Zealand of thin surfaces granular pavements where the pavements had not been subjected to traffic during construction. In all cases the construction appeared to comply with the contract specifications. A research project was initiated that investigated, at an accelerated pavement testing facility (CAPTIF), the effect of constructing the basecourse to a range of densities (88% to 95% of Maximum Dry Density). The stress distribution under a vibratory roller was also modelled and compared with stress distribution from a standard 40 kN wheel. The conclusion of the research is when using conventional New Zealand construction techniques and specifications some post construction deformation of "greenfield" pavements appears to be inevitable. However, the rut depth should not approach the levels (20 mm) that prompted the initiation of this research.

1 INTRODUCTION

The majority of pavements in New Zealand consist of a granular base with a chipseal surfacing. In construction of these pavements the specifications call for the compaction level to be closely controlled and monitored using nuclear density meters.

There have been a number of high profile early failures in New Zealand (rut depths greater than 20 mm observed early in the pavement life) associated with "greenfield" pavements (pavements not subjected to traffic during construction) where the granular layers were greater than approximately 400 mm thick. It is uncertain if the rutting was due to poor construction control, difficulties of measuring density of thick layers with a meter operating in backscatter mode, or a function of lack of traffic on the pavement. In response a research project was initiated that investigated, at the Transit New Zealand (TNZ) accelerated pavement testing facility (CAPTIF), the effect of constructing the basecourse and subbase to a range of densities (88% to 95% of Maximum Dry Density, MDD).

The conclusion of the research is when using condentinal New Zealand construction techniques and specifications some post construction deformation of "greenfield" pavements appears to be inveightable. The permanent strain developed will manifest itself into a larger rut depth as the granular thickness increases. However, the rut depth should not approach the levels (20 mm) that prompted the initiation of this research.

2 FIELD EXPERIENCE

A summary of the details of three pavements that rutted early in their lives is given in Table 1.

Each pavement was on the State Highway network and constructed to the TNZ B/2 specification. The basecourse materials all complied with the TNZ M/4 specification which specifies a 40 mm maximum particle size, low to zero plasticity, and at least 70% broken faces in the coarse aggregate. The sub-base materials were all a 65 mm maximum particle size, and partly to fully crushed. The construction specification TNZ B/2 specifies a compaction level in terms of percentage of MDD. MDD is determined in the laboratory using a vibratory hammer compaction method. The

Table 1. Details of three premature rutting failures.

	Site 1	Site 2	Site 3
Surfacing layer depth (mm)	30 (OGPA)	30 (OGPA)	Chipseal
Basecourse layer depth (mm)	150	180	150
Sub-base layer depth (mm)	450	390	200
Traffic volume (vehicle/lane/day)	9000	5000	6000
% heavy commercial vehicles	12	11	10
Rut depth (mm)	20	20	15
≈ Equivalent Standard Axles at six months	2.3×10^5	1.2×10^5	1.3×10^5

Table 2. TNZ B/2 Specification compaction requirements.

% of MDD	Sub-base	Basecourse
Mean value	≥95	≥98
Minimum value	≥92	≥95

New Zealand method is similar to the EU/BS method. The compaction requirements are given in Table 2. The necessity to achieve 95% of MDD compaction was established in a previous project.

The rut depths described in Table 1 were measured on the sites within six months of each site opening to traffic. On each site construction was completed before the pavement was opened to traffic.

A review of the construction quality records and post-mortems testing found that all sites complied with the specification requirements.

3 CAPTIF TESTING

3.1 *Description*

TNZ's accelerated pavement testing facility CAPTIF was used to explore the effect of the compaction level on the subsequent granular basecourse behaviour.

CAPTIF is located in Christchurch, New Zealand. It consists of a circular track, 58 m long (on the centerline), contained within a 1.5 m deep, 4 m wide concrete tank so that the moisture content of the pavement materials can be controlled and the boundary conditions are known. A centre platform carries the machinery and electronics needed to drive the system. Mounted on this platform is a sliding frame that can move horizontally by 1 m. This radial movement enables the wheel paths to be varied laterally and can be used to have the two "vehicles" operating in independent wheel paths.

Table 3. Basecourse densities.

	Compaction level, % of MDD		
	1	2	3
0–150 mm deep	88.0	95.0	98.4
150–300 mm deep	89.0	93.2	97.3
300–450 mm deep	93.8	94.4	94.3
Average	90.6	94.2	96.7

At the ends of this frame, two radial arms connect to the Simulated Loading and Vehicle Emulator (SLAVE) units. These arms are hinged in the vertical plane, so that the SLAVEs can be removed from the track during pavement construction, profile measurement etc., and the arms are hinged in the horizontal plane to allow vehicle bounce. Full details can be found in Pidwerbesky 1995; it basically consists of two wheel assemblies that can each apply a dual-tired half axle load of 40 kN.

For this research three test sections were constructed on a prepared subgrade that had a CBR of 3. The test pavement consisted of 450 mm of a 40 mm maximum particle size basecourse complying with the TNZ M/4 specification. This material would be regarded as a premium material. The intention was to construct three sections to 85%, 90%, and 95% of MDD.

The 450 mm layer was constructed in three 150 mm lifts which were compacted using a vibratory plate compactor. Due to the relatively tight radius of the test track it has been found very difficult to use a full sized vibratory roller and the plate compactor is the usual compaction system used.

It was found impossible to construct each layer to the specific density. This was especially noticeable with the target 85% MDD. After spreading and lightly compacting the layer the achieved percentage of MDD was approximately 88%.

The compaction level obtained within each layer of each test section is given in Table 3.

The three test sections will be referred to as 88% MDD, 95% MDD and 98% MDD reflecting the compaction level achieved on the top layer. The final surfacing was a 25 mm layer of dense graded asphaltic concrete. The pavement was then loaded and a transverse profile measured at regular intervals.

For each section at nine cross sections the vertical surface deformation (VSD) was measured at 25 mm intervals.

At CAPTIF, VSD is more conveniently measured with the transverse profilometer rather than measuring rut depth under a straight edge. VSD is defined as the vertical difference between the start reference level of the pavement and that measured. VSD and rut depth are very similar where there is no lateral heave of the

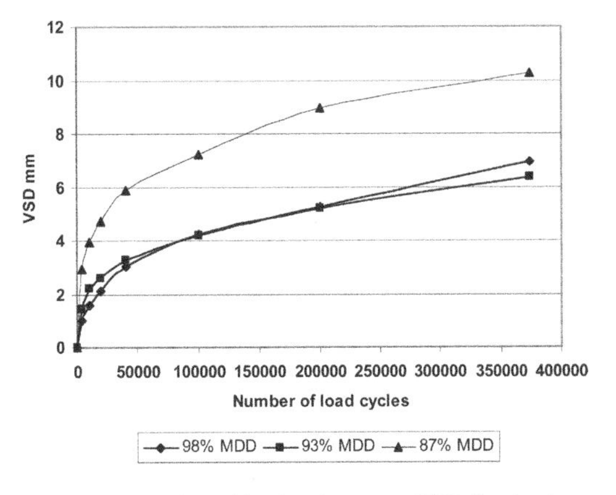

Figure 1. Number of load cycles versus VSD for the three test pavements.

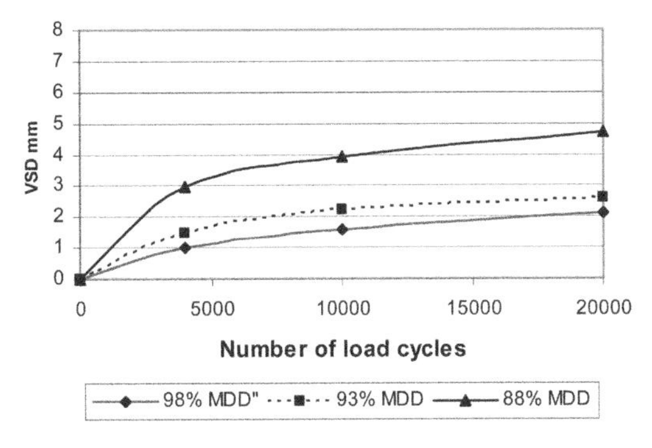

Figure 2. Number of load cycles (LC 0-20,000) versus VSD for the three test pavements.

pavement. The results are shown in Figure 1. In the 88% MDD section an area of rutting of 16 to 18 mm developed at 374,000 load repetitions cycles and the section was then resurfaced.

The VSD relationships for the 95% MDD and 98% MDD are nearly identical but the 88% has a higher level of deformation.

In Figure 2, the VSD up to 20,000 loading cycles is plotted and it can be seen the majority of the difference between the 88% MDD and the other two sections occurs in the first 2,000 loading cycles. The curves from 2,000 cycles are not parallel and the 88% MDD material is deforming at a slightly faster rate than the other two.

4 MODELLING COMPACTION STRESS

For a detailed investigation of pavement stress and strains, a 3-dimensional computational model is required. ReFEM, a finite element (FE) program was used to carry out this investigation. A

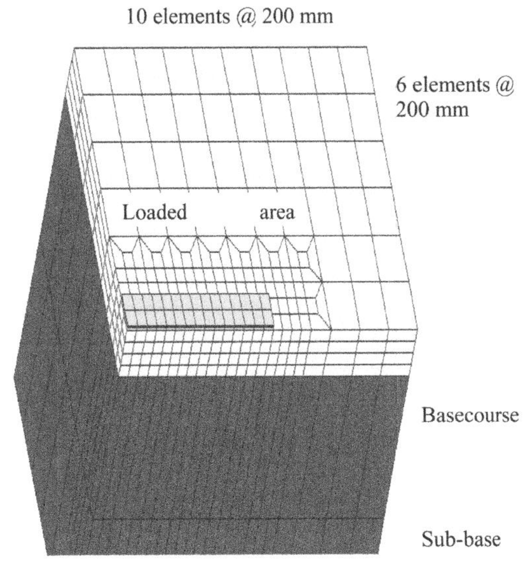

Figure 3. FE mesh for a pavement with a 200 mm thick basecourse and loaded by a 3 T roller drum.

rectangular model was developed to simulate the pavement. By making use of symmetry (quarter model) in the geometry the computational effort was reduced. The length of the FE section was 2.4 m (2 × 1.2 m) and the width of the section was 4.0 m (2 × 2.0 m). A 200 mm thick granular layer over a subgrade was modelled to explore the difference in stress distribution under a dual wheel compared to a vibratory roller.

The subgrade was modelled by 2 elements (loaded by a roller drum, each element 600 mm high) or 4 elements (loaded by a tire, each element 300,mm high). Figure 3 shows the FE meshes for a pavement with a 200 mm thick basecourse layer.

The basecourse was modelled by 4 elements. The asphalt layer was modelled by 2 elements, each 12.5 mm high (for the pavements loaded by a wheel). The mesh used in the analysis was finest at the contact zone (element size 66.6 × 66.6 mm × height) where large stress and strain gradients were expected. The bottom of the subgrade was prevented from axial movement in the three directions.

4.1 *Dual wheel*

In order to determine the contact area of the tire for use in the pavement modeling stage of this research, the tire footprints measured at CAPTIF under a 40 kN loaded dual wheel were used. The dimension of the tire footprint was approximately rectangular, 230 mm long and 215 mm wide (each wheel) when loaded to 40 kN. The loaded area was 0.0495 m^2 (each wheel). However, this area could not be modeled identically in the FE program that was used in this analysis. The loaded area of the truck was approximated by three 2 × 2 elements (each wheel 0.0533 m^2) as shown in Figure 2. The contact pressure of 0.338 MPa was assumed to be uniformly distributed over the contact area.

4.2 *Roller drum*

Vibratory rollers are typically used in granular pavement construction in New Zealand. Rollers normally have one or two different amplitude settings (see Table 4) and therefore represent a compromise regarding the compaction efficiently under various soil and basecourse conditions.

The system of a vibratory roller compacting a basecourse layer is fairly complex. A smooth drum roller, type Caterpillar CS 563C was selected to calculate the load under a typical 3 T vibratory roller drum. The width of the contact area is dependant on the soil/basecourse stiffness and was assumed as 266 mm for a high amplitude dynamic load and 133 mm for a low amplitude dynamic load (Hamm 2007). The roller specification is listed in Table 4.

The loaded area of the roller drum was modelled by fifteen 2 × 2 elements for a high amplitude dynamic load (loading area 0.532 m^2 = 266 mm × 2,000 mm) as shown in Figure 3 and modelled by fifteen 1 × 2 elements for a low amplitude load (loading area 0.266 m^2 = 133 mm × 2,000 mm). Because of the high stiffness of the roller drum it is assumed that the roller drum does not deform significantly in the transverse and vertical direction during the compaction process.

Hence, the roller drum was modelled as a rigid plate with a resilient modulus of 320,000 MPa in the transverse and vertical direction.

The elastic deformation of the base-course/subgrade is assumed to be governed by a parabolic load (Sandström 2006).

In order to reproduce this parabolic load, the roller drum was modelled with a very low E-Modulus value of 50 MPa in the longitudinal direction. The peak stresses and vertical elastic strains within the pavement due to high-amplitude and low-amplitude vibratory compaction load at conventional travel speed (4.4 to 6.6 km/h) were calculated.

Table 4. Roller specification (Ping et al. 2002).

Roller type : Caterpillar CS 563C	
Drum diameter	152.4 mm
Drum width	2,000 mm
Drum module weight (static load)	2.96 T
High amplitude dynamic load	15.195 T
High amplitude loaded area	0.532 m^2
Low amplitude dynamic load	4.65 T
Low amplitude loaded area	0.266 m^2

5 MATERIAL MODELS

5.1 *The asphalt layer*

The asphalt layer was treated as linearly elastic with an assumed Poisson's ratio of 0.35 and a resilient modulus of 3,000 MPa.

In order to determine the parameter for the elastic and plastic model, Repeated Load Triaxial (RLT) tests were conducted on the basecourse material used at CAPTIF. Resilient modulus tests as well as permanent strain tests, according to the draft TNZ RLT Standard (T/15), were conducted at two different Degrees of Compaction (DOC) (87% and 95%) and at a moisture content of 70% of the Optimum Moisture Content (OMC).

5.2 *Nonlinear Dresden Model for the basecourse*

On the basis of RLT testing an empirical non-linear elastic-plastic deformation design model (Dresden Model) was formulated. This model has been implemented into the ReFEM program. Further details are available elsewhere (Werkmeister 2003, Oeser 2004). This non-linear elastic model is expressed in terms of resilient modulus, E, and Poisson's ratio, ν, as follows:

$$E = p_a\left(Q + C\cdot\left(\frac{\sigma_3}{p_a}\right)^{Q_1}\right)\cdot\left(\frac{\sigma_1}{p_a}\right)^{Q_2} + D \quad (1)$$

$$\nu = R\cdot\frac{\sigma_1}{\sigma_3} + A\cdot\frac{\sigma_1}{p_a} + B \quad (2)$$

where σ_3 (kPa) minor principal stress (absolute value); σ_1 (kPa) major principal stress (absolute value); D (kPa) constant in term of modulus of elasticity; Q, C, Q_1, Q_2, R, A, and B are model parameters. On the basis of the results of multi-stage RLT tests it is possible to determine the parameters of the elastic model. Table 5 shows the parameters for the Dresden Model used for the FE calculation process.

Table 5. Parameter for the elastic Dresden-Model.

Material Elastic DRESDEN-Model	CAPTIF 1 Greywacke 0/40 95% DOC	87% DOC
Q	14,004	14,003
C	6,540	1,227
Q_1	0.346	0.555
Q_2	0.333	0.333
D(kPa)	65,000	65,000
R	0.056	0.056
A	−0.0006	−0.0006
B	0.483	0.483
p_a(kPa)	1	1

5.3 *Subgrade*

To overcome the limitation of the depth of the FE model the subgrade was modelled as a linearly elastic material with a resilient modulus of 70 MPa and a Poisson's ratio value of 0.4.

6 RESULTS

6.1 *Stress distribution (high amplitude dynamic load + static drum weight)*

Figure 4 shows the pattern of vertical compressive stress developed in the pavement structure when a 40 kN loaded dual wheel and a 3 T roller drum (high amplitude dynamic load + static drum weight) travels over the pavement layers.

The maximum compressive stress under a dual wheel naturally occurs in the middle of the wheel path next to the surface. In contrast the results of the FE calculation illustrate that the highest vertical stress under a rigid 3 T steel roller drum occurs next to the drum edges.

The stress values shown in Figure 5 demonstrate that the vertical stress in the upper two thirds of the basecourse under a 40 kN wheel load is significantly higher compared to the vertical stress under a 3 T roller drum (drum centre; high amplitude dynamic load + static drum weight). Only the vertical stress under the roller drum edge is higher compared to the stress under a 40 kN loaded dual wheel. The vertical compressive stress induced at a pavement depth of 175 mm and deeper is considerably higher than the vertical stress induced by a 40 kN loaded dual wheel. It can be concluded that the roller drum induces higher stress in the subgrade when compacting a 200 mm thick basecourse layer compared to the stress induced by a 40 kN loaded wheel load. Hence, this roller should be able to apply sufficient compaction effort on the lower part of the base-course and the subgrade only.

6.2 *Strain Distribution (high amplitude dynamic load + static drum weight)*

The vertical elastic strains determined under a 40 kN loaded dual wheel and under a 3 T roller drum (high amplitude dynamic load + static drum weight) are shown in Figure 5. They follow a similar pattern to the stress analysis.

Figure 5 illustrates that the vertical elastic strains in the top half of the basecourse under a 40 kN loaded dual wheel are significantly higher than those induced by a 3 T-vibratory roller drum (roller drum centre). However, under the edge of the roller drum higher vertical elastic strains are induced in the pavement compared to the 40 kN loaded dual wheel. In addition, the vertical elastic strains occurring under a 3 T roller drum (drum centre) in the sub-grade are on average as twice as high compared to the strains occurring under a 40 kN loaded dual wheel.

Furthermore, Figure 5 clearly illustrates that the highest vertical elastic stains due to a 3 T roller drum (drum centre) appear in the subgrade. In contrast, the highest stains due to a 40 kN loaded dual wheel can be observed in the basecourse.

7 CONCLUSIONS

This research has demonstrated that the initial post construction deformation of a thin surfaced granular

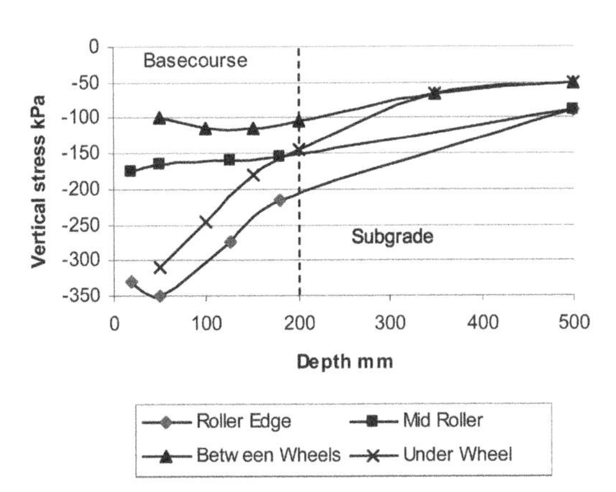

Figure 4. Vertical compressive stress under a 3 T roller drum and a 40 kN loaded dual wheel. (Basecourse thickness 200 mm, DOC 95%, high amplitude dynamic load + static drum weight.

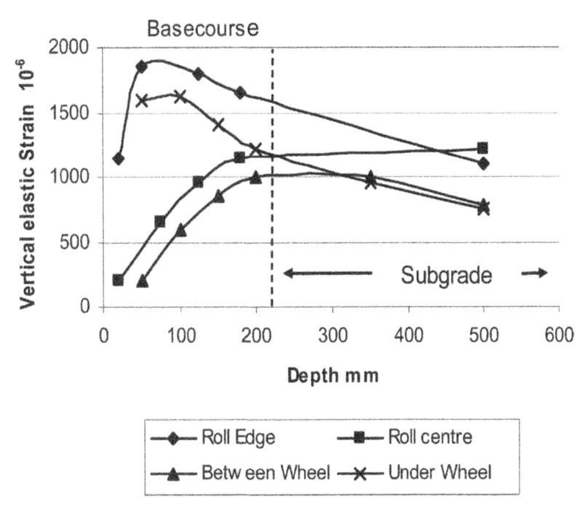

Figure 5. Vertical elastic strain under a 3 T roller drum and a 40 kN loaded dual wheel. (Basecourse thickness 200 mm, DOC 95%, high amplitude dynamic load + static drum weight.

pavement is affected by the compaction level achieved in the field it has also demonstrated that there is probably a contribution from the compaction equipment used. The FEM has shown that the stress distribution under the centre of a roller is lower in the top of a basecourse than that generated by a standard wheel load.

It is postulated that the shear deformation that has been found in the field is associated not just with post construction rutting but this is exacerbated by water being forced through the surface under traffic tyre forces.

Research into the influence of the compaction methodology and the waterproofing ability of thin surfacings on the performance of unbound granular materials is continuing.

REFERENCES

Hamm, AG, (2007) *Information from the Company*, Tirschenreuth, Germany.

New Zealand Standard NZS 4402 Test 4.1.3 *Determination of the Dry density /Water Content Relationship, NZ Vibrating Hammer Test.*

Oeser, M Numeric *Simulation of the non-linear behaviour of flexible pavements (in German: Numerische Simulation des nichtlinearen Verhaltens flexibler mehrschichtiger Verkehrswegebefestigungen),* PhD Thesis, University of Technology, Dresden, Germany.

Patrick, J.E., Alabaster, D. J., 1998. Pavement density. *Transfund NZ Research Report 100.*

Pidwerbesky, B.D. 1995. Accelerated dynamic loading of flexible pavements at the Canterbury accelerated pavement testing indoor facility *Transportation Research Record 1482*. pp. 79–86.

Sandström, A, Numerical Simulation of a vibratory roller on cohesive soil, *Geodynamic Research Report 2007.*

Ping Yang, Z, Leonhard, M & Puchta, S, Laboratory Simulation of Field Compaction characteristics on sandy soils, *Transportation Research Record 1808* paper No. 02-3164.

Transit New Zealand. 2005. *Specification for Construction of Unbound Granular Pavement* TNZ B/2:2005.

Transit New Zealand 2006*Specification for Basecourse Aggregate,* TNZ M/4.

Werkmeister, S. *Permanent deformation behaviour of unbound granular materials*. PhD thesis, University of Technology, Dresden, Germany.

Werkmeister, S, Steven, B.D, Alabaster, D.J, Arnold, G & Oeser M 3D Finite Element analysis of accelerated pavement test results from New Zealand's CAPTIF Facility, *Proceedings of the 7th International Symposium on the Bearing Capacity of Roads and Airfields (BCRA),* Trondheim (N), 24–26. June 2005.

Advances in Transportation Geotechnics – Ellis, Yu, McDowell, Dawson & Thom (eds)
© 2008 Taylor & Francis Group, London, ISBN 978-0-415-47590-7

Laboratory characterization and influence of mineralogy on the performance of treated and untreated granular materials for unpaved roads

P. Pierre & J.-P. Bilodeau
Department of Civil Engineering, Laval University, Quebec, Canada

G. Légère
FPInnovations – Feric division, Montreal, Quebec, Canada

ABSTRACT: A laboratory study on the influence of granular mineralogy sources (basalt and limestone) on the performance of treated and untreated granular materials used as surface pavements in unpaved roads is presented. Results of CBR, resilient modulus and water drop test are discussed. Calcium chloride, natural brine, Portland cement and a polymer were tested and compared to untreated samples. Results are also compared to earlier work conducted on a gneiss source material. The potential to increase bearing capacity, increase strength and reduce dust is discussed.

1 INTRODUCTION

The performance of unpaved roads is directly affected by environmental stresses such as wind, rain, water, frost, as well as mechanical stresses from vehicle tires, heavy axle loads and traffic volumes. The characteristics and mineralogy of granular materials used as surfacing material also play a key role in road performance. A combination of these factors may cause surface distress such as rutting, corrugation, potholes and erosion amongst others, which can affect user comfort and safety. Therefore, maintenance work such as grading must be carried out on a periodic schedule to maintain the pavement serviceability to an acceptable level. These maintenance costs represent a major expense for those who manage unpaved road networks.

Because of their rapid deterioration rate, most unbound granular materials used as surfacing materials for unpaved roads are treated in order to provide a better ride quality for users and to prolong the life of the material. There is an economical advantage to unpaved road surface treatments, which is costly but usually reduces maintenance costs (Sanders 1997). Many types of products are used to increase the general performance of unpaved roads. Dust suppressant agents such as alkali-chloride solutions, polymer solutions, lime and resins are used to reduce excessive dust problems causing safety and respiratory hazards. On the other hand, stabilization agents like cement, bituminous and polymer emulsions are used to reduce traffic related pavement performance problems such as rutting.

The effect of various products has been widely studied and reported by the scientific community. While it is recognized that most dust suppressant agents do not significantly improve the bearing capacity (Doré et al. 2005, Pelletier 2007, Pierre et al. 2007a, b), stabilization agents generally improve both bearing capacity and dust lifting resistance due to the strong cohesion forces brought to the unbound granular material skeleton. A recent study by Pelletier (2007) demonstrated with direct shear tests that shear resistance tends to increase with compaction water content (3 to 6%) for calcium chloride, natural brine, cement (1.5 and 3%) and polymer solution. In addition, Pierre et al. (2007b) showed that, within the suggested application rates recommended by suppliers, performance measured with CBR, unconfined compression and resilient modulus tests increase with natural brine, polymer solution and cement contents for equivalent compaction water content while it decreases for the calcium chloride. In addition, the results demonstrated that high polymer or natural brine content samples presented similar performance to low cement content samples. The studies from Hoover (1987) and Doré et al. (2005) previously demonstrated the mechanical advantage of using polymer or natural brine for mechanical stabilization while the work reported Bergeron (2000), Pouliot (2004) and Santoni et al. (2001) showed the mechanical advantages of cement and polymer for unpaved roads.

The objectives of this study were twofold; evaluate the influence of granular mineralogy (basalt and limestone) on the performance of treated and untreated granular materials and to compare the performance of various dust suppressants and stabilization agents. It should also be noted that some results from phase 1 of this study conducted by Pierre et al. (2007b) on a granitic (granite) material are also discussed and compared. This report presents the results of a series of laboratory tests (CBR, resilient modulus and water drop test) conducted with calcium chloride, polymer, natural brine and Portland cement at various concentration rates. Untreated specimens were also tested and used as a control to perform comparative analysis.

2 TEST SPECIMENS

As part of the CARRLo project (Pierre 2005), this present study is a continuation of the laboratory investigation described in Pierre et al. (2007b) for different granular material mineralogy classes, which are crushed basalt and crushed limestone. Standard geotechnical characterization and grain-size distribution of the aggregate sources are presented in Figure 1. For each aggregate source, the gradation and the compaction water content were kept constant while the treating product concentration rate varies. The tests were also performed on untreated granular materials to obtain the relative effect of the products on the performance. The granular materials were treated with two families of products which are chemical additives and cement products. These treating agents are a natural brine (60% of $CaCl_2$, 37% of NaCl, 2% of $MgCl_2$, and 1% of KCl), calcium chloride (94% of $CaCl_2$, 3.7% of NaCl, 0.04% of $MgCl_2$, and other impurities), aqueous acrylic vinyl acetate polymer emulsion and Portland cement (type GU). The polymer emulsion was dissolved in water at a 6.8:1 proportion as specified by the manufacturer. Table 1 presents the test matrix including the cure period and the various concentrations tested per product, which are mostly based on the suppliers' specifications and experience. It should be noted that each test was repeated three times in order to obtain more representative results by using average values.

The samples were compacted in 4 layers of approximately 25.4 mm with a vibrating hammer inside ABS plastic moulds with an internal diameter of 152.4 mm, a wall thickness of 7.658 mm and a height of 152.4 mm. While the bottom of the sample is level with the mould bottom, a compacted material free space of approximately 55.1 mm is found in the upper end of the mould (Figure 2). The samples were all compacted near the same optimum water content of 4.5% which includes the water content of the product. The optimum compaction water content of the untreated materials was 5% for both the limestone and basalt. A cure of 7 days at room temperature was applied to all samples.

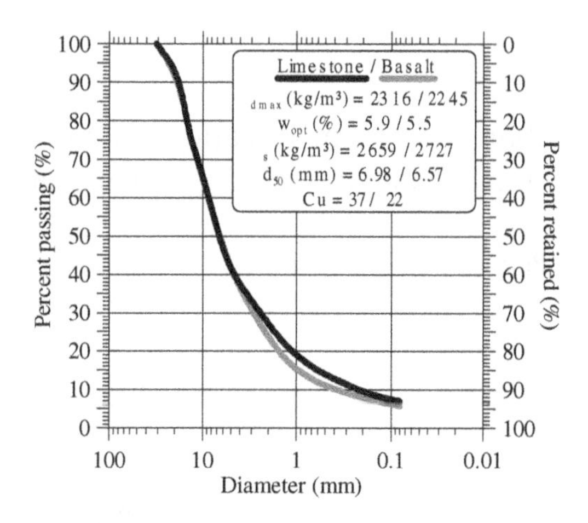

Figure 1. Aggregate sources main geotechnical properties.

Table 1. Tests matrix.

Product	CBR, M_R and Water drop
Brine (L/m^2)	1.2/1.5/1.8 for the 3 tests
Brine cure (days)	7, 7 and 14
Brine #samples	3/conc. for the 3 tests
$CaCl_2$ (L/m^2)	1.3/1.8/2.3 for the 3 tests
$CaCl_2$ Cure (days)	7, 7 and 14
$CaCl_2$ #samples	3/conc. for the 3 tests
Polymer (L/m^2)	1.4/2.4/3.4 for the 3 tests
Polymer cure (days)	7, 7 and 14
Polymer #samples	3/conc. for the 3 tests
Cement (%)*	1.5/3/4.5/6 for the two tests
Cement cure (days)	7 and 14
Cement #samples	3/conc. for the 2 tests
Reference	–
Reference cure (days)	7, 7 and 14
Reference #samples	3 for the 3 tests

* M_R tests were not performed for cement samples.

3 TESTING PROGRAM

The mechanical performance of treated granular materials is measured with CBR and resilient modulus tests slightly modified from the original standards ASTM D1805 (ASTM 1993) and AASHTO T-307 (AASHTO 2000). In addition, the surface sealing potential is measured with a test that is inspired from a European standard (RILEM 1980) which basically consists of measuring the average surface area of a 5 mL water drop (WD) on the sample surface. Figure 3 presents a scheme of the three tests. Because several specimens needed to be prepared the same day to respect the same cure time, ABS moulds were used rather than ASTM standard steel moulds for both CBR and resilient modulus tests.

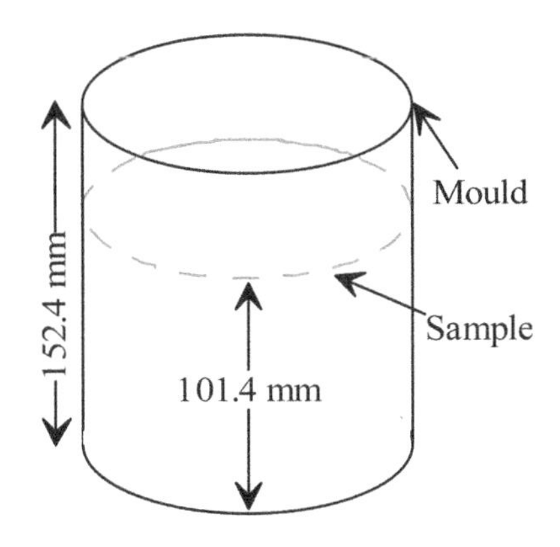

Figure 2. Scheme of the compacted samples.

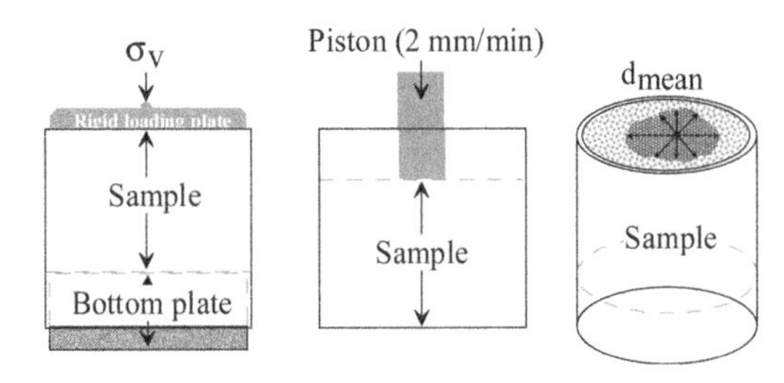

Figure 3. Scheme of the three performed tests.

The resilient behaviour was measured on the bottom end of the sample, with the top end resting on an appropriate support. Since the samples are inside the ABS moulds, the resilient behaviour can be compared to an oedometric resilient behaviour. The samples were submitted to 300 cycles at a vertical stress of 280 kPa which is applied in a cyclic haversine manner. To ensure a uniform contact with the samples, 10% of the total vertical stress is static while 90% is cyclic. This vertical stress level corresponds to ½ ESAL. The resilient modulus at 300 cycles is then recorded as the resilient behaviour performance parameter since most of the resilient deformation is found to be mostly stabilized after 150 to 200 cycles as stated by Pierre et al. (2007b). These tests were not performed on cement treated samples.

The CBR tests were conducted on the sample top end at a penetration rate of 2 mm/min to reach a maximum penetration of 10 mm. The force recorded at a corrected 5 mm of penetration is used as the reference force since statistical analysis showed that this value presents the highest correlation strength with the product concentration (Pierre et al. 2007b). Finally, the water drop test is performed on the undisturbed surface of the specimen, which is the one used for the M_R tests after a cure of 14 days. In order to quantify the surface sealing potential, 5 mL of water is dropped on the surface sample with no elevation energy. The diameter of the wet surface is then measured in four axes after 5 minutes. The average diameter (d_{mean}) of the wet surface is calculated and the corresponding surface is computed in order to compare the pore sealing efficiency of each product for each concentration. The first results reported by Pierre et al. (2007b) presented conclusive results on the potential to use this test method to quantify the pore sealing potential of various products commonly used on unpaved road surfaces for mechanical stabilization and dust suppression.

Table 2. Relative test results.

		Limestone/Basalt		
Product	Dosage	M_R	CBR	WD
Brine	1.2	1.02/1.07	0.92/0.88	0.99/1.37
Brine	1.5	0.91/1.06	0.82/0.86	1.01/1.48
Brine	1.8	0.93/1.02	0.80/0.95	1.32/1.14
$CaCl_2$	1.3	1.14/1.06	0.73/0.83	1.28/1.06
$CaCl_2$	1.8	1.18/1.08	0.79/0.82	0.99/1.55
$CaCl_2$	2.3	1.15/1.23	0.82/0.70	0.90/1.29
Polymer	1.4	1.06/1.18	1.02/1.03	0.68/0.93
Polymer	2.4	1.00/1.19	1.00/1.04	0.57/1.03
Polymer	3.4	1.00/0.99	1.03/1.13	0.68/1.31
Cement	1.5	–/–	1.24/1.62	0.91/0.86
Cement	3.0	–/–	2.03/2.15	1.17/0.99
Cement	4.5	–/–	2.04/2.49	0.88/1.07
Cement	6.0	–/–	2.57/2.77	1.09/0.99
Ref.	Rel.	1.0/1.0	1.0/1.0	1.0/1.0
Ref.	Abs.*	147/133	30/22	38/30

* MPa for M_R tests, kN for CBR tests and cm^2 for water drop tests.

4 TESTS RESULTS

Since the primary objective of this research project is to determine the efficiency of various products on the performance of granular materials, the results are presented in Table 2 as relative results (Rel.) which corresponds to the test result obtained for the treated granular material (treated granular material divided by the test result). The absolute test values (Abs.) obtained for the untreated granular materials are also presented in order to put the results into perspective. It should be pointed out that each test was performed three times for a given product and concentration in order to obtain a more representative average value as presented in the test matrix. Therefore, the relative results presented in Table 2 are average values. Presentation of relative results allows for comparing performance gain/loss caused by various products. CBR and resilient modulus both presented, on average, a range of 25 to −25% (gain/loss) for the natural brine, calcium chloride and polymer whereas higher values were obtained for the water drop test. CBR test results show higher values

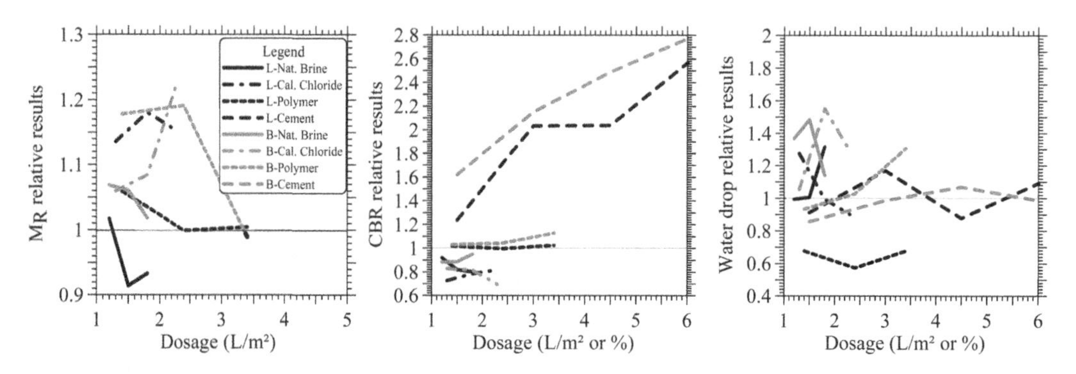

Figure 4. Relative test results.

for cement stabilized granular materials. The relative results introduced in Table 2 are shown graphically in Figure 4 for the basalt (B) and the limestone (L). From these figures, the important effect of material mineralogy and gradation can be observed since the CBR relative results are generally lower than those obtained by Pierre et al. (2007) for a granitic gneiss granular material.

5 DISCUSSION

The relative results presented in Table 2 show the different behavior of dust suppressant (natural brine, calcium chloride and polymer) and stabilizing agents (polymer and cement). When looking at the affect of natural brine on the performance of the granular material, the general trend indicates a decrease in mechanical performance (CBR and M_R) with an increase in concentration for the limestone while the opposite trend was found for the basalt CBR results. However, the addition of natural brine to the limestone does not improve the mechanical characteristics over the unstabilized material. An equivalent resilient modulus to the reference material is found at low concentrations for the limestone (1.02). For the basalt, the M_R relative results are higher than unity with performance gains of 7 to 2% while CBR relative results showed a loss in performance of 0.883 to 0.949. As a general trend, it can be concluded that none to little mechanical improvement is measured when natural brine is added to the limestone or basalt granular materials tested in this study. For the natural brine, the general trend observed shows that the basalt material presents a higher mechanical performance (when the average values are considered) and that low brine concentrations yield higher strengths than high concentrations for both limestone and basalt. Opposite trends were reported by Pierre et al. (2007b) for a granitic gneiss material. The only exception measured in this study was found for the basalt CBR relative results where the peak value is at $1.8\,L/m^2$. This is also confirmed by the surface sealing potential measured with the water drop test, where the basalt shows higher relative results than the limestone, especially at lower product concentrations. This means that less water penetrates the basalt samples than the limestone samples.

The calcium chloride relative results for the limestone show an increase of CBR and M_R with an increase in product concentration and a peak M_R value is measured at $1.8\,L/m^2$. The M_R results are higher than unity while the CBR results are lower. This may be explained by the calcium chloride crystallization within the granular material voids which increase the cohesion forces in the elastic range but seems to weaken the general strength of the granular skeleton at high deformations. The calcium chloride may decrease the grain to grain frictional forces when materials are submitted to high shearing forces due to its water retention capacities but at the same time may reduce dust. The basalt granular material shows similar mechanical behavior as the limestone in terms of resilient modulus with slightly lower relative values at a lower product dosage. However, the CBR relative results decrease with product concentration and are still lower than unity (0.827 to 0.695). The peak values measured in this study suggests that higher calcium chloride dosage should be used (except for the basalt CBR relative results) which is the opposite of what was found by Pierre et al. (2007b) and which demonstrates the important effect of grain-size distribution and mineralogy. For the limestone, the surface sealing potential is higher at a low product concentration (1.275) while the opposite is observed for the basalt (1.55 at $1.8\,L/m^2$). On average, the surface sealing potential of the natural brine is higher than the calcium chloride. It is also possible to conclude that even if no to little mechanical gains are observed when dust suppressant agents are used, calcium chloride increases the mechanical performance more than the natural brine, which was also found for the granitic gneiss with CBR tests. CBR relative results are similar for both mineralogy sources (lower than 1) while average M_R relative results are significantly higher than 1 for the calcium chloride (on average 1.124 for the basalt and 1.156 for the limestone for all the tested product concentrations).

The average M_R relative results for the polymer show higher values than the natural brine but lower M_R values than the calcium chloride. The previous phase of the study conducted on a granitic gneiss showed that the polymer and natural brine effect were equivalent in terms of resilient modulus. It can also be observed with M_R results that the mechanical performance is improved at a lower product concentration (1.4 L/m^2 for the limestone and 2.4 L/m^2 for the basalt). In addition, both mineralogy CBR relative results are slightly higher than 1 and are higher than the CBR relative results observed for the natural brine and the calcium chloride. However, the CBR relative results obtained for the basalt and the limestone are significantly lower than the ones previously obtained for the granitic gneiss. Since the general trend shows that both CBR and M_R relative results are higher than 1, the polymer is thus a better choice for maintaining or slightly improving the mechanical characteristics than the natural brine or the calcium chloride even though the performance gains are small (up to 19% for the basalt M_R results). Also, as a general trend, it is observed for both aggregate sources that an increase in product concentration causes a decrease in M_R and an almost neutral affect on CBR. For the limestone, the use of this type of polymer does not significantly influence the bearing capacity (on average 2 and 1% for M_R and CBR respectively) and does not improve the surface sealing potential. However, the polymer used with the basalt granular material causes an increase of bearing capacity and surface sealing potential. This finding may be due to the differences in gradation shape between the two aggregates.

The addition of cement to the granular materials tested in this study showed an increase of CBR relative results of 24 to 277%. The effect of cement is approximately 15% higher for the basalt when it is compared to the limestone. This is mainly because the cement particles are less dispersed within the basalt granular skeleton in comparison to the limestone granular skeleton. The cement mechanical stabilization, as measured in this study, is much higher than the polymer especially at high cement application rates. At low cement content for the limestone, the gap between high polymer dosages is the lowest and is approximately 20%. The limestone surface sealing potential is higher than 1 on average because the cement particles gradually fill the large material voids. For the basalt, the relative results are lower than 1 mostly because the voids are larger and the cement particles fill the granular skeleton more efficiently than the limestone voids. Material surface absorption may also play a more important role for the basalt and it explains the relative results lower than 1. This corroborates with previous findings on a granitic gneiss material. Nevertheless, the basalt relative results suggest that the cement percentage at which the absorption seems to play a more important role in the surface sealing potential results is at 4.5% where a peak value of 1.069 is measured. No such affect is observed for the limestone.

6 CONCLUSION

The performance of unpaved aggregate roads can be affected by the quality of the granular materials, the mineralogy of the material and the type of dust suppressant or stabilizer used. This paper presented the results of a laboratory study aimed at measuring the effects of different sources of granular material mineralogy on their performance as measured with CBR, resilient modulus and the water drop test. Basalt and limestone sources where documented and compared to previous work conducted by Pierre (2007b) on a gneiss source. Treated (stabilized) and untreated results were presented.

On average, both CBR and M_R results were higher for the untreated gneiss followed by limestone and basalt respectively. Overall similar trends were found for treated samples with calcium chloride, natural brine, polymer and Portland cement. The highest CBR's (within a given mineralogy class) were found with the cement followed by the polymer whereas both the brine and calcium chloride showed little to no gains. Increasing the cement content had a direct affect on increasing the CBR while this trend was not as obvious for the other products which showed variable results.

The water drop tests showed higher moisture retention on the gneiss, followed by the basalt granular material for the natural brine and calcium chloride while little to no gains was documented on the limestone. This may suggest that hygroscopic products are more suitable to reduce dust on gneiss and basalt materials. Both stabilizers tested (cement and polymer) yielded no benefits or even a negative affect in some cases for the water drop test.

The results of this study will help guide the authors for future full scale in situ testing which will later be reported as part of this project.

ACKNOWLEDGEMENTS

The authors wish to thank NSERC and all the partners involved in the CARRLo project for their financial support in this research project.

REFERENCES

American Association for State Highway and Transportation Officials. 2000. Determining the Resilient Modulus of Soils and Aggregate Materials. *In Standard Specifications for Transportation Materials and Methods of Sampling and Testing, 20th Edition*. AASHTO, Washington, D.C.

American Society for Testing and Materials. 2005. Standard Test Method for California bearing ratio of Laboratory-Compacted Soils. *Standard D1883-05, ASTM*, Philadelphia, Pa.

Bergeron, G. 2000. Retraitement en place à Transports Québec: Résultats des suivis de performance de 1991 à 2000.*Service des chaussées, Routes et Structures, Innovation Transport*, pp. 15–25.

Bureau de normalisation du Québec. 1982. Granulats. Analyse granulométrique par tamisage. *Standard BNQ 2560-040*, BNQ, Québec.

Bureau de normalisation du Québec. 1983. Granulats – Détermination de la densité et de l'absorptivité du gros granulat. *Standard BNQ 2560-067*, BNQ, Québec.

Bureau de normalisation du Québec. 1986. Sols – Détermination de la relation teneur en eau – masse volumique – essai proctor modifié. *Standard BNQ 2501-255*, Québec.

Bureau de normalisation du Québec. 1987. Granulats Sols – Analyse granulométrique des sols inorganiques. *Standard NQ 2501-025*, BNQ, Québec.

Bureau de normalisation du Québec. 1989. Granulats – Détermination de la densité et de l'absorptivité du granulat fin. *Standard BNQ 2560-065*, Québec.

Doré, G., Pierre, P., Juneau, S. and Stephani, E. 2005. Établissement de données techniques pour la saumure naturelle Solnat utilisée comme abat-poussière sur les chaussées et analyse comparative avec le chlorure de calcium. *Rapport GCT-2005-04. Département de Génie civil, Université Laval*, Québec.

Pelletier, L. 2007. Étude comparative de la performance en laboratoire de matériaux granulaires stabilisés utilisés comme surfaces de roulement pour les chaussées non revêtues, *Mémoire de maîtrise, Université Laval*, Québec.

Pierre, P. 2005. Amélioration de la qualité des chemins d'accès aux ressources et routes locales dans le contexte canadien, *Subvention de recherche et développement coopératrice*, CRSNG.

Pierre, P., Pelletier, L., Légère, G., Juneau, S. and Doré, G. 2007. Comparative laboratory study of shear behavior of granular materials stabilized with dust reducing products. *CSCE 2007 Annual General Meeting & Conference*, Yellowknife, Northwest Territories.

Pierre, P., Bilodeau, J.-P., Légère, G. and Doré, G. 2007. A laboratory study on the relative performance of treated granular materials used for unpaved roads. *Accepted for publication in the Canadian Journal of Civil Engineering.*

Pouliot, N. 2004. Traitement de surface sur routes gravelée, *Direction du laboratoire des chaussées (DLC), Bulletin d'information technique, vol. 9 no 3*, Ministère des Transports du Québec.

Sanders, T. G., Addo, J. Q., Ariniello, A., and Heiden, W. F. 1997. Relative effectiveness of road dust suppressants. *Journal of transportation engineering*, 123(6): 393–397.

Santoni, R.L., Tingle, J.S., and Webster, S.L. 2002. Stabilization of Silty Sand with Nontraditional Additives. *Transportation Research Record 1787*, TRB, National Research Council, Washington, D.C., pp. 61–70.

RILEM 1980. Water absorption under low pressure (pipe method), *Test No. 11.4, RILEM Commission 25-PEM*, Tentative Recommendations.

Advances in Transportation Geotechnics – Ellis, Yu, McDowell, Dawson & Thom (eds)
 ISBN 978-0-415-47590-7

Shakedown analysis of road pavement performance

P.S. Ravindra
University of Sydney, Australia

J.C. Small
School of Civil Engineering, University of Sydney, Australia

ABSTRACT: Shakedown behaviour of road pavements was investigated in laboratory controlled conditions using the Sydney University Pavement Testing Facility. Wheel loads lower than the shakedown load generated low permanent deformations for a larger number of load cycles in comparison with high permanent deformations for a lower number of load cycles for wheel loads higher than the shakedown load. Computer software was developed to calculate the shakedown limit using elastic stress distributions calculated with Sydney University's FLEA (Finite Layer Elastic Analysis) program.

1 GENERAL

1.1 *Introduction*

In modern day pavement design, decisions are primarily based on the pavement life cycle cost together with an acceptable pavement maintenance regime for the particular road under consideration. Functional failure (i.e. increase in roughness) will be an important factor in the design process although the pavement needs to be designed against structural failure in order to carry the design loads. The design inputs range from load and traffic analysis, environment factors, material properties, improvement of available material, properties of subgrade, properties of bases and sub bases, available construction standards and equipment, surfacing material and fundamental stress-strain analysis. The design process involves selecting the optimum pavement configuration based on a series of iterations on the assumed design layer thickness.

The current "mechanistic" pavement design procedure is based upon a failure mechanism identified with certain critical elastic strains reaching a critical level, whereas embankments or foundations are designed against failure by using plasticity theory. Pavement design procedures such as LR 1132 (1984), AASHTO (1994) and AUSTROAD (2000) are based on an elastic and fully recoverable stress- strain relationship. Therefore any attempt in utilizing plasticity theory in the development of a pavement model which predicts various types of pavement failures such as rutting, shoving and pushing, cracks etc. is a clear advancement in the pavement design process.

Many authors have detailed the nature of repeated application of load, which can be much more severe on a structure than simple loading to collapse. Application of a simple load beyond the first yield load but below the static collapse load will cause plastic deformation in the body. Repeated application of similar loading cycles will cause gradual accumulation of plastic deformation.

An analysis which incorporates the substantial strength existing prior to the point of static collapse has been suggested by (Sharp and Booker,1984) when they pioneered the application of Melan's (1936) Shakedown Theory to model "shakedown" behavior of road pavements. The shakedown model identified a critical load level below which shakedown occurs, but above which the permanent strains continue to occur as the "shakedown load".

Extensions for Melan's lower bound approach to calculate the shakedown load was presented by (Raad et al. 1988, 1989), (Hossain and Yu, 1996) and (Shiau and Yu, 2000). Upper estimates of the shakedown load were obtained by Collins and Cliffe (1987) who employed the dual kinematic theorem due to Koiter (1960). They have shown that in the two-dimensional case, results were identical with Sharp and Booker's lower bound approach. Whilst Sharp and Booker (1984) have shown convincing consistency of some AASHTO (1950) experimental results with their parametric study, they concluded that there is a clear need for further experimental work so that such design criteria could be validated.

In this paper, it is intended to discuss the detailed laboratory experiments performed at the University

Figure 1. Spring Loading Arrangement and the Loading Wheel.

of Sydney Pavement Testing Facility in Australia to examine the application of shakedown theory by means of measuring accumulated plastic deformation after applying traffic loads below and above the shakedown load.

2 SYDNEY UNIVERSITY PAVEMENT TESTING FACILITY

2.1 *Introduction*

The testing facility shown in Fig. 1 was initially developed by Wong and Small (1994) in order to test model pavements and was modified in order to change the position of the tyre across the pavement randomly by Moghaddas-Nejad and Small (1996). The facility consists of four main structural components, namely, the test tank, the overhead track and the loading carriage. The support and guidance for the moving loading carriage is provided by the overhead rails. The test section of pavement is constructed inside the test tank and positioned below one of the straight sections of the overhead track. The test wheel runs on a plywood track when outside of the test section of pavement. A conductor rail system supplies power to the motor which drives the wheel. The wheel passes over the test section of pavement once during each revolution around the track and triggers a micro switch that starts a micro computer recording data.

2.2 *Instrumentation and data acquisition*

All subsurface settlements are measured by buried Perspex discs (30 mm diameter) that are connected to a wire that passes through the base of the tank and is connected to a transducer. Data acquisition from the transducer output is carried out only during the passage of the wheel across the test section of pavement.

2.3 *Loading wheel*

The loading wheel is supported by a rotating arm which is pivoted at the top by means of roller bearings. The position of the spring loading system is almost in the middle and the loading wheel is at the end of the horizontal portion of the rotating arm. By means of a hinged connection, the spring loading system is connected to the bottom bogie plate. To adjust the spring load on the rotating arm, a set of two adjusting nuts has been provided above and below the rotating arm so that compression in the springs can be changed by adjustment of the upper nut. Four identical springs with guide rods placed concentrically inside them make up the loading unit. To satisfy different loading requirements, springs with different stiffnesses can be set up in the system. A LVDT transducer fitted between two reaction plates is used to monitor the variation of compression in the springs. There are two aluminium boxes supported by the bogie connection frame. The smaller box contains a power supply for the transducer and the larger one contains the analog to digital signal converter, for transmission of data from the wheel carriage.

2.4 *Measurement of surface settlement*

In order to cater for the large number of cross section measurements envisaged in the project, a new laser transducer based surface deformation measurement system was designed and developed to measure and record the surface deformations. With this new surface deformation measurement system, measuring of cross-sections was made easier and all measurements were recorded directly to a database.

3 MATERIAL PROPERTIES

3.1 *Sand subgrade*

The grading of the sand used for the subgrade in the tests is given in Table 1 below:

Table 1. Particle Size Distribution of Loose Silica Sand (Test Specification AS 1289 3.6.1).

Sieve size (mm)	4.75	2.36	1.18	0.600	0.425	0.300	0.15
% Passing	100	99	99	98	89	54	2

Table 2. Soil Strength Test Results for Base Material.

Texas Triaxial	Direct Shear Box
$\phi = 46°$ c = 45 kPa	$\phi = 51^0$ c = 56 kPa

Sieve size (mm)	37.5	19	13.2	9.5	6.7	4.75	2.36
%Passing	100	93	77	65	58	51	44
Sieve size (mm)	1.18	0.600	0.425	0.300	0.15		
%Passing	38	32	27	6	4		

3.2 *Base material*

Base material used in the experiment was made up of recycled crushed concrete aggregate obtained from the Randwick Yarra Bay Material stockpile site. This material mainly consisted of crushed concrete obtained from council building demolition sites. This material is used as a base material in road rehabilitation works in the City Council area and is found to be easy to handle at the site with respect to compaction and spreading.

All material tests carried out were compared with current AUSPEC specifications and relevant ARRB (Australian Road Research Board) recommended specifications for recycled crushed concrete. Direct Shear Box tests and Texas Triaxial tests were carried out to determine the angle of internal friction and cohesion values and values are given in Table 2, while its grading is shown in Table 3.

4 CONSTRUCTION OF PAVEMENT

4.1 *Pavement*

A Standard Proctor hammer was used to compact the pavement layers. Moisture content was maintained at OMC and the number of blows applied were evenly spread across the layer and were selected so as to give the same energy level to the base material as in the Standard Compaction Test. The pavement surface was allowed to dry at 20°C room temperature before lightly brushing off the loose fines in order to apply the bitumen emulsion seal coat with 5 mm single sized cover aggregate. Emulsion was applied with a roller brush and spread evenly across the pavement. After application of the cover aggregate, a light compaction pass was applied to embed the cover aggregate into the emulsion layer as a single layer of thickness equal to the aggregate's least dimension. A further 24 to 48 hours drying period was allowed before brushing the loose cover aggregate off the surface. To date, three different test pavements have been tested in the experiment. Wheel entry and exit sections were kept at 350 mm thickness for all tests and the 700 mm length mid section thickness was varied to obtain different theoretical shakedown loads.

5 CALCULATION OF SHAKEDOWN THICKNESS

5.1 *General*

The shakedown theorem due to Melan (1936) as presented by Maier (1969), described the main limitations of the classical theory as:

"(a) *Inviscid perfectly-plastic* (non-hardening) laws govern the local deformability, and involve *convex* yield surfaces, *associative* flow rules and *constant* elastic moduli (the term *inviscid* rules out time-effects, such as creep and rate-sensitivity); (b) geometric changes do not significantly affect the equilibrium relations; (c) temperature changes have negligible influence on the material properties; (d) external agencies act so slowly that the system behaves in a *quasi-static* way (with negligible inertia and viscous forces); (e) adaptation guarantees the survival of the structure, i.e., rules out structural crises by excessive deformation or local failure." The question of the validity of the shakedown theorems for materials with non-associated flow rules has been examined by Maier (1973). He showed that the bounds given by Koiter's theorem are still upper bounds to the shakedown load, even though the real material has a non-associated flow rule.

Melan's theorem can be used to obtain a lower bound to the shakedown limit for a non-standard material, but the yield-surface must be replaced by a "potential surface" which lies inside the yield surface.

Sharp and Booker (1984) applied the linear programming technique adopted by Maier (1969). They assumed a plane strain model with a trapezoidal pressure distribution under a roller. The material of the half space was assumed to be isotropic and homogeneous and the resulting permanent deformation and residual stress distribution were assumed independent of horizontal distance and dependent on the depth. The tangential shear load was taken as a trapezoidal distribution.

The failure criterion was Mohr-Coulomb and the material properties were assumed to be linear elastic – perfectly plastic.

5.2 *Calculation of shakedown limit*

Melan's static shakedown theorem (lower bound) states that "if any time-independent distribution of residual stresses can be found which, together with

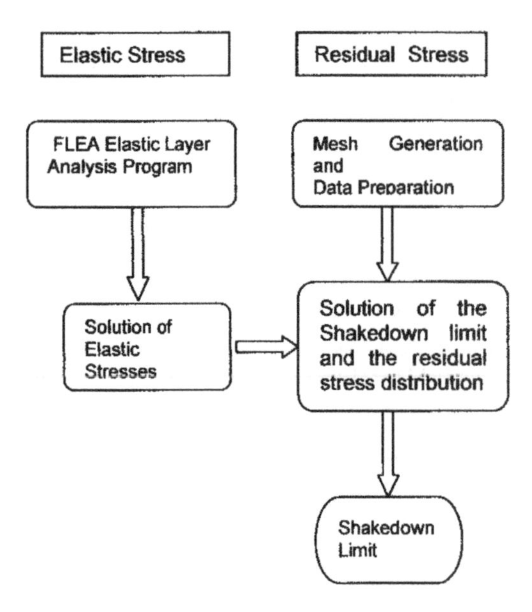

Figure 2. Calculation of shakedown limit.

the elastic stresses due to the load, constitutes a system of stresses within the elastic limit, then the system will shakedown". In other words, the elastic stresses associated with the maximum load, together with any distribution of residual stress, which just touches the yield surface will give a lower bound to the shakedown limit.

Calculation of Shakedown limits for the various pavement configurations tested in this experiment is based on the lower bound calculation procedure indicated in Figure 2. This procedure assumes that both elastic stresses and residual stresses required by the lower bound shakedown analysis are linearly distributed across the continua and the resulting deformation is plane strain by replacing the wheel load as an infinitely wide roller. A trapezoidal load distribution was selected as the contact load distribution. A stand alone windows based computer software package was developed to calculate the shakedown limit for this experiment. Output from Sydney University's FLEA (Finite Layer Elastic Analysis) program was directly read to provide the elastic stress distributions in the continua and the residual stress field was determined using the linear residual stress finite elements as indicated in S.H.Shiau (2001) in his PhD thesis. The Active Set Linear programming technique indicated by Best and Ritter (1985) and Sloan (1988a and 1988b) was used to calculate the optimum shakedown limit. The shakedown factor is defined as λ where

$$\sigma^t = \lambda\sigma^e + \sigma^r$$

and σ^t, σ^e, and σ^r are the total, elastic and residual stresses in the pavement. It is the factor that the elastic stress field can be multiplied by to keep the total stresses below the (Mohr-Coulomb) failure criterion. The calculated shakedown limits for the test pavements of 50 mm and 200 mm thickness for the wheel load at 80 N (44 kN/M^2) were 0.82 and 1.9 respectively. Elastic Moduli for recycled base material and sand sub grade were used as 350,000 kPa and 200,000 kPa respectively.

6 APPLICATION OF SHAKEDOWN METHOD

6.1 *Introduction*

Current methods of pavement performance prediction depend on traffic loadings, (increment in analysis year by damage functions), surface distress (increment in analysis year by mode: cracking, ravelling, potholes, and rut depth by pavement type classification by probability) and surface roughness (increment in year by components: traffic, surface distress, age/environment). Elastic or plastic behaviour of a road pavement under repeated loading condition will depend on its shakedown parameters.

6.2 *Pavement distress parameters*

Road pavements deteriorate over time under the combined effects of traffic and weather. Traffic axle loads induce levels of stress and strain within the pavement which are functions of elastic properties of pavement materials and layer thicknesses. Under repeated loading, a pavement will build up residual stresses and shakedown depending on material yield condition and the magnitude of the applied traffic loading. This may proceed to rapid accrual of permanent deformation of all materials or initiation of cracking through fatigue depending on the shakedown state. Weathering causes bituminous surfacing material to oxidize, become brittle and more susceptible to cracking and to disintegrate. This may initiate ravelling, spalling, and edge-breaking. Once initiated, cracking extends in area, increases in intensity (closer spacing) and increases in severity (or width of crack) to the point where spalling and ultimately potholes develop. Open cracks on the surface and poorly maintained drainage systems allow excess water to enter the pavement, speeding up the process of disintegration, reducing the shear strength of unbound materials (lowering the yield limit of the material) and thus increasing the rate of deformation under the stresses induced by traffic loading. Depending on the shakedown state, a pavement may move into a state of "alternating plasticity" or "ratcheting". The cumulative deformation throughout the pavement appears in the wheel paths as ruts and more generally on the surface as an unevenness or distortion of the profile termed roughness.

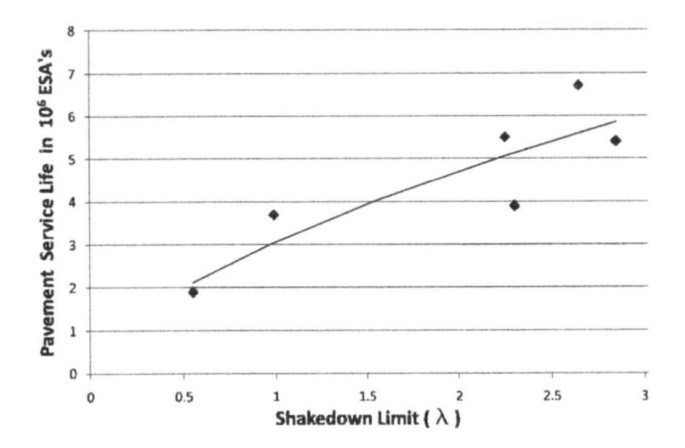

Figure 3. Pavement Service Life (ESA) plotted against calculated Shakedown Limit of the pavement.

Table 3. Pavement Parameters from (Sharp and Booker, 1983).

Road	Pavement Layers Base(M) Sub-Base(M)	E (kPa)	ν	C kPa	ϕ^0	(ESA) Life (ESA)
F4 Free	0.12	225000	0.4	600	25	6.7
Way	0.15	61000	0.3	65	51	
Windsor	0.15	302000	0.4	900	38	5.4
CastleHill	0.32	33000	0.3	108	37	
Windsor	0.075	882000	0.4	3000	35	3.7
Kellyville	0.125	39000	0.3	25	48	
Old	0.15	500000	0.4	1600	40	1.9
Windsor	0.2	75000	0.3	75	54	
Kissing	0.19	400000	0.4	700	30	5.5
Point	0.15	75000	0.3	79	49	
Rock	0.15	287000	0.4	850	33	3.9
Wood	0.1	64000	0.3	80	50	

Wheel Pressure 760 kPa

6.3 *Pavement service life*

Pavement service life can be defined as the cumulative traffic loading, expressed in ESAs, (Equivalent Standard Axles) from the immediate post construction or rehabilitation/replacement condition to a predetermined threshold value of a range of distresses beyond which the pavement is no longer acceptable for use. Similarly pavement service traffic can be defined as the cumulative traffic loading, expressed in ESAs, from the immediate post construction or rehabilitation/replacement condition to a predetermined threshold value of a range of distresses beyond which the pavement is no longer acceptable for use. Pavement service lives for six different pavements in NSW were compared with their shakedown limits in the Figure 3. The pavement parameters selected were as in (Sharp and Booker, 1983) Tables 8.3 and 8.4 are indicated in Table 3.

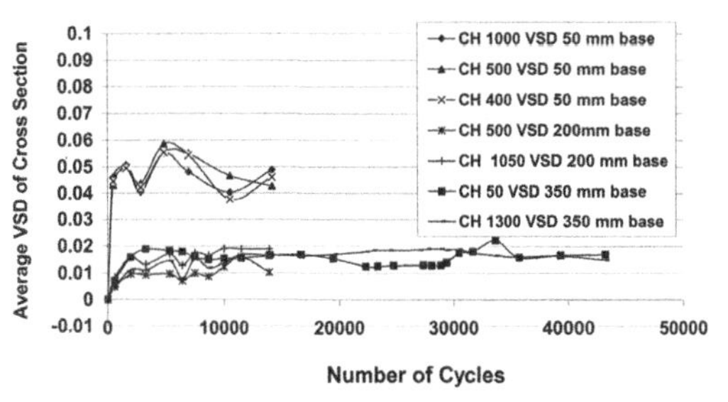

Figure 4. Average VSD (laser reading) of the particular cross section plotted against number of cycles.

7 TEST RESULTS AND DISCUSSION

7.1 *Database*

Data acquired was comprised of subsurface pavement settlement data from LVDTs, the time of reading, the lateral location of the wheel (as the wheel can be moved laterally relative to the pavement), number of test cycles, spring load monitoring data, subsurface permanent deformation details and transducer calibration data. Results were written to a relational database enabling online analysis and processing (OLAP) of test data.

Results indicate that larger deformations occur for fewer load cycles when the wheel load is more than the shakedown load.

7.2 *Test results*

Vertical surface deformation (VSD) of the pavement surface was measured at 9 cross sections initially. The number of cross sections was increased to cover the progression of VSD and to produce 3D mesh images of the pavement surface. This made it possible to visualize the settlement pattern of the total pavement rather than individual cross sections. Standard measures such as pavement rutting and roughness are related to VSD but do not have the same degree of reliability. Usually rutting is determined by calculating or measuring the depth of a rut from a straight edge placed across the wheel path and is affected by any heaving at the edges.

Roughness represents the variation of VSD along the wheel path. Results obtained by this experiment indicated that there was a rapid increase in VSD in the case of wheel loads more than the shakedown load calculated by the two dimensional shakedown load for the test pavement (see Figure 4). In these tests, trafficking was terminated when the pavement reached a shakedown state or when VSD increased more than 10 mm in a particular cross section (this corresponds to a 40 mm depth in the prototype). More tests are needed to verify the long term behavior of the pavement.

8 CONCLUSIONS

Preliminary testing of pavements with the University of Sydney Pavement Testing Facility have indicated that the shakedown limits predicted by 2-D shakedown theory analyses are a good indicator of whether a pavement will undergo continued deformation under cyclic wheel loading. Shakedown limits predicted appear to correlate with the long term the pavement performance which is an indicator of pavement service life.

REFERENCES

AASHTO 1994. *Guide for Design of Pavement Structures*, American Association of State Highway and Transportation Officials.

AUSTROADS 2000. *Pavement Design: A Guide to the Structural Design of Road Pavements*, Sydney, Australia.

Best, M.J. and Ritter, K. 1985. *Linear Programming: Active Set Analysis and Computer Programs*, Prentice-Hall, New Jersey.

Collins, I.F. and Cliffe, P.F. 1987. Shakedown in frictional materials under moving surface loads. *International Journal for Numerical and Analytical Methods in Geomechanics*:11, 409–420.

Hossain, M.Z. and Yu, H.S. 1996. Shakedown analysis of multi layer pavements using finite element and linear programming. *Proc. 7th Australia-New Zealand Conference on Geomechanics*, Adelaide: 512–520.

Koiter, W.T. 1960. General Theorem for Elastic-Plastic Solids, In: *Progress in Solid Mechanics*, Eds. Sneddon, J.N. and Hill, R.,1, North Holland, Amesterdam, The Netherlands:165–221.

LR 1132 (1984). *The Design of Bituminous Roads*, Transport and Road Research Laboratory, UK.

Maier, G.1969. Shakedown theory in perfect elastoplasticity with associated and non-associated flow-laws: A finite element, linear programming approach, *Meccanica* 4: 250–260.

Melan, E. 1936. Theorie statisch unbestimmer Systeme, Prelim. Publ. *2nd Congress of International Association of Bridge and Structure Engineering*, Berlin, 43.

Moghaddas-Nejad, F. and Small, J.C. 1996. Effect of Geogrid Reinforcement in Model Track Tests on Pavements, *Journal of Transportation Engineering* 127(6): 468–474.

Raad, L.,Weichert, D. and Najm, W.1988. Stability of multilayer systems under repeated loads, *Transportation Research Record*,1207: 181–186.

Raad, L.,Weichert, D. and Haider, A. 1989a. Analysis of full-depth asphalt concrete pavements using shakedown theory, *Transportation Research Record* 1227: 53–65.

Sharp, R.W. and Booker, J.R. (1983). Shakedown Analysis and Design of Pavements, *PhD Thesis*, University of Sydney, April 1983.

Sharp, R.W. and Booker, J.R. (1984). Shakedown of Pavements under moving surface loads, *Journal of Transportation Engineering*, ASCE110(1): 1–14.

Shiau, S.H and Yu, H.S. 2000. Shakedown analysis of flexible pavements, *Developments in Theoretical Geomechanics*: Eds. Smith & Cater, Balkema.

Shiau, S.H. 2001. *Numerical Methods for Shakedown Analysis of Pavements under Moving Loads*, PhD thesis, University of Newcastle, Australia.

Sloan, S.W.1988a. Lower bound limits analysis using finite element and linear programming, *International Journal for Numerical Methods in Geomechanics* 12: 61–67.

Sloan, S.W. 1988b. A steepest edge active set algorithm for solving sparse linear programming problems, *International Journal for Numerical Methods in Engineering* 26: 2671–2685.

Wong, H.K.W. and Small, J.C. 1994. Effect of Orientation of Approach Slabs on Pavement Deformation, *Journal of Transportation Engineering* 120(4): 590–602.

Advances in Transportation Geotechnics – Ellis, Yu, McDowell, Dawson & Thom (eds)
© 2008 Taylor & Francis Group, London, ISBN 978-0-415-47590-7

Anisotropy level prediction model of unbound aggregate systems

R. Salehi, D.N. Little & E. Masad
Texas A&M University, College Station, Texas, USA

ABSTRACT: This study establishes a procedure to determine the level of anisotropy of unbound aggregate systems based on particle geometry, mechanical response and physio-chemical properties of the fine aggregate portion (particles smaller than 75 μ) of the aggregate system. Stress induced directional dependency of material properties based on multiple variable dynamic confining pressure (MVDCP) stress path tests was determined for ten aggregate sources. Anisotropic responses for various gradations and saturation levels were determined for each aggregate source. The cumulative Weibull distribution function is used to describe aggregate size and aggregate geometrical characteristics. The fine portion of the gradation was characterized by the Rigden voids test and methylene blue test to account for fine particle shape properties and deleterious effect of plastic fines, respectively. Cross-anisotropic modular ratios are used as indicators of the level of anisotropy. The anisotropy model is developed based on a comprehensive aggregate matrix consisting of twenty seven aggregate features for sixty three aggregate systems. The sensitivity analysis of the model reveals the significant impact of particle geometry on level of anisotropy and orthogonal load distribution capacity of unbound aggregate systems. The model is also shown to be sensitive to the level of bulk stress and shear stress as reflected by the k_2 and k_3 parameters.

1 INTRODUCTION

Several researchers have studied anisotropy or directional dependency of material properties of aggregate systems (Tutumluer, et al. 1997), (Adu-Osei et al. 2001), (Kim et al. 2005). Tutumluer suggested the use of anisotropic constitutive models to characterize the response of aggregate base layers in flexible pavements (Tutumluer et al. 1997). He reported that the use of a cross-anisotropic model results in a drastic reduction or in some cases elimination of tensile stresses at the bottom of an unbound layer that are generally calculated using isotropic layered elastic solutions. He also found that compressive stresses throughout the aggregate layer and at the top of the subgrade are generally higher when the aggregate layer is considered as nonlinear and anisotropic. This conforms to results reported by Adu-Osei and Kim.

Adu-Osei developed a testing protocol and employed an error minimization technique called system identification to determine cross-anisotropic material properties. He utilized state of the art testing equipment (Rapid Triaxial Tester) to apply variable dynamic confining pressure in order to realistically simulate stresses induced in the aggregate base by moving wheel loads (Adu-Osei et al. 2001).

Oh (Oh et al. 2006) used Multi-Depth Deflectometers (MDD) to measure deflections of several flexible pavement sections subjected to passenger cars as well as overweight truck traffic (gross vehicle weight up to 556 KN). He compared the deflections measured with MDDs with responses calculated using different material constitutive models. He reported that the best matches between field measurements and finite element solutions were achieved when the aggregate base layer and subgrade were considered to be nonlinear and cross-anisotropic. A similar conclusion was reported by (Tutumluer et al. 2003) based on the analysis of Georgia Tech pavement sections.

Kim developed a first generation model for calculating anisotropy level in unbound aggregate systems and discussed the influence of aggregate geometry on the level of anisotropy. Kim's equation predicts the cross-anisotropic model parameters as a function of the response of an aggregate system to laboratory cyclic loading, aggregate geometry, and gradation parameters (Kim et al. 2005).

Seyhan studied the directional dependency of aggregate stiffness and used shear modular ratios as performance indicators of unbound aggregate systems. He established correlations between shear strength ratios and anisotropic modular ratios and reported that "good quality" aggregate systems have higher modular ratios (horizontal modulus/vertical modulus) or are less anisotropic and "poor quality" materials have lower modular ratios or are more

Table 1. Particle size distribution.

Sieve Opening (mm)	Percent Passing (%)		
	Fine	Intermediate	Coarse
25	100	100	100
19	85	85	85
12.5	74	74	72
9.5	70	66	62
4.75	67	54	40
2.36	62	41	25
1.18	52	30	18
0.6	42	23	14
0.3	34	18	10
0.15	28	14	8
0.075	20	10	7

anisotropic (Seyhan et al. 2002). The study conducted by Salehi on high fine content aggregate systems emphasized that the analysis should be based on considering all anisotropic material properties and not only modular ratios. He showed that despite the fact that unstabilized high fine systems are less anisotropic; they performed poorly and developed significant plastic strains when subjected to repeated loading at elevated saturation levels (Salehi et al. in press).

Thompson argued that granular layer performance (based on rutting resistance) is primarily related to aggregate shear strength ratio (ratio of deviator stress at the top of subgrade to unconfined compressive strength) not resilient modulus as utilized in the AASHTO design guide. He reported that shear strength ratios for natural subgrade soils under pavements are generally in the range of 0.4 to 0.75, where lower shear strength ratios indicate better performance and less rutting (Thompson 1993).

The objective of this study is to develop a new model to predict the level of anisotropy of aggregate systems based on properties measured using simple and routine tests.

2 MATERIALS AND TESTING METHODS

Elastic cross-anisotropic material properties were determined for 10 aggregate sources and considering three different gradations and at three different moisture contents for each aggregate in this study. Table-1 summarizes particle size distributions of the aggregate samples.

2.1 *Determination of anisotropic material properties*

In order to assess the directional dependency of the response of aggregate bases under a moving wheel load, aggregate samples were tested following the International Center for Aggregate Research (ICAR) loading protocol. Figure 1 schematically illustrates the stress paths applied to aggregate samples using the small strain ICAR protocol.

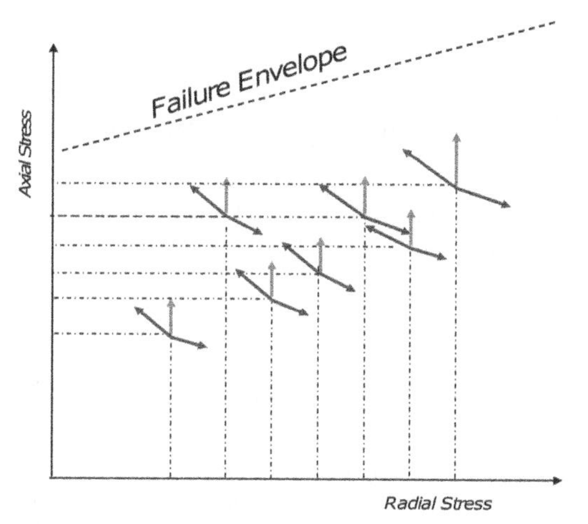

Figure 1. Schematic representation of variable confining pressure stress path protocol.

This protocol provides a means to determine cross-anisotropic material properties: E_x and E_y (elastic modulus in the horizontal and vertical directions, respectively), ν_{xy} and ν_{xx} (Poisson's ratio in the horizontal direction due to vertical loading and Poisson's ratio in the horizontal direction due to horizontal loading, respectively) and G_{xy}, shear modulus. Applied stresses and measured strains were used as input into an iterative error minimization technique called the system identification method to simultaneously solve for four of the five anisotropic material properties (E_x, E_y, ν_{xx} and ν_{xy}).

$$\begin{bmatrix} \frac{1}{E_x} & -\frac{\nu_{xy}}{E_x} & -\frac{\nu_{xx}}{E_x} \\ -\frac{\nu_{xy}}{E_x} & \frac{1}{E_y} & -\frac{\nu_{xy}}{E_x} \end{bmatrix} \begin{bmatrix} \sigma_x \\ \sigma_y \\ \sigma_x \end{bmatrix} = \begin{bmatrix} \varepsilon_x \\ \varepsilon_y \end{bmatrix} \quad (1)$$

The fifth material property, G_{xy}, is directly determined using elastic work potential relationships derived specifically for the shear stress regime. Details regarding the derivation of equation (2) can be found in (Adu-Osei et al. 2001).

$$G_{xy} = \frac{3}{4} \frac{\Delta\sigma_y}{(\Delta\varepsilon_y - \Delta\varepsilon_x)} \quad (2)$$

In cross-anisotropic materials, a plane of isotropy exists such that material properties in the x and z directions are equal. Since the horizontal plane is the plane of isotropy, the term G_{xx} is related to v_{xx} and E_x by equation (3).

$$G_{xx} = \frac{E_x}{2\,(1+v_{xx})} \tag{3}$$

The values of vertical (E_y), horizontal (E_x) and shear (G_{xy}) moduli are fitted to the bulk stress (θ) and octahedral shear (τ_{oct}) stress using nonlinear functions shown in equations (4–6).

$$E_y = k_1\left(\frac{\theta}{Pa}\right)^{k_2}\left(\frac{\tau_{oct}}{Pa}+1\right)^{k_3} \tag{4}$$

$$E_x = k_4\left(\frac{\theta}{Pa}\right)^{k_5}\left(\frac{\tau_{oct}}{Pa}+1\right)^{k_6} \tag{5}$$

$$G_{xy} = k_7\left(\frac{\theta}{Pa}\right)^{k_8}\left(\frac{\tau_{oct}}{Pa}+1\right)^{k_9} \tag{6}$$

Where k_1 through k_9 are fitting parameters and P_a is the atmospheric pressure (101 kPa).

2.2 *Aggregate geometry*

Aggregate geometry was characterized in terms of particle form, angularity and texture using the Aggregate Imaging System (AIMS). Aggregate form defines the flat or elongated nature of aggregate particles. Angularity refers to the sharpness or degree of roundness of aggregate corners. Finally, texture relates to small asperities at the surface of particles that defines surface roughness. More details regarding the AIMS device and testing method can be found in the work of Kim et al. (Kim et al. 2005). Fifty-six (56) aggregate particles from three aggregate sizes of each source were tested with the AIMS device. It is intuitive that aggregate shape, angularity and texture should affect level of isotropy as these geometric properties impact the interaction among aggregate particles, especially under compaction and loading induced stresses.

2.2.1 *Distribution functions*

Various two parameter and three parameter mathematical models have been used to describe aggregate particle size and shape distributions *(3, 9)*. However, widespread applications of some of these functions have been limited due to complexity in physical interpretations regarding each distribution parameter.

In developing the first generation anisotropy model Kim et al. (Kim et al. 2005) used the three parameter Fredlund model presented in equation 7 to fit the cumulative distribution functions of gradation and shape properties of aggregates.

$$P_p = \frac{100}{\ln\left[\exp(1) + (\frac{g_a}{d})^{g_n}\right]^{g_m}} \tag{7}$$

Where Pp is the percent passing a particular sieve; d is sieve opening; and ga ,gn and gm are fitting parameters that corresponds to initial break in the curve, maximum slope and curvature of the distribution function, respectively.

The Rosin-Rammler distribution function described by Djamarani has long been used to describe the particle size distribution of powders of various natures and sizes. The function is particularly suited to representing particles generated by grinding, milling and crushing operations (Djamarani et al. 1997). The Rosin-Rammler function is represented by two parameters: mean particle size (Dm) and n that is a measure of the spread of particle size distribution. Rosin-Rammler function is presented in equation 8.

$$Q(D) = 1 - \exp\left[-\left(\frac{D}{D_m}\right)^n\right] \tag{8}$$

The two parameter Weibull cumulative distribution function is very similar to the generally accepted Rosin-Rammler distribution commonly used by researchers working in the area of powder technology and cement industry.

Several two parameter and three parameter distributions were fitted to the data and the goodness of the fit was determined through least square error criteria. It was observed that the two parameter Weibull distribution provides a reasonable fit to both particle size and shape properties data at a 95% confidence level. Equation 9 presents the general form of the two parameter Weibull cumulative distribution function.

$$Q(d) = 1 - \exp\left[-\left(\frac{d}{\alpha}\right)\right]^{\beta} \tag{9}$$

Where d is the aggregate size, α is the scale parameter and β is the shape parameter.

Figure 2 and figure 3 show the impact of scale parameter and shape parameter on the overall shape function of the two parameter cumulative Weibull distribution function. Figure 2 shows that as the scale parameter, α, increases the distribution becomes more spread out and small values of the scale parameter α correspond to more condensed distributions. Gradations with larger particle sizes have higher α values. In terms of angularity distribution, a higher α value indicates a higher number of angular particles when compared with a sample with a lower α. The same analogy is valid for aggregate texture properties; an increase in the α value in texture distribution corresponds to an increase in the number of particles with a rougher microstructure at the surface of particle while distributions with a lower α value correspond to more smooth and polished particles.

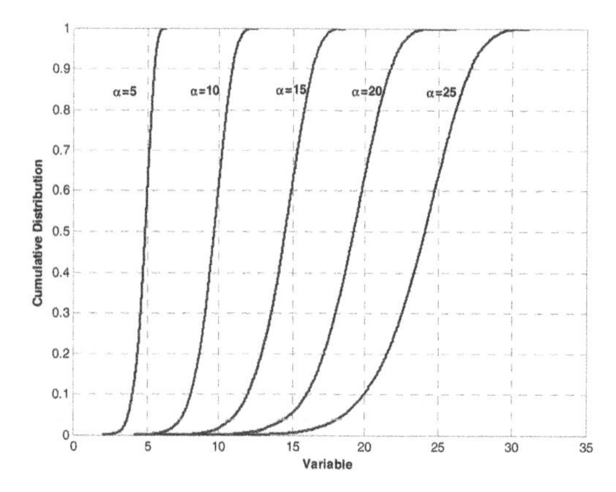

Figure 2. Effect of scale parameter (α) on the shape function of cumulative Weibull distribution.

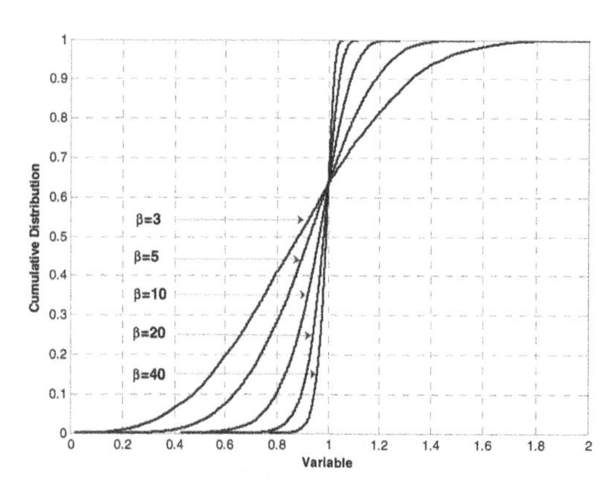

Figure 3. Effect of shape parameter (β) on the shape function of cumulative Weibull distribution.

Figure 3 shows that the shape parameter, β, relates to the uniformity of the distribution. For instance, a lower β value corresponds to an aggregate gradation that spreads over a wide range of particle sizes. This figure also suggests that finer gradations- with longer tails- exhibit a lower β value.

The fact that the Weibull distribution is a well established statistical function and the sensitivity of the distribution parameters fits the physical characteristics of aggregates such as particle size and geometry made it our popular choice of fitting distribution.

The distribution parameters determined from fitting the particle size distribution and aggregate shape properties data to the Weibull distribution were later used as input to the aggregate database for the purpose of developing the anisotropy model.

2.3 *Methylene blue test*

Since the model used in this study to assess anisotropy considers the impact of moisture, the authors felt that

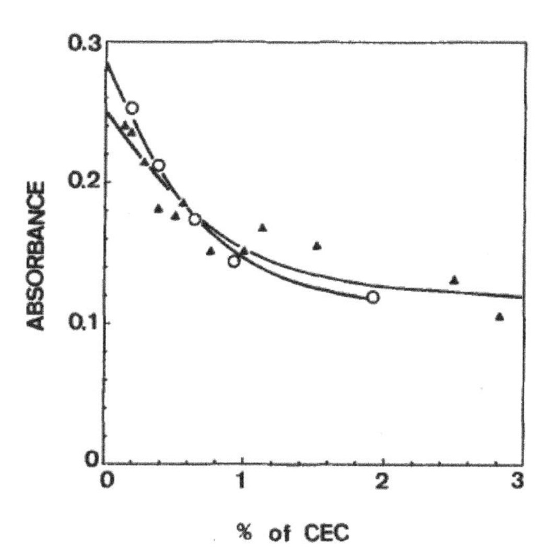

Figure 4. Variation of methylene blue absorbsion with cation exchange capacity (CEC) for two clay sources (After Cenens & Shoonheydt, 1988).

the activity (or sensitivity to moisture) of the fines fraction should be quantified. This can be done through a quantification of soil suction or by some, simpler surrogate test. The authors sought to find a simpler index test for this purpose and selected the methylene blue test. This test (ASTM C832-2003) was adopted as an indicator of the activity of fine particles in the mix. Several studies showed that the deleterious effect of plastic fines particularly shrink-swell potential is strongly correlated with the methylene blue value (Cokca et al. 1993).

Methylene blue is a large polar organic molecule that is adsorbed onto the negatively charged surfaces of clay minerals. The amount of methylene blue adsorbed by a given mass of clay depends on the relative concentration of negatively charged sites on the clay particle surfaces as well as surface area of the clay per unit mass.

Several researchers reported that the cation exchange capacity of fine particles can be measured by the absorption of methylene blue dye from an aqueous solution (Fityus et al. 2000), (Nevins et al. 1967) and (Hills et al. 1985). Nevins found a strong relationship between methylene blue value and cation exchange capacity as a measure of deleterious activity of fine aggregate particles in terms of volumetric expansion of fine grained soils.

According to Gouy-Chapman equation presented in equation 10, thickness of water layer absorbed to clay surface is proportional to the cation exchange capacity of the clay particles present in the aggregate mix. The thickness of the absorbed water layer controls the swell potential of aggregates and their susceptibility to attract and hold moisture. In general, the thicker the

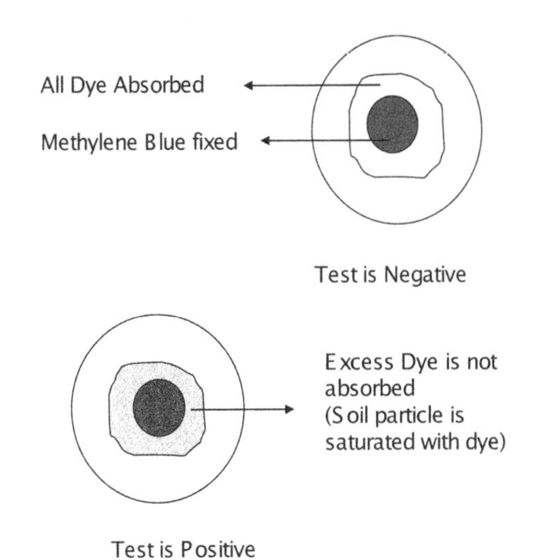

Figure 5. Schematic Representation of Methylene Blue Absorption Test.

diffused water layer, the less the tendency for particles to folliculate and the higher the swelling pressure in expansive soils.

$$\frac{1}{K}=\left(\frac{D\,k\,T}{8\pi n_o e^2 v^2}\right)^{0.5} \tag{10}$$

Where $\frac{1}{K}$ = the thickness of adsorbed water later to the clay surface; n_o = electrolyte concentration; ν is cation valance; D = dielectric constant of the medium; T = temperature; E = charge of an electron ($1.60217646 \times 10^{-19}$ c); and k = Boltzmann's constant (1.3807×10^{-16} erg/K).

Since Methylene blue molecules are preferentially adsorbed onto the negatively charged sites on clay surface, titration with Methylene blue can be considered to provide a good indication of the cation exchange capacity of clay particle. (Hills et al. 1985). Hence, in this study methylene blue test was adopted as a measure of moisture susceptibility and swell potential in aggregates systems.

The ASTM C832-2003 procedure was used to determine the methylene blue value for 10 aggregate sources. The concept behind the methylene blue test is that a certain amount of materials smaller than 75 microns titrated with Methylene blue dye and a spot is tested on a filter paper. The addition of more dye to the solution continues until the spot of material absorbs no more dye. This could be evidenced by a lighter blue ring around the spot tested. Figure 5 shows a schematic representation of dye absorption by fine particles.

Initially, as seen in Figure 5 a faint blue spot of solids was observed surrounded by a transparent ring of clear solution. As the process of addition of methylene blue solution was continued, the color of the inner circle becomes darker due to the fact that the fine particles absorb more dye. The surrounding solution remained distinct at this stage. Eventually, the end point is reached when the outline boundary of the inner blue spot breaks down into a light blue-green circle.

Once the blue-green circle was appeared, titration was stopped and the drop test on the filter paper was repeated for five times. If the green-blue circle persists, the end point has been reached; otherwise more methylene blue was added to the solution and titration continued. Once the end point has reached, equation 11 was used to calculate the methylene blue value for each aggregate sample.

$$MBV(gr/100gr)=\frac{\left[V_{cc}(ml)\times 10(gr)\times 100(gr)\right]}{1000(ml)\times Ws(gr)} \tag{11}$$

Where V_{cc} = the volume of methylene blue injected to the soil solution (ml) and W_s = the dry weight of fine particles used (gr).

2.4 *Dry compacted fines*

As with the methylene blue test, the authors sought to find a relatively simple and reliable test to assess the impact of the compacted fines on anisotropy. It is intuitive that the level of frictional interaction among the particles of the fines matrix will impact larger aggregate particle interaction and thus anisotropy. In order to characterize the shape properties of the fine portion of the aggregate matrix modified Rigden test was performed on particles smaller than 75 μ in the aggregate mix. The dry compacted fines test (Rigden voids tests) provides a measure to relate the maximum packing of fine particles to the geometry of particles as well as the uniformity of particle size distribution in the fines portion of the aggregate systems. It is assumed that higher density and tighter packing between fine particles will be achieved in samples with more uniform size distribution.

MODOT-T73 Volume of Voids in Compacted Filler test procedure was followed in this study to calculate the density and void contents of fine particles of ten aggregate sources.

The test method is based upon the assumption that the densest packing (maximum bulk density) of fines can be obtained by compacting the dry fines in a mold by a 100 grams compaction hammer. More details on sample mold and hammer presented in figure 6. Equations 12 through 15 were then used to determine the dry compacted air void content in fine aggregate samples. The calculated air void contents were used as inputs to the anisotropy level prediction model.

$$Vfb=\frac{\pi D^2 t}{4} \tag{12}$$

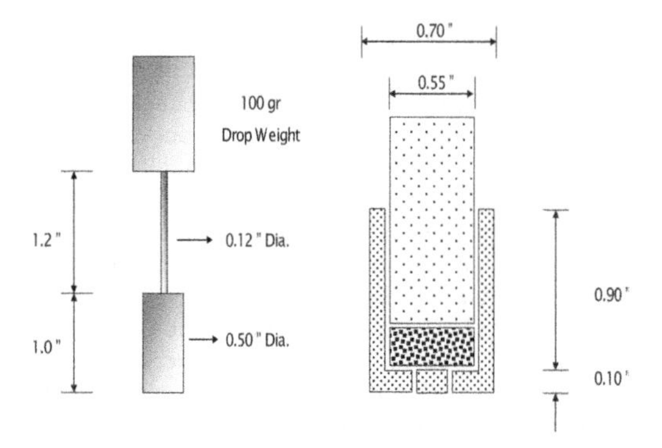

Figure 6. Compaction Hammer and Sample Mold Used in Determination of Dry Compacted Air Voids of Fine Aggregates.

$$V_{fs} = \frac{W_{fs}}{\gamma_w \, Gs} \tag{13}$$

$$\%DCF = \frac{V_{fb} - V_{fs}}{V_{fb}} \times 100 \tag{14}$$

Where V_{fb} = bulk volume of compacted fines (gr/cm^3); V_{fs} = volume of the fine solids (cm^3); t = change in the thickness of the sample (t_1-t_2)(cm); d = diameter of the mold (cm); W_{fs} = weight of compacted fines (gr); G_s = specific gravity of solids (gr/cm^3) and %DCF = percentage of voids in dry compacted fines.

3 ANALYSIS OF AGGREGATE DATABASE

Anisotropic material properties of ten (10) aggregate sources were determined using multiple variable dynamic confining pressure stress paths. Then, Equations 4–6 were fitted to the stress levels and measured strains in order to calculate the k parameters. The k parameters capture the anisotropic, stress sensitive and nonlinear response of unbound aggregate systems in the lab.

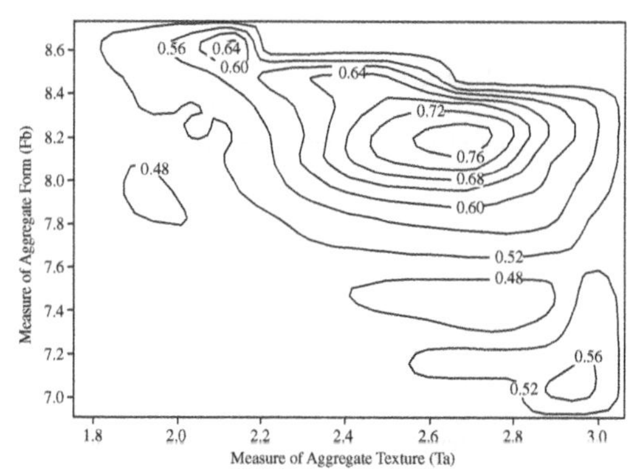

Figure 7. Impact of aggregate texture and form on anisotropy level assessed using modular Ratio (G_{xx}/G_{xy}).

As discussed earlier, aggregate shape parameters were determined using the Aggregate Imaging System (AIMS). The measured shape features of particles were then fitted to the Weibull distribution function presented in equation (9). Statistical parameters of cumulative Weibull distribution namely the shape parameter and scale parameter were then used as input data for the aggregate database. Particle size distributions were also fitted to the Weibull distribution equation and the corresponding parameters were determined and used in the aggregate database.

The moisture content at the time of testing, dry density, dry compacted air voids in the fines fraction and the methylene blue value of the aggregates were also determined and used as input to the aggregate database.

Figures 7 and 8 show the relationship between degree of anisotropy and aggregate shape characteristics measured using AIMS. The results clearly demonstrate the influence of aggregate geometry features on the degree of anisotropy in unbound aggregate systems. Figure 7 shows the effect of aggregate texture and aggregate form on the level of anisotropy as characterized by the shear modular ratio (G_{xx}/G_{xy}). Aggregate sources with more cubical particles and

$$\begin{bmatrix} \frac{G_{xx}}{G_{xy}} \\ \frac{E_x}{E_y} \\ \frac{G_{xy}}{E_y} \end{bmatrix} = \begin{bmatrix} -2.4 \\ -4.29 \\ -0.51 \end{bmatrix} + \begin{bmatrix} -0.00003 & 0.221 & 0.369 & -0.042 & 0.042 & 0.023 & 0.100 & -0.0022 & 0.24 & -0.005 & 0.016 \\ 0.00006 & 0.316 & -0.362 & -0.084 & 0.050 & 0.022 & 0.156 & -0.0023 & 0.18 & 0.017 & -0.011 \\ 0.00002 & 0.0721 & -0.137 & -0.028 & 0.017 & 0.008 & 0.042 & -0.0006 & 0.05 & 0.004 & -0.003 \end{bmatrix} \begin{bmatrix} k_1 \\ k_2 \\ k_3 \\ w \\ A_\alpha \\ F_\beta \\ T_\alpha \\ T_\beta \\ G_\beta \\ DCF \\ MBV \end{bmatrix} \quad \begin{bmatrix} R^2 = 0.95 \\ R^2 = 0.75 \\ R^2 = 0.79 \end{bmatrix} \tag{15}$$

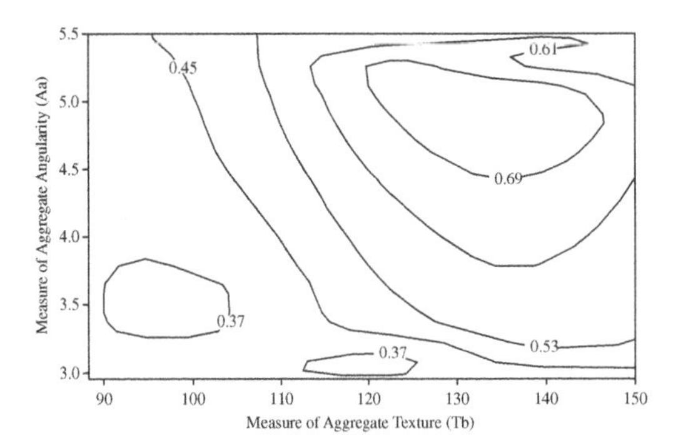

Figure 8. Impact of aggregate angularity and texture on anisotropy level assessed using modular ratio (E_x/E_y).

rougher texture demonstrated higher shear modular ratios (G_{xx}/G_{xy}), which are synonymous with less anisotropic systems as evidenced in figure 7. Less anisotropic unbound systems perform better in terms of load distribution characteristics throughout the aggregate layer.

Figure 8 demonstrates the impact of particle texture and aggregate angularity on the level of anisotropy characterized by modular ratios (E_x/E_y). Figure 8 shows that aggregate systems containing particles with rougher texture and more crushed surfaces (more angular) result in less anisotropic systems. Particle surface texture and angularity contribute to inter-particle frictional forces and affect aggregate interlock. Aggregate systems with rougher texture and more angular particles result in systems that more efficiently distribute load and are less prone to plastic deformation under traffic.

Figure 7 and Figure 8 emphasize the importance of lithology as well as rock crushing techniques used by the aggregate producers in the pavement industry. These graphs indicate that because the aggregate properties of angularity and texture impact anisotropy, which in turn impacts performance, they can be used as quality control/quality assurance tools for aggregate producers.

4 ANISOTROPY MODEL

Aggregate parameters discussed in the materials testing section of this paper were used as input data to an aggregate database that was in turn used to develop a model to predict the level of anisotropy in unbound aggregate systems. In this study modular ratios (G_{xx}/G_{xy}), (E_x/E_y) and (G_{xy}/E_y) were used to characterize the level of anisotropy for sixty three (63) aggregate systems. The inputs to the model are:

- The stiffness properties in the vertical direction represented in terms of k_1, k_2 and k_3.
- The scale parameter (α) and shape parameter (β) of the Weibull distribution equation fitted to aggregate form, angularity and texture.
- The shape parameters and scale parameters of the Weibull distribution fitted to aggregate size distributions.
- The moisture content at the time of testing.
- Dry density.
- Percent air voids in dry compacted fines.
- The methylene blue value.

Equation 12 presents the regression model that was established using a stepwise regression analysis with a 95% confidence level. The stepwise regression performs variable selection by adding or deleting predictors from the existing model based on an F-test. This method is a combination of a forward and backward feature selection and elimination process.

The shear modular ratio (G_{xx}/G_{xy}) was found to have a superior goodness of the fit ($R^2 = 95\%$) when compared to the other two anisotropy level characterizers, i.e., (E_x/E_y) and (G_{xy}/E_y), that had coefficients of correlation (R^2) of 0.75 and 0.79 respectively.

Parameters selected for the anisotropy model as presented in equation (12) are:
k_1, k_2 and k_3 = fitting parameters presented in equation (4); w: percent moisture content at the time of testing; A_α = shape parameter of angularity cumulative distribution function; F_β = scale parameter of form cumulative distribution function; T_α = shape parameter of texture cumulative distribution function; T_β: scale parameter of texture cumulative distribution function; G_β = scale parameter of gradation cumulative distribution function; MB = methylene blue value and DCF = air voids in dry compacted fines (percent).

Improvements over the first generation anisotropy model developed by Kim (Kim et al. 2005) can be summarized as:

- The number of specimens that populate the database was increased from thirty six (36) (in the previous model) to sixty three (63) aggregate samples from ten aggregate sources with three gradations and tested at various moistures states.
- Representations of the level of anisotropy are now based on three modular ratios (G_{xx}/G_{xy}), (E_x/E_y) and (G_{xy}/E_y).
- The Weibull distribution was used to fit the distributions of both characterizers of aggregate geometry and gradations in lieu of the Fredlund fitting equation used by Kim.
- The representative sieve concept was used with the motivation that particle shape properties are functions of particle size as well as crushing techniques and the mineralogy of the parent rock. Furthermore, based on the analysis of the distribution functions of aggregate shape properties for several sieve sizes of the same source, it was observed that

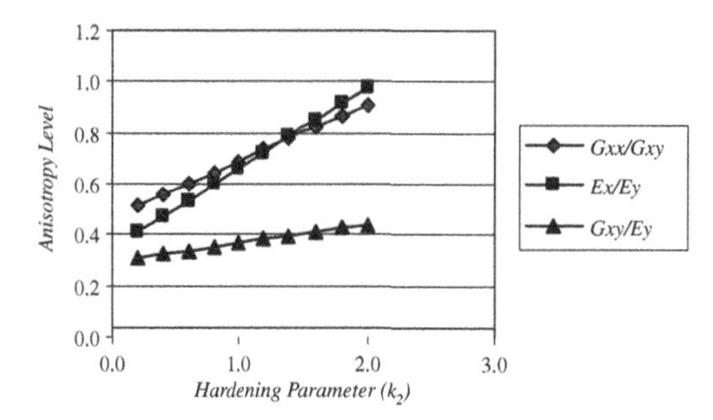

Figure 9. Influence of hardening parameter K_2 on modular ratios.

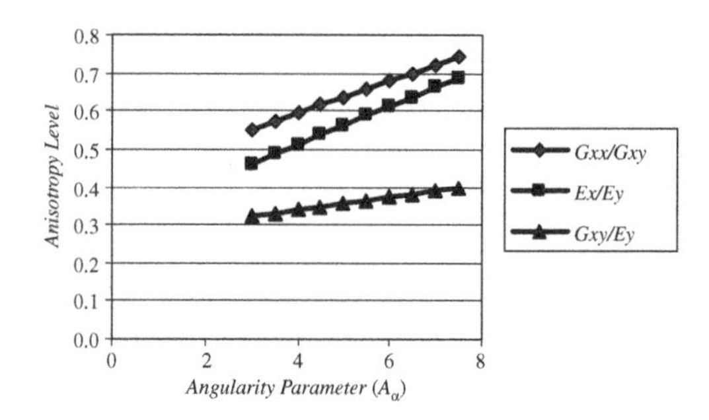

Figure 10. Influence of angularity parameter A_α on modular ratios.

different particle sizes exhibit different shape properties. Therefore, averaging shape parameters over different sieve sizes will induce systematic error in developing an anisotropy prediction model. Hence instead of reporting only one set of shape parameters for an aggregate source regardless of gradation, the representative sieve concept allocates a set of shape parameters for each gradation variant. In other words, aggregate shape parameters pertaining to the #3/8 sieve were for coarse gradation, values pertaining to the #1/4 sieve were used for intermediate gradation and values pertaining to the No. 4 sieve are used for fine gradation.
- The Rigden voids test was used as a measure of fine particles (particles smaller than 75 μ) shape features.
- The methylene blue value (MBV) test was used as a measure of the deleterious effect of plastic fines (particles smaller than 75 μ) in terms of moisture susceptibility of aggregate systems.

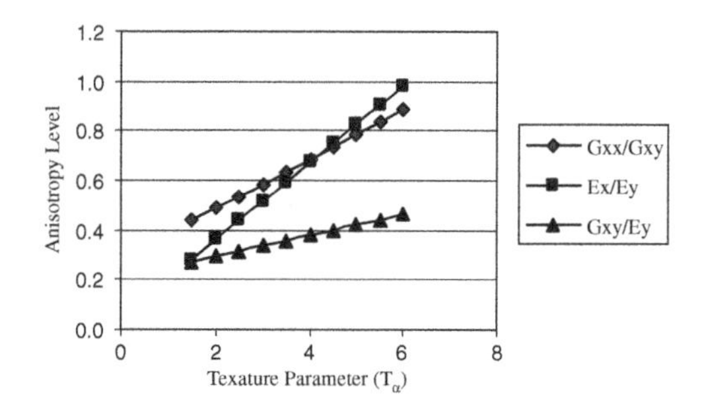

Figure 11. Influence of Texture parameter T_α on modular ratios.

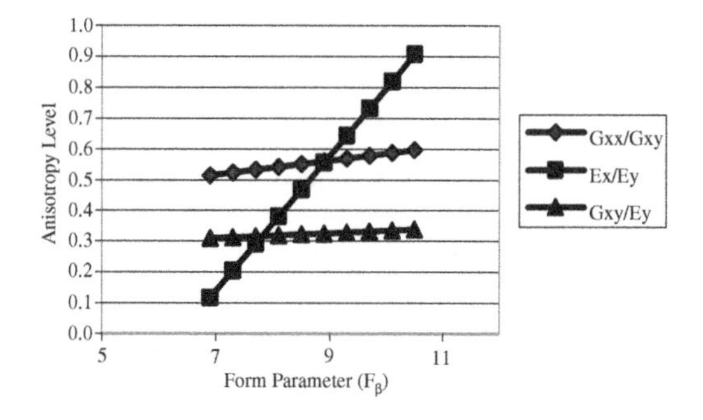

Figure 12. Influence of form parameter F_β on modular ratios.

5 SENSITIVITY ANALYSIS OF ANISOTROPY MODEL

A parametric analysis was conducted to evaluate the contribution of aggregate features to anisotropy as reflected by the model presented in equation 12. As with most regression-based models, it is not necessarily realistic to hold selected parameters within the model constant and arbitrarily change only one parameter. For instance, change in the gradation parameter will result in a change in the optimum moisture content, dry density and mechanical responses of the aggregate system (k parameters). However, despite the inter-correlation of aggregate parameters, it is instructive to monitor the impact of each variable on the level of anisotropy of aggregate systems.

Figures 9 through 12 illustrate the sensitivity of the model with respect to aggregate parameters. Figure 9 shows the impact of hardening parameter (k_2) in equation 4 on the level of anisotropy as quantified using the modular ratio (E_x/E_y). An increase in k_2 results in a higher modular ratio, which is synonymous with less anisotropic behavior. This can be explained by the fact that an increase in k_2 means that the modulus is more sensitive to confinement level. An increase in confinement level improves particle interlock and load distribution in orthogonal directions and reduces the anisotropy of the system.

Figures 10 through 12 illustrate the impact of aggregate geometry on the level of anisotropy of unbound aggregate systems. Figure 10 shows the impact of aggregate angularity on the degree of

Table 2. Sensitivity of Modular Ratios to Aggregate Features.

Modular ratio	Slope			
	k_2	T_a	A_a	F_b
G_{xx}/G_{xy}	0.22	0.10	0.04	0.02
E_x/E_y	0.32	0.16	0.05	0.22
G_{xy}/E_y	0.07	0.04	0.02	0.01

anisotropy. This plot demonstrates that aggregate systems with more angular particles have higher modular ratios or less anisotropy. Figure 11 shows the effect of aggregate texture on anisotropy. Aggregate particles with higher surface texture are less anisotropic. The trend in Figure 12 indicates that systems that consist of more cubical particles (lower form index) exhibit less anisotropy when compared with systems with flat and elongated aggregate particles.

The slopes of the lines plotted in Figure 9–12 are used to assess the sensitivity of modular ratios to the selected features. These slopes are summarized in Table 3.

A higher slope (absolute value) indicates more sensitivity of the level of anisotropy to the selected aggregate feature. Among the particle geometry features in the aggregate database, modular ratio (E_x/E_y) was found to be most sensitive to the degree of elongation of the aggregate particles or how cubical the aggregate particles are. This is in conformity with the fact that upon field compaction, elongated materials tend to re-orient to the horizontal plane which results in significant differences in material properties in orthogonal directions. The modular ratio (G_{xy}/E_y) was found to be less sensitive to aggregate features compared to the other two anisotropy level characterizers (G_{xx}/G_{xy}) and (E_x/E_y).

6 CONCLUSION

The level of anisotropy can be determined with a high level of reliability based aggregate properties that can be readily determined. These include: aggregate geometry features, fines shape and size distribution as measured using the Rigden voids test and fines reactivity potential as measured by the methylene blue test. The sensitivity analysis of the anisotropy model indicates that modular ratios (G_{xx}/G_{xy}) and (E_x/E_y) are highly sensitive to particle geometry. Aggregates with higher angularity and texture are less anisotropic. The hypothesis is that an increase in angularity and texture increases aggregate friction and interlock in all directions leading to a reduction in anisotropy. On the other hand, more elongated particles tend to re-orient themselves under loading in the direction of the horizontal plane leading to more anisotropic systems.

Finite element-based structural models of unbound aggregate bases can be used to model the cross-anisotropic, stress sensitive nature of unbound aggregate systems. Therefore, routine and efficient tests to determine the level of anisotropy can be effectively used to assess the performance of aggregate systems and to investigate the impact of aggregate features on the performance of unbound aggregate systems. This approach can be used as a quality control tool for aggregate producers and pavement material engineers.

REFERENCES

Adu-Osei, A., Lytton, R.L. and Little D.N. 2001,Cross-Anisotropic Characterization of Unbound Granular Materials. In *Transportation research Record: Journal of the Transportation Research Board*, No. 1757, TRB, National Research Council, Washington, D.C., pp. 82–91.

Cenens, J. and R.A. Schoonheydt, 1988, Visible Spectroscopy of Methylene Blue on Hectorite, Laponite B, and Barasym in *Aqueous Suspension. Journal of Clays and Clay Minerals*, Vol. 36; No. 3, June pp. 214–224.

Cokca, E. and A. Birand. 1993, Determination of Cation Exchange Capacity of Clayey Soils by the Methylene Blue Test, *Geotechnical Testing Journal*, Vol. 16 Issue 4, pp. 518–524.

Djamarani, K.M. and I.M. Clark, 1997, *International Journal of Powder Metallurgy and Powder Technology*, Vol.8, No 2, pp. 101–108.

Fityus, S., D.W Smith, and A.M Jennar. 2000, Surface Area Using Methylene Blue Adsorption as a Measure of Soil Expansivity, Geo2000 Conference, Australia.

Hills, J.F. and G.S. Pettifer, 1985, The Clay Mineral Content of Various Rock Types Compared with the Methylene Blue Value", *Journal of Chemical Technology* Vol. 35A, pp.168–180.

Kim, S., D. N. Little and E. Masad, 2005, Simple Methods to Estimate Inherent and Stress-Induced Anisotropy of Aggregate Base In Transportation Research Record: *Journal of the transportation Research Board*, No. 1931, TRB, National Research Council, Washington, D.C., pp. 24–31.

Nevins, M.J. and D.J Weintritt. 1967, Determination of Cation Exchange Capacity by Methylene Blue Adsorption, *Ceramic Bulletin*, Vol.46, Issue 6, pp. 587–592.

Oh, J.R.L. Lytton, and G.E. Fernando, June 2006, Modeling of Pavement Response Using Nonlinear Cross-Anisotropic Approach, *Journal of Transportation Engineering*, Vol. 132 Issue 6, pp.458–468.

Thompson, M.R. and D. Nauman, 1993, Strength and Deformation Characteristics of Pavement Structures. In Transportation Research Record: *Journal of the Transportation Research Board*, No. 1384. TRB, National Research Council, Washington, D.C. pp. 36–48.

Tutumluer, E. and M. R. Thompson. 1997 Anisotropic Modeling of Granular Bases in Flexible Pavements. In *Transportation Research Record: Journal of the Transportation Research Board,* No. 1577. TRB, National Research Council, Washington, D.C., pp.18–26.

Tutumluer, E., D. N. Little, and S. H. Kim. 2003, A Validated Model for Predicting Field Performance of Aggregate Base Courses. In *Transportation Research Record: Journal of the Transportation Research Board*, No. 1837, TRB, National Research Council, Washington, DC, , pp. 41–49.

Salehi R., D.N. Little, and E. Masad, In Press, Evaluation of Impact of Fines on the Performance of Lightly Cement Stabilized Aggregate Systems. Presented at 86th Annual Meeting of Transportation Research Board, Washington D.C.

Seyhan, U. and E. Tutumluer. 2002, Anisotropic Modular Ratios as Unbound Aggregate Performance Indicators. *Journal of Materials in Civil Engineering*, Vol.14, pp. 409–416.

Advances in Transportation Geotechnics – Ellis, Yu, McDowell, Dawson & Thom (eds)
 ISBN 978-0-415-47590-7

Utilization of waste foundry sand (WFS) as impermeable layer (drainage blanket) for pavement structures

P. Solmaz
Foster Wheeler, Bimaş, Istanbul, Turkey

A.G. Gedik
General Directorate of Highways, Istanbul, Turkey

M.A. Lav & A.H. Lav
Istanbul Technical University, Istanbul, Turkey

ABSTRACT: Pavement structures are susceptible to water and always been a major concern of pavement engineers. Water may enter the pavement structure from high water table and reduce the strength of granular pavement layers, causing pumping of structure under wheel loadings, resulting diminished pavement support. In this study waste foundry sand is considered as impermeable layer to prevent water penetration. Waste foundry sand samples were collected from different foundries operating in and around Istanbul. Index properties were determined. Optimum moisture contents and corresponding maximum dry densities of samples then obtained applying standard compaction effort. Hydraulic conductivity of samples prepared at maximum dry unit weight and optimum water content were then determined. The effects of compaction energy and bentonite content on the hydraulic conductivity were also investigated. The hydraulic conductivity values determined were in the range between 1×10^{-5} m/s and 1×10^{-7} m/s for bentonite content higher than 5%. According to the results obtained, the material may be used as drainage blanket.

1 INTRODUCTION

Road pavements become useless in a very short time when it is saturated. In other words, saturated layers under traffic loads results in pumping and reduce design life of pavements significantly. Water may enter the pavement structure from a high groundwater table by means of capillary forces or surface water may flow from the pavement edges and side ditches.

Although a proper drainage may prevent this adverse effect, to be on the safe side, there are several ways such as installation of geomembrans may be utilized to prevent high water table to penetrate the pavement structure. However, in this study waste foundry sand is considered as an impermeable layer to prevent water penetration. Waste foundry sand is a by product of molding industry where sand is mixed with bentonite and some other particles to improve its binding. As the material is stockpiled after the molding process, utilization of waste foundry sand in this way not only reduces the stockpiles but also an economic alternative to geomembrans. For this experimental study, waste foundry sand is obtained from different sources as each manufacturer mixed the sand with different quality and quantity of bentonite and grading, amount of bentonite, Atterberg limits, unit weight of samples were determined and compared in laboratory. Optimum moisture contents and corresponding maximum dry densities of samples then obtained by different compaction effort. Hydraulic conductivity of samples for each group of compaction effort were then determined. Since increasing amount of bentonite in sand change the hydraulic conductivity, additional bentonite is added in two of the WFS samples to observe the effect of increasing bentonite on hydraulic conductivity.

1.1 *Foundry sand facts*

A foundry is a facility that produces metal parts by pouring liquid metal (melted at extreme temperatures) into a pre-shaped mold. In order to produce metal castings, foundry operators require high quality silica sand. The sand makes the outside shell of the mold cavity in which molten metal poured and it should be bonded to keep its form inside the mold. Foundry sand is consumed large amounts by manufacturing industries such as car manufacturing, iron-steel industries, alloy production and various branches of metallurgy industry. Generally two types of binder systems are used in the industry as follows:

- Bentonite is a type of clay and 4% to 10% by weight is blended with sand for foundry operations. Majority of foundry sand used by the industry is the clay-bonded sand called "green sand".
- Various resins are also used when high strength of the foundry sand is necessary to withstand the heat of poured metal, however this type of sand, namely "resin sands" are used to produce roughly 10% of casting volume.
- During the process, foundry sands are reused several times until the material reach to the point that it can no longer be used for the purpose and called as "Waste Foundry Sand" (WFS).

1.2 *Foundry sand utilization*

At present, approximately 450,000 tons of WFS is produced annually in Turkey. Unfortunately, virtually no WFS is utilized and the material is dump stockpiles. Majority of metal casting industry is concentrated in Istanbul and surroundings where dumping grounds are being exhausted. Increasing amount of landfilled WFS becoming an environmental problem and a public concern. The industry is desperate to find a way of utilization. Consequently, it is a significant problem for the country.

Road construction and geotechnical fill applications being two primier field of civil engineering where large amount of WFS can be utilized, (Abichou et al., 2004a, Partridge, 1998, American Foundry Society, 2002). However, the following areas were found to have the potential to utilize WFS, (FHWA, 2004):

- Pavement base, subbase or subgrade fill material in highway constructions,
- Building bases in construction industry,
- Cement kiln feedstock,
- Aggregate in hot mix asphalt,
- Grouts and mortars,
- Ready mix concrete,
- Hydraulic barrier in landfills,

Particularly, after compacting WFS presence of bentonite in the material satisfy the hydraulic conductivity requirement to be utilized as hydraulic barrier (Abichou et al. 1998, 2000, 2002, 2004b, Kunes et al. 1983). Among the several potential use of the hydraulic barrier, it may be placed between subbase and subgrade in a pavement structure to prevent water penetration, as mentioned previously, (Guide for Mechanistic-Empirical Design of New and Rehabilitated Pavement Structures, 2001). It should be noted that the aim of this study is to investigate the suitability of the material as drainage blanket. Consideration of WFS as a pavement layer is beyond the scope of this paper.

2 FOUNDRY SAND AS HYDRAULIC BARRIER

An impermeable layer of earthwork is considered a hydraulic barrier. The performance of hydraulic barriers constructed using compacted WFS, both in laboratory and in the field have been the subject of some detailed studies in the literature as follows, (Abichou et al. 1998, 2000, 2002, 2004b, Kunes et al. 1983).

2.1 *Laboratory studies*

Kunes et al. (1983) carried out grain size distribution, compaction and hydraulic conductivity measurements on a typical sample from compacted WFS during their research on the use of the WFS as landfill cover material. The sample they tested contained 0.7% gravel, 88.3% sand, 7.5% silt and 3.6% clay. They determined the hydraulic conductivity of compacted WFS between 1.3×10^{-6} and 5×0^{-6} cm/s. They concluded that compacted WFS can manifest similar properties to those materials that are used to built impermeable layers.

Abichou et al. (2000) carried out a study on utilization of WFS as hydraulic barriers. In their study, WFS samples were compacted at a series of water content using three levels of compactive effort. Then they conducted permeameter tests on the samples using rijid wall and flexible wall testing devices to investigate relationships between hydraulic conductivity, compaction water content and dry unit weight. Besides, they investigated the effect of wet and dry cycles, freze and thaw cycles and the presence of chemical liquids on the hydraulic conductivity of compacted WFS. They show that the hydraulic conductivity of WFS samples diminishes with an increase in liquid limit, plastic limit and bentonite content. They also show that, if liquid limit is greater than 20, plasticity index is greater than 2, or bentonite content is greater than 6, then the hydraulic conductivity of compacted WFS samples can be determined as smaller than 10^{-7} cm/s. Besides, according to their test results, hydraulic conductivity values are not affected by freeze-thaw and wet-dry cycles. They tested the variation of hydraulic conductivity of WFS samples under the presence of four different liquid as permeant (deionized water, tap water, 0.1 N $CaCl_2$ solution and municipal solid waste leachate) for short (75 days) and long period (433 days) to check the performance of hydraulic barrier. They concluded that the hydraulic conductivity values were not changed significantly over time.

2.2 *Field studies*

Grede Foundries in Wisconsin supported a series of field performance tests to show the feasibility of utilization of WFS as hydraulic barrier in landfill liner and cover system, (Abichou et al. 1998, Freber, 1996).

The study consisted of five full scale lesimeter tests. Clay was used as barrier layer in two of these tests and WFS was used in the remainings. Clay barrier layers were compacted to at least 90% modified Proctor density. On the other hand, WFS barrier layers were compacted to at least 95% standard Proctor density.

Data were collected for four years and it has been concluded that, the test sections made of WFS performed better than those made of clay. When clay is used, due to freeze-thaw and wet-dry cycles, significant cracks were developed inside the barrier layers, which is why hydraulic conductivity is found greater for clay sections. On the other hand, the test sections made of WFS remained still intact even after many years, (Abichou et al. 2002).

3 INDEX PROPERTIES

In this study, geotechnical laboratory tests were conducted on waste foundry sand collected from 15 foundry companies operating in and around Istanbul. Of 15 group of material collected, 14 of which were green sand and the remaining was binded by resin. In order to determine the index properties, liquid and plastic limit and grain size distribution tests were carried out along with tests for determination of bentonite content and loss on ignition.

3.1 *Atterberg limits*

Liquid and plastic limit tests were conducted in accordance with ASTM D 4318-05. According to the results all samples are classified as nonplastic soils as shown in Table 1.

3.2 *Grain size distribution*

Sieve and hydrometer analyses to determine grain size distribution of the samples were conducted in accordance with ASTM D 422. Methilen Blue Tests (ASTM C 837) were carried out to determine bentonite content of the samples. Grain size distribution curves are given in Figure 1. As can be seen in Figure 1 the amount of fines vary between 6% and 28.3% for all samples except sample No 10 which is resin binded WFS having a fine content 2.2%. Green sands are classified as SM or SM-SC according to Unified Soil Classification System. WFS with resin is classified as SP. Bentonite contents were determined (ASTM C 837) as varying between 2.7% and 12.5%.

3.3 *Specific unit weight*

Pycnometer tests were conducted to determine specific unit weight of samples. The test results for specific unit weight, bentonite content and USCS classifications are given in Table 1.

Table 1. Index properties.

Sample No	Specific Unit Weight (gr/cm^3)	Plasticity Index	USCS Class	Binder	Fines (%)	BC* (%)
1	2,53	NP	SM	B*	28.3	6.79
2	2,46	NP	SM	B	13.6	4.08
3	2,48	NP	SM	B	19.4	4.08
4	2,48	NP	SM	B	16.7	2.72
5	2,45	NP	SM	B	16.0	6.9
6	2,45	NP	SM	B	16.0	12.5
7	2,56	NP	SM	B	12.8	5.43
8	2,53	NP	SM	B	13.5	6.11
9	2,58	NP	SM	B	12.1	4.07
10	2,61	NP	SP	Resin	2.2	–
11	2,45	NP	SM	B	22.9	8.15
12	2,59	NP	SM	B	15.8	5.43
13	2.62	NP	SP-SM	B	6.0	2.71
14	2.60	NP	SM	B	17.2	6.8
15	2.49	NP	SM	B	19.2	11.2

*B denotes Bentonite
*BC denotes Bentonite Content

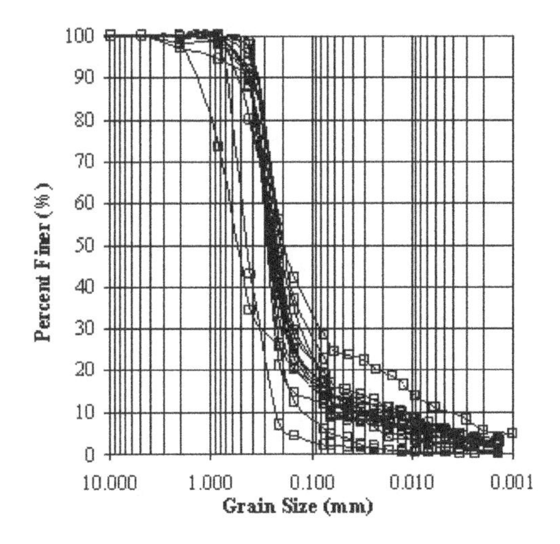

Figure 1. Grain size distribution curves.

3.4 *Loss on ignition*

Tests were conducted in accordance with TS 3245 to determine the amount of organic matter (coal dust) in the WFS. According to the results the content of organic matter was varying between 4% and 23%.

4 COMPACTION TESTS

4.1 *Standard compaction tests*

Before compaction tests were begun, metal parts in WFS were collected and removed by hand. Standard

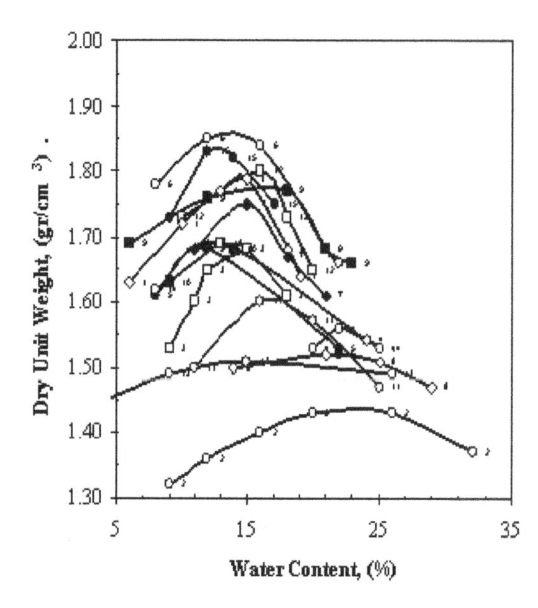

Figure 2. Standard compaction curves for all samples.

Table 2. Reduced compaction test results.

No	Bentonite content %	Optimum water content %	Dry Unit Weight gr/cm^3	Hydraulic conductivity m/s
7	5.43	17	1.66	1.63E-06
8	6.11	21	1.49	1.48E-05

compaction tests of the samples were carried out in accordance with ASTM D 698. Test results for all of the standard compaction tests are given in Figure 2. As can be seen from the figure, maximum dry unit weight is varying between 1.44 gr/cm^3 and 1.86 gr/cm^3. Optimum water contents in the figure is varying between 12% and 22%.

4.2 *Reduced and modified compaction tests*

Modified Proctor and reduced Proctor tests were conducted on two samples to compare different compaction effort. Modified Proctor tests were conducted in accordance with ASTM D 1557. The only difference between reduced Proctor and standard Proctor is that the number of blows is 15 in reduced Proctor instead of 25 that applies in standard Proctor tests (Abichou et al. 2000, Daniel et al. 1990). The test results of reduced, standard and modified Proctor are given in Table 2, Table 3 and Table 4, respectively. As can be seen from the Tables 2,3 and 4, as the compaction effort is increased higher dry unit weights were obtained at

Table 3. Standard compaction test results.

No	Bentonite content %	Optimum water content %	Dry Unit Weight gr/cm^3	Hydraulic conductivity m/s
7	5.43	15	1.75	1.44E-06
8	6.11	22	1.52	1.20E-05

Table 4. Modified compaction test results.

No	Bentonite content %	Optimum water content %	Dry Unit Weight gr/cm^3	Hydraulic conductivity m/s
7	5.43	16	1.77	1.23E-06
8	6.11	17	1.66	5.43E-05

lower values of water contents, in other words, a better compaction is obtained. This leads to lower values of hydraulic conductivities which will be explained in detail in the next section. It should be noted that, the effect of compaction effort on hydraulic conductivity is low as Abichou et al. (2000) mentioned in a detailed study.

4.3 *Optimum bentonite content*

Figure 3 shows the maximum dry unit weight versus bentonite content. As can be seen from the figure, the values of maximum dry unit weight tend to increase as bentonite content increases, yet after a certain value of bentonite content which is around 10%, the relation is reversed. This behavior was reported by other researchers, Kenney et al. (1992) and Abichou et al. (2000).

The variation of optimum water content with bentonite content is given in Figure 4. As can be seen in Figure 4 which is similar to Figure 3, that is, at bentonite content of 10% the trend of the curve is reversed.

Among the samples, the effect of bentonite content on compaction characteristics was investigated using Sample No 8 by increasing bentonite content to 10% and 15% by weight. Standard Proctor curves for bentonite contents of 6.11%(initial bentonite content), 10% and 15% for Sample No 8 are given in Figure 5. It can be seen in Figure 5, as bentonite content increased from 6.11% to 10%, dry unit weights are increased and the curve became more sensitive to water content increase. But, as bentonite content is increased from 10% to 15%, lower values of dry unit

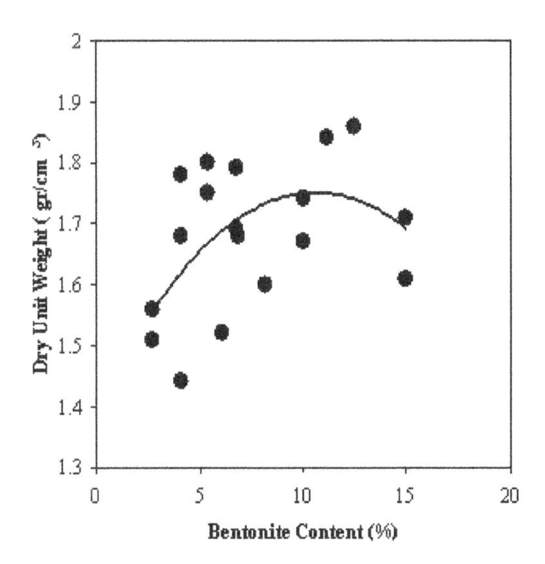

Figure 3. Maximum dry unit weight versus bentonite content.

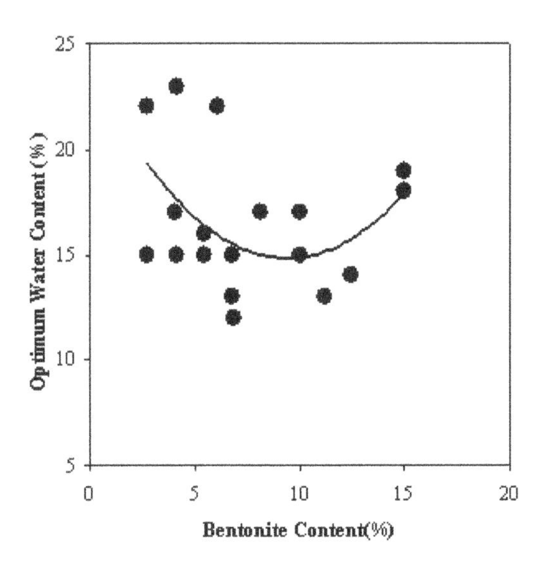

Figure 4. Optimum water content versus bentonite content.

weights are obtained. Similarly, optimum water content is 16% for bentonite content of 10% and increases 19% for bentonite content of 15%. The optimum value of bentonite content shown in Figure 3, depends on the grain size distribution of foundry sand and swell potential of bentonite, (Abichou et al. 2000, Howel et al. 1997a,b). The test results obtained for samples with bentonite content 10% and 15% are also included in Figure 3 and Figure 4.

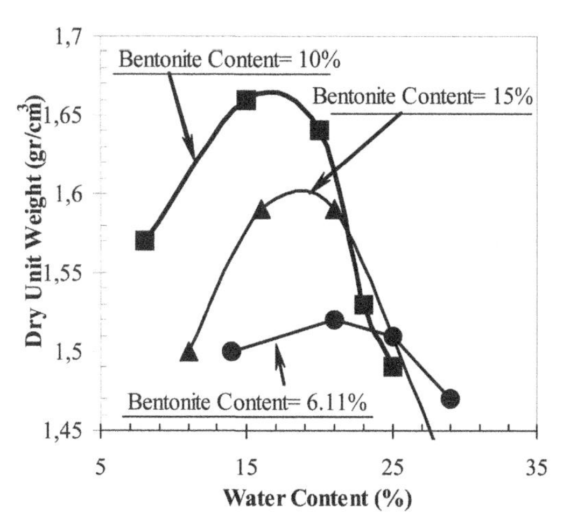

Figure 5. Standard Proctor curves for Sample No 8 with bentonite contents %6.11 (as it is), %10 and %15.

5 HYDRAULIC CONDUCTIVITY MEASUREMENTS

Samples prepared at optimum water content and maximum dry unit weight for hydraulic conductivity measurements. Hydraulic conductivity (k) were determined by falling head permeability device. Tap water is used as permeant liquid. Tests were terminated when consecutive readings are similar. According to the test results, the variation of hydraulic conductivity with bentonite content is given in Figure 6. As can be seen in Figure 6, hydraulic conductivity values are measured as low as 1×10^{-7} m/s. Hydraulic conductivity values reduce quickly as bentonite content increases to 4% – 5%. At higher values of bentonite content, the reduction of hydraulic conductivity slowed down and stayed almost constant between 1×10^{-5} and 1×10^{-7} m/s. No lower values than 1×10^{-7} m/s was measured thereafter even at higher bentonite contents. The variation of hydraulic conductivity with bentonite content is parallel with the results of similar research (Abichou et al. 2000, Kenney et al.(1992). As the voids are filled by bentonite, the hydraulic conductivity should be measured close to the hydraulic conductivity of bentonite itself beyond a value of bentonite content which is %7 according to Abichou et al. (2000) and %12 according to Kenney et al. (1992) . In Figure 6, for bentonite content of 15% the hydraulic conductivity converged were in between 1×10^{-5} and 1×10^{-7} m/s which is higher than hydraulic conductivity of bentonite itself. It seems that it is necessary more than 24 hours should be allowed for bentonite to hydrate (according to Abichou et al. (2000) the hydradation period is at least one week). In this study, however, 24 hours or less is allowed as hydradation period.

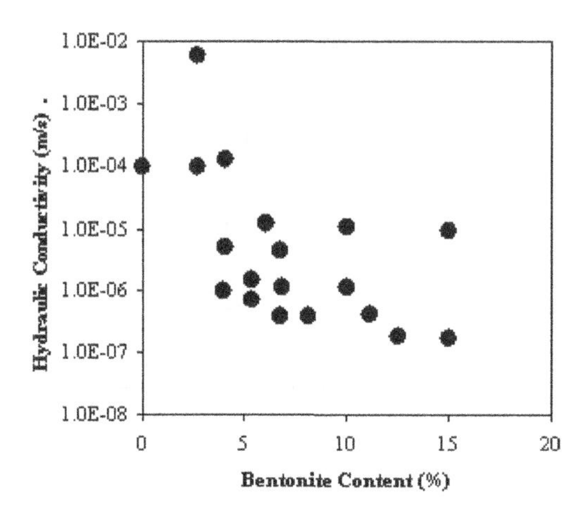

Figure 6. Bentonite content versus hydraulic conductivity.

6 CONCLUSIONS

One of the most suitable area of utilization for WFS is construction of highways as high amounts of material may be used. Considering this, initial results of an experimental project on suitability of WFS as impermeable layer for permeable structures is presented in this paper.

For the purpose of the study, index properties, bentonite content, compaction properties and hydraulic conductivity of the samples are determined for 15 waste foundry sand samples collected from foundries operating in and around Istanbul. According to the test results the following conclusions can be drawn:

1) The hydraulic conductivity of the compacted WFS were determined as low as 1×10^{-7} m/s which suitable for the purpose.
2) Optimum bentonite content was found to be 10% .
3) Duration of hydradation should be longer than 24 hours.
4) Effects of the quality of bentonite and organic substances need to be investigated.
5) Further tests should be carried out by means of a flexible wall hydraulic conductivity testing device to reconstruct in situ conditions.

ACKNOWLEDGEMENTS

The authors would like to acknowledge the following persons and companies who either sent or gave samples of WFS by which this research was initiated especially Nizamettin Özdemir, Can Akbaşoğlu, Bülent Nilüfer, Yusuf Tarhan, Arif Oktay Şimşir. Döktaş Dökümcülük Tic.San.A.Ş., Emin Döküm, Bıyıklı Döküm, Arı Metal, Ferro Döküm Sanayii Tic. A.Ş., Körfez Döküm San.Tic.A.Ş., Ata Döküm, Ümit Döküm, Akmetal Metalurji End.A.Ş., Haytaş Döküm San.A.Ş., Çemaş Döküm San.A.Ş., Anadolu Döküm San.A.Ş., Acarer Döküm San.Tic.Ltd.Şti., Hisar Çelik Döküm.San.Tic.A.Ş., Demrad Döküm Ürünleri Sınai Tic.A.Ş., Kutes, Trakya Döküm San.Tic.A.Ş., Erkunt San. A.Ş., Entil End. Yat.Tic.A.Ş., DemisaŞ Döküm Em. Mam.San.A.Ş.

REFERENCES

Abichou, T., T.B., Edil, Benson, Craig H.B. & Bahia, H., 2004 "Beneficial Use of Foundry by-Products in Highway Construction", *ASCE, GeoTrans* 2004, 154, 58.

Partridge, B., 1998, "Field Demonstration of Highway Embankment Constructed Using Waste Foundry Sand" *PhD Thesis, Purdue University*, Indiana.

American Foundry Society 2002, Benefical Reuse of Fondry Sand:A Review of State Practices and Regulations, Sector Strategies Division Office of Policy, Economics, and Innovation U.S. Environmental Protection Agency Washington, DC.

Foundry Sand Facts for Civil Engineers, 2004, FHWA-IF-04-004, *US Federal Highway Administration*.

Abichou, T., Benson, C. & Edil, T., 1998 Using waste foundry sand for hydraulic barriers. *Recycled Materials in Geotech. Applications, GSP79, ASCE*, Reston, Va., 86–99.

Abichou, T., Benson, C. & Edil, T., 2000, Foundry Green Sands As Hydraulic Barriers: Laboratory Study, *J. Geotech. and Geoenvir.Engrg.*, 126(12), 1175–1183.

Abichou, T., Benson, C. & Edil, T., 2002, Foundry Green Sands As Hydraulic Barriers: Field Study 2002, *J. Geotech. and Geoenvir.Engrg.*, 128 (3), 206–215.

Abichou, T., Edil, T.B., Benson, Craig H.B. & Tawfiq, K. 2004, Hydraulic Conductivity of Foundry Sands and Their Use as Hydraulic Barriers, *ASCE, Recycled Materials in Geotechnics,* 149, 13.

Kunes. T.&Smith, M. 1983, Waste Disposal Considerations for Green Sand Use in the Foundry Industry. AFS-CMI Conference on Green Sand-Productivity for the 801s.

Guide for Mechanistic-Empirical Design of New and Rehabilitated Pavement Structures, Appendix SS: Hydraulic Design, Maintenance, and Construction Details of Subsurface Drainage Systems, NCHRP, 2001.

Freber, B., 1996, Beneficial reuse of selected foundry waste material, *Proc. 19th Int. Madison Waste Conf., Dept. Of Engrg. Professional Dev.*, Univ. Of Wisconsin, Madison, Wis., 246–257.

Daniel, D.E. & Benson, C.H., 1990, Water content-density criteria for compacted soil liners, *J. Geotech.Engrg., ASCE*, 116(12),1811–1830.

Kenney, T., Van Veen, W., Swallow, M. A. & Sungaila, M. A. 1992, Hydraulic Conductivity of Sand-Bentonitee Mixtures, *Canadian Geotechnical Joumal*, Vol. 29, NO. 3, pp. 364–374.

Howell,J. & Shackelford, C. 1997, Hydraulic conductivity of sand admixed with processed clay mixtures, *Proc. 14th Int. Conf. On Soil Mech. And Found. Engrg.*, Balkema, Rotterdam, The Netherlands, 307–310.

Howel, J. & Shackelford, C., Amer, N., and Stern, R. 1997, Compaction of sand-processed clay soil mixtures,*Geotech. Testing J.*, 20(4), 443–458.

Advances in Transportation Geotechnics – Ellis, Yu, McDowell, Dawson & Thom (eds)
© 2008 Taylor & Francis Group, London, ISBN 978-0-415-47590-7

Resilient behaviour of paving materials and the change in deflection level with time

G. Trichês
Federal University of Santa Catarina, Florianópolis, Santa Catarina, Brazil

G.P. Simm Júnior
Prosul Ltda., Florianópolis, Santa Catarina, Brazil

ABSTRACT: The aim of this paper is to show the results obtained with the adoption of deflectometric control during the construction process of two experimental sections located in the state of Santa Catarina, Brazil. Deflection basins were measured in all layers of the pavement structure built, using Benkelman beam. Dynamic triaxial testing was conducted to determine the resilient behavior of the materials used in the construction. With the use of the Kenlayer program, load application was simulated directly upon the subgrade, the base course and the surface course, thus simulating the measure of deflections in each layer built during the construction process. Such modeling verified that the modulus values and the deflections calculated were identical to the field values. The work conducted shows that deflection control during the construction process, associated with mechanistic analysis, constituted a valuable procedure for the analysis and certification of the pavement structure constructed.

1 INTRODUCTION

One of the ways encountered to increase the probability of good performance of a new paved road is to control the resilient deformation of the structure during the construction process, through the systematic measure of the rebound deflections. The deflections measurement enables an indirect evaluation of the magnitude of the tensile strains to which the asphalt layer will be submitted at the beginning of the life cycle.

The work hereby presented seeks to demonstrate the efficiency and the advantages of the adoption of deflection control during the construction process of two experimental sections implanted on a highway of medium traffic volume (a rural collector-road) located in the western portions of the state of Santa Catarina, Brazil.

2 CHARACTERIZATION OF THE STUDY AREA

The studies were conducted during the "Works of Implementation of Highway SC-469, Section: Campo Erê – Saltinho," located in western Santa Catarina. The stretch of highway is part of the Santa Catarina state road network and is under the jurisdiction of the state's Department of Infrastructure. The construction started in September 2003 and the road was opened to traffic on March 19, 2005. The pavement structure was designed for a traffic load of 3,44 million cycles and a design california baring ratio of 11%. The pavement structure dimensioned for the highway consisted of:

- dry macadam sub-base, 16 cm thick.
- crushed stone base, 15 cm thick.
- hot mix asphalt course, 5 cm thick.

3 CHARACTERISTICS OF THE EXPERIMENTAL SECTIONS

With the objective of reducing paving costs by reducing the total thickness of the pavement structure, two experimental sections were built on the highway, with an extension of 200 meters each. On experimental section 01, the upper subgrade layer (60 cm thick) was constructed with residual basalt soil, whose tactile-visual description is that of a dark reddish silty clay (lateritic). In the first 40 cm, the clay was compacted at 100% of the normal Proctor energy and the last 20 cm of the layer was compacted at 100% of the intermediate Proctor energy. Compactation at the intermediate Proctor elevated the design CBR from 11% to 17%, permitting a reduction in pavement thickness. The base course was constructed in crushed stone with a thickness of 20 cm. The surface course was constructed with hot mix asphalt concrete (HMA), with a thickness of 5 cm.

Due to the large quantity of excavated rock (around 20% of the total earthwork volume), on experimental section 02 the fill was constructed with dynamited rock. The last 90 cm of fill was constructed with stones at a maximum diameter of 20 cm and the surface was regularized with fine crushed stone. A design CBR of 20% was adopted for this course. The base course was built in crushed stone with a thickness of 16 cm and the surface course is HMA concrete, with a thickness of 5 cm.

4 CHARACTERIZATION OF THE RESILIENT BEHAVIOR OF THE MATERIALS

4.1 *Subgrade layer – Section 01*

Dynamic triaxial testing was conducted in order to characterize the resilient behavior of the soil used during construction. It was adopted the Brazilian specification DNER-ME 131/94 for the testing procedures. For cohesive soils, the specification establishes a constant confining stress of 21 kPa and repeated deviatoric stresses varying from 21 to 210 kPa, with load application frequency of 20 cycles per minute and impact duration of 0,1 second. The resilient modulus is calculated for each stress level, after 200 applications of the repeated loading. The models obtained were of the bi-linear type, and the following equations were found for the soil compacted at normal Proctor energy:

$$M_R = 81.820 + 930 \times (54 - \sigma_d) \text{ for } \sigma_d < 54 \quad (1)$$

$$M_R = 81.820 - 80 \times (\sigma_d - 54) \text{ for } \sigma_d > 54 \quad (2)$$

where M_R = resilient modulus, in kPa; and σ_d = deviatoric stress, in kPa.

For the top 200 mm of the subgrade layer, compacted at the intermediate Proctor energy, the following equations were obtained:

$$M_R = 104.625 + 545 \times (55 - \sigma_d) \text{ for } \sigma_d < 55 \quad (3)$$

$$M_R = 104.625 - 77 \times (\sigma_d - 55) \text{ for } \sigma_d > 55 \quad (4)$$

where M_R and σ_d = defined previously, in kPa.

4.2 *Crushed stone base*

For granular soils, the specification establishes the use of confining stresses within a range of 21 to 140 kPa. For each level of confinement three different levels of repeated deviatoric stresses are applied (i.e. one, two and three times the value of the confining stress, thus with a maximum deviatoric stress of 420 kPa). The resilient modulus is registered after 200 applications of the deviatoric stress, for each pair of loading. The results were expressed by the K-θ Biarrez model, as per the following equation:

$$M_R = 16.816 \times \theta^{0,57} \quad (5)$$

where M_R = defined previously; and θ = bulk stress. Sum of the principal stresses σ_1, σ_2 and σ_3, in kPa.

4.3 *Subgrade layer – Section 02*

Regarding Section 02, it was obviously impossible to determine, in laboratory, the model of resilient behavior of the final layer constructed in detonated rock, due the large dimensions of the granular material. It was supposed that the material would present a behavior similar to that of the crushed stone. Based on consultations of the available literature and on simulations performed, it was pondered that the behavior of this material could be modeled by the following equation:

$$M_R = 7.000 \times \theta^{0,6} \quad (6)$$

where M_R and θ = previously defined, in kPa.

4.4 *Asphalt concrete surface course*

The determination of the resilient modulus of the HMA mix used on the asphalt surface course was done through dynamic diametrical compression testing. For such, four samples were collected from each section. An average modulus value of 5.523 MPa was obtained for section 01 and an average modulus value of 5.159 MPa was obtained for section 02.

5 DEFLECTOMETRIC CONTROL RESULTS

During the entire construction process, deflection control was performed with a Benkelman beam, with the measure of deflection basins on all of the layers, including the subgrade. The basins were measured every 20 m, at the position corresponding to the path of outer wheel, in both lanes of traffic. Temperature measures of the asphalt surface course were not taken into account during the deflection measures. This type of procedure is not common in Brazil and it is not part of the official specifications. However, numerical simulations showed that, for a pavement structure with a 5 cm thick asphalt layer and an average modulus value of 5.000 MPa, variations of 2.000 MPa in the modulus value of the asphalt concrete course would cause a variation of less than 1×10^{-2} mm in the calculated deflection. Therefore, it was considered that modulus variation of the asphalt concrete course caused by temperature variation would not have a significant impact on the measured deflections. Table 1 presents

Table 1. Average deflections obtained in each layer.

Layer	Section 1 0,01 mm	Section 2 0,01 mm	Highway 0,01 mm
Detonated rock subgrade	–	60	–
Cohesive soil subgrade	119	–	106
Dry macadam sub-base	–	–	83
Crushed stone base	72	62	75
Asphalt concrete course	49	58	50

the average deflections obtained on each layer on both sections and those obtained for a 2 km stretch of the highway with normal pavement structure.

5.1 *Analysis of the results of deflection control*

Analysis of Table 1 indicates that the experimental sections presented a deflectometric behavior similar to the one observed on the highway section constructed with a normal pavement structure. It was verified that an average deflection of 119×10^{-2} mm was obtained at the top of the subgrade layer of section 01 (built upon clay soil), much higher than the deflection that was obtained at the top of the subgrade layer of section 02 (built upon dynamited rock), which was of 60×10^{-2} mm. However, on the asphalt surface course of section 01, an average deflection of 49×10^{-2} mm was obtained, lower than the average deflection obtained on section 02, which was 58×10^{-2} mm. In other words, with the construction of the base course and asphalt surface course, there occurred a rather significant reduction in the deflectometric level in the structure of section 01.

For section 02, however, the deflectometric behavior remained practically constant, hovering around the same order of magnitude as the value measured at the top of the subgrade layer.

This apparently contradictory behavior can be explained by the Pavement Mechanics, through the analysis of the resilient behavior of the granular and cohesive soils inside the designed structure under the action of the load applied. According to Medina (1997), in the non-cohesive material, the resilient modulus is a function primarily of the confining stress applied. As such, the higher the level of stresses acting upon the granular layer, the higher the modular value obtained will be. In the case of cohesive soils, the resilient modulus is an inverse function of the deviatoric stress applied, i.e. the greater this stress is, the lower the material's resilient modulus will be. In this way, as the construction of the pavement layers progresses, there is a reduction of the stress level acting upon the lower layers due to the external load applied by the truck used for measuring the deflections. It is to be expected that there will be a positive variation in the modular values in case of the clay material and a negative variation in the case of the granular material. As such, there is a consequential change in the deflectometric level.

This effect can be better understood by means of a computer-based analysis with the Kenlayer program. For such, the same standard load used in the deflection control in the field was adopted in the numerical simulation, i.e. 80 kN equivalent single axle loading, distance between tire centers of 32 cm, contact radius of 10,8 cm and contact pressure of 560 kPa. A Poisson coefficient of 0,3 was adopted for the asphalt concrete, 0,35 for the non-cohesive material, and 0,45 for the cohesive soils.

The effect caused by the application of the standard load on each layer of the pavement was simulated, i.e., the measure of the deflections during the construction process was simulated. The moduli were calculated at the center of each layer (half of the thickness) and between the two wheels, and the modulus of the subgrade layer was determined 2,5 cm below the top, as recommended by Huang (2001). In the case of the load applied directly on the subgrade layer, we opted to subdivide it into 3 distinct sub-layers (two of 20 cm and one semi-infinite), thus refining the analysis.

5.1.1 *Results of the analysis of section 01*

Table 2 presents the results obtained the numerical simulation of experimental section 01. It was verified that, in the case of the load applied on the subgrade, a calculated deflection of 99×10^{-2} mm was obtained, slightly lower than the average on-site deflection of 119×10^{-2} mm. In the case of the load applied on the base, the calculated deflection was 66×10^{-2} mm, a value somewhat near the average deflection of 72×10^{-2} mm obtained on-site. Regarding the load applied on the asphalt surface course, a rather satisfactory field/laboratory correlation was obtained, with a calculated deflection of 52×10^{-2} mm, a value much closer to the average on-site deflection, on the order of 49×10^{-2} mm.

In the table one can also observe that, as the construction of the pavement layers moves forward, variations also occur at the level of stress acting on each layer and, consequently, variations occur in the modular values. Analyzing the layer of crushed stone, it can be seen that, with the load applied directly on the base, there is $\theta = 276$ kPa and an average modular value of 412 mpa for the crushed stone. With the execution of the asphalt surface course, the value of θ is reduced to 176 kPa and the resilient modulus falls to 318 MPa. In other words, the addition of the surface course brought about a reduction in the level of stresses acting on the underlying layers, due to the greater spreading of the pressure bulb. With this reduction in the value of the stress invariant, there also occurs a decrease in the modular value of the granular layer. This modular reduction leads to an increase in

Table 2. Results of the structural analysis of experimental section 01.

Layer	Depth* cm	σ_d kPa	θ kPa	M_R MPa
Loading applied on the subgrade:				
First (subgrade)	10	51	–	107
Second (subgrade)	30	50	–	86
Third (subgrade)	42,5	53	–	83
Calculated deflection: 99×10^{-2} mm				
Measured Deflection: 119×10^{-2} mm				
Loading applied on the base:				
Base	10	–	276	412
First (subgrade)	30	52 –	106	
Second (subgrade)	42,5	52	–	84
Calculated deflection: 66×10^{-2} mm				
Measured Deflection: 72×10^{-2} mm				
Load applied on the asphalt surface course:				
Base**	15	–	176	318
First (subgrade)	35	54	–	105
Second (subgrade)	47,5	39	–	96
Calculated deflection: 52×10^{-2} mm				
Measured Deflection: 49×10^{-2} mm				

* Depth at which the stresses were calculated.
** Corresponds to the 5 cm of the surface plus 10 cm corresponding to the center of the base course.

Table 3. Results of the structural analysis of experimental section 02.

Layer	Depth* cm	σ_d kPa	θ kPa	M_R MPa
Loading applied on the subgrade:				
First (subgrade)	10	–	314	219
Second (subgrade)	30	–	148	141
Third (subgrade)	42,5	–	95	109
Calculated deflection: 62×10^{-2} mm				
Measured Deflection: 60×10^{-2} mm				
Loading applied on the base:				
Base	10	–	314	442
First (subgrade)	30	–	146	139
Second (subgrade)	42,5	–	95	108
Calculated deflection: 58×10^{-2} mm				
Measured Deflection: 62×10^{-2} mm				
Load applied on the asphalt surface course:				
Base**	13	–	181	325
First (subgrade)	31	–	92	106
Second (subgrade)	43,5	–	62	84
Calculated deflection: 58×10^{-2} mm				
Measured Deflection: 58×10^{-2} mm				

* Depth at which the stresses were calculated.
** Corresponds to the 5cm of the surface plus 10 cm corresponding to the center of the base course.

deflection. However, with the execution of the asphalt concrete, a slightly increase occurred in the modular value of the subgrade, which went from 84 MPa to 96 MPa. This, allied with elevated rigidity of the concrete asphalt layer and, consequently, the higher deflection-reducing power, brought about a reduction in the final deflection level of the structure, which went from 66×10^{-2} mm to 52×10^{-2} mm with the execution of the asphalt concrete layer.

5.1.2 *Results of the analysis of section 02*

Table 3 presents the results of the analyses executed for experimental section 02. It was verified that the calculated deflections presented an excellent agreement with the deflections measured on-site, remaining practically constant during the entire construction process. With the execution of the upper layers of pavement, occurs the reduction of the level of stresses acting upon the lower layers. Because the resilient modulus of the granular material is a direct function of the level of the confining stress applied, there also occurs a reduction in the modular value.

This effect is very clear if we observe the modular variation presented by the layer of crushed stone: with the load applied directly on the base course, there is an average modular value of 442 MPa for this layer. With the load applied on the surface course layer, there is an average modular value of 325 MPa for the crushed stone. Such variation occurs due to the reduction in the value of the stresses invariant θ that—with the execution of the asphalt surface course—was reduced from 314 kPa to 181 kPa. The same thing occurred with the detonated rock subgrade. This reduction in the moduli of the granular layers was partially compensated by the elevated rigidity presented by the asphalt surface course layer, which permitted the maintenance of the structure's deflection level. It was obtained an excellent correlation between the field values and the calculated ones.

It is worth mentioning that the K-θ model leads to some slightly odd consequences, i.e. the decrease of stiffness with depth within the subgrade. Which means that, for an almost infinite depth, the stiffness would be near zero, which is absurd. The model does not take into account numerous factors that might interfere with the material stiffness. It is also worth mentioning that, during the use of the K-θ model, only the stresses generated by external loading are taken into account. In the computerized simulations, the gravitational stresses

Table 4. Evolution of the average deflections over time.

Period 0,01 mm	Section 01 0,01 mm	Section 02
Traffic opening	49	58
Month 1	46	45
Month 6	44	39
Month 18	43	32
Month 27	36	32

were not considered. According to Medina (1997), the use of gravitational stresses with the K-θ model leads to unrealistic low deflections, therefore, they are not used in the computations. So, though the numerical simulations showed a good correlation between field and laboratory testing, it must be kept in mind that this is an approximate relationship, whose field of application is valid only under certain conditions, such as a limited depth of analysis and stress level compatible to the ones used in the laboratory testing.

6 EVOLUTION OF THE DEFLECTION LEVEL OVER TIME

Deflection measures were performed 1, 6, 18 and 27 months after traffic opening. Table 4 presents the average deflections obtained for each section in the periods analyzed. Section 01 was the one that presented the smallest total reduction in the deflection, 26% after 27 months. Section 02 was the one that presented the largest reduction in deflection, 45%. Considering that both sections were subjected to same traffic loading and that there is no field or laboratory testing for determining the compaction level of detonated rock, one can ponder that this kind of material is subjected to greater post-compaction effect than cohesive soils when used as subgrade material. It is interesting to see that, after two years, section 02 presented an average deflection inferior to the one observed in section 01. It also can be noticed in both sections a tendency towards stabilization of the deflection level after six months.

The deflection basins were also used to estimate the tensile strain acting on the bottom of the asphalt layer. In order to estimate them, it was used an expression that allies geometry and solid mechanic concepts as follows (Momm et al. 2003):

$$\varepsilon_t = E_R \cdot \arccos\left(1 - \frac{F}{R}\right) \tag{7}$$

where ε_t = tensile strain acting on the bottom of the asphalt layer; E_R = thickness of the asphalt layer, in meters; F = difference between maximum deflection D_0 and deflection D_{25} (deflection measured at a 25cm distance from the load application point), in meters; R = radius of curvature, calculated by the expression $(D_0 - D_{25})/3125$, in meters; and arccos = the inverse cosine of the term in parenthesis, in radians.

Table 5. Evolution of the average tensile strain estimated.

Period	Section 01 microstrains	Section 02 microstrains
Traffic opening	120	108
Month 1	52	56
Month 6	64	69
Month 18	91	68
Month 27	43	43

Table 5 shows the average tensile strains estimated for each period. It can be seen that, after 27 months, there was a reduction of approximately 60% in the estimated values. It also can be noticed that most of the reduction occurred during the first month and that there was no continuous reduction over time. This may be attributed to the fact that weather conditions and temperature were not taken into account during deflection measures, thus affecting the quality of the field surveys.

Analysis of table 5 also shows that the post-compaction effect of the pavement layers due to traffic loading can significantly affect the fatigue life of a newly-constructed pavement structure. For the sake of illustration, if we use the following equation to estimate the maximum admissible tensile strain on the asphalt layer (Momm 1998):

$$N = 0{,}0045 \cdot \left(\frac{1}{\varepsilon_t \cdot 10^{-3}}\right)^{8{,}3035} \tag{8}$$

for a N number of 3,44 millions cycles, it is obtained a maximum tensile strain of 85 microstrains. Table 5 shows that the tensile strains obtained for the first period are superior to the limit value. However, the tensile strains estimated for the other periods are significantly inferior to the adopted maximum value, thus indicating that the post-compaction effect during early years has a very positive impact on the performance of a pavement structure. It can mean the difference between acceptance or rejection of a recently built pavement structure.

7 CONCLUSIONS

Based on the results obtained from the research, one can conclude that:

- The cohesive soil subgrade presented an almost elastic behavior, being almost unaffected by stress level variations.

- The resilient behavior of unbound material (i.e. the detonated rock subgrade and the crushed stone base) is highly dependent of the stress level, being very susceptible to stress variations.
- In an analysis of the deflectometric behavior of a pavement structure, it is extremely important to consider the non-linear behavior of the materials used in the paving process. Numerical simulations permitted us to explain rather satisfactorily the deflectometric behavior observed in the field, indicating a high degree of coherence between the results of laboratory and computacional testing and the field structural behavior.
- The post-compaction effect due to traffic loading may significantly improve the performance of a newly-constructed highway, which can be observed by means of deflectometric surveys after traffic opening.

ACKNOWLEDGMENTS

The authors wish to thank the company PROSUL and the Santa Catarina State Department of Infrastructure (DEINFRA/SC) for providing support and the data necessary for the development of this work, as well as the National Council for Technological and Scientific Development (CNPq) for the resources provided.

REFERENCES

DNER. 1994. *Procedure DNER-ME 131/94 – Determination of the resilient modulus*. Rio de Janeiro, RJ.

Huang, Y. H. 2001. *Manual of the Kenlayer Program*.

Medina, J. 1997. *Pavement Mechanics*. Rio de Janeiro: COPPE/UFRJ

Momm, L. 1998. *Study of the effects of granulometry over the superficial macrotexture of asphalt concrete specimens and its mechanical behavior. PhD Thesis*. São Paulo: University of São Paulo.

Momm, L. Kryckyj, P. R. Santana, W. C. & Oliveira, A. 2001. Correlation between Deformation of Concrete Asphalt Layer and Curvature of Deflexion. In: W. Uddin, R. M. Fortes & J.V Merighi, J.V. (eds), *Second international symposium on maintenance and rehabilitation of pavements and technological control*, vol.1: 140–154. Auburn: Proceedings.

3 Slope instability, stabilisation, and asset management

Advances in Transportation Geotechnics – Ellis, Yu, McDowell, Dawson & Thom (eds)
© 2008 Taylor & Francis Group, London, ISBN 978-0-415-47590-7

Installation of optical fibre strain sensors on soil nails used for stabilising a steep highway cut slope

B.L. Amatya, K. Soga & P.J. Bennett
University of Cambridge, Cambridge, UK

T. Uchimura
University of Tokyo, Tokyo, Japan

P. Ball
Systems Geotechnique Ltd, St. Helens, UK

R. Lung
Highways Agency, UK

ABSTRACT: A distributed optical fibre monitoring system based on Brillouin Optical Time Domain Reflectometry has been used to monitor the strain profile of soil nails installed on a steep highway cut slope along the A2 near Dartford, UK. Soil nails of the required length were assembled in the laboratory and twelve-core ribbon optical fibre sensor cable was attached using adhesive at frequent intervals along the length. At the site, the instrumented nails were installed into drilled holes in the cut slope. The holes were then filled with a neat cement grout. Finally a rock-fall netting was placed over the slope surface and head plates were tightened onto the soil nails. This paper presents a detailed description of the instrumentation, installation and some preliminary results of strain profile along the nails. It is expected that the outcome of this research will contribute in upgrading the design technique/methodology of soil nails for cut slopes.

1 INTRODUCTION

Recently soil nailing has become a commonly used method of earth reinforcement for stabilisation of existing retaining walls, existing unstable slopes and the construction of new steep slopes. The need to widen motorways and make improvements to the national railways networks in the UK economically demands steeper side slopes. Stabilisation of such slopes by soil nails is gaining acceptance. Soil nailing is a sustainable solution as it aims to strengthen existing earth structures rather than to replace or reconstruct the soil mass. This method has technical and economical benefits over more traditional methods and now possesses a substantial market share in the UK. However experience of the application of soil nailing is relatively limited in the UK. A recent report (CIRIA, 2005) gives guidance about the design, installation and testing of soil nails in the context of the UK. However current codes of practice, particularly design, could be further improved if more was known about the actual performance of the elements in the ground. Thus the main objective of this research is to generate site based information on *in situ* strains of the steel elements and perhaps allow improved confidence and economical design in the future.

The performance of soil nails installed in a real construction site is being investigated using optical fibre sensors (OFS). These sensors provide distributed strain measurement, which is based on Brillouin Optical Time Domain Reflectometry (BOTDR). The concept behind such technology for civil engineering application is now available in various publications (Ohno et al. 2001, Shi et al. 2004 etc.). In this study, strain distribution was measured along the whole length of various nails installed in a stretch of cut slope using a BOTDR analyser (Yokogawa AQ8603). This unit allows the measurement of the strain distribution along a standard single-mode optical fibre using a reflective technique, requiring access to only one end of the cable (e.g. Klar et al., 2006). In this paper, the detailed description of instrumentation, installation procedures of OFS and some preliminary results of strain profile along the length of the soil nails are presented.

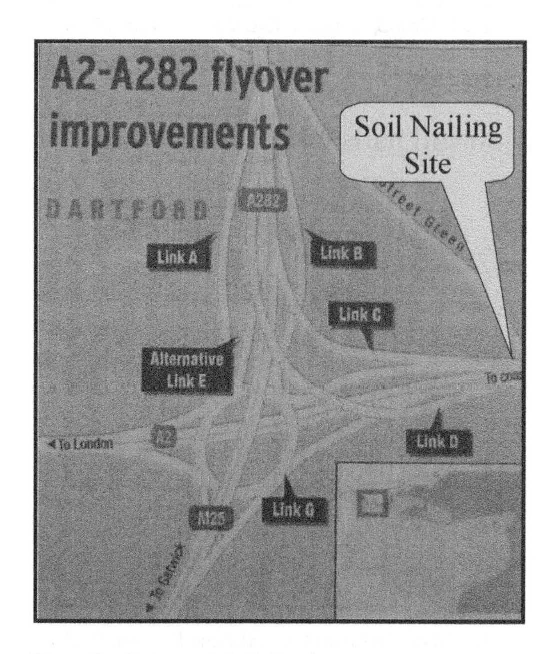

Figure 1. Instrumented site location.

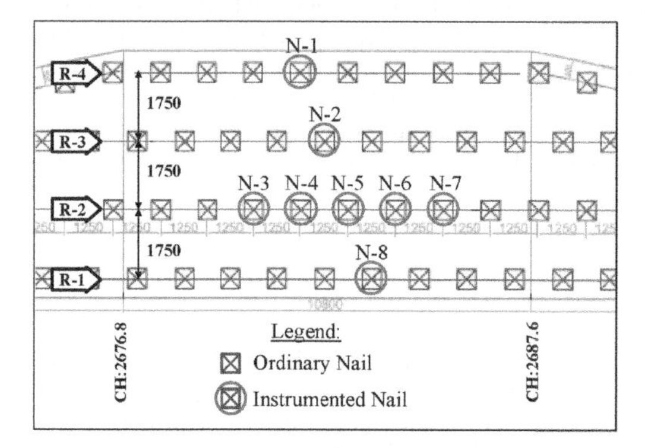

Figure 2. Arrangement of instrumented nails on the site.

2 INSTRUMENTED SITE LOCATION DETAILS

The instrumented site is part of the A2/A282 widening project near Dartford, London (Figure 1) and is located approximately at chainage 2680 along the eastbound carriageway of the A2 highway at gantry sign G19. At this location, the cut slope angle is 60° and the height is ~6 m. The back slope, above the cut slope, lies at about 26° angle and stands approximately 2m high. There are four rows of nails, at a vertical spacing of 1.75 m, and in each row nails are separated horizontally by 1.25 m, as shown in Figure 2.

Five nails in a row (R-2) and one nail in each of the other rows were instrumented. A typical cross section of the ground is shown in Figure 3.

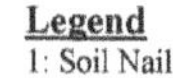

1: Soil Nail

2: Rockfall netting over Jute layer

3: Geocell filled with top soil

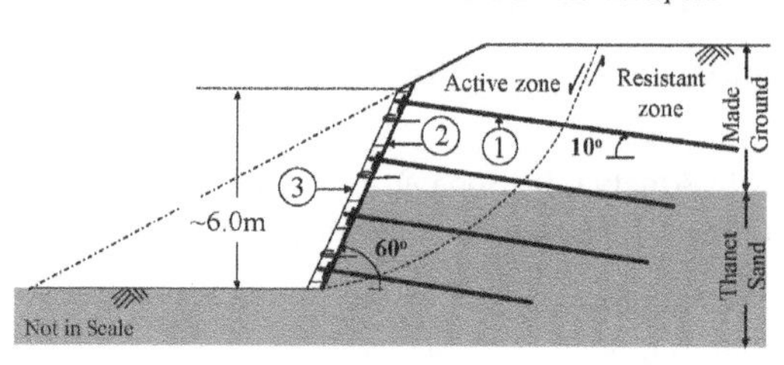

Figure 3. Typical cross section of ground at the site.

Table 1. Properties of steel hollow bars.

Bar Designation	Nominal Outer Diameter (mm)	Cross Sectional Area (mm^2)	Yield Strength (kN)	Ultimate Strength (kN)	Nominal Weight (kg/m)
R25N	25	300	150	200	2.6

3 MATERIALS

3.1 *Steel hollow bars*

Dywidag Systems International's R25N 'Mai' hollow bars were used for reinforcement. These are ribbed bars to improve nail-grout adhesion. The properties of such bars are presented in Table 1. Galvanised steel bar was used for uppermost part exposed to atmosphere or top soil.

3.2 *Optical fibre sensors*

Two types of OFS were used. A twelve-core ribbon fibre: SM12 (produced by Fujikura) and a loose-tube internal/external grade fibre 'Universal Unitube': Singlemode 9/125 OS1(produced by Excel) were used. The ribbon fibre was used for direct strain measurement. The response of this OFS is affected by temperature of the surrounding ground/bar. The Universal Unitube cable was therefore used for temperature compensation as the optical fibres are contained in a gel-filled tube and respond only to temperature changes, not to applied strain.

3.3 *Adhesive*

Araldite 2021 adhesive was used to attach the ribbon OFS to the steel nails. This product is suitable for metal and plastic surfaces. In addition to this a silicone sealant was also used to provide a protection layer over the fragile fibre in some locations; this will be discussed in the next section.

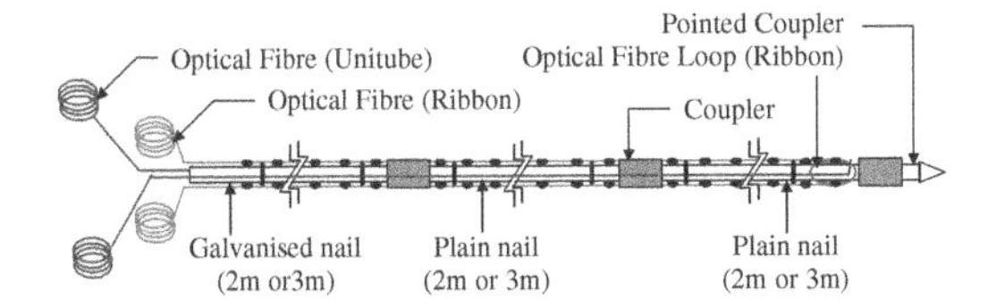

Figure 4. Instrumented nail details.

Figure 5. A loop of OFS at toe end of nail.

4 LABORATORY PREPARATION WORKS

Steel nails of the required length (e.g. 9 m, 8 m and 6 m) were assembled in the laboratory at Systems Geotechnique Ltd, St. Helens, UK. The steel rebar is available in 2 m and 3 m lengths. Couplers were used to connect individual lengths to make the required total lengths. The couplers were modified with the addition of flanges which were used to centralise the nail in the borehole during the installation phase on site. The flanges served to provide space for sensor cables to pass over the coupling without being in contact with the soil during field installation. The surface of each rebar was cleaned using an alcohol based thinner to ensure that they were free of dirt and grease thus ensuring a good bond with the adhesive.

A typical sketch of an instrumented nail is shown in Figure 4. Ribbon OFS cable was stretched along the length of the nail from top to toe and back again with an extra length of cable looped at both top ends for connection to the BOTDR analyser or for future inter-connection of ribbon between nails. About 2 to 3 m of ribbon was arranged in a small loop at toe end of nail (see Figure 5). The fibre was pretensioned to about 0.1% strain before attaching it on the nail surface using the adhesive (Araldite 2021). Small nodes of adhesive were placed on the bar at about 15cm intervals as shown in Figure 6. Silicone sealant was applied liberally over the loop to protect the fibre from possible damage during the field installation phase (Figure 5). To provide temperature compensation, as described above, Universal Unitube OFS cable was installed on selected nails as shown in 4. Finally, readings were taken for all nails to ensure the assembly was working. Each nail was wrapped in multiple layers of bubble plastic sheet for safe handling during storage and transportation to the construction site (Figure 7).

Figure 6. Ribbon fibre attachment in nail.

Figure 7. Nails with bubble wrap.

5 FIELD INSTALLATION

Field installation work was carried out in May 2007. Eight instrumented nails were installed in the slope. Table 2 shows details of the nails, their location, instrumentation and present conditions (Figure 2).

A typical cross section of a borehole with an instrumented nail is shown in Figure 8. To ensure the instrumented nails were not damaged, they were installed after the other ordinary soil nails in the vicinity were placed. Boreholes of the required length were drilled at an inclination of 10° below horizontal using temporary drilling tools as shown in Figure 9. The holes were

Table 2. Details of nail, their location and their present status.

Nail	Row	Length (m)	OFS (Unitube)	OFS (Ribbon)	Continuity	Present Status
N-1	4th	9	Yes	Yes	OK	Reading
N-2	3rd	8	No	Yes	OK	Reading
N-3	2nd	8	No	Yes	Broken	Reading
N-4	2nd	8	No	Yes	Broken	Unreadable
N-5	2nd	8	No	Yes	OK	Reading
N-6	2nd	8	Yes	Yes	OK	Reading
N-7	2nd	8	Yes	Yes	Broken	Reading
N-8	1st	6	No	Yes	OK	Reading

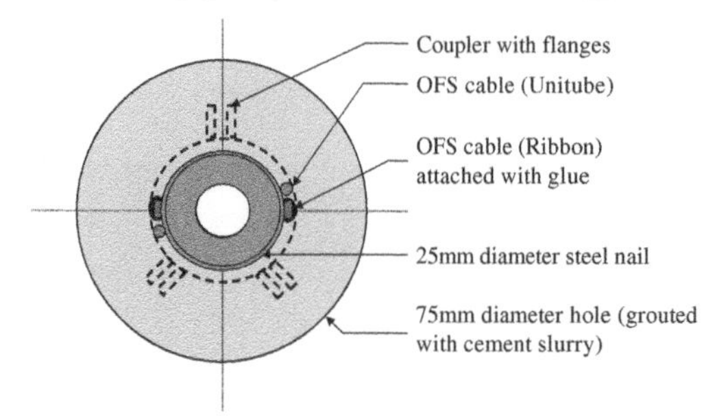

Figure 8. Typical cross section of borehole with instrumented nail.

Figure 9. Drilling borehole.

water flushed to avoid grout setting before installation of the instrumented nails could be completed. It was not possible to drill the instrumented nails directly into the soil as the exposed OFS cables would be destroyed. The instrumented nails were inserted carefully into the drilled holes in one piece using two mobile access platforms (MEWP) (Figure 10). The holes were subsequently backfilled using neat cement grout. Before positioning the nail head plate and then retaining fixed nuts, the cables were passed through holes in the plates. Figure 11 shows a loose plate mounted over a nail before the fixing nut was positioned.

Figure 10. Instrumented nail installation.

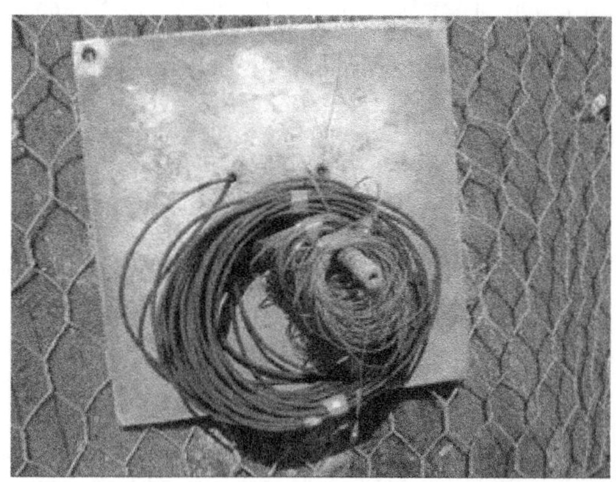

Figure 11. Loose head plate mounted in nail and OFS cables for access to analyser or interconnection.

After tightening the head plates onto the soil nails, the cables from each nail were arranged across the slope and finally passed to a junction box for access and connection to the BOTDR analyser. Cables were protected using sections of PVC pipes and expanding foam (see Figure 12). The nail heads were buried below geocell mattress filled with top soil so that observers are unaware of their existence and for aesthetic

Figure 12. Cable protection measures on the slope.

Figure 13. The nailed slope after covering by geocell and top soil.

purposes. The view of the site during September 2007 is shown in Figure 13 – note the gantry base where the junction box is located.

6 SOME PROBLEMS AND LIMITATIONS

As the assembly of the nails with OFS is time consuming as well as a delicate job, it was preferred to do the fabrication in laboratory conditions rather than at a construction site. This required a large space for the assembly area and a long vehicle for transportation to the site. Despite the utmost care, some damage to the cables occurred during handling and transportation (see Figure 14). In this case strain measurements are limited to the sections of nail that can be access from each of the top connections; sections of nail between breaks in the cable cannot be measured. Although two extra nails were fabricated during laboratory preparation work, only five nails out of eight were safely and securely installed in the slope. Some damage to the sensor during field installation can not be ruled out, but it was observed that most of the damage found resulted

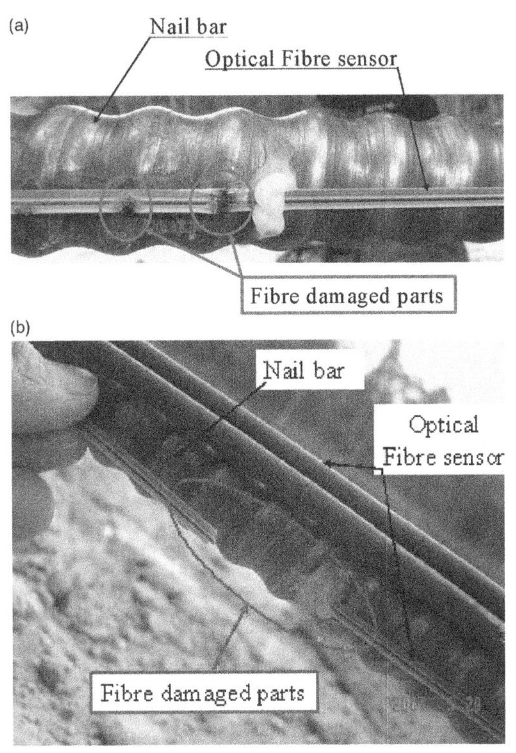

Figure 14. Damage to OFS cable during handling and transportation. (a) Damage due to friction (b) Damage due to cuts.

from handling and transportation. Securing the load sufficiently to make it safe to travel is potentially detrimental to the exposed cables.

The use of the twelve-core ribbon fibre means that eleven cores are redundant. As a result, in spite of damage to some cores, others continued to function satisfactorily enough for measurement. Table 2 shows which nails experienced problems and which nails are still reading in the ground. Thus it is recommended that more reliable techniques are devised for safe handling, storing and transporting of the nails; simply bubble wrapping the nails was not enough. Rigid steel troughs were used to handle the complete nails but these added to the manual handling of each nail.

7 PRELIMINARY READINGS

The readings from only one nail (N-6) are presented in this paper as an example of the preliminary results from the initial phase of monitoring. Figure 15 shows strain profiles measured from the ribbon OFS cable installed on the nail (N-6) at three different times. The two prestrained sections of optical fibre can be seen for the two sections from top to toe and back. The first set of reading was taken during June 2007 more than

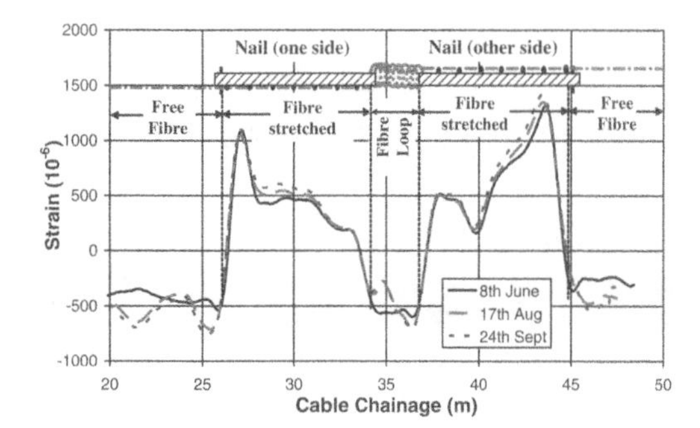

Figure 15. Raw strain profile along OFS cable installed in N-6 at three different times.

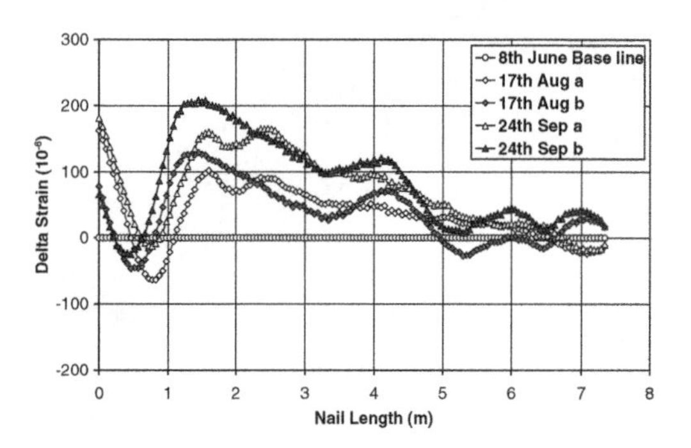

Figure 16. Profile of change in strain along length in N-7 at different times.

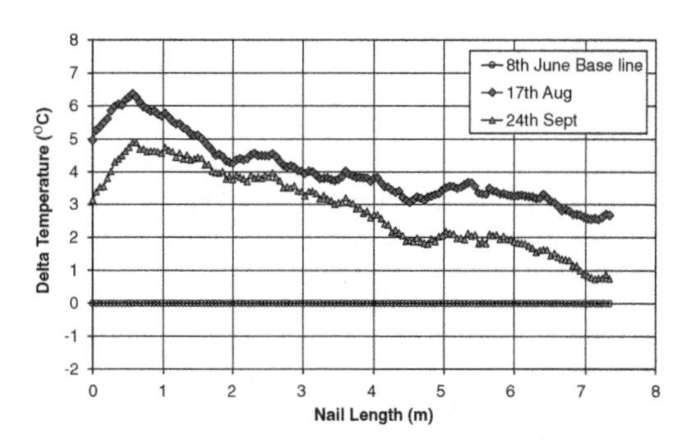

Figure 17. Profile of change in temperature along Nail-6 at different times.

one week after the soil nails were installed. This reading is a baseline and subsequent changes of strain can be calculated by subtracting this from further measurements. Two sets of further measurements during August and September show that strain in the nail was increasing gradually indicating the process of slope relaxation. The change in strain along the nail is shown in Figure 16. Since the response of OFS is influenced by induced strain both due to mechanical loading and thermal change in the vicinity, the above results are due to the combined effects.

To have a proper understanding of the effect due to seasonal change in ambient ground temperature in the results shown above, the response of a separate OFS cable (Universal Unitube) for temperature measurement is shown in Figure 17. Considering the reading of June as a baseline, the temperature of the ground along the nail increased by 1°C to 6°C by the time of four months of observation and the temperature along the nail during August was higher than that on September. However despite the fluctuation in the ambient ground temperature in the mentioned time period, the total strain in the nail is increasing gradually. This indicates that the nail is in tension and being active for stabilising the slope. These results indicate that the impact of thermal strain in the overall strain measurement should not be ignored and further work is currently conducted to separate these two effects from the total measured value.

8 CONCLUDING REMARKS

It has been demonstrated that it is possible to install optical fibre sensors on soil nails and take measurements of the strain profile. It is anticipated that readings will be possible for a considerable period (e.g. 10 years or more) providing *in situ* strain measurements to assess its long term performance. The difficulties of transportation and installation have been highlighted by the loss of some strands. The data are currently gathered and interpreted, which will be reported in near future. It is expected that the ultimate outcome of this research will contribute in improving design technique/methodology of soil nails for cut slopes.

ACKNOWLEDGEMENT

Taro Uchimura thanks Kajima Foundation for providing financial support of his stay in Cambridge as an academic visiting researcher.

REFERENCES

CIRIA Report C637: *Soil Nailing-best practice guidance*, Phear, A., Dew, C. Ozsoy, B., Wharmby, N. J., Judge, J., Barley, A. D.(2005).

Klar, A., Bennett, P.J., Soga, K., Mair, R.J., Tester, P., Fernie, R., St John, H. & Torp-Peterson, G. (2006) "Distributed strain measurement for pile foundations." *Proceeding of the Institution of Civil Engineers-Geotechnical Engineering*, 159, pp.135–144.

Ohno, H., narasu, H., Kihara, M., & shimada, A. (2001) "Industrial application of the BOTDR Optical Fibre Strain Sensor", *Optical Fibre Technology*, Vol.7, No.1, pp. 45–64.

Shi, B., Zhang, D., Ding, Y., Xu, H.Z. & Cui, H.L. (2004) "A new distributed optical fibre sensor for structural health monitoring-BOTDR", *Proceeding of the third international conference on earthquake engineering*, Ch.5, paper no.8.

Advances in Transportation Geotechnics – Ellis, Yu, McDowell, Dawson & Thom (eds)
© 2008 Taylor & Francis Group, London, ISBN 978-0-415-47590-7

Behaviour of the reinforced cuts in flysch rock mass

Ž. Arbanas
Civil Engineering Institute of Croatia, Rijeka Regional Unit, Faculty of Civil Engineering, University of Rijeka, Croatia

M. Grošić
Civil Engineering Institute of Croatia, Rijeka Regional Unit, Croatia

S. Dugonjić
Faculty of Civil Engineering, University of Rijeka, Croatia

ABSTRACT: The paper presents some experiences during the reinforced cut construction in the flysch deposits on the Adriatic motorway near city of Rijeka, Croatia. The major part of the motorway was constructed by cutting in flysch rock mass. Stabilities of the cuts were ensured by reinforcing of rock mass with rock bolts and appropriate supporting systems. The self boring rock bolts in combination with multilayered sprayed concrete or reinforced mat construction were used. Shallow bored drains were installed for dissipation of ground water collected at the contact of permeable cover and impermeable flysch rock mass. Interactive rock mass cutting design, based on the observational methods, was introduced during the construction. The active design approach has allowed the designer, based on the rock mass conditions and monitoring results to change the support system at some unfavourable locations. The measured values and back analysis has enabled the establishment of real rock mass strength parameters and deformability modulus.

1 INTRODUCTION

The geotechnically most dependent part of the Adriatic motorway was constructed in the Draga valley near Rijeka in past few years. The geological fabric of the Draga valley is very complex. The Cretaceous and the Paleogene limestones are situated on the top of the slope, while the Paleogene flysch crops out at the lower slope, and in the bottom of Draga valley where the motorway is located. Unlike limestone rocks at the top of the slope, flysch rock mass is almost completely covered by weathering zone material and talus breccia.

The major part of the motorway was constructed by cutting in flysch rock mass. Stabilities of the cuts were established by reinforcing of rock mass with rock bolts and appropriate supporting systems. As appropriate solutions, the self boring rock bolts in combination with multilayered sprayed concrete as a first stage and reinforced concrete net construction were used. The shallow bored drains were installed for dissipation of ground water collected at contact of permeable cover and impermeable flysch rock mass. Due to low value of flysch rock mass strength, extra attention was focused at the interaction between rock mass and rock bolts.

Interactive design of the rock mass cutting, based on the observational methods, was introduced during the construction. The appropriate measured equipment was installed before and during the construction – vertical inclinometers, horizontal deformeters, piezometers and geodetic marks. This active design approach has allowed the designer, based on the rock mass conditions and monitoring results, changing the support system at some unfavourable locations. The measured values and the back analysis enabled establishing real rock mass strength parameters and deformability modulus.

2 GEOTECHNICAL PROPERTIES OF FLYSCH ROCK MASS

During the period from 2004 to 2006, segments of the Adriatic motorway through the Draga valley near Rijeka were made. This segment of the Adriatic motorway with a length of 6.8 km showed to be very demanding in geotechnical terms, because of the significant number of structures (3 junctions, 2 tunnels and several viaducts) and therefore was expensive. The steep slopes of Draga valley are made of limestone rock mass. At the bottom of the valley, there are deposits of paleogene flysch mainly made of siltstones with rare layers of sand, marl, and breccia. Flysch rock mass is covered with slope formations, which tend to slide and denude (Arbanas et al. 1994). Usual geotechnical profile consists of three layers: clay cover made after disintegration of flysch rock mass (residual soil) or brought by gravitation from higher parts of the slope, layer of weathered flysch deposits with variable weathered characteristics, which decrease with depth and the fresh flysch zone.

The rock mass is mainly made of siltstones which exhibit visual transfer from the completely weathered zone with yellow color through highly weathered, moderately weathered and slightly weathered deposits all the way to completely fresh rock mass colored gray and blue. With completely weathered siltstones, the rock mass is completely disintegrated, but the original structure of the rock mass has stayed intact (ISRM 1978). The layer of fresh siltstone rock has no visible weathering marks, except for color change on the main discontinuity surfaces. During decomposition of singular weathering zones of the flysch rock mass, along with visual check of the material from test drills, significant contribution came from results from geophysical measurements with surface seismic refraction methods and down-hole seismic method (Arbanas et al. 2007a).

Determination of geotechnical properties of the flysch rock mass, during geotechnical examination works, was prevented because of the flysch rock mass behavior. During boring, it is difficult to get undisturbed samples, because the rock mass damages in high to moderate weathered siltstones and sudden degradation and disintegration of slightly weathered to fresh siltstones after geostatic loads are disturbed and exposure to air and water during examination boring. The consequence of these processes in siltstones is a very small number of usable test results of geotechnical characteristics. The main number of tests is made with the Point Load Test (PLT) method, where samples, obtained by boring, can be used without further processing and almost immediately after sampling (ISRM 1985). Disadvantage of Point Load Test is mainly the large dispersion of the results, which especially occurs with weak rock masses like flysch. The results dispersion is mainly influenced by rock mass layers, layers orientation during sampling, sample size, as well as weathering of the flysch rock mass. But, regardless of given disadvantages, used Point Load Test method is recommended in case of lack of more reliable testing, lack of appropriate representative samples, and with detailed description of tested samples of flysch rock mass. Test results with Point Load Test method on tested samples of fresh siltstones showed that uniaxial strength of these materials is from 10 to 15 MPa, and in extreme cases to 20 MPa.

Because there are no more reliable laboratory results of strength parameters and deformability, for determination of the behavior the GSI classification of rock mass (Marinos & Hoek 2000), and findings of flysch mass behavior (Marinos & Hoek 2001, Marinos et al. 2005) are used. The classification of fresh siltstone rock mass is placed in group E (Weak siltstone or clayey shale with sandstone layers) to H (Tectonically deformed silty or clayey shale forming a chaotic structure with pockets of clays. Thin layers of sandstone are transformed into small rock pieces.), with GSI values from 30 to 10 (Marinos et al. 2005). A significant parameters decrease is found with increase of weathering of flysch rock mass. This effect points out the need of further elaboration of GSI classification for various weathering categories of rock mass. For determination of strength laws the Hoek-Brown failure criterion for rock mass is used (Hoek et al. 2002) with uniaxial strength values of siltstone rock mass of $\sigma_c = 10$ MPa and disturbance factor $D = 0.7$, which corresponds to machine excavation.

Deformation characteristics of siltstone are even harder to determine than strength parameters. By using suggested relations and values (Marinos & Hoek 2000, Marinos & Hoek 2001, Hoek et al. 2002, Marinos et al. 2005), deformability parameters of flysch rock mass, which come from back stress-strain analysis of *in situ* measurement results show relatively low values of elasticity modulus, which ranges from $E = 80$ to 200 MPa (Arbanas et al. 2007a).

3 REINFORCING AND SUPPORT SYSTEMS APPLIED ON CUTS IN FLYSCH ROCK MASS

During the construction of the motorway, on the major part of the road, the cutting in flysch rock mass is designed and executed, and the cut stability is ensured with rock bolts and appropriate supporting system. According to executed geotechnical investigation works, use of rock mass reinforcement system was specified in slightly weathered to fresh rock mass, while in weaker parts of rock mass the change of geometry is specified, with cut's construction with appropriate stable slopes. Flysch rock mass reinforcing works showed for the need of additional analysis of interaction between bolts and flysch rock mass.

Because of steep slopes it wasn't possible to select a stable geometry on most cuts in the flysch rock mass without additional reinforcement or support system. The support system was designed in two phases. The first phase was predicted rockbolt reinforcement system with multilayer sprayed concrete to enable stable excavation of the cuts with relatively low factors of safety. In the second phase, stiff concrete retaining construction was applied to fix the relatively soft reinforcement system.

The primary reinforcement systems were performed by excavation in the working stages, in

Figure 1. The photo of the cut with applied primary support system.

Figure 2. The photo of the cut with applied primary support system, drainage and strip footing.

longitudinal stories of 3.0 m height and a successive construction of a three-layered sprayed concrete support system reinforced by a self-boring rockbolts from top to bottom of the excavation. The stability of the cut in the flysch rock mass without applied support system is time dependent, so the working stages are very short. Simultaneously with reinforcement, the shallow drainage boreholes were drilled to allow dissipation and lowering of ground water in the cuts, Figure 1.

The secondary support systems included construction of a stiff strip footing, concrete pillars and concrete beam on the toe of the cut, Figure 2. That stiff reinforced concrete construction is connected with the rockbolts from the primary support system to enable coupled action. The fronts of the constructions were closed with prefabricated concrete elements that connected the strip footing with head concrete beam, Figure 3. The secondary supporting construction also represents an element of final esthetic forming of the cut.

Figure 3. The photo of the cut with applied secondary support system.

During the cut construction, a measuring, observing, and monitoring system was established. The monitoring system included deformation measuring in horizontal deformeters and vertical inclinometers, and observations of geodetic marks, like load measuring in rockbolts. This enabled obtaining the necessary data for the stress-strain back analysis of the real behaviour of excavated and reinforced rock mass. An active design procedure was established which made possible the changes required in the rock mass reinforcement system in cuts. The measurements were performed after any construction stage so as temporary in the long period after applying of the complete support system.

4 NUMERICAL ANALYSIS OF REINFORCED CUTS

As to approve the stability of the reinforced flysch rock mass limit state analyses and stress strain analyses of the reinforced slopes were carried out. Combination of these methods enables understanding of reinforcement system behaviour (Arbanas 2002; 2004; Arbanas et al. 2006a; 2006b; Arbanas et al. 2007a; 2007b). Overview of the stress strain analyses is given for higher cuts through excavation stages and reinforcement of flysch rock mass. Each stage is simulated as removing of existing material and installing appropriate rock mass reinforcement system. From engineering-geological cross-sections of the rock mass and soil classifications was made and numerical model was established to predict behaviour of the slope cut during executing of work.

Geotechnical cross-section is made of cover and bedrock. The cover is made of colluvial clayey deposits thickness from 0 to 4.0 m, weathered zone of flysch rock mass thickness from 2.0 to 4.5 m, and the bedrock is made of fresh grey-blue coloured flysch rock mass. The layer of fresh siltstone bedrock has no visible weathering, except colour change on the main discontinuity surfaces. During decomposition of singular weathering zones of the flysch rock mass, along with visual check of the material from test drills, significant contribution came from results from geophysical measurements with surface seismic refraction and seismic down-hole methods. Soil and rock parameters used in stress strain analysis are given in Table 1.

Geotechnical profile of soil is generated using quadrilateral and triangular elements with four and three nodes on edges. Behaviour of the rock mass and soil is reproduced with elastic-plastic model. In finite element model rockbolts are replaced by spring elements acting at cut face. Spring element stiffness is obtained from pull out test on installed trial rockbolts. Finite element model used for calculation is shown in Figure 4.

Stress strain analyses were carried out with Sigma/W, Geo-Slope (Geo-Slope, 1998) whose predict behaviour of the reinforced rock mass during executing of works. The results of measured horizontal displacement on the control-measuring cross-section are shown in Figure 5. The displacement of cut face must correspond with activating forces in the rockbolts. To calculate bearing capacity and behaviour of rockbolts during excavation stages and reinforced system construction displacements of the cut face are observed Figure 6.

Table 1. Geotechnical properties of the materials in the model.

Layer	Material	Model	Young modulus (MPa)	Poisson ratio
1.	Slope formations	Elastic-plastic	3.50	0.31
2.	Weathered flysch	Elastic-plastic	8.50	0.31
3.	Fresh flysch	Elastic-plastic	85.0	0.33

The maximum values are located in cover layer of slope (colluvial slope deposits and weathered flysch rock mass) until displacements in bedrock are significantly lower.

An interactive rock mass excavation cut design, based on the observation method (Terzaghi & Peck 1967; Peck 1969; Nicholson et al. 1999, Szavits-Nossan 2006; Arbanas et al. 2006a), was introduced in the phase of the construction. Rock mass cut design methodologies were shown by Hoek & Bray (1977; see also Wyllie & Mah 2004) and have been amended to include the selecting of support structures (Arbanas 2002; 2004; Arbanas et al. 2006a) and the appropriate rockbolts (Stillborg 1994). Measured displacements are compared to displacements from numerical model. This allow the control of actualized force in rockbolt and possibility of extra reinforced of the rock mass (Arbanas et al. 2006a; 2006b; Arbanas et al. 2007b).

The results of measured horizontal displacement on the control-measuring cross-section are shown in Figure 7. Measured strain in the vertical inclinometers-extensometers (deformeters) and horizontal extensometers (deformeters) showed a good match with the predicted calculated strain for all excavation stages. Changes in reinforcement systems were rare and mostly caused with differences between predicted and actual geological conditions.

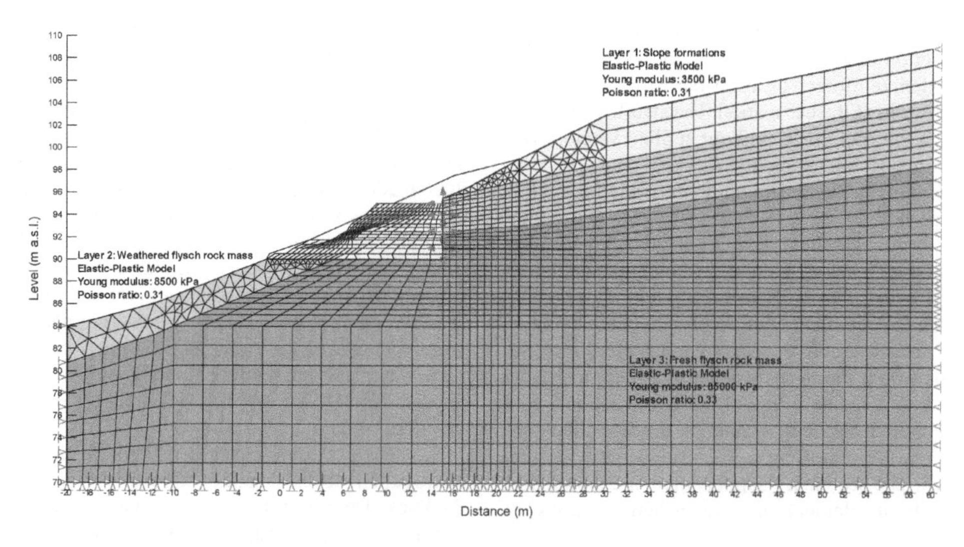

Figure 4. Finite elements model for stress strain analyses.

The reinforced of the flysch rock mass with designed systems were completed successfully according predicted design solutions. In part of the slope cut local instability were occurred mostly caused with unfavourable geological conditions or come late installation of the rock mass reinforcement system.

5 CONCLUSIONS

On the geotechnically most dependent part of the Adriatic motorway was constructed in the Draga valley near Rijeka in past few years. The geological fabric of the Draga valley is very complex. The Cretaceous and the Paleogene limestones are situated on the top of the slope, while the Paleogene flysch crops out at the lower slope, and in the bottom of Draga valley where the motorway is located. Unlike limestone rocks at the top of the slope, flysch rock mass is almost completely covered by weathering zone material and talus breccia. Determination of geotechnical properties of the flysch rock mass, during geotechnical examination works, was disabled because of the flysch rock mass behaviour. During boring, it is difficult to get undisturbed samples, because rock mass damages in high to moderate weathered siltstones and sudden degradation and disintegration of slightly weathered to fresh siltstones after geostatic loads are disturbed and exposure to air and water during examination boring. The main number of tests is made with Point Load Test method, where samples can be used without further processing and almost immediately after sampling. Test results with Point Load Test method on

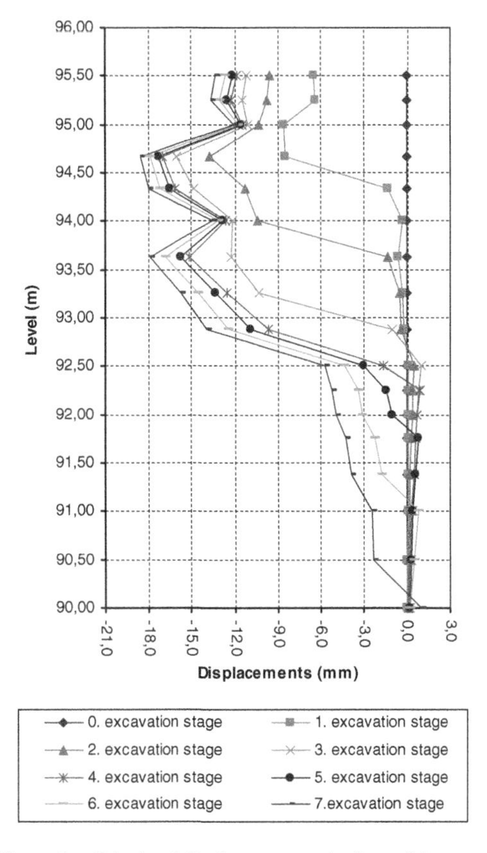

Figure 6. Calculated displacement on the face of the cut.

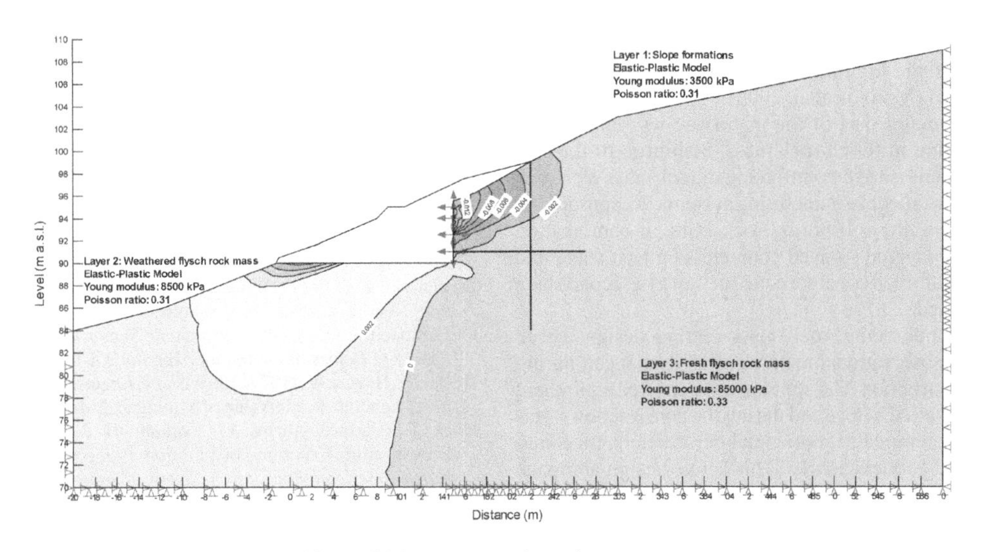

Figure 5. Horizontal displacements of the model from stress strain analyses.

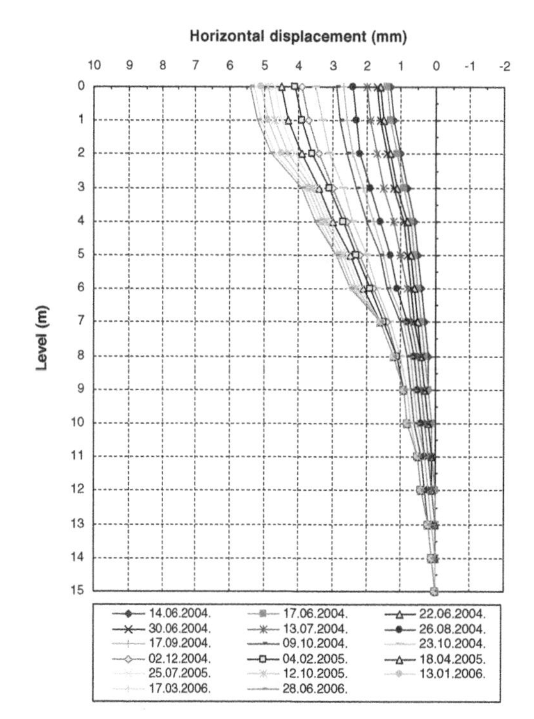

Figure 7. Measured displacement on the vertical inclinometer.

tested samples of fresh siltstones showed that uniaxial strength of these materials is from 10 to 15 MPa, and in extreme cases to 20 MPa. Because there are no more reliable laboratory results of strength parameters and deformability, for determination of the behaviour the Geological Strength Index classification of rock mass. The classification of fresh siltstone rock mass is placed in group E (Weak siltstone or clayey shale with sandstone layers) to H (Tectonically deformed silty or clayey shale forming a chaotic structure with pockets of clays) (Marinos et al. 2005).

The major part of the motorway was constructed by cutting in flysch rock mass. Stabilities of the cuts were established by reinforcing of rock mass with rock bolts and adequate supporting systems. As appropriate solutions, the self-boring rock bolts in combination of multilayered sprayed concrete as a first stage and stiff reinforced concrete construction as a second stage were used.

The interactive rock mass cutting design, based on the observational methods, was introduced during the construction. The appropriate measured equipment was installed before and during the construction – vertical inclinometers, horizontal deformeters, piezometers and geodetic marks. This active design approach has allowed the designer, based on the rock mass conditions and monitoring results, changing the support system at some unfavourable locations. The measured values and the back analysis enabled establishing real rock mass strength parameters and deformability modules. Executed measurements are indicated on the long-term deformation in the flysch rock mass caused by cut unloading typical for weak rocks.

The ensuring of slope stability in flysch rock masses by designed reinforcement and support system were generally successful. Only on the few cuts the local instability were observed caused by unfavourable then design predicted geological fabric or late installation of the rock mass reinforcement system works.

REFERENCES

Arbanas, Ž., Benac, Č., Andrić, M., Jardas, B. 1994. Geotechnical Properties of Flysch on The Adriatic Motorway from Orehovica to St. Kuzam. *Geotechnical Engineering in Transportation Projects, Proc. symp., Novigrad, 5–8 October 1994.*: 181–190. Zagreb (in Croatian).

Arbanas, Ž. 2002. *The influence of rockbolts on the rock mass behavior during excavation of deep cuts, MS Thesis*. Zagreb: Faculty of Civil Engineering, University of Zagreb (in Croatian).

Arbanas, Ž. 2003. Construction of open pit Zagrad in Rijeka, *Građevinar, Vol. 55, No. 10*: 591–597 (in Croatian).

Arbanas, Ž. 2004. *Prediction of supported rock mass behavior by analyzing results of monitoring of constructed structures, Ph.D. Thesis*. Zagreb: Faculty of Civil Engineering, University of Zagreb (in Croatian).

Arbanas, Ž., Kovačević, M.-S. and Szavits-Nossan, V. 2006 Interactive design for deep excavations, *Active Geotechnical Design in Infrastructure Development, Proceeding of XIII, Ljubljana, May 29–31, Vol. 2,*: 411–416.

Arbanas, Ž., Grošić, M., Jurić-Kaćunić, D. 2006. Influence of grouting and grout mass properties on reinforced rock mass behaviour. *Soil and Rock Improvement, Proc. 4th Conference of the Croatian Geotech. Society, Opatija, 5–7 October 2006.*: 55–64. Zagreb (in Croatian).

Arbanas, Ž., Grošić, M., Jurić-Kaćunić, D. 2007a. Experiences on flysch rock mass reinforcing in engineered slopes. *The Second Half Century of Rock Mechanics, Proc.11th Congress of the Int. Society for Rock Mechanics, Lisbon, Portugal, 9–13 July 2007.*: 597–600. London: Taylor and Francis Group.

Arbanas, Ž., Grošić, M., Kovačević, M.-S. 2007b. Rock mass reinforcement systems in open pit excavtions in urban areas. *Slope stability 2007, Proc.2007 Int. Symp. on Rock Slope Stability in Open Mining and Civil Engineering, Perth, Australia, 12–14 September 2007.*: 171–185. Perth: Australian Centre for Geomechanics.

Bieniawski, Z.T. 1989. *Engineering Rock Mass Classification*. New York: John Wiley & Sons.

GEO-Slope Int. Ltd. 1998. *User's Guide Sigma/W for Finite Element / Deformation Analysis, Version 4*. Calgary.

Hoek, E., Bray, J.W. (1977. *Rock Slope Engineering, 2nd. Edn.*, London: The Institute of Mining and Metallurgy.

Hoek, E., Carranza-Torres, C., Corkum, B. 2002. Hoek-Brown Failure Criterion-2002 Edition. *Proceedings of 5th North American Rock Mech. Symp., Toronto, Canada, Dept. Civ. Engineering, University of Toronto*: 267–273.

Hoek, E., Marinos, P., Benissi, M. 1998. Applicability of the Geological Strength Index (GSI) Classification for

Very Weak and Sheared Rock Masses. The Case of the Athens Shist Formation. *Bull. Eng. Geol. Env., No. 57*: 151–160.

ISRM, Commission on Standardization of Laboratory and Field Test 1978. ISRM Suggested Methods for the Quantitative Description of Discontinuities in Rock Masses. *Int. Jour. Rock Mech. Min. Sci. & Geomech. Abstr., Vol. 15, No. 6*: 319–368.

ISRM, Commission on Standardization of Laboratory and Field Test 1985. ISRM Suggested Methods for Determining Point Load Strength. *Int. Jour. Rock Mech. Min. Sci. & Geomech. Abstr., Vol. 22, No. 2*: 51–60.

Kilic, A.; Yasar, E.; Celik, A.G. 2002. Effect of Grout Properties on the Pull-out Load Capacity of Fully Grouted Rock Bolt. *Tunnelling and Underground Space Technology, No. 17*: 355–362.

Marinos, P., Hoek, E. 2000. GSI-A Geologically Friendly Tool for Rock Mass Strength Estimation. *Proc. GeoEng 2000 Conference, Melbourne.*

Marinos, P., Hoek, E. 2001. Estimating the Geotechnical Properties of Heterogeneous Rock Masses such as Flysch. *Bull. Eng. Geol. Env., 60*: 85–92.

Marinos, P., Marinos, P., Hoek, E. 2005. The Geological Strength Index: Applications and Limitations. *Bull. Eng. Geol. Env., 64*: 55–65.

Nicholson, D.P., Tse, C.M., Penny, C. 1999. *The Observational Method in Ground Engineering: Principles and Applications, Report 185*, London: CIRIA.

Peck, R. B. 1969. Advantages and limitations of the observational method in applied soil mechanics. *Géotechnique*, 19(2): 171–187.

Stillborg, B. 1994. *Professional Users Handbook for Rock Bolting*. Trans Tech Publications, Series on Rock and Soil Mechanics, Vol. 18, 2nd Edn. Clausthal-Zellerfeld.

Szavits-Nossan, A. (2006) Observations on the observational Methods, *Active Geotechnical Design in Infrastructure Development, Proceeding of XIII, Ljubljana, May 29–31, Vol. 2,*:.171–178.

Terzaghi, K. and Peck, R. B. 1967. *Soil Mechanics in Engineering Practice*. New York: John Wiley.

Windsor, C.R. 1996. Rock Reinforcement Systems, 1996 Schlumberger Award – Special Lecture. *Proceeding of EUROCK '96, Torino, Italy,* http://www.roctec.com.au/papers.html.

Windsor, C.R., Thompson, A.G. 1996. Terminology in Rock Reinforced Practice. *Proc. 2nd North American Rock Mechanics Conference NARMS'96 – Tools and Techniques, Montreal:* 225–232. Rotterdam: Balkema.

Windsor, C.R., Thompson, A.G. 1997. Reinforced Systems Characteristics, *Proceeding of Int. Symp. on Rock Support, Lillehammer*, http: //ww.roctec.com.au/papers.html.

Wyllie, D.C. and Mah, C.W. 2004. *Rock Slope Engineering,* Civil and Mining, 4th. Edn., Spon Press, New York: Taylor & Francis.

Advances in Transportation Geotechnics – Ellis, Yu, McDowell, Dawson & Thom (eds)
© 2008 Taylor & Francis Group, London, ISBN 978-0-415-47590-7

Numerical modelling of lateral pile-soil interaction for a row of piles in a frictional soil

I.K. Durrani, E.A. Ellis & D. J. Reddish
Nottingham Centre for Geomechanics, School of Civil Engineering, University of Nottingham, UK

ABSTRACT: Lateral pile-soil interaction arising from relative pile-soil movement can be modeled using a horizontal 2-d 'plan' section of the pile translating through the soil (or vice-versa). The ultimate lateral pressure on such a pile is well understood for an 'undrained' Tresca soil. However, behavior is not so firmly established for a purely frictional ('cohesionless') soil strength, or in any 'drained' situation. A series of *FLAC* analyses has been undertaken to determine if and how this 2-d situation can be modeled. The results demonstrate the effect of restraint on the section in the 'out of plane' direction, ultimately leading to the use of a 3-d analysis. The effect of pile spacing along a row (normal to the pile-soil movement), and in particular the phenomenon of 'arching' between adjacent piles, are also considered.

KEY NOTATION

d_p = pile diameter
p_p = 'equivalent pressure' on pile, defined as the nett load per unit length divided by the diameter (see Equation 1).
$p_w = p_p(d_p/s_p)$ = equivalent (average) pressure on a *row* of piles.
s_p = pile spacing parallel to (along) the row (see Fig. 3)
s_n = separation of 'remote' boundaries normal to the pile row (see Fig. 3)
δ_r = relative pile-soil displacement defined along a pile row (see Fig. 3)
σ'_{v0} = nominal overburden stress on horizontal section of soil below a level ground surface, corresponding to the weight of overlying material.

1 INTRODUCTION

The interaction of piles in rows where there is relative pile-soil lateral movement has been the subject of many studies (e.g. as summarized by Fleming et al. 1992). The relative movement may occur as a result of movement of the pile, sometimes referred to as 'active' loading. Alternatively the piles may be required to impede the potential movement of soil, for instance where 'discrete' piles are used to form a 'wall' in a slope that intersects a potential slip surface (e.g. Poulos 1999). This is normally called 'passive' loading.

This paper is related mainly to the specific example of a 'discrete pile wall' used to stabilize a slope, considering the generic lateral interaction for the long-term 'drained' condition, which is generally most critical in slope stability. The piles may typically be spaced at 3–4 diameter centers along the row, but it is necessary to ensure that soil will not 'flow' between the piles, but rather will 'arch' across the gap so that the pile row does actually behave as a continuous wall (Fig. 1). As the pile spacing increases the economic advantages are tempered by the increasing risk of soil flowing between the piles.

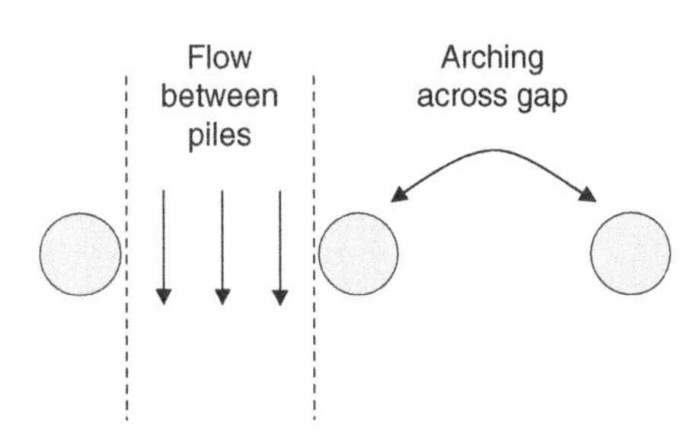

Figure 1. Flow and arching between piles for a discrete pile wall (passive loading).

2 EXISTING THEORETICAL SOLUTIONS FOR ISOLATED PILES

2.1 *Elastic response*

Analysis of lateral pile-soil interaction considers a (2-d) plane strain section normal to the axis of the

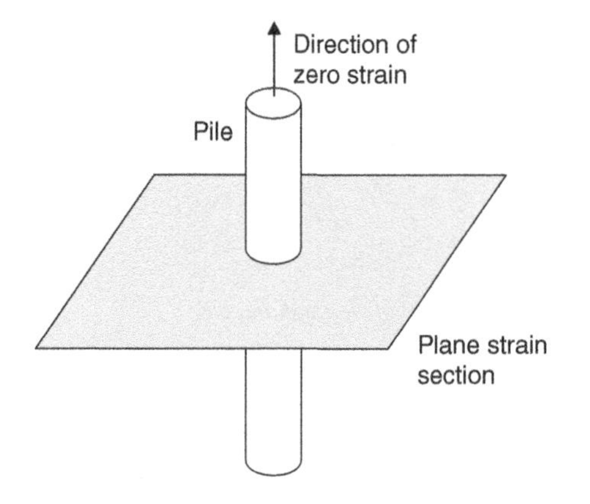

Figure 2. Plane strain section normally used for consideration of lateral pile-soil interaction.

pile as shown in Figure 2. Since the axis of the pile is generally vertical the plane strain section is horizontal.

Baguelin et al. (1977) derived an analytical solution for the force required to displace a circular section (representing the pile cross section) through uniform elastic material (representing the soil) with a rigid circular boundary:

$$\frac{p_p}{G} = 16\pi(1-\upsilon)\left\{2(3-4\upsilon)\ln\frac{D}{d_p} - \frac{2}{3-4\upsilon}\right\}^{-1}\left(\frac{\delta}{d_p}\right) \quad (1)$$

where
p_p = 'equivalent pressure' on pile, defined as the nett load per unit length in the direction of relative pile-soil movement (ie. the difference between load on the 'front' and 'back' faces), divided by the diameter;
G = shear modulus for the soil;
υ = Poisson's Ratio for the soil;
D = diameter of rigid boundary;
d_p = pile diameter; and
δ = displacement of the pile relative to the rigid boundary.

The value of (D/d_p) is somewhat arbitrary in practice, although it must be finite for there to be any resistance to the movement. It is now known that using a more realistic high stiffness at small strain would reduce this dependency on far field effects. A value of 30 has been used by previous researchers to consider an 'isolated' pile (i.e. one which is not significantly affected by any neighboring piles) in conjunction with the linear elastic response.

2.2 *Plastic response*

The 'ultimate' pressure on an isolated pile when yielding of the soil occurs has also been considered theoretically.

For a soil exhibiting undrained (Tresca) shear strength theoretical solutions exist (Randolph & Houlsby 1984). However, a range of semi-empirical results has been proposed for the drained (purely frictional) failure criterion, which will be considered (see Fleming et al. 1992). Here the following expression will be used:

$$p_{p,ult} = K_p^{\ 2}\, \sigma'_{v0} \quad (2)$$

where
$p_{p,ult}$ = ultimate equivalent pressure on the pile;
K_p = the passive earth pressure coefficient; and
σ'_{v0} = the nominal vertical effective stress in the ground at the depth considered.

3 2-D PLANE STRAIN DRAINED ANALYSIS OF PILES IN A SINGLE ROW

3.1 *Method*

Bransby & Springman (1999) demonstrated that 2-d plane strain finite element analysis could be used to replicate the results for elastic and plastic response for a Tresca soil described above for an isolated pile. The results were also extended to multiple rows of piles.

Similar analyses have been undertaken here using *FLAC*. Figure 3 shows a typical grid. s_p is the pile spacing parallel to (along) the row. $(s_n/2)$ is the distance to a 'remote' boundary (where movement is assumed to be restrained normal to the pile row), hence these boundaries are separated by a distance s_n. Lines of symmetry are exploited with corresponding boundaries through the centerline of the pile and at the midpoint between piles along the row. Only one row of piles is considered. The effect of varying s_p was to move the right hand boundary in the analysis. The diameter of the pile (d_p) was arbitrarily set at 1.0 m (the normalized spacings and pile-soil spacing relative to d_p are of consequence).

The perimeter of the (half) pile was modeled using beam elements so that it effectively formed a (hollow) rigid inclusion in the soil. These elements were translated normal to the pile row at a velocity of 2×10^{-7} m/step. This approach is described in Itasca 2002. As the analysis proceeded displacement (normal to the pile row) at point A (on the pile) and at point B (in the soil at the mid-point along the row) was recorded. Here, the relative pile-soil displacement (δ_r) is defined as the difference between these values – corresponding to a measure of 'local' deformation along the row. For an 'isolated' pile there is no movement at point B, and hence δ_r is equivalent to δ in Equation 1.

Drained soil response was modeled for a purely frictional soil with friction angle (φ') of 30°. The pre-yield material response was assumed to be linear with Young's Modulus 10 MN/m^2 and Poisson's Ratio 0.25. The kinematic dilation angle was assumed to be zero.

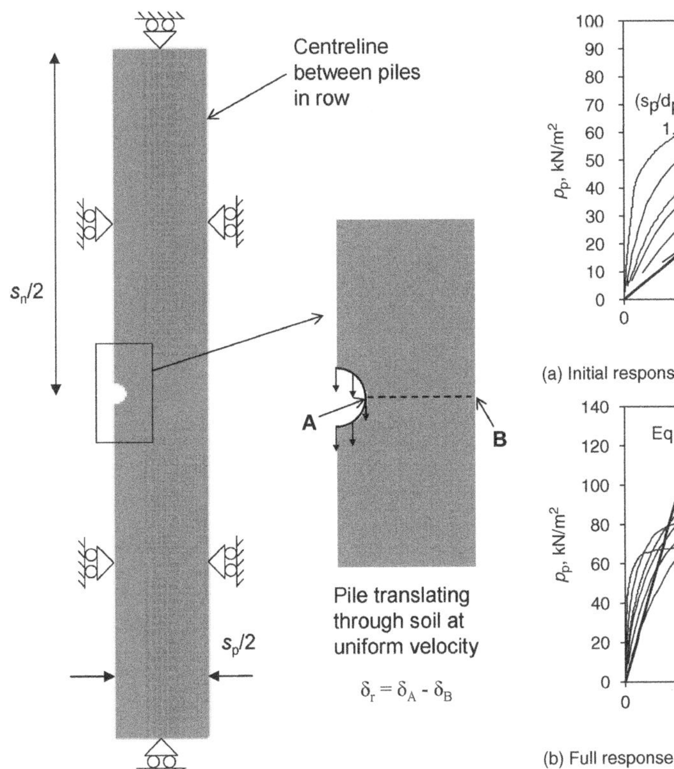

Figure 3. Typical *FLAC* mesh for 2-d plane strain analyses: spacing $(s_p/d_p) = 8$ (parallel to row), $(s_n/d_p) = 30$ (separation of boundaries normal to row).

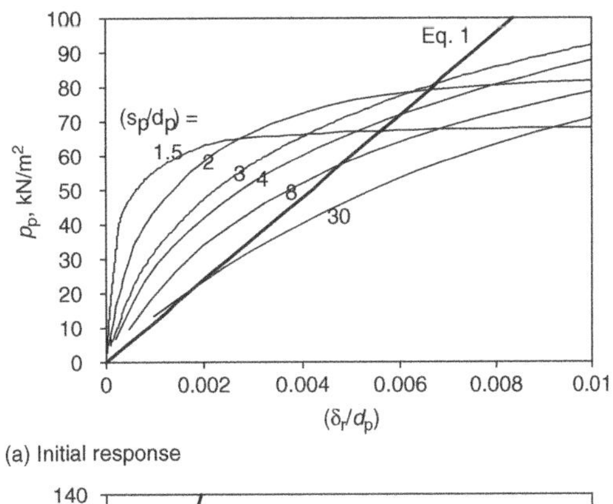

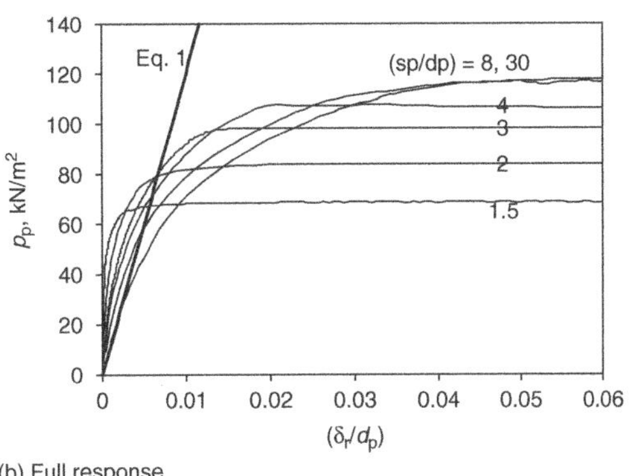

Figure 4. Variation of equivalent pressure on pile with normalized pile-soil displacement for 2-d plane strain analyses at 1.5 m depth in soil ($\sigma'_{v0} = 27\,\text{kN/m}^2$, $K_0 = 0.5$, $\varphi' = 30°$) and various spacing ratios along row.

Interface elements were incorporated between the pile and soil. These elements had high normal and shear stiffness, with identical strength properties to the soil. Hence they would allow shearing at the pile-soil interface, assuming it to be 'perfectly rough'. They also allowed the total force on the pile to be deduced.

The in situ stress state in the soil prior to translation was specified corresponding to 1.5 m depth in a soil with unit weight 18 kN/m^3, earth pressure coefficient $K_0 = (1\text{-sin } \varphi') = 0.5$, and zero pore water pressure. The effect of a sloping ground surface with varying depth to a horizontal plane was not considered at this stage, and hence the in situ stresses were uniform. Thus σ'_v (the 'out of plane' stress) was assigned an initial value of 27 kN/m^2 whilst σ'_h was 13.5 kN/m^2 for both the in plane directions. Since the plane is horizontal the acceleration due to gravity was zero in the analysis.

3.2 *Results*

The results for equivalent pressure on the pile (p_p) are plotted showing variation with normalized pile soil displacement (δ_r/d_p) in Figure 4 for a variety of pile spacings. The elastic response predicted by Equation 1 is also shown. As anticipated the 'isolated' pile initially shows good correspondence with this prediction. As (s_p/d_p) reduces the initial response becomes stiffer. Although this does not conform to the accepted variation of '$p-y$' response (normally used for active loading), it is anticipated when using δ_r, since as (s_p/d_p) reduces the smaller gaps present more resistance to relative pile-soil movement, emphasizing the logical advantage (in initial response) of reduced pile spacing for passive loading.

As the soil begins to yield (initially near the pile, and spreading through the soil) the response becomes less stiff and finally reaches an ultimate value, where relative displacement increases without further increase in the extent of the zone of yielding. The ultimate pressure tends to increase as (s_p/d_p) increases. This reflects the influence of the corresponding boundary in limiting the amount of soil that is available to yield and resist pile-soil movement.

Referring to Equation 2 an ultimate resistance of about 240 kN/m^2 would be expected. However, it can be seen that even for an isolated pile the value is only half this. This was thought to be because stresses in the entire area 'behind' the pile were tending towards zero (including the out of plane direction) at large pile displacements (see Fig. 7 later). In retrospect it is not surprising that this was happening since movement throughout this area is one-dimensional extension (except locally behind the pile).

Resistance to relative pile-soil movement is principally provided by normal stress acting on the 'front' of the pile. However, for a purely frictional soil the stresses on either side of the pile are strongly related (consider the bearing capacity equation), and hence reduction in stress behind the pile was affecting the resistance, which could be mobilized in front. It would be expected that such a reduction would occur, although it is not realistic that the out of plane stress (σ'_v) should drop significantly below the nominal overburden stress over a widespread area. Therefore further analyses were conducted in an attempt to examine the influence of this effect further.

4 3-D 'CONSTANT OVERBURDEN' ANALYSIS OF PILES IN A SINGLE ROW

Although the plane-strain idealization for a horizontal section seems appropriate for a Tresca soil (e.g. Bransby & Springman 1999), as discussed above it does not appear suitable for a purely frictional soil since variation of the out of plane stress (σ'_v) is unrealistically large and widespread when compared to the nominal overburden stress.

4.1 *Method*

To overcome this problem a 3-d mesh was used as shown in Figure 5. The top surface of the mesh was restrained by application of the nominal overburden stress (σ'_{v0}) rather than a restraint against movement. Initially the grid was arbitrarily given a height (or 'thickness', t) of 1.0 m. The pile diameter (d_p) was also 1.0 m. Other boundary conditions, soil properties and initial in situ stresses were identical to the plane strain section. Gravity was again assigned a value of zero so that σ'_v was nominally constant for the horizontal section.

The pile was now comprised of solid elements (which were effectively rigid compared to the soil), which were translated in the same way as for the plane strain section. The pile was restrained from movement in the vertical direction and δ_r was recorded as for the plane strain section. Interface elements were again used between the pile and soil, having similar properties as in the 2-d analyses. The interface nodes were used to derive shear and normal forces exerted on the perimeter of the pile, which were in turn used to derive p_p.

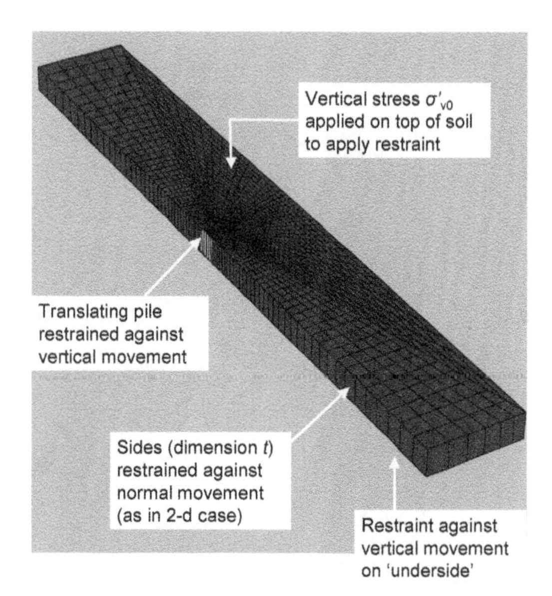

Figure 5. Typical *FLAC* mesh for 3d 'constant overburden' analyses: spacing $(s_p/d_p) = 8$ (parallel to row), $(s_n/d_p) = 30$ (separation of boundaries normal to row).

4.2 *Comparison with 2-d plane strain section*

Figure 6 is analogous to Figure 4 for the new results and shows the predictions from Equations 1 & 2. The results at small relative displacement show similar trends to the previous analyses. However, it can be seen that at larger displacements, the resistance is considerably greater and approaches the prediction from Equation 2. Also note that nearly all the results appear to tend towards a unique ultimate pile pressure, although as (s_p/d_p) increases larger relative displacements are required to mobilize resistance.

For $(s_p/d_p) = 2$ the response is very stiff but the ultimate resistance is lower. The results only extend to small δ_r because after this points A and B (Fig. 3) moved at the same rate so that δ_r did not increase further.

Figure 7 demonstrates the reason for the difference in behavior at ultimate conditions. In the 2-d plane strain analysis σ'_v (the out of plane stress) has dropped to a value of 5 to 10 kN/m^2 over the entire area behind the pile. It is clear that this is not realistic for a horizontal section at constant depth where the nominal overburden stress (σ'_{v0}) is 27 kN/m^2. The 3-d constant overburden analysis shows that the corresponding stress is in the range 25 to 30 kN/m^2 over the majority of the grid. σ'_v is locally reduced behind the pile and increased in front of it. In front of the pile $\sigma'_v \approx K_p\sigma'_{v0}$ and is the minor principal stress – this observation is consistent with Equation 2.

These local variations in stress are clearly more realistic. However, there is some question as to why they

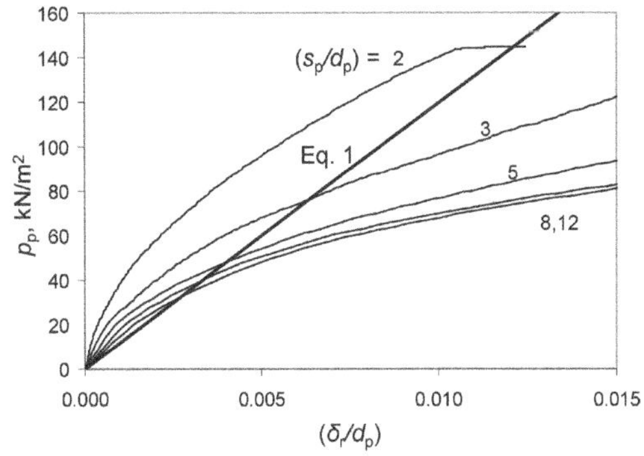

(a) Initial response

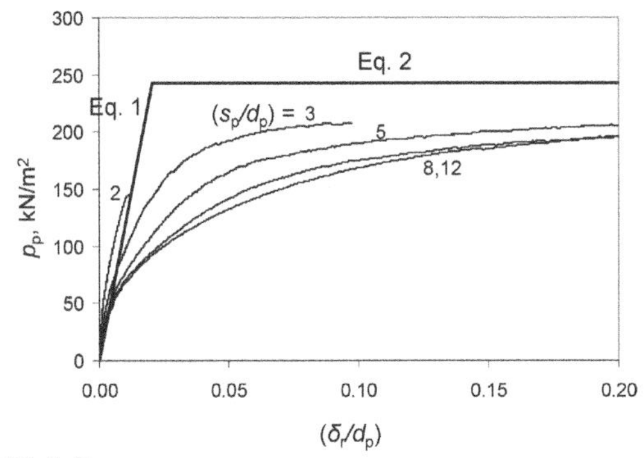

(b) Full response

Figure 6. Variation of equivalent pressure on pile with normalized pile-soil displacement for 3-d constant overburden analyses at 1.5 m depth in soil ($\sigma'_{v0} = 27\,\text{kN/m}^2$, $K_0 = 0.5$, $\varphi' = 30°$) and various spacing ratios along row.

arise at all when a 'confining' stress of σ'_{v0} has been applied at the top of the mesh. In fact the 3-d section is able to redistribute σ'_v by straining in this direction, and hence shearing in a vertical plane as this strain varied. Such shearing also occurred on the face of the pile. Hence σ'_v could vary by re-distribution to adjacent soil and the pile (implying a relatively small axial load in the pile). The tendency for the soil to 'heave' in front of the pile and 'settle' behind it led to local increase and reduction in σ'_v respectively.

4.3 *Effect of 3-d section thickness*

It seemed likely that the amount of stress redistribution would be affected by the thickness (t) of the 3-d section, which had been set arbitrarily since self-weight had not been considered in the analyses. Figure 8 shows the effect of doing this with $(s_p/d_p) = 3$. As expected the ultimate resistance reduced with the thickness of the section, and tended towards the result for the 2-d section as $t \rightarrow 0$. However, it is not clear if this is because the two situations were *directly* analogous. At thickness greater than 1.0 m the ultimate

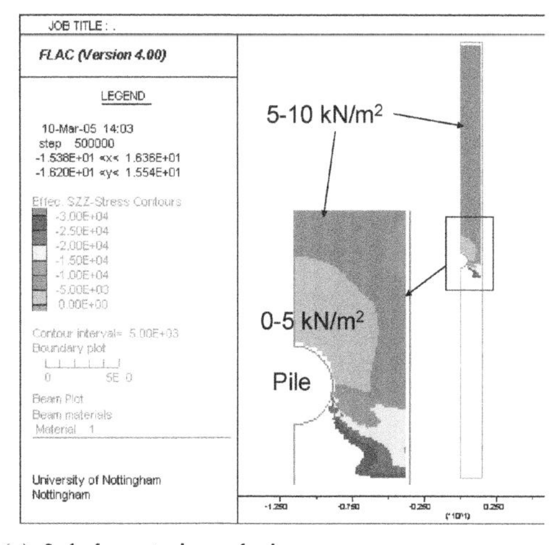

(a) 2-d plane strain analysis

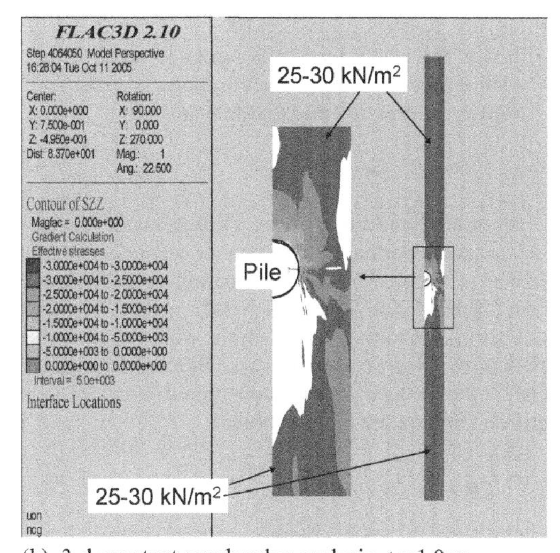

(b) 3-d constant overburden analysis, $t = 1.0$ m

Figure 7. Variation of σ'_v 'behind' pile for alternative analyses at 1.5 m depth in soil ($\sigma'_{v0} = 27\,\text{kN/m}^2$, $K_0 = 0.5$, $\varphi' = 30°$), $(s_p/d_p) = 3$.

resistance was virtually constant, implying that this was the maximum redistribution that would occur.

Three 'limits' are indicated on the figure, based on the following concepts of behavior for the pile:

- It is an 'isolated pile', with ultimate resistance predicted by Equation 2:

$$p_{p,\text{ult}} = K_p^{\,2}\,\sigma'_{v0}$$

- It is part of a 'continuous wall' (i.e. arching is effective, Fig. 1), with passive and active earth pressures in front of and behind it (respectively). Hence the equivalent pressure on the pile itself is given by

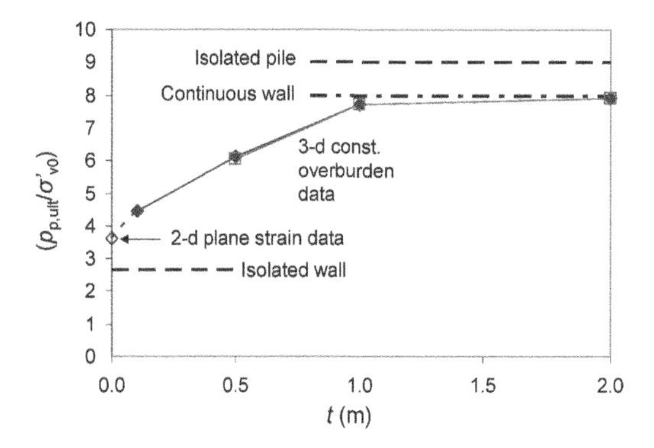

Figure 8. Effect of 'thickness' (t) of 3-d section on ultimate equivalent pressure on pile at 1.5 m and 3.0 m depth in soil ($s_p/d_p = 3$, $\sigma'_{v0} = 27$ and $54\,\text{kN/m}^2$, $K_0 = 0.5$, $\varphi' = 30°$). Theoretical limits also shown.

$$p_{p,ult} = (K_p\text{-}K_a)\,\sigma'_{v0}\,(s_p/d_p) \tag{3}$$

– It is a very short section of 'isolated wall' (rather than a pile behaving in accordance with Eq. 2). Hence the pressure on the pile is given by

$$p_{p,ult} = (K_p\text{-}K_a)\,\sigma'_{v0} \tag{4}$$

Since all the limits are proportional to σ'_{v0} this has been used to normalize the data. It can be seen that data at 3 m depth (and all other parameters the same) then follows exactly the same trend.

Specific conditions in the field would dictate the amount by which σ'_{v0} could vary, although it appears that for the specific situation considered here $t = 1.0$ m gives an approximate upper bound.

4.4 *Effect of arching*

Figures 9 & 10 show contours of stress parallel and normal to the pile row (respectively) for $(s_p/d_p) = 2$, 3 and 5. Figure 9 shows stress 'along' the row where it significantly exceeds σ'_{v0}. There is some evidence of arching of stresses between the piles (Fig. 1) in all cases, although $(s_p/d_p) = 3$ is perhaps most convincing.

Stresses approximately in the range σ'_{v0} to $K_p\sigma'_{v0}$ are shown by the contours in Figure 10. The analyses for $(s_p/d_p) = 2$ and 3 show considerably more evidence of increased normal stress in front the pile row across the full width ($s_p/2$). This implies that arching is no longer particularly effective in this sense at $(s_p/d_p) = 5$.

5 INTERPRETATION OF BEHAVIOR

From the perspective of discrete piles acting as a continuous wall (e.g. stabilizing a slope), the equivalent pressure on the pile (p_p) is of less interest than the equivalent pressure on the 'wall' (i.e. the average on

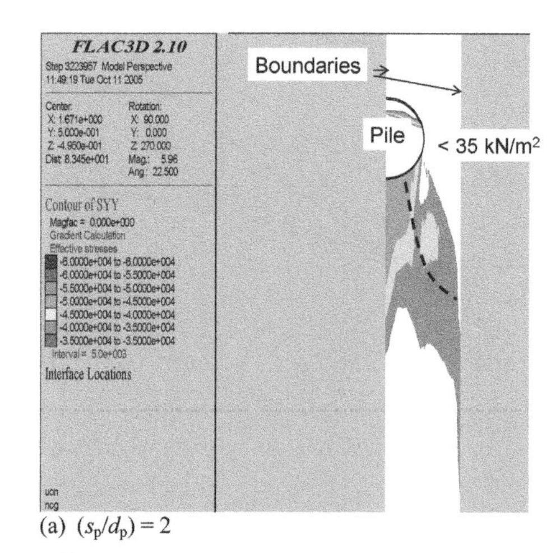

(a) $(s_p/d_p) = 2$

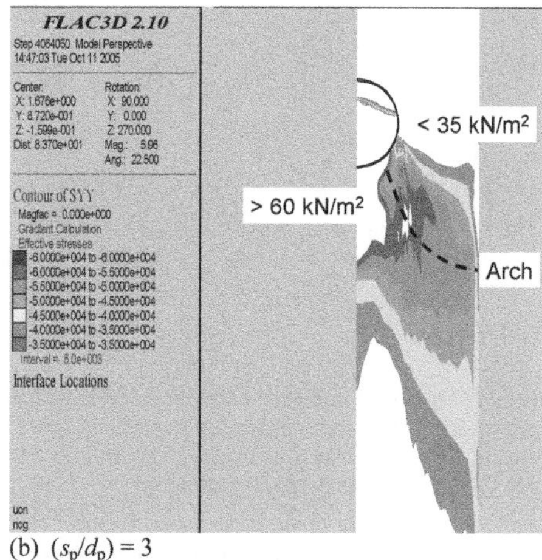

(b) $(s_p/d_p) = 3$

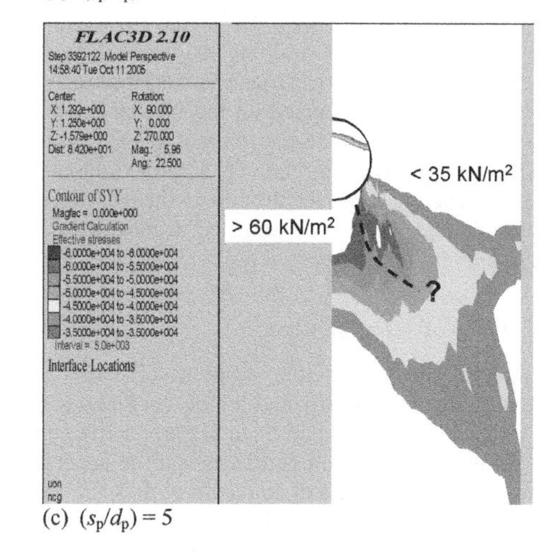

(c) $(s_p/d_p) = 5$

Figure 9. Stress parallel to pile row at various spacings parallel to row: constant overburden analysis at 1.5 m depth in soil ($\sigma'_{v0} = 27\,\text{kN/m}^2$, $K_0 = 0.5$, $\varphi' = 30°$, $t = 1.0$ m).

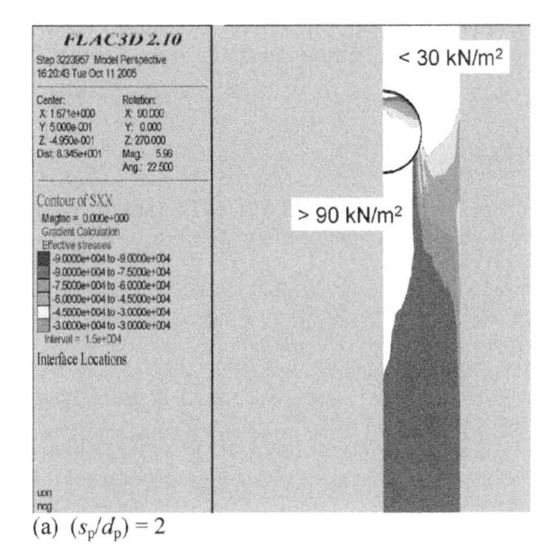

(a) $(s_p/d_p) = 2$

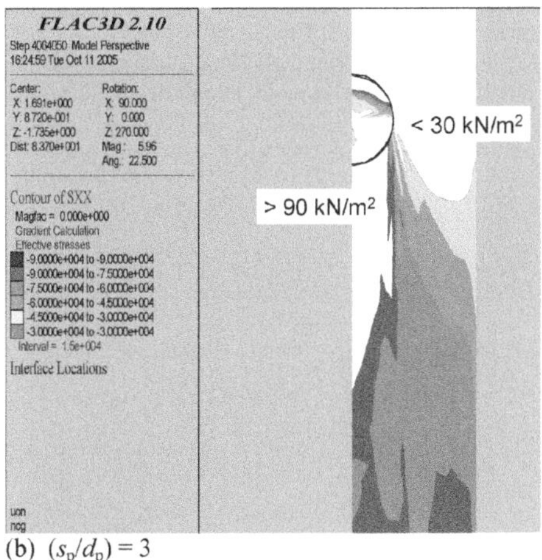

(b) $(s_p/d_p) = 3$

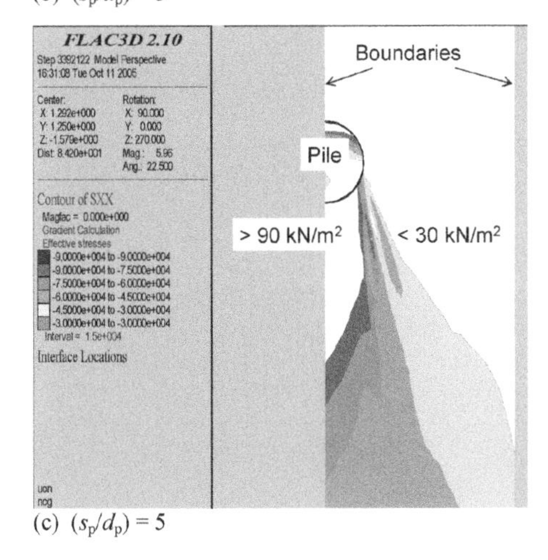

(c) $(s_p/d_p) = 5$

Figure 10. Stress normal to pile row at various spacings parallel to row: constant overburden analysis at 1.5 m depth in soil ($\sigma'_{v0} = 27$ kN/m², $K_0 = 0.5$, $\varphi' = 30°$, t = 1.0 m).

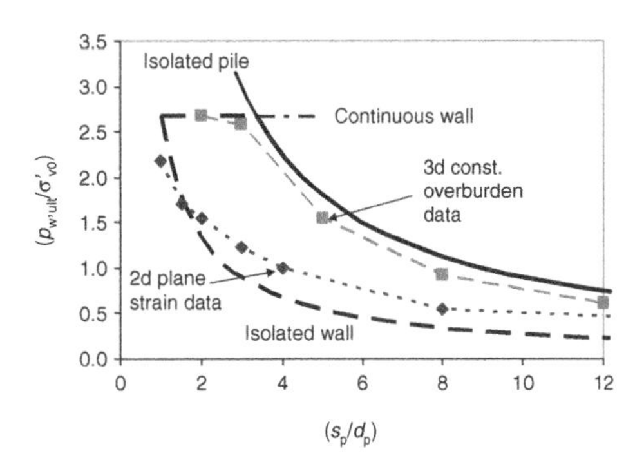

Figure 11. Ultimate equivalent uniform pressure on a pile row, showing variation with pile spacing along row. Theoretical limits also shown.

the row): $p_w = p_p(d_p/s_p)$. Figure 11 shows variation of $p_{w,ult}$ (normalized by σ'_v) with (s_p/d_p) for a variety of analyses. The limits described previously are also shown, again adjusted by the factor (d_p/s_p). Note that as would be expected the 'continuous wall' limit is then independent of s_p.

The data from the 3-d constant overburden analyses corresponds to the 'continuous wall' at small spacing, before dropping to follow the 'isolated pile' line at larger spacings. This is logical, and there is also evidence that the intersection of these 2 lines marks the point where arching between the piles is most 'effective', i.e. when $(s_p/d_p) = K_p^2/(K_p - K_a)$, from combination of Equations 2 & 3.

This value increases from 3.4 to 6.0 as φ' increases from 30° to 45°. The former value is consistent with existing practice, and confirms that $(s_p/d_p) = 3$ is close to the critical ratio for the analyses reported previously. However, the latter value is somewhat larger than would normally be considered. Limited *FLAC* analyses undertaken indicate that arching at wide spacings may be feasible if the material strength is high enough and there is sufficient pile-soil movement.

The data for the 3-d constant overburden analysis has also been shown to be independent of σ'_{v0} and K_0 when plotted in this way. Other analyses undertaken with $\phi' = 30°$ indicate that (unsurprisingly) the resistance to pile movement drops as the friction angle mobilized at the pile-soil interface reduces relative to the internal friction angle of the soil. Likewise, increasing the kinematic dilation angle increases resistance for the 'isolated pile' case, and hence can extend arching to wider spacings.

Finally, it can be noted that the 2-d plane strain analyses show reasonable correspondence with the 'isolated wall' limit. These analyses actually showed significant dependence on K_0, with $K_0 = 1.0$ giving increased resistance, but still below the 3-d constant overburden data. Thus it seems reasonable (and logical) to propose that the 3 limits bound $p_{w,ult}$, as shown in Figure 11. The distinction between the 'isolated pile' and 'isolated wall' response depends on the ability of the horizontal section of soil considered to redistribute σ'_v (increasing it locally behind the pile and increasing it locally in front) to maximize the differential normal stress on the pile (Figure 8).

6 CONCLUSIONS

The paper has demonstrated that:

- 2-d plane strain analyses and 3-d 'constant overburden' analyses give significantly different results when considering relative pile-soil lateral movement using a horizontal section for a purely frictional soil.
- Ultimate resistance to the pile-soil movement can be bounded by 3 simple and logical limits. The intersection of 2 of these limits seems to indicate the limit on pile spacing along a row where 'arching' of soil between adjacent piles is maintained when there is relative pile-soil movement. Such a limit would be of significant practical impact in (for instance) the design of 'discrete pile walls' used to prevent soil movement in potentially unstable slopes.

REFERENCES

Baguelin, R., Frank, R. & Said, Y.H. 1977. Theoretical study of lateral reaction mechanism of piles. *Géotechnique* 27(3): 405–434.

Bransby, M.F. & Springman, S.M. 1999. Selection of load-transfer functions for passive lateral loading of pile groups. *Computers & Geotechnics* 24: 155–184.

Fleming, W.G.K., Weltman, A.J., Randolph, M.F. & Elson, W.K. 1992. *Piling Engineering*, 2nd ed. Blackie, Glasgow (ISBN 0-216-93176-2).

Itasca Consulting Group, Inc. 2002. *FLAC – Fast Lagrangian Analysis of Continua, Version 4.0 User's Manual*. Minneapolis: Itasca.

Poulos, H.G. 1999. Design of slope stabilising piles. In Yagi, Yamagami & Jiang (eds), *Slope Stability Engineering*. Rotterdam: Balkema (ISBN 90 5809 079 5).

Randolph, M.F. & Houlsby, G.T. 1984. Limiting pressure on a circular pile loaded laterally in cohesive soil. *Géotechnique* 34(4): 613–623.

Advances in Transportation Geotechnics – Ellis, Yu, McDowell, Dawson & Thom (eds)
© 2008 British Geological Survey, Nottingham, ISBN 978-0-415-47590-7

New geophysical and geotechnical approaches to characterise under utilised earthworks

D.A. Gunn, H.J. Reeves & J.E. Chambers
British Geological Survey, Keyworth, UK

G.S. Ghataora, M.P.N. Burrow, & P. Weston
Birmingham University, Birmingham, UK

J.M. Lovell
Soil Mechanics Ltd., Leamington Spa, UK

L. Nelder
Scott Wilson Ltd., Nottingham, UK

D. Ward
Lankelma Ltd., East Sussex, UK

R. Tilden Smith
Great Central Railway (Nottingham) Ltd., UK

ABSTRACT: Transferring the freight burden from road to rail would bring about many environmental benefits. The Rail Contribution to the Energy Review (Dept. for Transport 2006) indicated that rail freight produces eight times less CO_2 per tonne.km than road freight. Implementing this strategy successfully will require further development of rail infrastructure to cope with additional capacity. Many new proposals, such as the EuroRail Freight Route, would utilise redundant and under-used infrastructure, much of which was constructed during the latter part of the nineteenth century. Earthworks of this age should be regarded as unique if they are to be improved or upgraded. Such upgrades would require investigations into the condition of the existing earthworks to assess the materials, variability in the geotechnical properties and engineering performance along the proposed route. This paper presents a section of embankment from the Great Central Railway as a case history that demonstrates the integration of a number of geophysical and geotechnical data to assess the condition of an embankment in relation to fill materials and track geometry. It emerged that embankment structure and strength information can be provided via combined use of non-intrusive mechanical and electrical techniques such as continuous surface wave profiling and resistivity surveying. It is envisaged that this information can be used to strategically plan intrusive investigations and works to improve the infrastructure.

1 INTRODUCTION

The Rail Contribution to the Energy Review (Dept. for Transport 2006) indicated that rail freight produces eight times less CO_2 per tonne.km than road freight. Successfully transferring freight from road to rail will require additional rail infrastructure to cope with additional capacity. As an example, the EuroRail Freight Route (2006) project proposes a network through the central spine of the UK linking the Channel Tunnel to Glasgow. It would utilise redundant and under-used infrastructure, much of which was constructed during the latter part of the nineteenth century. In those pioneering times, standards in track gauge and geometry, and cutting and embankment profiles were developed empirically based upon observations made by engineers during construction. Observations of the behaviour of embankment materials aided the construction of earthworks well before the scientific fundamentals of soil mechanics were developed. Consequently, many earthworks, especially those constructed in Victorian times, should be regarded as unique if they are to be improved or upgraded. Such processes would require investigations into the condition of the existing earthworks to assess the materials and variability in the geotechnical properties, and engineering performance along the proposed route.

Standard site investigation techniques such as drilling and pitting can be augmented by combined geophysical and geotechnical surveys to identify fill materials, and the scale of variability in their properties and distribution. Material property information can be provided via several means, such as: direct installation of probes (Gunn & Stirling 2004; Gunn *et al.* 2004; Nelder *et al.* 2006), portable, non-intrusive methods (Sussman *et al.* 2003; Clark *et al.* 2003; Gunn *et al.* 2005; Gunn *et al.* 2006a; Gunn *et al.* 2007), and from track-recording vehicles gathering related data (McAnaw 2001).

This paper presents a section of embankment from the Great Central Railway as a case history that demonstrates the integration of geophysical and geotechnical data to assess the condition of an embankment in relation to fill materials and track geometry. The approach shows how traditional techniques can be augmented with developing techniques such as continuous surface wave and resistivity surveying to provide information through the complete earthworks. It is envisaged that the techniques used can also provide information for the design of works required to improve the infrastructure.

2 EAST LEAKE EMBANKMENT RESEARCH SITE (ELERS)

The Great Central Railway was originally constructed in the 1890s as a link from the Manchester, Sheffield and Lincolnshire railway to London (Bidder 1900, Fox 1900). The section of line used in this study serves as a goods link from the mainline at Loughborough to the gypsum works at East Leake. Daily traffic includes two freight locomotives operated by EWS Railway Ltd. and GB Rail Freight Ltd. The line is managed by the Great Central Railway (Nottingham) Ltd., the Nottingham Transport Heritage Centre and the Mainline Steam Trust. The current investigation focuses on a section of earthworks SW of East Leake, Nottinghamshire (Fig. 1), which was constructed using local materials excavated from adjacent cuttings to the SW and NE of the site. This material was tipped and then compacted by subsequent movement of shunting locomotives and tipping wagons across the tipped material. The tipping method used along this section of the line was not stated explicitly by Bidder (1900), but has been deduced to have been end tipped from current observations and the information recorded by the engineers practicing at the time.

The section of embankment from the East Leake Tunnel (bridge 314) in the SW to the overbridge (bridge 313) in the NE forms the East Leake Embankment Research Site (ELERS). The site has been the subject of field investigations since September 2005.

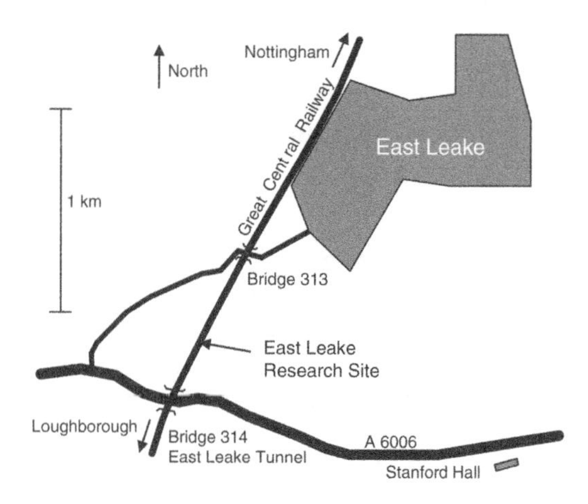

Figure 1. Site location: East Leake, South Nottinghamshire.

Victorian earthworks are generally very heterogeneous because of the techniques used in their construction. Data gathered from the field investigations at the ELERS has been integrated to provide an assessment of embankment condition in terms of the variability of fill materials. The scale of apparent heterogeneity observed is affected by the volume of ground sampled or sampling interval, (eg. probe spacing etc.). This study shows how a suite of rapid geophysical and geotechnical surveys can be combined to investigate potential larger scale features and lateral continuity of structure within the embankment, identify changes in fill regime and show the effect on track geometry. On the basis of these results recommendations are made for applying these survey techniques to evaluate the condition of existing earthworks as part of an upgrade scheme for re-introduction into the network.

3 SITE INVESTIGATION TECHNIQUES

Information on the physical and geotechnical property distribution within earthworks can be gathered from geophysical and geotechnical survey data respectively. A number of well understood relationships exist between geophysical properties such as resistivity (Rhoads *et al.* 1976; Mualem & Friedman 1991; Jackson *et al.* 2002, Chambers *et al.* 2004; Friedel *et al.* 2006) and dielectric constant (a key parameter in radar surveys) (Davis & Annan 1989; Baker 1991; Gallagher *et al.* 2000; Sussmann *et al.* 2003; Neal 2004) to lithology and moisture content. By combining geotechnical and geophysical data, it is possible to relate mechanical and engineering properties to lithological and moisture variation. This can be used as a basis for assignation or classification of the condition of earthworks.

The investigation at East Leake has included a series of intrusive and non-intrusive techniques comprising:

Intrusive

i. static cone penetration resistance tests (sCPT) to a depth of 10 m, W side of embankment,
ii. dynamic cone penetration resistance tests (dCPT) to 3 m; W, E sides and between rails,

Non-intrusive

iii. continuous surface wave surveys (CSW) to ascertain shear wave velocity and small strain stiffness logs to 8 m depth, W, E sides and between rails,
iv. Falling weight deflectometer (FWD) tests, profile on W side of rails and at depth in trial pits,
v. ground penetrating radar surveys (GPR), profiles along E, W sides, between rails and transects across embankment,
vi. resistivity surveys, profiles along W side and transects across embankment.

The sCPT technique (Meigh 1987, Soil Mechanics 2007) uses a cylindrical cone, pushed vertically from a rig into the ground at a constant rate of penetration of 20 mm per second. During penetration, measurements are made of the cone resistance, the side friction against the cylindrical shaft and, in piezocone tests, the pore water pressure generated at penetration by the cone. The dCPT technique uses a cone of known area driven into the ground with blows from a standard hammer onto the head of a piston or anvil attached to the cone by a set of steel rods. Commercial equipment is instrumented such that speed of impact and the penetration per blow are measured and used to calculate dynamic cone resistance using the Dutch formula (Langton 1999).

The CSW technique uses sinusoidal surface waves generated by an electromagnetic vertical vibrator seated on the ground surface. In practice, this produces a series of finite duration pulses, each at a single frequency over a range of frequencies, for example from 5 Hz to 100 Hz in increments of 0.5 Hz or 1 Hz. Field data acquisition at each frequency is synchronized with the control to the vibrator. Field dispersion curves are generated from the recorded signals at two or more receivers using a method based on the steady state Rayleigh method described by Viktorov (1967) and Richart *et al.* (1970). The CSW technique is particularly suited to railway sites where ambient noise levels are high, and the field data are interpreted to provide a stiffness-depth profile (Zagypan & Farifield 2000, Gunn *et al.* 2006a).

FWD systems have been mounted on trolleys for use on the railways and are operated by companies such as Scott Wilson Ltd. More recently, lightweight devices, such as the Prima 100 have been operated by companies such as Soil Mechanics Ltd., which allow single man operation without the need for a rail mounted vehicle. They provide a means of simulating axle loads on a single sleeper and of measuring the strain within the ballast with sub-millimetric accuracy (Grainger *et al.* 2001, Brough *et al.* 2003). Results can be presented either as measured deflection data or processed to provide performance indicators such as ballast strain factor or trackbed compression, the latter being used to calculate the effective stiffness of the trackbed at a given sleeper location.

GPR uses a transmitting antenna to provide a short pulse of high frequency (25–1000 MHz) electromagnetic energy into the ground. Variations in the electrical impedance within the ground generate reflections that are detected at the ground surface by the same or another antenna attached to a receiver unit (Davis & Annan, 1989). Variations in electrical impedance are largely due to variations in the relative permittivity or dielectric constant of the ground, and thus respond well to water distribution, fill materials and layered structure (Neal 2004).

The resistivity technique uses an array of four electrodes (Telford et al. 1976), where two electrodes pass a direct or a low frequency alternating current into the earth while a potential difference is measured between the other two. The ratio of the voltage between any two voltage electrodes and the current flowing from a current source electrode to a current sink electrode is measured by the field resistivity equipment to provide an apparent resistivity of the ground.

4 ASSESSMENT OF EMBANKMENT CONDITION

4.1 *Earthworks Materials*

Figure 2 shows an example of sCPT (15 cm^2 cone area) and dCPT (2 cm^2 cone area) profiles through the earthworks plotted with a log of the materials encountered. sCPT q_{net} values up to 30 MPa in the upper ballast coincide with the upper layer of the original ballast pavement described by Bidder (1900). Generally, values range from 2 to 3 MPa in the underlying embankment fill with the occasional peak, e.g. up to 18 MPa at 4.6 m depth in the mudstone fill derived from the Westbury Formation. Siltstone layers (e.g. between 2.25 and 2.5 m) are likely to be fragments of material, broken by the excavation and tipping process, rather than intact layers. The dCPT profile penetrated to a shallower depth due to manual deployment. Generally, the two CPT logs are similar, but the dCPT with a smaller cone area is more responsive to a finer scale of soil heterogeneity. This can be observed as greater detail on the log plot, for example, by the series of small peaks to 7.5 MPa in the mudstone fill from 1–1.5 m and the 10 MPa peak in the gravelly clay fill (reworked Westbury Formation) at 1.9 m. The materials from 0.4–0.8 m form the original engineered ballast pavement described by Bidder (1900) and GPR surveys show this pavement to be extensive across the embankment (Gunn *et al.* 2007). It appears as a strong continuous reflector at approximately 0.5 m depth on the W side of the embankment, but as a highly disrupted reflector between the rails. This disruption is possibly a recent phenomenon related to the line taking very high train loads. The original ballast comprises 'hand pitched stone' overlain by 'granite' chippings', Bidder (1900), which on inspection, are coated with a layer of soft, red-brown clay. This clay-coated interface was coincident with high moisture content in contrast to the above materials and produced a strong reflection on the GPR profiles.

Figure 3 shows a field and inverted small-strain stiffness profile on the embankment fill predominately of reworked Westbury Formation from about 0.5 m depth. The stiffness is the product of the square of the shear wave velocity and the density where the density at all depths is estimated at 2.0 Mgm^{-3}. Below the upper layers of fill from 0.8 m depth, the earthworks fill mainly comprises a re-worked mudstone gravel of angular lithoclasts of the Westbury Formation with sporadic cobbles of Blue Anchor Formation. The field data indicate that the ballast stiffnesses were around 100 MPa, while the subgrade was around 50 MPa at 1 m rising to around 80 MPa at 5.0 m below the top of the sleepers. Generally, the inverted stiffness profile shows a good match to the field data except at the near surface where the field data are sparse. The inversion was allowed to ascribe high stiffness values to the ballast on the basis that very high penetration resistances were recorded in the materials of this layer during both static and dynamic cone penetration surveys. Gunn *et al.* (2006b) showed a correlation between higher shear wave velocities and higher dynamic cone penetration resistance values associated with gravel-rich lenses in cohesionless deposits at coastal sites. A nearby dynamic cone penetration resistance profile is shown for comparison, where high penetration resistances are associated with the ballast and particularly with the original ballast pavement.

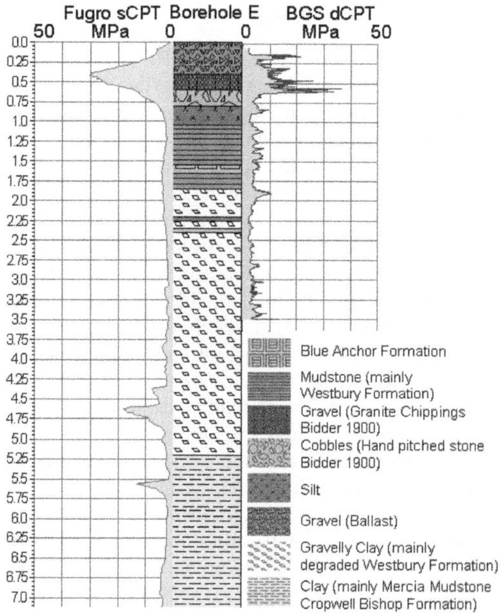

4.2 *Embankment Profiles*

Nelder *et al.* (2006) and Gunn *et al.* (2007) presented 2D sections of the sCPT, dCPT, friction ratio and the CSW which were created by infilling a grid between successive depth logs using an anisotropic inverse distance weighting between neighbouring grid nodes. These were used to investigate potential larger scale lateral continuity of structure within the embankment. For example, features formed as exposed surfaces by trafficking or drying out, or, as a natural accumulation of large intact blocks in the early phase of embankment construction appear as bands with high penetration resistances up to 20 MPa, which extend laterally for up to 30 m at around 5 m depth just above the top surface of the bedrock.

During construction of Victorian embankments the fill related directly to the material taken from adjacent cuttings (cut and fill). Consequently, Victorian embankments contain many interfaces between

Figure 2. Comparison of a static and dynamic CPT through the embankment.

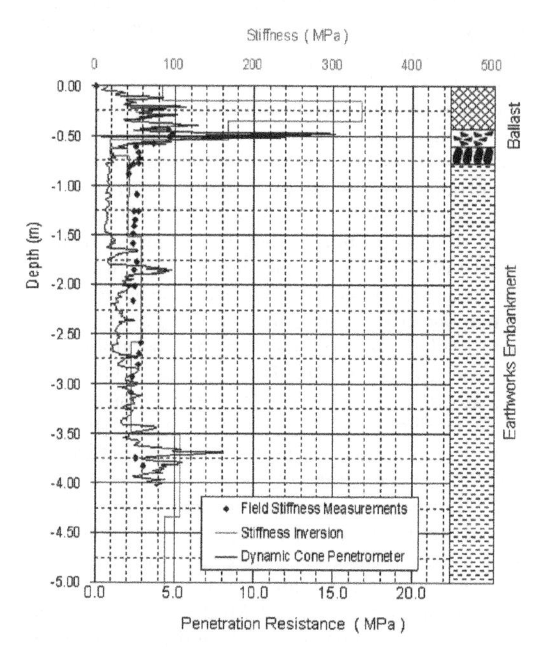

Figure 3. Example of a field and inverted small strain stiffness profiles with comparison to dCPT.

materials of differing engineering performance, which are very often associated with poor track geometry. Figure 4 shows the effect on penetration resistance within the embankment across an interface between clay (BH E: Clay Fill) and sand and gravel fill (BH G: Sand/Gravel Fill). The clay comprises re-worked mudstone gravel derived from the Westbury formation and the sand and gravel comprises an upper zone to 2 m of sand and rounded gravels of glaciofluvial origin overlying a gravel of angular lithoclasts of Blue Anchor Formation, which extends to 4 m depth.

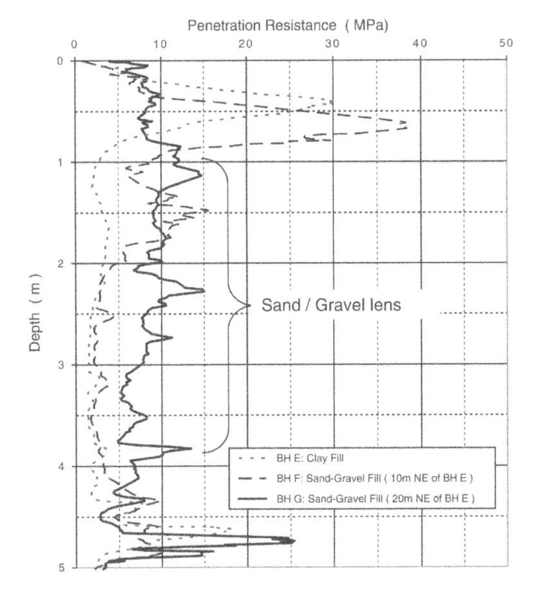

Figure 4. sCPT (qnet) profiles across interface between clay fill and sand/gravel lens.

Stiffness values (MPa) from surface FWD tests (Prima 100 with 300 mm plate dia.) were 43.9, 41.2 and 47.1 at E, F and G respectively. Further over the sand / gravel surface stiffness values were 26.3, 26.8 and 30.0 at 30, 40 and 50 m NE of BH E respectively. It is suspected that these surface FWD tests provide a measure of the stiffness of the ballast at the near surface and that deeper information can be gained either by using a larger plate or by testing in trial pits. At BH E a stiffness value of 80.3 MPa, measured at a depth of 0.4 m was considered to be representative of the original ballast layers, and a value of 36.8 MPa at 0.8 m representative of the silt and mudstone at the top of the underlying earthworks. FWD tests at successive depths during pitting provide stiffness profiles to limited depths. However, prior to pitting or where pitting is restricted, a single CSW test from the surface will provide a stiffness profile through the whole earthworks; an insight that is useful for planning remedial works.

Very poor track geometry occurred over a 20 m interface zone across the clay-to-sand/gravel fill boundary. The development of poor geometry on the NE side is coincident with a thinning of the sand/gravel lens (Fig. 5). Resistivity measurements (Chambers *et al.* 2007) respond well to lithology and moisture content, and resistivity images aid site mapping of fill materials. The lens of fill comprising sand, gravel and siltstone produces a wedge shaped zone with resistivities above 150 Ohm.m. The wedge develops from the surface at about the 40 m station (BH E) and thickens to about the 50 m station (BH F) such that it extends from the surface to 4 m depth. This high resistivity wedge

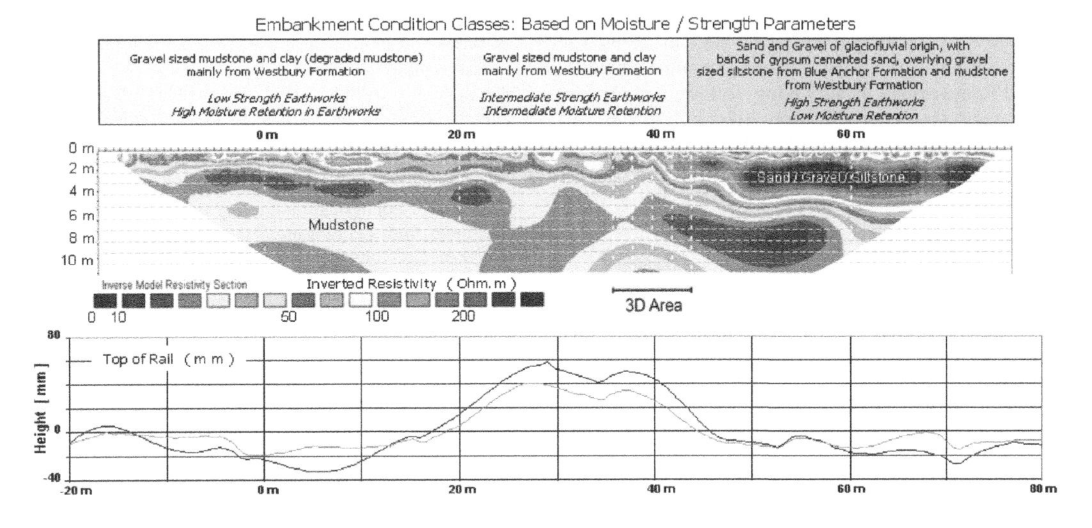

Figure 5. Longitudinal resistivity image compared to the track geometry and embankment condition.

persists laterally over this depth interval towards the 80 m station.

5 CONCLUSIONS

An investigation at the East Leake Embankment Research Site was undertaken to provide an assessment of the structure, the geotechnical properties and the variability in a Victorian Embankment. This involved the gathering of geotechnical data via probing, drilling, core sampling and laboratory testing and geophysical data from surface geophysical surveys. This is of relevance to the evaluation of the potential engineering performance of abandoned or regenerated rail routes in future programmes, such as the Euro-Rail Freight Route. This study indicated poor track performance at engineering interfaces, and again, highlighted that the problem is related to the materials, condition, engineering properties and performance of the subgrade and underlying embankment. It is essential that such problems are identified and remedied as part of the process to upgrade or improve infrastructure to cope with increased traffic and load schedules.

It is envisaged that there are many interfaces between different engineering materials throughout the UK network where track performance will be poor. Geophysical and geotechnical surveys can be combined in an embankment condition assessment scheme that will assist in their location and characterisation, and also, aid the design of geotechnical remedial or improvement measures. All the techniques discussed in this case history can be applied to provide rapid appraisal and data collection of several hundred metres of track per day. The recommendations from this paper would be to plan the surveying of the network in a modular fashion to maximise the information gained and its use downstream. Such plans should include phases of reconnoitring / locating interfaces / subgrade problems using non-intrusive surface survey techniques including GPR and resistivity surveying, followed by phases to characterise materials, properties and variability using combined techniques like CPT, CSW and FWD surveys in areas identified by large physical property contrasts.

6 ACKNOWLEDGEMENTS

This paper is published with the permission of the Executive Director of the British Geological Survey (NERC).

REFERENCES

Baker, P.L., 1991. Response of ground penetrating radar to bounding surfaces and lithofacies variations in sand barrier sequences. *Exploration Geophys*. 22, pp19–22.

Bidder, F.W., 1900. The Great Central Railway Extension: Northern Division. *ICE*, Vol. CXLII, Session 1899-1900, Part IV, Paper 3227, pp 1–22.

Brough, M., Stirling, A., Ghataora, G. & Madelin, K. 2003. Evaluation of railway trackbed and formation: a case study. *NDT&E Int*. 36, pp 145–56.

Chambers, J.E., Loke, M.H., Ogilvy, R.D. & Meldrum, P.I. 2004. Noninvasive monitoring of DNAPL migration through a saturated porous medium using electrical impedance tomography. *Jour. Contaminant Hydrology*, (68), pp 1–22.

Chambers, JE, Wilkinson, PB, Gunn, DA, Ogilvy, RD, Ghataora, GS, Burrow, MPN & Tilden Smith, R. 2007. Non-Invasive Characterization and Monitoring of Earth Embankments Using Electrical Resistivity Tomography (ERT). *Proc. 9th Int. Con. Railway Engineering*, London.

Clark, M., Gordon, M. & Forde, M. 2003. Issues over high-speed non-invasive monitoring of railway trackbed. *NDT & E Int*. 37, (2), pp 131–139.

Davis, J.L. & Annan, A.P., (1989). Ground-penetrating radar for high-resolution mapping of soil and rock stratigraphy. Geophysical Prospecting, 37, 531–51.

Department for Transport, 2006. *Rail Contribution to Energy Review*.

EuroRail Freight Route. 2006. http://www.blugman.freeserve.co.uk/efr/index2.html

Fox, F.D. 1900. The Great Central Railway Extension: Southern Division. *ICE*, Vol. CXLII, Session 1899-1900, Part IV, Paper 3209, pp 23–48.

Friedel, S., Thielen, A. & Sringman, S.M. 2006. Investigation of a sloe endangered by rainfall-induced landslides using 3D resistivity tomography and geotechnical testing. *Jour. Appl. Geophys*., (60), pp 100–114.

Gallagher, G., Leiper, Q., Clark, M. & Forde, M. 2000. Ballast evaluation using ground penetrating radar. *Railway Gazaette Int*., 2000, pp 101–102.

Grainger, P.S., Sharpe, P. & Collop, A.C. 2001. Predicting the effect of stiffness on track quality. *Proc. 4th Int. Con. Railway Engineering*, London.

Gunn, D.A, Nelder, L.M., Jackson, P.D. & Entwisle, D.C, Stirling, A.B., Konstantelias, S. & Lewis, R.W., Kingham, P. 2004. Geophysical inspection of the trackbed-subgrade stiffness and performance. *Proc. 7th Int. Con. Railway Engineering*, London.

Gunn, D.A. & Stirling, A.B. 2004. Geophysical trackbed and subgrade monitoring at Leominster Station. *Rail Technology Magazine*, 4, (6), pp 34–37.

Gunn, D.A, Nelder, L.M., Chambers, J.E., Reeves, H., Freeborough, K., Jackson, P., Stirling, A.B. & Brough, M., 2005. Geophysical monitoring of the subgrade with examples from Leominster. *Proc. 8th Int. Conf. Railway Engineering*, London.

Gunn, D.A., L.M. Nelder, J.E. Chambers, M.R. Raines, H.J. Reeves, D. Boon, S. Pearson, E. Haslam, J. Carney, A.B. Stirling, G. Ghataora, M. Burrow, R.D. Tinsley, W.H. Tinsley, R. & Tilden-Smith. 2006a. Assessment of railway embankment stiffness using continuous surface waves. *Proc. 1st Int. Conf. Railway Foundations*, Birmingham, September 2006, pp94–106.

Gunn, D.A., S.G. Pearson, J.E. Chambers, L.M. Nelder, J.R. Lee, D. Beamish and J.P. Busby, R.D. Tinsley & W.H. Tinsley. 2006b. An evaluation of combined geophysical and geotechnical methods to characterise beach thickness. *Q.J.E.G.H*. 39, 339–355.

Gunn, D.A, Reeves, H., Chambers, J.E., Pearson, S.G., Haslam, E., Raines, M.R., Tragheim, D., Ghataora, G., Burrow, M, Weston, P., Thomas, A., Lovell, J.M., Tilden Smith, R. & Nelder, L.M. 2007. Assessment of embankment condition using combined geophysical and geotechnical surveys. *Proc. 9th Int. Conf. Railway Engineering*, London.

Jackson, P.D., Northmore, K.J., Meldrum, P.I., Gunn, D.A., Hallam, J.,Wambura, J.,Wangusi, B. & Ogutu, G. 2002. Non-invasive moisture monitoring within an earth embankment – a presursor to failure. *NDT & E Int.*, 35, pp107–115.

Langton, D.D. 1999. The Panda lightweight penetrometer for soil investigation and monitoring material compaction. *Ground Engineering*, September, pp 33–37.

McAnaw, H.E. 2001. System that measures the system. *Proc. 4th Int. Con. Railway Engineering*, London.

Meigh, A.C. 1987. Cone penetration testing. Methods and interpretation. *CIRIA Ground Engineering Report: In situ testing*, Butterworths, London, 141p.

Mualem, Y. & Friedman, S.P. 1991. Theoretical prediction of electrical conductivity in saturated and unsaturated soil. *Water Resources Res.*, 27, (10), 2771–2777.

Neal, A. 2004. Ground-penetrating radar and its use in sedimentology: principles, problems and progress. *Earth-Science Reviews*, Vol. 66, pp261–330.

Nelder, L.M., D. A. Gunn & H. Reeves. 2006. Investigation of the Geotechnical Properties of a Victorian Railway Embankment. *Proc. 1st Int. Conf. Railway Foundations*, Birmingham, September 2006, pp 34–47.

Rhoads, J.D., Ratts, A.C. & Pather, R.J. 1976. Effects of liquid phase electrical conductivity, water content and surface conductivity on bulk soil electrical conductivity. *Jour. Soil Sci. America*, 40, 651–655.

Richart F.E. Jr., Wood R.D. & Hall J.R. Jr. 1970. *Vibration of soils and foundations*. Prentice-Hall, New Jersey.

Soil Mechanics Ltd. 2007. Geocone Homepage http://www.esgl.co.uk/soilmechanics/geocone/index.htm

Sussmann, T.R., Selig, E.T. & Hyslip, J.P. 2003. Railway track condition indicators from ground penetrating radar. *NDT & E Int.* 36, (3), pp 157–167.

Telford, W.M., Geldart, L.P., Sheriff, R.E. & Keys, D.A. 1976. *Applied Geophysics*. London, Cambridge University Press, 860p.

Viktorov, I.A. 1967. *Rayleigh and Lamb waves: physical theory and applications*. Plenum Press, New York. 154p.

Zagyapan, M. & Farifield, C.A. 2000. The use of continuous surface wave and impact techniques to measure the stiffness and density of trackbed materials. *Proc. 3rd Int. Con. Railway Engineering*, London.

Advances in Transportation Geotechnics – Ellis, Yu, McDowell, Dawson & Thom (eds)
© 2008 Taylor & Francis Group, London, ISBN 978-0-415-47590-7

Gault Clay embankment slopes on the A14 – Case studies of shallow and deep instability

O. Hamza & A. Bellis
Atkins Limited, Cambridge, England, UK

ABSTRACT: Highway embankment slopes constructed from locally won plastic clays have been regularly showing events of instability. This paper will present two cases in Gault Clay embankment slopes on the A14 in Cambridgeshire, where shallow and deep failures have been observed in addition to frequent cracking in the pavement at the same locations. The discussion is based on recent ground investigation and field observation along with slope stability analysis using standard methods. The results of the assessment help to identify mechanisms triggering failure of embankment slopes in Gault Clay fill and provide essential geotechnical inputs for the design of the remedial works.

1 INTRODUCTION

Instability of the Highway embankment slopes constructed from plastic over-consolidated clay fill is known to have occurred in the past, and observations suggest that instability is currently occurring at a number of identified sites (e.g. Perry, 1989; Greenwood, 1985; Stokes et al., 2004), producing costly maintenance problems

In order to design or evaluate an effective remediation that can prevent such instability, the performance of the embankments, the factors controlling the development of slip surface and stability degradation need to be understood.

In common with any typical earthwork, the geotechnical performance of highway embankments will depend on a combination of factors including: the mechanical characteristics of the fill material, its in-situ state (which will be governed largely by the initial compaction control, climatic conditions, drainage system and the effects of vegetation), the subsoil properties and the age of the embankment as well as its geometry.

In particular, when the fill material is won from a local borrow pit of high plastic cohesive soil and so is expected to have similar material characteristics to the subsoil, the shear strength and permeability play a key role in the engineering behaviour of the embankments. Indeed, a reduction in long-term stability due to the over-consolidation and plastic nature of clay fill, has been attributed to four reasons (e.g. Davies et al., 2004; Aubeny et al., 2004): they are likely to swell with time after placement because the over-consolidated clay fabric is unlikely to have been erased on excavation; highly plastic clays will change volume more significantly with seasonal variations in pore pressure (causing soil creep); poorly drained clays are likely to sustain higher pore pressures for longer following precipitation events (or block in ingress of water past a certain depth); their post-rupture drained angle of friction is likely to be in the range 20–25° (i.e. less than typical slope angles).

The surface of the embankment is likely to have undergone the most significant swelling cycles due to seasonal evapo-transpiration cycles or perhaps plant root ingress. Consequently, it may be aggravated with a significant number of tension cracks. This will increase the permeability of the surface layers significantly, whilst reducing the intrinsic shear strength parameters (c' and ϕ'). This may lead to a shallow failure event as the reduced shear strength in the weathered surface layers may be not sufficient to prevent sliding.

In a similar mechanism, failure events may consist of deep-seated slips with rigid body rotation of a mass of soil. For this type of failure, the shear plane will need to form within the main body of the fill material and may use pre-existing failure planes, include the tension crack at the crest i.e. highway level, emerge below the toe of the embankment. However, significant variation of depth and radius of curvature of the slip surface is likely depending particularly on the shear strength distribution within the slope.

In this paper, the mechanisms suggested above were examined on two cases of instability identified from the A14 between Girton and the River Cam (see the map in Figure 1). The A14 Cambridge Northern

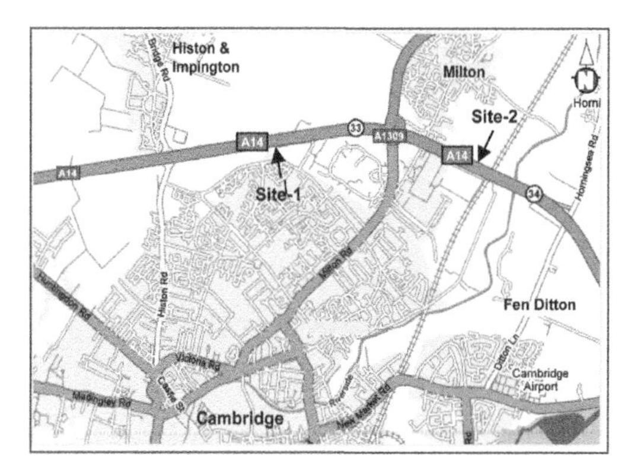

Figure 1. Map showing the location of the two case studies.

Bypass (CNB) is a two lane dual carriageway which is, for the most part, supported by 8 m high embankments constructed typically from locally won Gault Clay with slope angle of 1V:2H. Over this length of the carriageway, the embankment slopes have suffered a number of instability events comprising shallow and deep-seated failures; one case from each type is presented herein.

The studies were based on recent ground investigation and visual field observations together with back analysis of the failed slopes using limit equilibrium methods to determine the in-situ soil parameters at the time when the failure occurred. In the slope stability analysis, the possible effect of the drainage blockage (i.e. elevated pore water pressure) and the influence of the vegetation have been considered.

To help provide guidance for the geotechnical design of the maintenance works, the effectiveness of a repair method (comprising a combination of Geogrid reinforcement and lime modification) was evaluated in a slope stability analysis using the parameters obtained from the back analysis of the failed slopes.

2 FAILURE OBSERVATIONS AND HISTORY OF EMBANKMENT

2.1 *Failure observations*

The two case studies of unstable embankments presented in this paper have been assessed at different times; they are both located on the A14 carriageway between junction 32 and 34, north east of Cambridge as shown in Figure 1. Site-1 is approximately 300m long situated halfway between the Milton and Histon junctions westbound, while Site-2 is a 2.1 km long stretch of the eastbound between Milton and Fen Ditton.

The failure in Site-1 was (one of two at the same site) reported in December 2001, however the exact date of the event is unknown. The failure (shown in Figure 2 & 3) affected the full height of the 8m high embankment over a 35 m length with a backscar measured as 1.5m deep indicating to a deep-seated failure (Atkins, 2003). Drainage comprised a gravel drain at the toe of the slope and within the backscar. A fractured gully connection has been visible. At a close by location, a section of the embankment exhibiting a 30 m long tension crack at the crest has been observed. Temporary remedial works on the carriageway embankment were carried out using duckbill anchors with Geogrid pinned to the slope over a height of 3m to provide some face protection. Full depth slope repair was

Figure 2. Photos of the embankment slip of Site-1: (a) photo taken at the crest showing backscar, (b) the toe bulge of the slip, (c) photo taken from beyond the toe showing backscar and toe bulge of the slip. The vegetation comprises grass and newly planted trees.

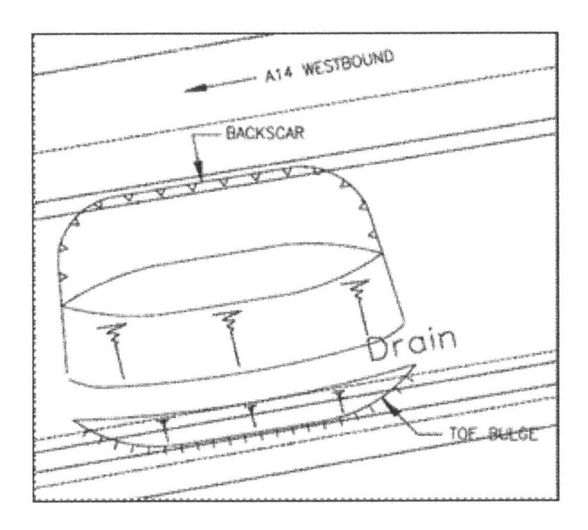

Figure 3. Schematic drawing shows the failure of the embankment of Site-1.

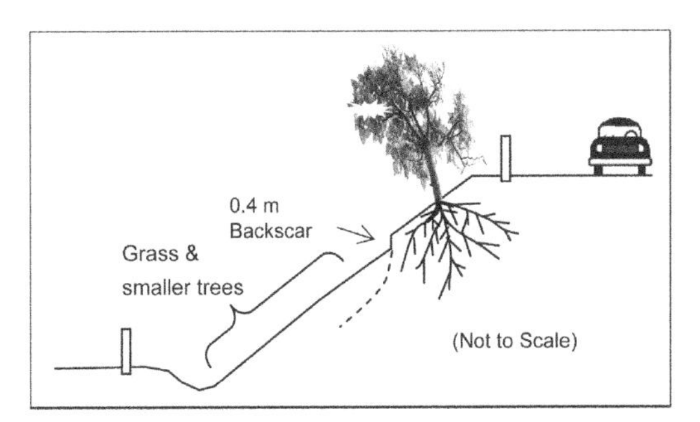

Figure 4. Basic sketch showing a cross section of the embankment fill at the identified shallow failure of Site-2.

then conducted as a permanent remedial work (Atkins, 2004).

A less severe event was reported in Site-2 after a walkover survey conducted in October 2005 (Atkins, 2006). A localized backscar of approximately 0.4m was identified at the mid height of the embankment suggesting a shallow failure as schematically illustrated in Figure 4. This soil movement may have indirectly contributed to transverse cracks observed in the past across the carriageway at regular intervals, and some longitudinal cracking in places. The height of the embankments is approximately 8m with slope angle of about 1V:2H, which are typical of the many clay slopes on the highway network (Greenwood, 1985). Much of the embankment slope had significant tree planting but not uniformly distributed. The trees were generally well established and appeared to be thriving, however some of them had been removed (chopped down) and those at the top of the slope were leaning down slope as shown in Figure 5. It was observed that concrete drainage channels had been provided from Milton Junction to the east. Gullies were provided

Figure 5. Photo taken for the embankment at Site-2, where Willow trees are leaning significantly.

beyond this point. It is believed that the outfall from the drainage occurs at the stream crossing towards the eastern end of this site.

2.2 *History of the embankments*

Prior to the construction of this section of the bypass arable farming was the primary land use within the subject sites (Atkins, 2003). The 1889 Ordnance Survey map of the area shows that the surface drains present today (with the exception of the ditches at the toe of the embankments) existed at that time. The *in situ* ground conditions in the area of the north east of Cambridge is often Gault Clay overlain by River Terrace Deposit, sometimes with a thin layer of alluvium (BGS, 1981). Fill of Gault Clay, which had been won from local borrow pits (according to an anecdotal evidence from staff present at the time of construction), was then compacted on the surface of the subsoil (natural ground) until reaching the final height of about 8 m of the recent embankments with gradient of 1:2.

Since the construction of the A14 (previously known as the A45) in the mid to late 1970s, the embankments are likely to have been deformed – typically due to consolidation, water pressure cycles and soil creep, induced cyclic mechanical loading from traffic and slope failures. The earliest known failures to have occurred within or near the subject sites are dated 1980 and 1983, with causes attributed to the over-consolidated clay of the embankment and an over-steep batter, possibly aggravated with planted trees, fractured gully connections or fractured pipes (Atkins, 2003). To ensure the integrity of the carriageway, the defects in the slope embankment as well as in the pavement have been inspected and repaired as part of the regular maintenance carried out for the carriageway. Further details on the surveys and maintenance at the subjects sites are discussed in the Preliminary Sources Study reported for the subject sites (Atkins, 2003 & 2006).

The effectiveness of different techniques for repair and prevention of slope failures (including excavation & granular replacement, lime stabilisation, geogrid reinforcement, Gabion wall, anchored tyre wall, ground anchors, rock ribs, and grout injection) have been investigated by Transport Research Laboratory TRL (Boden, 1995) in a field trial carried out on a 250 m long section of the embankment on the east-bound carriageway, west of the River Cam during the winter of 1983/84. This embankment trial was monitored until November 1994, when the performance of the different techniques was summarised. Following the TRL trial, it appears that most of the repairs to failed embankment slopes of the A14 have been based on a standard design geogrid reinforced re-profiling of the slope, occasionally lime has been added to the fill material.

The standard geogrid repair has been generally successful; however the method did not prevent repeated failures where the remedial programme did not include replacement of the damaged drains. The faulty drainage system is likely to have been leaking water into the embankment, creating a localised perched water table, softening the clay material until the slope failed, irrespective of the geogrid reinforcement. Therefore, a repair method comprising lime stabilisation in addition to geogrid reinforcement- has been recommended for the design as discussed in Section 4.3. This permanent repair method was conducted for the failed embankment on Site-1 involving the excavation of part of the embankment fill down to original ground level. The excavated material was then treated with lime and recompacted within the embankment, being also reinforced with layers of geogrid.

3 GROUND CONDITIONS

3.1 *Geology*

According to British Geological Survey map (BGS, 1981) and the accompanying memoir (Worssam & Taylor, 1969), the geological profile across the site consists of Terrace River Gravels underlain by Gault Clay and recent geological units. The horizon between the River Gravels and the underlying Gault Clay, is expected to be located between 1.8 m and 2.3 m below ground level. The Gault formation outcrops over the majority of the site except at the eastern end of the site where the Gault is overlain by the Lower Chalk, as a result of the solid geology dipping gently towards the southeast. The Gault formation varies from 27 to 43 m thick and consists of grey clay or marl.

Since the slope failures have been entirely within the embankment fill, specific ground investigation into this stratum and the natural ground below the base of the embankment was conducted as discussed in Section 3.3.

3.2 *Hydrology*

In order to predict the pore pressure distribution in a section of the embankment at any time instant (from which the overall stability of the embankment could be ascertained), the detailed recent hydrological input and the water contents and flow conditions in the embankment must be known (e.g. Davies et al., 2004). Infiltration into the slope which will vary spatially depends on the permeability of the soil surface. Soil water content of the slope will depend on precipitation minus runoff, evapo-transpiration, and drainage. Clearly, all these factors change with time as soil state near the surface changes. The current degree of saturation of any part of the soil will then reflect the net infiltration onto the soil surface and the hydrological conditions within the embankment (which also change with time).

From the Preliminary Sources Studies carried out for the subject sites (Atkins, 2003 & 2006), the original construction of the embankment has included kerb and gully drainage. Water from the carriageway would flow into the gullies, down the slope via piped connections to the toe drainage. The toe drainage comprises a French drain that lies within the embankment toe. However, no evidence of drainage for the slope within the subject sites has been observed suggesting that runoff is the predominant mode of groundwater movement in the embankment fill.

3.3 *Site investigation*

Along the A14 line between the M11 and A1303, various ground investigations have been conducted (Morgan, 1993). Accordingly, it has been established that the embankment fill at the subject sites (between Girton and the River Cam) is constructed from Gault Clay with similar engineering properties. However, to ensure that the remedial work design used correct parameters, further exploratory holes with in-situ and laboratory testing) were conducted (May Gurney, 2004; ScottWilson, 2006).

The embankment fill and the natural ground beneath it in the area adjacent to the shallow (Site-2) and deep-seated (Site-1) slope failures were investigated in a total of five cable percussion boreholes (from the crest to a maximum depth of 12 m) and fifteen window samples (at mid-slope, up to 8 m deep), in addition to a few 3m deep mechanically excavated trial trenches (at the foot of the embankments).

According to the data obtained from the site investigation, the pavement consisted of a 0.4 m thick layer of Rolled Bituminous Macadam over hardcore and a 0.7 m thick layer of brown medium dense silty sand with gravel. Approximate depths and general descriptions of materials encountered below the pavement are presented on the logs in Figure 6. Since no significant

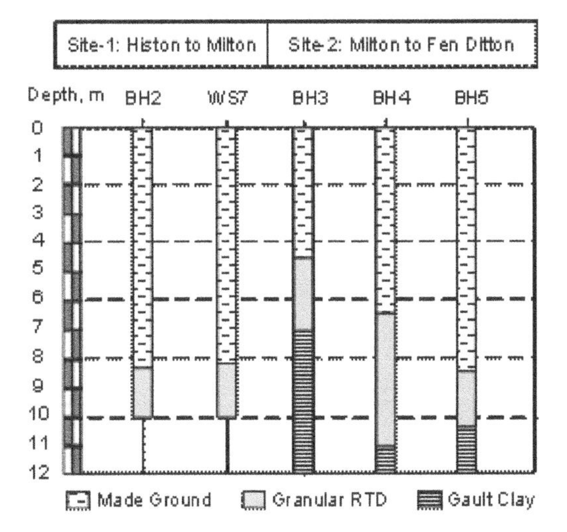

Figure 6. Logs of the boreholes excavated in the carriageway showing the soils and their depth below the top of each hole.

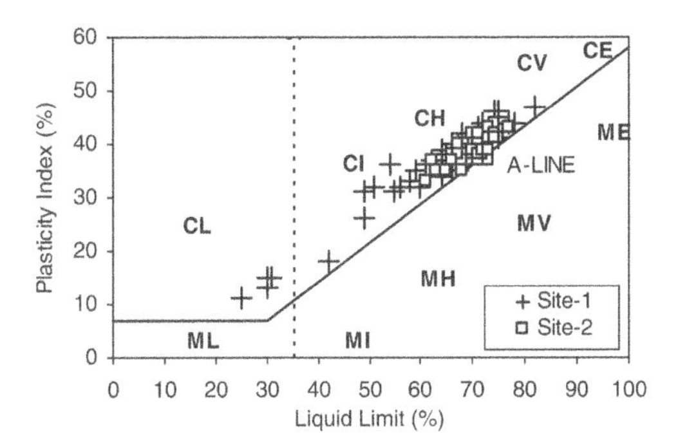

Figure 7. Plasticity Chart of the embankment fill.

difference was observed between the exploratory holes dug along the carriageway, the ground investigations confirmed the geology of the area comprising Made Ground of clay (cohesive fill) overlying natural ground of granular River Terrace Deposit underlain by Gault Clay. The Made Ground varies in thickness approximately from 4.5 to 8.5 m along the route corresponding to the engineered fill of the embankment construction for the A14.

The Made Ground was generally described as firm to stiff grey brown fissured clay with occasional medium subangular flint gravel. With depth, more granular material was encountered in the Made Ground i.e. becomes gravelly, sandy, or with occasional pockets of fine sand.

Below the Made Ground, up to 4.6 m thick of River Terrace Deposits were encountered. This material was typically described as dense light brown fine

Table 1. Summary of test results of the embankment fill.

Parameter (Unit)	Range (Typical value)	
	Site-1	Site-2
SPT 'N'	11–29 (20)	5–17 (9)
Natural Moisture Content (%)	9–34 (21)	12–38 (30)
Bulk Density (Mg/m^3)	1.8–2.0 (1.9)	
Liquid Limit (%)	25–82 (50)	38–77 (69)
Plastic Limit (%)	14–35 (25)	19–36 (30)
Plasticity index (%)	11–47 (35)	25–44 (39)
Vane Shear test, Hand Penetrometer, SPT correlation, Cu (kPa)	35–250	62–149
	Increases with depth, Fig 9	
Con. Undrained Triaxial c' (kPa)	2*–8**	–
test ϕ' (degree)	20*–24**	–
Organic Content (%)	–	0.1–0.5

* Post-rupture ** Peak

to medium sand with occasional fine to coarse subangular flint gravel. The amount of gravel content becomes more significant in some locations of this stratum, and was described as orange brown very sandy fine to medium angular flint gravel. The stratum may also have a slight content of silt and clay (up to 10%).

The Gault Clay was found to underlie the sites at between 7–11 m below the surface of the carriageway and extends to the base of the exploratory holes whenever it was reached. The Gault Clay was generally described as firm to stiff dark grey fissured clay with occasional fine to medium silt nodules. The Gault Clay is likely to be over-consolidated and the upper section of the stratum has been subjected to weathering by periglacial processes.

Several disturbed and undisturbed samples were obtained for laboratory testing to determine moisture content, soil classification, and shear strength parameters according to BS 1377. The results of these tests as well as the field tests are summarised in Table 1 and Table 2 for the Made Ground and Gault Clay respectively.

Groundwater strikes were noted within the River Terrace Deposits encountered beneath the embankment fill.

3.4 Discussion of geotechnical parameters

The plasticity chart in Figure 7 indicates that the embankment fills has a high to very high plasticity. The chart shows that the results are plotted above the A-Line indicating that the material consists predominately of clays, which is believed to have been sourced from borrow pits located adjacent to the line of the route, and to be Gault Clay. This was particularly confirmed by the test conducted on Gault Clay

Table 2. Summary of test results of the Gault Clay.

Parameter (Unit)	Range (Typical value)	
	Site-1	Site-2
SPT 'N'	–	22–27(24)
Natural Moisture Content (%)	26–77 (44)	24–31(27)
Bulk Density (Mg/m^3)	1.8–1.9 (1.85)	
Liquid Limit (%)	60–92(76)	50–81(69)
Plastic Limit (%)	21–35(27)	22–34(28)
Plasticity index (%)	35–63 (49)	28–47(40)
Vane Shear test, Hand Penetrometer, SPT correlation, Cu, (kPa)	25–430	81–112
Con. Undrained c′ (kPa)	6 (peak)	–
Triaxial test ϕ' (degree)	23 (Peak)	–

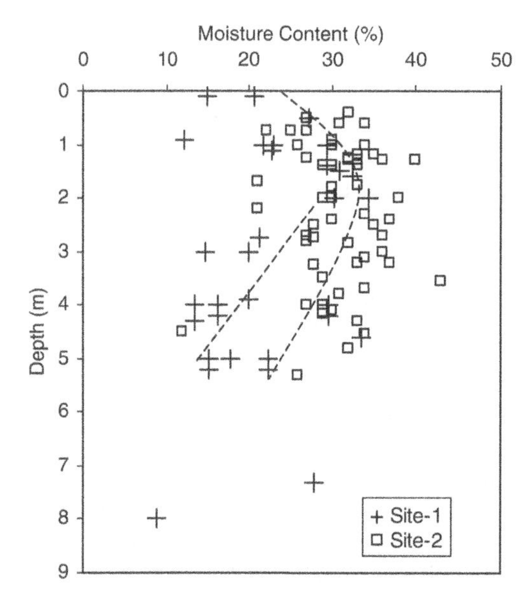

Figure 8. Moisture content of the embankment fill shown with depth below ground level.

(as presented in Table 2), where similar characteristic of plasticity was found.

In Figure 8 the moisture content is plotted against depth, showing a wide scatter of values and generally a trend slightly increasing with depth and then decreasing below 2 m depth. Within the upper 1 m of the embankment fill, the moisture content values typically ranged between 25 to 35%, while below 2 m values generally ranged between 10 and 20% for Site-1 and between 20 and 39% for Site-2; the overall average value is 29%.

The assessment of the typical values for the undrained shear strength was based on Vane Shear test, Hand Penetrometer, and SPT correlation, where the

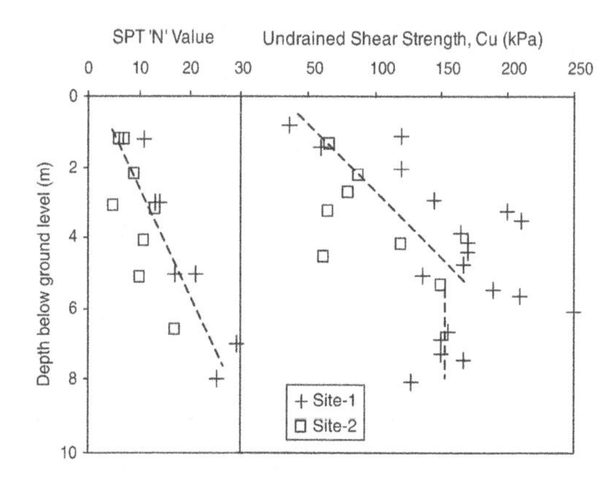

Figure 9. The results obtained from Standard Penetration Test (N value) and the undrained Triaxial test (Cu) plotted against the depth of the embankment fill.

undrained cohesion Cu was found to be increasing with depth as shown in Figure 9 with average value equal to 35 kN/m^2 near the surface increasing to 150 kN/m^2 at 3 m depth. Below 3 m, the undrained cohesion may be limited to this value (for the design purpose) because of the data variation.

The results obtained from Consolidated Undrained Triaxial tests performed on the embankment fill shows a typical value of effective cohesion of approximately 8kPa and a typical effective angle of shear resistance of 24°.

4 SLOPE STABILITY ANALYSIS

4.1 *Methods*

Because of the similarity in geometries and geologies of the two embankments at the subject sites, a typical model of 8m high and 1V:2H slope was chosen for the slope stability analysis. Particular details found in each site, such as the drainage conditions and the effect of vegetation, have been addressed in this analysis. The stability of the embankments was analysed using Slope/W (GEO-SLOPE, 2001), which is based on limit equilibrium theory (slice moment and force equilibrium). The slope stability in SLOPE/W is analysed by calculating the factor of safety, FOS, along slip surfaces – generated in a pre-defined area – by using a variety of standard methods including Bishop's, Janbu's, and Morgensten-Price method.

Standard methods of slope stability analysis seek to calculate the forces causing slope failure (the weight of the soil body plus any traffic loading) and compare this with calculated resistance forces (due to the mobilised shear stresses in the body and any possible root/mechanical reinforcement). This is done for every conceivable rotational failure mechanism and the factor of safety of the slope is given as the resistance force divided by the destabilisation force for the least stable failure plane. The key components are the weight of the soil zone contained by the shear surface (which will depend on the volume of the zone and the unit weight of the soil) and the shear resistance of the soil. The shear resistance is the area integral of the mobilised shear strength of the soil on the shear plane at failure. The shear stress mobilised will depend on the effective stresses in the soil on the shear surface, the soil properties (c′ and ϕ') and also the displacement magnitude (and possibly velocity). Standard slope stability analysis methods ignore the latter two components (displacement magnitude and velocity) and use the peak shear stress, which is assumed to act along the whole length of the slip plane simultaneously. Nevertheless, the standard methods provide simple approach and are believed to be adequate for the scope of the study since the mass movement and the progressive failure mechanism have not been quantitively considered.

The shear stresses developed on the postulated shear plane are critical. For consolidated undrained conditions it is often considered that the soil can be described with a Mohr-Coulomb failure criterion, $\tau = c' + \sigma' \tan \phi'$, where the shear stress, τ will depend on the effective stress acting on the shear plane. Clearly with almost a fixed total normal stress (due only to the weight of the soil above) it is the pore pressure changes that will change the effective stresses and alter the shear strength in the soil. Hence, knowledge of the pore pressure regime and hence drainage condition is vital.

4.2 *The effect of changing drainage conditions on slope stability – (Site-1)*

Slope stability analysis is likely to be particularly sensitive to the value of the pore water pressure ratio r_u, which was difficult to assess in the field with adequate certainty. Placement and compaction of embankment fill might have generated relatively high pore water suctions and the current (long-term) equilibrium pore pressures are likely to be higher than those at the end of construction. The strength of the fill may have then decreased with time as the pore pressures equilibrate (e.g. Potts & Zdravkovic, 1999).

The stability analysis used to assess the geotechnical parameters pertinent at the time of the failure was performed by reducing the peak shear strength parameters of the embankment fill and changing r_u to reproduce two different slope conditions. The first condition, which corresponds to the in-service condition where drains are assumed fully functional, was modelled by assuming a typical value of pore pressure ratio (r_u) of 0.2. The second condition, which corresponds to the condition of drainage blockage and elevated pore water pressure, was modelled by assuming r_u value within the embankment was increased to more than 0.2.

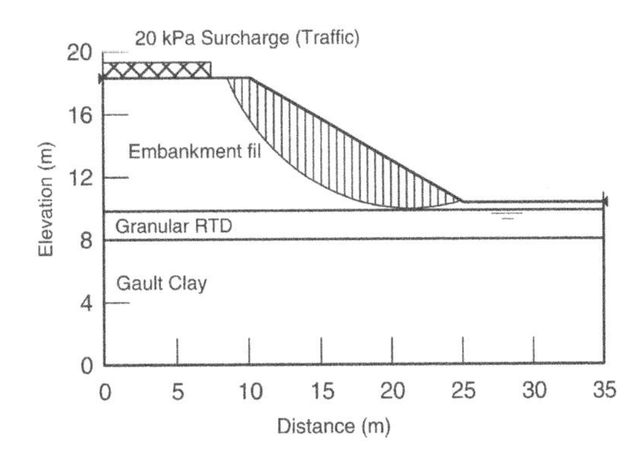

Figure 10. Rotational (deep-seated) slip surface with pore water pressure ratio, $r_u = 0.4$ assumed for the condition of drainage blockage (i.e. elevated pore water pressure).

In the analysis, the ground water was assumed to be at the top of the granular River Terrace Deposit in accordance with the finding of the ground investigation (Scott Wilson, 2006; May Gurney, 2004). The effective friction angle $\phi'$of the granular RTD was assumed equal to 32° (with no cohesion).

The analyses performed on the typical section of the embankment (Figure 10), showed that the slope model will have a marginal factor of safety of 0.99 when the embankment fill has $r_u = 0.4$, $c' = 5\,kN/m^2$, and $\phi' = 22°$. These results validate the hypothesis that it is very likely that the slope failures triggered when the embankment fill was wet and its pore water pressure was high (reducing the effective stresses in the soil as well as the shear strength parameters). The blocked or faulty drainage system is likely to have been one of the main causes of this increase in the pore water pressure within the embankment, by discharging the water load at localised points, rather than into the toe drain. The results confirmed (Figure 10) that the embankment model has developed a deep-seated rotational slip surface, which has been the case observed in Site-1 (see Figure 2).

4.3 *Geogrid reinforcement and lime modification (Site-1)*

Once the sources of the problem had been assessed, a series of possible solutions to repair the slope failures on Site-1 were considered for a generic design slope of 1V:2H and 8 m high. Geogrid reinforcement with and without lime modification were considered. The benefit of inclusion of tensile elements such as geogrid will provide a tensile capacity within the backfill on the slope and acts as a gravity structure to prevent deep failure within the embankment. The inclusion of the geogrids particularly at crest will help to mitigate against the risk of development of tension cracks.

Table 3. Summary of the factor of safety (FOS).

	Slope condition	
Reinforcement	Drains fully functional ($r_u = 0.2$)	Drains blocked ($r_u = 0.4$)
(i) Not used	1.23	0.99
(ii) Geogrid only	1.35	1.12
(iii) Geogrid & Lime	1.5	1.38

The major advantage of lime treatment is that it provides the opportunity to reuse the site material, largely independent of the weather conditions (Perry et al., 2003). The lime addition reduces the moisture content, reduces the PI, and allows a higher strength to be achieved.

The excavation of the fill started 2.5 m back from the kerbline and continued down slope in benches 1H:3V with a maximum depth of 1 m. The lime stabilised fill extended 0.5 m below the existing ground surface (below the embankment). Four layers of geogrid reinforcement were placed at a maximum 2 m vertical spacing. The geogrid type was designed to provide a minimum long term creep rupture strength ULS of 21 kN/m and be compatible with lime modified material.

Based on shear box tests, the addition of lime to the Gault Clay (from which the embankment fill is constructed) is assumed to increase the shear strength from a minimum values of $c' = 5\,kN/m^2$ and $\phi' = 22°$ (found in the failure back analysis) to a maximum achievable of $15\,kN/m^2$ and 25°.

A drainage blanket was placed at the base of the embankment in order to control the water regime in this region and the existing gully pipes were disconnected and replaced by a new surface water drainage system. However, the effect of the drainage blanket was intentionally ignored for analysis purposes.

The embankment was analysed considering the two typical values of $r_u = 0.2$ and 0.4, for both of which the results (FOS) are shown in Table 3 for three options: (i) without any reinforcement (ii) geogrid reinforcement only, and (iii) geogrid with lime. According to these results, even though the use of geogrid reinforcement alone may provide sufficient tensile strength to prevent any future slope failure, the addition of lime (which enhanced the FOS by up to 25%) is beneficial in conditioning the fill and so aiding compaction during construction.

4.4 *Stability analysis of Site-2 – The effect of vegetation*

According to the site visit survey conducted at Site-2, the vegetation on the embankment is well established

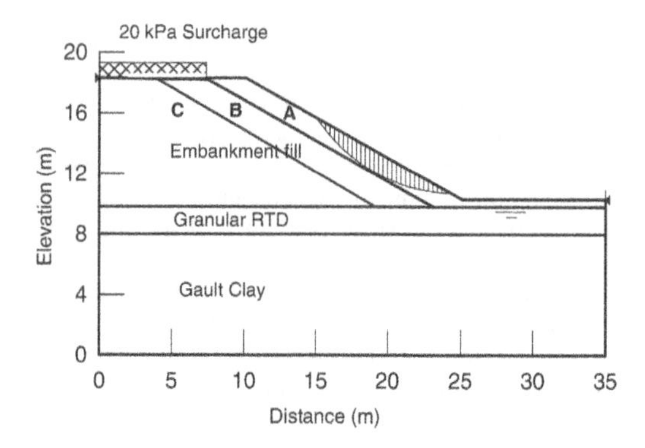

Figure 11. Shallow slip surface with pore water pressure ratio, $r_u = 0.2$ and three areas (A, B, C) assumed for the embankment fill to encounter the effect of the vegetation.

(unlike the vegetation on Site-1, which is small and newly planted and therefore it was not considered in the analysis discussed above). The mechanical influence of the vegetation on slope stability can not be ignored and may be modeled by incorporating root reinforcement using adjusted effective cohesion, c'veg (Wu et al., 1979) larger than the actual value used for the soil (c'). However, the near surface the soil is likely to have tension cracks due to the desiccation effect of the plants. Consequently, this will increase the permeability of the surface layers significantly, whilst reducing the intrinsic shear strength parameters (c' and ϕ'). The embankment slope may be subdivided therefore into three main areas: A, B and C (as shown in Figure 11) with different shear strength parameters corresponding to the vegetation effect. The soil parameters of layer A and B were adjusted to consider the effect of vegetation, while layer C (the core of the embankment) is assumed having the typical parameters obtained by the ground investigation without any adjustment (see Table 1).

The mechanical influence of vegetation may also include the negative effect of their weight. This was considered in the model slope by adjusting the unit weight of layer A and B to include the weight of the trees. Although the weight of trees surcharges the slope, increasing normal and downhill force components, tree weight in some situations is beneficial to slope stability (Greenway, 1987).

In addition to these mechanical effects, the hydrological effect of vegetation will alter the stability of the slope. As discussed previously, the performance/stability of embankments will depend on the pore pressures within. Increases in pore pressure (reduction of suction) due to inundation events may lead to failure events; seasonal fluctuations of pore pressure may lead to 'creep' serviceability displacements. Plants will reduce the likely maximum pore pressure through transpiration and rainfall

Table 4. Summary of the soil parameters considered for Site-2.

Layer ID (see Fig. 11)	Soil Parameters c' kN/m^2	ϕ' Degree	γ_b kN/m^3	r_u
A	2	20	22	0.2
B	10	24	23	0.2
C	8	24	19	0.2

interception, although it is likely that reduced transpiration due to seasonal cycles coincides with the worst inundation conditions. Vegetation will however increase the seasonal pore pressure fluctuations (Davies et al., 2004).

The slope stability analysis incorporating the mechanical effects of vegetation can be adjusted to include also the hydrological effects by adopting suitable value for the pore water pressure ratio $r_u = 0.2$ (Greenway, 1987; Morgan & Rickson, 1995).

Figure 11 shows the model slope arrangements, and Table 4 presents the soil parameters considered in the analysis, where the factor of safety approaches FOS ≈ 1 with potentially unstable conditions within the upper layer (layer A). These results are consistent with the general observation of the actual slope failure as shown in Figure 4.

5 CONCLUSION AND RECOMMENDATIONS

Two case studies of slope instabilities (shallow and deep) have been presented in this paper. In both cases, the embankments (which are 8 m high with slope angle of 1V:2H) have been typically constructed from overconsolidated clay, which has a capacity to absorb large amounts of water in a swelling and softening process.

Site investigation has been conducted to confirm the geology and geotechnical parameters of the embankment fill in addition to the underling natural ground. Although the local drift geology provided valuable information about the original materials that the embankments have been constructed from, the natural ground has a limited impact on the stability since the slope failures are entirely within the embankment fill.

In conjunction with the characteristics of the embankment fill material, its geometries, and seasonal fluctuations of pore pressure, the faulty drainage system and inadequate subgrade drainage are believed to be the principal causes of the deep slope failure that has occurred in the first case study. This was confirmed by standard slope stability analysis (using Slope/W), where a typical mode of failure was predicted when the embankment material had a reduced shear strength parameters as well as high ratio of pore water pressure.

The second site, which has thriving vegetation, had shown a less severe instability event of shallow failure. Although plants may reduce shear strength parameters near surface, the overall stability is likely to be improved because of the positive effect of the root reinforcement and the reduction of the likely maximum pore pressure. It may be concluded that the overall mechanical and hydrological influence of vegetation on slope stability may become more positive than negative if the vegetation has been well established and managed.

For the deep-seated failure case, full slope repair of geogrid reinforcement combined with lime modification was analyzed and adopted. Although geogrid reinforcement has been found to be the most effective and economic method of remediation (Boden, 1995) however the addition of lime is essential to condition the reworked fill for better compaction, and perhaps will minimise the risk of local soil softening during periods of wet weather.

The analysis can be improved by considering a progressive failure mechanism based on pore water pressure fluctuation and associated with systematic change in stiffness and strength of the materials. For this purpose, an advanced numerical modelling may be used, which has more potential for improving the understanding of the failure mechanism. Nevertheless, the approach presented herein (using standard methods) is found to be sufficient to satisfy the scope of the design.

ACKNOWLEDGEMENT

The authors gratefully acknowledge the support of the Highways Agency.

REFERENCES

Atkins. 2003. A14 Histon to Milton. *Preliminary Sources Study Report.* Ref. N. 4425082/GTG.2001527/003/R.002.

Atkins. 2004. A14 Milton-Histon Slope Failures. *Geotechnical Report.* Ref. N. 4425082/GTG.2001527/R.008.

Atkins. 2006. A14 Milton to Fen Ditton Eastbound. *Preliminary Sources Study Report.* Ref. N. 4425082/GTG. 2001527/038/R.002.

Aubeny, C.P. ASCE, M. Lytton, R.L. & ASCE, F. 2005. Shallow Slides in Compacted High Plasticity Clay Slope, *J. Geotech. and Geoenvir. Engrg.* Volume 130, Issue 7, pp. 717–727. July 2004.

Boden, D.G. 1995. Performance of slope maintenance and repair techniques on the Cambridge Northern Bypass. *Report prepared by Transport Research Laboratory.* Project Report PR/CE/52/95 E022A/HG.

British Geological Survey BGS. 1981. *Cambridge solid and drift edition. 1:50,000 scale, sheet* 188.

Davies, M. Hallett, P., Bengough, G. Bransby, F. Wilson, A. 2004. Vegetated embankment performance in the N-W

zone. *Desk study Report prepared by University of Dundee and SCRI for Network Rail.*
Geo-slope International Ltd. 2001. Slope/W v.5 Users Guide.
Greenway, D.R. 1987. Vegetation and Slope Stability. *In Slope Stability*. Wiley and Sons, New York.
Greenwoood, J.R. Holt, D.A. & Herrick, W. 1985. Shallow slip in highway embankments constructed of overconsolidated clay, *ICE Sym*. Thomas Telford Ltd.
May Gurney. 2004. Cambridge A14: Histon to Milton, Ground Investigation. *Factual report.*
Morgan, R.P.C. & Rickson, R.J. 1995. *Slope Stabilisation and Erosion Control: A Bioengineering Approach.* E and FN Spon.
Morgan, T. 1993. A45 Cambridge Bypass Widening. *Geotechnical Desk Study.*
Ordnance Survey Map. 1889. Ordnance survey map of Kings Hedges provided by *www.old-maps.co.uk.*
Perry, J. 1989. A survey of slope condition on motorway earthworks in England and Wales. *TRL Research Report RR199.* Crowthorne.
Perry, J. Pedley, M. & Reid, M. 2003. *Infrastructure embankments – condition appraisal and remedial treatment.* 2nd edition (C592), CIRIA.
Potts, D.M. & Zdravkoviæ, L. 1999. Finite Element Analysis in Geotechnical Engineering: Theory. *American Society of Civil Engineers.* Thomas Telford Ltd.
ScottWilson. 2006. A14 Milton to Fen Ditton: Eastbound. Geotechnical Ground Investigation. *Factual Report.* December.
Stokes, A. Spanos, I. Norris, J.E. & Cammeraat E. 2004. Eco- and Ground Bio-Engineering: The Use of Vegetation to Improve Slope Stability. *Proc. First intern. Conf. on Eco-Engineering 13–17 September.*
Worssam, B.C. & Taylor, J.H. 1969. Geology of the Country Around Cambridge. *Explanation of One-inch Geological Sheet 188, New Series, Institute of Geological Sciences.*
Wu, T.H. McKinnell, W.P. & Swanston, D.N. 1979. Strength of tree roots and landslides on Prince of Wales Island, Alaska. *Can. Geotech. J.* 16: 19–33.

Advances in Transportation Geotechnics – Ellis, Yu, McDowell, Dawson & Thom (eds)
© 2008 Taylor & Francis Group, London, ISBN 978-0-415-47590-7

Performance of a very deep soil nailed wall in the Istanbul subway

G. Icoz
Bogazici University, Istanbul, Turkey

T. Karadayilar
Zemin Teknolojisi A.S., Istanbul, Turkey

H.T. Durgunoglu & H.B. Keskin
Bogazici University, Istanbul, Turkey

ABSTRACT: Istanbul is a heavily populated city and the demand for mass transportation systems are increasing rapidly. Deep excavations and retaining structures are constructed in the city of Istanbul at different locations due to recent demand for the construction of subway lines within the underground transportation systems. Additionally, any construction project adversely effects the traffic load in the surrounding area, thus the construction time should be kept minimal. Excavation using soil nailed walls is relatively faster as compared to other alternatives leading to their widely usage in the metropolitan area of Istanbul. The project described in this paper consists of a station and crossover structure of Basak Residence-Kirazlı subway line. This cut and cover type construction requires excavations as deep as 40 meters. This is so far the deepest soil nailed wall constructed in the city. The plan area is approximately 10,000 square meters and the total area of the retaining structure is 22,500 square meters. The performance of these walls is monitored by means of inclinometers and surveying. The performance of this very deep soil nailed wall based on the observed lateral displacements at various steps of excavation is presented together with the overall safety of the wall under various loading conditions and stages of excavation as a case study.

1 INTRODUCTION

The city of Istanbul due to its recent growth in economy and population caused a great demand for additional mass transportation systems such as subway, light rail etc. Because of the existing structures and the varying topography of Istanbul, deep excavations are compulsory to allow the trains to travel safely and obeying grading and curvature limits set by train manufacturing companies. The subject subway line consisted of eight stations between Basak Residence-4 and Kirazlı stations. The stations are planned to be constructed by cut and cover construction technique and the tubes between stations are planned to be constructed by TBM working in both directions.

Considering the subway train limitations and topographical changes, very deep excavations were required to construct cut and cover type stations due to the topographical changes along the subway line in order to meet the four per cent maximum gradient. Consequently, the depths of the excavations have reached to 40+ meters below the existing ground surface.

Figure 1. Soil Nail Retaining Wall and TBM Entering Tunnel.

The excavation for Basak Residence-4 station site consists of 180 m long station and 165 m long crossover structure. The existing ground surface elevations vary between +135.0 m and

+150.0 m. The bottom of excavation is approximately +105.0 m, resulting in a total excavation as deep as 45.0 m.

Within the lateral influence zone of the excavation, no structures and/or utilities exist thus the lateral displacements and consequent vertical settlement as a result of excavations are not very strictly limited in terms of performance requirements of the retaining structures, thus employment of a soil nailed system is feasible and favorable.

2 SUBSURFACE CONDITIONS

The encountered subsoil formation is a thin layer of fill underlained by soft rock greywacke locally known as the Trace Formation, which is lithologically alternating sandstone, siltstone and claystone with various degrees of weathering and fracturing. The lithological unit is extensively fractured and non-homogeneous. In general, the subsurface mechanical properties of graywacke formation improve with increasing depth. On the other hand, the presence of a side valley at the project site was observed during excavation resulting in a localized shear zone and an attraction for groundwater flow. Obviously, the extend of weathering and fracturing controls the mechanical properties and in fact geological observations do agree well with the results of measurements reflecting mechanical properties of the formation. The geotechnical modeling of graywacke formation in the city, weathered zones, extend of fracturing and compressibility modulus of formation are usually obtained by means of integrated seismic survey and Menard pressuremeter testing performed within the boreholes at various locations and depths (Durgunoglu and Yilmaz, 2007a). No groundwater level is encountered in the completed borings or during excavation, however trapped water and the water flowing through the shear zone created by the side valley is observed. It is a well-known fact that the graywacke formation looses its strength significantly when it is in contact with water upon unloading during excavation. Therefore, a retaining system should be constructed considering minimum exposure of the formation to water during construction. Therefore, full coverage of excavation face with shotcrete employed in soil nailing offers a great advantage over other retaining systems.

3 SEISMICITY OF THE REGION

The city of Istanbul is potentially under the influence area of the Marmara Fault System, located at the south, in the Marmara sea, which is the western end of the North Anatolian Fault (NAF) of Turkey. In Istanbul, earthquake records spanning two millennia indicate that, on average one medium intensity (Io = VII–VIII) earthquake has effected the city every 50 years. The average return period for high intensity (Io = VIII–IX) events has been 300 years.

After the 1999 Kocaeli and Duzce earthquakes occurred on NAF within the Marmara Region in approximately 100–150 kilometers from the city of Istanbul, the structure of NAF system in Marmara sea attracted worldwide scientific attention.

Recent studies conducted after 1999 Kocaeli ($M_w = 7.4$) and Duzce ($M_w = 7.2$) earthquakes indicated about 65% probability for the occurrence of a $M_w > 7.0$ earthquake effecting Istanbul within the next 30 years due to the existence of potential seismic gaps (Parsons et al., 2000).

A site specific seismic risk assessment is carried out for the subject site. As a result, the peak ground acceleration (pga) is given as 0.185 g. According to the specification for the structures to be built in disaster areas by the Ministry of Public Works, the pga value for the structures in contact with the ground should be multiplied by a factor of 0.6. Consequently, a resulting effective horizontal earthquake acceleration coefficient of 0.11 g is obtained for design. This horizontal earthquake acceleration coefficient is to be used for pseudo-static analysis to assess seismic stability of the retaining system adopted.

4 TYPE OF RETAINING STRUCTURE

Among all options for the retaining wall systems for the subject site, the choice of the retaining structure is done based on pre-set criteria such as applicability, cost, speed, excavation safety and safety of surrounding structures. Since there is no surrounding structure within the influence zone of excavation, other remaining criteria dominated the choice of the system. Other than applicability, cost and safety, the speed of excavation is also a key parameter for this station, because this station is the entrance station for the TBM which is the major item effecting the schedule for the timely completion of the subway line, thus minimizing the duration of the excavation is a crucial criteria. For this reason, the excavation should be completed as soon as possible allowing the tunnel boring machine to be setup and start boring.

In addition, a cost comparison was done between the various retaining system alternatives considered. The two most economical systems were found to be soil nailed wall and the tied back bored piles with implementation of prestressed anchors. Eventually, it was estimated that soil nailed wall is about 30% more economical than the anchored tied back retaining system. In addition, considering the subsoil conditions and the depth of the excavation there is a potential for large diameter bored piles to deviate from vertical

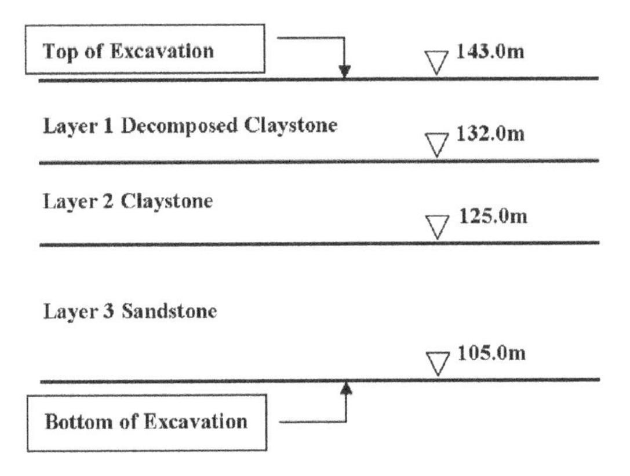

Figure 2. Idealized Soil Profile for H = 38.0 m of Excavation.

causing additional stability and safety problems during construction. Furthermore, based on the previous positive records of flexible earth retaining structures during earthquakes in Turkey Mitchell et al. (2000) and Durgunoglu et al. (2003a), soil nailed walls in such excavations performed within the city offer great advantage especially for the encountered subsoil and seismic conditions even as temporary structures. Considering all the factors described above, the soil nailed wall was chosen as the retaining system for the subject excavation.

5 DESIGN

5.1 *Soil modeling*

The initial step in the soil nailed wall design is the soil modeling using the available subsurface investigation data. Based on the information obtained from subsurface investigations and the laboratory analysis, a three layered model was established and the thin fill layer was not included in the design. The soil model is a 10 m-thick weathered claystone of Trace formation underlained by 7 m of weak claystone. Further, weak claystone was underlained by medium hard, moderately fractured sandstone. The idealized soil profile is presented in Figure 2 and the design soil parameters are summarized in Table 1.

The excavation plan was employed to minimize the wall height in such a way that the existing ground level will be excavated by 1:1 slope until approximate elevation of +143.0. Since it is known from the previous studies that a small inclination from the vertical has a great advantage in the performance of the nailed walls (French National Research Project Clouterre, 1993; Elias and Juran, 1989), the slope angle of $\beta = 79°$ (5:1 slope) was utilized for the remaining stages of the excavation.

Table 1. Soil Parameters Used in the Design.

	Layer 1	Layer 2	Layer 3
Ø′ (degrees)	30	35	38
c′ (kPa)	5	5	10
Unit Weight (kN/m3)	18	19	21
Nail Skin Friction (kPa)	200	250	250
Elasticity Modulus* (MPa)	70	100	150
Poisson Ratio	0.15	0.10	0.10

* For unloading, this value is multiplied by 3.0 for graywacke.

After elevation of +143.0 m, excavation will be completed by 5:1 (5 vertical by 1 horizontal) by soil nailed wall in three stages. For the first stage, excavation is planned from approximate elevation of +143.0 m to elevation of +132.0 m. At this level, a 5-meter wide platform is established. The second stage of excavation is planned from approximate elevations of +132.0 m to an elevation of +118.5 m. At this level, a 2.0 m wide another working and service platform platform is established. For the final stage, the excavation is planned from the bottom of the second stage to the final excavation level which is approximately at elevation of 105.0 m. Considering the 5.0 m and 2.0 m wide platforms, the corresponding effective slope angle for the complete soil nail wall (β_e) becomes equal to 69 degrees. Nail orientation with the horizontal were adopted as $\omega = 10°$ together with a nail diameter of Ø = 32 mm for the first stage of excavation and Ø = 40 mm for the other two stages. The common grouted drill hole diameter is adopted as D = 110 mm.

5.2 *Software*

Using the above listed parameters and excavation plan, a soil nail pattern is established and diameter and length of the nails are optimized to minimize excavation cost by means of minimizing nail density, i.e. average length of nail per square meter of excavation face. The detailed design cross section is presented in Figure 3.

Two computer programs were utilized to establish and optimize soil nail system. Talren 4, a stability analysis software for geotechnical structures with or without reinforcement used for global stability check for the retaining wall system using Bishop, Perturbations or Fellenius method. Plaxis V8, a 2D finite element software intended for the two dimensional analysis of displacements and stability.

5.3 *Results of the analysis*

The design criteria for the wall is to have a maximum performance ratio of 0.3% for the lateral displacements at the top of the wall. In addition, the general

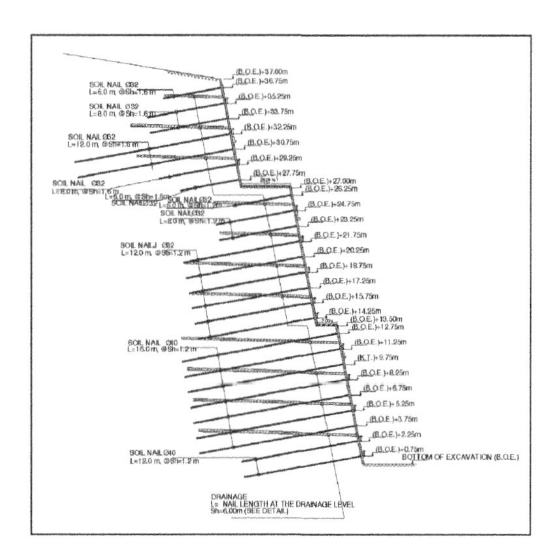

Figure 3. Detailed Design Cross Section.

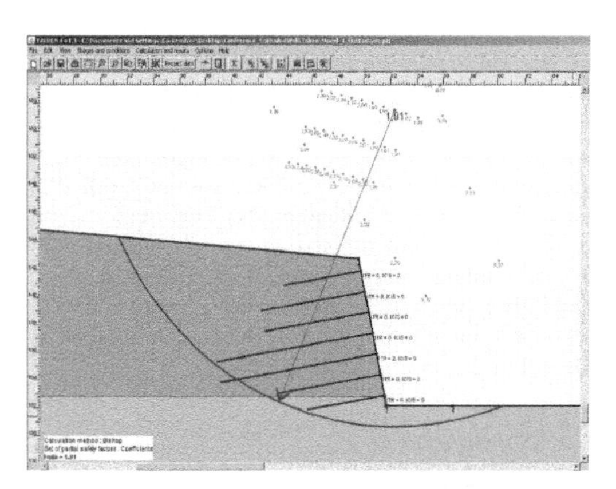

Figure 4. Result of the Static Overall Stability for the First Stage Excavation Under Static Conditions (F.S. = 1.91).

practice is to have a minimum factor of safety for overall stability of 1.3 for temporary structures under static loading and 1.1 under earthquake loading. Furthermore, in terms of internal stability, soil nails should not exceed allowable load or bond capacity. The analyses are carried out for both static and earthquake loadings. The results for the global stability analysis using Talren4 software for the static case is presented in Figure 4 for the first stage excavation, in Figure 5 for the second stage excavation and in Figure 6 for the final stage excavation. The minimum factor of safety for the static case was found as 1.48. On the other hand, the result of the stability analysis using Talren4 software under earthquake loading is presented in Figure 7, leading to a factor of safety of 1.24.

Further, using PlaxisV8 software, the general system stability and the horizontal wall displacements

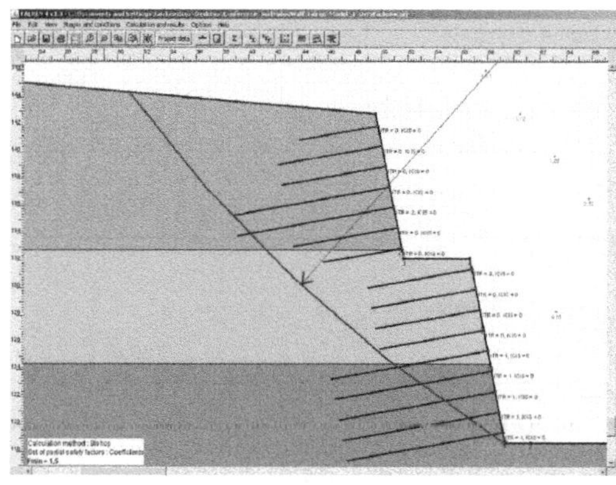

Figure 5. Result of the Static Overall Stability Analysis for the Second Stage Excavation Under Static Conditions (F.S. = 1.50).

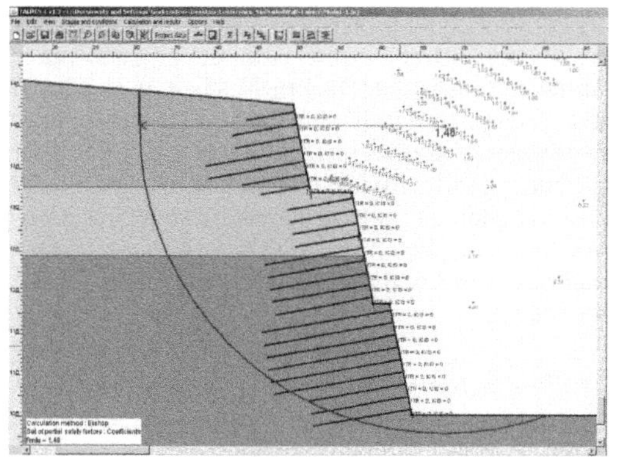

Figure 6. Result of the Overall Stability Analysis for the Final Stage Excavation Under Static Conditions (F.S. = 1.48).

are evaluated. The factor of safety for the system stability for static conditions is found to be 1.37. The failure mechanism is presented in Figure 8, the factor of safety for the system stability under earthquake loading is found to be 1.15. The failure mechanism under earthquake loading is presented in Figure 9. The result for the horizontal wall deformations is presented in Figure 10 and maximum lateral displacement is estimated as $\delta_{lmax} = 11.9$ mm corresponding to performance ratio of $P_r = 0.03\%$.

The wall stability and deformations are within the allowable design criteria as set by the design documents and project specifications.

6 PERFORMANCE OF THE NAILED WALL

The design of the wall is done using available subsurface information as outlined in "Subsurface Conditions" section of this paper. Where the subsurface

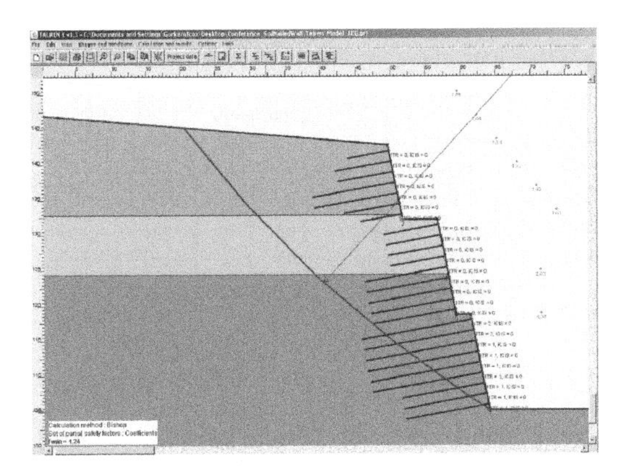

Figure 7. Result of the Overall Stability Analysis for the Final Stage Excavation Under Earthquake Loading (F.S. = 1.24).

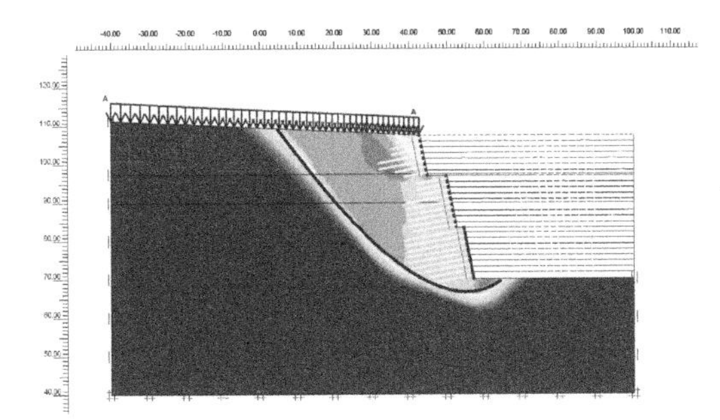

Figure 8. Failure Mechanism for the Static Conditions – Plaxis.

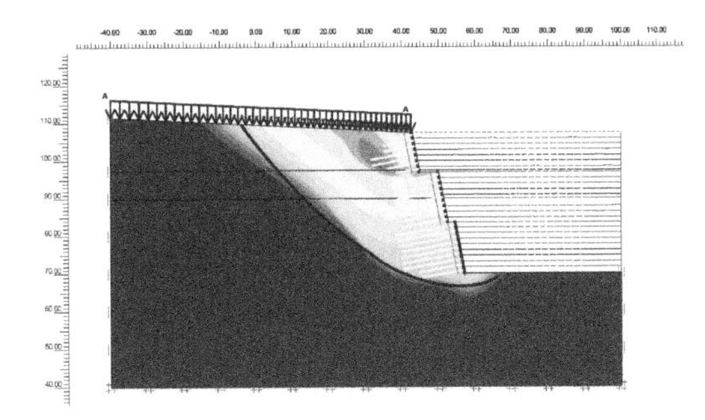

Figure 9. Failure Mechanism for the Earthquake Conditions – Plaxis.

conditions are similar to those summarized above, the excavation was carried out with no problems, however variations from initial design due to unexpected subsurface and geological conditions at some locations and stages have occurred. These variations are described in the next section.

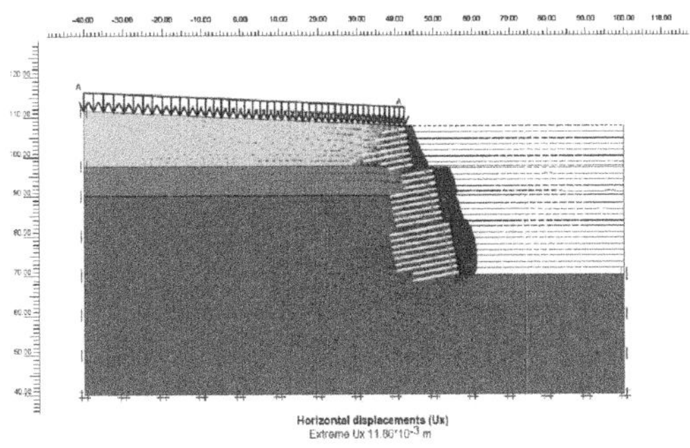

Figure 10. Horizontal Wall Deformations ($\Delta\delta_{hmax}$ = 11.9 mm).

Table 2. Summary of the Design and Performance of the Soil Nailed Wall For Each Layer.

	Layer 1	Layer 2	Layer 3
Vert. Spacing, S_v (m)	1.5	1.5	1.5
Hor. Spacing, S_h (m)	1.6	1.2	1.2
Area per nail, S (m^2)	2.4	1.8	1.8
Nail diameter, Ø (mm)	32	32	40
Avg. nail length, L (m)	8.6	10.2	12.9
Nail density, η (m/m^2)	3.5	5.7	7.2

According to recent compilation by Zetas (2007) about 185,000 m^2 of wall had been constructed in different projects and the performances of some of these soil nailed wall structures have been reported previously by Durgunoglu et al. (1997), Ozsoy (1996) and Yilmaz (2000).

The design and actual displacement and normalized displacement (i.e. performance ratio, P_r) data are presented together with some basic parameters of soil nailed walls (Phear et al., 2005) such as, height of wall (H), area per nail (S), average nail length (L), nail density (η = L/S). In table 2, area per nail (S), average nail length (L), nail density (η = L/S) for each layer of the three layered model are presented in Table 2.

In this study, the performance of the wall is monitored by means of electronic inclinometer recordings taken at certain time intervals in parallel to the excavation where excavation took place in locally well-known Trace Formation of greywacke. The two inclinometer recordings that are located at the crest of the deepest excavation points of the wall are presented in Figures 11 and 12.

During the design process, the maximum horizontal displacement at inclinometer locations is estimated as approximately 9 mm. As a summary, the height of the wall (H), average nail length (L), nail density (η), design and actual performance of the wall (P_r) at two inclinometer locations are presented in Table 3.

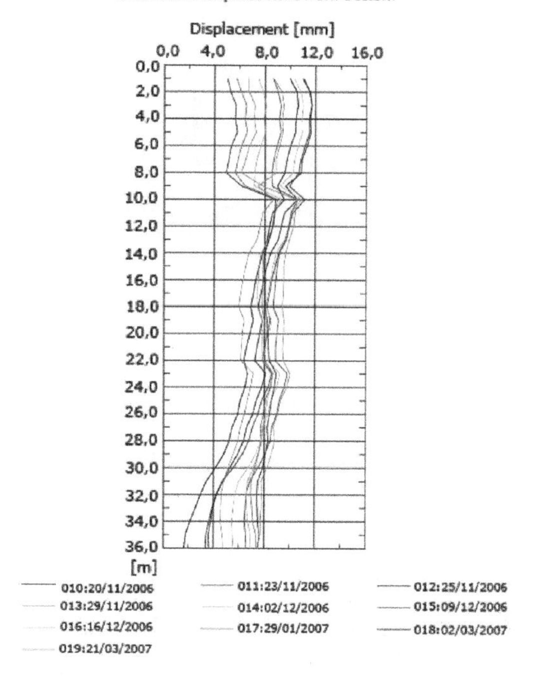

Figure 11. Inclinometer-8 Readings.

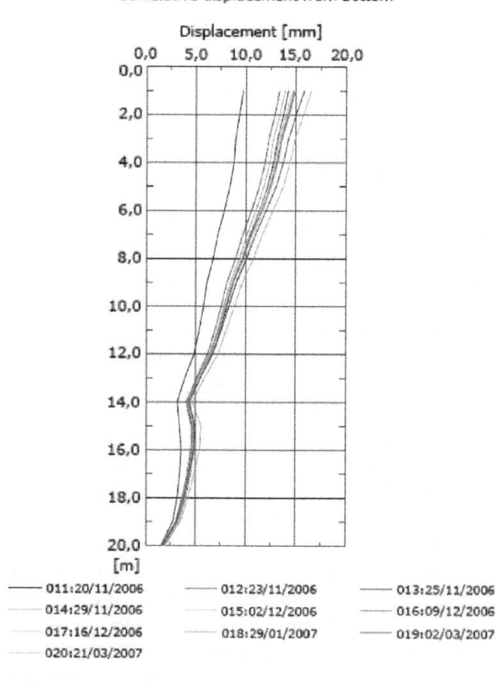

Figure 12. Inclinometer-9 Readings.

Table 3. Summary of the Design and Performance of the Soil Nailed Wall at two Inclinometer Locations.

	inclinometer-8	inclinometer-9
Wall Height, H (m)	37.1	37.5
Nail density, η (m/m^2)	5.7	5.6
Wall perf., P_r(10^{-3}, design)	0.23	0.23
Wall perf., P_r(10^{-3}, actual)	0.45	0.32

By the results of the data given in Tables 2 and 3, the following observations and evaluations are done for the soil nailed wall constructed in Basak Residence-4 subway station and cross-over structure:

A linear increase in lateral displacement with the height of the wall is observed until reaching to a certain height.

- The increase in slope occurs at various heights where the formation changes from sandstone to weak claystone to even weaker decomposed claystone. Similar observation was made previously by Durgunoglu et al. (2003b).
- The performance ratio, P_r, for the greywacke formation is within the range of $(0.32–0.45) \times 10^{-3}$.
- Average nail length, L, increases linearly with the height of soil nailed wall. The average nail length that was utilized is about $L = 12.9$ m for $H = 37.5$ m.
- Nail density, L/S (m/m^2) also increases linearly with the height of the soil nailed wall. It is about 5.7 m/m^2 for $H = 37.5$ m.
- For the similar graywacke formation, although the average nail length and density are similar to the values reported earlier by Durgunoglu et al. (2007b), performance ratio recorded for the subject project is much smaller. This is basically due to low effective slope of the wall for this case, $\beta_e = 69°$ in comparison to 85° reported in case studies of Durgunoglu et al (2007b). In other words, the lateral displacements do decrease considerably with decreasing effective slope angle from vertical.
- The performance of the wall as opposed to actual design shows variations. The observed displacements are generally higher than the values estimated at the design stage. This is believed to be because of the non-homogeneous structure of the graywacke formation, geological alterations and the existence of pocket and/or trapped water and possibly due to the higher modulus values used than the real one in the modeling of subsoil in numerical analysis.

7 LOCAL VARIATIONS FROM ORIGINAL DESIGN

The initially suggested soil nail pattern, length and diameter was applied for the big part of the project, however local variations from design occurred due to unexpected change in subsurface conditions.

A major revision needed to be made due to the existence of a side valley that could not have been foreseen before the excavation was initiated. Due to this geological formation, along the borderline of the side valley, a localized shear zone was observed. The graywacke was extremely weathered and became a natural water flow path along the sheared zone. It is a well-known problem that the trace formation looses its strength significantly as it is in contact with water upon excavation. Consequently, stress relief in horizontal direction as a result of excavations is the major potential hazard. Thus additional stability measures were taken for this part of the excavation by means of implementing prestressed anchors. The ability of the soil nailing system to adapt to such changes and additions during construction based on the site observations is another great advantage of the system.

8 CONCLUSIONS

Soil nailing is a very versatile excavation retaining system for deep excavations in urban areas surrounded by major structures and infrastructures provided that limiting lateral displacements are not exceeded.

Using conventional methods of design, Federal Highway Administration (2003) and previously developed charts for estimating lateral displacements or performance ratio may be misleading in deep soil nailing applications

Modern numerical analysis based design, application and construction monitoring for such deep soil nailed walls are of primary importance. Monitoring should be done by an experienced geologist/geotechnical engineer to observe changes of subsurface and geological conditions and by means of electronic inclinometers where recordings taken at certain time intervals in parallel to the excavation. The design should be revised in cases where adverse subsurface and/or geological conditions are observed or displacements are above certain limits as set by engineering practice, local codes or project specifications. This case study has shown that this could be achieved with success following the above outlined algorithm during design and construction.

ACKNOWLEDGEMENT

The soil-nailed wall for the Basak Residence-4 station and cross-over structure was constructed by main contractor Gulermak Dogus A.S. Design and performance monitoring during construction have been performed by Zemin Etud ve Tasarim A.S. We would like to thank both of these companies for their cooperation during construction and providing the pertinent data.

REFERENCES

Brinkgreve, R.B.J. & Broere, W., 2004. Plaxis 2D V8 Manual, Delft University of Technology & PLAXIS b.v.

Bureaux d'Ingenieurs-Conseils en Geotechnique, 2005. Talren4 Design Manual, Terrasol.

Durgunoglu, H.T., Kulac, H.F. & Olgun, C.G. 1997. Flexible Earth Retaining Structures – Soil Nailing, Engineering and Education, Symposium Honoring Vedat A. Yerlici, May 22, Istanbul, Turkey pp. 287–296.

Durgunoglu, H.T., Tari, T. & Catana, C. 2003a. Esnek Istinat Yapılarının Depremde Davranışı (in Turkish), 5. Ulusal Deprem Mühendisliği Konferansı, 26–30 May, Istanbul.

Durgunoglu, H.T., Kulac, H.F. & Arkun, B. 2003b. A Deep Retaining System Construction with Soil Nailing in Soft Rocks in Istanbul, Turkey, SARA-2003 Conference, June 22–26, MIT Cambridge, USA.

Durgunoglu, H.T. & Yilmaz, O. 2007a. An Integrated Approach for Estimation of Modulus Degradation in Soft Rocks, Proceedings of 4th ICEGE, Paper no. 1174, June 25–28, Thessaloniki, Greece.

Durgunoglu, H.T., Keskin, H.B., Kulac, H.F., Ikiz, S., & Karadayilar, T. 2007b. Performance of Soil Nailed Walls Based on Case Studies, Proceedings of the 14th ECSMGE, Vol. 2, pp. 559–564, Madrid.

Elias, V. & Juran, I. 1989. Soil Nailing for Stabilization of Highway Slopes and Excavations, Federal Highway Administration Report, FHWA-RD-89-198, Washington DC, USA.

Federal Highway Administration. 2003. Geotechnical Circular No: 7, Soil Nail Walls, US Department of Transportation, FHWA-AO-IF-OS-017, Washington DC, USA.

French National Research Project Clouterre. 1993. Recommendations Clouterre 1991, English Translation, Federal Highway Administration, FHWA-SA-93-026, Washington, USA.

Mitchell, T.K., Martin, J.R., Olgun, C.G., Emrem, C., Durgunoglu, H.T., Cetin, K.O. & Karadayilar, T. 2000. Chapter 9 – Performance of Improved Ground and Earth Structures, Earthquake Spectra, Supplement A to Volume 16, December, pp. 191–225.

Ozsoy, M.B. 1996. Soil Nailing Design Construction and Monitoring, M.S. Thesis, Bogazici University, Istanbul, Turkey.

Parsons, T., Toda, S., Stein, K.S., Barka, A. & Dietrich, J.H. 2000. Heightened Odds of Large Earthquake Near Istanbul: An Interaction Based Probability Calculation, Science, 288, pp. 661–665.

Phear, A., Dew, C., Ozsoy, B., Wharmby, N.J., Judge, K. & Barley, A.D. 2005. Soil Nailing – Best Practice Guidance, CIRIA C637, London.

Yilmaz, S. 2000. Behavior of Soil Nailed Walls in Different Soil Conditions, M.S. Thesis, Bogazici University, Istanbul, Turkey.

Advances in Transportation Geotechnics – Ellis, Yu, McDowell, Dawson & Thom (eds)
© 2008 Taylor & Francis Group, London, ISBN 978-0-415-47590-7

Landslide risk factors of geotechnical systems

S.I. Matsiy & Ph.N. Derevenets
Kuban State Agrarian University, Krasnodar, Russia

ABSTRACT: The main factors, which determinate reliability and risk of "soil mass-construction" geotechnical system, have been considered. The risk of exploratory design of an engineering protection scheme for a landslide section on a highway has been assessed. A set of the anti-landslide measures has been recommended and devised on the grounds of a colligative risk degree index. Anti-landslide retaining structures have been designed taking into consideration the landslide pressures between the tiers.

1 RELIABILITY AND RISK AS PROPERTIES OF GEOTECHNICAL INSTALLATIONS

Geotechnical installations should be considered from the point of view of the "soil mass-construction" united system under the particular engineering geological conditions. Solid mass reliability means its property to receive a whole aggregate of external impacts within the specified term with a provision of a normal operation of the constructions put in place (Ermolaev & Mikheev 1976). In its turn, such property depends on probability and intensity of negative influences of the hazardous geological and natural processes (HGNP) (Ragozin 1997).

A specified reliability level should be provided by a design taking into consideration an interconvertible reliability-risk index. The risk is determined as a product of probability of hazard (or frequency of event) by an expected damage:

$$R = P(A) \times Y(A) \quad (1)$$

where R = risk; $P(A)$ = frequency of event, probability of hazard; $Y(A)$ = expected damage from the event.

In other words, risk R is expressed in the form of (Einstein & Karim 2001):

$$R = P(A) \times P(Y) \times P(o) \quad (2)$$

where $P(Y)$ is probability of expected damage; $P(o)$ is the damage results, which can be expressed by the economic or environmental indices.

It means that risk can be described by "hazard-result" diagram and can be estimated with the help of several categories, for example: very high, high, middle, low, very low, optimal, and acceptable (Chowdhury & Flentje 2003). There are three approaches to risk assessment: qualitative, semi-qualitative and quantitative.

The specialists can assess the risk qualitatively on the grounds of their experience, slope investigation result analysis and the survey material being available. For the semi-qualitative assessment, each of many factors, which exert influence on stability (slope geometry, stability violation history and others), is investigated. Here, it is very important to assess a degree of the influence of each factor on the landslide process development and, as a result, on a possible damage.

The quantitative approach requires a quantitative assessment of the risk factors being expressed by the probabilities: an annual activation of motions; a capture of the existing buildings into the landslide process; a landslide displacement genesis when there are people and mechanisms in the section; damage and others.

Landslide risk is determined by two indices: probability and extent of risk (Tikhvinsky 2000). Probability of risk is probability of occurrence of a loss (economic, environmental, social loss) in case of an impact of the landslide process on a specific installation. The extent of risk is assessed by the caused damage, i.e. it is measured by the value or physical units. For the specific installations, a "risk with actions" relative index, which is expressed in a graphic form, is determined by a summation of the "initial risk" and "cost of measures" values.

2 RELIABILITY FACTORS OF THE GEOTECHNICAL SYSTEMS

The main factors, which determine reliability, are as follows (Ermolaev & Mikheev 1976):

- conformity of the accepted diagrams and a soil mass stability design method with the actual conditions of its operation;
- integrity of a description of the engineering-geological conditions of construction of the structures;
- integrity of initial materials concerning physicomechanical properties of soils taking into consideration the conditions of their folding and the structural-textured properties;
- integrity of the data concerning the loads and the impacts in the structure operation process;
- design decision implementation correctness control in the construction process.

The engineering geological conditions for the building projects situated in the highlands are ravelled by development of the unfavourable processes: landslides, erosion, denudation, river activities, etc. The physicomechanical properties of soils will be changed in the course of time depending on an activation of the hazardous engineering geological processes.

Design decision implementation correctness control should be combined with monitoring. In accordance with the building code (SNiP 2003) "... it is necessary to envisage an instrumentation equipment installation ... for a supervision of hazardous process development and an engineering protection structure operation within the period of construction and operation ... for a timely detection of an activation of the hazardous geological processes and taking the necessary steps for protection of the buildings and the structures and a provision of safety of people".

The factors, which improve "soil mass-construction" system reliability, are as follows: integrity of survey materials, rational design of the landslide measures and an implementation of monitoring. The factors, which abate reliability, are as follows: time, variability of the engineering geological conditions and soil properties, uncertainties (for example, assumptions in the design diagrams and the soil test methods).

3 RISK FACTOR ASSESSMENT AT THE INSTALLATION

A soil mass, which is represented by an ancient landslide section situated on the southern slope of the North-West Caucasus between the Malaya and the Bolshaya Khosta rivers, has been analysed. In January, 2006, a more recent landslide was formed here, its volume being nearly half a million cubic metres (Fig. 1).

Figure 1. Landslide at the section of Khosta-Verkhnyaya Khosta highway.

The displacement has involved a section of the highway of considerable length, which connects the city of Sochi with three communities: settlements Khleborob, Kalinovoe Ozero and Vorontsovka as well as Samurskaya-Sochi gas pipeline.

The engineering protection envisages several variants of risk and reliability analysis using the qualitative and quantitative approaches.

Variants 1a...e. The highway is laid on the top of the anti-landslide installation in the form of a solid pile structure (1a) or a trestle bridge employing separate supports (1b ... 1e) connected by the bay highway structures. The supports are made as the pile clusters employing cast-in-place piles being coupled by a reinforced concrete foundation cross. Practically, an axis of the installation coincides with the old alignment of the road.

Such solution will make the landslide bypass the supports without its stabilization and without violation of stiffness of the trestle itself. But a probability of soil displacement during construction is great. That's why in this case the risk belongs to "high" category.

Variant 2. It envisages a restoration of the initial location of the highway section by means of an arrangement of a high embankment on the body of the more recent landslide.

But in this case, the head of the landslide will be loaded. That's why it is necessary to erect an anti-landslide structure, which reduces the risk. In order to protect the section, a gravity structure, which involves considerable costs, will be required.

The rest variants given below envisage a set of the anti-landslide measures, including an erection of the pile retaining structures. They differ in an arrangement and a configuration of the structures. All variants include an arrangement of a double-row retaining installation employing cast-in-place piles in the landslide head (on the top of the disruption wall) as well as a single-row installation in the tongue across the main soil displacement vector.

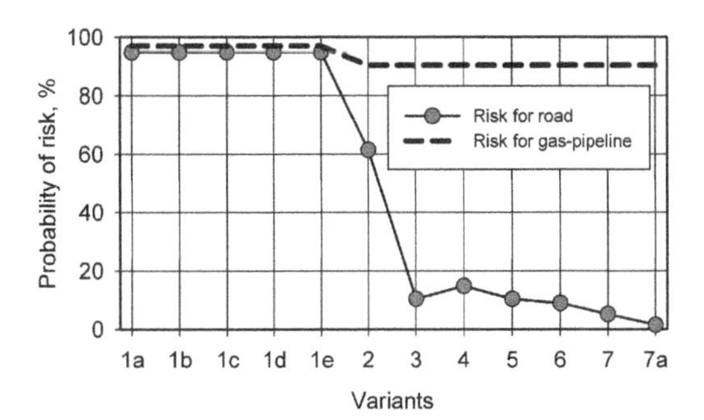

Figure 2. Anti-landslide measure risk probabilities at the section of Khosta-Verkhnyaya Khosta highway.

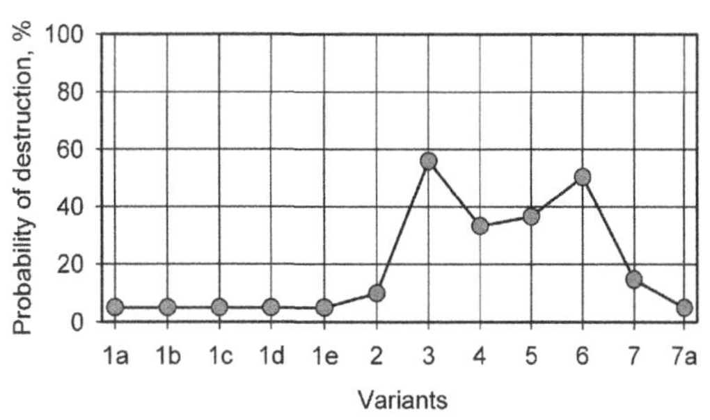

Figure 3. Cost of anti-landslide measures.

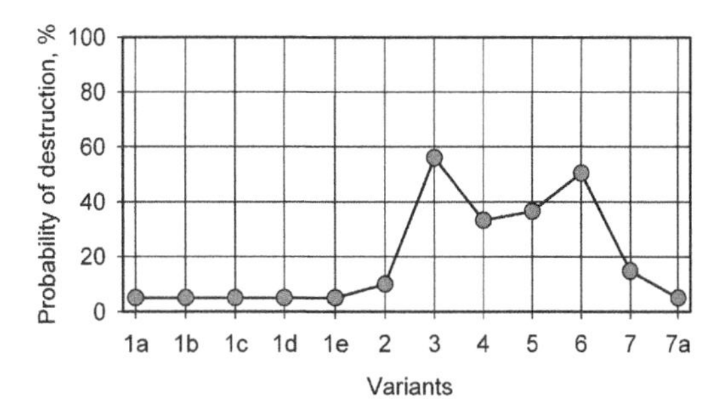

Figure 4. Prospective failure probability of the restored section of the highway depending on the engineering protection variants.

Variant 3. It envisages an arrangement of three double-row pile retaining installations along the base of the highway on the part being adjacent to the landslide. Two installations are arranged along an upper curb and one installation is arranged along the base of the left-hand edge of the modern landslide.

Variant 4. It is similar to variant 3, except the installation in the base of the left-hand edge of the landslide. Instead of it, a reinforced concrete retaining wall is arranged on the pile base employing two rows of cast-in-place piles. A backfill behind the wall plays the role of a counter-bank. The installation is arranged across the landslide in its middle part.

Variant 5. It is similar to variant 4. The landslide installation in the middle part of the landslide has U-shaped form on the plan. Due to it, a volume of the backfill behind the retaining wall is reduced.

Variant 6. It is similar to variant 3. It differs in a form of the installation being arranged on the top of the left edge close to the landslide head. It has a crinkle-crankle form on the plan.

Variant 7. It is similar to variant 3. It differs in an increased length of the installation in the area of the upper part of the landslide situated along the upper curb of the edge of the landslide in the part being adjacent to the highway. The retaining wall along the highway at the toe of the left-hand edge of the landslide has L-shaped form.

For a transition to the quantitative assessment of the risk, the most important components of reliability and risk have been specified: risk probability, cost of measures and failure probability. In accordance with the slope stability design results taking into consideration a complexity of the engineering-geological conditions, a "risk probability" diagram has been plotted (Fig. 2), i.e. probability of occurrence of a loss (economic, environmental, social loss) in case of the impact of the landslide on a line structure. The cost of the engineering protection measures has been determined according to the enlarged indices (Fig. 3). As it is clear from the diagram, some trestle variants are relatively inexpensive. But the risk probability is very high threat.

Highways section failure probability (Fig. 4) was determined for each variant taking into consideration a range of variation of the values of the soil physicomechanical indices being calculated in accordance with the laboratory test results.

A final diagram of the degree of risk of each variant of the developmental work has been plotted according to the summary relative indices of risk (Fig. 5). Variant 7a (variant 7+ monitoring) is an optimal variant of engineering protection, its "summary degree of risk" being 0.9.

4 ANTI-LANDSLIDE STRUCTURE CALCULATION

On the grounds of an assessment of the degree of risk, the most optimal variant of engineering protection of the section of Khosta-Verkhnyaya Khosta highway, which has been destroyed by the landslide motions, was chosen. The landslide pressures on the anti-landslide structures were calculated taking into consideration a pressure distribution between the tiers of the structures.

It is necessary to determine landslide load taking into account soil reaction below the structure along

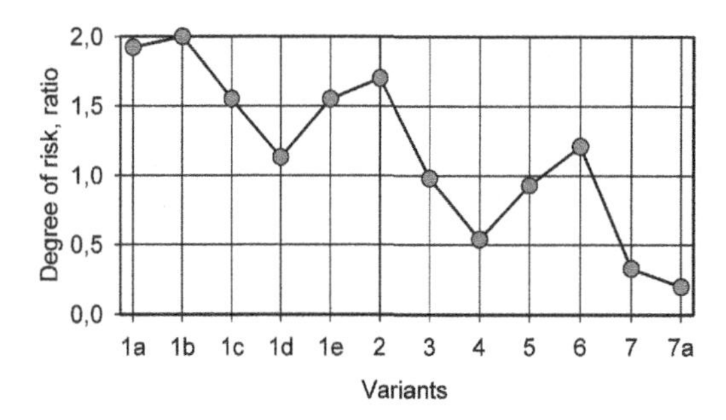

Figure 5. Summary diagram of degree of risk of the variants of engineering protection of the section of Khosta-Verkhnyaya Khosta highway.

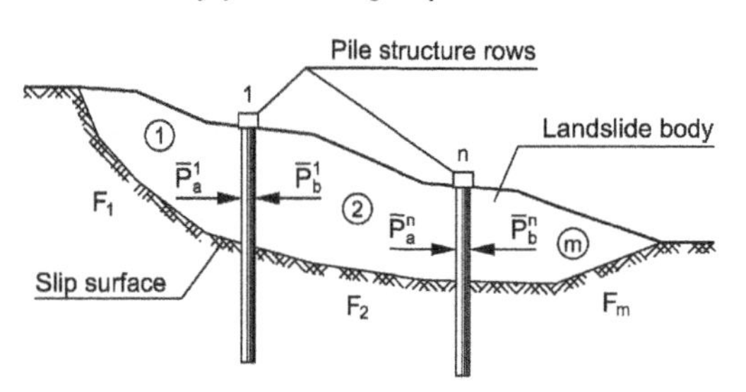

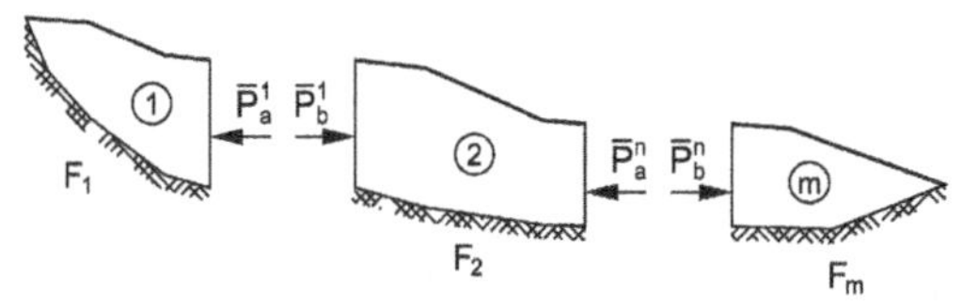

Figure 6. Basic diagram of the landslide load calculation on the tiers of pile constructions taking into account the soil reaction.

the slope (Yamagami et al. 1991) – as a difference of reactions of the anti-landslide pile structure to the parts of the landslide mass being arranged above (P_a) and below (P_b = soil reaction) the structure along the slope (Fig. 6):

$$P = P_a - P_b \geq 0,\ P_a > 0,\ P_b \geq 0 \tag{3}$$

Within the framework of limit equilibrium, the slope stability degree is usually determined by a ratio of a sum of all retaining forces (or moments) along the slip surface to a sum of all shearing power factors. Taking into account a construction of one or several tiers of the structures, landslide body stability is determined from a condition of equilibrium of all shearing and retaining forces in the following way (Matsiy & Derevenets 2006):

$$F_f = \frac{\sum_{i=1}^{m} R_i^f + \sum_{j=1}^{n} P_{af}^j}{\sum_{i=1}^{m} T_i^f + \sum_{j=1}^{n} P_{bf}^j} \tag{4}$$

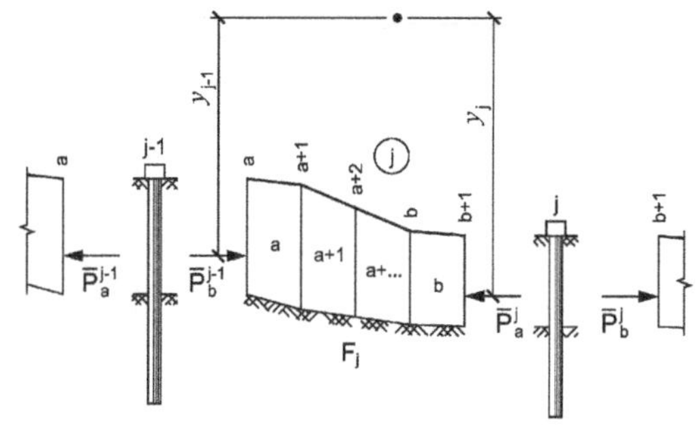

Figure 7. Design diagram of the section method for a calculation of the landslide loads on the multi-tier pile constructions.

where ΣR^f = sum of retaining forces along the slip surface; ΣT^f = sum of shearing forces; i = numbers of the slices of the landslide model; m = total quantity of the slices of the design diagram; and j = numbers of the cross-sections of an arrangement of the tiers of the retaining structures.

Respectively, the components of landslide load are equal to:

$$P_{af}^j = \left(\sum_{i=a}^{b} T_i^f + P_{bf}^{j-1} \right) F_j - \sum_{i=a}^{b} R_i^f \tag{5}$$

$$P_{bf}^{j-1} = \left(\sum_{i=a}^{b} R_i^f + P_{af}^j \right) \frac{1}{F_j} - \sum_{i=a}^{b} T_i^f \tag{6}$$

where a and b = the numbers of the first and the last slices of j part of the landslide body between $j-1$ and j tiers of the structures (Fig. 7); and F_j = safety factor for j part of the landslide body.

Similarly, it results from the condition of moment equilibrium of all shearing and retaining forces:

$$F_m = \frac{\sum_{i=1}^{m} R_i^m + \sum_{j=1}^{n} P_{am}^j y_j}{\sum_{i=1}^{m} T_i^m + \sum_{j=1}^{n} P_{bm}^j y_j} \tag{7}$$

$$P_{am}^j = \left[\left(\sum_{i=a}^{b} T_i^m + P_{bm}^{j-1} y_{j-1} \right) F_j - \sum_{i=a}^{b} R_i^m \right] \frac{1}{y_j} \tag{8}$$

$$P_{bm}^{j-1} = \left[\left(\sum_{i=a}^{b} R_i^m + P_{am}^j y_j \right) \frac{1}{F_j} - \sum_{i=a}^{b} T_i^m \right] \frac{1}{y_{j-1}} \tag{9}$$

where ΣR^m = sum of retaining moments along the slip surface; ΣT^m = sum of shearing moments; y = arm of a landslide load component in the design section (a distance from a point of force application $P_{a(b)}$ to a point of rotation).

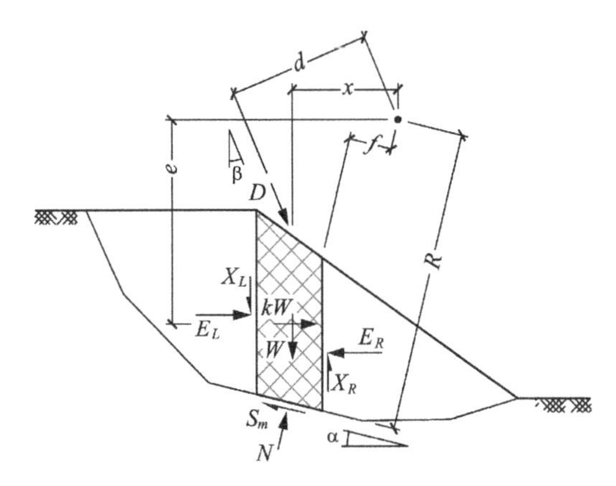

Figure 8. Design diagram of the combined method.

The end formulas of the calculation of landslide load have been obtained on the basis of the combined method (Krahn 2004), which takes into account an influence of the interslice forces E (Fig. 8).

$$E_i^R = E_i^L + N_i\left(\sin\alpha_i - \frac{\tan\varphi_i \cos\alpha_i}{F_j}\right) + \frac{u_i \tan\varphi_i - c_i}{F_j} l_i \cos\alpha_i + kW_i + D_i \sin\beta_i \tag{10}$$

$$N_i = \left[W_i - \left(X_i^R - X_i^L\right) + \frac{u_i \tan\varphi_i - c_i}{F_j} l_i \sin\alpha_i + D_i \cos\beta_i\right] \cdot \frac{1}{\cos\alpha_i + \frac{\sin\alpha_i \tan\varphi_i}{F_j}} \tag{11}$$

where W = soil weight in the slice; N = normal force at the base of the slice; E = horizontal component of the interslice forces. Indices L and R mean the left and right sides of the slice, respectively; X = vertical component of the interslice forces; D = resultant of external load; k = seismicity ratio; l = length of the base of the slice; α = angle of inclination of the base of the slice to a horizon; β = angle of inclination of force D to a vertical.

$$P_{af}^j = \left(\sum_{i=a}^{b} N_i \sin\alpha_i + k\sum_{i=a}^{b} W_i + \sum_{i=a}^{b} D_i \sin\beta_i + P_{bf}^{j-1}\right) F_j - \sum_{i=a}^{b}\left[c_i l_i + \left(N_i - u_i l_i\right)\tan\varphi_i\right]\cos\alpha_i \tag{12}$$

$$P_{bf}^{j-1} = \frac{\sum_{i=a}^{b}\left[c_i l_i + \left(N_i - u_i l_i\right)\tan\varphi_i\right]\cos\alpha_i + P_{af}^j}{F_j} - \left(\sum_{i=a}^{b} N_i \sin\alpha_i + k\sum_{i=a}^{b} W_i + \sum_{i=a}^{b} D_i \sin\beta_i\right) \tag{13}$$

$$P_{am}^j = \left\{\left(\sum_{i=a}^{b} N_i f_i + \sum_{i=a}^{b} W_i x_i + k\sum_{i=a}^{b} W_i e_i + \sum_{i=a}^{b} D_i d_i + P_{bm}^{j-1} y_{j-1}\right) F_j - \sum_{i=a}^{b}\left[c_i l_i + \left(N_i - u_i l_i\right)\tan\varphi_i\right] R_i\right\} \frac{1}{y_j} \tag{14}$$

$$P_{bm}^{j-1} = \left\{\frac{\sum_{i=a}^{b}\left[c_i l_i + \left(N_i - u_i l_i\right)\tan\varphi_i\right] R_i + P_{am}^j y_j}{F_j} - \left(\sum_{i=a}^{b} N_i f_i + \sum_{i=a}^{b} W_i x_i + k\sum_{i=a}^{b} W_i e_i + \sum_{i=a}^{b} D_i d_i\right)\right\} \frac{1}{y_{j-1}} \tag{15}$$

where R = arm of retaining force S_m; f = arm of normal force N; x = arm of gravity force of the slice W; e = arm of seismic load kW; d = arm of a resultant of external load D.

When a set of equations (10–15) is solved, an assumption concerning a pattern of the distribution of the interslice forces in the form of a functional dependence $f(x)$ is used.

$$X = E\lambda f(x) \tag{16}$$

Thus, according to the formulas (10–16) with the use of (2), it is possible to get the values of landslide load P on each tier of the structures.

5 DESIGN DECISIONS OF ENGINEERING PROTECTION

A set of engineering protection of the highway envisages a restoration of the alignment of the road situated on the top of the right-hand flange of the landslide. A new roadbed is arranged under a protection of two tiers of pile retaining structures situated along the top (I tier) and the base (II tier) of the landslide flange slope (Fig. 9). The physicomechanical properties of soils are given in Table 1.

According to the landslide pressure results taking into consideration a distribution between the tiers of the anti-landslide structures, a landslide pressure value on I tier was 746 *kN/m*; on II tier, it was 364 *kN/m*.

Thus, on the grounds of the calculations aimed at the engineering protection of the highway, an arrangement of the following structures is envisaged:

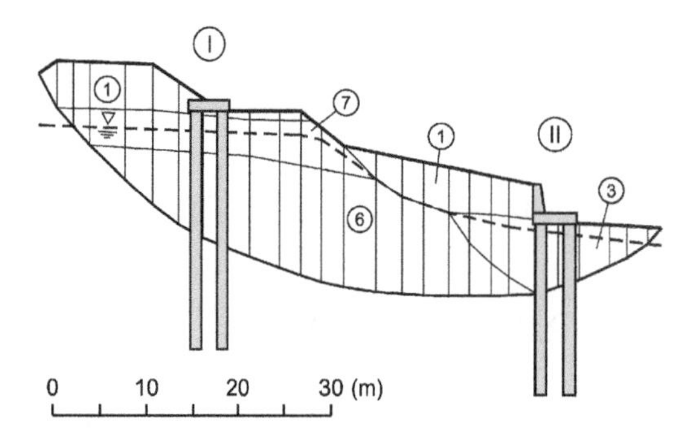

Figure 9. Design diagram in the characteristic section.

Table 1. Physical and mechanical characteristics of the landslide soils.

Soil	Unit weight γ, kN/m^3	Cohesion c, kPa	Angle of friction φ, deg
1	19.0	10.0	30.0
3	20.3	18.3	11.9
6	19.4	30.2	6.4
7	18.2	27.8	7.2

- the first tier: a structure employing two rows of the cast-in-place piles with Ø1200 *mm* and the length of 20 *m*, which are united by a monolith reinforced concrete raft foundation;
- the second tier: a retaining wall with the height of 4 *m* being erected on a pile footing employing two rows of the cast-in-place piles Ø1200 *mm* and the length of 20 *m*, which are united by a monolith reinforced concrete raft foundation.

The piles of both tiers of constructions has been placed with spacing of 3.0 *m* in a row and between rows.

6 CONCLUSIONS

1. Provision of the optimal conditions of the operation of the geotechnical systems requires an assessment of the factors, which determine their reliability.
2. Factor analysis of the landslide risk on the grounds of the qualitative and quantitative approaches makes it possible to choose a balanced design decision of the engineering protection.
3. According to the minimal value of the summary relative index of the degree of risk, the anti-landslide measures at the section of Khosta-Verkhnyaya Khosta highway have been worked out.
4. The pile retaining structures have been designed taking into consideration the distribution of the landslide pressures between the tiers of the structures.

REFERENCES

Chowdhury, R. & Flentje, P. 2003. Role of slope reliability analysis in landslide risk management. *Bull. Eng. Geol. Env* 62: 41–46.

Einstein, H.H. & Karam, K. S. 2001. Risk assessment and uncertainties. *International conference on "Landslides – Causes, Impacts and Countermeasures"*: 457–488. Davos (Switzerland).

Ermolaev, N.N. & Mikheev, V.V. 1976. Structure foundation reliability. Leningrad: Stroyizdat.

Krahn, J. 2004. *Stability modeling with Slope/W. An Engineering methodology*. Geo-Slope International Ltd., Calgary: Alberta.

Matsiy, S. & Derevenets, Ph. 2006. Assessment of landslide load on pile construction according to GLE method. J. Logar, A. Gaberc, B. Majes (eds), *Active geotechnical design in infrastructure development; Proc. Danube-European conf., Ljubljana, 29–21 May 2006* 2: 645–650.

Ragozin, A.L. 1997. Theory and practice of geological risk assessment. *Dissertation in the form of a scientific report for a competition of the scientific degree of Doctor of geological and mineralogical sciences.* Moscow: PNIIIS.

SNiP 22-02-2003 2004. *Engineering protection of the areas, buildings and constructions from the hazardous geological processes. Basic provisions.* Moscow.

Tikhvinsky, I.O. 2000. Landslide risk assessment on the regional and local level. *Assessment and natural risk management: proceedings of the All-Russian conference "Risk 2000"*: 242–246. Moscow: Ankil.

Yamagami, T. et al. 1991. A simplified estimation of the stabilizing effect of piles in landslide slope. Applying the Janbe method. *Landslides, Bell (ed.)*: 613–618. Rotterdam: Balkema.

Advances in Transportation Geotechnics – Ellis, Yu, McDowell, Dawson & Thom (eds)
© 2008 Taylor & Francis Group, London, ISBN 978-0-415-47590-7

Landslide investigation of roads in Sochi on the basis of the optimum risk method

S.I. Matsiy & D.V. Pleshakov
Kuban State Agrarian University, Krasnodar, Russia

ABSTRACT: Results of landslide investigation of motorways in Sochi are presented. Calculations have been accomplished on the basis of an optimum risk method. Using results of investigations conducted in the minimum time and with a minimum of engineering-geological survey data, the landslide risk level has been evaluated on the basis of risk factors. The approach has allowed quantitative risk assessment, and prioritisation of works and potential losses.

1 INTRODUCTION

For a risk estimate qualitative, semiquantitative and quantitative approaches (Chowdhury & Flentje 2003) are used nowadays. The qualitative and semiquantitative risk estimate, as a rule, is based on judgement, and also on qualifications and experience of experts and is applied in conditions of data deficiency. The landslide risk qualitative analysis needs accordingly a quantitative assessment of the factors promoting origin of hazard (Postoev 2003).

In some events absolute quantitative risk assessment is less important, than the qualitative (Fell, Ho, Lacasse, Leroi 2005). To such events it is possible to refer measures on planning and priority definition operations, and also at the substantiation of investments.

It is necessary to note absence of the uniform designed technique of hazard interpretation nowadays. In those few countries which began application of probabilistic approach to an slope stability estimation, it will be realized differently.

The main problem of hazard interpretation consists in select of measure of its applicability (Varga 2002). In geotechnical construction two types of criterion are scheduled: 1) normative – allowable and 2) acceptable, that is voluntary accepted by all faces and organizations which exposes potential hazard.

At voluntary accepted measure frequently it is required to define optimum balance between increase of profitability and efficiency of object, on the one hand, and the relevant growth of hazard with other side. Special difficulties originates if there is one only qualitative assessments of hazard.

Comparison of the calculated hazards with accepted criterion is finished by selection of a final design solution: the ratification of the calculated hazard or acceptance of organizational or technical measures on hazard reduction. In an event of voluntary accepted measure it is taken into account first of all magnitudes of eventual losses, cost of measures, charges on management, etc.

In broadly fathoming risk management process represents the continuous supervisory control of a risk level on the basis of long-time landslide structures monitoring.

Use of a new dilated system model of probabilistic interaction of a structures with geological environment corresponds more actual engineering – geological conditions, normally described by the considerable complexity and uniqueness of a structure and the long-lived multistage development of potentially hazardous process scenario.

Replacement of the absolute slope reliability concept with the limiting equilibrium methods by the optimum risk concept is more progressive methodological solution promoting increase of estimation accuracy in conditions of impossibility of absolute complete study of environmental change and human factor.

2 DEVELOPED SITUATION ON LANDSLIDE AREAS OF MOTOR-ROADS IN SOCHI REGION

During landslide areas investigation (Fig. 1–3) in Sochi region more than sixty landslide sites have been inspected. Operations included the analysis of archival materials, routing survey of territory. During surveys it accomplished a reconnaissance survey of adjoining

Figure 1. Destroyed site of the motorway Veseloe – Nignyaya Shilovka, km 3 + 690.

Figure 2. Landslide deforation of the motorway Makopse – Nadjigo, km 5 + 300.

territory of motorways, the expansive photographic material gathered.

The geological environment of investigated territory characterized by rocks of cretaceous, paleogene and psychozoic era. The region orient to the Tuapse foothill sag which large part is incorporated on a shelf zone of Black sea. Transient structural character of a zone has found the reflecting in lithological complex of breeds and tectonic structures of region.

The investigated territory is characterized by development of erosive and landslide processes. On a landslide surfaces the bulwarks of a flow are marked. Thus, the complex of geomorphological attributes testifies the existence of the modern landslide process.

Within the limits of inspected motorways and on adjoining territories it is possible to select the following modern engineering – geological phenomena and processes: landslips, erosion, a weathering, a planar ablation.

Figure 3. Landslide deforation of the motorway Lazarevskoe – Kirova, km 2 + 800.

It is necessary to note, that landslide hazard on the investigated territory existed before construction of motorways.

By the analysis of conducted survey it is fixed: all sites of motorways practically throughout passes in cuts and semi-excavations, embankments and semi-embankments, that considerably increases the probability of landslides.

Major factors of landslide process development on the examined motorway sites are: exogenous – humidifying of breeds by atmospheric and underground waters, undercut of slopes by erosion, a weathering; anthropogeneus – a human activity.

Development of the landslide phenomena is hardly influenced with processes of a weathering. Most actively processes of a weathering proceed in argillaceous rocks (argillites, clay aleurolites, marls).

Human factor in development of landslides play completely important role. The amount of landslips, caused by a road construction on the investigated territory, has considerably increased. Activation of landslides is caused, first of all, by slope excavations, the construction of the embankment reducing landslide resistance of slopes, absence of the organized drainages.

As has shown survey, landslips on the indicated sites are active, rises to the considerable sizes and deform a road-bed of a motorway.

During sites investigation, for grading landslide areas the technique of an optimum risk assessment was applied.

3 LANDSLIDE SITES INVESTIGATION BY THE OPTIMUM RISK METHOD

The optimum risk method is most effective at research of a plenty landslide slopes with the purpose of revealing the most hazardous areas. The approach allows to forecast behaviour of a slopes by an assessment of quantity risk factors and does not require the fulfilment of a major complex of engineering – geological surveys.

The approach allows to perform a quantitative risk level analysis and to organize the priority of designing structures. The eventual result of such analysis consists in definition of an integrated parameter of risk which is calculated with use of personal experience and the stored knowledge. The disadvantage of an optimum risk method is lack accounting of the composite interaction processes of structures and ground, and also some simplifications of the complex technique.

Reliability and efficiency of landslide measures depends on the forecast of local slopes resistance and of landslide hazard. The landslide behaviour is largely influenced with grounds reclamation by the person, a water balance, variability of a nature, and also other factors (Lee & Jones 2004).

For an estimation of a risk class during landslide areas investigation in Sochi the following factors were considered:

- A class of a motorway;
- A damage rate of a traffic-bearing surface;
- Availability of cracks and magnitude of crack opening displacement;
- An expansion of the motorway sites;
- Height and incline of the slopes;
- Capacity and volume of hazardous geological processes;
- Erosion processes and water-saturation of a ground;
- Availability of vegetation on the slope;
- Availability of an engineering protection measures and a level of their damage;
- Threat to inhabited objects and population;
- The human factor.

For each factor of an landslide risk assessment the relative indexes from 1 up to 10 were appropriated. Magnitudes of indexes were determined individually for each investigated site.

The factor magnitudes are empirically obtained by the analysis of great number of slopes. Factors allows to take into account the engineering – geological and hydro-geological conditions of an investigated site, the historical data, the human factor and other parameters, rendering influence on a landslide risk level.

Figure 4 shows a distribution of risk magnitudes and costs of engineering protection measures.

On the basis of an integrated parameter of risk, sites were subdivided into classes (Fig. 5). The technique has allowed to reveal the most hazardous sites (risk class 1). As a result of conducted researches the complex of engineering protection measures on the slopes has been determined.

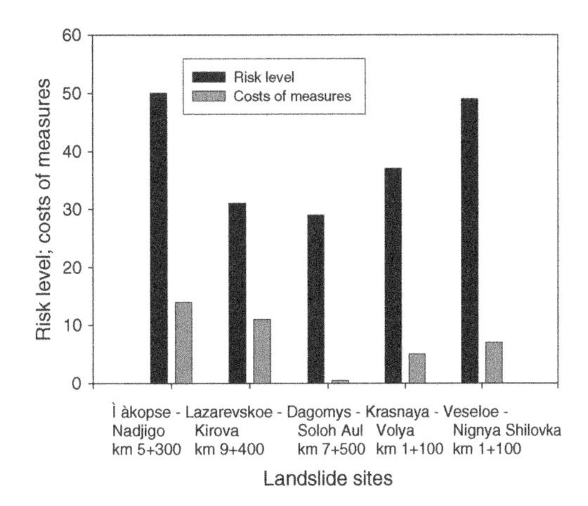

Figure 4. The diagram of risk level distribution and costs of engineering measures on the investigated landslide sites.

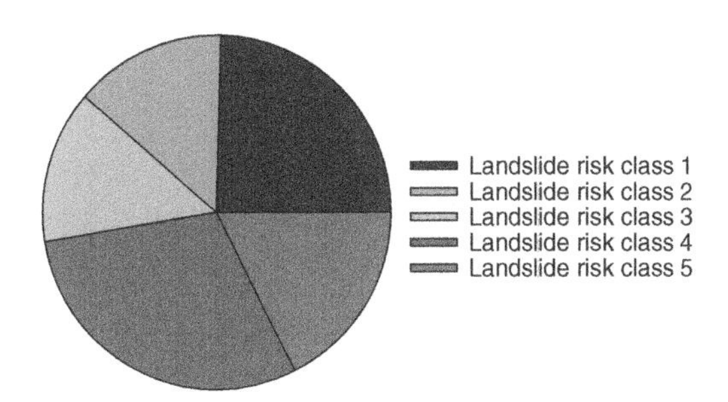

Figure 5. Percentage distribution of landslide sites to the risk classes.

4 CONCLUSION

1. With the help of optimum risk method in the shortest times and at a minimum of engineering geological survey data landslide risk level of the investigated sites has been estimated on the basis of risk factors. The given fact is extremely important on the projects, and also at the investments planning of construction measures.
2. The approach has allowed objectively, on the basis of quantity risk factors, to classify objects to the priority operations classes and to a level of potential losses.

REFERENCES

Chowdhury, R. & Flentje, P. 2003. Role of slope reliability analysis in landslide risk management. *Bull. Eng. Geol. Env*62: 41 – 46.

Postoev, G. 2003. About a quantitative landslide risk assessment. Natural risks assessment and management.

Proceedings of All – Russian conference «Risk – 2003»: 48–50. Moscow (Russia).
Fell R., Ho K. K. S., Lacasse S. Leroi E. 2005. A framework for landslide risk assessment and management. *Proceedings of international conference «Landslide risk management»*: 3–27. Vancouver (Canada).
Varga A. A. 2002. The probabilistic analysis of hydraulic engineering constructions safety at interaction with the geological environment. Engineering geology. Hydrogeology: 99–111. Moscow (Russia).
Lee E. M., Jones D. K. C. 2004. Landslide risk assessment: 375–381. London (England).

Advances in Transportation Geotechnics – Ellis, Yu, McDowell, Dawson & Thom (eds)
© 2008 Taylor & Francis Group, London, ISBN 978-0-415-47590-7

Slope stabilization of an embankment on the West Coast Main Line, UK

B.C. O'Kelly
Trinity College Dublin, Ireland, and formerly Scott Wilson Limited, UK

P.N. Ward & M.J. Raybould
Scott Wilson Limited, UK

ABSTRACT: A site inspection of an embankment on the West Coast Main Line (UK) revealed a general deterioration of the embankment, significant movement of the trackside services and waterlogging and bulging along the embankment toe. The medium-height embankment, with side slopes at about 30 degrees to the horizontal, was constructed on natural sidelong ground and the site had been an area of previous instability (former ash and slag tip). Site investigation, ground monitoring and analytical modelling indicated that the embankment was at limiting equilibrium and that recent progressive movement was occurring along a rotational slip surface located within the embankment core (ash and slag materials overlying reworked glacial till). The upgrade works comprised the installation of a row of stabilizing piles along the mid-height of the embankment side slope (facilitating Green Zone working); regrading to prevent shallower slips near the embankment crest and the installation of a toe drain that discharged to a nearby stream.

1 INTRODUCTION

The site area is part of an embankment (Figure 1) located at Wreay (about 5 km south-southeast of Carlisle, Cumbria) along a north-south aligned section of the Lancaster to Carlisle railway line (West Coast Mainline, UK) at Ordinance Survey national grid reference NY 439 482. The Lancaster to Carlisle line was constructed during the mid 1840s and currently comprises two tracks (Up-line and Down-line) supplied with 25 kV overhead electrification. Figure 2 shows some pertinent rail terminology as it relates to this particular site.

The embankment has an overall length of about 300 m and had been constructed on natural sidelong ground, which generally sloped in a north eastward direction. The embankment slope on the Up-line side was up to 13 m in height whereas the side slope on the Down-line side was less than 4 m in height (typically between 2 and 3 m). The side slopes were densely vegetated with grass, shrubs and mixed wood (Figure 3).

Two previous slope failures were known to have occurred on the Up-line side, adjacent to the site area behind Gill House farm (Figure 1).

Network Rail instructed Edmund Nuttall Limited to design and construct the necessary upgrade works, which were instigated following a site inspection (revealed leaning gantry at 63 mls 1556 yds, waterlogged ground conditions and bulging along the Up-line toe) and track monitoring data (indicated adverse movement of the rail tracks between 63 mls 1534 yds and 63 mls 1610 yds). Note that distances and locations along the rail lines are traditionally given in miles and yards in the industry (one mile equals 1.609 km and one yard equals 0.914 m).

Scott Wilson Limited was appointed by Edmund Nuttall Limited to undertake a desk study, site inspections, ground investigation and to design the necessary upgrade works to reduce future movement of the track and improve the stability of the critical embankment section identified. This paper presents the geotechnical design and construction of the upgrade works.

2 GROUND CONDITIONS

2.1 *Ground investigation*

The ground investigation works were detailed by Scott Wilson and carried out in stages by Ritchies Limited (UK) between April and October 2001. The works comprised 21 cable percussive boreholes, three trial pits and the installation of three standpipe-piezometer and seven inclinometer instruments. The locations of the exploratory holes and ground instrumentation are shown in Figure 1. Table 1 lists the inferred sequence of the different strata encountered. Figure 4 shows a typical geological cross-section for the site area. The undrained shear strength profile with depth was determined using empirical correlations from the N blow count values (Figure 5) that were measured from

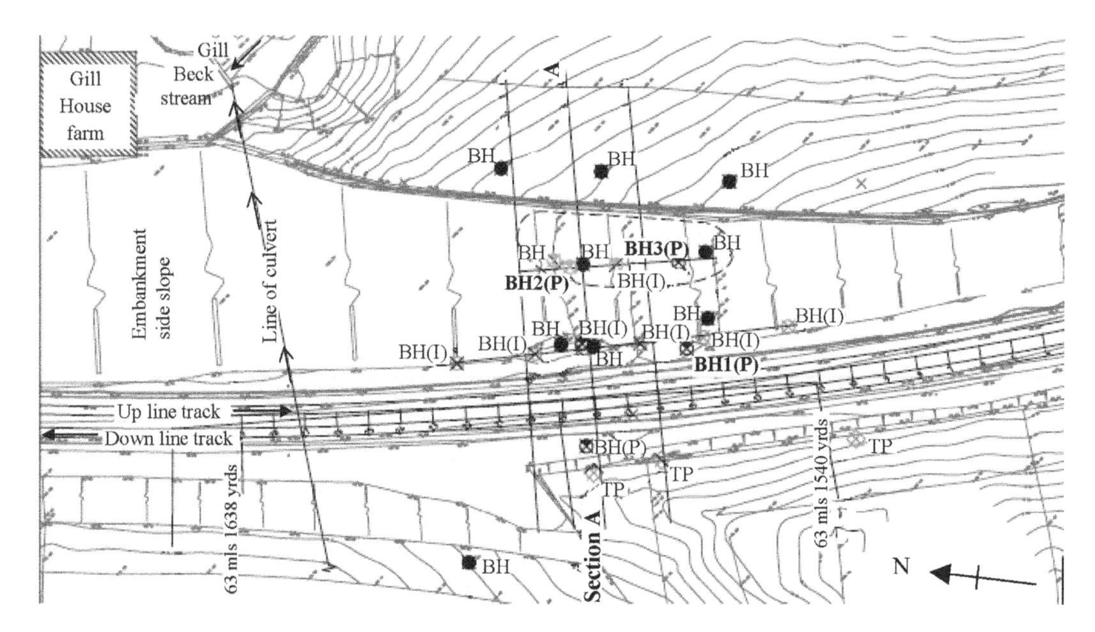

Figure 1. Site location plan. Note: BH, borehold; TP, trial pit; (I), includes inclinometer; (P), includes piezometer.

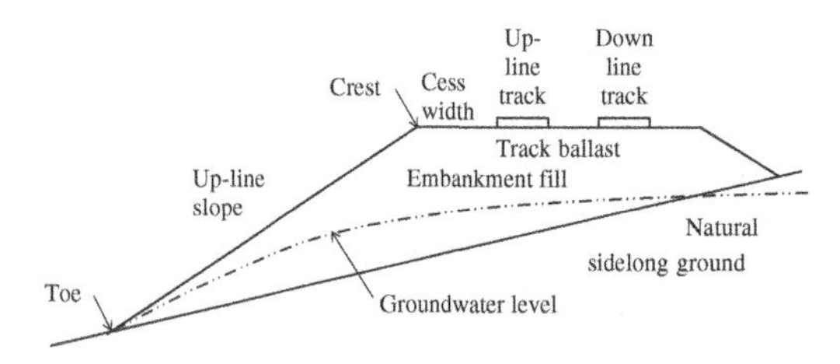

Figure 2. Terminology associated with twin-track railway.

Figure 3. View southwards along Up-line slope.

Standard Penetration tests (SPT) in the exploratory boreholes.

The Network Rail Hazard directory (issue 3rd January 2002) listed the site area as a former railway tip (Wreay ash tip), which explained the significant

Table 1. Stratigraphy beneath embankment centreline.

Stratum	Description	Thickness (m)
Ballast		0.3 to 0.5
Made ground	Loose to medium-dense ash, slag and cinders	3.8 to 4.4
Embankment fill	Very soft to firm silty sandy clay with some loose silty clayey sand	2.7 to 6.4
Glacial till	Firm to stiff glacial till with fine sand lens (ground foundation)	Not proven

depths of ash and slag materials that constituted the upper section of the embankment. Tipping began on the Down-line side in the 1900s with the level of the ash reaching track level by around 1914. Much of the ash material has since been removed from the Downside field for use in the manufacture of breezeblocks.

The lower section of the embankment was most likely constructed using locally-sourced glacial till material. The soils are generally of low plasticity and the SPT N-blow values (Figure 5) indicated frequent loose zones present.

The British Geological Survey map (Penrith: Sheet 24, 1:50,000 scale) and Geological Memoir (Appleby, Ulleswater and Haweswater) indicated that the solid geology at the site is Penrith Sandstone (poorly silicified red-brown sandstone).

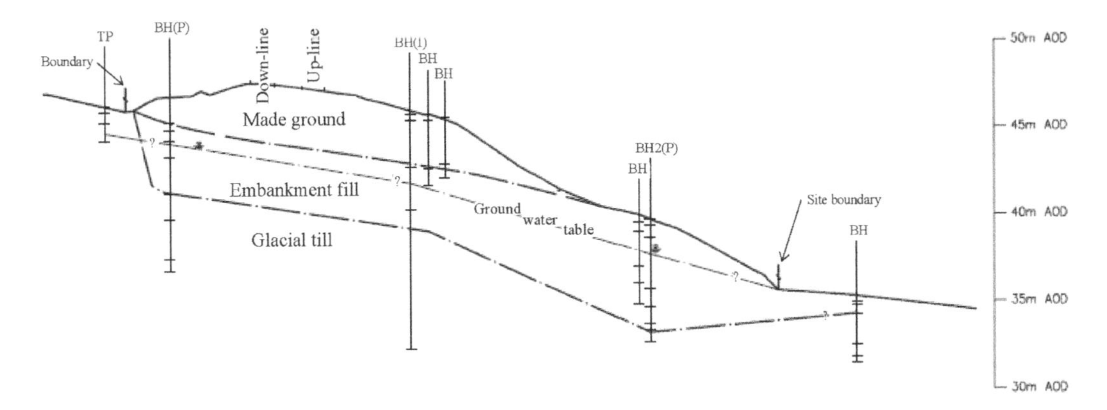

Figure 4. Geological profile at section AA in Figure 1. Note: BH, borehole; TP, trial pit; (P), includes piezometer.

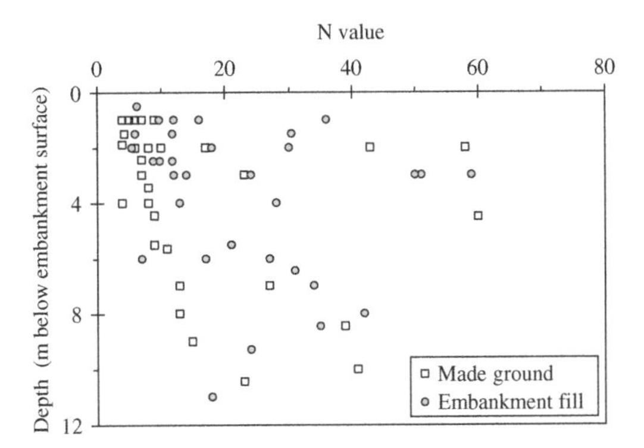

Figure 5. Standard Penetration test data.

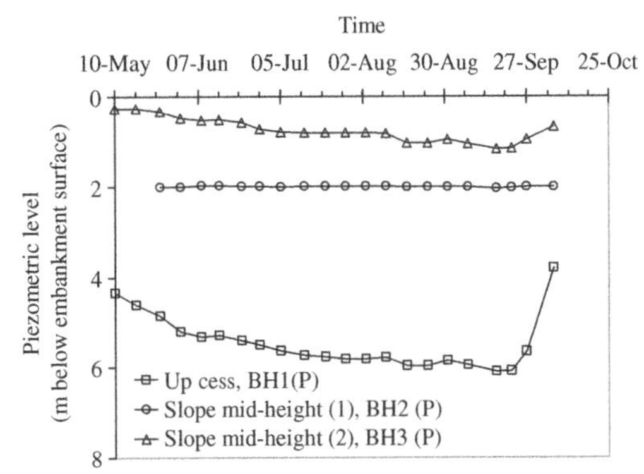

Figure 6. Groundwater levels within the embankment core.

2.2 *Groundwater*

The river Petterill is located about 200 m to the north-east of the site. Gill Beck, one of its tributaries, is culverted beneath the north end of the site area and exits at the embankment toe on the Up-line side, just south of Gill House farm (Figure 1).

Site inspections revealed seepage, water logging and bulging along the Up-line toe. However, there was no evidence of gullies or pre-wash slope failure.

The piezometer and inclinometer instruments were located along the Up-cess and at the mid-height of the embankment slope within the length of the toe bulge area. The response zone (sand cell) of the three piezometers at borehole locations BH1(P), BH2(P) and BH3(P) (Figure 1) were over the full depth of the embankment fill material and the instruments were periodically monitored between April 2001 and May 2002.

The data shown in Figure 6 indicated a relatively high groundwater table that fluctuated seasonally (substantial rise in groundwater levels between September and October 2001). The groundwater table was generally located between 3.7 and 6.0 m below the embankment crest level (within the reworked glacial till material) and between 0.3 and 2.0 m below the mid-height of the slope face. The groundwater table was almost coincident with the ground surface at the embankment toe. However, no seepage had been observed from the slope face itself.

2.3 *Ground movement*

Two inclinometers were initially installed to detect ground movement. Another five inclinometers were later installed to measure and define the extent of that movement. The lower section of the inclinometer tubes, 50 mm in diameter, were anchored to a depth of at least 3.0 m in the underlying glacial till foundation.

The instruments indicated that the embankment was moving laterally along a shear zone located at a depth of between 2.0 and 5.0 m below track level (within made ground and embankment fill layers) with up to 30 mm of movement recorded between April and September 2001. Monitoring of the track distortion including the Up-line and Down-line rail levels, cants and twists, and the Six-foot widths (refer to Figure 2) along with 22 target-monitoring points was undertaken on a weekly basis from 63 mls 1368 yds to 63 mls 1720 yds between June 2001 and May 2002.

A settlement trough, which affected both the Up and Down-line tracks over the full length of the site area, was identified (up to 16 mm of vertical settlement recorded) although the degree of track distortion was still within acceptable limits. Level monitoring of two gantry stanchions located in the Up-cess indicated that settlement was ongoing with up to 12 mm settlement recorded between July and September 2001. Moreover, bulging of the Up-line toe and tilting of the existing boundary fence, which had been installed about four years previously, had occurred over a distance of about 60 m.

3 GROUND MODEL AND DESIGN PARAMETERS

The existing embankment profile was accurately determined from a topographical survey (5.0 m square grid) undertaken by Interactive Track Services Limited (UK) in June 2001. Along the Up-line side, the embankment ranged between 9.5 and 13.0 m in height, with the Up-line side sloping at an overall gradient of about one vertical to 1.7 horizontal (30 degrees). The ground model (typical section presented earlier in Figure 4) was based on the exploratory borehole data and the recorded groundwater levels (Figure 6).

From the site inspection, ground conditions and a review of the monitoring data, it was concluded that the upper embankment section was slowly but progressively moving along a rotational slip surface located within the embankment core. The gantry at 63 mls 1556 yds was most likely leaning due to a shallower rotational slip which had induced a bearing capacity type failure.

A slope stability back-analysis was carried out on the existing embankment profile using the Geosolve *SLOPE W* program and assuming a limiting equilibrium condition (factor of safety (FOS) value of unity) in order to determine the mobilized values of the effective stress shear strength parameters for the different strata (Table 2). The analysis was carried out using the Morgernstern and Price Method of Slices in *SLOPE W*. Data from ring shear (peak), shearbox and consolidated-undrained triaxial compression tests with pore water pressure measurement (sets of three 38 mm diameter test specimens) were used as the initial input values.

A sensitivity analysis considered the FOS values mobilized for the most probable and most unfavourable ground conditions and the effects of variations in the groundwater levels. High groundwater levels are unconservative for the purpose of back calculating the values of the shear strength parameters. It was concluded that the critical slip surface, which ran from the Up-line cess and day-lighted at the embankment toe, was activated during transient rises in the groundwater level within the embankment fill material during torrential rainstorm events.

Table 2. Design values from *SLOPE W* back analysis.

Stratum	Bulk unit weight (kN/m^3)	Effective cohesion c′ (kPa)	Effective friction angle, ϕ' (degree)
Made ground	18	0	27
Embankment fill	20	3	22
Glacial till	21	3	32

4 UPGRADE WORKS

4.1 *General*

The upgrade works comprised the construction of an embedded pile retaining wall (shear dowels) along the mid-height of the Up-line embankment slope, the re-grading of the slope face and the construction of a toe drain (Figure 7). The embankment slope was regraded to one vertical in 2.5 horizontal above the new retaining wall and to one vertical in 2.0 horizontal below. In addition, a 2.8 m wide cess walkway and shoulder were constructed along the Up-line. The upgrade works used best practice measures, through a process of value engineering, to reduce the hazard of further slope movement that would affect the rail track.

A row of reinforced-concrete piles (46 number in total) were installed through the full depth of the embankment and penetrated the underlying stable ground, thereby preventing further lateral slope movement and reducing future movements at the track-bed level. The location of the piles at the slope mid-height was the most efficient and caused the least disruption to rail operations (clear spacing of 17.2 m between the pile row and the nearest Up-line rail). The re-grading of the upper slope reduced the risk of shallower rotational slips, which had occurred near the embankment crest, and the possibility of lateral spreading of the track-bed material. The construction of the toe drain with outfall to Gill Beck stream reduced the high groundwater table/water-logging near the embankment toe.

A number of alternative upgrade solutions were considered (but ultimately rejected) as follows:

- Construction of a pile retaining wall located along the embankment crest. Rejected as works would have had to be carried out next to the overhead line electrification; restricted sighting distance and would have also required possessions and larger (diameter and length) piles. Possession of all or part of the rail track was not possible during the course of the upgrade works.

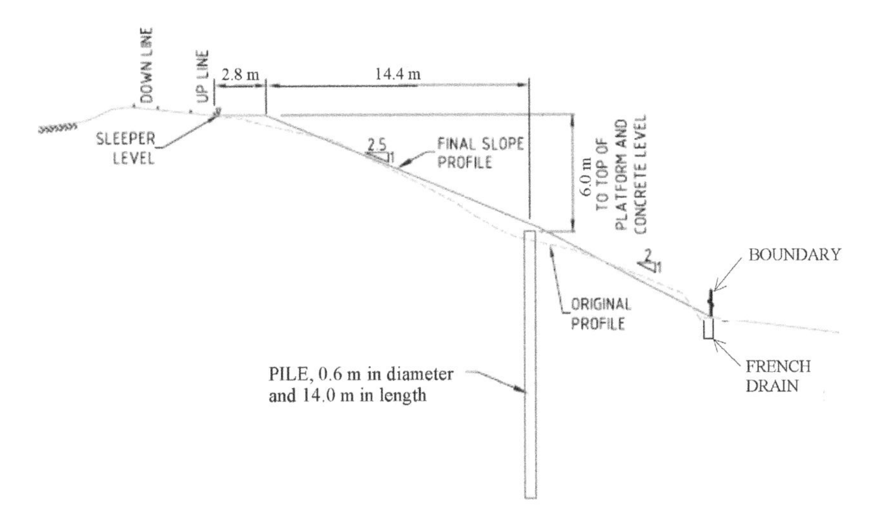

Figure 7. Upgraded embankment.

- Regrading the Up-line slope with a toe retaining-structure in place. The embedment requirement for the toe retaining structure led to concerns that excessive ground movement would have occurred during excavation works along the embankment toe;
- Regrading works over the full Up-line slope would have required land purchase next to the embankment toe with possible delays in reaching agreement for permanent land take;
- Counterfort drainage was rejected over concerns that excessive settlement would occur during the excavation works for the counterfort drains.

4.2 *Design*

The upgrade works were designed to improve the FOS value against slope instability as follows:

- The FOS value against a reactivation of the critical slip surface was increased from unity to 1.3;
- The FOS value against shallower secondary slips occurring within the upper section of the regraded slope was increased to at least 1.3.

As well as providing the necessary additional horizontal resistance to improve the global slope stability, the piles were also designed to support the upper section of the embankment in the event of a slip failure occurring below the location of the pile embedded retaining wall.

A *SLOPE W* analysis using the effective stress strength parameter values in Table 2, and with a vertical stress of 50 kPa applied along the twin tracks, indicated that a row of piles capable of exerting a horizontal resistance of 200 kN/m run at the mid-height of the embankment slope would increase the FOS value against a reactivation of the critical slip surface from unity to 1.3. The analysis also indicated that all potential shallower slips that day-lighted just upslope of the retaining wall would have FOS values greater than 1.3.

The pile design was based on a plane-strain analysis of the retaining wall using *WALLAP* geotechnical software assuming a loss of passive support of up to 1.0 m in depth in the event of a slip failure occurring further down the embankment slope. Modified coefficient of earth pressure values were used to accurately represent the sloping ground. A single row of cantilevered, reinforced concrete piles (600 mm in diameter and between 12.4 and 14.0 m in length) was capable of providing the necessary horizontal resistance of 200 kN/m run. The piles were spaced at 1.5 m centres along the row. The glacial till foundation provided the lateral resistance to the piles (FOS value of 2.0 against a rotation failure).

The steel reinforcement (10 number T40 mm diameter main steel bars with a 75 mm cover) was designed in accordance with BS5400 to provide a structural section capable of resisting the system of working loads and bending moments (applying a FOS value of 1.65) and for a design life of 120 years. The design calculations indicated that the bending moments and shear forces reduced to zero at a distance of about 13 m below the pile head level. The geometrical arrangement of the piles (46-number at 1.5 m centre spacing) also permitted the easy migration of groundwater through the embedded wall.

4.3 *Construction*

The engineering works were carried out between March and April 2002 and at no stage was temporary

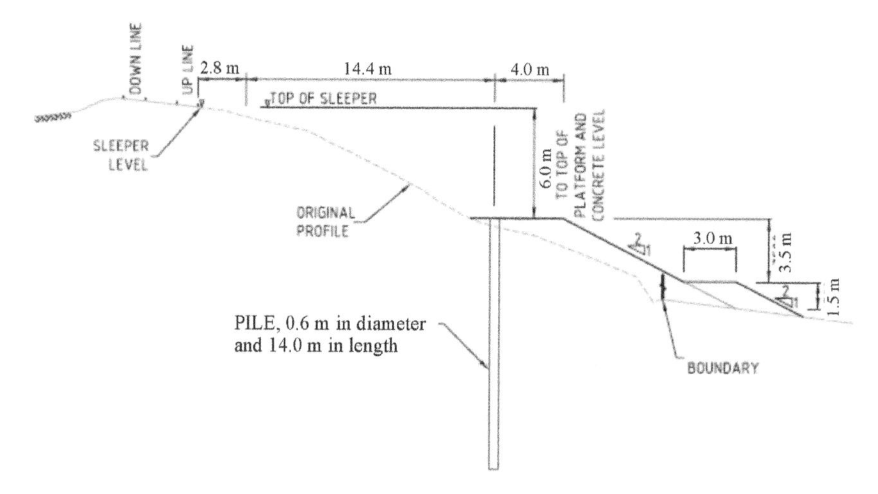

Figure 8. Temporary works.

possession of the rail track necessary (works sufficiently far away for Green Zone working).

A temporary haul road was constructed to access the base of the slope, which had been cleared of all vegetation. Then a piling platform up to 5.0 m in width was constructed along the length of the upgrade works (70 m in overall length between 63 mls 1534 yds and 63 mls 1610 yds) at the mid-height of the embankment slope (Figure 8). The platform was constructed using granular fill, progressively benching into the lower embankment slope to reach a distance of 1.3 m above the pile cut-off level. The benches were excavated in bays (typically 5.0 m in length, 600 mm in height and with a slope angle of about 60° to the horizontal) parallel to the track.

The bored piles were installed through the platform by Cementation Skanska using the continuous flight auger technique. The design included a slope stability analysis for the set up with the piling rig working above the platform with berm (FOS value on slope instability of 1.2). The spoil from the base of the boreholes indicated that the piles had penetrated the full depth of the glacial till layer and were founded in the underlying glacial sands. The piles were cast using 40 N concrete (28 day strength). After hardening, the pile heads were trimmed back to sound concrete. Integrity testing indicated that the integrity of all of the piles was sound.

The upper section of the embankment slope was regraded to the finished profile (one vertical to 2.5 horizontal). The unsuitable embankment fill was removed from the slope face in benches and replaced with compacted Highways Agency specification Class 6B granular fill material. The slope face was covered with a 300 mm deep Class 6F2 capping layer.

The piling platform was subsequently removed followed by regrading of the lower section of the embankment slope to the finished profile (one vertical to 2.0 horizontal). A drainage layer (2.0 m in width and 0.5 m in depth), which was wrapped in a geotextile membrane, was constructed at the interface between the ash/slag layer and the underlying embankment fill. The drainage layer was necessary since the Class 6F2 material (well graded limestone, three inch size down to dust) that had been used in the capping layer tends to hydrate when moist creating a very low permeable surface. A French drain was constructed along the line of the boundary fence near to the embankment toe with an outfall to a headwall structure at Gill Beck stream. The drain (0.5 m in width and 1.0 m in depth) was excavated in short lengths and backfilled without delay with Highways Agency specification Type B filter material. Finally, geo-matting and seeding were placed over the entire embankment slope to provide erosion protection and for the rapid reestablishment of vegetation.

5 POST CONSTRUCTION

Periodic monitoring of the track along the length of the site continued until June 2002 (three months after the substantial completion of the works). The data indicated that negligible movement had occurred at track-bed level during the construction works and the following three-month period.

A six-month review of the site in December 2002 indicated no discernable deficiencies or defects in the tracks. Interestingly, all of the inclinometers located above the embedded retaining wall had recorded ground movements of up to 10 mm up the slope, towards the track. It is postulated that the ground movements up the slope that had been recorded by the inclinometers were as a result of the stress relief that occurred following the removal of the pile platform and the slope regrading works.

6 SUMMARY

The ground investigation, monitoring and numerical analysis indicated that the upper section of the embankment was slowly but progressively moving (limiting equilibrium) along a rotational slip surface located within the embankment core (ash and slag materials overlying reworked glacial till). The groundwater level within the embankment core was relatively high due to the site topography (steeply sloping sidelong ground). The upgrade works comprised the construction of a row of stabilising piles (shear dowels) at the mid-height of the embankment slope, regrading of the slope face to prevent secondary shallow slips near the embankment crest and the construction of a toe drain.

The effective stress shear strength parameters for the design were determined from a slope stability back-analysis of the critical embankment section which was at limiting equilibrium. The critical slip surface, which ran from the Up-line cess and day-lighted at the embankment toe, was activated by transient rises in the groundwater table within the embankment core during torrential rainstorm events.

The 600-mm diameter stabilising piles, which provided the necessary horizontal resistance to increase the FOS value against slope instability from unity to at least 1.3, had the desired effect of preventing lateral movement of the embankment core, thereby reduced future movement at track bed level to an acceptable amount.

Advances in Transportation Geotechnics – Ellis, Yu, McDowell, Dawson & Thom (eds)
© 2008 Taylor & Francis Group, London, ISBN 978-0-415-47590-7

Some developments in rock slope hazard assessments for road and rail

D. M. Tonks, I. M. Nettleton & R. M. Denney
Coffey Geotechnics UK Ltd

ABSTRACT: The paper considers developments in the systematic appraisal through to the ongoing management of rock slopes affecting significant infrastructure. The Rock Hazard Index (RHI) system was originally developed for rapid appraisal of rock slopes affecting Scottish Roads. It has been developed further in connection with slopes appraisal under the Quarry Regulations and for rock slopes affecting rail, other infrastructure and property. A notable feature is the differences in targets/receptors and the owners' perceptions of these. It is particularly important to engage with the various interested parties in this.

This has led to development of the STAR (Stability Appraisal and Risk) system, following a source – pathway – receptor approach analogous to approaches used for contaminated land. This considerably assists communication and risk management planning, for instance distinguishing slopes with high hazard but for very little receptors at risk from those that may have low hazard but major receptors at risk. Remedial measures can be systematically studied in the light of this. Options can be readily assessed and costed ranging from improving the source (e.g. scaling, bolting), breaking the pathway, e.g. rock trap, catch fences, vegetation barrier, through to modifying the receptors (e.g. controlling development).

1 INTRODUCTION

1.1 *Nature of the hazards and experience to date*

Rock Slopes pose a wide range of risks to Roads, Rail and other Infrastructure located in their vicinity. These vary in nature and scale from localised degradation of only minor, nuisance value, through block falls which can pose significant risks, to large scale instabilities of major consequence. Since most rock slopes pose some risk, it is of considerable importance to be able to assess these in a structured and reasonably consistent manner, to enable appropriate actions and use of resources

The Rock Hazard Index (RHI) system was originally developed for rapid appraisal of hard rock slopes affecting Scottish Roads (TRL, 1995) following several major events. The method has been developed further in connection with rock slopes under the Quarry Regulations (DETR 1999) and for those affecting rail, infrastructure and property.

Slopes assessed as being of potential concern are then subjected to more detailed study, following the methods described by such workers as Hoek (Hoek 1976 & 2000) and others. These allow the hazards to be suitably assessed and prioritised.

Remedial measures can then be systematically developed in the light of this. Options can be assessed and costed ranging from reducing the hazards at source (e.g. scaling, bolting, netting), through to providing protection measures such as rock traps and catch-fences. Alternatively, or as an interim measure, where appropriate, the risks may be managed by removing or suitably managing the receptor. This may include traffic controls, management or diversions.

1.2 *Ongoing developments*

There is now more than 12 years experience of using the RHI and related schemes. It has been applied to more than 1500 slopes for a wide range of conditions and circumstances. This paper makes recommendations for further developments in the light of this experience.

There have also been similar developments elsewhere. It is noteworthy that some countries have taken a particularly pro-active approach to landslides and unstable land, often as a result of severe events. Hong Kong introduced a major system of Geotechnical Control, following several disastrous landslides in the 1970s. Australia has developed its Landslide Guidance, not least following significant loss of life at Thredbo in 1998, leading to the recent updated (Australian Geomechanics Society 2007).

Scotland has instituted significant further studies following several events in 2005, fortunately without loss of life. (Winter *et al* 2006, Nettleton et al 2005).

2 RHI SCHEME

2.1 *General*

In 1995 the Transport Research Laboratory (TRL) developed a new rock slope risk assessment system for roads, the Road Rock Slope Hazard Index System. The development was funded by the then Scottish Office (now Scottish Executive). The new approach was proactive, to seek to manage the risk to road users by identifying potential risks before incidents occurred, and to allow the effective prioritisation of available maintenance budget funds. The Road Slope Hazard Index (RHI) System, acts as a "coarse sift" eliminating slopes with a low risk potential from a later, detailed assessment stage (McMillan 1998).

More than 1500 rock slopes have been assessed using the RHI System in Scotland for the Scottish Executive and elsewhere in the United Kingdom for the Highways Agency. All major failure events that have occurred on slopes that have been assessed have so far been on slopes in the highest risk categories. No significant failure events have occurred on slopes in the lower risk categories.

The potential hazards present and the likelihood of them reaching the receptor at risk are determined and the RHI value is evaluated, taking into account the exposure to the potential hazards.

The system is based around rapid, standard field data collection in which estimates of influential geotechnical, geometric and remedial work factors are recorded on a standard form. There are a number of options for each factor and the relevant option is selected based on visual assessment of field conditions (TRL, 1995). Parameter values are derived for each input factor option to reflect the influence that the factors have on the rock slope hazard as a whole.

The RHI is derived by following a standard calculation procedure, using parameter values as input. The calculation process follows a logical route dictated by the influence of parameters on rock slope instability and rockfall hazard. The RHI values derived from these calculations are used to prioritise future action through classification of slopes into four action categories as follows:-

Slopes with an Index value of less than 1 do not present a significant hazard, fall into the No Action category and are unlikely to require maintenance in the foreseeable future. An index of between 1 and 10 indicates that conditions at a rock slope are such that hazards may develop. These slopes are normally scheduled for Review in Five Years. Slopes in this category require only minimal maintenance commitment and the review is similar to the initial RHI, but allowing assessment of change.

An index of 10 or more indicates that conditions at a rock slope may present a significant hazard, increasing with index number. These slopes therefore require action to investigate the nature and severity of the hazard. Prioritisation of action is achieved by grouping these slopes into two categories of Detailed Inspection and Urgent Detailed Inspection. Slopes in these categories may require significant maintenance commitment. Firstly they require a detailed inspection and secondly this detailed inspection may reveal the need for remedial action to reduce hazards to an acceptable level.

The RHI takes account of such remediation works as may be present. Where appropriate these may be matters for the detailed inspection stage.

Detailed inspections of the slope aim to record the location and type of failures present, the structural characteristics of the failures (including dimensions and discontinuity characteristics) and potential remedial work options. A local chainage system is established using temporary paint marks. Photomontages are taken of the slope and referenced to the chainage system.

Outcomes may be:-

Low hazard/no action. The slopes are shown to have acceptably low hazards and are signed off.
Manageable. Rock slopes showing significant hazards, but not sufficient to warrant immediate action. They are thus managed under the regime, normally scheduled for reassessment in 5 years, with such comments as may be appropriate.
Action. High hazard slopes are addressed on a priority basis. The effectiveness of remedial options can be assessed by comparing the anticipated post-remediation RHIs. Design would normally aim to reduce RHI to < 10.

Table 1. RHI Interpretation.

Action category	Road rock slope	Hazard index value
1.	No Action	< 1
2.	Review in Five Years	1–<10
3.	Detailed Inspection	10–<100
4.	Urgent Detailed Inspection	100 or greater

2.2 *Scottish roads*

The RHI system was originally trialled on 179 slopes on the A830 (Marshall 1995) and shown to give a good match with comparable studies by a range of experienced engineers and geologists. Over 50% were identified as needing no action; about 35% for 5 year review and 14% detailed inspections, of which 3 slopes were identified as urgent. It was then applied to about 1500 slopes around the Scottish Roads network (Nettleton et al 2000).

Figure 1. A890 Stromeferry Bypass.

2.3 *A890 stromeferry bypass*

The slopes at Stromeferry warrant particular mention (Davies & Lovett 1992). The Highland Rail line was constructed in the 1890s, adjacent to Loch Carron, along the toe of very major slopes for about 7 km. A single track road was constructed in the late 1960s on the inside of the rail line (Fig. 1), largely by blasting very steep cuttings. There were three reported landslides during construction, some very substantial, and the resulting route has been subject to considerable ongoing rockfall events.

The slopes have remained the responsibility of the Highland Council (rather than being part of the trunk road network). Use of the RHI has greatly assisted structured management of these slopes, with a programme of improvements spread over some 10 years. This has finally brought the rock slope hazards to levels comparable with other parts of the Highland roads system, albeit needing considerable ongoing management. The slopes are subject to regular maintenance and annual geotechnical inspections, with provision for significant ongoing remedial works.

A feature of the system is an algorithm relating the hazards to the road usage. The risks may thus be considered acceptable whilst the road is single track and with very low usage. However, it is clear that very substantial works would be needed to accommodate even quite limited increase in traffic, to the extent that alternative routes may need to be found.

Figure 2. M5 Wynhol Viaduct rock slopes.

2.4 *M5 wynhol*

The rock slopes within the Highways Agency Area 2 Network (south west England) were assessed using the TRL RHI system. The system identified three slopes requiring detailed inspections between Junctions 19 and 20 of the M5 (CIRIA C591 2003).

The total length of the rock slopes was in excess of 2.5 km (Fig. 2). A walk by reconnaissance of each slope was initially undertaken, with the purpose to identify areas of high potential hazard or areas of concern/uncertainty. These areas were then subject to a more thorough inspection from ground level or by rope access depending on the height and location in relation to the motorway.

This method of slope assessment allowed for the recording of data that would be required during the design phase of the remedial works.

3 SOME OTHER ROCK SLOPE APPLICATIONS

3.1 *Rail*

A similar system has also recently been used for rock slopes in the Railtrack (now Network Rail) NW area of England. The area extends from Carlisle in the north to Crew in the south, and from the outskirts of Sheffield in the east to Holyhead in the west.

The work involved the inspection of some 1800 individual rock slopes. The work was undertaken in a 10-week period by a team of 6 field geologists/engineers. The slopes ranged in height up to 30 m and extended up to a couple of km in length and include many major rock cuttings formed over the history of the rail network. The slopes reflect the extensive variety of geologies occurring in North West England and North Wales, including extensive areas of the Pennines, Lake District, Lancashire, Cheshire, Yorkshire and Derbyshire (Fig. 3).

Figure 3. 1 One of over 1800 rock slopes assessed.

Table 2. Geotechnical Appraisals and Assessment under the 1999 Quarries Regulations.

1999 Quarries Regulations require:
• Design
• Plans
• Site Investigation Information
• Records of materials tipped and placed
• Regular Hazard Appraisals and Geotechnical Assessments
Ongoing through life of quarry – enables an "Observational Approach" to reclamation planning.

3.2 *Quarries*

The 1999 Quarry Regulations (DETR, 1999) include a system of Geotechnical Appraisals and Assessments for rock slopes (Table 2). The RHI system has been adapted to create the Quarry Hazard Index (QHI) (Butler *et al*, 2000).

Geotechnical Appraisals required for rock excavations are carried out under the QHI system. This is particular useful for speed and introduces a valuable degree of consistency within and between quarries. These then form the basis of more detailed Geotechnical Assessments, which follow a rather similar format to that applied for roads, only allowing for the considerably different risk management approaches of experienced quarry operators.

Applications include a number of major limestone quarries, one with over 200 rock slopes, and many with 10–20 slopes. The authors have also contributed to a recent Mineral Industry Research Organisation (MIRO) study of reclamation of hard rock quarries (Cripps 2004). The QHI approach provides a useful tool to work towards safe and sustainable final faces.

3.3 *Infrastructure and property*

The RHI system is equally applicable to other facilities at risk from rock slopes. The system has been adapted for use with various other infrastructure and property.

A good example is for Hastings, where the local authority manages some 1.2 km of cliffs up to about 20 m high. Receptors at the toe of the cliffs include gardens, back yards or access passages as well as the extensive associated property. For this study specific action categories and appropriate comments have been developed and are presented in Table 3 below.

Table 3. Rock Hazard Index/Actions for Properties.

RHI value	Action category	Comment
<1	Not a Significant Hazard, no action	Risk normally accepted
1 to 10	Not a Significant Hazard, re appraise in two years	Risk commonly accepted but opportunities may occur to improve
10 to 100	Significant Hazard, Geotechnical Assessment	Inspection/remediation on a cost benefit basis
>100	Significant Hazard, Urgent Geotechnical Assessment	Urgent inspection/works Interim risk management/ remedials may be required

4 STAR – RISK EVALUATION SYSTEM

4.1 *General*

The RHI/QHI is intended to be a rapid means of assessing the hazards due to rock slopes in a zone and as such deals with all the hazards and potential hazards within that zone. It does not enable individual hazards to be assessed. Hence, a hazard specific Stability Appraisal and Risk (STAR) system has been developed which can also be used to assess soil slopes, and details are given below.

The STAR system, follows a source – pathway – receptor approach analogous to approaches used for contaminated land. This considerably assists communication and risk management planning, for instance to distinguish between those slopes identified as having a high hazard but low overall receptor risk from slopes with a low hazard but major receptor risk. Remedial measures can be systematically studied in the light of this. Options can be readily assessed and costed, ranging from removing the source (e.g. scaling, bolting); breaking the pathway, e.g. rock trap, catch fences, vegetation barrier; through to modifying the receptors (e.g. controlling development).

The system takes account of extensive recent work by many parties and applies knowledge built up during the development of the RHI/QHI systems. It is based upon relative risk assessment and subjective

Table 4. STAR system factors.

Hazard	Receptor vulnerability	Receptor value
1. Minor failure/erosion/weathering – typically slopes <25°.	Little or no effect	Undeveloped land
2. Moderate slope failure – typically slopes <25°. Small rock fall individual blocks <0.1 m^3	Nuisance or minor damage	Unoccupied/Infrequently visited building/Public rights of way
3. Substantial slope failure – typically slopes at 25° to 40°. Moderate rock fall <10 m^3	Significant damage	Roads/Footpaths
4. Large slope failure, typically slopes 25°–40°. Large rock fall 10–50 m^3	Major damage/Major Injury	Residential property/Commercial buildings (1 to few lives)
5. Major slope failure – typically slopes at > 40°. Major rock fall >50 m^3	Total loss/Loss of life	Major public buildings (many lives)

quantification of engineering judgement. The system assesses the risk as shown below:

Risk = Hazard x Pathway x Receptor

The Hazard factor is an estimation of the significance of the slope feature with the potential to do harm. The Pathway factor is an estimation of the percentage of the Hazard that is likely to interact with the Receptor. The Receptor factor is an estimate of the likely consequence should the Hazard interact with the Receptor. The Receptor factor is a combination of the vulnerability of the Receptor and the value of the Receptor. Definitions of the factors and their corresponding values are given in Table 4, the final STAR values and the Corresponding Risk Values are given in Table 5.

5 SUGGESTIONS FOR DEVELOPMENT

5.1 *General*

In the light of the foregoing the authors consider there are a number of particular areas where further progress can now be made.

There would be benefit in further developing guidelines for rock slope hazards. Some new works come under the UK Planning framework, in PPG14, Development on Unstable Land, Annex 1 (DOE 1996). This

Table 5. STAR values and the corresponding risk values.

STAR value	Risk	Comment
<10	Low	Normally accepted.
10 to 25	Low to Moderate	Significant Risk probably tolerable – client needs to be made aware of hazards and monitor
25 to 75	Moderate to High	Significant Risk requiring remedial measures/risk management actions
75 to 100	High	Significant Risk requiring major remedial measures
>100	Very High	Significant Risk requiring urgent action e.g. evacuation or emergency works then remedial measures

would benefit in updating in the light of comments herein and the considerable advances in assessment and rock engineering in recent years.

However, PPG14 only addresses certain new works. In our experience, most issues arise from existing slopes in proximity to property and infrastructure. Such situations can become unduly complex, especially where several parties are involved. A number of owners take pro-active risk management-based approaches to their slopes. CIRIA C591 & C592 (CIRIA 2003 and 2003) give useful information on some of these approaches for infrastructure cuttings and embankments, respectively, and quotes various examples of use of the RHI approach.

5.2 *Other landslide guidelines*

There would be benefit in development with other comparable systems. There is considerable consensus building around the various approaches. One of the most interesting recent developments is the updated Australian Geomechanics 2007 Framework (Australian Geomechanics Society 2007). The interested reader is referred to the full documents. Note that this attempts to cover all landslide risks and thus is a very substantial undertaking. Space precludes further comment here; it is hoped to address these issues in a further paper shortly.

5.3 *Low maintenance slopes & sustainability*

In the pursuit of developing interest in better understanding of Engineering Sustainability the authors are working on a scheme to index sustainability of slopes. Some early ideas were presented by Tonks & Law, 2007 (Table 6). It is planned to further develop such ideas in conjunction with ongoing research and development into sustainability in geotechnics.

Table 6. Sustainability of rock slopes.

Sustainability	RHI	Maintenance whole life cost	Risks >1 in
High	<5	Negligible	10^5
	5–10	Slight	10^4
Medium	10–20	Some	10^3
	20–50	High	10^2
Low	>50	Essential	10

In Scotland, such concepts are explicitly used in requirements for low maintenance rock slopes (McMillan 2000), and have been successfully used on the recent phases of the A830 Arisaig road improvements schemes and is currently being applied to a number of further schemes including the next stage of the A830 Arisiag improvements.

6 CONCLUSIONS

For many rock slopes, risks cannot be entirely eradicated, but they can and should be suitably and proactively managed appropriate to risks and costs. There would benefits to slope owners and other affected parties from further guidance in this.

The RHI system has provided a useful tool to appraise rock slope hazards. This has already shown considerable value for areas of road, rail and quarry slopes. The system is well suited to use for other infrastructure, including property and developments adjacent to cliffs.

More than 12 years experience of using the RHI and related systems has identified a number of areas where progress can be made, particularly as follows

- Integration/development with other comparable systems. There is considerable consensus around the various approaches.
- Widening to some applications for weaker rocks – e.g. chalk, mudrocks.
- Facilities to include/and adjust for ongoing records, and for cost-effectiveness of actions.
- Development of sustainability indicators.

There is now very substantial knowledge and experience, which should be suitably captured to develop further proportionate and helpful guidance.

ACKNOWLEDGEMENTS

The authors are very grateful to numerous parties who have supported this work, particularly the Highlands Council; Scottish Executive and numerous colleagues at EDGE Consultants UK Ltd and Coffey Geotechnics who have contributed to these projects.

REFERENCES

Australian Geomechanics Society. 2007. A National Landslide Risk Management Framework for Australia. *Australian Geomechanics.*

Butler A.J., Harber A.J., Nettleton I.M. & Terente V.A. 2000. Rock Slope Risk Management and the 1999 Quarries Regulations. In: *Proceedings of the 11th Extractive Industry Geology Conference and 36th Forum on the Geology of Industrial Minerals, May 8th–11th 2000. Bath, UK.* The Geological Society of London:

Perry, J., Pedley, M. & Brady, K. 2003. CIRIA C591 Infrastructure Cuttings. Condition Appraisal and remedial Treatment.

Perry, J., Pedley, M. & Brady, K. 2003. CIRIA C592. Infrastructure Embankments Condition Appraisal and remedial Treatment 2nd Edition.

Cripps, J.C, Nettleton I.M, et al. 2004. Reclamation Planning in Hard Rock Quarries: A Guide to Good Practice. MIRO.

Davies, T.P. & Lovett, N. 1992. Slope Stabilisation at Stromeferry Bypass. *Highways and Transportation, January 1992.*

DETR. 1999. The Quarries Regulations 1999. No. 2024. Department of the Environment, Transport and the Regions (DETR 1595). The Stationary Office Limited, London, UK.

DOE. 1990. Planning Policy Guidance 14: Development on Unstable Ground. Department of the Environment. HMSO, London, UK.

DOE. 1996. Planning Policy Guidance 14: Development on Unstable Land – Annex 1: Landslides and Planning. Department of the Environment. HMSO, London, UK.

Hoek, E. & Bray J.E, 1976. Rock Slope Engineering. IMM.

Hoek, E. 2000. Practical Rock Engineering. Hoek's Notes. Rockscience web page (http://www.rockscience.com)

Marshall, G.S. 1995. A830 Rock Slope Hazard Inventory Trail. Report on TRL's Project Report PR/SC/17/94. Transport Research Laboratory, Crowthorne, Berkshire, UK.

McMillan, P. Harber, A.J. & Nettleton, I.M. 2000. Rock Engineering Guides to Good Practice Road Rock Slope Remedial and Maintenance Works. Transport Research Laboratory, Crowthorne, Berkshire: UK. *Unpublished Report for the Scottish Executive.*

McMillan, P., Nettleton, I.M. & Harber, A.J. 1998. Rock Slope Risk Assessment and Maintenance Management. Transport Research Laboratory, Crowthorne, Berkshire: UK. *Annual Research Review 1998.*

Nettleton I.M. & McMillan, P. 2000. Ranking Risk Before the Fall. *Ground Engineering*, July 2000.

Nettleton I.M., Tonks, D.M., Low B., MacNaughton S. & Winter M.G. 2005. Debris Flows from perspective of the Scottish Highlands. *Landslides and avalanches.* ICFL Norway, 2005.

Tonks, D.M. & Law K.C. 2007. Developing On Unstable Land. ICE, Slope Engineering 2007.

TRL. 1995. Rock Slope Risk Assessment, Project Report PR/SC/22/95. Unpublished project report. Transport Research Laboratory, Crowthorne, Berkshire, UK.

Winter, M., Heald, A.P., Parsons, J.A., Shackman, L., & MacGregor, F. 2006. Scottish Debris Flow Events of August 2004. *Quarterly Journal of Engineering Geology and Hydrogeology, 39. Pp 73–78.*

Advances in Transportation Geotechnics – Ellis, Yu, McDowell, Dawson & Thom (eds)
© 2008 Taylor & Francis Group, London, ISBN 978-0-415-47590-7

Centrifuge modelling of slope stabilisation using a discrete pile row

B.S. Yoon & E.A. Ellis
Nottingham Centre for Geomechanics, School of Civil Engineering, University of Nottingham, UK

ABSTRACT: Discrete pile rows are an established technique for improving the stability of unstable infrastructure slopes. However the impact of pile spacing along the row is not clearly understood from a current design perspective. A series of centrifuge model tests were undertaken for a plane strain model with the upper layer of the slope tending to fail and interact with a discrete pile row. The results were interpreted to clarify the effects of pile row interaction with the slope.

1 INTRODUCTION

Discrete pile 'walls' are a popular method of improving the stability of slopes. The piles are typically spaced at 3–4 diameters centre-to-centre along the row.

The piles act in shear and bending to resist 'passive' lateral loading from a potentially unstable ('failing') soil mass, and resist this load a greater depth by mobilising active resistance (Figure 1(a)).

A key feature of this approach is that it is more economical at large pile spacing (provided sufficient moment capacity along the row can be provided), but there is increased risk of 'flow' of the unstable soil between the piles Figure 1(b). The geotechnical centrifuge tests reported consider interaction of a pile row with an unstable soil layer.

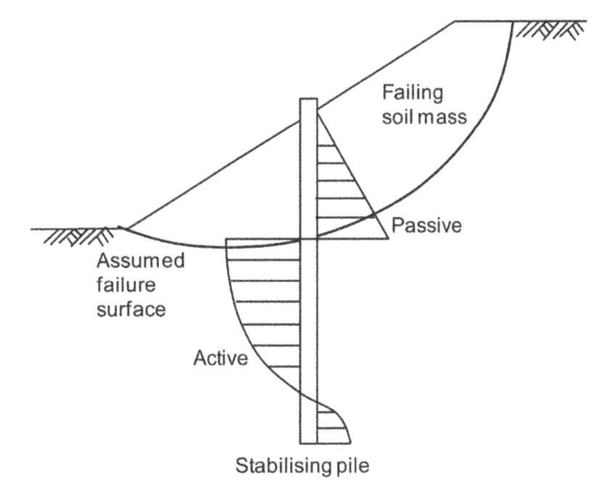

(a) Contribution to slope stability

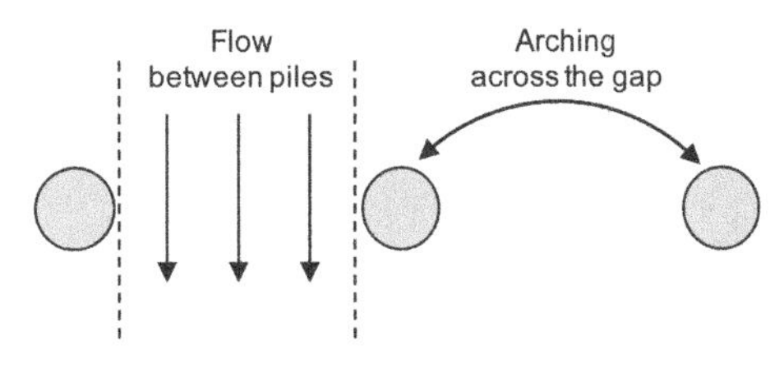

(b) Flow and arching between piles

Figure 1. Slope stabilisation using a discrete bored pile wall.

2 EXPERIMENTAL TECHNIQUE

2.1 *Nottingham Centre for Geomechanics centrifuge*

Geotechnical Centrifuge modelling is an established technique for study of soil-structure interaction. The principles of the technique have been described by a number of authors (eg. Taylor, 1995). In essence a $1/N$th scale model is subjected to N times normal gravity (g) in a geotechnical centrifuge. Hence full-scale ('prototype') stresses are generated in the model, and soil in the model exhibits behaviour representative of the full-scale prototype.

The Nottingham Centre for Geomechanics centrifuge is a 50 g-T machine, with 2.0 m platform radius – for further details see Ellis et al (2006).

2.2 *Plane strain model*

A plane strain package was used, with internal dimensions 700 mm (length) × 400 mm (height) × 200 mm

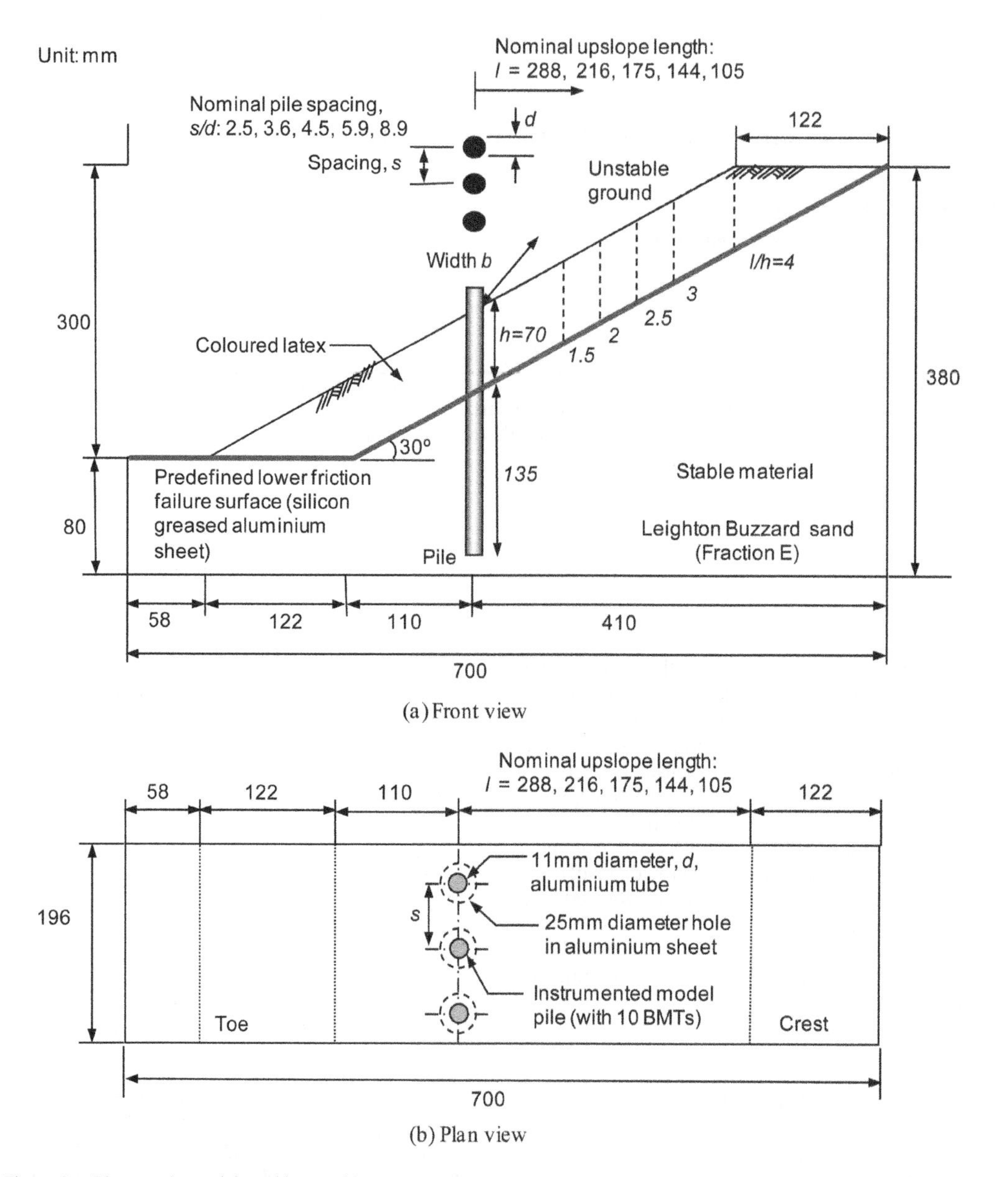

Figure 2. Plane strain model used in centrifuge tests.

(plane strain direction). Figure 2 shows the dimensions of the model slope tested.

The model is somewhat idealised. A failure interface separating the unstable and stable soil has a very low sliding resistance, increasing the tendency for the upper portion of the slope to cause passive loading on the pile row over a known depth, so that this aspect of behaviour could be studied in detail.

The model is intended to generically study this mechanism rather than to represent a specific prototype. Design procedures often do not exactly represent real life, and thus idealised generic approaches such as this can be at least as valuable in informing design as more 'realistic' (and hence specific) studies. Likewise, dry sand (Leighton Buzzard Fraction E, $d_{50} = 0.1$ mm) was used, but this is intended to represent any frictional soil (eg. drained behaviour of clay, although the friction angle would be lower). The sand was pluviated through air into the model container at 1g (before the centrifuge test), giving a relative density of 70–80%.

The failure interface was incorporated using aluminium sheet. It was intended that there should be significant tendency for the 'unstable' soil above this interface to move downslope during the test. Hence the upper surface of the aluminium sheet was coated with silicon grease. Latex sheet was then placed on the grease to separate if from the sand above. Modified shear box tests indicated that the resistance to sliding on this interface was less than 5 ° for a wide range of normal stress.

At the plane strain boundaries (vertical 'front' and 'back' faces) it is important that any effects of friction are minimized, and hence the latex-grease interface was also used at the sides above the interface, again to allow the 'unstable' soil to move downslope freely.

The model piles were installed in a row at discrete intervals across the slope before the test. As required, the edges of the plane strain section are lines of symmetry at the mid-point between piles. Between 2 and 7 piles were used, corresponding to $(s/d) = 8.9$ to 2.5 respectively. The piles were modelled using 9.5 mm OD aluminium tube with 0.91 mm wall thickness, giving a flexural stiffness of 16 Nm^2. At 30g this would correspond approximately to a concrete pile of 300 mm diameter.

One pile in each test was instrumented to measure bending moment at 10 locations along it's length, so that a bending moment profile could be established. This pile was positioned at the middle of the row so that any effects due to the side boundaries would be minimised. The strain gauges and associated 0.3 mm wires were positioned on the exterior of the tube. Hence these elements were protected with plastic 'heat shrink' around the tube, which also served to give a 'uniform' surface profile. The external diameter was then slightly increased to approximately 11 mm. Heat shrink was applied to all piles so that they remained similar in diameter.

At the failure interface the piles passed through oversized holes in the aluminium plate which would allow them to move in response to loading. Separate pieces of latex sheet were used on the interface above and below the pile row so that the unstable soil could move independently in these areas of the slope. For the same reason, separate pieces of latex sheet were also used on the sides for the unstable layer above and below the pile row.

Cameras mounted on the model package were used to observe the front and plan views (Figure 2) during the test. GeoPIV (White et al., 2003) was used to 'track' the movement of soil through a series of images, and this data was then converted to movement in the model using the known location of control points.

2.3 *Test procedure*

During the test models were subjected to increasing g-levels in the range 5 to 50g in a series of increments increasing from 1g initially to 4g at higher g-level. This approach is slightly unconventional since no single 'equivalent prototype' is modelled, but this again reflects the generic nature of the study. As g-level increases the 'weight' of the unstable soil above the interface tending to slide downslope increases, as does the stress in the soil governing interaction with the pile row for the frictional soil. At 30g the 70 mm thick unstable layer corresponded to 2.1 m, and hence soil behaviour broadly representative of a full-scale slope would have been exhibited.

At 1g the soil above the interface proved to be stable (although vibration could cause 'unraveling' at the surface). This was attributed to slight 'adhesion' in the grease used at the interface which was significant at 1g but not at higher g-levels, where the 30° inclination of the interface was several times higher than the friction angle at the greased interface. Thus during the test there was (as desired) significant tendency for the unstable material to move downslope and interact with the pile row.

Apart from the pile spacing, the other parameter which was varied in the tests was the length of the layer of unstable soil upslope of the piles, l. Expressed relative to the vertical thickness of the unstable layer, h, (l/h) varied from 1.5 to 4 (Figure 2). Increasing this value essentially increased the load acting on the pile row to maintain stability of the 'unstable' layer upslope of the pile row.

3 EXPERIMENTAL RESULTS

3.1 *Pile bending moment data*

Data from the 10 transducers was fitted with seperate polynomials above and below the failure interface, maintaining continuity of shear and moment across this interface. It was found that the moment data above the interface generally implied a triangular distribution of pressure with depth in the passive loading region. Hence for the free-headed piles the shear force at the interface (S_i) could be easily derived in terms of the moment at the interface (M_i) and thickness of the unstable layer (h):

$$\tau_p = \sigma_n \tan\left[JRCLog\left(\frac{JCS}{\sigma_n}\right) + \phi_b \right] \quad (1)$$

Assuming the load on the instrumented pile to be representative of all piles in the row, inspection of variation of S_i throughout the tests confirmed that the load on the row increased in proportion to both g-level and (l/h).

A variable B_{mob} was used to express the equivalent pressure mobilized on the pile (p = load per unit length/ diameter) due to interaction, relative to the

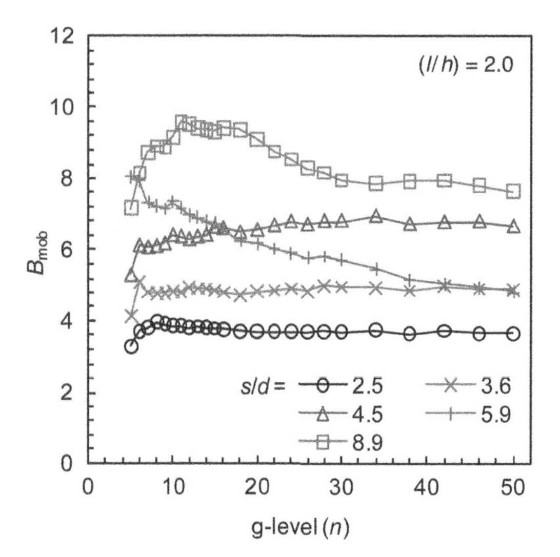

Figure 3. Variation of B_{mob} throughout a test ($l/h = 2$).

nominal overburden stress at a given depth in the unstable soil layer ($\sigma'_v = \gamma z$):

$$JRC = \frac{\arctan(\tau/\sigma_n) - \phi_b}{Log_{10}(JCS/\sigma_n)} \quad (2)$$

However, since the distribution of passive stress was approximately triangular with depth B_{mob} was assumed to be constant with depth. Hence

$$JRC = \frac{\alpha_s - \phi_r}{Log_{10}(JCS/\sigma_{n0})} \quad (3)$$

Since S_i increased approximately in proportion to g-level and γ increased directly in proportion to g-level it was found that B_{mob} was approximately constant throughout a test – eg. Figure 3 shows data for all tests with $(l/h) = 2.0$.

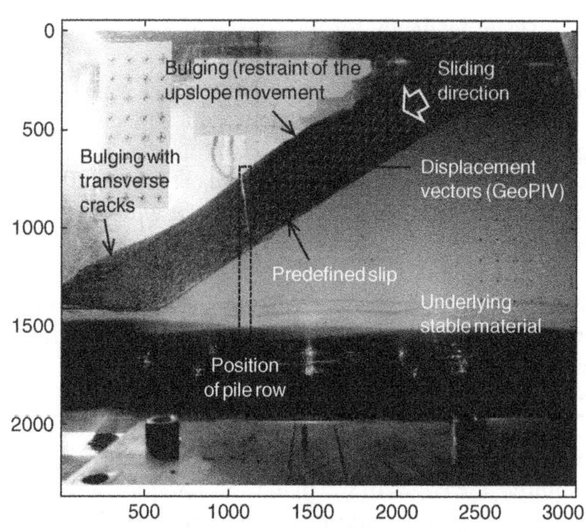

Figure 4. Front view of test ($s/d = 2.5$, $l/h = 4$) at $50g$.

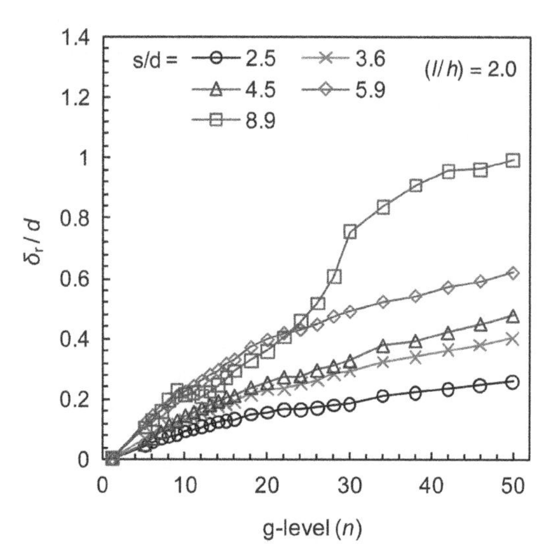

Figure 5. Relative soil-pile displacement throughout a test ($l/h = 2$).

3.2 *Relative soil-pile displacement*

Figure 4 shows the 'front' view of a model with $(s/d) = 2.5$ and $(l/h) = 4.0$ at $50g$ (the final stage of the test). The unstable layer is clearly visible due to the presence of the black latex sheet against the window. In fact the latex was spray-painted to give a visual 'texture' for GeoPIV to track (vectors of movement are shown on Figure 4).

Bulging can be clearly seen at the toe of the slope, and also just upslope of the pile row – indicating that the row had impeded downslope soil movement.

The camera with 'plan' view allowed general movement of the soil upslope of the pile row and movement of the pile heads to be observed. The relative soil-pile movement (δ_r) could then be deduced from images at each g-level.

Figure 5 shows how this movement (presented normalised by the pile diameter, d) increases with g-level for all tests with $(l/h) = 2.0$.

As anticipated, δ_r increases with g-level and (s/d), although the rate of increase with g-level tends to reduce as g-level increases. The test with widest pile spacing ($s/d = 8.9$) shows a 'jump' in the data at about $30g$, corresponding a significant movement event. Unfortunately the movement cannot be interpreted

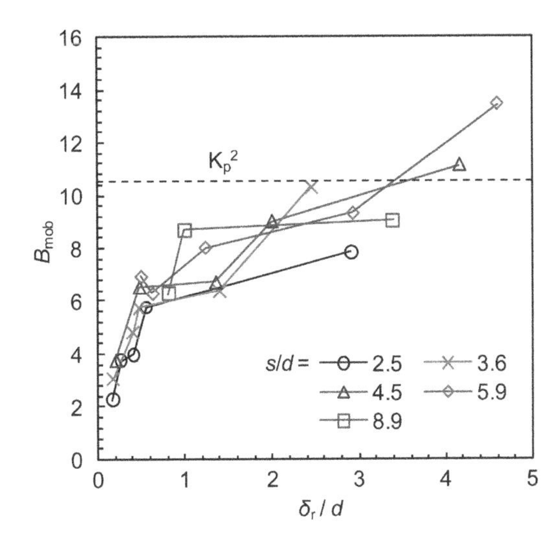

Figure 6. Variation of B_{mob} with relative soil-pile displacement at 50 g showing results from all tests.

directly as equivalent prototype values since it accumulated at a variety of g-levels. However, the results from the various tests can be compared.

3.3 *Limiting interaction pressure*

Figure 6 plots the normalised relative displacement at 50g in all tests against the (approximately constant) value of B_{mob} in the test. Separate data series show each value of (s/d) used in the tests. Moving along a data series from left to right shows increase of (l/h) in various tests, and hence increasing load on the pile row.

With the exception of one value at very large relative displacement B_{mob} tends to a maximum value of approximately 10. The critical state friction angle of the Leighton Buzzard Fraction E sand was found from triaxial tests to be 32°. This implies a passive earth pressure coefficient K_p of 3.25. Hence the limiting value of B_{mob} is approximately equal to K_p^2.

This value of limiting interaction for an 'isolated' pile in ground with a horizontal surface has previously been proposed by Fleming et al (1994). It could be argued that the limiting interaction pressure would be increased due to the inclination of the upslope soil loading the piles. However the limiting value does not appear to be significantly increased, and it is in any case conservative to ignore this potential effect in design.

3.4 *Effect of pile spacing*

According to numerical analyses conducted by Durrani et al (2008) the K_p^2 limit on interaction (for an isolated pile) could be used to propose a limit on centre-to-centre spacing (s) where behaviour of the pile row ceases to be a 'wall' and becomes more like individual piles at larger spacing. This limit is

$$\sigma_{v0} = \frac{W Cos\beta_s}{A} \tag{4}$$

For $K_p = 3.25$ (and $K_a = 1/K_p$), $(s/d)_{crit} = 3.6$. Based on this hypothesis when (s/d) is less than 3.6 the value of limiting interaction is reduced in proportion to (s/d) and hence for $(s/d) = 2.5$ the limiting value would be $(2.5/3.6) \times 3.25^2 = 7.3$. This is close to the maximum value observed in the tests for $(s/d) = 2.5$. However, it could be argued that a larger value would have been observed at higher δ_r. This would have required a test with $(l/h) > 4$. However, the test package would not allow this, and in any case there would have been very significant tendency for failure upslope of the piles, thus limiting interaction with the piles.

For a given magnitude of B_{mob} less than the limiting value it can be seen that the value of (s/d) does not have significant impact on the corresponding magnitude of δ_r. This supports another observation by Durrani (2006) that piles spaced wider than the critical value still offer support to the upslope material – it is just that the equivalent value per metre along the slope drops in proportion to $(1/s)$ for a given value of B_{mob}.

4 CONCLUSIONS

Using the photogrammetric and structural loading results a number of mechanisms of behaviour for the slope and pile row were identified ranging from a successfully stabilised slope (small relative soil-pile displacement), to slips which occurred upslope of the pile row and thus did not interact with it.

The mobilised interaction pressure B_{mob} was found to be approximately constant with depth and throughout a test (as g-level increased). Except at pile spacing less than a critical value proposed by Durrani et al (2008) B_{mob} tended to a maximum value of approximately K_p^2, corresponding to an isolated pile. The maximum value of B_{mob} does not appear to be significantly increased due to the inclination of the upslope soil loading the piles, and it is in any case conservative to ignore this potential effect in design.

The results have verified that the K_p^2 limit on interaction for an isolated pile can be used to propose the critical pile spacing where arching is effective.

REFERENCES

Durrani, I.K. 2006 *Numerical modelling of discrete pile rows to stabilise slopes*. PhD thesis, Nottingham University.

Durrani, I.K., Ellis E.A. & Reddish D.R. 2008 Numerical modelling of lateral pile-soil interaction for a row of piles in a frictional soil. *Proc. 1st Int. Conf. on Transportation Geotechnics*, Nottingham, UK.

Ellis, EA; Cox, C; Yu, HS; Ainsworth, A & Baker, N. 2006 A new geotechnical centrifuge at the University of Nottingham, UK. *ISSMGE International Conference on Physical Modelling in Geotechnics*, Hong Kong.

Fleming, W.G.K., Weltman, A.J., Randolph, M.F. and Elson, W.K. 1994 *Piling engineering*. Taylor & Francis Ltd.

White, D.J., Take, W.A. & Bolton, M.D. 2003 Soil deformation measurement using particle image velocimetry (PIV) and photogrammetry. *Geotechnique*, 53 (7), pp. 619–631.

Advances in Transportation Geotechnics – Ellis, Yu, McDowell, Dawson & Thom (eds)
ISBN 978-0-415-47590-7

The effect of moisture on the stability of rock slopes: An experimental study on the rock slopes of Khosh Yeylagh Main Road, Iran

M. Zare & S.R. Torabi
Department of Mining, Petroleum & Geophysics, Shahrood University of Technology, Shahrood, Iran

ABSTRACT: Instability of the rock slopes in mountainous areas renders the roads and railroads potentially hazardous. The presence and the behavior of discontinuities in the rocks, on the other hand, play a significant role in the instability of rocks and rock slopes. The Khosh Yeylagh Main Road is situated in north-east of Iran and connects Semnan province to the Northern provinces. This region contains numerous rock slopes with high potential of rock fall risk. These slopes are subject to much seasonal rain that results in a great change in strength components of discontinuities followed by increased risk of rock failure. In this paper, the effect of moisture on the properties of discontinuities is experimentally investigated. For this purpose, large numbers of rock samples containing natural discontinuities were collected from slope faces and then large and small scale tilt tests and also multistage direct shear tests were performed under both dry and wet conditions. It was observed that in low levels of normal load, rock joints are more reactive to presence of water. This condition is more expected in slope faces where mostly the normal load on the joint is only the weight of the upper side. It was shown that the results could be helpful in slope stability analysis and roadside trench design.

1 INTRODUCTION

The Increase in the population and necessity of the construction and development of roadways and highways ensuring safety and economy calls for the application of different technologies including the science of geotechnics. Evaluation of the stability of rock slopes adjacent to the roadways is an important part of the planning and design. Slope failure can cause loss of lives, disruption to the traffic and damage to the vehicles, equipment and structures. As a safety and economic measure, in such areas, evaluation of the stability of adjacent rock slopes should be carried out as part of roadways design.

The presence, nature, pattern and mechanical properties of discontinuities in the rock mass dictate the stability of the rock slope. Due to the porous nature of rock, moisture is potentially absorbed into the voids and cracks because of the surface free energy. As a result, a thin water film is formed on the crack surface, which in turn facilitates further crack propagation and deteriorates the material strength (West, 1994). Saturated rocks can as a result reach a failure state at a relatively low stress level compared to dry rock (Vasarhelyi, 2005). With respect to rock joints or fractures, which are quite commonly found in subsided strata post mining, water exists not only in the pore voids of intact parts but also on the failure surfaces in the form of a thin water film. Hence, the mechanical and geometrical parameters of the failure surface such as Joint Roughness Coefficient (JRC), dilation angle and basic friction angle etc are inevitably influenced. Accordingly, not only the intact rock but also the disturbed rock mass is at the risk of influence by water.

The detrimental effect of water has been observed on intact rock by many researchers (Ballivy, et al. 1976; Broch, 1979; Colback & Wiid 1965; Dube & Singh, 1972; Masuda, 2001; Parate, 1973; Van Eeckhout, 1976). Many relevant tests have been conducted on different types of rocks. Previous results showed that up to 90% of the unconfined compressive strength could be lost at approximately one third of the moisture content at saturation (Colback & Wiid, 1965; Hawkins & McConnel, 1992). In addition, it was found that the triaxial compressive strength of rocks decreases after the adsorption of water and the yielding strength varies almost linearly with the water content, so do the other strength parameters such as stiffness, cohesive strength and internal friction angle etc. Most of these analyses, however, are qualitative descriptions based on laboratory observations rather than quantitative expressions. Meanwhile, although various hypotheses have been put forward such as fracture energy reduction, capillary tension decrease, frictional reduction, chemical and corrosive deterioration and effective

Figure 1. A photograph of the Khosh Yeylagh main road.

Figure 2. Joint pattern in part of the Khosh Yeylagh region.

stress decrease due to the pore pressure increase (Van Eeckhout, 1976) in attempts to interpret the water effects, none of them provides a reliable and general quantitative measure of the problem.

Additionally, little attention has been paid to reduction of the rock slopes' stability under the consideration of moisture effects even though those accept great influence from water content and its damaging effects. The shear strength of rock joints and fractures of rock slopes will control their stability capacity and deformation of the disturbed rock. In addition, under saturated conditions, rock joints fail more easily than intact rock because the main resistance force of failed rocks is from the friction and cohesion of the already existing failure surface, which is more sensitive to the effects of water. Consequently, in order to assess the effect of moisture on stability of rock slopes, research needs to be focused on the rock joints or fractures.

In view of the above, a series of shear tests were carried out on sandstones as broken samples, under various normal load conditions. Corresponding results are presented and discussed. The variation of the strength property parameters due to the saturation is analyzed subsequently.

2 SAMPLING

The Khosh Yeylagh mountainous area is situated approximately 90 km north of Shahrood city, North-East of Iran, through which the main road connecting Semnan province and the northern provinces passes. Figure 1 is a photo of the area.

Rock samples containing dominant discontinuities from over fifty locations in the Khosh Yeylagh region were collected and transferred to the laboratory in separate packs ensuring that the couples of rock forming the two sides of the joint sets come together. The samples were of different shapes and sizes picked from the slope face next to the roadway. In an attempt to follow the linear scanlining procedure, during sampling, special attention was paid to ensure that the condition of joints is recorded such as spacing, orientation, kind of filling materials if any, aperture, persistence, orientation of major discontinuities. Figure 2 shows the joint pattern in part of the region.

3 EXPERIMENTAL INVESTIGATIONS

3.1 *Preparation of samples*

In the laboratory, the collected samples were cut to the size about 10 cm by 10 cm by 14 cm including about 7 cm each side of the discontinuity sets based on the suggested methods of ISRM (1978). The idea was to prepare the samples containing a single discontinuity suitable for testing. These samples were prepared to be used in direct shear test. Some samples of different dimensions were prepared to be used in the tilt tests.

3.2 *Laboratory tests*

Laboratory testing generally involves a constant normal load and an increasing shear load being applied to the dry and wet samples by the portable shear box which can accept specimens up to a size about a 140 cm cube. Both shear and normal loads are monitored, as well as displacements parallel and normal to the plane of shearing. Once the peak strength is reached, displacement is continued, usually at a significantly lower shear load, reflecting the residual strength.

Multistage testing refers to the case where several tests are undertaken at different normal loads. The peak and residual shear strength can be estimated for each test from plots of the shear displacement versus shear stress. A plot of normal stress versus shear stress can then be drawn allowing estimates to be made of the shear strength characteristics of the discontinuity. The same discontinuity sample is often used for multistage testing due to the difficulties in obtaining a sufficient number of identical samples, which allows the maximum information to be gained from each sample. However, Barton reported that when the same sample is used, only those tests performed at low normal stress would provide reliable information on the peak strength characteristics of the discontinuity. Repeated shearing of the specimen will cause the remainder of the test results to fall somewhere between the peak and the residual values.

3.3 *Barton's criterion*

After recognizing the inadequacy of the linear Coulomb model, Barton (1973) proposed an empirical criterion for the shear strength of rock joints, as follows:

$$\tau_p = \sigma_n \tan\left[JRC Log\left(\frac{JCS}{\sigma_n}\right) + \phi_b \right] \tag{1}$$

Where τ_p = the peak shear strength of the unfilled joint (that is, there is rock-to-rock contact across the plane), σ_n = the effective normal stress acting on the joint, JRC = the joint roughness coefficient (ranging from 0 to 20), JCS = the unconfined compressive strength of the rock immediately surrounding the joint and ϕ_b = the basic friction angle. The basic friction angle is a measurement of the shear strength along an artificially planar saw cut surface and is characteristic of the rock mineralogy (Giani 1992). The techniques available for estimating the input parameters to Barton's criterion are discussed below.

The measurement of JCS parameter is of fundamental importance in rock engineering since it is largely the thin layers of rock adjacent to joint walls that control the strength and deformation properties of the rack mass as a whole. The Schmidt Hammer provides the ideal measurement. This is a simple device for recording the rebound of a spring loaded plunger after its impact with a surface. The L-hammer used here (impact energy = 0.075 mkg) is described by the manufactures as being "suitable for testing small and impact-sensitive parts of concrete or artificial stone". It is suitable for measuring JCS values down to about 20 MN/m^2 and up to at least 300 MN/m^2 (Barton & Choubey 1977).

Barton and Choubey (1977) report that JRC could be estimated through the back analysis of shear tests, where Eq. (1) is rearranged into the following form:

$$JRC = \frac{\arctan(\tau / \sigma_n) - \phi_b}{Log_{10}(JCS / \sigma_n)} \tag{2}$$

They also described a residual tilt test in which pairs of flat sawn surfaces are mated and the pairs of blocks are tilted until sliding occurs. If as is the tilt angle at which sliding starts to occur, σ_{n0} is the normal stress acting on the joint when sliding begins to occur and ϕ_r is the residual friction angle, the JRC value can be estimated from the following equation:

$$JRC = \frac{\alpha_s - \phi_r}{Log_{10}(JCS / \sigma_{n0})} \tag{3}$$

The normal stress is calculated using

$$\sigma_{v0} = \frac{W Cos \beta_s}{A} \tag{4}$$

Where W = the weight of the upper block, A = the gross contact area, and β_s = the inclination angle.

The basic friction angle can be estimated from direct shear testing on smooth rock surfaces that have been prepared by means of a smooth, clean diamond saw cut (Hoek & Bray 1981). Barton and Choubey (1977) reported that the basic friction angle for most smooth unweathered rock surfaces lies between 25° and 35°. Stimpson (1981) suggested the use of tilt testing of diamond core samples for the estimation of the basic friction angle. He observed that the core surfaces produced by typical core drilling procedures are precut and smooth and therefore not dissimilar to a saw cut rock surface. The suggested tilt tests involve attaching two pieces of core to a horizontal base, ensuring that the core samples are in contact with one another and are not free to slide. A third piece of core is then placed on top of the first two pieces and the base is rotated about a horizontal axis until sliding of the upper piece of core along the two line contacts with the lower pieces of core begins. The following equation can then be used to estimate the basic friction angle.

$$\phi_A = \arctan(1.155 \tan \alpha_s) \tag{5}$$

Where ϕ_A is the basic friction angle for the upper piece of core and α_s is the angle at which sliding commences.

4 RESULTS AND DISCUSSION

4.1 *Effect of moisture on the friction angle of rock joint*

When the normal stress reaches a certain level, the surface asperities will undergo large deformation and eventually fail. Under this state the reduced surface asperities are not easily interlocked. Accordingly, the failure surface tends to be smoother, resulting in a smaller friction angle. Regarding the wet sample, the corresponding deformation is larger than that in the dry sample at the same normal stress levels (Li et al. 2005). Therefore, the relevant apparent friction angles are smaller, as demonstrated in figure 3.

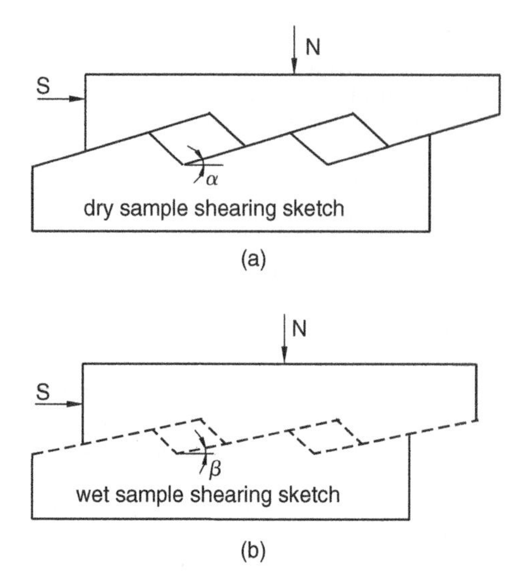

Figure 3. Schematic graph showing the deformation of surface asperities and different sliding surface because of saturation, note: β<α. (Li et al. 2005)

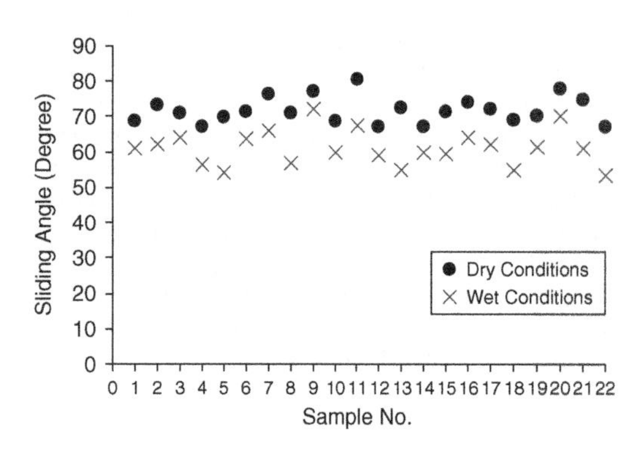

Figure 4. The sliding angle of wet and dry samples using tilt test approach.

For study of the weakening effects of moisture on the strength components and thereupon on the stability in rock slopes, the tilt tests were done on prepared samples. Also, Schmidt Hammer was used for measuring the strength capability of them.

Figure 4 shows the sliding angle in tilt test for 22 samples prepared for the laboratory testing.

In figure 5, data for Schmidt hammer rebound are presented for both wet and dry conditions. Finally, figure 6 demonstrates the reduction of residual friction angle caused by the moisture.

It has been approved that the above-mentioned types of testing are completely helpful in the slope stability analysis (Zare et al. 2007).

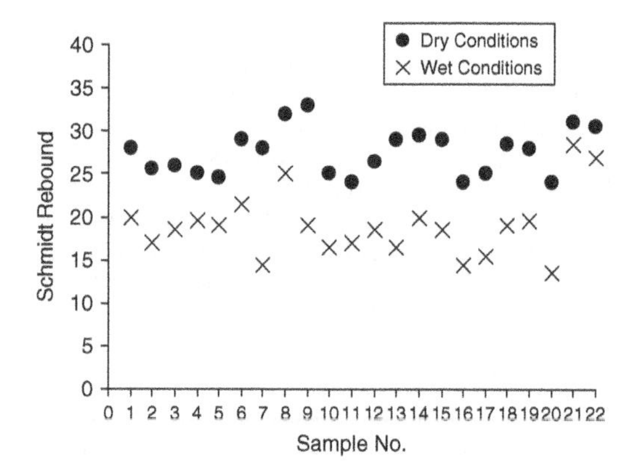

Figure 5. The Schmidt Hammer Rebound for wet and dry samples.

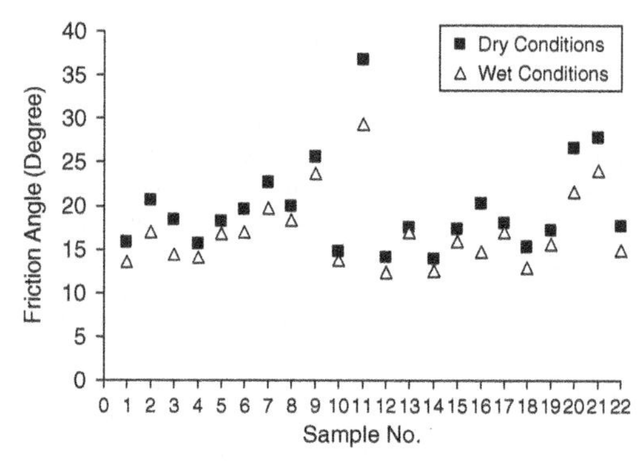

Figure 6. The effect of moisture on the residual friction angle.

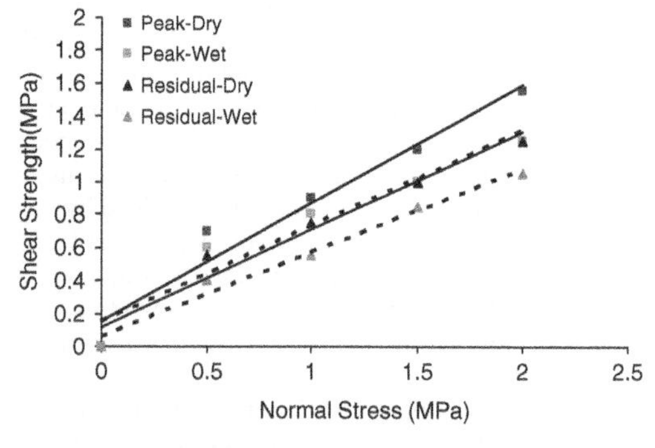

Figure 7. Graph showing the effect of moisture on the peak and residual shear strength.

4.2 *Effect of moisture on the peak and residual shear strength*

After testing and analyses of collected data, the graphs showing the detrimental influence of moisture on the strength components can be plotted. Figure 7 shows the reduction of peak and residual shear strength of rock joints caused by saturation. It is observed that the strength envelopes of wet samples are underneath the dry samples for both peak and residual strengths, which means the shear strength decreases when the rock mass is saturated.

Because of the existence of irregularities on joint planes, non-linear relationship normally exists between shear strength and normal stress at low normal stress levels. Generally, the gradient of the shear strength curve is increasing with decreasing normal stress. Hence, in particular at low normal stress levels, as relevant at most slope stability cases, it is of the greatest importance that the selection of shear strength parameters is adjusted to the normal stress level in question.

5 CONCLUSIONS

The experimental results and analyses presented in this paper show that the moisture can has a great influence on strength characteristics and subsequently on the stability of slopes containing jointed and fractured rocks. It is demonstrated that the shear strength reduction of rock joints is mainly from the water's serious deterioration of the residual friction angle. The main purpose of this paper has been to put emphasis on the importance of reliable determination of shear strength parameters, not in detail to discuss testing-method for achieving reliable results. However, as previously underlined, results from small scale shear tests will often be uncertain due to scale effects. At important slopes, the selection of shear strength parameters should preferably be based on large scale shear tests (in-situ or lab). Back analyses based on previously failed rock slopes may also be extremely valuable for evaluating shear strength parameters.

ACKNOWLEDGEMENT

The authors gratefully acknowledge the supports of Shahrood University of Technology and providing the laboratory equipment during the period that this work has been undertaken.

REFERENCES

Ballivy, G., Ladanyi, B. & Gill, D.E. 1976. Effect of water saturation history on the strength of low-porosity rocks, Soil specimen preparation for laboratory testing. *ASTM, American Society for Testing and Materials,* vol. 599, pp. 4–20

Barton, N. 1973. Review of a new shear strength criterion for rock joints. *Eng. Geol.* 7: 287–332.

Barton N. & Choubey, V. 1977. The shear strength of rock joints in theory and practice. *Rock. Mech.* 10. pp. 1–54.

Broch, E. 1979. Changes in rock strength by water. *Proceedings of the IV. International Society of Rock Mechanics, Montreux,* vol. 1, pp. 71–75

Colback, P.S. & Wiid, B.S. 1965. The influence of moisture content on the compressive strength of rocks. *Proceedings of the 3rd Canadian Symposium on Rock Mechanics, Toronto (Canada),* pp. 65–83.

Dube, A.K. & Singh, B. 1972. Effect of humidity on tensile strength of sandstone, *Journal of Mines, metals and fuels,* vol. 20(1), pp. 8–10.

Giani, G.P. 1992. Rock slope stability analysis. *Rotterdam A. A Balkema Publishers.*

Hawkins, A.B. & McConnel, B.J. 1992. Sensitivity of sandstone strength and deformability to changes in moisture content, *Journal of Engineering Geology,* vol. 25, pp. 115–130.

Hoek, E, & Bray, J.W. 1981. Rock slope engineering, *3rd ed. London: Institute of Mining and Metallurgy,* 358 pages.

ISRM, 1978. Suggested methods for the quantitative description of discontinuities in rock masses. *Int. J. Rock Mechanics Min. Sci. & Geomech. Abstr;* 15 (6) pp. 319–68.

Li Z., Sheng Y. & Reddish D.J. 2005. Rock strength reduction and its potential environmental consequences as a result of groundwater rebound. *9th International Mine Water Congress,* pp. 513–519.

Masuda, K. 2001. Effects of water on rock strength in a brittle regime, *Journal of Structural Geology,* vol. 23(11), pp. 1653–1657.

Parate, N.S. 1973. Influence of water on the strength of limestone, *Transaction of Society of mining engineers, AIME,* vol. 254, pp. 127–131.

Stimpson B. 1981. A suggested technique for determining the basic friction angle of rock surfaces using core. *Int J Rock. Mech. Min. Sci Geomech Abstr* 18:63–5.

Van Eeckhout E.M., 1976. The mechanism of strength reduction due to moisture in coal mine shales, *International Journal of Rock Mechanics, Mining Sciencs & Geomechanical Abstract,* vol. 13(2), pp. 61–67.

Vasarhelyi B. 2005. Statistical Analysis of the Influence of Water Content on the Strength of the Miocene Limestone *Rock Mech. Rock Engng.* 38 (1) pp. 69–76.

West, G. 1994. Effect of suction on the strength of rock, *Quarterly Journal of Engineering Geology,* vol. 27, 51–56

Zare, M., Torabi, S.R. & KhaloKakaie, R. 2007. Studying the Effect of Mechanical Properties of Discontinuities on the Slope Stability in Part of Angouran Open Pit Mine, IRAN. *Proceedings of the 7th International Scientific Conference SGEM. Bulgaria,* p. 58.

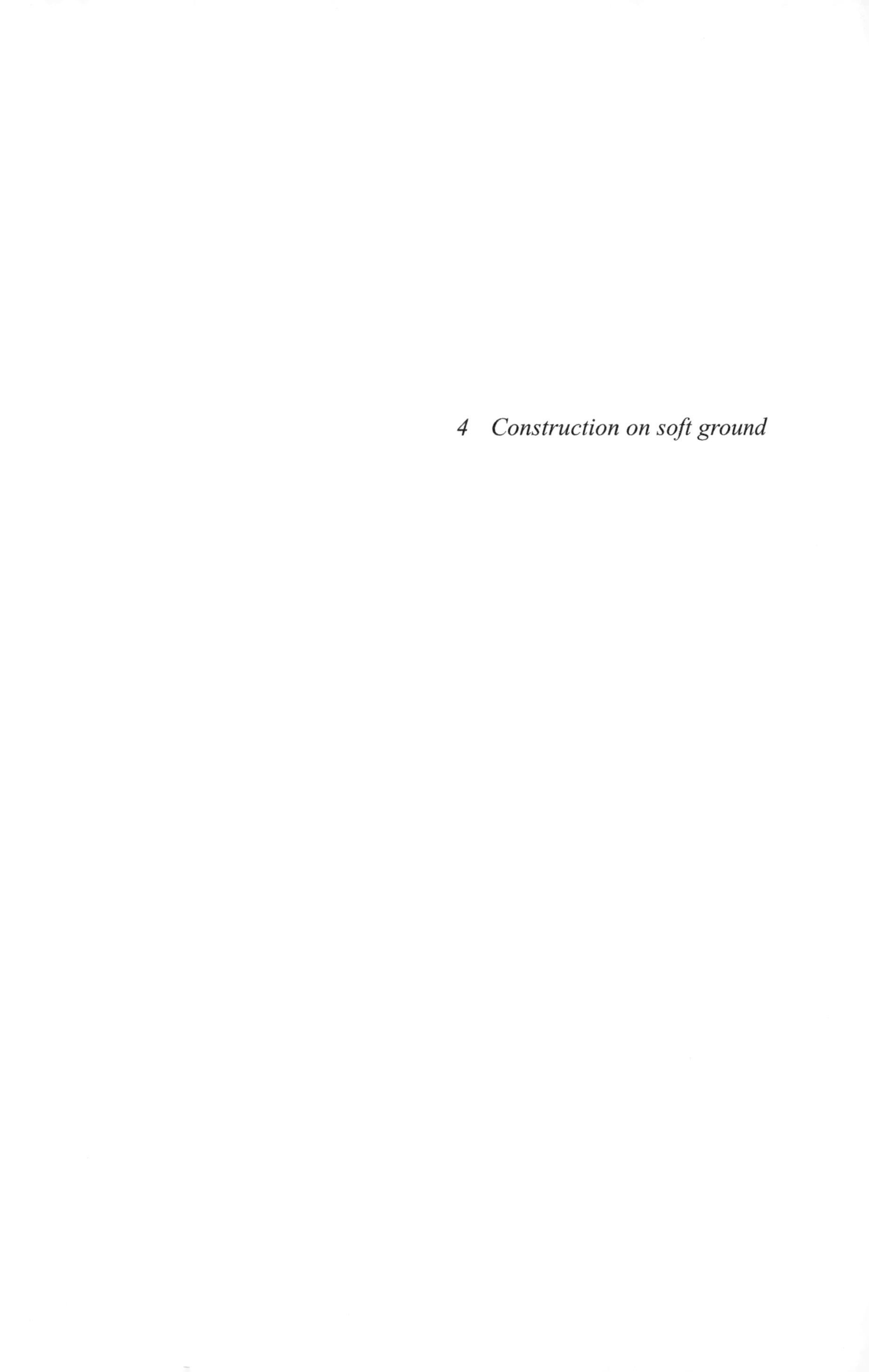

4 Construction on soft ground

Advances in Transportation Geotechnics – Ellis, Yu, McDowell, Dawson & Thom (eds)
© 2008 Taylor & Francis Group, London, ISBN 978-0-415-47590-7

Centrifuge modelling of piled embankments

R. Aslam & E.A. Ellis
Nottingham Centre for Geomechanics, School of Civil Engineering, University of Nottingham, UK

ABSTRACT: This paper presents a series of centrifuge tests examining the performance of unreinforced piled embankments constructed over a soft subsoil in terms of stress acting on the subsoil, and differential movement at the surface of the embankment. The effect of a 'working platform' below pile cap level and thus directly loading the subsoil is also considered. The normalised embankment height appears to be a critical parameter. Furthermore, a 'Ground Reaction Curve' concept can potentially be used to quantify variation of arching in the embankment with settlement. This would then allow the compatible contribution of the subsoil (and any geogrid) to also be considered in design.

1 INTRODUCTION

The concept of 'arching' of granular soil over an area where there is partial loss of support from an underlying strata has long been recognised in the study of soil mechanics (eg. Terzaghi, 1943). The 'trapdoor' approach has found some application in the analysis of piled embankments (Russell & Pierpoint, 1997), where the pile caps act as rigid supports, and the underlying soft subsoil as the 'trapdoor'. Other approaches for the design of such structures include a semicircular arch in the granular fill, as initially proposed by Hewlett & Randolph (1988) (and developed by Low et al., 1994; and Kempfert et al., 2004), and analogy with backfill over a buried pipe (BS 8006, 1995). Notably there is not one generally accepted approach for design (Love & Milligan, 2003).

The theories do not normally inherently consider the inclusion of a single or multiple layers of geogrid reinforcement in the embankment. (One exception is the so-called 'Guido' method).

This paper presents a series of centrifuge tests examining the performance of unreinforced piled embankments constructed over a soft subsoil in terms of stress acting on the subsoil, and differential movement at the surface of the embankment. A large range of embankment heights are considered. The effect of a 'working platform' below pile cap level and thus directly loading the subsoil is also considered.

2 EXPERIMENTAL METHOD

2.1 *Nottingham Centre for Geomechanics centrifuge*

Geotechnical Centrifuge modeling is an established technique for study of soil-structure interaction. The principles of the technique have been described by a number of authors (eg. Taylor, 1995). In essence a $1/N$th scale model is subjected to N times normal gravity (g) in a geotechnical centrifuge. Hence full-scale ('prototype') stresses are generated in the model, and soil in the model exhibits behaviour representative of the full-scale prototype.

The Nottingham Centre for Geomechanics centrifuge is a 50g-T machine, with 2.0 m platform radius – for further details see Ellis et al. (2006).

2.2 *Plane strain model*

Plane strain or axisymmetric packages are most commonly used in centrifuge modelling. However, in the tests described here a square pile layout was considered, and hence a 'unit cell' of the type shown in Figure 1(a) was used, requiring square vertical boundaries. A model which was 300 mm square in plan and 500 mm high was formed inside a 500 mm diameter axisymmetric tub as shown in Figure 1(b). The model contained a grid of 3 × 3 or 4 × 4 piles (ie. 100 or 75 mm centres), with the vertical boundaries located at lines of symmetry halfway between adjacent pile centres.

Wooden spacers were used at the bottom of the model to maintain the height of the surface as the thickness of other components varied. A 'model floor' was positioned above the spacers to allow the passage of wires from instrumentation.

The soft subsoil was positioned on the floor, and was actually modeled using Expanded Polystyrene Styrofoam (EPS), cut from blocks more normally used in the construction of lightweight embankments. The EPS block was 300 mm × 300 mm and 180 mm thick. Holes were drilled at the pile locations to allow the model piles to pass through the block.

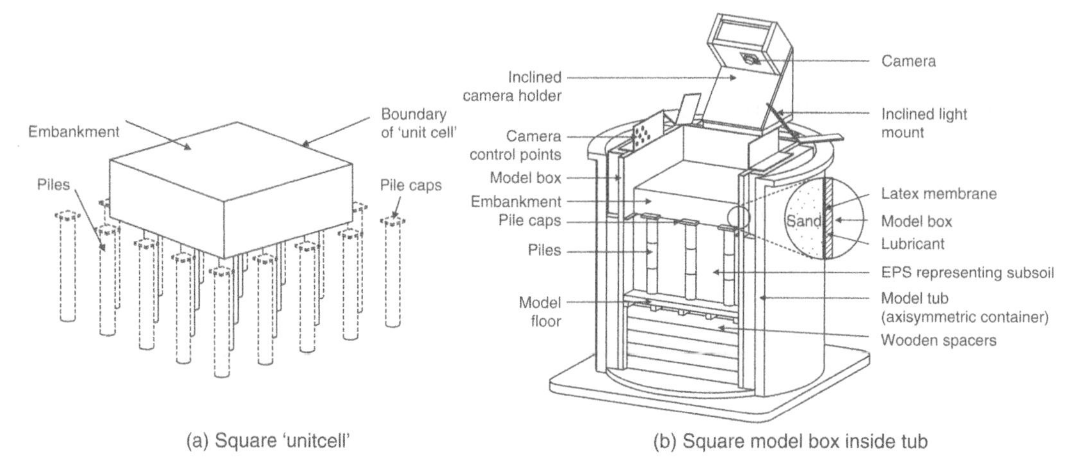

Figure 1. Centrifuge test model.

The use of EPS is an idealisation, but offered considerable convenience whilst still adequately modeling the presence of a soft underlying layer. Two grades of material were used: EPS 70 and 200, with one-dimensional stiffness approximately 2 and 8 MN/m^2 respectively, as measured in oedometer tests. These values were considered to be representative of a range of materials encountered in the field for soft subsoil. The stiffness was nominally constant up to the point of yield for the material, which generally was not reached in the tests.

The model piles were manufactured from 25 mm OD aluminium tube with 1.8 mm thickness. The piles were virtually incompressible compared to the surrounding EPS, and a number of piles in each test were instrumented to measure axial load. 30 mm square pile caps were fitted to the tops of the piles.

The embankment was modelled using Leighton Buzzard Fraction C sand ($d_{50} = 0.5$ mm). The sand was pluviated through air at 1*g* (before the centrifuge test) to give a high relative density of 90–95%. Embankment thicknesses in the range 35 to 210 mm were used.

As shown in the detail on Figure 1(b) silicone grease was used to reduce friction at the vertical side boundaries, separated from the sand by a latex membrane. The interface friction angle is then reduced to a small value of less than 5°.

A camera was mounted on the package, with an inclined view of the top surface of the model embankment during the test. GeoPIV (White, 2003) was used to 'track' the movement of soil through a series of images, and this data was then converted to actual movement in the model using the known location of control points, and assuming that movement was purely vertical (which is likely to be reasonable).

2.3 *Test procedure*

During the test models were subjected to increasing g-levels in the range 10 to 60*g* in 10*g* increments. This approach is slightly unconventional since no single 'equivalent prototype' is modelled. However, interpretation of the results generally indicated that behaviour was consistent throughout the g-levels. This is supported by the observation that the various calculation methods for piled embankments depend on the ratio of various dimensions rather than the absolute magnitude of the values.

At 30*g* the equivalent prototype dimensions were probably most plausible, corresponding to 0.9 m square pile caps at centre-to-centre spacing of 2.25 or 3.0 m. Equivalent embankment heights were in the range 1.0 to 6.3 m at this g-level. Throughout the range of g-levels stress in the model would be high enough for the soil behaviour to be broadly representative of a full-scale structure.

3 EXPERIMENTAL RESULTS

3.1 *Pile load data*

Most tests contained 4 piles with load cells. In general the results were consistent for the various devices, and also showed increase in load in direct proportion to g-level in a test. This indicates that 'efficiacy' (the proportion of the embankment weight carried by the piles rather than subsoil) was independent of g-level. As previously noted, the available analytical methods would also predict this.

From the load cell data it was possible to calculate both efficacy and the corresponding average normalised stress on the subsoil (EPS), with

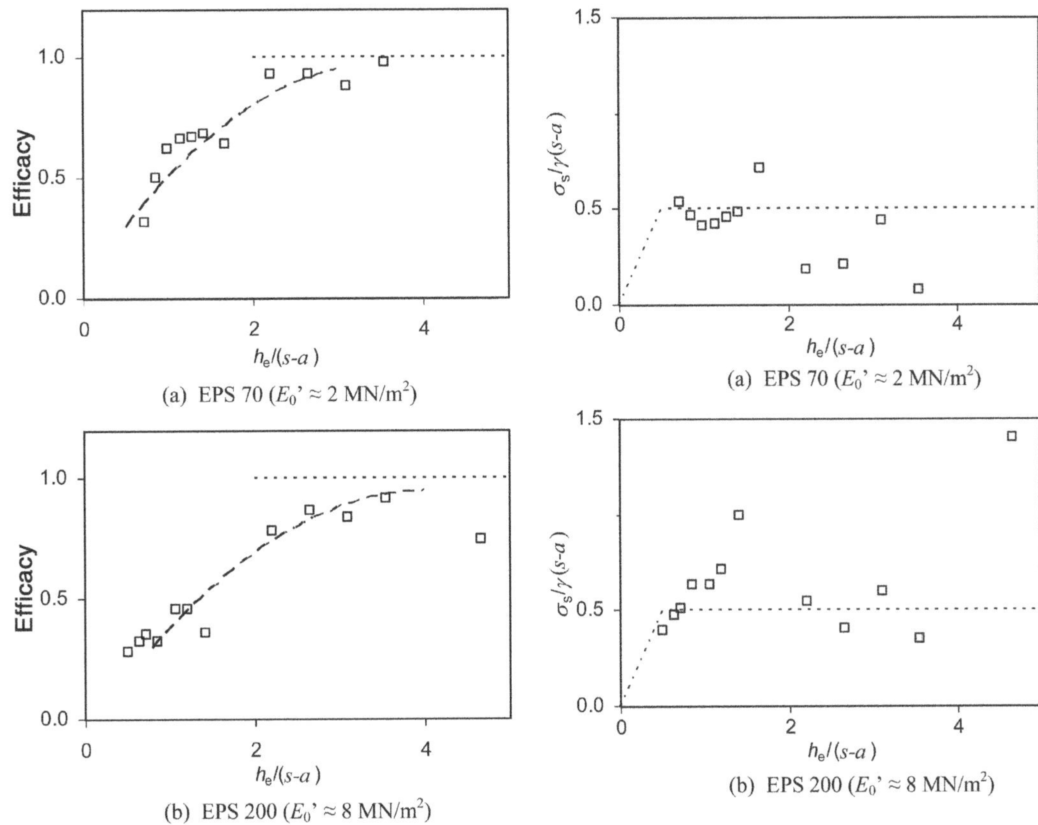

Figure 2. Efficacy showing variation with $h_e/(s-a)$.

Figure 3. $\sigma_s/\gamma(s-a)$ showing variation with $h_e/(s-a)$.

approximately unique values throughout each test. Hence Figure 2 shows a single data point for each test, plotting efficacy against the embankment height (h_e) normalised by the clear spacing between adjacent pile caps ($s-a$), where s is the centre-to-centre spacing and a is the pile cap dimension. The two subplots show the different subsoil (EPS) stiffnesses.

As anticipated, efficiacy increases with the normalised embankment height, tending to a value of 1.0 for $h_e/(s-a)$ in excess of about 2, irrespective of the subsoil stiffness.

The data can alternatively be interpreted in terms of the average vertical stress deduced to act on the subsoil (σ_s). In this case σ_s is plotted normalised by the value $\gamma(s-a)$, showing variation with $h_e/(s-a)$, Figure 3. On this plot a line from (0, 0) to (0.5, 0.5) indicates 'no arching' (since $\sigma_s = \gamma h_e$). It can be seen that for low embankment heights $\sigma_s/\gamma(s-a) \approx 0.5$. As the efficacy tends towards 1.0 small errors in pile load have a disproportionately large impact on estimation of σ_s, and hence the results appear erratic and are probably unreliable.

$\sigma_s \approx 0.5\gamma(s-a)$ shows good correspondence with the formula for conditions at the 'crown' of the arch proposed by Hewlett & Randolph (1988). For practical values of K_p applicable to granular embankments it an be shown that according to this method the vertical stress at the crown of the arch is relatively small, and the majority of stress arises from the 'infilling' material below the arch, which is assumed to contribute a value $0.7\gamma(s-a)$ to σ_s. For higher embankments conditions at the pile cap tend to be critical and thus this value is likely to be exceeded.

3.2 *Differential embankment displacement*

Figure 4(a) shows a typical 'surface plot' from a test with embankment thickness 70 mm and centre-to-centre pile spacing 100 mm (equivalent to 2.1 m and 3.0 m at 30g). The plot shows a 3-d view of the exaggerated deformed shape of the surface of the embankment derived from the image data. The horizontal axes show plan location in the 300 × 300 mm model box, and the vertical axis shows 'differential settlement' relative to a nominal zero value. As anticipated, the plot shows 9 'humps', corresponding to 'imprints' of the pile caps at

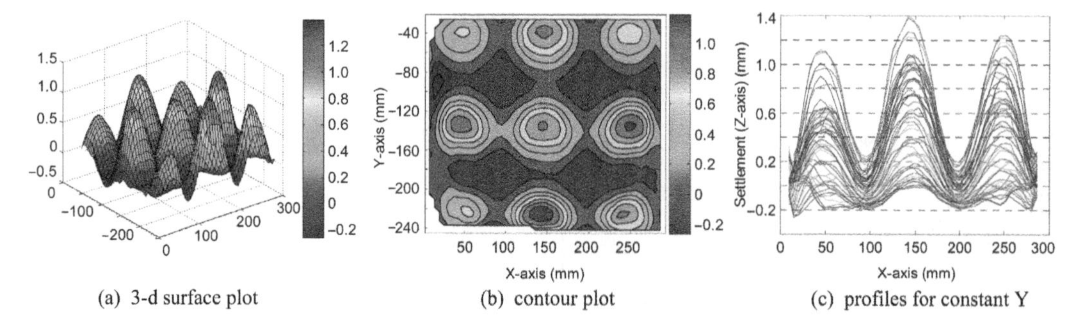

(a) 3-d surface plot (b) contour plot (c) profiles for constant Y

Figure 4. Typical data for differential settlement (model scale) at the embankment surface.

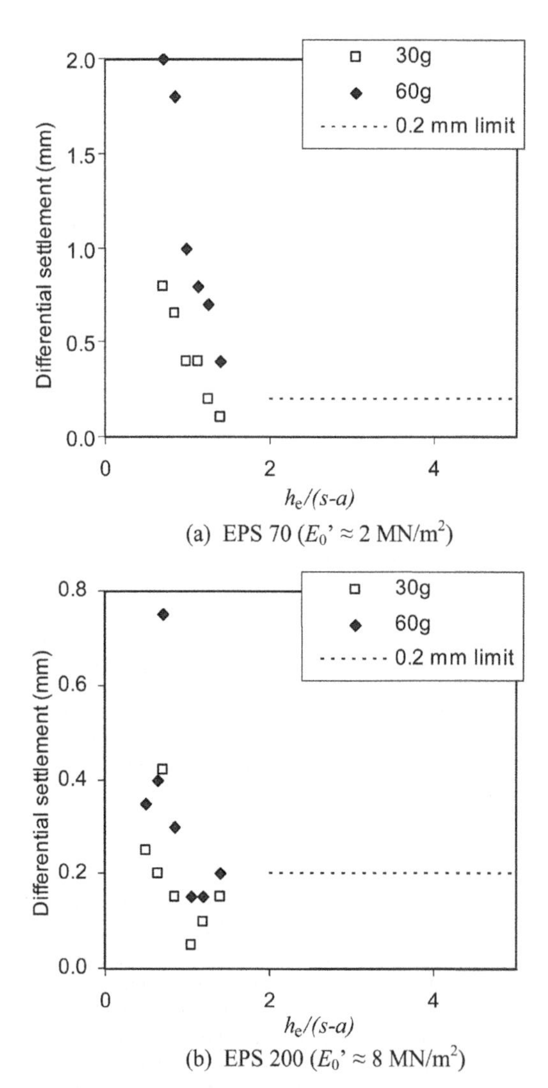

(a) EPS 70 ($E_0' \approx 2$ MN/m^2)

(b) EPS 200 ($E_0' \approx 8$ MN/m^2)

Figure 5. Differential settlement (model scale) at embankment surface showing variation with $h_e/(s - a)$.

the surface of the embankment in red. Areas between the piles (in blue) have settled more.

Figure 4(b) shows a corresponding contour plot of differential settlement. The 3×3 grid of pile cap imprints is again clear in red/yellow. Figure 4(c) shows corresponding profiles along lines of constant Y. The profiles again show 3 clear maxima at the pile cap locations. The magnitude of the maxima is largest for profiles passing 'over' pile caps, but significantly smaller for profiles at locations between the pile caps. These plots were used to manually assess the typical magnitude of differential settlement (δ) from the largest amplitude of variation. In this case a value of 1.0 mm was chosen.

It was generally observed that differential displacement increased in proportion with g-level in a test. Figure 5 shows variation of differential settlement at 30 and 60g with $h_e/(s - a)$ throughout the series of tests. Using the image data it was not possible to reliably discern differential settlement less than about 0.2 mm (6 mm at 30g), and this limit is shown by a dotted line.

For very low embankments the amount of differential settlement is higher for the softer subsoil (EPS), but in both cases as $h_e/(s - a)$ increases to about 2 the amount of differential settlement reduces to approximately zero (less than 0.2 mm at model scale).

3.3 *Effect of a working platform*

A number of tests were also undertaken with the upper portion of the subsoil replaced with granular material (like the embankment), representing a working platform initially placed on the soft subsoil to allow the piles to be installed or constructed. One such test will be reported here – the behaviour described below was consistently observed in all tests with a working platform.

The thickness of the embankment (h_e) was 270 mm, the thickness of the working platform (h_w) was 60 mm, and hence the thickness of the subsoil (EPS) was reduced from 180 mm to 120 mm so that the pile length remained at 180 mm. Critically, the working platform material is below the elevation of the pile caps, and hence is *not* supported by the pile caps but loads the subsoil directly. Centre-to-centre pile cap spacing (s) was 100 mm, and pile cap dimension (a) was 30 mm. At $30g$, $h_e = 8.1$ m, $h_w = 1.8$ m, $s = 3.0$ m and $a = 0.9$ m. The value of h_w is larger than most practical situations, but ensured that the effect of the working platform would be clearly evident in the test. EPS 70 was used for the subsoil.

The test initially proceeded in a routine manner, with constant efficiacy close to 1.0 up to $30g$, as would be expected for $h_e/(s - a) = 3.9$. However, for subsequent g-levels (up to $60g$) there was a marked drop in efficiacy. This was accompanied by a rapid increase in uniform settlement of the embankment surface with relatively little differential settlement, implying that the pile caps were 'punching' into the base of the embankment.

At $40g$ the stress on the subsoil due to the 60 mm thick working platform would have been $16 \times 0.06 \times 40 = 38\ \mathrm{kN/m^2}$. Thus making allowance for some additional loading from the embankment above the working platform, it seems probable that the EPS reached it's yield stress of $70\ \mathrm{kN/m^2}$ at $40g$ (rather like a soft subsoil reaching it's preconsolidation stress). Following yield the EPS would have deformed very significantly, and by the end of the test uniform settlement of the embankment had reached approximately 50 mm at model scale.

Iglesia et al. (1999) proposed a 'Ground Reaction Curve' (GRC) to consider the load on underground structures as they deform and soil arches above them (see also Zhuang et al., 2008). It was possible to plot a similar curve for the centrifuge test data. Stress on the 'subsoil' (in fact any material below pile cap elevation) was derived from the pile load and thickness of embankment material remaining above the elevation of the pile caps. This was plotted showing variation with uniform settlement at the surface of the embankment (which would reflect settlement of the subsoil), as this increased with g-level. Thus there were 6 data points (1 for each g-level 10 to $60g$).

Figure 6 shows this plot, noting that it is not precisely analogous to the original form of GRC proposed by Iglesia et al. (1999). The 'subsoil stress' deduced at pile cap elevation (σ_s) is normalised by the current nominal overburden stress γh_e, and hence is 1.0 before there has been any settlement of the embankment (at 1g prior to the test). Uniform settlement at the surface of the embankment is normalised by the clear gap $(s - a)$.

The GRC in Figure 6 shows a characteristically 'brittle' response (Iglesia et al, 1999), where a point of 'maximum arching' is reached, but then stress on the subsoil increases with subsequent settlement. Here maximum arching is reached at a relative settlement of about 5% – a similar value to that reported by Iglesia et al (1999). The subsequent behaviour illustrates reduction in efficiacy as the EPS yielded. During this process there were large deformations in the model, but the data does serve to illustrate the general brittle response.

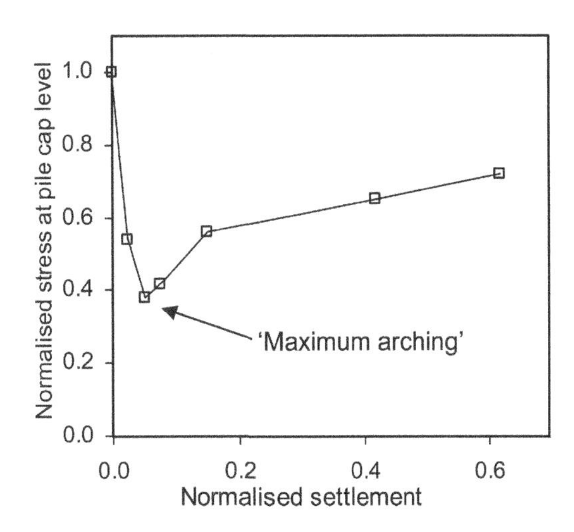

Figure 6. Normalised plot similar to GRC for centrifuge test data.

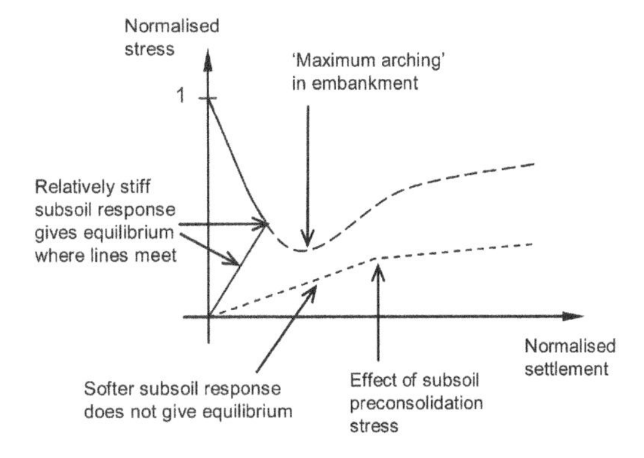

Figure 7. Potential use of GRC concept to assess settlement to give equilibrium.

3.4 *Potential use of GRC concept in design*

It seems possible that a GRC for a piled embankment could be used in design. Figure 7 schematically shows the normalised load on the subsoil, which in this case would be calculated to explicitly include the weight of any working platform. The settlement response of the subsoil to stress acting on it could be readily determined from one-dimensional compression characteristics, potentially including the preconsolidation stress. The effect of additional load capacity from geogrid layer(s) (also increasing with settlement) could also be added.

If the stress from the embankment GRC can be carried by the subsoil (and any geogrid) at a compatible settlement (ie. the lines intersect) stability should be ensured. However, if the lines do not intersect there will not be equilibrium – this effect is shown in Figure 7 for the softer subsoil. The probability of this is increased by the 'brittle' nature of the GRC. Likewise any working platform material beneath the elevation of the pile caps will increase the stress at the point of maximum arching, also meaning the line are less likely to intersect at a point of equilibrium.

The magnitude of the stress at the point of maximum arching can potentially be established using existing methods. However, the brittle response is likely to prove more problematic. Nevertheless the method adds to current understanding of the situation.

4 CONCLUSIONS

From the tests on unreinforced piled embankments the normalised embankment height appears to be a critical parameter:

- $h_e/(s-a) < 0.5$: stress on the subsoil is not reduced by arching, and there is significant differential settlement at the surface of the embankment
- $0.5 < h_e/(s-a) < 2.0$: there is increasing evidence of arching as h_e increases – the efficacy increases (tending towards 1.0), and differential settlement at the surface of the embankment reduces to a small value
- $2.0 < h_e/(s-a)$: there is 'full' arching with efficiacy close to 1.0 and little or no differential settlement at the surface of the embankment

Note that for typical values of (a/s), $h_e/(s-a) = 2.0$ corresponds to $h_e/s \approx 1.5$, and thus these results from physical modelling show good correspondence with numerical modelling reported by Zhuang et al (2008).

The Ground Reaction Curve concept can potentially be used to quantify variation of arching in the embankment with settlement. This would then allow the compatible contribution of the subsoil (and any geogrid) to also be considered in design.

ACKNOWLEDGEMENTS

Raveed Aslam was funded as a PhD student by the EPSRC (via the University of Nottingham School of Civil Engineering). This support is gratefully acknowledged.

REFERENCES

BS8006 1995. Code of Practice for strengthened/reinforced soils and other fills. British Standards Institution.

Ellis, EA; Cox, C; Yu, HS; Ainsworth, A & Baker, N. 2006. A new geotechnical centrifuge at the University of Nottingham, UK. *ISSMGE International Conference on Physical Modelling in Geotechnics*, Hong Kong.

Hewlett W.J. & Randolph M.F. 1988. Analysis of piled embankments. *Ground Engineering*, April 1988, pp. 12–18.

Iglesia G.R., Einstein H.H., & Whitman R.V. 1999. Determination of vertical loading on underground structures based on an arching evolution concept. *Proc. 3rd National Conf on Geo-Engineering for Underground Facilities.*

Kempfert H.G., Gobel C., Alexiew D. & Heitz C. 2004. German recommendations for reinforced embankments on pile-similar elements. *Proc. 3rd European Geosynthetics Conf*, 279–284.

Love J. & Milligan G. 2003. Design methods for basally reinforced pile-supported embankments over soft ground. *Ground Engineering*, March 2003, pp. 39–43.

Low B.K., Tang S.K. & Choa, V. 1994. Arching in piled embankments. *ASCE Journal of Geotechnical Engineering*, 120(11) 1917–1938.

Russell D. & Pierpoint N. 1997. An assessment of design methods for piled embankments. *Ground Engineering*, November 1997, pp. 39–44.

Taylor, R.N. (1995). *Geotechnical Centrifuge Technology*. Blackie Academic & Professional.

Terzaghi K. 1943. *Theoretical Soil Mechanics*. John Wiley & Sons, New York.

White, D.J., Take, W.A. and Bolton, M.D. (2003). Soil deformation measurement using particle image velocimetry (PIV) and photogrammetry. *Geotechnique*, 53 (7), pp. 619–631.

Zhuang, J., Ellis E.A. & Yu H.S. (2008). The effect of subsoil support in plane strain finite element analysis of arching in a piled embankment. *Proc. 1st Int. Conf. on Transportation Geotechnics*, Nottingham, UK.

Advances in Transportation Geotechnics – Ellis, Yu, McDowell, Dawson & Thom (eds)
© 2008 Taylor & Francis Group, London, ISBN 978-0-415-47590-7

Performance of embankments on soft ground: A1033 Hedon Road Improvement Scheme, UK

S.P. Beales
Scott Wilson Limited, UK

B.C. O'Kelly
Trinity College Dublin, Ireland, and formerly Scott Wilson Limited, UK

ABSTRACT: This paper presents the response of the ground foundation during the construction of the approach embankments to a new flyover at Salt End junction, A1033 Hedon Road Improvement Scheme, UK. The site area was underlain by about 8.0 m depth of soft alluvial deposits and the design specification required, amongst other factors, 100% consolidation of the alluvium under the higher embankments before the construction of the pavement layers could commence. The design solution incorporated ground improvement (prefabricated vertical drains and temporary surcharge) and a basal reinforcement layer in the transition zones next to the bridge abutments. An array of ground instrumentation monitored in real time the pore water pressure and deformation responses of the different strata comprising the ground foundation. The actual values of the primary consolidation parameters were determined from a back analysis of the settlement data.

1 INTRODUCTION

The A1033 Hedon Road Improvement Scheme involved the construction of a new stretch of dual carriageway, 6.7 km in length, that ran alongside the existing two-lane carriageway (west to east alignment) in close proximity to Hull estuary between Mount Pleasant junction, Kingston-upon-Hull, and Salt End junction, Hedon. The main construction on the £45 M (2001 data) design and build project commenced in autumn 2001 and was completed by summer 2003. The client was the Highways Agency (UK); the designer Scott Wilson Limited; the main contractor Alfred McAlpine and the client's representative Babtie. The alignment of the new carriageway traversed both greenfield and brownfield areas, which were generally underlain by about 8.0 m depth of soft alluvial deposits and with a high groundwater table (within 0.5 m of ground surface).

This paper focuses on the instrumentation and the ground foundation response during the construction of the eastern approach embankment (up to 8.4 m in height) to a new seven-span flyover at Salt End junction. The embankments were reinforced and steepened to 60 degrees on approach to the flyover. The objectives were to limit ongoing settlements to those agreed with the Highways Agency, namely:

1. For the higher embankments, achieve 100% primary consolidation in the alluvial deposits and seek to over-consolidate it to some degree in order to limit secondary compression settlements.
2. Limit the total differential settlements to between 20 and 50 mm over the five-year defect correction period following completion of construction.
3. Restrict ground movements along the line of the embankment toe due to the close proximity of the existing carriageway.

The design solution adopted was the stage construction technique in conjunction with ground improvement (prefabricated vertical drains and temporary surcharge loading). The surcharge would remain in place for a sufficient period in order to reduce the total differential settlements to an acceptable level. For the higher embankments, the surcharge was to induce a degree of over-consolidation in the underlying alluvium.

2 GROUND PROFILE

The Salt End east construction area (Figure 1) was a flat greenfield site (pastureland/playing fields) with the ground surface located about 2.5 m above Ordnance datum (mOD). The embankment site between chainages 6080 and 6400 m was bounded to the north by the existing carriageway and to the south by an open drain.

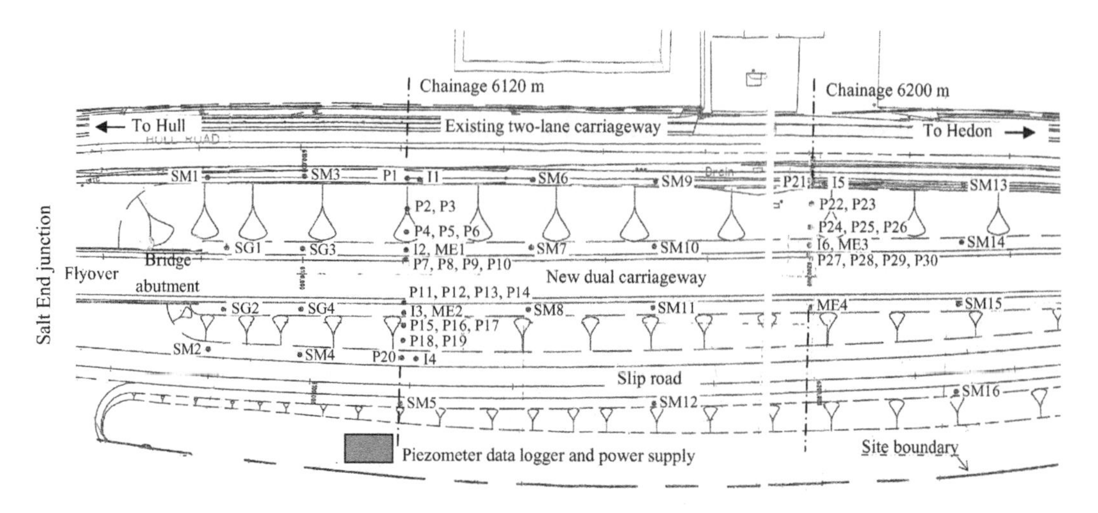

Figure 1. Ground instrumentation at Salt End junction (east). Note: P, piezometer; ME, magnetic extensometer; I, inclinometer; SG, rod and plate settlement gauge; SM, surface settlement marker.

Table 1. Stratigraphy.

Stratum	Chainage 6120 m Level (mOD)	Chainage 6120 m Thickness (m)	Chainage 6200 m Level (mOD)	Chainage 6200 m Thickness (m)
Ground surface level	+1.5		+1.9	
Crustal alluvium (firm silty clay)	−0.2	1.7	+0.1	1.8
Very soft alluvium	−5.3	5.1	−2.5	2.6
Soft, mottled organic clayey silt	−6.7	1.4	Not encountered	
Firm to very stiff glacial till	−16.2	9.4	−16.0	13.5
Glacial sandy gravel	−32.5	16.3	−32.5	16.5
Highly weathered chalk bedrock			Not proven	

The ground profile beneath the Salt End east site is shown in Table 1. The very soft alluvium comprised organic clay/silt and firm sandy gravelly clay. The glacial till layer included gravelly sand lenses (for example, encountered between −9.2 and −9.4 mOD at the location of instruments ME2 and I2, chainage 6120 m (Figure 1)) and was underlain by sandy gravel deposits under artesian pressure.

3 EMBANKMENT CONSTRUCTION

After erecting the site boundary and the removal of the top soil layer, a drainage mat comprising a 300- mm deep crushed stone layer sandwiched between Terram 1000 geo-membranes was constructed (Figure 2). With the working platform in place, Coffra Limited installed prefabricated drains (100 mm in width) vertically through the soft ground between chainages 6180 and 6300 m. The site investigation indicated that the thickness of the soft soil deposits varied over the site area. Hence, the drains were inserted through the full depth of the soft soil deposits (using a mandrel advanced under a set driving force) and penetrated up to about 1.0 m into the top of the underlying glacial till layer. Based on the ground model (interpolated from the ground investigation data) and the consolidation parameter values (determined from geotechnical laboratory tests on undisturbed specimens), a network of drains were designed in conjunction with a temporary surcharge to ensure that the geotechnical specification was achieved within the construction programme. The drains were arranged in a triangular grid pattern (plan view) at a centre spacing that varied between 1.0 and 2.0 m along the length of the Salt End site.

Figure 2. Installing ground instrumentation with drainage mat in place (view eastwards).

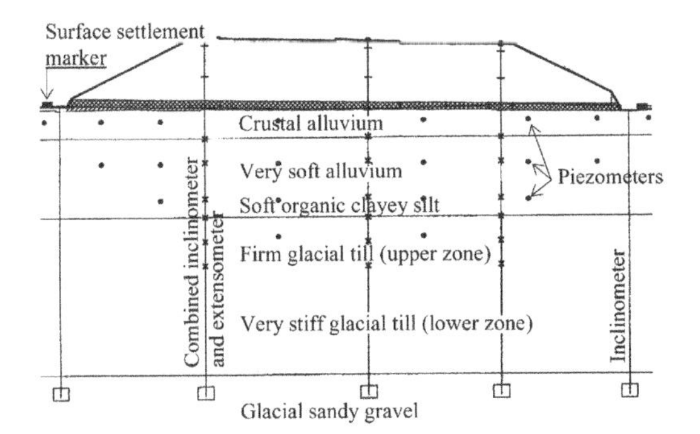

Figure 3. Layout of instrumentation at chainage 6120 m.

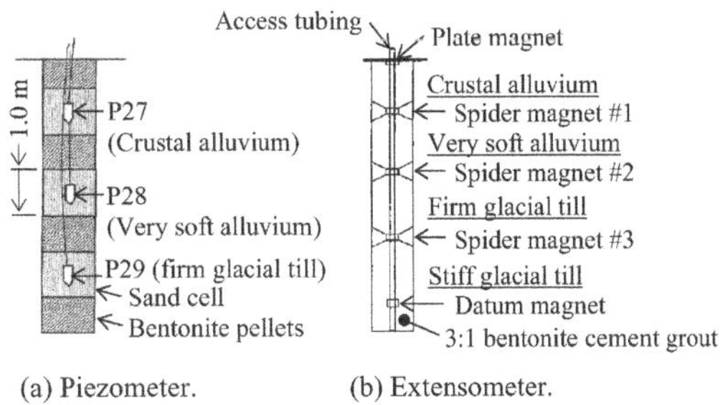

Figure 4. Installation details.

A basal reinforcement layer (high-tensile woven geotextile) was also included across the full embankment width in the transition zone to the bridge abutments (founded on pile groups end-bearing in the glacial sandy gravel stratum). The twin aims were to limit differential settlement in the transition along the new carriageway and reduce the lateral movements in the ground beneath the existing carriageway.

4 GROUND INSTRUMENTATION

4.1 *Overview*

An array of ground instrumentation monitored in real time the ground response during construction. Figure 1 shows the general plan layout of the instrumentation. Figure 3 shows in cross-section one of the two more heavily instrumented sections located at chainages 6120 and 6200 m.

4.2 *Pore water pressure*

An array of vibrating wire piezometer (P) instruments automatically recorded the pore water pressure response at the mid-height of the strata from which the consolidation responses of the different layers were determined.

Nests of between one and four piezometers were installed in a single borehole, 150 mm in diameter, which had been formed using a light cable-percussive rig (Figures 2, 4a). In plan, the boreholes were located equidistant from the three neighbouring prefabricated drains. The piezometers recorded the pore water pressure over a 1.0 m deep response zone (sand cell) that was centred at the mid-height of the strata. Piezometers were also located remote from the ground foundation to record variations in the natural groundwater level.

The electrical cables from the instruments were connected to an over-ground power supply and data logger unit that were located next to the embankment toe (see Figure 1). The cables were laid in a shallow trench that had been excavated in the drainage mat and backfilled with sand for protection. The site office dialled up the datalogger unit and downloaded the piezometer data in real time to the office PC for analysis.

4.3 *Settlement*

The settlement response was recorded using an array of magnetic extensometer (ME), rod and plate settlement gauge (SG) and surface settlement marker (SM) instruments (Figure 1). The extensometer boreholes (Figure 4b) were backfilled with 3:1 bentonite-cement grout. The datum magnet was located a distance of 0.5 m above the base of the borehole within the stable ground (glacial till). A series of spider sprung magnets were located at the interfaces between the overlying strata and a plate-mounted magnet was placed above the drainage mat.

The initial instrumentation scheme anticipated locating the datum magnet at depth within the sandy gravel stratum thereby also facilitating measurement of the settlement response over the full thickness of the glacial till layer. However, boiling sand filled the lower section of the open borehole once the borehole installation penetrated the top of the sandy gravel layer (artesian conditions). Hence, the datum magnets were actually located a distance of about 1.0 m above the base of the glacial till layer.

Rows of surface settlement markers (vertical steel pin fixed in a concrete footing) were located along the line of the embankment toe (Figure 1). The steel plates of the rod and plate settlement gauges were in contact with the top of the drainage mat.

4.4 *Lateral movement*

The lateral movement of the ground foundation was recorded using inclinometer instruments that were located beneath the embankment crest and toe (Figures 1, 3). The bottom section of the inclinometer tube was secured (acting as a reference) by penetrating the top of the glacial till layer by a distance of about 8.0 m. The upper section of the inclinometer tube recorded the lateral movement of the overlying alluvial deposits. Again, the boreholes were backfilled with a 3:1 bentonite-cement grout. In some instances, combined inclinometer and extensometer instruments were installed in the same borehole (for example, beneath the embankment crest, Figure 3).

5 GROUND MONITORING

All of the instrumentation was calibrated and the initial (reference) readings were recorded. The piezometer readings were initially recorded at 10 min intervals over a two-week period to study the variation in the natural groundwater level due to evapo-transpiration and tidal action (close proximity to Hull estuary).

The embankment was constructed in a series of short bursts of activity (for practical reasons) and included the use of some locally-available marginal fill material that had been improved by the addition of cement. Different surcharges and periods were adopted along the embankments to fit into the contractor's program requirements.

The piezometers were automatically recorded at one-hour intervals and all of the other instruments were manually recorded at least twice weekly as the construction progressed. The top of the rod and plate settlement gauge and surface settlement marker instruments were surveyed relative to a topographical benchmark to a measurement accuracy of ±5.0 mm. The level of the fill material that had been placed above the instrument locations was also recorded following each construction stage. The instruments were shielded from the machine plant inside concrete manhole rings, which were placed above the surrounding ground. The access tubing of the extensometer and inclinometer instruments and the rod sections of the rod and plate settlement gauges were increased in length by the addition of 1.0 m sections as the embankment increased in height. The fill material was then hand compacted around the extended access tubing and rod sections.

The construction rate was controlled from a geotechnical standpoint based on the level of build up in the pore water pressures, the lateral deformations beyond the toe and guarding against a general shear failure. The potential for the latter was assessed from the inclinometer deformations and the toe heave recorded by the surface settlement markers.

The temporary surcharge of uncompacted fill material placed above the final embankment height (Figure 5) was insitu by April 2003. Two metres surcharge depth was placed between chainages 6080 and 6350 m. The surcharge ramped down from 2.0 to 0.0 m in depth between chainages 6350 and 6410 m. A second row of surface settlement markers was installed along the

Figure 5. Final embankment height plus temporary surcharge in place (view westwards).

line of the embankment crest at the full surcharge height. Monitoring continued but at a reduced frequency throughout the surcharge period; namely, the piezometers were recorded at three-hour intervals and all of the other instruments were recorded at least once a week.

The surcharge remained in place (typically two to four month period) until such time as a back analysis of the monitoring data (section 7) had proved that the geotechnical design specification had been achieved. The surcharge was then removed. The embankment side slopes were regraded to the finished profile and top soiled after which the construction of the pavement layers could commence.

6 GROUND RESPONSE

6.1 *Pore water pressure*

Figure 6 shows some sample pore water pressure data recorded beneath the north embankment crest at the mid-height of the different strata at chainage 6200 m. The data span over about a three month period leading up to and including the surcharge with about 4.0 m depth of fill already in place. Piezometer P21 in Figure 6 shows the natural groundwater level recorded remote from the stressed ground foundation (typically 1.5 mOD). Initially, the crustal alluvium (P27) consolidated more rapidly (both vertical upward and radial inward flow conditions to the drains) whereas the flow regime in the underlying strata was predominantly radial flow to the drains.

There was an immediate build up in the pore water pressure when the fill material was placed. However, the excess pressures dissipated relatively quickly and the rate of cumulative build up was low. For example, Figure 6 shows that at chainage 6200 m, the excess pore water pressure reached a maximum value of about 25 kPa (2.5-m head of water) for an increase in total vertical stress of about 135 kPa. By the end of a fill stage (1.0 m depth of material placed over a one-day

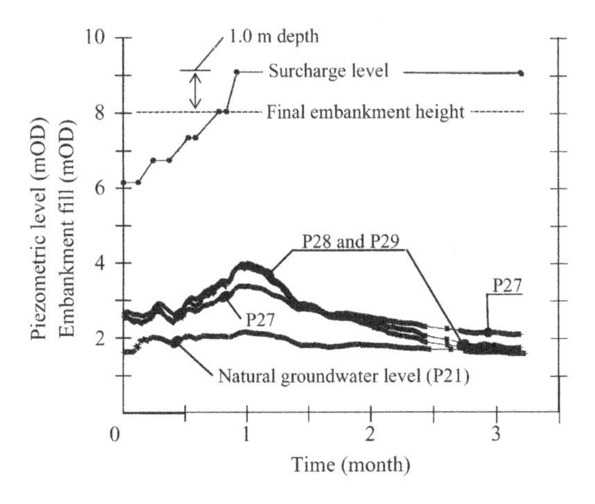

Figure 6. Pore water pressure response at chainage 6200 m.

period with consolidation ongoing), the ratio of the step increase in the pore water pressure to the increment in applied stress ranged between 0.25 (crustal alluvium) and 0.35 (alluvium and glacial till layers). At no stage did the recorded pore water pressures reach critical values from a geotechnical stability standpoint.

The excess pore water pressures had largely dissipated by the end of the two-month surcharge period used at chainage 6200 m. Complete dissipation occurred on removing the surcharge indicating that 100% consolidation of the alluvial deposits had been achieved under the final embankment height.

The groundwater level was found to fluctuate periodically with the tides, initially by up to 0.3 m in elevation in the soft alluvial deposits. However, the tidal effect reduced considerably with increasing effective stress (reducing hydraulic conductivity) and became masked by recurrent groundwater level changes arising due to evapo-transpiration (for example, see response of P27 in Figure 6).

6.2 *Settlement*

Figures 7 and 8 show the settlement response of the different soft soil layers beneath the embankment crest at chainages 6120 and 6200 m. Also included are the step increases in the vertical stress applied to the ground foundation by the embankment.

6.3 *Lateral movement*

Overall, the deformation response of the ground foundation was largely 1D compression. Figure 9 shows the lateral movements recorded at chainage 6120 m (within the area that included the basal reinforcement layer). Beneath the northern embankment crest and toe (next to existing carriageway in Figures 9b, c), maximum lateral movements of about 40 mm recorded

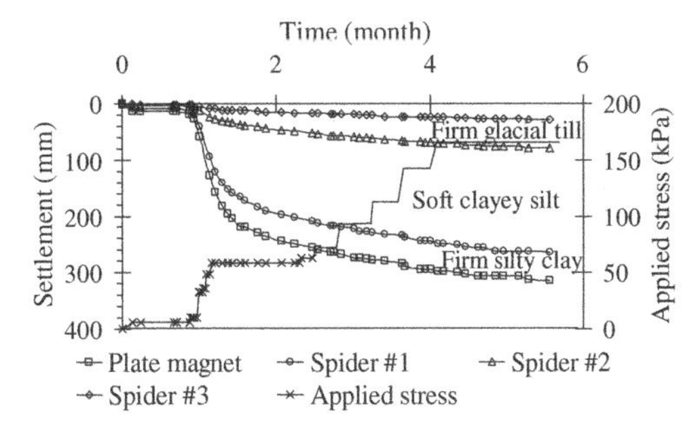

Figure 7. Settlement response from extensometer ME1 at chainage 6120 m.

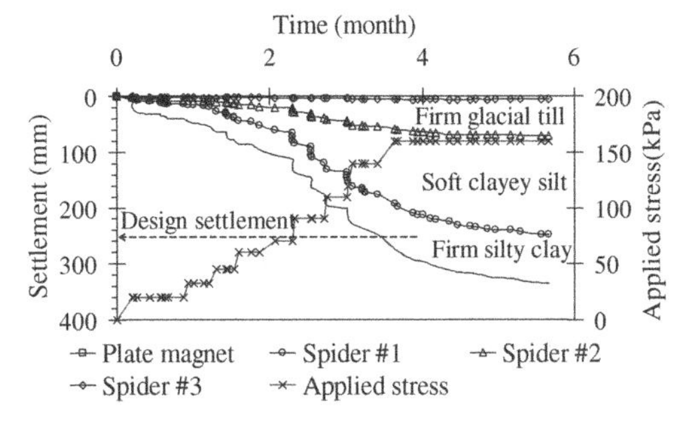

Figure 8. Settlement response from extensometer ME4 at chainage 6200 m.

within the soft clayey silt layer. The data from the surface settlement markers located along the embankment toe also indicated no significant heave (all recorded changes in elevation were within measurement accuracy) and there were no obvious signs of distress evident in the bituminous layer. Larger lateral movements of up to 70 mm (Figure 9a) were recorded beneath the side slope nearest the open drain (Figure 5). The maximum movement occurred at a shallower depth (although still within the soft clayey silt layer) and most likely occurred due to the reduction in the shear capacity of the ground foundation caused by the close proximity of the open drain. The lateral deformations of the embankment were also modeled using the finite element method. The recorded deformations were in good agreement with a PLAXIS analysis based on the stiffness values determined from pressuremeter tests.

7 SETTLEMENT BACK ANALYSIS

The actual values of the primary consolidation parameters (Table 2) were determined from a back analysis (curve fitting) of the recorded magnetic extensometer data and the applied load versus time history (Figures 7, 8). The parameter values for the different strata were refined until the theoretical consolidation curve

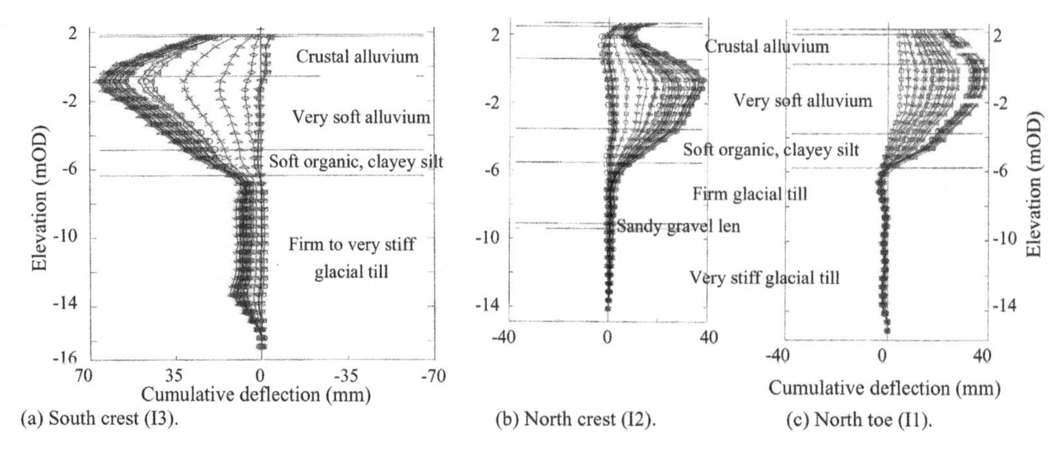

(a) South crest (I3). (b) North crest (I2). (c) North toe (I1).

Figure 9. Lateral ground movements at chainage 6120 m.

Table 2. Consolidation properties determined from settlement back analysis.

Stratum	Coefficient volume change (m^2/MN)	Coefficient of consolidation (m^2/year) Vertical direction Initial	Vertical direction Final	Horizontal direction Initial	Horizontal direction Final	Horizontal to vertical coefficient of consolidation ratio
Crustal alluvium	0.30 to 0.50	3.0 to 7.0	1.0 to 2.0	3.0 to 7.0	1.0 to 2.0	1.0 to 1.2
Very soft alluvium	0.40 to 0.70	4.0 to 12	1.0 to 2.0	8.0 to 24	2.0 to 3.0	2.0
Soft, mottled organic clayey silt	0.40 to 0.60	3.0 to 4.0	1.0 to 2.0	6.0 to 7.0	2.0 to 3.0	1.7 to 2.0
Glacial till (firm upper zone)*	0.05 to 0.15	3.0	2.5	3.0	2.5	1.0
(stiff lower zone)	0.01 to 0.05	3.0	—	3.0	—	—

* Penetrated over full depth by vertical drains and included sand lenses of up to 0.2 m in thickness.

matched the recorded settlement response, assuming 1D compression of the ground foundation.

Each layer was taken in sequence, commencing with the lowermost layer (glacial till) and using the laboratory-measured parameter values as a starting point. The computations were performed in Microsoft Excel taking into account the vertical drain arrays (1.0 and 2.0 m centre spacings at chainages 6120 and 6200 m, respectively) and the permeability anisotropy of the soft ground deposits. Table 2 lists the initial (greenfield) and final (surcharge) values of the coefficient of primary consolidation that were determined for the vertical and horizontal directions.

The back-analysed values were compared with the laboratory-measured values. Overall, both sets of coefficient of volume change values were in good agreement. However, the back-analysed coefficient of consolidation values were between two and four times greater than the values determined from standard oedometer test data with the discrepancy most likely arising due to scale effects. Moreover, the soft soil layers were cross-anisotropic with consolidation occurring up to twice as fast for horizontal (radial) flow rather than for vertical flow conditions.

The secondary compression index (C_α) values were estimated for the strata using the relationship: $C_\alpha = 0.04C_c$, where C_c is the compression index. For the calculation of the secondary compression settlements, $C_\alpha/(1+e)$ values of between 0.0035 and 0.0090 were used (where e is the void ratio).

The target settlements for the different strata under the final embankment height (achieving 100% consolidation of the alluvium) were calculated using the back-analysed parameter values in Table 2. The recorded settlements for each layer exceeded the target values proving that the geotechnical design specification had been achieved.

For example, Table 3 compares the settlements recorded by extensometer ME4 at the end of the surcharge period (applied stress of 160 kPa) with the

Table 3. Settlement response at chainage 6200 m.

Stratum	Final settlement recorded by ME4 (mm)	Target consolidation settlement (mm)
Crustal alluvium	86	65
Very soft alluvium	176	132
Firm to very (upper)	67	50
stiff glacial till (lower)	6	5
Total settlement (mm)	335	251

target values for the final embankment height (applied stress of 120 kPa).

The soft soil deposits were lightly over-consolidated (over consolidation ratio value of about 1.3) on removing the surcharge material. Settlements to date following construction of the embankments are in accordance with the design specification.

8 SUMMARY AND CONCLUSIONS

An array of ground instrumentation monitored the pore water pressure and deformation responses of the ground foundation (about 8.0 m in depth) during the construction of the approach embankments.

The degree of consolidation was assessed from the excess pore water pressure values recorded by the vibrating wire piezometers which were located at the mid-height of the different strata. The inclinometer data indicated that the ground foundation, aided by the basal reinforcement to the embankment, essentially deformed in 1D compression.

The actual values of the primary consolidation and secondary compression parameters for the different strata were determined from a back analysis of the magnetic extensometer data.

After a two-month surcharge period, the recorded settlements for the different strata exceeded the target values (100% consolidation of alluvium) set by the design specification for the final loading.

The back-analysed and laboratory-measured coefficient of volume change values were in good agreement. However, the actual coefficient of primary consolidation values were between two and four times greater than the values determined from standard oedometer test data, most likely due to scale effects. The soft soil layers were cross-anisotropic with consolidation occurring up to twice as fast for horizontal flow rather than for vertical flow conditions.

ACKNOWLEDGEMENTS

The authors would like to acknowledge the close collaboration of their colleagues in the Scott Wilson Limited - Alfred McAlpine project alliance.

Advances in Transportation Geotechnics – Ellis, Yu, McDowell, Dawson & Thom (eds)
© 2008 Taylor & Francis Group, London, ISBN 978-0-415-47590-7

The arching mechanism in piled embankments under road and rail infrastructure

E.J. Britton & P.J. Naughton
Institute of Technology Sligo, Ireland.

ABSTRACT: It is becoming increasingly necessary to construct road and rail networks on ground that was once considered unsuitable. Traditional methods of construction require prolonged construction periods and thus increased cost. An alternative method of constructing on soft ground is the use of piled embankments, which significantly reduces construction times as well as reducing differential settlements. Several design methods are available for estimating the magnitude of arching, however significant inconsistencies between these design methods have been found. A laboratory investigation, using a 1:3 scaled model was used to investigate the influence of pile and embankment geometry and embankment strength parameters on the magnitude of arching. The results from the experimental model have been compared to current design methods using the stress reduction ratio, and were found to be within the range predicted by the Terzaghi (1943) and Hewlett & Randolph (1988) arching theories.

1 INTRODUCTION

In many parts of the world, construction of road and rail networks is challenging due to marginal subsurface soils, such as soil with low bearing capacity or consolidation characteristics which could result in large differential settlements.

Designing structures, such as embankments, on soft foundation soils where the structure will impose a significant load over a large area, raises several concerns. These concerns are related to time constraints, excessive total and differential settlements, large lateral pressures and movement and slope stability. A variety of techniques can be used to address these concerns which include preloading or stage construction, using lightweight fill, over-excavation and replacement, geosynthetic soil reinforcement and piled embankments. The benefits of using piled embankments over other techniques are that superstructures can be built in a single stage without prolonged construction times and significant reduction in the total and differential settlements.

Theoretical studies on geosynthetic reinforced piled embankments have largely focused on the investigation of load transfer mechanisms including soil arching and tension developed along the geosynthetic. However, limited research has been carried out to investigate the true nature of these load transfer mechanisms and the factors which effect them such as dilatancy of the fill material and friction angle.

2 DESIGN METHODS

There are various methods available for the design of piled embankments. It is generally assumed in all cases that the total vertical load of the embankment is transferred to the piles by either soil arching within the fill material or by basal reinforcement spanning between adjacent piles. Six of the most popular design methods are reviewed in this paper.

The Stress Reduction Ratio, (S_{3D}), the ratio of the average vertical stress carried by the reinforcement to the average stress due to the embankment fill, first proposed by Low *et al.* (1994), is used to compare the output from the design methods.

2.1 *BS8006 (1995)*

BS8006 (1995) has adopted an empirical method initially developed by Jones *et al.* (1990), which is based on Marston's equation for positively projecting conduits. In the BS8006 method the stress concentration on the piles and consequently the stress remaining to be carried by the geosynthetic, depends on the pile type and the pile support condition. BS8006 identifies a critical height concept whereby the depth of fill is sufficient for the full arch to be deemed to have developed and any additional overburden or surcharge loads do not influence the tensioned membrane, but distribute to the boundary supports, i.e. the pile caps.

2.2 *Hewlett & Randolph (1988)*

The Hewlett & Randolph method is based on data observed from experimental tests carried out on free draining granular soil. It was observed that the region of sand between the pile caps comprised of a series of hemispherical domes. The stress reduction ratio is calculated assuming limited plastic stress in the arch. It was also found that there are two critical locations within these domes, which were shown to be at either the crown of the arch or at the pile cap. The higher stress reduction ratio is to be used in design.

2.3 *Jenner et al. (1998)*

The Jenner *et al.* (1998) design method was developed from the results of plate loading tests on samples of reinforced granular material in a confined rigid box reported by Guido *et al.* (1987). The method assumes that the arching mechanism in the fill above the pile caps is increased with the inclusion of geogrid and therefore the tensile force within the geogrid is lower than that assumed by the other design methods. The difficulty with the Jenner *et al.* method is that gravity acts in the opposite direction to that used in the laboratory trials on which the method is based (Love & Milligan, 2003).

2.4 *Terzaghi (1943)*

Terzaghi (1943) examined arching in sand directly above a yielding trap door. When the trapdoor was lowered the load in the sand was redistributed to the non yielding surrounds, i.e. the load on the trapdoor reduced while the load on the non yielding supports increased. The Terzaghi equations for the rectangular trapdoor problem were extended by Russell & Pierpoint (1997) to take account of the typical cruciform shape of a piled embankment.

2.5 *Russell et al. (2003)*

Russell *et al.* (2003) found from numerical analysis of the piled embankment problem that reinforcement tension was concentrated in the area directly between the pile caps. Because of this the geosynthetic reinforcement was divided into two types; primary, which spans between the pile caps and secondary, which covers the entire piled area. Also this design method allows for support from the subsoil, the magnitude of which can be determined from compatibility checks on the deformations of the reinforcement and subsoil.

2.6 *Kempfert et al. (2004)*

The Kempfert *et al.* (2004) method which was derived from 1:3 laboratory models of piled embankment problems. The magnitude of load on the soft soil, without reinforcement, is first calculated before the tension in the reinforcement is estimated. It was observed from the laboratory study that in the reinforcement between adjacent piles a higher tension was generated. The tension in the reinforcement is estimated based on the theory of elastically embedded membranes.

Figure 1. Experimental model.

3 EXPERIMENTAL MODEL

A 1:3 laboratory model of the piled embankment problem was developed as part of this study, Figure 1. The model consisted of a 1 m^3 box with a movable base. Four pile caps in a unit cell of a piled embankment are represented in the model by blocks of plywood. Load cells are located beneath the pile caps to measure the change in load due to arching as the base between the pile caps is lowered.

Sand samples, with homogeneous densities, were formed in the apparatus using a raining deposition technique similar to that described by Schnaid (1991). The target sample densities were achieved using a combination of different shutter plates and diffuser sieves. Dense samples were obtained by passing the sand through perforated plates having 6 mm holes on a 80 mm triangular grid and raining through 2 No. 6 mm sieves located 150 mm and 250 mm respectively from the base if the hopper. Loose samples were obtained by passing the sand through perforated plates

Figure 2. Electron microscope view of Sand at a magnitude (×75).

Table 1. Properties of sand investigated in this study.

Specific Gravity, Gs	2.66
Coefficient of uniformity, C_u	1.49
Coefficient of Curvature, C_c	1.10
Maximum Void Ratio, e_{max}	0.84 ± 0.01
Minimum void ratio, e_{min}	0.39 ± 0.006
Maximun particle size D_{max}	2 mm

having 20 mm holes on the same triangular grid, and omitting the diffuser sieves. The model was filled in four lifts and densities were measured at each lift to check the homogeneity of the sample density.

4 MATERIAL PROPERTIES

The sand investigated in this study, Figure 2, was uniformly graded, rounded, medium sand which was recovered from excavations (close to the ocean) at Ballyshannon, Co. Donegal, Ireland. The sand properties, Table 1, were determined in accordance with BS 1377 (1990).

The limiting densities and density indexes are presented in Table 2. Due to the shape and gradation of the sand used in this experiment it was not possible to obtain a truly loose sample using the raining apparatus. Therefore the sand samples tested can be classified as medium dense and dense samples.

The shear strength and dilatancy characteristics of the sand were obtained by direct shear tests. The samples were tested under normal stresses ranging from 123kPa to 368kPa. The angle of internal friction was found to be 43° for $I_D = 0.949$ and 36° for $I_D = 0.798$, with the angle of dilation of $10° \pm 1.5\%$.

Table 2. Limiting Dry Densities and Density Index's.

Sand State A			Sand State B		
Max density, ($\rho_{D,max,}$) kg/m^3	Sample density, (ρ_D) kg/m^3	Density Index, (I_D)	Min density, ($\rho_{D,min}$) kg/m^3	Sample density, (ρ_D) kg/m^3	Density Index, (I_D)
1.550	1.537	0.949	1.33	1.500	0.798

5 EXPERIMENTAL TESTING

The trapdoor at the base of the model was cruciform in shape and occupied 51% of the 1 m^2 base. Each pile cap area had plane area of 0.1225 m^2. The height of the sand above the pile caps was 1 m. The pile spacing and pile cap size in the experimental model was chosen so that the full arching mechanism was developed within the 1 m high sample. Naughton (2007) showed that full arching would develop at a height of less than 2.5 times the clear spacing between adjacent pile caps, corresponding to a maximum clear spacing of 0.4 for the model used in this study.

The sand was placed evenly on the pile caps and the trapdoor in 250 mm lifts using the sand raining technique. For the tests presented in this paper, the samples were placed at a density of 1400 kg/m^3.

After placing the sand the trapdoor was lowered and the sand allowed to yield. The output from the load cells was monitored every 3 seconds.

6 ANALYSIS OF RESULTS

When the trapdoor was released it dropped instantaneously and the sand directly on top of the trapdoor was allowed to fall freely out of the model. It was noted that the load on the pile caps increased dramatically in all cases and then remained relatively constant. The increase in load on each pile cap ranged from 100 kg to 120 kg, Figure 3. Good agreement was found between the outputs from the functioning load cells in any test. It was found that one or two load cells would not function properly. Work is currently under way to modify the pile cap design to ensure data is retrieved from all four load cells in future tests.

The load transfer is due entirely to arching within the fill material as geosynthetic reinforcement was not incorporated into the experimental model. Friction between the sand and the side of the box was not considered in the current analysis. It was assumed that

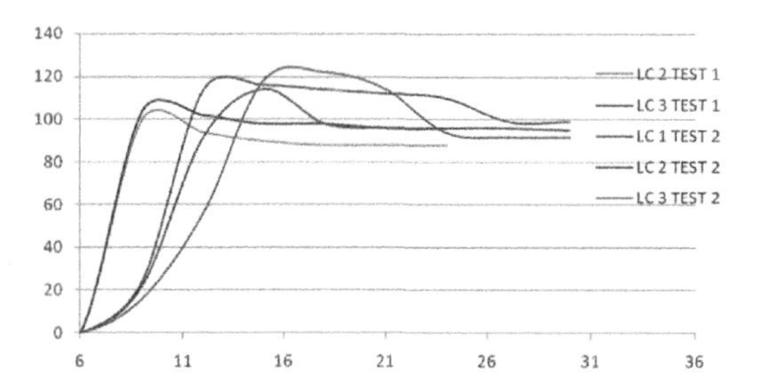

Figure 3. Load increase on pile caps after trapdoor is released ($\rho = 1400\,kg/m^3$).

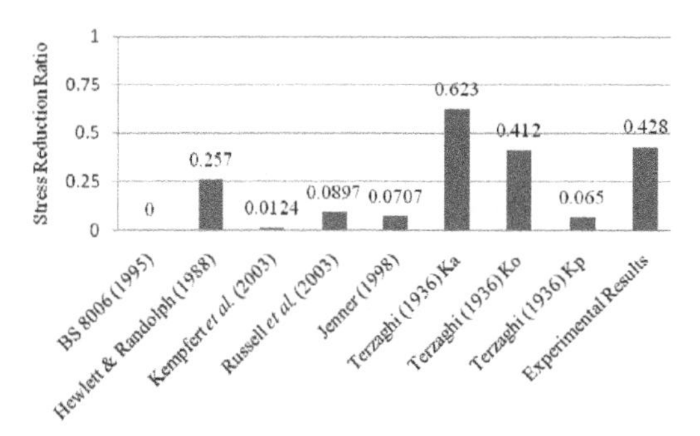

Figure 4. Comparison of Piled Embankment Design Methods.

friction could be ignored as the sand either yielded and fell out of the model or remained stationary. While the sides of the box were made as smooth as possible some frictional load may have been generated. However the analysis of the friction in this model is complex.

The stress reduction ratio of the experimental data was calculated based on the initial weight of sand over the pile cap and the increase in load recorded once the trap door was dropped. Comparisons of the stress reduction ratio from the design method, calculated using the dimensions of the experimental model, are presented in Figure 4. As Terzaghi's arching theory does not specify which earth pressure coefficient is to be used in calculation of the stress reduction ratio, values were calculated using K_a, K_p and K_0. The experimental models stress reduction ratio was similar to the value obtained from the Terzaghi Method (1943) when using K_0 as the earth pressure coefficient. The Jenner *et al.* (1998), Russell *et al.* (2003), BS 8006 (1995) and Kempfert *et al.* (2004) all predicted significantly lower stress reduction ratios and are therefore less conservative, underestimating the load at the base of the embankments between pile caps.

A possible explanation for the discrepancy in the stress reduction ratio may have been due to the cruciform shape of the trapdoor. The majority of previous experimental studies on piled embankments used a square or rectangular trapdoor as opposed to the cruciform shape which occurs in an actual piled embankment, (Russell *et al.* 2003).

7 CONCLUSIONS

Piled embankments are becoming increasingly popular in construction of road and rail networks on marginal subsurface soils as they are often the only practical and economic method available. The piled embankment application is truly a three dimensional problem and therefore should be modeled as such.

A series of model tests have been carried out to investigate arching in piled embankments. The results from the experimental model have been compared to current design methods using the stress reduction ratio, and were found to be within the range predicted by the Terzaghi (1943) and Hewlett & Randolph (1988) arching theories. The Jenner *et al.* (1998), Russell *et al.* (2003), BS 8006 (1995) and Kempfert *et al.* (2004) all predicted significantly lower stress reduction ratios and are therefore less conservative, underestimating the load at the base of the embankments between pile caps.

The experimental work discussed in this paper is currently ongoing and samples of various densities over different pile cap sizes are to be tested using the model, so as to determine whether the strength and dilatancy of the fill material has an effect on the arching behavior.

REFERENCES

BS8006, 1995. Code of practice for strengthened/reinforced soils and other fills, British Standards Institution, London.

Guido, V.A., Kneuppel, J.D. & Sweeney, M.A. 1987. Plate loading tests on geogrid-reinforced earth slabs. Proc. Goesynthetics '87 Conf., New Orleans.

Kempfert, H.G., Gobel, C., Alexiew, D. & Heitz, C. 2004. German recommendations for reinforced embankments on pile-similar elements. Proc. 3rd Eur. Conf. on Geosynthetics, Munich, Germany, 279–284.

Hewlett, W.J. & Randolph, M.F. 1988. Analysis of piled embankments. Ground Engineering, 21(3), 12–18.

Jones, C.J.P.F., Lawson, C.R. & Ayres, D.J. 1990. Geotextile reinforced piled embankments. Geotextiles, Geomembranes and related products, Den Hoedt (Ed), Balkema, Rotterdam.

Kempton, G.T. & Naughton, P.J. 2003. Piled Embankments with basal reinforcement: development of design methods and state of the Art. L'Ingegnere e l'Architetto, Italy. Volume 1–12/2002.

Kempton, G., Russell, D., Pierpoint, N.D. & Jones C.J.F.P. 1998. Two and Three Dimensional Numerical Analysis of the performance of piled Embankments. Proc. 6th Int. Conf. on Geosynthetics, Atlanta, Georgia.

Love, J. & Milligan, G. 2003. Design methods for basally reinforced pile supported embankments over soft ground. Ground Engineering, March 2003, 39–43.

Low, B.K., Tang, S.K. & Choa, V. 1994. Arching in Piled Embankments. Journal of Geotechnical Engineering (ASCE), Vol. 120, no.11: 1917–1938.
Naughton, P.J. 2007. The significance of critical height in the design of piled embankments. Proc. GeoDenver 2007.
Naughton, P.J. & Kempton, G.T. 2004. Comparison of analytical and Numerical Analysis Design Methods for piled Embankments. Proceedings Contemporary issues in foundation Engineering, GSP 131, ASCE, GeoFrontiers, Austin, Texas.
Quigley, P. & Naughton P.J. 2007. Design of piled Embankments. Proc conf. on soft ground engineering, Geotechnical Society of Ireland, 15–16 February, Portlaoise, pp 2.3.
Russell, D. & Pierpoint, N. 1997. An assessment of design methods for piled embankments. Ground Engineering 30, No. 11, November 1997, pp.39–44.
Russell, D., Naughton, P.J. & Kempton, G. 2003. A new design procedure for piled embankments. Proceedings of the 56th Canadian Geotechnical Conference and the NAGS Conference, pp.858–865.
Schnaid, F. 1991. Study of Cone Pressuremeter test in sand. PhD Thesis, University of Oxford.
Terzaghi, K., 1936 Stress distribution in dry and in saturated sand above a yielding trap door. Proc. 1st International Conference on soil mechanics and foundation Engineering, Harvard University, Cambridge Mass., Vol. 1, pp.307–311.

Advances in Transportation Geotechnics – Ellis, Yu, McDowell, Dawson & Thom (eds)
 ISBN 978-0-415-47590-7

FE analysis of the long-term settlement and maintenance of a road on peaty ground

H. Hayashi & S. Nishimoto
Civil Engineering Research Institute for Cold Region, Sapporo, Japan

ABSTRACT: Peaty ground commonly found in Hokkaido, Japan, is an extremely soft ground, which is highly organic and has special engineering properties. Finite element analysis (FE analysis) with consideration of the overlay pavement load as a measure against residual settlement was thus conducted concerning the long-term settlement of a road on peaty soft ground, and the relationship between the residual settlement and repair cost was studied. It was found that FE analysis using the Sekiguchi-Ohta model can express the long-term settlement of peaty soft ground. It was also revealed that the amount of residual settlement greatly affects the repair cost.

1 INTRODUCTION

Peaty ground distributed widely in Hokkaido, Japan, is an extremely soft ground, which is highly organic and has special engineering properties. Roads on peaty ground are usually designed to allow for a certain degree of settlement and are repaired and maintained during their service, since considerable settlement occurs over a long period of time in such roads. The Manual for Peaty Soft Ground Countermeasures (CERI, 2002) stipulates the allowable residual settlement as between 10 to 30 cm in three years after an expressway is placed in service. These values were, however, determined empirically and should be reconsidered for the minimization of the life cycle cost (LCC).

FE analysis with consideration to the overlay pavement load as a measure against residual settlement was thus conducted concerning the long-term settlement of a road on peaty soft ground, and the relationship between the residual settlement and repair cost was studied.

2 SETTLEMENT BEHAVIOR OF PEATY GROUND AND THE CONCEPT OF THE LIFE CYCLE COST

Figure 1 is a conceptual diagram of the settlement of peaty ground. A characteristic of the settlement of peaty ground is the long-term settlement called secondary consolidation, which occurs linearly to the logarithm of time and causes continuous settlement after the placement of roads in service. The type of ground, where soft clay layers often accumulate at the bottom of peat, is called peaty soft ground. In addition to the peculiarity of peat, the existence of thick clay layers makes the problem of long-term settlement even more complex.

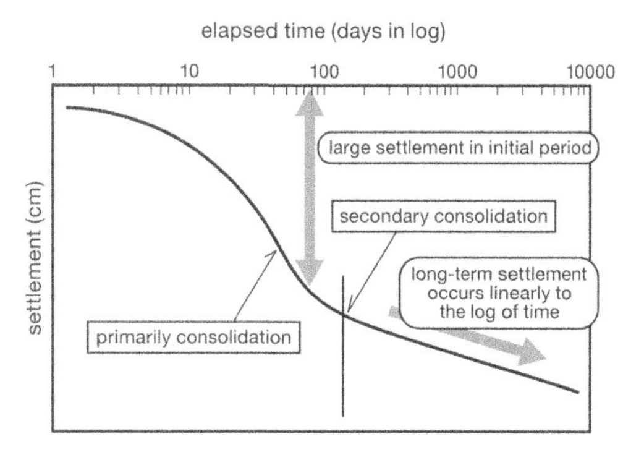

Figure 1. A typical settlement curve of peaty ground.

The standard method in the Peaty Soft Ground Countermeasures Manual for the construction of a road embankment on peaty soft ground is the method of allowing for a certain degree of residual settlement after the placement of the road in service and conducting maintenance and repair works during its service life. This method is used for reducing road construction costs, as well as for ensuring the flatness of the road surface and controlling road repair costs depending on the importance level of the road. It means that the concept of LCC shown in Figure 2 has been introduced. There are, however, some problems that must be solved for rationalization, such as the empirical setting of allowable residual settlement.

3 OVERVIEW OF FE ANALYSIS

3.1 *Overview of the section of analyzed*

The subject of the analysis, Mihara Bypass (1.4 km long), is an expressway on the outskirts of Sapporo, which is the largest city in Hokkaido. It is a two-lane road constructed on peaty soft ground. To ensure stability and reduce residual settlement, the preload method was adopted in combination with the counterweight fill method, which is the use of loading berms at the sides of the main embankment (Figure 3). Measurements of settlement have been continued since this expressway was placed in service in March 2005.

3.2 *Analysis conditions*

FE analysis was conducted using the Sekiguchi-Ohta model (Sekiguchi & Ohta, 1977), which can express long-term settlement. Table 1 presents the analysis cases. A comparison between measured settlement and analysis results was first made as a preliminary study, and the applicability of the constitutive model and validity of parameters were verified.

Next, in the main study, virtual simulation was conducted for three cases with different amounts of residual settlement three years after their placement in service. At that time, the ground and other conditions were assumed to be uniform and the specific amounts of settlement were set by varying the period of waiting time for decreasing settlement. Concerning the bumps caused by settlement, it was assumed that repairs (overlay pavement) were repeated whenever settlement reached 10 cm, based on the results of the fact-finding survey of repairs conducted by Nishimoto and Hayashi (2007). As the loading condition at that time, the load of the pavement thickness of 10 cm was uniformly distributed over the embankment crest, thus causing a slight increase in load from the embankment at this point. The traffic load generated by vehicles passing over the bumps was not taken into account because the embankment was high.

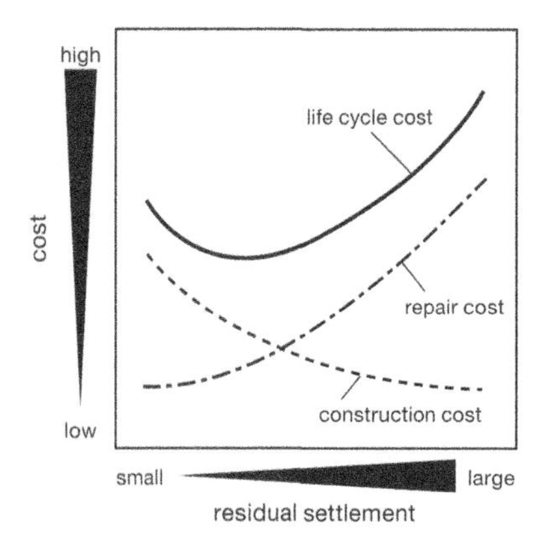

Figure 2. The concept of the life cycle cost.

3.3 *Determination of soil parameters*

When analyzing the long-term settlement of peat, it is important to determine the coefficient of permeability and coefficient of secondary consolidation. Hayashi

Table 1. Analysis cases and conditions.

Case		Analysis conditions: Overlay pavement load	Residual settlement (cm)
Preliminary case		no loading	–
Main case	1	loading	30
	2	loading	20
	3	loading	10

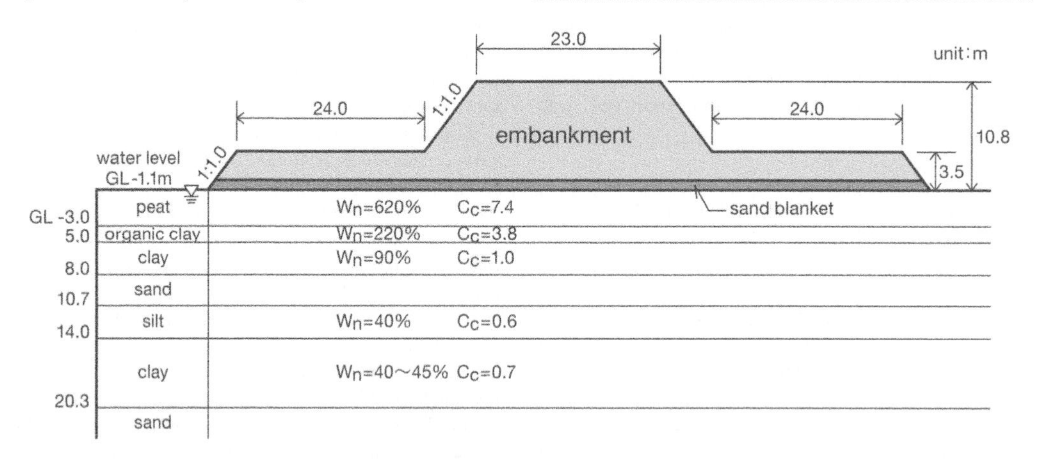

Figure 3. The cross section of the embankment.

et al. (2007a) have revealed that the coefficient of permeability of peat decreases considerably with consolidation and that the actual coefficient of permeability is 10 to 30 times as large as the values found by the oedometer test. These are engineering properties peculiar to peat. In this analysis, therefore, the initial coefficient of permeability was determined by Eqs. (1) and (2). The changes in the coefficient of permeability associated with consolidation were given by Eq. (3).

$$k_{oed} = C_v\, m_v\, \gamma_w \,/\, (8.64\times10^6) \quad (1)$$

$$k_0 = 10\, k_{oed\text{-}P0} \quad (2)$$

$$k = k_0 \exp((e-e_0)\,/C_k) \quad (3)$$

Where, k_{oed} is the coefficient of permeability found by the oedometer test (cm/s), C_v is the coefficient of consolidation (cm^2/day), m_v is the coefficient of volume compressibility (m^2/kN), γ_w is the unit weight of water (=9.81 kN/m^3), k_0 is the initial coefficient of permeability (cm/s), k_{oed-p0} is the coefficient of permeability at the time of effective overburden pressure

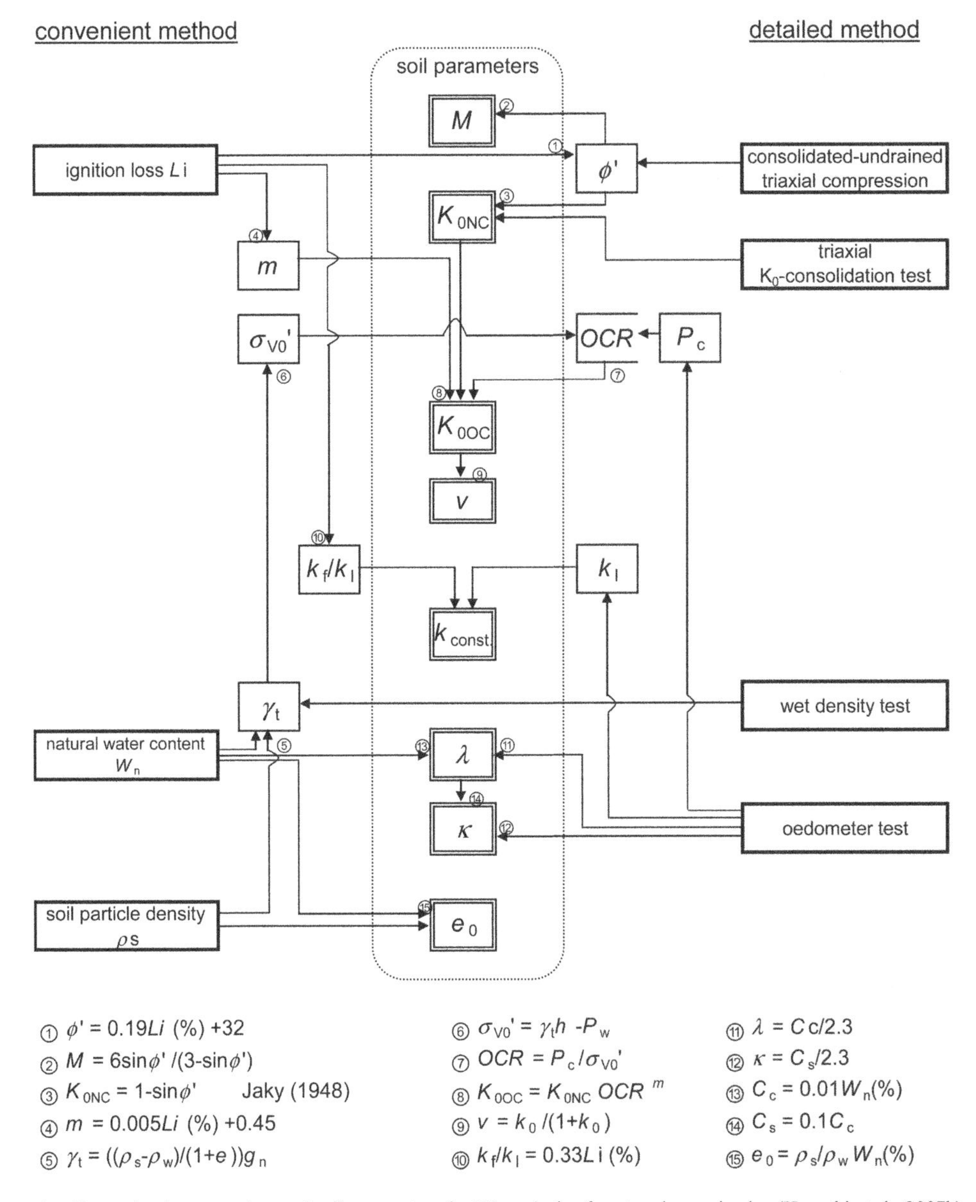

Figure 4. Determination procedures of soil parameters for FE analysis of peat and organic clay (Hayashi et al., 2007b).

found by the oedometer test (cm/s), e is the void ratio at the end of consolidation and e_0 is the initial void ratio. The constant C_k in Eq. (3) was determined from the relationship between the void ratio and the coefficient of permeability found by the oedometer test.

Noto (1990) related the coefficient of the secondary consolidation C_s (%) of peat with the natural water content W_n and expressed the relation by Eq. (4). The coefficient of secondary consolitaion in this analysis was determined using Eq. (4).

$$C_s = 0.033+0.000043W_n(\%) \quad (4)$$

Other soil parameters of peat were determined from the results of laboratory soil tests using the flow chart (Figure 4) proposed by Hayashi et al. (2007b). The parameters of clay were determined using the method of Iizuka & Ohta (1987). Table 2 lists the soil parameters used for analysis.

4 RESULTS OF PRELIMINARY STUDY

Figure 5 displays the changes in measured and analyzed ground surface settlement at the center of the embankment with the passage of time. The embankment constructed at the section was 10.8 m in thickness, and the measured settlement at the start of service was approximately 4 m, indicating the high compressibility of the ground. Measured values have been obtained for 2,075 days (684 days in service) so far, and almost correspond with the analysis results. This means that FE analysis using the Sekiguchi-Ohta model is effective in the examination of the long-term settlement of peaty ground. It also shows that the procedure to determine soil parameter presented in this paper was valid.

5 RESULTS OF THE MAIN STUDY

Figure 6 presents the analysis results of Case 1 (30 cm residual settlement in three years after placement in service). According to the analysis results, settlement will continue after placement in service and the residual settlement in 50 years is expected to be approximately 92 cm only in the case with the embankment load and 114 cm in the case where the overlay load is also taken into account. While the overlay load is rarely considered in normal settlement analysis, it is worth noting that, in the analysis results of this study, the amount of residual settlement was 22 cm greater with the overlay load than in the case with only the embankment load. Considering that residual settlement is often discussed on the order of around 10 cm, the overlay factor can be regarded as an important factor in the examination of residual settlement.

Table 2. The list of soil parameters used.

Soil layer	Coefficient of dilatancy D	Irreversibility ratio Λ	Critical state parameter M	Compression index λ	Initial void ratio e_0	Coefficient of secondary consolidation α	Initial volumetric strain rate V_0 (day/1)	Pisson's ratio V	Coefficient of in-situ earth pressure at rest K_{0OC}	Coefficient of earth pressure at rest K_0	Coefficient of permeability: Horizontal k_H (m/day)	Vertical k_V (m/day)	Change with void ratio λ_k
Peat	0.10	0.92	1.94	2.69	12.36	0.03	2.2×10^{-5}	0.21	0.30	0.26	8.6×10^{-2}	1.7×10^{-2}	1.16
Organic clay	0.15	0.90	1.48	1.62	5.44	0.02	3.3×10^{-5}	0.30	0.43	0.43	2.6×10^{-2}	5.2×10^{-3}	0.84
Clay	0.11	0.80	1.08	0.39	1.63	0.01	2.9×10^{-6}	0.37	0.78	0.58	8.6×10^{-4}	8.6×10^{-4}	0.43
Silt	0.07	0.82	1.16	0.20	0.91	0.01	2.3×10^{-5}	0.35	0.71	0.54	1.6×10^{-3}	1.6×10^{-3}	0.24
Clay	0.09	0.86	1.10	0.24	1.06	0.01	1.4×10^{-6}	0.36	0.59	0.57	9.9×10^{-4}	9.9×10^{-4}	0.31

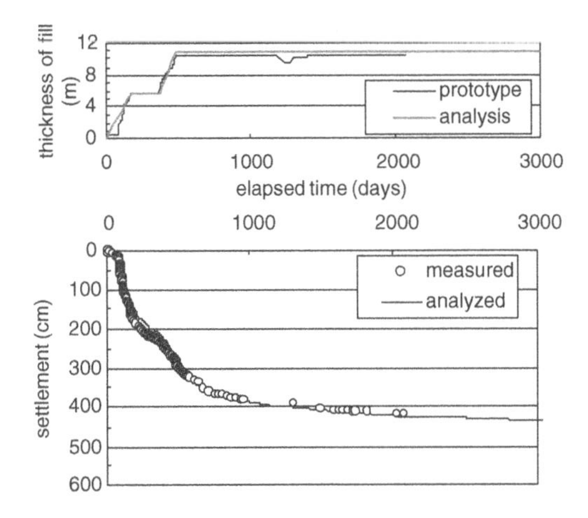

Figure 5. The changes in measured and analyzed ground surface settlement at the center of the embankment.

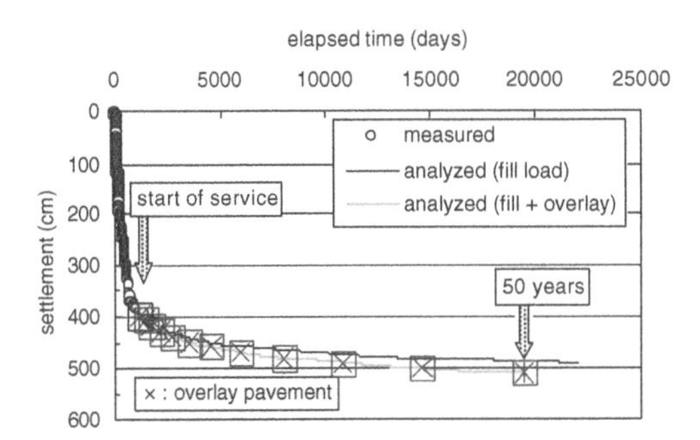

Figure 6. The effect of additional overlay pavement from repairs in increasing embankment load and settlement.

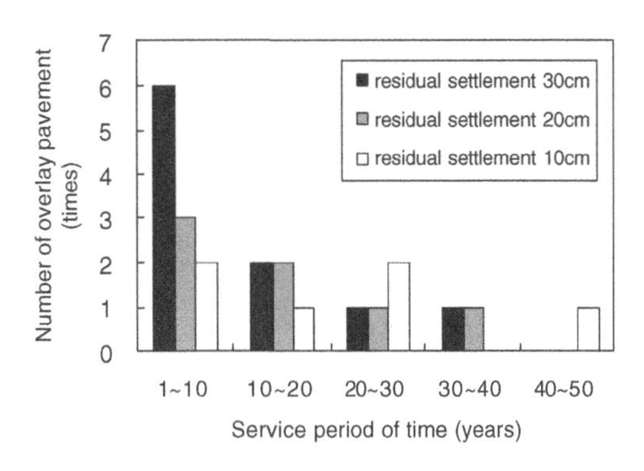

Figure 7. The number of overlay pavement repairs for respective cases.

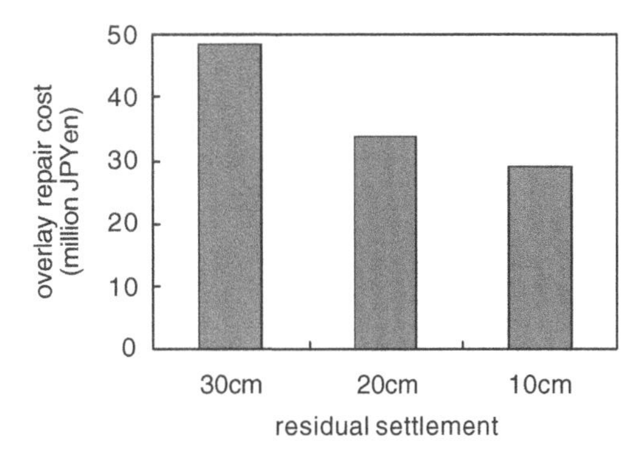

Figure 8. The relationship between residual settlement and the overlay repair cost.

Analysis was conducted for other two cases using the above method, and the schedule and number of overlay repairs for respective cases were found (Figure 7). In the case where residual settlement was 30 cm three years after placement in service, six repairs were necessary in 10 years. In the first four years, in particular, repairs had to be conducted every year. In the cases where residual settlement was 10 and 20 cm, however, the number of repairs was half or less. In the case of a two-lane road, closure is necessary even for a simple repair like overlaying. It is considered difficult to gain the understanding of users and related organizations if repairs requiring road closure must be made frequently soon after the opening of the road.

On the Mihara Bypass (1.4 km long), box culverts are placed at average intervals of 170 m. On the assumption that residual settlement would cause bumps before and after these culverts, the relationship between residual settlement and the overlay repair cost for the length of 1.4 km in 50 years was estimated (Figure 8). The conditions for the estimation were set using the actual repair cost found by Nishimoto & Hayashi (2007) as a reference. Compared with the case of 30 cm residual settlement, the repair cost was approximately 72 and 59% in the cases of 20 and 10 cm settlement, respectively. It can be seen that the setting of the amount of residual settlement greatly affects the repair cost.

6 CONCLUSION

It was found that FE analysis using the Sekiguchi-Ohta model can express the long-term settlement of peaty soft ground. It was also revealed that the amount of residual settlement greatly affects the repair cost.

The authors intend to conduct further studies using the method applied in this study as an analysis tool and present the rational amount of residual settlement, repair schedule and other guidelines.

REFERENCES

Civil Engineering Research Institute of Hokkaido (CERI). 2002. *Manual for Countermeasure against Peat Soft Ground*: 42–44. (in Japanese)

Hayashi, H., Mitachi, T. & Nishimoto S. 2007a. Permeability Characteristics of Peaty Ground. *Technical reports of Hokkaido branch, JGS* 46: 31–36. (in Japanese)

Hayashi, H., Mitachi, T. & Nishimoto S. 2007b. Determination Procedure of Soil Parameters for Elasto-plastic FE Analysis of Peat Ground, *Proceedings of the 13th Asian Regional Conference on SMFE*: 145–148.

Hayashi, H. & Nishimoto S. 2008. A Case Study on The Long-term Settlement and Life Cycle Cost of A Road on Peaty Ground, *Proceedings of 1st International Conference on Transportation Geotechnics*: (in Printing)

Iizuka, A. & Ohta, H. 1987. A determination procedure of input parameters in elasto-viscoplastic finite element analysis, *Soils and Foundations*. 27(3): 71–87.

Noto, S. 1990. Revised Formula for Prediction of Settlement on Soft Peaty Deposits. *Monthly Report of Civil Engineering Research Institute* 446: 2–9 (in Japanese)

Sekiguti, H. & Ohta, H. 1977. Induced anisotropy and time dependency in clays. *Proceedings of the 9th ICSMFE*: 229–239.

Advances in Transportation Geotechnics – Ellis, Yu, McDowell, Dawson & Thom (eds)
 ISBN 978-0-415-47590-7

Numerical modelling for life cycle planning of highway embankments on soft foundations

T. Ishigaki & S. Omoto
Nippo Corporation Research Institute, Tokyo, Japan

A. Iizuka
Kobe University, Kobe, Japan

T. Takeyama & H. Ohta
Tokyo Institute of Technology, Tokyo, Japan

ABSTRACT: This paper presents the applicability of soil/water coupled finite element modelling (S/W FEM) as a tool for supporting life cycle planning of highway embankments on soft foundations. To begin with, as a case study, the S/W FEM of a 30 years old existing highway embankment is conducted and its prediction of long-term settlement and deformation is verified. The constitutive model mainly employed in this modelling is an elasto-viscoplastic model developed by Sekiguchi and Ohta (1977). Secondly, the life cycle planning of highway embankments based on the FE simulation of long-term performance is trialled. The geotechnical performance-related indicators (GPRIs) relevant to geotechnical indicators (GIs) of highway embankments are investigated for life cycle planning. The case study with numerical modelling and life cycle planning of highway embankments described in this paper may be found to be applicable and efficient for use in geotechnical asset management practices.

1 INTRODUCTION

Excessive long-term deformation due to the result of consolidation settlement is usually observed in highway embankments on soft foundations. In many case, these reduce highway safety and serviceability. Prediction of the long-term deformation of fill body by appropriate numerical modelling may effectively be applicable to life cycle planning for use in geotechnical asset management practices. In this paper, the applicability of the soil / water coupled finite element modelling (S/W FEM) as a tool for supporting life cycle planning of highway embankments on soft foundations is presented.

To begin with, as a case study, S/W FEM of a 30 years old existing highway embankment in Hokkaido expressway is conducted and its prediction of long-term settlement and deformation is verified. It is one of the post mortem analyses during construction in a form of the so-called Class B prediction defined by Lambe (1973). The S/W FE code used in this study is DACSAR coded by Iizuka et al. (1987) and Takeyama (2007). The constitutive model mainly employed in the modelling is an elasto-viscoplastic model developed by Sekiguchi and Ohta (1977). The input soil parameters are determined by flow charts proposed by Iizuka et al. (1987) and Ishigaki et al. (2008). The flow charts are based on the laboratory and field tests together with a set of correlations proposed by many research workers. Secondly, the geotechnical performance-related indicators (GPRIs) relevant to settlement and deformation as geotechnical indicators (GIs) of highway embankment are investigated for life cycle planning. The life cycle planning of highway embankment based on the simulation of long-term performance is trialled. One of the results of this case study is compared with actual life cycle of maintenance works.

2 CASE STUDY

2.1 *Analyzed site and subsoil properties*

The Ebetsu trial embankment in Hokkaido Expressway between Sapporo and Iwamizawa, which placed 20 km away from Sapporo, was started to construct in 1977 and opened to traffic in 1983. The embankment as the analyzed site in this paper is the sand drain-treated test filling as a 7 m high embankment on a plane covered with highly compressive peat layer of 2 m thickness on

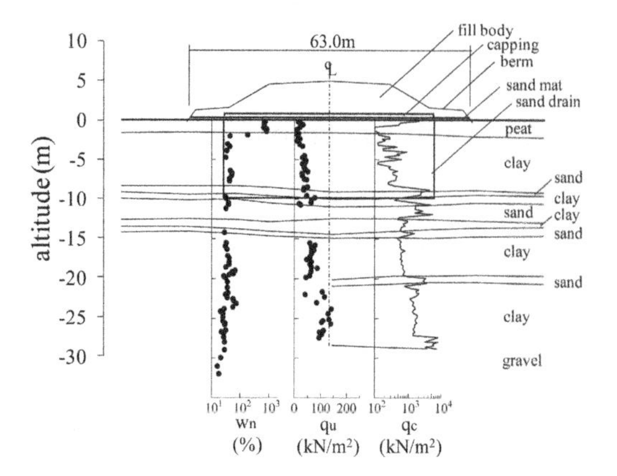

Figure 1. Cross section and subsoil properties of Ebetsu trial embankment.

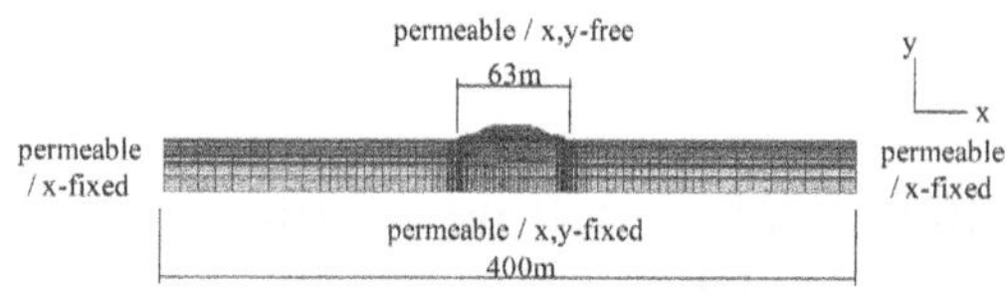

Figure 2. Two dimensional finite element model of Ebetsu trial embankment.

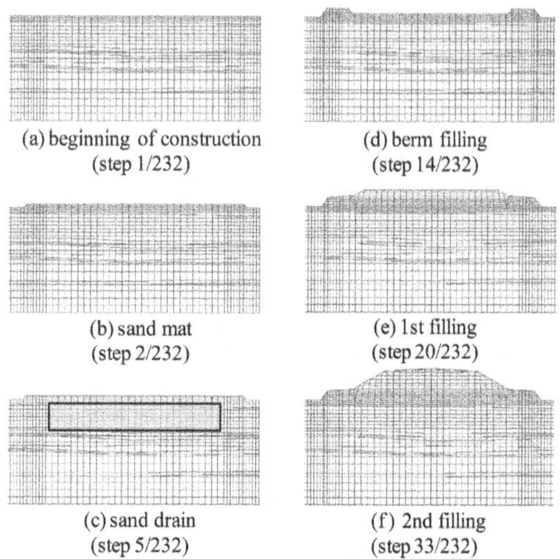

Figure 3. Finite element modelling of embankment construction.

alternation of soft clay layers and sand layers of 27 m thickness. Long-term settlement of embankment had occurred in the past 30 years.

The cross section and subsoil properties of the test fill are shown in Figure 1. The subsoil properties of natural water content, unconfined compressive strength and cone bearing capacity at the centre of embankment were investigated. The very soft peat layer with 700% of natural water content had deposited this part on the surface by a thickness of 1.8 m on the peaty clay of 7.3 m thickness. As countermeasures, the 10.1 m length and 0.4 m diameter sand drains were installed in triangle pattern in plan with a 1.8 m centre-to-centre pitch to accelerate the consolidation of this peat and peaty clay layers. The 1 m thick sand mat was placed on the ground surface. Drainpipes of 0.1 m diameter were placed from centre to toe of the embankment inside the sand mat at every 4 m interval along the longitudinal direction. The 0.5 m thickness of capping layer of clayey soil was performed between fill body and sand mat. The counterweight berms, pre-loading fill and extra fill were trialled. Slow filling by staged loading was also performed with an observational method.

2.2 *Soil/water coupled finite element modelling*

The soil-water coupled finite element modelling (S/W FEM) is conducted in one of the post mortem analysis during the embankment construction in a form of the so-called Class B prediction classified by Lambe (1973). The application of the Class B prediction may be expected to improve the accuracy of the long-term performance prediction of highway embankment. Because soil parameters and construction sequence of the highway embankment represented by boundary conditions of the numerical model are modified by comparing the results of numerical modelling with monitored data during construction works. It is considered that these calibrations are one of the condition assessments of highway embankments on soft foundations for the uncertainty of ground information in design. The authors assume that if the analysis can successfully simulate the deformation behaviour during construction, the analysis may successfully predict the long-term settlement and deformation of the highway embankment during operation by extending the boundary conditions to longer time.

The soil-water coupled FE code used in this analysis is DACSAR coded by Iizuka et al. (1987) and Takeyama (2007) , which is based on the consolidation theory proposed by Biot (1951) together with the formulation developed by Akai and Tamura (1978) . The formulation is adopted the technique proposed by Christian (1968) and Christian and Boehmer (1970). Two-dimensional FE modelling used in this analysis is shown in Figure 2. This model is considered that layers are the slight variation in thickness and level of the layers shown in Figure 1. Finite element modelling of embankment construction is shown in Figure 3. In this Class B prediction during construction period, embankment construction is modelled by adding elements to the mesh and the loading rate and the thickness of fill are assumed identical with those in the actual staged construction works. The level of underground water is set 0.6 m below ground surface by investigated data. Therefore, the buoyancy to act under the fill body is also considered in this modelling.

The constitutive models employed in this analysis are incremental elasto-plastic and elasto visco-plastic models developed by Sekiguchi and Ohta (1977). The yield function of elasto-plastic model is defined by:

$$f = MD \ln \frac{p'}{p'_0} + D\eta^* - \varepsilon_v^p = 0 \quad (1)$$

The yield function of elasto visco-plastic model is defined by:

$$F = \alpha \ln \left\{ 1 + \frac{t}{t_0} \exp\left(\frac{f}{\alpha} \right) \right\} - \varepsilon_v^p = 0 \quad (2)$$

where M = critical state parameter; D = coefficient of dilatancy proposed by Shibata (1963); p' = mean effective stress; p'_0 = initial mean effective stress; η^* = generalized stress ratio proposed by Sekiguchi and Ohta (1977); ε_v^p = viscoplastic part of volumetric strain; α = coefficient of secondary compression proposed by Sekiguchi and Ohta (1977); t = time; t_0 = initial time. Sekiguchi-Ohta model (The S-O model) was developed based on a set of assumptions very different from the original cam-clay model (The OCC model) by Roscoe et al (1963), but the final mathematical form is essentially the same as the OCC model. The S-O model can also describe the induced anisotropy, creep and relaxation characteristics of soils. In this analysis, peat layer is assumed an elasto-plastic model. However the subsurface of 0.6 m is assumed a linearly elastic model as surface crust. Because the value of cone bearing capacity is very high, (see Figure 1). Clay layers are assumed elasto-visco-plastic models. Compacted fill material, capping layer, sand mat and sand layer are assumed linearly elastic models.

The input parameters needed in the S-O model should primarily be determined through the triaxial test, oedometer test and permeability test. However, these laboratory tests were not performed prior to the actual construction works. Consequently, the procedure of input parameter determination of clay and peat follow the flow charts shown in Figure 4 (a), (b) proposed by Iizuka et al. (1987) and Ishigaki et al. (2008). The soil parameters in square are needed in the S-O model. The procedures are based on laboratory and field tests together with a set of correlations proposed by many research workers. Some local empirical correlations between natural water content and various soil properties in Hokkaido Expressway between Sapporo and Iwamizawa are developed and the input parameters were determined using Figure 4.

The input parameters of subsurface crust are determined by using correlation between the cone bearing capacity and deformation parameters after Sanglerat (1972). The deformation parameters and permeability coefficients of sand layers are estimated after Lunne and Chistophersen (1983) and Creager et al. (1945). The deformation parameters of compacted fill material, capping layer, and sand mat are assumed by reference to the author's past-analyzed experiences and permeability coefficients of these materials are estimated after Creager et al. (1945). The permeability coefficient of sand layers are calibrated by trial fitting with monitored performance. The permeability of peat and clay layer is changed during the process of consolidation by the relationship between permeability coefficient and void ratio after Taylor (1948). The estimated value of permeability coefficients in sand drain area shown in Figure 3(c) is modified by multiplying the factor calculated from the theory by Barron (1948) and calibrated by trial fitting with monitored performance. In this analysis, estimated permeability used in sand drained area is about 10 times of the original permeability coefficients of peat and clay layers.

2.3 *Computed results*

The computed results of settlement and compression under the centre of embankment being compared with the monitored performance are shown in Figure 5 and Figure 6. The computed results successfully agree well with the observed one. Figure 7 shows the computed pore water pressure of each clay layer under the centre of embankment. The computed results are relatively good agreement with the observed one, but the computed result of peat layer is quite lower than the observed values.

The computed results of lateral displacement under the toe of embankment and lateral displacement of ground surface at the toe of embankment are shown in Figure 8 and Figure 9. The deformation modes of computed results are relatively good agreement with the observed one, especially the difference between left and right side. This reason may be the efficiency of modelling to consider the subsoil layers precisely. However, both of the values of the computed results are lower than the observed one. The observed data in Figure 9 are suddenly getting high at 30days from the beginning of construction. This is the period of the installing the sand drains. Therefore, this sudden lateral deformation may be due to the effect of works of installing sand drain. The modelling of sand drain is not considered this effect of the works, only considered the effect of sand drains as permeability modification. Nevertheless, if taking no notice of this effect, the computed displacement curves are good agreement with the observed one.

The computed long-term settlement under the centre of embankment being compared with the past inspection data after opening to traffic is shown in Figure 10. The beginning of the both settlement are

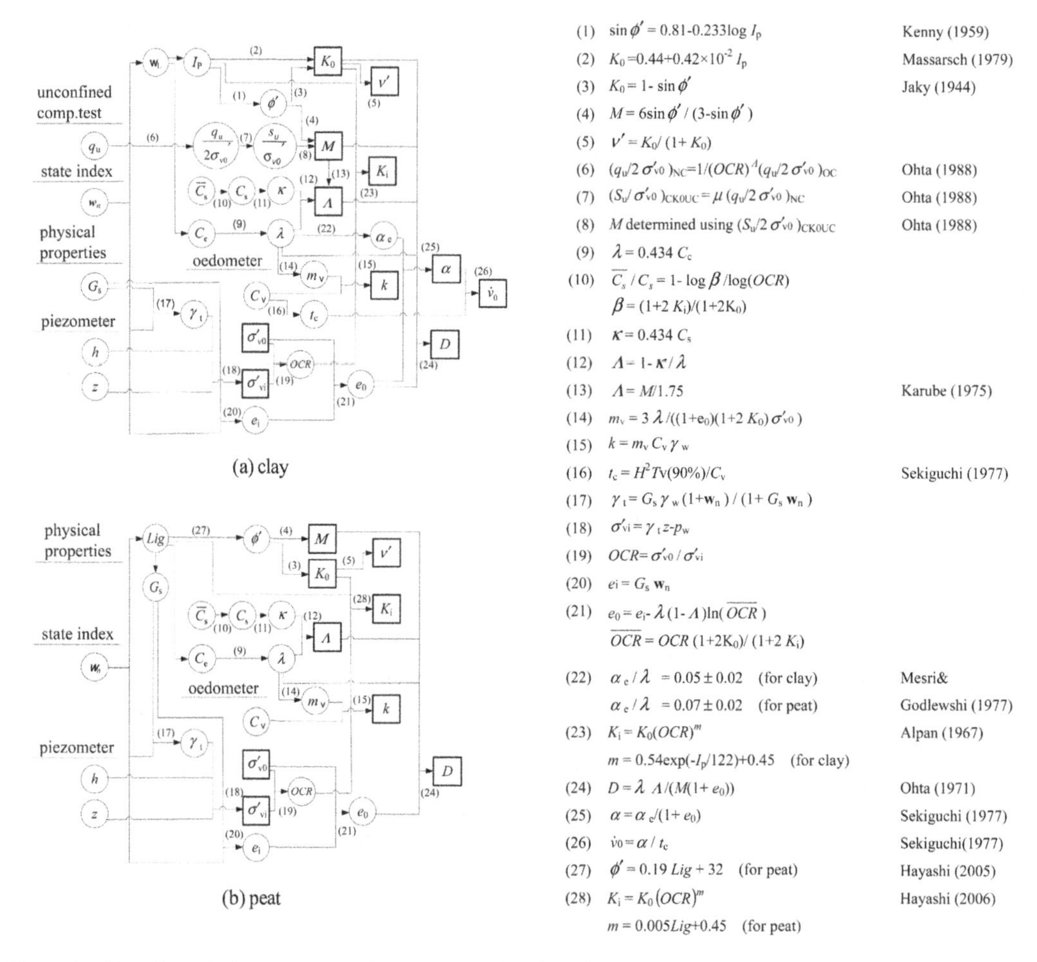

Figure 4. Flow charts for input parameter determination of Sekiguchi-Ohta model.

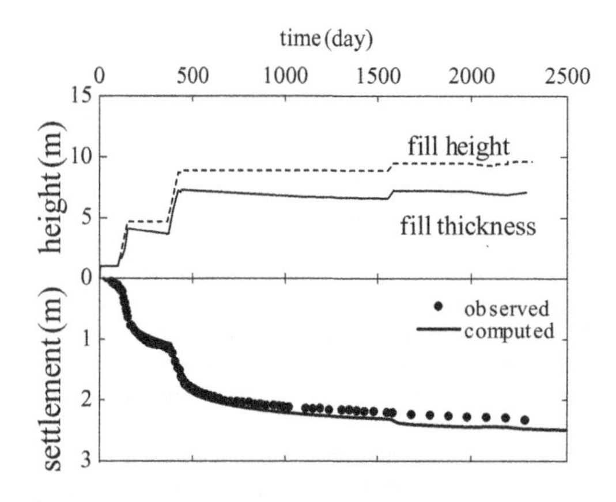

Figure 5. Settlement of the ground surface under the centre of embankment during construction period.

set to zero. The computed results successfully agree well with the observed one. The authors consider that Figure 10 evidence reasonably trustful predictability of Class B prediction made by the use of the integrated technique of modelling, parameter determination and computer simulation introduced in this paper.

3 LIFE-CYCLE PLANNINNG OF HIGHWAY EMBANKMENT

3.1 *Performance indicator of highway embankment*

For road asset management, it is required to define a set of performance indicators to evaluate the degree to which public agency's goals and objectives have been achieved (OECD, 2003). Nevertheless, geotechnical assets are unique in the supportive role that the assets

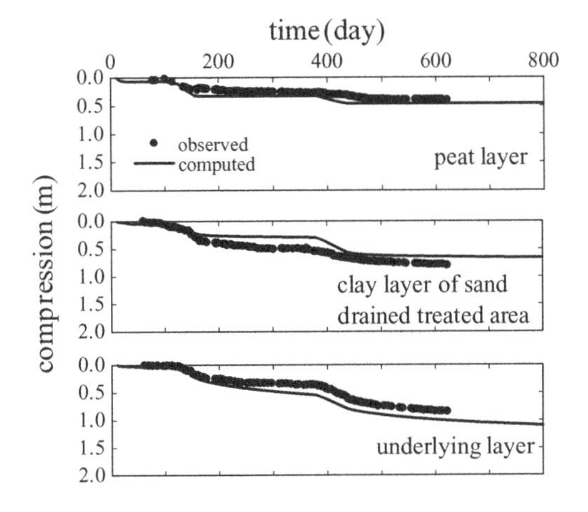

Figure 6. Compression of the ground surface under the centre of embankment during construction period.

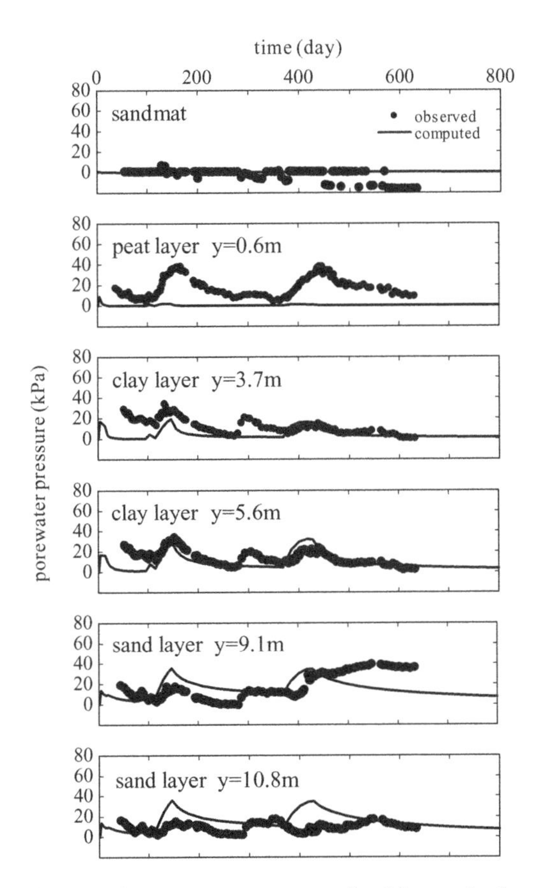

Figure 7. Excess porewater pressure of each layer under the centre of embankment during construction period.

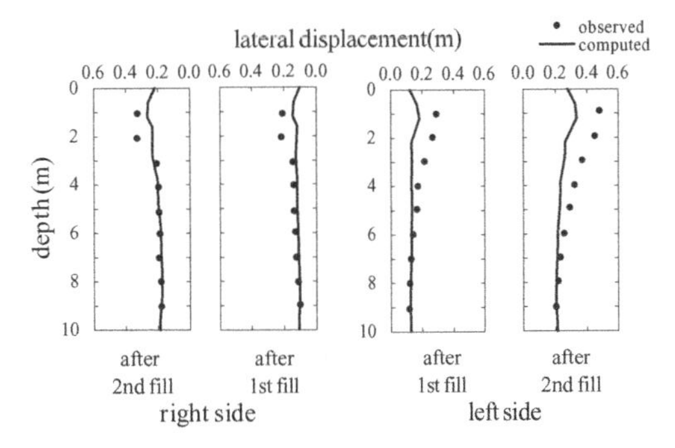

Figure 8. Excess porewater pressure of each layer under the centre of embankment during construction period.

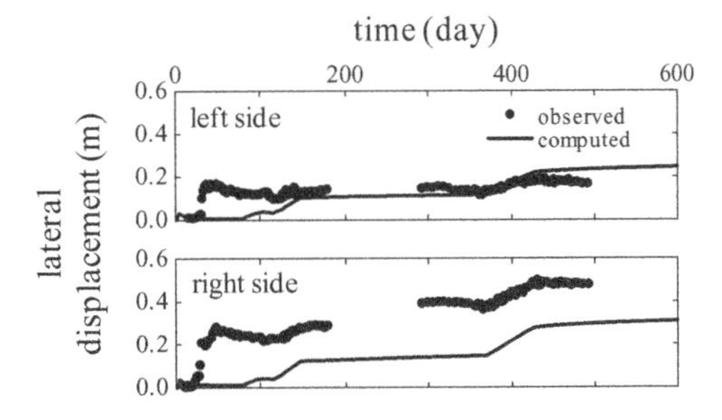

Figure 9. Lateral displacement of the ground surface at the toe of embankment during construction period.

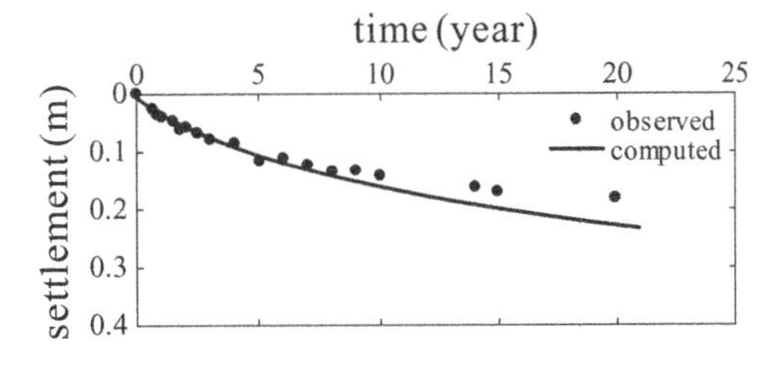

Figure 10. Long-term settlement of the ground surface under the centre of embankment after construction.

play for pavement and other road assets (Bernhardt et al. 2003). Therefore, the difficulty of developing geotechnical performance indicators (GPIs) is due to the requirements of considering the various interactions and integrations with other road assets. One of the authors suggested the geotechnical performance-related indicators (GPRIs) as mid-level indicator for use in life cycle planning of highway maintenance works (Ishigaki et al., 2008). Table 1 shows the relationship between the geotechnical indicators (GIs) and GPRIs of highway embankment on soft foundation related to road performance. The GIs are not only able to inspect at site but to predict by S/W FEM.

Table 1. Relationship between the geotechnical indicators (GIs) and the geotechnical performance-related indicators (GPRIs) of highway embankment on soft foundation.

GIs	GPRIs	Related performance
settlement of fill body	vertical alignment	safety
	faulting	user satisfaction
	cross slope	user costs
		travel time
	barrier height	safety
	drainage	
	stability	safety
lateral disp. of toe/surrounding ground	stability	safety
	ground inclination	surroundings
vertical disp. of toe/surrounding ground	ground inclination	surroundings
	ground height	

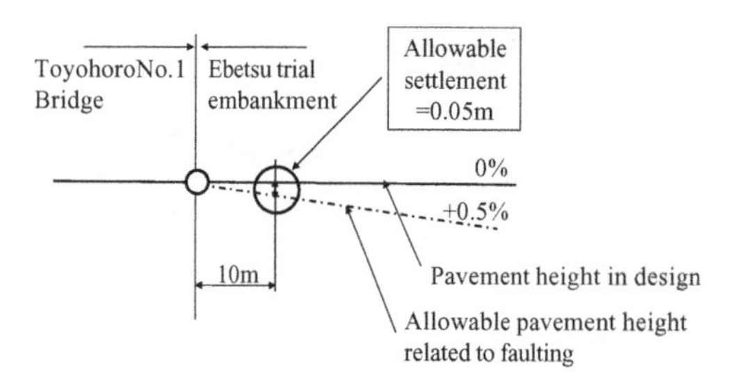

Figure 11. Settlement as indicator related to faulting between bridge abutment and adjacent pavement.

The value of GPRIs can be easily estimated from the value of related GIs. The authors assume the development of GPRIs may be applicable and efficient for development of geotechnical performance indicators (GPIs).

3.2 *Life cycle planning of highway embankment*

The life cycle planning of highway embankment based on the FE simulation of long-term performance is trialled. One of the actual maintenance criteria in this section is the faulting between a bridge abutment and adjacent pavement set to vehicle speed at 100 km/h. Figure 11 shows the allowable settlement as GCI related to faulting between abutment and adjacent pavement. In this section, the rate of change of the vertical slope is adopted as a definition of the faulting, and it is considered as 0.5% or more of the vertical slope of the bridge, which does not settle significantly. Figure 12 shows life cycle planning of asphalt overlay, which is predicted by the computed settlement. The maintenance criterion to the required service level is an allowable settlement of 0.05 m. The number of

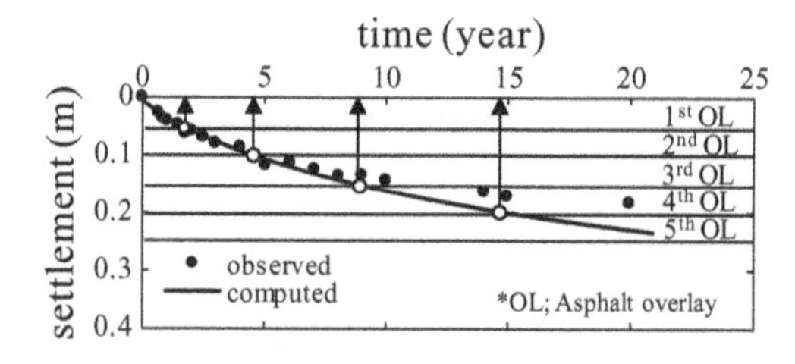

Figure 12. Life cycle planning of asphalt overlay for countermeasure of faulting between bridge abutment and adjacent pavement.

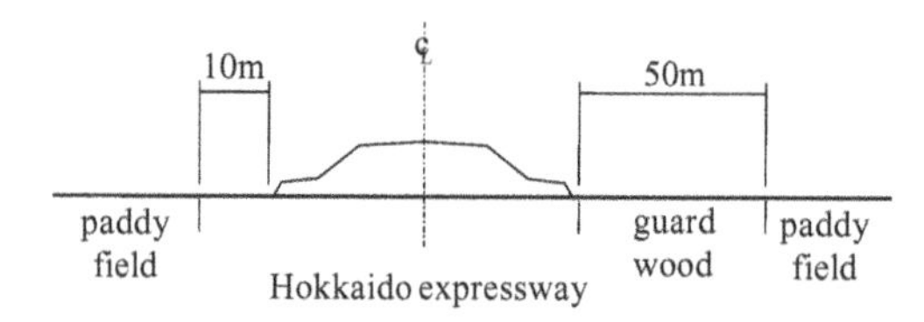

Figure 13. Perimeter of Ebetsu trial embankment.

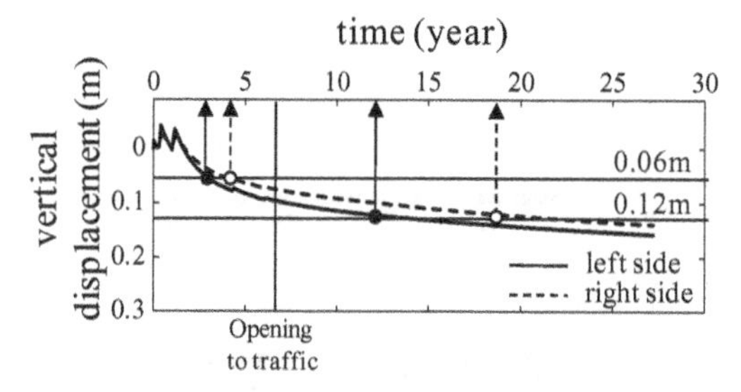

Figure 14. Life cycle planning of filling for countermeasure of vertical displacement of paddy fields.

times and timing of asphalt overlay are shown, when allowable settlement is exceeded. From this predicted settlement curve, the asphalt overlays are planned 4 times and timing is predicted to be 2nd year, 4th year, 8th year, and 14th year after opening to traffic. Actual asphalt overlays were performed 3 times and timing was 2nd year, 5th year, 9th year for 10years operating since opening to traffic. The life cycle planning based on computed settlement curves is successfully good agreement with the actual one.

Figure 13 shows the perimeter of Ebetsu trial embankment. Generally, filling is needed when settlement of paddy field exceed 0.06 m related to the height of young rice plant. Life cycle planning of filling for countermeasure of vertical displacement of surrounding paddy fields is shown in Figure 14. The number of times and timing of borrow material works are shown, when allowable displacement of 0.06 m is exceeded. From this predicted displacement curves, the works are planned both of 2 times and timing is simulated to be 3rd year and 12th year of left side and 4th year and 18th year of right side since starting construction.

4 CONCLUSION

The results of soil/water coupled finite element modelling of a 30 years old existing highway embankment evidence reasonably trustful predictability of Class B prediction made by the use of the integrated technique of modelling, parameter determination and computer simulation introduced in this paper. The life cycle planning of highway embankment based on the numerical modelling may be found to be applicable and efficient for use in geotechnical asset management practices.

ACKNOWLEDGEMENT

The authors are greatly indebted to Mr. M. Yoshimura, Mr. K. Ohkubo, and Dr. S. Yokota at Nippon Expressway Research Institute, Mr. K. Toyota at East Nippon Expressway Company Limited, for providing precious data and helpful criticisms and suggestion.

REFERENCES

Iizuka, A. and Ohta, H. 1987. A determination procedure of input parameters in elasto-viscoplastic finite element analysis. *Soils & Foundation*, Vol. 27, No. 3:71–87.

Ishigaki, T. Toyota, K. and Ohta, H. 2008. Long-term deformation used as indicator representative of highway embankment on soft foundation, *Proc, of 23rd World Road Congress*: PIARC

Sekiguchi, H. and Ohta, H. 1977. Induced anisotropy and time dependency in clays, *Constitutive Equation of Soils, Proc. of 9th International Conference on Soil Mechanics and Foundation Engineering, Specialty Session 9,* 305–315.

Advances in Transportation Geotechnics – Ellis, Yu, McDowell, Dawson & Thom (eds)
 ISBN 978-0-415-47590-7

A case study of the long-term settlement and life cycle cost of a road on peaty ground

S. Nishimoto & H. Hayashi
Civil Engineering Research Institute for Cold Region, Sapporo, Japan

ABSTRACT: Peaty Ground, which is found in Hokkaido, Japan, is very soft and problematic ground. When a road is constructed on peaty ground, long-term settlement caused by secondary consolidation occurs even after the road is put into service. In this paper, the relationship between the engineering properties of peaty ground at the study site and its long-term settlement is first considered. Then, the history of maintenance work conducted as a long-term settlement countermeasure, as well as its costs and negative impact on traffic, were studied. As a result, the relationship between the status of long-term settlement and life-cycle costs of expressways on peaty ground was clarified.

1 INTRODUCTION

Peaty ground distributed widely in Hokkaido, Japan, is an extremely soft ground, which is highly organic and has special engineering properties. Roads on peaty ground are usually designed to allow for a certain degree of settlement and are repaired and maintained during their service, since considerable settlement occurs over a long period of time in such roads. The Manual for Countermeasure against Peat Soft Ground (CERI, 2002) stipulates the allowable residual settlement as between 10 to 30 cm in three years after an expressway is placed in service. These values were, however, determined empirically and should be reconsidered for the minimization of the life cycle cost (LCC).

This paper outlines the status of the settlement of an expressway on peaty ground and repair history after its placement in service, and presents an evaluation including an estimation of LCC.

2 STATUS OF THE SETTLEMENT OF THE EXPRESSWAY AND ITS HISTORY OF REPAIR WORKS

2.1 *Overview of the route*

The subject of this study is an expressway in Hokkaido, Japan, which was placed in service in July 1998. Table 1 displays the history of the construction and repairs of the road. After the road was placed in service, bumps were observed before and after the bridge

Table 1. The history of the construction and repairs of the expressway.

construction stage	1996	1997	1998	1999	2000	2001	2002	2003
embankment construction	●———	——→						
waiting for reduction of settlement		●———	→					
pavement construction			●→					
in service of the expressway			start of service in July 1998 ———	———	———	———	———	——→
small-scale repair works					●	●	●	
large-scale repair works								●

and box culverts. As mentioned before, however, road surface settlement after its placement in service was allowed for in the design. Repair works were commenced in 2000, the third year of service of the road. However, since the actual settlement was greater than expected, it could not be controlled by normal repairs and a large-scale repair work had to be conducted in 2003, the sixth year of service.

2.2 *Overview of the ground and the status of settlement*

Figure 1 presents the ground profile and measurement results of road surface settlement in the fourth year of service. The ground can be roughly divided into two parts by the bridge at KP 8.3 km. The part between the interchange and bridge (3.7-km long) is soft ground consisting mainly of clay, while peat has accumulated thickly on the part between the bridge and box culvert

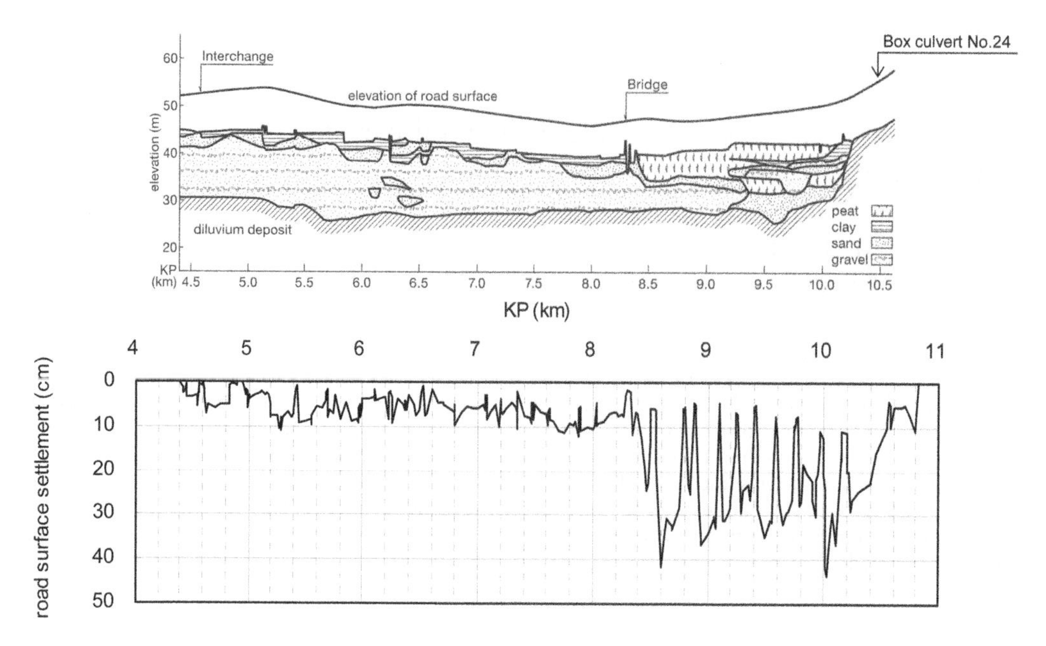

Figure 1. The ground profile and measurement results of road surface settlement in the fourth year of service.

No. 24 (2.2-km long). Even though the road embankment height is almost uniform throughout the route, the difference in soil profiles is clearly visible in the road surface settlement. While the road surface settlement on clay ground ranged from several to 10 cm, significant settlement exceeding 40 cm was observed in peaty ground.

2.3 *Small-scale repair works*

Measures against residual settlement taken between 2000 and 2002 included overlay pavement for the purpose of eliminating bumps before and after structures, the excavation of drain gutters to remove surface water, which had accumulated because the predetermined drain slope could not be secured, and the repair of cracks in the pavement near box culverts.

Figure 2 shows the costs of those small-scale repairs. While the repair cost for the section between the bridge and box culvert No. 24 (peat deposit), where settlement of 15 to 40 cm occurred up to the fourth year in service, was approximately 7.2 million JPYen per kilometer in three years, the cost for the section (clay deposit) where residual settlement was several to 10 cm was only 10% of that. The current Manual for Countermeasure against Peat Soft Ground (CERI, 2002) stipulates the allowable residual settlement (target values in three years after placement in service) as between 10 to 30 cm. Although a precise comparison cannot be made because the residual settlement in this study was measured in the fourth year of service, it can

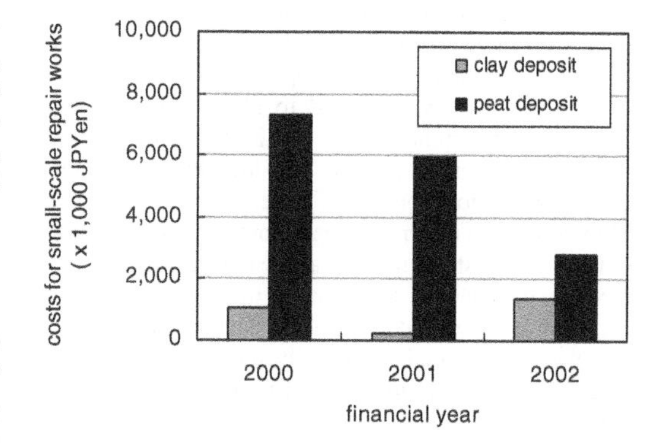

Figure 2. The costs of small-scale repairs of the expressway.

be seen that there is a considerable difference in repair cost between the minimum (10 cm) and maximum (30 cm) values of the allowable residual settlement.

The road had to be completely closed at night between 9:00 p.m. and 6:00 a.m. for 2 to 5 days for repair works. Although it is unavoidable to involve certain traffic regulations since the road was designed to allow for residual settlement, complete closure is necessary in the case of a two-lane road and the understanding of road users and related organizations must be gained. This is a point that should be considered in the future along with the maintenance of road functions and the LCC.

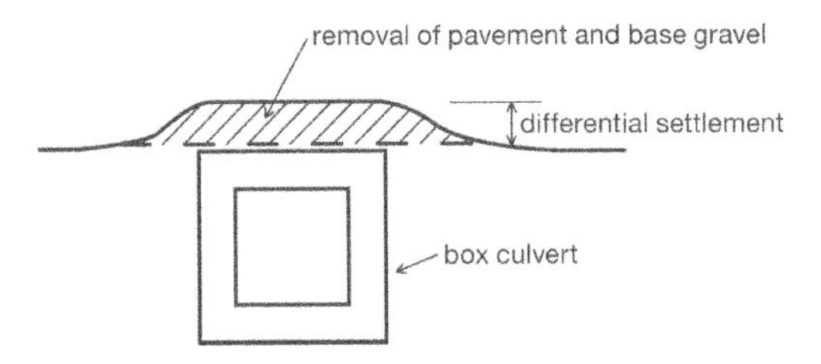

Figure 3. The handling in large-scale repair works.

Table 2. A summary of the large-scale repair works.

section	repair costs (×1,000 JPYen)	loss by delay in travelint time: time (days)	loss by delay in travelint time: costs (×1,000 JPYen)
interchange and bridge (clay deposit, L = 3.7 km)	248,273	50	9,800
bridge and No. 24 box (peat deposit, L = 2.7 km)	411,222		
total	659,495		

2.4 *Large-scale repair works*

Although settlement had been handled with small-scale repair works until 2002, it was becoming difficult because residual settlement was more significant than expected. In addition, there was a risk that the longitudinal gradient of the road and the height of guard fences might not satisfy the specified values. A large-scale repair of the entire road was thus conducted instead of the local elimination of bumps. After examining different repair methods, the method of lowering the road height with the bridge height as the control point was adopted to minimize re-settlement caused by load increase. While it was necessary to lower the road height on box culverts at that time, it was handled by removing gravel for base material (Fig. 3).

Table 2 shows a summary of the large-scale repair works. Although it was also considered possible to handle the section between the interchange and bridge by repeating small-scale repairs, it was judged more advantageous to conduct a large-scale repair considering the fact that the closure of the entire route was unavoidable anyway because of the access points with ordinary roads and the necessity of future maintenance and repair works. The large-scale repair required a cost of approximately 660 million JPYen and road closure at night for 50 days. Since cars would have to detour to a national highway during the closure, the loss by the delay in traveling time was estimated to be approximately 10 million JPYen.

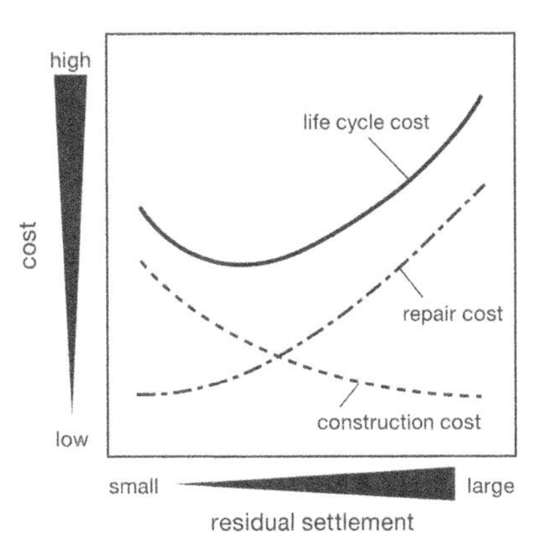

Figure 4. The concept of the life cycle cost.

3 EVALUATION OF COUNTERMEASURES AND THE ESTIMATION OF THE LIFE CYCLE COST

3.1 *Current evaluation*

Settlement that occurred after the placement of the route in service was thought to be due to a combination of the following two causes:

(1) Residual primary consolidation in the peat and clay layers
(2) Occurrence of secondary consolidation in the peat layer

The route was designed in 1996 and 1997. In those days, a method for the accurate prediction of settlement behavior peculiar to peat, which is described in the current Peaty Soft Ground Countermeasures Manual, had not been developed yet. It is thus understandable that such residual settlement could not be predicted although methods of the highest possible technological level at that time were selected for geotechnical investigations, settlement prediction and countermeasures for the route.

The following sections will present evaluation of the effect on residual settlement and the LCC (Fig. 4) for the route, based on the current technological level.

3.2 *Effects of new technologies*

The preload method is used in combination with the counter weight fill method as an actual measure against the soft ground of the route. The embankment was constructed in two separate years to ensure its stability. This case was compared with a virtual case where the prefabricated vertical drain method (PVD), which is a type of accelerated consolidation method, was used in combination. At the time of the construction of the route, it was generally believed that the accelerated consolidation method had very little effect on peaty ground. After that, however, Hayashi

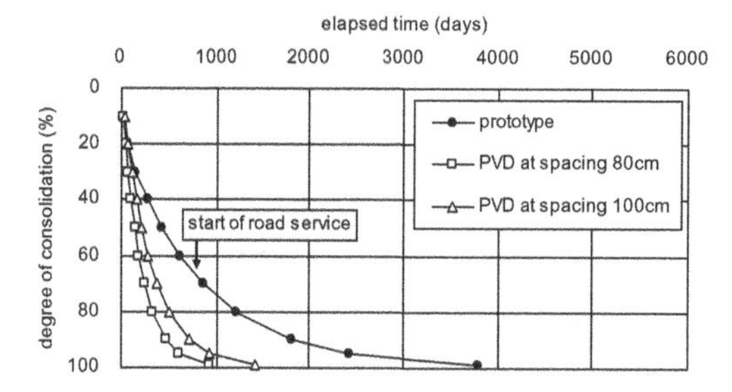

Figure 5. The analysis results of the degree of consolidation for each case.

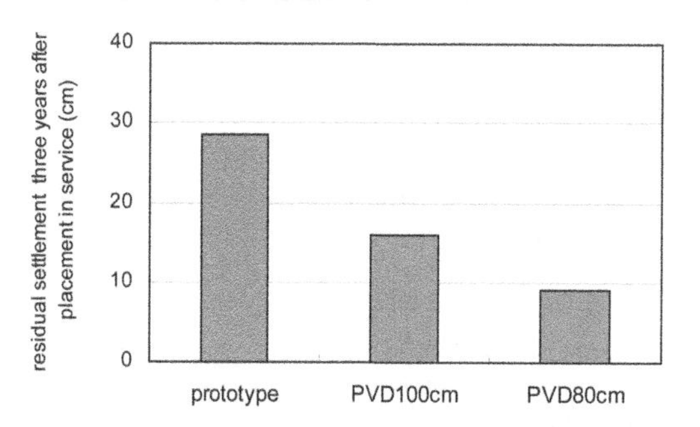

Figure 6. The results of residual settlement calculation summarized from Fig. 5.

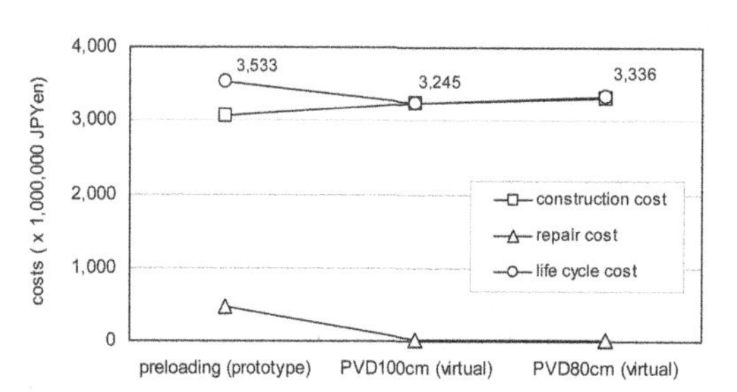

Figure 7. The results of the LCC simulation.

et al (2002) revealed the effect of the PVD method on peaty ground and presented the settlement prediction method using it.

Figure 5 displays the analysis results of the degree of consolidation for each case under the above-mentioned peaty ground condition (between the bridge and box culvert No. 24). The PVD cases of the analysis were performed using Yoshikuni's theory (Yoshikuni, 1979). Figure 6 shows the results of residual settlement calculation summarized from Fig. 5. When PVD was applied at spacing intervals of 80 cm, residual settlement three years after placement in service could be reduced to 9 cm. This is equivalent to the amount of residual settlement (minimum value of allowable residual settlement) measured in clay ground (between the interchange and bridge).

3.3 *Simulation of the LCC*

The LCC up to the 50th year after the placement of the road in service was estimated for the above three cases. Here, the above-mentioned status is reflected in the maintenance and repair costs. Figure 7 presents the results of the LCC estimate for the section between the bridge and box culvert No. 24 (2.2-km long). The construction cost of the section was kept down by selecting the most economical preload method. However, large-scale repairs were necessary due to significant residual settlement after placement in service and resulted in greater repair costs. Although the construction cost will be slightly higher when PVD is used in combination, it will be possible to control residual settlement and small-scale repairs will be sufficient. The LCC will thus be reduced by 6 to 8%. If it is possible to deal with settlement using peaty soft ground countermeasures at the level of PVD, it is rational to select 10 cm as the target design value out of 10 to 30 cm of the current allowable settlement three years after the placement of an expressway in service.

The estimation here was made under a variety of hypothetical conditions. It should thus be noted that the estimation results here may not be applicable to other site conditions.

4 CONCLUSION

In this study, the significant influence of the setting of residual settlement on the repair cost was quantitatively clarified. It was also found that, under some conditions, the minimization of the LCC could not be achieved only by a reduction in construction cost.

The authors intend to develop a method for the prediction of long-term settlement, in which the repair history can be taken into account, to minimize the LCC of peaty ground countermeasures in the future, and to present methods for setting the appropriate allowable residual settlement and determining the time of maintenance and repair using the prediction method as an analysis tool.

REFERENCES

Civil Engineering Research Institute of Hokkaido (CERI). 2002. *Manual for Countermeasure against Peat Soft Ground*: 42–44. (in Japanese)

Hayashi, H., Nishikawa, J. and Egawa, T. 2002. Improvement Effect of Prefabricated Vertical Drain to Peat Ground. *Proceedings of 4th International Conference on Ground Improvement Techniques. KL, Malaysia*: 391–398.

Yoshikuni, H. 1979. Design and Execution Management of the Vertical Drain Method. Gihodo Shuppan publication. Tokyo: 49–58. (in Japanese)

Advances in Transportation Geotechnics – Ellis, Yu, McDowell, Dawson & Thom (eds)
© 2008 Taylor & Francis Group, London, ISBN 978-0-415-47590-7

Pile-supported embankments on soft ground for a high speed railway: Load transfer, distribution and concentration by different construction methods

M. Raithel & A. Kirchner
Kempfert + Partner Geotechnik, Wuerzburg, Germany

H.-G. Kempfert
University of Kassel, Germany

ABSTRACT: For the foundation of traffic infrastructure in areas with soft subsoil often embankments supported by piles are necessary due to the settlement requirements. For sufficient load transfer into the piles different construction methods are possible. For the construction of a high speed railway line in China reinforced concrete slabs, a cement stabilization of the embankment material and geogrid reinforcement on top of the piles were used. The different methods are described and important field monitoring results are presented.

1 INTRODUCTION

In areas with soft subsoil embankments have often to be founded on piles or columns, especially if restricted settlement requirement exists. Thereby a sufficient load transfer into the piles is necessary.

Without any constructive elements for load transfer and distribution an arching effect in the embankment will be present, due to the higher stiffness of the piles compared to the soft soil. The mechanism of stress redistribution can be modelled by a system consisting of several arching shells. In many cases this is not sufficient for settlement reduction, especially if there is a risk of high cyclic/dynamic loading.

For a high speed railway line in China therefore three different construction methods for a better and safe load transfer, distribution and concentration were carried out:

- Reinforced concrete slab on top of the piles
- Horizontal geogrid reinforcement on top of the piles (so-called "geosynthetic-reinforced pile-supported embankment")
- Cement stabilization of the embankment material

In this contribution the design and construction of the pile-supported embankments will be presented.

2 PROJECT

Between Beijing and Tianjin 115 kilometres of high speed railway are being built by order of the Chinese Railway Department. The railway line will go into service at the Olympic Games 2008. With a design speed of 350 km/h, the superstructure of the line will be constructed with the so-called slab track system "System Bögl" developed by Max Bögl GmbH & Co. KG – Germany. The Bögl slab track system consists of laterally tensioned, prefabricated slabs with mounted rail fastenings.

About 80% of the railway line will run on a structure consisting of single-span and multi-span bridges with

Figure 1. High speed railway between Beijing and Tianjin.

standard span lengths of 24 and 33 m and in some cases between 60 and 120 m. The remaining length of the track will run on an embankment with piled foundations.

Due to subsoil conditions characterized normally by soft to stiff silt and clay down to depths of 50–80 m and a groundwater level just beneath the ground surface, all structures were founded on piles with lengths up to 70 m.

Considering the speed, the requirements concerning settlements of the foundation with 15 mm after construction of the slab track are very high. In order to determine the actual bearing capacity of the pile foundations pile load tests were conducted on special test piles (see Raithel et al. 2006). To guarantee the necessary quality, working piles (auger- and driven piles) were also tested.

3 PILESYSTEM – AUGER- AND DRIVEN PILES

For the foundation of the embankments two different pile systems are used.

The CFA-piles are drilled with continuous flight augers up to depth of 28 m. The diameter of the auger is about 40 cm and the diameter of the hollow stem is about 10 cm.

Additional precast concrete driven piles (prestressed) with diameter 45 cm and a total length up to 35 m were used. According to our experiences this system provides very good bearing capacities if the piles are properly produced and installed. The coupling was achieved by welded joints.

In order to determine the actual bearing capacity of the pile foundation, pile load tests on driven piles as well as on auger piles with lengths of 25 – 35 m each were conducted (Raithel et al. 2006).

A problem is the group effect of the pile foundation, due to the small separation of the piles (distance of 1.2 to 2 m). Therefore the settlements of the pile group are increased compared to the result of single-pile tests. The group effect of the piles was analyzed by singe pile tests and special calculation programs for definition of the group effect. Further information can be found in Raithel et al. (2006).

Figure 2. Continuous flight augers and centrifugally precast concrete driven piles.

4 STANDARD CONSTRUCTION – REINFORCED CONCRETE SLAB

Normally a reinforced concrete slab with a thickness of about 50 cm on top of the piles for a safe load transfer, distribution and concentration was used. On both sides a cantilever retaining wall was constructed. A schematic view of the pile supported embankment is shown in Figure 3 and a picture is given in Figure 4.

5 HORIZONTAL GEOGRID REINFORCEMENT ON TOP OF THE PILES

5.1 *General*

In recent years a new kind of foundation, the so-called "geosynthetic-reinforced and pile-supported embankment" (GPE) was established. Above the pile heads, the reinforcement of one or more layers of geosynthetics (mostly geogrids) is placed, see Kempfert & Raithel (2005). The application of such solutions is recently developed in Germany, see Alexiew & Vogel

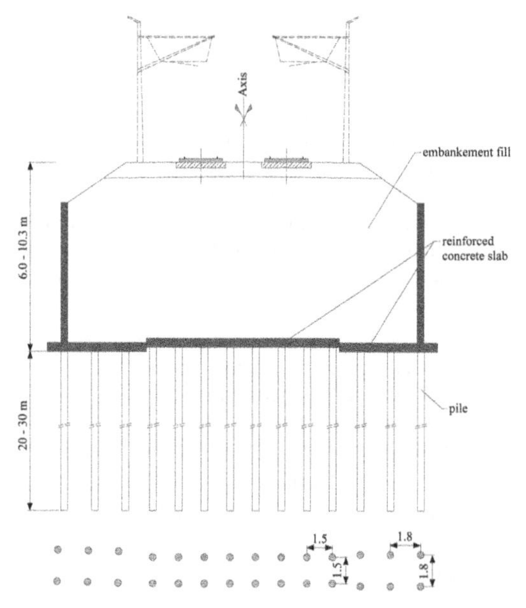

Figure 3. Pile supported embankment with concrete slab on top of the piles.

(2001). In the following the main principles of calculation and design and afterwards the construction in Beijing are shown.

5.2 *Calculation and design*

The stress relief from the soft soil results from an arching effect in the reinforced embankment over the pile heads and a membrane effect of the geosynthetic reinforcement, see Figure 5 (Kempfert et al. 2004).

Due to the higher stiffness of the columns in relation to the surrounding soft soil, the vertical stresses from the embankment are concentrated on the piles. Simultaneously soil arching develops as a result of differential settlements between the stiff column heads and the surrounding soft soil.

The 3D-arches span the soft soil and the applied load is transferred onto the piles and down to the bearing stratum.

The stress distribution can be modelled in various ways. Figure 6 shows, for example, a system consisting of several arching shells (Zeaske 2001, Zaeske & Kempfert 2002).

Figure 4. Construction of retaining walls.

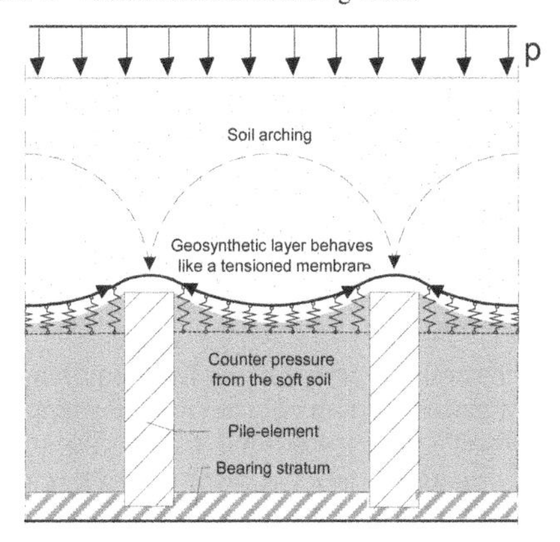

Figure 5. Mechanisms of load transfer and interaction.

This model leads to a differential equation, which is a function of the described vertical stresses σ_z [z] in the arching system (Zaeske 2001):

$$-\sigma_z \cdot dA_u + (\sigma_z + d\sigma_z) \cdot dA_o - 4 \cdot \sigma_\Phi \cdot dA_s \cdot \sin\left(\frac{\delta\Phi_m}{2}\right) + \gamma \cdot dV = 0 \tag{1}$$

For the areas above the arches a load depending stress distribution is assumed. The effective stress on the soft soil stratum σ_{zo} results from the limiting value consideration $z \rightarrow 0$ with $t =$ height of the load depending arch, so the following equation can be formulated. Simplified σ_{zo} can also be derived from dimensionless diagrams (DGGT, 2003).

$$\sigma_{zo} = \lambda_1^{\chi} \cdot \left(\gamma + \frac{p}{h}\right) \cdot \left(h \cdot (\lambda_1 + t^2 \cdot \lambda_2)^{-\chi} + t \cdot \left(\left(\lambda_1 + \frac{t^2 \cdot \lambda_2}{4}\right)^{-\chi} - (\lambda_1 + t^2 \cdot \lambda_2)^{-\chi}\right)\right) \tag{2}$$

with:

$$\chi = \frac{d \cdot (K_{krit} - 1)}{\lambda_2 \cdot s_d}, \qquad \lambda_2 = \frac{s_d^2 + 2 \cdot d \cdot s_d - d^2}{2 \cdot s_d^2}$$

$$\lambda_1 = \frac{1}{8} \cdot (s_d - d)^2, \qquad K_{krit} = \tan^2\left[45° + \frac{\varphi'}{2}\right]$$

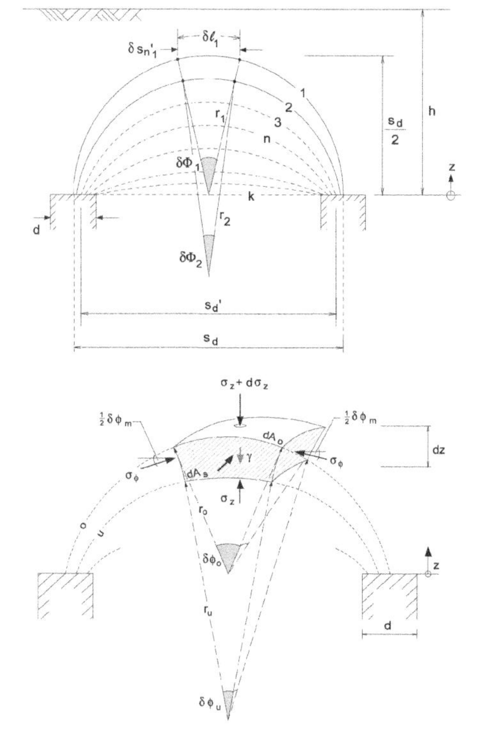

Figure 6. Theoretical arching model.

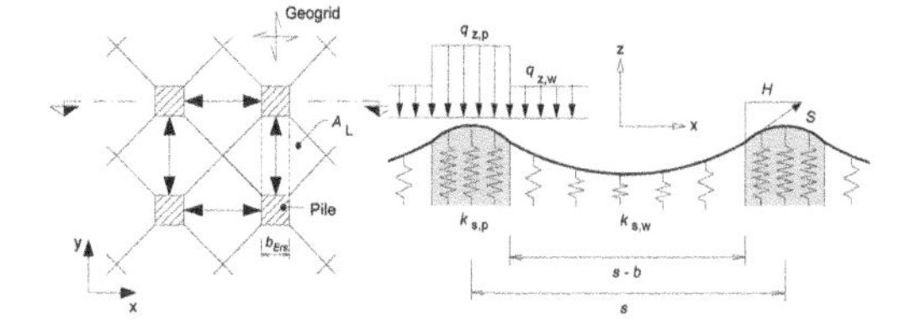

Figure 7. Loading of the reinforcement.

The loading of the reinforcement is expressed by the differential equation of the elastic supported cable, in which the vertical displacement z_W over the soft soil and the horizontal force H, (Zaeske & Kempfert 2002) are the unknown variables, see Figure 7.

$$\frac{d^2z}{dx^2} = \frac{q_{z,w}}{H} + \frac{k_{s,w} \cdot z_w}{H} \tag{3}$$

with:

$$H = \frac{2 \cdot \int_0^i \sqrt{1 + z_W'^2}\,dx + 2 \cdot \int_i^j \sqrt{1 + z_P'^2}\,dx - l_0}{2 \cdot \int_0^i \left(1 + z_W'^2\right)dx + 2 \cdot \int_i^j \left(1 + z_P'^2\right)dx} \cdot J$$

Finally the loading of the reinforcement S can be calculated directly as a function of the elongation ε ($J =$ stiffness) of the geosynthetic (for dimensionless diagrams see DGGT 2003):

$$S[x] = \varepsilon\,[x] \cdot J = H \cdot \sqrt{1 + z'^2[x]} \tag{4}$$

5.3 *Soil arching under cyclic loading*

Whereas the system behaviour (soil arching and membrane effect in geosynthetic reinforcement) under static loading is well-known, the bearing behaviour and the settlements expected under cyclic loading are not yet fully understood. Under cyclic loading the arching effect can only be formed in a very limited extent and part of the load carried directly by the piles can decrease remarkably, which results in an increase of the load on the soft soil and on the reinforcement. Due to the reduction of the soil arching, the strains in the geogrid and the surface settlements increase considerably. Based on the results of model tests and numerical investigations, the main parameters which cause a reduction of the arching effect have been identified (Heitz 2006). Critical geometrical and loading limits were worked out, see Table 1. For system values beyond these limits a modified calculation procedure is proposed which takes a soil arching reduction and an increase of the geosynthetic strains into account (Heitz 2006).

Table 1. Simplified requirements for consideration of cyclic effects on arching.

Cyclic loading	Geometry[1]	Reduction of arching
Large (railway – about 30 kPa)	>1.5	negligible
	≤1.5	partly up to complete
Middle (roads – about 15 kPa)	>0.7	negligible
	≤0.7	partly

[1] see Figure 15

Figure 8. Embedded geogrid.

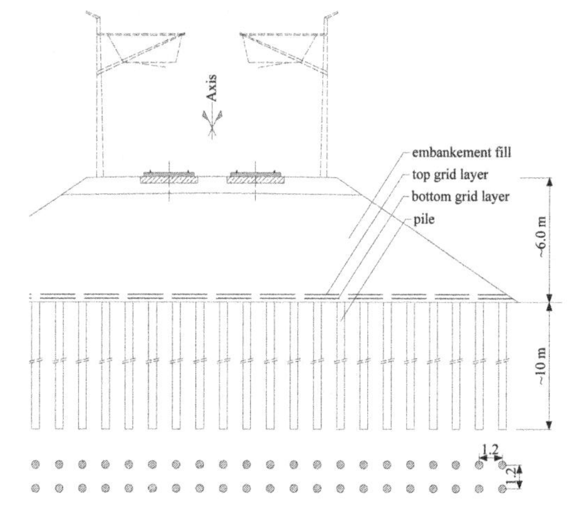

Figure 9. Geosynthetic-reinforced and pile-supported embankment Beijing.

5.4 *Construction*

For the construction of the railway line from Beijing to Tianjin a geosynthetic-reinforced and pile-supported embankment" (GPE) was built in the city area of Beijing. Figure 8 shows a picture of the embedded geogrid. A schematic view of the pile supported embankment with the geogrid reinforcement is shown in Figure 9.

Figure 10. Situation before filling.

6 CEMENT STABILIZATION OF THE EMBANKMENT MATERIAL

6.1 *Calculation and design*

In sections with very low embankments a cement stabilization of the embankment material instead of the geosynthetic reinforcement is more reasonable, due to the reduction of the arching effect and the increase of the forces in the geogrid reinforcement by a high cyclic loading.

The cement stabilized embankment acts similar to a slab over the pile heads, whereas the expensive reinforcement can be saved. For the design of the cement stabilized embankment a Finite Element Model FEM was used, considering a so called unit cell, which means the consideration of one pile with the surrounding soil (infinite pile grid) in an axially symmetric model.

Figure 10 shows the situation before filling. A schematic view of the pile supported embankment with the cement stabilization is shown in Figure 11.

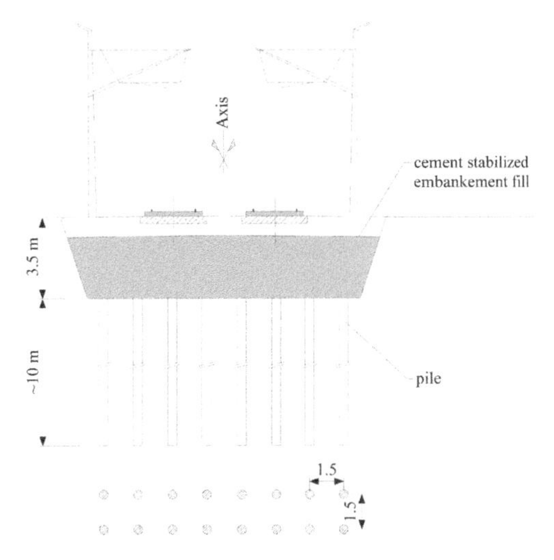

Figure 11. Cement stabilization in Beijing.

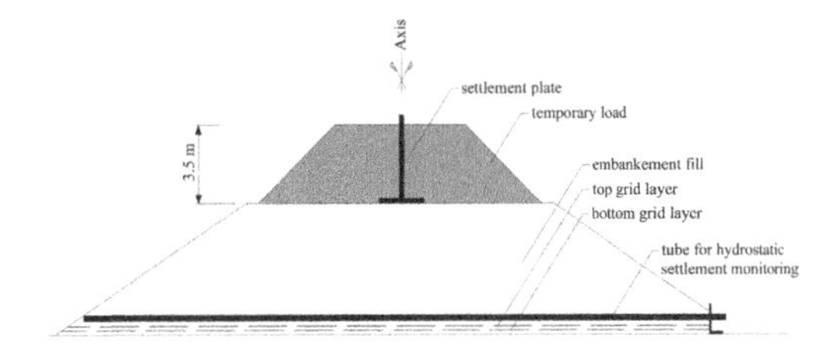

Figure 12. Typical measurement cross sections.

7 MONITORING

7.1 *Monitoring concept and measurement cross sections*

Due to the settlement requirements a monitoring program for verification of the design and certification of stability and serviceability was installed. The different monitored cross sections have distances between 5 and 50 m. A large quantity of vertical and horizontal inclinometers and geophones were installed. Additionally, the settlements of the rails were measured.

In a typical measurement cross sections one horizontal inclinometer and up to four settlement plates are used for the examination of the deformation behaviour, see Figure 12.

In Figure 12 additionally the temporary overload for minimization of the settlements after the construction time can be seen.

7.2 *Measurement results*

In January 2008, measurements had been performed for about 10 to 14 months. The long-term monitoring has confirmed the stability and serviceability. In December 2007 the slab track was installed in all parts of the railway line.

Figure 13 shows typical results for the settlements at different heights in a cross section with a reinforced concrete slab and Figure 14 the settlements of a section with a horizontal geogrid reinforcement on top of the piles instead of the concrete slab. In these sections the embankment is ultimately about 8.5 and 5.0 m high (following removal of the 3.5 m temporary load), respectively.

The settlements in the section with a cement stabilization of the embankment material instead of the geosynthetic reinforcement or a concrete slab are very low (up to 2 mm) due to the very low embankment. Therefore these measurements are not comparable to the settlements shown above.

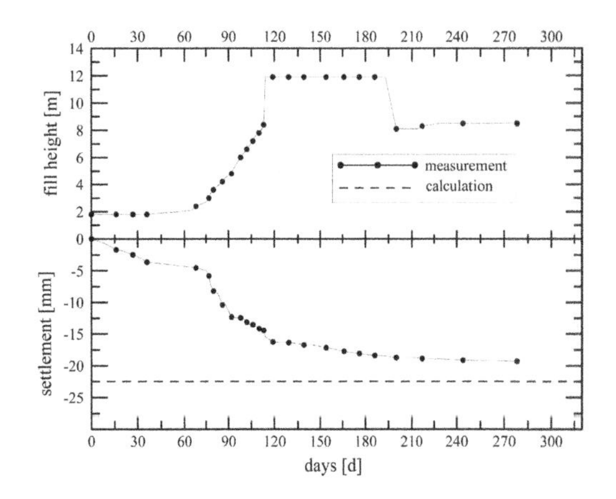

Figure 13. Settlements in a section with a concrete slab on top of the piles.

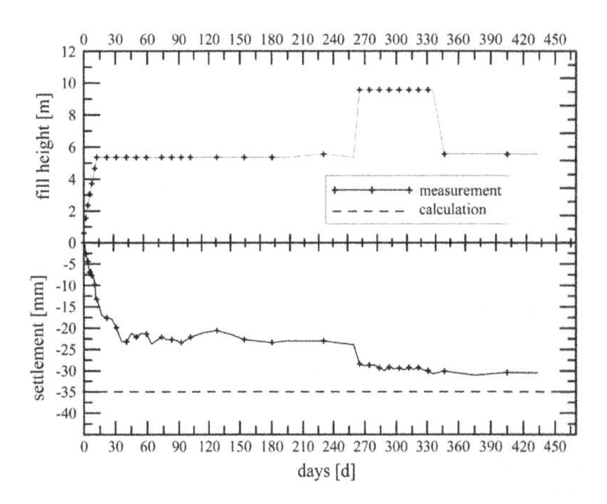

Figure 14. Settlements in a section with horizontal geogrid reinforcement on top of the piles.

8 CONCLUSION

According to the experiences during the construction of the railway line Beijing-Tianjin and the experiences in Germany a geosynthetic reinforcement as well as a cement stabilization of the embankment material can be used instead of a concrete slab to guarantee a sufficient load transfer and distribution.

Based on German and international experiences with geogrid-reinforced pile-supported embankments, practical reasons, experimental results and the validity of the analytical model the following recommendations are established:

The center-to-center distance s (see Figure 15) of the piles and the pile diameter d (pile caps) should be chosen as follows:

- $(s - d) \leq 3.0$ m: in the case of static loads
- $(s - d) \leq 2.5$ m: in the case of heavy live loads

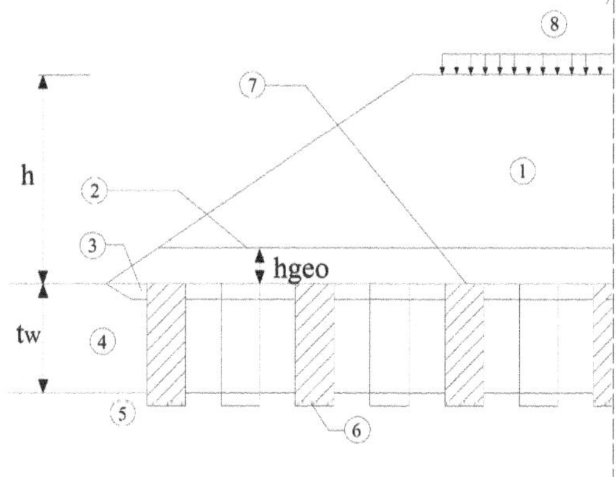

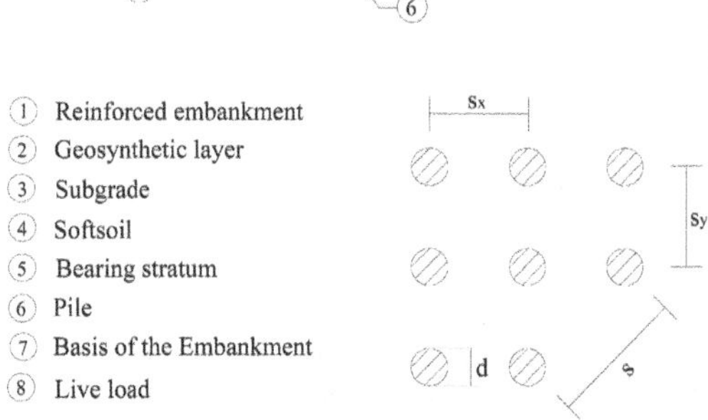

Figure 15. Geosynthetic-reinforced pile supported embankment (GPE) – Definition of parameters.

- $d/s \geq 0.15$
- $(s - d) \leq 1.4\,(h - z)$

The distance between the reinforcement layer and the plane of the pile/column/wall heads should be as small as possible, in order to achieve maximum efficiency of the geogrid membrane. However, it is recommended to have a safe distance z (interlayer) between the lowest reinforcement and the pile heads in order to prevent a structural damage of the reinforcement because of shearing at the edge of the pile heads:

- maximum two reinforcement layers
- $h_{geo} \leq 0.15$ m for single layer reinforcement
- $h_{geo} \leq 0.30$ m for two layers reinforcement
- for two layers the distance between the geogrid layers should be 15 to 30 cm
- design value of the tensile strength $R_{Bd} \geq 30$ kN/m; ultimate strain $\leq 12\%$.
- overlapping is generally allowed, but only just above the pile (caps) and only in the secondary bearing direction; length of overlapping $\geq$ d.

REFERENCES

Alexiew, D. & Vogel, W. 2001. Railroads on piled embankments in Germany: Milestone projects, Landmarks in

Earth Reinforcements. In Ochiai et al. (eds), Swets & Zeitlinger: 185–190.

Empfehlung 6.9 2003. Bewehrte Erdkörper auf punkt- oder linienförmigen Traggliedern. *Kapitel 6.9 für die Empfehlungen für Bewehrungen aus Geokunststoffen, EBGEO, DGGT (German Geotechnical Society)*, draft.

Heitz, C. & Kempfert, H.-G. 2007. Bewehrte Erdkörper über Pfählen unter ruhender und nichtruhender Belastung. *Bauingenieur*, Band 82: 380 – 387.

Kempfert, H.-G., Goebel, C., Alexiew, D. & Heitz, C. 2004. German recommendations for reinforced embankments on pile-similar elements. *Third European Geosynthetic Conference, DGGT (German Geotechnical Society)*, Volume 1: 279–285.

Kempfert, H.-G. & Raithel, M. 2005. Soil Improvement and Foundation Systems with Encased Columns and Reinforced Bearing Layers. In "Ground Improvement Case Histories" Ed. Buddhima Indraratna and Chu Jian. Elsevier B.V.

Raithel, M., Kirchner, A. & Kneissl, A. 2006.Gründung der Hochgeschwindigkeits-Teststrecke Beijing-Tianjin auf Grundlage von vergleichenden Probebelastungen an Großbohrpfählen, Rammpfählen und SOB-Pfählen. In Vorträge der Baugrundtagung 2006 in Bremen, *DGGT (German Geotechnical Society)*.

Raithel, M., Kirchner, A. & Kempfert, H.-G. 2008. German Recommendations for reinforced Embankments on Pile-similar elements. *Proceeding of the 4th Asian Regional Conference on Geosynthetics*, June 17–20, Shanghai.

Zaeske, D. 2001. Zur Wirkungsweise von unbewehrten und bewehrten mineralischen Tragschichten über pfahlartigen Gründungselementen. *Schriftenreihe Geotechnik*, Universität Kassel, Heft 10.

Zaeske, D. & Kempfert, H.-G. 2002. Berechnung und Wirkungsweise von unbewehrten und bewehrten mineralischen Tragschichten auf punkt- und linienförmigen Traggliedern. *Bauingenieur*, Band 77.

Advances in Transportation Geotechnics – Ellis, Yu, McDowell, Dawson & Thom (eds)
 ISBN 978-0-415-47590-7

Geotechnical challenges faced in the construction of the A55 expressway across Anglesey

S. Solera
Arup, Wrexham, UK (formerly Hyder Consulting)

J. Baple
Hyder Consulting, Guildford, UK

ABSTRACT: The new 31 km dual carriageway extension to the A55 in Anglesey, North Wales, runs from the Menai Straight to the Port of Holyhead. A wide variety of geotechnical challenges were encountered during the design and construction of the road. Amongst others, these included instrumented embankments, stone columns in very silty soils, piles in variable soils and rocks, and the procurement of granular backfill for integral bridges. This paper concentrates on the design and construction of an instrumented embankment over soft clays, and briefly discusses how other technical challenges were resolved. As a design and build contract, the paper also presents how the interaction between the designer and contractor under this type of contract helped enormously to resolve these problems during the works, something that would have been more difficult in a traditionally procured contract.

1 INTRODUCTION

The A55 Expressway across the island of Anglesey, North Wales (Figure 1) was constructed at a cost of £100 m and opened in March 2001. The scheme was designed by Hyder Consulting and constructed by Carillion-Laing under a Design, Build, Finance and Operate Contract for The National Assembly for Wales.

The geological setting of this road is unique in the United Kingdom, as it was constructed in parts through very old rocks of the Precambrian Mona Complex (schists, gneisses and some granitic intrusion), overlain by relatively thin glacial till and thick alluvial or marine deposits in places.

There were a variety of geotechnical challenges which the designer and contractor resolved by working together following an observational method approach. Geotechnical challenges included the construction of embankments over soft clays, selection of backfill material for integral bridges, construction of stone columns in deep alluvial deposits, resilient ("bouncy") embankments as well as the choice of adequate pile penetration to achieve the required design load. This paper concentrates in the design and construction of instrumented embankments over soft clays, silts and peats, and briefly discusses how other challenges were resolved. The designer provided three full time geotechnical engineers on site during construction to assist during construction. The interaction between these engineers and the contractor played a significant role in the implementation of design intent and the ability to respond to the observed behaviour of the embankments.

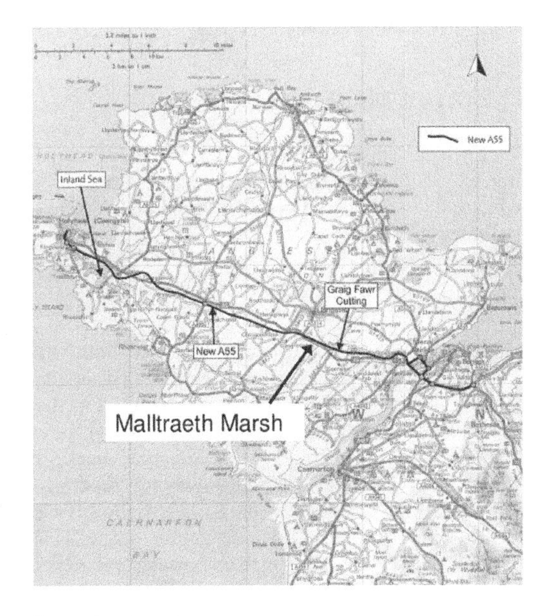

Figure 1. Location.

2 THE SCHEME

The construction of the A55 Expressway required 31 km of new dual two-lane carriageway and 34 bridges and other structures, and completes a dual carriageway trunk road from Chester to Holyhead.

The route of the scheme (see Figure 1) runs, generally, east to west, parallel to the old A5 constructed by Thomas Telford in the 19th Century. The road runs from the end of the Llanfairpwllgwyn east of the Island, crossing a body of water known as the Inland Sea on a new causeway adjacent to the Stanley Embankment (constructed by Telford in 1822). When it reaches Holy Island, it runs adjacent to the railway line to Holyhead.

3 GEOLOGY

Drift deposits are relatively thin glacial till and thick alluvial or marine deposits. The notable exception to this is Malltraeth Marsh, where soft clays, silts and peats overlie rockhead. Bedrock consists of Pre-cambrian metamorphic rocks of the Mona Complex comprising intensely folded schists and basic gneiss with a strong foliation, mica schists and granite intrusions. Faults separate the Pre-cambrian rocks from Carboniferous rocks, which comprise mudstones and limestones in the vicinity of Malltraeth Marsh.

4 EARTHWORKS

A very brief discussion is given below of some problems encountered during construction of earthworks and lessons learnt, before proceeding to discuss Malltraeth Marsh.

4.1 *Backfill to integral bridges*

All except one bridge were integral (i.e. without any bearings). The scheme was one of the first in the country where integral bridges were designed to the requirements of HA42/96 (UK's Highways Agency, 1996). The design required that granular backfill to structures had friction angles varying generally between 35° for wall abutments and 42° (shallow bank seats). Higher angles of friction than the design values could have led to higher than expected moments induced in the abutments, as a result of thermal expansion and contraction of the bridge deck. Resourcing compliant materials on the island was a challenge, as naturally occurring rounded deposits were not generally available in Anglesey. Some fills that met the angle of friction specification had too great a fines content and did not meet the permeability requirements. Some slight relaxation of the permeability was initially allowed, but was later found to drain poorly and had to be removed.

In addition, those bridges founded on rock required a sliding layer of granular fill, approximately 300 mm thick, with an angle of friction of 33°, to permit movement of the base. The only material to meet this angle of friction was wind-blown sand, which was to difficult to contain in place.

Lessons learnt from this experience indicate that it would have been better to design the bridges with a knowledge of locally won material, and to limit the required angle of friction to one or two different design values. Changes to the design guidelines for integral bridges (HA42/96) in relation to the determination of K* now allows the design of integral bridges to accommodate slightly higher friction angles for backfill.

It should also be said that it is extremely important that the granular backfill/embankment interface be compacted adequately to ensure that settlement does not occur in this area, as occurred in some cases at the site.

4.2 *Capping*

Within the area to the east of the Inland Sea a resilient ("bouncy") behaviour was observed in the majority of the clay embankments constructed from two of the borrow pits. This phenomenon could be described as a "wave" moving along the embankments under the weight of construction plant. This was attributed to a high moisture contents in the sandy silty clay soils, which under traffic build up of an excess pore-water pressure. In these areas it was clear that the intended 500 mm of granular capping was going to be insufficient to stop this resilient movement despite the compliance of the measured CBR values with the design values. To resolve this problem, lime was considered initially as an option to reduce the water content and increase the MCV values, but it was cheaper to place a thicker capping by bringing rockfill from nearby Gwalchmai Quarry. Trials were carried out using fully loaded 40 tonnes lorries over varying thicknesses of capping. Based on these trials, it was established that 1000 mm of Class 1C fill would be sufficient to reduce the dynamic movement ("bounciness") to acceptable levels and allow construction during inclement weather. This pragmatic approach developed by the close cooperation between designer and contractor would probably not have happened on a traditional contract.

5 MALLTRAETH MARSH

The new road crosses on a 5 m high embankment over a length of 1.6 km Malltraeth Marsh immediately to the west of Graig Fawr cutting (Figure 1). In this area there are approximately 12 m of soft alluvial clays, silts and peats associated with Afon Cefni River, underlain by glacial tills over carboniferous rocks. A simplified geological section is shown in Figure 2.

Afon Cefni required crossing by a new underbridge on piled foundations with two corrugated steel culverts in the embankments behind. The design of the embankment required a smooth transition between the piled foundation of the bridge, expected to be relatively rigid, and the adjacent approach embankments over clays and peats. In addition, the settlement of the two culverts needed to be reduced to acceptable levels. Instrumentation was used to monitor the embankment construction. A comparison of predicted versus measured settlements together with back analyses of consolidation properties are given below.

5.1 *Embankment design*

The 10–12 m thick alluvium in Malltraeth Marsh comprises a soft to very soft silty clay/clayey silt with zones of sand (see Figure 2). A virtually continuous band of silty sand about 1.5 m thick exists at between 1 m to 2 m depth on the east side of Afon Cefni. On the west side this sand thickens significantly and dominates the alluvial deposits.

Undrained shear strength results were obtained from various in situ (CPTs, SPTs and vanes) and laboratory tests (triaxial tests and laboratory vane). There is an upper crust of firm clay at the surface with an undrained shear strength of 30 kPa, below which the strength reduces to 10 kPa at 1.6 m where it remains constant to 3.5 m depth. Below this depth the undrained shear strength increases gradually to 20 kPa at 11 m depth. The following settlement design parameters were adopted, based on initial test data:

- Coefficient of consolidation (vertical and horizontal): $3\,m^2/yr$ for $H>3\,m$ and $4.5\,m^2/yr$ for $H<3\,m$ where $H =$ embankment height (oedometer results vary from 0.5 to $7\,m^2/yr$ with a mean value of $2.7\,m^2/yr$),
- Compressibility : $c_c/(1+e_0) = 0.16$ (laboratory values 0.05 to 0.17), where c_c is the compression index and e_o is the initial void ratio,
- Moisture content: 15% and 60%
- Creep Coefficient c_{sec}: 0.006

5.2 *Embankment construction*

A staged construction approach was adopted, which included surcharges, temporary berms and geotechnical instrumentation.

The embankments were fully instrumented with rod settlement gauges at the base of the embankments, magnetic extensometers, inclinometers and hydraulic piezometers. One of the on-site geotechnical engineers plotted and reviewed the instrumentation data during embankment construction to determine observed settlement behaviour and confirm the required timescale for each load stage.

Band drains were used to accelerate the settlement beneath the embankment, with the exception of the 30 m of embankment nearest to the piled abutments on either side of the bridge, where stone columns were installed to provide a transition between the band drain section of the embankment and the bridge. The stone columns accelerated the consolidation of the marsh, and provided foundation support to the two, 3 m diameter, corrugated drainage culverts just behind the bridge foundations. These also caused the lateral movement of the soft clays to be completed rapidly, which permitted piling of the bridge abutments. A summary of the ground improvement measures is given in Figure 2.

Construction of the embankment started with the placement of a geotextile separator directly over the existing topsoil, and the placement of a 300 mm thick granular drainage blanket. This blanket was used as a platform for the installation of the stone columns, band drains and instrumentation.

5.2.1 *Band drains*

The band drains were 100 mm × 3 mm arranged in a triangular spacing at 1.4 m to 1.9 m centres, installed by a closed-ended mandrel. They extended 4 m to 12 m depth through the soft clays and peats onto the glacial till.

5.2.2 *Stone columns*

Stone columns, 800 mm diameter, and 11 m to 12 m long, were installed using the dry bottom-feed method, terminating in the underlying glacial till. A triangular arrangement was used, varying from 1.5 m to 2.6 m, the closer spacing being nearest to the abutments. The rigs were fully instrumented to read the volume of stone used per metre depth and the penetration/extraction rate of the vibroflot.

Acceptance criteria comprised the visual inspection of the site operations and the review and interpretation of the on-board computer plots for each stone column. Concerns were raised during the early stages of column construction, as the vibroflot appeared to be withdrawn at relatively faster rates than expected from the method statement. This was also apparent from the monitoring records. It was feared that a rapid withdrawal could produce a suction effect (similar to a piston), drawing the soft clays up and leading to liquefaction of silts present at shallow depths.

To investigate and solve these problems, a number of stone columns were exhumed to a depth of 3–4 m. The excavation indicated a reduction of the diameter of the stone column from 800 mm at the

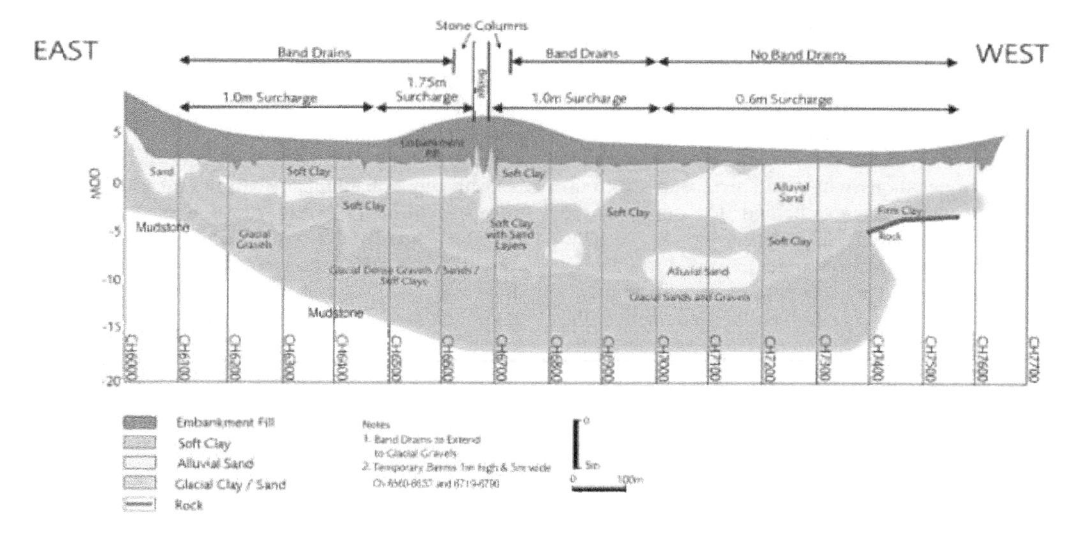

Figure 2. Geological cross section at Malltraeth Marsh.

surface to virtually nil at 1–2 m depth (within a zone where a number of silt layers were present), gradually re-appearing at greater depth. The computer plots were used to confirm that these defects coincided with areas of rapid vibroflot withdrawal and/or insufficient stone feed. A revised, slower (and uniform) withdrawal was agreed on site, to guarantee that a continuous column be formed in this type of ground.

The effectiveness of the stone columns in reducing the settlement measured is discussed below.

5.3 *Embankment instrumentation: performance and back analysis*

Inclinometers, hydraulic piezometers, extensometers and rod settlement gauges were installed to monitor the staged construction of the embankment. The reasons for the instrumentation were in order to:

- Monitor the performance of the embankment and surcharge.
- Monitor the rate of settlement and stability of the underlying soils in the short term following placement of fill, to allow a potential acceleration (or extension if required) of the construction programme.
- To assess progress of primary consolidation and dissipation of pore water pressure of the underlying soils for stability purposes
- Predict long term creep.

Temporary berms 1 m high and 5 m wide were required in the vicinity of Afon Cefni for stability purposes where the embankments were high.

At the highest point, in the vicinity of Afon Cefni, up to 5 m of fill was required with surcharges between 0.6 m and approximately 20 m above finished road level. The design required a Stage 1 embankment construction at a rate of 1 m/week to a certain level (for stability purposes) followed by a rest period of between 3 to 4 weeks. Stage 2 then would continue up to full height of the surcharge at a rate of 0.5 m/week, with rest periods between 3 to 4 weeks after each 0.5 m lift. The contemporaneous plotting and review of the instrumentation data by Hyder's geotechnical engineer allowed the programme to be modified as appropriate. The criteria followed were:

- Maximum horizontal settlement: 2.5 mm/day
- Maximum incremental ratio horizontal/vertical displacement: 0.5
- Overall horizontal/vertical settlement ratio: 0.25
- Maximum pore water pressure at any single stage to ensure slope stability

The limit to the horizontal/vertical displacement ratio to 0.5 was based on empirical observations of embankment failures when this ratio reaches 0.5 (Matsuo and Kawamura, 1977). Other criteria were based on experience of Hyder Consulting in other jobs.

The settlement field data and pore pressure data were back-analysed in order to establish a value of the horizontal consolidation in the area with band drains and stone columns. Settlement readings from the rod settlement gauges, magnetic extensometers and pore-water pressure measurements were also plotted using the method suggested by Asaoka (in Holtz et al, 1991) to estimate the magnitude and time for the end of primary consolidation. In this method, settlements or

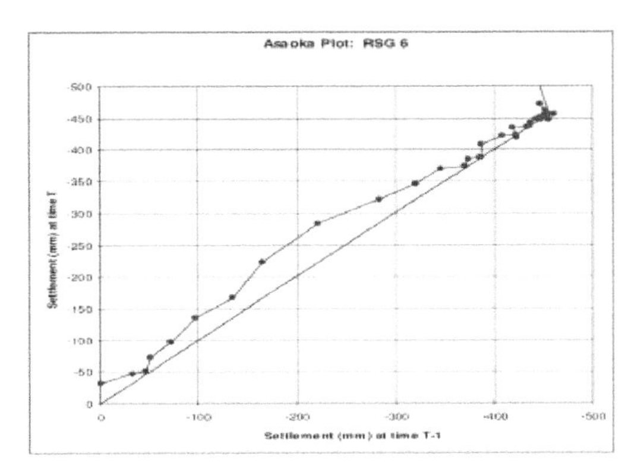

Figure 3a. Typical Asaoka Plots (settlement).

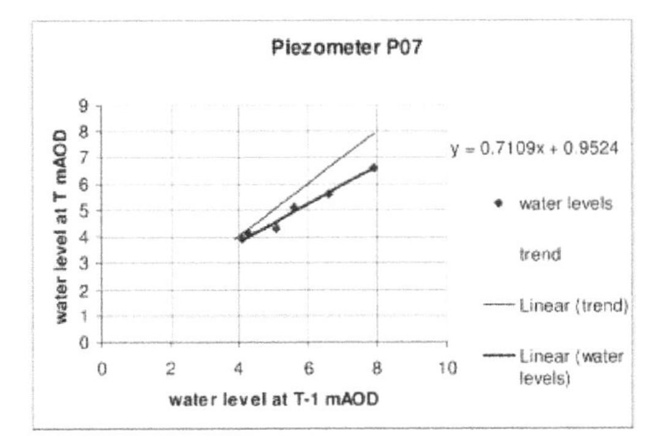

Figure 3b. Typical Asaoka Plots (water level).

porewater pressures were monitored at fixed intervals and plotted against the previous reading, with the data eventually converging towards a slope with a gradient 1:1 when the primary consolidation was complete. Creep settlement in peaty soils was shown on the plots as a sub-parallel line to the 1:1 gradient, indicating a slow on going movement at a diminishing rate. The coefficient of consolidation was calculated using a formula based on the slope angle of the straight line trend in the Asaoka plot, drain spacing and time interval used. For reason of space this formula is not reproduced here (see Holts et al, 1991 for further details). Examples of Asaoka plots used for the back analyses are shown on Figures 3a and 3b.

The results indicate values of the horizontal coefficient of consolidation (c_h) of between 10 and 20 m^2/year for the section of embankment with band drains. Similar values were obtained using the relationship of c_h with the Time Factor T_h for 90% consolidation and the spacing of drains. The back analysis of five piezometers using an Asaoka plot for the water level readings with time indicated field values of between 9 and 17 m^2/year. These field values are up to three times greater than the value used for design, and up to five times the mean value from laboratory tests. This difference can be attributed to the horizontal microfabric of the alluvium with several sand layers within the matrix, which permits greater horizontal drainage than assumed in a design based on oedometer samples. The area without band drains indicated a field coefficient of consolidation of 11 m^2/year.

The back analysed coefficient of consolidation in the stone column area is between 5 and 10 m^2/year, lower than in the band drain area. The reason for this lower value could be attributed to remoulding produced by the installation of the columns, which may destroy the fabric.

The main purpose of the stone columns was to reduce settlement in the vicinity of the bridge abutments, rather than accelerating consolidation. Total settlements measured within the stone column area were of the order of 300 mm, within the band drain area 300–760 mm, and 25–350 mm in areas of no treatment. A comparison of settlement between a zone of embankments with stone columns adjacent to a zone with band drains is shown on Figure 4. The settlement within the stone column area was approximately 50% of that of the adjoining band drain area, indicating the effectiveness of the stone columns in reducing the settlement significantly, providing a transition between the band drain zone and the piled bridge abutments.

Horizontal deflections measured in inclinometers ranged between 30 mm and 170 mm, measured within the zone of soft clay and silt at about 2 m to 5 m below the base of the embankment. Where band drains were installed, the average horizontal to vertical ratio when comparing inclinometer readings against nearby settlement gauges was 0.19, whereas in areas of no band drains this ratio increased to 0.34. In the stone column area, a ratio of 0.12 was observed, based on average readings.

The hydraulic piezometers, installed in groups of three at different levels at each location, proved to be an extremely useful and reliable tool for ensuring that the porewater pressure assumed in the design was not exceeded during construction. Significant acceleration of the programme was allowed where no increase of pore water pressure was observed. However, at certain locations during embankment construction, one piezometer would show a rapid build up of water pressure whereas the other two of the same group, probably installed in more permeable layers of the alluvial soils, would show very little change in water level (see Figure 5). Placement of additional fill was therefore delayed until an acceptable pore pressure had been obtained to guarantee an adequate factor of safety for the next lift (in excess of 1.1).

The surcharge was maintained over a period of between 4 to 9 months, depending on its location along the embankment. The recorded settlements of between

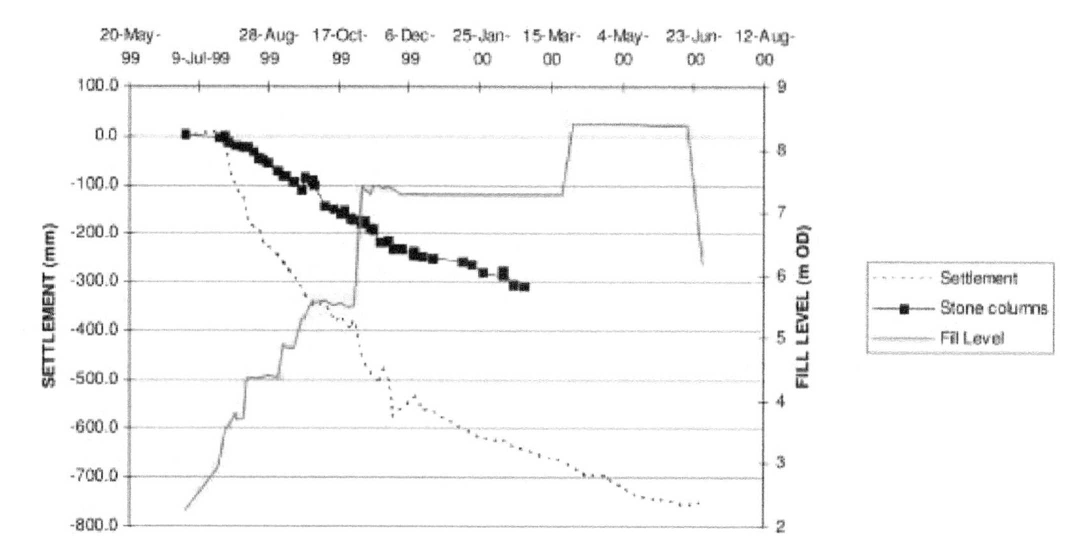

Figure 4. Comparison between settlement under stone column and band drain areas.

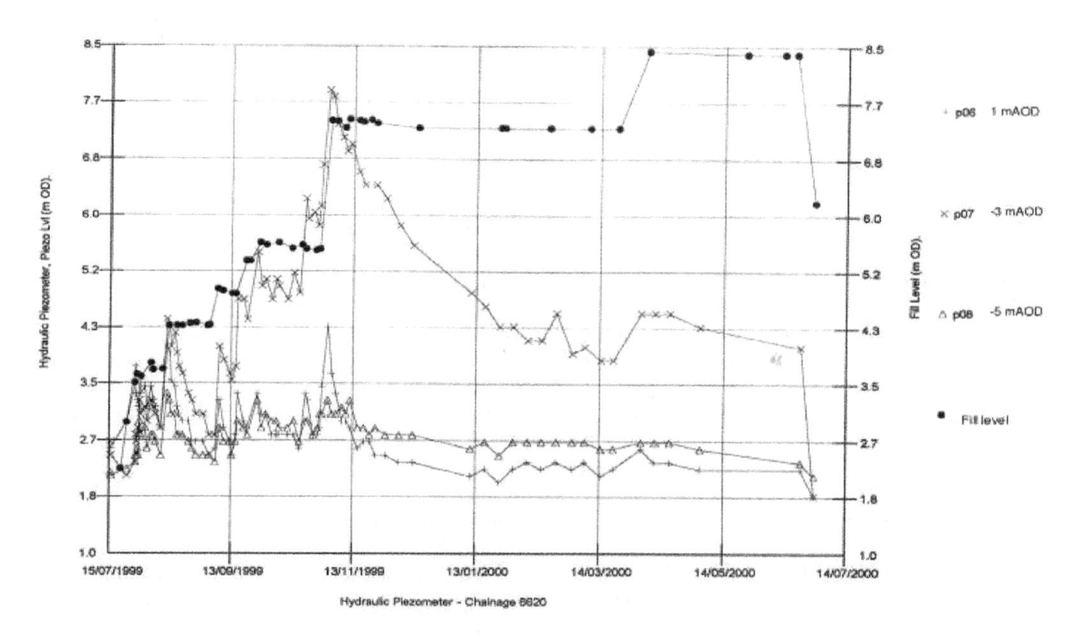

Figure 5. Example of hydraulic piezometer readings.

300 and 760 mm, were generally close to predicted values. In particular, long-term creep was of concern within an area east of Afon Cefni where the silts and clays were significantly softer than expected. To solve this problem, an additional 800 mm surcharge was placed over an extended period of 2.5 months (within the 4 to 9 month period stated above). In this way, creep that would occur in the future could be reduced by ensuring that it took place within a timescale that would fit the pavement maintenance schedule. The placement of this additional surcharge caused sufficient overconsolidation to reduce the long-term creep to acceptable limits within the intervention period for resurfacing of 20 years.

5.4 *Bridge piling*

No piling was allowed to commence until the following criteria were met:

- horizontal movement measured by the inclinometers was substantially complete (taken as less than 10 mm projected long term movement)
- vertical consolidation settlement of the surcharged embankment immediately adjacent the proposed piles was 90% complete

This was required to avoid excessive bending moments in the piles due to horizontal soil displacement from primary consolidation of the compressive soils.

The bridge is underlain by 10–12 m of alluvium followed by glacial till to approximately 21 m depth, and carboniferous mudstones beneath. The original design was to use continuous flight auger (CFA) piles into the glacial till. However, following the experience with the stone columns as described above, some concern was raised about the possibility of necking of CFA piles during construction. For this reason, precast piles, 305 mm × 305 mm were chosen, to carry the 800 kN design load plus 110 kN of downdrag.

It had been anticipated that the precast piles would encounter refusal within the dense sands and gravels in the glacial till, where the SPT values recorded in boreholes were high. However, during the trial drives it became apparent that no reasonable set (better than 2 mm/blow) could be achieved within the till. This was confirmed by the pile driving analyser (PDA) used during the driving of the trial piles, which indicated there was insufficient ultimate dynamic capacity for the proposed loading. For this reason, the piles were driven to a depth of approximately 21–23 m into the carboniferous mudstones (−18 mOD to −20 mOD). The pile capacity was confirmed by static load tests on the preliminary trial piles, and CAPWAP analyses on both the preliminary piles and 19 working piles. Working piles were also monitored for uplift, but no movement was observed.

6 CONCLUDING REMARKS

The instrumentation of the embankment at Malltraeth Marsh proved to be a valuable tool to manage the staged construction of the embankment. Back analyses of the data indicated that the horizontal fabric of the alluvial clays provided a higher horizontal drainage than assumed in design.

The use of stone columns at the interface with the piled abutment foundations provided a smooth transition with the adjacent area of embankment with band drains.

The procurement of granular backfill with low angle of friction for integral bridges needs to be addressed at design stage, to ensure that the material exists in the vicinity of the site.

For future projects, better quality material should be reserved for the top of the embankments to reduce movement problems at sub-formation. Other methods such as the use of lime could also be considered.

The work was continuously reviewed by geotechnical engineers on site, which allowed a successful interaction between designer and contractor on site. Innovative solutions such as the use of trials with fully loaded trucks on varying thicknesses of capping would probably have never happened on a traditional contract.

ACKNOWLEDGEMENTS

Thanks are due to the Transport Directorate of the Welsh Assembly for permission to publish this paper.

REFERENCES

Highways Agency. 1996. HA42/96: *The Design of Integral Bridges. Amendment No 1. Design Manual for Roads and Bridges. Volume 1A. 1996.*

Holtz, R.D., Jamiolkowski, M.B., Lancellotta, R., & Pedroni, R. 1991. *Prefabricated Vertical Drains: Design and Performance. CIRIA Ground Engineering Report: Ground Improvement. Butterworth- Heinemann.*

Matsuo M. & and Kawamura K. 1977. *Diagram for construction control of embankments on soft ground. Soils and Foundations. Japanese Society of Soil Mechanics and Foundation Engineering. Vol. 17 no 3. September.*

Advances in Transportation Geotechnics – Ellis, Yu, McDowell, Dawson & Thom (eds)
 ISBN 978-0-415-47590-7

The effect of subsoil support in plane strain finite element analysis of arching in a piled embankment

Y. Zhuang, E.A. Ellis & H.S. Yu

Nottingham Centre for Geomechanics, School of Civil Engineering, University of Nottingham, UK

ABSTRACT: The results of a series of linearly elastic-perfectly plastic plane strain finite element analyses to investigate the arching of a granular embankment supported by pile caps over a soft subsoil are reported. The analyses demonstrate that the ratio of the embankment height to the centre-to-centre pile spacing is a key parameter.

1 INTRODUCTION

The concept of 'arching' of granular soil over an area where there is partial loss of support from an underlying stratum has long been recognised in the study of soil mechanics (eg. Terzaghi, 1943). Terzaghi considered shear stress on the vertical interfaces originating from the rigid supports at either side of a 'trapdoor' where support was reduced. This approach has found some application in the analysis of piled embankments (Russell & Pierpoint, 1997), where the pile caps act as rigid supports, and the underlying soft subsoil as the 'trapdoor', Figure 1. Other approaches for the design of such structures include a semicircular arch in the granular fill, as initially proposed by Hewlett & Randolph (1988) (and developed by Low et. al., 1994; and Kempfert et. al., 2004), and analogy with backfill over a buried pipe (BS 8006, 1995). Notably there is not one generally accepted approach for design (Love & Milligan, 2003). The theories tend to initially consider plane strain conditions, which are then modified to account for the 3-dimensional nature of most piled embankments which use a square (or triangular) layout of pile caps in plan. This paper will consider a plane strain analysis, as the logical starting point which can be later extended to 3-d. The possible inclusion of a single or multiple layers of geogrid reinforcement in the embankment is not considered at this stage.

The above methods consider that there is sufficient tendency for the soft subsoil to settle that arching of the embankment material will occur, but they do not specifically link arching with the amount of support from the subsoil. An interesting contrast to this is the concept of a 'ground reaction curve' (GRC) used to determine the load on a plane strain underground structure such as a tunnel, Figure 2(a). Figure 2(b) shows the form of GRC proposed by Iglesia et. al. (1999) based on geotechnical centrifuge tests. p is the support pressure from the roof of the underground structure to the soil above, presented in non-dimensional form as

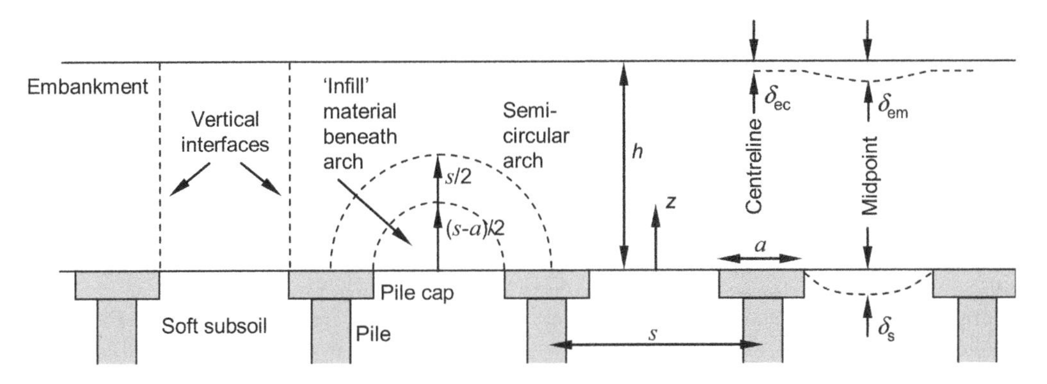

Figure 1. Piled embankment showing potential arching mechanisms, and notation for geometry and settlement (δ) used in this paper.

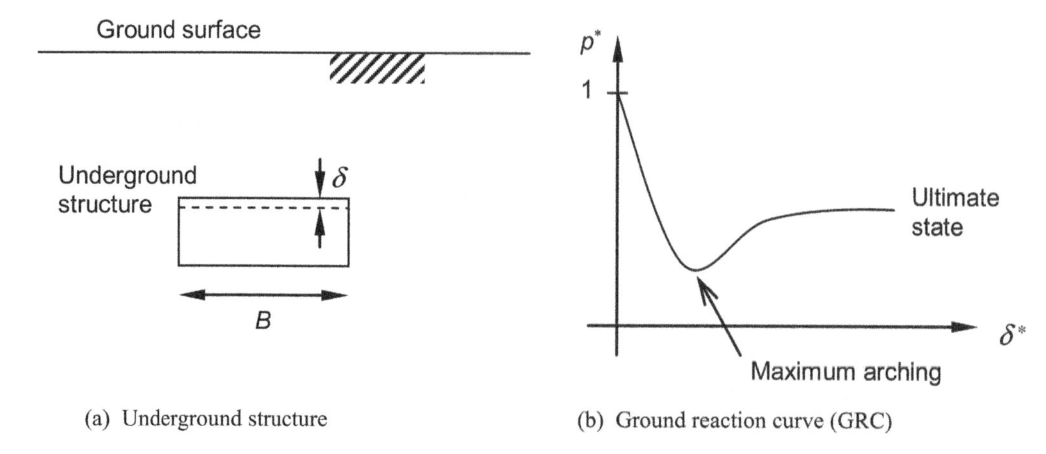

Figure 2. Ground reaction curve for underground tunnel (Iglesia et. al., 1999).

$p^* = p/p_0$ where p_0 is the nominal overburden total stress at the elevation of the roof derived from the thickness of overlying soil (and any surcharge at the ground surface). Variation of p^* is plotted against $\delta^* = \delta/\mathrm{B}$, where δ is the settlement of the roof and B is the total width of the underground structure.

Initially $p^* = 1$, but as δ^* increases arching develops in the soil above the underground structure. A point of 'maximum arching' (minimum stress on the structure) is reached where Iglesia et. al. consider a parabolic (similar to semicircular) arch to form in the soil above the structure. However, as δ^* increases further 'brittle' response is observed as p^* increases again to an ultimate state where shear on vertical planes at the edge of the structure is used to determine p, giving a higher load than the parabolic arch. The value of δ^* to reach the point of maximum arching is stated to be about 2 to 6 %.

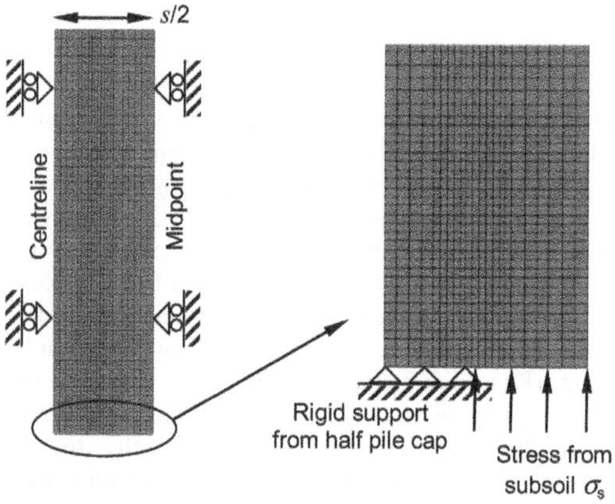

Figure 3. Typical finite element mesh ($h = 5$ m, $s = 2.5$ m) and boundary conditions.

2 ANALYSES PRESENTED

The analyses reported here were undertaken in plane strain using Abaqus Version 6.6. Figure 3 shows a typical mesh for the embankment, whose vertical boundaries represent lines of symmetry at the centreline of a support (pile cap), and the midpoint between supports. Hence there is a restraint on horizontal (but not vertical) movement at these boundaries. No boundary conditions are imposed at the top (embankment) surface, and no surcharge is considered to act here. The bottom boundary represents the base of the embankment, which is underlain by a half pile cap (width $a/2$) on the left, and subsoil (width $(s-a)/2$) at the right. The pile cap is assumed to provide rigid restraint to the embankment, whilst the vertical stress in the subsoil supporting the embankment (σ_s) is used to control the analysis – the subsoil itself was not actually modelled. The pile cap width (a) was fixed at 1 m

Table 1. Standard material parameters for granular embankment fill.

Young's Modulus (MN/m^2)	Poisson's Ratio	c' (kN/m^2)	ϕ' (deg)	Kinematic dilation angle (ψ) (deg)
25	0.2	1	30	0

and the centre-to-centre spacing (s) was 2.0, 2.5 or 3.5 m. The embankment height (h) was varied using values in the range 1.0 to 10 m.

The embankment material was assumed to be granular (and hence with predominantly frictional strength), and modelled using the linear elastic and Mohr Coulomb (c', ϕ') parameters shown in Table 1. For $s = 2.5$ m the effect of increasing ϕ' to 40°, or

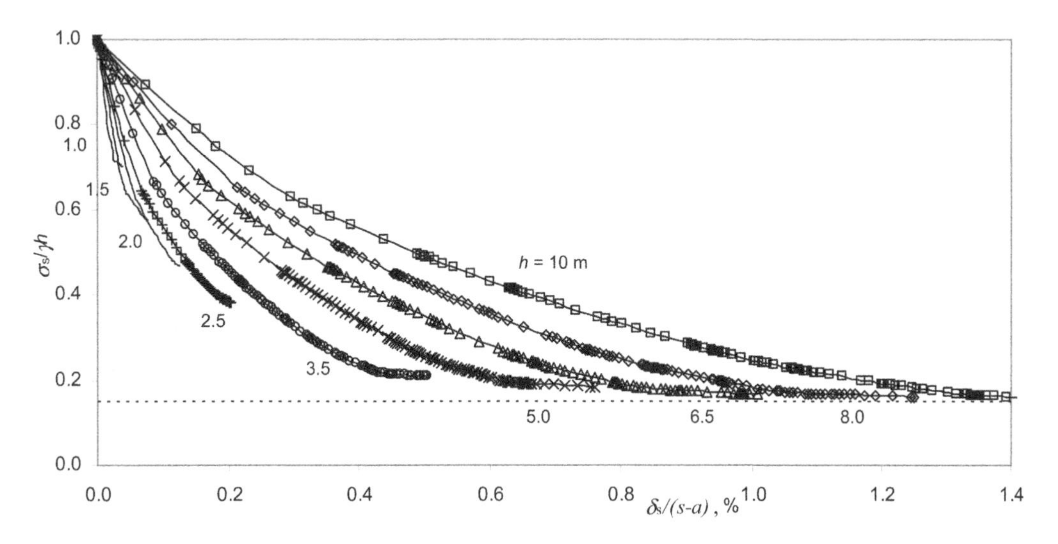

Figure 4. Ground Reaction Curves for the standard soil parameters (Table 1), $s = 2.5$ m, and a variety of embankment heights (h).

increasing the kinematic dilation angle (ψ) to 22° was also considered.

The sequence of analysis was straightforward. First the in-situ stresses were specified (based on a unit weight of 17 kN/m^3 and a K_0 value of 0.5) using the 'Geostatic' command in Abaqus. Initially σ_s was specified as the nominal vertical stress at the base of the embankment to give equilibrium with the in situ stresses, but this value was then reduced (generally allowing Abaqus to determine increment size automatically) to mimic loss of support from the subsoil. The subsoil in question is generally of low permeability, and thus this process has direct analogy with consolidation of the subsoil, which causes arching of the embankment material onto the pile caps.

3 RESULTS

3.1 *Ground reaction curves*

Figure 4 shows normalised ground reaction curves (GRC) equivalent to Figure 2(b) but using slightly different notation. Because the analysis was controlled by reducing σ_s, corresponding settlement of the subsoil (at the base of the embankment) increased from zero at the edge of the pile cap to a maximum value at the midpoint between pile caps (Figure 1). The maximum value at the midpoint will now be referred to as δ_s, and is thus slightly different to the definition in Figure 2(a) (where δ is constant). Data points are shown at the values of σ_s generated by the automatic incrementation in Abaqus, and thus become more dense towards the end of the analysis as plasticity is more prevalent and there is more difficulty in achieving convergence.

The GRC curve is modelled up to the point of maximum arching. Using displacement (rather than stress) controlled analyses it was found that at larger displacements a constant value of σ_s was observed, rather than the subsequent increase exhibited in Figure 2(b). It was concluded that the post-maximum stage of the GRC would only be observed in the finite element analysis if brittle soil behaviour was introduced, and it was chosen not to introduce such complexity at this stage. Nevertheless, the analysis does give the stress at maximum arching, and the displacement required to reach this point.

Figure 4 shows results for the standard soil parameters (Table 1), $s = 2.5$ m, and a variety of embankment heights (h). The highest (10 m) embankment requires the largest displacement to reach the point of maximum arching, but even here the normalised displacement is only slightly larger than 1%. However, this value is directly related to the soil stiffness which has been chosen – the value was doubled for an analysis with half the soil stiffness.

The ultimate normalised stress is in the range 16 to 20% for $h \geq 3.5$ m, but tends to increase rapidly as h reduces below this value. These results will be discussed further in a broader context below.

3.2 *Midpoint profile of earth pressure coefficient*

It was found that the earth pressure coefficient (K) plotted on a vertical profile at the midpoint between piles (the right-hand boundary of the mesh in Figure 3) gave a good 'illustration' of arching behaviour. Figure 5 shows the profiles plotted with z – vertical distance upwards from the base of the embankment (Figure 1), normalised by s. The profiles as plotted

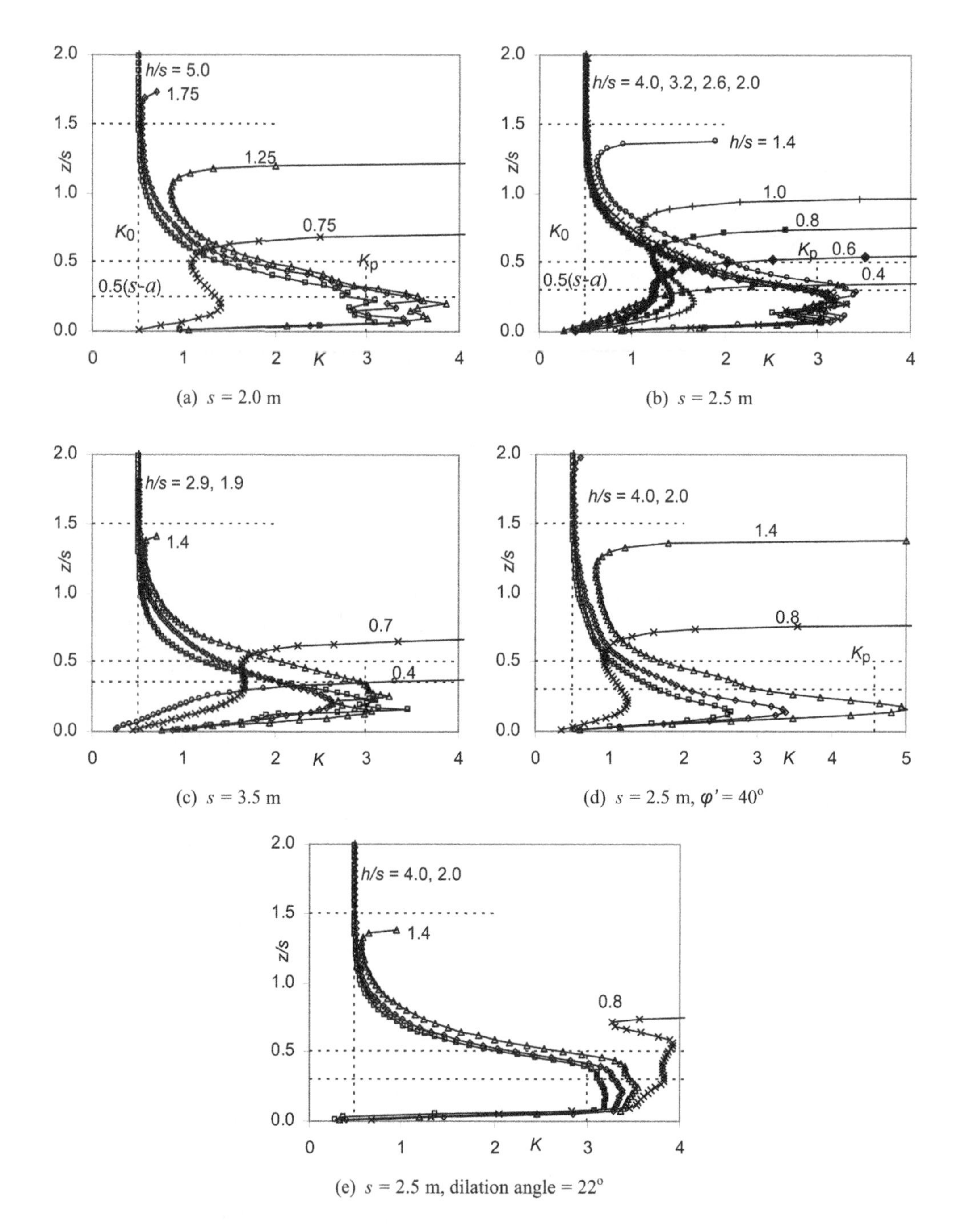

Figure 5. Profiles of earth pressure coefficient (K) on a vertical profile at the midpoint between piles (z measured upwards from base of the embankment, Figure 1), showing variety of embankment heights (h).

do not extend to the top of the embankment for the higher embankments. Values of $0.5(s-a)$, $0.5s$ and $1.5s$ are highlighted on the z axis; and $K = K_0$ and $K = K_p = 3.0$ (taking the standard Rankine value and ignoring the small cohesive element of strength) on the K axis. Subplots (a) to (c) show variation of s whilst (d) and (e) show the effect of increased friction angle and non-zero dilation angle respectively.

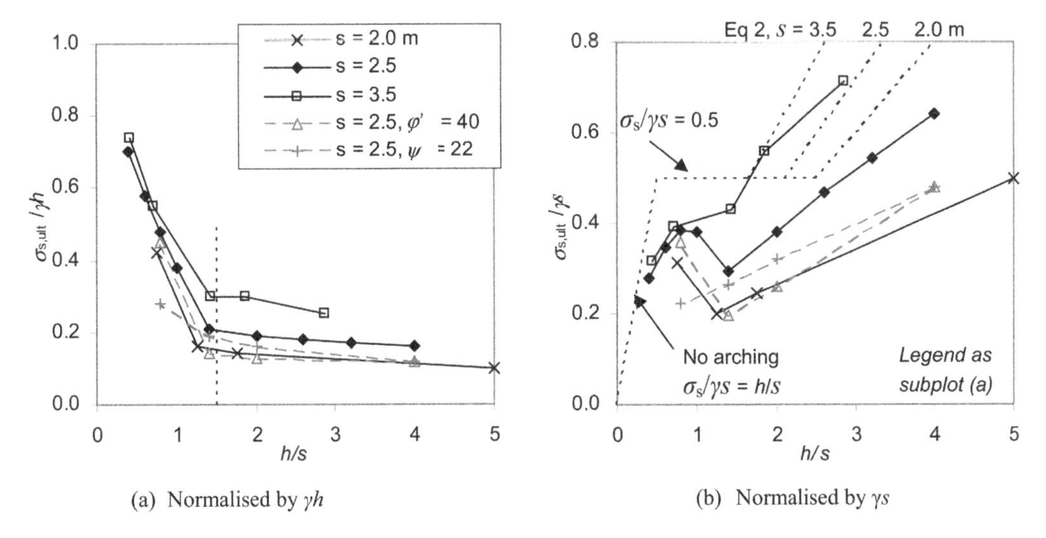

(a) Normalised by γh

(b) Normalised by γs

Figure 6. Normalised stress on subsoil at ultimate conditions ($\sigma_{s,ult}$) showing variation with (h/s).

Referring to Figure 5(b) for $(z/s) > 1.5$, $K = K_0$, and thus has not been modified by the formation of the arch. For embankments where $(h/s) > 1.5$, K increases with depth for $z/s < 1.5$, reaching K_p when $z \approx 0.5(s - a)$. Comparing this with a semicircular arch (Figure 1), the upper limit of the effect of arching is about 3 times higher, but the passive limit is only reached at the inner radius (and below) the arch, where the 'infill' material is evidently in a plastic state.

For embankments where $(h/s) < 1.5$ there is increasing tendency for the highest value of K to occur at the surface of the embankment, initially giving an 'S-shaped' profile, and then monotonic reduction in K with depth in the embankment for the lowest h. In fact K_p as indicated on the plots neglects the small cohesion intercept, and thus can be exceeded, particularly when stress is small (eg. near the surface of the embankment or immediately above the subsoil).

Subplots (a) and (c) ($s = 2.0$ and 3.5 m respectively) show trends of behaviour which are similar to (b). When $s = 3.5$ m there is some reduction in K at $z = 0.5(s - a)$ for the largest h, perhaps reflecting an increased tendency for failure of the arch at the pile cap rather than the 'crown' (top of arch) for large s (Hewlett & Randolph, 1988). This trend is supported for $s = 2.0$ m, where there would be increased tendency for failure at the crown, and where K is high at and below $z = 0.5(s - a)$.

Subplots (d) and (e) show the effect of increased friction angle and non-zero dilation angle respectively. The data again show similar trends. The higher K_p for the increased friction angle is only fully mobilised when h is close to the 'critical value' of $1.5s$, and K is generally quite considerably less than K_p for $z = 0.5(s - a)$, particularly for the higher embankments. The non-zero dilation angle slightly promotes the tendency for K_p (for the standard friction angle) to be mobilised when $z = 0.5(s - a)$ compared to subplot (b), and the data shows less fluctuation with depth in the plastic infill zone for $z < 0.5(s - a)$. This probably reflects improved numerical stability in the analysis when plastic strains show a greater degree of normality.

3.3 *Ultimate stress on subsoil*

Figure 6 shows the ultimate stress on the subsoil ($\sigma_{s,ult}$), illustrating variation with (h/s). Subplot (a) shows normalisation of $\sigma_{s,ult}$ by γh (as in the GRC), whilst (b) shows normalisation by γs.

Figure 6(a) shows that for $(h/s) > 1.5$, $(\sigma_{s,ult}/\gamma h)$ reduces slowly as (h/s) increases, but when $(h/s) < 1.5$, $(\sigma_{s,ult}/\gamma h)$ increases rapidly, tending towards 1.0. This behaviour was previously noted in Figure 4. The additional data shown here (variation of s and friction and dilation angles) only show two significant differences: for $s = 3.5$ m, the value at large (h/s) is slightly larger, and the non-zero dilation angle shows less dramatic increase when $(h/s) < 1.5$.

Figure 6(b) shows lines $\sigma_s = \gamma h$ (ie. 'no arching'), and $\sigma_{s,ult} = 0.5\gamma s$. Also shown is a simplified version of the condition for failure of the arch at the pile cap proposed by Hewlett & Radolph (1988). The equation of vertical equilibrium for the 2-d situation, assuming σ_s and σ_c (the vertical stress on the subsoil and pile cap respectively) to be constant is:

$$\sigma_c a + \sigma_s (s - a) = \gamma h s \quad (1)$$

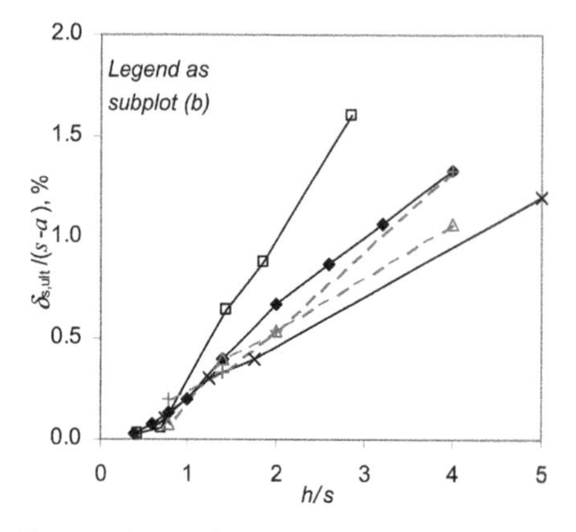

(a) Ultimate settlement of the subsoil at the midpoint between piles ($\delta_{s,ult}$) normalised by the clear gap between pile caps (s-a).

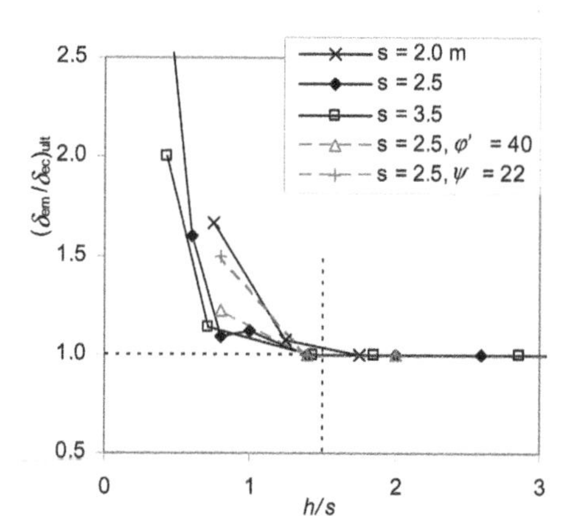

(b) Ratio of the settlement at the top of the embankment at the midpoint between piles (δ_{em}) to the equivalent value at the centreline above the pile cap (δ_{ec}) at ultimate conditions.

Figure 7. Settlement results at the subsoil and surface of the embankment.

It is then assumed (from analogy with bearing capacity) that $\sigma_c = K_p^2 \sigma_s$, to give:

$$\frac{\sigma_s}{\gamma s} = \frac{h}{s} \frac{1}{K_p^2(a/s) + (1 - a/s)} \tag{2}$$

This result is plotted for the 3 values of s.

For small (h/s), $(\sigma_{s,ult}/\gamma s)$ is less than 0.5, and when $(h/s) \approx 0.5$ the data converge with the 'no arching' line. At larger h Equation 2 shows the correct trend of behaviour, but tends to overestimate $\sigma_{s,ult}$, particularly as s reduces.

As would be anticipated, in Figure 6(b) increased friction angle or dilation angle tends to reduce $\sigma_{s,ult}$, for a given value of s.

3.4 *Settlement at the subsoil and surface of the embankment*

Figure 7(a) shows the maximum value of subsoil settlement at the midpoint between piles (Figure 1) required to reach ultimate conditions: $\sigma_{s,ult}$ (this value has been estimated 'by eye' from plots such as Figure 4, and thus is somewhat subjective). The value has been normalised by the clear gap between pile caps $(s - a)$ so that it is analogous to δ^* (Figure 2(b)). Variation with (h/s) is shown.

The clearest trend is that the normalised displacement to reach ultimate conditions increases with (h/s), tending to zero when $(h/s) \approx 0.5$, also corresponding to the point of convergence with the 'no arching' line in Figure 6(b). If there is no arching then no displacement is required to reach this 'ultimate' condition. This is also evident in Figure 4, where there is less tendency for an ultimate 'plateau' at lower h. As h increases arching occurs, and the amount of stress redistribution from the subsoil to the pile cap increases with h, thus it is not surprising that the amount of displacement required to achieve ultimate arching conditions also increases. This observation is also consistent with the variation with s, which indicates more displacement as s increases (for a given h) since this also implies increased redistribution of load from the subsoil to the pile cap.

The absolute magnitude of $\delta_{s,ult}/(s - a)$ is somewhat smaller than the value of δ^* of 2 to 6% quoted for 'maximum arching' by Iglesia et al (1999), which appears to be relevant to $(h/s) \approx 2$ to 5 in the reference. However, the finite element analyses reported here are linear elastic, and the initial gradient of the GRC (Figure 2(b)) implies that the ultimate arching conditions would be reached at a lower value of about 1%. Furthermore, the values shown in Figure 7(a) would vary directly in inverse proportion to the value of Young's Modulus used in these analyses (eg. if the Young's modulus had been reduced by a factor of 2 to better simulate the secant modulus to failure the normalised displacement would be doubled).

The results showed that the ratio of settlement at the top of the embankment at the midpoint between piles (δ_{em}, Figure 1) to the equivalent value in the subsoil (δ_s) at the point where ultimate conditions are reached: $(\delta_{em}/\delta_s)_{ult}$, was typically in the range 0.4–0.6. These values seem reasonable: the settlement at the surface of the embankment is less than beneath the arch.

Figure 7(b) shows the ratio of δ_{em} to the equivalent value at the centreline above the pile cap (δ_{ec}, Figure 1) at the point where ultimate conditions are reached: $(\delta_{em}/\delta_{ec})_{ult}$, showing variation with (h/s). This is a measure of differential settlement at the surface of the embankment, which is of considerable practical importance in terms of piled embankments. For $(h/s) > 1.5$ the value is 1.0, indicating no differential settlement. As (h/s) reduces differential settlement increases, dramatically so for (h/s) less than about 0.75.

4 CONCLUSIONS

The results of a series of linearly elastic-perfectly plastic plane strain finite element analyses to investigate the arching of a granular embankment supported by pile caps over a soft subsoil have been reported. The analyses demonstrated that the ratio of the embankment height to the centre-to-centre pile spacing (h/s) is a key parameter:

- $(h/s) \leq 0.5$ there is virtually no effect of arching: 'ultimate' conditions are reached almost immediately (with very small displacement) in the analysis, relative differential settlement at the surface of the embankment is very large, and the stress acting on the subsoil is virtually unmodified from the nominal overburden stress.
- $0.5 \leq (h/s) \leq 1.5$ there is increasing evidence of arching: as (h/s) increases the displacement required to reach 'ultimate' conditions increases, relative differential displacement at the surface of the embankment reduces, and the stress acting on the subsoil reduces compared to the nominal overburden stress.
- $1.5 \leq (h/s)$ 'full' arching is observed: the displacement required to reach ultimate conditions continues to increase and a clearly defined ultimate state is maintained at large displacement. There is no differential displacement at the surface of the embankment, and the stress acting on the subsoil is considerably reduced compared to the nominal overburden stress. For a high embankment the stress state is not significantly affected above a height of $1.5s$ in the embankment.

Furthermore, it has been shown (Figure 6(b)) that up to a critical value of (h/s) the stress on the subsoil is less than $0.5\gamma s$, approximately representing the effect of the infill material below the arch. At higher values of (h/s) conditions at the pile cap are critical and Equation 2 can be used to conservatively estimate the stress on the subsoil.

ACKNOWLEDGEMENTS

Yan Zhuang is financially supported by a Dorothy Hodgkin Postgraduate Award at the University of Nottingham. This support is gratefully acknowledged.

REFERENCES

BS8006 1995. Code of Practice for strengthened/reinforced soils and other fills. British Standards Institution.

Hewlett W.J. & Randolph M.F. 1988. Analysis of piled embankments. *Ground Engineering*, April 1988, pp. 12–18.

Iglesia G.R., Einstein H.H., & Whitman R.V. 1999. Determination of vertical loading on underground structures based on an arching evolution concept. *Proc. 3rd National Conference on Geo-Engineering for Underground Facilities.*

Kempfert H.G., Gobel C., Alexiew D. & Heitz C. 2004. German recommendations for reinforced embankments on pile-similar elements. *Proc. 3rd European Geosynthetics Conf*, 279–284.

Love J. & Milligan G. 2003. Design methods for basally reinforced pile-supported embankments over soft ground. *Ground Engineering*, March 2003, pp39–43.

Low B.K., Tang S.K. & Choa, V. 1994. Arching in piled embankments. *ASCE Journal of Geotechnical Engineering*, 120(11) 1917–1938.

Russell D. & Pierpoint N. 1997. An assessment of design methods for piled embankments. *Ground Engineering*, November 1997, pp39–44.

Terzaghi K. 1943. *Theoretical Soil Mechanics*. John Wiley & Sons, New York.

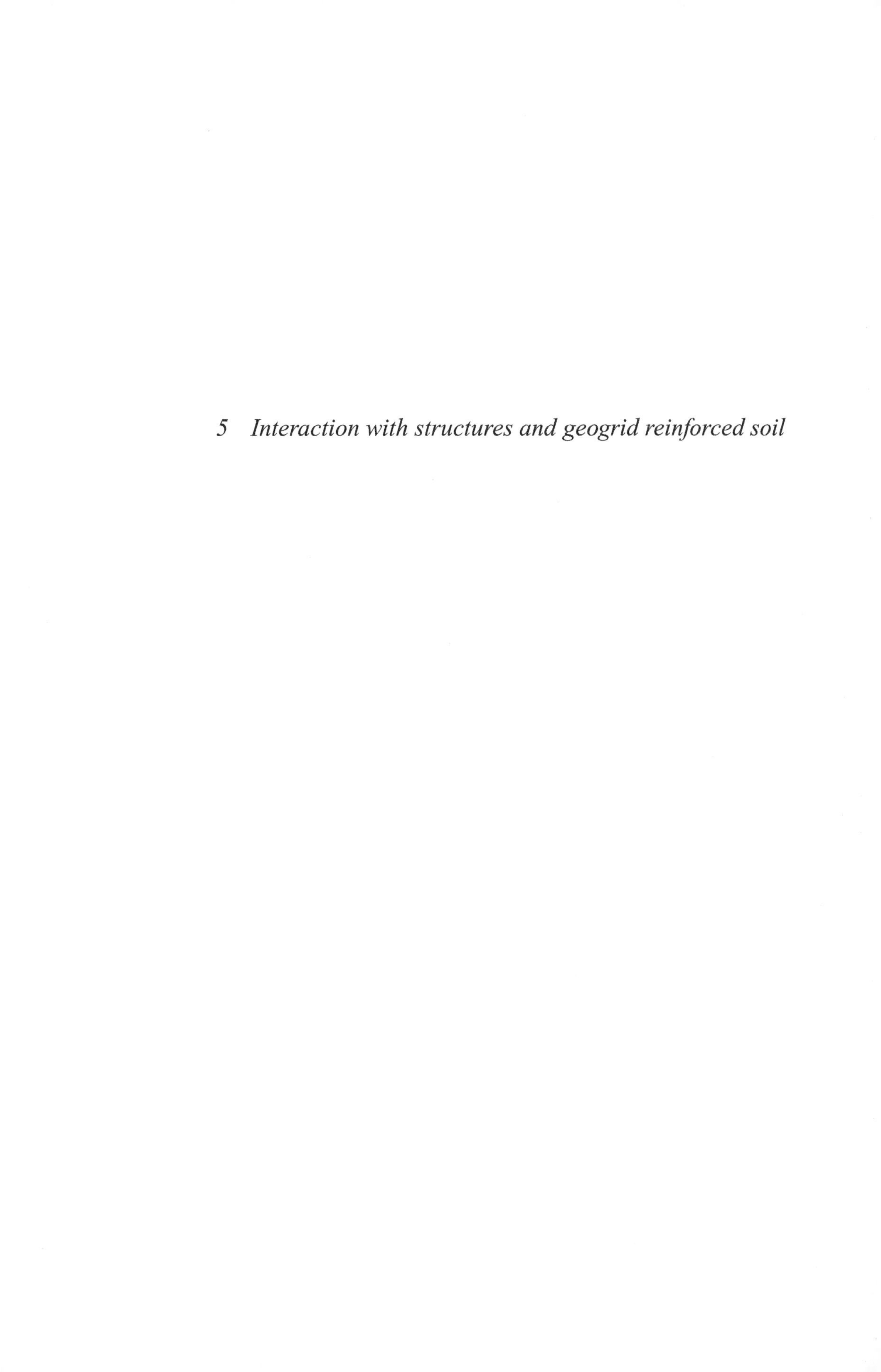

5 *Interaction with structures and geogrid reinforced soil*

Advances in Transportation Geotechnics – Ellis, Yu, McDowell, Dawson & Thom (eds)
 ISBN 978-0-415-47590-7

Design of reinforced soil structures using fine grained fill types

J.M. Clancy & P.J. Naughton
Institute of Technology, Sligo, Ireland

ABSTRACT: Large volumes of fine grained fill, which are not generally considered suitable for construction, are generated on construction sites. It is estimated that 500 million tonnes of construction and demolition (C&D) waste is generated annually in the European Union (EU), with over 11 million tonnes generated in Ireland in 2004. It was estimated that in 2001, 38% of the total C&D waste produced was excavated soil. Current design codes do not preclude the use of fine grained fill containing high proportions of fines (<63 μm) in reinforced soil structures. This material can be successfully reused in reinforced soil structures, which can tolerate some deformation of the face and settlement at the crest. The deformation and settlement of the structure can be reduced by incorporating a drainage component in the body of the structure. The properties of four typical soils, which would not normally be considered as backfill material, are reviewed in terms of their classification, strength and consolidation characteristics. Design guidance on utilizing fine grained fills in traditional reinforced soil design processes is presented. Finally possible applications of this soil, using the modified design procedures, are identified.

1 INTRODUCTION

Traditionally free draining granular fills are used in the construction of reinforced soil structures. This is due to their high strength and ability to prevent the development of excess pore water pressures. Zeynep (1992) reported the use of granular fill to be the most expensive component of a reinforced soil retaining system, usually corresponding to about 40% of the total construction cost.

The main concerns with using fine grained fills in reinforced soil structures is the potential build up of pore water pressure in the reinforced block results in lower shear strength than granular fill and reduced bond between the soil and the reinforcement which may result in deformation and settlement of the structure. Poor draining soils are also more difficult to compact when the moisture content is high, resulting in longer construction periods (Zornberg & Mitchell, 1994).

Large volumes of fine grained soils are disposed of each year as there are little uses for them on site. Annon. (2007a) has reported that an estimated 500 million tonnes of construction and demolition (C&D) waste is generated annually in the European Union, with over 11 million tonnes generated in Ireland in 2004. It was further estimated that in 2001, 38% of the total C&D waste produced that year was excavated soil (Annon, 2007a). Total construction and demolition waste for England was estimated at 89.6 million tonnes in 2005. 46 million tonnes were recycled and a further 15 million tonnes were spread on exempt sites with the remaining 28 million tonnes were sent to landfill as waste (Annon, 2007b).

Many studies have focused on the use of fine grained, cohesive fills in the construction of reinforced soil structures. The Transport Research Laboratory (TRL) investigated the use of cohesive soil in a soil wall with various forms of impermeable reinforcement such as plastic and steel strips. The wall itself consisted of upper and lower layer of sandy clay with a middle layer of granular fill. Its early performance was described by Murray and Boden (1979). High excess pore water pressures developed in the clay during construction, resulting in large deformations occurring.

Liu et al (1994), reported on a 12 m test embankment constructed of cohesive fill. The embankment was divided into four sections, one unreinforced, and the other three with differing geogrid reinforcements. The development of pore water pressures was linked to placement of the fill material. The pore pressures increased as fill was placed and dissipated when this activity ceased. It was also shown that as fill height increased, the rate of pore water dissipation decreased. Significant settlement of the embankment was only observed when the embankment was fully constructed to its final height.

The effectiveness of non-woven geotextiles in the reinforcement of steep clay slopes was assessed by

Table 1. Treatment of cohesive fills by various codes of practice.

Code of Practice	Requirements
B.S. 8006 (1996) U.K.	Cohesive fills may be used in new or reinstated slopes in combination with the appropriate reinforcement
HA 68/94 (1994) U.K.	Does not prohibit the use of cohesive fills
FHWA (2001) U.S.A.	Permits the use of soils with up to 15% passing the No. 200 sieve (0.075 mm)
Geoguide 6 (2002) H.K.	Permits the use of soils with up to 30% passing the No. 200 sieve (0.075 mm)

Tatsuoka & Yamouchi (1986). Two large scale embankments were constructed using Kanto loam, volcanic ash silty clay, which is common in Japan. Embankment I used the geotextile reinforcements at different vertical spacing on each side while Embankment II used the same vertical spacings but different reinforcement lengths. In Embankment I, larger horizontal and vertical deflection was noted on the slope with the bigger vertical spacings. The same result was noted for the slope with shorter reinforcement at Embankment II. The performance of the geotextile as reinforcement was assessed when the embankments were demolished and a cross-section of the structure was visible. Large cracks were evident only in the unreinforced sections while only minor hair cracks were observed in the top soil layers of the reinforced zones.

2 APPROACH TO FINE GRAINED FILLS ADOPTED BY CODES OF PRACTICE

Design codes treat the coarse and fine grained soils used in backfills differently. Both U.K. documents, BS 8006 (1995) & HA 68/94 (1994), offer no gradation limits for the material used, with BS 8006 (1995) stating that cohesive fills are permitted providing adequate reinforcement is used. The Federal Highway Administration (FHWA, 2001) document does provide a gradation limit for a maximum proportion of fines, as does Geoguide 6 (2002) although its limit is not as stringent, as illustrated in Table 1.

3 RESEARCH ON THE USE OF A DRAINAGE COMPONENT WITH POORER SOILS

Much research has been undertaken into the inclusion of a drainage component in reinforced soil structures. These components usually take the form of a novel geocomposite which has the dual functions of reinforcement and drainage. The geocomposite is designed with in-plane drainage which dissipates excess pore water pressure resulting in improved strength and reduced settlement of the structure.

Kempton et al (2000) reported on dissipation and pull-out testing on a geosynthetic in English China Clay, this material being chosen due to its low permeability. Included in the soil mass was the geocomposite with combined drainage and reinforcement capabilities. The study revealed that the geocomposite dissipated excess pore pressure to 20% of its original value in 32 hours. Increased pull-out resistance was recorded in comparison to a similar geosynthetic, but without the drainage component, both after full and partial dissipation of the excess pore water pressures.

Lopez et al (2005) also compared the performance of a geosynthetic with in-plane drainage (Paradrain) to one without (Paragrid). The efficiency of the Paradrain was defined as the difference in pullout strength achieved by the former at the same initial pore pressure. The results showed that Paradrain was more efficient at higher initial pore pressure values.

Zornberg & Kang (2005) showed that the use of in-plane drainage with a geogrid increased the pull-out resistance by approximately 30% compared to a geogrid without drainage capacity for the soil placement and loading conditions used in the testing program.

Boardman (1998) studied the change in the rate of consolidation associated with the Paradrain geocomposite using a modified Rowe cell apparatus. It was shown that the inclusion of the composite geotextile resulted in reducing the drainage path by half, in turn increasing the rate of consolidation. For a soil sample consolidated at $50\,kN/m^2$ the time required for consolidation was reduced from approximately 100 to just 40 hours. A smaller reduction was noted for $100\,kN/m^2$ consolidation pressure. A small improvement in pullout resistance was recorded for the smaller consolidation pressure with larger improvements after consolidation at $100\,kN/m^2$.

Naughton and Kempton (2004) reported on the use of the geocomposite in the reinstatement of a failed slope in Taiwan. The original slope had failed due to a combination of pore water pressure build-up during typhoon season, poor drainage and soil conditions. For the reinstatement the silty clay from the failed slope was reused in combination with the geocomposite meaning considerable cost saving as expensive granular fill did not have to be imported. The geocomposite allowed rapid dissipation of the pore water pressures and permitted the work to be carried out in only three weeks during the typhoon season.

Heshmati (1993) reported on how the geocomposite improves the physical properties of fine grained cohesive soils. The study concentrated on the shear strength

Table 2. Classification properties of soils included in testing program.

Soil	Plastic limit (%)	Liquid limit (%)	Plasticity index (%)	Unit weight (kN/m^3)*	Percentage passing 63 μm sieve
A	21.3	43.2	22.1	17.505	31
B	48.1	69.2	21.1	13.190	50
C	23.0	42.5	19.5	16.083	42
D	25.7	52.5	26.8	15.495	58

*Corresponding to > 95 % of maximum dry density.

Table 3. Shear strength and consolidation properties of soils included in testing program.

Soil	$(\varphi)'$	c'	Coefficient of consolidation, C_v (m^2/year)	Coefficient of volume compressibility, m_v (m^2/MN)
A	33.0	9.7	2.12–25.64	0.02–2.25
B	26.3	35.8	0.97–30.19	0.02–0.74
C	27.7	10.6	0.30–27.65	0.01–0.15
D	18.6	21.5	1.37–28.83	0.02–1.88

properties of kaolin provided by the drainage function and separately by the reinforcing function of a number of geotextiles. The interaction between the cohesive soil and the geosynthetics were monitored using a electron scanning microscope. The use of geosynthetics improved both the cohesion and shear strength of the clay, with cohesion being improved five-fold. Unexpectedly it was found that the use of a composite geosynthetic did not improve the shear strength which was possibly due to water being maintained between the drainage and reinforcing layers with failure probably along this plane. It was proposed that a geocomposite in which the reinforcement is embedded within the drainage geotextile would produce a positive result instead.

4 PROPERTIES OF FINE GRAINED SOILS

Four fine grained soils were examined in this study. The classification, strength and consolidation characteristics are presented in Tables 2 & 3. All testing was carried out in accordance with BS 1377 (1990). The strength and consolidation characteristics were determined from incremental consolidated direct shear and oedometer testing on recompacted samples prepared at optimum moisture content using standard compaction. The soils were selected randomly from construction sites in the Northwest of Ireland.

Soil A was taken from a site which was being infilled by surplus excavation material and is classified as a clay of intermediate plasticity. Soil B, a silt of high plasticity, was extracted from approximately 2 m below ground level from a site on which a local authority machinery yard is being constructed. The origin of Soil C is similar to Soil A, being surplus excavation material deposited on infilled land. It was classified as a clay of intermediate plasticity. Finally, Soil D was excavated from an area where development was taking place and the soil was not suitable for reuse there. Its classification is a clay of high plasticity. The four soils selected are considered representative of fine grained excavated soil waste available on construction sites. The values of coefficients of consolidation and volume compressibility, determined from incremental oedometer testing, show significant scatter which could be attributed to the sample preparation technique and gradation characteristics of the soils.

5 DESIGN CONSIDERATIONS FOR REINFORCED SOIL STRUCTURES USING FINE GRAINED SOIL

Designers generally specify granular material as backfill due to its excellent strength and drainage characteristics. Particular consideration needs to be paid to the drainage conditions in both the short and long term when using fine grained fills. Christopher et al (1998) recognised that there were different design criteria when considering reinforcement-drainage geocomposites in marginal fills. Three adverse conditions were identified in the design of reinforced soil structures using poorly draining material.

- Condition 1: *Generation of pore water pressures within the reinforced fill.* Excess pore water pressures can build up in poorly draining soils during construction, particularly during the placement of load on the soil, e.g. compaction. A permeable reinforcement can dissipate these pressures owing to their secondary function as a lateral drain.
- Condition 2: *Wetting front advancing into the reinforced fill.* Post-construction infiltration into the backfill soil may result in the loss of shear strength of the soil. This infiltration is possible due to the formation of tension cracks on the surface of the soil. A geocomposite with in-plane drainage capability can drain these cracks when they reach down to the first reinforcement layer.
- Condition 3: *Seepage Configuration within the reinforced soil.* Seepage into the reinforced fill can come from adjacent ground or from fluctuations in the water table. Again this adverse condition could be countered by the lateral drainage ability of the geosynthetic.

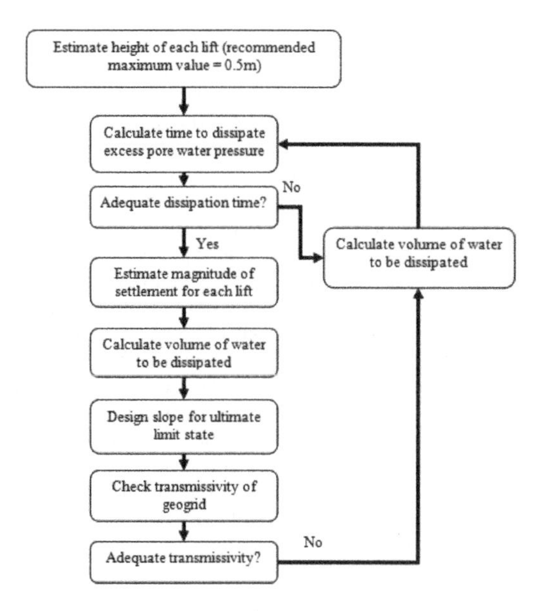

Figure 1. Flow chart for design of steep slopes using fine grained backfill, after Naughton et al (2001).

The design philosophy for reinforced soil structures suggested by Christopher et al (1998) was for a two-phase analysis. Firstly, a total stress analysis was performed for each of the three conditions ignoring the drainage contribution given by the geocomposite, and secondly, an effective stress analysis was performed for each condition taking the contribution to drainage into account. Conditions 2 and 3 can be addressed by the inclusion of adequately designed drainage system underneath and behind the reinforced soil block.

A design method for steep slopes constructed from cohesive fills and an innovative geocomposite was proposed by Naughton et al (2001). This design method aims to dissipate any excess pore water pressure present in the slope during the construction stage which will result in an increase in shear strength of the fill and enhanced bond between the reinforcement and fill. By dissipating the pore water pressures during construction required adjustments to the slope due to vertical and horizontal displacements can be made as construction proceeds. This approach also results in a one stage stability analysis, an effective stress analysis, as the excess pore pressure is dissipated fully before construction of subsequent layers.

The authors proposed a simple flow chart which set out the steps to be taken in designing the slope, illustrated in Figure 1.

Naughton et al (2001) proposed a limit of 0.5 m of the height of each lift to control short term stability of the slope face. The authors calculated the dissipation time based on the coefficient of consolidation and applied an appropriate factor of safety to account

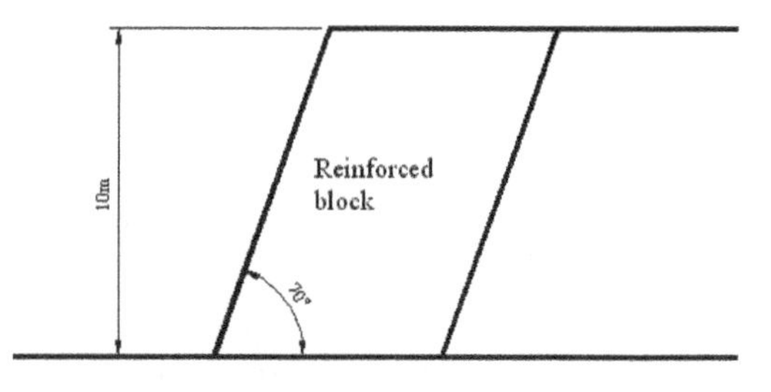

Figure 2. Slope used in design examples.

for unforeseen events. The settlement of each lift was shown to be related to initial height of the lift, the coefficient of volume compressibility and the change in the vertical effective stress. The volume of water to be dissipated could then be determined from the magnitude of settlement assuming a saturated soil. The slope was then designed using an effective stress analysis for the ultimate limit state. The required transmissivity of the geogrid could be calculated once the time for consolidation and volume of water leaving the soil are known. This can then be compared to the available transmissivity. If the transmissivity provided by the geogrid is insufficient, the height of the lift should be reduced and the design procedure repeated.

6 STABILITY OF REINFORCED SLOPES USING FINE GRAINED FILLS

The method proposed by Naughton et al (2001) was used to design steep reinforced slopes using each of the four soil types investigated in this study. The critical element in the design was to select a lift height that would facilitate construction of a single lift in a period of approximately 24 hours. A construction time of 24 hours per layer is considered appropriate for steep slopes of short to medium length.

Figure 2 illustrates the slope dimensions used in the designs, a 10 m high steep slope inclined at 70° to the horizontal. The bulk unit weight of the foundation fill was taken as 18 kN/m^3.

Once the height of lift was selected to give the required dissipation time the stability was checked in accordance with BS 8006 (1995) using a limit equilibrium approach using Oasys Slope 18.2 (Oasys, 2007). The reinforcement used for all slopes was a polyethylene/polypropylene geogrid, with initial short term strength of 100 kN/m. The partial material factors, corresponding to a design life of 120 years, for the geogrid are presented in Table 4. The analysis assumed that adequate drainage was provided at each reinforcement lift to facilitate double drainage during dissipation of excess pore pressures. The interaction between the geogrid and the soil was based on the angle of internal friction and the coefficients given in

Table 4. Partial material factors for the geogrid used in analysis.

Friction interaction	0.6
Adhesion interaction	0.6
Creep reduction	0.27
Manufacture	1.2
Damage	1.1
Environmental	1.0
Extrapolation of test data	1.3

Table 5. Dimensions and results from design procedure.

Soil	Suitable lift height (m)	Longest dissipation time (hours)	Total dissipation time (hours)	Reinforcement length (m)	Factor of safety
A	0.40	28.5	668	6.0	1.201
B	0.40	20.2	462	6.0	1.171
C	0.30	19.7	556	7.0	1.001
D	0.25	16.0	483	10.0	1.031

Table 4. No account was taken of the cohesion of the soil in assessing the geogrid-soil interaction. This is a conservative approach. Table 5 shows the final design of the slope for each soil type.

The lift height suggested by Naughton et al (2001) of 0.5 m was greater than that calculated for any of the soil types examined. Soil A had the longest dissipation time, 28.5 hours, on the lower layers, which overall increased the total dissipation time for the slope, 668 hours, although this could have been shortened by using shallower lifts. Soil B had the shortest length of reinforcement, 6 m, and also the shortest dissipation time, 462 hours. This is due to relatively high C_V values established the materials low unit weight. Soil C required increased reinforcement and a lower lift height to achieve adequate dissipation time and stability, owing to the materials low strength and C_V values at low confining stresses. Soil D proved to be the most unsuitable soil for use in the slope. A much greater length of geosynthetic and a reduced vertical spacing was required to stabilise the structure. Analysis showed that achieving adequate stability was the critical design element for this slope, not the dissipation time. This was possibly due to the soil properties, with the material having a low angle of shearing resistance and the largest proportion of fines, 58% passing 63μm. The properties of this soil would be the most typical for fine grained fills. The large quantity of reinforcement used highlights the problems associated with using fine grained fills in reinforced soil structures.

7 APPLICATIONS OF FINE GRAINED FILLS IN THE CONSTRUCTION SECTOR

Based on the current available knowledge on the use of fine grained soil it is not possible to recommend their use in all steep reinforced slope applications. Care needs to be taken in design to adequately predict the magnitude of vertical settlement and deformation at the face of the slope to meet serviceability limit state requirements (BS 8006, 1995).

Possible applications include non critical structures, structures which can tolerate some deformation without affecting their performance. These types of structures would include noise bunds, landscaping features and other non trafficked structures.

8 CONCLUSIONS

Large volume of fine grained soils are produced on construction sites each year (Annon, 2007a & b). Most of this fill is considered waste and is not routinely reused for the construction of reinforced steep slopes and other soil reinforcement applications.

Research (Kempton et al, 2000, Lopez et al, 2005, Zornberg & Kang, 2005, Boardman, 1998 & Heshmati, 1993) has shown that fine grained soil can be successfully used as backfill material provided adequate drainage is proceed in the body of the structure. Excess pore water pressures, generated using construction, can be rapidly dissipated resulting in increased strength and deformations occurring during the construction period (Naughton et al, 2001). Dual function geosynthetics, combining reinforcement and drainage components offer a practical means of utilising fine grained soils as backfill materials. Existing design method needs to be modified to take account of the properties and problems associated with fine grained fills, especially the need to dissipate excess pore pressure during construction.

The properties of four fine grained soils, typical of the waste materials generated on many construction site, were presented and were shown to have a wide range of strength and consolidation characteristics.

A steep slope was designed using each of the fine grained soil examined as backfill material. The design presented combines a method for determine the maximum height of each lift to allow dissipation of excess pore pressures in a 24 hour period. The stability of the slope was checked using effective stress analysis, as excess pore pressures generated during construction have been dissipated, using conventional slope stability software. Fine grained fills with high percentage of fines and low angles of shearing resistance require longer lengths of reinforcement placed at closer vertical spacing's than that expected from the use of granular free draining fills.

REFERENCES

Annon. 2007a. Department of the Environment, Heritage and Local Government. www.environ.ie retrieved 20/11/07

Annon. 2007b. Department for Environment, Food and Rural Affairs. www.defra.gov.uk retrieved 04/12/07

Boardman, D.I. 1998. Investigation of the Consolidation and Pull Out Resistance Characteristics of the Paradrain Geotextile, Unpublished research report, University of Newcastle, U.K.

BS 1377, 1990. British Standard Method of Test for Soil for Civil Engineering Purposes, British Standard Institution.

BS 8006, 1995. Code of Practice for strengthened/reinforced soil and other fills, British Standards Institution, London

Christopher, B.R., Zornberg, J.G., and Mitchell, J.K. (1998). Design Guidance for Reinforced Soil Structures with Marginal Soil Backfills. *Proceedings of the Sixth International Conference on Geosynthetics, Atlanta, Georgia, March, Vol. 2, pp. 797–804*

Federal Highway Administration (FHWA) 2001. Mechanically Stabilized Earth Walls and Reinforced Soil Slopes. Design and Construction Guidelines, Publication No. FHWA-NHI-00-043 , Washington DC

Geoguide 6, 2002. Reinforced Fill Structure and Slope Design (Geoguide 6), Geotechnical Engineering Office, Hong Kong.

HA 68/94, 1994. Design methods for the reinforcement of highway slopes by reinforced soil and soil nailing techniques, Design Manual for roads and bridges, HMSO, London

Heshmati, S. 1993. The action of geotextiles in providing combined drainage and reinforcement to cohesive soil, University of Newcastle upon Tyne

Kempton, G.T., Jones, C.J.F.P., Jewell, R.A. & Naughton, P.J. 2000. Construction of slopes using cohesive fills and a new innovative geosynthetic material. *EuroGeo 2, Bologna*: 825–828

Liu, Y., Scott, D.J. and Sego, D.C., 1994. Geogrid Reinforced Clay Slopes in a Test Embankment, *Geosynthetics International*, Vol. 1, No. 1, pp 67–91

Lopez, R.F., Kang, Y.C. & Zornberg, J. 2005. Geosynthetic With In-Plane Drainage As Reinforcement In Poorly Draining Soil, *Inter American Conference on Non-Conventional Materials and Technologies in Ecological and Sustainable Construction IAC-NOCMAT 2005 – Rio de Janeiro – Brazil, November 11–15th*

Murray, R.T. & Boden, J.B. 1979. Reinforced earth wall constructed with cohesive fill, *Colloque International sur le Renforcement des Sols*, Vol. 2, Paris, France, pp. 569–577

Naughton, P. & Kempten, G. 2004. Construction of steep slopes using cohesive fill and an innovative geogrid, *International Converence on Geosynthetic and Geoenvironmental Engineering (ICGGE), Bombay, India, December 2004*

Naughton, P.J., Jewell, R.A. & Kempten, G.T. 2001. The design of steep slopes constructed from cohesive fills and a geogrid, *Proceeding of the International Symposium on Soil Reinforcement, IS Kyushu, Japan, November 2001*

Oasys, 2007. Slope 18.2. Oasys Ltd., U.K.

Tatsuoka, F. and Yamouchi, H. 1986, A Reinforcing Method for Steep Clay Slopes using a Non-woven Geotextile, *Geotextiles and Geomembranes*, Vol. 4, pp. 241–268

Zeynep, D. & Tezcan, S. 1992. Cost Analysis of Reinforced Soil Walls. *Geotextiles and Geomembranes*, Vol. 11, No. 1, pp. 29–43

Zornberg, J.G. & Kang, Y. 2005. Pullout of Geosynthetic Reinforcement with In-plane Drainage Capability. *Geosynthetics Research and Development in Progress,Eighteenth Geosynthetic Research Institute Conference (GRI-18), January 26, Austin, Texas*

Zornberg, J.G. and Mitchell, J.K. 1994. Reinforced soil structures with poorly draining backfills. Part I: Reinforcement interactions and functions. *Geosynthetics International*, Vol. 1, No. 2, pp. 103–148

Advances in Transportation Geotechnics – Ellis, Yu, McDowell, Dawson & Thom (eds)
© 2008 Taylor & Francis Group, London, ISBN 978-0-415-47590-7

Back-analysis of a tunnelling case history: A numerical approach

F.M. El-Nahhas & A.M. Ewais
Faculty of Engineering, Ain-Shams University, Cairo, Egypt

Tamer M. Elkateb
Mott MacDonald Limited, Dubai, UAE/Faculty of Engineering Ain-Shams University, Cairo, Egypt

ABSTRACT: This paper presents the results of numerical back analysis carried out for a case history of twin tunnels constructed in downtown Edmonton, Canada. The tunnels were 6.2 m in diameter, with clear spacing of about 5 m, and were constructed in clay till using open face TBM. The analyses were carried out under plane strain conditions using the FLAC 5.0 software. Field measurements were used to calibrate the numerical model used in the analysis. A reduction factor for tunnel lining was used to take into account the effect of lining joints on its stiffness. The analysis results are presented in terms of stress and displacement fields, and surface settlement. The study results indicated that numerical models calibrated using field monitoring records can be effectively used to mimic tunnel behavior and provide insight on the behavior of twin tunnels. The comparison between behavior of twin tunnels and a single tunnel indicated that interaction between twin tunnels has the greatest impact on surface settlement and the smallest influence on stress fields around the tunnels.

1 INTRODUCTION

The use of tunnels in urban areas has increased in recent years, especially for infrastructure projects in heavily populated cities (El-Nahhas, 2006). Tunneling is a relatively complicated engineering problem as it involves soil-structure interaction in addition to the significant influence of the selected construction technique. The complexity gets higher when two tunnels are closely spaced (twin tunnels), a common case in transportation projects.

Several studies have been carried out to study the behavior of twin tunnels. Hoyaux & Ladanyi (1970) conducted a numerical study to determine the stress distribution around twin tunnels in soft soils to help identify the major factors affecting the development of plastic zones. Cording & Hansmire (1985) back analyzed the results of the field instrumentation for twin tunnels of Washington, D.C metro and developed a correlation between additional surface settlement, attributed to construction of the second tunnel, and tunnels separation distance. Chapman et al. (2003) developed an empirical method to determine surface settlement troughs for twin tunnels constructed in stiff soils.

This paper presents the results of a numerical back analysis carried out for twin tunnels constructed through stiff clay till in Downtown Edmonton, Canada. The analyses were conducted in a numerical framework using the FLAC 5.0 software. The numerical model used in the analyses is described together with the method used to calibrate the model using field instrumentation records. Stress and displacement fields, and surface settlement troughs predicted by the model are presented. A comparison is made between the behavior of twin tunnels and that of single tunnel.

2 BACKGROUND ON THE CASE HISTORY

A rapid Light Rail Transit (LRT) system was constructed in the City of Edmonton, Canada to connect the downtown area to the northeast suburbs in the 1970s and later to the south side. The stretch of the LRT passing through the heavily developed downtown area was constructed underground. The underground stations were constructed using the cut-and-cover technique and were connected with two twin circular tunnels about 6.2 m in diameter, and spaced at 11 m centerline to centerline. The tunnels were constructed using a Lovat open face Tunnel Boring Machine (TBM) with the west-bound tunnel, referred to herein as the first tunnel, constructed first followed by the east-bound tunnel, referred to as the second tunnel. The primary lining system consisted of steel ribs spaced at 1.22 to 1.53 m center to centre, and 100 × 150 mm lagging placed between the webs of successive ribs. The secondary lining consisted of cast-in-place reinforced concrete. Detailed information on the lining activation

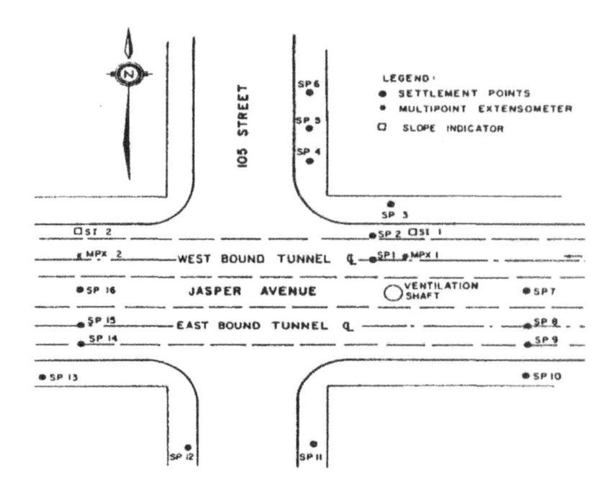

Figure 1. Site plan showing the area under study of Edmonton LRT and instrumentation layout.

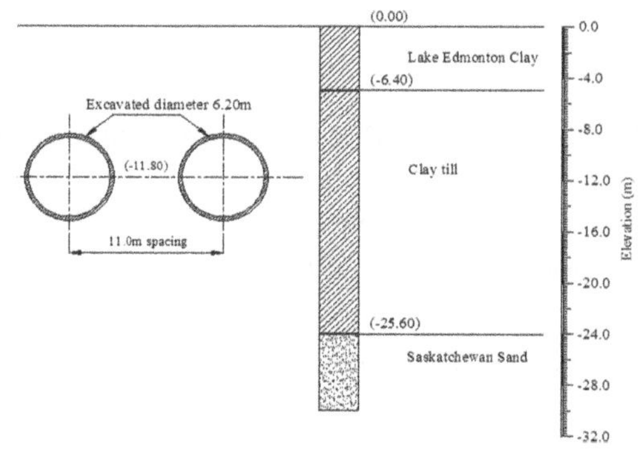

Figure 2. Idealized stratigraphic cross section at the study area.

method can be found elsewhere, e.g. Eisenstein & Thomson (1978) and El-Nahhas (1980).

The geology of downtown Edmonton was described in detail by Kathol & McPherson (1975). The bedrock consists of successive layers of clay shale and sandstone with occasional coal layers. The bedrock surface was eroded resulting in a pre-glacial broad valley with gently sloping sides. Parts of these pre-glacial valleys were later filled with granular deposits, termed Saskatchewan Sand. The advance of ice into the Edmonton area during the late Pleistocene age laid down two relatively similar clay till sheets above the Saskatchewan Sand. The clay till was overlain, later, by glacial clay sediments called Lake Edmonton Clay. This study is concerned with the heavily instrumented section of the LRT located to the east of the intersection of the 105 Street and Jasper Avenue on the North valley wall, as shown in Figure 1.

An idealized stratigraphic cross section at the study area, based on the results of a detailed geotechnical site investigation programs, is shown in Figure 2. The following distinctive layers were identified:

- High plastic clay (Lake Edmonton clay) from ground surface to a depth of about 6.4 m;
- Glacial clay till with some randomly oriented water bearing sand lenses (intra-till sand) from below the Lake Edmonton clay to a depth of about 26 m; and
- Uniformly graded dense sands (Saskatchewan sand).

It can be concluded from Figure 2, that the tunnel was constructed mainly through clay till. Piezometers installed at the study area indicated that groundwater was located well below the tunnel invert. A comprehensive field instrumentation program was implemented at the study area, as shown in Figure 1 (Thurber Consultants, Ltd. 1982).

3 ANALYSIS METHODOLOGY

Numerical back analysis of the twin tunnel behavior in the study area was carried out using the FLAC (Fast Lagrangian Analysis of Continua) 5.0 software (Itasca Consulting Group Inc. 2005). A typical mesh used in the analyses is shown in Figure 3. The model boundaries were taken at considerable distances (two tunnel diameter or greater) from the tunnel openings to minimize their effect on displacements and stress fields. The tunnel lining was modeled using linear elastic beam elements.

Soil was assumed to exhibit linear elastic perfectly plastic behavior, under plane strain conditions, applying the Mohr-Coulomb failure criterion. This simplifying assumption implies that soil was given constant elastic properties that did not vary with stress level or stress path (loading vs. unloading). Soil was also assumed to be under drained conditions, as field records indicated that the behavior of clay till was time-independent since stabilized ground deformations were recorded 3 days after construction (Eisenstein & Thomson 1978). Typical ranges for the engineering properties of the main soil layers were obtained from various studies conducted in the project area. Comprehensive numerical sensitivity analyses were carried out to refine these ranges and to obtain single-valued representative soil parameters together with Lining Reduction Factor (LRF), as outlined below, that would reproduce vertical settlement profile above the first tunnel centerline similar to the one captured in the instrumentation program. The results of the sensitivity analyses are shown in Figure 4.

A summary of the representative soil parameters is provided in Table 1. The elastic modulus of clay till was assumed to be stress dependent following the correlation of Janbu (1963) and hence the value presented in Figure 4 represents the value of the elastic modulus at the depth of the tunnel centerline.

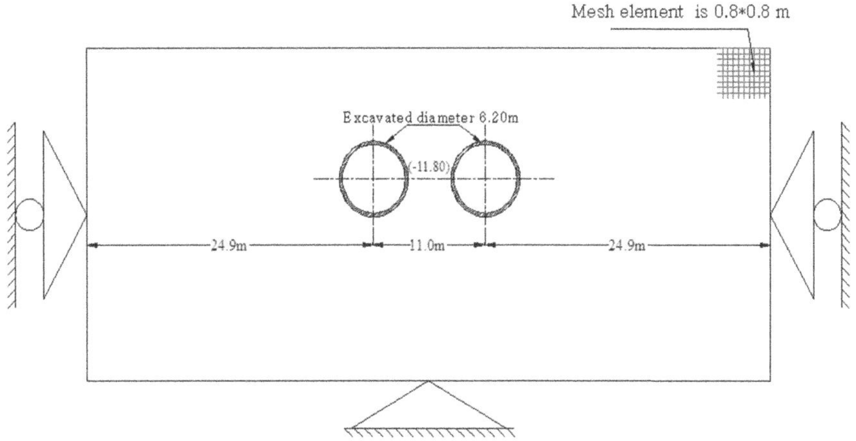

Mesh element is 0.8*0.8 m

Excavated diameter 6.20m

(-11.80)

24.9m

11.0m

24.9m

Figure 3. Schematic diagram showing the typical model used in the numerical analyses.

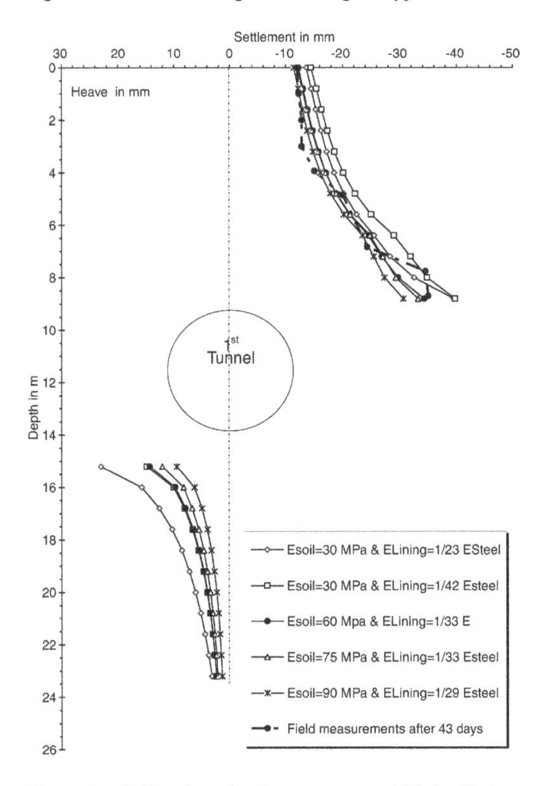

Figure 4. Calibration of soil parameters and Lining Reduction Factor using vertical displacement profiles at the location of the first (westbound) tunnel centerline.

The numerical analyses were performed in four stages to simulate the construction of the twin tunnels. These stages were,

- Stage 1: after excavation of the first (West-bound) tunnel prior to lining installation.

Table 1. A summary of the geotechnical properties for different soil layers.

	Lake edmonton clay	Clay till	Saskatchewan sand
γ(kN/m3)	20	21	19
c (kPa)	5	10	0
ϕ	25	35	40
E (MPa)	15	60*	150
	0.4	0.35	0.3
Ko	1	1	0.25

- Stage 2: after installation of the lining system of the first (West-bound) tunnel.
- Stage 3: after excavation of the second (East-bound) tunnel prior to lining installation.
- Stage 4: after installation of the lining system of the second (East-bound) tunnel.

The stress release method was used in the analysis where a stress release factor (λ) was applied to in-situ stresses to count for soil arching and ground relaxation prior to lining installation. A stress release factor of 50% was selected in this study, which is in close agreement with the finding of Heinz (1984) and Branco & Eisenstein (1985).

A Lining Reduction Factor (LRF) was applied to the lining stiffness used in the analyses to account for the effect of connections, with minimal flexural stiffness, between steel ribs and to avoid numerical inaccuracies that may result from the high contrast between the stiffness of the lining and the surrounding soil. A value of the LRF equal to 0.033 was obtained from the above noted sensitivity analyses. This is in reasonable agreement with the findings of Negro (1988) in a previous numerical tunneling study carried out for tunnels constructed through clay till in downtown Edmonton.

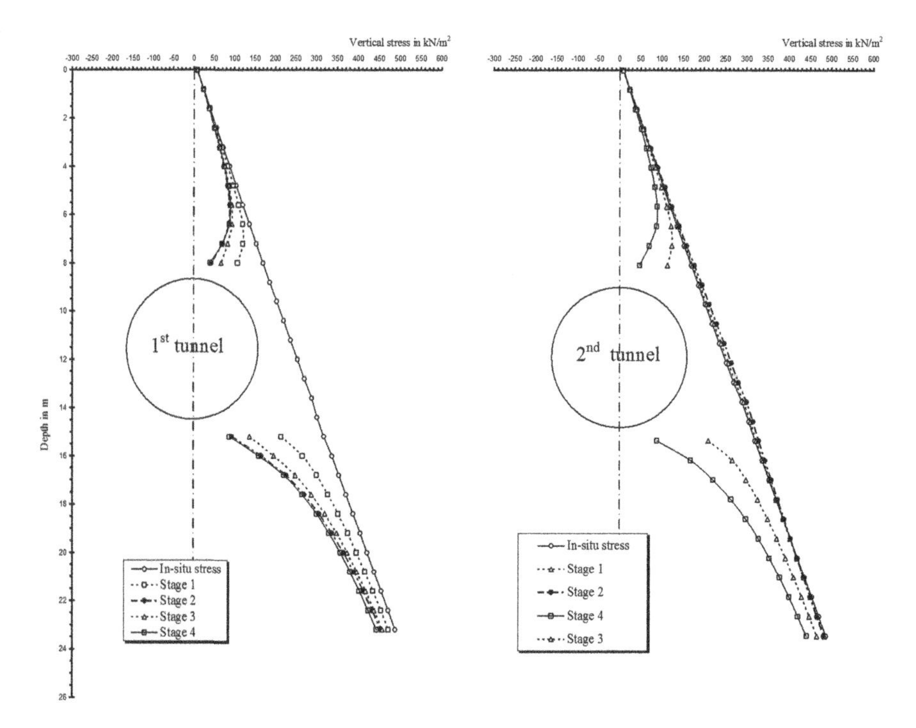

Figure 5. Vertical profiles of vertical stresses at the centerlines of the first and second tunnels.

4 ANALYTICAL RESULTS

4.1 *Stresses field*

Vertical profiles of the calculated vertical stresses at the location of the first and second tunnel centerlines are shown in Figure 5 for the four stages defined in Section 3. As expected, vertical stresses decreased at tunnel crown and invert and increased at tunnel springline.

For the first tunnel, vertical stresses decreased at the crown to about 63%, 25%, 40% and 23% of the initial in-situ vertical stress for stages 1 to 4, respectively. Vertical stresses increased gradually towards ground surface to reach the initial in-situ stress at a distance of about one tunnel diameter above tunnel crown. Vertical stress at the invert decreased to about 67%, 29%, 43% and 28% of the initial in-situ vertical stress for stages 1 to 4, respectively. Vertical stresses increased with depth to reach the initial in-situ vertical stresses approximately at a distance of about one tunnel diameter below tunnel invert. For the second tunnel, vertical stress at the crown and the invert decreased to 67 and 27.5% of their initial in-situ vertical stress for construction stages 3 and 4, respectively.

The above results indicate that, for the conditions of this case study, interaction between twin tunnels would have a minor effect on the vertical stress fields around tunnels compared to those developed from single tunnel construction. The effect of twin tunnel construction on the horizontal stress fields can be found elsewhere, e.g. Ewais (2007).

4.2 *Displacement field*

Vertical profiles of the calculated vertical ground displacements at the centerlines of the two tunnels and mid distance between them are presented in Figure 6. As expected, the maximum vertical displacements occurred immediately above tunnel crown and decrease gradually towards ground surface while the maximum heave occurred at tunnel invert and decreased gradually with depth. The ratio between the maximum vertical displacements at the first tunnel crown and tunnel diameter was equal to 0.17%, 0.51%, 0.5% and 0.61% for stages 1 to 4, respectively. The ratio between the maximum heave at the first tunnel invert and tunnel diameter was equal to 0.09%, 0.18%, 0.178% and 0.21%, for stages 1 to 4, respectively. The ratio between the maximum vertical displacements at the second tunnel crown and tunnel diameter was equal to 0.22% and 0.6% for stages 3 and 4, respectively. The ratio between the maximum soil heave at the second tunnel invert and tunnel diameter was equal to 0.09% and 0.21% for stages 3 and 4, respectively. The above results indicate that mutual interaction between the

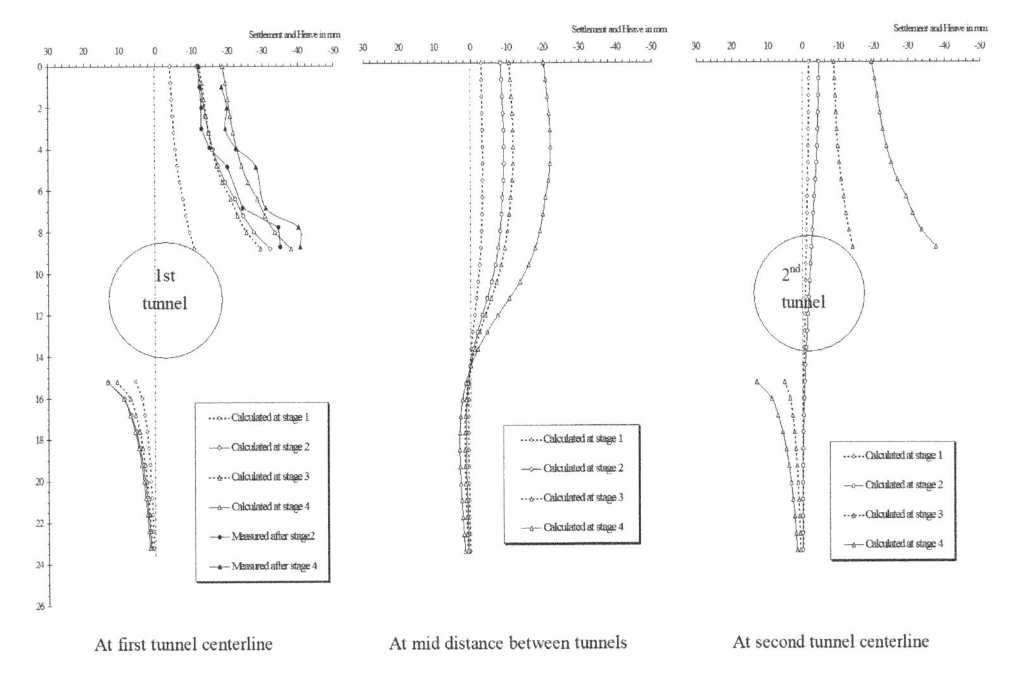

Figure 6. Vertical profiles of vertical displacement at the centerlines of the two tunnels and mid distance between them.

two tunnels resulted in an increase in the final ground settlement and heave at the crown and invert of each tunnel of about 20%, compared to the case of single tunnel.

For horizontal displacement, the ratio between the maximum displacements, occurred at the first tunnel springline, and tunnel diameter was equal to 0.11%, 0.29%, 0.245% and 0.31% for stages 1 to 4, respectively. The ratio between the maximum horizontal displacements and tunnel diameter for the second tunnel were equal to 0.15% and 0.31% for construction stages 3 and 4, respectively.

At mid distance between the two tunnels, horizontal displacement increased with the construction of the first tunnel and then decreased with the construction of the second tunnel to reach a value close to zero at stage 4. This indicates that the effect of construction of the second tunnel on horizontal displacements in the area located between the two tunnels was almost in opposite direction and of equal magnitude to that of the first tunnel.

4.3 *Surface settlement*

Surface settlement troughs were determined for the different stages defined in Section 3. For stage 2, the calculated maximum settlement above the first tunnel centerline was in the order of 11.6 mm (about 0.2% of tunnel diameter) which is in close agreement with the measured field values (12 mm). The settlement trough width was about 7.2 m, which is relatively greater than the measured value (5.1 m). For stage 4, the maximum calculated surface settlement was equal to 20.4 mm (0.33% of tunnel diameter) and was in close agreement with the measured values (17.5 mm). The maximum settlement was located between the two tunnels at distance of about 7.0 m (2/3 the spacing between tunnels) measured from the first tunnel centerline. These results indicate that the surface settlements predicted in this study are generally in good agreement with the field values while the width of settlement trough overestimated by about 40%. It should be noted, however, that overestimation of settlement trough widths have been encountered in several numerical studies of tunneling problems, e.g. Eisenstein (1982) and El-Nahhas et. al, (1997). More attention should be given in future studies to techniques that can be followed to improve the ability of numerical tunneling models to predict settlement trough widths.

The ratio between the calculated surface settlements above the first tunnel centerline and tunnel diameter was about 0.06%, 0.19%, 0.20% and 0.3% for stages 1 to 4, respectively. The ratio between the maximum calculated surface settlements above the second tunnel centerline and tunnel diameter was about 0.03%, 0.07%, 0.14% and 0.31% for construction stages 1 to 4, respectively. The above results indicate that the mutual interaction between the two tunnels resulted in an increase in maximum surface settlement above each tunnel centerline by 60 to 65%, with slightly more increase above the centerline of the second tunnel, compared to the case of single tunnel construction. On the other hand, the absolute maximum surface settlement, between the two tunnels, increased by 75% compared to the case of single tunnel construction.

5 CONCLUSIONS

A numerical back analysis of a tunneling case study that involved the construction of two twin tunnels in stiff clay till was presented. The analysis results were discussed in terms of stress and displacement fields, and surface settlement. The following conclusions can be drawn based on the study results:

1 Numerical analyses calibrated with field measurements can be efficiently used to back analyze tunneling case histories and to predict tunnel behavior in relatively similar construction and geological conditions.
2 A reduction factor of 0.03 should be applied to the stiffness of the tunnel lining to account for the connections (with minimal stiffness) between steel ribs (primary lining) and to avoid numerical inaccuracies that might result from high stiffness contrast between the lining and the surrounding soil. Effort is needed in future studies to provide theoretically sound basis for the selection of the Lining Reduction Factors for different soil and lining types.
3 Tunnel construction in clay till disturbs vertical stresses within a zone of about one tunnel diameter above and below the tunnel opening.
4 Interaction between the twin tunnels results in an increase in vertical ground displacement at tunnel crown and in surface settlement above tunnel centerlines equal to 20 and 60%, respectively, compared to single tunnel construction. Interaction between the twin tunnels has minor impact on stress field around the tunnels.
5 Settlement trough width calculated from numerical analysis could be relatively overestimated. Attention should be given to investigate different techniques that can be adopted to effectively predict settlement trough widths using numerical models.
6 The maximum calculated surface settlement after the twin tunnels construction was located between the two tunnels at distance of about 2/3 the spacing between the two tunnels measured from the first tunnel centerline. Construction of the twin tunnels resulted in an increase in maximum surface settlement by about 75% compared to single tunnels.
7 Attention should be given in future studies to investigate the effect of other constitutive models, which take into account stiffness variation with stress level and stress path, on the predicted tunnel behavior.

REFERENCES

Branco, P. and Eisenstein, Z. 1985. Convergence-Confinement Method in Shallow Tunnels. Proceedings of 11th International Conference. on Soil Mechanics and Foundation Engineering, San-Francisco, 4: 2067–2072.

Chapman, D.N., Rogers, C.D. and Hunt, D.V. (2004), "Predicting the settlements above twin tunnel constructed in soft ground". – AITES World Tunneling Congress, Singapore, Tunnels and Underground Space Technology 19, 378 (CD-ROM).

Cording, E.J. and Hansmire, W.H. (1985), "Soil Test Section: Case History Summary", Journal of Geotechnical Engineering, ASCE, 111: 1301–1320.

Eisenstein, Z. 1982. The Contribution of Numerical Analysis to Design of Shallow Tunnels. Proceedings International Symposium on Numerical Models in Geo-Mechanics, Zurich.

Eisenstein, Z. and Thomson, S. 1978. Geotechnical Performance of a Tunnel in Till. Canadian Geotechnical Journal, 15: 332–345.

El-Nahhas, F. 1980. Behaviour of Tunnels in Stiff Soils. Ph. D. Thesis, Department of Civil Engineering, University of Alberta, Canada, 305 p.

El-Nahhas, F.M. (2006) Tunnelling and Supported Deep Excavations in the Greater Cairo. Int. Symposium on Utilization of Underground Space in Urban Areas, Keynote Lecture, Egyptian Tunnelling Society & International Tunnelling Association, Egypt, pp. 27–56.

El-Nahhas, F.M., Ahmed, A.A. and Esmail, K.A. 1997. Prediction of Ground Subsidence above Tunnels in Cairo. Proceedings of 14th International Conference. on Soil Mechanics and Foundation Engineering, Hamburg, 3: 1453–1456.

Ewais, A.M.R. 2007. Numerical modeling of twin tunnels constructed using open face TBM in stiff clayey soil. M.Sc. Thesis, Department of Structural Engineering, Ain Shams University, Egypt, 171 p.

FLAC (Fast Lagrangian Analysis of Continua) User's Guide (2001), Itasca Consulting Group Inc.

Heinz, H.K. 1984. Applications of the New Austrian Tunneling Method in Urban Areas. M.Sc. Thesis, Department of Civil Engineering, University of Alberta, Canada, 323 p.

Hoyaux, B. and Ladanyi, B. (1970), "Gravitational Stress Field around a Tunnel in Soft Ground". Canadian Geotechnical Journal, 7: 54–61.

Janbu, N. 1963. Soil Compressibility as Determined by the Oedometer and the Triaxial Tests. Proceedings of European Conference on SMFE, Wiesbaden, 1: 19–25.

Kathol, C.P. and McPherson, R.A. 1975. Urban geology of Edmonton. Alberta Research Council Bulletin 32, Edmonton, Alta., 61 p. plus maps.

Negro, A.J. 1988. Design of Shallow Tunnels in Soft Ground. Ph. D. Thesis, Department of Civil Engineering, University of Alberta, Edmonton, 1480p.

Thurber Consultants, Ltd. 1982. South LRT Extension, Geotechnical report No. 14 and No. 15. Instrumentation Test Section at 105th Street. File 14-31-9.

Advances in Transportation Geotechnics – Ellis, Yu, McDowell, Dawson & Thom (eds)
© 2008 Taylor & Francis Group, London, ISBN 978-0-415-47590-7

Numerical modelling of the stability of shallow mine workings

P.R. Helm, C.T. Davie & S. Glendinning
School of Civil Engineering and Geosciences, Newcastle University, Newcastle upon Tyne, UK

ABSTRACT: Numerical modelling work using FLAC 3D has been undertaken to establish the key parameters in the initiation of roof collapse leading to void migration in shallow abandoned workings. The results give an indication of the rockmass strengths required for stability with differing excavation sizes and over-burden loads. Results for varying rock mass strength and stiffness properties indicate that increases in pore pressure, due to increases in the level of the groundwater table, are a significant factor in the initiation of collapse of the excavation roof of shallow mine workings.

1 BACKGROUND

During May and June 2001 (Donaldson Associates, 2002) a number of crown holes (surface subsidence features related to the migration of sub-surface voids) were found near the East Coast Main Line (ECML) track at a site near Dolphingstone. The presence of these crown holes and the possibility of further subsidence at this site prompted Network Rail to divert the ECML over a distance of approximately 1 mile. The diversion required the construction of a continuous reinforced concrete raft supported on end bearing piles and the whole operation cost Network Rail approximately £52 Million. During the site investigation carried out by Donaldson Associates Ltd (2002), it was found that the most likely cause of this subsidence was the collapse of abandoned mine workings beneath the site, leading to the process of void migration. Currently the ground conditions that lead to mine roof collapse and void migration are poorly understood (Healy & Head, 1984).

The instability at Dolphingstone was caused by roof collapse of pillar and stall mine workings. These are mine workings where pillars of the mineral deposit (in this case coal) are left in place to support the roof of the mine (Attewell & Taylor, 1984).

The most common type of pillar and stall workings comprised regular square pillars (Healy & Head, 1984, Waltham, 1989) however the geometry may vary as differing regions developed varying geometries depending on varying local conditions.

1.1 *Void migration and crown hole formation*

Roof failure within pillar and stall workings most commonly leads to surface subsidence in the form of crown holes or depressions. These failures most commonly occur at roadway crossings due to the span of the roof being at a maximum at these points (Attewell & Taylor, 1984).

This collapse may initiate as strata begun to delaminate due to tensile splitting along bedding or lamination planes (Diederichs & Kaiser, 1999). These strata then sag into the excavation.

Ultimately this would result in brittle failure of the roof strata which would collapse into the mine working resulting in the upwards migration of the void. This migration will continue to the surface unless halted by natural arching, choking of the void by collapse debris or by the presence of a high strata within the rock mass above the void. If the void reaches the surface, a crown hole will form.

2 NUMERICAL MODELLING

To investigate the parameters affecting excavation stability, numerical modelling work has been undertaken using the Finite Difference geomechanics code FLAC3D (Cundall, P. A. 2003).

2.1 *Methodology*

A numerical modelling capability has been created to allow the automatic variation of parameters that may influence the initiation and progression of collapse and void migration. In all the modelling discussed here, the models are created with a mesh composed of cubic elements 0.1 m^3 in size. Model boundaries are fixed displacement boundaries. The left and right boundaries are free to displace vertically but not horizontally and the model base is fixed in all directions.

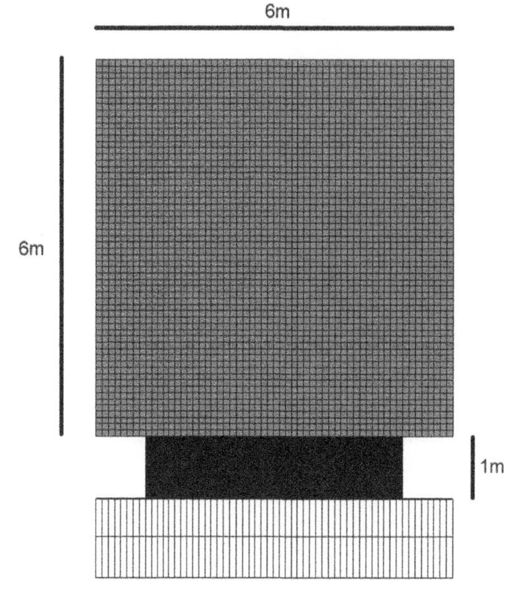

Figure 1. Descritisation of the model over the excavation. Zone sizes are 0.1 m^3.

Figure 1 shows the mesh discretisation for the rockmass overlying the excavation. The grid also extends 10 m below the excavation base. Bulk density and elastic properties are applied to the mesh, and the model brought to equilibrium under gravity. The force ratio limit for equilibrium was set to 1×10^{-5}. This helps to ensure unbalanced forces do not influence the results. Mohr coulomb properties are applied during excavation phases and elastic properties are reapplied during model changes (such as increasing groundwater level) and the model brought back to an equilibrium state to prevent unbalanced forces due to model changes initiating failure.

2.2 *Rockmass strength and stiffness parameters*

The strength and stiffness properties of coal measures strata were derived using ROCLAB software from the company Rocscience which uses the Geological Strength Index (GSI) parameter to estimate rockmass strength and stiffness and the Mohr-Coulomb parameters for use in FLAC. The GSI system was developed to allow reliable rockmass properties to be derived for numerical analysis (Marinos et al 2005). The system allows the modeling properties to be varied to account for variations in the rockmass which can greatly affect stability. This value is based upon assessment of the lithology, the rock structure and the condition of joint surfaces in the rockmass. This GSI value is then input into the software along with estimates of intact strength and stiffness properties. Corrected strength and stiffness parameters are then derived to account for variations in the rockmass.

Table 1. Strength and stiffness rockmass properties..

GSI	100	90	80	70	
K	2.3×10^{10}	2.2×10^{10}	2.0×10^{10}	1.7×10^{10}	Pa
G	7.6×10^{9}	7.3×10^{9}	6.7×10^{9}	5.6×10^{9}	Pa
c	1.1×10^{7}	5.5×10^{6}	2.9×10^{6}	1.5×10^{6}	Pa
Ø	60	62	64	64	
σt	4.4×106	2.1×106	9.8×105	4.56×105	Pa
GSI	60	50	40	30	
K	1.2×10^{10}	$7.0 \times 10^{9}0$	$3.7 \times 10^{9}0$	$1.9 \times 10^{9}0$	Pa
G	3.9×10^{9}	2.4×10^{9}	1.2×10^{9}	6.2×10^{8}	Pa
c	8.5×10^{5}	5.1×10^{5}	3.3×10^{5}	2.4×10^{5}	Pa
Ø	64	63	61	59	
σt	2.2×10^{5}	1.0×10^{5}	4.8×10^{5}	2.3×10^{5}	Pa
GSI	20	10			
K	$1.1 \times 10^{9}0$	$7.0 \times 10^{8}0$	Pa		
G	3.5×10^{8}	2.3×10^{8}	Pa		
c	1.7×10^{5}	1.1×10^{5}	Pa		
Ø	55	49			
σt	1.1×10^{4}	5.0×10^{3}	Pa		

The values derived for coal measures strata are summarised in table 1.

It is important to note that use of the Mohr-Coulomb model along with the GSI value is applicable to bedded strata where failure behavior is not controlled by steeply dipping joint planes as the Mohr-Coulomb model. In cases where there are steeply dipping weakness planes these parameters may over estimate stability.

2.3 *Roof yield behaviour in FLAC*

The initial tensile yielding and resulting sagging described by Diederichs & Kaiser (1999) can be reproduced in FLAC and is demonstrated here by plotting volumetric strain as seen in Figure 2 which represents a 1 m by 1 m excavation with an overburden height off 22 m, rock mass with a GSI of 20. The zone of positive volumetric strains (0.02%) above the excavation represent the initiation of bedding or lamina separation within the rockmass.

As the groundwater level in the model is increased the rate of yielding increases and the roof strata deform to a larger extent. Figure 3 shows the same excavation as Figure 2 with the water table level increased to 5 m above the excavation roof. This results in significant yielding and displacement and is equivalent to the roof strata sagging into the excavation. Total displacement seen in Figure 3 equal to 0.089 m or 89% Strain.

2.4 *Overburden load*

Coal measures strata are typically composed of interbedded sandstones, siltstones and shales. Dry

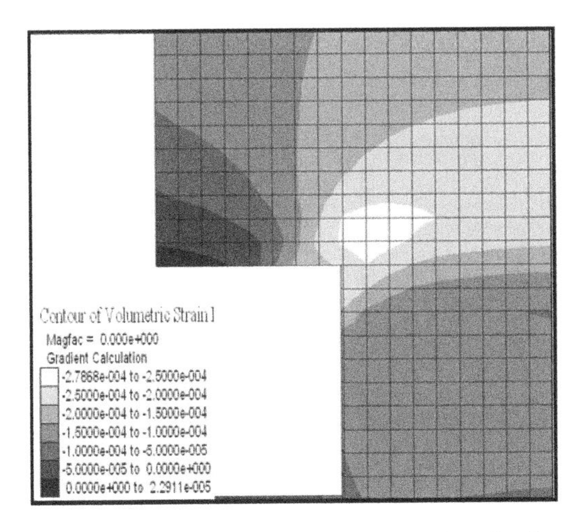

Figure 2. Volumetric strain over the excavation showing tension in the rockmass.

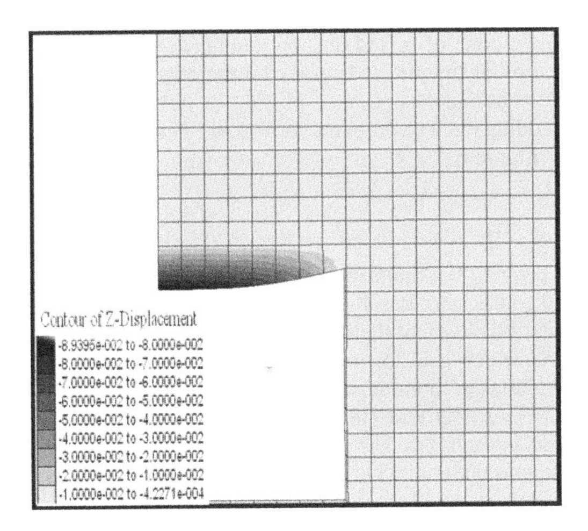

Figure 3. Sagging of the roof strata due to tensile failure within the roof of the excavation.

densities of the theses materials can vary from around 1800 kgm^{-3} up to 2650 kgm^{-3} and with porosity values from 0.05 to 0.3 (Bell, 1992).

As the overburden load on the roof of the excavation is a sum of the bulk density and the overburden thickness, the effect of both the overburden density and the overburden thickness on excavation stability have been investigated.

2.4.1 *Effect on stability of density variations*

To test the effect of density variations of the overburden on the excavation, a series of models were run at varying rock mass strengths, and water levels at a constant overburden load (30 m). The properties used were porosity: 0.3; Dry density: 2000–2700 (kgm^{-3}); Resultant bulk density range: 2300–3000 (kgm^{-3}). The variation in displacements for the range of densities is very low at all groundwater levels. At a GSI of 10 (the lowest rockmass strength tested), the maximum variation occurs at the highest level of groundwater table, where the difference in displacements from an overburden density of 2000 to 3000 kgm^{-3} was equal to 4×10^{-2} m. This is less than 1% of the total displacement and is not considered significant.

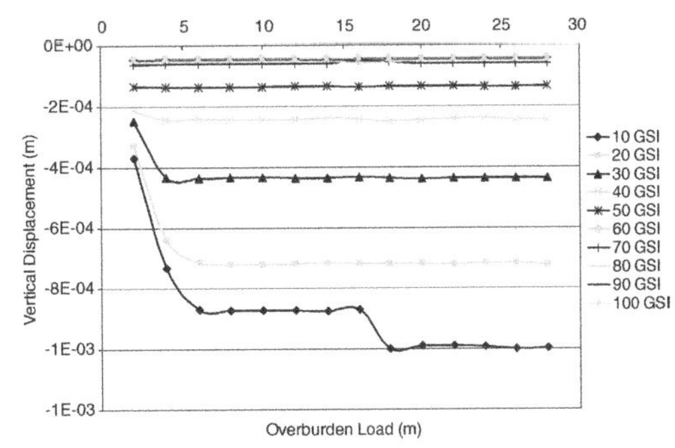

Figure 4. Variations in displacement for varying rockmass strengths and groundwater below the excavation base.

2.4.2 *Effect of overburden thickness variation at varying rockmass strengths*

Excavations with lowered groundwater display uniform displacement within the roof centreline at ranges of rockmass strength from 100 down to 60 GSI and at all overburden thicknesses as can be seen in Figure 4. At GSI values below 60, progressively increased displacements are observed as the overburden thickness increases.

At GSI values between 40 and 10, it can be seen that the initial 4 to 6 m of additional overburden increases the displacement within the roof strata for a given rock mass strength. However additional increases in overburden beyond the first 6 m do not effect roof displacements. The one exception to this is for the rock mass with a GSI of 10 where there is a sudden change in displacements at around 16 m. This corresponds with a limited occurrence of tensile yielding within the models with overburden greater than 16 m thick. Excavations at this depth of overburden with such low rockmass strength would require additional support for stability.

Figure 4 indicates the overburden thickness has virtually no effect on excavation stability above a GSI value of 60. Below this value the influence of overburden load increases progressively as the strength of the rockmass decreases, but only the first 6 m have any effect. Even at the lowest value of rockmass strength tested, the overburden thickness has a limited effect on excavation stability with very low excavation roof displacements ranging from 1×10^{-4} m

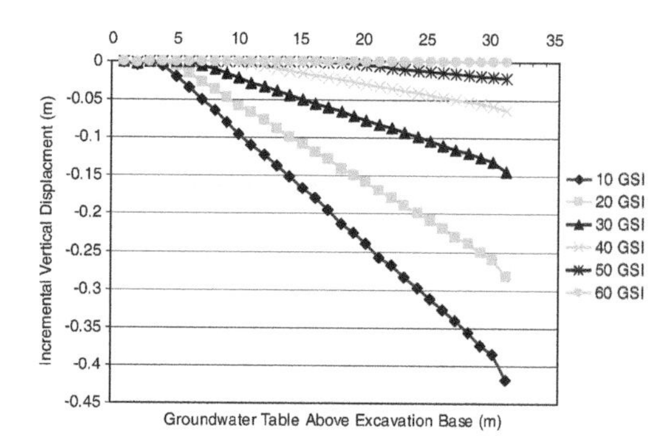

Figure 5. Plot of roof displacement for varying rockmass strengths and water levels.

(2 m overburden) up to 1×10^{-3} m (28 m overburden). It is doubtful that such small displacements would contribute significantly to roof failure or collapse.

2.5 *Effect of groundwater on stability at varying rockmass strengths*

Where excavations (in this case, coal mine workings) are below the water table, pumps are used to lower the groundwater to a level below the excavation to enable mining to be undertaken. Once the mine is abandoned, water extraction is stopped and the water table will return to its original level. This rise will be dependent on the hydrogeologic properties of the local geology. This rise in some cases may be sufficient in itself to promote instability and collapse of excavations particularly if the rockmass strength is low. However the groundwater level may also vary due to seasonal fluctuations in rainfall and due to extreme rainfall events.

In order to assess the effect of a rising groundwater table on the rockmass, a hydrostatic pore water pressure was applied to the model in one metre increments from 1 m below the base of the excavation to the upper boundary of the model grid representing the surface. The density of the zones was also increased to account for the body forces generated by the weight of the water. This was repeated for varying rockmass strengths (GSI 10 to 100) and for varying thicknesses of overburden material (2 to 30 m).

It is readily apparent from Figure 5 that strength values above a GSI of 60 do not result in significant roof displacements at any water level. At rock mass strengths below this value, the excavation may be initially stable (with roof displacements in the millimetre range even for the lowest GSI value). However even small increases in the groundwater table can have a significant impact on the stability of the roof strata, as can be seen by the very large incremental displacement values for 1 m changes of groundwater level.

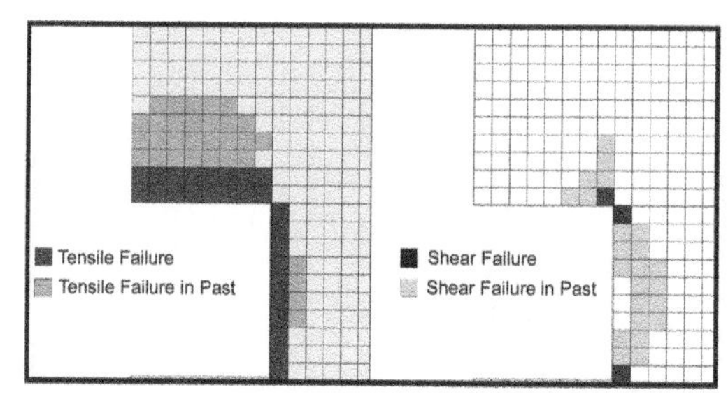

Figure 6. Plasticity state plot of zones around excavation.

For example an excavation in a rockmass with GSI = 20, a change of groundwater level from 20 to 21 m results in an additional 0.15 m of displacement.

This instability in the model takes the form of plastic yielding due to the tensile stress within the excavation roof exceeding the tensile strength of the material. A plot of the failure state of zones in the roof of the model shows the zone of tensile failure within the roof, and a zone of mixed tension and shear failure above the pillar margin. It is assumed the tensile failure in this case represents the splitting or delamination of bedding planes within the roof strata. This can be seen in Figure 6.

2.6 *Numerical modelling of void migration*

Currently the majority of the modelling work undertaken has been based on the identification of parameters which affect the initial stability of an excavation and which lead to roof collapse. Work is also being undertaken to model void migration to the surface.

FLAC 3D is a finite difference continuum code, and as such only has limited ability to model discontinuous materials. Therefore a custom function is being written that will remove failed elements from the excavation roof to simulate the roof collapse and allow void migration to the surface.

In this case the failure or collapse criterion is based initially around detection of tensile yielding of the elements comprising the roof of the excavation. To do this FLAC uses a Mohr-Coulomb constitutive model with tensile cut off (Cundall, 2003) to assess shear and tensile yielding within the rockmass. Once tensile yield is detected within an element, the customised code assesses the grid point displacement at the base and top of the element. The difference between these two displacements is used to calculate the elements vertical strain. If this exceeds a set parameter the element is removed from the model. Currently work is being undertaken to assess the influence of this strain parameter on the modelling of void migration. An initial trial of this function based around the site geology encountered at Dolphingstone produced a collapse progression resulting in a sinkhole chimney.

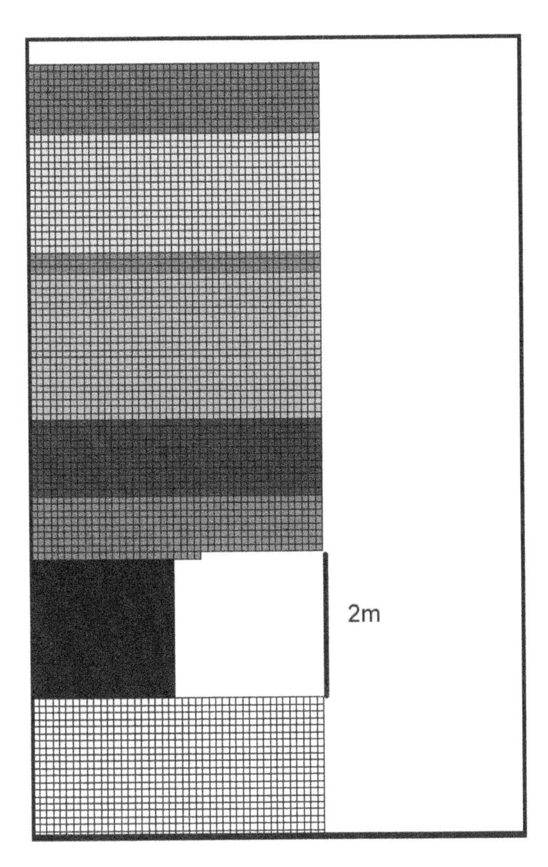

Figure 7. Dolphingstone model prior to modeling of collapse progression.

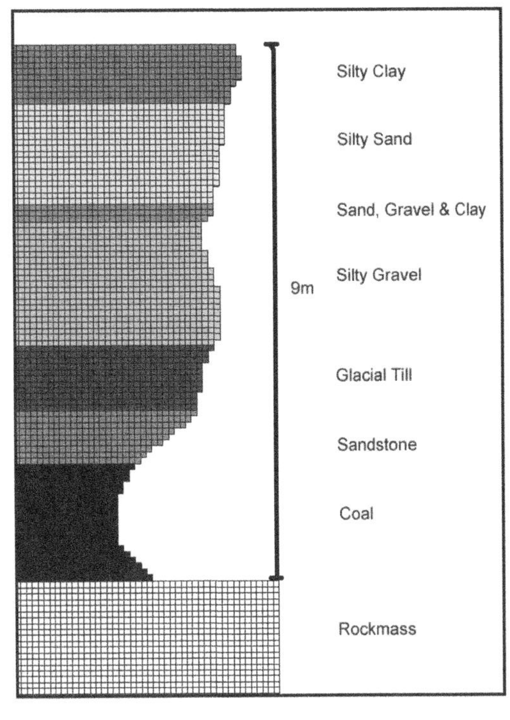

Figure 8. Plot of Dolphingstone geology and collapse chimney.

In this case the model represented a 2 m extraction height with 2 m wide rooms and equal sized pillars (75% extraction ratio). The initial model geometry can be seen in Figure 7 and the final collapse geometry is shown in Figure 8.

It can be seen in Figure 8 that the angle of caving varies depending on the strata through which the void migrates. With the shallowest angle within the sandstone layer with the caving angle increasing as the void migrates into the superficial deposits above the sandstone.

An empirical relationship between caving angle and the uniaxial compressive strength of the rockmass has been suggested by Das (2000). It may be possible to use this model to investigate this relationship further and use it as a tool to estimate the height of caving before the formation of a natural arch occurs in varying strength materials. Further to the above, the diameter of the crown hole produced in the model (approx 1 m) very closely matches those encountered at the site in Dolphingstone.

Along with the parametric study to assess the influence of the strain criterion on the collapse progression and ultimate geometry, work is being undertaken to allow the emplacement of elements into the excavation to represent the collapse roof material. It hoped that this will represent an effective method of modeling void migration using continuum geomechanics software.

3 CONCLUSIONS

Numerical modelling work on the influence of varying parameters on the stability of shallow abandoned mine workings indicates that the influence of overburden density and thickness have a limited influence on the stability of excavations at shallow depths. By far the largest influence on roof stability is the level of the groundwater table combined with the rockmass strength. For values of GSI greater than 60, shallow excavations are apparently very stable. However below this value they become significantly less stable with increasing pore water pressures due to rising groundwater. Failure of shallow mine workings may be induced by rising water levels after mine abandonment or by small seasonal variations in the groundwater table where the stability of the excavation roof is marginal.

ACKNOWLEDGEMENTS

The authors wish to acknowledge Network Rail for providing funding for this project.

REFERENCES

Attewell, P.B. and R.K. Taylor, Eds. (1984). *Ground Movements and their Effects on Structures*. London, Blackie & Son Ltd.

Bell, F.G. (1992). *Engineering Properties of Soils and Rocks*. London. Butterworth-Heinemann Ltd.

Cundall, P.A. (2003) *FLAC3D: Fast Lagrangian Analysis of Continua in 3 Dimensions*, Version 3.0, Itasca Consulting Group, Inc., Minneapolis, MN.

Das, S.K. (2000), Observations and classification of roof strata behaviour over longwall coal mining panels in India. *International Journal of Rock Mechanics and Mining Sciences*. 37(4): 585–597.

Diederichs, M.S. and Kaiser, P.K. (1999), Stability of large excavations in laminated hard rock masses: the voussoir analogue revisited. *International Journal of Rock Mechanics and Mining Sciences* 36(1): 97–117.

Donaldson Associates (2002) *ECML Dolphingstone Remediation Works Geotechnical Interpretive Report*. Unpublished Report.

Healy, P.R. and Head, J.M. *Construction Over Abandoned Mine Workings*. CIRIA, London, 1984. (CIRIA Special Publication 32).

Marinos, V. Marinos, P. & Hoek E. (2005) The geological strength index: applications. *Bulletin of Engineering Geology and the Environment* 64: 55–65.

Waltham, A.C. (1989) *Ground Subsidence*, London, Blackie & Son Ltd.

Advances in Transportation Geotechnics – Ellis, Yu, McDowell, Dawson & Thom (eds)
© 2008 Taylor & Francis Group, London, ISBN 978-0-415-47590-7

Analytical solution of multi-step reinforced soil slopes due to static and seismic loading

E. Kapogianni & M. Sakellariou
National Technical University of Athens, Athens, Greece

ABSTRACT: The purpose of this paper is to present an analytical solution of multi-step reinforced- soil slopes due to static and seismic loading. This solution is based on the kinematic theorem of limit analysis and on the quasi-static approach and concerns homogenous cohesionless soils that are expected to deform plastically, following the Coulomb yield criterion. For the implementation of this methodology, a computer programme has been designed and representative examples are shown. With this programme, different types of potential failure mechanisms are studied and the most critical ones are used for the final design. Finally, multi-step and one-step reinforced slopes with the same height and soil mechanical characteristics are analysed and compared and conclusions are derived concerning their stability and the amount of the reinforcement that is necessary to prevent failure.

1 INTRODUCTION

Soil is ideal for use in construction, since it is a relatively inexpensive and abundant construction material. Moreover, soil is capable of providing very high strength in compression, but virtually no strength in tension. Like other construction materials with limited strength, soil can be reinforced with materials such as strips, grids, sheets, rods and fibers in order to form a composite material that has better mechanical characteristics. The need to reinforce soil is even higher in seismically active areas. For these reasons, the construction of reinforced-soil slopes with geosynthetics has expanded in the last 20 years.

The stability of a reinforced slope is related to the geometrical and mechanical characteristics of the construction. Height and inclination of the slope along with static and seismic loading conditions determine the amount of the reinforcement that is necessary to prevent failure. Slopes with small height and gentle inclination demand relatively low reinforcement. On the other hand, high and steep slopes are vulnerable to earthquake loading and therefore require long reinforcement with large tensile strength. Moreover, the erosion due to the water flow is another possible problem that affects high and steep slopes. The surface water covers larger distance along high and steep slopes with high velocity, leading to erosion. For these reasons, there is need for another approach for the design and construction of high slopes in order to limit these problems.

The design of high reinforced-soil slopes with steps has many advantages. Soil weight loading and inertial force induced by the earthquake loading are reduced and therefore the necessary tensile strength and length of the reinforcement is much lower. In addition, the surface water is collected segmentally in every one step limiting this way the potential erosion. Reinforced, steep embankments with irregular, curved groundplan have been constructed for the Egnatia Highway in Greece.

The analysis method for multi-step reinforced slopes with geosynthetics that is presented here and emphasizes to the former advantages is based on the kinematic theorem of limit analysis and on the quasi-static approach and concerns homogenous cohesionless soils that are expected to deform plastically, following the Coulomb yield criterion.

2 ANALYSIS METHODS

Various methods of analysis have been used in order to define the behaviour of reinforced slopes subjected to seismic and static loading.

The limit equilibrium method is traditionally applied to obtain approximate solutions in soil stability problems and entails assumed failure surfaces of various simple shapes such as plane, circular and log-spiral (Terzaghi 1943). With this assumption, each of the stability problems can be reduced to one of finding the most critical position for the failure surface of the shape chosen. In this method, an overall equation of equilibrium, in terms of stress resultants can be written for a given problem. This makes it possible to solve various problems by simple statics.

In contrast to limit equilibrium method, the limit analysis method considers the stress-strain relationship of the soil in an idealized manner. This idealization (expressed by the flow rule) establishes the limit theorems on which limit analysis is based. Within this framework, the approach is rigorous and the techniques are in some cases much simpler than those of limit equilibrium. The plastic limit theorems of Drucker et al (1952) can be employed to obtain upper and lower bounds of the collapse load for stability problems.

If the same collapse mechanisms are used, the results of the limit equilibrium method and the limit analysis method are identical.

The finite element analysis method is a very comprehensive approach and stress-strain analysis can be performed. Moreover, soil-reinforcement interface can be taken into account along with the material properties of the soil structure. However, an analytical solution is an accurate, closed form solution with small numerical cost and therefore can be used as a fast tool for the design of soil structures.

The analysis method used in the current study is based on the kinematic theorem of limit analysis and on the quasi-static approach.

2.1 *Kinematic theorem of limit analysis*

The kinematic theorem of limit analysis is based on the upper – bound theory of plasticity and states that the slope will collapse if the rate of work done by external loads and body forces exceeds the energy dissipation rate in any kinematically admissible failure mechanism.

$$\int_V \sigma_{ij}^* \varepsilon_{ij}^* dV \geq \int_S T_i \, v_i dS + \int_V X_i v_i^* dV, \; i,j = 1,2,3 \qquad (1)$$

where ε_{ij}^* is the strain rate in a kinematically admissible velocity field, σ_{ij}^* is the stress tensor associated with ε_{ij}^*, T_i the stress vectors on boundary S, velocity $v_i^* = v_i$ on boundary S(given kinematic boundary condition), X_i is the vector of body forces (unit weight and the distributed quasi-static inertial force), and S and V are the loaded boundary and the volume, respectively.

In this study, pore water pressure and potential liquefaction are not considered. The rate of external work is due to soil weight and inertia force induced by the seismic loading and the only contribution to the energy dissipation is provided by the reinforcement. In addition, it is assumed that the energy dissipation is performed only by the tensile strength of the geosynthetics, while resistance to shear, bending and compression are ignored.

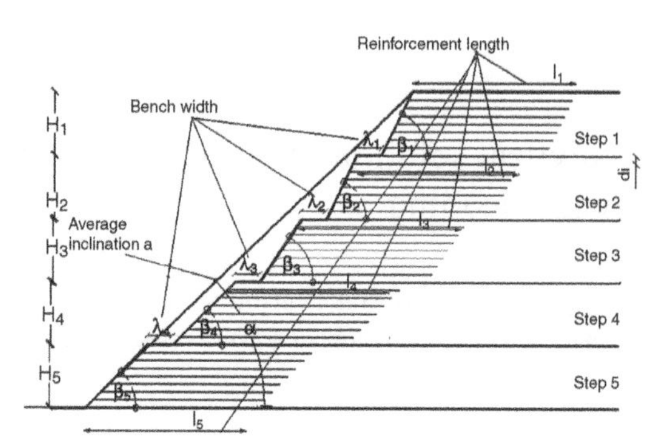

Figure 1. Multi-Step Reinforced Slope.

2.2 *Quasi- static approach and seismic coefficient*

According to the quasi-static approach, a static force with horizontal direction represents the seismic influence on the failure soil mass. This force is estimated by the product of seismic intensity coefficient and weight of the potential sliding soil mass. This approach is a widely accepted method, despite the fact that it neglects the acceleration history.

The evaluation of the seismic coefficient can be accomplished with various empirical predictive relations based on the seismotectonic environment of each region.

Ambraseys (1995) predictive relations for peak horizontal ground accelerations generated by earthquakes in the European area:

$$\log(a_h) = -1.09 + 0.238 M_s - 0.0005 r - \log(r) + 0.28 P \qquad (2)$$

if no account is taken of the focal depth, where, $r = (d^2 + 6^2)^{0.5}$, d is the source distance in km, M is the surface wave magnitude, P is 0 for 50 percentile values and 1 for 84 percentile.

If the effect of the focal depth h (km) is allowed for, the equation becomes:

$$\log(a_h) = -0.87 + 0.217 M_s - 0.00117 r - \log(r) + 0.26 P$$
$$\text{where, } r = (d^2 + h^2)^{0.5} \qquad (3)$$

where, $r = (d^2 + h^2)^{0.5}$

2.3 *Analytical solution of multi-step reinforced soil slopes*

According to this method, a high and steep reinforced slope is divided into more slopes with smaller height and scaled inclination. For example, a slope with 50 meters height and inclination 3:2 is divided into 5 slopes with 10 meters height each and scaled inclination as shown in Figure 1.

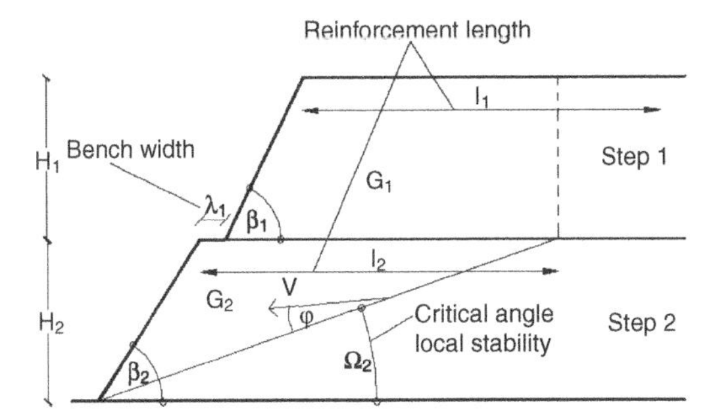

Figure 2. Local Stability Failure Mechanism- Step 2.

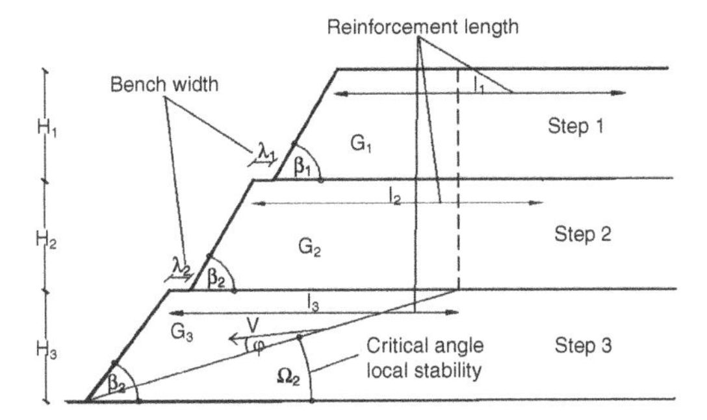

Figure 3. Local Stability Failure Mechanism- Step 3.

In order to determine the amount of the reinforcement that is necessary to prevent failure, different failure modes are examined and the most critical ones are used for the final design. The failure mechanism presented here is the plane failure mechanism.

In this failure mechanism, it is assumed that the reinforced soil mass translates as a rigid body with velocity V (Figure 3). Height $H$of the slope and angle Ω that the failure plane forms with the horizontal, specify the mechanism. Main goal is to determine the critical value of Ω (for a slope with given height) and therefore define the critical failure mechanism. Once the critical failure mechanism is defined, the amount of the necessary reinforcement can be calculated.

The plane failure mode is applied twice, once in order to ensure that the tensile strength and length of the reinforcement are adequate against local stability for each step separately (Figures 2, 3) and then against global stability for the whole slope (Figure 4). It is necessary that both failure modes are applied, since the failure mode that concerns local stability gives critical results for the upper layers of the reinforcement and the failure mode for global stability for the lower layers. The assumption of sliding at an angle φ on the base of the mechanism (corresponding to normality) accounts for the strength of the soil on this surface.

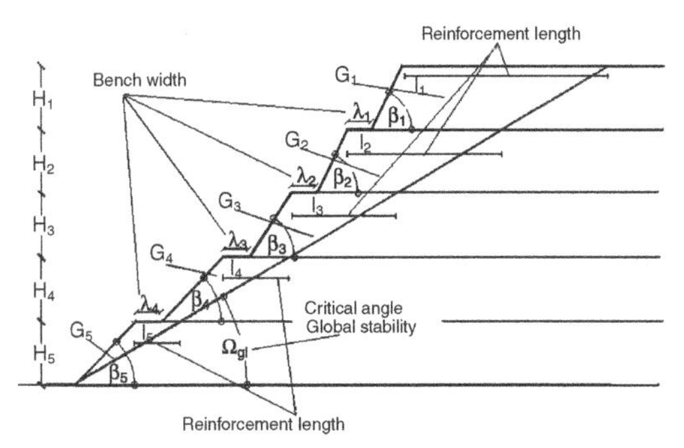

Figure 4. Global Stability Failure Mechanism.

In particular, the rate of external work done by soil weight and inertial force is:

$$\dot{W} = (G_1 + G_2 + ... + G_i)V \sin(\Omega - \phi) + k_h(G_1 + G_2 + ... + G_i)V \cos(\Omega - \phi) \quad (5)$$

where $G_1, G_2 \dots G_i$ indicate the weight of the soil wedge for every one step and have different expressions for local and global failure modes and k_h the seismic coefficient. In the current study, five slopes can be designed at maximum.

According to Ausilio et al (2000), the energy dissipation is

$$\dot{D} = V \cos(\Omega - \phi)\sum_{i=1}^{n} T_i \quad (6)$$

Moreover, Ling et al (1997) suggested that the tensile strength (T_i) can be calculated approximately by the following equation:

$$T_i = K \gamma z_i d_i \quad (7)$$

where K represents the total reinforcement in a normalized form with the following expression:

$$K = \frac{\sum_{i=1}^{n} T_i}{(1/2)\gamma H^2} \quad (8)$$

and $d_i$is the tributary area of layer i.

Eq. (6) owing to Eq. (7) becomes:

$$\dot{D} = \frac{1}{2} V \cos(\Omega - \varphi) K \gamma H^2 \quad (9)$$

Equating the rate of external work done by soil weight and inertial force to the energy dissipation, the total reinforcement in a normalized form K can be calculated:

a) Local Stability:

Step 1

$$K_1 = \frac{2G_1 \tan(\Omega_1 - \varphi_1) + 2k_h G_1}{\gamma_1 H_1^2} \quad (10)$$

Step 2

$$K_2=\frac{2(G_1+G_2)\tan(\Omega_2-\varphi_2)+2k_h(G_1+G_2)}{\gamma_2 H_2^2} \quad (11)$$

Step 3

$$K_3=\frac{2(G_1+G_2+G_3)\tan(\Omega_3-\varphi_3)+2k_h(G_1+G_2+G_3)}{\gamma_3 H_3^2} \quad (12)$$

and so on for steps 4 and 5.

By substituting the maximum values of K_i to equation (7), the maximum necessary tensile strength can be calculated for every one step.

The necessary length of reinforcement (l_i, Figures 2 and 3) based on geometry of the critical local mechanisms without specific consideration of the pull-out resistance is:

$$l_i=\frac{H_i\sin(\beta_i-\Omega_i)}{\sin(\beta_5)\sin(\Omega_i)} \quad (13)$$

(The necessary length cannot be lower than $0.7H_i$)

b) Global Stability

$$K_{gl}=\frac{2(G_1+G_2+...+G_5)\tan(\Omega-\varphi)+2k_h(G_1+G_2+...+G_5)}{\gamma(H_1+H_2+...+H_5)^2} \quad (14)$$

The necessary tensile strength of the reinforcement for each layer derives by substituting again K_{gl} to Eq. (7).

The necessary length of the reinforcement for global stability can be calculated by the following equations:

$$l_1=\frac{H_1+H_2+...+H_5}{\tan(\Omega)}-\frac{H_1}{\tan(\beta_1)}-\frac{H_2}{\tan(\beta_2)}-...-\frac{H_5}{\tan(\beta_5)}-\lambda_1-...-\lambda_4 \quad (15)$$

$$l_2=\frac{H_2+H_3+H_4+H_5}{\tan(\Omega)}-\frac{H_2}{\tan(\beta_2)}-\frac{H_3}{\tan(\beta_3)}-...-\frac{H_5}{\tan(\beta_5)}-\lambda_2-\lambda_3-\lambda_4 \quad (16)$$

$$l_3=\frac{H_3+H_4+H_5}{\tan(\Omega)}-\frac{H_3}{\tan(\beta_3)}-\frac{H_4}{\tan(\beta_4)}-\frac{H_5}{\tan(\beta_5)}-\lambda_3-\lambda_4 \quad (17)$$

$$l_4=\left(\frac{H_4+H_5}{\tan(\Omega)}-\frac{H_4}{\tan(\beta_4)}-\frac{H_5}{\tan(\beta_5)}-\lambda_4\right) \quad (18)$$

$$l_5=\frac{H_5}{\tan(\Omega)}-\frac{H_5}{\tan(\beta_5)} \quad (19)$$

(The necessary length cannot be lower than $0.7H$)

For the final design, the maximum values of the necessary tensile strength and length for each one layer are chosen. As mentioned before, local stability calculations give longer reinforcements with larger tensile strength for the upper layers and global stability calculations for the lower layers. By applying both analyses, the multi step reinforced soil slope can be designed in a more accurate and safe way.

Similarly, the critical acceleration factor can be obtained:

a) Local Stability:

$$k_{y1}=\frac{K_1\gamma_1 H_1^2-2G_1\tan(\Omega_1-\varphi)}{2G_1} \quad (20)$$

$$k_{y2}=\frac{K_2\gamma_2 H_2^2-2(G_1+G_2)\tan(\Omega_2-\varphi)}{2(G_1+G_2)} \quad (21)$$

and so on for k_{y3}, k_{y4}, k_{y5}

b) Global Stability:

$$k_{ygl}=\frac{K_{gl}\gamma H_{total}^2-2(G_1+G_2+G_3+G_4+G_5)\tan(\Omega_{gl}-\varphi)}{2(G_1+G_2+G_3+G_4+G_5)} \quad (22)$$

The minimum value of k_y is the critical yield acceleration of the slope.

Similarly, equations for the design with static loading can be obtained:

a) Local Stability:

$$K_1=\frac{2(FS)G_1\tan(\Omega_1-\varphi_1)}{\gamma_1 H_1^2} \quad (23)$$

$$K_2=\frac{2(FS)(G_1+G_2)\tan(\Omega_2-\varphi_2)}{\gamma_2 H_2^2} \quad (24)$$

$$K_3=\frac{2(FS)(G_1+G_2+G_3)\tan(\Omega_3-\varphi_3)}{\gamma_3 H_3^2} \quad (25)$$

and so on for step 4 and step 5

b) Global Stability:

$$K_{gl}=\frac{2(FS)(G_1+G_2+...+G_5)\tan(\Omega-\varphi)}{\gamma(H_1+H_2+...+H_5)^2} \quad (26)$$

where F.S. is the desirable Factor of Safety.

3 SOFTWARE IMPLEMENTATION

For the implementation of the former methodology, a computer programme has been designed. This software has been developed in Borland Delphi and has the following features.

1) Calculation of the Horizontal Peak Ground Acceleration (HPGA), based on the empirical predictive relation mentioned before. Av independent value can also be chosen.

2) The next step is to define the soil mechanical characteristics (φ, γ), the height H and inclination β of each step, the appropriate bench width λ and the number of the reinforcement layers n. It is assumed that the layers of the reinforcement have equal length for each step and the same spacing for the whole slope (d_i). Moreover, an appropriate value of the HPGA must be chosen based on step (1).
3) Afterwards, the dynamic analysis for the plane failure mechanism takes place for local and global failure modes. The results of the analysis are shown and specifically: the maximum total reinforcement in a normalized form (K_i and K_{gl}), the maximum tensile strength for each layer (T_i and T_{gl}), the necessary length of the reinforcement (l_i and l_{gl}), the final length of the reinforcement (lf_i and lf_{gl}), since as mentioned before the necessary length cannot be less than $0.7H_i$, and finally the angle that specifies the critical failure mechanism (Ω_i and Ω_{gl}). In addition, the necessary tensile strength and length of the reinforcement for local and global stability analysis are compared and the maximum ones are used for the final design of the multi step reinforced slope due to seismic loading.
4) Similarly, static analysis is performed using local and global failure mechanisms. Apparently, in this case the HPGA is zero and the user must choose an appropriate value for the Factor of Safety of the construction. As before, the calculated values of the necessary tensile strength and length of the reinforcement for local and global failure modes are compared and the maximum ones are chosen for the final design.

4 EXAMPLES OF MULTI-STEP AND ONE-STEP REINFORCED SOIL SLOPES

In this section, calculations are carried out in order to demonstrate how the program works and how geometry, seismic loading and soil properties affect the necessary tensile strength and length of the reinforcement.

In the first example it is assumed that soil is cohesionless with unit weight $\gamma = 18\,kN/m^3$ and angle of soil shearing resistance $\varphi = 35^0$. The slope angles for the 5 steps are: $\beta_1 = 2{:}1$, $\beta_2 = 2{:}1$, $\beta_3 = 3{:}2$, $\beta_4 = 1{:}1$ and $\beta_5 = 1{:}1$(vertical: horizontal). The height of each step is 10 m and the reinforcement consists of 20 equally spaced layers for each slope ($d_i = 0.5$ m). The plane failure mechanism is applied for local and global stability and the critical mechanism is defined, initially for every one step and then for the whole slope. In general, for steeper slopes and at lower seismic coefficient, the critical mechanism is located at $\Omega > \varphi$ and for slopes with gentle inclination and higher seismic

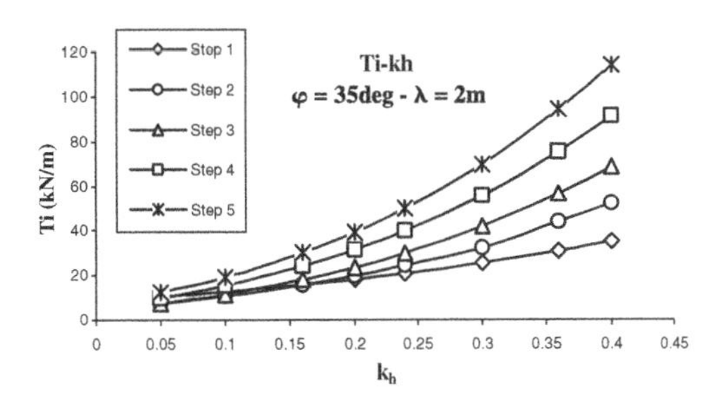

Figure 5. Seismic influence on the necessary tensile strength for $\lambda = 2$ m.

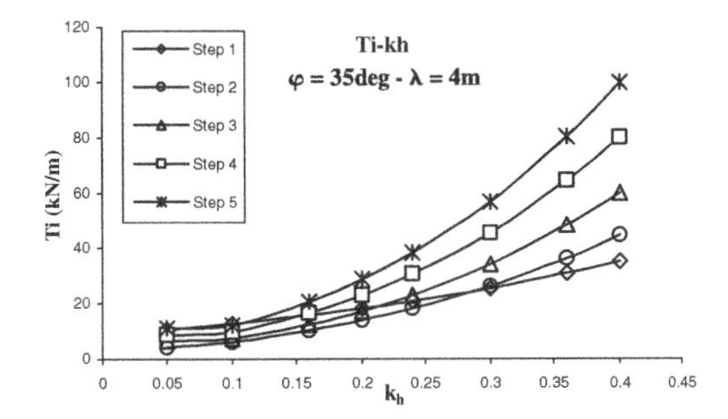

Figure 6. Seismic influence on the necessary tensile strength for $\lambda = 4$ m.

coefficient at $\Omega < \varphi$. For example, for a one-step slope with $\varphi = 35^0$, $\beta = 45^0$ and $k_h = 0.16$, the critical mechanism is at $\Omega = 34^0$. On the other hand, for a one-step slope with $\varphi = 35^0$, $\beta = 65^0$ and $k_h = 0.16$, the critical mechanism is at $\Omega = 42^0$ and for $k_h = 0.36$, Ω is smaller than φ ($\Omega = 33^0$).

The seismic influence on the amount of necessary reinforcement of every one step is illustrated. Moreover the effect of the bench width λ between the 5 steps is shown. Finally, a one step and a multi step slope are compared.

Figures 5 and 6 show the required tensile strength for different values of the seismic coefficient k_h for every one step. As can be expected, the required tensile strength increases with increasing seismic coefficient k_h. In addition, Step 5 requires the most tensile strength due to the influence of the weight of the upper steps. In Figure 6, where bench width λ between the 5 Steps is larger, it can be seen that the necessary tensile strength is reduced and especially for the lower layers of the slope and for higher seismic coefficient. Specifically, Figures 7 and 8 illustrate how λ affects the necessary tensile strength for every one step. Steps 2, 3, 4 and 5, require reduced tensile strength as distance λ increases, while Step 1, as expected requires the same tensile strength.

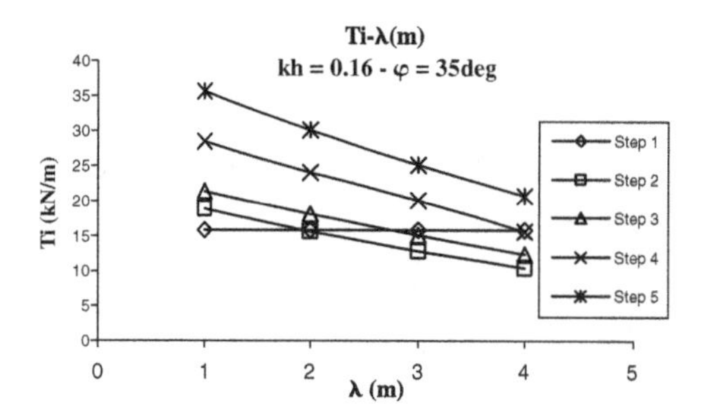

Figure 7. Effect of distance λ between the 5 steps, for kh = 0.16.

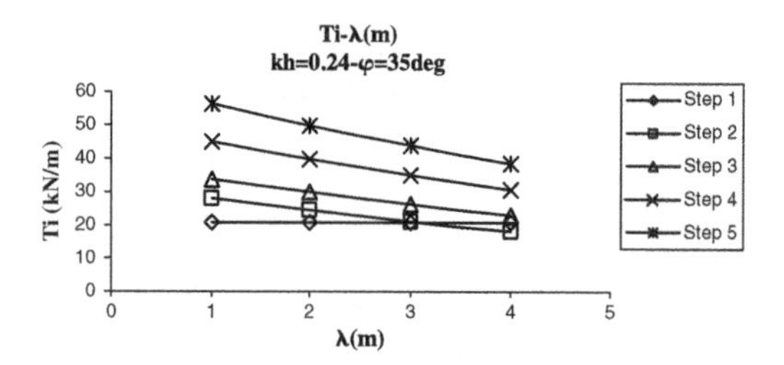

Figure 8. Effect of distance λ between the 5 steps, for kh = 0.24.

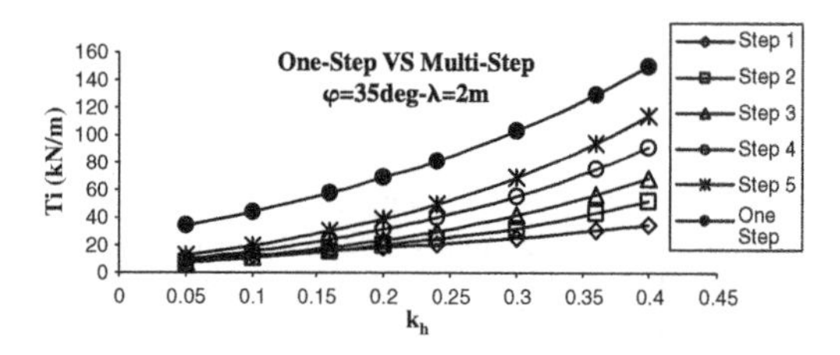

Figure 9. One-Step (α = 3:2) versus Multi-Step (λ = 2 m), at different seismic coefficients.

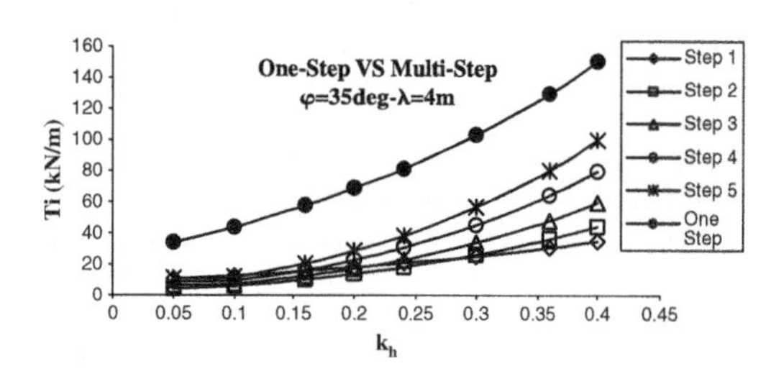

Figure 10. One-Step (α = 3:2) versus Multi-Step (λ = 4 m), at different seismic coefficients.

In addition, Figures 9 and 10, illustrate the comparison of a one-step reinforced slope with a multi-step reinforced slope for different values of the seismic coefficient. It is assumed that the one-step has average inclination *a* (Figure 1), associated with the geometrical characteristics of the multi-step slope through the following equation:

$$\tan(a) = \frac{(H_1 + H_2 + \ldots + H_5)}{\frac{H_1}{\tan(\beta_1)} + \frac{H_2}{\tan(\beta_2)} + \ldots + \frac{H_5}{\tan(\beta_5)} + \lambda_1 + \lambda_2 + \lambda_3 + \lambda_4} \quad (27)$$

Specifically, for the case of the multi-step mentioned before ($\beta_1 = 2:1$, $\beta_2 = 2:1$, $\beta_3 = 3:2$, $\beta_4 = 1:1$ and $\beta_5 = 1:1$), the corresponding average inclination for $\lambda = 0$ is slightly less than 3:2 and the height of the one-step slope is $H_{total} = 50\,m$. The soil mechanical characteristics are the same. As shown in Figures 9 and 10, the necessary tensile strength of the multi-step slope is significantly lower than the necessary tensile strength of the one-step slope. The inclination is taken as 56° (3:2) for the single step slope, whilst the average inclination *a* for $\lambda = 2$ and 4 is 48° and 46° respectively, hence accounting for the reduction in reinforcement.

5 CONCLUDING REMARKS

The expressions derived in this study based on the kinematic theorem of limit analysis can be conveniently used for the design of high reinforced slopes with steps. The amount of the reinforcement that is necessary to prevent failure due to static and seismic loading can be calculated for both local and global failure modes and the most critical ones can be used for the final design. In general, the critical mechanism can be defined for $\Omega > \varphi$ for steeper slopes or/and at lower seismic coefficient and for $\Omega < \varphi$ for slopes with gentle inclination or/and at higher seismic coefficient. The necessary tensile strength of the multi-step compared to that of the one-step is significantly reduced, while the potential erosion due to the water flow is limited. Moreover the critical acceleration factor k_y can be obtained and static analysis with the desirable Factor of Safety can be performed.

The calculations carried out indicate that reinforcement increases with an increased seismic force and reduces as bench width λ increases. Finally, the inclination of each step affects the amount of the reinforcement.

This paper is supported by the project PYTHAGORAS. The project is co-funded by the European Union-European Social Fund & National Resources-EPEAEK II.

REFERENCES

Ambraseys, N.N. 1995. The prediction of earthquake peak ground acceleration in Europe. In: *Earthquake Engineering and Structural Dynamics, 24, pp. 467–490.* John Wiley & Sons, Ltd.

Ausilio, E. & Conte, E. & Dente, G. January 2000. Seismic stability analysis of reinforced slopes. In: *Soil Dynamics and Earthquake Engineering, 19, pp. 159–172, Elsevier.*

Brandl, H. 2004. Innovative methods and technologies in earthworks. In: Gomez Correia, A. & Loizos, A. (eds), *Geotechnics in pavement and railway construction; Proc. intern. seminar, Athens, 16–17 December 2005.* ISSMGE.

Chen, W.F. 1975. Limit analysis and soil plasticity; *Developments in Geotechnical Engineering,* Elsevier.

Ling, H.I. & Leshchinsky, D. & Perry, E.B. 1997. Seismic design and performance of geosynthetic-reinforced soil structures. *Geotechnique* 47, 5, pp. 933–52.

Ling, H.I. & Leshchinsky, D. & Chou, N.S.C. January 2001. Post-earthquake investigation on several geosynthetics-reinforced soil retaining walls and slopes during the Ji-Ji earthquake of Taiwan. In: Soil Dynamics and Earthquake Engineering, 21, 4, pp.297–313, Publisher: Elsevier.

Michalowski, R.L. August 1998. Soil reinforcement for seismic design of geotechnical structures. In: *Computers and Geotechnics, 23. pp. 1–17, Elsevier.*

Mitchell, J.K & Villet, W.C.B. 1987. Reinforcement of earth slopes and embankments; *National Cooperative Highway Research Program Report 290.*

Roessing, L.N. & Sitar, N. March 2006. Centrifuge model studies of the seismic response of reinforced soil slopes. In: *Journal of Geotechnical and Geoenvironmental Engineering, 132, 3, pp. 388–400.* ASCE.

Advances in Transportation Geotechnics – Ellis, Yu, McDowell, Dawson & Thom (eds)
© 2008 Taylor & Francis Group, London, ISBN 978-0-415-47590-7

Piled foundations for the M25 Motorway near Heathrow Airport

Mark Pennington & John Spence
Stent Foundations Ltd, Basingstoke, Hampshire, UK

ABSTRACT: Ever since opening, the M25 motorway to the west of London has carried a rapidly increasing volume of traffic. The section of M25 between the junctions with the M3 and M4 motorways, which is in the near vicinity of Heathrow airport, is the most heavily trafficked section. Piling and ground treatment was required for motorway widening works over this route, which is 7 miles long. The widening scheme had to fit within the existing highway boundary. The foundation works were undertaken next to the live motorway and beneath the flightpath for aircraft landing and taking off at the airport. The paper describes the unique method of foundation construction techniques used for the piled retaining walls, and the how the methods were affected by the live motorway and the height restrictions imposed by the airport.

1 INTRODUCTION

Since its completion in 1986 the M25 London Orbital Motorway has carried a rapidly increasing volume of traffic. The section of M25 between the junctions with the M3 and M4 motorways, which is in the near vicinity of Heathrow airport, is the most heavily trafficked section. The project route which is 7 miles long is shown in Figure 1.

Between Junctions 12 (M3 motorway) and 13 (Staines interchange) the M25 was originally constructed as dual three lane motorway. Between Junctions 13 and 15 (M4 motorway) the M25 was originally constructed as four lane dual carriageway, reducing to three lanes at the junctions with the fourth lanes forming the slip roads.

In 1988 this four lane design, with the same lane reduction arrangement at the junctions, was constructed between Junctions 12 and 13. This was achieved without any new earthworks.

Between 1990 and 1998 plans were made to further increase the motorway capacity over this section. Widening was to be achieved entirely within the existing motorway boundary, retaining all existing bridges, as follows;

- five lane dual carriageway between Junctions 12 to 14 (Heathrow Terminal 4)
- six lane dual carriageway between Junctions 14 to 15
- four lane dual carriageway locally through Junctions 13 and 14.
- construction of a new spur road to the new Heathrow Terminal 5.

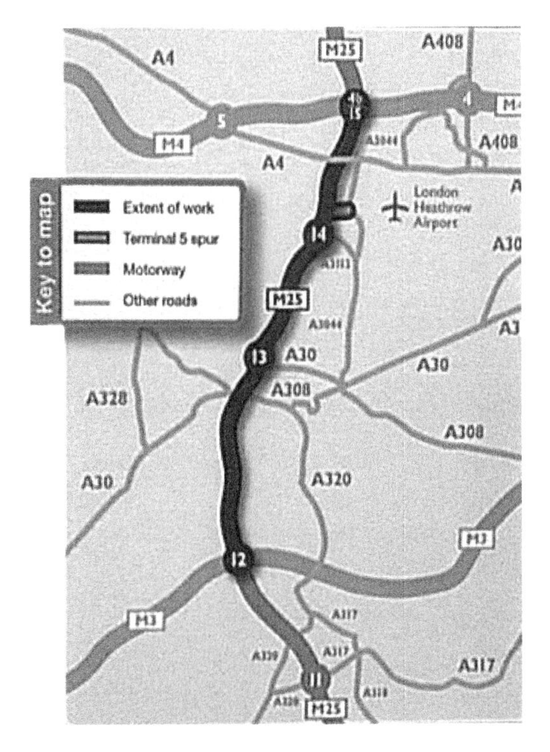

Figure 1. Project Route.

The vicinity of the project contains urban developments, including many villages, and both industrial and agricultural land. Notable features are the large number of reservoirs and gravel pits. The scheme lies within the Metropolitan Green Belt and north of Junction 13 is the Colne Valley Regional Park.

During 2004 to 2005 Stent completed a £10million piling contract for the works. The piling comprised 2000No. piles including 2.5 miles of piled retaining wall and piling for 45No. new gantries. The foundation works were undertaken next to the live motorway, see Figure 2, and beneath the flightpath for aircraft landing and taking off at Heathrow Airport, see Figure 3.

Figure 2. Piling Rig next to Live Motorway.

Figure 3. Piling Rig beneath Flight Path.

Piling was carried out over a 16 month period with up to 7No. piling rigs working. Rotary piling methods were used by both crane mounted and hydraulic machines. Unusually the crane mounted units were extended so that piles could be installed up to 8.5 m in front of the tracks.

This paper will describe the foundation construction techniques used for the piled retaining walls, and the how the methods were affected by the live motorway and the height restrictions imposed by the airport.

2 SITE GEOLOGY

The project is underlain by solid geology of the Lower Eocene Series, almost exclusively London Clay Formation. Sediments of the Bagshot Formation have been encountered at Junction 12.

Much of the route is underlain by River Terrace Gravel deposited during the Pleistocene Ice Age about 10,000 years ago. The gravel outcrops over most of the area between Junctions 12 and 13, but further north the gravel is overlain by Alluvium of Holocene Age deposited within the flood plains of the rivers Thames and Colne.

Before 1999 over 800No. exploratory boreholes had been installed over the project route. At Junction 15 the London Clay is about 30 m thick, increasing to 131 m thick at Chertsey, south-east of Junction 12. At Junction 14 the London Clay was proved to be 42 m thick.

The London Clay formation consists mainly of dark blue to brownish grey clay, containing variable amounts of fine grained sand and silt. The silt is particularly abundant at the base and near to the top of the formation. Horizontal beds of calcareous claystone concretions occur sporadically.

Sandy horizons are also occasionally present, becoming more common near to the top of the deposit towards the boundary with the Claygate Beds, which were found to be up to 15 m thick near to Junction 12. The Claygate Beds are a subdivision of the London Clay, comprising laminated orange sands interbedded with pale grey to lilac clays.

The maximum thickness of the River Terrace Gravel is about 4 m. Alluvial deposits are up to 4 m thick as well, but much has been previously removed as a result of the sand and gravel workings.

Many of the landfill sites were infilled prior to the Control of Pollution Act 1974 and records here are scarce. A wide variety of landfill materials are present.

Two aquifers are present; a shallow River Terrace Deposits aquifer and a deep Upper Chalk aquifer.

3 SITE TOPOGRAHY

The topography of the project route is flat and relatively low lying, with the M25 motorway crossing the valleys of the river Thames and its southward flowing tributaries associated with the river Colne.

At the northern end of the route, Junction 15 is situated between several disused water-filled sand and gravel pits. South of Junction 15 the M25 motorway is on low embankment constructed over areas of old landfill. The landfill has been either excavated and backfilled with London Clay fill or was treated using dynamic compaction.

South of Junction 14 the motorway is still on low embankment, with a 1.5 mile wide old landfill area to the east. The river Wraysbury was diverted for the original M25 construction.

Between Junctions 13 and 14 Wraysbury Reservoir dominates the western side of the motorway. The motorway is at grade on the west but on embankment up to 3 m high on the east.

Towards Junction 13 the embankment rises to 10 m high. This interchange was located on a spit of land between two water-filled gravel pits.

South of Junction 13 the motorway crosses the river Thames and the embankment reduces in height to only 1 m high. Further south Long-side Lake, which is a disused gravel pit now used for recreation, runs for 700 m to the west of the motorway which is constructed on a 7 m high embankment.

4 PILED RETAINING WALLS

In order to avoid additional land take the widening of the M25 was mostly built on the shoulders of the current embankment. This meant that piled retaining walls were required part way down the existing embankment shoulder.

This type of widening is becoming increasingly more common on motorway widening schemes in the UK. It provides a cost effective method of meeting the increased demand for the highway. However, this does provide a design and logistical problem to the construction of the piled retaining walls.

Many of the existing embankments are constructed at or beyond their theoretical long term angle of failure. Careful consideration needed to be given to the active and passive resistances on the retaining walls constructed through the shoulder of the embankment. In cases where the existing embankment was constructed from clay fill, allowance needed to be made for the possibility of future long term slips in front of the retaining wall. This led to larger and deeper piles than may be expected from the apparently small retained heights. Piles were typically 750 and 900 mm diameter, 10 to 18 m long.

In order to construct piles of this size relatively large and heavy plant was required. This in turn led to logistical problems of operating this plant on the edge of a live motorway and in the vicinity of an embankment.

Installing piles part way down an embankment slope can be carried out from three locations:

- from the top of the embankment, reaching out to the pile position,
- from the toe of the embankment, reaching up to the pile position,
- from a berm or platform constructed to get the piling rig to the pile location.

All of the above options were considered for this stretch of motorway widening. In this instance, the most appropriate method was to construct the piles from the top of the embankment. A lack of working space at the toe of the embankment meant that piling from the embankment toe was not an option. The size and weight of the piling plant meant that berms or platforms on the embankment shoulder were not practical for the main piling plant.

The width of the existing carriageway meant that it was just about possible to operate the large piling plant on the edge of the carriageway alongside a workable traffic management scheme. A linear working site was set up which also enabled other construction traffic to pass by the site. Piling had to be carried out from the hard shoulder lane of the motorway meaning that the piling rigs would need to reach out up to 8.5 m.

The ground conditions found would normally be ideal for CFA piling. This technique couldn't have coped with the large boulders, timber sleepers, and metal objects encountered during boring within the fill materials. In addition it is not currently possible to construct CFA piles at a reach of up to 8 m from the piling rigs. Rotary bored piling was thus the chosen technique.

To do this Stent produced extended frames, see Figure 7, to carry the Watson 5000 drilling units that are ideally suited to this type of piling work. A considerable amount of design work had to be carried out to ensure that the working platform, base crane unit, and frame could withstand the additional lever arm forces produced.

A unique system was developed to produce the retaining walls with an exposed aggregate finish. Firstly the retaining wall piles 750 mm and 900 mm diameter 10 to 18 m deep were installed, see Figure 4. The piles were cast through a guide wall to ensure accuracy of position and cast 50 to 100 mm above the guide wall using a circular steel former, as shown in Figure 5. The concrete mix design ensured that there was no bleed or laitance and therefore virtually no breaking down of piles was required. This was essential to ensure that the construction of the retaining wall could progress quickly behind the piling operation.

Tests were carried out to ensure that the integrity of the concrete was acceptable with no or little breaking down of the concrete.

Figure 4. Constructing Retaining Wall Piles.

Figure 5. Use of Circular Former.

The guide wall was not then removed as it also served as the base for positioning the precast concrete wall units, which varied in height from 1 to 4.5 m. The wall units were temporarily propped and this is shown in Figure 6. Once the unit was positioned a shutter was erected on the retained side of the pile and in-situ concrete poured to tie the precast unit to the piles through a capping beam. Drainage was then installed and the void behind the wall units backfilled.

In addition, specially adapted low headroom piling rigs were used in areas with high tension overhead

Figure 6. Precast Concrete Retaining Wall Units Being Positioned.

Figure 7. Piling Rig with Extended Frame.

wires, under bridges and where there were height restrictions under the approaches to the airport runways, see Figure 8. These specially adapted rigs meant that the piling works could continue without having to work around airport restrictions which would have affected the programme significantly.

5 CONCLUSIONS

Ever since opening, the M25 motorway to the west of London has carried a rapidly increasing volume of traffic. The section of M25 between the junctions with

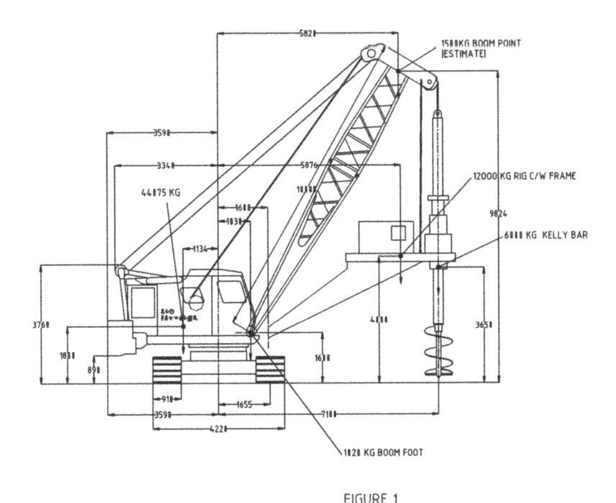

Figure 8. Low Headroom Piling.

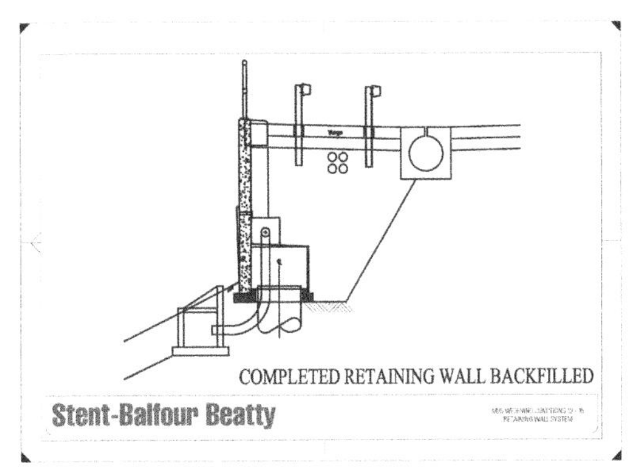

Figure 9. Finished Retaining Structure.

the M3 and M4 motorways, which is in the near vicinity of Heathrow airport, is the most heavily trafficked section.

Piling and ground treatment was required for motorway widening works over this route, which is 7 miles long. The foundation works were undertaken next to the live motorway and beneath the flightpath for aircraft landing and taking off at the airport.

As is becoming increasingly popular with many motorway widening projects currently proposed for the UK, the widening had to be carried out within the existing highway boundary. This led to logistical restrictions and limited the construction methodology that could be used.

This paper describes the foundation construction techniques used for the piled retaining wall, and the how the methods were affected by the live motorway and the height restrictions imposed by the airport. The paper covers the reason for the construction methods adopted and how they were modified to suit the linear construction site.

Over 200,000 man hours were required for the piling works and there was not a single notifiable accident.

A unique method of constructing a piled retaining wall is described, along with the special adaptations to the piling rigs to enable them to reach out 8.5 m and install piles at the toe of the existing carriageway embankment. The sequence of wall construction was;

- regrade existing slope
- install guide wall
- construct piles with piling rig at road level
- erect and prop precast wall panels
- construct capping beam
- install drainage
- backfill behind wall panels

The finished product is illustrated in Figure 9.

Advances in Transportation Geotechnics – Ellis, Yu, McDowell, Dawson & Thom (eds)
© 2008 Taylor & Francis Group, London, ISBN 978-0-415-47590-7

Geotechnical aspects of masonry arch bridges

G.M. Swift, C. Melbourne & J. Wang
School of Computing, Science and Engineering, University of Salford, Greater Manchester UK

M. Gilbert & C. Smith
Department of Civil Engineering, University of Sheffield, UK

ABSTRACT: The existing European railway infrastructure is in general between 60 and 100 years old and as such, has not been designed to support the increased loads generated by modern railway traffic; in general, freight and passenger train traffic has increased significantly in terms of frequency and speed, but also in terms of weight (axle loads). For the existing infrastructure, in particular bridges, these increased loads could have a significant impact in terms of stability.

Ongoing research at the Universities of Salford and Sheffield, in collaboration with other UK and European partners, has been focussed primarily on the structural controls on railway bridge stability, with a view to improving the efficacy of existing assessment procedures. More recently, the focus has shifted towards the soil-structure interaction issues that impact on the overall stability, in particular issues such as stress distribution and redistribution through the backfill around a typical arch, prior to, during and subsequent to, failure. Assessment has taken the form of physical modelling, using large scale models to simulate the behaviour of the arch system under varying surface loads, using different types of soil backfill. The results from these physical models, coupled with Particle Image Velocimetry (PIV) techniques have been used by the University of Sheffield to validate the output from an in-house 2D analytical tool, as well as to allow the development of more sophisticated finite element modelling at both Sheffield and Salford.

This paper centres on the development of this research programme and documents the primary outputs to-date, as well as giving an overview of the future directions of the research, for example, the implications of climate change, dynamic loading, three-dimensional effects and soil reinforcement.

1 INTRODUCTION

1.1 *Background*

Masonry arch bridges have been in use for at least 4000 years (Ford et al, 2003) and it is believed that there are almost a million masonry arch bridges worldwide, and that approximately 40% of the European railway bridge stock comprises masonry arch bridges, and that in excess of 60% of these are more than 100 years old, (Melbourne et al, 2007). The maintenance and assessment of these bridges is a constant concern for the bridge owners, and it is therefore imperative that the existing bridge stock is not adversely affected by changes in the loading regime and bridge condition and that appropriate assessment, modelling, repair and strengthening techniques are available.

Research at the University of Salford and in more recent years, at the University of Sheffield, has focused on two main issues that can impact upon the bridge's capacity and on its service life, and these are:

- fatigue performance of multi-ring brickwork arches; and
- soil-structure interaction in masonry arch bridges

The aim of this paper is to provide an overview of this research.

1.2 *Masonry arch behaviour*

Railway bridges are carrying increasing volumes of traffic and carrying loads significantly in excess of those envisaged by the bridge designers. Experience over the last 30–40 years has indicated that long term cyclic loading from the repeated application of heavy traffic loads, can accelerate the deterioration of masonry arches, however, much of the existing research has focused on the behaviour of the arch under

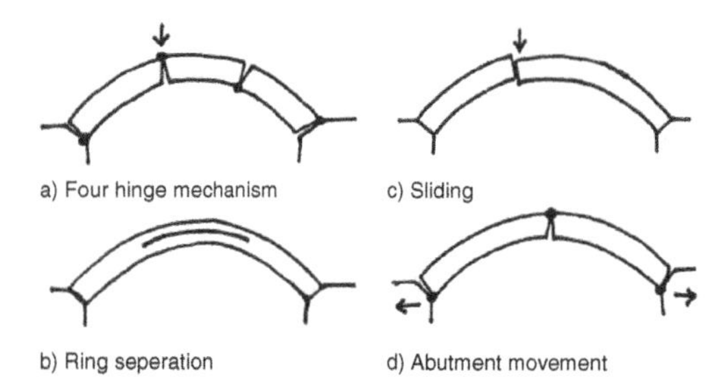

Figure 1. Masonry arch failure modes.

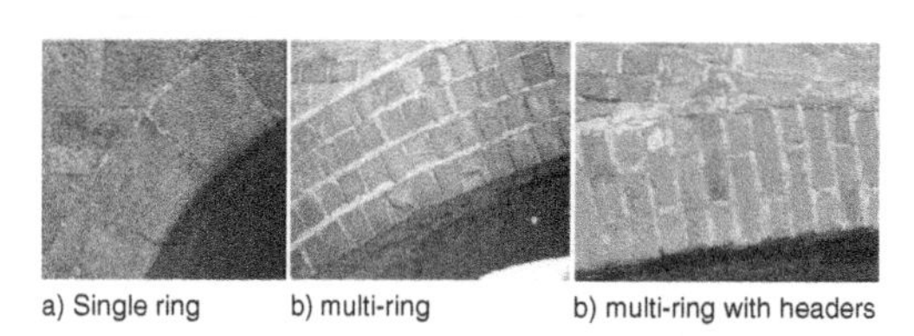

Figure 2. Different degree of shear connection.

monotonic loading conditions. The failure of masonry arch bridges is typically associated with three modes:

- formation of a hinge mechanism;
- snap through failure prior to the full formation of hinges; and
- crushing failure

These failure modes generally relate to single span, single ring square arches without taking the effects of soil-structure interaction, possible abutment movements, mortar washout, long-term fatigue deterioration, etc. into account. In addition to the classical failure modes, masonry arches may also fail by a number of alternative failure mechanisms, such as ring separation, sliding, abutment movement, etc. Figure 1 shows typical arch failure modes.

Railway bridges in Europe almost invariably contain multiple arch rings. These rings are either simply connected by uninterrupted mortar joints or crossed by headers which provide an interlocking effect and greatly increases the shear strength of the connection between rings, as shown in Figure 2.

The type of arch construction shows certain regional consistencies around Europe, for example arch bridges in Britain are generally built without headers, while in Southern Europe arches with interlocking headers are more common.

The shear resistance of arch rings is also influenced by the quality of mortar and brick units. Aged arch bridges often suffer from weak shear connections from washed out mortar joints and/or weak interlocking bricks that can result in unexpected ring separation and may significantly reduce the load carrying capacity of the bridge.

2 ASSESSMENT METHODS

2.1 *Overview*

The assessment of masonry arch bridges is difficult as there is no widely accepted reliable structural assessment procedure. Current assessment methods across Europe fall broadly in to three categories (after Melbourne et al, 2007), and these are:

- the semi-empirical MEXE method;
- limit analysis methods (eg. RING 2.0 (Gilbert, 2007); and
- solid mechanics approaches (including finite element and discrete element analyses)

In addition, a new approach has been developed at the University of Salford, termed the Sustainable Masonry Arch Resistance Technique (SMART), which will be discussed in the next section.

Of the aforementioned methods, it is recognized that many are too conservative in their estimates of ultimate failure loads, requiring uneconomical and often unnecessary mitigation measures, whilst numerical modelling approaches tend to be impractical as an assessment tool as they require input parameters that cannot be readily determined.

The MEXE method as described by the UIC Code 778-3R (1994), is founded on principles of elasticity, in which a two-pinned parabolic arch static system is assumed to have a limited compressive strength of circa $1.4\,N/mm^2$. Empirical formulae are then used to calculate the load capacity using subjectively estimated modifying factors to allow for geometry and material condition of the bridge. The real behaviour of the arch is not represented by the MEXE formulations, but the approach does provide reasonable approximations given the simplicity and speed of calculation, and is therefore widely used in the estimation of railway and highway masonry arches. Harvey (2007) provides a comprehensive review of this method, describing the basic analysis and the limitations.

Limit analysis techniques, such as the RING software, estimate the ultimate load carrying capacity of bridges by modelling the constituent masonry blocks explicitly. The blocks are assumed to be rigid and separated by contact surfaces at which rocking, crushing and sliding may take place. The analysis can also allow, in a simplistic way, for the presence of backfill material surrounding the arch; the soil is not modelled explicitly, and therefore does not represent an attempt to examine the soil-structure interaction.

The basic algorithm followed by any structural assessment method however, can be summarised in Figure 3.

2.2 *SMART*

The SMART assessment method takes an holistic approach, incorporating such factors as the form of

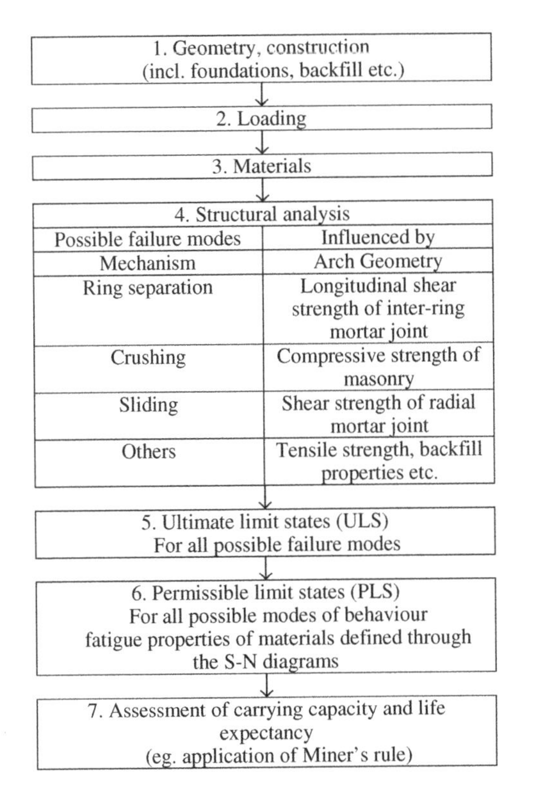

Figure 3. Masonry arch assessment procedure.

construction, the materials, the loading conditions, limit states, actions (both current and historic), analysis and modes of failure. In general, more traditional assessment methods tend to treat the arch barrel in isolation, and therefore tend to ignore the interaction between the barrel and other elements of the bridge.

The method gives an assessment of the bridge's working and ultimate load capacity and an insight in to its residual life.

3 SOIL-STRUCTURE INTERACTION

3.1 *Overview*

Although it is well accepted that soil-arch interaction has a significant influence on the load carrying capacity of many masonry arch bridges, the complex nature of the soil-arch interaction is not well understood. In the past decades comprehensive research has been carried out on masonry arch bridges. However, the performance of the backfill has generally not been the focal point and almost all laboratory bridges tested to date have been backfilled with granular material bridges (e.g. Page 1993, Fairfield and Ponniah 1993, Melbourne and Gilbert 1995). Additionally, rigid abutments have been adopted in most laboratory tests which are less representative of those found in practice.

Recent intrusive investigations performed on local authority owned bridges in the United Kingdom (UK) have frequently identified that abutments are relatively insubstantial (e.g. of comparable thickness to the arch barrel), and that clay backfill is present. At present no experimental studies of the performance of clay backfilled arches seems to exist.

Additionally, it has frequently been found that abutments are of comparable thickness to the arch barrel (i.e. are relatively insubstantial). Unfortunately conventional arch assessment methods have not been calibrated using clay backfill materials, or using abutments which are not rigid. This makes realistic assessment difficult at the present time.

The principal aim of the investigations described in the subsequent sections is to gain an improved understanding of the soil-arch interaction. This is being achieved by conducting a series of carefully controlled laboratory soil-arch tests to obtain high quality data-sets, which can be used to calibrate numerical models.

3.2 *Small-scale experiments*

Since full-scale tests are comparatively time-consuming and expensive to set up, simple small-scale tests are being undertaken at the University of Sheffield alongside the full-scale tests being carried out at the University of Salford. The small-scale apparatus comprises an arch barrel comprising Perspex voussoirs, constructed inside a clear-sided Perspex tank. Simple instrumentation is employed: load cells to measure the applied load and a digital camera to capture structural and soil movements. With this apparatus tests can be performed rapidly, enabling parametric studies to be performed, thereby guiding the full-scale test programme.

Additionally, the influence of test variables which may be difficult to investigate at full-scale can be investigated. For example, the load capacity of an arch bridge subjected to flooding up to road level has been investigated. In the case of cohesionless fill materials, Archimedes principle of buoyancy implies that the capacity will be significantly reduced if the arch is flooded up to road level. This was confirmed experimentally using the small-scale apparatus.

The small-scale test apparatus is shown diagrammatically in Figure 4. The arch in Figure 4 has a span of 375 mm and the model is therefore a 1:8 scale of the full-scale model used for the experiments described in section 3.3. The test methodology incorporates supplementary characterisation tests for the soils, including soil density and shear strength.

Recent application of Particle Image Velocimetry (PIV) technology to geotechnical testing allows the

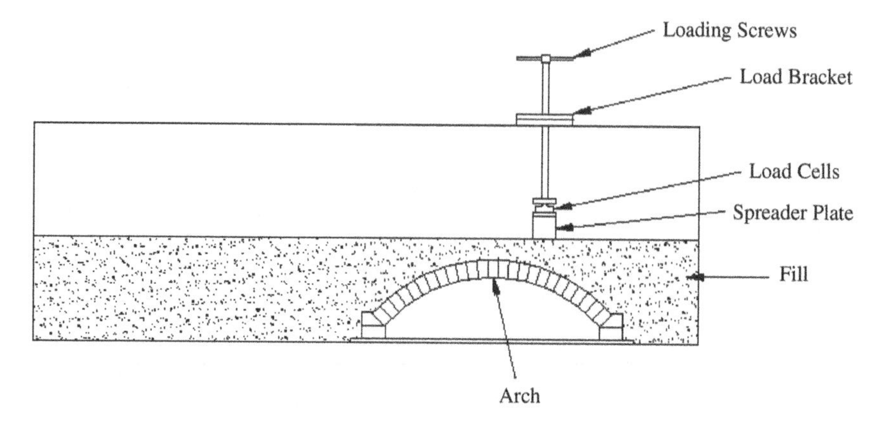

Figure 4. Small-scale test apparatus.

Figure 5. Sample GeoPIV image of the sand test using a fine mesh.

Figure 6. Sample GeoPIV image of crushed limestone test using a coarse mesh.

rapid investigation of soil-structure interaction mechanisms. By using two dimensional plane strain soil models tested in transparent walled chambers, images of the progressive soil-structure deformation can be captured using digital cameras. Utilization of the PIV techniques can then be used to generate the displacement field from the movements of individual clusters of soil grains. The GeoPIV computer program enables the displacements and 2D movement vectors of soil particles from two related but slightly different digital camera images to be plotted out. The software places a grid on the first image and allocates a series of targets at the centre of the grid squares. The first image is then compared with the second image. The Particle Image Velocimetry (PIV) analysis enables the conversion between image space and object space, and it is this that enables the creation of the displacement field.

Small-scale tests have been conducted on bridges with sand, crushed limestone and clay backfill, and typical PIV output is presented in Figure 5, 6 and 7 for these tests, respectively.

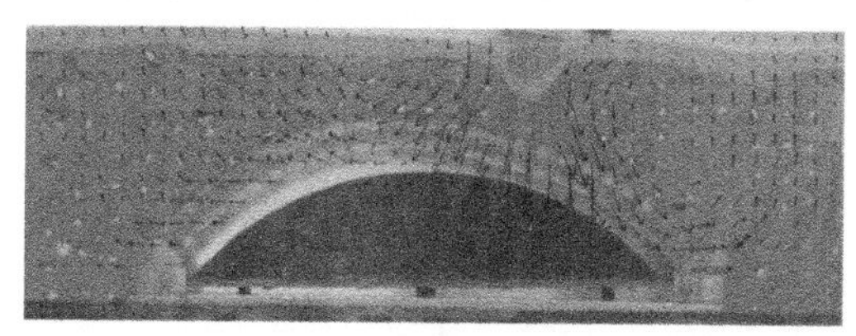

Figure 7. Sample GeoPIV image of clay test using a coarse mesh.

The tests on the sand indicate the relative movement of the sand grains clearly, in response to the surface loading and in response to the movement of the arch barrel. Movement of the abutments in to the sand backfill can also be observed.

As with the sand test, the extent of the failure surface is governed by the angle of friction of the backfill material, in this case, and to a lesser degree with the sand, the friction angle is high, and therefore the failure mechanisms extend out beyond the arch abutments.

The observed failure mechanisms for the clay test are markedly different to those for the granular material. With the latter, the failure surface was oriented approximately 45° to the barrel, however, for the clay, the failure surface trends towards the surface at a much steeper angle; indeed, above the abutments, the clay is effectively moving vertically upwards. The clay backfill used for this test was a soft to firm material.

3.3 *Large-scale experiments*

Three bridges have been tested to failure to date, with the following objectives:

- To test the performance of the apparatus (tank, imaging and instrumentation).
- To determine the influence of abutment fixity on arch behaviour.
- To provide benchmark test results.

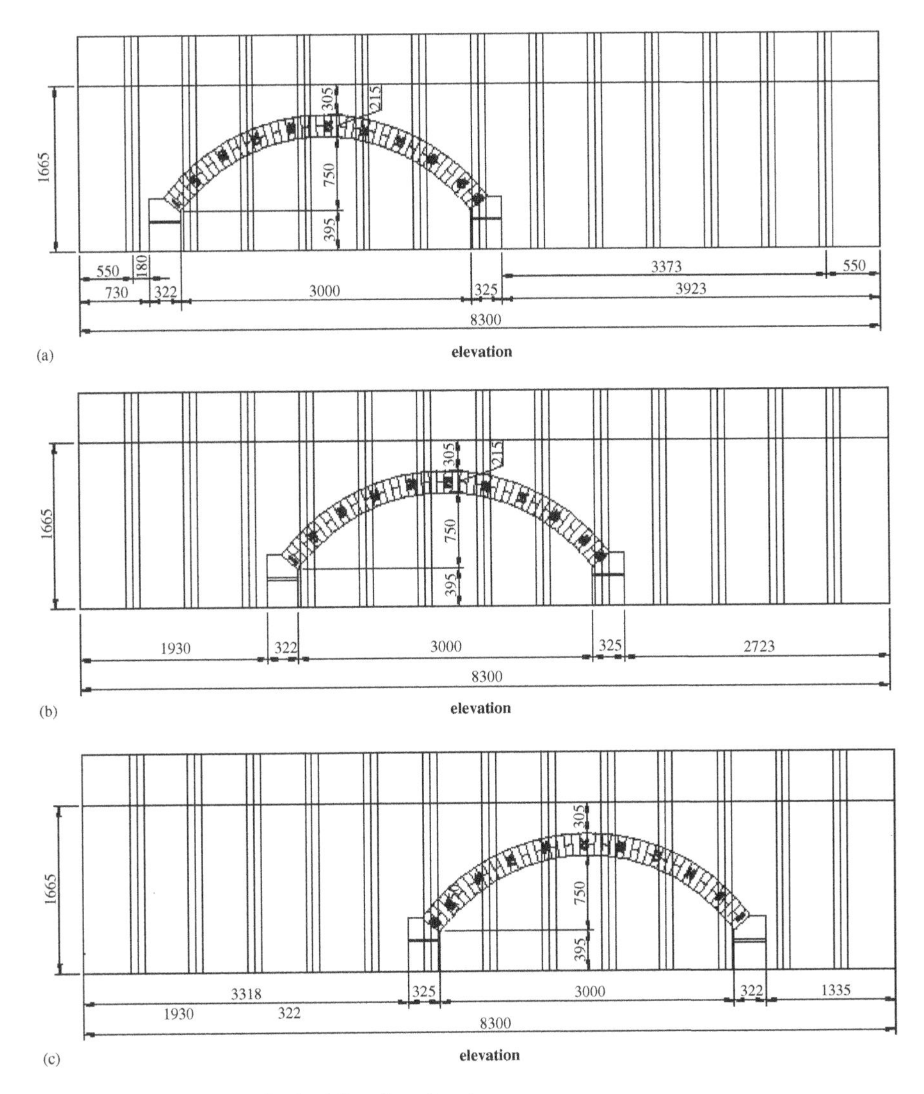

Figure 8. General test arrangement for the full-scale arch tests.

- To better understand the behaviour of masonry arch bridges with clay backfill.

The first test bridge (Figure 8a) was designed to be similar to the 3 m span bridges tested at Bolton in the 1990s (Melbourne & Gilbert 1995), thereby permitting direct comparison. However, unlike the Bolton bridges, which had been constructed between rigid abutments potentially movable abutments were specified and furthermore the walls of the plane-strain testing tank marked the edges of the bridge, rather than the brickwork spandrel walls used previously.

The second test bridge (Figure 8b) was designed to be identical to the first with the exception that fill material below the level of the crown of the arch was replaced with soft clay, representative of that found in some local authority owned bridges in the UK. Crushed limestone was used above the crown to reflect the fact that competent near surface road/sub-base material is normally present in real bridges. Additionally, had this not been used then a local failure of the soil in the vicinity of the applied load would have been likely.

The third test bridge (Figure 8c) was successfully tested recently and the results are currently being processed and should be completed shortly. The third test bridge was filled with the same limestone used for the first bridge test. Unlike Bridge 1, twelve pressure cells were incorporated in this bridge, enabling valuable additional information to be obtained.

Figure 8 shows the general arrangement for these tests which were carried out in a purpose built tank 8.3 m long by 2.1 m high. Transparent sides (50 mm in thickness) were provided for the tank to allow image analysis of the tests. The arches tested were segmental and had a 3 m span and a nominal span to rise ratio of 4:1. The backfills employed were compacted to a depth of 305 mm above the crown of the arch.

The midspan point of the arches varied according to the different anticipated soil-arch failure mechanisms.

For the first and third tests, the backfill used was an MOT Type 1 crushed limestone, whilst in the second test, a clay soil was used. Average properties are given in Table 1.

The fills were compacted in layers using vibrating compaction plates. Sensitive areas adjacent to the walls and the arches were compacted with a hand rammer. For the fill comprised MOT Type 1 graded crushed limestone (Bridge1 and Bridge3), the required weight for each layer was loaded into a crane mounted hopper for transfer to the rig and spread evenly using a shovel to the required thickness, and then compacted to the required specification using 10 kN (1t) whacker plate.

Filling and compacting of clay on such a large scale is challenging. Trial compactions were carried out using a Proctor compaction test and compaction on a small scale was assessed using a 1 m^2 box and a vibrotamper at the north end of the fully assembled tank. Guided by the experience obtained from these tests, a filling methodology was established. The clay was first wetted up in batches as appropriate to the required consistency and thoroughly mixed using the backacting arm of an excavator. It was then transferred to the test rig using the excavator bucket and spread evenly using a shovel to the required thickness. Each layer was then compacted using a 10 kN vibrating compaction plate suspended from a crane to ease handling.

Table 1. Material properties (Gilbert et al, 2007).

Material	Type	
Masonry units	Compressive strength (N/mm^2)	154
	Unit weight (kN/m^3)	23.2
Mortar	Compressive strength (N/mm^2)	1.9
	Unit weight (kN/m^3)	14.4–15.4
Masonry	Compressive strength (N/mm^2)	24.5
	Unit weight (kN/m^3)	22
Backfill	Unit weight (kN/m^3)	Limestone: 19.1 Clay: 22.1
	Angle of friction (deg)	Limestone: 46.4
	Cohesion (kN/mm^2)	Limestone: 22.4 Clay: 78

Sensitive areas adjacent to the walls and arch itself were compacted with a hand rammer. Each pressure cell was hand covered using clay. No compaction was employed directly adjacent or over the cells.

During clay filling, small soil samples were taken at regular intervals for moisture content testing. A total of 95 samples were collected and the average moisture content was 13.4%. Readings from a pocket penetrometer were also taken at regular distances after compaction of each layer during the whole filling process.

To protect the clay from drying out after compaction into the tank, several cloth and plastic sheets were used to cover the clay during and after filling.

Linear Variable Differential Transformer (LVDT) type displacement transducers were placed beneath the intrados (interior face) of the arch barrel and the top parts of the abutments to measure its movement. Additionally, Wall-Surface Soil Pressure Transducers (200 kPa and 500 kPa) were used to monitor earth pressure. A total of four pressure cells were embedded into the arch extrados (external face) in test 2 and twelve in test 3.

As with the small-scale tests, the relative movement of the arch barrel and the soil backfill was observed and analysed using PIV techniques, as discussed previously.

Finally, acoustic emission equipment was also installed, with 8 acoustic gauges installed in the intrados of the arch barrel to monitor real-time fracture activity.

A load was applied to the upper soil surface in increments of between 5–10 kN until collapse occurred. Figure 9 shows the collapse mechanisms in the three tests. The limestone filled bridge (test 1) failed at a load of approximately 125 kN, the clay filled bridge (test 2) failed at a load of approximately 90 kN, and Bridge 3 failed at the load of approximately 148 kN. The load-deflection response is presented in Figure 10, whilst Figure 11 shows a typical output from the GeoPIV analysis.

3.4 *Numerical modelling*

While the Finite Element Method is widely and successfully used when analyzing both steel and reinforced concrete structures, its utilization in the analysis of masonry still provokes some dispute in the scientific community. Several earlier developed FE based computer programs do not adequately describe the complex mechanical behaviour inherent to masonry elements. In the last decade significant effort has been made in the numerical modelling of masonry, but current knowledge about masonry mechanics is underdeveloped, in comparison with other fields such as concrete or steel. It is still very difficult to introduce correct constitutive laws to describe the basic material

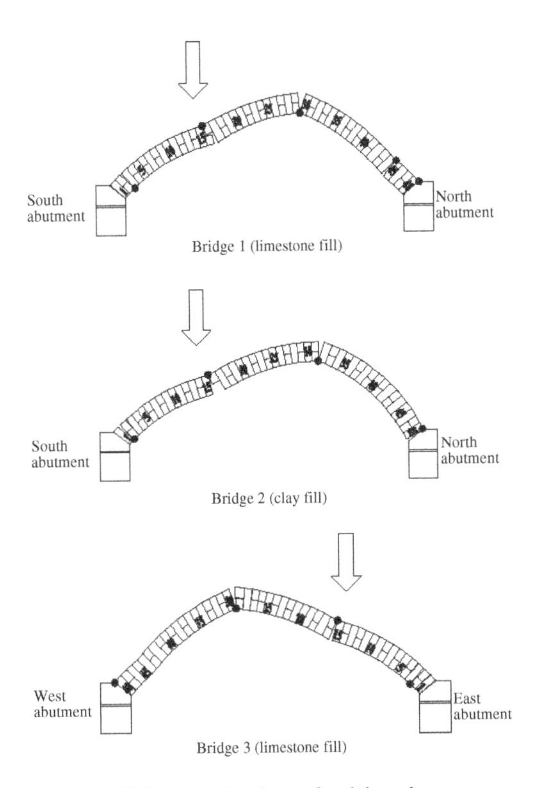

Figure 9. Collapse mechanisms of arch barrel.

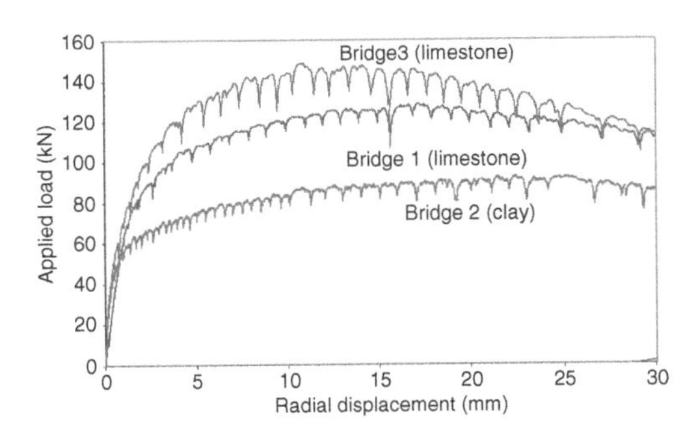

Figure 10. Load deflection response at quarter span position.

and to ensure the convergence of such highly non-linear problems. Special care must be taken, therefore, when using the results from finite element analysis of masonry structures.

A simple soil-arch interaction model and a full bridge model were created using the commercially available finite element package ANSYS 9.0. The simple soil-arch interaction model is mainly used for parametric studies within the service loading range. The material for both the arch and soil in the simple soil-arch model is assumed to be within the elastic range and the interface between the arch and soil is characterised as a friction contact interface (Wang & Melbourne, 2007).

The full bridge model, however, has been created to attempt to predict the failure of the corresponding full-scale test. In the full bridge model, the smeared cracking approach is adopted for modelling the masonry behaviour, while the non-linear behaviour of the soil is simulated with a Drucker-Prager constitutive material model. The same friction contact interface is adopted for the full bridge model.

The finite element mesh indicating the geometry of the model analysed, is presented in Figure 12. The boundary conditions were such that the abutments were fixed, and plane strain conditions were ensured. Typical output is shown in Figure 13 from the simulation of the full-scale test.

The finite element models have allowed a number of variables to be examined in the context of soil-arch interaction, including soil depth above the arch crown, the relative stiffness of arch barrel and soil and the contact stiffness at the arch-soil interface, mesh density, interface angle of friction and abutment fixity.

4 CONCLUSIONS

The paper has provided a review of the current state of research into the assessment of the stability of masonry arch bridges. The authors have sought to emphasise the importance of further studies in this area, in particular examining the role played by backfill in terms of soil-arch interaction. The experimental programme

Figure 11. Typical soil displacement around deforming arch barrel.

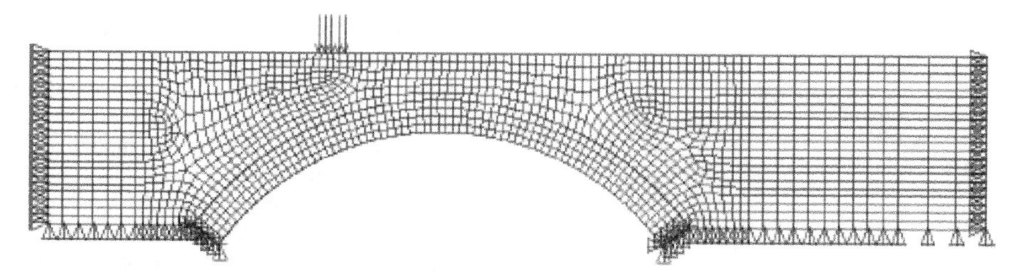

Figure 12. Finite element mesh, loading and boundary condition for simple soil-arch interaction model.

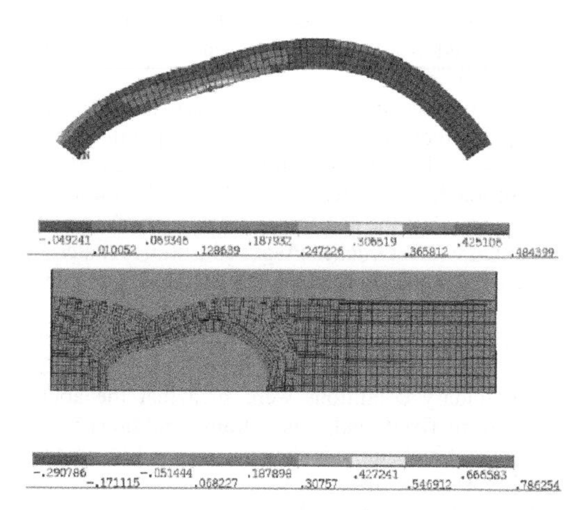

Figure 13. Major principal stress contours in the arch barrel and the abutment under a surface load of circa 60 kN.

has indicated that the limestone filled arch bridges proved capable of carrying more load than the clay filled bridges, indicating the importance of fill type on bridge capacity.

Additionally, this also demonstrates the importance of performing intrusive investigations to ascertain fill type in the case of bridges with 'borderline' load carrying capacity.

The test programme has allowed the suitability of the particle image velocimetry (PIV) technique to be confirmed and has highlighted the value of such techniques in order to help better understand the nature of the soil-arch interaction.

The numerical modelling programme is still in its infancy, but a number of important conclusions can be drawn from the results to date, such as the importance of the relative stiffness of the arch and the backfill soil. It was also observed that the fixity of the abutments can affect the load at which the first crack appears in the arch, and will in turn affect the load redistribution in the arch and abutments.

It is also evident that numerical modelling can simulate the full-scale tests, with further refinements.

REFERENCES

Fairfield, C. & Ponniah, D. 1993. Model tests to determine the effect of fill on buried arches. *Proc. ICE Structures and Buildings*, 104(4), pp 471–482.

Ford, T., Augarde, C. & Tuxford, S. 2003 Modelling masonry arches using commercial finite element software. In *Proc. 9th International Conference on Civil and Structural Engineering Computing*, Egmond aan Zee, The Netherlands, 2–4 September 2003.

Gilbert, M. 2006. *Guide to the use of RING 2.0 for the assessment for railway masonry arches*. UIC Report.

Gilbert, M., Smith, C.C. Wang, J., Callaway, P.A. and Melbourne, C. 2007. Small and large-scale experimental studies of soil-arch interaction in masonry bridges. *ARCH'07 – 5th international Conference on Arch Bridges*, Portugal. pp. 381–388.

Harvey, W. 2007. *Review of the MEXE method*. UIC Report (draft).

Melbourne, C. & Gilbert, M. 1995. The behaviour of multiring brickwork arch bridges. *The Structural Engineer*. 73(3), pp 39–47.

Melbourne, C., Wang, J. & Tomor, A. 2007. The analysis and assessment of masonry arch bridges. In Jan Bien et al (eds), *Sustainable Bridges: Assessment for Future Traffic Demands and Longer Lives*, Wroclaw, Poland 10–11 October 2007.

Page, J. 1993. *Masonry Arch Bridges*. London, HMSO.

UIC Code 778-3R. 1994. *Recommendations for the assessment of the load carrying capacity of existing masonry and mass-concrete arch bridges*, Paris.

Wang, J. & Melbourne, C. 2007. Finite Element Analyses of Soil-Structure Interaction in Masonry Arch Bridges. *ARCH'07 – 5th international Conference on Arch Bridges*, Portugal. pp. 515–523.

Advances in Transportation Geotechnics – Ellis, Yu, McDowell, Dawson & Thom (eds)
© 2008 Taylor & Francis Group, London, ISBN 978-0-415-47590-7

Reinforced soil structures under the influence of superstructure loads

I.E. Zevgolis
Geotechnical Engineer, Athens, Greece

P.L. Bourdeau
School of Civil Engineering, Purdue University, West Lafayette, IN, USA

W.-C. Huang
Fugro Consultants Inc., Houston, TX, USA

ABSTRACT: Reinforced soil structures combined with pile foundations are often used as bridge abutment – retaining walls systems. However, eliminating piles and using reinforced soil structures for direct support of bridge decks, would be a significant simplification in the design and construction of such systems. This would require special design considerations regarding the reinforcement elements, which would have to sustain the concentrated superstructure loads, in addition to the earth pressures from the approach embankment. The present study presents the preliminary results of an investigation using finite elements, about the performance of reinforced soil structures under the direct influence of concentrated loads. Analysis is performed in plane strain conditions. An elastic perfectly plastic Mohr-Coulomb model is used to capture the behavior of the reinforced granular material. The interaction between soil and reinforcement is modeled using elastoplastic interface elements. Analysis is performed taking into account stage construction by using sequential calculation phases. For the conditions examined in the present study, the results indicate that superstructure concentrated loads increase the magnitude of the induced tensile forces and influence the locus of points of maximum tensile forces within the reinforced soil mass. In addition, these loads increase the magnitude of the induced horizontal deformations of the facing panels, particularly on the upper side of the wall.

1 INTRODUCTION

Soil reinforcement technologies have found numerous applications in transportation systems all over the world (Goughnour and DiMaggio, 1978; Mitchell and Christopher, 1990; Schlosser, 1990; Jones, 1996). Soon after the introduction of these technologies in civil engineering, reinforced soil structures combined with piles, soon became a reliable solution of choice for bridge abutment retaining walls. However, eliminating piles and using reinforced soil for direct structural support of bridge decks is a significant simplification in the design and construction of such systems. It results in faster construction of highway bridge infrastructures, and in significant cost savings in construction, maintenance and retrofitting. On the other hand, the above configuration requires special design considerations, since the reinforced soil wall shall be designed to sustain not only the earth pressure from the approach embankment, but also the concentrated loads from the bridge superstructure (Elias et al., 2001).

The objective of the present study is to investigate the performance of reinforced soil structures under the influence of superstructure loads. In this context, a case example of a typical reinforced soil wall that supports a single span bridge is analyzed, using the finite element method. Results are provided in terms of induced tensile forces on the reinforcement elements and horizontal deformations of the wall's facing. In the light of these data, the particular effect of the bridge loads on the above quantities is discussed.

2 NUMERICAL MODELING

2.1 *Finite element program*

A software package, *Plaxis v.8.2.*, was used for the analysis (Brinkgreve, 2002). This software, developed at the Technical University of Delft specifically for analyzing geomechanics and soil-structure interaction problems using the finite element method, has been extensively used in academia and industry. Its capabilities to simulate stage construction and interface

Table 1. Basic features of the analyzed example.

Properties	Values
Visible height of the structure, H	5 m
Embedment depth, D	1 m
Width of the structure, W	7 m
Bridge seat width, b	2 m
Bridge seat height, h	2.4 m
Length of bridge span, L	30 m
Dead concentrated vertical load, F_V	215 kN/m

response between soil and other materials are critical features for the application considered in this study.

2.2 *Basic features*

Analysis was performed in plane strain conditions. Horizontal fixities ($u_x = 0$) and full fixities ($u_x = u_y = 0$) were imposed on vertical and bottom boundaries of the mesh, respectively, as displacement boundary conditions. Basic geometric and loading conditions are tabulated in Table 1. The construction sequence of reinforced soil structures followed in practice was taken into consideration in the numerical simulations by using sequential calculation phases. Initial conditions were calculated using K_0 conditions, where K_0 was calculated based on Jaky's formula. Groundwater conditions were simulated by assigning a phreatic level at the ground surface (i.e. at the top of the foundation soil) and using γ_w equal to 10 kN/m^3.

Based on the results of limit equilibrium analysis and conventional design methods, horizontal and vertical spacing of the reinforcement layers was set at 0.50 m, with the first layer being placed 0.30 m above the leveling pad of the wall. Safety with respect to ultimate limit states was verified using conventional methods of design.

The soil profiles were modeled with 15-node triangular elements. The Mohr-Coulomb model was used to capture the behavior of the reinforced soil, the approach embankment, and the foundation soil. Constitutive model (effective) parameters are provided in Table 2.

Steel strip reinforcement elements were modeled using line elements that can sustain tensile forces, but cannot sustain compressive forces. Facing panels and EPDM (Ethylene Propylene Diene Monomer) bearing pads were modeled using elastic plates. Values of properties of the above are tabulated in Table 3 (scaled to equivalent values for plane strain conditions). Finally, the interaction between soil and reinforcement elements, as well as between soil and facing panels, was captured by interface elements, whose behavior was described by an elastic – perfectly plastic model and a Coulomb failure criterion.

Note that the presence of discrete strips and of the resulting interaction between soil and strips, involves

Table 2. Mohr – Coulomb model parameters.

Parameter	Reinforced soil mass	Approach embankment	Foundation soil
γ_{unsat} (kN/m^3)	20	19	15.5
γ_{sat} (kN/m^3)	21	20	17
E_{ref} (kN/m^2)	50,000	40,000	50,000
c_{ref} (kN/m^2)	0	0	5
φ (°)	34	30	30
ψ (°)	7	6	0
ν (–)	0.306	0.320	0.350

γ_{unsat}: unsaturated unit weight, γ_{sat}: saturated unit weight, E_{ref}: Young modulus, c_{ref}: cohesion, φ: friction angle, ψ: dilatancy angle, ν: Poisson's ratio.

Table 3. Reinforcement, facing panels, and bearing pads parameters (scaled for plane strain conditions).

Parameter	Reinforcement elements	Facing panels	Bearing pads
EA (kN/m)	80,000	3,500,000	533.3
EI (kN·m^2/m)	–	5,716.7	0.321
d(m)	–	0.14	0.085
w (kN/m/m)	–	3.29	0
ν (–)	0.30	0.20	0.495

EA: axial stiffness, EI: bending stiffness, d: thickness, w: weight, ν: Poisson's ratio.

a 3-D mechanism that cannot be modeled in plane strain analysis, unless an approximation is made. This approximation consists on replacing strips and the corresponding contact areas with soils, with a plate extended to the full width and breadth of the structure (Figure 1). So, strips and the corresponding interaction are considered continuous in the out-of-plane direction (z direction). Further details on the mathematical formulation of this approximation are provided by Zevgolis (2007).

3 RESULTS AND DISCUSSION

3.1 *Construction stages of interest*

As already mentioned, the construction sequence that is followed in practice was taken into account in the simulations by using sequential calculation phases (of approximately 0.25 to 0.30 m each). In total, simulation was completed in 36 calculation phases. Results here are provided for three construction stages of interest. The first is the stage where the erection of the reinforced soil wall has just been completed. So, in this case it is only the self-weight of the wall that is accounted for in the calculation of the induced tensile stresses (Figure 2a). The second stage is the one

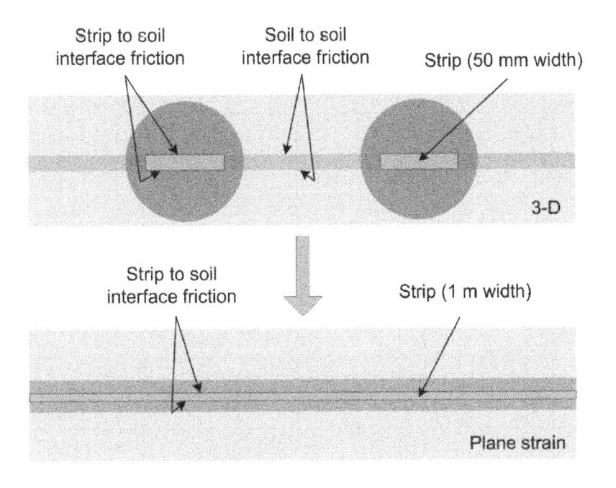

Figure 1. Idealization of 3-D interaction mechanism in plane strain conditions.

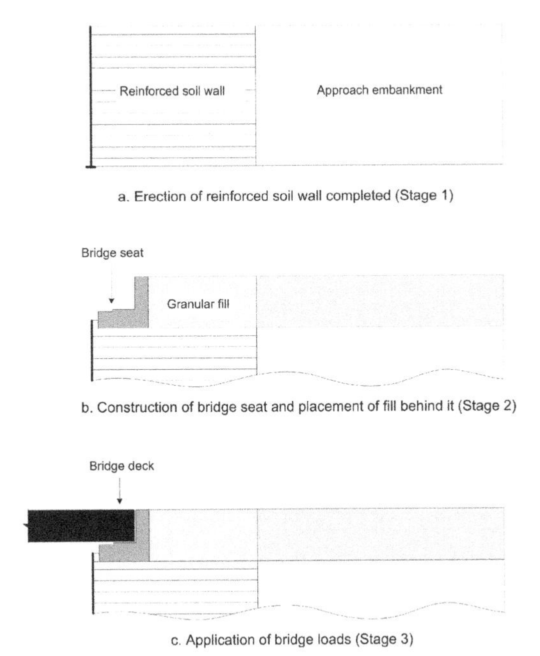

Figure 2. Geometry of different construction stages.

where the cast-in-place (CIP) bridge seat has been constructed and the fill behind it, has already been put in place (Figure 2b). Finally, the third stage corresponds the time immediately after the application of the concentrated bridge loads (Figure 2c).

3.2 *Influence on the magnitude of induced tensile stresses*

Figure 3 shows the calculated tension on some representative steel strips, for the three construction stages

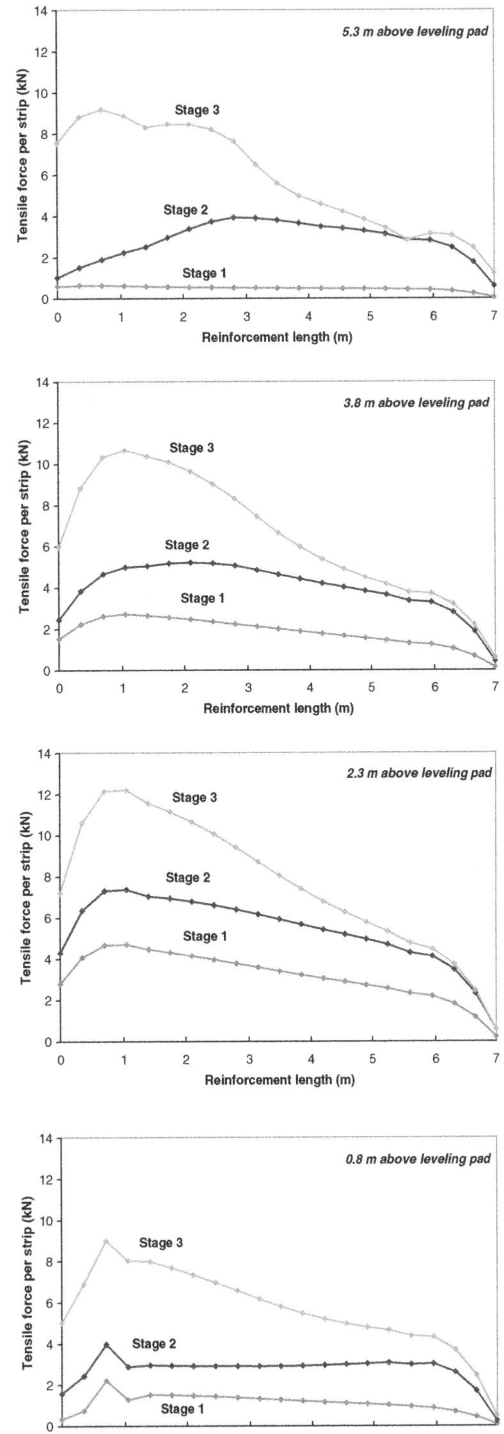

Figure 3. Tensile forces along the reinforcement elements.

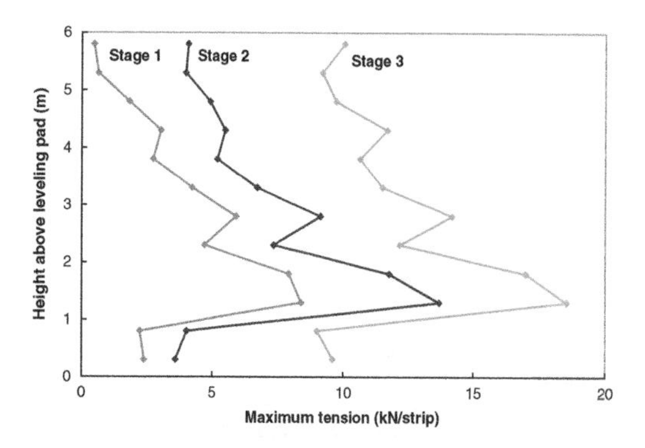

Figure 4. Values of maximum induced tension over depth.

that were described above. The figure shows that tension on the right-hand end of the reinforcement is zero. It also indicates the general trend that is often observed in these type of structures: tension gradually increases as we move from the rear to the front side of the strips, reaches a maximum at a short distance from the facing of the wall, and then slightly decreases at the connection with the facing panels. As far as the different construction stages are concerned, the following comments can be made: The construction of the CIP footing and the placement of the fill behind it (Stage 1 to Stage 2), causes a relatively uniform increase in the induced tensile forces. Note that this is not the case for the strip located 5.3 m above the leveling pad, since the left-hand side of this strip is only subjected to the weight of the CIP footing, and not the weight of the fill behind it. When the bridge deck is constructed and the concentrated loads are being applied (Stage 2 to Stage 3), tension increases significantly in the front part of the reinforcements. This increase naturally decays as we move towards the rear side of the inclusions.

Figure 4 shows the values of maximum tension that is induced on each reinforcement strip. As shown in the figure, tension generally increases with depth for all three construction stages. This is reasonable, due to the overburden pressure that increases with depth. Particularly in reference to Stage 3, it is indicated that the concentrated loads from the bridge superstructure influence the magnitude of maximum tension on the strips located in the proximity of the bridge seat.

3.3 *Influence on the locus of maximum tensile stresses*

Concentrated bridge loads were found to influence the location of points of maximum tension, as well. More specifically, as illustrated in Figure 5, when the erection of the reinforced soil wall is completed (Stage 1), the loci of maximum tension points is at maximum 1 m away from the facing panels. When the fill behind the footing is placed, the loci of maximum tension

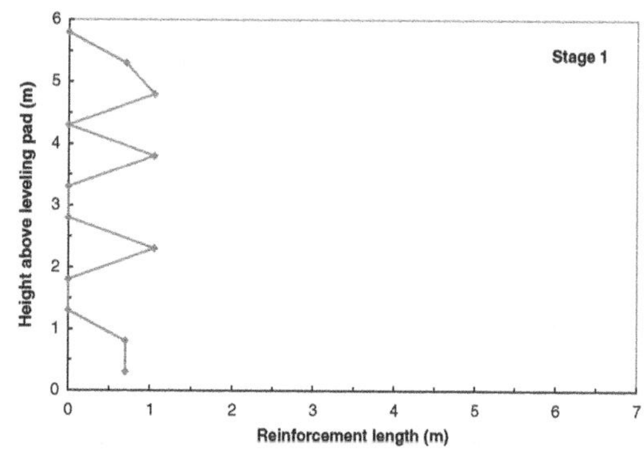

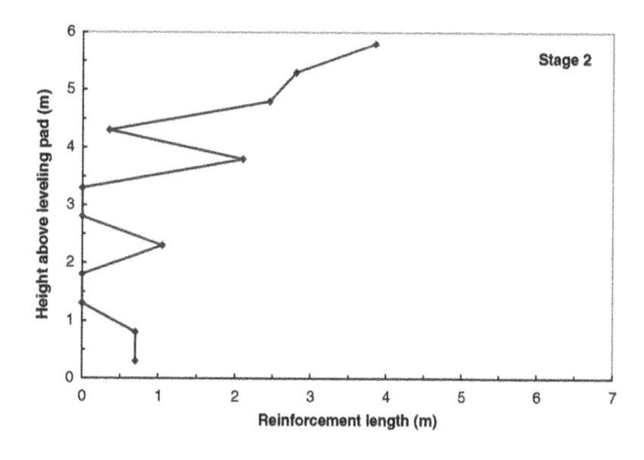

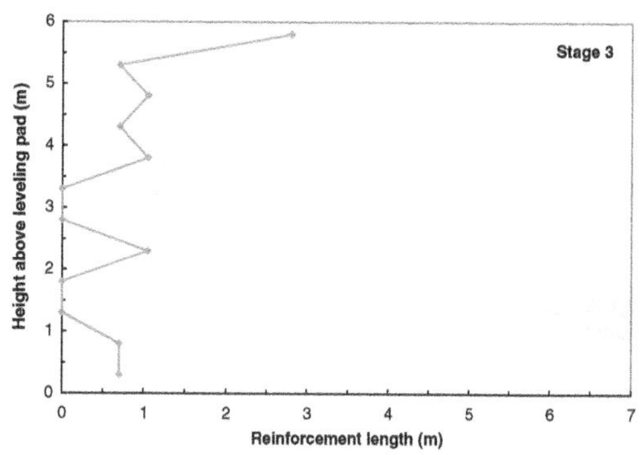

Figure 5. Location of points of maximum tension.

points gradually extend further towards the rear side of the inclusions. When the concentrated bridge loads are applied, the loci is almost identical with that of the first stage, with the exception of the layer immediately below the bridge seat.

3.4 *Influence on the deformations of the facing panels*

Issues relative to deformations and serviceability limit states of structures that are used as direct bridge abutments are addressed elsewhere (Zevgolis and

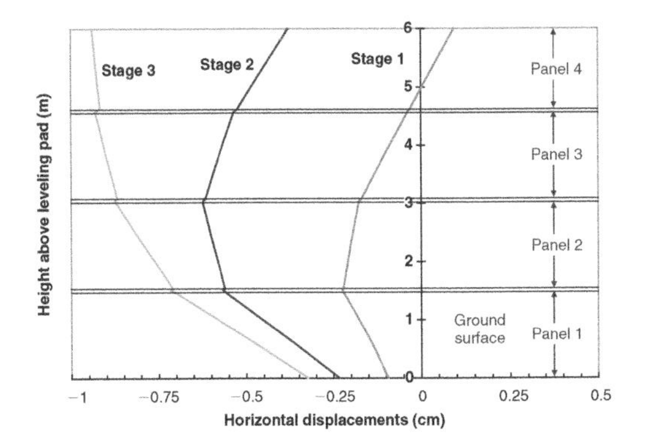

Figure 6. Horizontal deformations of facing panels.

Bourdeau, 2008). However, in connection to the forces developed on the reinforcement elements, Figure 6 shows the induced horizontal deformations of the facing panels (negative values indicate movement towards the left side). As shown in the figure, the construction of the CIP footing and especially the placement of the fill behind it (Stage 2), causes a relatively uniform increase in the horizontal displacements. On the other hand, the influence of application of the bridge loads is mostly concentrated on the upper part of the wall.

4 CONCLUSIONS

Using reinforced soil structures for direct structural support of bridges, i.e. eliminating the use of deep foundations, would result in faster construction of bridge infrastructures and in significant cost savings.

In order to investigate the performance of such systems under the influence of superstructure loads, a finite element analysis was performed for a typical example case. For the assumptions used in this study, it was shown that concentrated loads applied by the superstructure increase the magnitude of induced tensile forces, influence the locus of points of maximum tensile forces within the reinforced soil, and increase the horizontal deformation of the wall's facing. For an in-depth study of the issue, further analyses that would cover a wide range of cases are necessary.

REFERENCES

Brinkgreve, R.B.J. (2002). *Manual for Plaxis 2D – Version 8.* Balkema, Plaxis b.v., Netherlands.

Elias, V., Christopher, B.R., and Berg, R.R. (2001). "Mechanically stabilized earth walls and reinforced soil slopes – Design & construction guidelines." FHWA-NHI-00-043, US Department of Transportation, Federal Highway Administration, Washington D.C., USA.

Goughnour, R.D., and DiMaggio, J.A. (1978). "Soil-reinforcement methods on highway projects." *Symposium on Earth Reinforcement*, April 27 1978, Pittsburgh, USA, p. 371–399.

Jones, C.J.F.P. (1996). *Earth reinforcement & soil structures.* Thomas Telford Ltd. (new edition).

Mitchell, J.K., and Christopher, B.R. (1990). "North American practice in reinforced soil systems." *Design and Performance of Earth Retaining Structures (Geotechnical Special Publication No. 25)*, 18–21 June 1990, Ithaca, USA, p. 332–346.

Schlosser, F. (1990). "Mechanically stabilized earth retaining structures in Europe." *Design and Performance of Earth Retaining Structures (Geotechnical Special Publication No. 25)*, 18–21 June 1990, Ithaca, USA, p. 347–378.

Zevgolis, I. (2007). "*Numerical and Probabilistic Analysis of Reinforced Soil Structures.*" PhD Dissertation, Purdue University, West Lafayette, IN, USA.

Zevgolis, I.E., and Bourdeau, P.L. (2008). "Assessment of settlements of reinforced soil structures used as direct bridge abutments." *Proceedings of the 2nd BGA International Conference on Foundations (ICOF 2008)*, IHS BRE Press, 24–27 June 2008, Dundee, Scotland.

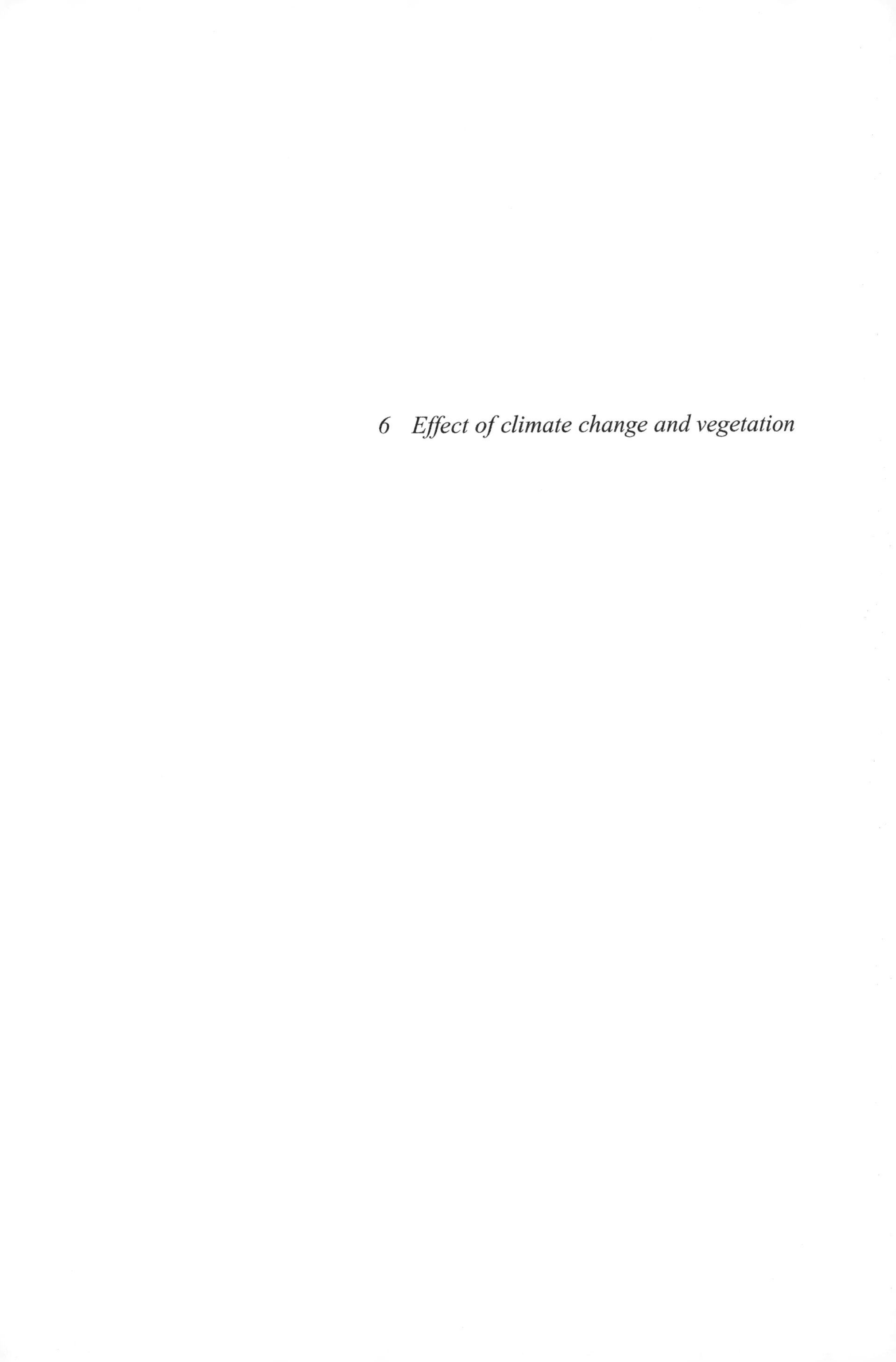

6 *Effect of climate change and vegetation*

Advances in Transportation Geotechnics – Ellis, Yu, McDowell, Dawson & Thom (eds)
© 2008 British Geological Survey, Nottingham, ISBN 978-0-415-47590-7

Non-invasive time-lapse imaging of moisture content changes in earth embankments using electrical resistivity tomography (ERT)

J.E. Chambers, D.A. Gunn, P.B. Wilkinson & R.D. Ogilvy
British Geological Survey, Nottingham, UK

G.S. Ghataora & M.P.N. Burrow
Department of Civil Engineering, University of Birmingham, Birmingham, UK

R. Tilden Smith
Great Central Railway (Nottingham) Limited, Nottingham, UK

ABSTRACT: Earth structures, such as embankments, require ongoing monitoring and maintenance to identify potential failure zones and to compensate for the effects of settlement. Extreme weather events leading to prolonged periods of desiccation or saturation are becoming more frequent and threaten embankment stability. In this paper the development of electrical resistivity tomography (ERT) as a non-invasive tool for characterizing and monitoring earth embankments is described.

A study is presented in which ERT was applied alongside conventional intrusive techniques to investigate and monitor a section of Victorian era embankment on the Great Central Railway. ERT electrodes were permanently installed as a series of linear 2D arrays, both parallel and perpendicular to the long-axis of the embankment. The resulting ERT images, when calibrated using intrusive sampling methods, revealed the spatial variability of the embankment soils and were used to identify major discontinuities between material types at locations associated with poor track geometry. Subsequently time-lapse ERT images were used to monitor moisture content changes in the embankment; these images revealed both the spatial extent and magnitude of water content variations, and were used to assess the effect of an exceptionally prolonged and heavy period of rainfall during the summer of 2007.

1 ELECTRICAL SURVEY DESIGN

The test area (Figure 1) covers a 100 m length of embankment ($y = -20$ to 80 m), which includes the crest and the two flanks ($x = 0$ to 31 m). Geotechnical tests (trial pits, boreholes & cone penetration tests), described by Gunn et al. (2007), were restricted primarily to the crest of embankment ($x = 8$ to 24 m). However, the resistivity images extended the full length and width of the area (Figure 2).

Prior to the field survey, synthetic modelling studies were undertaken to quantify the effects of a conductive metal rail on electrical resistivity tomography (ERT) measurements made along a line running parallel to it for a range of standard ERT electrode configurations (e.g. Dahlin and Zhou, 2004). It was considered that conductive rails could potentially short-circuit the injected current, thereby distorting the resistivity images. These studies revealed that the dipole-dipole array configuration was less affected by the

Figure 1. Test site location. Ordnance Survey ©Crown copyright. All rights reserved. Licence number 100017897/2008.

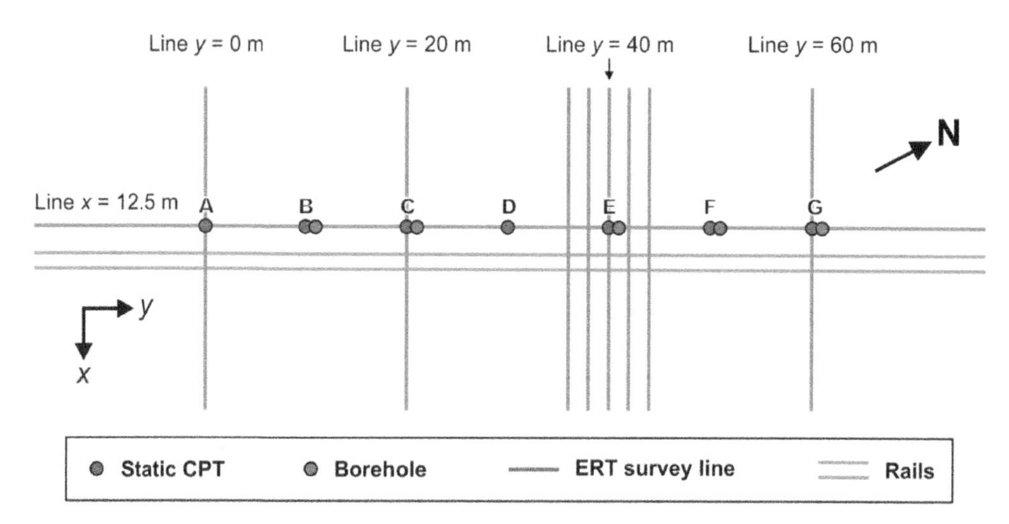

Figure 2. ERT survey line and intrusive sample positions.

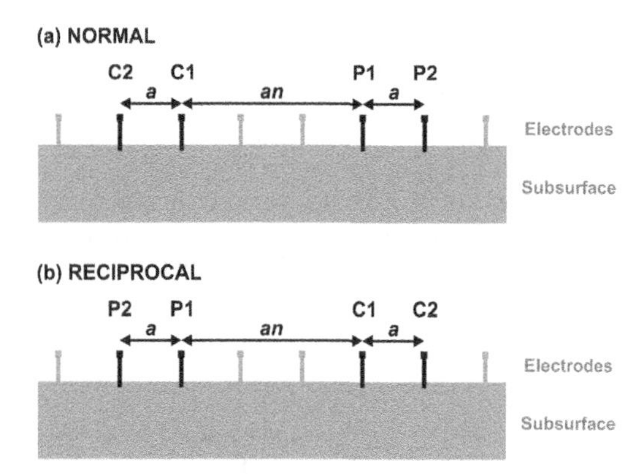

Figure 3. Normal and reciprocal dipole-dipole measurement configurations.

conductive rails than other standard array types; moreover, it was found that given a sufficiently high contact resistance between the rail and the ground, the influence of the conductive rails on the resistivity image would be small; in this case assessment of the field data showed that rail-ground contact resistances were sufficiently high to have effectively eliminated the problem of current short circuiting through the rails.

All resistivity data were collected using the dipole-dipole array configuration (Figure 3), and were inverted using Res2DInv software (Loke, 2006) to produce images of the resistivity distribution within the embankment. The dipole-dipole command sequences comprised both normal and a full set of reciprocal measurements. Reciprocal measurements provide a robust means of assessing ERT data quality and determining reliable and quantitative data editing criteria. Data collected parallel to the embankment long-axis (y) used electrode arrays with dipoles (a) of 1.5, 3, 4.5 and 6 m, and unit dipole separations (n) of 1 to 8, whilst measurements collected on the perpendicular sections employed dipoles of 1, 2, 3, and 4 m, and unit dipole separations of 1 to 8. Resistivity images collected during an initial reconnaissance survey of the site during September 2005 are shown in Figure 4.

2 EMBANKMENT CHARACTERISTICS

Intrusive investigations within the test area have shown that the embankment is approximately 5.5 m high, and is underlain by Mercia Mudstone bedrock. The embankment comprises weathered mudstone material between $y = -20$ and 40 m, which was excavated from a nearby cutting and is characterized by relatively low resistivities of less than 100 Ωm (Figure 4, ERT line $x = 12.5$ m). Beyond $y = 40$ m the composition of the embankment changes to gravel, sand and silt, and is characterized by resistivities significantly in excess of 100 Ωm. At the interface between these two distinct lithologies significant distortions in track geometry can be observed, which are likely to be the result of differential settlement. The compositional changes shown in the resistivity section parallel to the embankment long-axis are also seen in the perpendicular sections at $y = 0$, 20, 40 & 60 m (Figure 4). However, these sections also show significant features associated with variations in moisture content

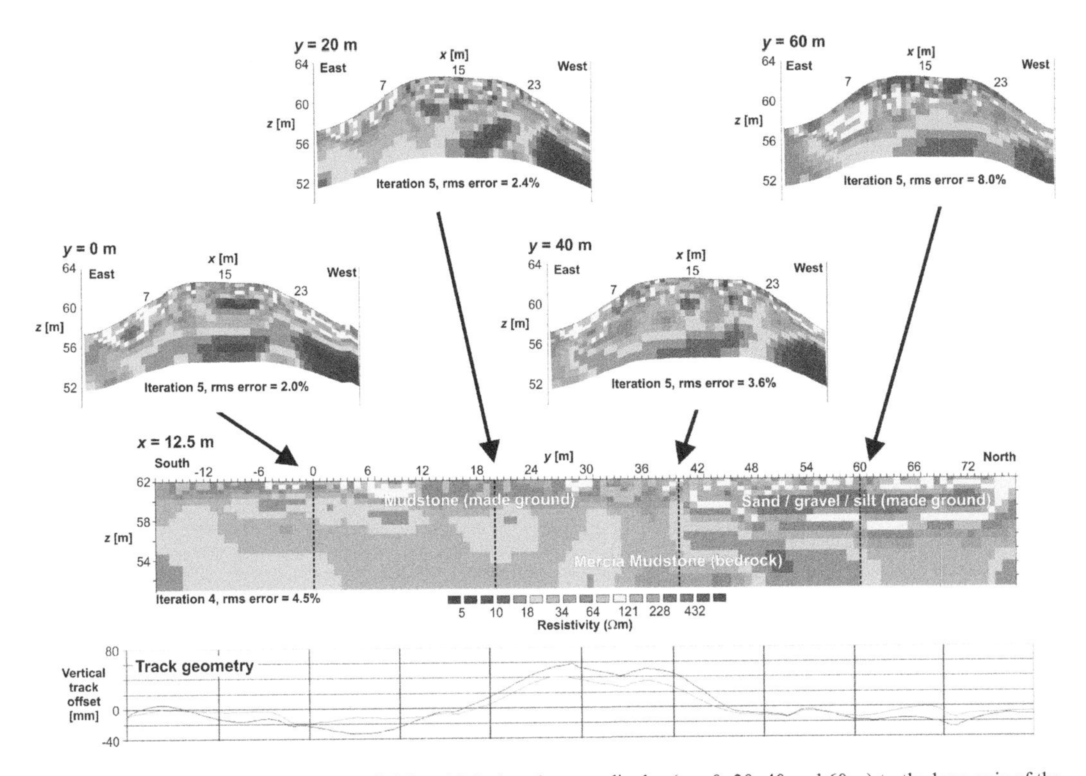

Figure 4. Resistivity models (top) parallel (x = 12.5 m) and perpendicular (y = 0, 20, 40 and 60 m) to the long axis of the embankment (September 2005). Track geometry profiles (bottom); blue trace – west rail, pink trace – east rail. Track geometry data provided by the Railway Research Centre, University of Birmingham, UK.

within the embankment. In particular, the flanks are dominated by relatively high resistivities due to moisture loss from evaporation and transpiration during the preceding summer months.

3 SEASONAL MONITORING

Permanent ERT imaging arrays were installed to better understand the spatial extent and magnitude of moisture content changes within the embankment in the area of poor track geometry. The results from one of these sections, located at $y = 40$ m are described here. Datasets have been collected at intervals of approximately 6 weeks in order to monitor seasonal changes. To date, datasets have been collected for the period July 2006 to November 2007. These time-lapse data were inverted using Res2DInv. The July dataset was inverted with a smoothness-constrained (L_2-norm) least-squares method to produce a reference model (Figure 5). Then each of the subsequent datasets was inverted using the July 2006 image as an initial model. A spatial L_2-norm constraint was applied to the resistivity differences between the reference model and the model being generated to ensure that these differences

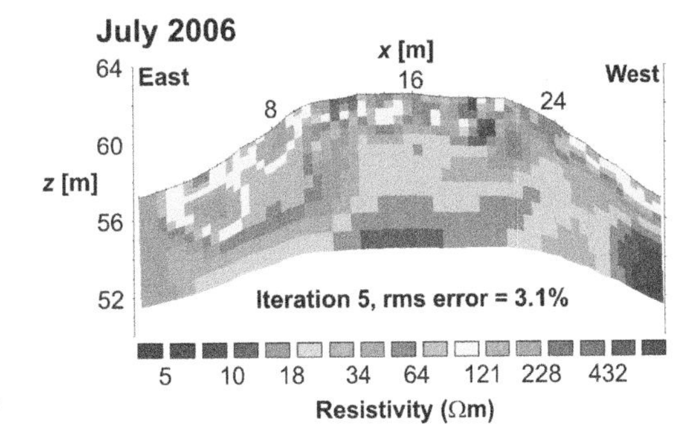

Figure 5. July 2006 ERT reference model at $y = 40$ m.

varied smoothly with position. This type of cross-model constraint has been shown to be effective in reducing artefacts that can occur if time-lapse data are inverted independently (Loke, 1999).

The time-lapse resistivity models from August 2006 to November 2007 have been plotted as differential images, which show the percentage change in resistivity relative to the July 2006 reference model; these differential time-lapse resistivity models are shown in Figure 6. This figure shows that drying of the

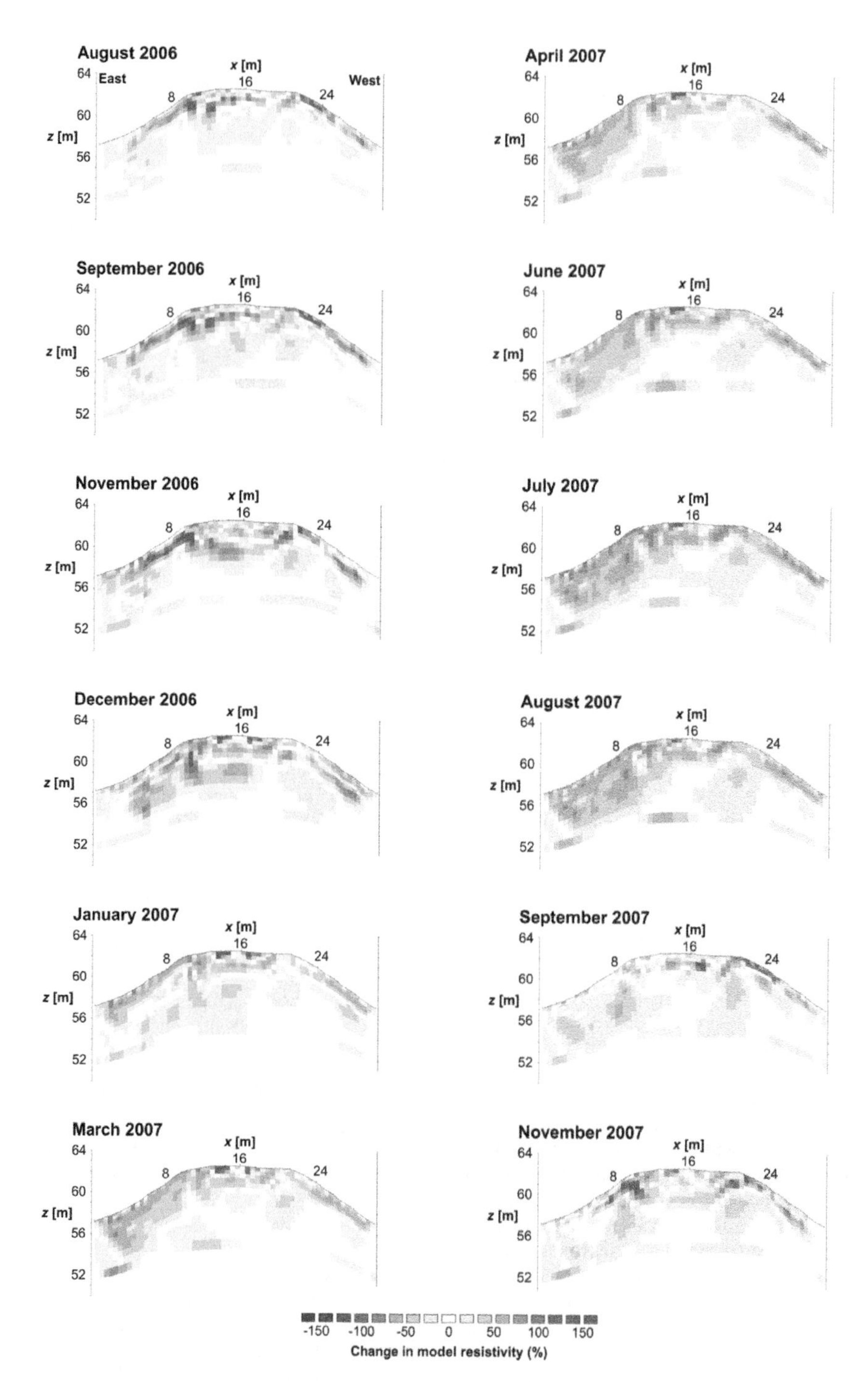

Figure 6. Differential resistivity images at y = 40 m showing percentage change in resistivities from the baseline model (July 2006).

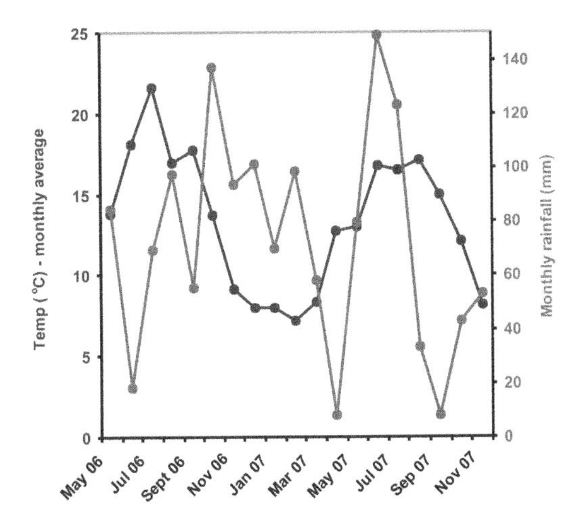

Figure 7. Rainfall and temperature data recorded from a nearby weather station located in Loughborough.

embankment flanks continued until November 2006, after which time resistivities throughout the near surface of the embankment decreased. By March 2007 elevated moisture levels, as identified by decreases in resistivity, are shown to extend to more than 4 m below ground level. Greater infiltration of moisture appears to have occurred on the eastern flank, from $x = 0$ to 10 m, than is seen on the western flank; this may be a function of the prevailing wind direction resulting in uneven evaporation across the embankment, or differences in the density and type of vegetation on the two flanks. The general decrease seen in near surface resistivities over the winter months does not appear to be a function of increased rainfall, which is relatively consistent over the monitoring period (Figure 7). Instead it is likely to be primarily due to decreased vegetation cover and lower temperatures during the winter, which reduced evaporative loss of moisture from the embankment allowing water to penetrate deeper into the subsurface.

Between March and August 2007 the embankment continued to become wetter despite increasing temperature and evapotranspiration. The cause was the extraordinarily high rainfall in June, July and August (Figure 7). The effect of the wet summer on the moisture distribution in the embankment can be seen clearly by comparing the August 2006 timelapse image with that from August 2007. The climatic conditions leading up to August 2006 were far more typical of what would normally be expect during the summer months, and hence the embankment was significantly drier than at the same point in 2007.

The period between September and November 2007 was characterized by much lower rainfall than the preceding months. Consequently, the embankment begins drying, with resistive zones appearing on the crest and flanks of the embankment during September to November.

4 CONCLUSIONS AND FUTURE WORK

A study is described in which 2D track-parallel ERT has been used to identify significant compositional variations associated with embankment instability. Further work is required to prove the technique at other sites, and to model and assess the effects of topography on ERT images generated from measurements collected parallel to the long axes of embankments.

Qualitative seasonal changes in embankment moisture content have been monitored using 2D ERT. The monitoring period has included the heavy rainfall that occurred from June to August 2007, which prevented the embankment from drying out over the summer. Monitoring will continue for at least an additional 10-month period to allow the collection of data over two entire yearly cycles. Additional ERT monitoring arrays have recently been installed at other locations within the test area to study the response of different embankment materials (e.g. gravel, sand and silt) to drying and wetting cycles. Direct in-situ and laboratory-based moisture content and resistivity measurements of embankment materials will be carried out to calibrate the resistivity models and to move towards a quantitative assessment of moisture content changes using ERT data.

The investigation of the relationship between moisture content and the instability of the embankment is a key element of our ongoing research; in particular the shrink/swell effects associated with clay minerals, the mobilization and precipitation of soluble constituents, and changes in strength caused by varying water content are being considered.

Given its low cost and relative rapidity compared to many intrusive methods, ERT may be suitable as a reconnaissance and monitoring tool for identifying potential areas of instability in old earth structures.

ACKNOWLEDGEMENTS

This paper is published with the permission of the Executive Director of the British Geological Survey (NERC).

REFERENCES

Dahlin, T., and Zhou, B. 2004. A numerical comparison of 2D resistivity imaging with 10 electrode arrays. *Geophysical Prospecting* 52: 379–398.

Gunn, D.A., Reeves, H., Chambers, J.E., Pearson S.G., Haslam, E., Raines, M., Tragheim, D., Ghataora, G.S., Burrow, M.P.N., Weston, P., Thomas, A., Nelder, L.M.,

Lovell J.M. & Tilden Smith, R. 2007. Assessment of embankment condition using combined geophysical and geotechnical surveys. *Proc. 9th International Con. Railway Engineering, University of Westminster, London, 20–21 June.*
Jackson, P.D., Northmore, K.J., Meldrum, P.I., Gunn, D.A., Hallam, J.R., Wambura, J., Wangusi, B., and Ogutu, G. 2002. Non-invasive moisture monitoring within an earth embankment - a precursor to failure. *NDT & E International* 35: 107–115.
Loke, M.H. 2006. RES2DINV Rapid 2-D resistivity and IP inversion using the least-squares method. *Manual, Geotomo Software, Penang, Malaysia.*
Loke, M.H., 1999. Time-lapse resistivity imaging inversion. *5th Meeting of the Environmental and Engineering Geophysical Society European Section Proceedings.* Em1.
Sjodahl, P., Dahlin, T. and Zhou, B. 2006. 2.5D resistivity modeling of embankment dams to assess influence from geometry and material properties. *Geophysics* 71: G107–G114.

Advances in Transportation Geotechnics – Ellis, Yu, McDowell, Dawson & Thom (eds)
© 2008 Taylor & Francis Group, London, ISBN 978-0-415-47590-7

Predicting seasonal shrink swell cycles within a clay cutting

O. Davies, M. Rouainia, S. Glendinning & S.J. Birkinshaw
Dept. of Civil Engineering and Geosciences, Newcastle University, England, UK

ABSTRACT: A numerical method for predicting the seasonal pore water pressure changes and consequently the shrinking and swelling of a clay cutting is proposed. The cutting is located on the A34 Newbury bypass and has been extensively monitored by Smethurst et al (2006) since January 2003. A hydrological finite difference model, SHETRAN, is used to accurately predict the pore water pressure changes within the cutting using meteorological data together with soil data and vegetation properties. The daily surface pore water pressure changes are then imported into a geotechnical finite difference program FLAC tp flow which is used to predict the mechanical response of the cutting. This paper details the comparability of the hydraulic simulation with 1 year observed data and further predicts the hydraulic and mechanical response over the next 3 years.

1 INTRODUCTION

Observations and numerical modeling of infraearthwork embankments and cuttings have revealed that these earthworks shrink and swell seasonally. Such shrink swell cycles can cause irrecoverable plastic strains and possibly some softening of the earthwork material. This softening could eventually lead to ultimate failure of the earthwork. It is therefore important to be able to accurately model these seasonal shrink swell cycles to determine the magnitudes of movements and so also the potential strain softening of the embankment or cutting material. In order to model the movements caused by seasonal pore pressure fluctuations the modeler must first be able to accurately predict the seasonal pore pressures. This paper outlines a method of first predicting the pore water fluctuations within a cutting using a 3D hydrological model SHETRAN. The surface pore pressure predictions obtained by SHETRAN were then transferred to a geotechnical numerical model FLAC two phase (tp) flow which is capable of modeling the earthworks response to these fluctuations.

1.1 *SHETRAN*

SHETRAN is a 3D coupled surface/subsurface physically based spatially distributed finite difference model for coupled water flow together with sediment and solute transport modeling capabilities (Ewen et al., 2000). For the purpose of this modeling only the water flow component is considered. The SHETRAN flow model for an embankment (Fig. 1) can be thought to consist of 3 components:

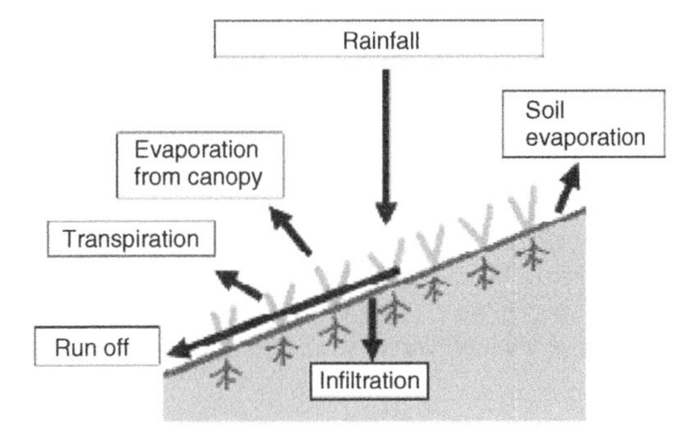

Figure 1. SHETRAN surface model.

1 Interception and evapotranspiration
2 Run off
3 Subsurface flow

Data requirements for the model are: meteorological data, soils data, vegetation properties and run off data together with boundary and initial condition settings. SHETRAN automatically outputs pore pressures for each cell within the grid.

At the surface boundary interception of precipitation is modeled by a modified Rutter model (Rutter et al., 1971) allowing the calculation of net rainfall reaching the ground together with the amount of stored water on the vegetation canopy and evaporation from the canopy. Evapotranspiration, the movement of water from the soil and within plants, is modeled within SHETRAN using the Penman-Montieth equation for actual evapotranspiration (Monteith, 1965). This is calculated as a loss term to describe the uptake of water through plant roots. Run off is also calculated within the program. The amount of runoff water is determined from the available water from the interception evapotranspiration component and the rate of infiltration into the subsurface. Flow resistance parameters are then used to model the run off using approximations of the St. Venant equations of continuity and momentum.

The subsurface is assumed to consist of a variably saturated porous medium. Flow through the medium is calculated by solving the non-linear partial differential Richard's equation.

1.2 *FLAC tp flow*

The FLAC finite difference code allows the numerical modeling of earthworks built of soil and rock. The two-phase flow option within the FLAC program is able to model two immiscible fluids within a porous medium. This allows the modeling of an unsaturated soil with the fluids present being water and gas. FLAC tp flow is capable of solving a fluid only calculation, a mechanical only calculation and a fully coupled fluid-mechanical calculation. A fully coupled calculation with a Mohr-Coulomb constitutive model is presented within this paper. FLAC tp flow requires soil properties, water properties and boundary and initial conditions to be specified. Water is able to enter the grid by specifying either a discharge or a pore water pressure at the boundary.

2 SITE DESCRIPTION

The site to be modeled is a cutting on the Newbury bypass in Southern England. The site has been extensively monitored by Smethurst et al (2006). The cutting height is 8 m and its length 28 m. It was cut in London clay in 1997. The London clay in the area is about 20 m thick and the top 2.5 m (from original ground level) has been extensively weathered. The cutting material consists of predominately stiff grey clay although there are several bands of silty clay and also bands of large flints present within the slope. The weathered clay is a highly variable material changing from a stiff orange brown clay to a clayey silt over small distances and depths. Vegetation on the slope consists of mainly rough grass and herbs with some small shrubs of less than 0.5 m high (Smethurst et al., 2006).

The site was monitored by Smethurst et al., to distinguish soil water content, pore water pressure, soil temperature, free water surface, rainfall and runoff.

Table 1. Soil hydraulic properties.

	London clay	Weathered London Clay
Residual volumetric moisture content	0.28	0.28
Saturated volumetric moisture content	0.45	0.45
Van Genutchen parameter n	1.443	1.443
Van Genutchen parameter α cm^{-1}	0.458	0.458

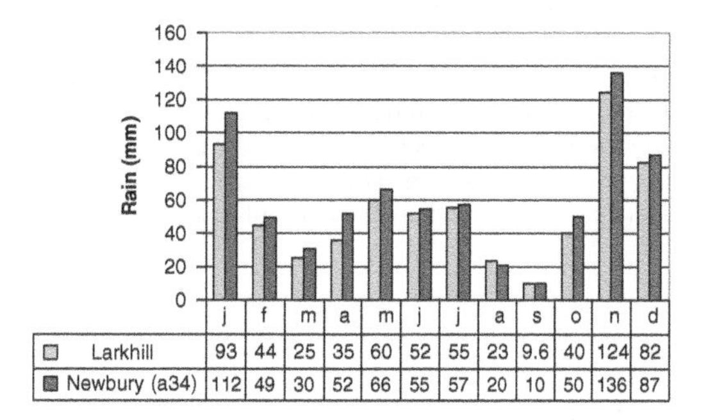

Figure 2. Comparison of monthly rain data from the Newbury site and summed hourly rain data from the Larkhill weather station.

3 SHETRAN MODELLING

3.1 *Soils data*

Soil water characteristics used for the modeling were derived from Croney (Croney, 1977) and initial permeabilities from tests carried out on site by Smethurst et al (2006). The soil hydraulic properties are outlined in Table 1.

3.2 *Weather data*

SHETRAN requires hourly rain data. The weather data used for the simulations was obtained from Met Office weather station at Larkhill (UK meteorology cal Office). Fig. 2 compares the monthly rainfall data recorded on the Newbury site and the summed data from the Larkhill weather station for 2003.

Fig. 2 shows that the measured monthly rainfall differs slightly but overall there is a good comparison. Fig. 3 shows the potential evapotranspiration (PET) measured on the Newbury by-pass and the calculated PET from the weather data recorded at Larkhill. The calculated Larkhill PET is consistently higher than the PET measured on the A34 site. This will lead to higher levels of evapotranspiration and lower overall pore pressures within the model. Smethurst et al

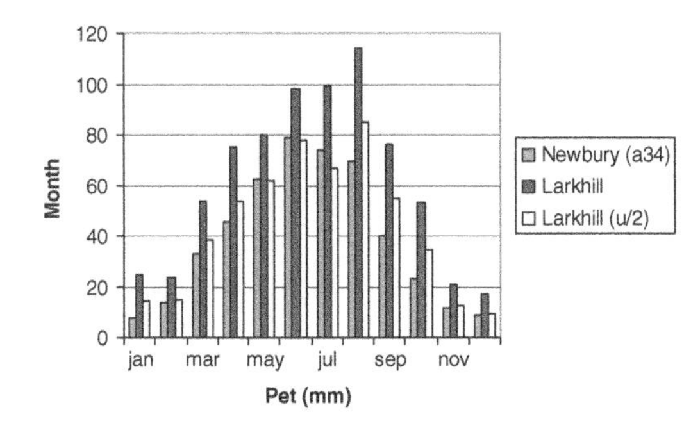

Figure 3. Calculated PET from the weather data recorded at the Nebury site and PET calculated from the weather data from the Larkhill weather station for maximum recorded windspeed (u) and a reduced wind speed.

Table 2. Vegetation properties.

	grassland
Evapotranspiration factor at −0.7 m	0.44
Canopy storage capacity (m)	1e−4
Maximum canopy drainage rate (m^{-1})	14×10^{-9}
Fractional rate of change of canopy drainage storage water (m^{-1})	5.1×10^{-3}
Canopy resistance factor ($s\,m^{-1}$)	100
Vegetation height (m)	0.3
Leaf area fraction	1
Fraction of energy absorbed by the canopy	0.9
Maximum rooting depth (m)	0.3

(2006) noted that the average monthly measured site PET was generally 20–30% lower than the long term data for Southampton (Southampton approximately 35 miles South of Newbury site). The reason for the discrepancy between the PET levels is the aspect of the slope. The majority of winds in the U.K. are Westerly and the cutting is East facing and sheltered by trees above. The wind speed (u) for the site within the calculated data has therefore been reduced by a factor of 2 so the calculated PET becomes more comparable with the site measured PET. The resulting PET is also shown in Fig. 3 which shows that there is generally a good comparison although calculated PET is generally overestimated for the months of August through to October.

3.3 *Vegetation data*

The cutting has a covering of mainly grass. This vegetation type was modeled at the surface of the SHETRAN grid. The properties assumed for this vegetation are shown in Table 2.

A vegetation time series file was also used within the simulation which allowed the variation of the ratio of leaf area to ground patch area with time. This time

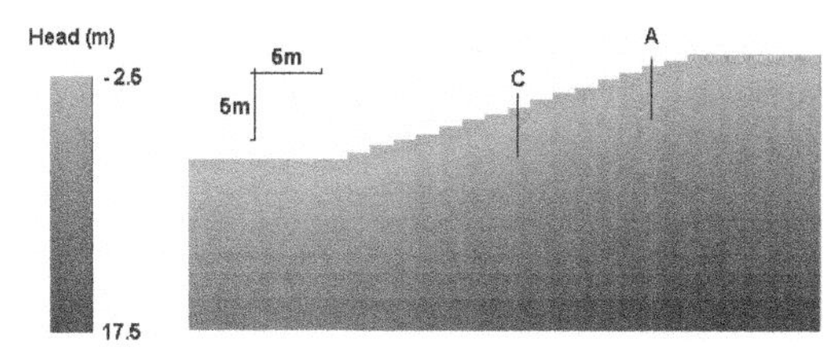

Figure 4. SHETRAN grid showing initial pore pressures.

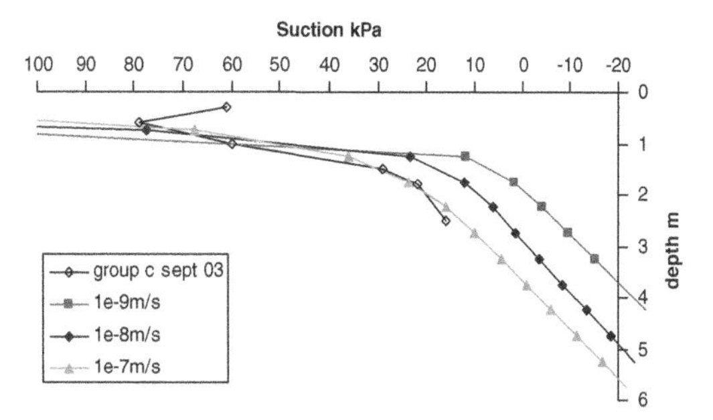

Figure 5. Calculated and recorded pore pressures for instrument group C, end of Sept 2003.

series file essentially simulated the thinning out of vegetation during the winter months and thickening during the summer months.

3.4 *SHETRAN grid*

A representative slope was built within the SHETRAN code and the soils given the relevant properties as detailed in Tables 1 and 2. Figure 4 shows the SHETRAN grid with initial pore pressures. Each cell within the model is 0.5 m deep and 1.75 m long. A horizontal surface of just over 10 m was modeled at the crest and also along from the toe of the cutting, a 12 m foundation was also modeled beneath the cutting. This is consistent with the depth of London clay in this area. The pore pressures were assumed to be hydrostatic with a surface pore pressure of −50 kPa. This is consistent with the site measured data after the wet winter of 2002.

3.5 *Modeling results*

The soil properties for the modeling were those specified within Table 2 with an assumed permeability of 1.0×10^{-9} m/s for the whole cutting. This value was comparable to the average permeability from triaxial tests (2.3×10^{-10} m/s) and in situ bail out tests (3.7×10^{-9} m/s) for the London clay. From these models it became clear that the permeability was too low. Although there was good comparison at the surface (Figs. 5, 6, 7 and 8), the pore pressure could not respond as fast as the site measured pore pressures

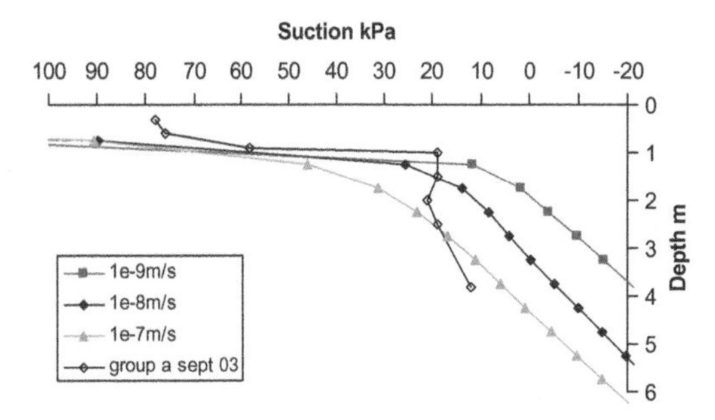

Figure 6. Calculated and recorded pore pressures for instrument group A, end of Sept 2003.

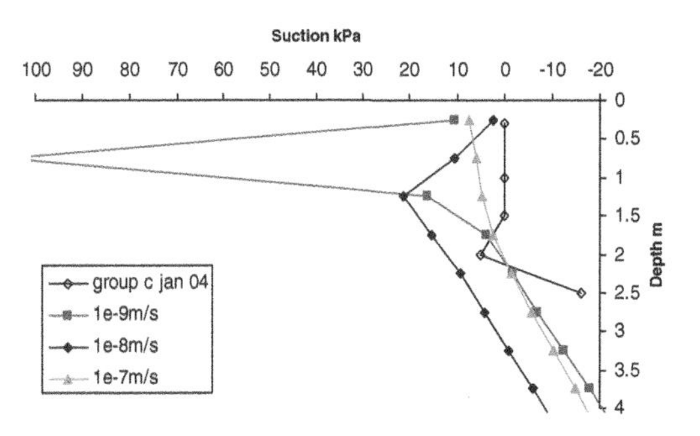

Figure 7. Calculated and recorded pore pressures for instrument group C, beginning of Jan 2004.

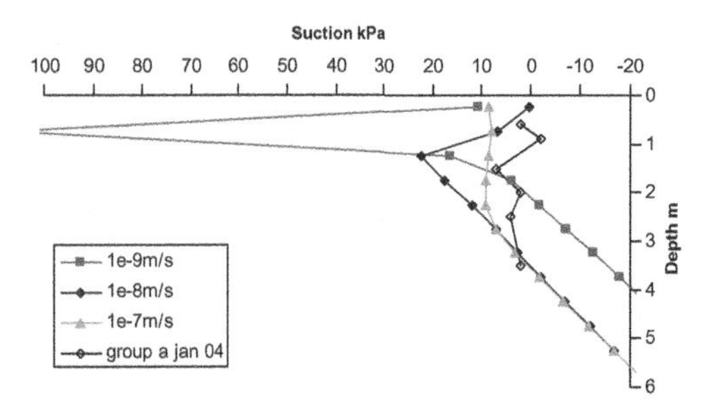

Figure 8. Calculated and recorded pore pressures for instrument group A, beginning of Jan 2004.

deeper within the slope. Water could not penetrate the slope to replenish the loss of moisture during the summer and the slope dried out. The permeability of the model was therefore increased to give a better comparison with the recorded instrument data. Permeability was increased first to 1.0×10^{-8} m/s and then to 1.0×10^{-7} m/s. All other parameters remained constant. Figs. 5, 6, 7 and 8 show the effects of increasing the mass permeability on two vertical profiles within the slope at the end of September 2003 and at the beginning of January 2004. These profiles are located at the points of instrument group C and at instrument group

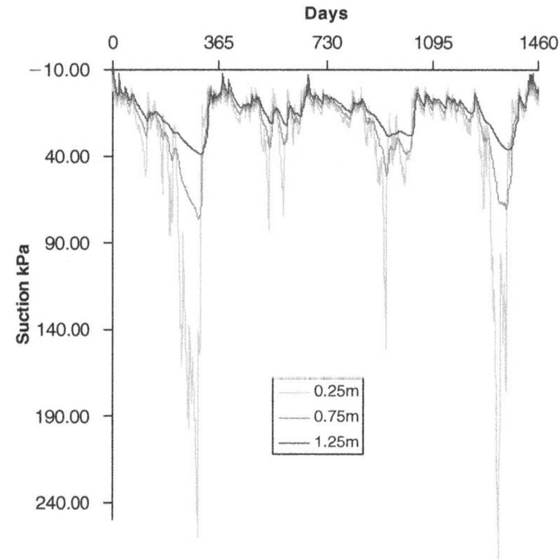

Figure 9. Calculated variation of pore pressure within the Newbury cutting from Jan 2003 to Dec 2007.

A (positions shown on Fig. 4). The site measured pore pressure profiles are also shown for reference. It is clear from these Figs. that comparison with the site measured data improves with permeability increase. A mass permeability of 1×10^{-7} m/s gave a very reasonable comparison to the site measured data. The measured data at instrument group C shows a maximum suction of 80 kPa, this maximum suction was most likely exceeded as this is close to the limit of the tensiometer instrument used. Equitensiometer data at this location shows that this suction increased to a maximum of 440 kPa. At this location SHETRAN captures the variation of pore water pressure with depth very well. The maximum suction of 440 kPa has not been reproduced yet a very high suction of 250 kPa was simulated (omitted from figs for clarity). The comparison at the same permeability for the location at instrument group A is not as close as that at instrument group C. SHETRAN has, however, managed to model the pore pressures at the top of the profile (suctions too high for measurement) and at the bottom of the profile reasonably well. This profile is within the weathered London clay material which is highly spatially variable. The soil within the SHETRAN simulation was assumed to be homogenous. This would explain why SHETRAN has not managed to fully capture this complex pore pressure profile.

The SHETRAN modeling was continued with a mass permeability of 1×10^{-7} m/s for a further 4 years. Fig. 9 shows the variation of pore pressures within the slope at instrument group C over this period. The winter of 2002 was very wet which gave rise to very high pore pressures at the beginning of 2003. The summer of 2003 was also exceptionally dry and that

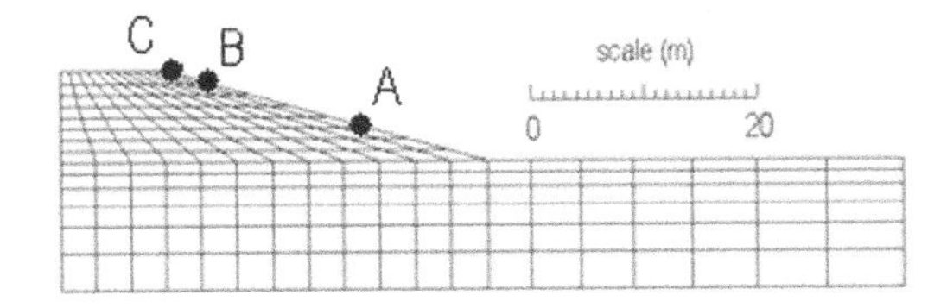

Figure 10. The FLAC tp flow grid used for the calculation of shrink swell cycles.

resulted in the high negative pore pressures within the slope at the end of September 2003. The pore pressures at the end of 2003 have also not recovered to their end of 2002 maximum owing to the dry summer. Also the suctions at the end of September 2004 were not as high as the previous year. 2005 was similar to 2004 but 2006 was once again a dry summer and high suctions similar to those of Sept 2003 were once again calculated within the cutting. This dry summer was then followed by a wet period at the end of 2007 which resulted in the wetting up of the cutting once more with maximum pore pressures comparable to those of January 2003.

4 FLAC TP FLOW MODELLING

4.1 *Introduction*

For this part of the modeling the surface pore pressures calculated by SHETRAN were transferred to the FLAC tp flow numerical model. This transfer allowed the modeling of the response of the cutting to the changing pore pressures.

4.2 *FLAC tp flow grid*

The FLAC tp flow grid used for the analysis is shown in Fig. 10. The vertical dimension of each cell within the cutting is 1 m. The horizontal dimension varies within the grid from an average of 0.1 m at the crest to 3 m above the foundation. The foundation cell size increases with distance away from the cutting. Mechanical boundary conditions were imposed at the base of the foundation to restrict vertical and horizontal movement and at the sides of the model to restrict horizontal movement only.

4.3 *Soils data*

The constitutive model used for the mechanical simulation was a Mohr-Coulomb model. Table 3 details soil properties for this model. The bulk unit weight was that recorded by Smethurst et al (2006) from testing on material from the Newbury site. The remainder of the properties were assumed to be similar to the properties used by Potts et al (Potts et al., 1997) who have done extensive numerical modeling on cuttings within the London clay.

Table 3. Soil parameters.

Bulk unit weight kN/m^3	14.2
c′ kPa	7
θ'°	20
Poisson's ratio	0.2
Young's modulus (kPa)	2500(p′ + 100) Min 4000
Coefficient of earth pressure at rest	1.5

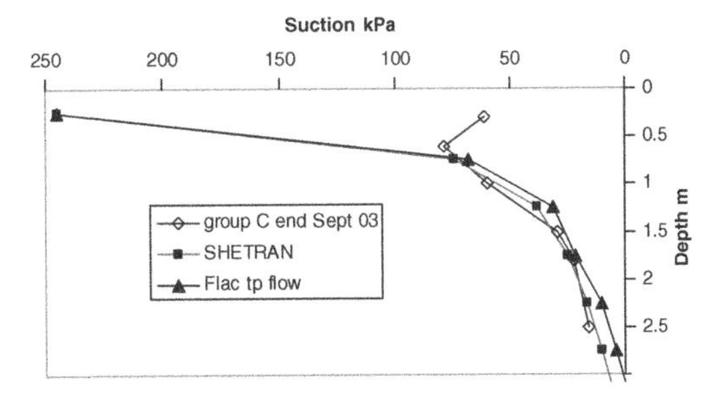

Figure 11. Pore pressure profile for instrument group C calculated by FLAC tp flow compared to the SHETRAN calculation and recorded data.

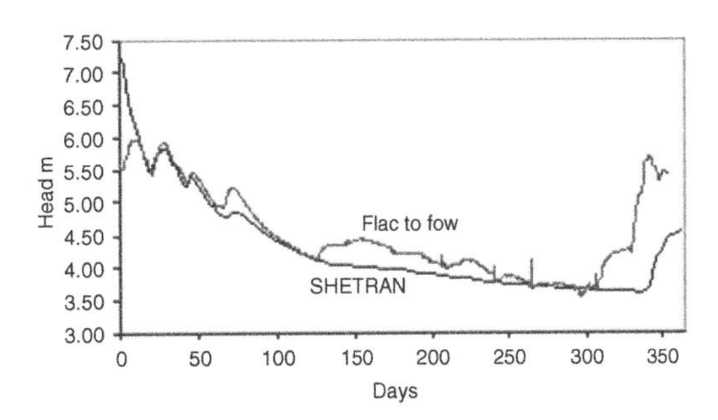

Figure 12. Variation of pore pressure over the first year (Jan 2003–Jan 2004) of the simulation at 7.5 m depth below cutting crest.

4.4 *Modeling results*

The surface pore pressures were transferred onto the FLAC tp flow boundary and a plot made of the pore pressure profile at instrument group C for the end of summer 2003 as a check that the transfer was successful. Fig. 11 shows this profile together with the profile from the SHETRAN simulation and recorded results.

Fig. 12 shows a time series for the first year which was also plotted to show the difference between the SHETRAN calculation and the FLAC tp flow calculation for a point 7.5 m below the crest of the cutting. SHETRAN is a single phase model which calculates flow by solving the non-linear partial differential Richards equation. The FLAC tp flow model is a two

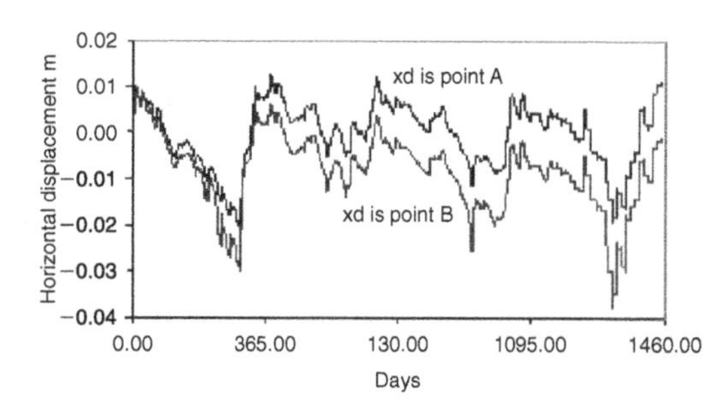

Figure 13. Horizontal displacement on slope surface for the period Jan 2003–Dec 2006.

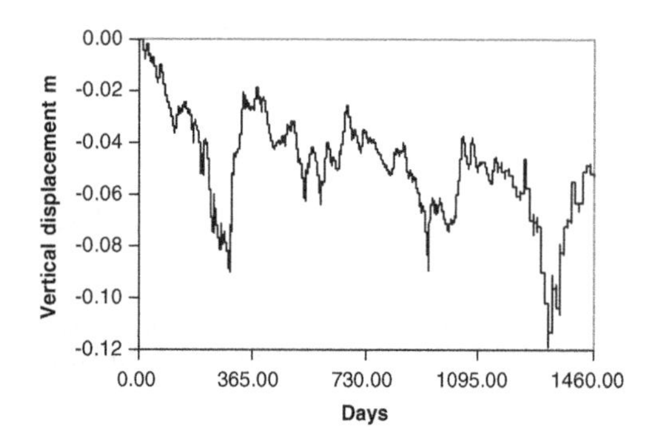

Figure 14. Vertical displacement on slope crest (point C Fig. 10) for the period Jan 2003–Dec 2006.

phase calculation which approximates to the Richards equation assuming the air within the calculation is at atmospheric pressure, the air density is zero and the porous medium cannot deform. The air within the FLAC calculation has been given the set parameters and no mechanical generation of pore pressure was permitted. These approximations are validated when observing the time series in Fig. 12, there is some deviation between the two profiles although this is deemed to be a small variance and have little overall effect on the results. Figs. 13 and 14 show the horizontal and vertical displacement at the points indicated by A, B and C on Fig. 10. The cutting shows a maximum horizontal cyclic movement of under 0.04 m and vertical cyclic movement of 0.06 m for the first year of the calculation. This was the period of exceptionally wet winter and dry summer. The cyclic displacements for the next two years then reduces to a horizontal average of 0.03 m and vertical average of 0.04 m before once again there is a dry summer and a wet winter. The model also calculates a net downward movement of the crest of about 0.04 m.

5 CONCLUSIONS

The pore pressure calculated by SHETRAN for a material with a mass permeability of 1×10^{-7}m/s were found to correlate reasonably well with those recorded at the Newbury site. The model managed to simulate the pore pressures at instrument group C very well and also capture the key features of the pore pressure profile at instrument group A despite the highly variable nature of the material at this location. The high suctions measured on site were not captured by the SHETRAN calculation. This was most likely due to the material in which these suctions were recorded being intact and having a low permeability similar to that recorded within the laboratory on samples from the site. The higher mass permeability used within the SHETRAN calculation could not generate such high suctions. It has become clear from this exercise, however, that the mass permeability of the material is the important factor when determining the pore pressures within a cutting. The fact that the permeability SHETRAN required to produce these results was three orders of magnitude greater than the permeabilities recorded by Smethurst et al., within triaxial cells and two orders of magnitude greater than estimates on site from bail out tests highlights the importance of accurately measuring this property.

Minimal accuracy was lost due to the transfer of the pore pressure from SHETRAN to FLAC tp flow. The cyclic movements of the Newbury slope show the dependency of movement on the climate, particularly extreme seasonal variations. Correctly calculating the seasonal pore pressure variations within a cutting is critical in order to accurately model the rate of strain softening which may occur.

REFERENCES

UK Meteorological Office. MIDAS Land Surface Stations data (1853-current) [Online]. Available at: http://badc.nerc.ac.uk/data/ukmo-midas (Accessed: 2007).

Croney, D. (1977). *The Design and Performance of Road Pavements.* Her Majesty's Stationary Office: London.

Ewen, J., Parkin, G. and O'Connell, P. E. (2000). 'SHETRAN: Distributed River Basin Flow and Transport Modeling System', *Journal of hydrologic engineering,* 5, (3), pp. 250–258.

Monteith, J.L. (1965). *Proc. 15th Symposium Society for Experimental Biology.* Swansea:Cambridge University Press, London.

Potts, D.M., Kovacevic, K. and Vaughan, P.R. (1997). 'Delayed collapse of cut slopes in stiff clay', *Geotechnique,* 47, (5), pp. 953–982.

Rutter, A.J., Kershaw, K.A., Robins, P.C. and Morton, A.J. (1971). 'A predictive model of rainfall interception in forests, 1. Derivation of the model from observations in a plantation of Corsican pine.' *Agricultural Meteorology,* 9, pp. 367–384.

Smethurst, J.A., Clarke, D. and Powrie, W. (2006). 'Seasonal changes in pore water pressure in a grass-covered cut slope in London Clay', *Geotechnique,* 56, (8), pp. 523–537.

Advances in Transportation Geotechnics – Ellis, Yu, McDowell, Dawson & Thom (eds)
© 2008 Taylor & Francis Group, London, ISBN 978-0-415-47590-7

Centrifuge modelling of embankments subject to seasonal moisture changes

P. Hudacsek & M.F. Bransby
University of Dundee, Dundee, UK

ABSTRACT: Embankments experience seasonal, periodic infiltration/evapo-transpiration conditions. This causes wetting-drying cycles which may lead to serviceability movement of an embankment both producing shrink-swell behaviour and generating gradual down-slope movements. In addition, slope failure may be triggered after a number of cycles due to cyclic softening of the soil. The response of these embankments is particularly relevant in light of global climate change which is likely to exacerbate the climate input conditions. This paper presents a study of the above effects for a compacted clay embankment by centrifuge model testing. This paper describes the apparatus developed and presents results from one example embankment test. Results show the nature of the soil displacements which occur periodically during wetting and drying events and demonstrate the efficiency of the apparatus to investigate the long-term response of embankments.

1 INTRODUCTION

There are thousands of kilometres of railway and highway embankments in the UK alone. The conditions of these assets have to be maintained so that they remain fit for purpose. One emerging threat to their performance is caused by global climate change.

Current climate change models generally predict drier summers and more precipitation in the winter in the UK. This poses a dual threat to performance: (i) slope failures due to excessive inundation/infiltration events; and (ii) increased seasonal serviceability slope movements because of moisture differences between wet winters and dry summers.

Whilst this increasing annual variation leaves embankments constructed of permeable, granular material unaffected, it has a more significant impact on other soils, particularly for intermediate permeability, expansive clay embankments (O'Brien et al., 2007). Annual moisture variations (caused by water balance changes due to summer evapo-transpiration against winter infiltration-dominant periods) will cause seasonal swelling and shrinkage of an earthwork. This will lead to track and pavement displacements with consequences of various severity from travel inconvenience, through disruptions and excessive maintenance expenses of the vehicles to accidents. As well as serviceability issues due to these seasonal displacements, these moisture changes may lead to 'ratcheting' failure as pointed out by Take & Bolton (2004).

Seasonal behaviour was observed in field conditions by (Standing et al., 2001) at Cannon Park railway embankment where 30–35 mm of rail displacement was reported within the 16 months of observation. Similar observations are reported by Smethurst et al (2006), O'Brien (2007) and Scott et al (2007) for other railway embankments generally constructed of London clay.

Permanent slope movements are caused because of the irreversibility of the strains in the embankment during the cycles of drying and wetting. The gradual accumulation of plastic strains either cause crest movements exceeding serviceability limits or bring the slope into its ultimate state of equilibrium. Numerical models (Kovacevic et al., 2001) and centrifuge tests (Take, 2002) proved the theory of accumulation of small plastic strains due to the seasonal movements. Take (2002) used saturated over-consolidated kaolin as the slope material. This soil state may not truly replicate the response of variably compacted infrastructure embankments, but showed clearly one mechanism that could occur. The over-consolidated kaolin model embankment had relatively high permeability and the relatively small drop in the critical state shear strength to residual were considered to produce conservative results. In order to obtain more realistic data about real earthworks such as Victorian end-tipped railway embankments, compacted clay models should be investigated.

The aim of paper is to present a test methodology to study the effect of seasonal climate changes on compacted clay embankment performance. This was achieved using centrifuge model testing. A brief description of the apparatus, model preparation methods, and soil properties will be given first. Finally, preliminary results from an example experiment will

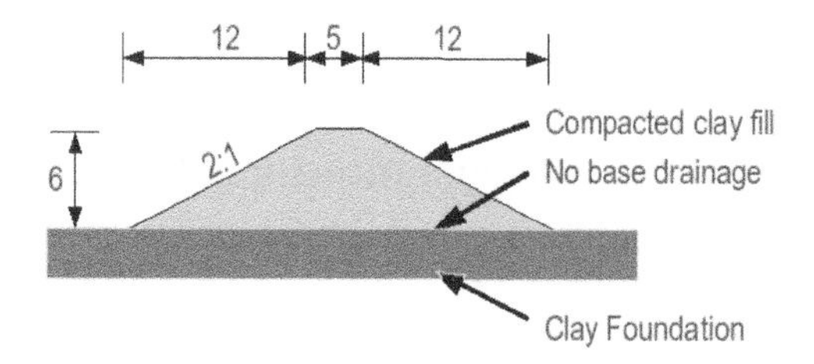

Figure 1. The embankment geometry modelled in the test programme (full scale; dimensions in meters).

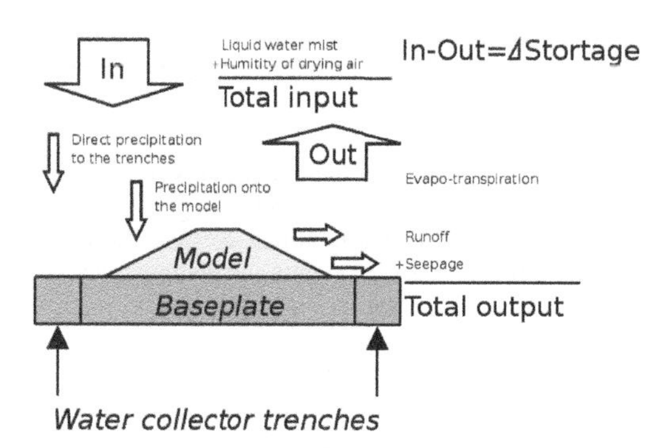

Figure 2. Schematic of the overall hydrological modelling strategy.

be presented, focusing mainly on displacements during three consecutive years of simulated climate.

2 EXPERIMENTAL APPARATUS AND INSTRUMENTATION

This section gives a brief description of the apparatus used to model an embankment with different hydraulic boundary conditions. The embankment geometry modelled is shown in Figure 1 and was that used in full-scale and numerical modeling in the UK EPSRC-funded BIONICS project (Hughes et al., 2008; Toll et al., 2008; Davies et al., 2008). Model tests were conducted at a scaling factor of 60.

The 1/60th-scale model slope was contained within a sealed environmental chamber which was placed on the arm of the centrifuge and spun to give a centripetal acceleration 60 times Earth's gravity. Different climate scenarios were simulated by alternating wet and dry periods. The components of the hydrological cycle being modelled are shown in Figure 2. More details of the modelling are given in this and the following section.

2.1 *Centrifuge apparatus*

The Dundee geotechnical centrifuge is a 3 m effective radius beam centrifuge. Data acquisition and the process control is done through an on-board PC which is equipped with four multi-function data acquisition cards. Process control operations are done through a relay box interface controlled by digital input/output ports on the DAQ cards. The compressed air and the water used in the test series is supplied to the package through the rotary joints (slip-rings) of the centrifuge.

2.2 *Water delivery system*

Water is delivered to the model package by an array of 12, variable position, fine misting nozzles placed in 3 rows. By adjusting the nozzle positions it was possible to compensate for the Coriolis-effect and ensure uniform spray distribution over the soil surface in the model. The water flow rate through the nozzles is a linear function of the supply pressure. Hence, the flow rate can be calculated at any instant during a test through measurement of the water pressures at the nozzle head following calibration of the pressure-flow relationship. To simulate realistic precipitation rates and times, the nozzles have to be operated for short periods of time only. Consequently, an air water interface tank was used to provide rapid build up of the nozzle supply pressure and a quick discharge system of the nozzle feed line was introduced to reduce the shut-off time. As a further improvement, a valve is used upstream of each nozzle to ensure nozzle supply pressure always exceeds the dripping pressure.

The system was controlled by using the Labview data acquisition software to switch on or off the water pumps and solenoids required to start and stop water flow. Maximum flow rate corresponds to an inundation rate of 1.4 mm/h over the whole soil surface at full scale.

2.3 *Liquid water output measurement*

To measure the mass of the discharged water (i.e. the sum of the surface run off along the embankment slopes and the water seeping out of the soil) a water collection trench was placed parallel to each toe of the model embankment (shown schematically in Figure 2). The water leaving the model was collected in these trenches while drainage taps at the base of the trenches were kept closed. As the trenches filled up, the hydrostatic pressure of the impounded water was measured with a pair of pressure transducers placed under the base plate connected to the deepest points of the trenches. By calibrating the relationship between the voltage signal from the pressure transducers and the collected mass, this allowed the net volume collected in each trench to be calculated at any time. By monitoring the change of the water volume over time, the net flow rate of water into the trench could be deduced. To prevent over-topping of the overflow trenches, when a certain level of water is exceeded, the

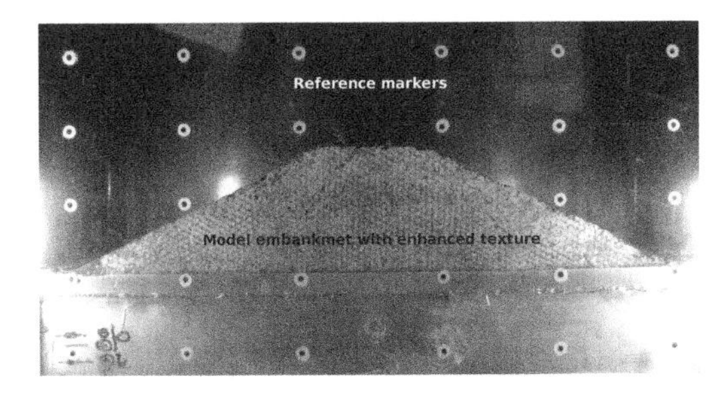

Figure 3. Typical digital image of the embankment model taken 'in-flight'.

water egress lines are automatically opened, releasing the collected water and allowing water levels in the trenches to rise again.

2.4 *Dry air delivery, humidity, temperature and air volume measurement*

To reduce the humidity of the air in the environmental chamber, dry air was supplied to the chamber at the start of each dry period (summer). The air was supplied by a compressor outside the centrifuge chamber through a desiccant dryer unit. The air input speed was selected in combination with the output values from the box to ensure that the transient air pressure changes in the box did not exceed 4 kPa during air input.

To facilitate the calculation of the evaporation related water losses of the system the temperature and the humidity of the entering dry air and the leaving moist air was measured. The input air volume was measured by a purpose built orifice plate device. An additional humidity sensor is installed within the environmental chamber to measure the actual climate conditions in the box.

The humidity sensors were commercial, DC output components, calibrated to allow measurement of relative humidity (RH%). For the air temperature measurement, quarter-bridge thermistor circuits were designed and calibrated by means of a hot and cold air source using a high-resolution electronic thermometer as the reference.

2.5 *Image capture*

Embankment displacement data was captured using digital image analysis of photographs taken 'in-flight' during each test. The quality of these images had to be high due to the small seasonal soil displacements which were measured.

Digital images were captured with a Cannon S50 digital camera through the observation window of the centrifuge strongbox. A typical image is shown in Figure 3. On the internal side of the observation window, reference markers were applied in order to facilitate later translation of the data from image to model space. The reference markers were placed with known positions by means of a perforated steel plate, to an estimated precision of 0.1 mm. A light dusting of kaolin powder was placed on the surface of the soil adjacent to the observation window to enhance the texture on the image side of the model to facilitate image analysis.

3 EXPERIMENTAL PROCEDURES AND SOIL PROPERTIES

3.1 *Soil preparation method*

The compacted soil in the full-scale embankment comprised of heavily over-consolidated glacial till which was removed from a borrow pit and compacted in place. Its macro-fabric would reflect the original properties of the borrow pit soil and 'clods' of this original soil fabric (probably with typical dimensions of the order of 10–30 cm) would remain following removal from the borrow pit. These 'clods' would then be partially re-worked during mechanical compaction (e.g. O'Brien, 2007). This overall process was re-created in a controlled manner in preparation of the centrifuge embankment model with the clod sizes reduced in size appropriately.

Glacial till soil used in the BIONICS full-scale embankment (Hughes et al., 2008) was first soaked in water, homogenised with a mixer and passed through a 1.2 mm sieve. The resulting slurry with moisture content, $w = 90\% - 100\%$ was then placed in a consolidometer of diameter 0.8 m and consolidated one-dimensionally in load increments to a maximum pre-consolidation pressure of 640 kPa. After full consolidation was achieved at $\sigma'_v = 640$ kPa, the vertical stress was reduced to 80 kPa and the sample equilibrated with free water drainage. Finally, the drainage lines were closed and the vertical total stress was removed to leave a saturated, overconsolidated block of soil under suction.

To mimic the macro-fabric of the prototype embankment, the homogeneous soil block was then cut into 'clods' of typical dimension 3 mm × 2 mm × 6 mm (180 × 120 × 360 mm at full-scale). These were placed into a mould attached the model base plate and compacted quasi-statically by applying quickly a vertical load to the top of the whole sample. During the compaction process, the applied force and displacement were logged to allow calculation of the compaction energy applied to the sample. By applying different peak compaction pressures the density of the soil can be controlled. To achieve low compaction levels, a maximum pressure of 120 kPa was applied.

The soil sample was removed from the mould and cut to shape. Finally, the centrifuge strongbox was re-assembled and sealed prior to centrifuge flight.

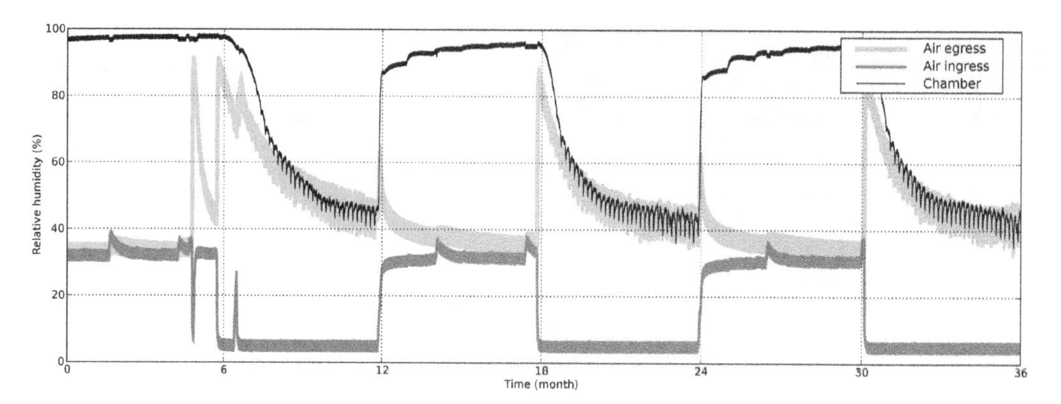

Figure 4. Measured relative humidity during three annual climate cycles.

3.2 *Soil properties*

The natural fill material used in the BIONICS project is a well graded glacial till with particle size ranges from clay size up to coarse gravel/cobbles (Toll et al, 2008; Hughes et al., 2008). The Atterberg limits are LL = 40% and PL = 22% so the soil can be classified as a clay of intermediate plasticity. The shear strength parameters of the intact soil are $\phi' = 18°$ and $c_u = 55$ kPa (shear vane). The bulk density (saturated) of the intact soil is 2.00 Mg/m^3.

The bulk density and the shear strength of the compacted soil depends on the applied compaction energy. For the example test presented later, the maximum compaction pressure was 120 kPa which produced a bulk density, $\rho = 1.85$ Mg/m^3 and an average undrained shear strength, $c_u = 32$ kPa (vane).

In order to obtain the drained shear strength parameters of the soil, undisturbed samples were taken from both the intact and compacted soil blocks for later soil testing.

3.3 *Model geometry and scaling laws*

The geometry of the centrifuge prototype is shown in Figure 1. The cross-sectional dimensions of the 1/60th scale model are height, h = 10 cm, breadth, w = 48.3 cm and the thickness out of the plane of the model was 50 cm. As a consequence of the scaling law for diffusion related processes in the centrifuge 10 years of prototype time in the centrifuge at 60 g can be modelled in 24 hours of model centrifuge time.

3.4 *Spin up and equilibration*

The centrifuge was started and the rotation rate increased to apply an acceleration of 60 g to the soil sample. Once the target acceleration level was achieved, the model was left for one year at prototype scale (2.4 hours of model time) to equilibrate. During this period some further volume changes may have occurred due to equilibration of pore fluid pressures.

3.5 *Climate modelling*

Following the equilibration period, the climate simulation was commenced. Each simulated year consists of two parts: a dry and a wet 'season'. By changing the duration and frequency of the simulated rainfall patterns and the ratio between the dry and the wet periods a wide variety of climate scenarios can be simulated.

In the example test presented in this paper, the ratio between the dry and the wet part of the year was 1:1, which means two 'seasons' of 72 minutes duration (half a year at prototype scale) were modelled. The wet period consists of six rainfall events of 60 hours duration and 1.4 mm/hour intensity rainfall (prototype scale) evenly distributed over the 6-month (prototype) wet 'season'.

4 RESULTS

Results are presented for a single test to demonstrate system behaviour and indicate likely embankment performance. The example test consisted of an embankment with a low compaction level (120 kPa maximum compaction stress) which was subjected to five years of climate cycles detailed above. Preliminary results are presented to indicate particularly the hydraulic conditions and the embankment displacements induced during the last three years of climate cycles. In this example test, two climate cycles were first applied to the model before a period of no climate input (corresponding to overnight centrifuge running) with the final three years of climate input. Future tests will contain continuous climate inputs during the whole duration of the model tests.

4.1 *Climate input data*

The relative humidity data measured during testing are shown in Figure 4. The graph shows data from

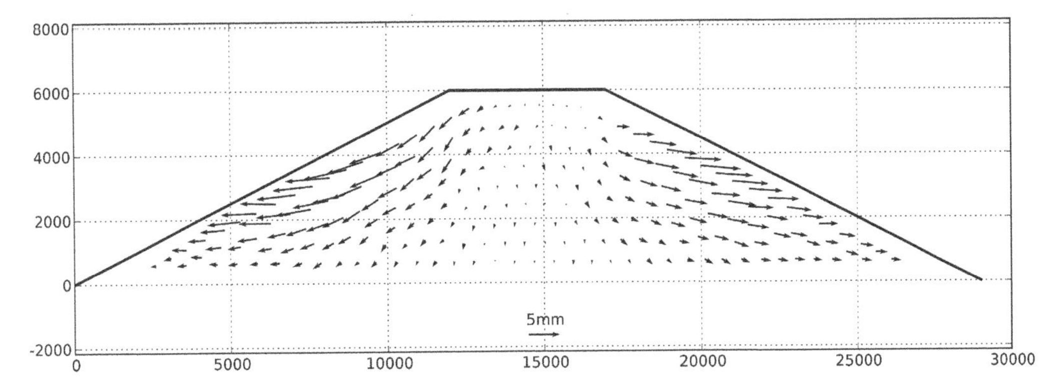

Figure 5. Soil displacements measured during the 5th winter 'season' (dimensions in mm at full-scale).

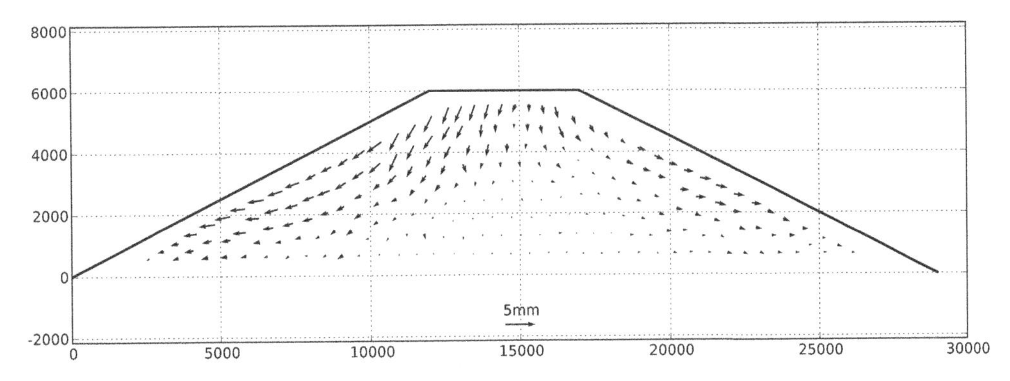

Figure 6. Soil displacements measured during the 5th summer 'season'.

years 3, 4 and 5 and presents the annual fluctuation of relative humidity in the air above the soil embankment ('Chamber') and in the air inlet and outlet sensors. The relative humidity within the chamber gives a good qualitative indicator of the hydraulic excitation of the model. It is seen clearly that high humidity wet seasons are followed by increasingly dry 'summer' seasons as expected.

4.2 *Calculated soil displacements*

Soil displacements were obtained by digital image analysis using a modified version of the PyPIV open source PIV software modified to reduce the 'pixel locking' phenomenon (Gui & Werely, 2002).

Figures 5 and 6 present calculated displacement vectors for the 5th winter and summer seasons. The vectors indicate the magnitude and direction of soil movement during the entire winter (Figure. 5) or summer (Figure. 6) season. Similar patterns of behaviour are observed but the magnitudes of displacement in the winter are larger due to the expected higher moisture contents in the soil. The largest full-scale equivalent displacement for the 5th winter season is in the order of 5 mm (0.08 mm at model scale).

Figures 5 and 6 reveal that the slope bulges near the middle. Furthermore, it is interesting that although plastic (i.e. irreversible) deformation takes place, no clear localisation can be seen. This is highlighted when the displacement data is re-plotted in terms of equivalent inclinometers (Fig. 7b).

Figure 7a shows data from pseudo extensometers. This reveals that there is continuous vertical compression of the embankment and that volumetric strain is concentrated in a layer 2 or 3 m below the surface.

Figure 8 shows the calculated trajectories of soil at selected points near the embankment surface. Displacements are shown enlarged 75 times to reveal the soil movement directions.

There is significantly different behaviour between the movement of soil at the crest and soil on the side slopes. Downward movement (settlement) of the crest is associated with volume reduction and is most active during the dry seasons. Horizontal displacements occur on the slope sides and are most active in the wet periods.

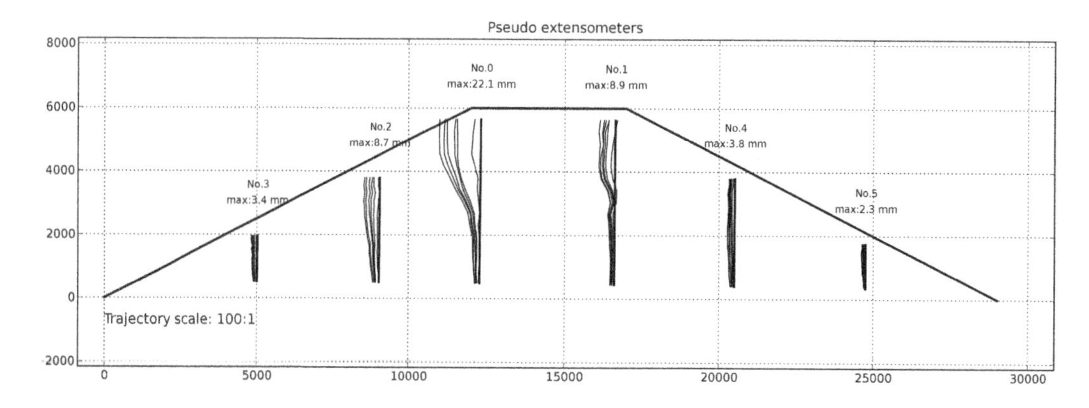

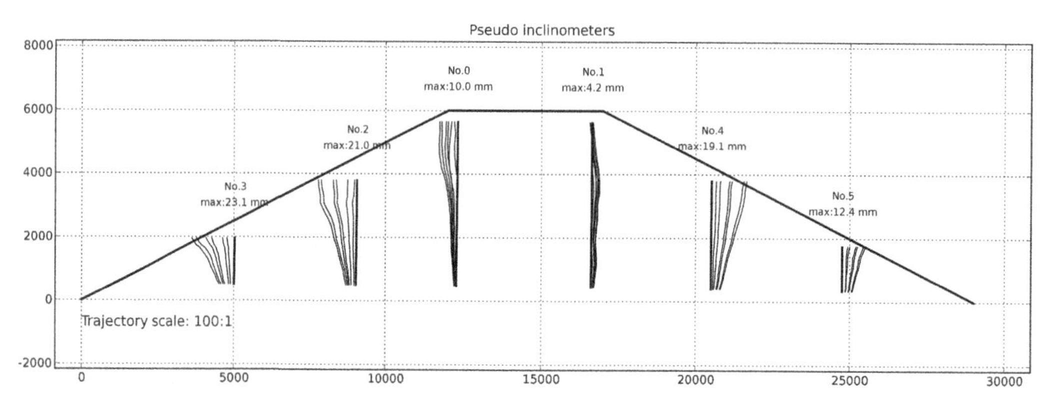

Figure 7. Soil displacement data of the 3 simulated years, re-plotted as: (a) equivalent extensometers; and (b) equivalent inclinometers.

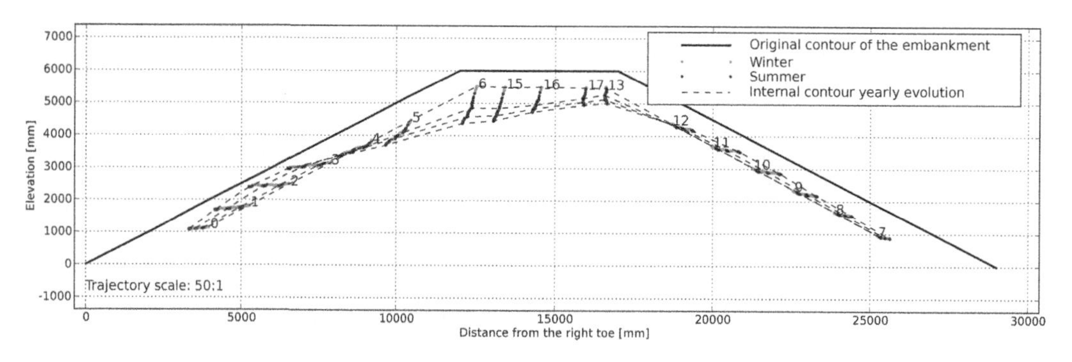

Figure 8. Calculated soil displacement trajectories for selected points in the embankment.

Figure 8 also reveals different behaviour between the two sides of the embankment. This is probably due to the imperfection in sample compaction or to non-uniform spray arrangements. The right side of the slope indicated by patches 18, 13, 17 (in a zone which may be more compacted) show minor (2–3 mm prototype scale) but measurable swelling in each winter period.

Finally, figure 9 reveals the temporal changes in vertical and horizontal displacement of each of the patches identified in Figure 8. Patches on the left hand side slope (numbered 0–5 in Figure. 8; Solid lines in Figure. 9) reveal accelerations of lateral displacement in the winter and slower movements during the dry periods with an average lateral permanent movement ranging from 4 to 9 mm per year. In contrast, patches

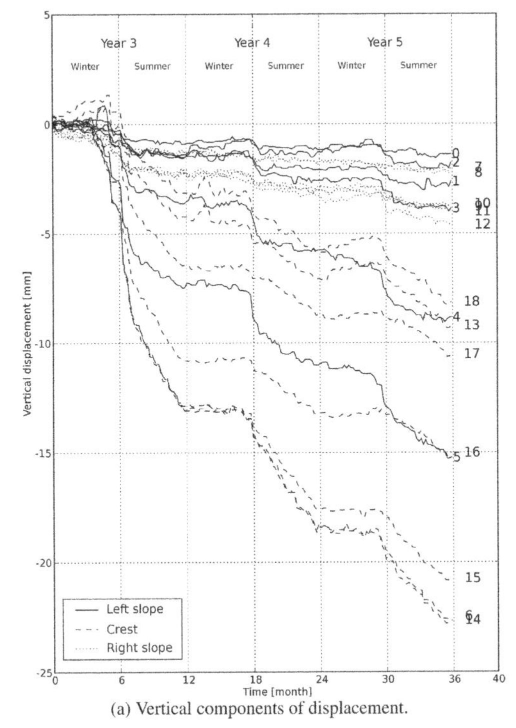

(a) Vertical components of displacement.

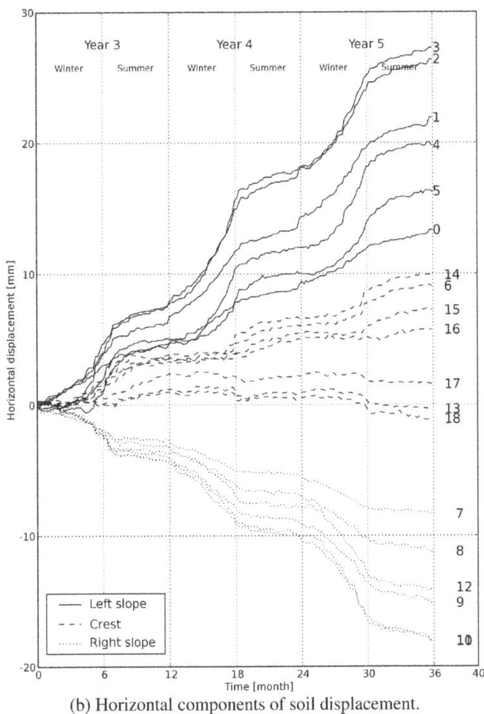

(b) Horizontal components of soil displacement.

Figure 9. Cumulative soil displacements for the final three years of excitation for selected points on the surface of the embankment (refer to Fig. 8 for numbering).

on the crest (dashed lines in Figure. 9) show small lateral movements and seasonally varying downward settlements. As discussed previously, settlements of the crest occur quickest in summer (dry) periods and slowest in the winter and fall in the range of 5–8 mm per annual cycle.

5 CONCLUSIONS

The paper reports the development of experimental apparatus to investigate the performance of a compacted embankment subject to seasonally induced hydraulic cycles. A new model preparation method was introduced in order to mimic the macro fabric of a real compacted clay embankment.

With the calibrated input and output control and measurement devices it is possible to control or measure the parameters of the hydraulic boundary conditions.

The measured preliminary displacement data shows that the slope reacts to the simulated hydraulic excitation. During the three years of simulated climate presented it was found that the wet seasons are mostly responsible for horizontal displacements of the slope surfaces, whilst most of the vertical crest movements happen during the dry spells. These down-slope movements were accumulating and resulted in a deformation field which resembles the deformations of poorly compacted old embankments.

Interesting, these results differed significantly to those from a previous study conducted on slopes constructed from saturated, overconsolidated kaolin (Take, 2002; Take & Bolton, 2004).Those results shown significant heave and shrink cycles while the compacted, medium plasticity clay embankment shown continuous downslope movements, with seasonally changing gradients in its vertical and horizontal components suggesting that the compacted nature of the soil and its plasticity may be critical to the seasonal and/or long-term response of transportation earthworks

Further tests will be carried out with different climate conditions and on model embankments with different compaction levels to investigate these phenomena further.

ACKNOWLEDGEMENTS

The work forms part of the BIONICS research project supported by the UK EPSRC, GR/S87430/01. We are grateful for this support and for interesting discussion with colleagues in the BIONICS consortium.

REFERENCES

Davies, O., Rouainia, M., Glendinning, S., Birkenshaw, S. 2008. Predicting seasonal shrink swell cycles within a clay cutting. *1st ISSMGE International Conference on Transportation Geotechnics, Nottingham, UK*, September 2008.

Gui, L. & Werely, S.T. 2002. A correlation-based continuous window-shift technique to reduce the peak-locking effect in digital PIV image evaluation. *Experiments in Fluids* 32, 506–517.

Hughes, P.N., Glendinning, S., Davies, O. & Mendes, J. 2008. Construction and monitoring of a test embankment for evaluation of the impacts of climate change on UK transport infrastructure. *1st ISSMGE International Conference on Transportation Geotechnics, Nottingham*, UK, September 2008.

Kovacevic, N., Potts, D.M., Vaughn, P.R. 2001. Progressive failure in clay embankments due to seasonal climate change. *Proc. 15th ICSMGE, Istanbul, Turkey*, Vol. 3: 2127–2130.

O'Brien, A.S. 2007. Rehabilitation of urban railway embankments: research, analysis and stabilization. *Proc. ECSMGE, Madrid:* 125–143.

Scott, J., Loveridge, F. & O'Brien, A.S. 2007. Influence of climate and vegetation on railway embankments. *Proc. ECSMGE, Madrid:* 659–664.

Smethurst, J.A., Clarke, D. & Powrie, W. 2006. Seasonal changes in pore water pressure in a grass covered cut slope in London clay, *Geotechnique* 56 (8): 523–537.

Standing, J et al. 2001.

Take, A. The influence of seasonal moisture cycles on clay slopes. PhD thesis. University of Cambridge.

Take, A. & Bolton, M.D. 2004. Identification of seasonal slope behaviour mechanisms from centrifuge case studies, *Proc. The Skempton Conf., London*: Thomas Telford, Vol. 2: 992–1004.

Toll, D.G., Mendes, J., Karthikeyan, M., Gallipoli, D., Augarde, C.E., Phoon, K.K. and Lin, K.Q. (2008) Effects of Climate Change on Slopes for Transportation Infrastructure, *1st ISSMGE International Conference on Transportation Geotechnics*, Nottingham, UK, September 2008.

Advances in Transportation Geotechnics – Ellis, Yu, McDowell, Dawson & Thom (eds)
© 2008 Taylor & Francis Group, London, ISBN 978-0-415-47590-7

Construction and monitoring of a test embankment for evaluation of the impacts of climate change on UK transport infrastructure

P.N. Hughes, S. Glendinning & O. Davies
Newcastle University, UK

J. Mendes
Durham University, UK

ABSTRACT: Our climate is set to change significantly over the next century. The Engineering and Physical Sciences Research Council (EPSRC) has recognised the importance of climate change and have introduced the Building Knowledge for a Climate Change (BKCC) programme, this has now been followed up by the Sustaining Knowledge for Climate Change (SKCC) programme. EPSRC and the UK Climate Impacts Programme (UKCIP) are working together to fund consortia (academia and stakeholders) to study the potential long term impacts of climate change in the UK on the built environment, transport and utilities. BIONICS is concerned with the establishment of a research facility (a full-scale embankment) which will be fundamental to improving the understanding of the long-term impact of climate change on infrastructure embankments and inform industry and stakeholders of the adaptation strategies required to mitigate the effects. The initial research programme utilising the BIONICS embankment is assessing the impact of climate change on embankment vegetation, developing a modelling capability for the prediction of long-term behaviour of embankments subjected to a changing climate and beginning to develop a methodology for identifying at-risk sections of the transport infrastructure.

This paper will describe the design, construction, and testing of the test embankment. Testing data demonstrates the difficulties encountered in trying to reproduce older less effective compaction techniques with modern plant and equipment and the effects of different compaction methods on construction generated pore water pressures.

1 BACKGROUND

Climate change has the potential to have a serious detrimental effect on huge parts of our infrastructure. Whilst this is generally accepted by stakeholders, there is as yet no strategy to facilitate the planning required to act upon it. BIONICS will enable the effects of climate change on infrastructure slopes to be deduced by establishing a unique facility consisting of a full-scale, instrumented soil embankment, planted with a variety of flora with controlled heating and rainfall at its surface. It is now establishing a database of high-quality embankment performance data to enable modeling of the interaction of climate, vegetation and engineering on the behaviour of infrastructure earthworks. By bringing together key stakeholders and academics, it has provided the focus for several "spin-off" projects including the development of new slope monitoring instrumentation (ALARMS, led by Professor Neil Dixon of Loughborough University) and remote sensing techniques (Linear Assets, led by Professor John Mills of Newcastle University). Further spin off projects are being developed with Southampton University, Queens University Belfast, Loughborough University and Durham University. In the longer term this strategy will enable advanced procedures for maintaining serviceability and safety of strategic embankments and cuttings in addition to advancing the science base.

Slopes make up a large proportion of UK transport networks (£20B of the estimated £60B asset value of major highway infrastructure is earthworks). Major (*ultimate limit state*) failure of these slopes causes disruption to network operation and frustration to the public. Minor (*serviceability state*) failure of rail slopes causes speed restrictions and daily commuter misery due to delays. Continuous maintenance of these slopes is essential and, according to Perry et al (2001), cost £50 million in the year 1998/9 for embankments alone. However, the cost of emergency repair is ten times greater than the cost of planned maintenance works (O'Brien, 2001) so the ability to predict the future effects of climate on the infrastructure would be of significant financial benefit. There is evidence that the scenario of more intense rainfall is already having an impact on the national transport infrastructure, including major landslides in Scotland (e.g. Stromeferry) and, in the winter of 2000/1, which was documented the wettest on record, over 100 slope failures in the Southern Region of Railtrack alone (O'Brien, 2001). More recently, the summer of 2007 was the wettest summer on record. This led to many slope failures at the time but, perhaps more importantly, did not allow the usual build up of suctions that maintain stability during the winter (Loveridge, 2007).

2 AIMS AND OBJECTIVES OF PROJECT

The aim of the project is to establish a unique facility for engineering and biological research to improve the fundamental understanding of the effects of climate change on slopes. The specific objectives of the project include:

1. Build and monitor an embankment representative of UK infrastructure subjected to different climates.
2. Plant and monitor representative vegetation subjected to different climates.
3. Create a controlled climate.

3 THE EMBANKMENT

After consultation with academic and industry stakeholders a final design for the embankment was selected to be representative of UK transport infrastructure. The embankment is 90 meters long and has been constructed in two distinct parts. Half of the embankment is constructed to modern highway specifications using modern compaction plant (0.3 m lifts, 18 passes with a vibrating roller), and half has been constructed to poorer specification using as little compaction as possible in order to simulate older rail embankments (1 m lifts, minimum tracking with construction plant). A diagram of the embankment is shown in Figure 1.

4 IN-SITU TESTING

Core cutters were used during construction to assess the levels of compaction being achieved in each of the test sections of the embankment. As can be seen in Table 1 higher densities were achieved in the "well" compacted sections of the embankment although the difference was smaller than had been anticipated from laboratory testing.
A soil suction probe developed by Durham University (Mendes et al, 2007) was also used to test compacted fill during construction. Recorded soil suction results are shown in Figures 2 and 3.

Site conditions did not allow the tests to be run until equilibrium had been reached but the results indicate that considerably higher suctions were being generated in the "well" compacted sections of the embankment (in excess of −400 kPa in places) than in the "poorly" compacted sections (between −40 kPa and −165 kPa).

5 MATERIALS TESTING

During the installation of the instrumentation on the BIONICS site undisturbed samples were recovered from instrumentation boreholes. These samples were used to characterize the properties of the embankment material. Testing has included, shear box, quick undrained and consolidated drained triaxial testing, atterberg limits, 2.5 kg and 4.5 kg compaction. Soils

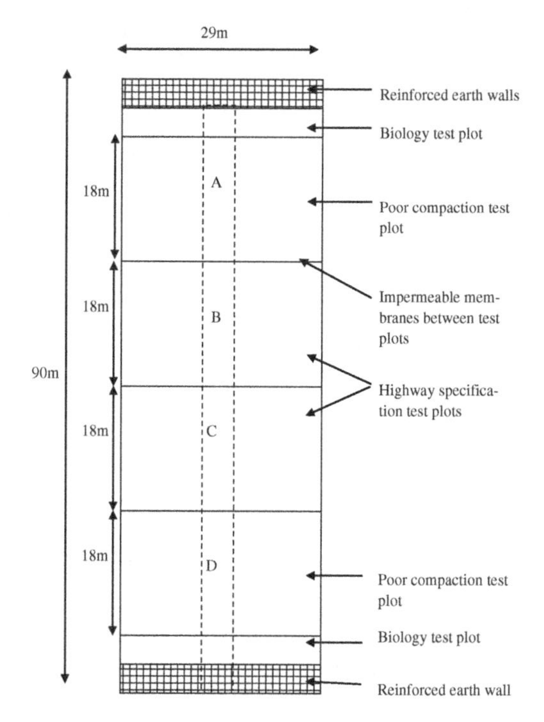

Figure 1. BIONICS embankment layout.

Table 1. Summary of core cutter density results.

	Bulk Density (Mg/m^3)	Water Content (%)	Dry Density (Mg/m^3)	Air voids (%)	Degree of saturation (%)
"Poor" compaction	1.93	20.7	1.6	6.0	85.3
Good compaction	2.01	20.1	1.7	3.2	91.4

testing data has been posted on the project website for use by the BIONICS researchers, a summary of which is shown in Table 2 and in Figure 4.

6 INSTRUMENTATION

Since the construction was completed the embankment has been extensively instrumented with extensometers, inclinometers, flushable piezometers, and soil suction tensiometers. The data from these instruments shows that both settlement and slope movement so far has been low. The pore pressure within the embankment has been more active however. As was shown in Figures 2 and 3, during construction negative pore pressures were measured in both the "well" and "poorly" compacted parts of the embankment. Figures 5 and 6 show pore pressures recorded in the two sections of the embankment 12 months

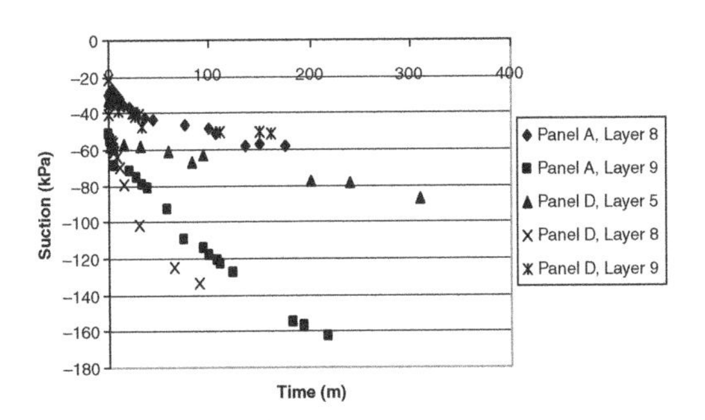

Figure 2. Soil suctions recorded in samples taken from the "poorly" compacted sections of the embankment (Panels A and D). Layer numbers indicate the compaction lift of tested specimen.

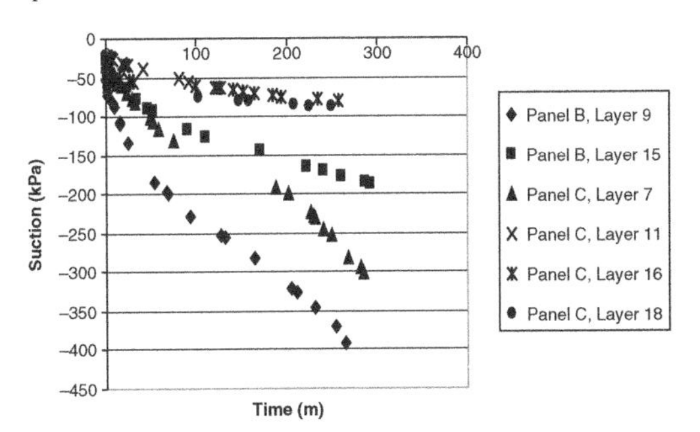

Figure 3. Soil suctions recorded in samples taken from the "well" compacted sections of the embankment (Panels B and C).

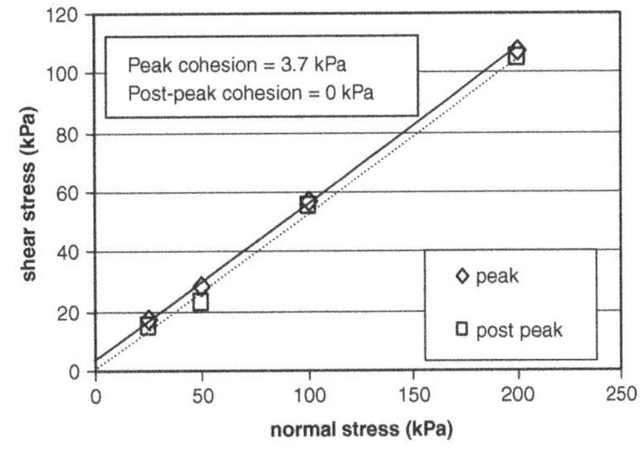

Figure 4. Shear box results from embankment soil.

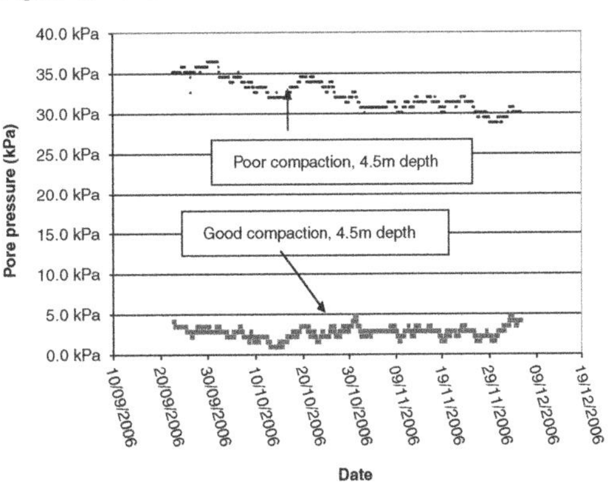

Figure 5. Pore pressures in embankment core Sept 06–Dec 06.

Table 2. Soils testing data summary.

		Bulk Density (Mg/m³)		Dry Density (Mg/m³)	
Compaction	2.5 kg 4.5 kg	2.11 at 15.5%WC 2.25 at 12.9%WC		1.82 at 15.5%WC 2.00 at 12.9%WC	
Atterberg limits	Average Liquid Limit (%) 41.7	Average Plastic Limit (%) 23.2	Average Plastic Index (%) 21.6	Average natural water content (%) 19.4	Soil is clay of intermediate plasticity (BS: 1377)
Laboratory Permeability		"poorly" compacted panels Well compacted panels		1.6×10^{-10} m/s 8.77×10^{-11} m/s	

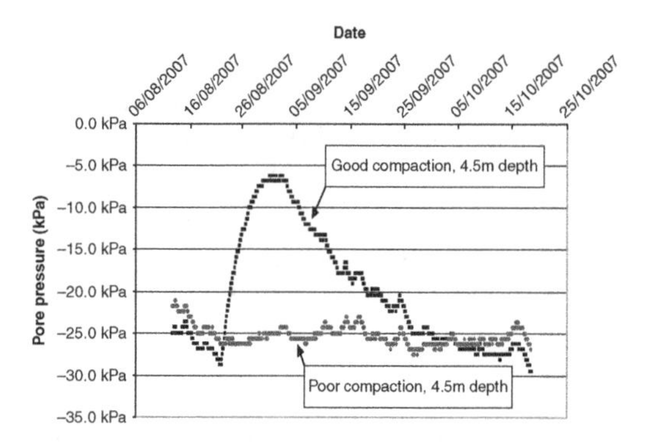

Figure 6. Pore pressures in embankment core Aug 07–Oct 07.

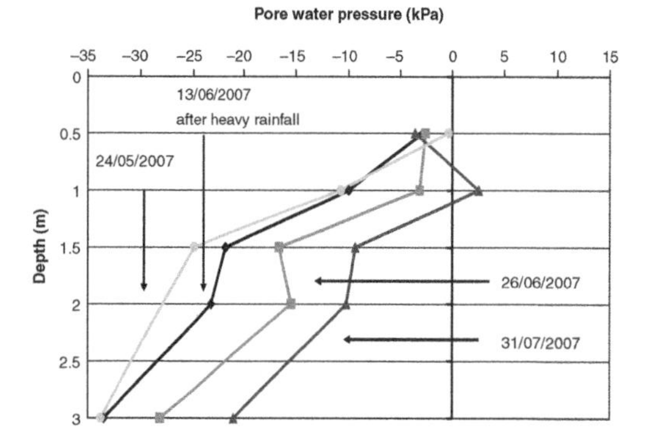

Figure 7. Well compacted panel pore water pressure profiles for different weather conditions (after Mendes et al 2008).

after the compaction was completed. As can be seen in Figure 5 the pore pressure recorded in the core of the embankment was positive 12 months after construction. Readings taken between August and October 2007 from the same positions show that the core of the embankment is now in suction. It is believed that the construction process and foundation conditions contributed to the dissipation of the initial high negative pore pressures. The establishment of vegetation has now begun to generate negative pressures once more, albeit much lower values.

Soil suction tensiometers installed by Durham University in the summer of 2007 are now also recording soil suctions in the 0–100 kPa range within the embankment (as discussed by Mendes et al, 2008). A summary of data from tensiometers installed in the "well" compacted panel A is shown in Figure 7.

7 REMOTE SENSING

In addition to the geotechnical instrumentation installed within the embankment BIONICS is also host to several spin off projects, including the Remote Asset Inspection for Transport Coridors project headed by John Mills at Newcastle University. As part if this project the embankment is routinely surveyed using GPS and Terrestrial Laser Scanning technology. These tools are being developed to be used in network scale assessments of transport infrastructure. A typical output from the terestrial laser scanner is shown in Figure 8.

Figure 8. 3-D model from terrestrial laser scan of BIONICS embankment.

Figure 9. Rainfall sprinkler component of climate control system.

8 NEXT STEPS

Additional instrumentation is being installed (principally standpipe piezometers and theta probes) to obtain further data on infiltration rates and soil permeability. These additions coincide with the recent installation of a climate control system which will be used to simulate extreme rainfall events over half the length of the embankment (shown in Figure 9).

Ground temperature and water content probes have also recently been installed along with 2 mini weather stations which will be used to calculate evapotransporation and determine the effects of slope aspect on water content and pore pressure. Figure 10 shows data from the first month of recording, the plot shows a

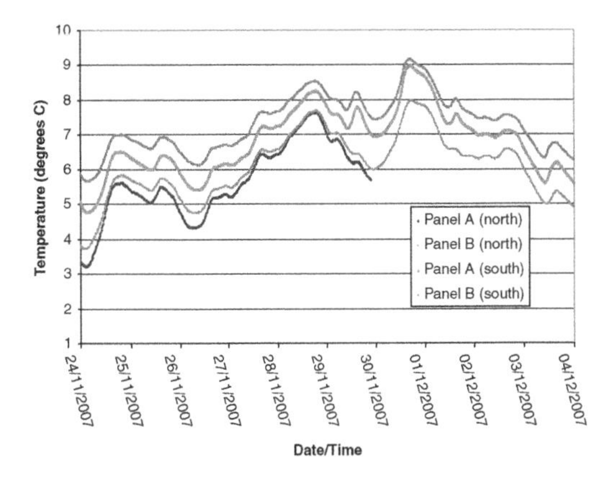

Figure 10. Ground temperature of north and south facing slopes.

clear positive difference in temperature between the North and South aspect.

Numerical analysis has shown that the permeabilities measured in the laboratory do not reflect the mass characteristics of the embankment therefore further analysis of field permeability will form part of the inundation tests. It is intended to repeat in-situ permeability tests throughout 2008 in order to investigate temporal variability. Also in 2008 a cover system will be in operation over part of the embankment length enabling drought conditions to be simulated. In the longer term it is hoped that the BIONICS embankment will be used for further spin-off projects.

9 CONCLUSIONS

Measurement of soil suctions during construction has demonstrated that significant negative pore pressures are generated during the construction process, strongly influencing slope stability. These negative pore pressures have been shown to be partially dependant upon compactive effort and water content.

Despite low permeabilities measured in the laboratory (see Table 2) early instrumentation datas now coming from the embankment indicates that water infiltration has dissipated the initial high negative pressures. One possible reason for this is that the in-situ permeability differs significantly from permeability measured in the laboratory caused by macro scale factors such as cracking and plant roots having a greater impact on soil permeability than the micro scale.

The development of numerical models from the data now being generated by the project will enable practitioners and infrastructure asset owners to target remediation works more effectively. It is estimated that early remediation costs one tenth of the cost of remediation post failure, this also prevents disruption to the transport network for the public and business users.

ACKNOWLEDGEMENTS

The authors wish to acknowledge the EPSRC, SRIF and the Railway Safety and Standards Board for funding the project. Additional thanks are extended to all researchers at Newcastle, Durham, Dundee, Bristol, Loughborough and Nottingham Trent Universities for their ongoing and significant input into the project and all members of the project Steering Group for their continued support. Further information on the project may be found at www.ncl.ac.uk/bionics.

REFERENCES

British Standards Institute (BSI) (1990). Methods of Test for Soils for Civil Engineering Purposes, BS 1377, London.

Mendes, J., Toll, D.G., Augarde, C.E. and Gallipoli, D. (2008). A System for Field Measurement of Suction using High Capacity Tensiometers, 1st European Conference on Unsaturated Soils, Durham, UK, July 2008.

O'Brien, A. (2001). Personal communication.

Perry, J. Pedley, M. and Reid, M. (2001). Infrastructure Embankments – condition appraisal and remedial treatment. CIRIA publication C550.

Loveridge, F. and Anderson, D. (2007). What to do with a vegetated clay embankment? Slope Engineering Conference, Thomas Telford, London, UK.

Advances in Transportation Geotechnics – Ellis, Yu, McDowell, Dawson & Thom (eds)
© 2008 Taylor & Francis Group, London, ISBN 978-0-415-47590-7

Field testing and modelling the effects of climate on geotechnical infrastructure – A collaborative approach

P.N. Hughes & S. Glendinning
Newcastle University, UK

N. Dixon & T.A. Dijkstra
Loughborough University, UK

D. Clarke, J. Smethurst & W. Powrie
Southampton University, UK

D.G. Toll & J. Mendes
Durham University, UK

D.A.B. Hughes
Queens University, Belfast, UK

ABSTRACT: Slopes make up a large proportion of UK transport networks and failure of these slopes causes disruption to network operation and frustration to the public. Our climate is predicted to change significantly over the next century; future change is likely to have a significant effect on much of our infrastructure. There is evidence that the scenario of more intense rainfall is already having an impact on the UK transport infrastructure including major landslides in Scotland (e.g. Stromeferry) and, in the winter of 2000/1, which was documented the wettest on record, over 100 slope failures in the Southern Region of Railtrack alone.

Newcastle, Southampton, Belfast, Durham and Loughborough Universities have all been carrying out research into the impacts of climate and vegetation on embankment and cut slope stability. These five Universities, along with international partners in Canada, Singapore, China, Thailand, Hong Kong, Spain, are conducting a collaboration programme the aim of which is to link research groups undertaking full-scale monitoring of slopes in order to ensure that there is an understanding of the resources required to maintain transport infrastructure in a changing global climate. This paper presents results of current full scale infrastructure slope monitoring and model development at the involved universities and plans for future collaborations.

1 INTRODUCTION

The UK's geotechnical transport infrastructure is generally well over a hundred years old, and requires continuous maintenance if serviceability problems and even slope failures are not to cause disruption and delays to the travelling public. Furthermore, the climate of the UK is set to change significantly over the next century, which is likely to have significant additional detrimental effects.

Slopes make up a large proportion of UK transport networks (£20 bn of the estimated £60 bn asset value of major highway infrastructure is earthworks). Major (ultimate limit state) failure of these slopes causes disruption to network operation and frustration to the public. Minor (serviceability limit state) failure of rail slopes causes speed restrictions and daily commuter misery due to delays. Continuous maintenance of these slopes is essential and, according to Perry et al (2001), cost £50 million in the UK in the year 1998/9 for embankments alone. However, the cost of emergency repair is ten times greater than the cost of planned maintenance works (O'Brien, 2001). This does not include the impact of failures of natural slopes adjacent to the transport network, or failures of other types of slopes such as those used for flood defence. There is evidence that the scenario of more intense rainfall is already having an impact on the UK transport infrastructure, including major landslides in Scotland (e.g. Stromeferry) and, in the winter of 2000/1, which was documented the wettest on record, with over 100 slope failures in the Southern Region of Railtrack alone.

2 COMMUNICATION AND NETWORKING

The nature of the problem of climate impacts on slopes is such that it affects many different stakeholders and end-users. The problem is also being approached from many different angles and with different objectives in mind. It thus forms a very broad multi-disciplinary field in which geographers, mathematicians, statisticians, physicists, engineers, ecologists, hydrologists, etc. try to work out their own particular problem angles and seek to forge links to provide a broader solution than would be achieved individually. This is not always easy as specialists speak different (scientific) languages and do not always share the same philosophical approach to problem solving.

With this in mind the network CLIFF*S* (climate impact forecasting for slopes) was funded by the UK Engineering and Physical Sciences Research Council (EPSRC) in 2005 to bring together academics, research and development agencies, stakeholders, consultants and climate specialists. The main aim of bringing these people together is to stimulate an integrated research response to address the intricately linked problem of forecasting, monitoring, design, management and remediation of climate change induced variations in slope instability. The size of the task and the complexity and multi-disciplinary nature of the problem requires active participation of a wide group to assess the magnitude of the resulting impact on UK society and to identify appropriate management and remediation strategies To achieve a better insight into the links between climate change and slope stability in the UK, firstly there is a need to determine the information requirements and, secondly, a need to focus research efforts on targeted assessments of long-term scenarios. Although detailed processes or individual site conditions are being addressed, general process-response issues are still not well understood or researched – a problem exacerbated by poor communication in this multi-disciplinary field (Dijkstra & Dixon 2007).

CLIFF*S* is managed by Loughborough University and is supported by a large core group of academic institutions (including the Authors') and stakeholders. It currently has more than 150 members, mainly from the UK. It operates by organizing multi-disciplinary themed workshops and by providing a web-based information exchange facility. Workshop themes have included issues of risk and uncertainty, and aspects of the responses of natural and constructed slopes to changes in climate. Details of these workshops can be accessed at the network's website on cliffs.lboro.ac.uk. Whilst current membership is mainly UK based, CLIFF*S* is being extended to include contacts with research groups in other regions around the world with the aim of sharing experience and data. The authors are collaborating on an EPSRC funded project (PICNIC – Potential Impacts of Climate on eNgineered Infrastructure Corridors) to communicate with and learn from researchers and practitioners who deal with slopes in a wide range of soils and vegetation, subjected to different climates. To date, the authors have held discussions with researchers from Spain, and through a study tour of Hong Kong and South East Asia have been instrumental in the formation of a South East Asia group studying climate slope processes (SEA CLIFFS). The activities of this group will be disseminated through CLIFFS.

3 CURRENT UK RESEARCH

Interest in the effects of climate on slopes has been generated by owners and operators of transport systems in the UK, as evidenced by their support of several ongoing research programmes. Five universities, Newcastle University, Queens University Belfast, University of Southampton, Loughborough University and Durham University have all been carrying out research into the impacts of climate and vegetation on embankment and cut slope stability. This has already included field instrumentation work to measure seasonal moisture and pore water pressure changes in a number of embankments and cut slopes, back analysis and numerical modelling.

The research has started to give a more detailed picture of embankment response (lateral and vertical deflections) to seasonal variations of both moisture content and pore water pressure. The behaviour of these embankments is complex, and in terms of trying to model their behaviour there are still many challenges to be overcome. Recent work has shown that the numerical models are very sensitive to the values and distributions of parameters such as permeability, which in a clay embankment can vary considerably as a result of summer desiccation and cracking close to surface, and the nature and compaction of the clay fill. It is possible that a very dry summer followed by a wet winter is most critical for stability, as the summer cracking allows a path for rainfall infiltration. However, this is still not well understood.

Future changes in climate in the UK are likely to lead to more extreme rainfall events with higher intensity storms. Such rainfall events are common in the tropical regions of the world and the team is drawing on collaborative work in Singapore, Thailand, China and Hong Kong.

The five universities recently received a major travel grant from the Engineering and Physical Sciences Research Council (EPSRC). This will allow the team to visit and build better links with both UK and overseas infrastructure owners and research organisations.

The Roads Service in Northern Ireland (Department for Regional Development in Northern Ireland) and

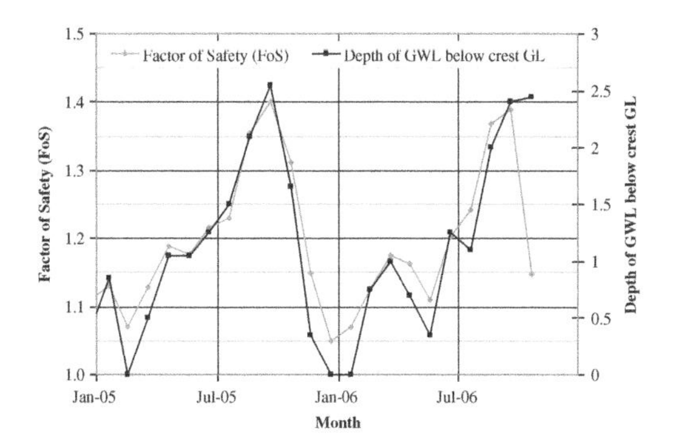

Figure 1. Monthly fluctuation of GWL.

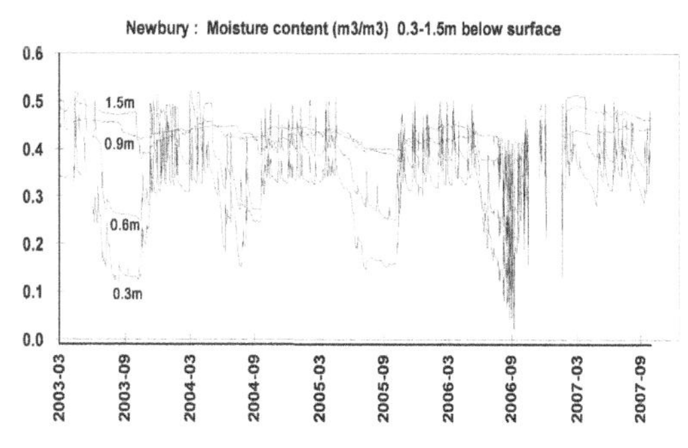

Figure 2. Moisture content (m^3/m^3) 0.3–1.5 m below surface.

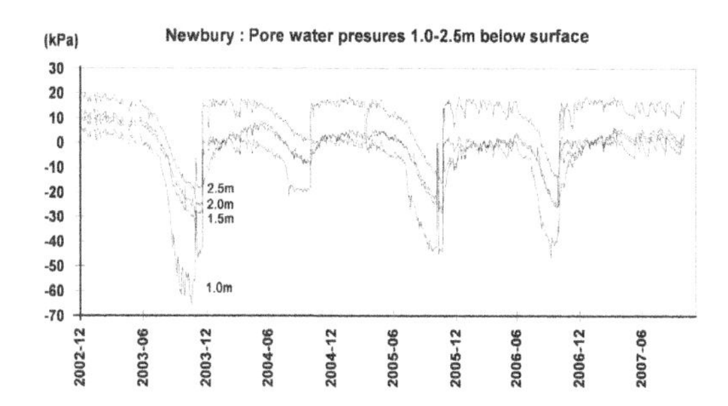

Figure 3. Pore water pressures 1.0–2.5 m below surface.

Northern Ireland Rail are funding Queens University Belfast to develop a risk based method of assessment of the geotechnical infrastructure on the Northern Ireland road network following a major slope failure on the road network in 2000 (Hughes et al 2007). As part of this research programme a cutting on the A1, 4 miles south of Dromore has been heavily instrumented (Clarke & Hughes et al).

Pore water changes were recorded during the excavation of the cutting and currently much data is being gathered on the pore water dynamics forced by rainfall and evapotranspiration effects (Figure 1). A transient predictive model incorporating climate events has been calibrated and verified against the field data using GeoStudio 2007. GeoSlope International (from Calgary, Alberta, Canada) have been supporting the project with technical assistance and the provision of the latest modelling software.

Two further sites are currently being instrumented to monitor deep and near surface pore water pressure changes. Craigmore cutting is a 23 m deep cutting on the rail network which is now 150 years old and experiencing some shallow failures. The cutting is heavily vegetated and it is intended to monitor the effects of the root zone and climate on shallow failures. Tullyhappy is a further 30 year old cutting in glacial till on the road network. A similar investigation ising been carried out here.

Southampton University has been carrying out intensive monitoring of soil moisture and pore water pressures at a Highways Agency owned road cutting near Newbury in Southern England since 2002. The climate is temperate with average annual rainfall of 850 mm, summer temperatures of +20°C and winter temperatures of 0–3°C. An array of 40 sensors were inserted in five groups along a cut slope in the London Clay, vegetated with a mixture of short grass and small bushes up to 0.5 m tall. Data were recorded for soil moisture content using Time Domain Reflectrometry (TDR) in the upper layers of the soil (0–2.5 m) below the surface. Pore water pressures were also monitored using Vibrating Wire Piezometers. Readings have been made every 10 minutes since 2002.

Hydrological inputs and losses at the site have been measured, including rainfall, surface runoff, depth to saturation together with climatic parameters to estimate potential evapotranspiration (temperature, humidity, wind speed, solar radiation). Soil characteristics have been determined using both field and laboratory approaches and the soil moisture sensors were calibrated using gravimetric methods and backed up with regular neutron probe measurements

Figure 2 shows the variation in volumetric soil moisture content (m^3/m^3) between 0.3 and 1.5 m depth This long term monitoring clearly shows the cyclical changes between summers (warm and relatively dry) and winters (cool and relatively wet). Most drying occurs in the upper 1.0 m of the soil profile, where the roots from the vegetation are most active. The maximum drying usually occurs at the end of summer in September (month 9), followed by a rapid re-wetting of the profiles in November-January. Also apparent is the effect of the climatic patterns of different years; 2003, 2005 and 2006 were relatively dry in the summer whereas 2004 and 2007 had higher than average rainfall.

Figure 3 shows the associated variations in pore water pressures for the same period. Near hydrostatic conditions occur in the winter months

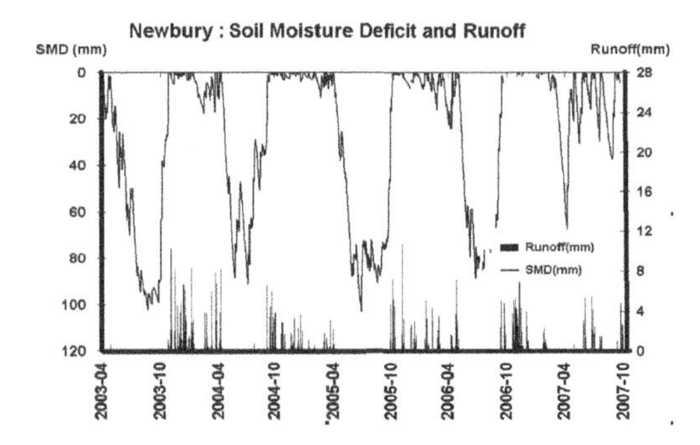

Figure 4. Soil moisture deficit and run-off.

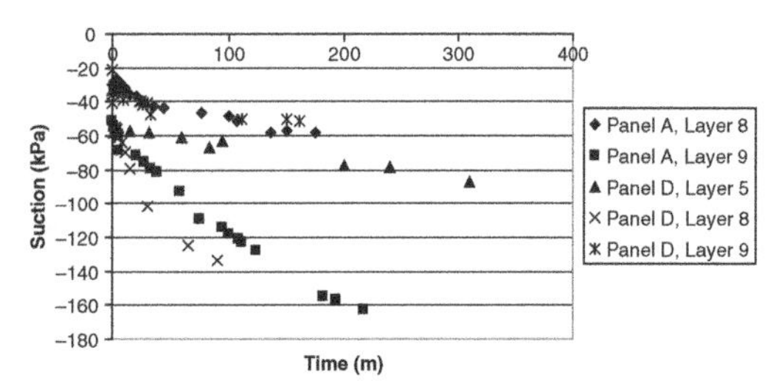

Figure 5. Suctions measured during construction of the poorly compacted sections.

Note: Section constructed in 1 m layers, beginning with layer 1.

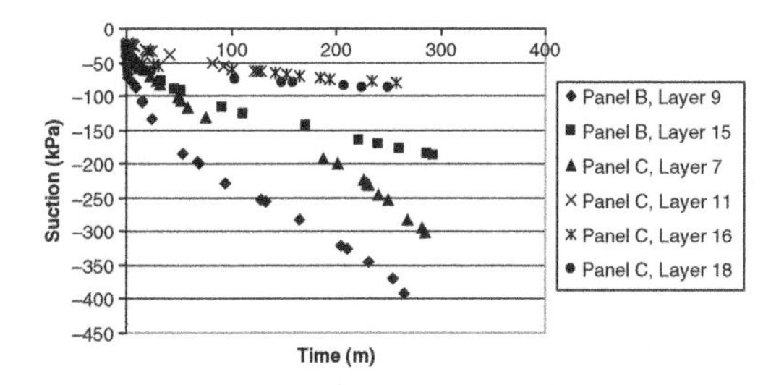

Figure 6. Suctions in the well compacted sections.

Note: Sections constructed in 0.3 m layers.

(November–March) but the seasonal growth of vegetation between April and September dries the soil and negative pore water pressures develop up to 2.5 below the surface. Suctions of up to −70 kPa are recorded at 1.0 m depth and suctions as high as −400 kPa have been recorded at 30 cm using some temporary instruments. The magnitude and duration of the negative pore water pressures varies from year to year, again depending on the climatic conditions experienced.

A series of hydrological and numerical models have been developed to describe and explain the behaviour of the processes at the site. Figure 4 shows the result of a Soil Moisture Deficit model based on the FAO CROPWAT methodology. The losses of moisture from the soil profile are calculated based on evapotranspiration and a root zone model and converted to an equivalent pore water pressure. These have been used to validate a FLAC model of the slope.

This work has demonstrated the use of hydrological models in describing the surface boundary conditions and their impacts on pore water pressures. These models are being used to explore the impact of many years repeated cycling of wetting and drying on the slope stability and UKCIP climate change scenarios are being using to investigate the long term performance of the slopes.

In addition to the monitoring of 'real' infrastructure slopes, subjected to 'real' UK climate, a consortium of asset owners has been put together by Newcastle University to oversee the BIONICS (Biological and Engineering Impacts of Climate change on Slopes; www.ncl.ac.uk/bionics) research project. This is a four year programme that aims to establish a unique facility for engineering and biological research. This facility is in the form of a full-scale, fully instrumented embankment, with climate control over part of its length. Thus, the facility allows the control of the climate necessary to study the effects of future climates, coupled with a fully characterized engineering soil and vegetative cover. A database of embankment performance data is being compiled of the results of testing and monitoring during both the construction and the climate experiments. This unique set of data, describing the full history of the embankment will be available for all future research based at the facility. The embankment is 90 meters long and has been constructed in two distinct parts. Half of the embankment is constructed to modern highway specifications using modern compaction plant, and half has been constructed to poorer specification using as little compaction as possible in order to simulate older rail embankments.

Data from the in-situ testing conducted on the embankment during its construction utilizing high suction tensiometers developed at Durham University (Lourenço et al, 2006) has demonstrated that high negative pore pressure were generated during the construction process. Figures 5 and 6 show suctions measured in the "poor" and "well" compacted zones respectively. These tests clearly indicate that modern construction techniques generate higher soil suctions than older methods.

Piezometers installed after construction have shown that the initial soil suctions had in part dissipated six months after construction was complete. Now that vegetation has become more established on the embankment soil suctions of up to −30 kPa have been measured in both poor and well compacted sections at 4.5 m depth using fully flushable piezometers. An additional system for measuring soil suctions, developed by Durham University has been installed in the BIONICS embankment using Wykeham Farrance – Durham University tensiometers (Mendes et al, 2008). The borehole locator system allows readings at different levels in a single borehole, permitting observations of the variation of suction with depth. It also allows tensiometers to be removed for re-saturation whenever necessary. The wide measuring range of the tensiometers (down to −2 MPa) allows usage of such a system in most natural and manmade earth structures.

Preliminary results (from three months of monitoring) show that there are different patterns of suction measurements from the tensiometers installed in the well compacted section of the embankment compared to those installed in the poorly compacted section (Mendes et al, 2008). It has been observed that tensiometers installed in the poorly compacted section of the embankment react rapidly to rainfall. The well compacted section instead shows a slower change of suction and does not respond rapidly to rainfall.

Currently a climate control system is being constructed, consisting of flexible, roofing sections that can be pulled over the embankment when required (similar to those used to cover sports facilities), and arrays of computer controlled rainfall sprinklers mounted on poles. Automatic weather stations will monitor wind speed, net radiation, temperature, relative humidity and atmospheric pressure, with tipping bucket and storage rain gauges to measure rainfall rates and totals. The performance of the proposed arrangements will be measured to ensure that it provides the climatic conditions required. In particular, the heating effect provided by covering (and leaving covers in place over night to prevent heat loss) will be compared to the temperatures predicted by climate change. This system will then be used to study the response of the embankment and the vegetation to controlled patterns of rainfall and heating.

Associated with the BIONICS project is an EPSRC funded research project at Loughborough University to develop novel instrumentation to detect slope instability (Dixon & Spriggs, 2007). A real-time continuous slope monitoring system based on detection and quantification of acoustic emission generated by slope deformations is currently being trialed on the BIONICS embankment. Performance of the acoustic system is being compared to traditional deformation measurement techniques including in place inclinometers. Acoustic emissions are related to slope deformation rates. The system is sensitive to both small magnitudes and rates of displacements and the technique has potential to provide an early warning of instability. This too will be monitored closely during the controlled climate experiments.

Durham University has been studying rainfall-induced slope failures in collaboration with Universities in Singapore and Thailand who currently experience more extreme patterns of rainfall and temperature. Experience in both countries is that there has been an increase in landslide activity associated with increased rainfall events (Toll et al, 2008; Jotisankasa et al, 2008).

Rainfall has been the dominant triggering event for landslides in Singapore and Thailand. Studies show spates of landslides occurring after unusually wet periods. Observations of past landslides in Singapore suggest that a total rainfall of 100 mm within a six day period is sufficient for minor landslides to take place (Toll, 2001). In Thailand, a total rainfall of 150–400 mm would tend to trigger major landslides (Jotisankasa et al, 2008).

Measurements of pore-water pressures in slopes in Singapore and Thailand show that rainfall infiltration produces changes in pore-water pressure near to the surface. However, at greater depths (around 3 m) the pore-water pressures do not change significantly (Tsaparas et al, 2003). Numerical modeling shows that this is because water tends to flow down the slope within the zone of higher saturation (which has higher permeability) that develops near the surface (Tsaparas and Toll, 2002). As a result, rainfall-induced failures tend to occur within the near surface zone and are not usually deep-seated.

Pore-water pressures measured within slopes in Singapore (Tsaparas et al, 2003) were, for a large part of the monitoring period, only slightly negative and at 3 m depth were generally positive. However, there were periods during the year when pore-water pressures reduced significantly following a drier period. Pore-water pressures dropped to as low as −70 kPa near the surface (0.5 m depth). Interestingly, this is similar to the values of suction measured at similar depths by Southampton University in very different climatic conditions in the UK. However, piezometer data in Singapore shows that there was little change in ground water table level (which remained below 15 m depth). Therefore, these suction changes were the result of infiltration and evapotranspiration occurring at the surface and were not due to changes in water table.

Therefore, it is important that when studying climate effects on slopes that we do not always assume that rainfall will produce a rise in water table level. Infiltration of rainfall at the surface can produce significant changes in pore-water pressure without a change in water table (although a perched water table may be induced at the surface).

Field measurements in Singapore suggest that pore-water pressures do approach the hydrostatic condition near the surface due to infiltration (Toll et al, 2001). However, pore-water pressures remain significantly below the hydrostatic line, even at the wettest time of the year. Therefore, assuming that pore-water pressures were hydrostatic throughout the slope (as would often be assumed in a saturated soil analysis) would be over-conservative.

Work is now underway with the National University of Singapore to investigate the impact that future climate change will have on Singapore, including slope stability problems (Toll et al, 2008).

4 CONCLUSIONS/SUMMARY

There is compelling evidence that our climate is changing and that this change will have a significant impact on the behaviour of earthworks infrastructure globally. In the UK, there is sufficient concern for the owners and operators of its transport networks to be actively funding research to investigate the problem. However, it has been recognized that the problem of climate influences on slopes is a sufficiently complex problem that a much greater understanding of the problem can be gained by sharing the existing knowledge from a wide range of disciplines. The CLIFF*S* network has been funded to facilitate such an exchange of ideas in the UK and the Authors are partners in the PICNIC project that aims to learn from international experience with researchers world wide who are determining the effects of a range of climates, vegetation and soil types. The rewards, in terms of shared experience and improved understanding are only just beginning to be realized, although it has already led to the formation of the SEA CLIFFS group in South East Asia, and it is anticipated that the shared experience will provide a more complete picture of the impacts of climate (and climate change) on slopes.

REFERENCES

Clarke, G.R.T., Hughes, D. A. B., Barbour, S.L. and Sivakumar, V. (2005) Field Monitoring of a Deep Cutting in Glacial Till: Changes in Hydrogeology. *GeoSask2005*, the 58th Canadian Geotechnical Conference and 6th CGS & IAH-CNC Joint Groundwater Specialty Conference, Saskatoon, Canada, September 18–21.

Dijkstra, T. and Dixon, N. (2007) Networking for the future – addressing climate change effects on slope stability. Proceedings Int. Conf. on Landslides and Climate Change: Challenges and Solutions, Isle of Wight, UK, May 2007 (CD).

Dixon, N. and Spriggs, M. (2007) Quantification of slope displacement rates using acoustic emission monitoring. Canadian Geotechnical Journal, 44 (8), 966–976.

Hughes, D., Sivakumar, V., Glynn, G., Clarke, (2007) A Case Study: Delayed Failure of a Deep Cutting in Lodgement Till, Journal of Geotechnical Engineering, ICE, Volume: 160 | Issue: 4 Cover date: October 2007 (accepted) Page(s): 193–202 Print ISSN: 1353–2618.

Jotisankasa, A., Kulsawan, B., Toll, D.G. and Rahardjo, H. (2008) Studies of Rainfall-induced Landslides in Thailand and Singapore, 1st European Conference on Unsaturated Soils, Durham, UK, July 2008.

Lourenço, S.D.N., Gallipoli, D., Toll, D.G. & Evans, F. D., 2006. Development of a commercial tensiometer for triaxial testing of unsaturated soils. Proc. 4th International Conference on Unsaturated Soils, Carefree, USA, Geotechnical Special Publication No. 147, ASCE, Reston. Vol. 2, 1875–1886.

Mendes, J., Toll, D.G., Augarde, C.E. and Gallipoli, D. (2008) A System for Field Measurement of Suction using High Capacity Tensiometers, 1st European Conference on Unsaturated Soils, Durham, UK, July 2008.

Toll, D.G. (2001) Rainfall-induced Landslides in Singapore, Proc. Institution of Civil Engineers: Geotechnical Engineering, Vol. 149, No. 4, pp. 211–216.

Toll, D.G., Mendes, J., Karthikeyan, M., Gallipoli, D., Augarde, C.E., Phoon, K.K. and Lin, K.Q. (2008) Effects of Climate Change on Slopes for Transportation Infrastructure, 1st ISSMGE International Conference on Transportation Geotechnics, Nottingham, UK, September 2008.

Toll, D.G., Tsaparas, I. and Rahardjo, H. (2001) The Influence of Rainfall Sequences on Negative Pore-water Pressures within Slopes, Proc. 15th International Conference on Soil Mechanics and Geotechnical Engineering, Istanbul, Rotterdam: Balkema, Vol. 2, pp. 1269–1272.

Tsaparas, I. and Toll, D.G. (2002) Numerical Analysis of Infiltration into Unsaturated Residual Soil Slopes, in Proc. 3rd International Conference on Unsaturated Soils, Recife, Brazil, Lisse: Swets & Zeitlinger, Vol. 2, pp. 755–762.

Tsaparas, I., Rahardjo, H., Toll, D. and Leong, E.C. (2003) Infiltration Characteristics of Two Instrumented Residual Soil Slopes, Canadian Geotechnical Journal, Vol. 40, No. 5, pp. 1012–1032.

Advances in Transportation Geotechnics – Ellis, Yu, McDowell, Dawson & Thom (eds)
© 2008 Taylor & Francis Group, London, ISBN 978-0-415-47590-7

Effects of climate change on slopes for transportation infrastructure

D.G. Toll, J. Mendes & C.E. Augarde
School of Engineering, Durham University, UK

M. Karthikeyan & K.K. Phoon
Department of Civil Engineering, National University of Singapore, Singapore

D. Gallipoli
Department of Civil Engineering, University of Glasgow, UK

K.Q. Lin
Land Transport Authority, Singapore

ABSTRACT: The paper reports on two studies of the expected effects of future climate change on slope stability in the UK and Singapore. In the UK the BIONICS project is investigating the role of climate in the stability of embankments for transportation infrastructure. The project involves a full-scale instrumented embankment that has been constructed with facilities for climate control. A novel instrumentation system is described that is being used to observe changes in pore water pressures (suctions) due to rainfall. In Singapore a national study is being carried out to investigate the impact of climate change, one aspect of which is slope stability. The mode of slope failure due to rainfall infiltration is reviewed and the need to use unsaturated soil mechanics to explain the observed behaviour is emphasised. Numerical modelling has been carried out using the finite element code SEEP/W to allow the effects of future climate events to be investigated. The identification of appropriate parameters for unsaturated flow modelling is described and limitations in commercial finite element codes are identified.

1 INTRODUCTION

The IPCC 4th Assessment Report (IPCC, 2007) provides convincing evidence of global warming as a result of increased greenhouse-gas production since the start of industrialisation in 1750. The implications of this, as the report states, are: "Continued greenhouse gas emissions at or above current rates would cause further warming and induce many changes in the global climate system during the 21st century that would *very likely* be larger than those observed during the 20th century".

We are already seeing the occurrence of more extreme climate events. In the UK, the winter of 2000/1 was the wettest on record and the period May-July 2007 was the wettest for 250 years (leading to extensive flooding in Gloucestershire, Worcestershire and Yorkshire). In 2006, Singapore had its wettest December (766 mm) since recording of rainfall started in 1869 (NEA, 2006). In December/January 2006/7 the maximum rainfall was 345 mm, lasting over 20 hours (one of the highest recorded in Singapore in the last 75 years) (Ng *et al.*, 2007). Whether or not these extreme events are attributable to global warming is not important; they are the realities of our current climate and we need to ensure that our infrastructures can cope with such events and possibly more extreme events in the future.

Since a large part of any transportation network comprises slopes (in embankments and cuttings) it is important that we understand the impact of climate change on this part of our infrastructure. There is already evidence that more intense rainfall is having an impact on the UK transport infrastructure, including 100 slope failures on the UK national rail network (Turner, 2001) and 60 on the road network (Ridley *et al.*, 2004) during the winter of 2000/1. Similarly in Singapore, the Building and Construction Authority (BCA, 2007) noted increased landsliding due to the exceptional rainfall in December/January 2006/7 and the Land Transport Authority recorded 17 landslides on the transportation infrastructure of Singapore during December 2006 to March 2007.

This paper reports on studies underway in the UK and Singapore to examine the effects of climate change on slope stability. To predict the implications of future climate change will inevitably require the use of numerical modelling based on anticipated future climate regimes. However, to achieve this it is vital that the models used are validated against measurements from the past and present. This means such studies must involve field measurements of responses to current climate as well as the development of modelling methodologies that can deal with the complexity of climate/soil interaction.

The BIONICS project in the UK involves a full-scale instrumented embankment that has been constructed with facilities for climate control (Glendinning *et al.*, 2006). This will provide experimental evidence for changes in pore water pressures (suctions) within a UK embankment. In Singapore the impact of climate change on slope stability is being investigated through numerical modelling as part of a national study of the impact of climate change. This involves modelling the unsaturated flow processes due to rainfall infiltration.

2 RAINFALL-INDUCED SLOPE FAILURES

To properly understand rainfall-induced slope failures we need to apply an understanding of unsaturated soil behaviour. For embankments above the water table, pore water pressures can be negative (i.e. suctions). Suctions can also exist in cuttings where the water table is deep. Therefore, we need to understand the role of suction in supporting the slope (increasing the strength of the soil) and how infiltration of rainwater causes changes in the pore water pressures (or suctions).

Toll *et al.* (2001) and Toll (2006) noted that field measurements in Singapore show that pore water pressures do approach the hydrostatic condition near the ground surface during rainfall infiltration. However at 2.5–3 m depth little change in pore water pressure was observed. The pore water pressures at this depth remained significantly below the hydrostatic line, even at the wettest time of the year. Ridley *et al.* (2004) came to the same conclusion for UK conditions. Therefore, assuming that pore water pressures were hydrostatic throughout the slope (as would often be assumed in a saturated soil analysis) could wrongly predict that more deep-seated failures might occur, whereas the important mode of failure might be shallow.

Pore water pressure responses to rainfall can be better understood by numerical modelling. A major factor in controlling the response is the changes in water permeability near the surface (Tsaparas and Toll, 2002). The surficial permeability is often significantly higher due to desiccation cracking and root passages; this increase can be two orders of magnitude (Chappell and Lancaster, 2007). In addition to this, when water infiltrates at the surface, a near-surface zone with a high degree of saturation is produced. This produces a zone of much higher permeability (this can be 3–4 orders of magnitude higher than the unsaturated permeability). Since the permeability at depth (2–3 m below the ground surface) will be so much lower, water is not encouraged to flow to greater depths (even though the hydraulic gradient might be greater in this direction); instead flow tends to take place down the slope within the near-saturated surface zone.

This means that most pore water pressure changes in clayey soils are likely to take place near the surface. Failures therefore tend to occur within the near-surface zone where pore water pressures increase close to hydrostatic levels. This is consistent with observations of failures on the UK highway network, where many failures are quite shallow (Perry, 1989). It also fits with the fact that many rainfall-induced landslides in Singapore are less than 2 m deep (Toll *et al.*, 1999).

3 MEASUREMENTS OF SUCTION (UK)

The objective of the BIONICS project is to investigate what could happen to infrastructure embankments in the UK when subjected to climate change. An experimental embankment has been built in four panels (Figure 1) separated by vertical impermeable membranes and constructed using different compaction efforts (Glendinning *et al.*, 2006). Panels A and D are poorly compacted (intended to represent old rail embankments constructed in Victorian times) while panels B and C are well compacted (representing modern embankments).

Current measurements of suction have been obtained during natural rainfall conditions. However, a climate control system will be used to impose expected future climate patterns on the embankment.

A novel instrumentation system was installed in the BIONICS embankment to allow continuous measurements of suction or pore water pressure (Mendes *et al.*, 2008). Ten high capacity tensiometers capable of direct measurement of pore water pressure to −2 MPa (Lourenco *et al.*, 2006) were installed in two borehole probe locators. Each locator includes five suction stations at depths of 0.5 m, 1 m, 1.5 m, 2 m and 3 m. The borehole probe locator consisted of a 3 m long PVC pipe with an outer diameter of 90 mm. Five guide tubes were inserted inside the borehole probe locator to individually connect each suction station to the surface and allow insertion of the tensiometers. This design allows the tensiometers to be removed individually whenever necessary. This is important as tensiometers can cavitate when measuring high suctions for a long period of time and then require removal for resaturation.

One borehole probe locator was installed in the poorly compacted panel A while the second was installed in the well compacted panel B (see Figure 1).

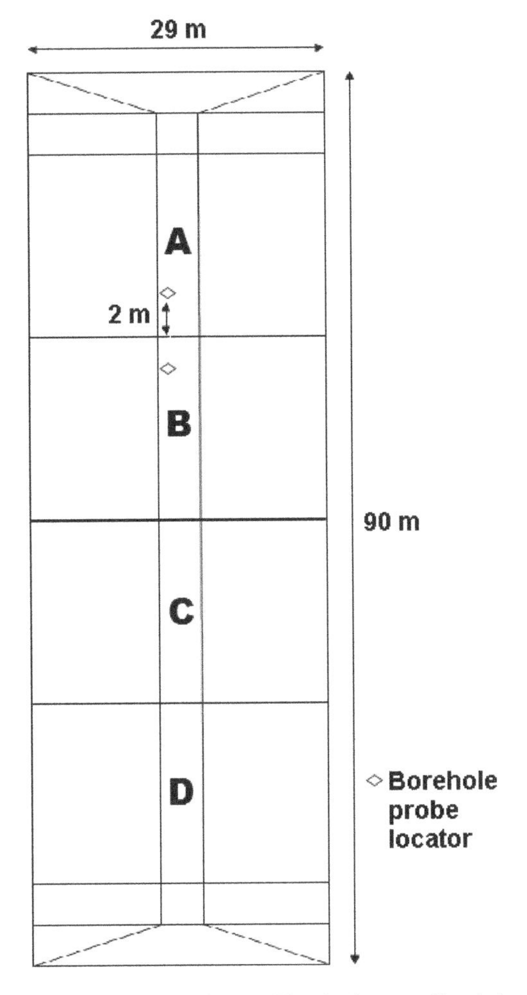

Figure 1. Plan view of BIONICS embankment and borehole probe locators (after Glendinning et al., 2006).

Both were located close to the south facing slope of the embankment at about 1 m from the edge of the crest.

Preliminary suction measurements in the embankment are available from May to July 2007. Data for the well compacted panel are reported here (Figures 2 and 3). Figure 2 also shows the rainfall recorded at the site.

Suctions measured during the construction of the embankment in 2005 (Hughes *et al.*, 2007) showed suctions in the well compacted panel ranging from 80 kPa to over 400 kPa. However, by the time the continuous suction monitoring system was installed in 2007 the suction readings were lower; close to zero near the ground surface (0.5 m) and with a maximum value of 35 kPa at 3 m depth.

The results show a progressive wetting up (reduction in suction) since monitoring started in May 2007.

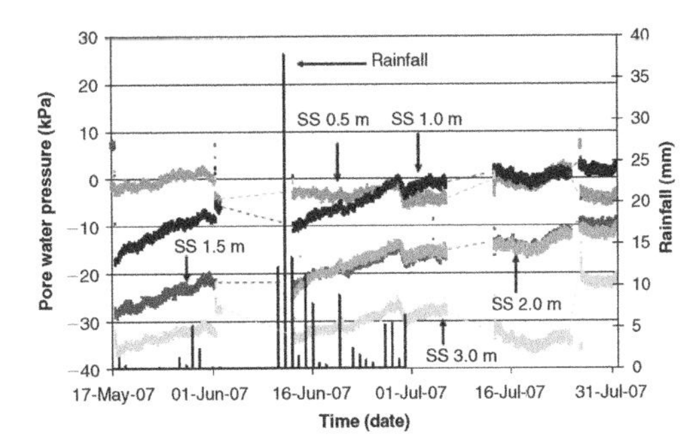

Figure 2. Pore water pressure records for the well compacted panel suction (SS indicates suction station at different depths). Vertical spikes show daily rainfall.

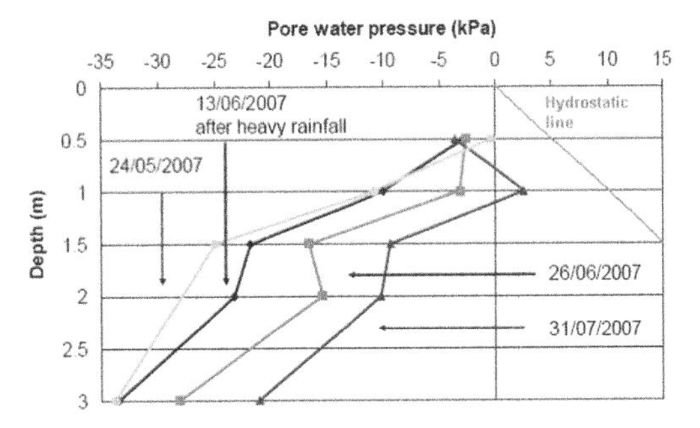

Figure 3. Well compacted panel pore water pressure profiles for different weather conditions.

There was heavy rainfall on 13th June 2007 followed by a continued wet period. At 1 m depth the initial pore water pressure was −20 kPa but this increased with time to reach small positive values. Below 1 m, pore water pressures reduced with depth but the values of pore water pressure have been gradually rising with time (see Figure 2).

The changes in pore water pressure profiles can be seen in Figure 3. This shows a progressive wetting up, since the initial readings in May, with pore water pressures approaching zero within the top 1 m (and becoming positive at 1 m). The results to date show pore water pressure profiles well below hydrostatic conditions, but further monitoring through the winter months are needed to confirm whether this remains true for worst case conditions.

In the well compacted material the tensiometers do not show rapid responses to rainfall events, rather a general increase in pore water pressure (reduction in suction) has occurred with time. Results for the poorly compacted panel reported by Mendes *et al.* (2008) show more rapid responses to rainfall but the response to wetting is less consistent than for the well compacted panel.

The suctions measured in the BIONICS embankment are consistent with the order of measurements recorded elsewhere in the UK. Ridley *et al.* (2004) show suctions of 50–100 kPa in an underdrained embankment. Smethurst *et al.* (2006) report somewhat higher values, with suctions of over 80 kPa near the surface in a cutting in London Clay (with some evidence for higher suctions greater than 400 kPa at 0.3 m depth).

4 NUMERICAL MODELLING (SINGAPORE)

A study is underway to provide predictions of the impact of climate change on slopes in Singapore. A commercial finite element code, Seep/W (Geo-Slope International, 2004) has been used for the numerical modelling of unsaturated flow, adopting a two-dimensional and transient seepage model in an infinite slope. For validation, the model has been compared with field measurements reported by Tsaparas *et al.* (2003).

The input parameters needed to undertake transient seepage analyses require an understanding of the relationship between matric suction and volumetric water content (the water retention curve) and the relationship between permeability and suction (the permeability function).

Figure 4 shows the water retention curve for a soil sample obtained from the slope at a depth of 0.4 m (Tsaparas, 2002). The saturated volumetric water content was 0.53 and at a suction of 200 kPa the volumetric water content reduced to 0.38.

Agus *et al.* (2001) reported an envelope for water retention curves established for Jurong residual soils (the slope is in the Jurong formation) and these are included in Figure 4 for comparison. The upper, average and lower water retention curves shown in Figure 4 were established using saturated volumetric water content of 0.53 and the curve fitting parameters reported in Agus *et al.* (2001).

Other water retention curves from the location of the slope reported by Agus *et al.* (2003) are also shown in Figure 4. These are for different depths (UP-1 from 5.60 m and UP-3 from 4.0 m). The figure shows that there is similar trend in the shape of the water retention curve for Jurong residual soils even though there is difference in the saturated volumetric water content. Agus *et al.* (2001) also examined the effect of weathering on the shape of the water retention curves and reported that there are no significant difference between the shape of these curves and the depth of weathering for Jurong sedimentary formation residual soils.

Measurement of permeability functions for unsaturated soils is a time consuming and labour-intensive process and there are often limited data. However, Agus *et al.* (2003; 2005) presented unsaturated permeability functions measured in the laboratory for

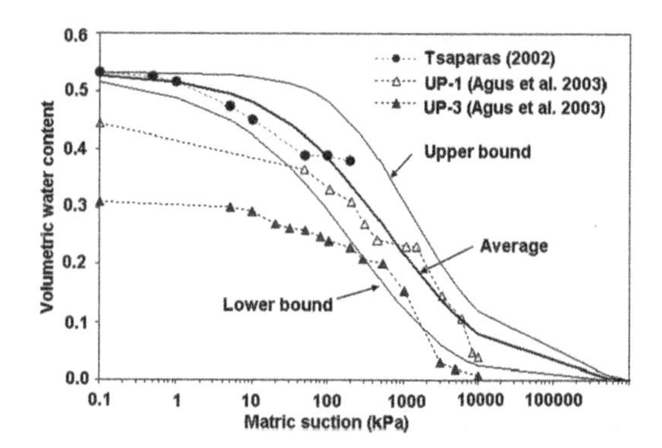

Figure 4. Water retention curves for Jurong formation residual soils.

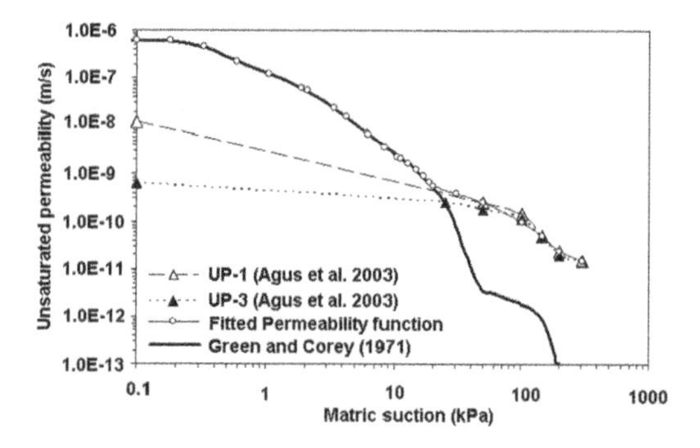

Figure 5. Unsaturated permeability functions for Jurong formation residual soils.

Singapore residual soils. Figure 5 shows the unsaturated permeability functions measured for Jurong residual soil samples.

From Figure 5, it can be seen that there is close agreement between these two soil samples for suctions above 20 kPa. Even though the saturated permeability values are different (due to differences in saturated volumetric water content as can be seen in Figure 4) the unsaturated permeability functions are almost identical in the higher suction range.

Karthikeyan *et al.* (2008) described the use of the commonly used integration functions for predicting the unsaturated permeability function from water retention data for the Jurong soil. The predicted curves showed significant differences between the experimental data reported by Agus *et al.* (2003) and the predictions. The Green and Corey (1971) method gave a curve closest to the experimental results but was still very different. This indicates the danger in using such expressions without an experimental confirmation of their validity.

In situ permeability measurements at the study area showed that the saturated coefficient of permeability with respect to water, k_s, for the soil is 6×10^{-7} m/s, measured at approximately 0.4 m deep using a

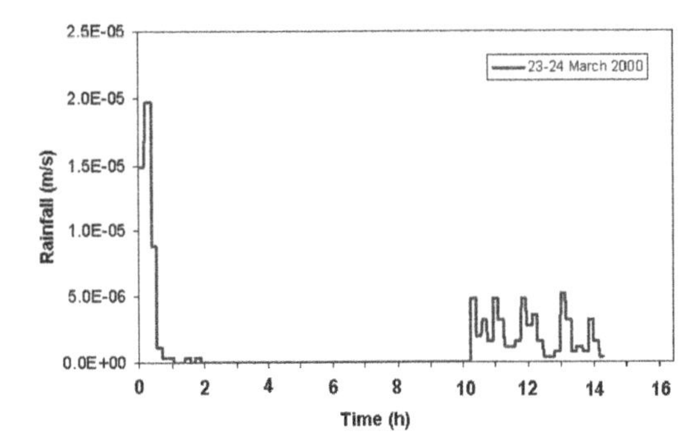

Figure 6. Recorded rainfall events from 23–24 March 2000.

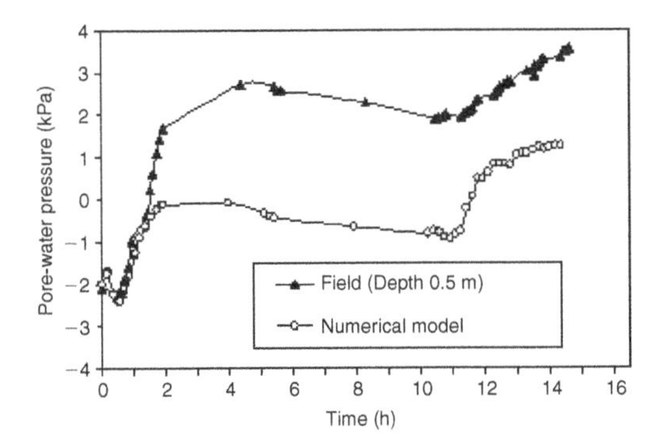

Figure 7. Comparison of pore water pressure variations from field observations and modelling.

Guelph Permeameter (Tsaparas *et al.*, 2003). Other measurements for k_s of Jurong soil reported by Agus *et al.* (2003) and others show that k_s can vary between 10^{-10} m/s and 10^{-6} m/s (Karthikeyan *et al.*, 2008). The field measurement by Tsaparas (2002) falls within the range of values, and is in agreement with data by Agus *et al.*, 2003.

It can be seen from Figure 5 that there is lack of information on experimental data near the air entry value of the soil (below a matric suction of 20 kPa). Therefore, the Green and Corey (1971) equation was used to estimate the unsaturated coefficient permeability values near the air entry value up to the matric suction of 20 kPa (using the water retention curve reported by Tsaparas, 2002 shown in Figure 4) with a saturated coefficient of permeability of 6×10^{-7} m/s. The fitted permeability function shown in Figure 5 is based on the Green and Corey equation up to 20 kPa suction and on experimental observations for higher suctions. This curve was used in the analyses.

Tsaparas and Toll (2002) recognized the effect of the higher permeability of the surficial layer due to desiccation cracking and root passages. They accounted for this by including a more permeable surface layer with a thickness of 0.25 m (the depth affected by rooting). Karthikeyan *et al.* (2008) also showed that modelling the behaviour without taking account of this higher permeability zone did not compare well with field measurements. Therefore, a 0.25 m layer was introduced at the surface, with highly anisotropic permeability; a higher permeability was used perpendicular to the surface ($k_y = 6 \times 10^{-4}$ m/s) (taking account of cracking and root passages in this direction), but the same value of permeability as the matrix soil was maintained for flow parallel to the ground surface ($k_x = 6 \times 10^{-7}$ m/s).

Figure 7 shows a comparison between the field observations and a numerical simulation of the variation of pore-water pressure based the rainfall data for 23–24 March 2000 (shown in Figure 6). It can be seen that the changes are quite small. The comparison of the results shows the trend in the results is the same as that shown by the field measured values. However, the numerical prediction still underestimates the magnitude of the pore water pressure change when compared with field measurement.

Further attempts have been made to introduce dual porosity and permeability models into the analysis to improve the predictions. However, Seep/W is unable to handle the steep changes in material properties required; the solution tends to diverge instead of converge and oscillates between two extreme solutions represented by the extremities of the permeability function.

Limitations of using commercial finite element codes for modelling unsaturated flow have been reported by Karthikeyan *et al.* (2001), Tan *et al.* (2004), Cheng *et al.* (2007) and Fredlund (2007). Some of these limitations can be overcome by correct selection of mesh size and time step intervals. Nevertheless, Cheng *et al.* (2007) show that numerical limitations can lead to an over prediction of the wetting front in infiltration analyses. This can have serious implications for slope stability calculations; in the example given by Cheng *et al.* the factor of safety was over predicted at 2.3, compared to a correct solution of 1.1. These shortcomings need to be addressed and this is now the focus of work at the National University of Singapore.

5 CONCLUSIONS

A large part of any transportation network comprises slopes (in embankments and cuttings) and so it is important that we understand the impact of climate change on this part of our infrastructure. To predict the effects of future climate change on slope stability requires the use of numerical modelling, based on anticipated future climate regimes. To achieve this it is vital that the models are validated against field measurements. Therefore, this paper reports on studies involving field measurements of pore water pressure responses to current climatic conditions and also discusses the development of modelling methodologies and parameter selection for modelling the complexity of climate/soil interaction.

Results of field monitoring in the BIONICS embankment in the UK shows a progressive wetting up over the period May to July 2007. The measurements show the pore water pressures approaching zero within the top 1 m (and becoming positive at 1 m). The results to date show pore water pressure profiles well below hydrostatic conditions, but further monitoring through the winter months are needed to confirm whether this remains true for worst case conditions.

The results from the numerical modelling for Singapore show that trends in observed pore water pressures can be predicted, but the magnitudes are not in full agreement. Further research is required to improve the accuracy of the numerical analysis. It is shown that some of the conventional assumptions about the critical input parameters (such as the use of integration functions for predicting the unsaturated permeability function from water retention data) can give results that are seriously in error. These discrepancies were only evident due to the significant research effort to obtain the water retention and flow properties for the Singapore residual soils. This emphasises the need for experimental data to provide these properties, as well as the need to validate the models against field observations.

REFERENCES

Agus, S.S., Leong, E.C. & Rahardjo, H. (2001). Soil-water Characteristic Curves of Singapore Residual Soils. Geotechnical and Geological Engineering, Vol. 19, pp. 285–309.

Agus, S.S., Leong, E.C. & Rahardjo, H. (2003). A flexible wall permeameter for measurements of water and air coefficients of permeability of residual soils, Canadian Geotechnical Journal, Vol. 40, pp. 559–574.

Agus, S.S., Leong, E.C. & Rahardjo, H. (2005). Estimating Permeability Functions of Singapore Residual Soils, Engineering Geology, Vol. 78, pp. 119–133.

BCA. 2007. Slope Stability for a Safe Built Environment, Pillars Issue 2, Building and Construction Authority, Singapore, p. 16.

Chappell, N.A. & Lancaster, J.W. 2007. Comparison of methodological uncertainties within permeability measurements. Hydrological Processes, 21, pp. 2504–2514.

Cheng, Y.G., Phoon, K.K. and Tan, T.S. 2007. Unsaturated soil seepage analysis using a rational transformation method with under-relaxation. Submitted to International Journal for Geomechanics.

Fredlund, D.G. 2007. Engineering Design protocols for unsaturated soils. Proceedings of the 3rd Asian Conference on Unsaturated Soils, edited by Yin, Z., Yuan, Z. and Chiu, A.C.F. Science Press, Beijing, China, pp. 27–45.

Geo-Studio (2004) User's manual for SEEP/W, Geo-Slope International Ltd., Canada.

Glendinning, S., Rouainia, M., Hughes, P. & Davies, O. 2006. Biological and engineering impacts of climate on slopes (BIONICS): The first 18 months, Proc. 10th IAEG Congress, Nottingham, Paper 348 (on CD).

Green, R.E. & Corey, J.C. 1971. Calculation of hydraulic conductivity: A further evaluation of some predictive methods, Proc. Soil Sci. Soc. Am., 35, pp. 3–8.

Hughes, P., Glendinning, S. & Mendes, J. 2007. Construction Testing and Instrumentation of an Infrastructure Testing Embankment, Proc. Expert Symposium on Climate Change: Modelling, Impacts & Adaptations (eds. Lion, Singapore.

IPCC. 2007. Summary for Policymakers. In: Climate Change 2007: The Physical Science Basis. Contribution of Working Group I to the Fourth Assessment Report of the Intergovernmental Panel on Climate Change [Solomon, S., D. Qin, M. Manning, Z. Chen, M. Marquis, K.B. Averyt, M. Tignor and H.L. Miller (eds.)]. Cambridge University Press, Cambridge, United Kingdom and New York, NY, USA. (http://www.ipcc.ch/pdf/assessment-report/ar4/wg1/ar4-wg1-spm.pdf)

Karthikeyan, M., Tan, T.S. & Phoon, K.K. 2001. Numerical Oscillation in Seepage Analysis of Unsaturated Soils. Canadian Geotech. Journal, 38, pp. 639–651.

Karthikeyan, M., Toll, D.G. & Phoon, K.K. 2008. Prediction of changes in pore-water pressure response due to rainfall events, 1st European Conf. Unsaturated Soils, Durham, July 2008.

Lourenço, S.D.N., Gallipoli, D., Toll, D.G. & Evans, F.D., 2006. Development of a commercial tensiometer for triaxial testing of unsaturated soils. In Proc. 4th International Conference on Unsaturated Soils, Carefree, USA, Geotechnical Special Publication No. 147, ASCE, Reston. Vol. 2, 1875–1886.

Mendes, J., Toll, D.G., Augarde, C.E. & Gallipoli, D. 2008. A System for Field Measurement of Suctions using High Capacity Tensiometers, 1st European Conf. Unsaturated Soils, Durham, July 2008.

NEA. 2006. Singapore experienced its wettest December since 1869, National Environment Agency, Singapore http://app.nea.gov.sg/cms/htdocs/article.asp?pid=2815

Ng, P.B., Poh, K.K. & Tenando, E. 2007. Slope Failure after Prolonged Rainfall, BCA Seminar – Approach to Structural Inspection and Slope Stability, Building and Construction Authority, Singapore.

Perry, J. 1989. A Survey of Slope Condition on Motorway Earthworks in England and Wales, Research Report 199, Transport and Road Research Laboratory, Crowthorne.

Ridley, A., McGinnity, B. & Vaughan, P. (2004) Role of pore water pressures in embankment stability, Geotechnical Engineering, Proceedings of the Institution of Civil Engineers, 157(GE4), pp. 193–198.

Smethurst, J. A., Clarke, D. & Powrie, W. 2006. Seasonal changes in pore water pressure in a grass-covered cut slope in London Clay, Géotechnique 56(8), pp. 523–537.

Tan, T.S., Phoon, K.K. & Chong, P.C. (2004). Numerical study of finite element method based solutions for propagation of wetting fronts in unsaturated soil. Jnl. Geotech. and Geoenvir. Engineering., 130(3), 254–263.

Toll, D.G. 2006. Landslides in Singapore, Ground Engineering, 39(4), pp. 35–36.

Toll, D.G., Rahardjo, H. and Leong, E.C. 1999. Landslides in Singapore, Proc. 2nd International Conference on Landslides, Slope Stability and the Safety of Infra-Structures, Singapore, pp. 269–276.

Toll, D.G., Tsaparas, I. & Rahardjo, H. 2001. The Influence of Rainfall Sequences on Negative Pore water Pressures within Slopes, Proc. 15th International Conference on Soil Mechanics and Geotechnical Engineering, Istanbul, Rotterdam: Balkema, Vol. 2, pp. 1269–1272.

Tsaparas, I. 2002. Field Measurements and Numerical Modelling of Infiltration and Matric suctions within Slopes, PhD Thesis, School of Engineering, Durham University, UK. 314p.

Tsaparas, I. & Toll, D.G. 2002. Numerical Analysis of Infiltration into Unsaturated Residual Soil Slopes, in Proc. 3rd International Conference on Unsaturated Soils, Recife, Brazil, Lisse: Swets & Zeitlinger, Vol. 2, pp. 755–762.

Tsaparas, I., Rahardjo, H., Toll, D.G. and Leong, E.C. 2003. Infiltration characteristics of two instrumented soil slopes. Canadian Geotechnical Journal, 40, pp. 1012–1032.

Turner, S. 2001. Climate Change Blamed as Landslip Incidents Treble. New Civil Engineer, 10 May 2001, p. 8.

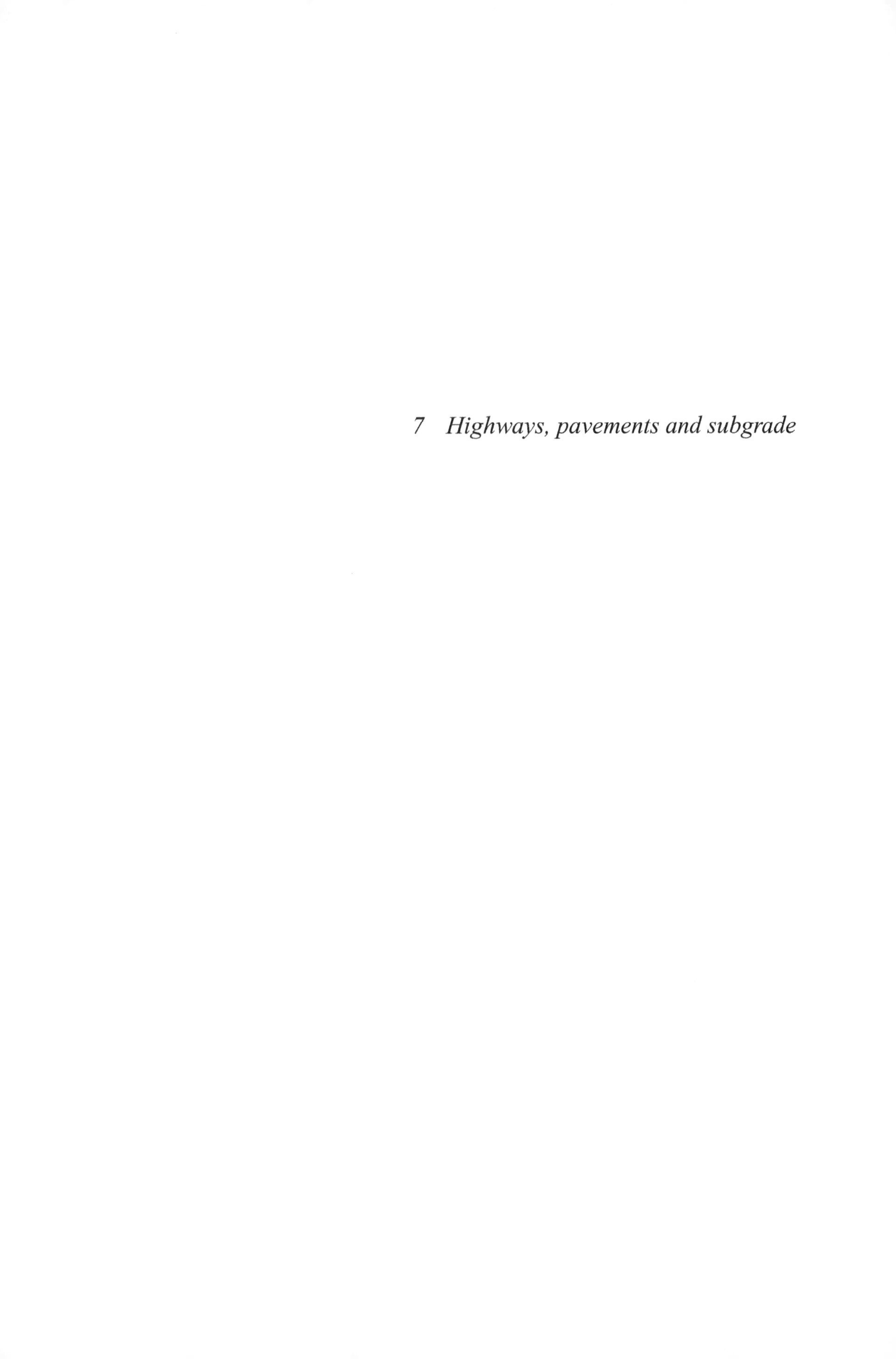

7 *Highways, pavements and subgrade*

Advances in Transportation Geotechnics – Ellis, Yu, McDowell, Dawson & Thom (eds)
© 2008 Taylor & Francis Group, London, ISBN 978-0-415-47590-7

Using soft clay modified with cement-agricultural wastes as road construction materials

C.M. Chan & K.A. Ibrahim
Research Centre for Soft Soils (RECESS), Faculty of Civil and Environmental Engineering, Universiti Tun Hussein Onn Malaysia, Malaysia

ABSTRACT: The southern region of Peninsular Malaysia has a large and widespread deposits of soft clay soil, frequently giving rise to problems in the construction of road embankments. Problems such as undulating pavement, severe potholes or even sinkholes were not uncommon. The usual practice was to discard and replace the original soft soil with suitable material prior to construction, but this mass replacement method could be costly and labour-intensive. This project was therefore conceived to explore the possibilities of utilising the in situ soil itself as the main road construction material, via modifications with stabilising additives. The stabilising agents experimented, apart from cement as the binding agent, were mainly natural wastes, including fibres from pineapples leaves, rubber chips as well as rice husks. These modified road construction materials were not only promoting sustainable and environmental-friendly engineering practice, but also opening up new opportunities of using in situ materials in road construction projects. A laboratory-based approach was adopted in this project, mainly aimed at identifying the suitable admixed materials and optimum proportion of mixes, as well as determining the engineering characteristics of the alternative materials in terms of their intended use within the road structure. Generally, it was shown that modified soft soils have huge potential to be used as site specific road construction materials, requiring minimal pre-construction trial tests to determine the suitable mixes.

1 INTRODUCTION

1.1 *Type area*

Roads are arguably the most widely used platform for travelling on land. With years of research and development, road construction technology and machinery are now well developed and advanced. The downside, however, is the relatively high costs incurred, which can sometimes hinder development in rural areas.

Malaysia has an extensive network of roads, primarily of asphalt and rigid pavements, while most rural areas in the country are still connected by *jalan kampung* or village roads, which are generally compacted but unpaved.

The country also has a widespread deposit of soft marine clay along the coasts, with an average depth of about 40 m (Ahmad Tajudin 2004). Subgrades made of soft clay is particularly unreliable to provide sufficient support, especially when soaked or saturated (Parsons and Kneebone 2005). This is due to low strength and high compressibility of the soft clay soil.

The soft clay is usually disposed of and replaced with other suitable backfill materials for road construction. However, instead of incurring more costs by transporting the materials, the soft clay can be reutilised via stabilisation. This is especially beneficial for coastal and rural areas with such problematic soils or limited resources. Besides the modified clay road has added 'green' value for reusing existing agricultural waste materials, such as pineapple leaves, rubberchips and rice husks. These wastes would have been disposed of in landfills or through open burning if not reutilised. With the addition of these materials, smaller quantities of cement may be required for improving the soil, giving the method an economical edge. The in situ mixing with materials available on-site also enhances the practicality of the stabilisation technique.

It is in this context that this study was executed to explore the possibilities of using soft clay modified with cement-agricultural wastes as road construction materials. The agricultural wastes were perceived to provide better bonding between soil-cement, where the soil particles and waste materials could be firmly bound by the hydrated cement, resulting in a stronger and stiffer soil matrix. Of course, the small bits of wastes also acted as reinforcement to the soil once the cement has hardened.

Table 1. Properties of clay (Chan 2007).

Average water content	74%
Bulk density	1.36 Mg/m^3
Specific gravity, G_s	2.66
Liquid limit, LL	77%
Plastic limit, PL	31%
Plasticity Index, PI	46%
Clay fraction (percentage by weight passing 2 μm sieve)	28%
Activity, A = PI/Clay fraction	1.64

2 MATERIALS AND METHODOLOGY

2.1 *Clay*

The soft clay used in this project was retrieved from the test site of the Research Centre for Soft Soils (RECESS) based in Universiti Tun Hussein Onn Malaysia, at a depth of approximately 1.5 m. The bulk, disturbed clay samples were wrapped in layers of cling film and plastic bags to prevent moisture loss during transportation and storage. Properties of the clay is given in Table 1.

2.2 *Ordinary Portland cement*

Ordinary Portland cement is a widely used stabiliser, whether on its own or admixed with other additives (e.g. Feng 2002, Kitazume 2005 and Hird and Chan 2006). The cement was first oven-dried at 105°C for 24 hours before being stored in airtight containers to maintain the consistency of cement used in the preparation of specimens.

2.3 *Pineapple leaf fibres (PF)*

Pineapple leaves are commonly discarded as waste upon harvest of the fruit, either disposed in landfill or burnt in open fires. In this study, to extract the fibres, surfaces of the leaves were first scarified to expose the inner layer. Next the leaves were boiled in a 5% solution of NaOH for 2 hours, after which the leaves were drained and dried at 105°C overnight. The dried leaves were then soaked in a bleach solution for 2 hours, before being dried again in the oven at 105°C for 12 hours. Details of the preparation of PF can be found in Ibrahim (2007).

2.4 *Rubber chips (RC)*

Rubber chips used in this study were retrieved from inner tyre tubes for bicycles. The thin rubber sheet was cut manually into chips passing the 2 mm sieve to avoid segregation in the stabilised material due to large chips. Actual RC used would be waste rubber (e.g. trim-off) collected from rubber processing plants. It was mainly for convenience that tyre tubes were used in this study.

Table 2. Test specimens.

Specimen ID	Cement (%)	Agricultural waste (%)
5-0	5	0
10-0	10	0
5-1.5PF	5	1.5
5-2.0PF	5	2.0
5-2.5PF	5	2.5
10-1.5PF	10	1.5
10-2.0PF	10	2.0
10-2.5PF	10	2.5
5-2.0RH	5	2
5-2.5RH	5	2.5
5-3.0RH	5	3
10-2.0RH	10	2
10-2.5RH	10	2.5
10-3.0RH	10	3
5-5RC	5	5
5-10RC	5	10
5-15RC	5	15
10-5RC	10	5
10-10RC	10	10
10-15RC	10	15

2.5 *Rice husks (RH)*

Raw rice husks were collected from a rice processing plant. The rice husks were used as they were without any processing.

2.6 *Preparation of specimens*

The agricultural wastes were admixed with cement prior to being mixed with the clay. Two quantities of cement were used, i.e. 5 and 10%, while the percentages of agricultural wastes varied accordingly (Table 2). These pre-determined percentages were calculated based on dry weight of the clay soil. Note that percentages of the agricultural waste added were related to the weight of the material itself, where light materials with large surface areas (i.e. RH and PF) were used in relatively small quantities.

The mixture was mixed thoroughly by hand to form a uniform paste, then compacted in a split mould to form specimens of 38 mm in diameter and 76 mm in height. A steel rod was used to compact the mixture in 4 layers, 40 blows each. The extruded specimens were then wrapped in cling film and stored for 14 days prior to testing.

2.7 *Unconfined compressive strength test*

The unconfined compressive strength test was conducted as prescribed in Part 7 of BS 1377 (1990), with

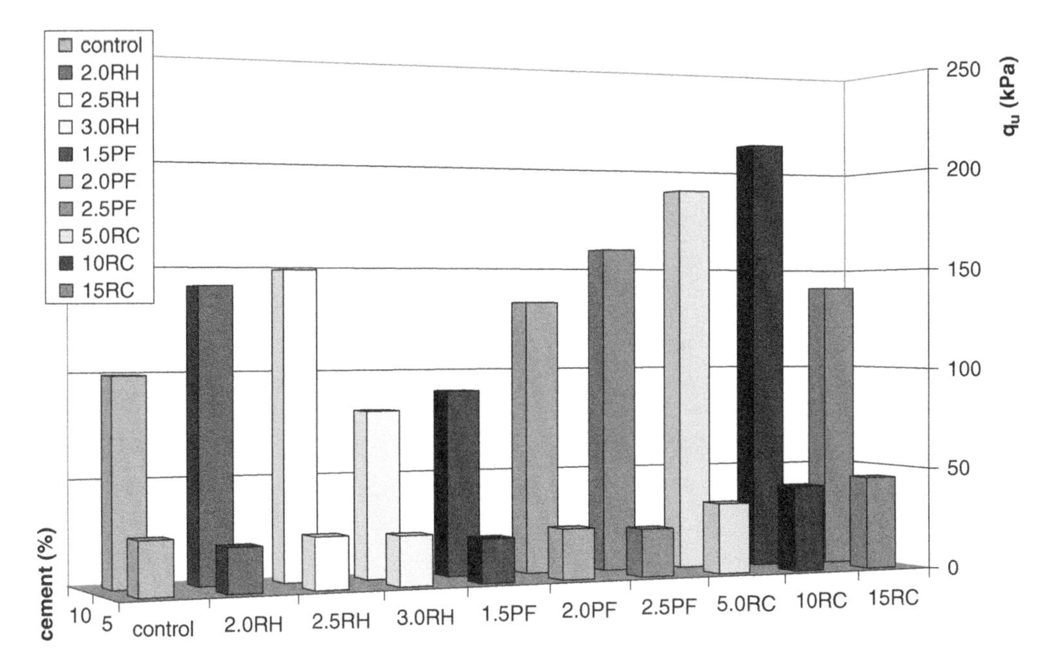

Figure 1. Unconfined compressive strength, q_u – stabiliser contents.

a conventional triaxial testing machine at a strain rate of 1.5 mm per minute. Care was taken to ensure that both ends of the specimen were as flat as possible to minimize bedding error during tests, especially with the stiff specimens.

3 RESULTS AND DISCUSSIONS

3.1 *Unconfined compressive strentgh*

The unconfined compressive strength (q_u) is plotted against stabilising agents in Figure 1. q_u data of the unstabilised specimens (i.e. clay with 0% cement) was unavailable as the soil was too soft to be tested in the compression apparatus. However previous work on the same clay reported the undrained shear strength measured with a shear vane to be a mere 4–5 kPa (Chan 2006).

Note that for the cement-modified specimens (labeled 'control'), 5% addition increased q_u of the clay by 25 kPa, while 10% addition produced approximately 100 kPa of strength increase.

From Figure 1, it is apparent that specimens with 5% cement were marginally improved, with q_u bordering at about 23 kPa for the cement-RH and cement-PF specimens, and 34–45 kPa for the cement-RC specimens, suggesting that cement was the dominant binding agent in the stabilised soil matrix. This was supported by the marked improvement in the specimens with 10% cement, where q_u was observed to increase dramatically by 130–180 kPa.

The optimum percentage of rice husks was found in specimen 10-2.5RH. A further 0.5% increment of rice husks drastically reduced q_u by almost 50 kPa. This was mainly attributed to the segregation effect excess rice husks had on the soil matrix. As for the cement-PF specimens, 1.5–2.5% of PF with 10% cement was obviously insufficient to achieve the highest strength. Further trials with higher PF content is necessary to ascertain the optimum quantity required. Specimen 10-10RC displayed the highest strength among all specimens. On contrary, 15% RC addition resulted in a significant drop in q_u, corresponding with an increase in settlement (see Section 3.2).

Referring to the correlation between q_u and consistency given in ASTM Standards (1992), 5% cement addition was insufficient to improve the clay's strength much, where the consistency lied between very soft to soft. All of the 10% cement-RH and cement-PF specimens fell under the category of medium to stiff, while the cement-RC ones were categorized as stiff to very stiff, corresponding with the highest strength achieved by 10-10RC.

3.2 *Vertical and radial deformation*

All specimens were observed to deform in both vertical and radial directions with compression to various

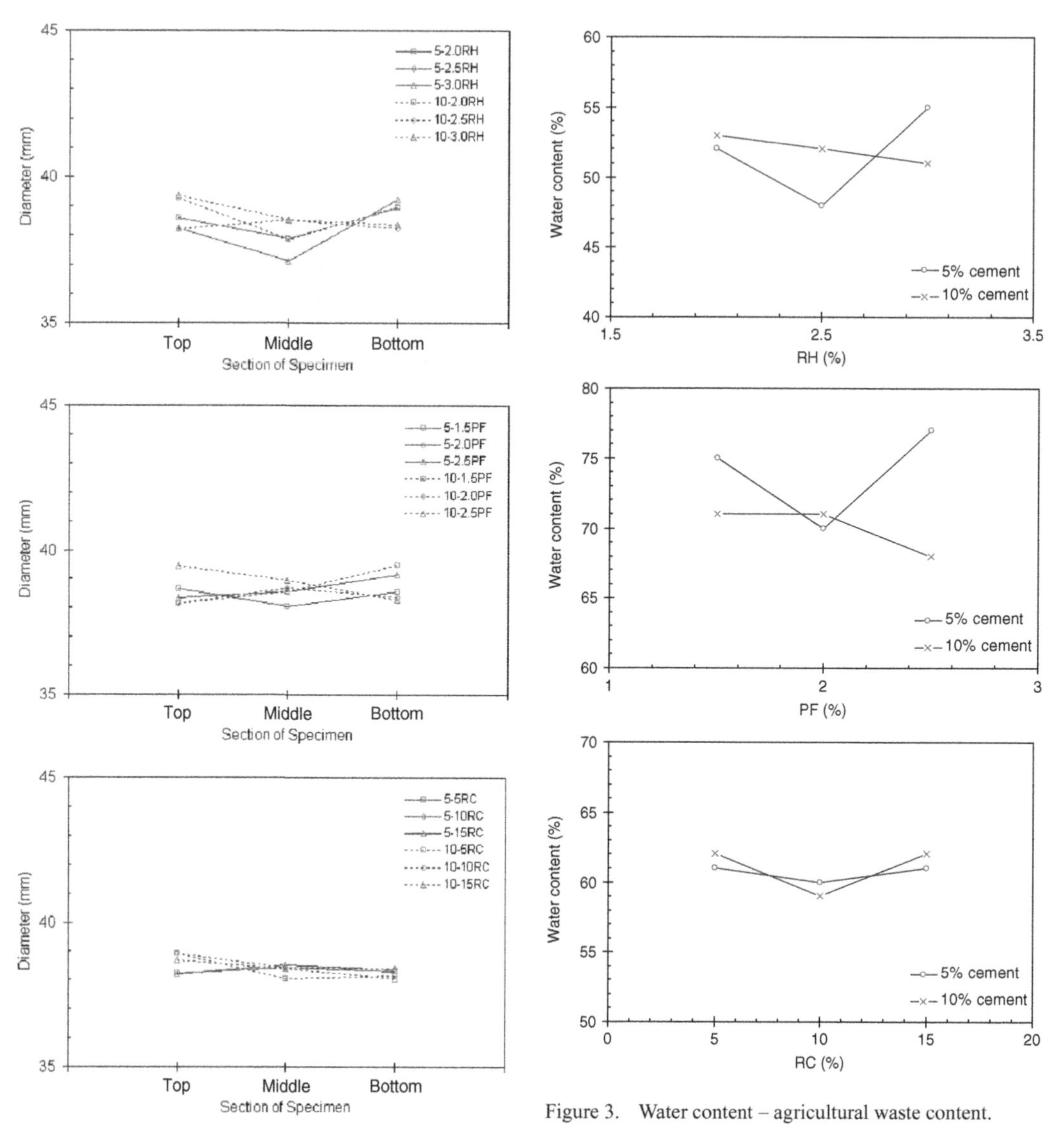

Figure 2. Diameter – section of specimen.

Figure 3. Water content – agricultural waste content.

degrees, despite the addition of stabilisers. It should however be reminded that the unconfined compressive strength test was a quick, undrained test which cause specimens to deform or distort but not change in volume.

Vertical deformation ranged between 3.3–5.1% for all cement-RH and cement-PF specimens with 5% cement, and below 2.0% for the cement-RC ones. This points to the elastic nature of rubberchips, providing additional stiffness to the specimens to resist compression. With 10% cement, all specimens showed less settlement with compression.

All specimens displayed radial expansion with bulging at the middle and bottom sections, except for 5-1.5RH, 5-2.0RH and the 10% cement-RC specimens, which seemed to be 'necking' in the middle section (Figure 2). Nevertheless it is interesting to note that 10-15RC deformed vertically by 2.4%, compared to 1.4% for 10-10RC. This observation corresponded with the sudden drop in q_u mentioned in Section 3.1.

3.3 *Water content*

The final water content of the specimens are shown in Figure 3. From the plots it can be readily perceived that higher cement content (i.e. 10%) kept the drop in water content marginal. As for the 5% cement specimens, loss of water from the clay was significant, more so in specimens with RH and PF. This was apparently due to water absorption by the RH and PF in the specimens. These observations are suggestive that cement, at a higher percentage, is the dominant binder in the soil matrix.

For the RH and PF specimens with 5% cement addition, the drop in water content at 2.5% RH and 2.0 PF corresponds with the highest strength recorded within the specimen group, i.e. coincidentally 25 kPa respectively (see Figure 1). This is indicative that the optimum additive content is closely related to the water content present in the stabilised soil matrix. In other words, 'drying' of the stabilised soil specimen seems to contribute to the strength increase.

The cement-RC specimens, on the other hand, showed relatively minimal change in water content, regardless of the cement or RC contents. This can probably be explained by the water-repellent nature of rubberchips, leaving cement hydration as the sole factor to water reduction in the specimens.

4 CONCLUSIONS

Following are the main conclusions drawn from this study:

- The consistency check suggests that cement-RC is perhaps the most effective stabiliser for the clay, corresponding with the highest strength achieved.
- Cement is the dominant binder in the stabilised soil matrix, as shown in the marginal change in strength and water content with the 5% cement specimens.
- The water content of the stabilised soils can be related to the unconfined compressive strength, where the stronger specimens displayed lower water contents.
- Generally, vertical and radial deformations decrease with increased stabiliser, but cement-rubberchips can provide extra stiffness due to the elastic property of rubber.

ACKNOWLEDGEMENT

This project was partially funded by a research grant from the Ministry of Science, Technology and Innovation (MOSTI), Malaysia.

REFERENCES

Ahmad Tajudin, S.A. 2004. *Engineering characteristics, mineralogy and microstructure of soft clay in Peninsular Malaysia (in Malay)*. Master's thesis. Universiti Teknologi Malaysia.

American Society for Testing and Materials. 1992. Soils stabilization with admixtures. Philadelphia: *Annual book of ASTM standards.*

British Standards Institution (BSI). 1990. *BS1377: British Standard Methods of Test for Soils for Civil Engineering Purposes*. Part 7: Shear strength tests (total stress).

Chan, C-M. 2006. *A laboratory investigation of shear wave velocity in stabilised soft soils.* PhD thesis. University of Sheffield, UK.

Chan, C-M. 2007. Improved compressibility of a cement-stabilised Malaysian soft clay. *Proceedings of the 2nd International Conference on Geotechnical Engineering* (Geo-Changsha). Changsha, China.

Feng, T.W. 2002. Effects of small cement content on consolidation behaviour of a Lacustrine clay. *ASTM Geotechnical Testing Journal,* Vol. 25, No. 1, pp. 53–60.

Hird, C.C. and Chan, C-M. 2005. Correlation of shear wave velocity with unconfined compressive strength of cement-stabilised clay. *Proceedings of the International Conference on Deep Mixing Best Practice and Recent Advances*, Stockholm, Sweden, Vol. 1, pp. 79–85.

Ibrahim, K.A. 2007. *Alternative road construction materials using clay admixed with natural wastes.* B. Eng thesis. Universiti Tun Hussein Onn Malaysia.

Kitazume, M. 2005. State of Practice Reports: Field and laboratory investigations, properties of binders and stabilised soils. *Proceedings of the International Conference on Deep Mixing Best Practice and Recent Advances,* Stockholm, Sweden, Vol. 2, pp. 660–684.

Parsons, R.L. and Kneebone, E. 2005. Field performance of fly ash stabilised subgrades. *Ground Improvement* Vol. 9, No. 1, pp. 33–38.

Advances in Transportation Geotechnics – Ellis, Yu, McDowell, Dawson & Thom (eds)
 ISBN 978-0-415-47590-7

Time dependant volumetric behaviour of asphalt pavements under heavy loading and high temperature

M.A. Kamal & I. Hafeez
Department of Civil Engineering, University of Engineering and Technology, Taxila, Pakistan

D.A.B. Hughes
School of Civil Engineering, Queen's University Belfast, UK

ABSTRACT: Permanent deformation and fatigue cracking are two common signs of failure in a flexible pavement. High ambient temperature and heavy loading are two major factors that cause early failure by reducing the percentage air voids in a pavement mix with time. This behaviour depends on the in situ air void content, bitumen type, aggregate quality and hot mix asphalt (HMA) properties. A field study on two pavement sections was carried out in Pakistan to investigate the volumetric behavior of HMA under severe loading conditions experienced in this region. Section 1 was paved at the start of the winter (i.e. the start of the 6 month cooler season) and Section 2 was paved at the start of the summer (ie at the start of the 6 month hot season). The initial air void content of the mix was close to 6.0%. Air void content was measured by taking cores every three months. The seasonal development of permanent deformation was also monitored on both sections over the 2 year test period. The study clearly illustrated that the rate of change in the development of air voids in the mix was different for each section and followed a seasonal trend related to temperature. The study also recorded the reduction in air void content in the pavement hot mix material with depth. Data is presented which correlates season (temperature) and depth (magnitude of applied stress) to the reduction in air void content in the bituminous mixes. Some conclusions are drawn on the efficient design and construction of bituminous pavements in Pakistan.

1 INTRODUCTION

The southbound pavement section between Turnol and Taxila (N-5) is one of the most critical sections of highway network in Pakistan. There is poor performance history and premature failure, exclusively due to wheel path rutting in the outer lane. Heavy traffic, generated from Margalla aggregate quarries, Lawrence Pur (sand source), and the steel and cement industries from the vicinity passes through this section and feed most of the cities of Southern Punjab. The pavement was constructed during the eighteenth century and its performance had never been satisfactory. Since the independence of Pakistan in 1947 ever-increasing axle loads have caused this section to be rehabilitated many times and the pavement has hardly survived more than one complete seasonal cycle. Maximum and minimum air temperature during the summer ranges from 23°C to 40°C and in winters from 3°C to 17°C. Heavy rainfall usually occurs from July to September and during the months of February and March.

The major form of distress along the section is rutting, which can be termed as non-structural due to the shear failure or lateral flow of the bituminous material. The rutting is driven by slow moving heavily loaded axles applying sustained loading at the high ambient pavement temperatures experienced in the region.

Rutting was observed on the outer lane of the carriageway causing the formation of raised shoulders along the wheel paths. These shoulders of asphalt have been milled off repeatedly during rehabilitation over the years. This has resulted in surface irregularities manifested in poor ride quality for the user. Polymer modified bitumen with Elvaloy Terpolymer was used to reconstruct a ten-kilometer long trial section in order to monitor the volumetric properties under actual traffic loading and environmental conditions.

Huber et al. (1987) examined 9 test sites in Saskatchewan to evaluate the mix design characteristics and performance and concluded that asphalt contents and voids filled with asphalt were the most basic parameters that effected rutting. Prithvi et al. (1995) demonstrated the difference between the volumetric properties of laboratory design and plant produced hot mix asphalt and Kandhal et al. (1996) reexamined minimum VMA requirements in the field and established optimum film thickness criteria for mix durability.

2 PAVEMENT DESIGN

Various design options were considered by National Highway Authority (NHA), taking into account the history of the section. They adopted the specified layer thickness method based on the AASHTO 93 Design Guide (MoC, NHA, 2002), for a period of 10 years using the following parameters;

Design ESAL's	=	120 Million
Design period	=	10 Years (2003–2012)
Reliability	=	90%
Standard Deviation	=	−1.282
Overall Standard Deviation	=	0.45
Initial Serviceability	=	4.2
Terminal Serviceability	=	2.5
Design CBR	=	10% (12,000 Psi)
Drainage Coefficient	=	1 (for unbound layers)

3 MATERIAL DESIGN

3.1 *Aggregate properties*

The results of various laboratory tests performed on the aggregates used in the asphaltic wearing course and base course in order to determine their physical properties have been tabulated in Table 1.

Table 1. Physical properties of aggregates.

Test description	Specification reference	Results
Sodium Sulphate Soundness value	AASHTO T104	3.32%
Aggregate crushing value (ACV)	BS 812, Part 1	22.5
Toughness Index (TI)	BS 812, Part 1	74.0
Ten percent fine value (TFV)	BS 812, Part 3	0.70
Aggregate Impact value (AIV)	BS 812, Part 3	13.5
Los Angles Abrasion value (LAA)	ASTM C 131	23%
Specific gravity (Gsb)	AASHTO T85 ASTM C 127	2.661
Elongation Index (EI)	BS 812, Part 1	11%
Porosity	AASHTO T 85	0.78%
Flakiness Index (FI)	BS 812, Part 1	4.75%
Sand equivalent	AASHTO T 176	72%
Polished Stone Value	BS 812-114	70%

3.2 *Polymer Modified Bitumen (PMB)*

The 60/70 penetration grade bitumen was modified with 1.6% (for wearing course) & 0.8% (for base course) Elvaloy®4170 and superphsophoric acid in Attock Laboratory, to produce a PG 78-23. The Specifications are tabulated in Table 2.

The desired performance grading was developed using Attock Refinery 60/70 grade bitumen and Elvaloy at Mathy Technology & Engineering Services, Australia (Gerald, 2001) where different trials were made to prepare a final blend to suit local climatic conditions for the specific project in Pakistan. Elvaloy was used to enhance elasticity and viscosity of the mix.

3.3 *Granular sub-base*

Crushed aggregate material (gradation shown in Table 3) was used as subbase material for this specific project. The selected material was free from any organic matter and other deleterious substances and

Table 2. Specifications limits of PMB.

Test type	Austroads specifications	Austroads test method	US equivalent
Elastic consistency @ 600C, pa.s	1500 min,	MBT 21	Not Known
Stiffness @ 25°C, kpa,	130 max,	MBT 21	Not Known
Brookfield Viscosity @ 165°C, pa.s	0.75 max,	MBT 11	Not Known
Flash point °C	250 min,	MBT 12	ASTM D92
Loss on Heating, % Mass	0.6 max,	MBT 03	ASTM D1754
Personal Recovery, @ 25°C %	12 min,	MBT 22	Not Known
Melting point°C	60 min,	MBT 31	ASTM D36

Table 3. Summary of grading for granular sub-base course.

Passing sieve size (mm)	Specifications % passing
60	–
50	100
25	55–85
9.5	40–70
4.75	30–60
2.0	20–50
0.425	10–30
0.075	5–15

compacted at optimum moisture content to form a firm and stable layer. The achieved compaction of the layer was 98% of the maximum dry density as specified by General Specification (MoC, NHA, 1998) for the sub-base and was determined according to AASHTO T-19.

3.4 *Water bound macadam base*

Aggregate gradations used for water bound macadam as given in Table 4 are in accordance with NHA specification (MoC, NHA, 1998). Material passing No. 40 sieve had a liquid limit less than 25% and a plasticity index of less than 6%. Fine aggregates (filler material or screening) consisted of pure crushed stone, free of clay lumps and dirt. Binding material to prevent raveling of water bound macadam was fine grained material 100% passing the 425 micron sieve with plasticity index value of less than 6%. Los Angeles Abrasion value and flakiness index values obtained through laboratory tests were 22% and 12% respectively.

Crushed stone was deposited and spread on the prepared surface to the proper depth and lightly compacted with a 10 tonne roller to establish the required grade and level of stones. Following the initial rolling, dry binding was applied uniformly over the surface while compaction was continued. When the interstices were filled with fine aggregate the surface was saturated by sprinkling water. The rolling, sprinkling and application of additional fine aggregate was continued until a grout was formed that filled all the voids and formed a wave of grout in front of the roller, as specified by general specification of NHA (MoC, NHA, 1998).

3.5 *Asphaltic base course*

Mineral aggregates for the bituminous base course consisted of coarse aggregates, fine aggregates and filler materials with at least two mechanically fractured faces. Asphalt binder used in the mix was Polymer modified bitumen with 0.8% Elvaloy Terploymer. Aggregate gradations and mix design properties have been tabulated in Table 5. Job Mix Formula (JMF) was prepared following Asphalt Institute Manual Series (Asphalt Institute, 1993).

Loss of Marshall Stability by immersion of specimen in water at 60°C for 24 hours as compared with stability measured after immersion in water at 60°C for 20 minutes was less than 25% (MoC, NHA, 1998). Grading used for asphaltic wearing course laid as per NHA requirement has been given in Table 6.

Optimum Asphalt Content, Marshal Stability, Bulk Specific Gravity (Gmb) and Design Air Voids (Va) were kept 3.83% , 1200 kg, 2.37 and 5.8% respectivily.

4 FIELD LAYING AND COMPACTION PROCEDURE

HMA using PMB (0.8% &1.6% Elavloy Terploymer) with specified gradation for asphaltic base course and

Table 4. Gradations of water bound macadam base.

Sieve design (mm)	Class B	Class C
102	–	–
89	–	–
76	100	–
63.5	90–100	100
50	25–75	90–100
37.5	0–15	35–70
25	–	0–15
19	0–5	0–5
12.5	–	–
102	–	–

Table 5. Summary of gradation and JMF of asphaltic base course.

Mix designation	Class B
U.S. standard sieve size	–
(Percent passing by weight)	–
50 mm	–
38 mm	100
25 mm	75–90
19 mm	65–80
12.5 mm	55–70
9.5 mm	45–60
4.75 mm	30–45
2.38 mm	15–35
0.300 mm	5–15
0.075 mm	2–7
Asphalt content (%)	3.35
Stability (Kg.)	1000
Bulk specific gravity (Gmb)	2.393
Maximum theoretical specific gravity (Gmm)	2.541
Design air voids (Va)	5.8

Table 6. Summary of grading for asphaltic wearing course.

Passing sieve size (mm)	Achieved gradation	Specifications % passing
25	100	100
19	90–100	90–100
12.50	–	–
9.50	56–69	56–70
4.75	38–46	35–50
2.36	25–33	23–35
0.300	5–12	5–12
0.075	3.4–5.4	2–8

Table 7. Details of pavement structure (sec. 01 & 02).

Asphaltic wearing course	60 mm
Asphaltic base course	100 mm
Water bound macadam (Class C)	150 mm
Water bound macadam (Class B)	150 mm
Granular sub base	150 mm

Table 8. Summary of loading.

Sr. no.	Month	Accumulative-ESAL'S (millions)
1	Sep-2003	0.399
2	March-2004	3.03
3	Sep-2004	6.14
4	March-2005	9.50
5	Sep-2005	13.01
6	March-2006	16.76
7	Sep-2006	20.50
8	March-2007	24.14
9	Sep-2007	28.00

wearing course was prepared using Parker (160 Ton/Hr Capacity), fully automated asphalt plant capable of producing 100% controlled mix. Mixing, paving and laying temperatures of HMA were kept approximately 165°C, 160°C & 158°C respectively. Initial break down rolling started at a temperature between 150°C to 160°C with a vibratory tandem steel wheel roller equipped with an arrangement of sprinkling water over the drum. Intermediate rolling with Pneumatic Tyre Rollers (PTR) with a tyre pressure of 65 psi was started until the final rolling completed and mat temperature fell down to 140°C–150°C. Thickness of various layers of pavement achieved at site has been given in Table 7.

5 ANALYSIS OF RESULTS

The accumulative axle load was monitored at the same section and extrapolated to estimate the applied load on the three lane wide pavement and reported in Table 8. The damaging factors for loaded trucks travelling at an average speed of 25 Km/hr were taken from Axle load study (1995) conducted by National Transportation Research Centre (NTRC). Damaging factor for empty trucks and other type of vehicles were taken from Axle load study conducted by M/s Associated Consulting Engineers, Pakistan (Kamal, 2006).

Cores were taken after 24 hours of rolling, ensuring pavement surface and ambient temperatures were the same. The volumetric parameters were calculated as per AASHTO & ASTM standards. Cores from the same sections were taken every three months period, volumetric properties have been reported in Table 9

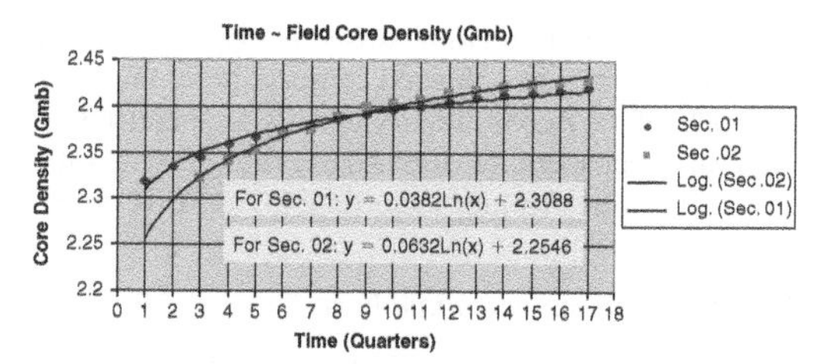

Figure 1. Relationship between time and air voids.

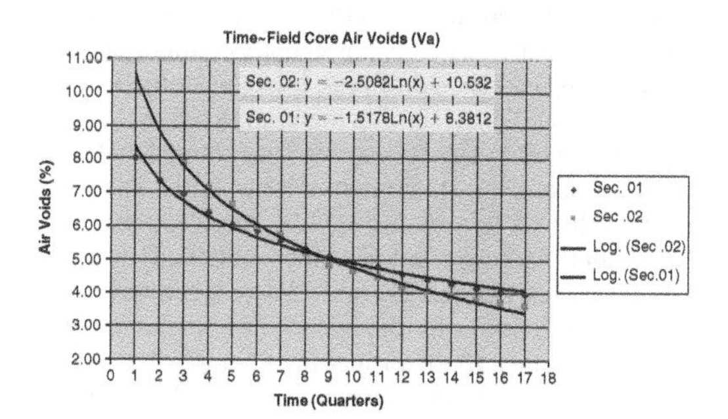

Figure 2. Relationship between time and core density.

and plotted graphically in Figures 1 & 2. From the figures it can be observed that the increase in core density or decrease in air voids of the asphaltic wearing course during the initial three years was abrupt and continued at a slower rate with time. Trend line equations for air voids and core densities have been developed based on the data collected from the field and are shown in Table 10.

6 CONCLUSIONS

Following Conclusions have been drawn:

- Change in asphalt density of field core is a function of accumulative axle load and time.
- Seasonal variations have significant impact on the overall performance of the mix during its initial stage.
- Mix laid after summer season showed better performance as compared from the one laid before.
- Chang of volumetric properties observed after three years is very low.

REFERENCES

Asphalt Institute, "Superpave Mix Design" Superpave series No. 02. 1993.

Gerald Reinke; "Elvaloy Formulation with Attock Asphalt" Mathy Technology & Engineering Services, Inc; October 2001.

Huber, G.A., and G.H. Heiman, Effect of Asphalt Concrete Parameters on Rutting Performance: A field Investigation, Proceedings of the Association of Asphalt Paving Technologists, Vol. 56, 1987.

Kandhal, Prithvi S., Sanjoy Chakraborty, "Evaluation of Voids in the Mineral Aggregate for HMA Paving Mixture, NCAT Report No. 96-4, National Center for Asphalt Technology, March, 1996.

Kamal, M.A., Hafeez, I; "Field and Laboratory Evaluation of Volumetric properties of Asphaltic concrete". 7th International Congress on Civil Engineering, May 8–10, 2006.

Ministry of Communication.; National Highway Authority, Pakistan, Central Design Cell, Pavement Design Methodology & Strengthening Recommendations. 2002.

Ministry of Communication., "National Highway Authority, Pakistan" General Specifications. 305-1, 1998.

Prithvi, S. Kandhal, Kee Y. Foo, Jhon A.D., "Angelo Field Management of Hot Mix Asphalt Volumetric Properties NCAT Report 95-04, December 1995.

Advances in Transportation Geotechnics – Ellis, Yu, McDowell, Dawson & Thom (eds)
© 2008 Taylor & Francis Group, London, ISBN 978-0-415-47590-7

Quantifying effects of lime stabilized subgrade on conventional flexible pavement responses

O. Pekcan, E. Tutumluer & M.R. Thompson
Department of Civil and Environmental Engineering, University of Illinois Urbana, Illinois, USA

ABSTRACT: Lime stabilization is commonly used to improve weak natural subgrade in Illinois. ILLI-PAVE nonlinear finite element (FE) program was utilized in this study as an advanced pavement structural analysis tool to quantify the improvement from the lime stabilized soil layer in conventional flexible pavements. Using ILLI-PAVE, many combinations of typical Illinois highway pavement layer thicknesses and material properties were analyzed to establish a database of surface deflections and critical pavement responses. Using this database, Artificial Neural Network (ANN) structural models were trained successfully as forward analysis tools, surrogate to ILLI-PAVE, to be used in the backcalculation of the Falling Weight Deflectometer data. The ANN model predictions were on the average within 1% of the ILLI-PAVE FE results. The developed ANN models were therefore quite accurate for the rapid analyses of conventional flexible pavements built on lime stabilized soils replicating the ILLI-PAVE results.

1 INTRODUCTION

Lime stabilization is a major improvement technique over weak soils for subgrade preparation and working platform construction. It can significantly enhance the stability and load-bearing capacity of the subgrade soil while decreasing its moisture sensitivity. However, structural improvements due to the addition of a lime stabilized soil (LSS) layer are often not taken into account in pavement analysis for the thickness design since this layer is considered as a working platform. Quantification of the changes in pavement responses, i.e., deflections, stresses and strains in the pavement profile, is especially important in pavement management and life cycle costing studies involving nondestructive evaluation, e.g., Falling Weight Deflectometer (FWD) testing, and backcalculation of pavement layer properties. Recent research has clearly indicated different responses obtained from full depth asphalt pavements on lime stabilized soils when compared to those built on natural soils (Pekcan et al. 2007). Separate analyses, therefore, need to be conducted to account for the contribution of the LSS layer on the responses and deflection profiles of the conventional flexible pavements built on lime stabilized soils (CFP-LSS).

The current state-of-the-art mechanistic-empirical pavement design requires proper estimation of critical pavement responses to link them to pavement failure through empirical distress models. Typically elastic layered solutions or more advanced numerical solution techniques, such as the finite element method, are utilized to analyze pavement structural behavior under wheel loading. Certain pavement responses estimated from the solutions and responsible for pavement deterioration are then used in empirical distress models to obtain performance predictions. Successful applications with numerical analyses usually involve accurately determining pavement responses, which is only possible with proper pavement layer geometry and material property inputs. Identification of field material properties, however, is often difficult or expensive due to complexity of many contributing factors and experimental routines. As a result, rapid and practical means of generating accurate solutions may require conducting sensitivity analyses of the many pavement input properties.

Artificial Neural Networks (ANNs) are computational tools that operate similar to biological neurons in human brain. They are mainly used for processing information, where inherent and complex patterns are available. They are capable of capturing the nonlinear relationship between the inputs and outputs of a dataset reliably (Haykin 1999). In the field of transportation geotechnics, ANNs have been usually implemented as powerful regression tools. For example, they were successfully used for pavement layer backcalculation by Ceylan et al. (2005), Meier & Rix (1995) and Pekcan et al. (2006). Previous studies also showed that ANNs were powerful and quite effective as forward analysis tools to replace or mimic advanced pavement analyses (Ceylan et al. 1999; Meier et al. 1997).

In these studies, the effects of nonlinear, stress dependent aggregate base and cohesive subgrade soils on the pavement responses were also successfully modeled using ANNs.

Building upon the findings from these recent research studies, this paper mainly focuses on the development of quick and robust structural analysis tools to quantify effects of the LSS layer on the behavior of conventional flexible pavements.

2 NUMERICAL MODELING STUDIES

The finite element (FE) numerical solution is an advanced state-of-the-art technique nowadays preferred to conduct realistic pavement analyses. ILLI-PAVE 2005 program was mainly used in this study as the advanced structural model for analyzing CFP-LSS pavements. ILLI-PAVE 2005, the most recent version of the well known ILLI-PAVE FE program, has been extensively tested and validated over almost four decades (Gomez-Ramirez et al. 2002; Raad & Figueroa 1980). The ILLI-PAVE flexible pavement FE solution incorporates stress dependent geomaterial characterization models and the Mohr-Coulomb failure criteria or the limiting strength of materials (Thompson & Elliott 1985).

2.1 *Constitutive modeling*

Proper implementation of the pavement layer behavior into the FE solution is a crucial step for accurate numerical modeling. Most of the pavement layer deformations under traffic loading are usually recoverable and thus considered elastic. In the ILLI-PAVE analyses conducted of the CFP-LSS, the asphalt concrete (AC) layer was modeled linear elastic with an elastic or resilient modulus (E_{AC}) and a constant Poisson's ratio of 0.35. Similarly, the behavior of LSS layer on top of the natural subgrade under compressive stresses was assumed to be linear elastic and represented by an elastic modulus, E_{LSS} (Little 1999). The LSS layer Poisson's ratio was taken to be 0.31 (TRB Circular 1987).

Both the unbound granular base (GB) and the subgrade soil layers are known to exhibit nonlinear, stress dependent modulus behavior (Tutumluer 1995). Essentially the GB layer with a stress-hardening type elastic or resilient modulus (M_R) is constructed to spread the wheel load to protect the weaker subgrade from overstressing and at the same time support the AC layer to prevent fatigue. In this study, the K-θ model (Hicks & Monismith 1971) was used to capture the nonlinearity of the GB layer modulus through the use of "K" and "n" model parameters. For modeling the stress dependency, "K" values typically range from 20.7 to 82.7 MPa based on a comprehensive

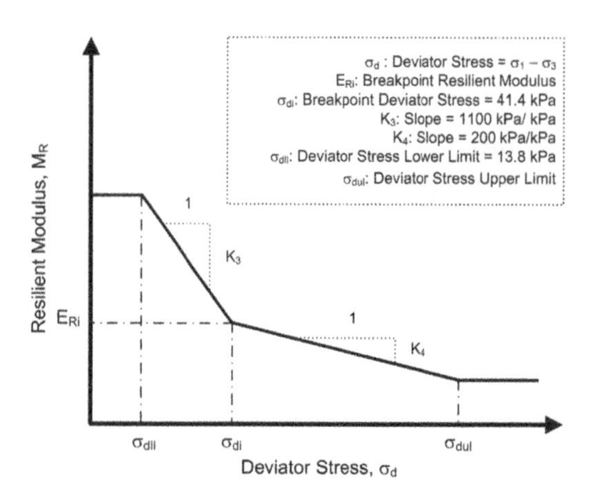

Figure 1. Bilinear modeling of the stress dependency of fine-grained soils (Thompson & Robnett 1979).

granular material database compiled by Rada & Witczak (1981) and parameter "n" is also correlated to K using Equation 1:

$$\log_{10}(K^*6.895) = 4.657 - 1.807 * n \tag{1}$$

where K is in MPa. This way, the unbound aggregate modulus characterization model requires only one parameter. Poisson's ratio was taken as 0.35 when $K \geq 34.5$ MPa, otherwise it was assumed 0.40.

In Illinois, subgrade soils are typically fine grained and therefore exhibit stress softening behavior with increasing applied stress levels. Thompson & Robnett (1979), using repeated load triaxial tests in the laboratory, showed that the elastic or resilient properties of fine grained soils could be represented by a bilinear nonlinear model as shown in Figure 1.

In addition, according to Thompson & Elliot (1985), the breakpoint resilient modulus (E_{Ri}) can be used to classify fine grained soils as soft, medium or stiff according to the breakpoint deviator stress (σ_{di}). Finally, the unconfined compressive strength (Q_u) of these soils can be related to both E_{Ri} value and the upper limit deviator stress (σ_{DUL}) (defined in Figure 1) using Equation 2:

$$\sigma_{DUL} * 6.895 = Q_U = \frac{E_{Ri} * 6.895 - 0.86}{0.307} \tag{2}$$

where σ_{DUL}, Q_u are in kPa and E_{Ri} is MPa. Finally, the subgrade Poison's ratio was taken to be 0.45.

2.2 *Falling weight deflectometer simulation*

In ILLI-PAVE, FWD testing was modeled with an axisymmetric FE mesh using the standard 40-kN equivalent single axle wheel loading applied as a

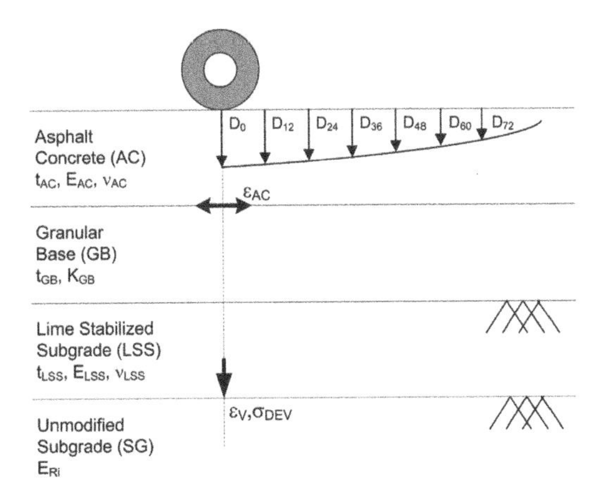

Figure 2. Falling weight deflectometer simulation of conventional flexible pavements on lime stabilized soils.

uniform tire pressure of 552 kPa over a circular area of 152-mm radius. The horizontal meshing was done according to the spacing of the FWD sensors as follows: 0, 305, 610, 915, 1219, 1524 and 1829 mm away from the center of FWD loading. The surface deflections corresponding to the locations of these FWD sensors were abbreviated as D_0, D_{12}, D_{24}, D_{36}, D_{48}, D_{60} and D_{72}, respectively. These deflections are in conformity with the uniform spacing commonly used by many state highway agencies in FWD testing (Ceylan et al. 2005). In addition to the deflections, the critical pavement responses, i.e., horizontal strain at the bottom of AC layer (ε_{AC}), vertical strain at the top of the subgrade (ε_V), and the deviator stress on top of the subgrade (σ_{DEV}) directly at the centerline of the FWD loading, were also extracted from ILLIPAVE results to ensure reliability of the results (see Figure 2). These critical pavement responses play a crucial role in the context of mechanistic-empirical pavement design procedures as they directly relate to major failure mechanisms in the form of excessive fatigue cracking and rutting in the wheel paths of asphalt pavements.

3 ARTIFICIAL NEURAL NETWORKS

ANNs are biologically inspired information processing tools. They are essential elements of Soft Computing field, which has grown extensively in the last two decades. ANNs are different than conventional mathematical approaches in that the solutions produced by them are non-universal and non-unique (Ghaboussi 2001; Ghaboussi & Wu 1998). Their ability to work with noisy data makes them very good candidates for solving complex engineering problems. By proper training, ANNs can be used to model various nature induced phenomena such as the solutions

Table 1. Ranges of pavement geometries and material properties used for Artificial Neural Network training.

Material type	Layer thickness (cm)	Material model	Layer modulus inputs (kPa)
Asphalt concrete	$t_{AC} = 7.6$ to 45.7	Linear elastic	$E_{AC} = 6.9 \times 10^5$ to 172.4×10^5
Granular base	$t_{GB} = 10.2$ to 55.9	Nonlinear K-θ	$K_{GB} = 20.6 \times 10^3$ to 110.3×10^3
Lime Stabilized subgrade	$t_{LSS} = 25.9$ to 50.8	Linear elastic	$E_{LSS} = 110.3 \times 10^3$ to 103.4×10^4
Fine-grained subgrade	$(76.2 - t_{AC} - t_{GB} - t_{LSS})$	Nonlinear bilinear	$E_{Ri} = 6.9 \times 10^3$ to 103.4×10^4

of inverse problems, where conventional mathematical approaches do not work successfully (Liu & Han 2003).

3.1 *Backpropagation neural networks*

Backpropagation is the most commonly used algorithm for training ANNs (Reed & Marks 1999). They are also known as multi layered feed forward neural networks. They can learn functional relationships when the available data are presented as input-output pairs (i.e. supervised learning). These types of neural networks consist of data processing elements, i.e., neurons, which are usually interconnected in layers: input layer, one or more hidden layers and an output layer. The information flow between layers is provided through weighted connections. The error calculated between the summed activation values and real outputs is minimized using an algorithm by sending signals forward and backward through the nodes. Each cycle of such minimization is called an "epoch." The network can obtain a finalized set of weights at a certain number of epochs or when the smallest allowable error level is reached. This process is referred to as proper learning or training.

3.2 *Forward models*

A fully trained ANN structural analysis model needed to be capable of replacing the ILLI-PAVE FE program for generating accurate solutions. For this purpose, typical values of thicknesses and material properties of CFP-LSS in Illinois were first identified. They were then extended ±20% for proper learning of ANNs in the field ranges. The training ranges used in this study are provided in Table 1. Approximately 30,000 combinations of randomly selected inputs were fed into ILLI-PAVE for FE analyses. The results were collected in a database of input-output pairs. 29,000 of the randomly chosen pairs were used for ANN training and while the remaining 1000 datasets were used for testing.

Two different ANN models were developed (see Table 2). The inputs to both models were geometry and design thickness data for the CFP-LSS. The outputs of Model 1 were the surface deflections under FWD loading while those of Model 2 were critical pavement responses. In both models, two hidden layers were used with 20 neurons in each hidden layer. In addition, both ANN models were trained for up to 10,000 epochs to properly minimize errors and to prevent overtraining.

The developed ANN models were also evaluated for accuracy using an independently chosen testing dataset. The model predictions were compared with the corresponding ILLI-PAVE FE results in the dataset. The performance of each model was reported using Average Absolute Error (AAE) values as shown in Figures 3(a) through 3(g) and in Figures 4(a) through 4(c) for Models 1 and 2, respectively. Model 1 could predict the surface deflection values with less than a maximum AAE value of 0.7%. For example, this error level corresponds to $\pm 5\,\mu m$ for D_0 deflection

Table 2. Artificial neural network forward models for conventional flexible pavements on lime stabilized soils.

ANN Models	Input variables	Output variables
Model 1	t_{AC}, t_{GB}, t_{LSS}, E_{AC}, K_{GB}, E_{LSS}, E_{RI}	D_0, D_{12}, D_{24}, D_{36}, D_{48}, D_{60}, D_{72}
Model 2	t_{AC}, t_{GB}, t_{LSS}, E_{AC}, K_{GB}, E_{LSS}, E_{RI}	ε_{AC}, ε_V, σ_{DEV}

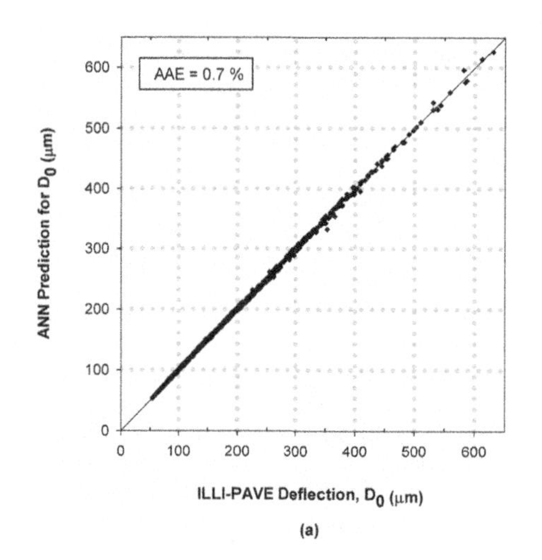

Figure 3. Artificial neural network model 1 prediction performances for surface deflections.

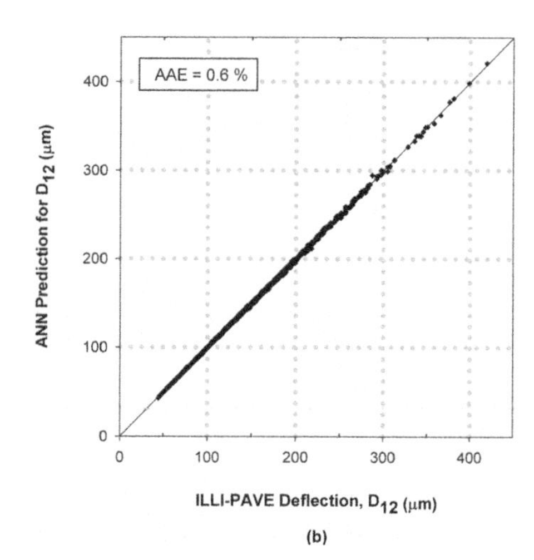

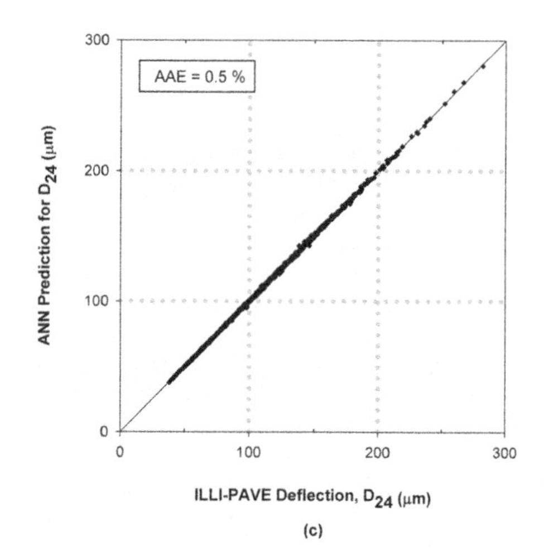

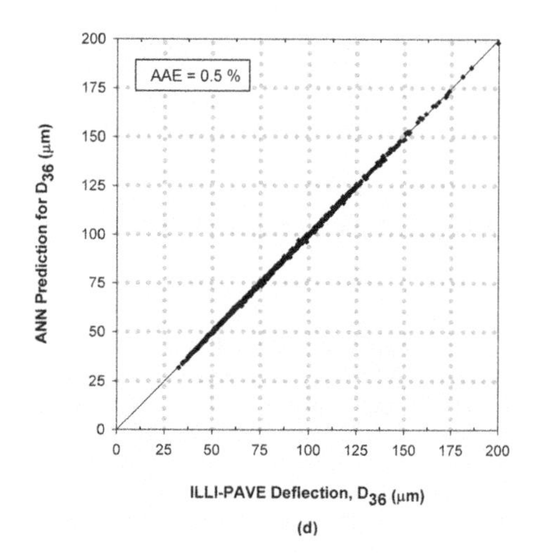

Figure 3. Continued

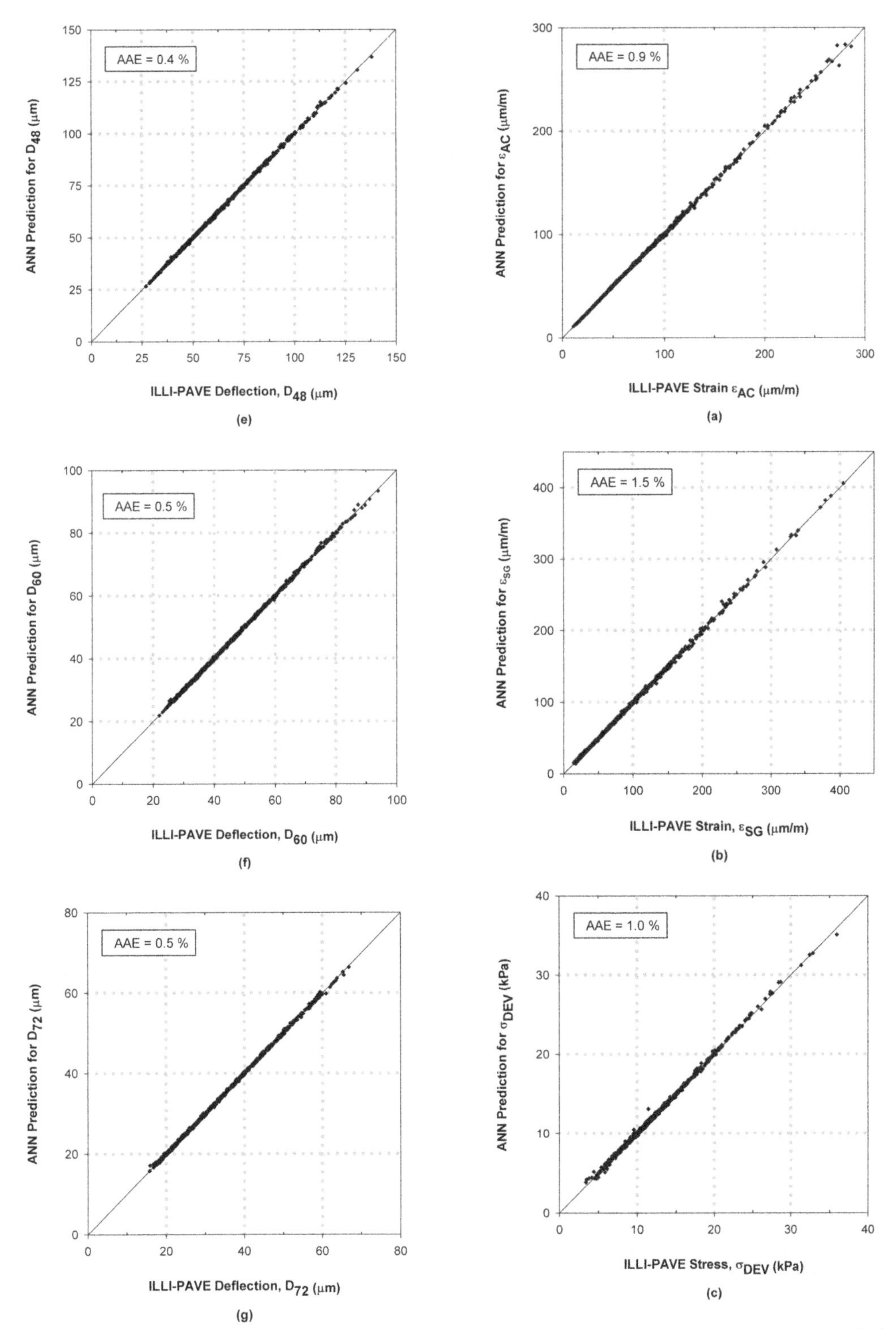

Figure 3. Continued

Figure 4. Artificial neural network model 2 prediction performances for critical pavement responses.

from FWD testing. Similarly, Model 2 was successful in predicting the critical pavement responses: ε_{AC}, ε_V and σ_{DEV} with AAE values of 0.9% (±3μm/m), 1.5% (±6 μm/m) and 1.0% (±0.4 kPa), respectively. These close results proved that solutions from ANN models were in very good agreement with those of the ILLI-PAVE. In other words, ANN models could successfully replace ILLI-PAVE analyses as surrogate forward analysis tools by eliminating the need for complex FE inputs. Consequently, these ANN models provide practical and accurate structural analyses of the CFP-LSS.

4 SUMMARY AND CONCLUSIONS

Conventional flexible pavements on lime stabilized soils were modeled in this study. A validated nonlinear finite element (FE) program, ILLI-PAVE, was used for structural analyses of these pavements. A number of different pavement geometries and pavement layer properties were analyzed for Falling Weight Deflectometer (FWD) simulations. The analyses also took into account the nonlinear aggregate base and subgrade behavior. Considering the stabilized layer as a subbase under a typical granular base, both the number of layers and the complexity of the forward problem had increased. The ILLI-PAVE FE results collected in a database were then used with the corresponding analysis inputs to train Artificial Neural Network (ANN) models for forward analyses. Since ANNs are capable of capturing nonlinear behavior in the granular base and unstabilized subgrade layers, the trained ANN models successfully predicted pavement responses to quantify the effects of lime stabilized subgrade soils. The ANN model predictions were on the average within 1% of the ILLI-PAVE FE results. For use with FWD based backcalculation and nondestructive pavement evaluation, these trained ANN models will serve as forward structural models to determine critical pavement responses in an accurate and rapid means.

REFERENCES

Ceylan, H., Guclu, A., Tutumluer, E. & Thompson, M.R. (2005). "Backcalculation of full-depth asphalt pavement layer moduli considering nonlinear stress-dependent subgrade behavior." International Journal of Pavement Engineering, 6(3), 171–182.

Ceylan, H., Tutumluer, E. & Barenberg, E.J. (1999). "Artificial neural networks for analyzing concrete airfield pavements serving the Boeing B-777 aircraft." Transportation Research Record, 1684, 110–117.

Ghaboussi, J. (2001). "Biologically inspired soft computing methods in structural mechanics and engineering." Structural Engineering and Mechanics, 11(5), 485–502.

Ghaboussi, J. & Wu, X. (1998). "Soft computing with neural networks for engineering applications: Fundamental issues and adaptive approaches." Structural Engineering and Mechanics, 6(8), 955–969.

Gomez-Ramirez, F., Thompson, M.R. & Bejarano, M. "ILLI-PAVE based flexible pavement design concepts for multiple wheel-heavy gear load aircraft." Proceedings of the 9th International Conference on Asphalt Pavements, Copenhagen, Denmark.

Haykin, S.S. (1999). *Neural networks: A comprehensive foundation*, Prentice Hall, Upper Saddle River, N.J.

Hicks, R.G. & Monismith, C.L. (1971). "Factors influencing the resilient response of granular materials." Highway Research Record, 345, 15–31.

Little, D.N. (1999). Evaluation of structural properties of lime stabilized soils and aggregates Vol. 1: Summary of findings, National Lime Association.

Liu, G.R. & Han, X. (2003). *Computational inverse techniques in nondestructive evaluation*, CRC Press, Boca Raton.

Meier, R., Alexander, D. & Freeman, R. (1997). "Using artificial neural networks as a forward approach to backcalculation." Transportation Research Record, 1570, 126–133.

Meier, R.W. & Rix, G.J. (1995). "Backcalculation of flexible pavement moduli from dynamic deflection basins using artificial neural networks." Transportation Research Record, 1473, 72–81.

Pekcan, O., Tutumluer, E. & Thompson, M.R. (2006). "Nondestructive flexible pavement evaluation using ILLI-PAVE based artificial neural network models." In CD-ROM Proceedings of the ASCE Geo-Institute Geo-Congress, Geotechnical Engineering in the Information Technology Age, Atlanta, Georgia, February 26-March 1, 2006.

Pekcan, O., Tutumluer, E. & Thompson, M.R. (2007). "Analyzing flexible pavements on lime-stabilized soils using artificial neural networks." Advanced Characterization of Pavement and Soil Engineering Materials, Proceedings of the International Conference on Advanced Characterisation of Pavement and Soil Engineering, 20–22 June 2007, Athens, Greece, 587–596.

Raad, L. & Figueroa, J.L. (1980). "Load response of transportation support systems." Journal of Transportation Engineering 106(1), 111–128.

Rada, G. & Witczak, M.W. (1981). "Comprehensive evaluation of laboratory resilient moduli results for granular material." Transportation Research Record, 810, 23–33.

Reed, R.D. & Marks, R.J. (1999). *Neural smithing: Supervised learning in feedforward artificial neural networks*, The MIT Press, Cambridge, Mass.

Thompson, M.R. & Elliott, R.P. (1985). "ILLI-PAVE based response algorithms for design of conventional flexible pavements." Transportation Research Record, 1043, 50–57.

Thompson, M.R. & Robnett, Q.L. (1979). "Resilient properties of subgrade soils." Journal of Transportation Engineering ASCE, 105(TE1), 71–89.

TRB Circular 1987. "Lime stabilization: Reactions, properties, design, and construction." State-of-the-art Report: 5, Transportation Research Board, National Research Council, Washington, DC

Tutumluer, E. (1995). "Predicting behavior of flexible pavements with granular bases." PhD Thesis, School of Civil and Environmental Engineering, Georgia Institute of Technology.

Advances in Transportation Geotechnics – Ellis, Yu, McDowell, Dawson & Thom (eds)

The capacity of improved Khon Kaen Loess as a road construction material

P. Punrattanasin
Department of Civil Engineering, Khon Kaen University, Khon Kaen, Thailand

ABSTRACT: The loess terrain exists in abundance in the Northeastern region of Thailand. The soil has a potential to collapse, which has been caused by wetting. This study reports the characteristics of Khon Kaen loess under various conditions. This study aims to investigate the suitability of this soil as a material for road construction. The undisturbed samples were excavated from the field and to be tested. The results show that the addition water can dramatically reduce the bearing capacity of the loess. The study also performed with the treated loess by static compaction. The results show the effectiveness of the additional energy from compaction in the increasing of bearing capacity of the soil. A series of trial slope embankments were also constructed in order to evaluate the behavior of the loess. The field observations showed that the testing embankment was stable under traffic loading and erosion of soil occurred by water infiltration from ground surface. The trial embankment reinforced with geosynthetic materials was also tested. The results showed the effectiveness of geosynthetic material as one of the remedial measures.

1 INTRODUCTION

Loess or wind-deposited soil is known as one of the major problematic soils. This soil has high potential to collapse which has caused severe settlement problems. The loess terrain exists and covers in many areas of the Northeastern region of Thailand (Fig. 1). The loess deposits can be classified into two colors, Red Loess and Yellow Loess (Phien-wej *et al.*, 1992). The minerals of the two units are very similar but different in oxidation states. The thickness found in Thailand ranges from a few meters to more than 6 m. Thick loess deposits are found in high elevation areas including Khon Kaen, which is the economic center of the Northeastern region. The soil was firstly considered as a good bearing layer at dry state but it was found that wetting of the soil by water can cause the lost in strength, the severe settlement and finally the damage of structures can be occured. In this study, the properties of Khon Kaen loess are presented from laboratory testing. The main objective of this research is to evaluate the potential of the stabilized loess by compaction as a material for road construction. Trial embankments are also tested to show the effectiveness of the energy from the compaction.

Figure 1. The loess terrain in Khon Kaen city.

2 SOIL PROPERTIES

Basic, index and engineering properties of Khon Kaen loess are summarized in Table 1. The soil is silty fine sand (SM). The loess consists of 65% sand, 30% silt and 5% clay. Udomchoke (1991) studied the microstructure of this soil and found that Khon Kaen loess consists of well-sorted fine sand grains but poorly sorted silt and clay particles. Sand grains have smooth, sub-rounded surfaces indicating eolian origin. The oedometer test is used for the determination of the consolidation characteristics of soil. From the tests it was found that the compression index of undisturbed loess is 0.0135 with maximum past pressure of 1334 kPa. A series of tests were then carried out to define the collapse potential of this soil. An undisturbed sample was cut and fit into a

Table 1. Properties of Khon Kaen loess.

Property	Khon Kaen Loess
Specific gravity	2.60
Natural dry unit weight (kN/m^3)	15.3
Natural water content during rainy season (%)	8–12
Liquid limit (%)	16.0
Plastic limit (%)	13.0
Plasticity index	3.0
Soil classification (USCS)	SM
Optimum moisture content (%)	9.7
Maximum dry density (kN/m^3)	21.1
Permeability coefficient (cm/s)	2.80×10^{-6}
Strength parameter, ϕ (degree)	38

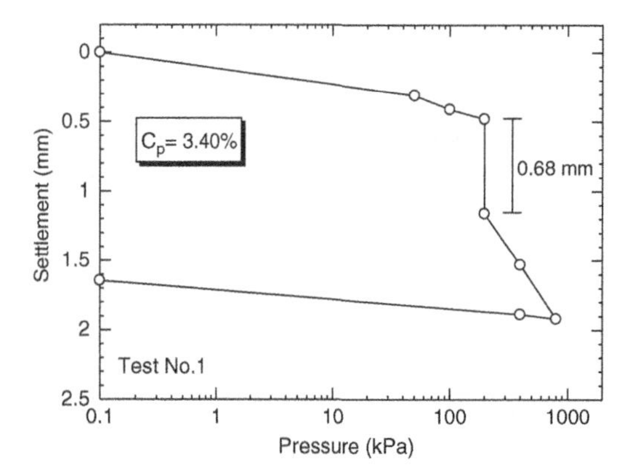

Figure 2. The collapse potential of the undisturbed loess sample.

consolidometer ring and pressures are then applied progressively until 200 kPa. At this pressure the specimen was flooded by water to make the sample in the saturation condition. The e-log p curve plotted from the data obtained can be used to calculate the collapse potential of the soil, C_p. The results show that the collapse potential value of undisturbed loess sample at a natural water content of 7% is expressed the severe trouble condition after getting wet (Fig. 2). It is also found that the compacted loess performed by adding more energy by the standard proctor method can reduce the collapse potential value to the stage of no problem condition (Fig. 3).

3 THE PHYSICAL MODELING TESTS

This part presents the results of from 1 g physical modeling test. It is aimed with investigating the performance and the suitability of the soil as a layer for road construction. All tests were conducted on the same initial conditions so that test results from

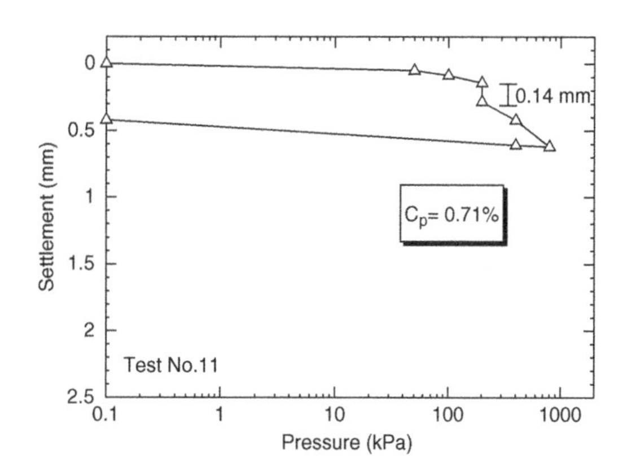

Figure 3. The collapse potential of the compacted loess sample.

Figure 4. Equipments used in 1 g physical modeling test.

different parametric studies can be properly compared and discussed.

3.1 *Loading apparatus*

Experimental testing has been conducted using a transparent tank with a loading system. A key feature of this apparatus is that the bearing capacity of the soil can be evaluated and the displacement of the soil beneath the foundation can be monitored. The loading apparatus consists of a system of jacks, control system, a system of transducers and data acquisition system. Figure 4 shows the overview of the apparatus at Khon Kaen University. The configuration allows the jacks to move up and down so that any designed vertical load can be applied.

3.2 *Model foundation and soil samples*

The model foundation used for the test program is made of stainless steel with a rectangular shape. The footing is a constant width, B, of 50 mm with a length (L) of 400 mm and can be considered this model as

Figure 5. An undisturbed block sample.

Figure 6. A model sample of compacted loess.

a striped foundation. The soil used for all tests was red Khon Kaen loess collected in Khon Kaen University area. The physical properties are already given in the Table 1. An undisturbed loess sample with a natural water content of 7% was carefully excavated and prepared from field (Fig. 5). The sample was cut to fit in the transparent container. This first sample was tested to find the bearing capacity of the soil under natural condition. It was found from the loading test that the bearing capacity of the soil was very high at the natural water content. The next test program consists of 3 more samples under the same 1 g loading test. The second sample was also excavated from the same excavation pit with same water content but before performing the loading test the water was added to the soil surface. This simulation represents the soaked condition of the soil. For the last two cases, the samples were prepared in laboratory by adding high energy to the remolded soil by the static compaction using the standard proctor method (Fig. 6). The compacted sample was then tested to investigate the bearing capacity and the suitability of soil. Sample No. 4 was prepared with the same procedures of compacted sample but before the

Table 2. Test conditions in 1 g physical modeling.

Test No.	Test condition
1	Undisturbed sample with natural water content
2	Undisturbed sample under soaked condition
3	Compacted sample with natural water content
4	Compacted sample under soaked condition

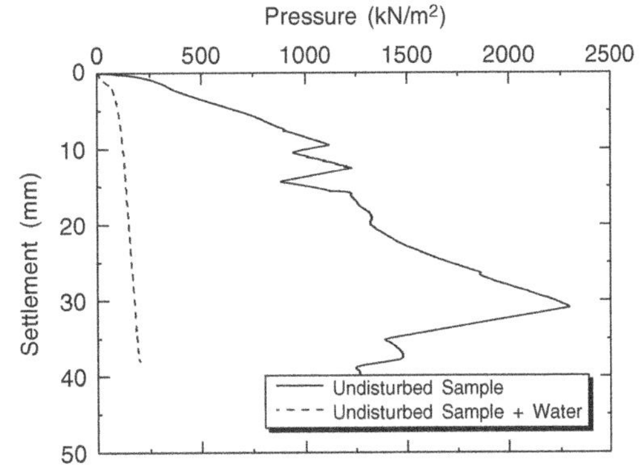

Figure 7. Pressure-settlement relationship of undisturbed samples under natural water content and soaked condition.

test, water was added with the same amount used the test No. 2. The objective of Test No. 4 is to observe the bearing capacity of the compacted soil under soaked condition. All tests are detailed in Table 2.

3.3 *1 g test results*

The ultimate bearing capacity of the footing in this study is defined at the peak load in pressure-settlement relationship. In the case where large settlement occurred, peak load was not selected but instead the load at a settlement of 10% of footing width (0.1 B or 5 mm) was chosen. Since the tests are displacement-controlled, the displacement can be continued after the peak load. An ultimate bearing capacity of 1125 kN/m^2 at a displacement of 9.0 mm was found in the pressure-settlement relationship (Fig. 7) for the footing resting on the undisturbed sample at natural water content. For the foundation on undisturbed sample at soaked condition, the ultimate bearing capacity of 160 kN/m^2 was found at a settlement of 6.0 mm. Fig. 7 clearly demonstrate that the additional water can dramatically decrease the ultimate bearing capacity of the footing on undisturbed loess. The ultimate bearing capacity of compacted loess sample under natural water content in the plot in Figure 8 is about 1300 kN/m^2 at a settlement of 4.0 mm. For the compacted sample under soaked condition, it is found that the ultimate bearing capacity of 215 kN/m^2 was observed at a displacement of 4.5 mm. Table 3 summarizes both the ultimate bearing

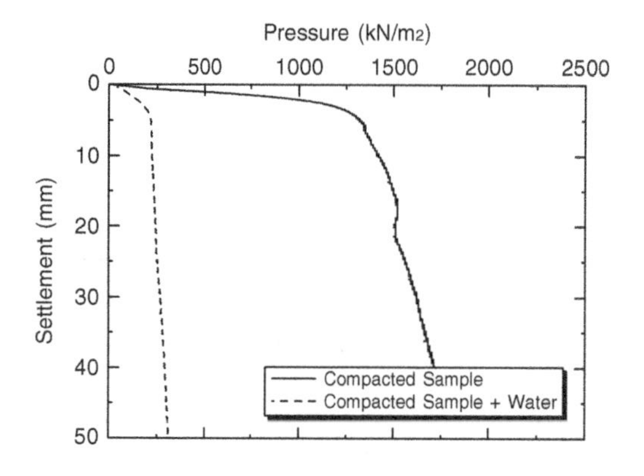

Figure 8. Pressure-settlement relationship of compacted samples under natural water content and soaked condition.

Table 3. Results from 1 g tests.

Test No.	Ultimate bearing capacity (kN/m^2)	Settlement at ultimate bearing capacity (mm)	Capacity at 0.1 B (kN/m^2)	Settlement at 0.1 B (mm)
1	1125	9.0	700	5.0
2	160	6.0	100	5.0
3	1290	4.1	1360	5.0
4	215	4.5	220	5.0

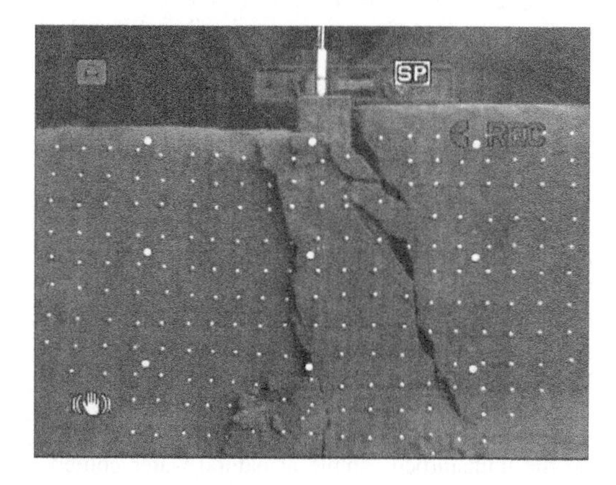

Figure 9. Picture after test of Test No. 1.

capacity and the bearing capacity of footings at a settlement of 0.1 B or 5.0 mm. It is expected from Test No. 3 and No. 4 that the higher bearing capacity of loess could be obviously improved by adding more energy by changing the method of compaction from standard to modified proctor.

Using the benefit of transparent tank and an image processing technique, the soil beneath the foundation can be observed by images taken from the video camera. Figures 9 to 12 were taken at the end of the

Figure 10. Picture after test of Test No. 2.

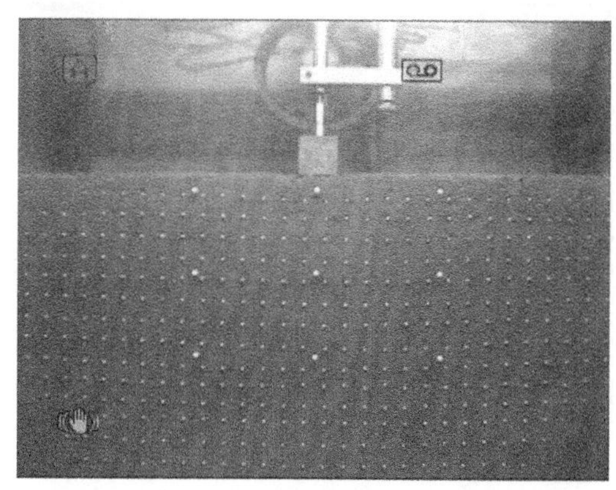

Figure 11. Picture after test of Test No. 3.

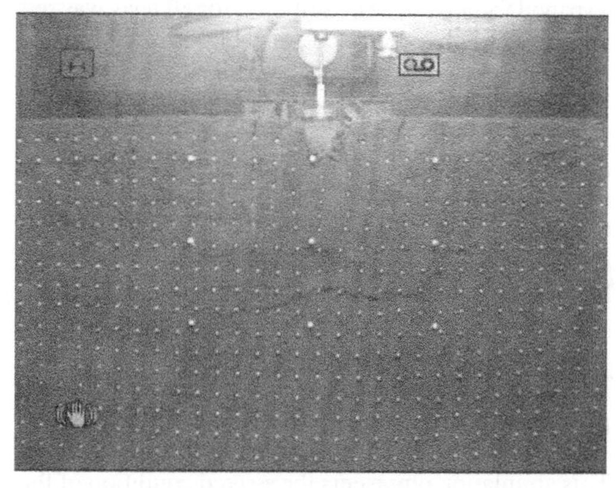

Figure 12. Picture after test of Test No. 4.

test from tests No. 1 to 4, respectively. The pictures can provide a good indicator to define the mode of failure of the soil by incorporating with the pressure-settlement relationship. It can be concluded the mode of failure in Table 4.

Table 4. Mode of failure for all tests.

Test No.	Mode of failure
1	Local shear failure
2	Punching shear failure
3	Local shear failure
4	Punching shear failure

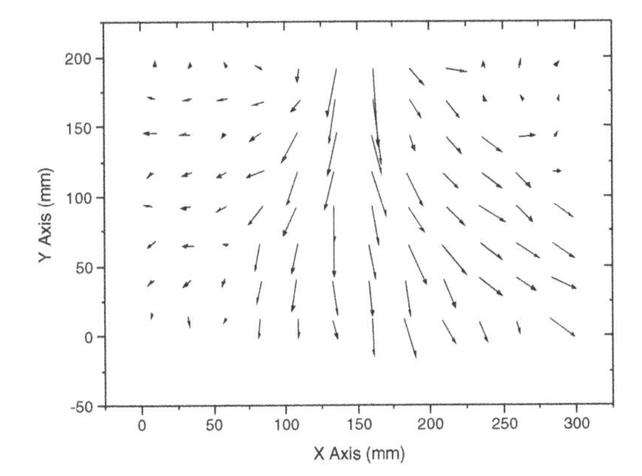

Figure 13. Displacement vectors of Test No. 4.

The targets placed on the soil can be used to calculate the displacement vector of the soil movement below the foundation. The sample analysis of the displacement vectors from pictures taken before and after tests No. 4 can be generated in Figures 13. After adding the water, the compacted sample became in the weak stage so that the size of displacement vectors at the center are very high and the size of vectors at the left and right sides of center are in the inclined direction.

4 FULL SCALE OF TRIAL EMBANKMENTS

The main objective of making Khon Kaen loess trial embankments is to assess the potential of the soil as a material for road construction. Natural slopes in soil are of interest to geotechnical engineers.

4.1 *Site conditions and design of embankment*

The test site is located in Khon Kaen University. The chosen construction area was flat for the most parts and located near the borrow pits of fill material. The design of the test embankment configuration was based on the typical section from the Department of Highway, the Kingdom of Thailand. The layout of the fill and the cross section of the embankment are schematically shown in Figures 14 and 15. The first test embankment is divided the area into two parts with the aim of simulating the general traffic load of 2 t/m^2 on the road and simulating water in the rainy season in the Northeastern region of Thailand. To facilitate

Figure 14. Trial slope embankment.

monitoring of the construction and to observe the performance of the embankment, field instruments were systematically installed. The instrumentation of unreinforced slope consists of piezometers, total pressure cells and inclinometer casings. Pneumatic piezometers were also installed to monitor the positive and negative pore pressures. The pneumatic-type total pressure cells were installed to measure the total pressure imposed by the fill on the bottom of the embankment. The applied total stresses measured from the monitoring agreed well with those deduced based on the thickness of fill and the measured unit weight of the fill. Horizontal movement of the ground was monitored by measuring the displacements from inclinometer.

4.2 *Compacted embankment*

The Khon Kaen loess was used as a fill material for the trial embankment. A series of soil layers were systematically constructed by the compaction with a thickness of 0.5 m in each layer. Field density measurements on the layers indicated an average unit weight of 18.5 kN/m^3. After reaching a designed thickness of 2.5 m, the first half of the trial embankment was statically loaded with a dead weight of 2 t/m^2 represented the traffic loads. The staged loading is 0.333 t/m^2 in steps. The second half of the embankment was soaked with water simulating the flood after a heavy rainfall in this region. The variations of horizontal displacement with depths obtained from inclinometers are measured. At the end of simulation processes, maximum horizontal displacement of about 1 mm and 8 mm were found from loading area and soaked area, respectively. The embankment was found to stand free at a slope of about 53°. Inclinometers reported the rapid movement when applied loads were changed from 2.5 to be 3.0 t/m^2. The displacement profile at 2.5 m- embankment thickness did not indicate any failure zone. The soils in the soaked area suddenly dropped after being soaked for 20 days as caused by the water infiltration. The erosion continued to occur

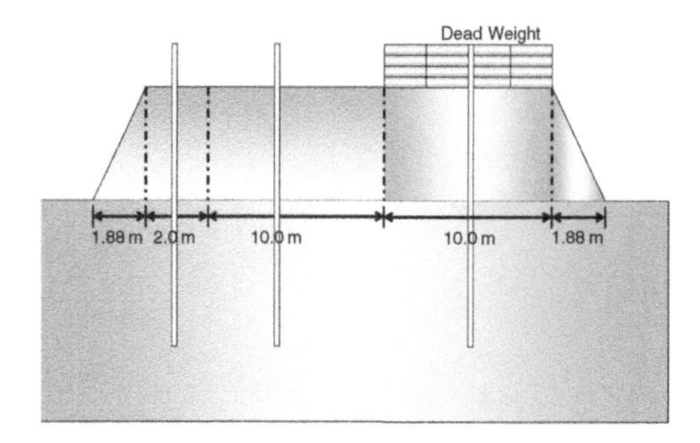

Figure 15. Cross section of the trial embankment.

Figure 16. The protection of erosion failure.

as more water was added. The test was terminated due to a safety reason.

4.3 *Reinforced embankment*

To enable the redesign of failed slope by additional water, the planning and redesign of preventive and remedial measures were necessary. The second phase of trial embankment was then constructed. The geogrid was selected and to be used for the reinforcement. The specification of compaction method is the same used in the first plain loess embankment. Concrete piles were placed on the top of reinforced embankment as a dead weight. Then, a uniform distributed load could be applied on the top of the test embankment. The test was run until the embankment showed its instability. After a uniform load of 2 t/m^2 was applied on the embankment, total pressure cells reported that the average pressure beneath the reinforced embankment with geogrid increased by about 0.7 t/m^2. If this embankment was constructed without geosynthetic materials, an increased pressure should be more than 1.0 t/m^2 according to the pressure bulb or pressure distribution with depths. The results show that the geosynthetic can transfer and make a good pressure distribution in loess and it was found the stable reinforced embankment after static loading test. The soaking condition of reinforced embankment was then tested with the same criteria used in the unreinforced embankment. The test result reported the effectiveness of geogrid and the soil compaction to prevent the slope failure. It was found only failure at the side slope and this problem can be solved by a simple and traditional method by growing the local glass on the side slope as shown in Figure 16.

5 CONCLUSIONS

The characteristics of Khon Kaen loess have been investigated by both laboratory and physical model testing. The results of basic, index and engineering properties of Khon Kaen loess under various conditions are reported. The research primarily focused on the bearing capacity of this soil under natural and soaked conditions. An undisturbed soil sample was tested and compared the result with compacted soil at natural water content and at the soaked condition. The results from the pressure-settlement relationship demonstrate that addition energy from a standard compaction can increase the bearing capacity of loess under both natural and soaked conditions. The pictures taken after test and the displacement vector provide a key indicator to define the mode of failure of the soil. Since this soil was found in abundance in the Northeastern part of Thailand, it was considered that the usage of this soil as a road construction material can reduce the construction cost. Due to its collapse potential, the compacted loess reinforced with a geosynthetic material can then be considered as an option to improve the soil properties. The effectiveness of compaction and reinforced material was found from the tests on two trial embankments.

ACKNOWLEDGEMENT

The research was developed under the cooperation between Khon Kean University, Thailand and the Tokyo Institute of Technology, Japan. The construction and monitoring of the test embankments were funded by the Department of Civil Engineering, Faculty of Engineering, Khon Kaen University and the Hitachi Scholarship Foundation, Japan. The authors wish to acknowledge the assistance of Mr.Nattapong Kowittayanant. The geosynthetic materials were supplied by Cofra (Thailand) Ltd. Author would like to extensively thank to Dr.Watcharin Gasaluck and Mr.Kosawat Changjaturat for his kind information and helps.

REFERENCES

Phien-wej, N., Pientong, T., Balasubramaniam, A.S., 1992. "Collapse and strength characteristics of loess in Thailand", Engineering Geology, Vol.32, pp.59–72.

Udomchoke, V., 1991. "Origin and engineering characteristic of problem soils in Khorat Basin, Northeastern Thailand", D.Tech.Diss, Asian Institute of Technology, Bangkok, Thailand.

Advances in Transportation Geotechnics – Ellis, Yu, McDowell, Dawson & Thom (eds)
© 2008 Taylor & Francis Group, London, ISBN 978-0-415-47590-7

Non-stationary temperature regime of the pavements

B.B. Teltayev
Kazakhstan Highway Research Institute, Almaty, Kazakhstan

ABSTRACT: The results of the complex (experimental and theoretical) research of the temperature regime of the flexible pavement and its subgrade in the non-stationary carrying-out in January (the coldest month of the year) and August (one of the hottest month of the year) in the climatic conditions of the Northern-Eastern Kazakhstan. The subgrade is represented by sandy clay. The experimental temperature measurement in the points of the pavement have been realized by thermo-electrical indicator of the resistance. The theoretical research of the non-stationary regime of the pavement and subgrade has been realized under the program ROTER designed by the author on the basis of the finite element method which takes into consideration phase transfer of the dampness in the subgrade. The comparison of the theoretical and experimental results showed their satisfactory coincidence.

1 INTRODUCTION

Mechanical and natural-climatic factors influence over the strength and longevity of the road constructions. One of the defining natural-climatic factors is temperature. Physic-mechanic and other indices of the asphalt-concrete change within the limits depending on the temperature. The temperature regime defines also the process of the freezing and melting of the subgrade and the layers of the pavement. Therefore, the complex (experimental and theoretical) research of the temperature regime of the flexible pavement and its subgrade in the non-stationary carrying-out has been considered in this paper.

2 EXPERIMENTAL RESEARCH

The experimental study of the temperature regime of the pavement and its subgrade in the conditions of the varying continental climate of the Northern-Eastern Kazakhstan has been realized on the specially chosen section of the motor road "Pavlodar-Barnaul", the 8th km. This section is in the pavement with the height 1.2–1.5 m. The subgrade is represented by plastic clay. The ground water is in the depth of more than 8–10 m. The pavement consists of two layers: the first one is the asphalt-concrete of 15 cm thickness, and the second one is the sand-gravel mixture of 20 cm thickness.

The thermo-electrical indicators of the resistance have been used for measuring the temperature in the points of the pavement and subgrade. They allow to carry out long measuring of the temperature within the range of −50°C and +180°C. The indicators have been placed in the depth: 0.07; 0.15; 0.35; 0.55; 0.75; 0.95; 1.15; 1.35; 1.55; 1.75; 1.95; 2.20; 2.70 and 3.20 m from the surface of the asphalt-concrete layer. Their ends have been taken out to the safe distance and put into the special crate. The temperature of the surface of the asphalt-concrete layer and air in the shadow has been also measured. The experimental measurements have been realized for a year.

The norm requirements for organic binders are stated with taking into account summer and winter temperature regimes of the roadbed in many countries of the world including Kazakhstan. Therefore the results of the measurement and stated peculiarities of the temperature regime of the pavement and its subgrade are given in the coldest month of the year – January and in one of the hottest month of the year – August.

As one can see from the Figure 1 the daily change of the air temperature causes the temperature change only of the pavement. Daily temperature is practically constant in the depth of 0.35 m from the surface of the asphalt-concrete layer, i.e. under the sand-gravel mixture. The surface temperature of the asphalt-concrete cover is always higher the air temperature. One can explain this phenomenon by the solar radiation influence. The temperature difference during various day periods is different according to its value. As one can expect this difference is quite low during the night period and quite high during the day time. Together with it the difference value between air temperature and surface temperature of the asphalt-concrete layer during the day time isn't constant. It is from 7°C

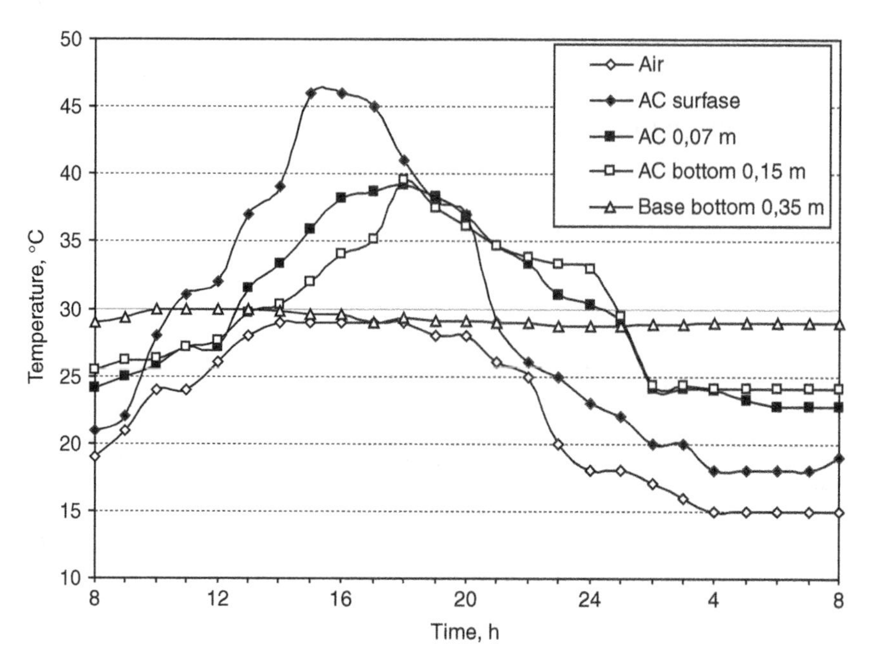

Figure 1. Temperature Change in Pavement Points from 8 a.m. the 18th of August till 8 a.m. the 19th of August, 2004.

to 17°C for 9 hours of the day time from 11 a.m. to 8 p.m. Together with it the maximum temperature difference equal to 16–17°C has been marked for 2 hours between 3 p.m. and 5 p.m. Maximum temperature meaning of the asphalt-concrete surface during this period of time is 46°C. Its minimum meaning is 18°C and has been marked between 4 a.m. and 7 a.m. This difference changes very little during the rest period of the day. It is 1–4°C in the morning till 10 a.m., and 3–5°C in the evening and at night after 9 p.m. We can accept this on the condition that the difference between asphalt-concrete temperature and air temperature is constant and equal to 2–4°C during the whole period between 9 p.m. and 10 a.m.

The temperature regime of the pavements has relatively simple consequences during the winter period compared with the summer one. As one can see from Figure 2, the air temperature change causes the temperature change only of the asphalt-concrete layer. The temperature doesn't practically change in all other depths from the surface of the asphalt-concrete layer during the day. One can accept precisely for practical calculations that the air temperature change during the winter period causes the synchronic change of the surface temperature of the asphalt-concrete layer and the absolute air temperature is always 2–3°C higher than the surface temperature of the asphalt-concrete layer. One can see that the air temperature and the surface temperature of the asphalt-concrete layer changes step by step during the winter period for a day. Together with it the surface temperature of the asphalt-concrete is – 22°C from 11 a.m. to 5 p.m., −24°C – from 6 p.m. to 1 a.m., and – 27°C – from 2 a.m. to 10 a.m.

The speed change of the temperature of the asphalt-concrete layer is one of the practically important indices of the temperature regime of the pavements. It is known that elasticity strength occurs with the temperature decrease in monolith asphalt-concrete layer. If the temperature decrease is essential and occurs quickly the amount elasticity strength appearing under the mutual influence of temperature and automobile load will be larger in value than the asphalt-concrete strength. It will result in appearance of new and increase of the existing cracks in asphalt-concrete layer of the pavement.

If the temperature decrease occurs slowly in the asphalt-concrete layer the amount elasticity strength will be lower than the asphalt-concrete strength because of its relaxation properties.

In our case the maximum meaning of the temperature decrease of the surface of the asphalt-concrete layer took place between 8 p.m. and 9 p.m. in August and was 8.0°C/h. The value of this index was 4.5°C/h on the low surface of the asphalt-concrete layer and took place between 6 p.m. and 7 p.m. The maximum speed decrease of the temperature was 4.6°C/h for the whole asphalt-concrete layer with the 15 cm thickness and it was between 8 p.m. and 9 p.m. The next largest value of the sped decrease of the temperature was

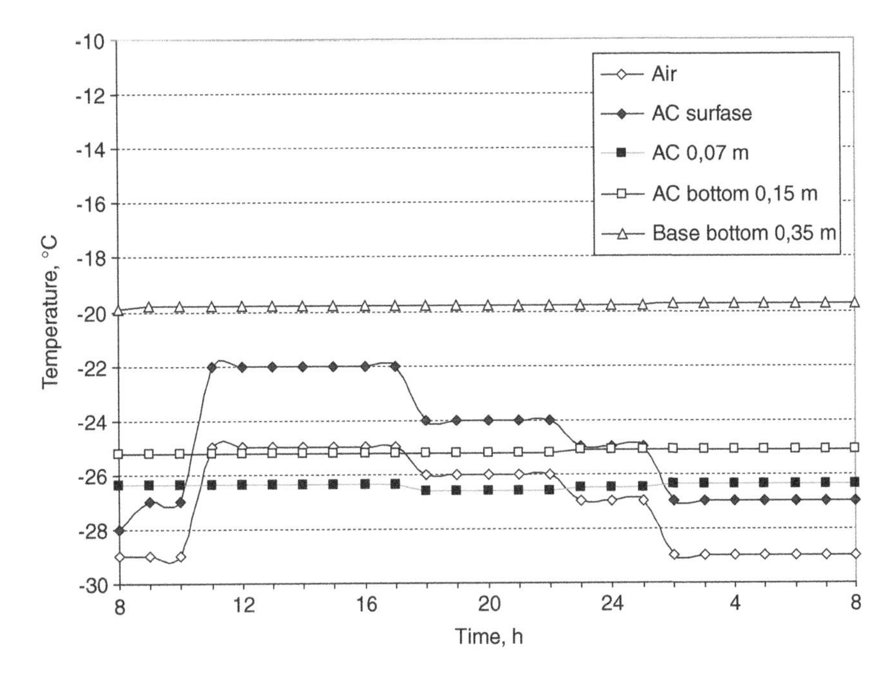

Figure 2. Temperature Change in Pavement Points from 8 a.m. the 28th of January till 8 a.m. the 29th of January, 2005.

4.0°C/h in the asphalt-concrete layer and took place between 1 a.m. and 2 a.m.

It was stated in January that the temperature changes only in the surface of the asphalt-concrete layer during the day time (it was mentioned above). The speed decrease of the surface temperature of the asphalt-concrete layer is maximum between 5 p.m. and 6 p.m. and between 1 a.m. and 3 a.m. and equal to 2.0°C/h.

It is stated that the temperature changes only within the pavement during day time in summer, and only on the surface of the asphalt-concrete layer during the winter period. It is stated that the temperature decreases in the subgrade in summer with the depth increase with gradient 3.7–5.2°C/m. The decrease of the absolute temperature meaning also occurs with the depth increase in winter. Therefore the gradient meaning from the depth 0.07 m until low surface of the asphalt-concrete layer is 23.2°C/m, and in the subgrade is 10.0–19.4°C/m.

3 MODELLING BY THE FINITE ELEMENT METHOD

The theoretical research of the temperature regime of the pavement and subgrade has been realized by finite element method. Taking into account that the motor road length is very large compared with its height and width and for any definite section of the road one can believe with the largest practical precise that geometrical and heat-physic parameters of the road and environment along the road length doesn't change. It's enough to consider the double-measured process of heat conduct.

Differential equation of double-measured non-stationary heat conduct for isotropy material is (Abramanovich & Levin 1969):

$$\lambda\left(\frac{\partial^2 T}{\partial x^2}+\frac{\partial^2 T}{\partial y^2}\right) = c\rho\frac{\partial T}{\partial t} \tag{1}$$

where T = temperature; t = time; λ = termal conductivity coefficient; c = specific heat; ρ = density; and x, y = components of the Cartesian coordinate system.

Taking into account the layers of the construction considered the problem has been solved by finite element method (Teltayev 1999). The finite-element analogue of the differential equation 1 has been shown as the system of the linear algebraic equations (Zienkiewicz 1971) and (Segerlind 1976):

$$[C]\cdot\left\{\dot{T}\right\}+[K]\cdot\{T\}+\{Q\}=0 \tag{2}$$

where [C] = matrix of specific heat of system; [K] = matrix of termal conductivity of system; [Q] = vector of joint heat flows at the time moment t + Δt; $\{\dot{T}\}$ = vector of velocity of joint temperature at the time moment t + Δt; and {T} = searching vector of joint temperature at the time moment t + Δt.

Table 1. Calculated Indices of the Pavement and Subgrade.

Materials of layers and ground	Density ρ (kg/m^3)	Termal conductivity coefficient λ (J/(m h°C))	Specific heat C (J/(kg°C))
Asphalt-concrete	2400	5040	1650
Sand-gravel mixture	1875	6876/7740	1000/900
Sandy clay	2000	5832/7092	1450/1250

Algorithm of the equation solution (2) has been used in this paper, according to which the heat balance in the joints of the finite-element system has been shown as the following:

$$[\overline{K}] \cdot \{T\} = \{\overline{q}\} \quad (3)$$

where $\{\overline{q}\}$ = efficient vector of the joint heat flows; and $[\overline{K}]$ = efficient matrix of the heat conduct of system.

The above-mentioned finite-element model doesn't take into account the phase transfers (freezing and melting) of moisture contained in the subgrade. To take them into account one must include vector of heat phase transfer of system $\{Q_f\}$, into the right part of the equation system additionally (3), which is calculated by means of summing up the respective vector elements of the heat phase transfer finite elements $\{Q_f\}$.

Calculations on the temperature regime definition of the pavement and subgrade have been made under the applied program ROTER, designed by the author according to the above mentioned methods. Cross-section of the road was discretized into 722 triangle finite elements with common number 400 joint points for calculations. Equation system 3 has been solved for each time step with the help Gauss-Zeidel's iteration method.

Calculated meanings of the material indices of the pavement layers and subgrade have been given according to (Teltayev 1999) which have been shown in Table 1.

The meanings of the heat conduct coefficient λ and heat capacity c have been shown for the sand-gravel mixture and sandy clay in the upper part of the fraction for the melting (non-frozen) condition, and in the lower part of the fraction for their freezing (frozen) condition in the Table 1.

4 NUMERICAL CALCULATIONS AND COMPARISON OF THEIR RESULTS WITH THE EXPERIMENTAL DATA

Numerical calculations have been made for all months of the year with the help of finite element method. Some results are given in Tables 2 and 3. Thus, the surface temperature of the asphalt-concrete layer has been given under the experimental data, and the horizon temperature in the depth of 3.2 m from the surface of subgrade has been accepted according to Pavlodar meteorological station and experiment.

Table 2. Comparison of Measured and Calculated Temperature Meanings in the Pavement and Subgrade Points, 12 a.m., 28th–29th January, 2005.

Depth, m	Measured temperature, °C	Calculated temperature, °C	Difference, %
0	−22,0	−22,0	0
0,07	−26,4	−25,1	−4,9
0,15	−25,2	−23,7	−6,0
0,35	−19,8	−21,0	+6,1
0,75	−15,7	−15,7	0
0,95	−13,5	−13,6	+0,7
1,15	−11,5	−11,5	0
1,35	−9,5	−9,8	+3,2
1,55	−8,5	−8,3	−2,4

Table 3. Comparison of Measured and Calculated Temperature Meanings in the Pavement and Subgrade Points, 12 a.m., 18th–19th August, 2004.

Depth, m	Measured temperature, °C	Calculated temperature, °C	Difference, %
0	32,0	32,0	0
0,07	27,2	25,9	−4,8
0,15	27,6	27,4	−0,7
0,35	30,0	30,2	−0,7
0,75	27,8	27,7	−0,4
0,95	27,4	26,9	−1,8
1,15	26,0	26,1	−0,3
1,35	24,9	25,2	+1,2
1,55	24,7	24,3	−1,6

It is stated that the experimental and theoretical results coincide satisfactory. The maximum difference of theoretical results and experimental ones, equal to 12.1%, occurs in asphalt-concrete layer in January at 12 p.m. Coincidence of theoretical and experimental results (difference is equal to zero) has been also marked in January in points located in subgrade in the depth of 0.75 m and 1.15 m from the surface of the pavement. Average meanings of this difference are equal to 3.6% for January and 2.6% in August.

5 CONCLUSION

One can make the following conclusions on the basis of the above mentioned material:

1. Daily temperature difference causes the temperature change only in pavement points in conditions of

sharp continental climate of the Northern-Eastern Kazakhstan in August (one of the hottest months of the year). The surface temperature of the asphalt-concrete layer is always higher than the air temperature. This difference is maximum for two hours between 3 p.m. and 5 p.m. and equal to 16–17°C. The maximum temperature meaning is equal to 46°C, and the minimum one is minus 18°C.
2. The air temperature difference causes the temperature change only in the surface of the asphalt-concrete layer in January (the coldest month of the year). The maximum negative temperature, equal to −27°C, occurs between 2 a.m. and 10 a.m., and the minimum negative temperature, equal to −22°C, has been marked between 11 a.m. and 5 p.m.
3. Finite-element model, taking into account the phase transfers in the subgrade, describes satisfactory the non-stationary temperature regime of the pavement. Comparison of the experimental and theoretical results showed that the average difference meaning had been 3.6% in January and 2.6% in August. It allows to use this finite-element model for the forecast of the temperature regime of various pavements and their subgrades in different climatic conditions.

REFERENCES

Abramanovich, I. G., & V. I. Levin. 1969. equations of mathematical physics. Moscow: Nauka.

Teltayev, B. B. 1999. Strains and stresses in flexible pavements. Almaty: KazATC.

Zienkiewicz, O. C. 1971. The finite element method in engineering science. New York: McGraw-Hill.

Segerlind, L. J. 1976. Applied finite element analysis. New York: John Wiley & Sons.

Advances in Transportation Geotechnics – Ellis, Yu, McDowell, Dawson & Thom (eds)
© 2008 Taylor & Francis Group, London, ISBN 978-0-415-47590-7

Contaminated land: Recent research and development for UK Highways Agency

D.M. Tonks, E.M.G. Gallagher & D.M. Smith
Coffey Geotechnics Ltd

D. Gwede
Highways Agency

ABSTRACT: This paper summarises the outcomes of recent research and development work for the UK Highways Agency (HA) into the investigation, risk assessment and treatment of contaminated land. There has been a substantial increase in legislation, guidance and understanding of this topic over recent years. This paper outlines the key issues to Highways Agency and how these are addressed under present and proposed guidance. A few case studies have been selected to illustrate the range of issues arising. The paper also considers planned developments to ensure the appropriate knowledge and technologies are suitably used and managed. This is particularly important in respect of the consistent management of risks and the sustainable use of resources. The HA has been supporting development of industry standards on contaminated land through the umbrella of Institution of Civil Engineers Site Investigation Steering Group.

1 INTRODUCTION

1.1 *Scope*

It is Highways Agency (HA) policy to pro-actively address contaminated land in an integrated and proportionate manner with other ground-related risks, in accordance with best practice and its position on sustainable construction as set out in Building Better Roads (HA, 2003).

Over the last few the years the authors have been involved with ongoing research and development studies of contaminated land issues affecting or potentially affecting HA assets (EDGE & NGI, 2005), with particular reference to the UK Contaminated Land Regime, Part IIA (TSO, 1995).

A sustainable approach to highway construction favours re-use of contaminated/derelict land where possible, to minimise taking greenfield land and give the opportunity to fund a proportionate clean-up. Remediation and re-use will increasingly be preferred to 'dig and dump'/landfilling – both for cost and sustainability. However, this approach can be technically and procedurally difficult and expensive. There is some need for more pro-active 'drivers', protocols & procedures, for example by introducing the measurement of sustainability in road building procurement processes such that contractors have a clear financial incentive to follow such a policy.

A consistent approach is needed for procedures, knowledge management, staged investigations and assessment of risks to users and wider stakeholders and to controlled waters. It can seem exceptionally difficult and require massive and disproportionate effort to satisfactorily resolve quite simple tests, viz:

- Is the land actually Part IIA 'Contaminated Land' – and if so, under which criteria;
- Who are the responsible parties (liability groups or 'appropriate persons'); and
- What are the appropriate remedial measures?

Added to these challenges, appropriate consideration must also be given to planning requirements (noting the planning regime is arguably the primary driver for contaminated land *investigations*) and which also requires an assessment of risk appropriate to the end use.

Perhaps the biggest difficulty with the new Part IIA regime is the combination of technical and legal uncertainty which can occur, allowing relatively simple 'issues' to become protracted. It is to the advantage of HA (and often the other parties affected) to take a pro-active approach to speedy and effective resolution.

The research review (EDGE & NGI, 2005) for HA takes account of:

- Existing HA procedures and guidance;
- Existing experience on highways;

- Existing experience on contaminated land and related matters; and
- More than 200 references, including numerous websites and databases.

From this basis, the research review:

- Identifies key documentation relevant to UK Highways Agency assets and projects;
- Discusses issues in the light of this for the various situations and project stages including:
  - Existing assets – i.e. non-project;
  - Ongoing maintenance and management;
  - New project feasibility, environmental impact assessment to planning approval;
  - New project procurement;
  - Investigations and design;
  - Construction; and
  - Completion and records.
- Considers existing HA guidance and makes recommendations for future development & research; and
- Gives guidance as to best practice – with a view to agreement with the Environment Agency (EA) and other stakeholders.

1.2 *Background*

Contaminated land is defined under the UK Environment Act (TSO, 1995) as:

78A (2) – any land which appears to the local authority in whose area it is situated to be in such a condition by reason of substances in, on or under that land, that-

(a) significant harm is being caused, or there is a significant possibility of such harm being caused, or

(b) pollution of controlled waters is being or is likely to be caused.

Significant potential problems and liabilities can be identified. However, relatively little has materialised so far for HA (or others) by way of consistent procedures to collate/unify knowledge or records. Issues tend to be addressed 'ad hoc', with varying success. No HA land has actually been determined as Part IIA Contaminated Land, but there are thought to be land holdings which could give rise to some issues. Some cases are going through investigation/voluntary actions.

There can be a need to act quickly, effectively and consistently when potential cases arise, although in practice actions are often triggered through planning requirements rather than under Part IIA.

1.3 *International position*

A rather full review of experiences elsewhere in the world may be summarised as showing:

- General consistency of thinking;
- No key points missing or inconsistencies in UK;
- Similar difficulties over liabilities of ownership/purchase/operational activities; and
- Some countries are more pro-active/have less procedural barriers to remediation.

On review, there are good reasons to prefer the UK 'suitable for use' and 'risk assessment' approaches as opposed to multi-functionality or more prescriptive, quasi-legalistic regulation. It is recognised there may be difficulties in detail. HA may assist as a key player promoting sustainable approaches.

As a postscript to this review, we share the concerns recently expressed by DEFRA (2007) regarding potentially prescriptive, disproportionate and expensive proposals highlighted in the consultation on the Soils Framework Directive. These include onerous new requirements for identifying 'contaminated sites', reduced site-specific, risk-based investigation (hence increasing potential risk of blight), also the move away from voluntary action by agreement towards actions driven by formal enforcement.

2 CURRENT SITUATION

2.1 *General*

Management of contaminated land is an important aspect of sustainable development. Highways constitute some of the least sensitive uses of land in this respect. The surfaced areas normally preclude access to the ground below and prevent direct pathways. Surfacing also provides controls on infiltration and hence risks to groundwater – but drainage arrangements and detailing warrant considerable thought and care. Vehicle users are generally at low risk with respect to contamination. Some areas which are not surfaced, including landscaping, roadside verges and some ancillary works may be more sensitive. Conversely, it is rare, but not impossible, for highways to contaminate adjacent land or controlled waters.

For the benefit of society and the environment it may be desirable and beneficial to locate highways on brownfield, contaminated and derelict ground rather than greenfield, where the opportunity arises.

Strategic-level responsibility for identifying Part IIA Contaminated Land rests with the local authority as primary Enforcing Authority. It appears rare for HA land to be considered under this, but there is clearly such potential and care is needed in addressing this. There is merit in developing guidance for a reasonably consistent and effective approach.

More commonly, new works may be planned on land which is, or may be, contaminated in some way. It will be important to ensure that the project procedures and design are appropriate, with particular reference to

Part IIA and other relevant regulations. These include investigation, risk assessment and design procedures for new works. Consideration should also be also given to related matters such as waste management and remediation.

Whilst there has been substantial concern and research effort through Europe and elsewhere, and many sites affected by contamination, there have been relatively few proven cases of actual harm to humans from contaminated land, either recently or in the past. There are, however, a number of celebrated and often-quoted cases (and rather more known to specialists) which certainly posed significant risks and have been remedied accordingly.

We have yet to find a case of significant harm to humans in a UK highways context. Whilst Part IIA Contaminated Land does not appear to have much adverse impact on highways to date, reasonably pro-active procedures may be warranted. This should be kept in context with the various other risks associated with highways, either to the public as users, or to construction and maintenance personnel.

It is proposed that relatively simple registration of risk assessments be developed for highways, consistent with the UK approach to contaminated land. This would allow prioritisation and tiered assessment on the basis of risk and need, for existing land and new works. It would also aim for consistency in approach. It could be most beneficial to use an existing Geotechnical (asset) Data Management System such as HA GDMS to collate these risk registers.

The research review also sets out the roles and issues relating to the various other stakeholders with which HA needs to liaise including:

- Government, Defra, EA,
- Local Authorities, and
- Public (and interest groups and advisors).

Recommendations are made for improved procedures and guidance in the light of existing HA documents where appropriate and new documents where necessary.

3 HA PROCEDURES AND GUIDANCE

3.1 *General*

Current HA guidance on contaminated land comes within the scope of the topics and documents reviewed below. Recommendations are made for updating these.

3.2 *Managing geotechnical risk – New works*

HD22/02 gives standard procedures for management of geotechnical matters on new works. Projects affected by contamination generally come in 'Geotechnical Category 3'.

The 'Preliminary Sources Study' should include all relevant information on contamination. Further information can be given via the 'Geotechnical Risk Register'.

Findings come under the 'Geotechnical Report':

- Appendix C, existing information, and
- Appendix D geotechnical feedback reports.

These are being updated in relation to contaminated land generally and particularly:

- Ground risk assessment;
- Sufficiency and management of data;
- Risk screening on available data;
- Conceptual models;
- Current or future contamination issues; and
- Plan/vertical extents of areas affected.

3.3 *Maintenance of highway geotechnical assets. HD41/03*

This gives brief reference to contaminated land. Key issues to develop/amplify include:

- Risk screening on available data;
- Could there be current or future contamination issues;
- Could site be Part IIA Contaminated Land;
- Plan/vertical extents of areas affected;
- Present and possibly future land uses; and
- Planning and management of maintenance activities.

Appendix C: Geotechnical principal inspection report & geotechnical maintenance form (GMF) could be developed for this purpose.

3.4 *Site investigation for highway works on contaminated land. HA73/95*

This has served well over many years and contains much still very useful advice. However, in keeping with HA current policy to support and work to UK industry standard documents where possible, HA is supporting the updating of the Site Investigation Steering Group (SISG) series of documents (SISG, 2008, in prep.). HA hopes these will effectively replace HA73/95 when completed or at least reduce this to minimum requirements. A major part of the revision of the SISG documents is specifically to incorporate elements directly related to site investigation work on contaminated and/or potentially contaminated land, including laboratory testing requirements etc. The documents will contain general guidance, a robust standard specification and specific health and safety guidance in this respect.

Parts of HA73/95 are also applicable to investigations for existing highways assets and may be usefully updated for this. For example, HA73/95 gives no particular guidance as to how field and laboratory contamination testing should be reported, noting the introduction of MCERTS, waste acceptance testing, leachability and availability becoming more significant issues, also developments in digital data management (HA GDMS, etc). Modern protocols are needed – cross linked to updated specifications. Some key references to draw upon in the next version of HA73/95 include *inter alia* BS10175: 2000 Code of practice: investigation of potentially contaminated sites and EA (2000) Technical Aspects of Site Investigation.

A key objective is to ensure investigations do not create new pathways enabling pollutant linkage which could create Part IIA Contaminated Land.

3.5 *Earthworks – design and preparation of contract documents. HA44/91*

Some key issues to update/develop include:-

- Management of waste;
- Remediation (brief update, pending more detailed separate guidance?); and
- Earthworks, disposal and classification of contaminated material – to ensure earthworks do not create new pathways/pollutant linkage(s), which could create Part IIA Contaminated Land.

3.6 *Foundations BD74/00*

It is important to ensure foundations do not create new pathways enabling pollutant linkage which could create Part IIA Contaminated Land. Updated advice should be included on piling and ground improvement works in contaminated soils. It is important to ensure contamination and ground conditions generally could not adversely affect materials in the ground e.g. buried concrete, culverts and suchlike.

3.7 *Suggested other guidance and procedures*

Issues of land ownership/purchasing/acquisition/ leasing/handover etc are significant. These need interdisciplinary work with asset managers, lawyers and planners together with the appropriate technical advisors. Aspects include due diligence, management of data and asset management more generally.

Classification and management of unacceptable material and waste needs an urgent update to comply with recent legislation, waste acceptance procedures and criteria for contaminated soils (and possibly other wastes).

Guidance is needed on procedures and Construction Quality Assurance (CQA) to validate remediation.

The Specification Series 600 Earthworks needs update to address contamination and waste issues. DMRB Chapter 11 Environmental Assessment defines what is necessary/desirable at planning stages in relation to contaminated land.

- Part 10 Water Quality and Drainage. Need to clarify and update risks to controlled waters under Part IIA.
- Part 11 Soils and geology. Contaminated land sections could usefully be updated.

4 PROJECT MANAGEMENT, CONSTRUCTION AND REMEDIATION

4.1 *Construction/Earthworks*

There is the opportunity to further address contaminated land in the context of current procurement methods such as design and build (D&B) and private finance initiative (PFI). It will be important to closely define minimum standards and suitable procedures for the HA role of client.

Highways may introduce barriers/prevent pathways, thus making the land not 'Contaminated Land'. These may not necessarily be suitable for other uses i.e. significant contamination may still be in place but there is no pathway and/or no target, hence not 'Contaminated Land' under the act. It is important for HA to keep adequate asset registers.

4.2 *Waste*

Management of waste has become a substantial issue in the last few years. European Landfill Directive and UK Regulations add substantial complexity and greatly affect construction wastes, and work on contaminated land. Wastes are now classified using 'waste acceptance criteria' (WAC) as inert, non-hazardous or hazardous. Each waste stream is further classified under the new and complex (6 digit) European Waste Catalogue (EWC). This is presently problematic with a need for clear guidance to users.

4.3 *Recycling and re-use*

The construction industry recycles some 60–70% of its wastes, and makes much use of waste from other industries – because this makes economic sense. Much is hardcore and suchlike, PFA, mine waste, and some other industrial by-products. Some can be usefully treated for use as various grades of fills etc. HA has taken a lead on aspects of this. There is scope for more research and updated guidance.

4.4 *Remediation*

Procedures for remediation are presently very involved. HA needs to spell out the roles of parties and there is substantial need for guidance.

Highway pavements are likely to constitute adequate cover in many instances. A recent report on cover schemes for housing (NHBC, 2004) concludes that 600 mm clean cover is frequently sufficient for gardens. Highways are normally far less sensitive end-uses (receptors), so normal highway construction thicknesses are likely to suffice. A parallel design / guidance document on this may be of assistance.

5 SOME CASE HISTORIES

5.1 *General*

A few case studies have been selected which show the range of issues arising Some details have been altered or omitted to illustrate potential matters which can arise. No project specific judgments are intended in the comments.

5.2 *Viaduct – Existing asset*

An existing viaduct was shot blasted in the early 1990s. Sheep were grazing below and there were some reported sheep deaths in late 1990s. Issues included:-

- Did the circumstances actually constitute significant harm or significant possibility of significant harm (SPOSH)? Several quite detailed studies failed to be conclusive on this point.
- Was the land contaminated before? If so, by whom? What prior studies are appropriate?
- Works under contract. What information should be obtained, and records kept – and by whom?
- Ongoing maintenance. What information should be obtained, and records kept – and by whom?
- What contamination may arise from run-off?
- Procedures are needed to design out / manage those activities which could cause contamination.

An interesting feature is that the problem only arose because of the use of the land. There is not normally grazing below viaducts, but it seems the shelter here was used as a feed area. An obvious solution is exclusion of grazing animals with a fence, or other change of use.

5.3 *Road over old landfill*

In the early 1990s a new highway was constructed across the central part of an old landfill. About 10 years later the adjacent part was developed for housing, introducing a new receptor. A further complication arose from leachates and potential risks to the aquifer. A cut-off wall was formed on the boundary. However, for various reasons it was suggested that the land could come under Part IIA.

Interestingly, the local authority was also an owner of some of the old landfill and would thus have to consider its own land as Enforcing Authority. Several Geotechnical Reports (after HD22/02) considered the issues in considerable detail, including extensive risk assessments, but were inconclusive: outcome indeterminate – more study needed.

Fortunately, it was recalled that the original scheme had taken account of the issues and indeed had received previous National Rivers Authority approval. Whilst this did not necessarily demonstrate compliance with subsequent law and standards, it proved very helpful in establishing the prior situation, and suitability of the previous design.

The importance of keeping and managing suitable records cannot be over-emphasised.

5.4 *Compulsory purchase land*

A plot of brownfield land was bought in preparation for an option future highways scheme. As discussed above, this can be highly desirable to make sustainable use of such land, and thus contribute to urban regeneration. It is therefore particularly important that such beneficial activities are not undermined by possible negative consequences.

In the event, this plot of land was not in the preferred route and the HA looked to sell the site. However, the contamination became an issue with regard to the sale and finding a suitable use compatible with the conditions and/or possible remediation.

Issues included:-

- Implications for purchase/sale/transfer;
- Purchase of land vs. transfer of liabilities;
- Waste management/disposal;
- What investigations needed/warranted;
- Ongoing site management;
- Risks to adjacent surface water; and
- Risks to groundwater.

6 CONCLUSIONS AND RECOMMENDATIONS

Technical knowledge in land contamination has advanced very rapidly in recent years and this is set to continue. This includes related matters such as waste and remediation.

There is much new and developing science. It is important to bear in mind that much is provisional and rapidly superseded by improvements. Likewise, there is much technical guidance, subject to regular revision and updating.

There have also been substantial developments in UK contaminated land law and procedures over recent years. Part IIA puts the duty for identification of potentially contaminated sites on to local authorities, not the landowner. However, it is recommended that HA should be pro-active where appropriate.

Work is in progress to suitably incorporate these developments into HA procedures and guidance, noting the prospects for ongoing progress and change.

Some key recent developments are being incorporated on a prioritized basis, with a view to needs and best value. Other matters will need more work.

Some topics are in a state of significant ongoing development and appropriate expert advice will be needed in risk assessments, developing solutions and in approvals procedures.

There is need for new or improved guidance on most contaminated land topics including:

- Procedural and management;
- Investigations;
- Field and laboratory testing;
- Risk assessments;
- Construction; and
- Remediation.

In the first instance, much may be best addressed by use of industry standard documents (e.g. BS 10175, CLR11) and others which are currently in preparation and supported by HA, notably SISG.

It is envisaged that HA procedural documents can then be refined, with inclusion of references to appropriate texts and websites.

Contaminated Land issues are addressed in various HA documents, notably:

- HA73/95 Site Investigation for Highway Works on Contaminated Land;
- HD22/02 Managing Geotechnical Risk;
- HD41/03 Maintenance of highway geotechnical assets;
- HA44/91 Earthworks;
- BD74/00 Foundations; and
- DMRB Chapter 11 Environmental Assessment.

These are best kept updated on contaminated land matters by reference to developing industry standard documents within their respective scopes, with particular further HA comment where necessary.

It is important to ensure that new highway works, do not create new pathways enabling pollutant linkage which could give rise to Part IIA Contaminated Land. Risks particularly apply to:

- New investigations and boreholes;
- Earthworks in or adjacent to contamination;
- Work in or adjacent to surface waters and groundwater;
- Foundations, especially those introducing pathways such as stone columns and some forms of piling; and
- Drainage (specific guidance is likely to be needed for this).

Procedures should include suitable quality assurance in construction (CQA), remediation and validation testing.

It will be useful to track selected new works projects from initiation/early stages, through to completion and ongoing maintenance.

Where contamination is identified or may be an issue for existing HA land or assets (i.e. not in a new project), procedures are being developed to establish if it may be Part IIA Contaminated Land and if so, to agree suitable actions. Voluntary remediation is likely to be the normal, preferred approach.

Due consideration also needs to be given to prevention of contamination from highways where they cross areas of special environmental interest or other sensitive uses. Consideration should be given to establishing registers of such sensitive adjacent sites, on a prioritised basis. Special procedures may be needed – possibly through BD41/03. This may be integrated with other environmental impact procedures (e.g. noise, air).

Records and knowledge management are of particular importance for contaminated land management and sustainability, both for present and future use.

Further study of case histories is planned to checklist and examine issues arising, assess where and what guidance is required and assist in prioritising future tasks. This will involve review of selected projects known to have addressed significant contaminated land, successfully or otherwise. Several projects of interest for this have been identified.

ACKNOWLEDGEMENTS

The authors are grateful to HA for commissioning and supporting this work. However, the findings and views expressed are their own and should not be construed as necessarily representing the views or policy of HA.

REFERENCES

BS 10175:2000 *Code of Practice for Investigation of Potentially Contaminated Sites*

Environment Agency, CLR No. 11 *Model procedures for the management of contaminated land.* Sept 2004

Defra July 2007. *Consultation on the proposed EU Soil Framework Directive*

EDGE Consultants UK Ltd., Norwegian Geotechnical Institute, NGI, 2005. *Contaminated Land, Review Report.* Published on www.highways.gov.uk

Environment Agency, 2000 *Technical Aspects of Site Investigation*: Vols 1 & 2, R&D Technical Report P5-065/TR

National House Building Council (NHBC) *Technical Standards 2004*

HA 2006 DMRB Chapter 11 Environmental Assessment

HA Dec 2003 *Building Better Roads: Towards Sustainable Construction*, Published on www.highways.gov.uk

HA 73/95 DMRB Vol 4, Sect 1, Pt 7, *Site Investigation for Highways Works on Contaminated Land*

HD 41/03 DMRB Vol 4, Sect 1, Pt 3, *Maintenance of Highway Geotechnical Assets*

HA 44/91 DMRB Vol 4, Sect 1, Pt 1, *Earthworks – Design and Preparation of Contract Documents*

BD 74/00 DMRB, Vol 2, Sect 1, Pt 8, *Foundations*

HD 22/02 DMRB Vol 4, Sect 1, Pt 2, *Managing Geotechnical Risk*

TSO, 2000. *Environment Act, 1995. Part IIA. The Contaminated Land Regulations SI 2000/227*

Site Investigation Steering Group (SISG) 2008 (in preparation), ICE, BGA,*Specifications and Advice for Site Investigations, including for Contaminated Land*

Advances in Transportation Geotechnics – Ellis, Yu, McDowell, Dawson & Thom (eds)
© 2008 Taylor & Francis Group, London, ISBN 978-0-415-47590-7

An assessment of the selected reinforcements of motorway pavement subgrade

D. Wanatowski
Nottingham Centre for Geomechanics, School of Civil Engineering, The University of Nottingham, UK

A. Florkiewicz & W. Grabowski
Institute of Civil Engineering, Faculty of Civil and Environmental Engineering, Poznan University of Technology, Poland

ABSTRACT: Thawing fine-grained soils are often saturated and have extremely low bearing capacity. In many countries with long-lasting winter seasons such weak soils may cause serious problems in the maintenance of road pavements. Consequently appropriate selection of the most efficient method of subgrade reinforcement is one of the most important factors in enhancing the performance and extending the service life of roads. In this paper, some of the most common methods, such as removal and replacement, cement stabilisation or the use of geotextile, used for reinforcement of motorway subgrade in Poland are analysed. Advantages and disadvantages of selected reinforcement examples with regard to environmental conditions are discussed.

1 INTRODUCTION

Thawing fine-grained soils are often fully saturated and have extremely low bearing capacity. In many countries with long-lasting winter seasons such weak soils may cause serious problems in the maintenance of road pavements. An appropriate selection of the most efficient method of subgrade reinforcement is therefore one of the most important factors in enhancing the performance and extending the service life of roads. This is particularly significant in the case of motorways, which are of the utmost importance to the development of any country. Owing, however, to systematic traffic of heavy vehicles and frost action in subgrade, motorway pavements can suffer serious structural damage during the spring thaw, and thus can last considerably less longer than expected.

In order to protect motorway pavements from damaging frost action and increased moisture content during the spring thaw, the subgrade must have sufficient bearing capacity. For example, in Poland, according to the criteria of the General Directorate for Public Roads (GDDP 1997), a frost-susceptible motorway subgrade is required to have an elastic modulus $E_2 \geq 120$ MPa and a relative compaction $RC \geq 103\%$. The elastic modulus E_2 is determined by the plate-loading test and the relative compaction by the standard Proctor test. If these requirements are not fulfilled the subgrade has to be reinforced by one of the available methods, e.g. removal and replacement, cement stabilisation or geosynthetic reinforcement.

In this paper selected methods of reinforcement of a motorway pavement subgrade are compared. The advantages and disadvantages of each method are pointed out. Practical aspects of the real technical solutions used in Poland with regard to environmental conditions are also discussed.

2 SELECTED REINFORCEMENT METHODS

It is well known that different reinforcement aspects need to be considered for roads built on embankments and in cuttings. These two cases are illustrated schematically in Figure 1. It can be observed from Figure 1 that in the case of embankment constructed on a weak subgrade the reinforcement is normally located at its base (Fig. 1a) whereas in the case of cutting the reinforcing layer is located immediately below the pavement (Fig. 1b). In addition, the natural subgrade beneath embankment is normally modified by one of the deep improvement methods such as precompression, grouting, or in situ soil mixing (Fig. 1a).

It should also be pointed out that the soil used as a fill in an embankment is generally remoulded, well compacted and has good drainage characteristics. On the other hand, the soil beneath a pavement of the

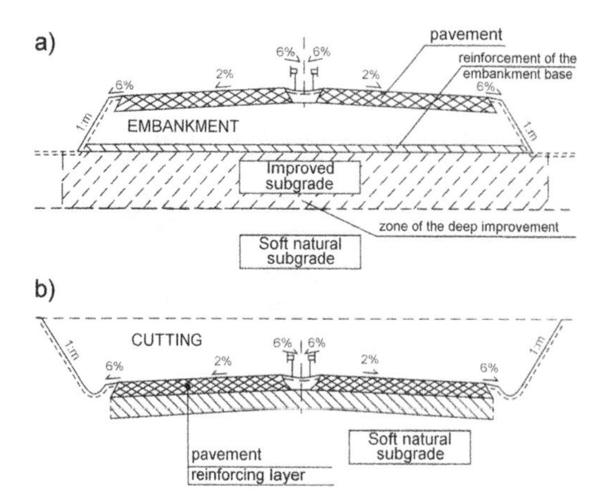

Figure 1. Reinforcement of a weak subgrade: (a) embankment; (b) cutting (modified after Florkiewicz & Grabowski 1999).

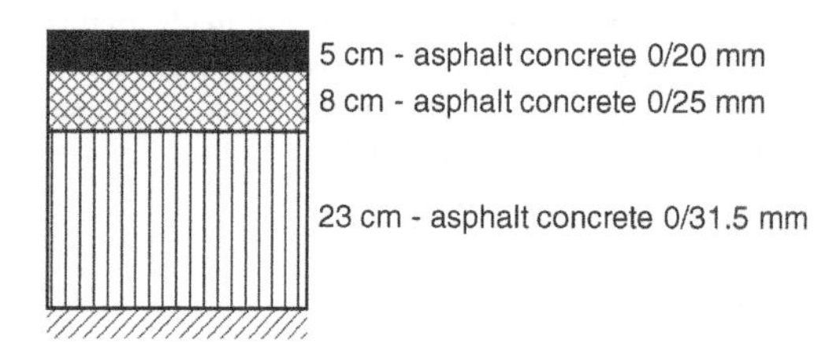

Figure 2. Flexible pavement structure considered in the study.

road in cutting is normally in its natural undisturbed state. Furthermore, the road in cutting often requires a more complex drainage system compared with the road on embankment. These two situations will therefore normally require separate consideration (Brown 1996, Florkiewicz & Grabowski 1999).

In this study, a flexible motorway pavement constructed in a 1 m deep cutting is considered. A standard pavement structure selected from the Polish catalogue (GDDP 1997) is analysed. The pavement consists of three asphalt concrete layers and has a total thickness of 36 cm, as shown schematically in Figure 2. According to the Polish requirement (GDDP 1997), such motorway pavement requires a foundation with an elastic modulus $E_2 \geq 120$ MPa and a relative compaction $RC \geq 103\%$. It is obvious that such parameters cannot normally be obtained for fine-grained soil subgrade so, in order to meet the required parameters, five different reinforcement techniques were used:

a) Removal and replacement of the weak soil,
b) Geotextile-reinforced well-graded gravel,
c) Mechanically-stabilised crushed aggregate,
d) Geogrid-reinforced crushed aggregate,
e) Cement stabilisation.

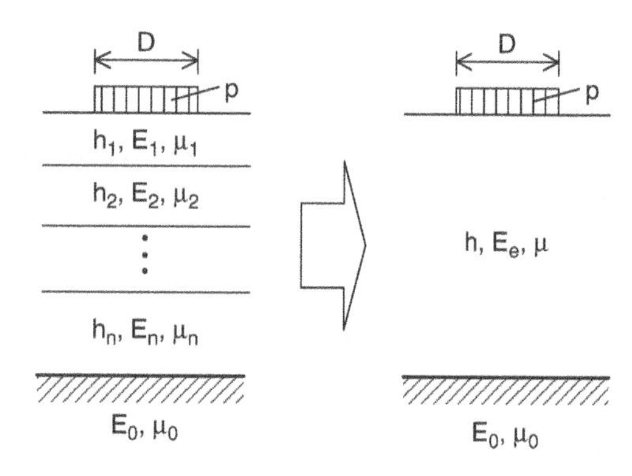

Figure 3. Transformation used for the pavement without geotextile reinforcement.

3 SUBGRADE

A very soft sandy clay (symbol CL, according to the Unified Soil Classification System) in a plastic state was chosen as an example of a weak, frost-susceptible subgrade. The clay has the liquidity index $I_L = 0.50$, the undrained shear strength $c_u = 37$ kPa, the California Bearing Ratio, CBR = 1.6%. The assumed elastic modulus $E_0 = 10 \times CBR = 16$ MPa.

The groundwater table at the depth of 1 m from the bottom of the pavement structure and the depth of frost penetration $h_z = 0.90$ m were assumed. According to the Polish standard (GDDP 1997), this situation is considered to be a bad groundwater condition. As a result, the total thickness of the pavement structure, including the reinforcing layer, must be at least $0.85h_z$, which in this case equals 0.76 m.

4 DESIGN METHODOLOGY

Three different methods were used to design the required thickness of reinforcing layers. In all three methods the same design criterion was used, that is, to obtain $E_2 \geq 120$ MPa at the top of the reinforcement. As mentioned earlier, according to the Polish requirement (GDDP 1997) this condition must be fulfilled for all motorway pavements.

All conventional solutions (i.e. without geotextile reinforcement), such as removal and replacement, mechanically-stabilised aggregate and cement stabilisation, were designed by the use of a method of equivalent modulus based on the elasticity theory (Yoder & Witczak 1975). The principle of this method is to transform a multi-layer elastic system into an equivalent two-layer system to which Burmister's equations (Burmister 1943) can be applied (Fig. 3). In the equivalent model, the pavement material is assumed to be homogeneous, isotropic and elastic, and is characterised by

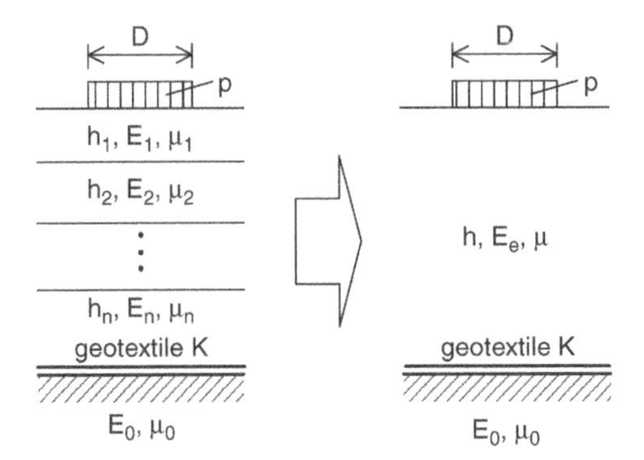

Figure 4. Transformation used for the geotextile reinforced pavement.

Poison's ratio, μ, and the equivalent elastic modulus, E_e, which can be obtained from the following equation:

$$E_e = \frac{\Sigma h_i E_i}{\Sigma h_i} \tag{1}$$

where h_i and E_i are the thickness and the elastic modulus of single layer, respectively.

The subgrade soil is also assumed to be homogeneous, isotropic and elastic and is characterised by Poisson's ratio, μ_0, and the elastic modulus, E_0, as shown in Figure 3.

A geotextile reinforcement of the weak subgrade was designed by the modified method of equivalent modulus described by Wojtowicz (1994) and Bugajski & Grabowski (1999). This method is based on the assumption that the elastic properties of the geotextile-reinforced pavement will not be lower than those of the conventionally-reinforced pavement (i.e. without geotextiles). The principle of this method is similar to that of the previous method, i.e. to transform a multi-layer elastic system into the equivalent two-layer system. The modified method, however, also accounts for the tensile properties of the geotextiles, in terms of the tensile modulus, K, as shown schematically in Figure 4.

A geogrid reinforcement of the subgrade was designed with Tensar geogrids characterised by a high tensile stiffness. A method developed at the University of Hannover in Germany (Golos 2005) was adopted. This method is based on Boussinesq's theory and full-scale experiments carried out by Kennepohl et al. (1985). The method accounts for the interlock created by Tensar geogrids and the compacted aggregate and is presented as a set of design curves for different values of the equivalent elastic modulus (E_e) required at the top of the reinforcement. In this study, the curves for the $E_e = 120$ MPa were used (Fig. 5).

All the materials used in the design of reinforcing layers are summarised in Table 1 (Rolla 1987). A circular tyre imprint with the diameter $D = 0.313$ m and

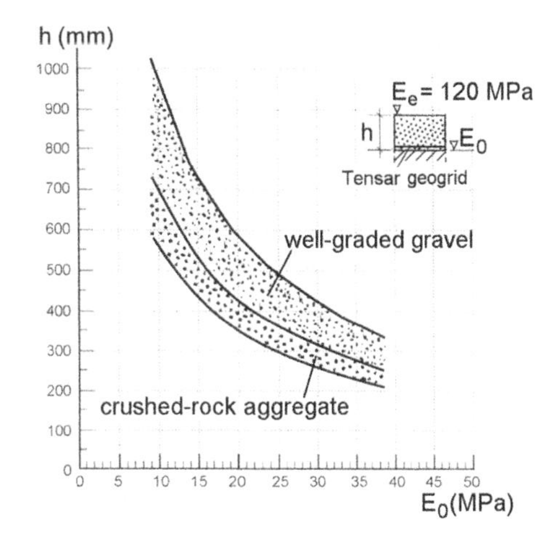

Figure 5. Design curves for reinforcement of aggregate layers using Tensar geogrids (modified after Golos 2005).

Table 1. Summary of the materials used in reinforcing layers.

Material	Elastic modulus E (MPa)	Symbol in Fig. 6
Asphalt concrete 0/20 mm	1500	
Asphalt concrete 0/25 mm	1500	
Asphalt concrete 0/31.5 mm	1500	
Cement-stabilised soil	600	
Crushed-stone aggregate	400	
Well-graded gravel	200	

a contact pressure $p = 0.65$ MPa were assumed in all the calculations. This corresponds to a 50 kN single wheel load (100 kN single axle load).

5 ASSESSMENT

5.1 *Proposed reinforcement solutions*

All the proposed reinforcements of the weak subgrade are presented in Figure 6. In addition to the materials summarised in Table 1, the following geosynthetics were considered in the design:

- A woven geotextile with a minimum tensile strength of 45 kN/m (Fig. 6b),
- Tensar SS30 geogrid with the tensile strength of 30 kN/m (Fig. 6d)
- An unwoven geomat acting as a separation layer at the interphase of the subgrade and the reinforcement (Figs 6a, c, d, e).

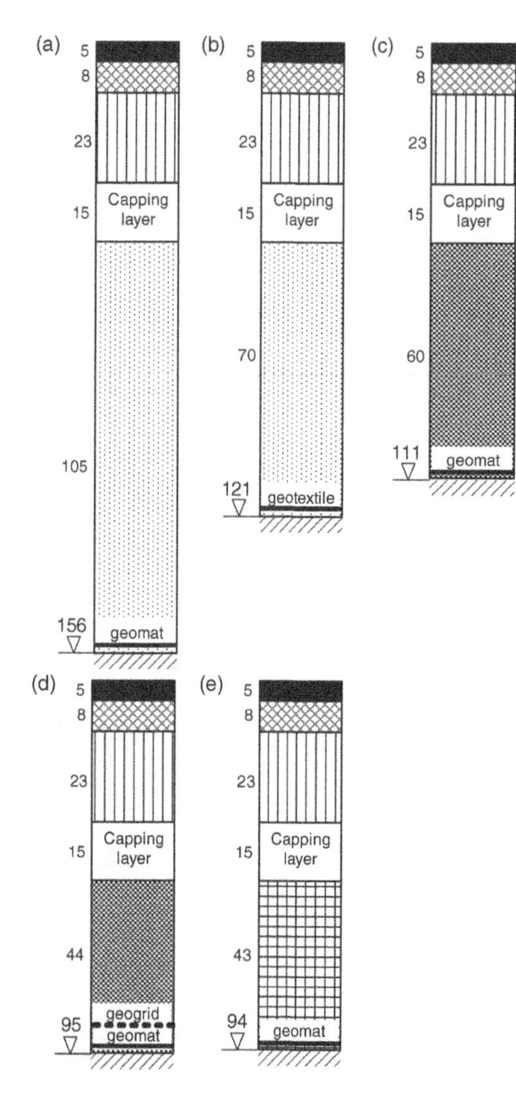

Figure 6. Proposed reinforcement solutions for the subgrade with CBR = 1.6% (all dimensions in cm): (a) removal and replacement; (b) geotexile reinforced well-graded gravel; (c) mechanically-stabilised crushed-stone aggregate; (d) geogrid-reinforced crushed-rock aggregate; (e) cement-stabilised soil.

Figure 6 shows that in addition to the reinforcing layer required by the soft subgrade, a capping layer of 15 cm was provided for all the structures. According to the Polish requirement (GDDP 1997) such a capping layer is required for all the roads that carry large volumes of heavy traffic in order to provide a working platform for heavy machinery during construction of the pavement layers. The Polish standard (GDDP 1997) specifies that the capping layer shall be at least 10 cm thick. The capping layer should be made of crushed aggregate or cement-stabilised soil of CBR ≥ 40%.

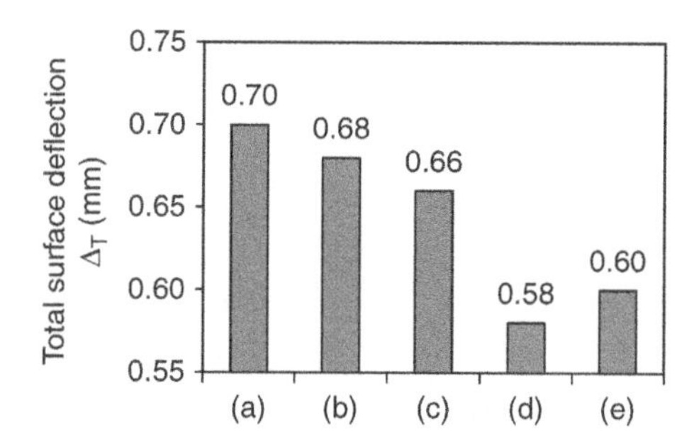

Figure 7. Total surface deflections.

5.2 *Total surface deflection*

It is well known that most of the total surface deflection is caused by the elastic compression of the subgrade layer (Yoder & Witczak 1975). Furthermore, deflections are simply the mathematical integration of the vertical strain with depth. It can, therefore, be assumed that the same factors that affect vertical strains in the subgrade also affect the surface deflection. As a result, the values of total surface deflection were used in this study to assess the effectiveness of the proposed reinforcement solutions.

Deflection values were calculated by use of the solution of the two-layer system proposed by Burmister (1943). In this paper, it is assumed that the surface layer represents all the pavement layers and is characterised by the equivalent modulus, E_e. The underlying layer represents the natural subgrade and is characterised by the elastic modulus, E_0 (Fig. 3).

Total surface deflection, Δ_T, for the two-layer system was calculated by means of equation:

$$\Delta_T = 1.5 \frac{pa}{E_0} F_2 \tag{2}$$

where p = contact pressure; a = radius of tire imprint; E_0 = elastic modulus of the subgrade; F_2 = dimensionless factor depending on the ratios of E_e/E_0 and h/a.

The total surface deflections calculated for each reinforcement method are compared in Figure 7. It should be noted that letters on the horizontal axis correspond to the solutions shown in Figure 6.

It can be observed from Figure 7 that the calculated values of total surface deflection are in the range of 0.58–0.70 with the lowest obtained for the removal and replacement method and the highest for geogrid-reinforced crushed-stone aggregate. Figure 7 also shows that the geotextile reinforcement of well-graded gravel causes a slight reduction in the deflection compared with the well-graded gravel without geotextile. On the other hand, the use of Tensar geogrid together with crushed aggregate reduces the pavement deflection significantly when compared with the layer of crushed aggregate without any geogrid. This is because the interlock mechanism between the granular material and the geotextile can only be created when the geogrid reinforcement is provided. When a typical woven geotextile is used, the mechanism of interlock cannot be created.

It can also be observed from Figure 7 that the deflection obtained for the cement stabilisation is much lower than that for the conventional removal and replacement method even though the thickness of the former layer is much smaller than that of the latter. This suggests that a greater reduction in the deflection will be obtained by increasing the modulus or rigidity of the reinforcing layer rather than by increasing its thickness.

5.3 *Shear stresses in the subgrade*

The preceding section showed that different values of the total surface deflection will be obtained for different reinforcement solutions. It was demonstrated that the pavement deflection could be reduced by incorporation of more rigid reinforcing layers and/or increasing their thickness. As the pavement layers, however, become stiffer and provide increased load spreading capability, shear stresses within the pavement and the subgrade will change. Therefore, it is important to verify whether the shear stresses developed in the subgrade do not exceed the shear strength of the subgrade soil.

Similarly to the total surface defections, the shear stresses in the natural subgrade were calculated by use of Burmister's theory. The maximum shear stress in the subgrade, τ_{max}, was calculated under the edge of the tyre. The value of τ_{max} was determined from nomographs based on the ratios E_e/E_0 and h/D, and the shear strength of the subgrade soil (Rolla 1987).

The maximum shear stresses in the subgrade obtained for each reinforcement are compared in Figure 8. Although all the maximum shear stresses shown in Figure 8 are smaller than the undrained shear strength of the subgrade soil, $c_u = 37$ kPa, differences between the values of τ_{max} obtained for each reinforcement are noticeable. The highest shear stress of 24.0 kPa was obtained for the removal and replacement, whereas the lowest shear stress of 16.6 kPa was obtained for the cement stabilisation.

It can also be observed from Figure 8 that an increase in the reinforcement stiffness causes a reduction in the shear stress in the subgrade. In other words, the effect of stiff reinforcing layers such as crushed aggregate with geogrid or the cement-stabilised soil is pronounced. Similar observations have also been made

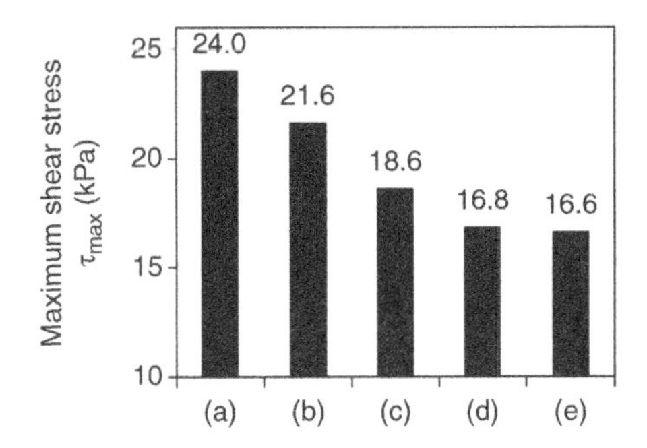

Figure 8. Maximum shear stresses in the subgrade.

for two stiffer subgrade soils with CBR of 2.5% and 3.9%. The differences between shear stresses in the subgrade obtained for different reinforcements were, however, much smaller (Wanatowski & Florkiewicz 2000).

From the values of total surface deflection and shear stresses developed in the subgrade it can be concluded that the design of geogrid reinforcement or cement stabilisation can guarantee the highest bearing capacity of the subgrade. The reinforcement with the use of geotextile or mechanically-stabilised aggregate can provide a good bearing capacity. Finally, the replacement of the soft soil with well-graded gravel provides a satisfactory bearing capacity of the subgrade.

5.4 *Drainage*

As mentioned earlier, the capping layer of 15 cm was provided for all proposed motorway pavement structures. According to the Polish standard (GDDP 1997), a crushed aggregate or cement-stabilised soil with CBR $\geq$ 40% can be used for such layers. In order to improve long-term drainage in the pavement, however, the material used for the capping must also have a sufficient permeability. Therefore, it is proposed that an aggregate with the coefficient of permeability k $\geq$ 0.01 m/s rather than a cement-stabilised soil should be used for the capping layer in all the proposed solutions. In addition, the capping layer should be connected to a deep drainage system in the form of drains installed under the ditches. Such a drainage system will provide an excellent way of removing all troublesome water from the pavement structure. An example of the motorway drainage system used in the case of removal and replacement of soft soil is shown in Figure 9.

It should be noted that although the capping layer provides an excellent drainage of the pavement, none of the proposed reinforcements can fully protect the natural subgrade from surface water that can infiltrate through cracks in the pavement or at the pavement edges. Nevertheless implementation of a deep

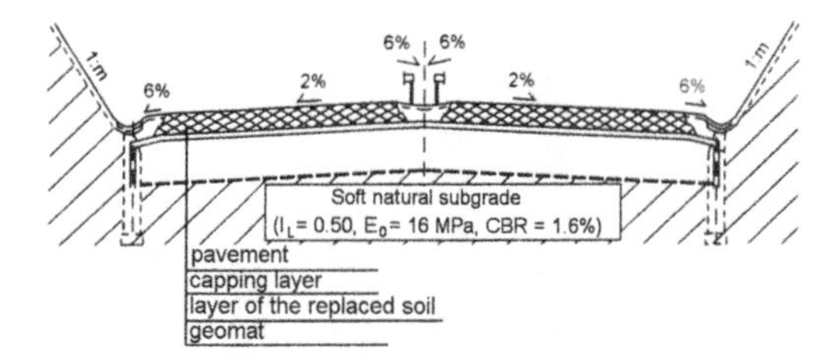

Figure 9. Example of deep drainage system of motorway pavement (modified after Florkiewicz & Grabowski 1999).

drainage system should provide a very efficient protection of the soft subgrade from further plasticising.

5.5 *Frost resistance*

It can be seen from Figure 6 that the total thickness of each pavement structure is greater than the required value of $0.85h_z = 0.76$ m. Therefore, all the proposed structures meets the design criterion with respect to the frost resistance (GDDP 1997). The motorway pavement reinforced with the use of Tensar geogrid and cement stabilisation have the smallest thicknesses of 0.95 and 0.94 m, respectively (Figs 6d, e). These values are very close to the depth of frost penetration $h_z = 0.90$ m. Therefore, it is possible that in the case of a severe winter, the two above-mentioned pavements may still be subjected to damaging frost action. The total thickness of the other three structures is noticeably larger (Figs 6a, b, c), which guarantees a very good frost resistance of the motorway pavement.

5.6 *Construction costs*

Adequate analysis of the selected reinforcement methods must include consideration of their construction costs. In this study, a simplified economic analysis was carried out. A normalised cost index (NCI) defined by Equation 4 was used in the assessment.

$$NCI = \frac{Cost\ of\ 1m^2\ of\ analysed\ reinforcement}{Cost\ of\ 1m^2\ of\ the\ cheapest\ reinforcement} \quad (3)$$

For the sake of simplicity it was assumed that all necessary materials are available within 15 km of the construction site. It should also be noted that maintenance costs were not considered in the analysis.

The normalised costs of the proposed reinforcement are summarised in Figure 10. It can be observed from Figure 10 that highest construction cost was obtained for the conventional removal and replacement method. On the other hand, the lowest cost was calculated for the layer of well-graded gravel reinforced with geotextile. Figure 10 also shows that the layer of Tensar geogrid-reinforced aggregate is more cost-effective compared with the layer of mechanically-stabilised

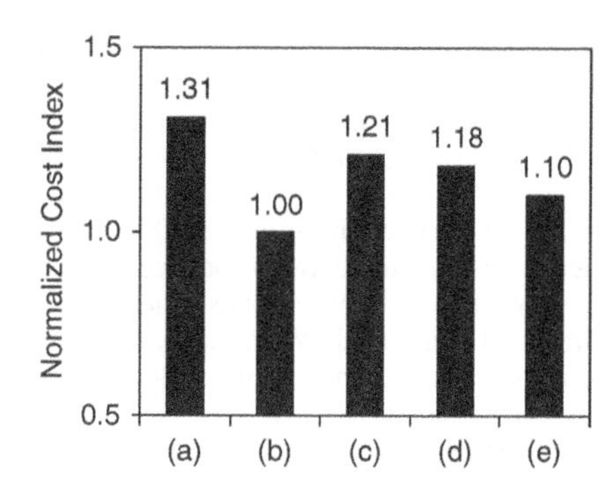

Figure 10. Normalised costs of the proposed reinforcements.

aggregate. This is because reinforcement of the aggregate layer with Tensar geogrid allows the overall construction depth to be reduced, saving on materials and excavation.

Similar conclusions have also been drawn for the subgrade with CBR of 2.5% and 3.9%. The differences between construction costs of different reinforcements were, however, smaller (Wanatowski 1999).

6 CONCLUSIONS

Rapid development of modern technologies and application of new materials in highway engineering make it possible to solve even very complicated reinforcement problems by a variety of methods. It can be noticed that geotextiles play a significant part in most of the modern reinforcement techniques. This is because they have proved to be among the most versatile and cost-effective ground modification materials.

Selected reinforcement methods of the subgrade presented in this paper show that the mode of operation of a geotextile in such applications is defined by three functions: separation, reinforcement and drainage. Depending on the application, the geotextile can perform two or more of these functions simultaneously, e.g. a woven geotextile used for the reinforcement of the layer of well-graded gravel (Fig. 6b) performs both reinforcement and separation functions.

The simplified assessment of the selected reinforcement techniques has shown that all solutions possess both advantages and disadvantages. For instance, the conventional removal and replacement method can provide a very good frost protection of the subgrade but it may be relatively expensive. On the other hand, the layer of well-graded gravel reinforced with geotextile can be very cost-effective but it does not provide the most efficient reinforcement for the natural subgrade.

In the authors' opinion, successful reinforcement of a weak subgrade of motorway pavement requires a complex geotechnical analysis which should be included in motorway design reports. Such analysis should consider the combined effect of technical and economical aspects of the proposed solutions and the actual environmental conditions on the durability and reliability of motorway pavement.

REFERENCES

Brown, S.F. 1996. Soil mechanics in pavement engineering. *Géotechnique* 64(3): 383–426.

Bugajski, M. & Grabowski, W. 1999. *Geosynthetics in highway engineering*. Poznan University of Technology, Poznan (in Polish).

Burmister, D.M. 1943. The theory of stresses and displacements in layered systems and applications to the design of airport runways. *Proceedings of Highway Research Board*, 23: 126–148.

Florkiewicz, A. & Grabowski, W. 1999. Stabilisation and drainage of the motorway pavement subgrade. *Drogownictwo* 54(12): 381–385 (in Polish).

GDDP, 1997. *Katalog typowych konstrukcji nawierzchni podatnych i polsztywnych*. Instytut Badawczy Dróg i Mostów, Warszawa, Poland (in Polish).

Golos, M. 2005. Methods of design of bases made of aggregate reinforced with geogrid on weak soil. *Drogownictwo* 60(7–8): 222–228 (in Polish).

Kennepohl, G., Kamel, N., Walls, J. & Haas, R. 1985. Geogrid reinforcement of flexible pavements: design basis and field trials. *Proceedings of the Association of Asphalt Paving Technologists* 54: 45–70.

Rolla, S. 1987. *Projektowanie nawierzchni*. WKiL, Warsaw (in Polish).

Wanatowski, D. 1999. *Technical and economical analysis of the selected methods of reinforcement and drainage of motorway subgrade.* M.Sc. Thesis, Poznan University of Technology, Poland (in Polish).

Wanatowski, D. & Florkiewicz, A. 2000. Influence of different improvements of the subgrade on its bearing capacity. *Drogownictwo* 55(11): 323–327 (in Polish).

Wojtowicz, J. 1994. Granular road bases reinforced by the non-woven geotextile. Part 1. Theoretical solution. *Drogownictwo* 49(10): 332–337 (in Polish).

Yoder, E.J. & Witczak, M.W. 1975. *Principles of pavement design*. Wiley & Sons, New York.

Advances in Transportation Geotechnics – Ellis, Yu, McDowell, Dawson & Thom (eds)
 ISBN 978-0-415-47590-7

Digital image based numerical approach for micromechanics of asphalt concrete

Z.Q. Yue
Department of Civil Engineering, The University of Hong Kong, Hong Kong, P.R. China

ABSTRACT: This paper presents a summary of the effort to develop a digital image based numerical approach for micromechanics of asphalt concrete. This approach incorporates the actual microstructure and heterogeneity of asphalt concrete into the conventional finite element or different methods for numerical mechanical prediction. It is a workable approach for the micromechanics of asphalt concrete and other geomaterials.

1 INTRODUCTION

Since the late 19th century, asphalt concrete (AC) has been widely used as the surface layer materials for the construction of roads, highways, runways and parking areas (Roberts et al. 2002). AC mainly consists of asphalt, fine and coarse aggregates. Sometimes, a small percentage of additives is added to the mix for special engineering properties. Different mix design formulae produce different asphalt concretes after hot compaction. The AC mix design is to determine a suitable asphalt content for selected aggregates so that a desired or optimal internal structure can be produced with roller or other compaction tools in the field. The key criterion for an optimal mix design is to allow the AC pavement have traffic serviceability over long time.

As shown in Figure 1, AC is a heterogeneous material and consists of asphalt cement, voids, fine particles, sand, and coarse aggregates. These individual materials and components have different physical and mechanical properties and behaviour. The mechanical behavior of different AC mixtures under traffic and climate loading can be influenced by many factors including quality of asphalt cement, aggregate gradation and shape, and quality of compaction. Most importantly, the AC microstructure (i.e., the internal distribution of aggregates in AC solids) can play a significant role in AC mechanical properties and in their resistance to major distress of AC pavements. These distresses include rutting, fatigue, thermal cracking, and low temperature cracking.

Since 1992, we have developed a workable approach to realize the quantitative characterization of AC microstructure and to predict AC mechanical behavior by taking into account their actual internal structure (Sepehr et al. 1994; Yue et al. 1995a, b, 1996, 1997,). We have found that digital image and processing techniques can be used to achieve the goals for quantitative examining and better predicting AC mixtures in pavements (Chen et al. 2004, 2005, 2007a; Yue et al. 2002a, b, 2003a, b, 2004) and Yue (2004, 2006 & 2007).

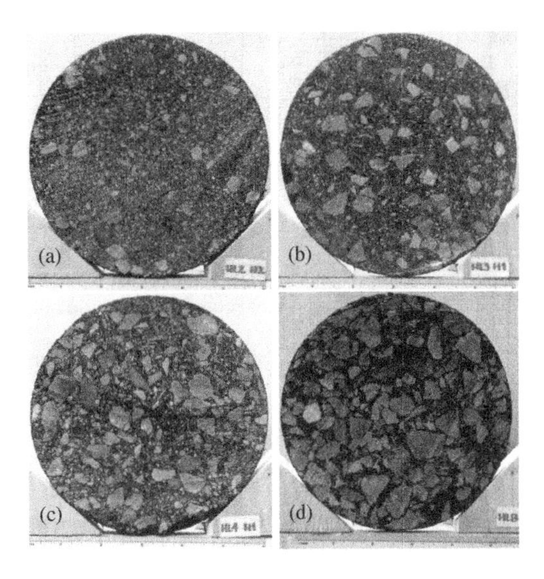

Figure 1. Digital images of four 150 mm diameter circular cross-sections of four asphalt concrete samples (a), (b), (c) and (d) for the mix design HL2, HL3, HL4 and HL8, respectively.

In this paper, we will present a brief summary of our efforts on the development of a workable approach to incorporate the AC microstructure into the mechanical analysis and numerical simulation. The approach is based on digital image and its process techniques. The digital image results are then further incorporated into conventional numerical methods such as finite element or difference methods for the numerical examination. Results of our studies have shown that this approach can be used to predict the micro-mechanical responses of AC materials. The actual spatial distribution of aggregates can be considered exactly.

2 DIGITAL IMAGE SURVEYING AND REPRESENTATION

2.1 *Digital image for surveying and representing*

As shown in Figure 1, AC internal microstructure can be classified at least two regions. One region is the area occupied by aggregates. The other is the area occupied by the matrix. The matrix is composed of the mixture of asphalt cement and fines and voids. Besides, the major differences among the four AC samples of different mix design formulae are the differences in the sizes and distribution of aggregates.

More importantly, each of the digital images in Figure 1 is not only an accurate topographical survey but also a digital representation of the actual spatial distribution of AC internal materials on that circular cross-section. There was an obstacle for surveying and representing the spatial distribution of individual materials such as aggregates consisting the AC microstructure because of lack of effective data acquiring and representation tools in the past. Consequently, the different individual components could not be taken into account in the prediction and analysis of AC mechanical response and failure process under loading.

Since 1995, the popularization of window-based personal computers and the developments of digital image processing techniques have provided a completely new way for accurate measurement and digital representation of AC internal structure.

Digital image techniques convert an actual AC section into its corresponding image via photographing and other tools and store the image in digital form in computer. When an AC section is converted into its digital image, the information at the section such as cracks, materials and texture and their positions are automatically and accurately represented as pixels assigned with grays or colors and fully distributed over the entire digitalized area. As shown in Figure 1, a digital image itself has become an efficient method for accurate measuring and digital representing of the internal structure and spatial distribution of an AC smooth section.

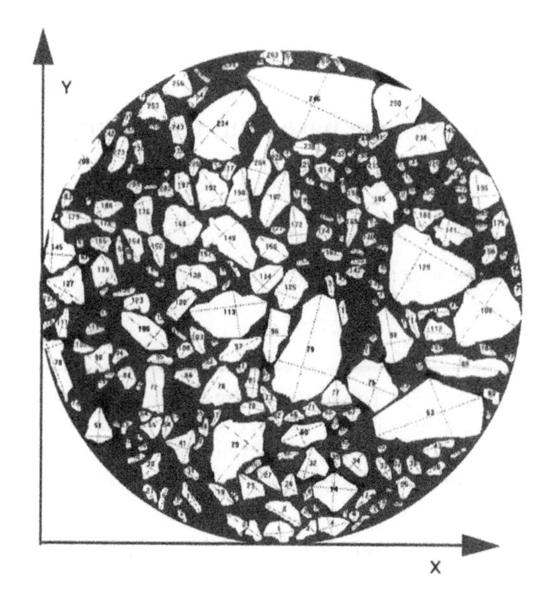

Figure 2. Digital representation of aggregate distribution on horizontal cross-section of a cylindrical asphalt concrete sample of 150 mm in diameter (Yue & Morin, 1996).

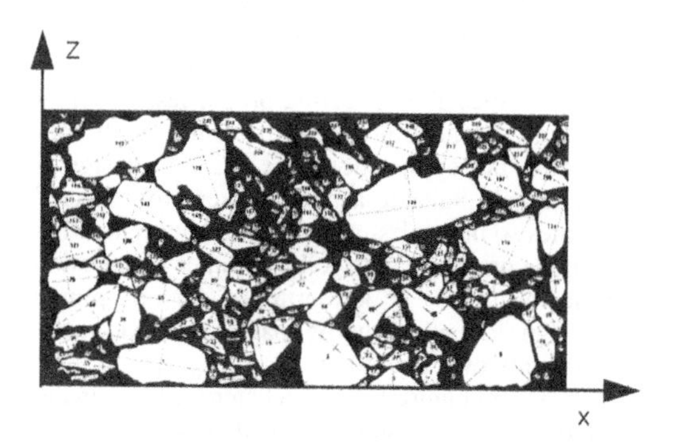

Figure 3. Digital representation of aggregate distribution on vertical cross-section of a cylindrical asphalt concrete sample of 150 mm in diameter (Yue & Morin, 1996).

2.2 *Digital image processing*

Digital image processing software for editing and statistics have been developed to assess, evaluate and identify the actual area occupied by each aggregate. Figures 2 and 3 show two processed digital images of the horizontal (XY plane) and vertical (XZ plane) smooth cross-sections for a cylindrical AC sample, respectively. In the processed digital image, each aggregate has an identifying number and its major and minor axes are also drawn with two perpendicular straight lines. The morphological parameters describing the geometry of an aggregate cross-section are also measured. These parameters include the perimeter, the area, the centroid, the major and minor axes, the Feret diameter and the shape factor. Details of the digital image processing can be found in Yue et al. (1995a) and Yue & Morin (1995b, 1996, 1997).

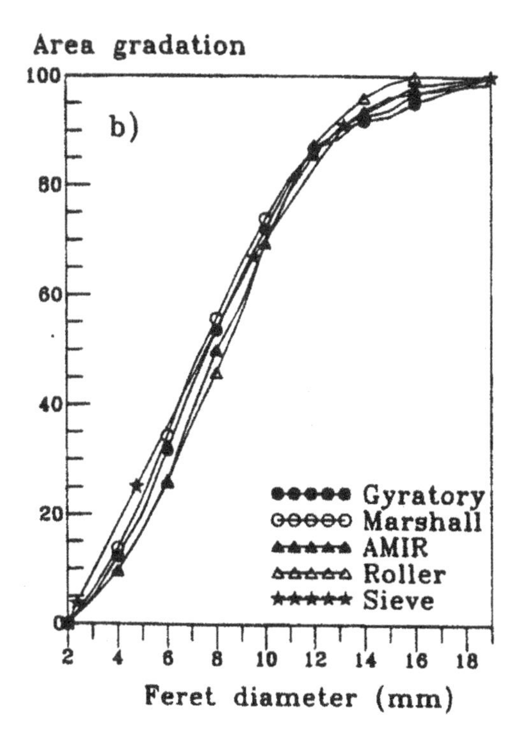

Figure 4. DIP based area-gradation of HL4 asphalt concrete samples prepared with gyratory, Marshall, AMIR or steel roller compactors (Yue et al., 1995).

2.3 *Image gradation analysis*

With the processed digital image data, image gradation analysis can be performed. The digital area or number gradations can be calculated with respect to the minor or major axis length, the Feret diameter, the shape factor and the orientation.

Figure 4 gives a result of the area gradation by the Feret diameter. It is defined as the summation of the areas of the aggregates whose Feret diameters are between 2 mm and a dimension d ($d\_ > 2$ mm) over the total aggregate area. It is expressed in percentage as a function of the dimension d (i.e., the Feret diameter, see Yue et al. 1995). Besides, the conventional sieve analysis result is also plotted. It is evident that the two-dimensional (2D) image area gradations follow the same patterns and are very similar to the three-dimensional (3D) sieve result. Similar results were also found in Kuo & Freeman (1998), Masad et al. (1999) and Obaidat et al. (1998).

2.4 *Statistical analysis of aggregate distribution*

Statistical analysis of the 3D spatial distribution of aggregates in cylindrical AC samples was performed from calculating and comparing the 2D spatial distribution of aggregates on horizontal and vertical AC cross-sections (Yue et al. 1995a, Yue & Morin 1995b, 1996, 1997).

The aggregates on the horizontal cross-sections have a substantial difference with those on the vertical cross-sections. The areas and shape factors of aggregates on the horizontal cross-sections are greater than those on the vertical cross-sections. This result indicated that the aggregates had their major cross-sections placed likely on horizontal plane in AC samples. It is consistent with the compaction formation process of AC samples in laboratory and at the field. The compactors such as steel roller or gyratory compactor forced the randomly oriented aggregates in the original hot mixtures to be re-oriented horizontally and permanently during the compaction and cooling.

3 MESH GENERATION AND RESULTS

3.1 *General*

In recent years, we have further developed digital image based numerical methods for simulating and predicting the mechanical behavior of AC samples by taking into account their actual internal microstructures. For this purpose, we have developed three mesh generation methods. Two are for the 2D analysis and the other for 3D analysis of AC micromechanics.

3.2 *Image pixels for 2D square mesh*

A digital image is comprised of rectangular array of image pixels. Each pixel is the intersection area of any horizontal scanning line with the vertical scanning line. All these lines have an equal width h. For gray level image, at each pixel, the image brightness is sensed and assigned with an integer value that is named as the gray level. For the mostly used 256 gray images and binary images, their gray levels have the integer interval from 0 to 255 and from 0 to 1 respectively. For color image, at each pixel, there are three integer values to represent the red, green and blue color respectively. As a result, the digital image can be expressed as a 2D discrete function $f(i,j)$ or 3D discrete function $f(i,j,gray)$ in the Cartesian coordinate system (i,j). An example is shown in Figure 5, where $h = 0.1960$ mm, $i = 1, 2, \ldots, 10$ and $j = 1, 2, \ldots, 12$.

Since it occupies a square area, a pixel can be treated as a square element. The four corners of a square pixel can be used as the four nodes of a corresponding square element. Accordingly, the rectangular array of image pixels can be automatically transferred to form a mesh with the same size and number of square elements. The material properties can be assigned to each element by its revised individual grey level or colour. A typical example is shown in Figure 6 for the rectangular array of image pixels in Figure 5.

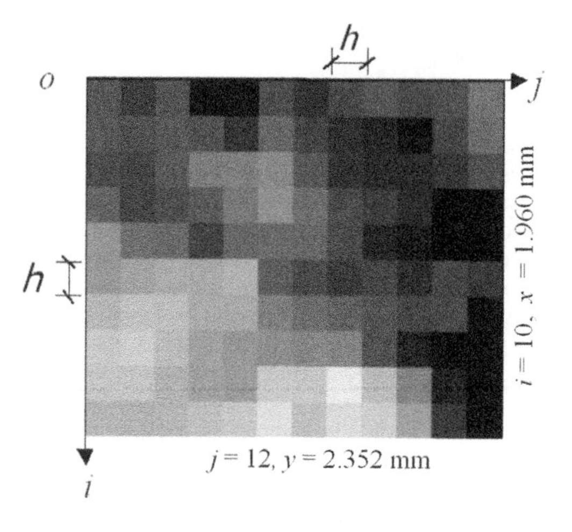

Figure 5. A rectangular array of 12×10 square pixels representing a digital image of a rectangular AC cross-section of 2.352 mm by 1.960 mm (Yue et al. 2003a, b, Chen et al. 2004).

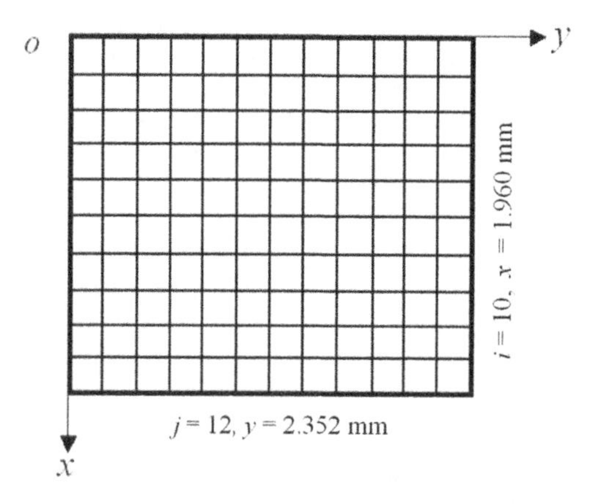

Figure 6. A total of 12 × 10 square grids representing a digital image of a rectangular AC cross-section of 2.352 mm by 1.960 mm (Yue et al. 2003b, Chen et al. 2004).

3.3 *Aggregate isolation for 2D Auto-mesh*

In this method, a polygon is used to represent an aggregate section. The polygon boundary is determined and isolated from their square pixel boundary on the digital image. After all the aggregates have been isolated and represented with polygons, auto-mesh generation algorithms are used to automatically produce meshes with triangular, quadrangular or mixed elements. An example is shown in Figure 7.

3.4 *Image pixel extension for 3D cubic mesh*

The square pixel array or element mesh in Figures 5 and 6 can be further used to form 3D cubic mesh for 3D analysis. The digital image for an AC section can be assumed to represent a thin layer of AC material. Hence, the identified geomaterial on each pixel can be vertically extended to form a cubic or rectangular solid. Consequently, a group of cubic solids with the width and height can form an AC layer. An example is shown in Figure 8.

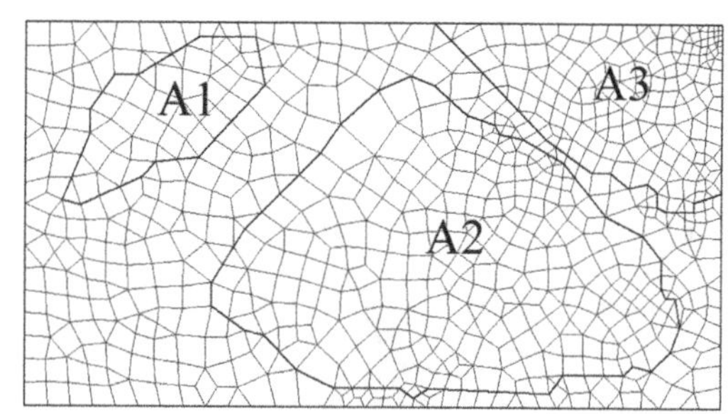

Figure 7. Automatic mesh with triangular and quadrangular elements for isolated aggregates (A1, A2 and A3) and their surrounding matrix of an AC section (Yue et al., 2003a).

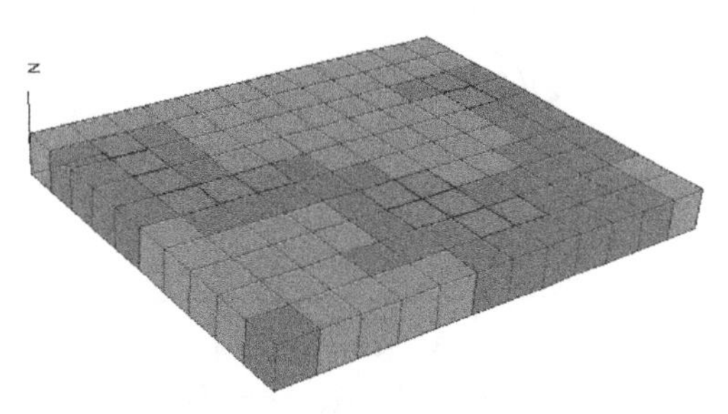

Figure 8. One layer of cubic mesh automatically generated from a vertical extension of a rectangular array of image pixels for an AC section (Chen et al. 2007a).

The cylindrical AC sample can be cut, milled and smoothened one seam by one seam. Digital image can be scanned or photographed for each smooth and fresh surface. So, a series of digital image can be obtained. Each digital image can be extended vertically to the milled thickness. A digital layer can be formed and constitute a group of cubic solids. Each layer can be further connected with its bottom or up layers. Consequently, a 3D mesh can be formed for the cylindrical AC sample. The 3D mesh is formed with cubic elements in a series of layers. An example is shown in Figure 9.

4 NUMERICAL ANALYSIS AND RESULTS

4.1 *Basis assumptions*

The conventional continuum mechanics adopt the phenomenological approach in establishing constitutive equations for geomaterials. Due to limitations in data

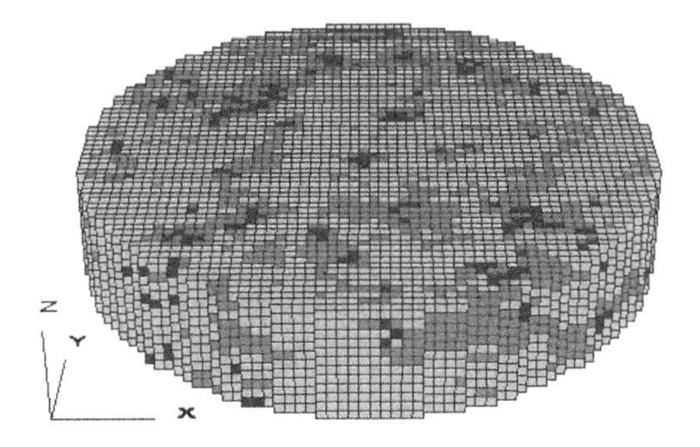

Figure 9. An actual 3D mesh of cubic elements in 12 layers for a HK granitic rock (Chen et al. 2007a).

acquiring and digital representation, the internal structure of AC solids is homogenized. The constitutive equations cannot completely take into account the microstructure of actual AC materials.

It should be noted that in general, the conventional continuum mechanics are valid and applicable to the micromechanics of asphalt concrete if AC microstructure can be considered. Therefore, the digital image techniques offer us a tool for data acquiring and digital representation of the AC microstructure. The developed 2D or 3D meshes can be further incorporated into existing numerical analysis software package. Numerical methods such as finite element or difference methods can be adopted to perform the numerical analysis and simulation and prediction of the mechanical behavior of AC materials under loading.

It is further assumed that AC materials consist of two or more different individual components according to certain natural laws. Each individual component such as air, water, matrix and aggregates is homogeneous and has relatively stable and simple constitutive relationship between stresses and strains. The contact between any two individual components can be either continuous or discontinuous.

4.2 *Numerical analysis and results*

Using the above assumptions and results, numerical analysis can be carried out. An example is shown in Figure 10. In Figure 10, a digital image AC section is transferred to a finite element mesh with triangular elements. The aggregates on the AC section are also identified as polygons with thick interface lines. The aggregates and the matrix are treated as two types of solids with different mechanical properties and parameter values.

As shown in Figure 11, finite element analysis was carried out to examine the elastic response of the cylindrical AC solid under the Brazilian split tension test with the consideration of the AC microstructure and the material differences between aggregates and matrix. Figures 12 and 13 shows the contour distributions of the minimum and maximum principal stresses on the non-homogeneous circular AC section. It is evident that the stress fields are strongly affected by the presence of the aggregates.

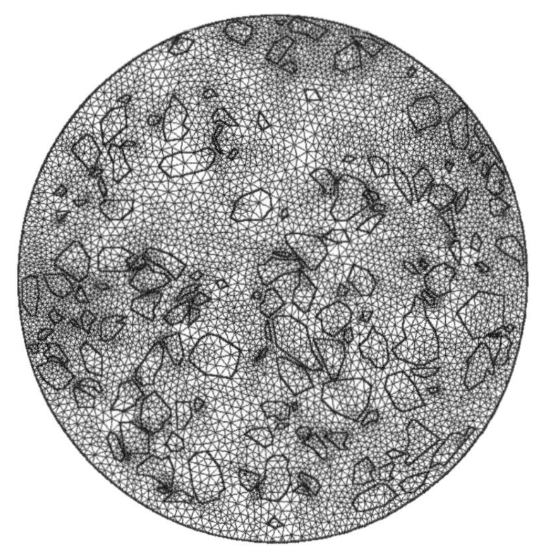

Figure 10. Automatic mesh with triangular elements for identified aggregate zones and their surrounding matrix of an AC section (Yue et al. 2003a).

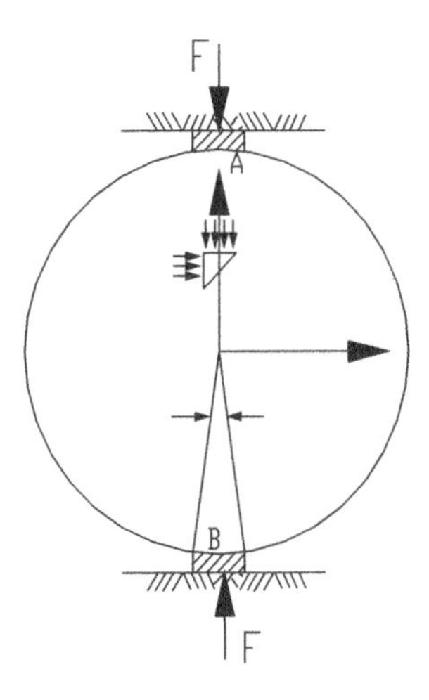

Figure 11. Numerical modelling of a cylindrical AC solid under the Brazilian split tension test (Yue et al. 2003a).

4.3 *Numerical findings*

Our numerical analysis and results have shown that the microstructure and material heterogeneity have a limited influence on the initial elastic response, but has

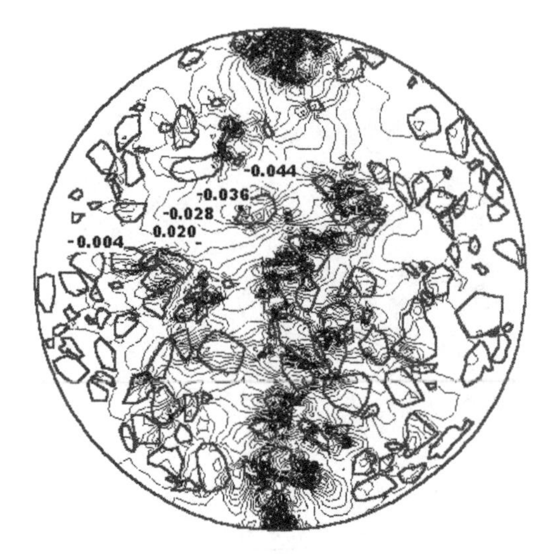

Figure 12. The contour distributions of the minimum principal stress on a non-homogeneous model for the asphalt concrete cross-section from the DIP-FEM analysis (Yue et al. 2003a).

Figure 13. The contour distributions of the maximum principal stress on a non-homogeneous model for the asphalt concrete cross-section from the DIP-FEM analysis (Yue et al. 2003a).

a great effect on the mechanical response before and during the failure stage.

In other words, the conventional phenomenological based approach in geomechanics can give a good prediction and analysis on the deformation and stress of AC materials under working loading conditions or at the elastic stage. The elastic or deformation modulus of an individual geomaterial group can be determined correctly. However, it has limitations to predict and analyze the mechanical response of geomaterials near the failure stage.

5 FURTHER INVESTIGATIONS

In recent years, further investigations have been conducted to use the digital image based approach for the examination of actual geomaterials where their microstructure and internal heterogeneity are taken into account (Yue 2007, Xu et al. 2007).

5.1 *Seepage simulation in heterogeneous soils*

Yue et al. (2003b) presented a digital image based finite element simulation of seepage in weathered soils by taking into account the actual soil internal structure and heterogeneity. Digital image techniques were used to capture the internal material distribution for the mesh formation. The numerical results showed that the soil inhomogeneity had significant effects on seepage flow.

5.2 *Failure prediction in heterogeneous rocks*

Chen et al. (2004, 2005, 2007a) and Yue et al. (2002b) presented digital image based 2D and 3D numerical prediction of failure process in heterogeneous rocks, where actual granite sections were used.

5.3 *Compaction of clayey fills*

Hu et al. (2005) developed a digital image based method to quantitatively examine the effect of dynamic compaction of clay fills. Microstructures of dry clay fill samples were obtained with a scanning electronic microscope (SEM). The clay SEM images were converted into ternary images representing particles, pores and contact zones.

The results revealed that the changes of the soil microstructure have a good correlation with the observed ground settlements during the dynamic compaction. The settlements were due to the size and shape changes and re-orientation of the grain particles and pores of the clay fills.

5.4 *Moving boundary of fluid in sandstone*

Zhou et al. (2003) presented a video microscope based examination of the shape of moving front boundary of fluid flow in sandstone. The front of the moving boundary of fluid flow in sandstone is a comprehensive result caused by the average tendency and the fluctuation tendency of fluid flow.

5.5 *Microstructure of cement concrete*

Yue (2007) noted a digital image based result that the aggregates on the horizontal and vertical cross-sections of a cubic cement concrete had a noticeable difference. The major sections of the aggregates were normally lain vertically in the concrete specimens which were formed with a mechanical vibration bar. The physical and mechanical properties of the concrete specimens were different along the horizontal and vertical directions.

Chen et al. (2007b) presented a digital image based numerical investigation on the micromechanics of cement concrete. Actual cement concrete section was used to examine the failure mechanism and to develop a global constitutive model.

6 CONCLUSIONS AND RECOMMENDATIONS

Over the more than 100 years, many researchers have developed many methods for mechanics of AC material. Due to limitations in technologies, however, it was difficult to include the actual internal structure of AC materials into the numerical and mechanical analysis.

At present, new technologies have provided us the tools for more realistically examine and predict the behavior of AC materials under loading.

The work and findings presented in the above have demonstrated that we can now measure the heterogeneous spatial distribution of AC materials and we can now take them into consideration in numerical analysis and simulation. A workable approach has been developed for the micromechanics of AC mix and other solid geomaterials.

In summary, the workable approach has the following three main tasks or steps.

- Step 1: Digital image representation of heterogeneous AC materials or other geomaterials.
- Step 2: Automatic mesh generation from the digital image representation
- Step 3: Numerical calculations using existing mechanical software packages to incorporate the material microstructure and heterogeneity.

It is noted that such research and application have just been started. It is believed that more and more results will be reported in the near future.

ACKNOWLEDGEMENTS

The author thanks National Research Council of Canada and the Research Grants Council of Hong Kong Special Administrative Region for financial supports. The author also thanks Mr. W. Bekking, Miss I. Morin, and Dr. S. Chen for their assistance in laboratory testing and numerical simulation.

REFERENCES

Chen, S. Yue, Z.Q. & Tham, L.G. 2004. Digital image-based numerical modeling method for prediction of inhomogeneous rock failure, *Int. J. Rock Mech. Min. Sci.*, 41, 939–957.

Chen, S., Yue, Z.Q. & Tham, L.G. 2005. Numerical simulations of fracture propagation in heterogeneous geomaterials based on digital image modelling method, *Compilation of Abstracts for the Third M.I.T. Conference on Computational Fluid and Solid Mechanics*, June 14–17, 2005, Cambridge, MA 02139, USA, pp.62.

Chen, S., Yue, Z.Q. & Tham, L.G. 2007a. Digital image based approach for three-dimensional mechanical analysis of heterogeneous rocks, *Rock Mechanics and Rock Engineering*, 40(2), 145–168.

Chen, S., Yue, Z.Q. & Kwan, A.K.H. 2007b. Actual microstructure-based numerical method for mesomechanics of concrete, *Magazine of Concrete Research*, submitted.

Kuo, C.Y. & Freeman, R.B. 1998. Image analysis evaluation of aggregates for asphalt concrete mixtures,*Transportation Research Record*, 1998, No. 1615: 65–71, National Research Council, Washington D.C., USA.

Hu, R.L., Yue, Z.Q. & Tham, L.G.. 2005. Digital image analysis of dynamic compaction effects on clayey fills. *J. of Geotech. Geoenviron. Eng, ASCE*, 131(11), 1411–1422, 2005.

Masad, E., Muhunthan, B., Shashidhar, N. & Harman, T. 1999. Internal structure characterization of asphalt concrete using image analysis. *J. Computing in Civil Eng.*, ASCE, 13(2): 88–95.

Obaidat, M.T., Al-Masaeid, H.R., Gharaybeh, F. & Khedaywi, T.S. 1998. An innovative digital image analysis approach to quantify the percentage of voids in mineral aggregates of bituminous mixtures. *Canadian J. of Civil Eng.*, 25: 1041–1049.

Roberts, F.L., Mohammad, L.N. & Wang, L.B. 2002. History of hot mix asphalt mixture design in the United States, *J. of Materials in Civil Engineering*, 14 (4): 279–293.

Sepehr, K., Svec, O.J. Yue, Z.Q. & El Hussein, H.M. 1994. Finite element modeling of asphalt concrete microstructure, *Proceedings, LOCALIZED DAMAGE 94, Computer Aided Assessment and Control*, 21–23 June, 1994, Udine, Italy. pp. 225–232.

Xu, W.J., Yue, Z.Q. & Hu, R.L. 2007, Current status of digital image based quantitative analysis of internal structures of soils, rocks and concretes and associated numerical simulation, *Journal of Engineering Geology*, 15(3): 289–313. (in Chinese).

Yue, Z.Q., Bekking, W. & Morin, I. 1995a. Application of digital image processing to quantitative study of asphalt concrete microstructure, *Transportation Research Record*, No. 1492: 53–60, USA.

Yue, Z.Q., & Morin, I. 1995a. An investigation of aggregate orientations in asphalt concrete using digital image processing, *Proceedings, 1995 Annual Conference of Canadian Society of Civil Engineering*, Ottawa, Canada, June 1–3, 1995. Vol.2, pp.59–69.

Yue, Z.Q., & Morin, I. 1996. Digital image processing for aggregate orientation in asphalt concrete mixtures, *Canadian J. of Civ. Eng.*, 23, 479–89.

Yue, Z.Q. & Morin, I. 1997. Digital image processing for aggregate orientations in asphalt concrete:reply to discussion by Ashraf M. Ghaly, *Canadian J. of Civil Eng.*, 24:334–334.

Yue, Z.Q., Chen, S. & Tham, L.G. 2002a. Digital image based finite element method and its applications in geotechnical engineering, *Proceedings of the 2002 Annual Conference of the Society of Mechanics of Guangdong Province*, September 28, Guangzhou, China. pp.31.

Yue, Z.Q., Chen, S. & Tham, L.G. 2002b. Digital image processing based finite element method for rock mechanics, *Proceedings of the 2nd International Conference on New Developments in Rock Mechanics and Rock Engineering, Supplement*, October 10–12, 2002. Shenyang, China, editors: Yunmei Li, Chun'an Tang, Xiating Feng, Shuhong Wang, Rinton Press, Inc. Princeton, New Jersey, USA. pp.609–615.

Yue, Z.Q., Chen, S. & Tham, L.G. 2003a. Finite element modeling of geomaterials using digital image processing, *Computers and Geotechnics*, 30, 375–97.

Yue, Z.Q., Chen, S. & Tham, L.G. 2003b. Seepage analysis in inhomogeneous geomaterials using digital image processing based finite element method, *Proceedings of the 12th Panamerican Conference for Soil Mechanics and Geotechnical Engineering and the 39th US Rock Mechanics Symposium, Soil and Rock America*, MIT, Boston, USA, June, 2003, pp.1297–1302.

Yue, Z.Q. 2004. Digital image processing based numerical methods for mechanical analysis of heterogeneous materials, *Proceedings of the 8th (2003–2004) Annual Meeting of Hong Kong Society of Theoretical and Applied Mechanics*, Hong Kong, China, March 6, 2004. pp.7.

Yue, Z.Q. 2006. Recent development of digital image processing based numerical methods for micromechanics of inhomogeneous geomaterials, *Proceedings of Abstract of the 15th U.S. Congress on Theoretical and Applied Mechanics*, Boulder, Colorado, U.S.A. between June 24 and 30, 2006, Session S13b-1.

Yue, Z.Q., 2007. Digital representation of meso-geomaterial spatial distribution and associated numerical analysis of geomechanics: methods, applications and developments, *Frontiers of Architecture and Civil Engineering in China, Selected Publications from Chinese Universities*, 1(1): 80–93.

Zhou, H.W., Yue, Z.Q., Tham, L.G. & Xie, H. 2003 Video microscopic investigation of the moving boundary of fluid flow in sandstone, *Transport in Porous Media*, 50(2003): 343–370.

8 Rail

Advances in Transportation Geotechnics – Ellis, Yu, McDowell, Dawson & Thom (eds)
© 2008 Taylor & Francis Group, London, ISBN 978-0-415-47590-7

Assessment of layer thickness and uniformity in railway embankments with ground penetrating radar

F.M. Fernandes
Department of Civil Engineering, University of Minho, Guimarães, Portugal

M. Pereira
Geotechnique Department, Laboratório Nacional de Engenharia Civil, Lisboa, Portugal

A. Gomes Correia & P.B. Lourenço
Department of Civil Engineering, University of Minho, Guimarães, Portugal

L. Caldeira
Geotechnique Department, Laboratório Nacional de Engenharia Civil, Lisboa, Portugal

ABSTRACT: In the aim of a national research project entitled, the knowledge concerning the methodology for the construction and quality control of the railway embankments and rail track layers for high speed trains was developed. One of the objectives is to establish a methodology for quality control of construction layers by different available test methods. Non-destructive testing (NDT) methods are currently. Within the NDT available, ground penetrating radar (GPR), a very fast and reliable technique, is very attractive due to their ability to provide information about layer thickness and state condition without causing damage or requiring the removal of material samples. For that, a trial embankment was constructed with different materials, layer's thicknesses, water contents and compaction energy levels. GPR was used in order to detect the thickness of the sub-ballast layer located over the compacted sand layer and its uniformity along the track, but also along the cross-section of the track. In order to control some parameters of the sub-ballast layer, like thickness and uniformity, several metallic plates had been used in the base of the sub-ballast layer, along an alignment. It shows clearly the ability of GPR to detect the sub-ballast layer and its thickness variations along the profile.

1 INTRODUCTION

In the aim of a national research project entitled "Interaction soil-rail track for high speed trains", it was established a protocol between the National Railway Network (REFER) and four national research institutions to develop the knowledge concerning methodologies for the construction and quality control of railway embankments and rail track layers for high speed trains. One of the objectives is to establish a methodology for quality control of construction layers by different existing test methods.

For this purpose, non-destructive testing (NDT) methods are currently very attractive due to their ability to provide information about layer thickness and state condition without causing damage or requiring the removal of material samples. Within the NDT available, ground penetrating radar (GPR) is a fast and reliable technique, whose advantage is the repeatability and capability of acquiring continuous.

2 GEORADAR

The georadar is a non-destructive and non intrusive inspection technique, which principle is based on the radiation of very short electromagnetic impulses (<10 ns) that are reflected at interfaces of materials with different dielectric properties (Figure 1).

For applications in civil engineering, usually frequencies from 500 MHz to 2.5 GHz are applied via bow tie antennas. The reflections recorded with the receiving antenna by moving both, transmitter and receiver along a predefined line on the surface can be visualized as 2D images or radargrams, where the intensity of the reflected impulses is displayed in a grey scale.

3 RAILWAY INVESTIGATION

The construction of new railway infrastructures and the maintenance and upgrade of existing ones for higher speeds and loads depends greatly on obtaining field information about construction quality, materials characteristics and conditions. For that purpose, ground penetrating radar (GPR) is being actively used and studied for several decades by diverse scientists (Clark et al. 2004, Hugenschmidt 2000, Gallagher et al. 1999, Jack & Jackson 1999, Roberts et al. 2006). This technique has become an accepted method for *in-situ* monitoring.

The georadar allows to obtain a significant number of parameters that are very relevant to analyse and assess the conditions of construction layers and embankments. It includes the assessment of the thickness, elastic modulus, dielectric and magnetic properties and the moisture condition of ballast, sub-ballast and base layers (Narayanan et al. 2004, Clark et al. 2001). It also can help in the detection of base or ballast settlements.

GPR has several key characteristics that made it very adequate for the referred purpose. It is a portable and compact equipment, which allows the continuous survey and data acquisition along very large distances and at rather fast speeds (up to 50 km/h). It allows the characterisation of the soil and ballast at large depths (up to 5 m), depending on the antenna frequency used. Furthermore, it is significantly faster, more reliable and precise than traditional techniques based on hole drilling, destructive methods or simple visual inspection.

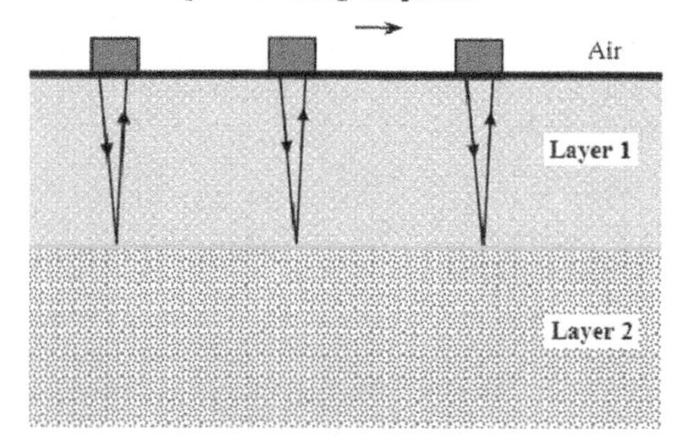

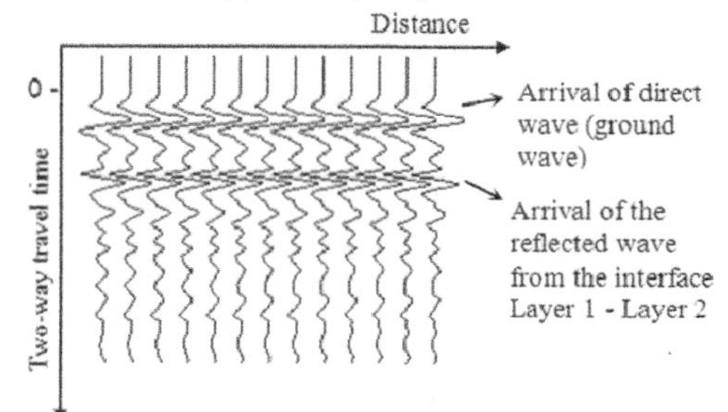

Figure 1. Working principle of georadar in reflection mode and resultant radargram.

4 TESTING SITE AND METHODOLOGY

To reach the proposed goal, a trial embankment was constructed with different materials, layer's thicknesses, water contents and compaction energy levels. Two tracks were built, one constituted by a compacted sand layer and a second one constituted by a sub-ballast layer over a compacted sand layer (see Figure 2). Additionally, on the first track, metallic plates were introduced in the base of the sub-ballast layer, along an alignment, which provides an indicator for thickness control. The surveys were carried out in order to detect the thickness of the sub-ballast layer located over the compacted sand layer and its uniformity along the track and the cross-section of the track.

To accomplish this, two GPR systems were used together with different antenna frequencies. One RAMAC system (MALA Geoscience, Inc.) was used together with three different antennas, each one with a different objective: 500 MHz antenna for deep sounding, 800 MHz antenna for optimum detection of the layers in the first half meter and 1600 MHz antenna for the survey of shallow characteristics. A SIR-10 system (Geophysical Survey Systems, Inc.) was used with an antenna of 900 MHz, as illustrated in Figure 3.

5 RESULTS

The radargrams obtained with the surveys carried out in the trial embankments showed that GPR provide very good results regarding obtaining information about layer thickness. Both the surveys with low (500 MHz) and medium (800 MHz) frequencies

Figure 2. Partial view of the trial embankment with execution of a GPR profile.

were successful in showing the thickness and evolution of the sub-ballast layer. However, the 800 MHz antenna showed more sensitivity. These radargrams are presented in Figure 4 and Figure 5, respectively.

The conditions on site during the survey were wet; therefore, the dielectric constant of the sub-ballast material was estimated to be around 11. With this value

Figure 3. Equipment: System SIR-10 – GSSI, and antennas of 900, 500 and 200 MHz.

it was possible to compute the thickness of the layer, which is presented in Figure 6.

The track was additionally surveyed transversally with the 800 MHz antenna, whose radargram is illustrated in Figure 7. This information complements what was already obtained in previous longitudinal radargrams by showing the thickness evolution through the width of the track section, but also pointing out areas where the contrast is weak, and could be due to the occurrence of fouling or moisture. In the same way, the resultant cross-section is illustrated in Figure 8, where it can be seen that the average thickness of the sub-ballast layer is 30 cm (in the test position) and is rather uniform. In consequence, with this technique, it is rather easy and fast to determine the thickness along a particular track and detect possible defects and anomalies.

Other example from a research study using GPR, in a trial embankment, with the system SIR-10 (Geophysical Survey Systems, Inc.) and antennas of 900 MHz is presented in Figure 9.

In order to control some parameters of the sub-ballast layer, like thickness and uniformity, several metallic plates had been used in the base of the sub-ballast layer, along an alignment. Each plate that was placed in the interface between the sub-ballast and the

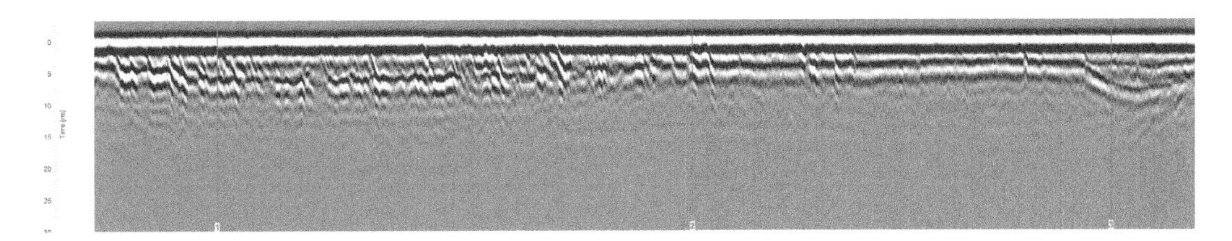

Figure 4. Partial radargram from the RAMAC system and the 500 MHz antenna.

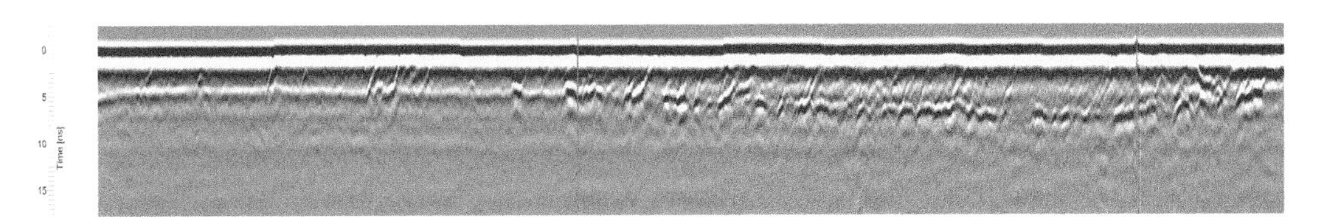

Figure 5. Partial radargram from the RAMAC system and the 800 MHz antenna.

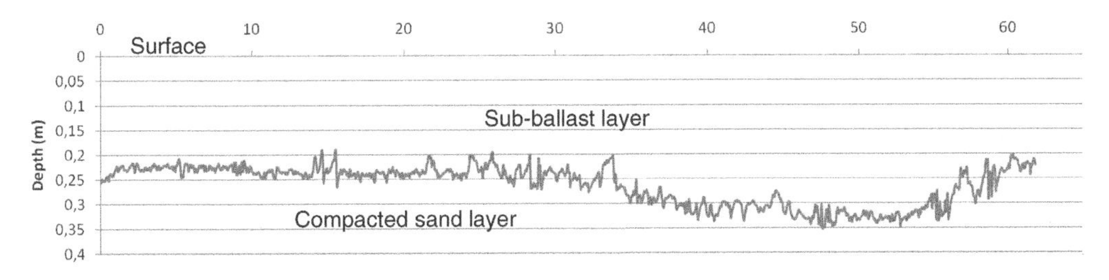

Figure 6. Evolution of the thickness of the sub-ballast layer obtained with the 800 MHz antenna.

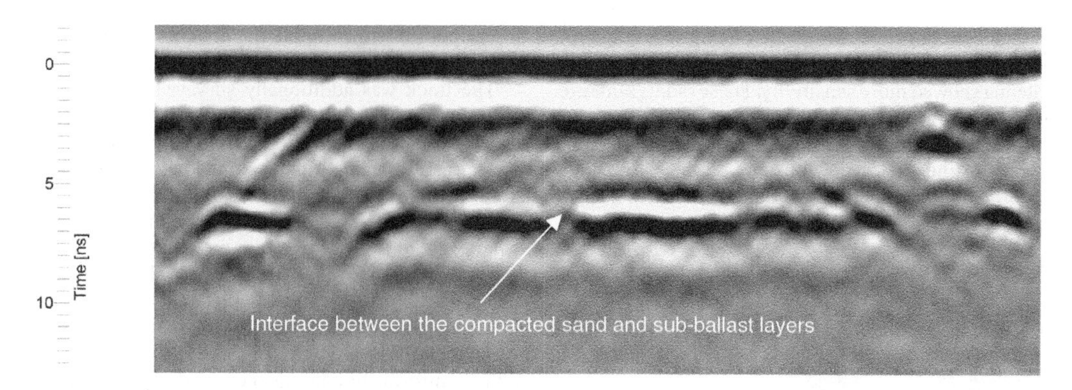

Figure 7. Transversal profile obtained with the 800 MHz antenna.

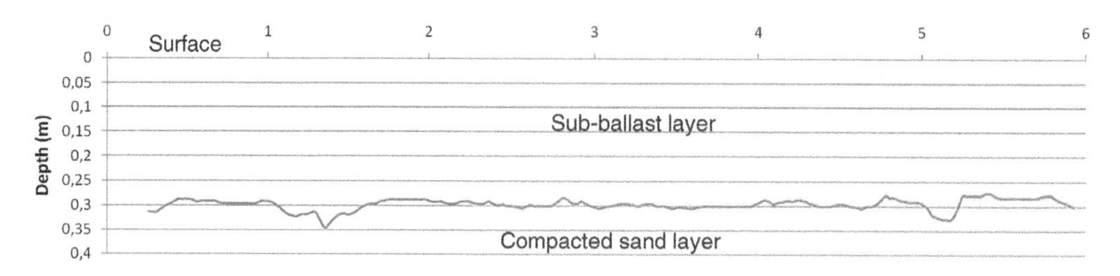

Figure 8. Cross-section of the embankment from the profile obtained with the 800 MHz antenna.

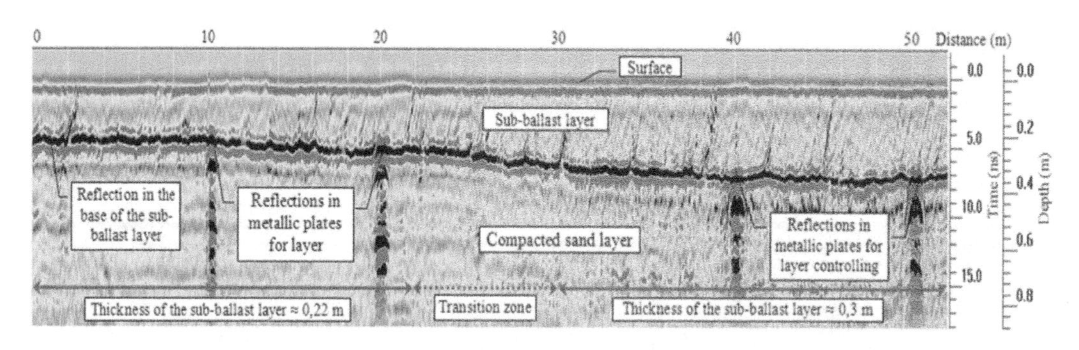

Figure 9. GPR image obtained with 900 MHz antennas and their interpretation.

compacted sand layer worked as a reference and was a point of control, due to the fact that metallic targets can mark GPR records with multiple and strong reflections. The image in Figure 6 shows clearly the ability of GPR to detect the sub-ballast layer and its thickness variations along the profile. A similar methodology was followed in Carpenter et al. (2004) but, instead of metallic plates, a specially designed high reflectivity polymer band was used instead, providing layer control and speed estimation of the base soil, allowing to control the moisture content evolution.

6 CONCLUSIONS

Georadar was applied in trial railway embankments to validate it's effectiveness in detecting the thickness and evolution of the construction layers. The

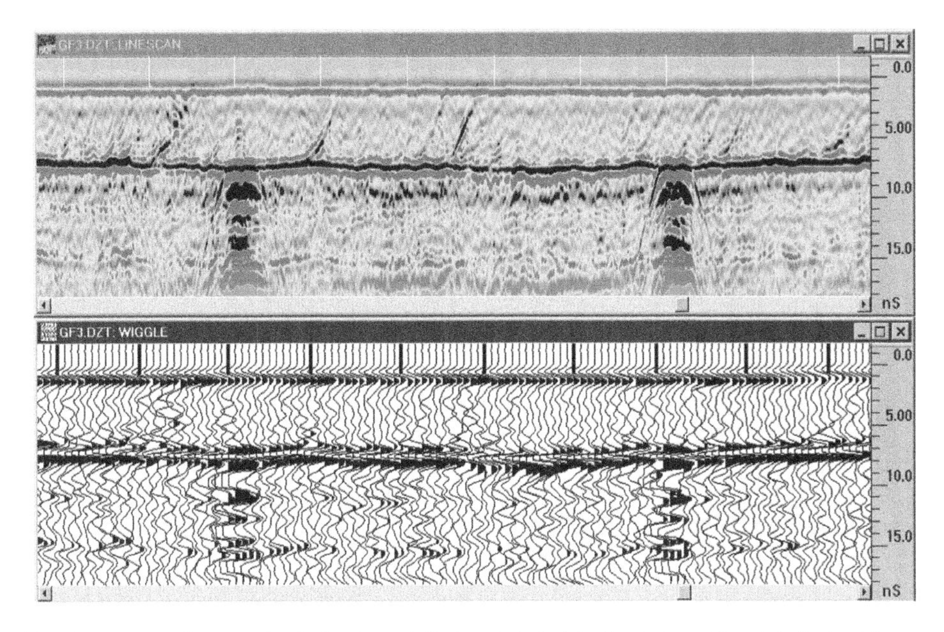

Figure 10. Radargram with strong signals due the metallic targets in the interface between the sub-ballast and the compacted sand layer.

results showed that the georadar is capable of providing information about the layers thickness, evolution and moisture content, all in a very fast and non-destructive way.

In this way, georadar can be applied during any phase of the construction to verify if the design project is being followed.

ACKNOWLEDGEMENTS

The authors gratefully acknowledge the National Railway Network (REFER) for building the trial embankments. The second author would like to acknowledge the partial support from FCT (Portuguese Foundation for the Science and Technology) by the grant number POCTI SFRH/BPD/26706/2005.

REFERENCES

Clark, M.R., Gillespie, R., Kemp, T., McCann, D.M. Forde, M.C. 2001. Electromagnetic properties of railway ballast. *NDT&E International*, 34(5):305–311.

Clark, M.R., Gordon, M., Forde, M.C. 2004. Issues over high-speed non-invasive monitoring of railway trackbed. *NDT&E International*, 37(2): 131–139.

Roger Roberts, R. Al-Qadi, I., Tutumluer, E., Boyle, J., Sussmann, T. 2006. Advances in railroad ballast evaluation using 2 GHz horn antennas. *Proc. 11th International Conference on Ground Penetrating Radar, Columbus Ohio, USA, 19–22 June 2006.*

Carpenter, D., Jackson, P.J., Jay, A. 2004. Enhancement of the GPR method of railway trackbed investigation by the installation of radar detectable geosynthetics. *NDT&E International*, 37(2): 95–103.

Gallagher, G.P., Leipera, Q., Williamson, R., Clark, M.R., Forde, M.C. 1999. The application of time domain ground penetrating radar to evaluate railway track ballast. *NDT&E International*, 32(8): 463–468.

Jack, R., Jackson, P. 1999. Imaging attributes of railway track formation and ballast using ground probing radar. *NDT&E International*, 32(8): 457–462.

Hugenschmidt, J. 2000. Railway track inspection using GPR. *Journal of Applied Geophysics*, 43(2–4): 147–155.

Narayanan, R.M., Jakub, J.W., Li, D., Elias S.E.G. 2004. Railroad track modulus estimation using ground penetrating radar measurements. *NDT&E International*, 37(2): 141–151.

Advances in Transportation Geotechnics – Ellis, Yu, McDowell, Dawson & Thom (eds)
© 2008 Taylor & Francis Group, London, ISBN 978-0-415-47590-7

Ballast stressing on a railway track and the behaviour of limestone ballast

K. Giannakos
University of Thessaly, Volos, Greece

A. Loizos
National Technical University of Athens, Athens, Greece

ABSTRACT: In this paper the results of the tests performed on the ballast used in the Greek network are presented. The tests were conducted in laboratories in France and Greece, and led to a change in the Greek technical specifications. Moreover, the need for a new method for the calculation of the actions on the railway superstructure is analyzed in terms of track maintenance and passenger's comfort. Finally, the influence of the limestone ballast on the track response and the need for new higher demands in the specifications is discussed.

1 INTRODUCTION

During the study for the dimensioning of a railway track and the selection of the individual materials constituting the track, the "weak links" are the ballast and the substructure. These are the elements of the track that develop residual deformations: subsidences and lateral displacements, directly connected to the deterioration of the so-called geometry of the track, which can be nevertheless described much more specifically as quality of the track. The smaller the residual deformations and the smaller their increase over time is, the better the quality of the track.

It is therefore imperative to reduce as much as possible the development of subsidence, primarily, as well as lateral displacements. In the Greek network during the 1970's and the 1980's appeared cracks on twin-block concrete sleepers and an extended investigation program begun. In the frame of this investigation, a new approach for the actions on sleepers and the ballast has been developed, by taking into account the real conditions of the track (maintenance etc.) which led to the increase of the demands of the specifications for railway ballast.

In this paper the new method is described and the results of the tests performed on the ballast used in the Greek network are presented. The tests were conducted in laboratories in France and Greece, and led to a change in the Greek technical specifications. Moreover, the need for a new method for the calculation of the actions on the railway superstructure is analyzed in terms of track maintenance and passenger's comfort. Finally, the influence of the limestone ballast on the track response and the need for new higher demands in the specifications is discussed.

2 SCOPE OF THE TEST PROGRAM

2.1 *General*

In the frame of a technical cooperation between the French Railways (SNCF) and the Hellenic Railways Organization (OSE)/Track Directorate, a test program was performed in the Test Center of the "Direction de l' Equipement" of the SNCF, to study the interaction between the ballast and different sleeper types. The scope of the experimental program was the simulation of the dynamic loads acting on the track in order to determine the influence of the sleeper's type on the "behaviour" of the ballast (limestone at that era) used in the Greek network.

These tests were performed with the application of a pulse load in the laboratories of SNCF. Measurements of the quality of the ballast material which come from different quarries in Greece were performed.

Quality tests were performed on Vagneux U31 twin-block concrete sleepers (produced according to the technical specifications of the SNCF in Thessaloniki, Greece and sent from Greece to France), and two sleepers, one twin-block Vagneux U41 type and one monoblock, manufactured. in France.

Investigation programs were also performed in Greece in cooperation of OSE with Universities Greek and foreign (Tasios et al., 1989, Abakoumkin et al., 1992–1993, etc.)

2.2 *Pulse load tests*

Tests were performed in a cylindrical bucket from lamina (60 cm diameter, 250 cm height) with two different loading mechanisms: (i) using a steel

Table 1. Ballast from Domokos quarry (Cooperation OSE/SNCF 1989).

Dimensions of the plate (cm)	Weight of ballast before the test (kg)	Subsidence Cycles × 10^3 250 mm	500 mm	750 mm	1000 mm	Weight of ballast after the test (kg) sieve > 25	sieve < 1.6	rest
31 × 31 steel	112	3.95	7.80	10.00	11.50	109.00	0.10	2.90
34.5 × 34.5 steel	112	6.00	7.50	8.20	9.00	110.60	0.08	1.32
31 × 31 wood	112	3.30	4.40	4.80	5.40	109.40	0.10	2.50

plate with dimensions 31 × 31 cm^2, and (ii) using a steel plate with dimensions 34.5 × 34.5 cm^2, that is 1.25 × (31 × 31) cm^2. The number of cycles of each test was 1.000.000 and trhe load was applied at a frequency of 4 Hz.

The load was calculated so that the pressure under the plate would be correspondent to the Region R1 for the concrete sleepers strength (see also Giannakos et al., 2007), i.e. 120 kN. The base of the bucket was replaced with a "pad" made of synthetic felt.

Only the ballast particles not passing through the 25 mm sieve were chosen and the weight of the ballast which was placed in the bucket was measured. The subsidence was measured every 250.000 cycles, with the first measurement performed after 5.000 cycles At the end of the tests, that is after 10^6 cycles, measurements were performed on the ballast quantity not passing through the 25 mm sieve as well as the ballast quantity that passed through the 1,6 mm sieve

2.3 *Cyclic load tests*

The device used for this type of tests is a unidirectional vibrator which applies cyclic load on the sleeper between +80 kN and 2.50 kN at a frequency of 50 Hz. The duration of each test was 50 hours.

In these tests only the ballast particles not passing through the 25 mm sieve were tested.

After 100 hours in the device the ballast was removed and the quantity not passing through the 25 mm sieve as well as the crushed ballast (during the test) that passed through the 1.6 mm sieve were measured.

Tests were also performed on:

(a) 1 twin-block concrete sleeper type Vagneux U31 produced in Greece
(b) 1 twin-block concrete sleeper type Vagneux U41 produced in France
(c) 1 wooden sleeper
(d) 1 monoblock sleeper of prestressed concrete produced in France
(e) 1 twin-block concrete sleeper type Vagneux U2 produced in Greece

3 LABORATORY TEST RESULTS

3.1 *Results from the pulse load tests*

The results are presented in Table 1. The tests were performed under a cyclic load between 10 and 59 kN. The ballast abrasion (wear) appears to be directly connected to the pressure acting on the ballast – sleeper interface, without any influence from the nature of the seating surface. The subsidence measured at the seating surface which is smaller than the one measured with the steel support surface is possibly due to a notable difference of the rheologic quality of the assembly for the test to the frequency imposed from the load, which reacts differently since the quantities of the abraded ballast are almost equivalent.

We should note that the behavior of a very polluted and very compacted ballast on track is practically quite the same like that of a slab track, with the disadvantage that the polluted ballasted track does not maintain the geometry of the track.

3.2 *Results from the cyclic load tests*

Results are presented in Table 2. It must be noted that the sleepers produced in France presented an abrasion not exceeding 0.9%. The ballast wear seems to be of less importance with the wooden sleeper even if the fractured pieces of ballast are sensibly equivalent to that ascertained with a U41 sleeper or a monoblock sleeper.

As far as the abrasion is concerned, the U31sleper produced at Thessaloniki presented a worse behavior than the sleepers produced in France. Very similar to that is the behavior of the U2 sleeper with octagonal blocks produced in Greece.

The U41 sleeper presents very good results. The deterioration of the ballast is sensibly worse than that of monoblock sleeper.

The ballast of a high Los Angeles (L. A.) coefficient (bad quality) does not improve its behavior when used with a wooden sleeper in comparison with its use with a concrete twin block sleeper with grater blocks or a

Table 2. Ballast from Domokos quarry (Cooperation OSE/SNCF 1989).

	Weight of ballast in kg					
		After test			Loss of weight of the sleeper	
Sleeper model	Before test Sieve > 25	Sieve > 25	Sieve < 1.6	Rest	Before (kg) after (kg)	Loss (kg) %
U31 OSE 68 cm block	1499.00	1423.00 95.0%	19.00 1.3%	57.00 3.7%	206.50 202.80	3.70 1.8%
U41 84 cm block	1413.00	1377.00 97.5%	10.30 0.7%	25.70 1.8%	262.00 261.10	0.90 0.34%
Wooden sleeper	1571.00	1528.50 97.0%	3.50 0.2%	39.50 2.8%		
Monoblock sleeper	1560.50	1500.60 96.2%	13.40 0.8%	46.50 3.0%	240.30 239.50	0.80 0.33%
U2 OSE block octagonal	1449.00	1365.00	19.10	65.00	184.30 180.70	3.60 1.95%

monoblock concrete sleeper. But the complete compaction of the support will be delayed with wooden sleepers if we take into consideration the fines that appeared in the test.

3.3 *First conclusions*

The results enable us to estimate from one hand the actions to be undertaken for the amelioration of the existing situation and from the other hand to make the relevant decisions for the choice of the appropriate way of the modernization of the network, such as the choice of ballast quarries or of the type of sleeper.

4 DYNAMIC LOAD ACTING ON THE TRACK SUPERSTRUCTURE

4.1 *Action on the sleeper*

During the investigation, a new method was developed (Giannakos 2004), for the calculation of the load acting on a sleeper as a combination of static and dynamic components (see also Giannakos et al., 2007 etc). According to this method, the dynamic loads acting on the track are calculated at the dominant frequencies of the track system, in contrast to the method presented in the German bibliography that assumes static load increased by a coefficient representing the additional dynamic component due to the vibration of the vehicle – rail system.

In high frequencies, the superstructure does not respond because of its low eigenfrequency, therefore, for safety reasons, we assume that dynamic loads (semi-statics are included) are not distributed to the adjacent sleepers, in contrast to static loads that are distributed. Therefore it is recommended that the service load should be considered as the static load increased by 3 times the standard deviation of the dynamic load (P = 99.7%):

$$R_{serv} = \bar{A}_{dyn} \cdot (Q_{wh} + Q_{\alpha}) + 3 \cdot \sqrt{\left[\sigma(\Delta R_{NSM})\right]^2 + \left[\sigma(\Delta R_{SM})\right]^2} \quad (1)$$

Giannakos et al., 2007) by (2)

$$\bar{A}_{dyn} = \frac{1}{2\sqrt{2}} \cdot \sqrt[4]{\frac{\ell^3 \cdot h}{E \cdot J}} \quad (2)$$

where: Q_{wh} wheel static load, Q_a load due to cant deficiency, $\sigma(\Delta R_{NSM})$ the standard deviation of the dynamic component due to non-suspended masses, $\sigma(\Delta R_{SM})$ the standard deviation of the dynamic component due to suspended masses and h dynamic track stiffness (Giannakos et al., 2007) derived by the following equation:

$$h = 2\sqrt{2} \cdot \sqrt[4]{\frac{E \cdot J \cdot \rho_{total}^3}{\ell^3}} \quad (3)$$

where E, J the elasticity modulus and moment of inertia of the rail, ℓ the distance of the sleepers and ρ_{total} the total static stiffness of the track as multi-layered structure. The coefficient 3 just before the square root in Eqn (1) covers a probability of 99,7% (Giannakos 2004) for the appearance of this value of action and it should be used for the calculation of sleepers.

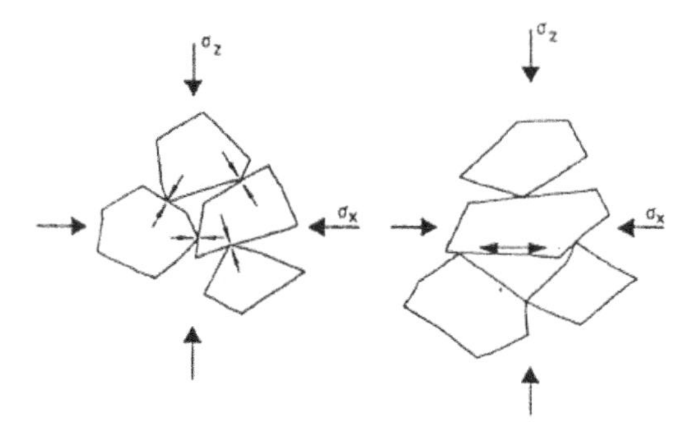

Figure 1. Ballast grains in ballast bed (Eisenmann et al., 1980).

4.2 *Actions on the ballast bed – Subsidence – Average pressure*

The ballast bed undertakes the actions from the sleeper's seating surface. It is clear that the French bibliography pays small attention – if any- to the pressure exerted on the ballast at the seating surface of the sleeper (Giannakos et al. 1989 – a, Giannakos et al. 1989 – b). The performance of the ballast grains (see Figure 1) in the ballast-bed -where the contacts between them are "points"- are not described or simulated with reliability by the average pressure, which in that case has rather "symbolic" mean. With this knowledge the pressure does not represent the reality but it is a "virtual" tool only for theoretical calculations.

There is no uniform support of the sleeper on the ballast, nor uniform compaction of the ballast and the ground and there are faults on the rail running table, imperfections on the wheels etc. A decisive coefficient in determining the dimensioning of the superstructure is the maximum value – which is calculated, based on the rules of probabilities, from the mean value and standard deviation. In order to calculate the value of the load that stresses the sleeper, the triple value of the standard deviation is taken. (P = 99.7%), as mentioned already above.

According to the German bibliography (Eisenmann 1980, Eisenmann et al., 1980) the average "pressure" under the sleeper is used as "measure of judgement" or as "the mean value of pressure is a *criterion* for the stressing of the ballast on track" (Eisenmann 1988).

In order to calculate the stress of the ballast-bed under the seating surface of the sleeper, we should calculate the subsidence y_{static}, $y_{dynamic}$, y_{total} and the mean value of pressure ($\overline{p}$), according to the dynamic analysis (Giannakos et al 1990, 1991, 1992 and also Abakoumkin et al., 1992–1993) of the phenomenon. Based on the above, as well as on the points made by Prof. J. Eisenmann (1981), it is suggested that the double standard deviation (P = 95.5%) should be taken in order to calculate the load exerted on the sleeper-ballast contact surface and the results are comparable to those of the German method. Therefore for the ballast according to Giannakos (2004):

$$R_2' = (Q_{wh} + Q_a) \cdot \overline{A}_{dyn} + 2 \cdot \sqrt{\sigma(\Delta R_{NSM})^2 + \sigma(\Delta R_{SM})^2} \tag{4}$$

The seating coefficient (or coefficient of ballast) is calculated by:

$$C = \frac{p_{total}}{F_{sleeper}}$$

The subsidences:

$$y_{stat} = (Q_{wh} + Q_a) \cdot \frac{1}{2\sqrt{2}} \cdot \sqrt[4]{\frac{\ell^3}{E \cdot J \cdot \rho^3}} \tag{5}$$

$$y_{dyn} = \frac{R_{dyn}}{h} \tag{6}$$

and the mean value of pressure:

$$\overline{p} = \frac{\overline{A} \cdot (Q_{wh} + Q_a)}{F_{sleeper}} + C \cdot y_{dyn} \tag{7}$$

where according to Zimmermann's theory $\overline{A}$ is given by Eqn (2) and therefore:

$$R_{dyn} = R_2' - \overline{A}_{dyn} \cdot (Q_{wh} + Q_\alpha) \tag{8}$$

where
ℓ = the distance of the sleepers (usually 600 mm)
h = the dynamic stiffness of the track given by Eqn (3)
F_{sleep} = semi-sleeper's active surface
E = the modulus of elasticity of the rail
J = the moment of inertia of the rail
C = the coefficient of the ground (bettungsmodul in kg/cm³)

Then respective values y and P -according to German bibliography- were calculated on the basis of the statistical analysis of the phenomenon (Eisenmann 1980, Eisenmann et al 1984):

$$L = \sqrt[4]{\frac{4 \cdot E \cdot J}{b \cdot C}} \qquad b = \frac{F_{sleeper}}{\ell} \tag{9}$$

Static subsidence

$$y_{stat} = \frac{Q}{2 \cdot b \cdot C \cdot L} \tag{10}$$

Subsidence y_{stat} is augmented by being multiplied by a coefficient:

$$\lambda = 1 + t \cdot \overline{s} \tag{11}$$

Table 3. Correlation between probability P (%) and coefficient t.

P	0%	68.3%	80.0%	90.0%	95.0%	95.5%	98.0%	99.7%
t	0	1.00	1.28	1.65	1.96	2.00	2.33	3.00

Table 4. Comparison of the results of mean stress $\bar{p}$ and subsidence y resulting from the formulas of this chapter (Giannakos et al., 1990).

		According to K. Giannakos				According to German bibliography			
		U31 sleeper		U41 sleeper		U31 sleeper		U41 sleeper	
k = 12 NSM = 15 kN		V = 200	V = 140	V = 200	V = 140	V = 200	V = 140	V = 200	V = 140
ρ = 250 kN/mm	y_m	1.47	1.29	1.47	1.29	1.22	1.32	1.22	1.32
	$\bar{p}$	0.6	0.52	0.5	0.44	0.5	0.54	0.42	0.45
ρ = 100 kN/mm	y_m	1.86	1.66	1.86	1.66	1.64	1.77	1.64	1.77
	$\bar{p}$	0.51	0.46	0.43	0.38	0.45	0.49	0.38	0.41
ρ = 80 kN/mm	y_m	2.01	1.80	2.01	1.80	1.80	1.95	1.80	1.95
	$\bar{p}$	0.49	0.44	0.41	0.37	0.44	0.47	0.37	0.40
ρ = 60 kN/mm	y_m	2.25	2.03	2.25	2.03	2.07	2.23	2.07	2.23
	$\bar{p}$	0.45	0.41	0.38	0.34	0.42	0.45	0.35	0.38
ρ = 40 kN/mm	y_m	2.69	2.45	2.69	2.45	2.56	2.77	2.56	2.76
	$\bar{p}$	0.41	0.37	0.34	0.31	0.39	0.42	0.33	0.35

where:

$$\bar{s} = \alpha \cdot \varphi \quad \text{and} \quad \varphi = 1 + \frac{(V-60)}{140} \qquad (12)$$

α = 0.1 for superstructure in very good condition
α = 0.2 for superstructure in good condition
α = 0.3 for superstructure in poor condition
t = 1 for P = 68.3% where P stands for probability
t = 2 for P = 95.5%
t = 3 for P = 99.7%

Analytically the probabilities are given (Eisenmann et al., 1984):

In the comparative study, the following data in the various formulas were used (Giannakos et al., 1990):

α = 0.3 for V ≤ 140 km/h,
α = 0.2 for V = 200 km/h and
t = 2 because we examine the pressure on the ballast.

It is not permitted to have poor superstructure condition for V > 140 km/h.

Next, $\bar{p}$ is calculated using the formulas:

$$y_{max} = \lambda \cdot y_{st} \quad \text{and} \quad \bar{p} = C \cdot y_{max} \qquad (13)$$

The derived results show that dynamic analysis gives slightly worse results, mainly because in the calculation of y_{dyn} the dynamic stiffness (h) of the track interferes, which is approximately two times the static stiffness (ρ), therefore the stress on the ballast is higher.

The comparison of the results for various types of sleepers can be found in the following table.

5 BALLAST STRESS AND DEFORMATIONS

As for the issue of ballast fatigue, the existing bibliography assumes a uniform distribution of stresses under the sleeper and without further details uses the mean value of pressure. Nevertheless, various researchers (UIC, SNCF, OSE) have questioned whether the mean value of pressure is representative. *Based on bibliography, the maximum moment measured actually on track results from parabolic stress distribution* (ORE D71). But in reality, the seating of the sleepers is support on discrete points (points of contact with the grains of the ballast[Giannakos et al., 1988]) and the resulting necessity to calculate the stress per grain of ballast cannot give comparative results to the rest of the bibliography. So it is possible to use the mean value of pressure not as an absolute quantity, but comparatively and in combination with the possibility it covers (Giannakos et al., 1990). There is no uniform support of the sleeper on the ballast, nor uniform compaction of the ballast and the ground and there are faults on the rail running table, imperfections on the wheels etc.

Undeflected (stiff) seating (e.g. in the case of a concrete bridge, rock at the bottom of a tunnel as substructure) with great axial load (e.g. 225 kN) leads to faster deterioration of the ballast and therefore, to

Table 5. Results of the performance of 4 types of sleepers (Giannakos 2004 p. 250).

Type of sleeper + fastening type	Region R2 kN	Average value of pressure MPa	Subsidence y mm	Surface mm^2
Wooden sleeper + «K»	261.9	0.505	1.166	275.000
Sleeper U3 + RN	264.7	0.751	1.044	185.800
Sleeper U31 + Nabla	228.3	0.598	1.468	197.200
Sleeper U41 + Nabla	228.3	0.503	1.468	243.600

Table 6. Results of the ballast from Domokos quarry on the Vibrogir at St Ouen (SNCF).

Type of sleeper	Fines <1.6	Intermediate >1.6 or <2.5	Total
Wooden sleeper	0.2%	2.8%	3.0%
U31 sleeper	1.0%	4.0%	5.0%
U41 sleeper	0.7%	1.9%	2.6%

Table 7. Results of the ballast from Domokos quarry on the Pulsateur at St Ouen (SNCF).

Type of sleeper	% Fines <1.6	% Intermediate >1.6 or <2.5	% Total
Wooden sleeper	0.08%	2.23%	2.31%
U31 sleeper	0.08%	2.59%	2.67%
U41 sleeper	0.07%	1.18%	1.25%

deterioration of the geometry of the track. In such cases the phenomenon can be prevented by placing rubber sub-mats in order to smooth out the great differences in the stifness of the substructure, during the transition from embankment into a tunnel or a concrete bridge. In the bibliography it is suggested (Eisenmann 1988) that, regarding the substructure, as load should be taken the sum of the mean load +1 standard deviation, and regarding the ballast from $1 \div 3$ ($P = 68.3\% \div 99.7\%$) standard deviations depending on the speed and the necessary maintenance work.

The most important issue, though, is that since the publication (ORE D117, Rp2, Rp4) of ORE's research, (Office des Recherches et Etudes of the U.I.C.), it has been established that the material of the sleepers (wood, concrete) gives almost identical values of settlement of the track.

The residual settlement is a percentage of the total subsidence during the passing of the loads (Hay 1982). Therefore, it can be extrapolated that we will have an almost identical performance in the deterioration of the geometry of the track. This fact is confirmed by a more recent publication in 1984 (F.I.P. 1984, see also more recent fib 2006).

This experimental confirmation, which has been also verified through calculations (Giannakos et al., 1990 a, b) means that in relation to the sustaining of the geometry of the track, the material of the sleeper has no significant influence. We will observe the same frequency of maintenance interventions, whether using a wooden sleeper or concrete sleeper, in relation to the sleeper-ballast contact surface. To verify the above experimental data with calculations, the calculation of the reaction of the sleeper has been done since 1990, as well as that of the mean value of pressure $\overline{p}$ and the subsidence y for four types of sleepers. Indicative results are included in the following table (for $V = 200$ km/h, $k_l = 12$, $\rho = 250$ kN/mm, N.S.M. = 1.5 t): R2 region has been calculated by using 3 standard deviations.

If the mean value of pressure is used as a criterion, it can be calculated that even though the surface of a wooden sleeper is about 13% greater than that of the U41, it bears about 3% higher pressure, evidently because of the different elastic pad, in undeflected seating.

For soft substructure ($\rho_{subgr.} = 40$ kN/mm) the U41 gives $y_{total} = 2.69$ mm and $\overline{p} = 0.343$ MPa and the wooden gives $y_{total} = 2.48$ mm and $\overline{p} = 0.307$ MPa.

Besides, during experiments on Greek ballast from the Domokos quarry (DRi = 9) conducted in the SNCF laboratory of St Ouen, it was established that the total deterioration caused to the ballast is shown in the following table for the cyclic load tests.

As it can be seen, the fracture caused to the ballast is similar to that of the wooden sleeper and for U41, which have almost similar surfaces.

On the contrary, there is less powder on the wooden sleeper, which shows that the ballast will be rendered non-apt for tamping and will demand replacement after a longer period of time.

For the above reasons, the fact that the wooden sleeper in the pulse tests produced a markedly lower value of subsidence than U31 and U41 can be interpreted as "a result of the rheology of the test's assembly, which responds differently to the frequency that the load imposes, since the quantities of ballast deterioration are equivalent" (Cooperation OSE/SNCF 1988, 1989).

Moreover this can be seen in Table 7 which refers to the pulse load test. Heavier concrete sleepers, in relation to the wooden ones, hinder the settlement of the track that is caused by vibrations (Dimitriadis et al., 1987). With those sleepers there arise no peaks, which characterize the amplitude of vibration in the resonance area, and whose creation leads to destabilization of the ballast. Moreover, the reduction of Non Suspended Masses and the use of a more elastic pad, i.e. pad with small ρ ($\rho < 100$ kN/mm and/or 80 kN/mm), leads to a reduction of the stressing of the ballast.

6 CONCLUSIONS

This paper presents an extension of the method of estimation of actions on sleepers (Giannakos, 2004, Giannakos et al., 2007) to the estimation of pressures under the seating surface of sleeper on the ballast. Experiments verify that the subsidence is independent of the material (wood, concrete).

Since the calculated pressure at the seating surface of the sleeper on ballast coincide with the average measured values they can safely be used as criterion for the behaviour of the ballast-sleeper system. The presented method gives results that justify the tests and the behaviour of the system. It also gives a quantifying reasoning for the experience of the real situation on track.

REFERENCES

Abakoumkin K., Trezos K., Loizos A., Lymberis K. 1992–1993. Normal gauge line monoblock sleepers made of pre-stressed concrete *NTUA/Sector of Transport & Communication Infrastructure, Athens.*

Cooperation OSE/SNCF, 6/1998. Comportement des traverses en relation avec les charges dymaniques.

Cooperation OSE/SNCF:3/1989. Programme d' essays realize au center d' essays de la Direction de l' Equipement.

Dimitriadis G., Papagiannopoulos T., Koroneos D., 1987. Notes from the training seminar in the French Railways, *OSE/Track Directorate.*

Eisenmann J., 1981. Verhaltensfunktion des Schotters, Folgerungen fur hohe Fahrgeschwindigkeiten, *Der Eisenbahningenieur*, 3/1981 p 100–103.

Eisenmann J., 1980. Qualitat des Oberbauzustandes, *Der Eisenbahningenieur*, 3/1980.

Eisenmann J., 1988. Schotteroberbau – Moglichkeiten und Perspektiven fur dieModerne Bahn.

Eisenmann J., Kaess G., 1980. Das Verhalten des Schotters unter Belastung, *ETR* (29) 3, Hestra-Verlag Darmstadt 1980

Eisenmann J., Mattner L., 1984. Auswirkung der Oberbauzustandes auf die Schotter und untergundbeanspruchung, *Der Eisenbahningenieur* 3/1984.

Federation Internationale du beton, 2006. Precast concrete railway track systems, June 2006.

Federation Internationale Precontrainte (F.I.P.), 1987. Concrete railway sleepers, London.

Giannakos K., 2004. Actions on the Railway Track, Papazissis publ., p. 250.

Giannakos K., Iordanidis P., Vlasopoulou I., 1989a. Brief Committee Report on the subject of twin-block sleepers after their second visit to the SNCF in March, 1989, *OSE/Track Directorate, Athens 1989.*

Giannakos K., Iordanidis P., Vlasopoulou I., 1989b. OSE Expert's Committee Report on their visit to SNCF from 15.5.1988–11.6.1988, *OSE/Track Directorate, Athens.*

Giannakos K., Loizos A., 2007. Loads on railway superstructure – Influence of high-resilient fastenings on sleepers loading, *Proceedings of Advanced Characterization of Pavement and Soil Eng. Materials, Athens, Greece, 20–22 June 2007*, page 1363.

Giannakos K., Vlasopoulou I., 1991. Problems and Solutions for the "sleeper-ballast" system in OSE», *OSE/Track Directorate,* Athens 1991.

Giannakos K., Vlasopoulou I., 1992. Research to determine the type of sleeper to be adopted by OSE – Comparative Presentation of the Various Types of Sleepers – Evaluation of Professor K. Riessberger's research, *OSE/Track Directorate*, Athens 1992.

Giannakos K., Vlasopoulou I., 1990b. Study of the fatigue of the ballast and the performance of the superstructure in relation to the type of sleepers used» *OSE/Track Directorate.*

Giannakos K., Vlasopoulou I., 1990a. Study of the system "sleeper-ballast", *OSE/Track Directorate*

Hay W., 1982. Railroad Engineering, John Wiley & Sons, Chapter 15 – Track Analysis.

Jenkins H., Stephenson J., Clayton G., Lyon D., 1974. Incidences des parametres caracteristiques de la voie et des vehicules sur les efforts dynamiques verticaux qui de developpent entre rai de roue, *Rail International* 10/1974, 682–702.

ORE, Question D 71. RP 9, page FIG 97.

Tasios Th. P., Trezos K. 1989. Laboratory tests and measurements on OSE sleepers, *National Technical University of Athens (NTUA)/Reinforced Concrete Laboratory*, Athens 20-10-1989.

Advances in Transportation Geotechnics – Ellis, Yu, McDowell, Dawson & Thom (eds)
 ISBN 978-0-415-47590-7

Design methodology of slab track systems

K. Giannakos
University of Thessaly, Volos, Greece

S. Tsoukantas
National Technical University of Athens, Athens, Greece

ABSTRACT: The substantial ballast wear in high-speed lines has led to the adoption of the slab track. In this paper a methodology for the calculation of the actions applied on the slab track and the design philosophy of such systems are presented. The role of the fastenings is also discussed.

1 INTRODUCTION: SLAB TRACK – THE NEED FOR ITS APPLICATION

After many years of international experience in high-speed lines (in Japan, Germany and France, et al), substantial ballast wear was observed. Ballast is can be literally crushed and compressed due to dynamic loads, breaking, etc., thus resulting in loss of elasticity, insufficient drainage of rainwater, etc. Under these circumstances, maintaining the line's geometry requires most frequent and expensive maintenance interventions, while the structural components of the superstructure (sleepers, rails, fastenings, etc.) incur unacceptable wear and subsequently need to be replaced long before their expected life-cycle. Costly interventions are also required on the infrastructure.The ballast must be entirely replaced at intervals of traffic, much more frequently than that used for lower speeds, even if it is comprised of granitic or basaltic rocks.

However, when the ballast is made of limestone, as is the case in Greece, then the dynamic phenomena of the «vehicle-track» system render the application of the ballasted track extremely costly, with increased maintenance costs, which lead rather to unacceptable number for high-speed lines of mixed traffic. These negative experiences of all the networks led to the adoption of the slab track (or ballast-less track) on an international level. With the Slab track method (ST), the disadvantages of the ballasted track are substantially removed (as proved in practice during the thirty years of the method's application and experience), since the ballast is entirely substituted by a reinforced concrete slab (plate) on a suitably formed blanket layer.

In Japan, due to Japanese Railways' negative experience of ballasted tracks during the operation of the first high-speed line Shinkansen, it was decided over twenty years ago to exclusively use the slab track technique for the construction of new high-speed lines. In Germany the railway company, Deutsche Bahn, likewise decided to specify a rather exclusive use of the slab track (Feste Fahrbahn) method for the construction of mixed traffic high-speed lines, following tests that were conducted for many years on various ST systems. Regarding tunnels specifically, almost all European networks have adopted the slab track technique. The primary reason for this (beyond the extensive benefits of ST) is that the substructure in tunnels, which is principally constructed by concrete, as a rigid body, multiplies the internal stresses incurred in the body of the ballast from the trains' motion, resulting in rapid loss of its elastoplastic behavior, and consequently greater width of ballast bed and more frequent maintenance intervention is required.

The slab track system (or ballastless track) is recommended for use in the infrastructure of high-traffic, high-speed lines, it offers increased passenger comfort, has a longer life cycle and it requires minimal maintenance over time (compared to the ballasted track) provided that fastening of high elasticity are used (Leykauf et al 2006). The initial construction cost of the slab track is approximately 30% to 40% higher than the cost of the ballasted track, according to Deutsche Bahn data (DB-AG). However, this difference in cost is depreciated drastically over time, given the fact that, comparatively, the cost of maintenance is almost non-existent. Moreover, the use of slab track in newly constructed tunnels may also allow the narrowing of its cross-section, resulting in the reduction of the project's total cost.

2 SLAB TRACK STRUCTURE

The slab track consists basically of a rigid bearing concrete structure and appropriately-shaped fastenings, which are affixed to it and hold the rails in the appropriate position.

The rigid bearing structure consists of a bearing concrete slab of appropriate thickness and width (e.g. 20 cm and 300 cm correspondingly) on a founded Cement Treated Base (CTB) of appropriate thickness ($30 \div 40$ cm) and width equal to the upper concrete slab increased by twice the thickness of the layer itself. This layer is based on appropriately formed and well compacted (stabilized) soil, with or without the interposition of an anti-frost layer. In Figure 1 the form of the rigid bearing structure of the slab track is schematically presented. The role of this rigid bearing structure is to undertake and distribute –securely– to the soil the actions resulting from the static and dynamic loads of the vehicles as well as the actions from temperature fluctuation.

The role of the fastening systems is to firmly fasten the rails in the correct position and to undertake, absorb (through special arrangements), and transfer the dynamic energy created by the vehicles movement, braking forces etc. The elastic pads of these resilient fastenings should be in the range of 18 kN/mm to 25 kN/mm, whereby for the time being mostly stiffness $c_{stat} = 22.5$ kN/mm $\pm$ 2.5 kN/mm will be installed. In Figure 2 the typical assembly of such a fastening is presented. High dynamic elasticity is indispensable for a low maintenance track with high availability, otherwise an increased amount of wear, wave corrugation and head checks must be expected (Eisenmann 2005).

3 ESTIMATION OF ACTIONS ON SLAB TRACK

3.1 *Introduction*

In order to calculate the forces acting on a slab track system we adopt the same methodology utilized for the calculation of actions on a ballasted track (see also Tsoukantas et al, 1999). A brief description of the method is presented herein while a more detailed description can also be found in Giannakos et al. (2006).

3.2 *Static load calculation*

The total load acting on the sleepers comprises of: (a) the static load, and (b) the dynamic load. The static load is further divided in two components: (i) the load caused by the wheel, and (ii) the load due to the cant deficiency.

In order to estimate the load due to the action of the wheel on the track, the track is simulated as a continuous beam on elastic support whose stressing

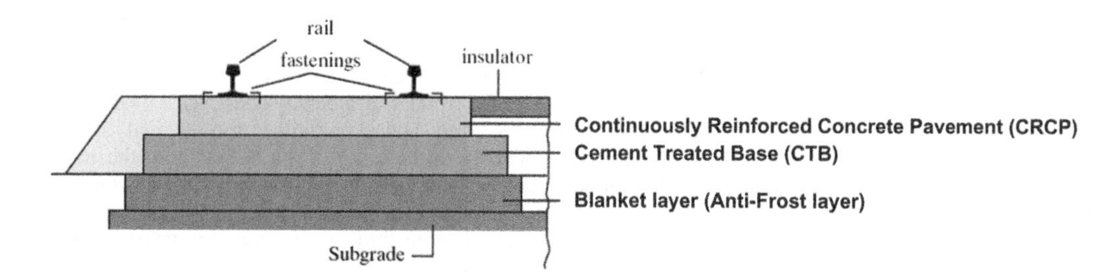

Figure 1. Schematical section of the un-ballasted track.

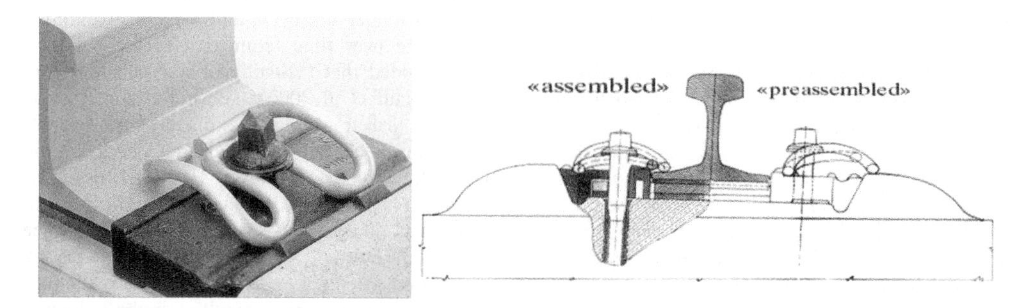

Figure 2. Fastening W15 of Vossloh for slab track.

is described by the following equation (see also Giannakos, 2004):

$$\frac{d^4 y}{dx^4} = -\frac{1}{E \cdot J} \cdot \frac{d^2 M}{dx^2} \quad (1)$$

where y is the deflection, M is the moment that stresses the beam, J is the moment of inertia of the rail and E is the modulus of elasticity of the rail.

From this formula it is concluded that the reaction R_{sl} of a sleeper reaches a value of:

$$R_{sl} = \frac{Q_{wh}}{2\sqrt{2}} \cdot \sqrt[4]{\frac{\ell^3 \cdot \rho}{E \cdot J}} \quad \Rightarrow \frac{R_{sl}}{Q_{wh}} = \bar{A} = \frac{1}{2\sqrt{2}} \cdot \sqrt[4]{\frac{\ell^3 \cdot \rho}{E \cdot J}} \quad (2)$$

where Q_{wh} the wheel load, ℓ the distance between the sleepers, ρ reaction coefficient of the sleeper which is the ratio R/y and is a quasi coefficient of track elasticity (stiffness).

In reality, the ballasted track consists of a sequence of materials –in the vertical axis– (rail, ballast, elastic pad, sleeper, substructure), and each one of them is characterized by its own coefficient ρ_i. The unballasted track consists correspondingly of a slightly different sequence of materials, rail, elastic pads, sleeper, concrete slab and substructure.

Hence, for each material it is:

$$\rho_i = \frac{R}{y_i} \Rightarrow y_i = \frac{R}{\rho_i} \Rightarrow y_{total} = \sum_{i=1}^{v} y_i = \sum_{i=1}^{v} \frac{R}{\rho_i} = R \cdot \sum_{i=1}^{v} \frac{1}{\rho_i}$$
$$\Rightarrow \frac{1}{\rho_{total}} = \sum_{i=1}^{v} \frac{1}{\rho_i} \quad (3)$$

where ν is the number of various layers of materials that exist under the rail –including the rail itself– elastic pad, sleeper, ballast etc.

According to Giannakos 2004, the recommended reaction coefficient (stiffness) of the slab track is $\rho_i = 100$ kN/mm (Giannakos 2004). A fluctuation between values of $\rho_i = 86, 100, 114$ kN/mm can be also found in the literature (Leykauf et al., 1990). Moreover, an average deflection of 0.26 mm is proposed for the slab (Eisenmann et al., 1979).

The load due to cant deficiency is not purely static since it is caused by the centrifugal acceleration exerted on the wheels of a vehicle that is running in a curve with cant deficiency. It is not, however, a dynamic load either, and hence it is often referred to as semi-static load. The following equation (see also Giannakos et al 1994).:

$$Q_\alpha = \frac{2 \cdot \alpha \cdot h_{CG}}{e^2} \cdot Q_{wh} \quad (4)$$

provides the accession of the vertical static load of the wheel that is obtained from equation (2), at curves with cant deficiency. In the above equation α is the cant deficiency, h_{CG} the height of the centre of gravity of the vehicle from the rail head, and e the track gauge.

3.3 *Dynamic load calculation*

3.3.1 *Calculation of the reaction of the sleeper*

According to the German bibliography, the load acting on the track Q_{total} is given by the following formula (Fastenrath 1981).:

$$Q_{total} = Q_{wh} \cdot (1 + t \cdot \bar{s}) \quad (5)$$

where $\bar{s} = 0.1\varphi \div 0.3\varphi$ dependent on the condition of the track, and

$$\varphi = 1 + \frac{V - 60}{140}$$

where V the speed and t coefficient dependent on the probabilistic certainty P (t = 1 for P = 68.3%, t = 2 for P = 95.5% and t = 3 for P = 99.7%). The reaction of each tie is calculated from the total load Q_{total} acting on the track. From the German bibliography (Eisenmann J., 2004) the action (or reaction) on each tie is calculated from the following equation:

$$\frac{R_{max}}{Q_{total}} = \frac{Q \cdot \ell}{2 \cdot L} = \frac{1}{2\sqrt{2}} \cdot \sqrt[4]{\frac{\ell^3 \cdot \rho}{E \cdot J}} = \bar{A}_{stat} \quad (6)$$

where L is the "elastic length" of the track.

As shown in Giannakos et al. (2006), equation (6) significantly underestimates the actions on sleepers. Therefore, the use of a dynamic coefficient of elasticity $\bar{A}_{dyn}$ is proposed instead of $\bar{A}_{stat}$ calculated by the following equation (Giannakos 2004):

$$\rho_{dyn} = h_{TR} = 2\sqrt{2} \cdot \sqrt[4]{E \cdot J \cdot \left(\frac{\rho}{\ell}\right)^3}$$
$$\Rightarrow \bar{A}_{dyn} = \frac{1}{2\sqrt{2}} \cdot \sqrt[4]{\frac{\ell^3 \cdot \rho_{dyn}}{E \cdot J}} \quad (7)$$

Alternatively, the maximum load acting on the track can be calculated by the following formula *(Alias 1984, Prud'home et al. 1976)*:

$$Q_{total\,max} = Q_{wh} \cdot (1 + \frac{Q_\alpha}{Q_{wh}} + 2 \cdot \frac{\sqrt{[\sigma^2(\Delta Q_{NSM})] + [\sigma^2(\Delta Q_{SM})]}}{Q_{wh}}) \quad (8)$$

where $\sigma(\Delta Q_{NSM})$ is the standard deviation of the dynamic component due to the non-suspended masses and $\sigma(\Delta Q_{NSM})$ is the standard deviation of the dynamic component due to the suspended masses of the railway vehicle.

In this case, in order to calculate the reaction R of the sleeper without underestimate the value of the actions, the following are recommended (Giannakos et al. 2007):

(a) use $\bar{A}_{stat}$ for the static and semi-static components of the load (i.e. load due to the wheel and cant deficiency, respectively), and
(b) do not use any distribution the dynamic component of the total load.

By applying the aforementioned methodology to equation (8) the following equation is derived for the reaction of the sleeper:

$$Q_{total\ max} = Q_{wh} \cdot (\bar{A}_{stat} \cdot \left(1 + \frac{Q_\alpha}{Q_{wh}}\right) + 2 \cdot \frac{\sqrt{\left[\sigma^2(\Delta Q_{NSM})\right] + \left[\sigma^2(\Delta Q_{SM})\right]}}{Q_{wh}}) \quad (9)$$

3.3.2 *Calculation of the load due to the non-suspended masses*

The theoretical analysis leads to the following equation for the standard deviation of the load that is caused by the non-suspended masses (Giannakos 2004):

$$\sigma(\Delta R_{NSM}) = k_\alpha \cdot V \cdot \sqrt{m_{NSM} \cdot h} \quad [t] \quad (10)$$

where R is a symbol of the load region (or Q) of the sleeper, k_a is a coefficient that depends on the rail running table geometry (Table 1), V the speed in km/hr, m_{NSM} in [t] the non suspended mass (NSM), and h the track stiffness in kN/mm.

By setting as reference the measured response of a track laid in the SNCF railway network with the following characteristics (Giannakos et al 1990, 1994, Ambakoumkin K., Loizos A., et al 1992): (i) $V = 200$ km/h, (ii) $(m_{tr} + m_{NSM}) = 1.7804$ t, and (iii) $h = 75$ kN/mm, the following equation is derived:

$$\sigma(\Delta R_{NSM}) = k'_\alpha \cdot \frac{V}{200} \cdot \sqrt{\frac{m}{1.7804}} \cdot \sqrt{\frac{h_{TR}}{75}} \quad [t] \quad (11)$$

where m, measured in tones, is the NSM of the vehicle and track, the coefficient k'_α is equal to 30% of k_l whose values are presented in Table 2, and h_{TR} is given by equation (7) (Giannakos et al 1994, Giannakos et al., 2007).

Table 1. Values for k_α (after Giannakos et al 1994).

	Non ground rail	Ground rail
Coefficient k_α	$0.00779 \div 0.01558$	$0.00389 \div 0.00584$

Table 2. Values of coefficient k'_α for the standard deviation of accelerations according to SNCF.

k_l values	Ground Rail running table	Non-ground Rail running table
maximum	$1.35 \div 1.8$	3.6
minimum	0.9	1.8

Hence, two ways are proposed (Giannakos et al 1991, 1994) in order to calculate the load due to the non-suspended (unsprung) masses:

(i) apply equation (10) with the values of k_α from Table 1, or
(ii) apply comparative equation (11) with the values of $k'_\alpha = 0.3k_l$ and use values for k_l provided in Table 2.

It must be noted that for the calculation of k_α the measurement data of the French network were used. Measurements gave a fluctuation amplitude of an "intermediate" coefficient k_l from 12 to 6 for a non-ground rail running table, and k_l from 4.5 to 3 for a ground surface of the rail running table.

3.3.3 *Load due to the suspended (sprung) masses*

The standard deviation of the load is given by the formula (SNCF 1981, Giannakos et al., 1994):

$$\sigma(\Delta R_{SM}) = \frac{V - 40}{1000} \cdot N_L \cdot Q_{wh} \quad (12)$$

where N_L is the mean standard deviation of the longitudinal level condition of the track. N_L is given as the standard deviation of the longitudinal level defects along the track, which for the SNCF tracks are in the order of 0.7 mm. For the Greek network, N_L has to be taken equal to 1 mm. According to the experts assessment of the French and Greek railways.

3.3.4 *Dynamic load distribution*

In high frequencies, the superstructure does not respond because of its low eigenfrequency, therefore, for safety reasons, we assume that dynamic loads (semi-statics are included) are not distributed to the adjacent sleepers, in contrast to static loads that are distributed. Therefore it is recommended that the service load should be considered as the static load increased by 3 times the standard deviation of the dynamic load (P = 99.7%):

$$R_{serv} = \bar{A}_{dyn} \cdot (Q_{wh} + Q_{\alpha}) + 3 \cdot \sqrt{\left[\sigma\left(\Delta R_{NSM}\right)\right]^2 + \left[\sigma\left(\Delta R_{SM}\right)\right]^2} \quad (13)$$

where $\bar{A}_{dyn}$ is calculated by (2) where instead of ρ_{total} we use the dynamic stiffness h_{TR} (or ρ_{dyn}) of the track, given by the equation (7).

3.3.5 *Slab track – distribution of reactions*

3.3.5.1 *German Bibliography*

The real seating surface of the sleeper on the slab is depicted in Figure 3a above. For calculation purposes we can consider the equivalent seating surface as depicted in Figure 3b.

Under the assumption that the acting seating surface of the sleepers (their reaction) under the static load (on rail) of the wheel Q_{stat}, is:

$$F = (d - m) \cdot b_1 \quad [in\ \ mm]$$
$$b = \frac{F}{2 \cdot \ell} \quad [in\ \ mm] \quad (14)$$

Where:

ℓ = sleeper's length

b_1 = average width of the sleeper at the loaded seating area in mm

m = the unloaded seating area in the middle of the sleeper

The design philosophy of slab-track's continuously reinforced concrete pavement (CRCP) is the same as that of "rigid pavements on elastic foundation".

For the calculation of the forces acting on the sleepers –and through them- on the bearing layer (B.L), in Germany the Zimmermann method is used. The basic equations of this method are given in the Eqns 15 and 16 that follow

Deflection of rail:

$$y = \frac{(\Sigma Q_i \cdot n_i)}{2 \cdot (b \cdot C) \cdot L} \quad (15)$$

Action on the sleeper [see (6) above]:

$$S = \frac{(\Sigma Q_i \cdot n_i) \cdot \ell}{2 \cdot L} = (b \cdot C) \cdot \ell \cdot y \quad (16)$$

where:

$$n = \frac{\sin\xi + \cos\xi}{e^{\xi}} \qquad \xi = \frac{X}{L},$$
$$L = \sqrt[4]{\frac{4 \cdot E \cdot J}{b \cdot C}}, \qquad b \cdot C = c / \ell \quad (17)$$

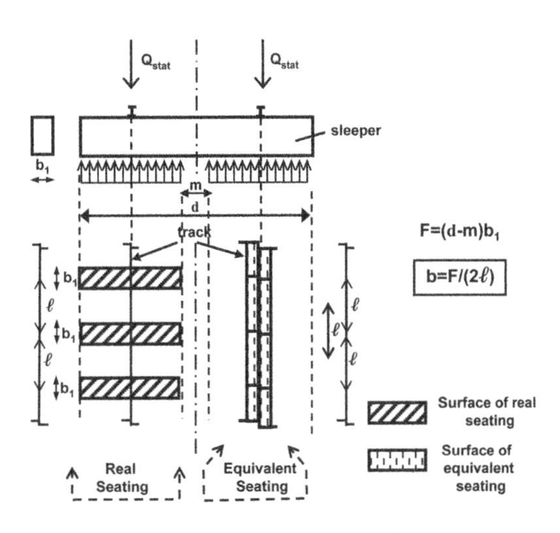

Figure 3. (a) Conventional distribution of pressures (reactions) of sleeper (b) Real and equivalent seating of the track.

and E modulus of elasticity of the rail, J moment of inertia of the rail , C ballast modulus, ℓ sleeper distance, b the theoretical seating width of the sleeper, X distance of the wheel load from the sleeper, $c = \rho$ reaction coefficient of the sleeper, Q wheel load, L elastic length of the track .

3.3.5.2 *Proposed methodology*

The philosophy behind Eqn (13) can also be applied in the case of a slab track system. Therefore, the authors propose the use of Equation (13) for the calculation of the forces that act on a sleeper instead of Equation (16).

4 DESIGN OF CRCP OF THE SLAB TRACK

The CRCP of the slab track is designed (as in pavements too, with a concrete slab):

(a) as a layer of appropriate thickness with contraction joints to concentrate the cracks:
   (i) without dowels (Figure 4)
   (ii) with dowels(Figure 5)

(b) as continuous reinforced concrete slab (with free cracking) and percentage of reinforcement $\rho_s = 0.8 \div 1.0\%$ in the middle of the concrete layer (Figure 6)

5 THE ROLE OF THE FASTENING SYSTEMS

The role of the fastening systems is to firmly fasten the rails in the correct position and to undertake, absorb (through special arrangements), and

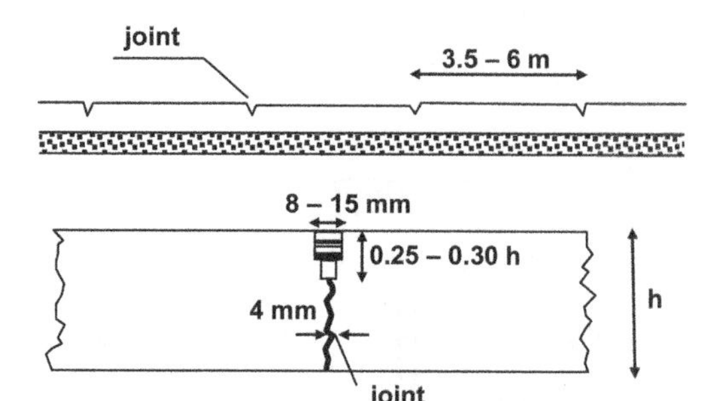

Figure 4. Concrete slab with contraction joints without dowels (upper part joints, lower part detail of a joint).

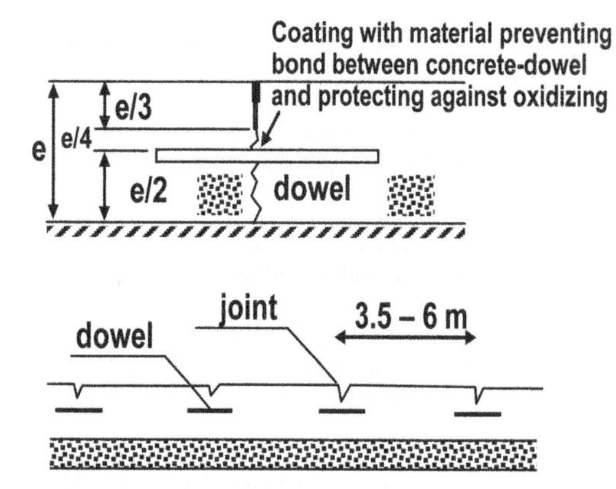

Figure 5. Bearing concrete slab with contraction joints with dowels (right joints in slab track, left detail of joint).

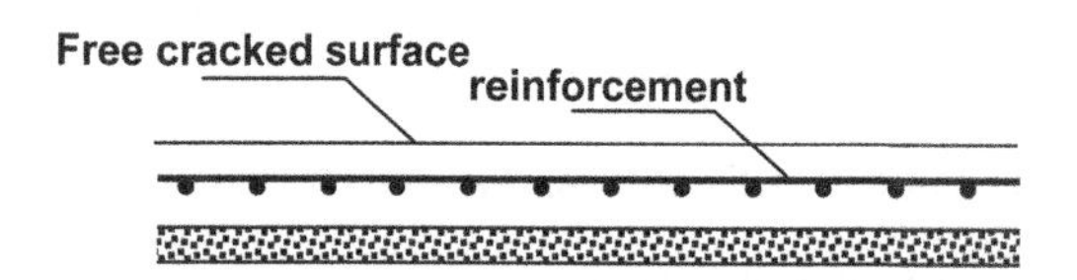

Figure 6. Continuous concrete slab with continuous reinforcement, $\rho = 0.8 \div 1.0\%$ (in the middle of the width).

transfer the dynamic energy created by the vehicles movement, braking forces etc. The elastic pads of these resilient fastenings must be of a stiffness $c_{stat} = 22.5$ kN/mm $\pm$ 2.5 kN/mm or even lower. In Figure 7 the typical assembly of such a fastening is presented. "*Important is that ratio of secondary deflection to rail deflection d / y is smaller than 3–4%. The secondary deflection is the additional rail deflection between the rail supports. At values higher than >7% corrugation and head checks can be expected above the rail supports and it can lead to buzzing in the vehicles.*" If the rail deflection y is increased by using fastening systems with lower stiffness the ratio d/y will be reduced. (Eisenmann, 2004)

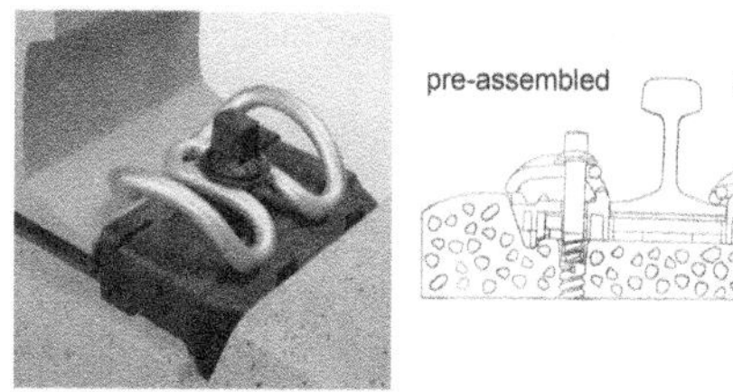

Figure 7. The fastening system 300 for slab track.

High dynamic elasticity is indispensable for a low maintenance track with high availability. Otherwise an increased amount of wear, wave corrugation and head checks must be expected.

Current developments have the target to go even lower than 22.5 kN/mm.

6 DESIGN PHILOSOPHY AND BEHAVIOR OF SLAB TRACK UNDER LOADS

The design philosophy and the behavior of the rigid bearing concrete layers (CRCP, CTB) of the slab track is the same as these of rigid pavements, supported on an elastic base.

Advantages of slab track in comparison to pavements are:

(a) Slab track has determined (small) width, hence, longitudinal joints in the slab are not required, and
(b) the acting loads are applied in determined constant positions (rails), away from the edges (the extremities of the slab) of the bearing structure; that is in this case they are acting on favorable places.

For the design of the slab track, after the choice of its formation (structure) the acting loads must be determined as well as the level of its resistance.

The state of actions of the slab track is derived from:

- The static and dynamic loads from the railway vehicles whose magnitude is estimated as in the sleepers on ballast
- The temperature fluctuation –mainly- in the CRCP

For the calculation of the actions on the sleepers and through them on the CRCP the Zimmerman method is applied. According to this method, the railway track is modeled as an infinite beam founded on an elastic medium, supported by Winkler springs at the position of the sleepers (see also Tsoukantas, et al. 1999, 2006).

For the calculation of the stresses on the bearing layers of the slab track, CRCP and CTB, the Eisenmann method is mainly used. This method, apart of its simplicity, gives the possibility to take into account the existence of the adhesion between the layers of the multilayered system (e.g. see Figure 8) or to ignore it.

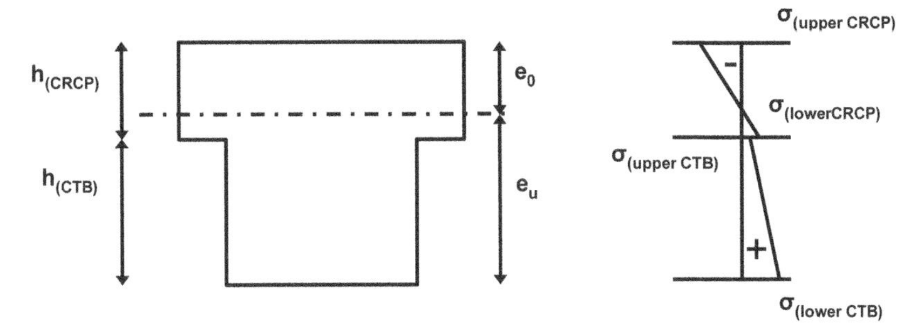

Figure 8. Geometric characteristic of multi-layered system.

In both cases and using the Zimmermann method, the bending moments on the equivalent system are derived, depending on the resistance modulus K of the bearing layer (according to Odemark), the equivalent modulus of elasticity E of the equivalent system, and the thicknesses of the layers. Based on the above data the stresses at the interfaces of the different layers are calculated.

For the concrete bearing layer the repetitive character of the external loads imposes the investigation of the reduction of the strength of concrete (mainly the tension strength). The fatigue of concrete is of decisive importance. The stresses owing to temperature are considered as permanently acting and they reflect the lower value of the (permanent) actions. The stresses from the loads of vehicles reflect the maximum stresses –of small duration- but non-foreseen magnitude every time and of repetitive character. Generally the Smith diagram (extended for $2*10^6$ cycles of loading) may be used, to estimate the reduced concrete tension strength (Figure 9).

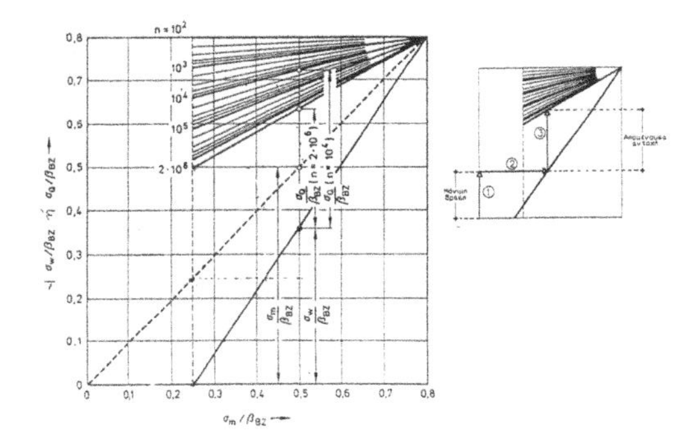

Figure 9. The extended Smith diagram for $2*10^6$ cycles.

7 SLAB TRACK APPLICATION

7.1 *General*

Slab track is applied to tunnels, bridges and to the plain track.

(a) In tunnels, the CRCP can be based directly on the concrete of the bottom of the tunnel.
(b) In bridges the connection of the "body" of slab track to the bridge must be carefully studied, in relation to the type of bearer (simply supported, continuous beam from reinforced concrete or prestressed concrete etc.) and the span of the bridge. Appaarently influences due to earthquakes demand particular, specific study.
(c) On plain track it is obligatory for the CRCP to be placed on the CTB layer.
(d) Specific formation of the slab track is demanded in transitional areas of adaptation e.g. in a case of transition from slab track to ballasted track etc.

7.2 *Application of slab track in Greece*

In Greece, with "soft" limestone ballast, a pilot project for the first application of slab track was materialized and since 2004 a section of 11.5 km has been in operation in the main corridor Athens – Thessaloniki, which is designed for speeds from 200 to 250 km/h mixed traffic (passenger and commercial trains). Today the operational speed is 160 km/h for mixed traffic. Hellenic Railways Organization (OSE) is the only user and manager of infrastructure. In order to (Tsoukantas et al., 2006):

(a) examine the usefulness and the financial influence (during time) of the application of slab track
(b) cover the need for acquainting know-how and deepening, in Greece, in the technology of slab track
(c) possibly adapt some of the applied slab track systems in Europe to the Greek conditions

OSE established a scientific team for the implementation of the first application of slab track in Greece, under the supervision (i) of the General Directorate of Infrastructure and (ii) of the General Directorate of ERGOSE, its subsidiary company for the co-funded projects by the EU and Greek Government. This scientific team consisted of Greek and German external Consultants and academics.

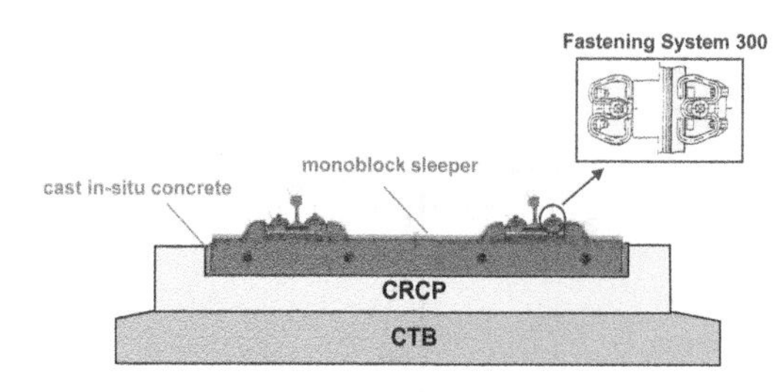

Figure 10. Schematic representation of Rheda – Sengeberg system of slab track.

The authors participated in this team. After a thorough investigation, OSE decided to adopt the Rheda – Sengeberg system of slab track because of its multiyear experience and its very good behavior (Figure 10). It is remarkable that up to now the Rheda – Sengeberg system has not demanded any substantial maintenance. Moreover, according to the experience gained in Germany during the last 30 years, the total annual maintenance cost is less that 25% of the correspondent cost of the ballasted track. In Greece, OSE decided to act conservatively for the first application of slab track and to adopt this system because it presents the longest period of use.

8 CONCLUSIONS

In this paper a methodology for the calculation of the actions on a slab track system is presented. The method is compared with others proposed in the French and German bibliography. Moreover, the design philosophy of slab track systems is analysed.

The main advantages of the slab track (or ballastless track) are that it is the recommended type of infrastructure for high-traffic, high-speed lines, it offers increased passenger comfort, has a longer life The results of the calculations with the proposed method were found to be in agreement with the observed in situ behavior of the sleepers (cracking, minimal maintenance compared to the ballasted track. etc), as well as with laboratory experiments.

Finally, the influence of the stiffness of the fastening on the load transmitted on the sleeper is briefly discussed.

REFERENCES

Abakoumkin K., Loizos A., Trezos K., and Lymberis K. 1992-1993. Normal gauge line monoblock sleepers made of pre-stressed concrete, *NTUA/ Sector of Transport & Communication Infrastructure, Athens.*

Alias J. 1984. La voie ferree, IIeme edition, Eyrolles, Paris.

Eisenmann J., 2005. Die Schiene als Fahrbahn. *Eisenbahningenieur 6/2005*

Eisenmann J., 2004. Die Schiene als Tragbalken, *Eisenbahningenieur 5/2004*

Eisenmann J., Duwe B., Lempe U., Leykauf G., Steinbeisser L., 1979. Entwinklung, Bemessung und Erforschung des schotterlosen Oberbaues, Rheda, *ETR (34)*

Fastenrath Fritz, 1981. *Railroad Track-Theory and Practice*, Frederic Ungar Pub.Co., New York, part 2, The Rail as support and Roadway, theoretical principles and practical examples, by J. Eisenmann

Giannakos K., Loizos A., 2007. Loads on railway superstructure – Influence of high-resilient fastenings on sleepers loading. *Proceedings of Advanced Characterization of Pavement and Soil Eng. Materials, Athens, Greece, 20–22 June 2007, 1363.*

Giannakos K., Tsoukantas S., 2006. Ballasted and Unballasted Tracks – Need for Fastenings With Soft Pads. *International Railway Symposium, Ankara – Turkey, 14–16 December*

Giannakos K., 2004. *Actions on the Railway Track*, Papazissis publications, Athens,

Giannakos K., Vlassopoulou I., 1994. Load of Concrete Sleepers and Application for Twin-Block Sleepers, *Technical Chronicles* (14), Scientific Journal of TCG.

Giannakos K., Vlassopoulou I., 1991. Problems and Solutions for the system "sleeper-ballast" at the OSE, *OSE/Track Directorate, Athens.*

Giannakos K., Vlassopoulou I., 1990. Study on the «sleeper-ballast» system, *OSE/Track Directorate, Athens.*

Leykauf G., Lechner B., Stahl W., 2006. Trends in the use of slab track/ballastless tracks, *RTR Special Issue*, September 2006.

Leykauf G., Mattner L., 1990. Elastisches Verformungsverhalten des Eisenbahnoberbaus – Eigenschaften und Anforderungen, *Eisenbahningenieur* 41(3)

Prud' Homme A., Erieau J., 1976. Les nouvelles traverses en beton de la SNCF, *RGCF* 2/1976.

SNCF/Direction de l' Equipement, 1981. Mechanique de la Voie, Octobre 3

Tsoukantas S., Giannakos K., Topintzis T., Zois H., 2006. System Rheda 2000 from the point of view of a Structural Engineer – the most modern evolution of superstructure technology in Railway projects, *Proceedings of Concrete Conference, TCG, Alexandroupolis, 24–27 October 2006.*

Tsoukantas S., Giannakos K., Kritharis A., Topintzis T., 1999. Examination of some factors that affect the actions in concrete sleepers, *Proceedings of Concrete Conference, TCG, Rethymnon, 25–27 October 1999, Volume I, 247–259.*

Tsoukantas S., 2000. Research program for the Investigation of problems for the application of slab track in Greece, OSE – ERGOSE, Athens

Tsoukantas S., Giannakos K., Topintzis T., Zois H., 2006. System Rheda 2000 from the point of view of a Structural Engineer – the most modern evolution of superstructure technology in Railway projects *Proceedings of Concrete Conference, TCG, Alexandroupolis, 24–27 October 2006.*

Advances in Transportation Geotechnics – Ellis, Yu, McDowell, Dawson & Thom (eds)
© 2008 Taylor & Francis Group, London, ISBN 978-0-415-47590-7

Geotechnical aspects of ballasted rail tracks and stabilising underlying soft soil formation

B. Indraratna
School of Civil, Mining and Environmental Engineering, Faculty of Engineering, University of Wollongong, Australia

H. Khabbaz
School of Infrastructure and the Environment, Faculty of Engineering, University of Technology Sydney (UTS), Australia

ABSTRACT: The necessity of keeping a competitive edge against other means of transportation has increased the pressure on the railway industry to improve its efficiency and decrease maintenance and infrastructure costs. A comprehensive research program was launched at the University of Wollongong (UoW), sponsored by the Australian Cooperative Research Centre for Railway Engineering and Technologies to investigate mechanical behaviour of ballast under cyclic loading as well as effectiveness of using geosynthetics in trackbed and various methods to stabilise tracks underlying soft soil formation. The research findings show that the bonded geogrids-geotextiles can improve the performance of ballasted tracks and the installation of prefabricated vertical drains in soft formation can reduce the preloading period significantly. According to experimental and field trial investigations it has been found that the ballast deformation and breakage can be reduced if appropriate ballast grading and track confining pressure are applied.

1 INTRODUCTION

Rail tracks are conventionally founded on ballast for several reasons, including economy, drainage and ease of maintenance. However, ballast breaks down and deteriorates progressively under heavy train cyclic loads, leading to costly rail track maintenance. Hence, an accurate quantification of mechanical behaviour of ballast, particularly at presence of underlying soft soil formation, is essential for stabilisation measures of railway tracks. On the other hand, the cost of substructure maintenance can be significantly reduced if a better understanding of the rail substructure behaviour is obtained.

This paper addresses a number of research programs, including theoretical and experimental investigations, carried out at the University of Wollongong (UoW) under the auspices of the Cooperative Research Center for Railway Engineering and Technologies (RAIL-CRC) and in collaboration with other railway organisations. Several large-scale triaxial and consolidometer facilities for testing ballast and soft soil have been designed and built at UoW to conduct the experimental side of this program. As expected, these testing rigs have provided more realistic information on the ballast stress-strain and degradation characteristics, using the prototype rock fragments. The following investigations have been accomplished at UoW, and the main findings of these studies are presented in this paper:

- Developing an elasto-plastic stress-strain constitutive model for ballast particles incorporating the degradation of particles;
- Studying the effects of ballast particle size distribution and confining pressure on ballast shear strength, settlement and degradation, based on monotonic and cyclic tests using large-scale cyclic triaxial apparatus;
- Investigating the prospective use of different types of geosynthetics for enhancing the performance of recycled and fresh ballast and quantifying the effect of their inclusion for reducing ballast degradation and excessive track deformation based on large-scale triaxial testing. A number of plane strain finite element analyses have been carried out to obtain the optimum location of geosynthetics in the trackbed.
- Optimising the accelerated primary consolidation of track soft formation using prefabricated drains vertical drains (PVDs). A number of analytical and numerical models have been developed and verified through several case studies. A large-scale consolidometer has also been employed to validate the models and examine the effect of the smear zone on the performance of the PVDs.

- Modelling and evaluating the effect of suction induced by native vegetation in the vicinity of the rail tracks;
- Conducting a field trial to validate the research findings.

2 DEVELOPMENT OF A CONSTITUTIVE MODEL FOR BALLAST PARTICLES INCORPORATING THE DEGRADATION OF PARTICLES

2.1 *Ballast degradation*

Ballast deforms and degrades progressively under heavy cyclic loading. The degradation of ballast particles can occur in several ways: grinding off of small-scale asperities (abrasion) in which resulting fines cause fouling and reduce drainage; breaking of fragments and angular projections, which influence the initial settlement; and fracturing or splitting of individual particles. This breakage is responsible for the long-term stability and safety of the track.

Generally, the main factors that affect ballast breakage can be divided into three categories: (1) ballast properties related to the characteristics of the parent rock (e.g. hardness, specific gravity, toughness, weathering, mineralogical composition, internal bonding and grain texture); (2) physical properties associated with individual particles (e.g. soundness, durability, particle shape, size, angularity and surface smoothness); and factors related to the assembly of particles, and (3) loading conditions (e.g. confining pressure, initial density or porosity, thickness of ballast layer, ballast gradation, presence of water or ballast moisture content, cyclic loading pattern including load amplitude and frequency).

2.2 *Constitutive modelling of ballast*

Research group at UoW (Salim & Indraratna 2004; Indraratna & Salim 2002, 2005) developed a continuum mechanics-based constitutive model incorporating dilatancy and plastic flow rule to predict ballast behaviour and particle breakage. The factors contributing in particle breakage, as explained in Section 2.1 were taken into account for the model development. The model uses a generalised 3-D system to define contact forces, stresses and strains in granular media including the plastic potential, hardening function and particle breakage parameters. The model is based on the critical state concept and the theory of plasticity with a kinematic-type yield locus (constant stress ratio). This model was verified using large-scale triaxial laboratory results at various confining pressure (Fig. 1). The developed constitutive model contains 11 parameters, which can be evaluated using drained

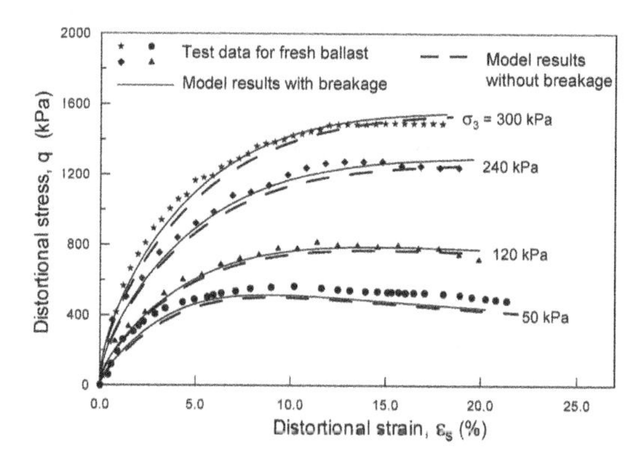

Figure 1. Analytical prediction of stress-strain of ballast with and without particle breakage compared to test data (modified after Salim & Indraratna 2005).

triaxial test results alongside the particle breakage measurements. The model was further extended for the more complex cyclic loading to embrace the number of load cycles, load amplitude and frequency. The details are given by Indraratna & Salim 2005.

3 EFFECTS OF BALLAST PARTICLE SIZE DISTRIBUTION AND CONFINING PRESSURE ON BALLAST DEGRADATION

According to research at UOW (Indraratna et al. 2005), the initial permeability of ballast would drop by approximately 50% if the moderately graded distribution were employed instead of the very uniform distribution. However, in the absence of fouling, the moderately graded ballast is still highly free draining, and in terms of degradation, settlement and strength, is far superior to current ballast grading specifications. The test results of ballast varying the gradation indicated that even modest changes in the coefficient of uniformity (C_u) can substantially affect strains and breakage behaviour, and that a distribution similar to the moderate grading would give improved track performance. Based on these findings, Indraratna et al. (2004) recommend using a modified particle size distribution range for railway ballast with C_u exceeding 2.2, but not more than 2.6, in comparison to very uniform (conventional) grading of $C_u = 1.4$–1.5. This recommended gradation, which is slightly more well-graded than the current Australian Standards (AS 2758.7 1996) of ballast grading, is shown in Figure 2.

It was determined that confining pressure should be considered as an important design parameter for railway lines. Significant reductions in settlement, lateral spreading and particle degradation were observed to occur with increased confining pressure, together with improved track resiliency and ride quality. The highly dilatant behaviour occurring at low confining

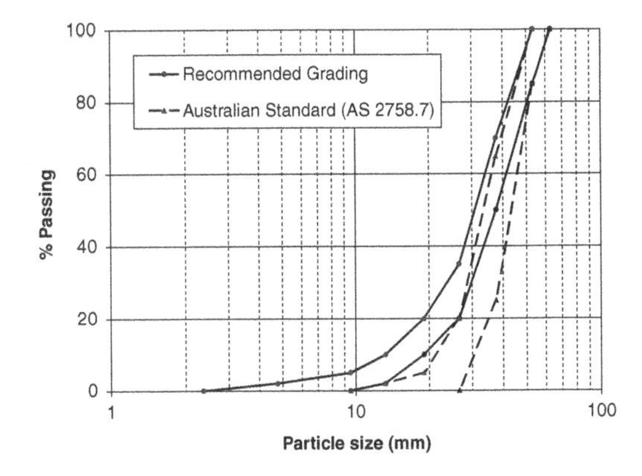

Figure 2. Recommended ballast gradation in comparison with current Australian Standard (after Indraratna et al. 2004).

pressure was acknowledged as particularly unfavorable in relation to permanent deformation and particle degradation, and should therefore be avoided in rail tracks. Simple, practical methods of increasing the in-situ level of lateral confinement were suggested by Indraratna et al. 2007.

Enhancement of the lateral track confining pressure could be achieved practically by utilizing one or more of the following methods:

- Incorporating lateral restraints such as barriers at the extremities of the sleepers or shoulder ballast;
- Increasing the sleeper-ballast frictional characteristics by changing the shape of the sleepers, reducing the sleeper spacing, or increasing the sleeper roughness;
- Placing geosynthetic layers within the railway substructure, most suitably at the ballast-subballast interface (Indraratna and Salim, 2003), to promote interlock between the geosynthetics and ballast;
- Increasing the effective overburden acting on the load bearing ballast by utilising greater ballast compaction during maintenance, compaction of the shoulder and crib ballast, or an increase in the height of the shoulder ballast.

4 EFFECTS OF GEOSYNTHETICS ON RAIL TRACK DEFORMATION

A wide range of geosynthetics with different properties have been developed to meet highly specific requirements corresponding to various uses in new rail tracks and track rehabilitation for more than three decades. Enhancing the performance of rail tracks by composite geosynthetics is now seriously considered by rail industry. Based on relatively low cost and the proven performance of geosynthetics in a number of railway applications, the UoW research team have conducted a comprehensive study to investigate the effects of the different types of geosynthetics on fresh ballast, recycled ballast, track drainage and stabilisation of railway formation.

It was expected that the use of geosynthetics would encourage the reuse of discarded ballast from stockpiles, reducing the need for further quarrying and getting rid of the unsightly spoil tips often occupying valuable land areas in the metropolitan areas. The fundamental and experimental studies (Indraratna et al. 2003, 2004) proved that a geogrid bonded with a drainage fabric (geotextile) will increase the load bearing capacity of the ballast bed while minimising the lateral movement of ballast and reducing degradation. Use of the composite geosynthetics also prevents the occurrence of liquefied soil (slurry) and its upwards pumping that would foul the ballast.

The feasibility of finite element analysis using the PLAXIS code to simulate the behaviour of fresh and recycled ballast with or without geosynthetics in a large scale triaxial rig was investigated and used to obtain the optimum location of geosynthetics in rail track substructure (Indraratna et al. 2007).

The results of the plane-strain finite element numerical modelling indicated that the optimum location of geosynthetics for improving rail track deformation might be taken 200 mm beneath the sleeper's bottom. However, if the design-required thickness of ballast is more than 200 mm, it will be more convenient to place the geosynthetics at the bottom of the ballast bed (i.e. at the ballast-capping interface), to allow for maintenance requirements such as ballast tamping and cleaning.

5 PERFORMANCE OF VERTICAL DRAINS SUBJECTED TO CYCLIC TRAIN LOADS

Low-lying areas with high volumes of plastic clays can sustain high excess pore water pressures during both static and cyclic loading. The effectiveness of prefabricated vertical drains (PVDs) for dissipating pore water pressures and factors influencing its efficiency (e.g. smear effect) was investigated at UoW (Indraratna et al. 2005). In poorly drained situations, the increase in pore pressures will decrease the effective load bearing capacity of the formation. Even if the rail tracks are well built structurally, undrained formation failures can adversely influence the train speeds apart from the inevitable operational delays.

The quality of a robust rail track construction is defeated, if the underlying soft soil is weak and compressible, thereby leading to unacceptable differential settlement or pumping of slurry soil (under heavy axle loads) causing ballast fouling. In this situation, the improvement of soft clays beneath rail tracks is imperative, and the use of PVDs prior to track construction is now encouraged in many coastal areas in Australia. Pre-construction consolidation of formation soil will eliminate excessive post-construction settlement of the track as well as increase the shear strength of the soil. Moreover, the PVDs will continue to function in the long-term to provide rapid pore pressure dissipation interfaces under cyclic load, especially in low-lying areas subjected to high annual rainfall.

If applying pre-consolidation with surcharge alone, the required time can be too long for busy rail track construction. Installation of vertical drains can reduce the preloading period significantly by decreasing the drainage path length in the radial direction (Indraratna & Redana 2000). When a higher surcharge load is required to meet the expected settlement and the cost of surcharge becomes costly, application of vacuum pressure with reduced surcharge loading can further accelerate consolidation (Rujikiatkamjorn & Indraratna 2007). The external negative load is applied to the soil surface in the form of vacuum through a sealed membrane system (Indraratna et al. 2005). The higher effective stress is achieved by rapidly decreasing the pore water pressure, while the total stress remains the same. Therefore, potential shear failure due to excess pore pressure can also be avoided using this method.

After rail track construction, it is well-known that the rail track structure including underlying soil formation usually subjects to cyclic load from heavy freight trains. Ballast fouling by local soil liquefaction occurs where drainage condition is poor. Researchers at UoW indicated the excess pore pressure dissipation expedited, as the existing high permeable PVDs could perform as an additional drainage hence, speeding up the dissipation of excess pore pressure built up due to cyclic loads.

As described earlier, PVDs accelerate consolidation and curtail lateral movements. The stability of rail tracks and highways built on soft saturated clays is often governed by the magnitude of lateral strains, even though consolidation facilitates a gain in shear strength and load bearing capacity. If excessive initial settlement of deep estuarine deposits cannot be tolerated in terms of maintenance practices (e.g. in new railway tracks where continuous ballast packing may be required), the rate of settlement can still be controlled by: (a) keeping the drain length relatively short, and (b) optimising the drain spacing and the drain installation pattern. In this way, while the settlements are acceptable, the reduction in lateral strains and gain in shear strength of the soil beneath the track improve its stability significantly.

5.1 *Laboratory testing*

An experimental study was conducted to investigate the influence of prefabricated vertical drains on the cyclic behaviour of soft clay, using large-scale triaxial equipment simulating typical cyclic loads encountered in railway environments. The excess pore water pressure ratio and the post-cyclic loading dissipation rate were considered in the assessment of the performance of PVDs.

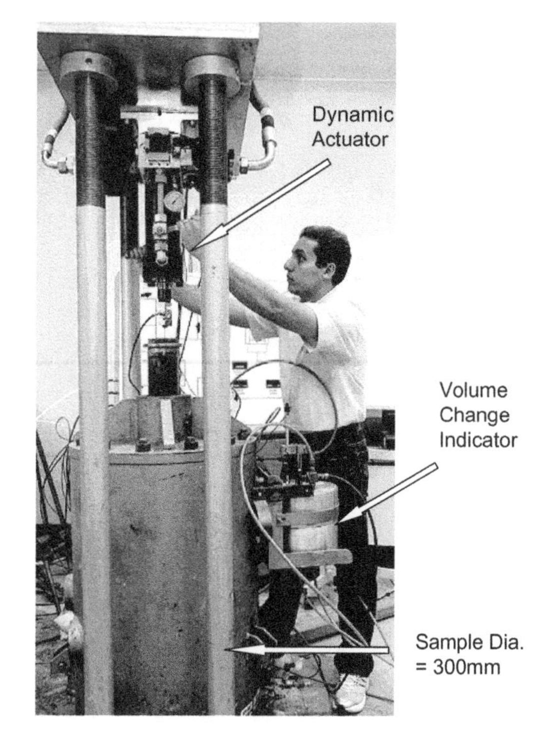

Figure 3. Large Scale Triaxial rig with a dynamic actuator designed and built at University of Wollongong.

The large-scale triaxial test was used to examine the effect of cyclic load on the radial drainage and consolidation by PVDs is shown in Figure 3. This testing chamber is capable of accommodating specimens of 300 mm in diameter and 600 mm in height. The excess pore water pressure is monitored via miniature pore pressure transducers.

During the application of cyclic loading, the PVDs reduced the rate of generation of excess pore water pressure, when compared to the case without PVD. Under the same cyclic stress ratio, the magnitude of the excess pore pressure generated was significantly less when PVDs were present (Attya et al. 2006). As expected, irrespective of the magnitude of the cyclic stress ratio and the number of cycles, the development of excess pore pressure was the least for the part of the soil specimen nearest to the central PVD, as monitored by the transducers located closest to the PVD.

While further testing is still ongoing to study in detail the cyclic behaviour of soft clay stabilised by PVD, the findings reported here clearly suggest that railway tracks will benefit considerably using PVDs installed in the soft subgrade, by reducing the risk of undrained failure and soil slurrying under high excess pore pressures. It was also shown that short prefabricated vertical drains may be used under rail tracks to dissipate cyclic excess pore pressure and to curtail lateral displacements to improve stability, if preloading is not used. However, where preloading can be applied, deeper soft formations can be stabilised using longer vertical drains, for more resilient soft soil foundations (Attya et al. 2006).

6 USE OF NATIVE VEGETATION FOR STABILISATION OF SOFT FORMATIONS

Tree roots provide three independent stabilising functions: (a) reinforcement of the soil, (b) dissipation of excess pore pressure and (c) establishing matric suction increasing the soil shear strength. The matric suction established in the root zone propagates radially and contributes in ground stabilisation near the root zone. Using native vegetation in semi-arid climates and coastal regions of Australia has become increasingly popular for stabilising railway corridors built over expansive clays and compressive soft soils. As a consequence of passage of heavy trains or ballast tamping to reshape and level the ballast, a ballast bowl (or ballast pocket), in which water accumulates and softens the ground, can be formed under the track granular layer. Using native vegetation can be a cost effective and environmentally friendly solution to remediate this problem.

Indraratna et al. (2006) have developed a mathematical model for the distribution of tree root water uptake within the root zone. This proposed model combines the effects of soil matric suction, root density and potential transpiration rate. Consequently, the rate of tree root water uptake, $S(x, y, z, t)$, can be formulated as:

$$S(x, y, z, t) = f(\psi)\, G(\beta)\, F(T_P) \tag{1}$$

where, $f(\psi)$ is the soil suction factor, $G(\beta)$ is the root density factor, and $F(T_P)$ is the potential transpiration factor. In this study, a fully coupled flow-deformation model was developed for unsaturated soils embracing the proposed root water uptake model. The deformation model used in the governing equations was based on the continuum mechanics theories and the modified effective stress concept for unsaturated soils. The flow model is based on Darcy's law, Henry's law and the conservation of mass. The deformation and flow models were coupled through the effective stress parameters as explained in detail by Fatahi (2007).

The proposed model for predicting the rate of root water uptake was included in a numerical analysis using the ABAQUS finite element code to examine the distribution of soil suction and the profile of the moisture content near a single black box tree located in western Victoria, Australia (Fig 4). The predicted results of a numerical analysis for the matric suction and the moisture content around that single black box tree were compared with the field data taken in May 2005. There were some uncertainties in some field and laboratory measurements of the soil parameters, the actual distribution of tree roots and atmospheric parameters. Nevertheless, a good agreement was generally obtained between the measured and simulated distribution of soil moisture (Fatahi, 2007).

Figure 4. Excavation around a black box tree in western Victoria, Australia for validation of the developed model based on field observations.

7 FIELD TRIAL

Instrumentation of a live track has been part of a field trial for the research into track, ballast and formation interaction (Fig 5). The objectives of the field trial have been as follows:

1. To verify the laboratory research findings through real track conditions and to assess the potential use of geosynthetics in enhancing performance of fresh and recycled ballast and decreasing maintenance cost; and
2. To validate the proposed constitutive and numerical rail track substructure models and to obtain the required dynamic parameters.

The trial track is located at Bulli site on the south coast line of New South Wales (NSW). Five adjacent sections have been instrumented:

- S1: A control section
- S2: Fresh ballast without geosynthetics
- S3: Fresh ballast with geosynthetics (Fig. 6)
- S4: Recycled ballast with geosynthetics
- S5: Recycled ballast without geosynthetics

Figure 5. Installation of displacement transducers at Bulli site.

Figure 6. Installation of geogrids and geotextiles at Bulli site.

According to UoW guidelines, RailCorp-NSW installed a set of pressure cells (21 off), displacement transducers (10 off) and settlement plates (12 off), into the track substructure.

Although the field data collection has not been completed yet, the preliminary comparisons indicates a good agreement between field measurements and previous results obtained at UoW using large scale laboratory rigs.

8 CONCLUSIONS

The research findings at University of Wollongong show that the bonded geogrids can improve the performance of ballasted tracks. It is also indicated that that prefabricated vertical drains (PVDs) can be used when accelerated rate of consolidation and improvement of soft formation of rail track is desired. However, the factors influencing the performance of PVDs, and particularly the disturbance of soil surrounding the drains due to the installation process, should be considered in design and practice procedures. Native vegetation can also improve the conditions of soft soil formation in the vicinity of rail tracks. The ballast deformation and breakage can be reduced if appropriate ballast grading and track confining pressure are applied. Recent field trials clearly have shown that the performance of the UoW recommend ballast grading is much better than the current standard for ballast grading. It is expected that railway organisations will adopt such modernisation in the future design of ballasted tracks.

ACKNOWLEDGMENTS

The authors acknowledge financial support from the Cooperative Research Centre for Railway Engineering and Technologies (RAIL–CRC). They express their special thanks to former and current research fellows and PhD students at University of Wollongong including Dr Wadud Salim, Dr Mohamed Shahin, Dr Joanne Lackenby, Dr Cholachat Rujikiatkamjorn, Dr I.W. Redana, Dr Rohan Walker, Dr Behzad Fatahi and Anass Attya for their assistance and contribution to this research project. The authors also wish to acknowledge the support and cooperation of industry colleagues: David Christie of RailCorp, NSW; Mike Martin of Queensland Rail; Julian Gerbino of Polyfabrics, Australia and Wayne Patter of ARTC.

REFERENCES

Australian Standards AS 2758-7. 1996. Aggregates and rock for engineering purposes: Part 7: Railway Ballast. Australia.

Attya, A., Indraratna, B. & Rujikiatkamjorn, C. 2007. Effectiveness of vertical drains in dissipating excess pore pressures induced by cyclic loads in clays. In Yee, K, et al. (eds), *Proceedings of the 16th Southeast Asian Geotechnical Conference*, Selangor, Malaysia, 8–11 May 2007: 447–451.

Fatahi, B. 2007. Modelling of influence of matric suction induced by native vegetation on sub-soil improvement", PhD Thesis, University of Wollongong, NSW, Australia

Indraratna, B., Fatahi, B. and Khabbaz, H. 2006. Numerical analysis of matric suction effects of tree roots. *Proceedings of the Institution of Civil Engineers Geotechnical Engineering*, 159(2): 77–90.

Indraratna B., Lackenby J. & Christie D. 2005. Effect of confining pressure on the degradation of ballast under cyclic loading. *Geotechnique*, 55(4): 325–328

Indraratna, B., Khabbaz, H., Salim, W., Lackenby, J. & Christie, D. 2004. Ballast characteristics and the effects of geosynthetics on rail track deformation. *Int. Conference on Geosynthetics and Geoenvironmental Engineering*, Mumbai, India: 3–12.

Indraratna, B., Khabbaz, H. & Salim, W. 2006. Geotechnical properties of ballast and the role of geosynthetics. *Journal of Ground Improvement*, ISSMGE, 10(3): 91–101.

Indraratna, B. & Redana, I. W. 2000. Numerical modeling of vertical drains with smear and well resistance installed in soft clay. Canadian Geotechnical Journal, 37: 132–145.

Indraratna, B. & Salim, W. 2002. Modelling of particle breakage of coarse aggregates incorporating strength and dilatancy. *Geotechnical Engineering*, Proc. Institution of Civil Engineers, London, 155(4), Issue 4, 2002, pp. 243–252.

Indraratna, B. & Salim, W. 2005. Mechanics of ballasted rail tracks – A geotechnical perspective. UK: A. A. Balkema – Taylor and Francis.

Indraratna, B., Rujikiatkamjiorn, C. & Sathananthan, I. 2005. Analytical and numerical solutions for a single vertical drain including the effects of vacuum preloading. *Canadian Geotechnical Journal*. 42(4): 994–1014.

Indraratna, B, Shahin, M. A., and Salim, W. 2007. Stabilising granular media and formation soil using geosynthetics with special reference to railway engineering. *Journal of Ground Improvement*. ISSMGE 11(1): 27–44.

Rujikiatkamjorn, C. & Indraratna, B. 2007. Analytical solutions and design curves for vacuum-assisted consolidation with both vertical and horizontal drainage. *Canadian Geotechnical Journal*, 44(2) : 188–200.

Salim, W. & Indraratna, B. 2004. A new elasto-plastic constitutive model for coarse granular aggregates incorporating particle breakage. *Canadian Geotechnical Journal*, 41(4): 657–671.

Advances in Transportation Geotechnics – Ellis, Yu, McDowell, Dawson & Thom (eds)
© 2008 Taylor & Francis Group, London, ISBN 978-0-415-47590-7

Effect of loading frequency on the settlement of granular layer

A. Kono
Railway Technical Research Institute, Japan

T. Matsushima
Tsukuba University, Japan

ABSTRACT: Cyclic loading tests were performed both experimentally and numerically to study the effect of impact loading velocity on the settlement of a shallow granular layer like a ballasted track. In the experiment equal-sized spherical balls were regularly stacked to make a granular layer to compare the result with 2D Discrete Element simulation. It was found that residual settlement after cyclic loading increased with increasing loading velocity when we used the unfinished pinballs with rough surface. On the other hand, the experiment with polished pinballs with smooth surface gave an opposite tendency. The results of DEM simulations imply that a volumetric increase due to the strong impact loading accelerates the deformation of the assemblage of unfinished pinballs, while it leads to considerable reversal of displacement during the off-loading process in the polished pinball assemblage.

1 INTRODUCTION

Differential settlement of the ballast causes a *hanging sleeper* which has a gap between sleeper and ballast layer. The *hanging sleeper* accelerates ballast deterioration and increase wayside vibration and noise. There are some methods for reducing the vibration around the *hanging sleeper*, but the mechanism of the gap formation has not been clarified yet.

This study is devoted to the *hanging sleeper* under the rail joint, which is thought to be affected by impact loading by running vehicle.

We developed an apparatus with a high-capacity stepping motor to apply high frequency cyclic loading. It can generate cyclic loadings of varying amplitude, pulse duration and inter-pulse interval. The cyclic loading tests employed two types of steel ball to avoid the unevenness found in irregularly shaped ballast grains.

Furthermore, a series of simulations by Discrete Element Method were performed to show in detail the behavior of particles during cyclic loading.

2 EXPERIMENT

Fig. 1 shows an apparatus developed for high frequency cyclic loading. These loadings are applied by using a stepping motor on the top of screw. Furthermore a box, 800 mm × 300 mm × 300 mm needs to be set up.

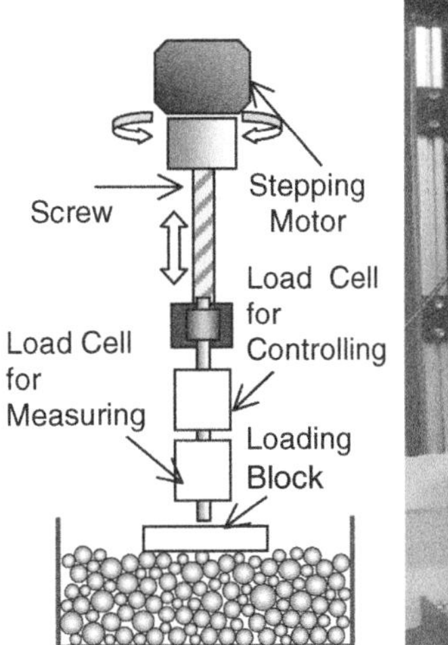

Figure 1. Apparatus.

2.1 *Material*

The box contained two types of steel balls for each case; pinballs and unfinished pinballs as shown in Fig. 2. These equal-sized pinballs have about 11 mm diameter but it is dispersed as shown in Fig. 3. These pinballs are stacked up in a simple stagger pattern only in one vertical face shown in Fig. 4, but not in the other vertical plane to imitate two-dimensional condition.

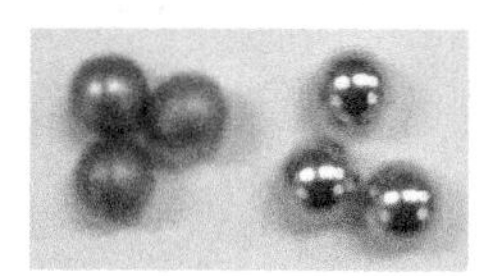

Figure 2. Two types of pinballs.

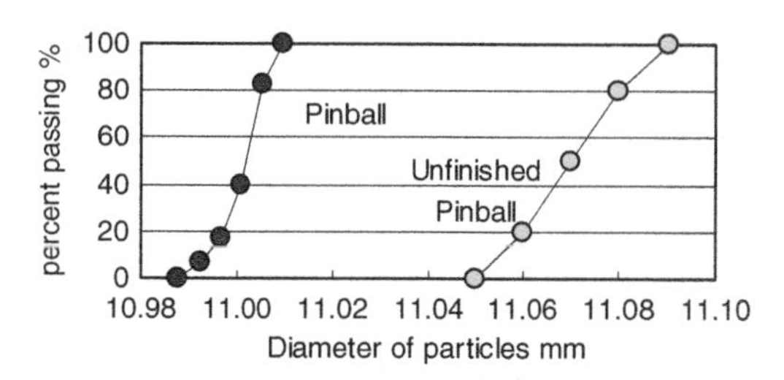

Figure 3. Grain Size Distribution of pinballs.

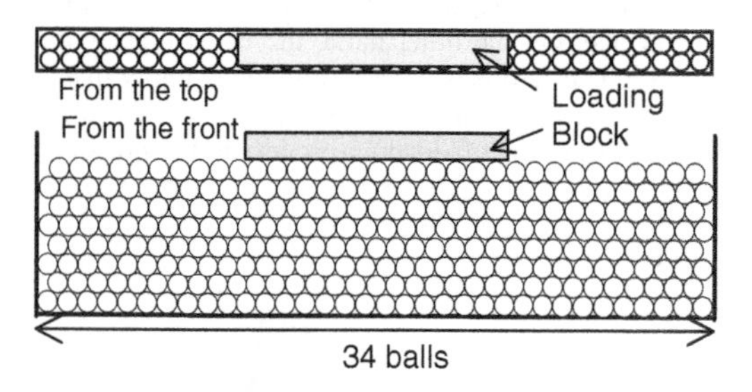

Figure 4. Stacked Up Pinballs.

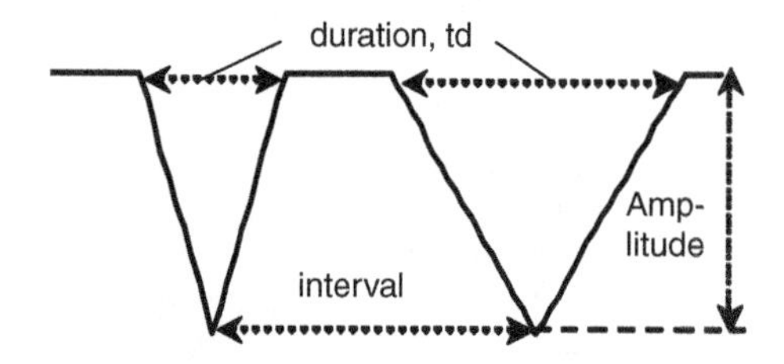

Figure 5. Loading Pattern.

2.2 *Loading conditions*

Various patterns of cyclic loadings can be applied by high-capacity stepping motor, which can control amplitude, duration and interval of cyclic loading independently as shown in Fig. 5.

In this study, the interval and amplitude of cyclic loadings are fixed for each tests of above mentioned materials. Only durations are set for three patterns, 0.2 s, 0.05 s and 0.02 s as shown in table 1. The amplitude is set to 1 kN.

2.3 *Results*

Fig. 6 shows an example of the displacement-time data from a test under 0.2 s duration loading. The displacement is represented by the settlement of the loading block shown in Fig. 1.

Table 1. Loading patterns.

	Materials	Duration, td	Amplitude	Interval
Case 1	Equal-sized Unfinished Pinballs	0.2 s 0.05 s 0.02 s	1 kN	0.5 s
Case 2	Equal-sized Polished Pinballs			

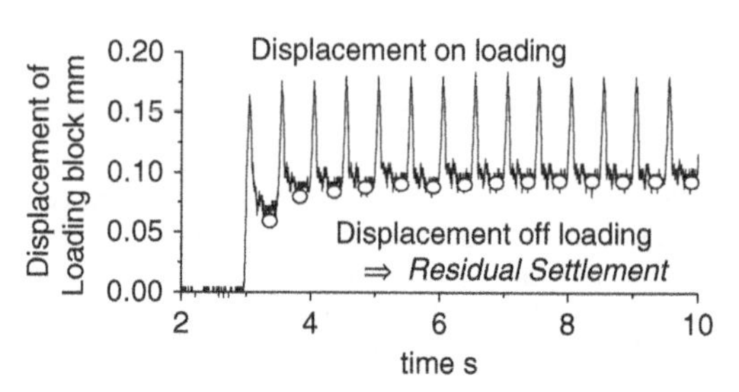

Figure 6. Residual Settlement.

Residual Settlement, the permanent deformation of the granular layer, will be discussed in this study, so the relationship between *Residual Settlements* and loading cycles is plotted as open circles in Fig. 6.

Fig. 7(1) shows the one of the results from three tests in case1 using unfinished pinballs. It expresses that *Residual Settlement* under cyclic loading is the smallest under the 0.2 s duration loading and is the largest under the 0.02 s duration loading. It means that the faster loading affect the larger deformation for unfinished pinballs, which has rough surface.

Fig. 7(2) shows the one of the results from three tests in case2 using well-polished pinballs. It expresses that the *Residual Settlement* is the smallest under the 0.02 s duration loading and is the largest under the 0.2 s duration loading. It means that the slower loading affect the larger deformation for smoothed pinballs. We carried out these tests 2 times for each case using those stacked-up pinballs.

3 DEM SIMULATIONS

3.1 *DEM parameters*

In DEM, grains are modeled as discrete rigid elements on which contact forces affected by other attached elements and gravity act as shown in Fig. 8. There are contact springs and dampers in the normal and tangential directions and sliders which control slippage at contact points. DEM defines that those elements satisfy the equations of motion which are solved by time steps.

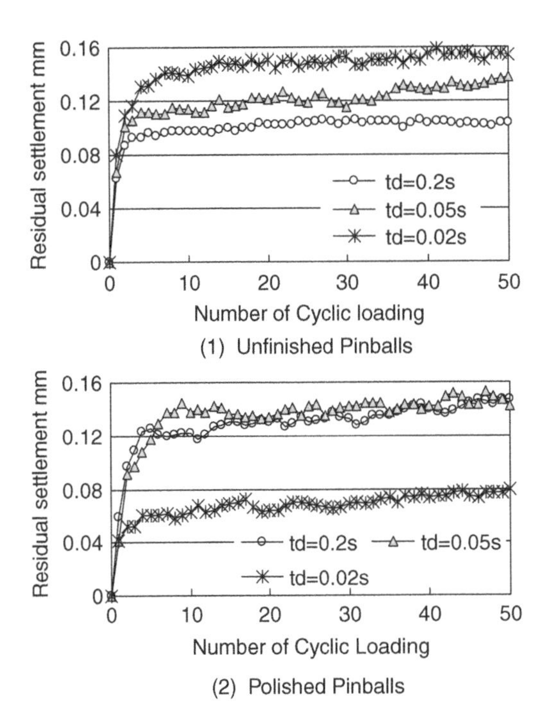

Figure 7. Residual Settlement from Experiments under various duration cyclic loadings.

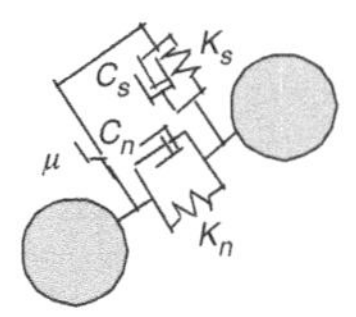

Figure 8. DEM model.

Table 2 shows the parameters for DEM in this study.

Kn and *Ks*, the contact spring coefficient between elements *i* and *j* is defined by the equation (1)[1], which is an approximate equation originating in Hertian Theory.

$$Kn = \frac{4b}{3\pi}\left(\frac{1}{\delta_i + \delta_j}\right) \qquad \text{----------------(1)}$$

$$\text{where, } \delta_i = \frac{1-\nu_i^2}{E_i \pi}, \delta_j = \frac{1-\nu_j^2}{E_j \pi}, b = \sqrt[3]{\frac{3}{2}\frac{(1-\nu^2)}{E}\frac{r_i r_j P}{r_i + r_j}}$$

In equation (1), *E*, Young's modulus of the pinball is set as 2.05×10^{11} N/m^2, and ν, the Poisson ratio, is set as 0.28[1]. The pressure between attached elements is presumed to be about 36 N assuming that 14 particles attached to the loading block receive 0.5 kN at maximum loading. *Ks* is defined from the ratio of Young's modulus for shear against Young's modulus.

The damping coefficient, *Cn* and *Cs*, are defined in equation (2), derived from the equation of motion of mass point with spring and damper.

$$e_b = \exp\left(-\frac{h}{\sqrt{1-h^2}}\pi\right), \quad h = \frac{c}{2\sqrt{k \cdot m}} \qquad \text{----(2)}$$

In equation (2), e_b, the coefficient of restitution of pinball is presumed to be 0.3 or 0.5 in this case. The value of *Kn* is represented by k to define *Cn*. As for *Cs*, it is also defined in equation (2) in which k's value is *Ks* as described above.

Friction coefficient is set to be 0.1, 0.3 and 0.6. As for steel balls, the friction coefficient is said to be about 0.3. The smaller value 0.1 is presumed for the polished pinball and the larger value 0.6 is presumed for the unfinished pinball here.

Table 2. Parameters for DEM.

Spring Coefficient (normal) N/m	1.48×10^7		
Spring Coefficient (shear) N/m	5.78×10^6		
Damping Coefficient (normal) N·	204.5	123	
Damping Coefficient (shear) N · s/m	127.8	76.8	
Friction Coefficient	0.1	0.3	0.6

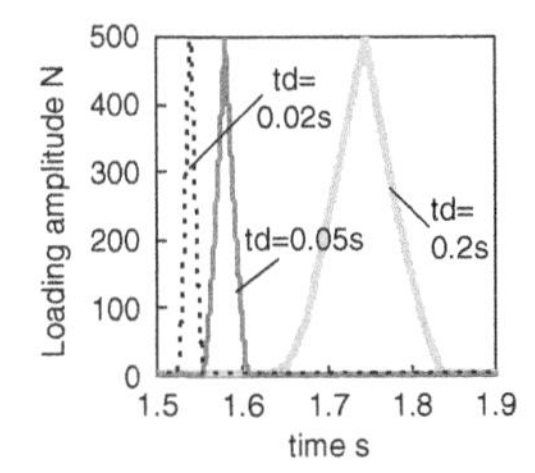

Figure 9. Example of loading-time data from DEM.

3.2 *Model preparation and loading patterns*

At the series in DEM simulations, circular elements having the same grain size distribution as real pinballs, as shown in Fig. 2, are stacked up the same way as the pinball in experiments as shown in Fig. 4.

During those simulations, mass and inertia of circle elements were calculated as spheres not cylinders to imitate experiments using pinballs.

The patterns of cyclic loadings are also applied as same as the experiments.

At these DEM simulations, the loading amplitude is set to be half value at the experiments because there are two vertical surfaces in depth at the experiments. Fig. 9 shows examples of the load-time data from the DEM simulations.

3.3 *Results*

3.3.1 *Residual settlement*

Fig. 10 shows the relationship between *Residual settlement* of the loading block and loading cycles from the above DEM simulations. Fig. 10-(1), the upper three graphs show the results of using the bigger damping coefficient and Fig. 10-(2), the lower three graphs show the results of using the smaller damping coefficient. Fig.-(a), the two graphs on the left, show the results using the friction coefficient 0.1, the two graphs (b) in the middle, show the results of using the friction coefficient 0.3, and the two graphs (c) on the right show the results of using the friction coefficient 0.6.

Those six graphs show that the circular assemblage having friction coefficient 0.6 deforms significantly under 0.02 s duration loading. Conversely, the same circular assemblage having the friction coefficient 0.1 deforms significantly under the 0.2 s duration loading.

This qualitative tendency is in good agreement with the results from the experiments.

As for the quantitative results, *Residual Settlement* from the DEM simulation is about 60–80 percent of that from the experiments. The difference seems to be caused from the conditions of pinball assemblage in the experiment being not completely two-dimensional.

3.3.2 *Displacement on-loading and off-loading*

Fig. 11 shows the displacement-time data from the DEM simulation. Fig. 11-(1) shows the result of using smooth circles with the friction coefficient 0.1 and Fig. 11-(2) shows that of rough circles with friction coefficient 0.6.

As for the smooth circle assemblage, the displacements at maximum loading the same as in the cases under 0.2 s or 0.02 s duration loading. After that, during off-loading, the deformation reverses significantly under 0.02 s duration loading. Then Residual settlement is small under the short duration loading for the case using smooth circles.

By contrast, the displacement of rough circle assemblage at maximum loading is large under 0.02 s

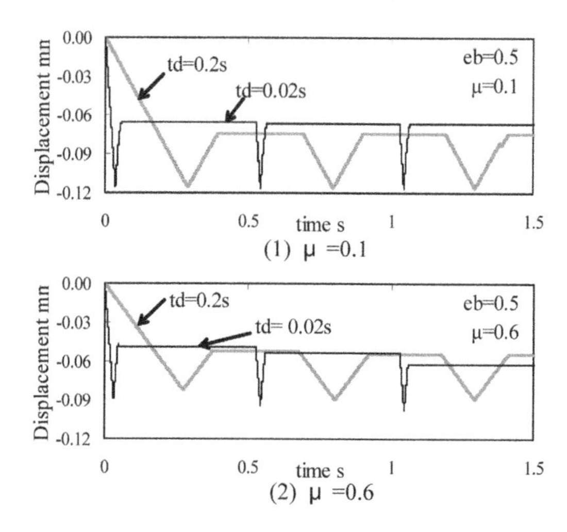

Figure 11. Displacement-time data from DEM.

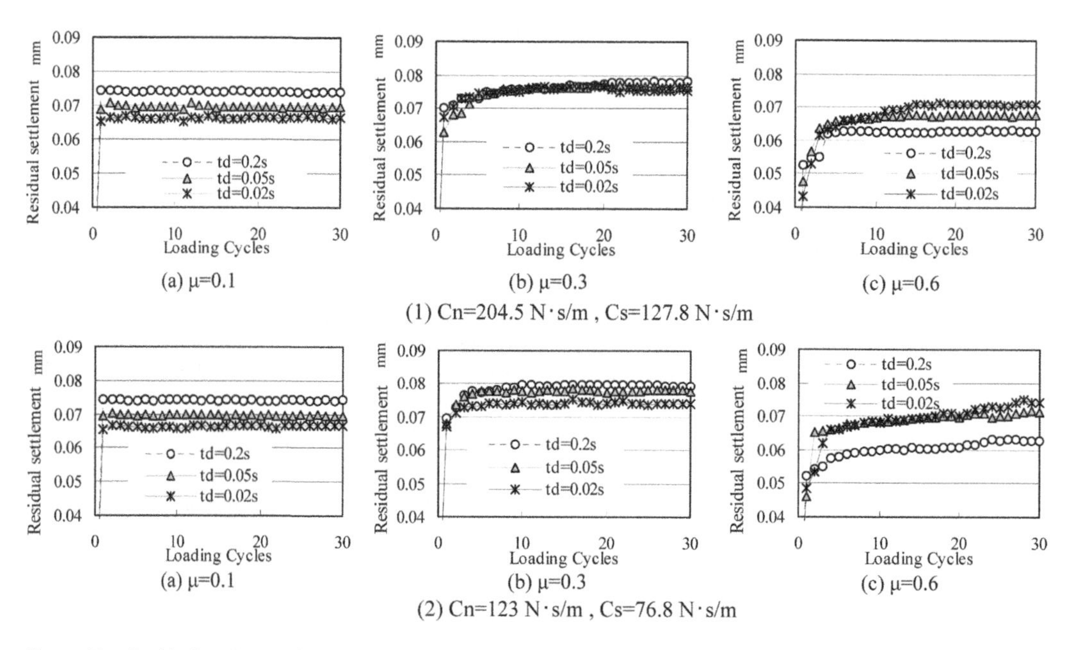

Figure 10. Residual settlement from DEM simulation under various duration cyclic loadings.

duration loading. The differences between those maximum displacements under 0.2 s and the 0.02 s duration loading are almost constant at every cycle. However, the reversible displacement under 0.02 s duration loading decreases more gradually. The reversible displacement is large in the 1st and the 2nd cycles, so *Residual settlement* is smaller than that under 0.2 s duration loading. Then the reversible displacement under the 0.02 s duration loading decreases with loading cycles. So *Residual settlement* under short duration loading increases gradually and exceeds that of long duration loading in the latter stages.

3.3.3 *Particle movement*

Fig. 12 (1) and (2) show particle movement before loading, at maximum loading and after loading in the 3rd cycle. Fig. 12-(1) shows the result under 0.02 s duration loading with smooth circles, Fig. 12-(1) shows the result of 0.02 s duration loading with rough circles and Fig.-(3) shows the result of 0.2 s duration loading with rough circles. The arrow length indicates 100 times of real displacement.

Fig. 12-(1) shows that the layer of smooth circles under short duration loading deforms and reverses dramatically. Fig. 12-(2) shows that the layer of rough circles under short duration loading deforms significantly on-loading, but reverses gradually off-loading, compared with the rough circles under long duration loading as shown in Fig. 12-(3) on loading.

3.3.4 *Contact and slipping point*

Fig. 13 shows the ratio of the number of contact points and slipping point on the initial contact points at the case of rough circles. It shows that the number of contact point decrease during on-loading or off-loading, especially under the 0.02 duration loading. The other hand, the numbers of slipping points increase during on-loading or off-loading.

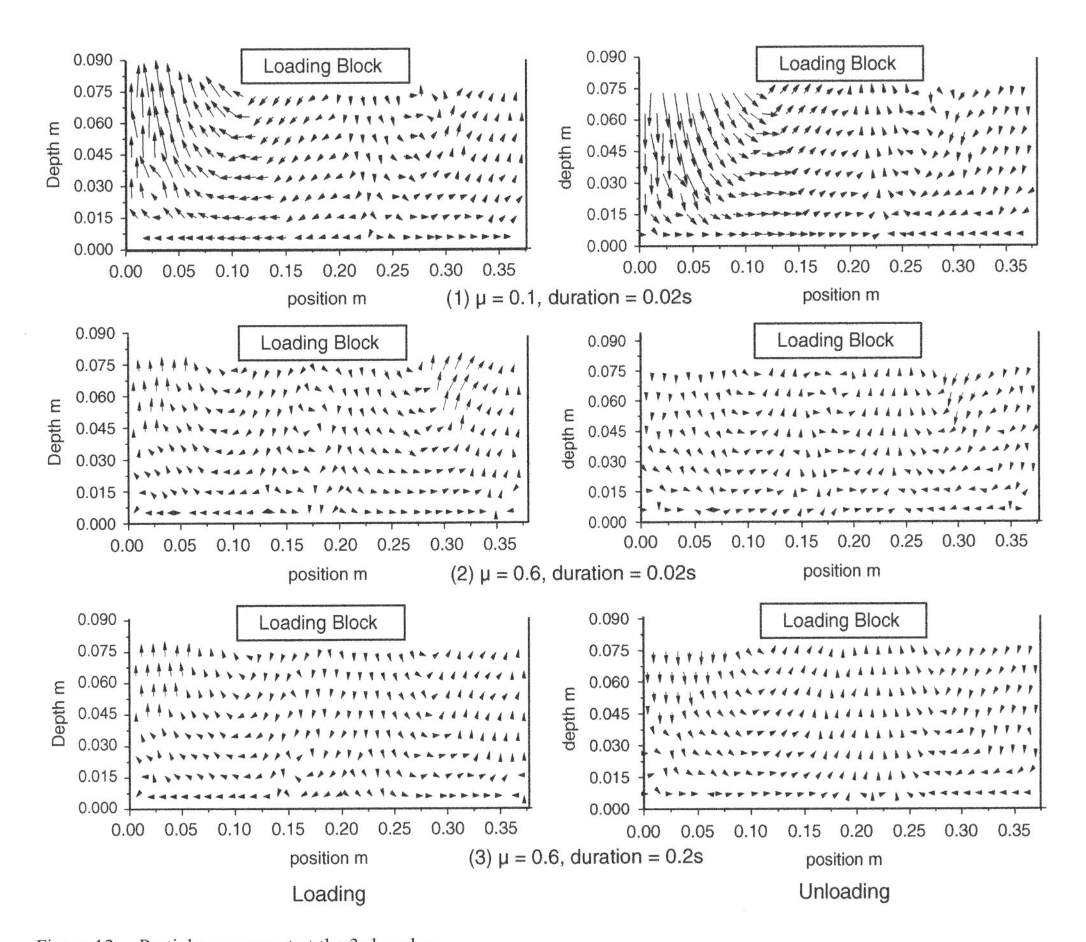

Figure 12. Particle movement at the 3rd cycles.

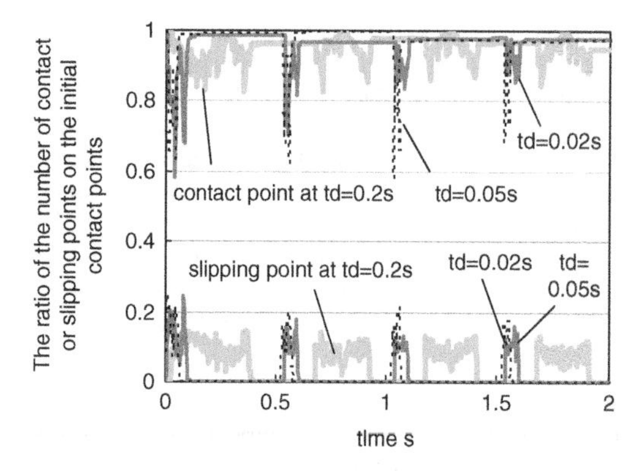

Figure 13. the ratio of the number of contact or slipping points on the initial contact points.

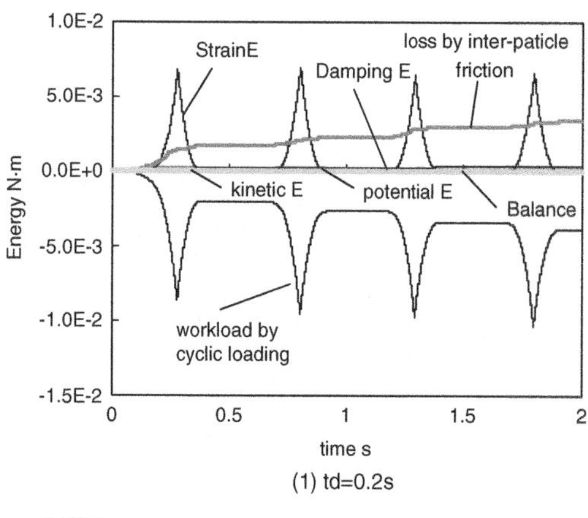

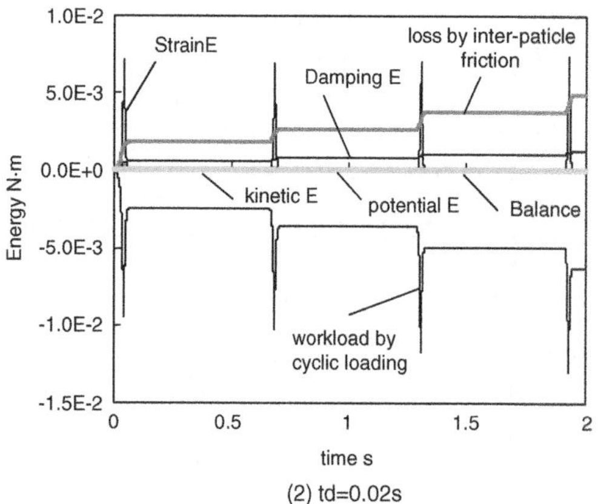

Figure 14. Balance of Energy.

3.3.5 *Energy balance*

Fig. 14 shows energy balance*2) during cyclic loading with the 0.2 s and the 0.02 s duration loading using rough circles. It is calculated by DEM simulation and the balance express error of the calculation.

At the case of the 0.02 s duration loading, the damping energy and potential energy are large compared with the case of the 0.2 s duration loading. From that results, volumetric deformation may be predominant at the grain assemblage under the 0.02 duration loading, fast loading in other word. As for the grain assemblage under the 0.2 s duration loading, slow loading, the shear deformation may be predominant.

4 CONCLUSION

Cyclic loading tests were performed experimentally using the equal-sized spherical balls and also simulated by DEM. The findings are as follows.

(1) The *Residual settlement* of railway ballast layers is rapid at first and decrease gradually under cyclic loading.
The residual settlement vs loading cycle curve given by experiment and simulation in this study shows the same tendency as ballast layers do.
(2) However, this tendency is slight in the case of granular layer composed of stacked-up equal-sized pinball, in this study, compared with real ballast layers.
(3) As for rough pinballs, the residual settlement is large at first under slow loading, but the later settlement increase minimally compared with the same under fast loading.
(4) The smooth pinball assemblage deforms on loading and reverse off-loading dramatically under fast loading, but rough pinballs doesn't reverse as much as smooth polished pinballs during off-loading.
(5) This tendency seems to relate to that volumetric deformation is predominant under fast loading and shear deformation is predominant under slow loading.

REFERENCES

The Society of Powder Technology; *Funtai simulation Nyumon* (in Japanese), 1998

T.Matsushima et. al; Grain-shape Effect on Peak Strength of Granular Materials, Computer method and Advances in Geomechanics, pp 361–366, 2001

Advances in Transportation Geotechnics – Ellis, Yu, McDowell, Dawson & Thom (eds)

 ISBN 978-0-415-47590-7

Testing the ultimate resistance at the sleeper/ballast interface

L. Le Pen & W. Powrie
University of Southampton, UK

ABSTRACT: Today's trains place ever greater and more complex loads on track infrastructure. In the UK, high speed tilting passenger trains have been introduced permitting enhanced speeds on curved sections of track. Maximum axle loads on freight trains may be permitted to increase from 25 to 30 tonnes; some sleepers such as the G44 have been designed with this in mind. Although vehicles undergo testing before entering service, the long term implications of the new loading regimes are not clear. In particular, the increased curving speeds of tilting trains introduce combinations of vertical lateral and moment loading onto the track which are not explicitly considered in the design or maintenance of railway track. The paper describes the development of a testing apparatus able to simulate combined vertical, horizontal and moment loading of a railway sleeper as may be experienced in service, and presents some preliminary data of sleeper and ballast behaviour.

1 INTRODUCTION

During curving, train/track interaction is at its most complex. Centrifugal and possibly wind forces give rise to lateral loads in addition to the normal vertical loads, leading to asymmetric loading of the track. Dynamic loading, caused by imperfections from the design geometry of the track, imperfections in the wheel profile and variation in load response of the track system, is also present. Modelling this complex interaction is difficult and most models incorporating dynamic effects focus on certain aspects of the interaction such as the wheel to rail interface (e.g. Vampire – DeltaRail, 2006) with sleeper to ballast interaction simplified to that of a linear spring and dashpot both vertically and laterally.

Recently, tilting trains (Figure 1) capable of travelling at up to 140 mph were introduced onto the West Coast Main Line (WCML), operating between London and Glasgow. The ability to tilt allows these trains to curve faster than conventional trains and maintain higher point to point mean speeds. These trains may apply more severe modes of loading to the track than conventional trains.

This paper reports preliminary data from tests designed to investigate the fundamental mechanisms and factors affecting the in service behaviour and ultimate resistance at the sleeper/ballast interface under the action of forces applied by high speed tilting trains. Results shown in this paper are limited to ultimate resistance, although further tests have been carried out investigating the in-service behaviour, and will be reported in future articles in due course.

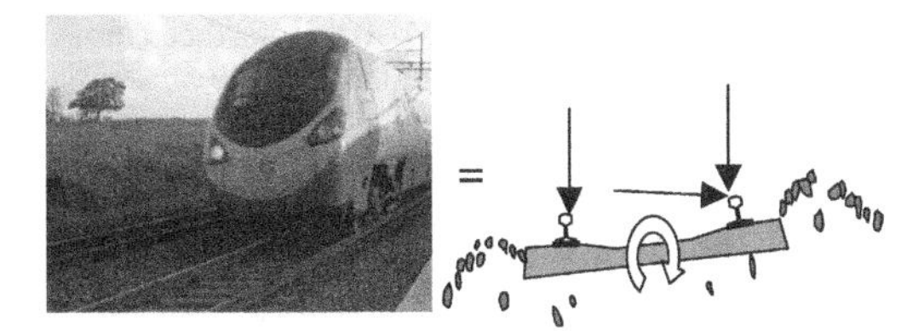

Figure 1. The loading under investigation.

2 PREVIOUS WORK

Trackbed design methods (e.g. Li & Selig, 1998a & b) throughout the world specify a depth of ballast on a suitable subgrade such that the track system is able to cope with the long term repeated vertical loading by trains without excessive deformation or failure.

For horizontal or lateral loading the track is then often assumed to remain structurally safe if the Prud'homme relation is obeyed. The relation has been reported variously in similar form but sometimes with different parameter values depending on whether wooden or concrete sleepers are used e.g. Prud'homme and Weber (1973). In the UK, Railway Group Standards (RGSs) specify a version of the Prud'homme relation such that rails and sleepers are deemed structurally safe provided lateral forces do not exceed $W/3 + 10$ over a sustained distance of 2 metres or more where W is the axle load in kN (RGS, 1993).

The resistance to lateral movement of the track system comes from the ability of the rails to spread the load over a number of sleepers which is governed

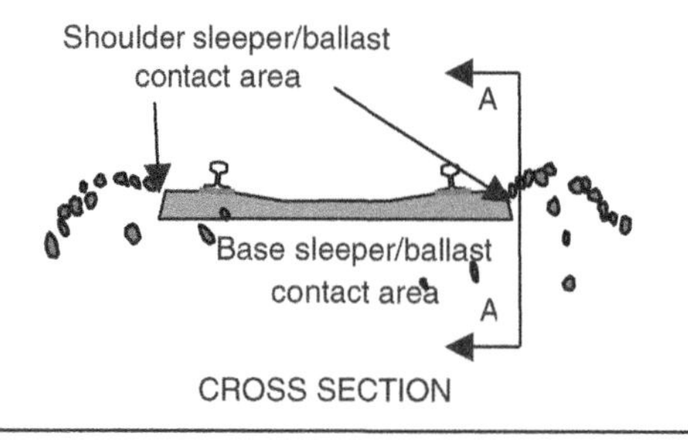

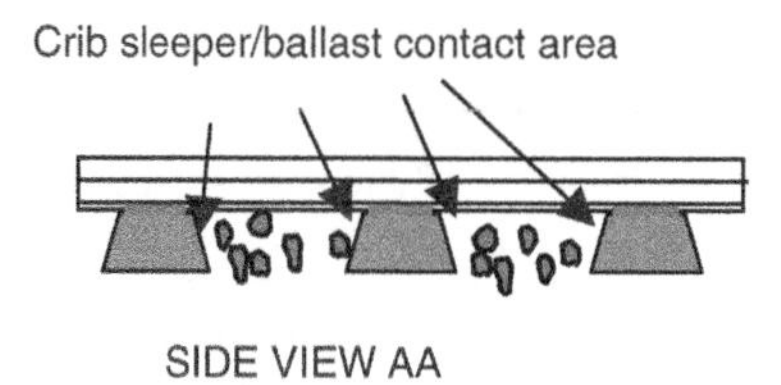

Figure 2. The three sleeper/ballast contact areas.

by the lateral and vertical relative stiffnesses of the track system. The load that is transferred into each sleeper must be resisted by the sleeper/ballast interface. Resistance at the sleeper/ballast interface comes from contact at the base, shoulder and crib as shown in Figure 2.

For consistency with the Prud'homme relation as used by NR, the track would be expected to have a fixed component of lateral resistance of 10 kN and a frictional component of W/3. However, previous tests have rarely reported resistance in this way. Commonly two types of test to measure the lateral resistance of the sleeper to ballast interface are carried out (ERRI, 1995b):

1. **Single sleeper push test**: A sleeper is detached from the rails pushed sideways by a machine attached to the rails, and the load/deflection response is recorded. E.g. see tests reported in Selig and Waters (1994).
2. **The panel pull method**: A section of in-service track is pulled sideways from the rail head and the load/deflection behaviour is recorded. From this, the individual sleeper resistance can be estimated. It can be carried out with the section either isolated (cut) or attached to the rest of the line (uncut). E.g. see tests reported in Esveld (2001).

Although various railway infrastructure companies and research institutions have carried out such tests, they are rarely reported in the literature and rail companies and consultancies are not always willing or able (due to fragmented and/or poor archiving) to share findings. Table 1 summarises the results of various lateral resistance tests accessed, but not necessarily carried out, by the European Rail Research Institute (ERRI). Lateral resistance is reported per sleeper, and is quoted for unloaded track in which the contribution from the three interfaces, crib, base and shoulder is typically one third each (ERRI, 1995a). Tests report the maximum or *peak* lateral resistance within a deflection of about 20 mm.

Table 1. Summary of lateral resistance data on unloaded track for concrete sleepers (ERRI, 1995a).

	Lateral resistance/sleeper (kN)			
	Minimum	20% less than	50% less than	Maximum
Loose tamped/relay	4.2	5.2	5.9	6.9
Just tamped (undisturbed)	5.9	7.1	8.3	11.8
Trafficked	5.4	8.1	10.3	15.7

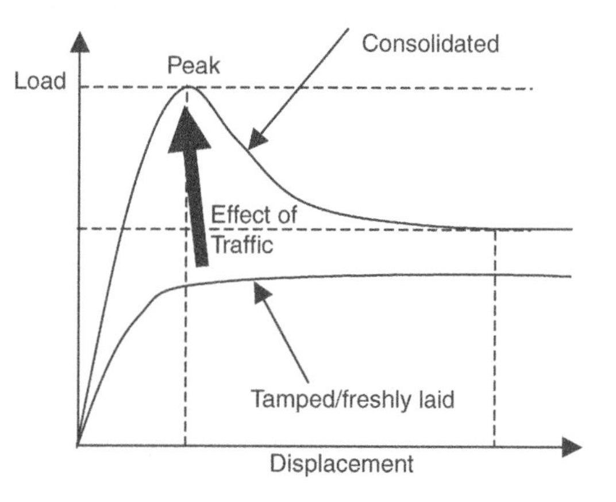

Figure 3. Effect of traffic on the lateral load/displacement behaviour of sleepers (ERRI, 1995b).

Differences between the minimum and maximum lateral resistance in each category could be due to differences between test sites, e.g. different specifications and levels of ballast and/or sleepers and fatigue. Some of the variation may be due to differences between panel or single sleeper tests and the type of sleeper, size of shoulder and crib ballast present. ERRI reported that the lateral resistance per sleeper was often less when testing a panel of sleepers; a result which ERRI attributed to interaction between sleepers. It is also possible that some of the panel tests incorporated hanging sleepers. Without access to the original test data it is difficult to assess the quality of the results. However, Table 1 indicates that the compaction of the ballast by tamping and trafficking has a large influence.

It should also be noted that the peak lateral resistance can be misleading. Lateral resistance at the sleeper/ballast interface can vary with deflection as shown in Figure 3.

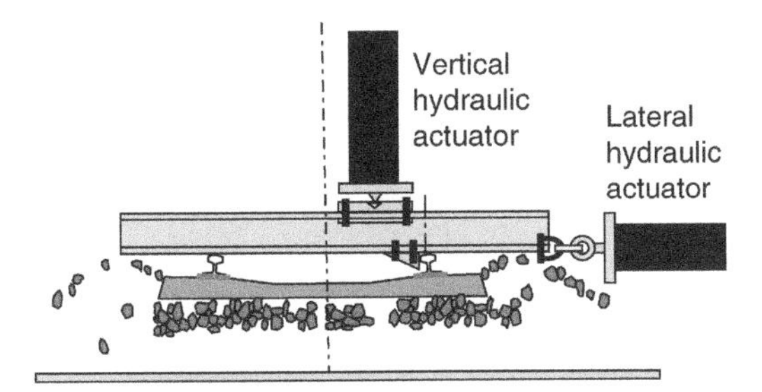

Figure 4. Laboratory schematic.

Figure 5. Photograph of laboratory set-up.

In the UK, moment loading is not considered from the point of view of the trackbed, although rollover of trains is considered in standards that specify safe operating speeds during high winds (RGS, 2000).

3 TESTING METHODS

To investigate the sleeper/ballast interface load resistance, laboratory experiments have been carried out which replicate aspects of service loading. Particular attention was given to replicating the moment loading that is caused as trains curve at high speed. To this end, one sleeper bay of track (650 mm wide) has been replicated in the laboratory. A bed of ballast 300 mm deep supports a type G44 sleeper (Tarmac, 2005) with short sections of BS113A rail (BSI, 1985). A loading beam across the railheads transfers applied load to rails and then to the sleeper from hydraulic actuators as shown in Figures 4 and 5. Moment loading can be applied by varying the position of the vertical hydraulic actuator. The ballast is seated onto a layer of softboard, representing a slightly compressible subgrade.

The laboratory tests reported in this paper are limited to the interaction between the sleeper base and the ballast. No shoulder or crib ballast was present: the effect of these is being investigated in subsequent tests.

Table 2. Key to tests reported.

Test	Vertical load (kN)	Position of vertical load
1A	75	Central
2A	45	Central
3A	15	Central
1B	45	0.5 m offset
2B	15	0.5 m offset
3B	30	0.5 m offset

Test preparation is as follows:

1. The ballast is placed and levelled.
2. The sleeper is placed into the testing rig and allowed to stabilize overnight.
3. The loading beam is connected and 100 vertical load cycles of 75 kN applied, corresponding to a likely load reaching one sleeper from a Pendolino axle immediately above.
4. 10 lateral load cycles are applied of not more than 1/3 of the applied vertical load to ensure the contact is stable.
5. The sleeper is pulled laterally under displacement control while the vertical load is maintained.
6. All is removed and replaced, including all the ballast prior to subsequent tests.

All the tests are on a bed of ballast that is intended to replicate an initial condition of tamped/freshly laid ballast with expected behaviour as shown in Figure 3.

4 RESULTS & DISCUSSION

Six tests are reported here. In each test the vertical load was held constant and the sleeper pulled at a constant rate (0.5 mm/s). Tests 1A, 2A and 3A were for a centrally placed and maintained vertical load and tests 1B, 2B and 3B were for a vertical load applied at an offset of 0.5 m from the centreline of the sleeper towards where the lateral actuator applied its pull from. The rail heads are 1.5 m apart, so that the vertical load in test run B was applied at 2/3 of the maximum possible offset. Also note that because the lateral load was applied onto the railhead there is an additional moment load in all the tests due to the height of the railhead above the sleeper/ballast interface. This is estimated to be at an eccentricity of 0.4 m.

Figures 6 to 8 show the lateral load/deflection graphs over different deflection ranges for each of the tests. Tables 3 and 4 show key values from these load/deflection graphs.

Over the full range of deflection it is possible to identify variability in each load/deflection line. Occasionally large drop-offs in the load occur followed by rapid return towards the previous value of load.

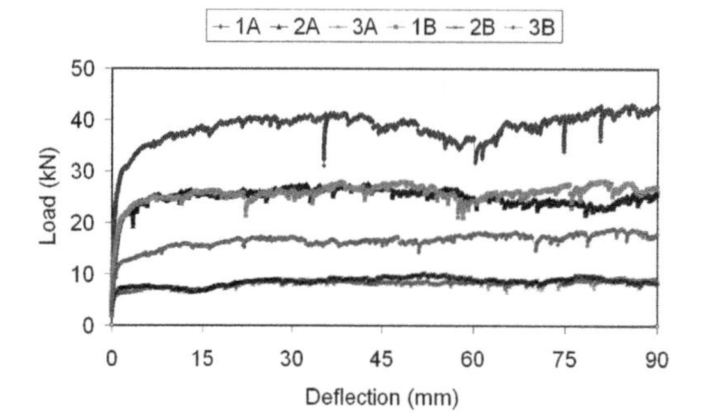

Figure 6. Load/deflection full scale.

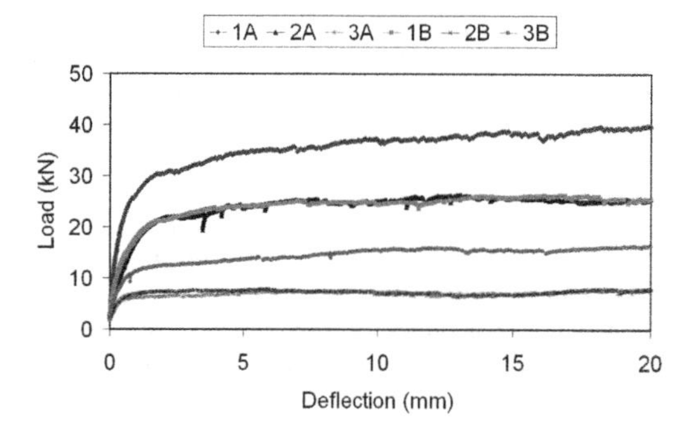

Figure 7. Load/deflection up to 20 mm.

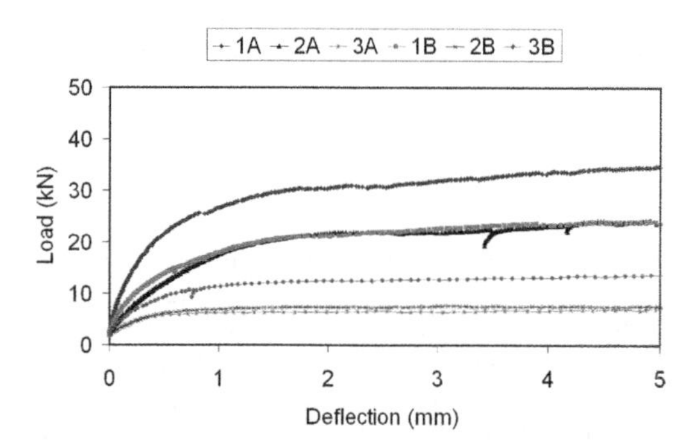

Figure 8. Load/deflection up to 5 mm.

These are thought to be due to ballast breakage or rearrangement events; that is, particles of ballast fracturing, crushing, rolling or sliding. During tests it was observed that noises likely to be associated with breakage/movement events accompanied the fall-offs in load.

Figures 7 & 8 show that, during the first 2 mm of deflection, the load/deflection plots are very smooth. This is because at low deflections the majority of movement is recoverable and little or no slip of the sleeper on ballast has yet occurred. This is confirmed by other non-failure magnitude cyclic lateral load tests

Table 3. Load at key deflections.

Test	Vertical load (kN)	Lateral load (kN) on sleeper at:			
		0.5 mm	1 mm	2 mm	3 mm
1A	75	21.4	26.5	30.4	34.5
2A	45	11.9	17.5	21.7	21.9
3A	15	5.5	6.1	6.4	6.5
1B	45	13.6	17.8	21.2	22.6
2B	15	6.0	6.8	7.4	7.5
3B	30	9.2	11.2	12.4	12.7

Table 4. Load at key deflections continued.

Test	Lateral load in kN on sleeper at:		
	Mean 2 to 10 mm	Peak up to 10 mm	Mean from 10 to 90 mm
1A	37.3	37.3	39.3
2A	23.8	25.4	25.4
3A	7.0	7.4	8.3
1B	23.9	25.0	26.0
2B	7.5	7.8	8.7
3B	20.0	15.5	16.8

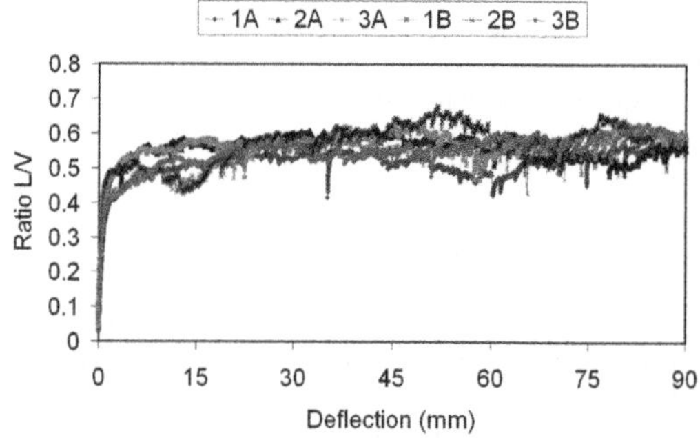

Figure 9. Deflection against ratio of lateral/vertical load.

not reported here. The magnitude of the recoverable deflection is related to the applied vertical load, with larger vertical loads giving rise to higher recoverable magnitudes of deflection.

Figure 9 shows that as the deflection moves beyond 2 mm all 6 tests move towards a limiting (failure) ratio of lateral to vertical load regardless of the magnitude of the constant vertical load and its eccentricity.

It had been thought that the moment loading could cause failure at lower lateral loads, a behaviour that granular soils are known to exhibit and which is taken into account when designing offshore structures subjected to moment inducing wave loading. However, these tests appear to indicate that the lateral load at failure is insensitive to the eccentricity of the vertical applied load, i.e. that the moment component of loading has a negligible effect on the ultimate lateral failure load within the range of load cases investigated.

This behaviour can be further described numerically using the equations proposed by Butterfield and Gottardi (1994) giving the failure of a shallow foundation under combined vertical (V), horizontal (H) and moment (M) loading. The main equations are summarised below and all the symbols used are described at the end of this paper:

$$\left[\frac{H/V_{max}}{t_h}\right]^2+\left[\frac{M/BV_{max}}{t_m}\right]^2-\left[\frac{2C\frac{M}{BV_{max}}\frac{H}{V_{max}}}{t_h t_m}\right]=\left[\frac{V}{V_{max}}\left(1-\frac{V}{V_{max}}\right)\right]^2 \quad (1)$$

$$V_{max}=\sigma'_f BL=N_\gamma s_\gamma(0.5\gamma B-\Delta u)BL \quad (2)$$

$$\frac{H}{t_h}=\frac{V(V_{max}-V)}{V_{max}} \quad (3)$$

$$\frac{M/B}{t_m}=\frac{V(V_{max}-V)}{V_{max}} \quad (4)$$

Where:

$$N_q=K_p e^{\pi\tan\phi} \quad (5)$$

$$K_p=\left(\frac{1+\sin\phi'}{1-\sin\phi'}\right) \quad (6)$$

$$t_h=\tan\delta \quad (7)$$

$$N_\gamma=(N_q-1)\tan(1.4\phi')\ \text{(Meyerhof, 1963)} \quad (8)$$

$$S_\gamma=1+0.1K_p(B/L)\ \text{(Meyerhof, 1963)} \quad (9)$$

By estimating the relevant material parameters, it is possible to plot the failure envelope for a G44 sleeper on Network Rail specification railway ballast. The main difficulty in using these equations is in deciding on a value for t_m which corresponds to the initial tangent to the failure surface on the graph of V against M/B (Figure 11). The difficulty arises because t_m can only be found experimentally and the tests so far carried out indicate that not enough moment loading has been applied to cause a moment loading failure and hence to determine the value of t_m. Therefore it has only been possible to indicate a minimum value for t_m, arrived at by ensuring that all test results reported here fall along the edge or within the failure envelope. In all likelihood this represents a significant underestimate of t_m.

Table 5. Values used in Butterfield's equations.

Symbol	Value adopted	Units
B	2.5	m
L	0.285	m
γ	16	kN/m^3
u	0	–
ϕ	0.698	radians
k_p	4.599	–
N_q	64.20	–
N_γ	93.69	–
s_γ	5.03	–
δ	0.44	radians
t_h	0.47	–
t_m	0.272	–
V_{max}	2688	kN

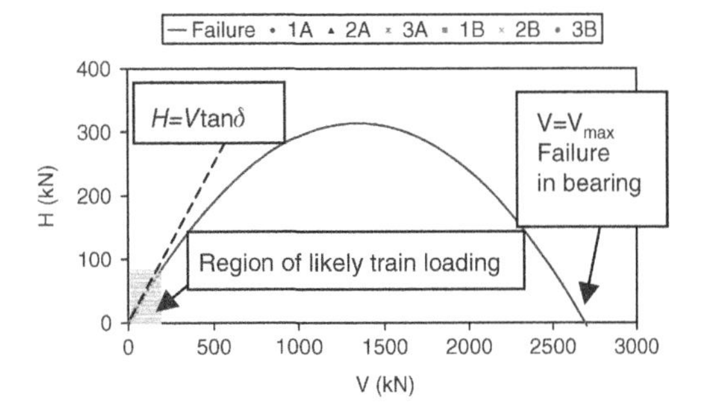

Figure 10. Vertical, horizontal loading failure envelope.

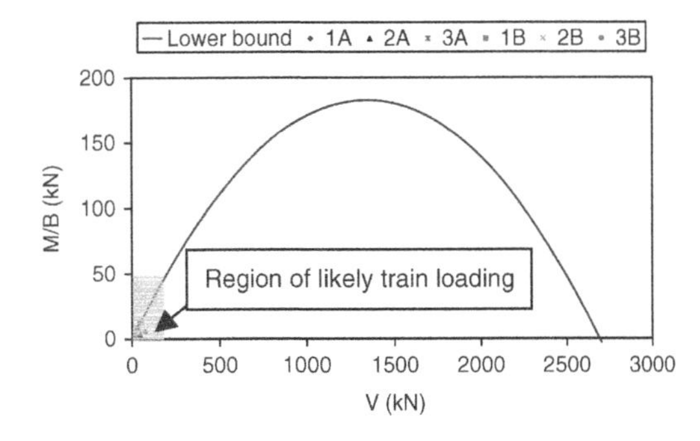

Figure 11. Vertical, moment loading failure envelope.

The calculations to plot the failure envelopes are not fully set out here but the parameter values used are summarised in Table 5 and the failure envelopes, which do not incorporate any factors of safety, are illustrated in Figures 10 & 11.

Figure 10 confirms that the vertical to lateral loading ratio for sliding failure remains more or less constant for any likely magnitude of train-applied vertical load. In Figure 11 it can be seen that even with the lower bound estimate for t_m from these tests, it is clear that the failure envelope remains close to linear in the likely range of train loading.

The likely range of values for t_m can also be assessed to provide a reality check on the minimum estimate of its value from these tests: A minimum value for t_m may be estimated by assuming that there is no effect from moment loading when the eccentricity of a vertical load V on a strip foundation of width B from the centre is less than $B/6$. This corresponds to the well known middle third rule where, provided a vertical load remains within the middle third, pressure is distributed across the full width B with no contact lost. It then follows that by replacing M with $VB/6$ in equation (4), at low values of V, the value for t_m is 0.167. A maximum value for t_m may then be estimated by assuming that the maximum eccentricity of a vertical load is $B/2$. Similarly M may be replaced with $VB/2$ in equation (4), hence, at low values of V, the value of t_m is 0.5. However, this simple analysis also assumes that the horizontal load has no eccentricity and applies no moment.

The lower bound value determined in these tests of 0.272 then places the true value of t_m in the range 0.272 to 0.5.

5 CONCLUSIONS

From the laboratory tests:

- The magnitude of recoverable lateral deflection increases with increasing vertical load, provided that the vertical load is held constant within a given test. This finding is confirmed by other cyclic lateral loading tests not reported here.
- The interface between sleeper and ballast at the base has surpassed a resilient recoverable region after a deflection of 2 mm for all the tests. This corresponds to a ratio of vertical to lateral load of about 0.45 (median value for all tests).
- The ratio of vertical to lateral load then tends to a limiting ratio at larger deflections of about 0.56 (median value of 10 to 90 mm mean, all tests).
- Sudden falls in the load/deflection plots appear to be due to breakage/rearrangement events with the overall trend quickly reasserting itself.
- In these tests, the ratio of vertical to lateral load of centrally and eccentrically loaded sleepers was not significantly different because the combinations of vertical, horizontal and moment loading tested fell well within the linear region of the Butterfield failure envelopes. However, care is needed in applying this finding to specific cases of train loading where, in addition to the exact combination of vertical, horizontal and moment loading to be considered arrangement and types of ballast and sleepers may also differ.
- The value of t_m was found to be a minimum of 0.272

6 FURTHER WORK

These tests have investigated the ultimate resistance available from the sleeper base to ballast contact area. Further work is ongoing to quantify the resistances from the shoulder and crib contacts and the factors governing them. Work is also being carried out to quantify the range of likely movement of the sleeper on the ballast during in-service loads, and, in collaboration with Manchester Metropolitan University, to incorporate realistic relations of resilient movement of sleeper on ballast into a dynamic vehicle/track interaction model.

7 DEFINITION OF SYMBOLS USED

B	Sleeper length	e	Eccentricity
L	Sleeper width	σ_f	Stress at failure
γ	Bulk unit weight of ballast		
u	Pore water pressure		
ϕ	Friction angle of ballast		
k_p	Passive pressure coefficient		
N_q	Bearing capacity factor		
N_γ	Analogous to the bearing capacity factor found from Meyerhof formula		
s_γ	Shape factor taken as the value for s_q from Meyerhof formula		
δ	Measured angle between soil and structure here taken as the 2 mm median ($\tan^{-1}0.46$ or 25°)		
t_h	Tangent to failure surface on graph of V against H when V = 0		
t_m	Tangent to failure surface on graph of V against M/B when V = 0 (Lower bound from these tests)		
V_{max}	Maximum bearing capacity		

For a fuller explanation the reader is referred to Powrie, (2004).

ACKNOWLEDGEMENTS

The authors are grateful for the financial support of the Engineering and Physical Sciences Research Council and to Network Rail, Tarmac, Pandrol, and Corus steel for their support in kind.

REFERENCES

BSI (1985). *British Standard: Specification for Railway Rails*, British Standards Institution, London, UK.

Butterfield, R. and Gottardi, G. (1994) A complete three-dimensional failure envelope for shallow footings on sand. *Géotechnique*, **44**(1), 181–184.

Deltarail (2006) *http://www.vampire-dynamics.com/.*
ERRI D202 (1995a). *Improved knowledge of forces in CWR track (including switches) RP 2, Review of existing experimental work in behaviour of CWR track.* European Rail Research Institute, Arthur van Schendelstraat 754, NL – 3511 MK UTRECHT.
ERRI D202 (1995b). *Improved knowledge of forces in CWR track (including switches) RP 3, Theory of CWR track stability.* European Rail Research Institute, Arthur van Schendelstraat 754, NL – 3511 MK UTRECHT.
Esveld, C. (2001) *Modern Railway Track,* Zaltbommel, MRT Productions.
Li, D. & Selig, E. T. (1998a) Method for railroad track foundation design. I: Development. *Journal of Geotechnical and Geoenvironmental Engineering,* Vol.124, pp.316–322.
Li, D. & Selig, E. T. (1998b) Method for railroad track foundation design. II: Applications. *Journal of Geotechnical and Geoenvironmental Engineering,* Vol.124, pp.323–329.
Meyerhof, G.G. (1963) some recent research on the bearing capacity of foundations. *Canadian Geotechnical Journal,* **1**(1), 16–26
Powrie, W. (2004) *Soil Mechanics: Concepts and Applications,* London, Spon.
Prud'homme, A. & Weber, O. (1973) L'aspect technique des grandes vitesses. la voie et les installations electriques des lignes a grande vitesse 1, 2. (Technical Aspects of High-Speed Trains. Track and Its Infrastructure. Electric Installations 1, 2). *Travaux,* pp.26–46.
GM/TT0088 (1993). *Railway Group Standard: Permissible Track Forces for Railway Vehicles.* Group Standards, Railway Technical Centre, Derby.
GM/RT2141 (2000). *Railway Group Standard: Resistance of Railway Vehicles to Derailment and Roll-Over.* Safety and Standards Directorate, Railtrack PLC, Railtrack House DP01, Euston Square, London, NW1 2EE.
Selig, E. T. & Waters, J. M. (1994) *Track Geotechnology and Substructure Management,* London, Telford.
Tarmac (2005) Rail Products Guide. Tarmac Precast Concrete Limited.

Advances in Transportation Geotechnics – Ellis, Yu, McDowell, Dawson & Thom (eds)
© 2008 Taylor & Francis Group, London, ISBN 978-0-415-47590-7

Settlement of concrete slab track considering principal stress axis rotation in subgrade

Y. Momoya, T. Takahashi & E. Sekine
Railway Technical Research Institute, Tokyo, Japan

ABSTRACT: Slab tracks on earth structures are widely constructed in Japanese high speed railways. Those slab tracks require significantly high performance against residual settlement under repeated train loads. In the design standard of earth structures in Japan, quality of soil materials and compaction control are strictly prescribed; however, construction cost becomes relatively high. If the prediction of residual settlement under repeated train load becomes possible, the design of earth structures for slab tracks will become more rational. In this study, to discuss residual settlement under repeated train road, scale model tests, FEM analysis and principal axis rotation tests were carried out. Based on the results of those tests and analysis, prediction method of residual settlement under moving wheel loading was discussed.

1 INTRODUCTION

Slab tracks (figure 1) are widely constructed in Japanese high speed railways. Those slab tracks were originally developed for viaducts and tunnels; however, in these 10 years, slab tracks are widely constructed on earth structures. Figure 2 shows standard cross section of slab track on the earth structure. Earth structures for slab tracks are required high performance against residual settlement under repeated train loads, because the restoration after excessive settlement is significantly difficult.

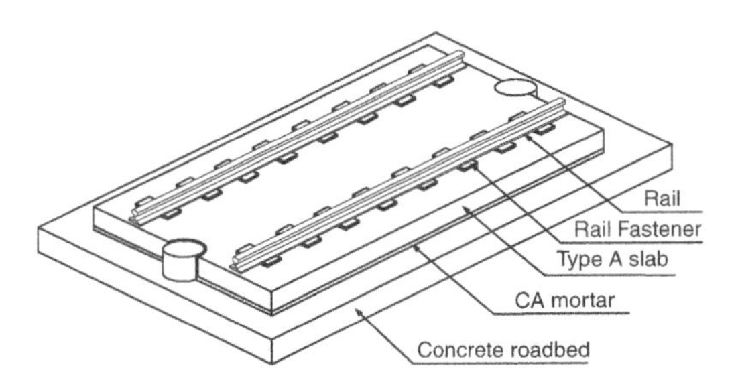

Figure 1. Slab track for Japanese Shinkansen bullet train.

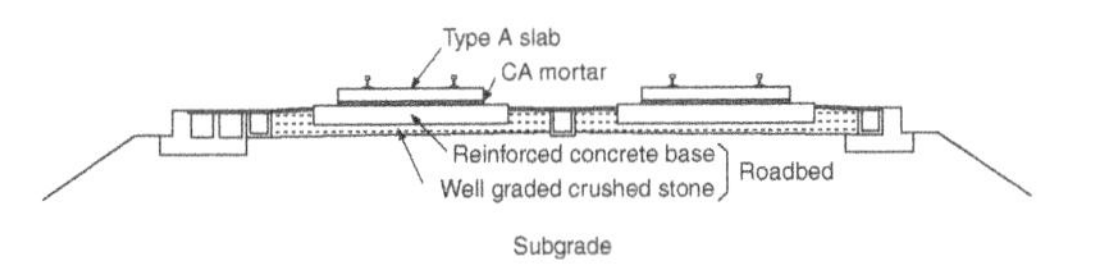

Figure 2. Cross section of slab track on earth structure.

It is difficult to predict residual settlement after the millions of repeated loading cycles, because the roadbeds and subgrades under traffic loads are affected by complicated stress condition including principal stress axis rotation. Therefore, the design standard prescribes the performance of earth structure based on empirical knowledge and real scale tests. In the present design standard, quality control of soil materials and compaction control are strictly regulated; however, construction cost becomes relatively high accordingly.

To develop more rational design method and to reduce the construction cost, it is necessary to predict residual settlement under repeated loading cycles. Hirakawa et al (2002) and Momoya et al (2005) pointed out the difference between moving wheel loading and fixed point loading based on the scale model tests. Towhata et al (1994) and Brown (2006) pointed out the effect of principal stress axis rotation under traffic loads based on the elemental tests.

In this study, to evaluate residual settlement of slab track on earth structure, moving wheel loading tests, FEM analysis and elemental principal stress axis rotation tests were carried out.

2 MOVING WHEEL LOADING TEST

2.1 *Model test method*

Figure 3 shows the moving wheel loading test apparatus, in which loading wheel with a given vertical load moves back and forth at 1000 mm/min repeatedly.

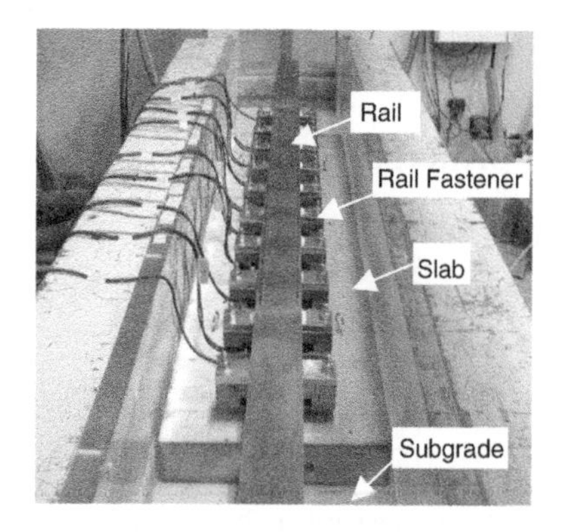

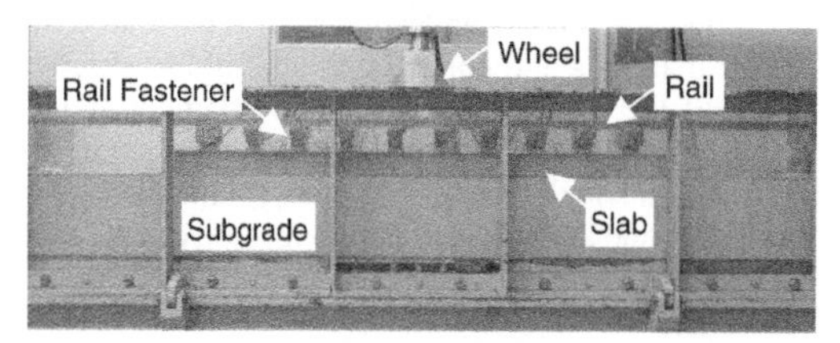

Figure 3. Moving wheel loading test model of slab track.

The sand box was 2000 mm in length and 300 mm in width, in which scale model tests were carried out in plane strain conditions. The model subgrade was made by air-dried Toyoura sand ($D_{50} = 0.26$ mm, $U_c = 1.91$, $G_s = 2.65$), which is widely used for model tests and elemental test in Japan. Model concrete slab was made of fiber (polyvinyl alcohol) reinforced concrete, of which the compressive strength (28 days) was 61.9 N/mm^2.

The scale of the model was 1/5 of typical slab track. To simplify the test condition, concrete slab (60 mm thick) was directly settled on Toyoura sand subgrade (200 mm thick). The length of concrete slab was 1200 mm. The second order moment of the model rail was equal to $1/5^4$ of the 60 kg rail used in Japan. In this test, two types of rail supporting system were compared. The first case is conventional fastening system, that the rail is connected to slab discontinuously. In the second case, rail was continuously supported by particular rubber track pad.

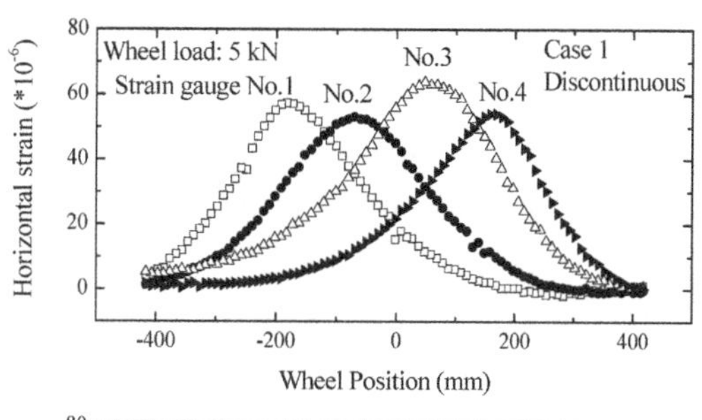

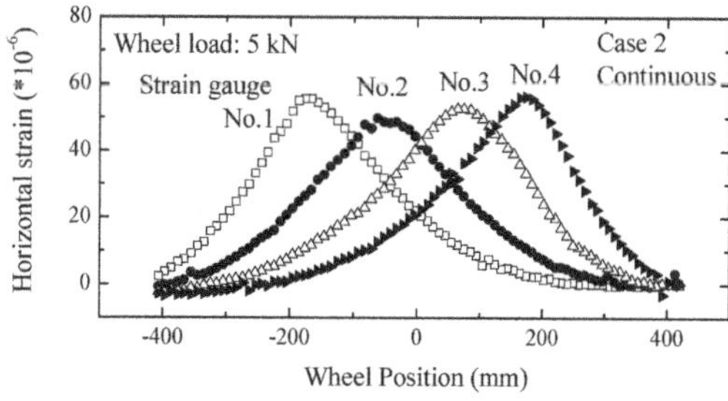

Figure 4. Strain at the bottom of concrete slab.

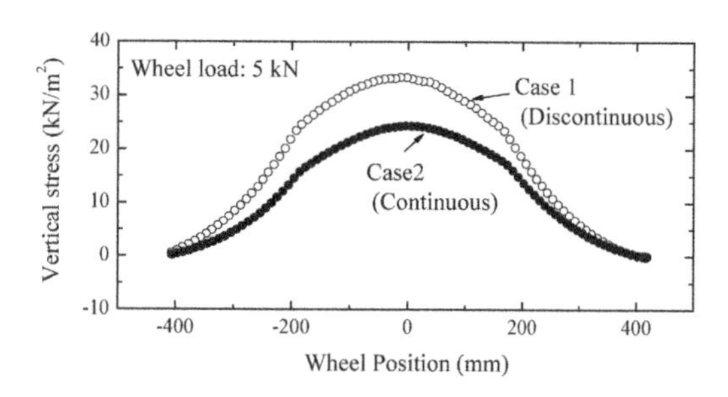

Figure 5. Vertical stress at the bottom of sand box.

2.2 *Model test results*

Figure 4 shows the strain at the bottom of concrete slab. Four strain gauges were fixed at the bottom of slab in 120 mm spacing. The strain was slightly smaller in case 2 (continuous rail support system) than in case 1 (discontinuous rail support system). Figure 5 shows vertical stress at the bottom of sand box. Vertical stress was smaller in case 2 than in case 1. Those results show that the continuous rail support system effectively decreases stress and strain applied on concrete slab and subgrade.

Figure 6 shows residual settlement of concrete slab. In case 2 (continuous rail support system), residual settlement became smaller than in case 1 (discontinuous rail support system); however the difference was not significant. Residual vertical strain was approximately 0.45% after 300 times loading with wheel load 5 kN.

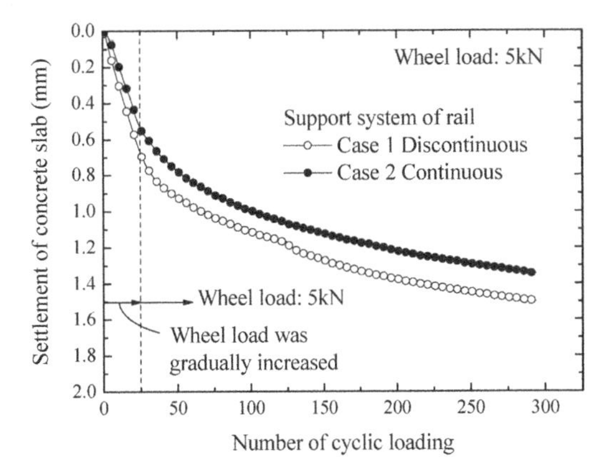

Figure 6. Residual settlement of concrete slab.

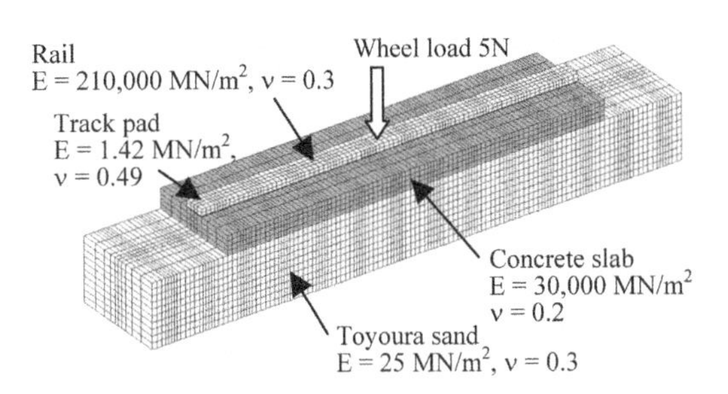

Figure 7. FEM analysis model of scale model.

3 FEM ANALYSIS OF MODEL TEST

Figure 7 shows the linear elastic FEM analysis model of continuous rail support system slab track by NASTRAN. Interfaces of different materials were continuous in this model. Horizontal spring elements were inserted beneath sbugrade to represent friction between base and subgrade. Wheel load of 5 kN was applied on the rail at the center of the model. In this chapter, strain of concrete slab and stress in subgrade were discussed based on the result of the FEM and model test.

In figures described below, horizontal axes for scale model test results were wheel position. On the other hand, horizontal axes for FEM were distance from loading point.

Figure 8 shows strain at the bottom of concrete slab. Result of FEM analysis corresponded well with the result of scale model test. Figure 9 shows vertical stress at the bottom of sand box. The peak stress in FEM analysis approximately equivalent with the peak stress in the scale model test, however, stress distribution area was larger in the FEM analysis than in the scale model test. This was due to limited distance of wheel moving in the scale model test. Due to the similar reason, peak shear stress in the FEM analysis was approximately

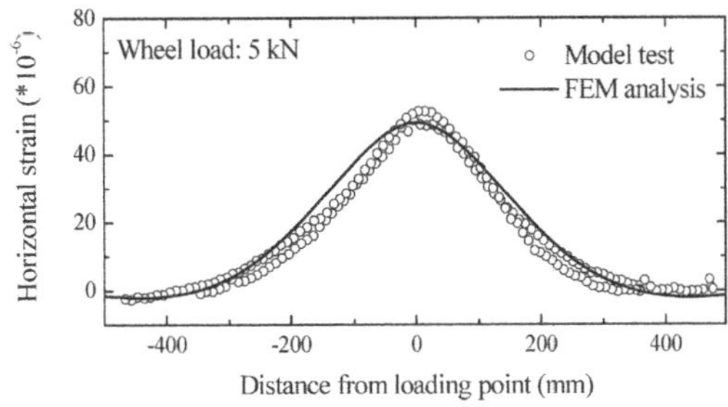

Figure 8. Horizontal strain at the bottom of concrete slab.

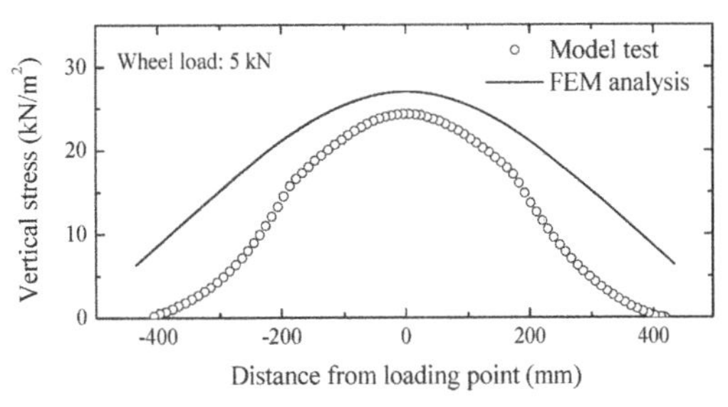

Figure 9. Vertical stress at the bottom of sand box.

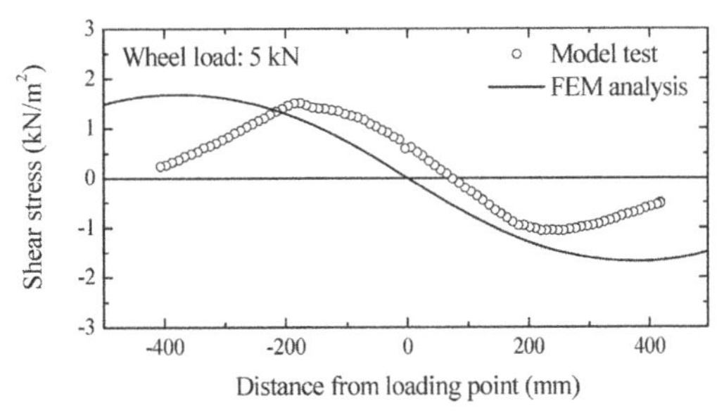

Figure 10. Cross section of slab track on earth structure.

equivalent with the peak stress in the scale model test as shown in figure 10; however, the shear stress distribution area was larger in the FEM analysis than in the scale model test.

Those results derive the conclusion that the elastic FEM analysis described here properly simulates the stress condition of scale model test under moving wheel load. Based on the stress condition obtained by FEM, elemental principal stress rotation tests were carried out as described in next chapter.

4 PRINCIPAL STRESS AXIS ROTATION TEST

To discuss residual settlement of concrete slab under moving wheel load in the scale model test, principal stress axis rotation tests were carried out. The test apparatus shown in figure 11 was developed to simulate principal stress axis rotation under traffic loads (Momoya et al (2006)).

The specimen (20 cm × 20 cm × 20 cm) was set in the laminar shear box, divided in 10 sub-layers in plane strain condition. The vertical normal and horizontal shear stresses, σ_z and τ_{zx}, on the top cap and the vertical shear stress, τ_{xz}, on the side walls are independently applied by using a set of air cylinders (i.e., Bellofram®cylinders). The horizontal normal stress, σ_x, on the side walls is applied by means of two air pressure bags, set immediately behind the side walls.

Figure 12 shows the time variation of principal stress axis rotation in the case of "railway" simulated in this test. Similarly, the stress conditions in the case of "roads" and in the case "without principal stress axis rotation" were carried out in the series of tests.

Figure 13 shows residual strain after 300 times loading cycles. In the cases "roads" and "railway", that with principal stress axis rotation, residual strain became larger than in the case "without principal stress axis rotation". In the case of "railway", residual stress was twice as large as the case "without principal stress axis rotation".

In the scale model test with moving wheel load described above, average residual strain of subgrade after 300 times loading cycles was 0.45%, where principal stress amplitude $\Delta\sigma_1$ was approximately 30 kN/m^2. In the principal stress axis rotation test with confining pressure $\sigma_c = 20$ kN/m^2, residual strain was approximately 0.3% in the case of "railway" with $\Delta\sigma_1 = 30$ kN/m^2, which was rather smaller than that in the scale model test.

Figure 14 shows the effects of confining pressure σ_c on the residual strain. Here, assuming $\sigma_c = 5$ kN/m^2 in the scale model test, residual strain became 1.4 times larger than in the test with $\sigma_c = 20$ kN/m^2.

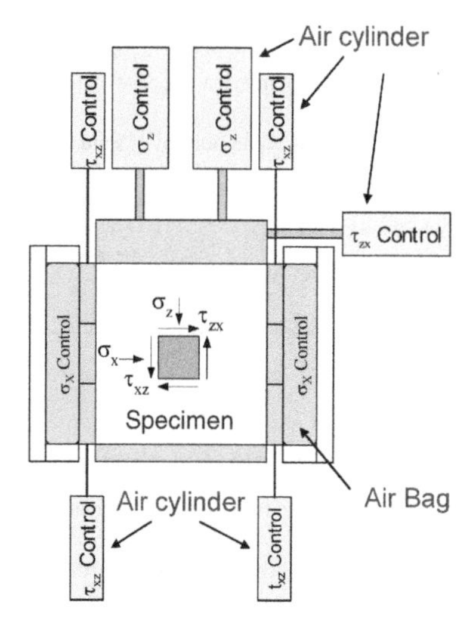

Figure 11. Principal stress axis rotation test apparatus.

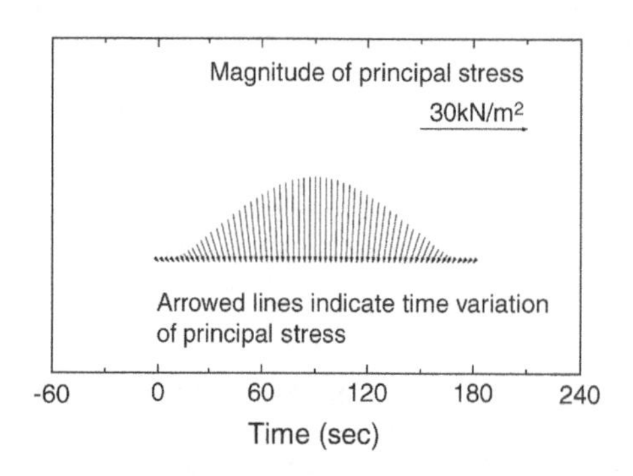

Figure 12. Time variation of principal stress axis rotation in the test (In the case of railway).

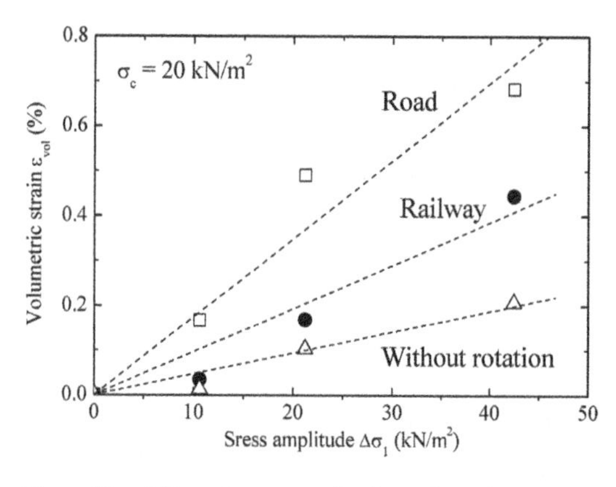

Figure 13. Effect of stress amplitude on the residual strain.

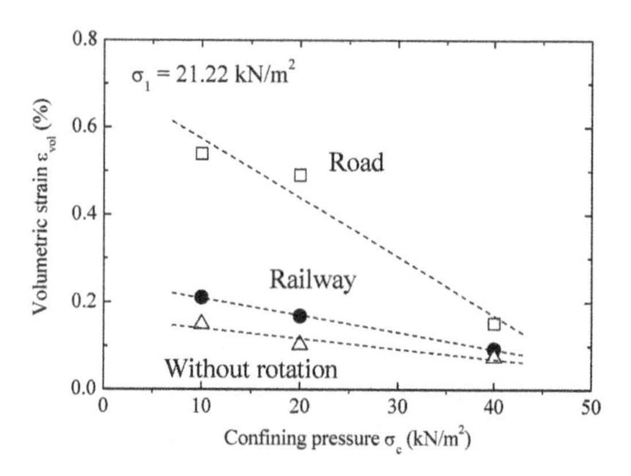

Figure 14. Effect of confining pressure on the residual strain.

In the test result of "railway" with $\sigma_c = 20\,kN/m^2$ and $\Delta\sigma_1 = 30\,kN/m^2$, residual strain was approximately 0.3%. To compensate the result with $\sigma_c = 20\,kN/m^2$ to $\sigma_c = 5\,kN/m^2$, it is necessary to multiply 1.4 to residual strain with $\sigma_c = 20\,kN/m^2$. Therefore, residual strain with $\sigma_c = 10\,kN/m^2$ in "railway" would be $0.3\%*1.4 = 0.42\%$, which approximately coincide with the residual strain of 0.45% in the scale model test.

Those discussions derive a conclusion that the result of principal stress axis rotation test corresponded well with the residual strain in moving wheel loading test.

5 CONCLUSIONS

In this study, moving wheel loading tests, FEM analysis and principal stress axis rotation tests were carried out. Those results showed that the residual strain in principal stress axis rotation test well simulated the residual strain in moving wheel loading test.

It is concluded that it is important to estimate the effect of principal stress axis rotation properly to predict residual settlement under traffic loads.

REFERENCES

Brown, S. F. 2007. The effects of shear stress reversal on the accumulation of plastic strain in granular materials under cyclic loading, *Design and Construction of Pavements and Rail Tracks-Geotechnical aspects and Processed Materials-*, pp. 89–108, Taylor & Francis.

Hirakawa,D., Kawasaki,H., Tatsuoka,F. & Momoya,Y. 2002. Effects of loading conditions on the behaviour of railway track in the laboratory model tests, *Proc. 6th Int. Conf. on the Bearing Capacity of Roads, Railways and Airfields*, Vol. 2, pp. 1295-1305, Lisbon: Balkema.

Towhata, I., Kawasaki, Y., Harada, N. & Sunaga, M. 1994. Contraction of soil subjected to traffic-type stress application, *Proc. of Inter. Sympo. on Pre-Failure Deformation Characteristics of Geomaterials*, pp. 305–310, Sapporo, Japan (Shibuya et al. eds.).

Momoya, Y., Sekine., E., & Tatsuoka, F. 2005. Deformation characteristics of railway roadbed and subgrade under moving wheel load, Soils and Foundations, Vol. 45, No. 4, pp. 99–118. Japanese Geotechnical Society.

Momoya, Y., Watanabe, K., Sekine, E., Tateyama, M., Shinoda, M., & Tatsuoka, F. 2007. Effects of continuous principal stress axis rotation on the deformation characteristics of sand under traffic loads, *Design and Construction of Pavements and Rail Tracks-Geotechnical aspects and Processed Materials-*, pp. 77–87, Taylor & Francis.

Advances in Transportation Geotechnics – Ellis, Yu, McDowell, Dawson & Thom (eds)
© 2008 Taylor & Francis Group, London, ISBN 978-0-415-47590-7

A comparison of trackbed design methodologies: A case study from a heavy haul freight railway

L.M. Nelder, C. England, R.J. Armitage & M.J. Brough
Scott Wilson Ltd, Nottingham, UK

P.R. Fleming & M.W. Frost
Loughborough University, UK

ABSTRACT: One of the major roles of railway trackbed layers is to reduce vehicle induced stresses applied to the underlying subgrade to a level that limits the progressive build up of permanent deformation. The ability of trackbed layers to satisfy this requirement is dependent upon the materials used for construction and their thickness. Numerous design methods, (both empirical and analytical), have been developed across the World to evaluate trackbed design thickness. However, where there is limited information or experience of previous trackbed design with the specific materials or site conditions under consideration, the choice of methodology becomes one of engineering judgment, in assessing the significance and reliability of the design input parameters.

This paper describes a number of design methods which were assessed in a recent project to design a new heavy haul freight railway trackbed, founded on moisture sensitive subgrades, using locally available materials for the track support layers. The produced design thicknesses for each of the methods are compared for differing subgrade conditions. The results show considerable variation of thicknesses from each method with little consistent pattern to the variation. Reasons for these variations are suggested and the choice of the final design used for specific subgrade conditions are presented together with appropriate justification. Concluding on these issues, recommendations are made for a more considered approach to trackbed design.

1 INTRODUCTION

The design of trackbed support layer thickness has historically been based on empirical solutions formed from observation and experience. These solutions have been developed within the constraints of the geology and traffic loading experienced in their geographical catchments. While these solutions will remain valid when used within the range of their original development, once the traffic or subgrade conditions are altered the confidence with which they can be applied will inevitably be affected.

Most of the standards for trackbed design have also been developed to stand alongside corresponding standards on ballast type, rail/sleeper configurations, maintenance methods and vehicle suspension systems. The combination of these standards serves to optimize the network usage and to protect the track and rolling stock from excessive damage. Implicit within these standards are acceptable levels of deterioration within which operations can continue normally.

Provision of rail infrastructure in areas not covered by pre-existing design codes can present significant problems to the designer. It is often the case that the individual components have been sourced from differing regions and will not easily be incorporated into another design code.

This was the case in a project to improve the rail access to a new mining development in the Caribbean which is expected to carry freight traffic of 11 million tonnes per annum at a line speed of 25 mph.This project involved designing trackbed layers for major upgrades to existing rail lines and the installation of new railway infrastructure. Extensive and detailed site and ground investigations where carried out as a preliminary phase of design (Brough et al. 2006) to determine the extent of subgrade variation. Additional laboratory tests were carried out to determine the mechanical behaviour of locally derived stone proposed as ballast for the construction as this differed considerably from the materials specified in the main design standards. The differences in subgrade and loading conditions envisioned in this design were significantly different from those encountered when using most standard design methods. For this reason it was decided to use a number of design methodologies to determine the depth of construction required to protect the subgrade and then to assess which method was most appropriate for each section of line. The methods used in the analysis are the most commonly applied national standards and a linear elastic model used by Scott Wilson for designing highway pavements, suitably modified to allow for rail wheel loading.

2 DESIGN METHODS

No suitable standard design code exists for the calculation of construction thickness for the axle loads proposed for this project in the Caribbean. The methods used in this project are the British Rail design chart standards (Heath and Shenton, 1972), UIC 719-R (UIC, 1994) the AREA Engineering Manual approach (Selig and Waters 1994) and a linear elastic model developed by Scott Wilson for use in pavement design.

The relevant parameters for design assume a static axle load of 31,000 kg, an estimated annual traffic load of 11 million gross metric tones per year a line speed of 25 mph, an axle spacing of 2.8 m and a driving wheel diameter of 33 inches. From these a dynamic load factor based on Clarkes equation (Clarke 1957) of 1.25 has been calculated, from which a maximum applied stress of 243 kN/m^2 was derived for the direct force applied by an individual sleeper to the underlying ballast. This stress assumed that the load applied by an axle over a sleeper is distributed 50% over the sleeper directly underneath the axle load and 25% over the two adjacent sleepers, which is a commonly used approximation that will vary depending on rail and trackbed relative stiffness.

One of the main aims of trackbed design is to provide a sufficient thickness of foundation material to reduce the stress applied to the top of the subgrade by loading to a level where there is negligible permanent deformation under each wheel load application; referred to as the threshold stress or threshold shear strength. The threshold shear strength in this design is assumed to be 50% of the shear strength determined from in-situ hand vane measurements obtained from trial pits. There was an additional requirement to consider the changes in shear strength within highly plastic clays due to seasonal moisture changes, as the site investigations were undertaken during relatively dry conditions. Where the plasticity index was greater than 65% the threshold values where reduce by a further 25% to allow for potential softening of materials based on the judgement of the senior design engineer.

2.1 *British Rail design approach*

This approach suggests that in order to reduce the stress level applied to below the threshold stress, granular thickness can be calculated from the following design chart which was based on laboratory studies undertaken on London Clay samples (Heath and Shenton, 1972).

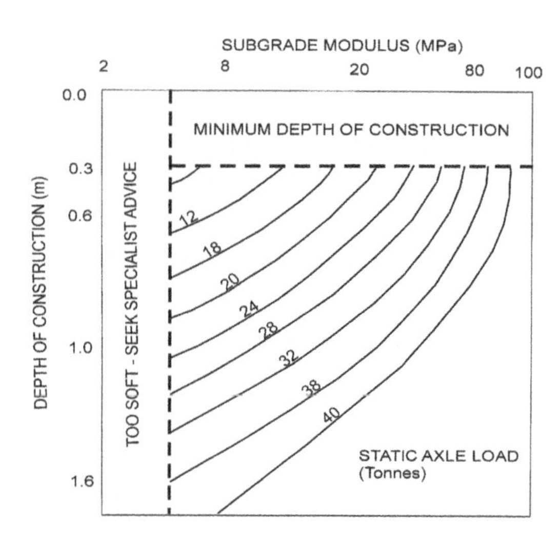

Figure 1. Britsh Rail trackbed thickness design chart (after Heath and Shenton, 1972).

For the purposes of this project the following assumptions were also made:

1 The depth of construction was based on a static axle load of 32 tonnes.
2 Subgrade resilient modulus, defined as the peak applied axial repeated stress divided by the samples recoverable axial strain was determined from either testing of recovered samples or estimated from the results of Dynamic Cone Penetrometer (DCP) testing, where samples could not be recovered.
3 Where existing layers of material are maintained in the new construction as granular sub-ballast, an equivalent foundation modulus (EFM) has been obtained from the equation:

$$\text{EFM} = 143 * \text{Surface Deflection}^{-1.0439}\ \text{MPa} \quad (1)$$

where the surface deflection was obtained from a proprietary linear elastic layer program using material properties obtained from the relevant site investigations.

2.2 *UIC 719R*

The UIC 719R approach, developed by the European Committee for Standardisation for Europe wide use, is an empirical method for calculating granular layer thickness based on soil descriptions and qualitative classification of the soil into bands applicable to European soils. For the purposes of this project soils were assigned to classifications equivalent to those in the UIC code based on observed ground conditions and material properties obtained from in-situ and laboratory testing.

It was assumed that existing granular ballast materials, where present and of suitable quality, could

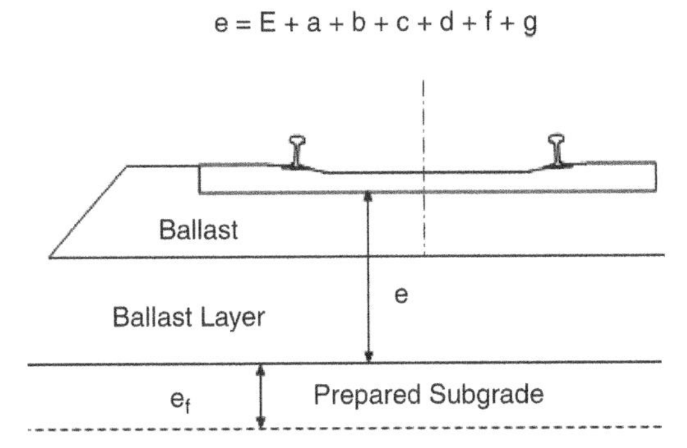

Figure 2. Calculation of Minimum Thickness of Trackbed (After UIC 1994).

be considered to be equivalent to sub-ballast for the purpose of the design; hence, where these materials remain in-situ their thickness needed to be subtracted from the calculated granular layer thickness. This approach demonstrates an advantage over other methods in that any existing suitable granular subgrade materials could be easily incorporated into the design to reduce the required new granular layer thickness.

For the purpose of this design the required bearing capacity of the sub-grade has been classified as Platform 2: 'average sub-grade' based on the required structural performance.

The overall thickness "e" is the sum of a number of factors (a–d, f–g etc on Figure 2) which relate to the subgrade conditions, proposed traffic speed, load and nature of traffic and trackbed structural requirements. The value of "E" depends on the required bearing capacity of the subgrade. The thickness of the prepared subgrade layer e_f will depend on the soil type and the required platform bearing capacity.

For the purposes of this design, values of a, b, c, f and g used in the calculation of granular thickness have been assumed to be zero in all calculations due to the low line speed and the lack of passenger comfort requirements. The value of d is taken to be 0.12 m due to the high axle loads proposed in the design.

2.3 *AREA Engineering manual approach*

This approach uses equations developed by Love (Love 1928) and Talbot (Talbot 1919).The Talbot equation is largely empirical and was developed from full scale laboratory tests using relatively low wheel loads. The Love equation is a development of the Boussinesq equation for elastic analysis, where the point load regime is replaced with a uniform pressure. The equations are:

TALBOT EQUATION:

$$P_c = \frac{16.8 P_m}{h^{1.25}} \quad (2)$$

LOVE EQUATION

$$P_c = P_m \left[1 - \left(\frac{1}{1 + r^2/h^2} \right)^{\frac{3}{2}} \right] \quad (3)$$

Where P_c = allowable subgrade pressure (psi); P_m = applied stress on ballast (psi); h = ballast depth (inches); r = radius of a circle whose area equals the sleeper bearing area (inches).

The Engineering Manual approach recommends a universal limit of 20 psi (140 kPa) for allowable subgrade pressure. However this approach was not used in this project as it was felt that it could lead to inadequate design over poor ground conditions and over design on more competent subgrades. Instead values of P_c were calculated using Terzagi's equation (Terzarghi 1943) for undrained soils (undrained conditions), working on the assumption that the railway acts as a rectangular pad foundation of dimensions equal to the sleeper areas rather than a continuous footing. This approach designs against individual sleeper punching which is assumed to be the worst case failure mechanism. Allowable subgrade pressure is then assumed to be 50% of the ultimate bearing capacity, as calculated from the Skempton equation (Skempton 1951). Allowable subgrade pressures for granular subgrades have been conservatively estimated from the results of DCP testing. Safe bearing pressures can be related to the density of the packing (Waltham 1994) of the granular material, which in turn is related to the SPT N values (BS 5930:1999). DCP testing was used in this project for which there is no standard correlation to SPT values, so engineering judgement was used to approximate suitable values.

2.4 *Analytical method using a multi-layer elastic model*

There is currently no standard analytical method to assess the thickness of railway ballast for a given subgrade condition and traffic loading. In order to develop an analytical model, knowledge of subgrade properties in terms of strength and stiffness, trackbed layer material properties, future traffic loading and relevant failure criteria are required. Two different failure criteria where considered to calculate ballast thickness:

- Limiting the shear stress applied to the subgrade under the locomotive axle load of 380 kN to 50% of the subgrade material shear strength.
- Limiting the maximum ballast surface deflection under 125 kN loading to 1.5 mm, which is associated with typical UK mainline poor track performance (Armitage and Sharpe 2000). The 125 kN

represents 50% of a typical UK axle load assumed to be applied on one sleeper.

This second criteria recognises the importance of limiting ballast elastic deformation under loading, since large movements can allow particle reorientation and ultimately loss of track quality, even where the subgrade is not overstressed. Using the design principles adopted in the field of Pavement Engineering a model was developed to test the trackbed against these failure criteria. The trackbed structure comprising subgrade, existing contaminated granular material and new ballast were modelled as a multi-layer linear elastic system where each layer is described by its stiffness, Poisson's ratio and thickness. An in-house multi-layer elastic program, used for pavement design, for nearly 20 years, was used to calculate the maximum shear stress at the top of the sub-grade and the deformation of the ballast surface under train loading for various assumed thicknesses of ballast. The calculated values were then compared with the subgrade material failure strength and the acceptable surface deflection of 1.5 mm, respectively. If either of the failure criteria was breached then the granular thickness was increased until a satisfactory design was achieved.

Subgrade and existing granular layer stiffnesses, shear strengths and thicknesses were predicted from a combination of laboratory and in-situ testing and observations.

3 DESIGN RESULTS

The results in Tables 1–6 show a selection of combined granular/ballast layer thicknesses for trackbed on a range of subgrade types found along the proposed route. Where the thickness is given as "less than" the requirement was below 300 mm, which has been assumed as the minimum for automated track maintenance by tamping.

3.1 *Rock sub-grade*

The design values of granular thickness for a rock subgrade are presented in Table 1.

The high values using the UIC approach are as a result of a localized area of weak rock which has been highly weathered and was assigned a lower soil classification compared to other rock subgrades. Within the range of values, the UIC design thickness comes out at around twice the value predicted by other methods. This is due to the automatic use of 550 mm of material specified for the P2 bearing classification.

Table 1. Typical design values for rock sub-grades.

Method	Stiffness (MPa)	Mode (mm)	High (mm)
Love eqn	200	<300	<300
Talbot eqn	200	<300	<300
BR	200	300	300
UIC	200	570	1200
LE Stress	200	–	–
LE def	200	300	300

The granular thickness specified in the design for rock subgrade was therefore set at 300 mm. This is in agreement with most methods and provides sufficient depth for future automated track maintenance.

3.2 *Granular sub-grade*

The granular material sub-grades are either densely or loosely compacted and the results for the two classes are separated and shown in Tables 2 and 3.

The granular material is typically layers of made ground to a depth of 1.2 m, which comprises gravelly sands with varying but small quantities of clay.

Table 2. Typical design values for loose granular sub-grades.

Method	Stiffness (MPa)	Mode (mm)	High (mm)
Love eqn	50	<300	600
Talbot eqn	50	<300	500
BR	50	400	400
UIC	50	670	670
LE Stress	50	–	–
LE Def	50	500	500

Table 3. Typical design values for dense granular sub-grades.

Method	Stiffness (MPa)	Mode (mm)	High (mm)
Love eqn	80	<300	600
Talbot eqn	80	<300	500
BR	80	300	300
UIC	80	670	670
LE Stress	80	–	–
LE Def	80	300	300

Table 4. Typical design values for very soft to soft clay sub-grades.

Method	Stiffness (MPa)	Mode (mm)
Love eqn	20	900
Talbot eqn	20	800
BR	20	900
UIC	20	–
LE Stress	20	>1000
LE Def	20	>1000

Once again the UIC method gives a higher construction thickness than the other methods, but does not distinguish between the different mechanical properties of the two subgrade density states. The BR and analytical methods show a slight increase in thickness for the less dense material reflecting the lower stiffness and higher susceptibility to deflection.

Within each density state there are is a range of acceptable threshold stresses related to the composition of the granular materials. This has an effect on the AREA Engineering Manual methods which are based on the mechanical bearing capacity of the materials, not the stiffness of the subgrade.

The granular layer thickness specified for these sections of trackbed are 300 mm for the 50 MPa stiffness and 400 mm for the 80 MPa stiffness in accordance with the BR design method.

3.3 *Clay Sub-Grade*

The clay subgrades were divided into three categories (20, 40, 60 MPa stiffness) in order to assess a range of conditions. The stiffness was related to the undrained shear strength as measured by a hand vane in trial pits using the correlation given in BS5930 1999 section 41.3.2.

At lowest sub-grade stiffness there is a reasonably consistent agreement between all methods that a granular thickness of around 1 m is required. The UIC method could not be used for this quality subgrade since it was considered too poor and should be improved. As the subgrade stiffness increases to 40 MPa the AREA and BR approaches show a decrease in required thickness, but the UIC and analytical methods both stipulate a thickness of around 1 m. Within each subgrade type there was a wide range of allowable bearing pressures and threshold shear strengths which contributed to variation in design thickness Each design method also uses different parameters as a basis for their results; in particular the approach used to allow for the presence of existing granular material (such as old trackbed). Since clay subgrades vary considerably, due to differences in their stiffness and strength as a result of moisture changes, design methodologies that reflect these differences are preferable to those that don't.

In general the soft clay subgrades all required in excess of 1 m depth of granular material, as calculated by the Linear Elastic Stress and Deflection design methods. Firm clay subgrades (either with or without existing sub-ballast) were felt to require a more conservative approach than that suggested by the AREA or BR manuals so the UIC method was stipulated, to allow for potential future softening. For stiff subgrades with no existing sub-ballast the UIC method was also stipulated, but where there was a thickness of sub-ballast up to 200 mm the BR method was used since there was a degree of subgrade protection already in place. These decisions were influenced by the results from the Linear Elastic analysis. Indeed, where there was a good thickness of sub-ballast, in excess of 200 mm, over a stiff clay subgrade the Linear Elastic analysis was used to check for allowable deflections when using the 300 mm minimum depth of granular material specified to allow for tamping.

Table 5. Typical design values for firm clay sub-grades.

Method	Stiffness (MPa)	Mode (mm)	High (mm)	Low (mm)
Love eqn	40	500	800	500
Talbot eqn	40	400	600	400
BR	40	500	500	500
UIC	40	1020	1020	1020
LE Stress	40	>1000	>1000	>1000
LE Def	40	900	900	300

Table 6. Typical design values for stiff to very stiff clay sub-grades.

Method	Stiffness (MPa)	Mode (mm)	High (mm)	Low (mm)
Love eqn	60	400	500	<300
Talbot eqn	60	300	400	<300
BR	60	300	300	300
UIC	60	1020	1020	1020
LE Stress	60	1000	>1000	400
LE Def	60	300	300	300

3.4 *General design considerations*

As can be seen from Tables 1–6 and Figure 3 the results from the differing design methodologies can vary significantly. It can also be seen that although there is some general trend in the differences (thicker granular trackbed for weaker subgrades) there is no simple relationship that allows for direct comparison. This is not unexpected as the methods use differing material parameters and assumptions to determine the thickness of granular material.

This lack of correlation highlights the problems faced by designers when specifying trackbed construction. The use of a design code in a situation outside of the limits upon which it was based may not produce adequate results when used elsewhere, especially when the method uses pre-prepared charts. If all methods are fundamentally based on providing adequate subgrade protection, then a more targeted design can be produced with increasing knowledge of the subgrade materials and properties. Otherwise, some methods may be unnecessarily conservative, or (perhaps worse) unable to deliver the level of performance required during the design life.

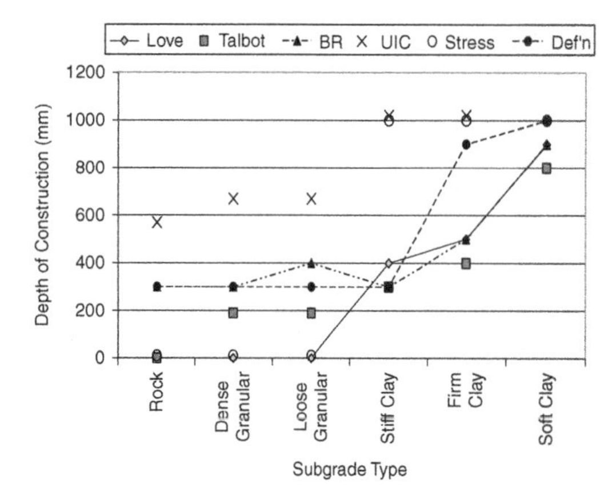

Figure 3. Summary Plot of Design Thicknesses.

There is always a balance to be made between initial cost and future maintenance cost. This balance is one that can only be made by understanding the client's perspective, so it is essential to involve the client by explaining the options and agreeing the preferred whole life value approach. The ability to maintain a track (especially when single line, such as the one considered in this paper) involves knowledge of available plant, local expertise, freight operations, economic cost of downtime, etc.

In recent years a number of finite element methods have been proposed Chang et al (1980), Rose et al (2003) some of which are used in conjunction with elastic models. As yet, due to the complexity of modelling and limits of time, these methods are not suited for routine use, especially for the design of trackbed renewals, although they may find application in the design of new high speed lines where further optimisation of trackbed design is financially beneficial.

A more practical solution to this problem is to use a method based on Linear Elastic analysis, validated using in service trackbed data. This would allow designers to develop more appropriate solutions, using site specific materials data which is now available, underpinned by a method proven for pavement design over nearly three decades.

4 CONCLUSIONS

- Results from different design methodologies produced varying results for the required depth of trackbed construction on a range of subgrade types.
- In this project a number of design approaches were used and compared, which enabled the designer to specify a suitable trackbed construction.
- A design method based on a Linear Elastic analysis appears to provide realistic trackbed thicknesses in most cases. This makes use of material information that can be captured relatively easily during a site investigation, so is not constrained by regionally developed empiricism.

5 FURTHER WORK

The authors are currently developing the linear elastic model approach for trackbed design under an Engineering Doctorate scheme. It is hoped to optimise future designs by using a greater range of material specific parameters and a more realistic approximation of trackbed loading. Notwithstanding the inelastic nature of granular layers it is felt that, for practical purposes, this approach would be suitable for bridging the gap between the current extremes of design method (complex finite element versus simple empirical) to provide more cost efficient solutions.

REFERENCES

Armitage, R.A., Sharpe P. (2000). Implementing Optimal Trackbed Evaluation to Predict and Reduce Track Degradation. Infrastructure Maintenance and Research, London January 18–20.

B.S. 5930 (1999). Code of Practice for Site Investigations. British Standards Institute.

Brough, M.J., England C., et al. (2006). Trackbed Design for Heavy Haul Freight Routes: Case Study: Carribean Bauxite Mine. RailFound, Birmingham, University of Birmingham Press.

Chang, C.S., Adegoke C.W., Selig E.T. (1980). The GEOTRACK Model for Railroad Track Performance. Journal of the Geotechnical Engineering Division, ASCE, 106 (11) pp 1201–1218.

Clarke, C.W. (1957). Track Loading Fundamentals. The Railway Gazette. 106. pp26.

Heath, D.L., Shenton, M.J. et al. (1972). "Design of Conventional Rail Track Foundations." Proc.Inst.Civ.Eng., 51, 251–267.

Love, A.E.H. (1928). The Stress Produced in a Semi-infinite Body by Pressure on Part of the Boundary. Philosophical Transactions of the Royal Society. London. 228: 377–420.

Rose, J.G., Bei S., Long W.B. (2003). Kentrack: A Railway Trackbed Structural Design and Analysis Program. Proceedings of the AREMA Annual Conference, Chicago.

Selig, E.T., Waters J.M. (1994). Track Geotechnology and Substructure Management. Thomas Telford, London.

Skempton, A.W. (1951). The Bearing Capacity of Clays. Proc. Building Research Congress. Vol 1, pp 180–189. London.

Talbot, A.N. (1919). Second Progress Report of the Special Committee on Stresses in Track. Proceedings of the American Railway Engineering Association.

Terzarghi, K. (1943). Theoretical Soil Mechanics. John Wiley and Sons. New York.

UIC (1994). Earthworks and Trackbed Layers for Railway Lines. UIC Code 719 R, International Union of Railways, Paris, France.

Waltham, A.C.W. (1994). Foundations of Engineering Geology. Blackie Academic and Professional, London.

Advances in Transportation Geotechnics – Ellis, Yu, McDowell, Dawson & Thom (eds)
 ISBN 978-0-415-47590-7

Performance of canted ballasted track during curving of high speed trains

J.A. Priest, L. Le Pen & W. Powrie
School of Civil Engineering and the Environment, University of Southampton, Southampton, UK

P. Mak & M. Burstow
Network Rail, London, UK

ABSTRACT: A train travelling along a curve applies both vertical and lateral forces to the track. The lateral forces are associated with the centripetal acceleration, and for a given radius increase with the square of the train speed. Traditionally, curved track is superelevated or canted to improve passenger comfort; however the level of cant chosen is a compromise between that required for the fastest and slowest trains expected. Therefore, passenger comfort is normally the limiting factor on maximum curving speed. Tilting trains enable faster curving and lead to an increase in the forces applied to the track. This paper reports the initial results of a study undertaken to assess the behaviour of canted track during the passage of high speed tilting trains, and compares the behaviour observed to that from traditional non-tilting trains.

1 INTRODUCTION

A train traveling round a curve experiences a centrifugal acceleration seeking to move it radially out from the curve. This acceleration is resisted by the track through the generation of lateral forces which are transmitted through the contact patches to the vehicles. Thus, curved railway track is subjected to lateral as well as vertical vehicle loading which contribute to track geometry degradation.

To improve passenger comfort, curved track is normally canted or superelevated (i.e. the outer rail is raised relative to the inner rail). This rotation of the train causes some of its vertical weight to act in a direction parallel to the (rotated) track plane, reducing the lateral centrifugal force experienced by the train. If the angle of cant is sufficient to tilt the body of the train so that the component of weight acting laterally towards the centre of the curve cancels the centrifugal force acting towards the outside of the curve then the vehicle is said to be at cant equilibrium. For a curve of a given radius, the angle of cant required to achieve cant equilibrium increases with train speed. Cant deficiency is used to describe the situation where the lateral forces are not balanced, and is defined as the additional cant which would be required to be applied to the track to attain cant equilibrium. For passenger comfort it is necessary for the train to operate with some cant deficiency, the maximum allowable value of which is specified in track design standards. On a mixed traffic railway, the speeds of trains will vary; thus the degree of superelevation will normally be a compromise between the highest and lowest expected train speeds.

Tilting trains, such as the Class 390 Pendolino introduced in 2003 on the West Coast Main Line (WCML), enable trains to travel at enhanced cant deficiency (i.e. faster) on a given curve while maintaining passenger comfort by rotating the vehicle body to compensate, at least in part, for the lack of superelevation. As a result of the increased speeds, the curving forces applied at the rail head are correspondingly increased, although there is no increase in the lateral forces experienced within the train.

The system of vertical and lateral forces applied to the rail by a train as it moves round a curve is indicated schematically in Figure 1a. The ratio of vertical forces applied to each rail from the train axles, and the magnitude of the lateral force, will depend on the train speed. The resultant equivalent forces imparted at the base of the sleeper are indicated in Figure 1b. Static tests carried out in the laboratory by Le Pen & Powrie (2008) on a single sleeper laid longitudinally on a ballast bed suggest that the sleeper/ballast interface should be capable of carrying the loads associated with high speed tilting trains. To investigate the actual behaviour of canted track, a programme of field monitoring of displacements has been carried out on a section of curved track on the WCML. This paper reports some of the initial results.

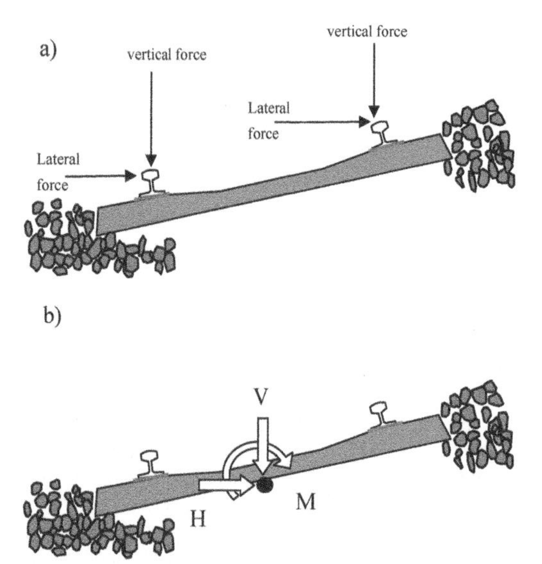

Figure 1. a) Force applied to the rails. b) Equivalent forces at the base of the sleeper.

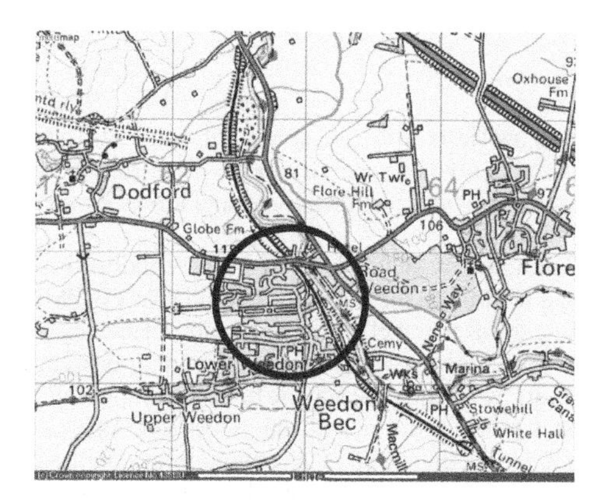

Figure 2. Site location at Weedon Bec, Northamptonshire (located in circle).

2 FIELD TESTING

2.1 *Site location*

To assess the response of canted track to different train loading conditions, a section of track on the WCML was chosen where both high speed Pendolino tilting trains and non-tilting trains are run. The site location was at Weedon Bec, just north of Northampton, where the track has a pair of reverse curves (Fig. 2). Measurements were obtained from a number of locations on a section of track which had a design cant of 150mm on a curve of radius 1230m. The section of track monitored had been recently tamped, as part of a track geometry maintenance programme, prior to the first monitoring visit in November 2006. A 2nd monitoring visit was undertaken in February 2007.

Figure 3. Instrumentation bracket attached to sleeper end showing PIV target with vertical geophone to the rear.

2.2 *Measurement techniques*

Two displacement monitoring systems were adopted. One system used low frequency geophones attached to the ends of sleepers to measure their velocity, from which dynamic displacements could be calculated. The other combined relatively simple remote video monitoring (Bowness et al., 2005) and Particle Image Velocimetry (PIV; Adrian, 1991) to assess sleeper displacements directly from the movement of a target attached to the sleepers. Figure 3 shows the PIV target attached to a sleeper with a vertical geophone fixed behind. Geophones in both vertical and horizontal orientations were used so that vertical, lateral and longitudinal displacements of individual sleepers could be measured. Figure 4 shows the typical layout of instrumentation at each measurement location. During the passage of a train, vertical motion of the sleeper at both the low rail end (1V & 2V) and the high rail end (3V, 6V & & 7V) of the sleepers were recorded; and horizontal motion along the track (4HL & 8HL) and normal to the track (5HT & 9HT) were measured.

2.3 *Data processing*

Figure 5a shows the typical output voltage obtained from a vertical geophone during the passage of a train. The output voltage of a geophone is generated by the motion of a spring-mounted core within a coil fixed to the sensor body. For the geophone used in this investigation (LF-24, supplied by Sensor Nederland) the sensitivity (i.e. the output voltage per unit velocity) is 15 V/m/s when the excitation frequency is above 2

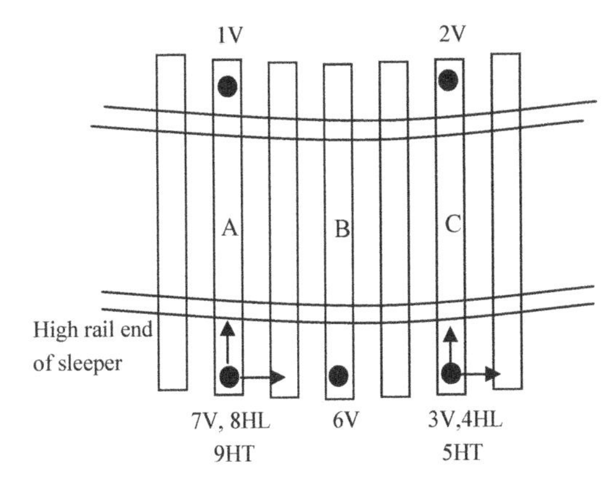

Figure 4. Layout of geophones on sleeper.

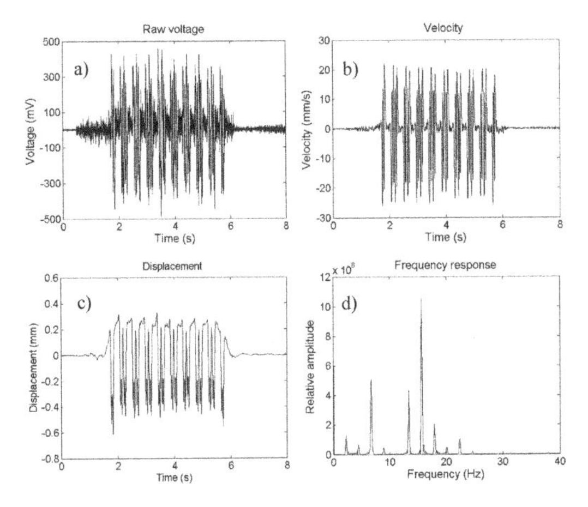

Figure 5. a) raw voltage output from geophone. b) Sleeper velocity obtained from geophone output. c) Sleeper displacement from integrating velocity data. d) Frequency response from velocity data.

Hz. For this study all train speeds were high enough for the dominant velocity frequency, related to passing individual bogies of the train, to be above this threshold.

Figure 5b shows the velocity data obtained from the raw voltage output on the basis of the geophone sensitivity. Sleeper displacements are then calculated by integrating the velocity data. Figure 5c shows the displacements calculated from the velocity data. A passband filter is applied to remove low frequency drift and high frequency noise from the dataset hence the apparent shift in datum (+ve displacement) and start-up transients are the result of signal processing. Figure 5d shows the power density spectrum from the velocity data. The peaks correspond to dominant frequencies relating to the passage of individual axles, bogies, carriages etc. and their multiples.

Table 1. Train classes and corresponding data.

Train Type	Train speed (km/h)	Nominal Axle load (tonne)
Class 390 Pendolino	155–180	12.9
Class 220 Voyager	153	14
Class 350 Desiro EMU	157	12
Class 87 Loco	165	21
Mk 3 Coaches	165	10

The remote video monitoring uses a high speed camera mounted on a telescopic lens to obtain individual images of the target during the passage of the train. The telescope allows the monitoring system to be positioned far enough away from the track not to be unduly affected by deflections and vibrations of the ground during the passage of a train.

A detailed description of these techniques and their validation is given by Bowness *et al.* (2007).

3 RESULTS AND DISCUSSION

A total of twenty five trains were monitored at different locations along the track. Of these, most were Class 390 Pendolinos with an average speed of ~180 km/h although two trains were recorded at ~155 km/h. Train speed is determined from the displacement time plot (Figure 5c) on the basis of the time difference between the first and last axle and the distance between them. Measurements were also made for one Class 87 pulling Mk3 coaches, two Class 221 Super Voyagers and one Class 350 Desiro EMU. Table 1 lists the different trains that were measured with their corresponding speeds and nominal axle loads.

Figures 6-9 show the displacements recorded for a 9-car Class 390 Pendolino travelling at 183 km/h for the right hand monitored sleeper (sleeper C) shown in Figure 4. The largest vertical displacement occurs at the high rail end of the sleeper The total displacement for the middle section of the train, in the absence of start-up transients, is about 0.86 mm with the low rail end exhibiting around 40 % less displacement (0.38 mm). The lateral displacement for this sleeper was 0.3 mm, which is ~35% of the maximum vertical displacement. Figures 6-8 show that the individual bogies are the dominant loading event, with individual axles being a minor event. In contrast, the longitudinal displacements (Figure 9) are dominated by the individual axles, with overall displacements of about 10% of the vertical displacements. Displacements of the left hand sleeper (sleeper A) shown in Figure 4 were similar, although the vertical displacement of the high rail was about 1.3 mm. It has been shown by Bowness *et al.* (2007) that vertical

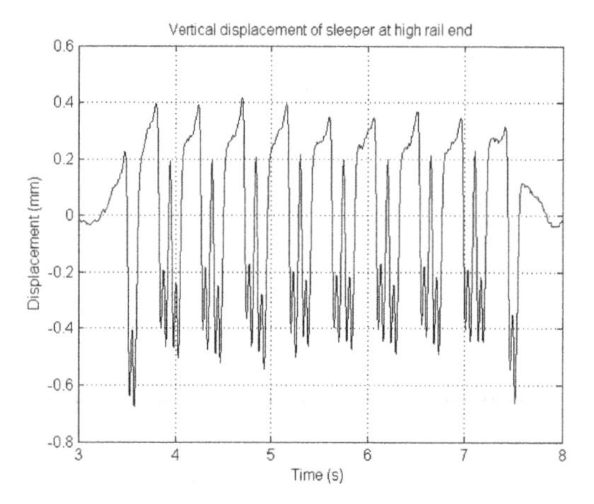

Figure 6. Vertical displacement of sleeper C at the high rail end for a Class 390 Pendolino.

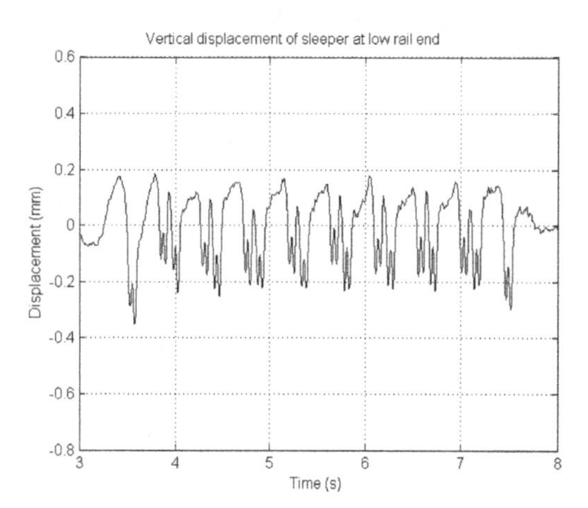

Figure 7. Vertical displacement of sleeper C at the low rail end for a Class 390 Pendolino.

sleeper displacements can be quite variable; thus this behaviour is not unexpected.

Figure 10 shows the vertical displacements of sleeper C obtained using the remote video monitoring and subsequent PIV analysis. Although the response derived using the PIV analysis is comparable to that in Figure 6, the video data are subject to appreciable ground vibration as seen by the apparent displacements around the zero datum between bogies. Although the telescope was positioned some 5m away from the track being monitored, it was still on the embankment supporting the two track railway.

Figure 11 shows the vertical displacements for the high rail end of sleeper A (at a different location along the track) for a range of trains and speeds. It can be seen the Class 350 train, which has the lowest axle load (nominal 12 tonne) and one of the lowest speeds,

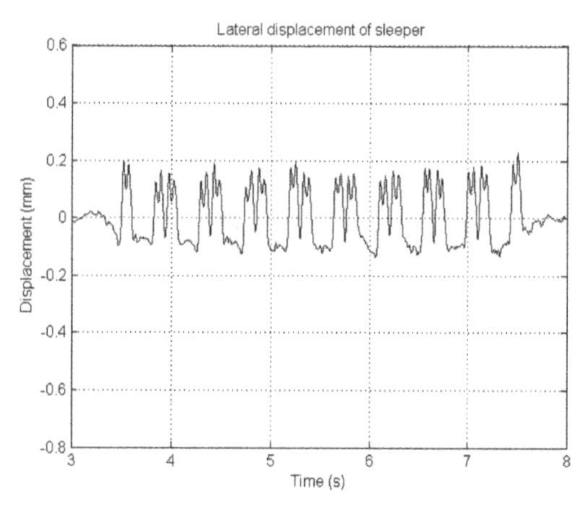

Figure 8. Lateral displacement of sleeper for a Class 390 Pendolino.

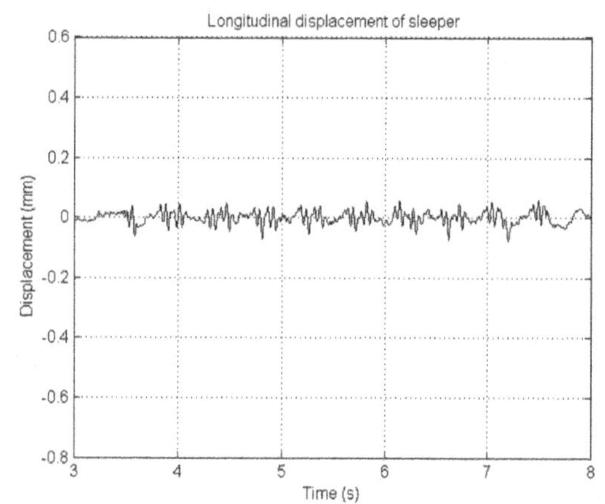

Figure 9. Longitudinal displacement of sleeper.

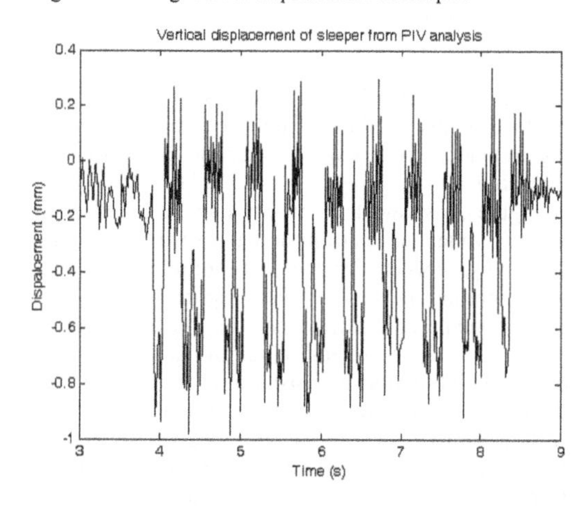

Figure 10. Vertical displacement of sleeper C from PIV analysis.

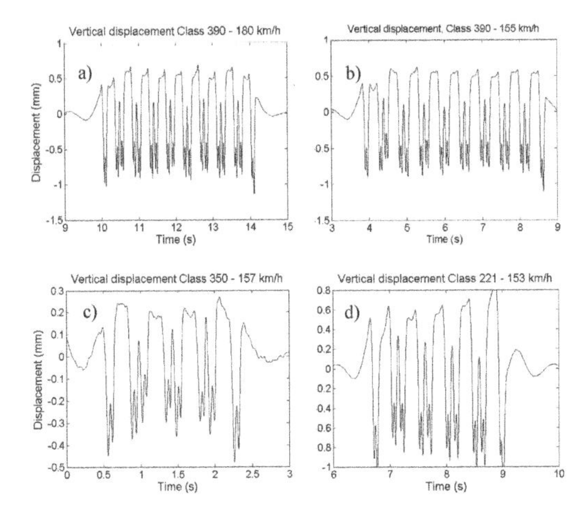

Figure 11. Vertical displacement of sleeper at the high rail end for a variety of train classes and speed. a) Class 390 at 155 km/h, b) Class 390 at 180 km/h, c) Class 350 at 157 km/h, d) Class 221 at 153 km/h.

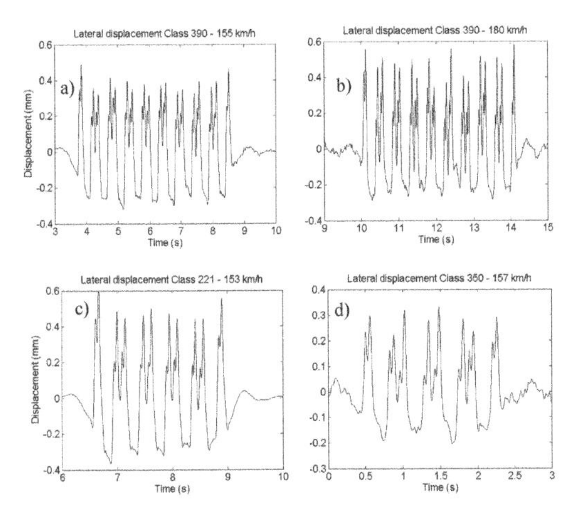

Figure 12. Lateral displacement of sleeper at the high rail end for a variety of train classes and speed. a) Class 390 at 155 km/h, b) Class 390 at 180 km/h, c) Class 350 at 157 km/h, d) Class 221 at 153 km/h.

produces the least displacement. The Class 221 travelling at the lowest speed has vertical displacements very similar to the Class 390. However, the Class 221 also has the largest nominal axle load of 14 tonne, which offsets the reduction in speed. Figure 12 shows the lateral displacement for the same sleeper. The difference in lateral displacements between the Class 350 and the Class 390 is much greater than observed for the vertical displacements owing to the effect of the greater speed on the lateral (centrifugal) force. The Class 390 train travelling at180 km/h produces the largest lateral displacement of the track.

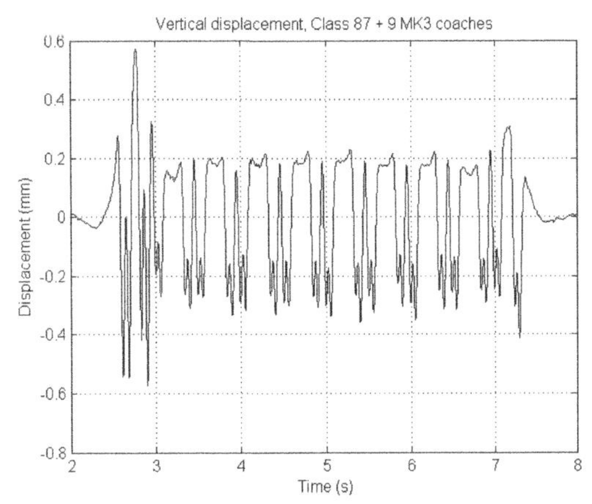

Figure 13. Vertical displacement of a sleeper at the high rail end for a Class 87 locomotive pulling 9 MK3 coaches.

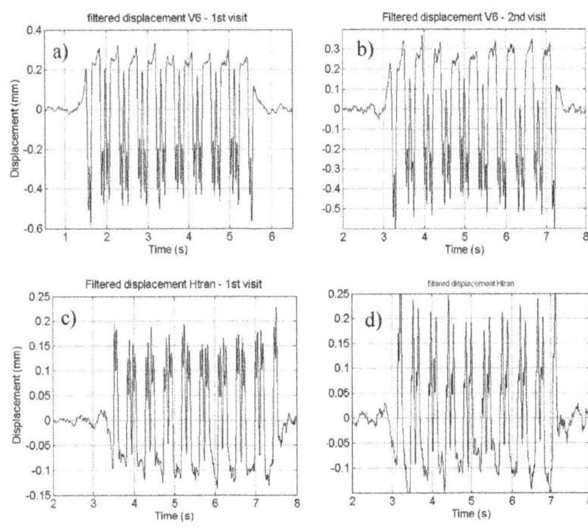

Figure 14. Change in response of the track during two monitoring visits for the passage of a Class 390 train. a) & b) comparison of vertical displacement, c) & d) comparison of lateral displacement.

Prior to the modernization of the WCML and the subsequent introduction of the Class 220, 221 and 390 Pendolinos, most trains were formed of Class 87 locomotives hauling Mk3 coaches. Figure 13 shows the vertical displacements obtained during the passage of this type of train. The influence of the locomotive, which has a nominal axle weight of 21 tonne, is clearly greater than that of the Mk3 coaches, which are considerably lighter. In comparing these displacements with those from Figure 6 it can be seen that while the locomotive produces vertical displacement comparable with the Class 390 train, the displacements from the Mk3 coaches are considerably less. The behaviour of the ballast and the underlying ground is influenced by both the magnitude of the load applied (vertical and

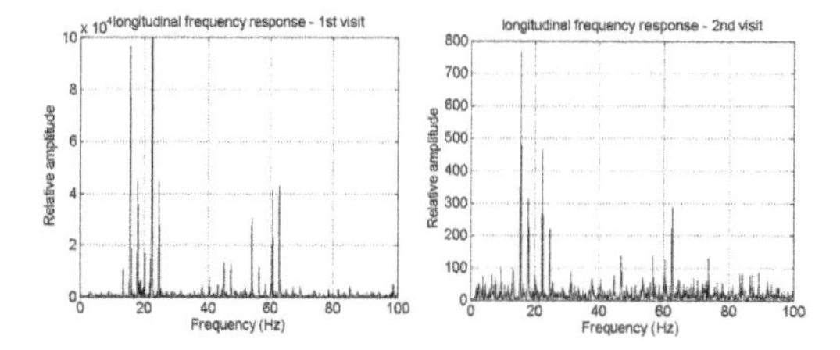

Figure 15. Comparison of frequency response for longitudinal motion on the high rail end of a sleeper for differing visits.

lateral) and on the number of loading cycles (Selig & Waters 1994). Therefore, the effect of the passage of a 9-car Class 390 train with 18 high displacement loading events is likely to be greater than a locomotive and coaches with 4 high displacement and 18 low displacement loading events.

Figure 14 shows the change in response of the track for both vertical and lateral displacements between the first visit in November 2006 and a subsequent visit in February 2007. It can be seen that vertical displacements at the high rail end of the sleeper have remained substantially unchanged, although a small minor reduction in the low rail end displacement was observed (not shown). However, the lateral displacements have noticeably increased, suggesting that the lateral track stiffness had reduced between visits. Figure 15 shows the frequency response obtained for the longitudinal velocity of the measured sleepers. Although the main response (peaks) is reasonably consistent between the first and second visit, there is a pronounced increase in general noise. This is thought to be related to a general loosening of the sleeper within the ballast.

4 SUMMARY

The introduction of high speed tilting trains on the West Coast Main Line allows higher curving speeds on canted track whilst maintaining passenger comfort. This gives rise to increased forces being applied to the track. An investigation was conducted to assess the deformations of canted ballasted track during high speed train passage.

Low frequency geophones were used to measure vertical and horizontal sleeper velocities, from which displacements were calculated. The data have shown that a canted sleeper is subjected to a complex set of displacements during a train passage. The vertical displacement of the high rail end of the sleeper is greater than that at the low rail end (~40% of high rail). Lateral sleeper displacement was up to ~35% of the maximum vertical displacement. Both the vertical and lateral displacements are dominated by the bogies and longitudinal motion by individual axles (10% of maximum vertical displacements). Lateral displacement was greatest for Class 390 Pendolino trains, although vertical displacements were comparable with those for a Class 221 Super Voyager and a Class 87 locomotive. The lowest displacements were recorded for the Mk3 coaches being hauled by the Class 87 locomotive, which was the common train configuration before the upgrade. Therefore, the introduction of Class 390 trains has given rise to a significant (order of magnitude) increase in the number of maximum displacement cycles over a given time. Thus frequent high speed trains may lead to a requirement for increased maintenance compared to that historically undertaken, owing to the increase in load and loading cycles on the track ballast and formation.

Comparisons of displacements immediately after track maintenance (tamping) and 3 months later showed that while vertical displacements were largely unchanged, lateral displacements had increased. Further work is required to fully understand the dynamic behaviour of the track system.

ACKNOWLEDGEMENTS

The authors are grateful for the financial support of the Engineering and Physical Sciences Research Council (grant no. GR/S12784/01) and to Network Rail for allowing and facilitating track access.

REFERENCES

Adrian, R. J. (1991). Particle imaging techniques for experimental fluid mechanics. *Ann. Rev. Fluid Mechanics,* Vol 23: 261–304.

Bowness, D., A. C. Lock, W. Powrie, J. A. Priest and D. J. Richards (2007). Monitoring the dynamic displacements of railway track. *Proc. IMechE Vol. 221. Part F: J. of Rail and Rapid Transport*: 13–22.

Bowness, D., A. C. Lock, D. J. Richards and W. Powrie (2005). Innovative remote video monitoring of railway track displacements. *Applied Mechanics and Materials, Vol* 3-4: 417–422.

Le Pen, L & W. Powrie (2008). Testing of track stability. *1st International Conference on Transportation Geotechnics*, Nottingham, UK.

Selig, E. T. and J. M. Waters (1994). *Track geotechnology and substructure management*, Thomas Telford, London.

9 Soil improvement

Advances in Transportation Geotechnics – Ellis, Yu, McDowell, Dawson & Thom (eds)
 ISBN 978-0-415-47590-7

Effects of freezing and thawing processes on the mechanical behavior of silty soils stabilized with fly ash

S. Altun
Ege University, Civil Engineering Department, Bornova, İzmir, Turkey

A.B. Goktepe
Kolin Construction Co., Cankaya, Ankara, Turkey

ABSTRACT: Effects of freezing and thawing on the mechanical behavior of silty soils stabilized with fly ash are investigated in this study. Laboratory based analyses required to obtain efficient fly ash addition rates and compaction procedure are discussed. Examples of laboratory evaluations are presented using silty soils. The soils are natural silts obtained from the vicinity of Izmir City, Turkey. Additionally, the effect of weakly cementation on the mechanical behavior of silt stabilized by fly ash is also studied experimentally. The cementing agents are cement and fly ash slurry, and the samples so formed were somewhat cemented.

Sity soils are stabilized by 10% and 30% percents of fly ash, as well as 2% and 8% percents of cement. After the compaction process in proctor densification, uniaxial tests are performed on the silty samples for various freezing and thawing cycles (one, three and seven) under different curing conditions. It is concluded that the influence of freezing and thawing on the mechanical behaviour of silty soils must be considered with the ratio of stabilization materials and curing conditions.

1 INTRODUCTION

Durability of pavement materials induced by environmental effects, namely, cyclic freezing/thawing action, can have a chief effect on the performance of a pavement structure. Freezing/thawing event is considered to be one of the most harmful actions that can make considerable damage to pavement. Therefore, the planning and analysis of pavements should take account for such seasonal effects. Besides, many studies recognized the importance of evaluating the durability of stabilized aggregate bases and suggested that such effects be carefully considered in the mixture design.

Coal burning utilities in the Aegean Region of Turkey are using sub-bituminous coal increasingly. These facilities typically produce fly ash, which may be classified as Class C fly ash, due to its high calcium oxide content. The ashes are characterized by their self-cementing property; therefore, they can be used for soil improvement as cement surrogates. Waste materials such as fly ash, may be utilized in several constructions, such as highways and concrete buildings, in order to save waste disposal costs.

Soil stabilization process is primarily made for maintaining the long-term soil stability. Over the last decades, the addition of fly ash and/or cement to fine soils is a widely used application performed for increasing the engineering properties, especially in highway and airport constructions. Soil type, application procedure, economy, curing time, and the type as well as the amount of the additive are crucial for a successful performance.

There are various studies on fly ash utilization in the literature. Turner (1997) compared fly ash stabilized foundation soil with sands reinforced with geosynthetics. Results denoted that fly ash addition exhibited at least same performance comparing with other reinforcements. Cokca (2001) conducted compaction tests on cement stabilized Soma fly ashes, and presented that increasing cement content decreases optimum water content and maximum dry density, but increases specific gravity. Yıldız et al. (2004), indicated that unconfined strength of highly plastic Aksaray clay is 15 times better after 28 days curing. On the other hand, for the same curing period, the strength of low plastic Doganhisar clay is approximately 2 times better than unstabilized samples. Besides, the strengths are decreased 10–15% after freezing and thawing cycles.

The strength of soil stabilized by self-cementing fly ash is usually determined by California Bearing Ratio (CBR) tests (Ferguson 1993, Ferguson & Leverson 1999, White & Bergeson 1998). In this investigation, laboratory tests are performed on fly ash (originated from Soma thermal plant) and Portland cement

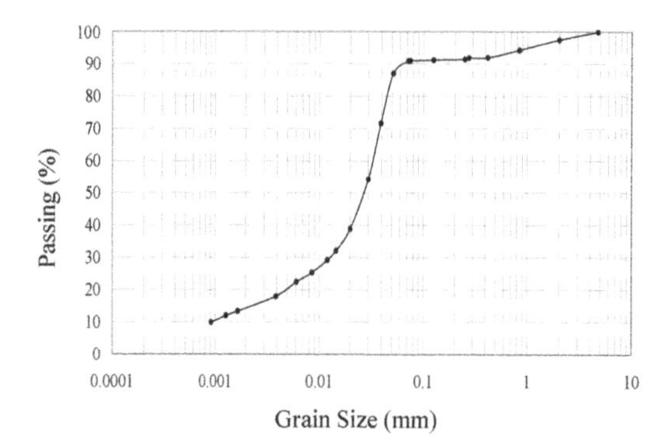

Figure 1. Granulometric distribution curve.

stabilized silty soils to observe strength gains with different mixing compositions. Moreover, the effect of fly ash and cement addition on freezing/thawing strength of silty soils is also studied within the content of this study. Consequently, economical and performance based evaluations are made for the determination of optimum contents.

2 MATERIALS AND TEST PROCEDURE

2.1 *Properties of testing materials*

Silt samples are collected from Turgutlu, Imir region. Soil classification is determined due to sieve analysis and Atterberg limit tests. Soil type is classified as A-4 according to AASHTO classification system. On the other hand it is determined as low plastic sandy and clayey silt, which is symbolized as ML due to unified classification system (UCS).

Granulometry curve of the soil is given in Figure Deney 1. The curve is plotted by the data obtained with dry/wet sieve and hidrometric analyses. The specific gravity (G_s) is found as 2.73 by applying Picnometer test. Finally, liquid limit, plastic limit, and plasticity index are obtained as 30.20, 24.28, and 5.92, respectively.

The fly ash used in this study is lignite originated and collected from Soma thermal power plant. The specific gravity and the specific area are 2.21 gr/cm^3 2351 cm^2/gr, respectively. The percents indicating the material over 0.045 mm and 0.090 mm are 33% and 14.3%, respectively. Chemical properties of the fly ash are given in Table 1.

Finally, cement used in laboratory analyses of this study is Portland cement CEM-I 42.5, which having specific gravity of 3.13 gr/cm^3 and specific area of 3670 cm^2/gr. Top-sieve values of the cement for the sieves 0.09 mm and 0.032 are 0.70% and 18.20%, respectively.

Table 1. Chemical properties of the fly ash.

Chemical Composition	Fly Ash (%)
Alumina (Al_2O_3)	23.55
Ferric Oxide (Fe_2O_3)	4.91
Calcium Oxide (CaO)	18.67
Magnesium oxide (MgO)	1.54
Silica (SiO_2)	45.98
Sulphur oxide (SO_3)	1.47
Na_2O	0.24
Free CaO	0.64

Table 2. Abbreviations and the properties of the test samples.

Sample	Fly Ash (%)	Cement (%)	LL	PL	PI
Natural soil	–	–	30.20	24.28	5.92
2C	–	2	37.82	26.98	10.84
5C	–	5	42.03	30.51	11.52
8C	–	8	42.50	33.12	9.38
10FA	10	–	36.90	26.66	10.24
20FA	20	–	36.51	29.78	6.73
30FA	30	–	37.17	30.59	6.58
10FA2C	10	2	38.87	29.47	9.40
20FA2C	20	2	35.42	29.31	6.13
30FA2C	30	2	35.10	29.79	4.63
10FA5C	10	5	44.29	32.84	11.45
20FA5C	20	5	38.80	30.08	8.72
30FA5C	30	5	36.81	31.13	5.68
10FA8C	10	8	41.56	33.81	7.75
20FA8C	20	8	38.52	30.59	7.93
30FA8C	30	8	39.85	32.60	7.25

2.2 *Laboratory tests and testing methodologies*

Within the context of this investigation, 16 different silt samples are prepared by adding 10, 20, and 30% of the fly ash as well as 2, 5, and 8% of the cement with different combinations. Symbols as well as properties of the samples are listed in Table 2. In Table 2, FA and C abbrevations are indicating the fly ash and the cement.

As mentioned before, particle size and hidrometric analyses as well as liquid limit and plastic limit tests are performed at the beginning. In order to determine optimum water content and maximum dry density, standard Proctor tets are conducted. Finally, freezing/thawing and unconfined pressure tets are carried out for the consideration of the strengths under cold weather conditions. It should be noted that unconfined pressure tests, which were performed in accordance with AASHTO T 208-82, are made under 1, 7, and 28 days of curing periods, as well as freezing/thawing tests are conducted for 1, 3, and 7 cycles in accordance with the ASTM D 560 test method. Consequently, the results are evaluated in terms of economical and performance based considerations; therefore, the optimum mixture is determined by comparing the outcomes

3 RESULTS AND DISCUSSION

3.1 *The influence of fly ash and cement additions on the stabilization*

For 1 day of curing, 10FA2C sample exhibited the best strength values; contrarily, the least performances were obtained for natural soils and 30FA5C samples. When adding 2% percent of the cement to the natural soil, the strength was increased. For the addition of 5% cement content, the ratio of the increase was decreased, but it was increased again for when 8% cement content was reached. In other words, the slope of the tendency curve is decreased for the gap indicating 5% cement content. On the other hand, fly ash additions of 10 and 20% to the natural soil increased the strength drastically, but if it is reahed to 30%, the strength value is less than that of 10% case. It should be mentioned that there is no linear relationship between fly ash/cement content and the strength. In addition, if fly ash ratio is increased for constant cement ratio is increased for the same fly ash content, it is possible to observe a decrease in unconfined strength. In summary, for 1 day of curing, there is neither direct nor inverse relationship between the strength and the additives.

For 7 days of curing, the maximum strength was observed for 10FA2C sample, and the least one was for the natural soil, as expected. If 10% and 20% of fly ash is mixed with the natural soil, there is consistent increase in the strength. However, the strength gain is significantly decreased for 30% percent of fly ash addition. Namely, for 7 days of curing, approximately same strength values were obtained for 20% and 30% fly ash additions. Apart, 2, 5, and 8% of cement addition 158%, 224% ve 286% of strength increases with respect to the natural soil were encountered, respectively. Similar to 1 day curing case, 10FA2C sample exhibited the maximum unfined strength value. The strength ratio increase in sample 10FA8C 346% with reference to the natural soil. If fly ash content is kept constant at 20% in weight, and cement ratio is 2, 5, and 8% increased, the increase in uncofined compression strength was experienced as 250, 306, and 452%, respectively. If cement content is varied as 2, 5, and 8% for constant fly ash content of 30%, the strengths are bigger than those of 20% cement content, namely percents are 448, 639, and 684%, respectively. It can be concluded with the help of these similar strength values that 5% cement addition may be preferred instead of 8% for 30% of fly ash content.

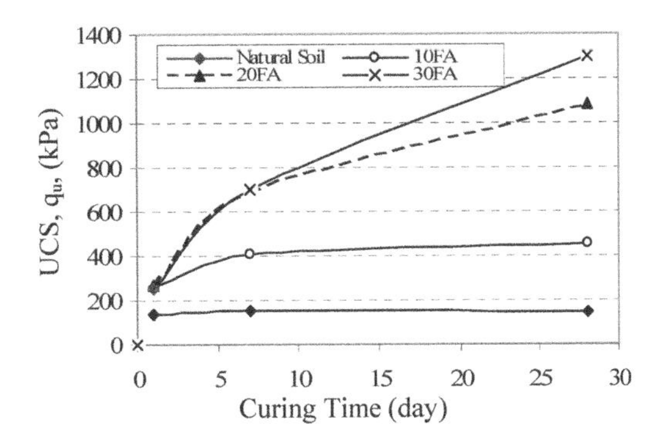

Figure 2. The variation in the 1, 7, and 28 curing days strength of samples including natural soil and the fly ash only.

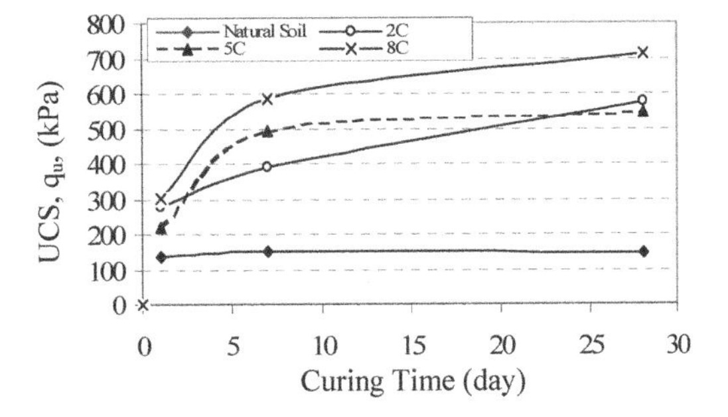

Figure 3. The variation in the 1, 7, and 28 curing days strength of samples including natural soil and the cement only.

Consequently, for 28 days of curing period, the highest strength performance was obtained for 30FA8C sample. The worst performance value was seen for the natural soil, as in the previous experiences. The strength of 10% fly ash added case is less than cases in which only cement is mixed. Nevertheless, the strength values of cases in which 20% and 30% content of fly ashes were added are much more than those of only cement were mixed. For mix designs in which both fly ash and cement were added, the increase in any component (fly ash or cement) increased the strength consistently. It should be noted that 10FA2C sample exhibited unexpected behavior, and relatively big compressive strength value was observed. The last point that should be underlined that all of 28 days cured samples exhibited crisp and brittle behaviors, and they were failed by blown up instantly. This behavior could not be seen for the samples cured 1 or 7 days (Figs. 2–4). All the results can be seen in Table 3.

Obviously, curing process increased the strengths of all samples. Minimum effect of curing was observed for natural soil sample. In other words, there is slight difference in the strength of natural soil sample related with the change in curing duration. When focusing

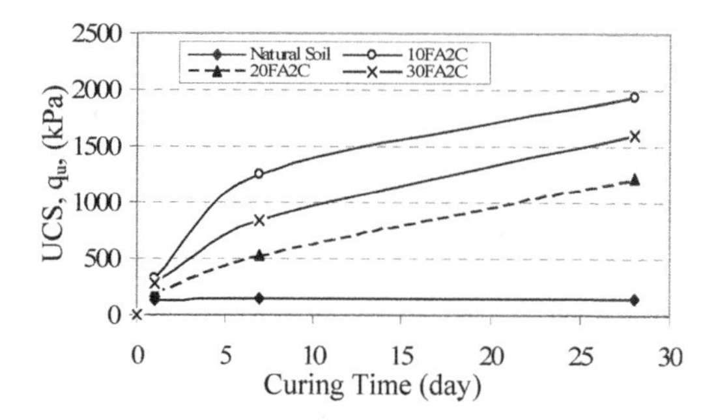

Figure 4. The variation in the 1, 7, and 28 curing days strength of samples including natural soil and 2% cement content.

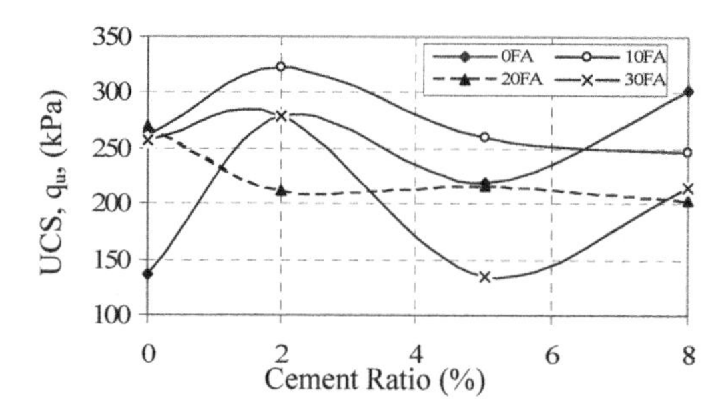

Figure 5. The variation in the 1 curing day strength of samples according to cement ratio with constant fly ash percent.

on cement added specimens, curing time increased the strength; nevrtheless, the rate of the increase is less than those of fly ash added samples. It can be concluded for cured samples that the addition of 10, 20, and 30% of fly ash is more advantageous than adding of 2, 5 and 8% cement into the silty soils. For cases in which only cement or fly ash content is generally increased, the strength values increased with the increase in the additive. In other words, a direct relationship between unconfined compressive strength and the additive parameters was observed during the experiments. The only exception to this inference is encountered in the test performed on 8C specimen (Figs. 5–6). In mixed designs including both of the additives, namely cement and fly ash, 10FA2C sample is worth considering in terms of the strength values. This can be explained with the existence of very low optimum water content. Furthermore, for soils with fly ash and cement mixture the rate of the strength increase was increased with the cement content for constant fly ash levels. The only exception to this evaluation is 30FA8C sample, whose rate of the strength increase is lower than that of 30FA5C sample. It should be mentioned that the most positive effect of curing in all samples is observed for 30FA5C sample.

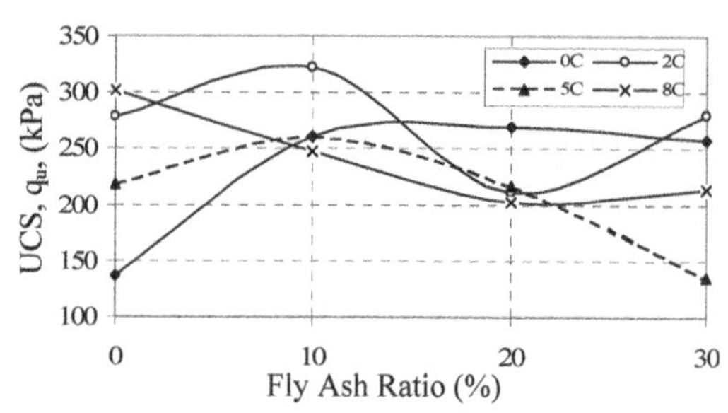

Figure 6. The variation in the 1 curing day strength of samples according to fly ash ratio with constant cement percent.

Table 3. Unconfined compressive strength values of all samples due to 1, 7, and 28 days of curing.

	Unconfined Compressive Strength (kPa)		
Sample	1 Day	7 Days	28 Days
Natural Soil	136.06	152.03	148.05
10FA	261.32	408.38	451.11
20FA	270.27	695.54	1085.05
30FA	257.35	703.49	1303.64
2C	278.22	391.49	547.32
5C	218.60	492.84	574.49
8C	302.06	587.24	710.45
10FA2C	322.93	1184.41	1939.57
10FA5C	260.33	447.13	1081.07
10FA8C	247.41	677.66	1569.94
20FA2C	211.64	532.59	1226.14
20FA5C	216.61	617.04	1114.85
20FA8C	202.70	838.62	2056.82
30FA2C	279.21	832.66	1599.75
30FA5C	135.13	1122.80	2066.75
30FA8C	214.62	1192.36	2533.76

3.2 *The effect of freezing-thawing behavior to the stabilization*

Geotechnical properties of natural and stabilized soils change with freezing/thawing cycles. Especially, strength parameters of saturated soils are much more sensitive to freezing/thawing effects than unsaturated soils. In this respect, freezing/thawing conditions are crucial for transportation infrastructures as well as geotechnical applications that are constructed in cold regions. As a result of cold regional effects, soil behaviors are fundamentally changed and crack propagation as well as stability failures is encountered after freezing/thawing cycles.

Cold regions are subject to freezing/thawing conditions every year. At the first step of freezing of fine grained soils, formation of initial glaciers becomes at the largest spaces (voids) which are exposed to lowest temperature and having the largest water potential.

Table 4. Unconfined compression strengths after 1, 3, and 7 cycles of freezing/thawing process.

Sample	Unconfined Comp. Strength (kPa)		
	1 cycle	7 cycles	28 cycles
Natural Soil	154.01	510.73	1164.54
10FA	514.70	472.97	487.87
30FA	1236.08	1806.42	1997.20
2C	596.18	555.44	667.72
10FA2C	1426.85	1862.06	2025.02
10FA8C	1212.23	1939.57	2122.40
30FA2C	1864.05	2348.94	2706.65

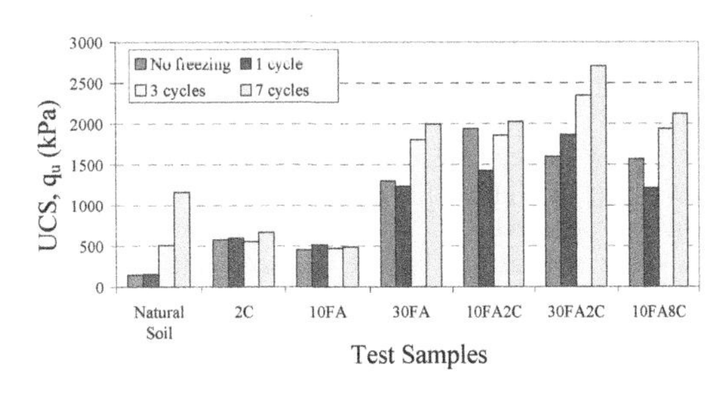

Figure 7. The variation of soil strength after different repetition of freezing-thawing action.

Subsequently, water existing in smaller voids starts to freeze. When water turns into glacier formation, volumetric increase is approximately 9%; therefore, unavoidable cracks occur in the soil (Yıldız et al. 2004). As a result of freezing/thawing cycles existing cracks are widened or new crack formations begin; therefore, strength of the soil decreases due to the sizes of the cracks. It should be noted that the thawing of a freezed layer commences at the same time from up and top surfaces.

It was shown in several studies that freezing/thawing cycles decrease the stability considerably (Simonsen & Isacsson 2002, Khoury & Zaman 2002, Kværnø & Øygarden 2006, Lee et al. 1995, Konrad 1989). Nevertheless, there are studies indicating the increase in the stability as a result of freezing (Lehrsch 1998). The reason of the cracks as well as the failure as a result of freezing/thawing conditions can be explained with the decrease in permeability of the soil (White, 2002). The shear strength and the stability of soils usually have the lowest values after winter seasons. Obviously, lower stability and shear strength are of primary reasons of the soil failure.

Sample preparation procedure for freezing/thawing tests is same as the procedure unconfined compression test. The samples are prepared at optimum water content and compacted in standard proctor molds. Subsequently, specimens are prepared by penetrating specific tubes into the mold. The height of the specimens is either 5 cm or 10 cm. Later, the specimens are covered with stretch films and put into special bags. In order to gain strength and the completion of the hydration, they are kept in curing room under special curing conditions for a certain period of time.

In this experimental study, freezing/thawing tests are applied to natural soil, 10FA, 30FA, 10FA2C, 30FA2C, and 2C samples. Samples were kept in 1 day cure condition, and then freezing/thawing cycles for 1, 3 and 7 times were performed on the samples. Subsequently, unconfined compression tests were performed on the specimens. The results of unconfined compression tests were compared with the outcomes of unconfined compression tests that were subject to freezing/thawing conditions. Therefore, positive and negative effects of the physical event were evaluated by considering natural soil conditions and different stabilization alternatives.

Axial and residual stresses can be obtained after freezing/thawing tests. In this respect, the results of unconfining compression tests that are performed on non-freezed samples are compared with the data gained from freezing/thawing tests. Therefore, the influence of freezing/thawing conditions can evaluated.

Due to the experimental program, prepared samples were exposed to 3 different freezing/thawing cycles, namely 1, 3, and 7 cycles. Results of the experiments are summarized in Table 4 and Figure 7.

The maximum strength value after 1 freezing/thawing cycle was observed for 30FA2C sample, which involves the highest additive contents. As expected, the least strength was of the natural soil sample. Another conspicuous issue on the behavior of these soils is that stabilized samples experienced instant ruptures after reahing the peak strength; therefore residual strengths were observed very low for these samples. In other words, after 28 days of curing, stabilized soils exhibited crisp behaviors after 1 freezing/thawing cycle.

As in the previous case, namely 1 freezing / thawing cycle case, the maximum strength value after 3 cycles was observed for 30FA2C sample. The least performance belongs to 10FA sample for this time, and the strengths of 2C as well as the control specimen are almost same. The strength values of 10FA2C, 10FA8C, and 30FA are similar. Samples which exhibited better performances with respect to the others, namely 30FA2C, 10FA2C, 10FA8C ve 30FA, were ruptured crisply and very low residual stress values were obtained.

The maximum performance in samples subjected to the maximum freezing/thawing conditions, i.e. 7 cycles, belongs to 30FA2C. Other noteworthy samples exhibiting high strength values are 10FA8C, 10FA2C, and 30FA. The least value is of 10FA sample. It should be noted that natural soil sample exhibited unexpected performance after 7 cycles. Obviously, crisp samples, which having higher unconfined compression strengths, exhibited extremely low residual strengths. Moreover, samples 10FA2C, 30FA2C, and 30FA ruptured instantly and approximately 0 residual strength observed.

When natural soil sample was subjected to 1, 3, and 7 freezing/thawing cycles, the strengths increased 4.03%, 244.97%, and 686.59%, respectively. 10FA sample exhibited inconsistent behavior, namely the strengths were increased 14.10%, 4.85%, and 8.15%, successively. On the other hand, the strength losses of 30FA sample were found to be 5.18%, 38.57%, and 53.20% for 1, 3, and 7 cycles, respectively. 28 days unconfined compressive strength of 10FA2C sample, which was 1939.57 kPa initially, decreased to 1426.85 kPa with a rate of 26.43%. Therefore, the maximum strength loss in all experiments belongs to 10FA2C sample for 1 cycle of freezing/thawing. For this sample, 4% and 4.41% strength losses were observed for 3 and 7 cycles, respectively. Sample 30FA2C exhibited consistent behaviors, and axial strengths increased 16.52%, 46.83%, and 69.19% for 1, 3, and 7 cycles. The strength loss of 10FA8C sample reached to 22.78% after cycle of the process, which equals to 1212.23 kPa. Losses of the sample were obtained as 23.54% and 35.19% for 3 and 7 cycles. Finally, sample 2C exhibited different trends, and the strength increased 3.81% for 1 cycle, decreased 3.29% for 3 cycles, and increased 16.26% for 7 cycles.

Unconfined compressive strengths of silty soils, in which 10% and 30% fly ashes were added, are higher than those of the natural samples. Additional 2% cement (in terms of dry weight) also increased the strength values considerably. In summary, after cycles of freezing/thawing, both natural and mixed soils exhibited crisp and brittle behavior with higher compressive strengths. It was considered that these results were based on the remarkably lose of water content of samples during the curing period. In the next stage of this study, it is aimed to investigate the freezing/thawing effect making the similar testing program performing on the samples protected the water content.

4 CONCLUSION

Within the content of this experimental study, class C fly ashes of Turgutlu region and Portland cement were mixed with silty soils at optimum water contents to increase unconfined compressive strengths. Consequently, more additive resulted in more compressive strength; therefore sample 30FA8C exhibited the best performance in terms of unconfined compressive strength. It should be noted that, the maximum performance was obtained by mixing both of the additives, not one of them. In addition, sample 10FA2C exhibited outstanding strength performance under the light of the total amount of the additives.

For the investigation the effects of freezing/thawing processes on the mechanical behavior of silty soils, a series of tests were carried out. In the results of the experiments, both natural and mixed soils exhibited crisp and brittle behavior with higher compressive strengths after cycles of freezing/thawing. It was considered that these results were based on the remarkably lose of water content of samples during the curing period.

As a result, the stabilization with fly ash and cement increased the strength considerably. Therefore, considering the waste properties of fly ash, it should be utilized for the stabilization of road bases in order to increase freezing/thawing strength of the soil. This is also quite beneficiary in terms of economical considerations.

ACKNOWLEDGEMENTS

The financial supports by TUBITAK with 105M336 and Ege University Research Funds with 2005-MUH-020 projects are highly appreciated and acknowledged.

REFERENCES

Cokca, E. 2001. Use of class C fly ashes for the stabilization of an expansive soil. *Journal of Geotechnical and Geoenvironmental Engineering*. 127(7): 568–573.

Ferguson, G. 1993. Use of Self-cementing Fly Ashes as a Soil Stabilization Agent. *Fly ash for soil improvement, Gotechnical Special Publication, No 36, ASCE, NewYork*, 1–14.

Ferguson, G. & Leverson, S.M. 1999. Soil and Pavement Base Stabilization with Self-Cementing Coal Fly Ash. *American Coal Association, Alexandria, VA*.

Khoury, N.N. & Zaman, M.M. 2002. Effect of wet-dry cycles on resilient modulus of class C fly ash stabilized aggregate base. *Transportation Research Record. 1787*. Transportation Research Board, Washington D.C., 13–21.

Konrad, J.M. 1989. Physical processes during freeze–thaw cycles in clayey silts. *Cold Regions Science and Technology*. 16(3): 291–303.

Kværnø S.H. & Øygarden, L. 2006. The influence of freeze-thaw cycles and soil moisture on aggregate stability of three soils in Norway, *Catena, Norwegian Institute for agricultural and environmental research (Bioforsk)*, Soil and Environment Division, 8 p.

Lee, W. Bohra, N.C. Altschaeffl, A.G. & White, T.D. 1995. Resilient modulus of cohesive soils and the effect of freeze–thaw. *Canadian Geotechnical Journal*. 32, 559–568.

Lehrsch, G.A. 1998. Freeze-Thaw Cycles Increase Near-Surface Aggregate Stability, *Soil Science*. 163(1): 63–71.

Simonsen, E. & Isacsson, U. 2001. Soil behaviour during freezing and thawing using variable and constant confining pressure triaxial tests. *Canadian Geotechnical Journal* 38, 863–875.

Turner, J.P. 1997. Evaluation of Western Coal Fly Ashes for Stabilization of Low-Volume Roads, *Testing Soil Mixed with Waste or Recycled Materials, ASTM STP 1275*, American Society of Testing and Materials.
White, D.W. 2002. Field Trials of Guidelines for Soil Stabilizers of Non-Uniform Subgrade Soils. Proposal. Iowa State University, Ames.
White, D.J. & Bergeson, K.L. 2001. Long-term strength and durability of hydrated fly-ash road bases, *Transportation Research Record, 1755*, 151–159
Yildiz, M. Sogancı, A.S. Demiroz, A. & Albayrak, V. 2004. The effect of freezing-thawing action on the strength and permeability behaviour of lime stabilized clay soil. Soil Mechanics and Foundation Engineering 10. National Congress. 227–236 (In Turkish).

Advances in Transportation Geotechnics – Ellis, Yu, McDowell, Dawson & Thom (eds)
© 2008 Taylor & Francis Group, London, ISBN 978-0-415-47590-7

Relationship between microstructure and hydraulic conductivity in compacted lime-treated soils

O. Cuisinier & T. Le Borgne
Laboratoire Central des Ponts et Chaussées, Centre de Nantes, France

ABSTRACT: The influence of lime addition on the hydraulic conductivity of soils is still an open question. There is no general agreement in the literature whether it consistently induces a strong increase of the hydraulic conductivity of soils or not. In this context, the paper presents some results to give insights into this question. The effects of water content, density and compaction technique were investigated. The results on lime-treated samples were compared to results obtained on untreated samples. These tests were followed by mercury intrusion porosimetry tests and scanning electron microscopy to determine the influence of lime treatment on soil microstructure. The hydraulic conductivity tests results showed that the hydraulic conductivity of the lime-treated soil can be of the same order of magnitude as for the untreated soil. It seems to be highly dependant on the compaction conditions. The observations of the microstructure of the sample demonstrated that lime is responsible for a dramatic reduction of the mean pore radius of the lime-treated samples. Hence, the density variation may be balanced by reduction of the mean pore radius in the lime-treated soil. That may explain the relative stability of the hydraulic conductivity of the lime-treated soil compared to the untreated one.

1 INTRODUCTION

Lime addition is a very wide spread technique to improve the engineering behaviour of fine soils since it could increase their workability as well as their mechanical behaviour after compaction. In addition to these well known effects, lime also alters several engineering properties like soil hydraulic properties. Hydraulic conductivity is a key issue in hydraulic infrastructures (earthdam, river levees, etc.) However there is no general agreement in the literature about the effects of lime on the hydraulic conductivity of soils.

The most common idea is that the hydraulic conductivity of the lime-treated soil is higher than the hydraulic conductivity of the corresponding untreated soil. That has been evidenced by some authors working with various types of soils, from highly expansive clays to low plasticity silts, using various samples preparation techniques (e.g. Ranganatham 1961, Brandl 1981, Nalbantoglu & Tuncer 2001). The most common hypothesis to explain this results is linked to the fact that, under a given compaction energy, the optimum dry density of the lime-treated soil is lower than the optimum dry density of the untreated soil. One may conclude that the lower the density, the higher the hydraulic conductivity. However, other studies evidenced that it is possible to obtain lower hydraulic conductivity after lime treatment. This has been shown by Fossberg (1965) and Leroueil et al. (1996), both working on samples prepared at high water content, near or above the liquid limit. Others authors evidenced that the hydraulic conductivity of a lime-treated soil is highly dependant on the compaction water content. El-Rawi & Awad (1981) evaluated the influence of the compaction water content and of the lime dosage on the hydraulic conductivity of a low-plasticity silt. They have shown that if the water content is below or equal to the optimum water content, the hydraulic conductivity of the lime-treated silt is almost one order of magnitude higher than the hydraulic conductivity of the untreated soil. However, on the wet side of the optimum of the lime-treated soil, the hydraulic conductivity of the lime-treated soil is of the same order of the magnitude as the untreated soil. Similar conclusions have been drawn by McCallister (1990) with three lime-treated expansive clays.

Therefore, there is no a general tendency about the influence of lime on hydraulic conductivity. It seems that permeability of lime-treated soil is dependant on the compaction conditions, and it is not systematically higher than this of the corresponding untreated soil. This might be related to the fact that lime treatment could not only dramatically influence the density of the compacted soil but also its microstructure of the soil, i.e. the geometrical organisation of soil pores (e.g. Cuisinier *et al.* 2007). Indeed, it is well known that hydraulic conductivity is highly dependant on soil microstructure (e.g. Garcia-Bengochea 1979).

Table 1. Identification properties of the untreated A34 clay.

Soil	w_L %	w_P %	Ip %	γ_s kN.m^{-3}	VBS g/100 g	<80 μm %
Chemillé silt	34.2	18.6	15.6	2.70	2.8	72.1
Clayey soil	51.0	39.8	11.2	2.68	5.8	98.0

Table 2. Optimum compaction references of the tested soils (normal Proctor compaction procedure).

	Untreated soil		Lime-treated soil	
	w %	γ_d kN.m^{-3}	w %	γ_d kN.m^{-3}
Chemillé silt	14.5	1.86	22.0	1.64
Clayey soil	19.0	1.58	26.0	1.52

In this context, it was decided to conduct an experimental program to evaluate (i) the influence of the compaction procedure on the hydraulic conductivity of lime-treated soil, (ii) the role of the water content, and (iii) the microstructural modification induced by lime treatment. In a first approach, two compaction techniques (static and dynamic), two water contents and two kinds of soil were considered. Hydraulic conductivity tests were performed with consolidation cells permeameter. Mercury intrusion porosimetry and scanning electron microscopy were used to assess the influence of lime treatment on compacted soil microstructure.

2 EXPERIMENTAL PROCEDURES

2.1 Properties of the materials

The main identification properties of the tested soils, determined following French standards, are given in Table 1. The silt has been taken in the vicinity of Chemillé, in the West part of France. The clayey soil has been sampled in North-East part of France. According to French technical recommendations, both soils are medium plasticity silts (A2). However, it can be seen from the data in Table 1 that the Chemillé silt has a lower VBS value that the clayey soil and that it is coarser than the clayey soil.

2.2 *Samples preparation for hydraulic tests*

The compaction characteristics of the soils considered in this study were determined according French standard for normal Proctor compaction procedure (Table 2). The lime dosage was 2 %, on a dry-weight basis, for both soils. This corresponds to an average lime dosage for this kind of soils in France.

In the study, consolidation cells permeameter were used. The height of the ring that contains the sample is 19 mm and its diameter 70 mm. A special procedure was selected to prepare dynamically compacted samples directly in the ring. A mini-compaction device similar to the one presented by Sridharan & Sivapullaiah (2005) was used. The samples were compacted in 2 layers. The mini-compaction procedure was adapted in order to obtain the same dry density as obtained with the standard Proctor procedure. The statically compacted samples were also prepared directly in the cell ring. The static compaction was performed with a displacement speed of 1.27 mm.min^{-1}.

For each soil, two water contents were considered, the optimum and a water content on the wet side of the optimum. It was 14.5% and 17.5% for the untreated Chemillé silt and 20.5% and 23.5% for the lime-treated Chemillé silt. It was 24.0% and 27.5% for the untreated clayey soil and 26.0% and 30.5% for the lime-treated clayey soil.

2.3 *Determination of the hydraulic conductivity*

After the compaction, the samples were inserted on the consolidation cell permeameter device. A vertical pressure of 10 kPa was immediately applied to the sample. The test began immediately after by the imposition of the desired water pressure. A hydraulic head of about 1 m was applied at the lower base of the sample. A reservoir of water surrounding the consolidation ring maintained atmospheric pressure at the effluent end of the specimen. The water flow was maintained several days. The hydraulic conductivity of the sample was the mean of several measurements made after the equalisation of water flow.

Then, the vertical stress was increased up to 60 kPa step by step. At each stress step, the hydraulic conductivity of the sample was determined with the same procedure. This allowed to increase the density of the samples and therefore to evaluate the influence of the mechanical load on the hydraulic conductivity measurements.

2.4 *Samples preparation for microstructural investigations*

Due to technical requirements, both Mercury intrusion porosimetry tests and Scanning Electron Microscope observations must be conducted on completely dried soil samples. Freeze-drying was selected for our study an alternative to oven drying to prevent the effects of shrinkage from water on drying on samples microstructure.

Soil samples were specially compacted for these investigations. Soil pieces of approximately 1 cm^3

were taken on the sample just after the sample compaction. These were quickly frozen with liquid nitrogen (temperature of 196°C) and then placed in a freeze-drier at least 72 hours for the sublimation of the water.

2.5 *Mercury intrusion porosimetry (MIP)*

In the MIP method, the mercury pressure is increased by steps and the intruded volume of mercury is monitored for each pressure increment. Assuming that soil pores are cylindrical flow channels, Jurin's equation is used to determine the pore radius associated with each mercury pressure increment:

$$r = \frac{2\,T_s \cos\alpha}{P} \qquad (1)$$

where r corresponds to the entrance pore radius, Ts to the surface tension of the liquid (0.485 N.m1 for mercury and 0.07275 N.m1 for water), α to the contact angle of fluid interface to solid (0° for air-water while 140° is an average value for mercury-air interface in soils), and P to the pressure difference between the two interfaces (Pa). A MIP test gives the cumulative mercury volume intruded as a function of the entrance pore radius. To further interpret MIP data, Juang & Hotz (1986) have defined the pore size density function (PSD) defined as follows:

$$f(\log r_i) = \frac{\Delta V_i}{\Delta(\log r)} \qquad (2)$$

where ΔV_i is the injected mercury volume at a given pressure increment corresponding to pores having a radius of $r_i \pm (\Delta \log r_i)/2$.

2.6 *Scanning electron microscope (SEM)*

Observations by SEM were made using a Hitachi S570 microscope. The freeze-dried samples selected for observation were fractured just before the coating in order to observe a fresh surface.

3 HYDRAULIC CONDUCTIVITY TESTS RESULTS

The results of the hydraulic conductivity tests are given on Figures 1 to 4. These figures give the hydraulic conductivity as a function of the dry density of the sample, and the higher the dry density, the higher the applied vertical stress to the sample in the consolidation cell permeameter. For a given compaction modality, the lowest density corresponds to a vertical stress of 10 kPa (initial state), and the highest to a vertical stress of 60 kPa.

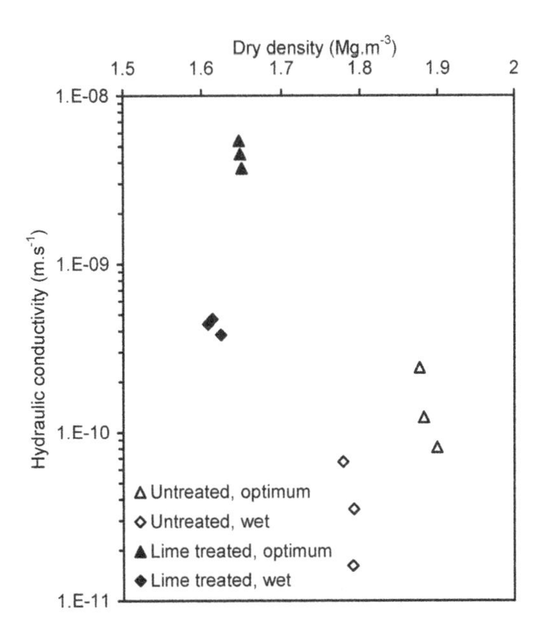

Figure 1. Influence of dry density of the water content on hydraulic conductivity: Chemillé silt – dynamic compaction.

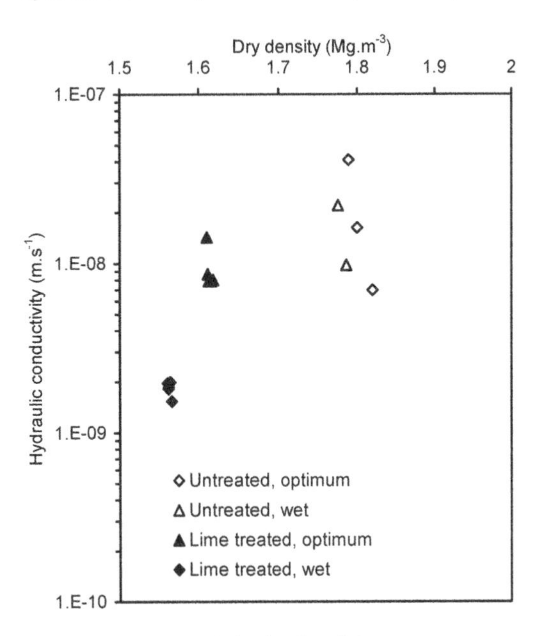

Figure 2. Influence of dry density of the water content on hydraulic conductivity: Chemillé silt – static compaction.

3.1 *Silt from Chemillé (Figures 1 and 2)*

It can be seen that the hydraulic conductivity determined on the wet side of the optimum is lower than the hydraulic conductivity at the optimum water content. This is true for both the untreated and the lime-treated

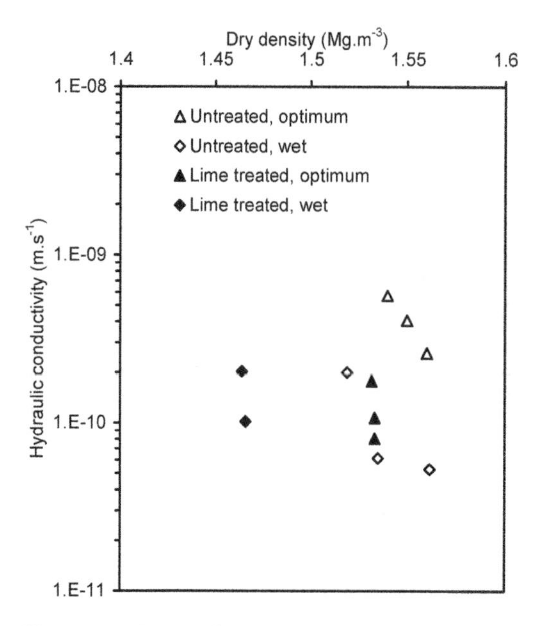

Figure 3. Influence of dry density of the water content on hydraulic conductivity: clayey soil – dynamic compaction.

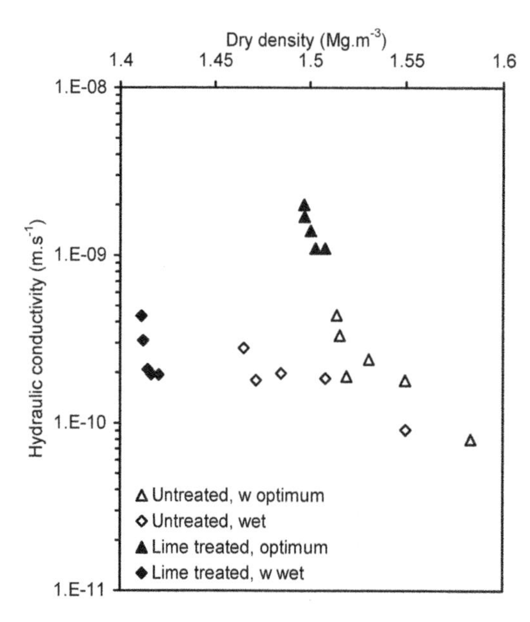

Figure 4. Influence of dry density of the water content on hydraulic conductivity: clayey soil – static compaction.

silt, and for both modes of compaction. This is a general trend observed in the case of untreated compacted soils (e.g. Daniel & Benson 1990).

When dynamic compaction is considered, there is one order of magnitude of difference between the lime-treated and the untreated soil hydraulic conductivity when the optimum water contents are considered. It is however interesting to note that the hydraulic conductivity of the lime-treated soil compacted on the wet side of the optimum is of the same order of magnitude as for the untreated soil at its optimum water content whereas the dry densities of these samples are dramatically different. In the case of the samples statically compacted, the lime treatment seems to less influence the hydraulic conductivity of the Chemillé silt. It is however interesting to observe that the lowest hydraulic conductivities were obtained with the lime-treated silt.

3.2 *Clayey soil (Figures 3 and 4)*

The results of the hydraulic conductivity tests performed on the clayey soil are given on Figures 3 and 4. It appears that, despite the differences between their initial densities, the hydraulic conductivity is approximately of the same order of magnitude for the clayey soil when dynamic compaction is considered.

For the static compaction (Figure 4), it can be observed that the hydraulic conductivity of the lime-treated clayey soil compacted at its optimum has the highest hydraulic conductivity. The others samples statically compacted have approximately the same hydraulic conductivity despite the dry densities are comprised between 1.45 and 1.57.

3.3 *Conclusion*

As a conclusion, it can be stated that the hydraulic conductivity of lime-treated soil is not systematically higher than the hydraulic conductivity of the corresponding untreated soil. This was observed with samples having significantly different dry densities. This clearly indicates that the dry density cannot be used as a criterion to evaluate the hydraulic conductivity of lime-treated soil.

As it is known in compacted soil, hydraulic conductivity of compacted soil is influenced by the microstructure. It is likely that microstructural modifications induced by lime treatment may explain the fact that the hydraulic conductivity of lime-treated soil is of the same order of magnitude than the untreated soil. In order to check this hypothesis, some microstructural investigations were carried on. The clayey soil was selected at its optimum water content. In a first approach, the static compaction was selected because it is easier to control the compaction conditions.

4 MICROSTRUCTURE OF THE CLAYEY SOIL

The main object of that part of the work is to highlight the impact in term of microstructure of the lime treatment. Then, to link these differences to the hydraulic conductivity.

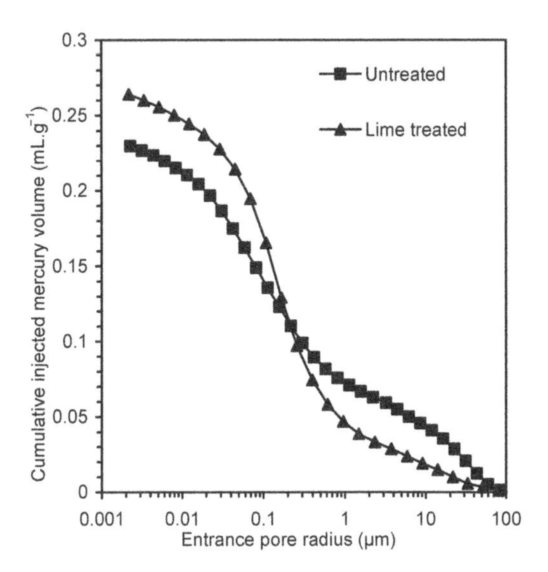

Figure 5. Microstructure of statically compacted clayey soil (optimum water content): cumulative injection curve.

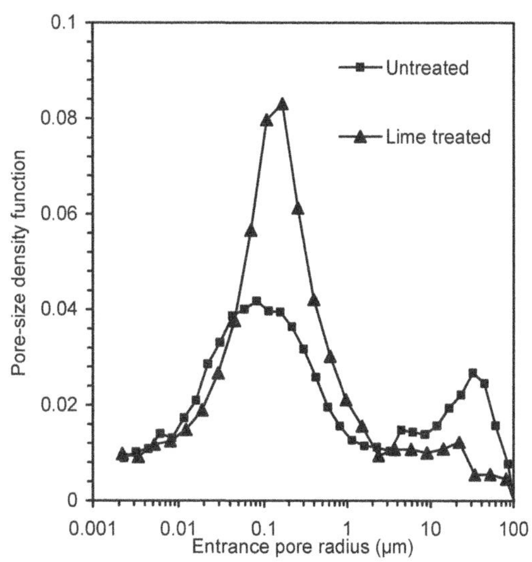

Figure 6. Microstructure of statically compacted clayey soil (optimum water content): pore size density function.

4.1 *Mercury intrusion porosimetry (MIP)*

It can be seen that the intruded volume of mercury is significantly lower for of the untreated sample than for the lime-treated sample (Figure 5). This is explained by the fact that the dry density of the untreated sample is higher, hence there is less void in the untreated sample.

The pore size density (PSD) function (Figure 6), gives more information about the repartition of the pore volume as a function of the pore radius. For the untreated clayey soil, the PSD function displays two pore classes. The common interpretation is that the smallest pores (micropores, below 3 μm) correspond to the pores inside the aggregates whereas the largest pores (macropores, above 3 μm) are the pores between these aggregates. This kind of structure has been termed "a double structure". It is possible to determine a micropore void ratio and a macropore void ratio from the intruded mercury volume. Then the micropore void ratio is 0.45 and the macropores void ratio is 0.16.

The results show that lime treatment significantly altered the microstructure of the clayey soil. It can be seen that there is no well-defined macropore class in the lime-treated soil. The decrease of the optimum density when the clayey soil is treated with lime seems to be mainly related to a strong increase of the micropore void ratio which increased from 0.45 in the untreated soil up to 0.62 in the lime-treated clayey soil.

4.2 *Scanning electron microscopy (SEM)*

SEM observations were done in order to better evaluate the impact of lime treatment in terms of microstructure.

Figure 7. Fabric of the untreated compacted clayey soil (×300).

Two typical pictures of the untreated clayey soil statically compacted at its optimum water content are given on Figure 7 and 8. Large aggregates are identified and the void spaces between these aggregates correspond to the macropores evidenced with MIP tests. The diameters of these macropores are of some 10 tens of μm from SEM pictures, that is the same order of magnitude as the macropore radius evaluated from MIP measurements. It is not possible from SEM pictures to observe the micropores.

The SEM pictures taken on the lime-treated clayey soil attested a dramatic alteration of the microstructure due to the lime treatment (Figure 9 and 10). There is no clear evidence of the presence of aggregates in the lime-treated sample. It is not possible to identify some

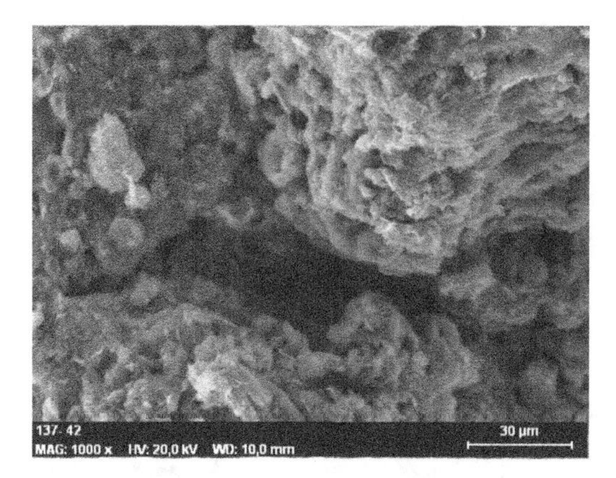

Figure 8. Fabric of the untreated compacted clayey soil (×1000).

Figure 9. Fabric of the 2 % lime-treated clayey soil (×250).

macropores on the Figure 9. The Figure 10 allowed the identification of some pores of diameter of only a few micrometers.

4.3 *Conclusion*

In addition to the dry density reduction, the lime treatment altered significantly the microstructure of the clayey soil statically compacted at its optimum water content. There is strong decrease of the macropore volume and an increase of the micropore volume in the lime-treated soil. The mean pore radius is lower in the case of the lime-treated soil. These conclusions are confirmed both by MIP tests and SEM observations.

5 DISCUSSION AND CONCLUSION

Considering a given compaction energy, the optimum dry density of the lime-treated soil is lower than this of the untreated soil. The common acceptation implies

Figure 10. Fabric of the 2% lime-treated clayey soil (×1300).

that the hydraulic conductivity of the lime-treated soil is significantly higher than this of the untreated soil. The hydraulic conductivity tests results presented in this paper demonstrate that this not systematic. It depends on the water content relatively at the optimum water content, on the type of soil and on the compaction procedure.

Several tests results showed that the hydraulic conductivity of the lime-treated soil can of the same order of magnitude as this of the untreated soil. This was observed beside large differences in term of dry density. In a first approach to interpret these results, SEM and MIP investigations were carried out. It appeared that there is much less macropores in the lime treated soil comparatively to the untreated soil. Even if the pore volume of the lime-treated soil is higher than this of the untreated soil, the radius of these pores is lower in the case of the lime-treated soil. Hence the density reduction impact on the hydraulic conductivity might be balanced by a reduction of the mean pore size in the lime-treated soil.

As a conclusion, it can be stated that the lime treatment dramatically alters the density at the optimum water content of a soil. This is not systematically associated to an increase of the hydraulic conductivity certainly because of the effect of lime treatment on the microstructure of the soil. This would be a function of the water content, the compaction procedure, the soil nature, etc. More investigations are needed to establish a sound framework to predict impact of lime treatment on soil hydraulic conductivity.

REFERENCES

Brandl, H. 1981. Alteration of soil parameters by stabilization with lime. 10th *Int. Conf. on Soil Mechanics and Foundation Engineering, Stockholm, Sweden*, vol. 3: 587–594.

Cuisinier., O., Masrouri,. F., Pelletier. M., Villiéras., F. & Mosser-Ruck, R. 2008. Microstructure of a compacted soil submitted to an alkaline plume. *Applied Clay Science, in press.*
Daniel, D.E. & Benson, C.H. 1990. Water content-density criteria for compacted soil liners. *J. of Geotechnical Engineering* 116: 1811-1830.
El-Rawi, N.M. & Awad, A.A.A. 1981. Permeability of lime stabilized soils. *Transportation Engineering J.* 107: 25–35.
Fossberg, P.E. 1965. Some fundamentals engineering properties of a lime-stabilized clay. 6th *Int. Conf. on Soil Mechanics and Foundation Engineering*, Montréal, vol. 1: 221–225.
Garcia-Bengochea, I., Lovell, C.W. & Altschaeffl, A.G. 1979. Pore distribution and permeability of silty clays. *J. Geotechnical Engineering Division* 105: 839–856.
Juang, C.H. & Holtz, R.D. 1986. A probabilistic permeability model and the pore-size density function. *Int. J. of Numerical and Analytical Methods in Geomechanics* 10: 543–553.
Locat, J., Tremblay, H. & Leroueil, S. 1996. Mechanical and hydraulic behaviour of a soft inorganic clay treated with lime. *Canadian Geotechnical J.* 33: 654–669.
McCallister, L.D. 1990. The effects of leaching on lime-treated expansive clays. PhD thesis. The university of Texas at Arlington, 417 p.
Nalbantoglu, Z. & Tuncer, E.R. 2001. Compressibility and hydraulic conductivity of chemically treated expansive clay. *Canadian Geotechnical J.* 38: 154–160.
Ranganatham, B. 1961. Soil structure and consolidation characteristics of black cotton clay. *Géotechnique* 11: 331–338.
Sridharan A. & Sivapullaiah P.V. 2005. Mini compaction test apparatus for fine grained soils. *Geotechnical Testing J.* 28: 240–246.

Advances in Transportation Geotechnics – Ellis, Yu, McDowell, Dawson & Thom (eds)
© 2008 Taylor & Francis Group, London, ISBN 978-0-415-47590-7

Characteristics of soft clay stabilized for construction purposes

M.M. Hassan, M. Lojander & O. Ravaska
Helsinki University of Technology, Espoo, Finland

ABSTRACT: Finnish post-glacial clay is very soft, sensitive and provides very little strength. Therefore it has been regarded as a waste and dumped as useless material so far. However, if it is mixed with relatively small amounts of a binding agent, the properties change considerably and can be used for construction purposes. The paper discusses the changes in index properties and particularly in the strength-stiffness properties of cement stabilized soft clay. Properties needed for construction purposes are also discussed.

1 INTRODUCTION

Ground improvement by treating soil with various types of binders is an attractive alternative and often economical compared to other ground improvement methods. Generally, typical Finnish clay in its natural water content with liquidity index more than unity is soft, structured, and sensitive and provides low strength. Addition of binder alters the engineering properties of the existing soil to create a new site material which is capable to meet certain material requirements instead of rejection, excavation and dumping as a waste. Deep stabilization or mass stabilization for various kinds of transport infrastructures is used throughout the world. The engineering characteristics of treated soil as well as the development in the machinery for homogenous mixing are extremely important to optimize the effect of stabilization. This study focuses on the characteristics of cement stabilized soft clay.

Table 1. General geotechnical properties of studied clay.

	Clay 1	Clay 2	Clay 3
	HUT	K	LV
Water content, w_0 (%)	74–76	63	210
Bulk unit weight, γ_t (kN/m^3)	15.4	16.2	12.3
Dry unit weight, γ_d (kN/m^3)	8.73	9.9	3.8
Liquid limit, LL (%)	71	51	185
Plastic limit, PL (%)	30	26	59
Plasticity index, PI (%)	41	25	–
Liquidity index, LI (%)	1.1	1.5	–
Specific gravity, G_s	2.8	2.75	2.5
Degree of saturation, Sr (%)	100	100	100
Grain size distribution (%)			
Clay (<0.002 mm)	81	60	40
Silt (<0.06 mm)	19	36	
Organic content (%)	<1	<1	8 ~ 9
pH (1:2.5)	7	5.2	4.4

2 EXPERIMENTAL INVESTIGATION

2.1 *General geotechnical properties of soil*

Three Finnish soils were selected for this study and their general geotechnical properties are presented in Table 1. The first soil studied (Clay 1) was taken from the Helsinki University of Technology (HUT) area and at the specific location where the clay deposit is approximately 4 m thick. The top 0.5 m is considered as weathered crust layer and the ground water table is located below 0.5 m. The natural water content of the remolded soil samples varied in the range 74 ~ 76%, whereas the liquid limit (LL) and the plastic limit (PL) are 71% and 30% respectively. Clay 1 is composed of 81% clay fraction with a low organic content and is considered as inorganic, highly plastic, fat clay.

Disturbed Kuopio clay (K), Clay 2, was collected near the city of Kuopio situated in the middle part of Finland. The natural water content of a remolded sample was 63% and the liquid limit (LL) and the plastic limit (PL) are 51% and 26% respectively. Clay 2 is inorganic, silty clay and acidic in nature with the pH value about 5.2.

The third soil (Clay 3) selected for this study was organic soil. Undisturbed and disturbed samples were collected using a piston sampler and an excavator respectively from Leppävaara (LV) situated near Helsinki. The initial water content, liquid limit (LL) and plastic limit (PL) of Leppävaara clay are 210%, 185% and 59% respectively. It is well known that the presence of organic matter has certain influences on the geotechnical characteristics of soil like increase in compressibility, reduction in the bearing capacity, and many researcher have observed that organic matter increases the liquid and plastic limits. Clay 3 was also found acidic with pH value about 4.4.

2.2 *Binder (cement) and preparation of specimens*

There are many types of binder materials that possess hardening properties and numerous studies have been published on the stabilization effects of binders either as a sole binder or as a combination with other binders. Cement is a hydraulic binder and addition of cement to any soil-water system results in several chemical reactions that cause profound alteration of the physicochemical properties of soil. The reactions that occur produce both immediate and long-term changes in the mechanical behavior of the soil. Cement as a hydraulic binder uses water to initiate a chemical reaction and produces hydrated calcium silicates (CSH), hydrated calcium aluminates (CAH) and hydrated lime (CH). Primary cementation products bind soil particles and increase the strength. Disassociation of hydrated lime increases the pH value of pore water and Ca^{+2} ions also initiate flocculation and react eventually with soil silica and alumina and produce more cementation products CS(A)H, which is called as a pozzolanic reaction.

In Finland, Portland-composite type (CEM-II 42.5 N) cement is generally used for soil stabilization. In this study, CEM II/A-M(S-LL) 42.5 N named Yleissementti (Finnsemetti, 2007) was used as dry powder. This is a Portland-composite cement in which the clinker percentage varies in the range 80–94% and the other constituents blast-furnace slag and limestone vary in the range 6–20%. The specific surface area of this type of cement is in the range 340–420 (m^2/kg) and according to the strength classification this corresponds to a compressive strength of 42.5 MPa in 28 days.

In this study, the cement amount C_w (kg/m^3) is defined as the mass of cement amount per unit volume of clay, whereas the cement content (A_w) is defined as the weight ratio of cement to clay in their dry state and W/C ratio is the water content of the clay to the cement content.

In the investigation of the cement treated material behaviour and effectiveness on stabilization cement amount C_w (kg/m^3) varied from 25 (kg/m^3) to 150 (kg/m^3). At first, the prescribed amount of remolded clay was deposited inside a plastic bag and then the required amount of dry cement powder was poured on it to simulate the dry mixing technique. Hand mixing was started immediately and continued up to 5 to 7 minutes to ensure a homogenous blend. Plastic tubes of 50 mm diameter and 140 mm height were used to prepare unconfined compression (UC) specimens. After mixing the soil-cement mixtures were placed in plastic tubes using a momentary compaction pressure of 100 kPa for Clay 1 (HUT) and

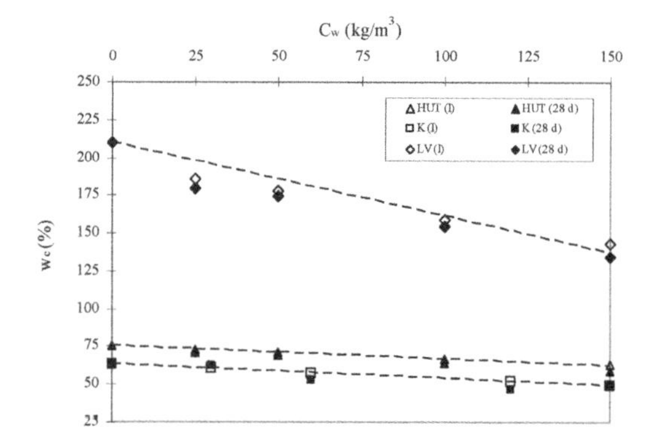

Figure 1. Water content w_c (%) of cement treated clays after mixing (I) and after 28 days.

Clay 2 (K) but for Clay 3 (LV) the soil-cement mixture was filled in the specimen tubes using hand pressure and tapping.

For unconfined compression tests, the specimens were cured for 7 and 28 days at $6 \pm 1°C$. After the specified curing time they were trimmed to the specified height (100 mm) with a height-to-diameter ratio of 2 and smoothed to form parallel end surfaces. Unconfined tests were performed at a rate of 1%/min. Water content, density, plastic limit and pH of the treated soils were also tested.

3 GENERAL GEOTECHNICAL PROPERTIES OF CEMENT TREATED CLAY

3.1 *Water content, unit weight and pH*

Figure 1 presents the water content of cement treated clay w_c(%) immediately after mixing (I) and after 28 days against the cement amount C_w (kg/m^3) for the three studied soils. In this study, the cement treated clay water content (w_c) is defined as the ratio of the weight of water to the total weight of solids including the soil and cement solids, as determined by heating the admixed clay in an oven at a temperature of 105°C for 24 h. Deviations were observed in the after mixing (I) water contents, but Figure 1 presents the mean values.

The addition of cement as dry powder to the soft soil decreases the soil water content because of the initial hydration reaction. Figure 1 shows that after mixing (I) the water content w_c (%) of cement treated clay decreases in all the studied clays from their initial values w_0 (%) and approximately follows a linear relationship with C_w (kg/m^3). The extent of the decrease in the water content mainly depends on the soil type, the amount of binder and the initial water content.

Further reduction (28 days) in water content of cement treated clay occurs mainly due to continuous hydration and pozzolanic reactions. Though the decrease or increase in the water content may

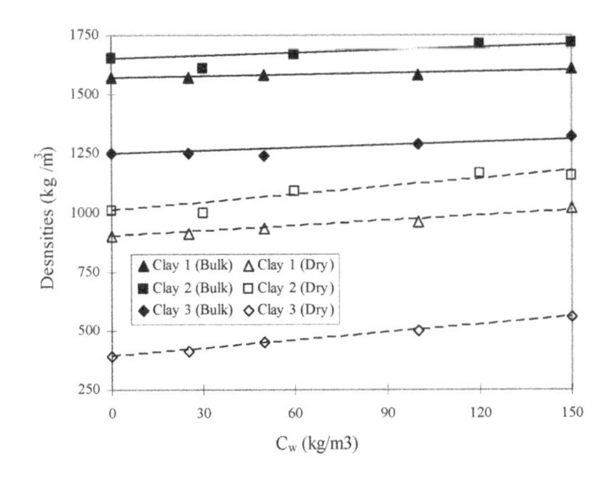

Figure 2. Bulk and dry densities of cement treated clay (28d).

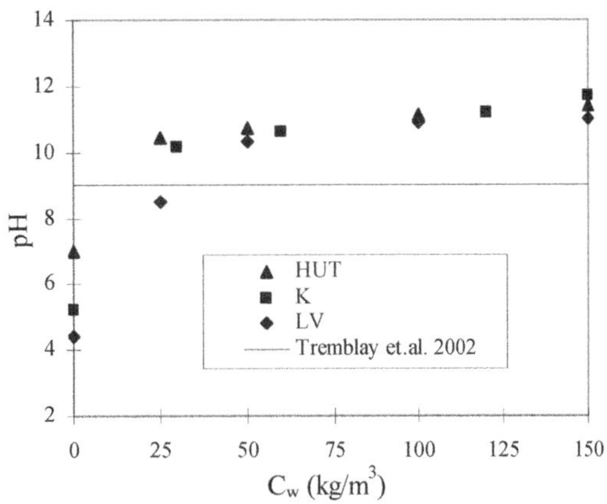

Figure 3. pH against cement amount (28d).

depend on the curing conditions, the presence of the groundwater table or availability/scarcity of water in other geo-hydrological sources, initial loading or compaction or the thickness of the admixed soil layer either in the field or in the laboratory, the binding process will go on using more water for chemical reactions.

Figure 2 presents the bulk and dry densities of specimens cured for 28 days against the cement amount C_w(kg/m^3). Generally, both the bulk and dry densities increase with the increasing cement amount C_w(kg/m^3) for all the three studied clays. However, with a lower cement amount the bulk densities remain as the same or decrease from their initial state. In spite of initial flocculation and structural change because of initial chemical reactions it can be argued that the bulk and dry densities increase approximately linearly with the cement amount C_w(kg/m^3) and also that the increase rates of the dry densities are higher than those of the bulk densities. The cement hydration and pozzolanic reactions reduce the water content (Figure 1) of treated soils and produce solid cement products which eventually increase the unit densities of treated soil.

It should be mentioned that after curing densities of cement treated soils cannot be attributed only to the cement amount C_w(kg/m^3) but also to mixing homogeneity, presence of air bubbles, initial loading or compaction and/or curing conditions.

The organic content and/or the initial acidity of soil also influence the characteristics of cement treated soil. It is well known that organic matter affects the cement stabilization either retarding it or preventing the hydration process and usually they demand larger quantities of binder in order to achieve satisfactory results. Tremblay et al. (2002) mentioned that not only the quantity but also the nature of the organic matter affects the cement stabilization process. A higher pH value of treated soil favours the cement reaction and the long term pozzolanic reaction. Cement hydration with soil pore water produces cementation gel and releases calcium hydroxide (CH) which disassociates and raises the pH value of stabilized soil.

In this study, Clay 1 (HUT) is inorganic clay with a neutral pH around 7, Clay 3 (LV) is organic acidic clay with a pH value of 4.4 whereas Clay 2 (K) is inorganic clay but acidic in nature (pH $\approx$ 5.2). To study the pH values of treated clay, 28 days cured, dried-pulverized samples were used with a soil to water ratio of 1:2.5 for all tests. A pH meter was calibrated by using a standard buffer solution and the pH values of soil suspensions were measured. The pH value increases considerably due to its alkaline nature from the initial value as cement is added. Above higher cement amounts ($C_w > 50$ (kg/m^3)) all the three treated soils show pH values around $10 \sim 12$. However, Clay 3 at a low cement amount 25 kg/m^3 also shows increase in pH though the strength development was not evident (Figure 5). Tremblay et al. (2002) reported that at the pH value below 9 no apparent increase in the strength was observed. Its acidic nature to neutralize the alkaline pH of cement treated soil and it can be argued that some cement will be used to neutralize the potential acidity of cement admixed clay.

3.2 *Plastic limit*

Binder reactions cause structural changes to soil particles and have an influence on the plastic limit. It is evident from earlier studies by Chew et. al. (2004) and Åhnberg (2006) that the plastic limit increases with increasing cement and with curing time. Bergado et. al (1996) also mentioned that the plastic limit increases for soft Bangkok clay with cement content and curing time. A well documented study by Brandl (1981) showed that the plastic limit increases with slaked lime both for clay and silt and also with time (7d and 270d).

Figure 4 shows the plastic limits of cement treated clays immediately (I) after mixing, after 1d and 7d. Plastic limits increase for all the studied clays

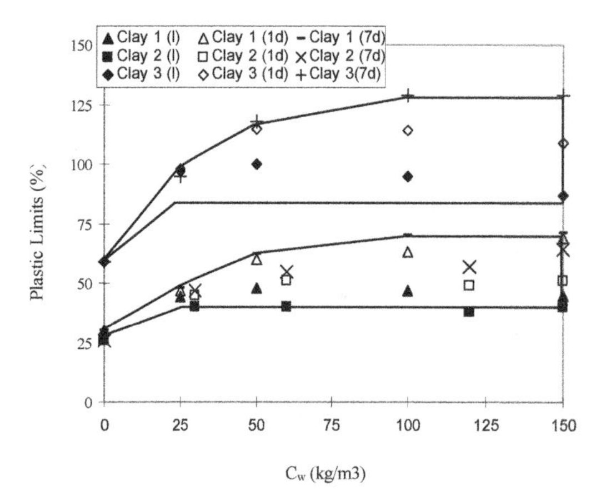

Figure 4. Plastic limits of cement treated soils.

immediately after mixing. The increase in the plastic limits is about 12 ~ 14% for inorganic clays (HUT and K), but the plastic limit increases considerably, about 28 ~ 40%, for organic clay (LV). No appreciable effect on the plastic limits for Clay 1 and Clay 2 with increasing cement amount C_w (25–150 kg/m^3) was observed with the exception of Clay 3, an organic clay, which showed decreasing tendency. It can be concluded that the apparent increase in the plastic limit just after mixing in any kind of clay is evident, and above a certain cement amount the effect of excess cement on the plastic limits (in this study C_W varies from 25 kg/m^3 to 150 kg/m^3) has a little or no appreciable effect. It should be mentioned that the plastic limit is not a property of cement, but the composite nature of clay-cement gel can hold more water than clay in the natural state.

Changes in the plastic limit after 24 h are also presented in Figure 4. With a rather small cement amount ($C_W = 25$–30 kg/m^3) the plastic limits (24 h) do not increase much or it can be stated that the plastic limit approximately remains as the same compared to the plastic limit after admixing (I). However, with a certain amount of cement (about 50 kg/m^3) the plastic limits increase considerably compared to the natural state and the state after mixing. Increases in plastic limits after 24 h for Clay 1, Clay 2 and Clay 3 are about 30 ~ 39%, 23 ~ 25% and 50 ~ 56% respectively compared to their natural states. With higher cement amounts the admixed soils behave like semi-hard to hard solid bodies because of initial cementation bonds resulting in aggregation of soil particles.

The plastic limits after 7d are also presented in Figure 4. Plastic limits do not increase considerably after 7 days compared to 1 day for Clay 1 and Clay 2 though an increasing trend can be seen for Clay 3 with higher cement amounts. However after 7 days the cement admixed soft clay becomes stabilized providing strength and starts to behave like a solid body. Evaluation of the plastic limit after 7 days may become erroneous because of the aggregation and cementation of the particles into lager size or as suggested by Locat et. al. (1996) 'the presence of intra-aggregate water increases the apparent water content without really affecting interaction between aggregates'.

4 STRENGTH AND STIFFNESS

4.1 *Unconfined compressive strength*

The effectiveness of cement on the strength development was evaluated with unconfined compressive strength (UCS) tests. Figure 5 presents the UCS against W/C ratio of Clay 1 (HUT), Clay 2 (K) and Clay 3 (LV) after 7 and 28 days curing. It is well established that a decreasing W/C ratio or an increasing C_W (kg/m^3) or A_W(%) will result in increase in strength. However, at a higher W/C ratio the strength development was marginal as very small amount of cement is available to bring any significant strength development. It can be seen from Figure 5 that the strength development (28d) of Clay 1 (HUT), which is fat clay, shows higher strength than Clay 2 (K) at the same W/C ratio. It should be noted that the initial water content of Clay 2 was lower than Clay 1 but Clay 2 is a silty clay and was inherently acidic, and therefore those initial conditions might also affect the strength development. The organic soil Clay 3 with a higher initial water content and acidity showed the lowest strength development at the same W/C ratio. Therefore a different soil with different initial conditions like soil type, organic content, acidity or after curing condition can also influence on the strength and that should be considered before stabilization. However, it can be argued that the strength rate development against W/C ratio rises rapidly after 200 kPa for all of the studied clays (Figure 5).

Time-strength development of the three studied clays is evaluated by normalizing the 7 day strength to the 28 day strength. However, the unconfined strength above a W/C ratio 20 in not considered in evaluating this ratio. The 7d strength ratio of the three studied clays is in between 0.4 $q_{u(28d)}$ ~ 0.8 $q_{u(28d)}$.

For field applications the selection of an appropriate C_w (kg/m^3) or W/C ratio depends on the specification regarding the strength and the factor of safety.

4.2 *Modulus of elasticity and strain at failure*

Figure 6 presents the modulus of elasticity of the tested specimens evaluated from stress-strain curves from UC tests. However, no internal measurements were performed and therefore the measured strains refer to the total measured strains. At low cement contents ductile stress-strain relationships were observed though brittle behavior appeared with increasing cement amounts. The secant modulus E_{50} was evaluated at

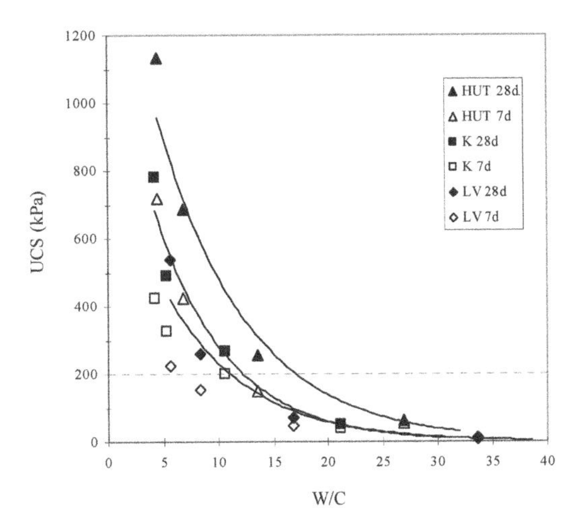

Figure 5. UCS against W/C ratio (7 and 28 days).

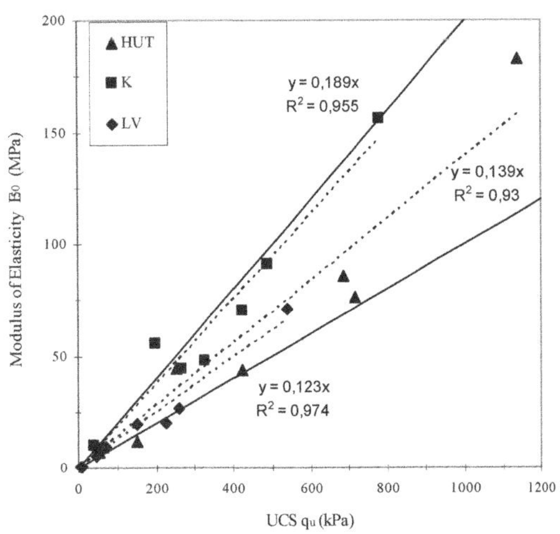

Figure 6. Modulus of elasticity (MPa) and UCS (kPa).

the stress level equal to 50% of $q_{u(max)}$ from the stress-strain curve. However, it was observed that a linear portion of the stress-strain curve often exceeds 50% of the unconfined compressive strength. Figure 6 shows that the modulus of elasticity E_{50} of the stabilized soils increases with increasing strength and an approximate relationship between E_{50} and $q_{u(max)}$of cement treated studied clays are in the range of $100\,q_{u(max)} \sim 200\,q_{u(max)}$. It can also be seen that the correlation of E_{50} varies depending on the clay type. Despite some scatter the correlations of E_{50} to $q_{u(max)}$of Clay 1, Clay 2 are found to be $139\,q_{u(max)}$ and $189\,q_{u(max)}$ respectively, whereas the relation of Clay 3 was about $123\,q_{u(max)}$.

Lorenzo & Bergado (2006) found the correlation of E_{50} to $q_{u(max)}$of Cement treated (Type I Portland) Bangkok clay to be $115\,q_u$–$150\,q_u$. EuroSoilStab (2001) reported that a fairly linear relationship exists between E_{50} and the UC strength and generally falls in the range $100\,q_{u(max)} \sim 200\,q_{u(max)}$. Lorenzo & Bergado (2006) also mentioned a study by Porbaha (1998) based on the extensive research of Saitoh et al. (1980). In that study E_{50} ranges from 105 q to 1000 q in soft clay on the port areas in Japan with Ordinary Portland cement and Boston Blue clay ranges from 50 q–150 q with Type II cement.

The measured strains at failure ε_f (total strain) are presented in Figure 7. Despite some scatter it can be seen that the measured strains at failure decrease with increasing strength for all of the three studied clays. At lower strengths the measured strains at failure are higher and the three studied clays exhibit a ductile behaviour. Stabilized organic clay (Clay 3) showed higher failure strains than the other two soft inorganic clays. Therefore the measured strains at failure varied with type and strength of the stabilized soil. An extensive study and comparison by Åhnberg (2006) showed that the strains at failure decreased rapidly

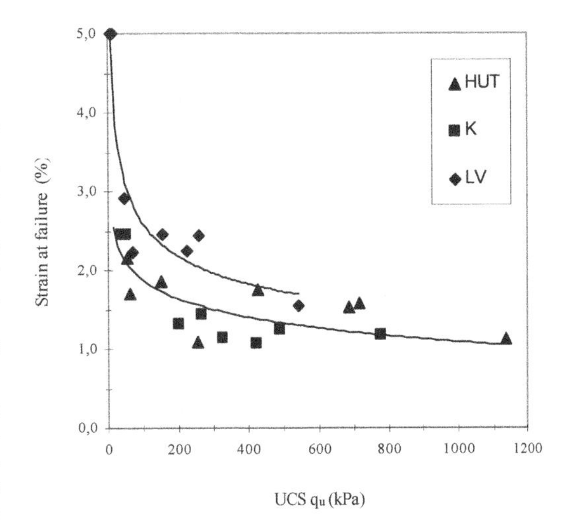

Figure 7. Measured strains at failure (%) of stabilized clays (7d and 28d).

above 200 kPa. The failure strains of inorganic clays are higher at low strength but above 200 kPa the failure strengths decreased rapidly and the strain at failure for Clay 1 and Clay 2 were found in between 1–1.7%. It should be noted that in the uniaxial compression tests, as the confining pressure is zero, ε_f will be higher when the confining pressure is applied.

5 FIELD ASPECTS

In general, soft clays have high water contents with the liquidity index more than unity. This kind of soft clay cannot be spread and compacted which is a prerequisite for soft clay to be used for construction.

Cement treatment reduces the water content after mixing and the plastic limits increase in clay soils. However, immediate workability depends mainly on the amount of cement used and initial water content of clay. If the amount of cement is high and the initial water content of clay is close to its natural plastic limit, any clay will be workable after mixing. It is shown in Figure 4 that at low cement contents the plastic limit does not increase much but around a certain cement amount an apparent plastic limit increases considerably from the mixing time to 1 day. Therefore it can be argued that highly plastic-fat clay (HUT) or silty clay (K) will be workable within 24 hour after mixing if at least a certain cement amount ($C_w \geq 50$ (kg/m^3)) is used despite the initial liquidity index of these clays is over unity.

According to the previous experience in Finland, 50 kPa is a rough estimate for the shear strength of clay needed for machinery handling. This value has not been verified by systematic experiments and therefore it has to be regarded as a directive value for further studies. The conditions in which this value is achieved depend on the amount of binder, time available for curing and other economical factors such as the volume of the stabilized soil in the whole construction project etc.

6 CONCLUSIONS

In this study general geotechnical and strength-stiffness properties of cement treated clay are discussed. The engineering parameters related to permeability and compressibility of cement treated clays will be published later. The following conclusion can be drawn from the investigation and analysis above.

Addition of cement as dry powder decreases the water content immediately from the natural water content due to chemical reactions of cement and after mixing the water content w_c (%) can be regarded as the initial water content of stabilized clay. The increase in bulk and dry densities are observed against the cement amount C_w (kg/m^3) though the densities will depend not only on the cement amount but also on the mixing homogeneity, presence of air bubbles and compaction etc.

The presence of organic content and the acidic nature should be taken into account before stabilization. Organic and/or acidic soils require more cement to achieve satisfactory stabilization results.

The plastic limit increases immediately after mixing in all the studied clays and a considerable increase takes place at a certain cement amount within 24 h. This increment in the plastic limit will ease the field compaction.

The strength increases with decreasing W/C ratio or with increasing C_w (kg/m^3) for all of the studied clays and different clays showed different strength characteristics against W/C ratio. However, the strength development rate against W/C ratio speeds up rapidly above 200 kPa for all clays.

The modulus of elasticity E_{50} of the stabilized soils increases with increasing strength. The modulus of elasticity E_{50} depends on the clay type but generally varies in the range $100\,q_{u(max)} \sim 200\,q_{u(max)}$ for all of the three studied clays. The measured strains at failure decrease with increasing strength for all clays. The strains at failure for a lower strength are higher and the material behavior is ductile. Organic soils showed higher strains at failure than inorganic clays.

ACKNOWLEDGEMENT

This paper was made with the aid of the research project 'Use of soft clay in protection barriers' financed by the Academy of Finland.

REFERENCES

Bergado, D. T., Anderson, L. R., Miura, N., Balasubramaniam, A. S. (1996), Soft ground Improvement in Lowland and other Environments, ASCE Press, pp. 253.

Brandl H. 1981. Alteration of Soil Parameters by Stabilization with lime. Proceedings of the Tenth International Conference on Soil Mechanics and Foundation Engineering., Vol 3, pp. 587–574, Stockholm, Sweden.

Chew S.H., Kamruzzaman, A.H.M. and Lee, F.H. 2004. Physicochemical and engineering behaviour of cement treated clays. Journal of Geotechnical and Geoenvironmental Engineering., ASCE, Vol. 130, No. 7, pp. 696–706.

EuroSoilStab. 2001. Development of design and construction methods to stabilize soft organic soils. Design Guide Soft Soil Stabilization. EC project BE 63–3177

Finnsementti Oy. 2007. http://www.finnsementti.fi, 30.11.2007

Locat, J., Tremblay, H. & Leroueil, S. (1996). Mechanical and hydraulic behaviour of soft inorganic clay treated with lime. Canadian Geotechnical Journal, 33, 654–669.

Lorenzo, G.A. and Bergado, D.T. 2006. Fundamental characteristics of Cement-Admixed clay in Deep mixing. Journal of Materials in Civil Engineering, ASCE, Vol. 18, No 2, pp. 161–174.

Porbaha, A. 1998. 'State of the art in Deep mixing Technology. Part I: Basic concept and overview'. Ground Improvement, 2, 81–92.

Saitoh, S., Kawasaki, T., Niina, S., Babaski, R., and Miyata, T. 1980. Research on DMM using cementitious agents (part 10)-Engineering properties of treated soils. Proc. 15th National Conference of the JSSMFE, Tokyo, pp. 717–720.

Tremblay, H., Duchesne, J., Locat, J. and Leroueil, S. 2002. Influence of the nature of organic compounds on fine soil stabilization with cement. Canadian Geotechnical Journal, Vol 39, pp. 535–546.

Åhnberg, H. 2006. Strength of stabilized soils-A laboratory study on clays and organic soils stabilized with different types of binder. Doctoral Thesis, Lund University, Sweden.

Advances in Transportation Geotechnics – Ellis, Yu, McDowell, Dawson & Thom (eds)
© 2008 Taylor & Francis Group, London, ISBN 978-0-415-47590-7

Effects of potential deleterious chemical compounds on soil stabilisation

T. Le Borgne, O. Cuisinier & D. Deneele
Laboratoire Central des Ponts et Chaussées, Centre de Nantes, France

F. Masrouri
Laboratoire Environnement, Géomécanique & Ouvrages, Nancy-Université, France

ABSTRACT: A study was undertaken in order to characterise the effects of several potential deleterious compounds on soil stabilisation. An original procedure was used: a silty soil was mixed with each compound at a given concentration. The suitability of the different mixtures was then determined by tests according to French standards: volumetric swelling, tensile and compression shear strength. For each compounds, the concentrations are representative of the concentration of that compound in French soil. Four different chemical compounds were considered in this study (sulphur, nitrates, phosphate and chloride). It appeared that sulphate induced strong swelling of the silt treated with lime and cement in accelerated curing conditions. In the case of nitrates and phosphate, a lowering of the unconfined compressive strength was observed, but the silt is still suitable for soil stabilisation according to French technical recommendations. The addition of sodium chloride doesn't alter significantly the mechanical behaviour of the silt treated with lime and cement. It is not possible to determine a threshold value for the considered compounds: the effect of a given compound is related to its concentration, but also to the type of cement and to the suitability test considered.

1 INTRODUCTION

The objectives of sustainable development impose an evolution of the French earthworks socio-economic context. One of the most remarkable impact is the requirement to exclusively use the materials located directly in the land reservation of a project to build the infrastructure (backfill, capping layer, etc.). This requirement concerns also the materials with very low engineering properties. One possibility to improve such materials is the soil stabilisation with lime and/or hydraulic binders. However, it is known that the treatment of these soils can be ineffective because of the presence of deleterious chemical compounds in the ground. Despite that lack of systematic information about their effects on soil stabilisation, several chemical compounds are reported as deleterious by the French technical guide for soil stabilisation (LCPC-SETRA, 2000). These compounds are: sulphate, organic matter, clay minerals, phosphate, nitrate, chloride, etc. In the literature, there is a lot of evidences that sulphate and organic matters can alter soil stabilisation. In the case of sulphur, Mitchell (1986) and Hunter (1988) have characterised its adverse effects for soil stabilisation with lime. Many other studies were carried out to evaluate the impact of sulphate on soil stabilisation (e.g. Rajasekaran *et al.* 1997, Wild *et al.* 1999). Questions arise however about the threshold concentration of the influence of sulphur on soil stabilisation. However, there are very few studies about the impact of other chemical compounds on soil stabilisation. Sherwood (1962) noticed that the compressive strength of 10% cement stabilized sandy soils was unaffected by 0.25–3% sulphate. But, in soils containing clay particles, the presence of low sulphate (0.2%) results in a strength loss of higher than 50%. There is one study presented by Guichard (2006) that evaluated the impact of nitrates on soil stabilisation. They do not evidenced significant effect of these compounds on the soil behaviour.

Hence, except for sulphur and to a lower extend for organic matters, it appears that the actual effect of supposed deleterious compounds on soil stabilisation is not well-known. In this context, a study was undertaken in order to identify the action and the threshold concentration of several potential deleterious agents.

Working on a natural soil can cause difficulties to underline the action of a particular element and to identify the exact causes of the disturbance. This has been clearly demonstrated by Cabane (2005). This is why an original procedure was used in that study. Instead of using natural soil containing deleterious compounds, a suitable soil for stabilisation was mixed with one deleterious compound at a given concentration. The suitability of that mixtures was then determined according to French standards. We studied the effects of four chemical compounds that are often met: calcium sulphate (use to validate our methods) sodium chloride, potassium phosphate, and ammonium nitrate, on a silty soil.

2 MATERIALS PROPERTIES

2.1 *Soil*

The selected soil has been sampled in Magny le Hongre, in the East of Paris. Soil stabilisation is common and is known to be efficient for this silt. The chemical analysis confirmed that it does not contain any substance that would be classified as deleterious compound for soil stabilization.

The main identification properties of the soil are given in Table 1. This soil is medium plasticity silt. The compaction characteristics of the treated soil considered in this study were determined according the normal Proctor compaction procedure.

2.2 *Selected chemical compounds*

Four different chemical compounds were considered in this study. Sulphur, in the form of calcium sulphate ($CaSO_4$, $2H_2O$), was selected as a reference since its effect on soil stabilisation is well characterised. Secondly, two fertilizers were chosen: ammonium nitrate (NH_4 NO_3) and potassium phosphate (KPO_4). Their use is very widespread in France and their potential effect on soil stabilisation is still unknown. The last compound considered is sodium chloride (NaCl).

A key point is the concentration of each compound in the soil. The adopted strategy, for each compound, was to select two concentrations. The first one, low value, was supposed to be the average value of that compound found in French soils. The second concentration, high value, was retained as the maximum concentration of each compound measured in French soils. The selected values (Table 2) were determined with the help of the soil database of the French institute for agricultural researches.

2.3 *Binders*

Typical soil treatment for capping layers with this kind of soil is a combination of quicklime and hydraulic binder. Typical dosages were selected, 1.5% for the quicklime and 6% for the hydraulic binders, both of the equivalent oven-dried soil mass.

One quicklime was considered. It has a high content in free lime, more than 91%.

Hydraulic binders are highly versatile regarding their composition since they can include several secondary constituents. To minimize the interactions

Table 1. Main identification properties of silty soil.

	Silt of Magny le Hongre
Particle size distribution	
Dmax (mm)	5
<2 mm (%)	99.9
<80 μm (%)	94.6
Physical properties	
Plastic limit (%)	20.8
Liquid limit (%)	36.5
Plasticity index (%)	15.7
VBS (–)	4.43
Specific gravity ($Mg.m^{-3}$)	2.7
Optimum compaction references	
w (%)	15.5
γ_d ($Mg.m^{-3}$)	1.80

Table 2. Concentration of chemical compounds.

Compound	Low concentration*	High concentration*
$CaSO_4$, $2H_2O$	0.62	6.2
NH_4 NO_3	0.16	1.5
KPO_4	0.29	0.85
NaCl	0.06	1.2

* Concentration: g per kg of dry soil.

between these secondary constituents and the added chemical compounds, binders having as less secondary constituents as possible were selected. The first hydraulic binder was pure Portland cement (CEM I 52.5 N) with more than 95 wt.% of clinker and the second one (CEM II 32,5 R) was a binder based on pure Portland cement (70 wt.% of clinker) containing limestone as secondary constituent.

3 PROCEDURE FOR ADDING CHEMICAL COMPOUNDS TO SOIL

One key issue is the preparation of the mixture of the soil and of the selected deleterious compound. In order to obtain homogenous sample, we adapted mixing procedure from ecotoxicology to the soil stabilisation. This procedure is recognized to be efficient for mixing soil and chemical compounds homogenously (ISO 11268-1).

3.1 *Ecotoxicology procedure*

This ecotoxicology procedure is an experimental procedure, designed to assess lethal or sub lethal toxic effects of chemical compounds on plants or animals in short time in soil system. Soils to be tested can be reference soils, artificial soils and site soils.

3.2 *Mixtures preparation*

The procedures from ecotoxicology were adapted to the question of soil stabilisation. The mixture of soil and of the desired chemical compound was prepared in five stages:

1. chemical compound was dissolved into distilled water;
2. soil and the solution were thoroughly mixed with mechanical mixer;
3. fifteen days equilibration phase was required to provide time for liquid/solid phase equilibration reactions;
4. soil was mixed again, and water content determined;
5. soils were wetted to the desired water content for compaction.

3.3 *Samples preparation*

The samples were prepared in order to determine the suitability of each mixture for soil stabilisation. They were all prepared by static axial compression in a cylindrical mould. Sample dry density was chosen at 98.5% of optimum dry density of the soil treated with quicklime and cement without any deleterious compound. It was 1.64 $Mg.m^{-3}$ with a mass water content of 20%. All the samples have the same compaction characteristics.

4 EXPERIMENTAL PROCEDURES

French technical standards, guides suggest several tests to determine soil suitability for treatment with lime and/or hydraulic binders. Three test procedures were used in our study.

4.1 *Soil suitability for stabilisation test*

In soil stabilization applications to the construction of capping layers, it is necessary to check the suitability of the soil for treatment. A suitability test is described in French standards. Its basic principle is to accelerate the setting reaction of the treated soil, and to measure the eventual swelling and the strength of the treated soil after the curing period.

First, samples are kept at constant water content until the workability period of the hydraulic binder is elapsed. Then samples, with a height of 50 mm and a diameter of 50 mm, are immersed in temperature controlled bath at 40°C, for seven days. After that period, the volumetric swelling is measured. The volumetric swelling is defined as the ratio between the initial and the final volume (after 7 days at 40°C). Moreover, tensile shear strength has been measured by Brazilian strength test (Rtb) on other samples which were submitted to the same maturing procedure. The criterion to determine soil suitability for lime and/or hydraulic binders is a combination of both test results (Table 3).

Table 3. Criteria for determining the soil suitability for treatment with hydraulic binders (LCPC-SETRA 2000).

Suitability of soil	Volume swelling Gv (%)	Brazilian strength Rtb (MPa)
Suitable	≤ 5	≥ 0.2
Marginal	$5 \leq Gv \leq 10$	$0.1 \leq Gv \leq 0.2$
Unsuitable	≥ 10	≤ 0.1

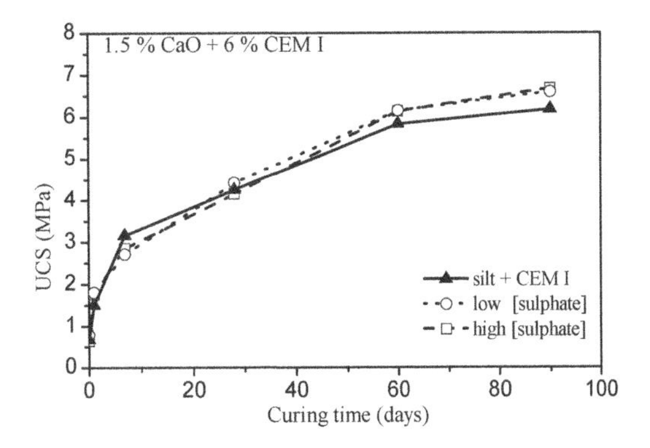

Figure 1. UCS of silt + sulphate + CEM I.

4.2 *Unconfined compression strength (UCS)*

UCS of treated soil has been measured on cylindrical samples, with a height of 100 mm and a diameter of 50 mm. The displacement speed of the press is constant, 1 mm/min.

Samples are stored in two different ways. First, samples are kept at constant water content. In that case, UCS has been determined at different curing time (3 h, 1, 7, 28, 60, 90 days). Secondly, some samples are kept at constant water constant at 20°C for 28 days, and then immersed in temperature controlled bath at 20°C for 32 days. Then the UCS is determined.

For non immersed sample, the results of UCS test are given on Figures 1 to 8. These figures give the UCS as a function of the curing time. Solids curves are given results of silts without deleterious compounds. Dotted curves correspond to the results of treated silt with compound at low concentration. Dashed curves correspond to the results of treated silts with compounds at high concentration.

For immersed sample, results of UCS test are given in text. The volumetric swelling of the samples was also evaluated on these samples after the immersion period. The criteria to evaluate the suitability for treatment is the ratio (Ri) between the immersed UCS and

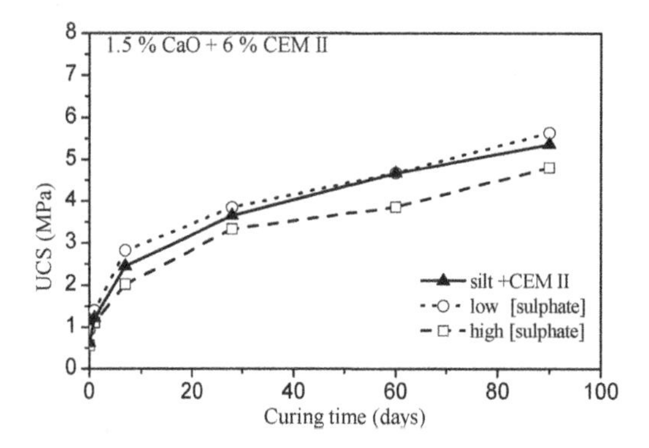

Figure 2. UCS of silt + sulphate + CEM II.

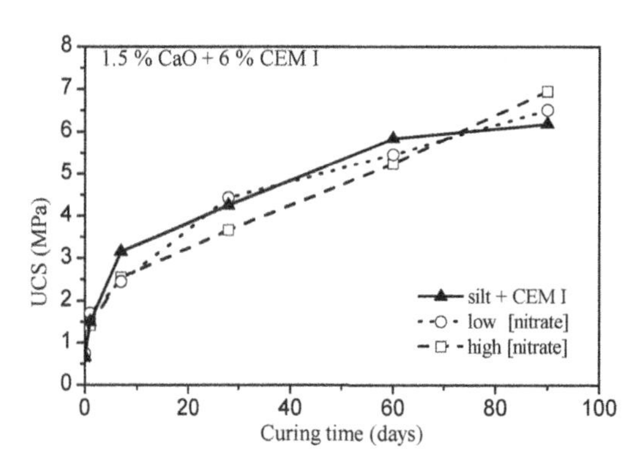

Figure 3. UCS of silt + nitrate + CEM I.

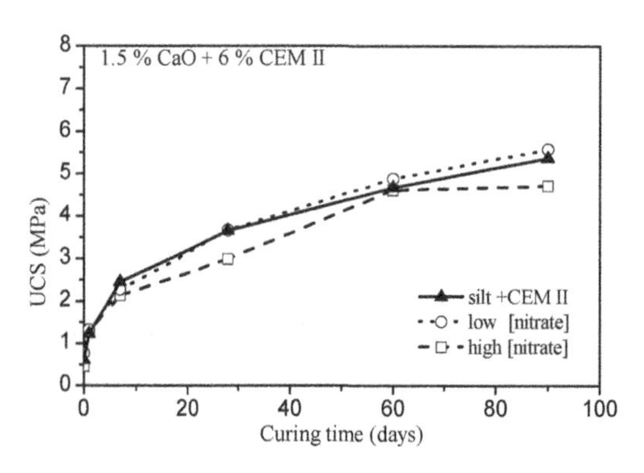

Figure 4. UCS of silt + nitrate + CEM II.

the no immersed UCS at 60 days and. That ratio Ri must be higher than 0.6.

5 EXPERIMENTAL RESULTS

For each soil mixture, the suitability test results are given in Tables 5 to 8, the results of UCS are given on

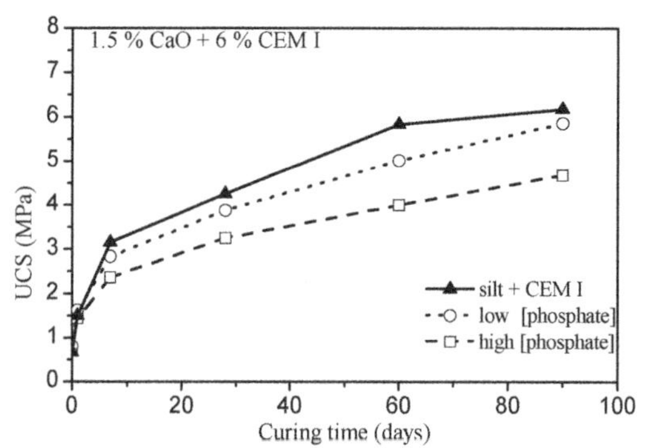

Figure 5. UCS of silt + phosphate + CEM I.

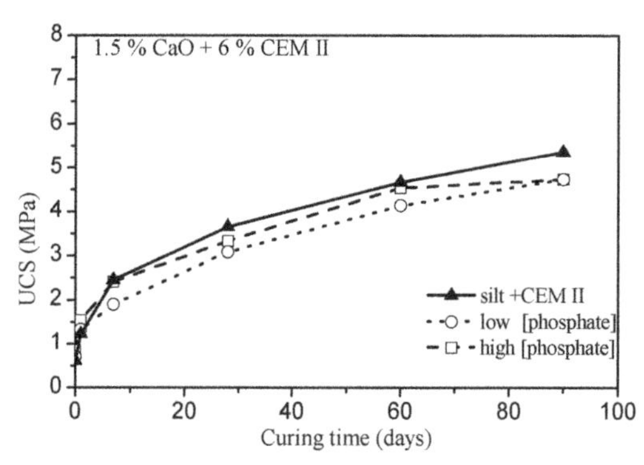

Figure 6. UCS of silt + phosphate + CEM II.

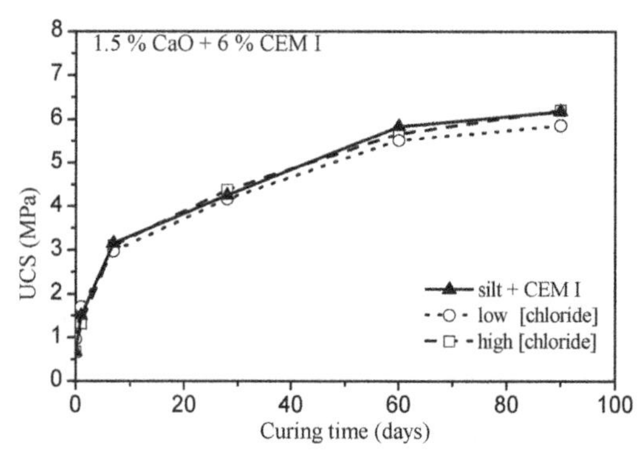

Figure 7. UCS of silt + chloride + CEM I.

Figures 1 to 8; and the results of immersed UCS are given in the text.

5.1 *Pure treated silt*

The average results of the suitability test for the pure silt are given in Table 4. The volumetric swelling is limited and similar for the two kinds of treatment, 0.6% for CEM I and 0.5% for CEM II. The average of Rtb

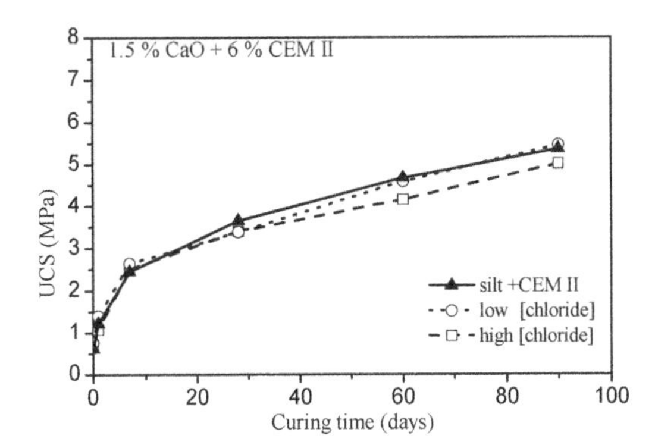

Figure 8. UCS of silt + chloride + CEM II.

Table 4. Results of pure treated silt suitability for treatment.

Cement	Volume swelling Gv (%)	Brazilian strength Rtb (MPa)
CEM I	0.6	0.511
CEM II	0.5	0.439

Table 5. Suitability test result for silt + sulphate mixture.

	Volumetric swelling (%)		Rtb (MPa)	
Cement	Low [C]	High [C]	Low [C]	High [C]
CEM I	0.52	3.55	0.674	0.550
	0.46	3.63	0.546	0.721
	0.49	3.13	0.602	
CEM II	0.260	18.88	0.566	0.377
	0.268	11.92	0.594	0.316
	0.409	13.56	0.437	0.364

measure for the silt treated with CEM I is 0.511 MPa. However the average of Rtb decreases when silt is treated with CEM II, (0.439 MPa).

The UCS results for the two treatments of the pure silt are given on Figures 1 and 2. It can be seen that UCS increases when curing time increase. This is a general trend observed in the case of treated soils. The UCS is lower when CEM II is used.

The ratios Ri are satisfactory for treatment with CEM II (0.87). The volumetric swelling is low (0,31%).

5.2 *Silt mixed with sulphate*

The suitability test results are given in Table 5. The soil behaviour is not significantly altered for the lowest concentration in sulphate. However, at high concentration it induced an increase of the volumetric

Table 6. Suitability test result for silt + nitrate mixture.

	Volumetric swelling (%)		Rtb (MPa)	
Cement	Low [C]	High [C]	Low [C]	High [C]
CEM I	0.12	0.25	0.514	0.597
	0.29	0.57	0.483	0.560
	0.27	0.28	0.480	0.582
CEM II	0.22	0.31	0.590	0.496
	0.30	0.19	0.517	0.483
	0.34	0.28	0.406	0.318

Table 7. Suitability test result for silt + phosphate mixture.

	Volumetric swelling (%)		Rtb (MPa)	
Cement	Low [C]	High [C]	Low [C]	High [C]
CEM I	0.55	0.10	0.668	0.631
	0.31	0.06	0.620	0.487
	0.51	0.11	0.390	0.434
CEM II	0.05	0.07	0.278	0.400
	0.19	0.24	0.333	0.311
	0.18	0.04	0.471	0.356

Table 8. Suitability test result for silt + chloride mixture.

	Volumetric swelling (%)		Rtb (MPa)	
Cement	Low [C]	High [C]	Low [C]	High [C]
CEM I	0.32	0.40	0.465	0.681
	0.03	0.29	0.670	0.387
	0.34	0.38	0.651	
CEM II	0.11	0.12	0.533	0.487
	0.06	0.16	0.497	0.436
	0.10	0.11	0.445	0.421

swelling. It can be seen that the type of cement has a dramatic effect since the silt with high sulphur concentration treated with CEM II is unsuitable for the treatment while it is still suitable if CEM I treatment is considered.

There is no significant influence of the sulphate on UCS (Figure 1). In the case of CEM II, the UCS of the silt mixed with sulphur is slightly lower than the pure silt treated with CEM II.

The ratios Ri are satisfactory for treatment with CEM I (low concentration: 0.86, high concentration: 0.82). The volumetric swelling is low for low sulphur concentration (0.11%), and is higher for high sulphur concentration (1.52%). The ratios Ri are satisfactory when CEM II is considered (low concentration: 0.86, high concentration: 0.93). The volumetric swelling is low for low sulphate concentration (0.19%), and is higher for high sulphate concentration (2.11%).

At the concentrations studied, sulphate can not be considered not systematically as a deleterious compound for soil stabilisation. It can be noted that the mixture suitability depends on the sulphate concentration, but also on the type of cement and the test procedure. Hence, when the silt mixed with sulphate at high concentration treated with CEM II is unsuitable for soil stabilisation if the suitability test is considered. However, UCS test immersed or not, indicate that the setting reaction is not altered by the presence of sulphate.

5.3 *Silt mixed with nitrate*

As it can be seen in Table 6, the addition of nitrates to the silt do not altered the suitability tests results, for both concentrations. The volumetric swelling are similar than this measured on the pure silt. The Rtb values are slightly higher than this determined on the pure silt.

The results of UCS test for silt treated with CEM I are given on Figure 3 and on Figure 4 for the CEM II. For both treatments, the results obtained with the silt mixed with nitrated at high concentration are slightly lower than the UCS determined with the pure silt.

The ratios Ri are satisfactory for treatment with CEM I (low concentration: 0.87, high concentration: 0,83). The ratios Ri are slightly lower the than pure silt treated with CEM II (low concentration: 0.71, high concentration: 0.71). The volumetric swellings are low for both binders.

As a conclusion, it can be stated that nitrates seems to slightly lower the efficiency of soil stabilisation when the nitrates concentration is high. All the tests procedures led to the same conclusion.

5.4 *Silt mixed with phosphate*

The suitability test results for the silt mixed with phosphate are given in Table 7. The volumetric swelling, as well as the Rtb, is not altered by the phosphate addition. One significant difference with the pure silt has been evidenced: Rtb is lowered by the addition of phosphate in the case of the CEM II.

The results of UCS test for silt treated with CEM I are given Figure 5. The UCS measures decreased when phosphate concentration increase. In the case of CEM II (Figure 6), the UCS of the silt mixed with phosphate is slightly lower than the pure silt treated with CEM II.

The ratios Ri are satisfactory for treatment with CEM I (low concentration: 0.73, high concentration: 0.86). The ratios Ri are also satisfactory for treatment with CEM II (low concentration: 0.82, high concentration: 0.88). The volumetric swellings are low for both binders.

As a conclusion, it appeared that phosphate lowered the efficiency of soil treatment. However, the silt is still suitable for soil stabilisation according to French standard.

5.5 *Silt mixed with chloride*

The suitability test results for the silt mixed with chloride are given in Table 8. The soil behaviour is not significantly altered. The addition of chloride, in low or high concentration doesn't alter the soil stabilisation.

The results of UCS test for silt treated with CEM I are given on Figure 7 and on Figure 8 for the CEM II. There is no significant influence of the chloride on UCS in that case.

The ratios Ri are satisfactory for treatment with CEM I (low concentration: 0.90, high concentration: 0,79). The ratios Ri are also satisfactory for treatment with CEM II (low concentration: 0.81, high concentration: 0.92). The volumetric swellings are low for both binders.

As a conclusion, with the concentration selected in that study, it can be seen that chloride can not be considered as a deleterious compound for soil stabilisation.

6 CONCLUSION

The results showed that fertilizers (phosphate and nitrate) tend to lower the mechanical characteristics of the silt, but, the soils containing the fertilizers are still suitable for soil stabilisation, and thus, usable for earthworks. In the case of sulphate, for each of binder, an influence was evidenced. However, the impact of sulphate on the mechanical behaviour appeared to be highly dependant on the kind of binder. In some cases, we observed large swelling of the silt, hence unsuitable for soil stabilisation, this kind of behaviour of sulphate containing soil has also been evidenced by several authors. In other cases, at the same sulphate concentration, only limited swelling and satisfactory mechanical properties were obtained. The study do not evidenced any impact of chloride on the mechanical behaviour of the tested soils.

This paper shows that some considered chemical compounds (sulphate and fertilizer) may alter silt stabilization process. However, the presence of a given chemical compound at a given concentration in the soil is not sufficient enough to determine the suitability of the soil for a treatment. As a concluding remark, it is difficult to determine whether a compound is deleterious for soil treatment or not and to define a concentration threshold. This study demonstrated that it is not an intrinsic property, even for sulphur, but it depends on the concentration, but also on the type of cement, and the type of suitability test considered.

ACKNOWLEDGEMENTS

This research was funded by the "Syndicat Professionnel des Terrassiers de France".

REFERENCES

Cabane, N., Nectoux, P., Gaudon, P. & Fouletier, M. 2005. Contribution to study of sulphur damages on traited soils. Proceedings of 2nd International Symposium of treatment and recycling of materials for transport infrastructure, Paris.

Guichard, C. 2006. Eléments perturbateurs de la prise dans les sols traités aux liants hydrauliques: définition, détection, seuils et remèdes. Master of sciences, INSA Strasbourg.

Hunter, D., 1988. Lime induced heave in sulphate bearing clay soils. *Journal of geotechnical engineering*, 114: 150–167.

LCPC – SETRA. 2000. Soil treatment with lime and/or hydraulic binders. Technical guide. ISSN: 1151-1516. 240 p.

Mitchell, J.K. 1986. Practical problems from surprising soil behavior. *J. Geot. Engineering Division*, 112: 259–289.

Ribera, D. & Saint Denis, M. 1999. The worm Eisenia fetida: interests and perspectives in terrestrial ecotoxicology. Bull Soc. Zool. Fr., 124(4): 411–420.

Sherwood, P.T. 1962. Effect of sulphate on cement and sulphate treated soils. Highways Res. Board Bul. 353, 98–107.

Wild, S., Kinuthia, J.M., Jones, G.I. & Higgins, D.D., 1999. Suppression of swelling associated with ettringite formation in lime stabilized bearing clay soils by partial substitution of lime with ground granulated blastfurnace slag. *Eng. Geol.* 51, 257–277.

Advances in Transportation Geotechnics – Ellis, Yu, McDowell, Dawson & Thom (eds)
© 2008 Taylor & Francis Group, London, ISBN 978-0-415-47590-7

Stability of the solidification process of a lime-treated silt under percolation conditions

B. Lequiller, V. Ferber, O. Cuisinier & D. Deneele
Laboratoire Central des Ponts et Chaussées, Centre de Nantes, France

Y.J. Cui
Soil Mechanics Teaching and Research Center, École Nationale des Ponts et Chaussées, Marne-la-Vallée, France

ABSTRACT: The long term stability of both geotechnical amelioration and physico-chemical processes brought by lime stabilization for earth structures exposed to long term water circulations remains an open question. In fact, there is no general tendency in the literature on the effect of leaching on the physico-chemical behavior of lime-treated soils and more precisely on the stability of the cementitious compounds formed. Hence, this paper examines the effects of long term water leaching from a physico-chemical point view on a silty soil treated at two different lime contents. Long term water leaching has been simulated using percolation equipment. Different types of chemical analyses have been conducted both on untreated and treated soils and collected leachates. The physico-chemical results show that the addition of lime leads relatively quickly during curing to the formation of cementitious compounds as well as carbonate compounds. The results also indicate that water leaching does not seem to have a negative impact on the cementations bonds formed during curing. In fact, they even suggest for specimens treated at the higher lime content that there could be formation of pozzolanic by-products during leaching.

1 INTRODUCTION

Lime stabilization is a widespread technique for the construction of embankments in the field of earthworks since it can both improve the workability and mechanical properties of wet clayey soils. However, the durability and stability of the improvements brought by this technique for the construction of hydraulic earth structures such as earthdams, dikes, or river levees exposed to long term water circulations remains an open question.

In fact, most of the research conducted over the last decade aimed at better understanding the lime treated soils behavior (e.g. Eades & Grim 1962, Diamond & Kinter 1965, Little 1995, Bell 1996) without taking into account the environmental stress factors to which earth structures can be exposed to: prolonged water contact, wetting-drying or freezing-thawing cycles, traffic, etc. It has been shown that two kinds of physico-chemical processes are initiated when adding lime to the soil. The first one, termed modification, occurs immediately after lime addition and is related to cation exchange and flocculation processes. The second one, termed solidification, occurs in the long term and corresponds to the development of the pozzolanic reactions leading to the formation of cementitious compounds. The geotechnical improvements brought by lime such as the decrease of soil plasticity, the increase in shear or unconfined strengths as well as the change in permeability can be directly attributed to these processes.

Only few studies were conducted to evaluate the durability – or permanence of these geotechnical improvements and therefore of the physico-chemical processes – of lime stabilized soils exposed to water circulation. A study by De Bel et al. (2005) shows that water circulations lead to a decrease of the unconfined compression strength of a lime-treated loam. However, this study does not conclude whether the decrease can be attributed to an increase of the specimens water content (i.e. saturation) or to a detrimental effect of the water circulations on the pozzolanic activity and therefore on the cementations bounds. In addition, a study of McCallister (1990) on three different clayey soils suggests that the extent of the decrease in unconfined compression strength entailed by water circulations is highly related to the added lime content. The general trend being the more lime used, the less decrease

Table 1. Geotechnical and physico-chemical identification properties of the Jossigny silt.

Geotechnical properties	
Liquid limit, w_L (%)	37.0
Plastic limit, w_P (%)	18.7
Plasticity index, I_p (%)	18.3
Methylene blue value (g/100 g)	3.2
Soil density, γ_s (kN.m^{-3})	2.69
Clay fraction, <2 μm (%)	29.4
Physico-chemical properties	
Soil pH	8.0
$CaCO_3$ content, (%)	1.30
Cation exchange capacity (meq/100 g)	15.1

in unconfined strength. The hypothesis given by the author to explain this result is that, when present in sufficient amount, the pozzolanic compounds are relatively stable to water circulations. Finally, the lack of studies in the literature on the behavior of lime-treated soil under percolation conditions does not allow concluding on their physico-chemical behavior and more precisely on the stability of the cementation compounds.

In this context, the present work aims at investigating the long term behavior of a soil treated with two different lime contents and exposed to water circulations, from a physico-chemical point of view. More precisely, the different objectives of the study are: (i) to develop an experimental procedure to simulate long term water circulations, (ii) to describe the physico-chemical behavior of a lime-treated soil under long term water leaching and (iii) to evaluate the influence of lime content on this behavior.

2 EXPERIMENTAL PROCEDURE

2.1 *Material – untreated soil and lime*

The soil tested in this study is a silty soil sampled in Jossigny, nearby Paris in France. This soil typically represents the soils that can be treated with lime in France. The main identification characteristics of the soil were determined according to the French standards and are given in Table 1. The mineralogical analysis from X-ray diffraction shows that the silt clay fraction is mainly composed of illite and kaolinite, with a significant amount of interstratified illite and smectite minerals. The other minerals are quartz and plagioclases (Cui 1993).

The differential thermal analysis obtained for the lime used for stabilization shows that it is constituted of around 80% of free calcium oxide, 15% of portlandite and 3% of carbonates.

Table 2. Optimum compaction characteristics of the tested materials.

Type of material	Water content %	Dry density kN.m^{-3}
Untreated material	16.0	1.76
Treated with 1% of lime	21.0	1.64
Treated with 3% of lime	23.0	1.58

2.2 *Specimens preparation*

Two different lime contents, on a dry-weight basis, have been tested in this study. The first one, 1% of lime corresponds to a usual lime content used in France for this kind of soils. The second one, 3% of lime, corresponds to an upper limit of lime content from an economical point of view.

It has also been chosen in this study to dynamically compact the specimens directly in cylinder molds of 100 mm height and 50 mm diameter using a mini compaction device similar to the one used by Sridharan & Sivapullaiah (2005). Besides, the specimens were compacted in 10 layers to ensure a homogenous compaction within the specimens. The compaction characteristics of the untreated and treated soils were determined from the French standard Proctor compaction curves of each material and chosen equal to their optimum compaction characteristics (Table 2).

After compaction, the specimens were sealed with wax to limit air carbonation and cured at ambient laboratory temperature (20 ± 1°C) for a period of 7 or 25 days.

2.3 *Percolation equipment and conditions*

Water circulations were simulated using percolation equipment per *ascensum*. Specimens were placed, after the curing period, in cells similar to flexible wall permeameters. Twelve cells were designed. Each cylinder cell was independently controlled so that cells can be connected or disconnected from the main control panel without affecting other leach tests.

Percolation cells were leached under a constant hydraulic head of 80 kPa. The resulting hydraulic gradient was 80. A confining pressure of 120 kPa was also applied to the cells in order to limit side leakages. Distilled water was used in order to have the same leach fluid throughout the test duration. Percolation tests were conducted at a temperature of 20 ± 1°C. The leachates were collected as they passed through the samples into 1 liter plastic bottle. Figure 1 provides a schematic diagram of a percolation cell.

One should note that the operating percolation conditions were chosen more severe than the one observed in the field in order to accelerate the long term degradation that may occur. In fact, the mean annual soil

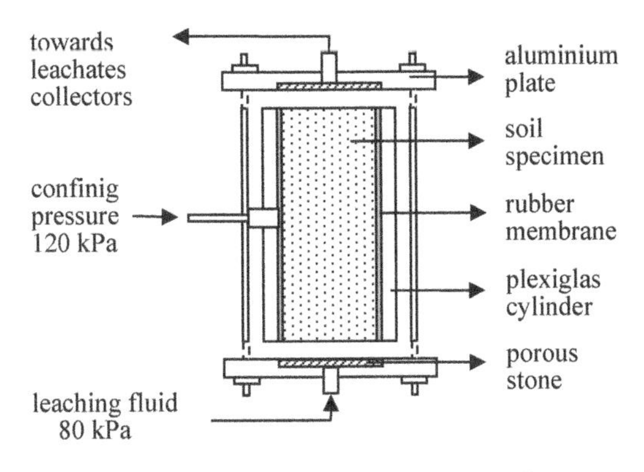

Figure 1. Schematic diagram of a percolation cell.

infiltration in France can be evaluated to 300 L.m^{-2}. With the percolation procedure used in this study, this value varies from 2700 L to 15000 L for the same surface and time period.

2.4 *Chemical analysis*

In order to describe the long term physico-chemical behavior of specimens under leaching conditions, several different types of chemical analyses have been conducted both on soil specimens and collected leachates. The analyses were made before, throughout and after percolation. The chemical analyses performed on the specimens are: pH, free lime content (Leduc method) and carbonates contents (NF ISO 10693, Dietrich-Frülhling method). Those conducted on the leachates are: pH and elementary element concentrations (ICP-AES) such as calcium, silicon and aluminum.

3 EFFECT OF LEACHING ON THE PHYSICO-CHEMICAL BEHAVIOUR OF SPECIMENS

3.1 *Leachates chemical analyses*

The chemical data presented in this section referred to the results obtained for untreated and treated specimens after a curing time of 25 days. Only one set of data is given for each lime content, the tendencies recorded for different specimens being the same for a given lime content.

Figures 2 to 5 show the evolution of pH and the concentrations in calcium, silicon and aluminium during the percolation according to the lime content used.

The leachates pH of each specimen can be considered as constant throughout the percolation (Fig. 2). However, the leachates pH values depend on the specimens lime content. Indeed, the leachates pH of the untreated specimens is equal to 8 whereas the leachates pHs of specimens treated at 1% and 3% of lime are respectively around 9 and 12.

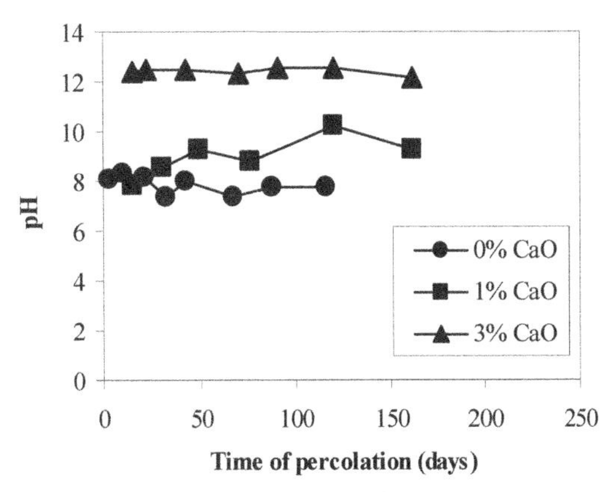

Figure 2. Evolution of the leachates pH with the time of percolation.

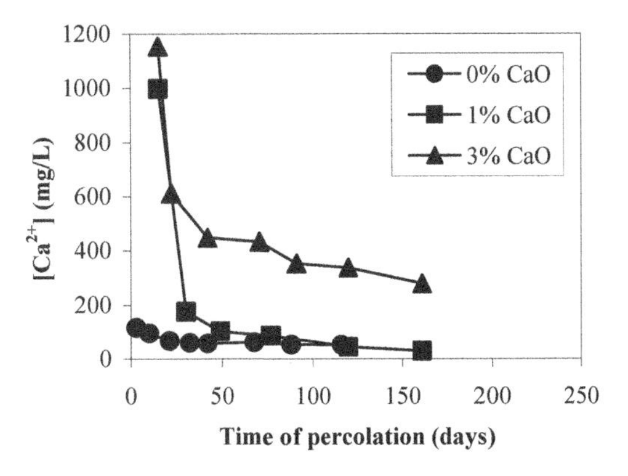

Figure 3. Evolution of the leachates calcium concentrations with the time of percolation.

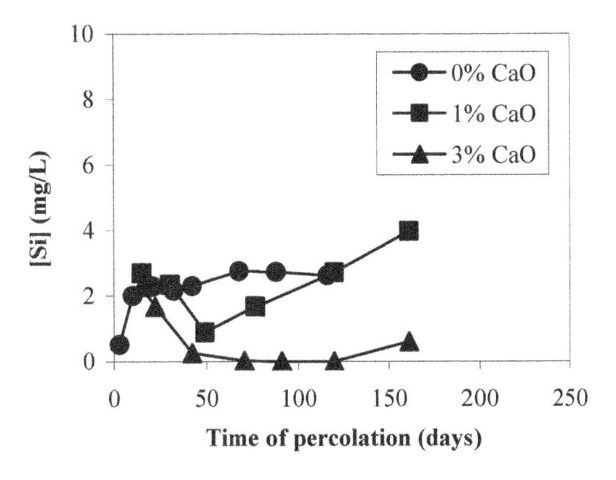

Figure 4. Evolution of the leachates silicon concentrations with the time of percolation.

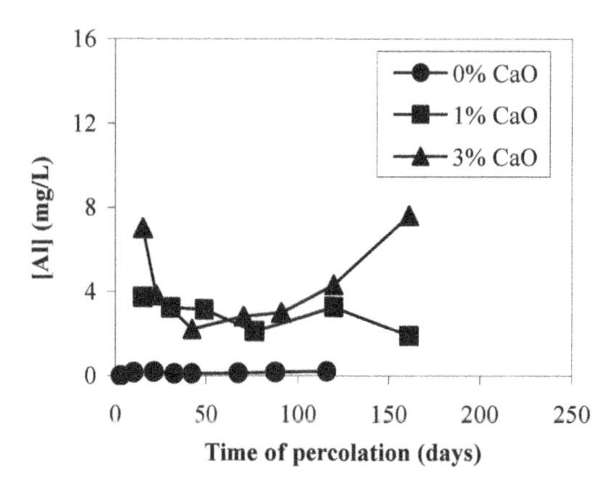

Figure 5. Evolution of the leachates aluminum concentrations with the time of percolation.

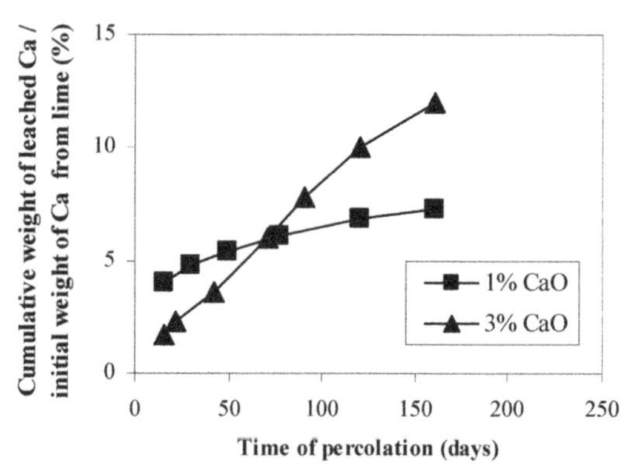

Figure 6. Variation of the cumulative weight of leached calcium over the initial weight of calcium from lime.

The leachates calcium concentrations decrease for each specimen with respect to the time of percolation (Fig. 3). The calcium concentrations of the untreated specimens decrease from 120 mg/L at the beginning of the percolation to 50 mg/L after 120 days of percolation. In the case of specimens treated at 1% and 3% of lime, the leachates calcium concentrations decrease respectively from 1000 mg/L and 1200 mg/L at the beginning of the percolation to 30 mg/L and 280 mg/L after 160 days of percolation. Furthermore, it can be noticed that the leachates calcium concentrations of lime treated specimens are higher than those of untreated specimens for a given time of percolation (especially in the case of specimens treated at 3% of lime).

The evolution of the leachates silicon concentrations varies during leaching with the amount of lime added (Fig. 4). In fact, the leachates silicon concentrations of untreated specimens increase until 50 days of percolation and remain thereafter close to 2 mg/L. Those of specimens treated with 1% of lime increase with the time of percolation to reach a value of 4 mg/L. The leachates silicon concentrations of specimens treated at 3% of lime decrease until 50 days of percolation and can be, from this point, considered as negligible.

The leachates aluminium concentrations are also a function of the amount of lime added (Fig. 5). The leachates aluminium concentrations of untreated specimens can be considered as zero throughout the percolation, whereas the leachates aluminium concentrations of specimens treated at 1% of lime are relatively constant to 3 mg/L with the time of percolation. In the case of specimens treated with 3% of lime, the leachates aluminium concentrations increase with the time of percolation to reach a value of around 8 mg/L.

Finally, Figure 6 shows the evolution during percolation of the cumulative weight of leached calcium compared to the weight of calcium coming from the lime initially added. One can notice that the percentages of leached calcium for specimens treated at 3% of lime are lower than those of specimens treated with 1% of lime during the first 75 days of leaching and that, thereafter, they become higher. The percentages of leached calcium compared to the initial weight of calcium provided by the lime added content reach, after 160 days of percolation, 7% and 12% for specimens treated respectively at 1% and 3% of lime.

3.2 *Soil chemical analyses*

Figures 7 and 8 show the free lime and carbonates contents of specimens treated respectively at 1% and 3% of lime for different types of stress (i.e. different curing times and different leaching periods).

One can note that after one day of curing the addition of the carbonates content and free lime content obtained for a given lime content corresponds more or less to the amount of lime initially added to the soil. Furthermore, it can be noticed that the carbonates content results recorded for specimens treated at 1% and 3% of lime are higher than the carbonates content of the untreated specimens: the difference being around 0.50% in the case of specimens treated at 1% of lime and 1.60% for specimens treated at 3% of lime.

The results also indicate, in the case of specimens treated at 1% of lime, that their free lime content decreases both with curing time and percolation time. In fact, it decreases from 0.45% after one day of curing to 0.20% after 25 days of curing and are equal to 0.1% after 200 days of percolation. On the other hand, their carbonates content remains relatively constant to 2% whatever the curing time or percolation time. The same kind of observations can be made for

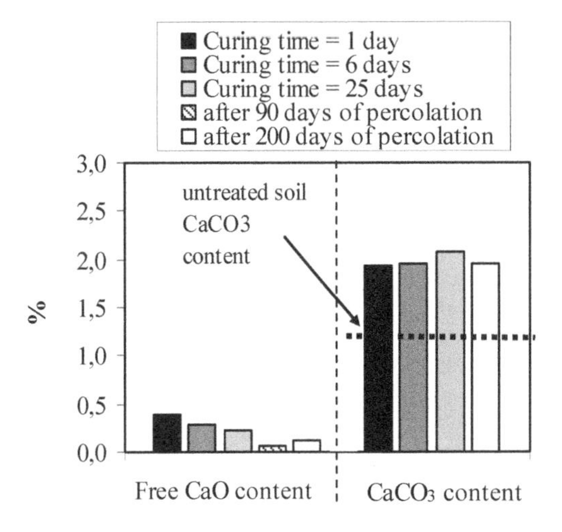

Figure 7. Free lime and carbonates contents of specimens treated with 1% of lime for different types of stress.

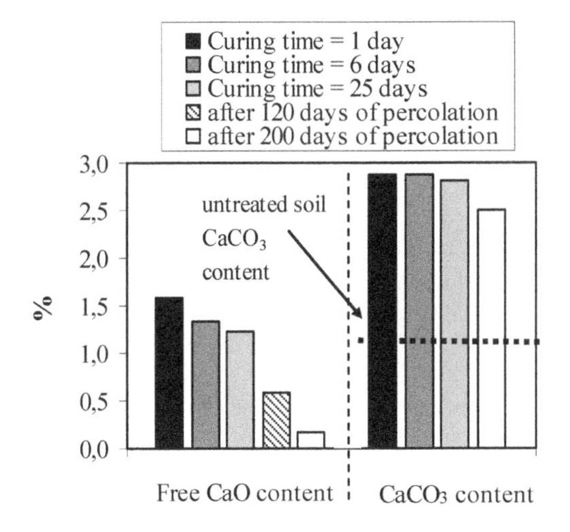

Figure 8. Free lime and carbonates contents of specimens treated with 3% of lime for different types of stress.

specimens treated at 3% of lime. Their free lime content decreases from 1.60% for a curing time of one day to 1.20% after 25 days of curing. Their carbonates content remains constant to 2.90% throughout the 25 days of curing. However, the carbonates content seems to decrease with percolation, since it reaches a value of 2.50% after 200 days of leaching.

4 DISCUSSION

4.1 *Before leaching*

The difference in carbonates contents between untreated and treated specimens shows, as mentioned by De Bel et al. (2005), that a large amount of lime-around 50% of the lime added for both lime dosages – could be directly precipitated as carbonates. This phenomenon could be caused either (i) by the carbon dioxide contained in the air during the mixing phase of the soil and the lime or (ii) by bicarbonates ions contained in the soil.

Furthermore, the decreases in free lime contents recorded for both lime dosages during the curing period combined to the stability of carbonates contents could be a sign of pozzolanic activity leading to the formation of cementitious by-products such as calcium silicate hydrates, CSH, or calcium aluminate hydrates, CAH (Malhotra & Baskar 1983).

4.2 *After leaching*

It can be noted from Figure 3 that the leached calcium does not seem to come from the soil itself since the calcium concentrations recorded for the untreated specimens are lower than those recorded for the treated specimens (especially in the case of specimens treated at 3% of lime). This observation suggests that the leached calcium must come from another source of calcium. Different assumptions can be made as for this source:

1. calcium coming from the decalcification of the cementitious products formed during curing;
2. calcium provided by the carbonates dissolution (carbonates either precipitated with the lime addition or initially present in the untreated soil);
3. calcium coming from the lime remaining in the soil and that has not yet been used (by carbonation and/or pozzolanic reactions).

Figure 9 gives the solubility of several soil minerals with regards to the soil pH. One can notice that silica is very soluble for pHs above 10, whereas for lower pHs, it shows a low solubility. The solubility of alumina is very high for pHs below 4 and above 9, whereas it is practically negligible for pHs between 4 and 9. The leachates pH recorded for untreated specimens -8- (Fig. 2) can therefore explain the absence of aluminum and the small amount of silicon collected in the leachates during leaching. The same type of observation can be made for specimens treated with 1% of lime. Their leachates silicon and aluminum concentrations are relatively constant throughout percolation (Fig. 4-5). However, these are higher than those of untreated specimens. This result could be explained by the fact that the pH recorded for specimens treated at 1% of lime -9- (Fig. 2) is slightly higher than the

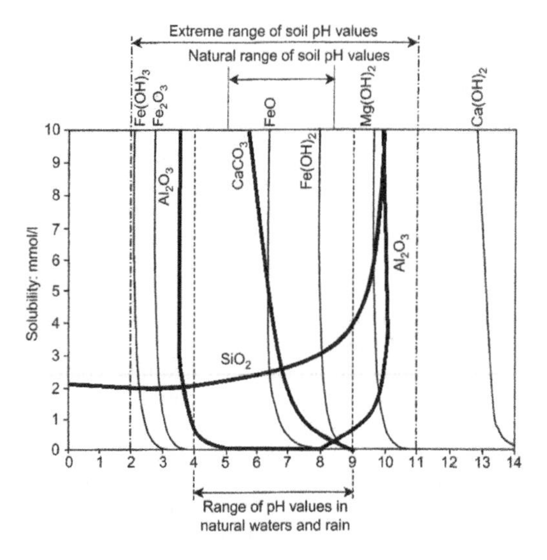

Figure 9. Solubility of some soil mineral species in relation to pH (after Loughnan, 1969).

one of untreated specimens -8-, hence favoring the dissolution of alumina and silica minerals (Fig. 9). In the case of specimens treated with 3% of lime, the pH recorded through the time of percolation -12- (Fig. 2) should favor the solubility of silica and alumina in the leachates (Fig. 9). As a result, a large amount of silicon and alumina should be observed in the leachates. Surprisingly, the results (Fig. 4) show leachates silicon concentrations almost equal to zero and leachates aluminum concentrations (Fig. 5) that remain low and close to the one recorded for specimens treated at 1% of lime. These results suggest, for specimens treated at 3% of lime, that the dissolved silica and alumina remain in the soil even during percolation. As mentioned by Boardman et al. (2001), this phenomenon could indicate the formation of cementitious compounds such as CSH or CAH. In addition, the fact that the aluminum concentrations are higher than the silicon concentrations could be a sign of a larger formation of CSH rather than CAH compounds. To conclude on the first assumption, the leached calcium does not seem to come from a decalcification of the cementitious compounds (no sign indicating that cementitious compounds could be soluble). On the contrary, there even seems to be signs, in the case of specimens treated at 3% of lime, of pozzolanic activity during leaching.

Assumption 2 can also be discarded for both specimens treated with 1% and 3% of lime. In fact, according to the method used to measure the carbonates content, the variations in carbonates contents observed during percolation (especially for specimens treated at 3% lime) are not significant. Hence, the carbonates contents of treated specimens during percolation can be considered as stable (Fig. 7 and 8).

Finally, the last assumption – "calcium coming from the lime remaining in the soil and that has not yet been used" – seems therefore the most plausible. Although the leached calcium concentrations of treated specimens are relatively high (Fig. 3), the ratios of the cumulative weight of leached calcium compared to the initial amount of calcium provided by lime are relatively low (Fig. 6): 7% and 12% for specimens treated respectively at 1% and 3% of lime. This result indicates that only a small amount of the lime added is susceptible to leach.

5 CONCLUSION

This study aimed at describing the physico-chemical behavior of a silty soil treated at two different lime contents under water leaching. The following conclusions can be made:

- A large part of the lime added for stabilization seems immediately precipitated as carbonates.
- There are signs of pozzolanic activity for both specimens treated at 1% and 3% of lime during curing.
- There is no sign of a decalcification of the cementitious by-products formed during leaching for both specimens treated at 1% and 3% of lime.
- The results indicate, in the case of specimens treated at 3% of lime, the existence of pozzolanic activity during leaching. This is not the case for specimens treated with 1% of lime.
- The amount of leached lime is relatively low compared to the amount of lime initially added (less than 12% for specimens treated at 3% of lime).

To conclude, although the severe leaching conditions used in this study, it seems that from a physico-chemical point of view, leaching does not reduce the efficiency of lime stabilization. Furthermore, the efficiency in long term stabilization is all the more since the lime content used is high.

REFERENCES

Bell, F.G. 1996. Lime stabilization of clay minerals and soil. *Engineering geology*, 42: 223–237.

Boardman, D.I., Glendinning, S., Rogers, C.D.F. 2001. Development of stabilisation and solidification in lime-clay mixes. *Géotechnique*, 50, No. 6: 533–543.

Cui, Y.J. 1993. *Etude du comportement d'un limon compacté non saturé et de sa modélisation dans un cadre élastoplastique.* PhD thesis, ENPC, Paris, 190 p.

De Bel, R., Bollens, Q., Duvigneaud, P.H., & Verbruge, J.C. 2005. Influence of curing time, percolation and temperature on the compressive strength of a loam treated with

lime. *International Symposium TREMTI, paper n° C022*: 10p.
Diamond, S. & Kinter, E.B. 1965. Mechanisms of soil-lime stabilization. *Highway Research Record*, 92: 83–102.
Eades, J.L. & Grim, R.E. 1962. Reaction of hydrated lime with pure clays minerals in soil stabilization. *Highway Research Board Bulletin*, 262: 51–63.
Loughnan, F.C. (ed.) 1969. *Chemical Weathering of silicate minerals*. New York: Elsevier.
Little, D.N. (ed.) 1995. *Stabilization of pavement subgrades and base courses with lime*. National Lime Association.
Malhotra, B. R. & Baskar, S.K. 1983. Leaching phenomenon in lime soil mix layers. *Highway Research Bulletin*, No. 22, New Delhi: 15–25.
McCallister, L.D. 1990. *The effects of leaching on lime-treated expansive clays*. PhD thesis, University of Texas, Arlington, 417 p.
Sridharan, A. & Sivapullaiah, P.V. 2005. Mini compaction test apparatus for fine grained soils. *Geotechnical Testing Journal*, 28: 240–246.

Advances in Transportation Geotechnics – Ellis, Yu, McDowell, Dawson & Thom (eds)
© 2008 Taylor & Francis Group, London, ISBN 978-0-415-47590-7

Development of composite improved soil with lightweight and deformability

K. Omine, H. Ochiai & N. Yasufuku
Department of Civil Engineering, Kyushu University, Fukuoka, Japan

ABSTRACT: Although ordinary improved soils such as cement-treated soil or air-formed lightweight soil has relatively high compression strength, the improved soil show high rigidity and brittle behavior. In this study, composite improved soil with lightweight and deformability was developed. This geomaterial is made by mixing granular material obtained from crushed lightweight soil and cement-stabilized soil in slurry as filling material. The cone penetration and triaxial compression tests were performed on the composite stabilized soil. Mechanical properties of composite improved soil were compared with those of air-formed lightweight soil or sand. From the test results, it was confirmed that the composite improved soil has well trafficability at the early curing period and is possessed of both properties of cohesive and frictional materials. It is also indicated that the composite improved soil is a geomaterial with lightweight, frictional property and deformability.

1 INTRODUCTION

Cement improvement method has been widely used for soft ground in Japan. One of the reasons is that compressive strength of the cement-treated soil can be controlled by amount of hardening agent. Recently, air-formed lightweight soil has also been developed and mechanical properties of the lightweight soil have been investigated by many researchers (Omine et al., 2001, Watanabe et al., 1999, Yasuhara, 2002). Air-formed lightweight soil is usually made by mixing sand or original soil with water, hardening agent and air bubbles. The main features of the soil are i) lightweight (wet density between 0.6 and 1.2 g/cm^3), ii) adjustable strength (unconfined compressive strength between 300 and 1000 kPa), iii) fluidity before hardening. However, those improved soils exhibit brittle behavior and low tensile strength. It is therefore that new geomaterial with toughness and frictional property is expected.

In this study, a composite improved soil with lightweight and deformability is developed. This geomaterial is made by mixing granular material obtained from crushing lightweight soil and cement-treated soil in slurry as filling material. The cone penetration and triaxial compression tests were performed on the composite improved soil. Mechanical properties of the composite improved soil are compared with those of air-formed lightweight soil and sand. Change of cone index for curing period, influence of confining pressure on the strength property and deformability are discussed based on the test results.

2 IDEA OF COMPOSITE IMPROVED SOIL

It is possible to control compression strength of ordinary improved soils such as cement-treated soil or air-formed lightweight soil by amount of hardening agent. However, it exhibits brittle behavior and low tensile strength. Figure 1 shows examples of stress-strain relationship on air-formed lightweight soil, granular material made by crushing the lightweight soil. The stress-strain relationship of Toyoura sand is also

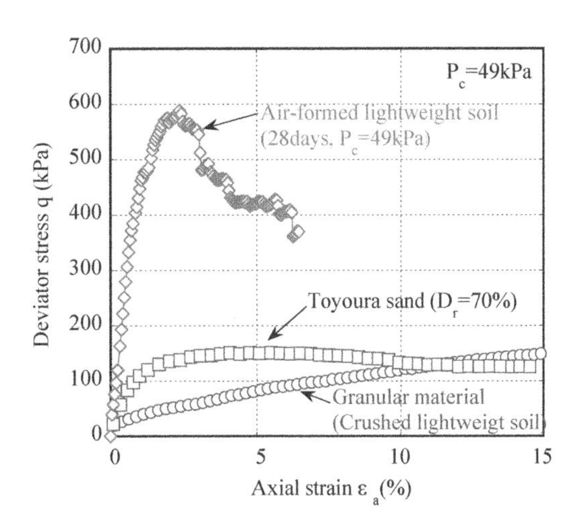

Figure 1. Typical stress-strain relationship on air-formed lightweight soil, granular material and sand.

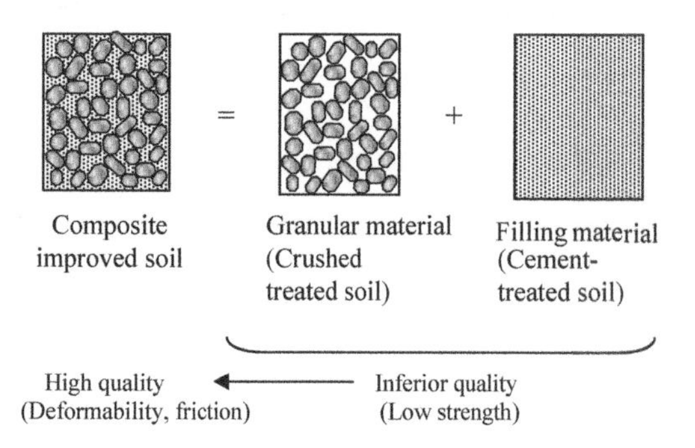

Figure 2. Image of composite improved soil.

shown for comparison. The lightweight soil reaches maximum stress at small strain of 1∼2%, and has high rigidity and low residual strength. This is a typical behavior of the improved soil. On the other hand, the granular material made by crushing the lightweight soil does not indicate such brittle behavior, but it becomes very compressible material with low strength. In the case of Toyoura sand with relative density of 70%, the difference of peak and residual strength is very small and the strain at failure is large. Fundamentally, it is expected that geomaterial has frictional property and high deformable performance.

Newly-developed geomaterial, namely composite improved soil, is made by mixing a granular material and cement-stabilized soil in slurry. In this study, crushed air-formed lightweight soil is used as the granular material. Cement-treated soil in slurry is made using soft soil such as surplus construction soil or dredged soil. Figure 2 shows an image of composite improved soil. Cement-treated soil as filling material exhibits low strength and shows brittle behavior. The granular material has very compressible and crushable property. Those are geomaterials with inferior quality. However, the composite improved soil comprising those materials is considered to possess high quality with frictional property and toughness.

3 TRAFFICABILITY OF COMPOSITE IMPROVED SOIL

3.1 *Samples and test method*

Granular material was prepared by crushing the air-formed light weight soil with curing period of 28 days. Wet density of the original lightweight soil is $0.60\,g/cm^3$. Grain size of the sample is 2.0∼9.5 mm and the shape is shown in Fig. 3. Conditions of samples and cone index test are shown in Table 1. Cement-treated soil with cement content of $100\,kg/m^3$ was used as filling material. Specimen of the composite improved soil was made by following procedure. The granular material was packed under relatively

Figure 3. Granular material made by crushing air-formed lightweight soil.

Table 1. Conditions of samples and cone index test.

Original sample		Ariake Clay
Hardening agent		Blast furnace cement (Type B)
Granular material	Type of material	Air-formed lightweight soil
	Cement content	$C_G = 200\,kg/m^3$
	Wet density	$0.60\,g/cm^3$
	Curing period	28 days
	Grain size	2.0∼9.5 mm
Filling material	Type of material	Cement-treated soil
	Cement content	$C_M = 100\,kg/m^3$
Size of speciment		Mold of 10 cm diameter
Curing conditions		Wet, 20°C, 1 & 3 days

dense condition into a mold with capacity of $1000\,cm^3$ and the cement-treated soil in slurry was poured into space between the grains. Wet density of the composite improved soil is approximately $1.0\,g/cm^3$. Cone index test on the composite improved soil was performed based on the Japanese standard (JGS A 1228).

3.2 *Cone index*

Relationship between cone index and curing period on various types of geomaterial is shown in Fig. 4. Cone index of the granular soil without the filling material is approximately 700 kPa. Cone index of the lightweight soil before hardening was also measured for comparison. It takes a curing period of 3 days at least for obtaining the lightweight soil with more than $q_c = 800\,kPa$. On the other hand, cone index of the composite improved soil exceeds 800 kPa at curing period of 1 day. Although the initial cone index of the composite improved soil without curing is 200 kPa and

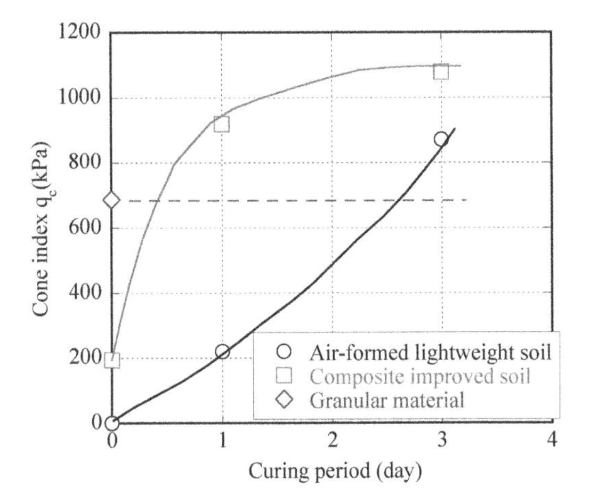

Figure 4. Relationship between cone index and curing period on various types of geomaterials.

less than that of the granular material, it may depend on the density or compaction condition.

Thus, the composite improved soil comprising granular material and cement-treated soil has well trafficability at the early curing period.

4 SHEAR PROPERTY OF COMPOSITE IMPROVED SOIL

4.1 *Samples and test method*

The same samples used in the cone index test were prepared in the triaxial compression test. The granular material was packed under relatively dense condition into a mold and the treated soil in slurry was poured into space between the grains. The test conditions are shown in Table 2. Specimen size is 50 mm in diameter and 100 mm in height.

Triaxial compression test on consolidation and unrained condition (CU) was performed on the composite improved soil. Pore water pressure was not measured because of unsaturated condition. Consolidation pressure is 49, 98 and 196 kPa, and strain rate is 0.5%/min.

4.2 *Stress-strain relationship*

Typical stress-strain relationship on various types of geomaterials is shown in Fig. 5. The test results of the granular soil and the filling material (cement-treated soil) were obtained from triaxial and unconfined compression tests, respectively. The cement-treated soil is cohesive material and reaches peak strength at small strain. The granular material exhibits very compressible property and does not have peak strength because

Table 2. Conditions of samples and triaxial test.

Original sample		Ariake Clay
Hardening agent		Blast furnace cement (Type B)
Granular material	Type of material	Air-formed lightweight soil
	Cement content	$C_G = 200\,kg/m^3$
	Wet density	$0.60\,g/cm^3$
	Curing period	28 days
	Grain size	2.0~9.5 mm
Filling material	Type of material	Cement-treated soil
	Cement content	$C_M = 100, 200\,kg/m^3$
Size of speciment		50 mm in diameter 100 mm in height
Curing conditions		Wet, 20°C, 7&28 days
Curing pressure		49, 98 & 147 kPa

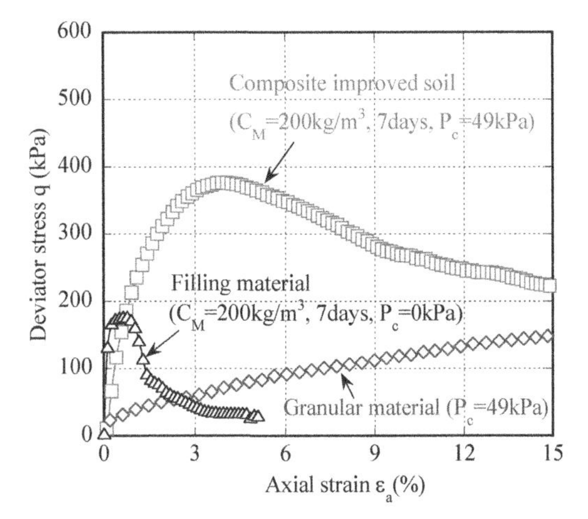

Figure 5. Typical stress-strain relationship on various types of geomateriasl.

of crushable material. On the other hand, the composite improved soil comprising the granular material and treated-soil has high strength by comparison with both materials.

Figure 6 (a) and (b) show the stress-strain relationship on the composite improved soil with cement content (filling material) of $C_M = 100\,kg/m^3$ for curing periods of 7 and 28 days, respectively. There is no clear peak strength and low residual strength in comparison with ordinary cement-treated soil. The maximum strength of the composite improved soil increases with increase in the consolidation pressure. This indicates that the composite improved soil has toughness and frictional property. The stress-strain relationship on the composite improved soil with cement content of $C_M = 200\,kg/m^3$ is shown in Fig. 7 (a) and (b). In this

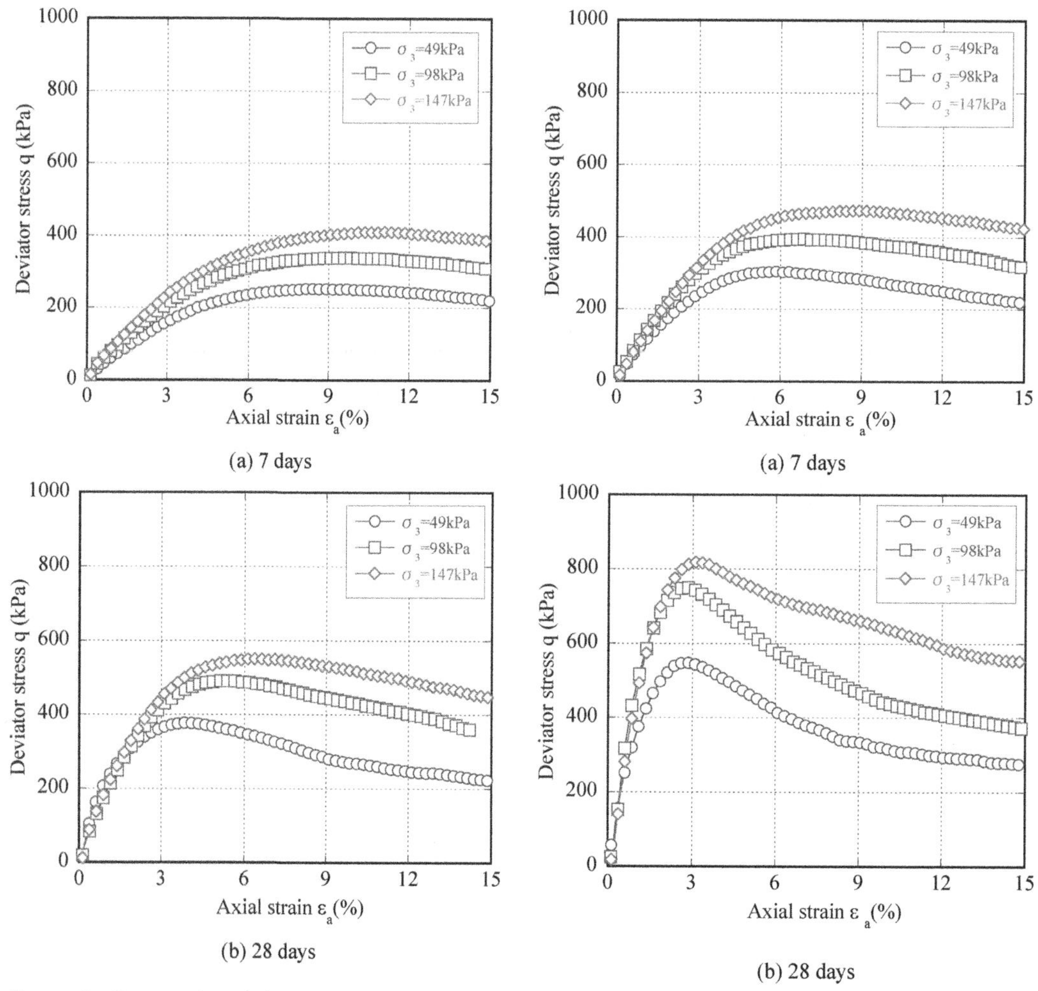

Figure 6. Stress-strain relationship on the composite improved soil with cement content (filling material) of $C_M = 100\,kg/m^3$.

Figure 7. Stress-strain relationship on the composite improved soil with cement content (filling material) of $C_M = 200\,kg/m^3$.

case, a similar result has been obtained. Thus, the composite improved soil is possessed of both properties of cohesive and frictional materials.

4.3 *Strength property*

Relationship between maximum deviator stress and consolidation pressure on each geomaterial is shown in Fig. 8 (a) and (b). Maximum deviator stress of the air-formed lightweight soil is almost constant for the consolidation pressure and this represents a property of cohesive material. Test result of Toyoura sand with relative density of 70% is also shown in this figure for comparison. The composite improved soil indicates significant increasing trend for increase of consolidation pressure. The value of maximum deviator stress at the same consolidation pressure becomes high as the cement content is large. These results suggest that the composite improved soil have not only cohesion but also frictional property such as sand.

In order to evaluate a deformability, the relationship between strain at failure and consolidation pressure is shown in Fig. 9 (a) and (b), where the strain at failure is defined as one at the maximum deviator stress. The strain at failure on the lightweight soil is less than that of 1% independent of consolidation pressure and the strain level at failure is very small. The strain at failure on Toyoura sand is in the range of 3~7%. Although the strain at failure on the composite improved soil depends on the consolidation pressure and cement content, the value is in the range of 3~10% and it is

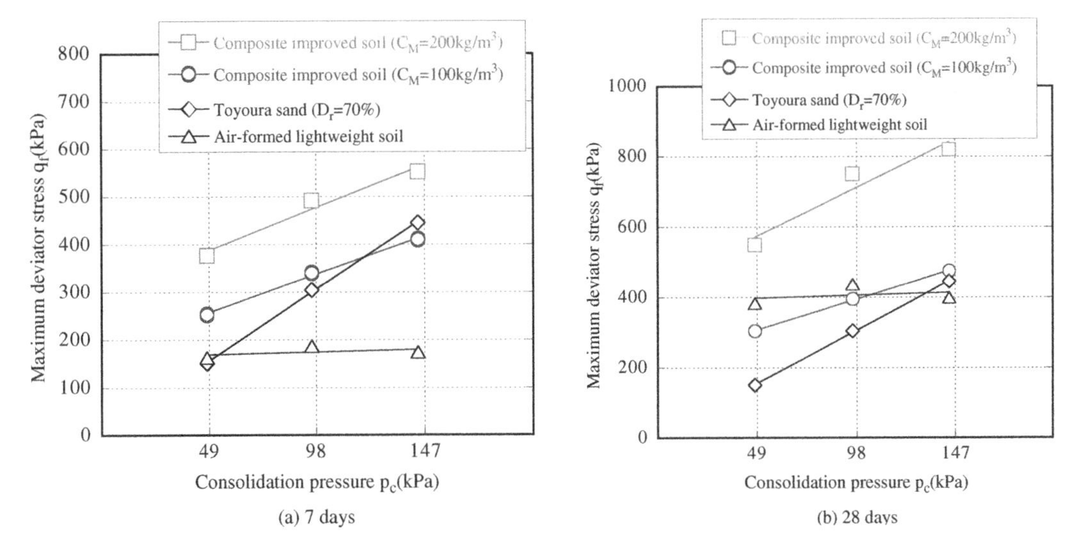

Figure 8. Relationship between maximum deviator stress and consolidation pressure on various type of geomaterials.

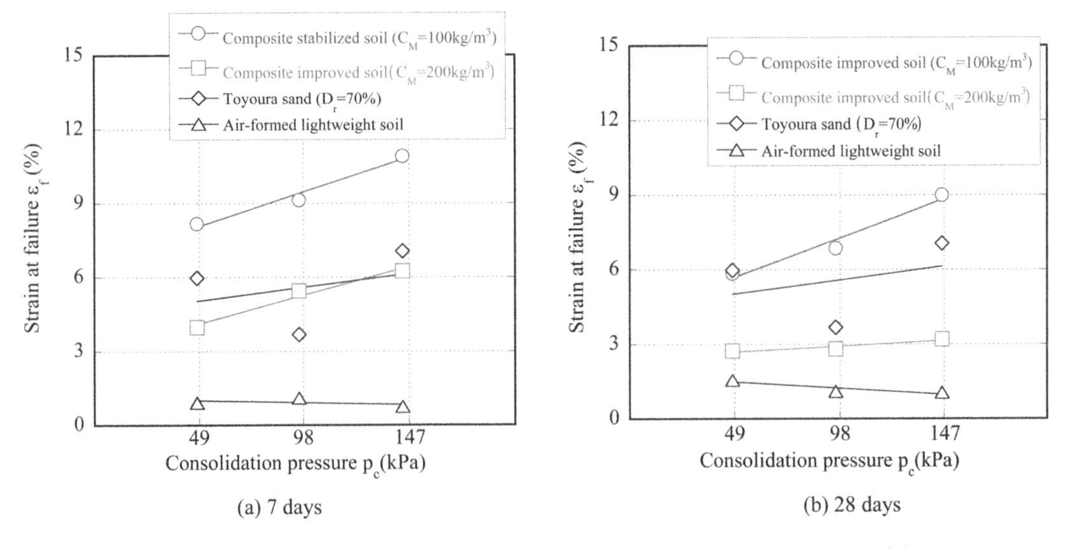

Figure 9. Relationship between strain at failure and consolidation pressure on various type of geomaterials.

the same level as that of Toyoura sand. These results suggest that the composite improved soil has high deformable performance.

The results of triaxial compression test are shown in Table 3. Wet density of the composite improved soil and the air-formed lightweight is approximately 1.0 and 0.6 g/cm^3, respectively, and these are lightweight as compared with general geomaterials. In viewpoint of strength constants, the internal friction angle on the composite improved soil is larger than 2° and the adequate cohesion exhibits. It is therefore said that the composite improved soil is a geomaterial with lightweight, frictional property and deformability. Furthermore, utilization of the composite improved soil as recycling material may be expected.

5 CONCLUSIONS

The main conclusions obtained from this study are as follows:

1) A new geomaterial comprising lightweight granular material and cement-treated soil as filling material, namely composite improved soil, has been developed.

Table 3. Results of triaxial compression test on the composite improved soil and lightweight soil.

(a) 7 days

	Cement content C_M (kg/m^3)	Wet density ρ_t (g/cm^3)	Consolidation pressure p_c (kPa)	Maximum deviator stress q_f (kPa)	Strain at failure ε_f (%)	Deformation modulus E_{50} (MPa)	Cohesion c_{cu} (kPa)	Internal friction angle ϕ_{cu} (°)
Composite improved soil	100	0.959	49	253.0	8.174	5.65	54.9	26.42
		0.964	98	340.2	9.133	7.40		
		0.968	147	410.2	10.899	8.10		
		1.082	49	377.0	3.972	24.86		
	200	1.053	98	492.0	5.438	18.64	89.5	28.13
		1.028	147	551.9	6.233	19.22		
Airformed lightweight soil		0.562	49	162.9	0.907	22.02	74.6	5.40
		0.569	98	186.5	1.092	31.12		
		0.564	147	173.3	0.742	27.33		

(b) 28 days

	Cement content C_M (kg/m^3)	Wet density ρ_t (g/cm^3)	Consolidation pressure p_c (kPa)	Maximum deviator stress q_f (kPa)	Strain at failure ε_f (%)	Deformation modulus E_{50} (MPa)	Cohesion c_{cu} (kPa)	Internal friction angle ϕ_{cu} (°)
Composite improved soil	100	0.956	49	304.1	5.842	9.74	66.6	27.74
		0.959	98	394.7	6.847	11.94		
		0.955	147	474.8	8.991	11.60		
	200	1.023	49	549.3	2.735	39.87	112.6	35.39
		1.063	98	749.9	2.809	55.15		
		1.066	147	819.1	3.197	46.05		
Air formed lightweight soil		0.569	49	382.5	1.551	31.88	168.7	8.14
		0.578	98	436.5	1.088	49.84		
		0.571	147	398.5	1.046	38.17		

2) The composite improved soil has well trafficability at the early curing period.
3) The composite improved soil is possessed of both properties of cohesive and frictional materials.
4) The composite improved soil is a geomaterial with lightweight, frictional property and deformability.

REFERENCES

1) Omine, K., Ochiai, H., Yasufuku, N., Kobayashi, T. & Takayama, E. 2006. Strength properties of air-formed lightweight soils with various types of short fibers, *Proc., of the 8th International Conference on Geosynthetics (8ICG)* vol.4, pp. 1659–1662.
2) Watanabe, T. & Kaino, T. 1999. "Shear strength of foam com-posite lightweight soil, *Proc., 11 Asian Regional Conf. SMFE*, pp. 47–48.
3) Yasuhara, K. 2002. "Recent Japanese experiences with light-weight geomaterials", *Proc., of International Work-shop on Lightweight Geo-Materials*, pp. 35–59.

Advances in Transportation Geotechnics – Ellis, Yu, McDowell, Dawson & Thom (eds)
© 2008 Taylor & Francis Group, London, ISBN 978-0-415-47590-7

Some results of laboratory tests on lime-stabilized soils for high speed railways in Italy

M. Schiavo & N. Moraci
University Mediterranea, Reggio Calabria, Italy

P. Simonini
University of Padova, Padova, Italy

ABSTRACT: Due to the limited availability of coarse grained soil the use of lime stabilization for the construction of embankments is being increased in Italy. This system of soil improvement is recently used also in the realization of new high speed railway embankments, which requires an excellent short and long-term mechanical performance of the materials supporting the rail lines. The paper presents and discusses some interesting results of laboratory and field geotechnical tests performed in one Italian site where large high speed railway embankments are under construction. In particular, the paper deals to the evaluation of the optimal quicklime content, taking into account basic soil characteristics, economical cost and long-term static and cyclic mechanical behavior.

1 INTRODUCTION

Due to the limited availability of coarse grained soil and to the increasing cost of material transportation, lime stabilization is being utilized for the first time in Italy to construct the embankments of the new high-speed railways and motorways. More particularly, this technique is used not only for the improvement of the natural subgrade but also for the embankment core (e.g. Roger *et al.*, 1996).

The Italian Railway Authority provides to the road/railway designers a typical cross section, that allows the lower part of the bank core to be built using fine grained soils mixed with quicklime (instead of granular material). The typical bank section is shown in Figure 1. Note that both subgrade and lower part of the bank are constructed using lime-stabilized soil.

The behaviour of artificially bonded soils, such as lime-stabilized soils, has been studied in the recent past by several researchers (e.g. Clought *et al.*, 1979, 1981; Leroueil & Vaughan, 1990; Chang & Woods, 1992; Gens & Nova, 1993; Consoli *et al.*, 1998, 2000; Balasubramaniam *et al.*, 2005; Liu & Shen, 2006; Åhnberg, 2007; Khattab *et al.*, 2007) who provided important explanations concerning the interaction among the artificially cemented soil particles.

Difference in the stabilized-material response from field to laboratory has been also considered and explained in terms of sample disturbance (Consoli *et al.*, 2000).

This study concerns the geotechnical characterization and performance of natural soils treated with quicklime, used to built up, between the second half of 2002 and 2004, the embankments of the new Milan-Bologna high speed railway line.

To this purpose, the natural soil mixed lime has been subjected to a comprehensive laboratory test programme, before and after the embankment construction.

The effect of lime stabilization on short and long term shear strength has been measured by using undrained triaxial, unconfined compression tests and direct shear tests carried out on compacted soil in dry and wet condition.

Within this programme, the Italian Authority required a new special test procedure for the samples taken from the bank core, to evaluate the behaviour of lime-stabilized material after several cycles of drying and wetting condition, used to increase artificially the curing time.

The paper presents and discusses some interesting results of the experimental tests carried out so far on field and laboratory samples, considering both short and long time behaviour and including the influence of wetting and drying cycles.

On the basis of these results, a new research has been recently undertaken to investigate the combined effect of cyclic loading and drying/wetting condition on the mechanical behaviour of lime stabilized-soil.

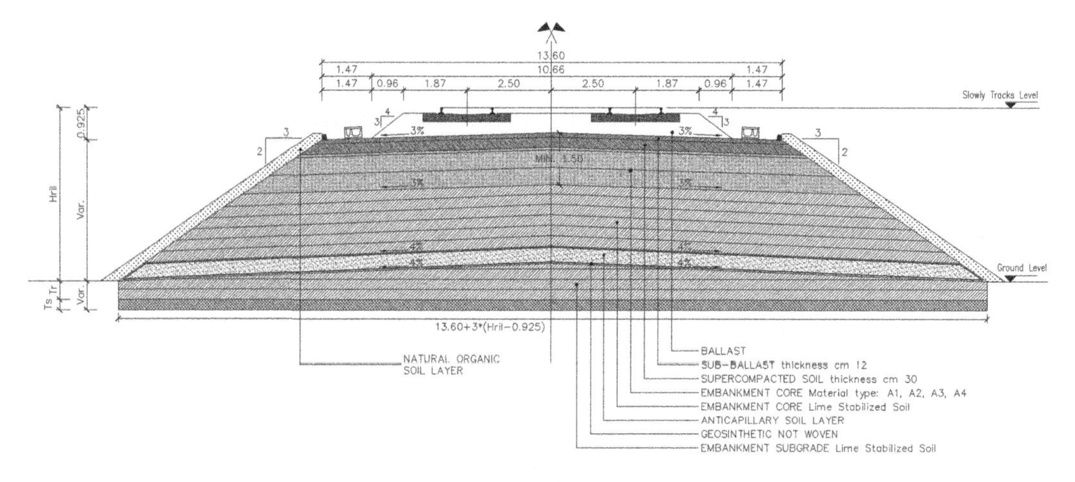

Figure 1. Typical embankment railways section with lime stabilized treated soil in subgrade and core.

2 GEOTECHNICAL CHARACTERIZATION

2.1 *Natural soil*

The natural soil utilized for the construction of embankment subgrade and core of the Milan-Bologna line comes from nearby excavations, necessary to realize shallow and deep foundations of the main railway viaducts and underpasses.

Referring to the classification AASHTO 145-82 (1986), the natural soil can be classified as A-7 (see plasticity chart, Fig. 2). According to the Italian Standards and Railways Authority Specifications, this soil can be used for treatment with lime.

2.2 *Compacted mixtures*

According to Eades and Grim (1966), preliminary tests were carried out to estimate the minimum quicklime content of the mixture necessary to stabilize the natural soil. For all the soils considered here it turned out a minimum value of 1.2%.

On the base of this result, three different lime percentages, that is 2.0%, 2.5% and 3.0%, were used in the preliminary laboratory tests.

Classical CBR, IPI, Unconfined Compression tests have been then performed on the different mixtures, to select a lime content of 2.5% as the best performing percentage to be used for the construction of the railway line embankments.

The Proctor/AASHTO Standard Compaction Test on this mixture provided a maximum dry unit weight equal to 16.16 kN/m^3 corresponding to an optimum moisture content of 19.15%.

On the compacted mixtures, several undrained triaxial compression tests and drained direct shear box tests were used to measure laboratory shear strength

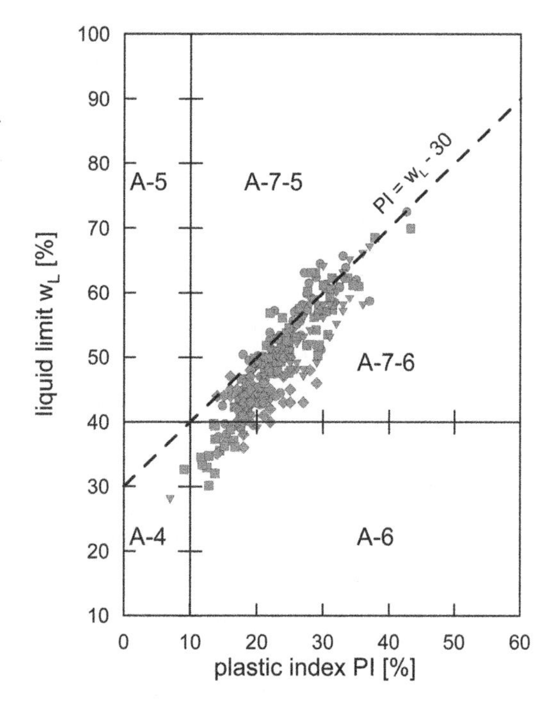

Figure 2. Plasticity Chart – ASSTHO Classification.

immediately after compaction. After a curing time of 28 days, the same tests were again carried out to evaluate time-effect of mixture strength.

Table 1 summarizes the main results of the laboratory tests, where s_u is the undrained shear strength, c is the cohesion and φ the friction angle.

Table 1 reports the minimum, mean and maximum value of all the three strength parameters. Note that short term resistance s_u was measured after a curing

Table 1. Shear Strength for natural and treated soil in short and long term behavior.

Soil Type		s_u [kPa]	c [kPa]	φ [°]
Natural	min	–	–	–
	mean	27.0	9.8	24
	max	–	–	–
Treated Soil	min	220.9	35.5	27
CaO 2.5%	mean	350.0	90.0	31
	max	437.2	127.1	35

Table 2. Shear resistance of an artificially aged samples.

Type of sample		s_u [kPa]	c [kPa]	φ [°]
NS	min	431.9	26.0	–
	mean	625.2	39.0	30
	max	784.6	52.0	–
WDS	min	555.4	41.0	–
	mean	741.8	61.0	34
	max	886.9	81.0	–

time of 7 days whereas long term strength parameters c and φ were estimated after 28 days.

Even if a large data scatter, probably due the natural soil variability and test procedure (i.e. direct shear) was measured, a very satisfactory increase of shear strength was obtained with the lime treatment.

On the basis of the above successfully test results, the Italian Railway Authority required further investigation concerning the effective conservation of such shear strength properties over long times.

To this end, a new testing procedure was proposed. An artificial aging of mixture was simulated by five cycles of wetting/drying, each of them taking 11 days (4 days wetting followed by 7 days drying). These samples are referred as to WDS samples in the next section to distinguish them from normal samples (NS). After the last WD cycle, 7 subsequent days of full saturation was imposed before starting with the new series of direct shear tests, which have been carried out on all the samples to evaluate the influence of multiple WD on s_u and c, φ parameters. The results are reported in Table 2.

It is worthwhile to note that the sequence of multiple WD effect a general increase of strength parameters, probably due to the acceleration of the pozzolanic reaction of the lime with the natural clayey soil. With respect to the data shown in Table 1, the cohesion increases significantly (around 33%), whereas friction seems to be substantially unaffected by the WD sequence. This seems to support the basic idea that lime treatment may form bonds in the fine soil particles, increasing the overall cohesion, but as a consequence of a higher degree of interlocking among larger aggregate particles.

To check the effectiveness of the lime treatment, the same testing procedure was applied to samples drown up from the embankment core. This will be discussed in the next section.

Table 3. Sample specification.

Code	Sampling Date	Tests		Testing Procedure	
E1	28.05.2004	TX.UU	DS	NFS	WDFS
E2	17.06.2004	TX.UU	DS	NFS	WDFS
E3	09.09.2004	TX.UU	DS	NFS	WDFS
E4	09.09.2004	TX.UU	DS	NFS	WDFS

Table 4. Water content and unit weight of the field stabilized soil before and after saturation.

Type of test	Condition		w [%]	γ [kN/m³]
TX.UU Undrained	NFS	min	7.1	15.7
		mean	16.8	18.1
		max	27.4	19.2
	WDFS	min	16.0	18.2
		mean	23.0	19.1
		max	31.5	20.1
DS Drained	NFS	min	16.3	16.9
		mean	21.6	17.8
		max	31.2	19.1
	WDFS	min	21.4	17.2
		mean	28.2	18.2
		max	33.1	19.1

3 FIELD SHEAR STRENGTH

Field samples (FS) of stabilized soil were taken from the shallowest layer of four embankments belong the Milano-Bologna high speed railway line.

The samples listed in table 3 were subjected to undrained triaxial compression using the same multiple WD procedure described above. These samples are referred as to WDFS (whereas NFS are normal field samples).

Table 4 reports the unit weight (γ) and water content (w) measured on the samples before compression and shear tests. Note that the moisture content in undrained tests corresponds to that of field condition. On the contrary, before direct shear tests, the stabilized soil was saturated, so that moisture content reached relatively high values.

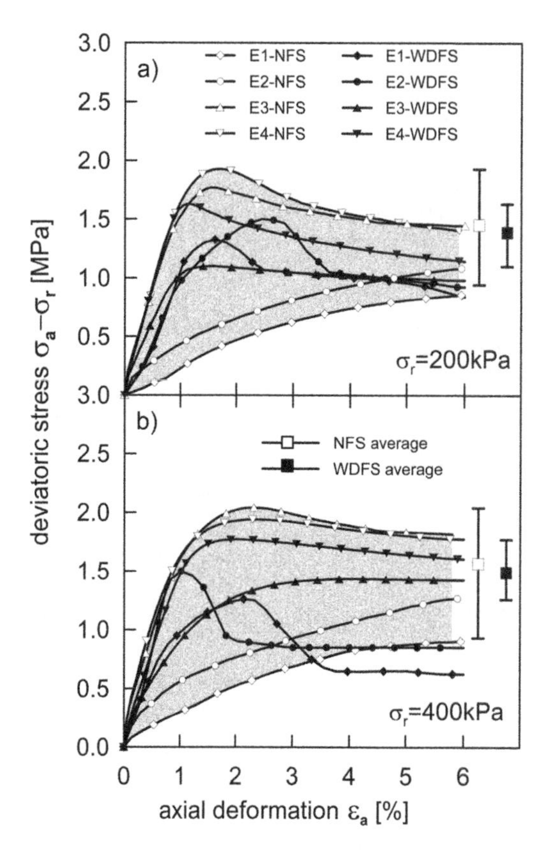

Figure 3. Stress-strain curves of undrained compression tests.

Noticeable variation of the water content can be observed. This may be due to the different weather and seasonal conditions occurring at the time of sampling (May / September 2004, see Tab. 3).

3.1 *Undrained stress-strain behaviour*

Stress-strain behaviour measured in undrained compression tests, for both normal and artificially wetted and dried samples, is shown in Figure 3, in terms of deviatoric stress ($\sigma_a - \sigma_r$) vs. axial strain ε_a. Figures 3a and 3b show the stress-strain response at confining pressure $\sigma_r = 200$ kPa and 400 kPa respectively (shaded areas envelope the response of NFS). The right part of the same Figure 3 and Table 5 reports the range of shear strength (maximum, mean and minimum value) for both NFS and WDFS.

Firstly, it is important to note the very different stress-strain behaviour measured on the Normal Samples and the related range of shear strength. In some cases, a brittle response with softening was observed.

In some others a continuously hardening behaviour was recorded at both levels of confining pressure. At $\varepsilon_a = 2\%$, the deviatoric stress presents a range of variation equal to around 1.6 MPa, which is an extremely large value for materials that, on the contrary, should provide similar strength. This may be related to the strong influence of seasonal condition at the time of sampling and testing operations. Samples E1 and E2 were taken in May and June, typically wet and rainy months in Italy, whereas samples E3 and E4 were collected in September, after a long dry summer period.

Table 5. Strength from triaxial compression tests.

Condition	σ_r [kPa]		$(\sigma_a - \sigma_r)$ [MPa]	ε_a [%]
NFS	200	min	0.94	9.94
		mean	1.45	–
		max	1.93	1.69
	400	min	0.93	7.50
		mean	1.57	–
		max	2.04	2.31
WDFS	200	min	1.10	1.51
		mean	1.39	–
		max	1.63	1.15
	400	min	1.26	2.14
		mean	1.49	–
		max	1.77	2.01

The multiple wetting/drying sequences reduce the range between maximum and minimum strength to 0.5 MPa, but the average strength remains substantially unaffected. The multiple wetting/drying sequences seem to minimize the seasonal effect during construction on both strength and stiffness, especially for the higher water content soils. In other words, the WD sequence reduces the heterogeneity in the response of the different samples toward similar stress-strain behaviour, typical of cemented soils. These results may support the idea of a general long term improvement of the mechanical behaviour of lime-stabilized materials.

3.2 *Drained stress-strain behaviour*

Mechanical behaviour was also measured in direct shear tests, for both normal and artificially wetted and dried samples. For both samples, Figures 4a,b,c report the shear stress vs. the horizontal displacement for three different stress levels namely $\sigma'_v = 50$, 100 and 200 kPa.

Again, a significant different stress-displacement behaviour can be observed for NF samples with a range between maximum and minimum strength decreasing as normal stress increases. In addition possible brittle response of treated material at $\sigma'_v = 50$ kPa vanishes completely at $\sigma'_v = 200$ kPa.

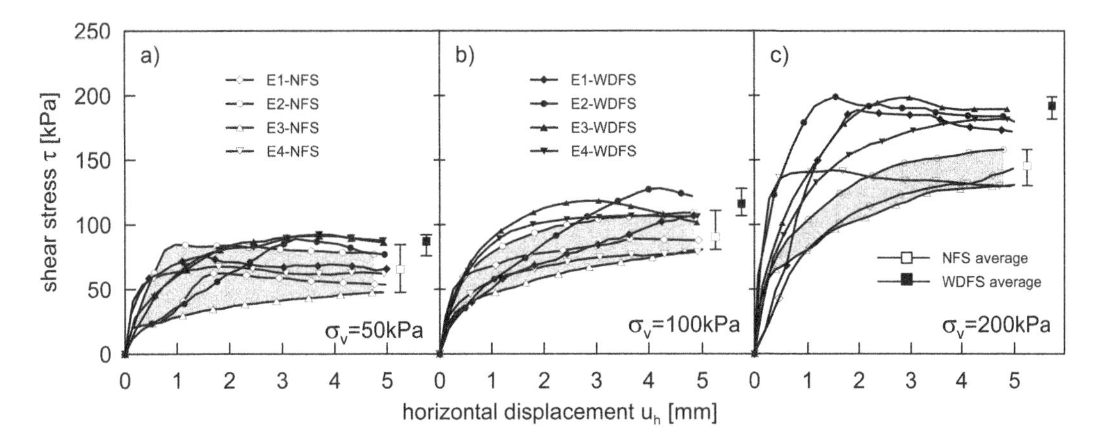

Figure 4. Stress-Strain curve for drained direct shear box.

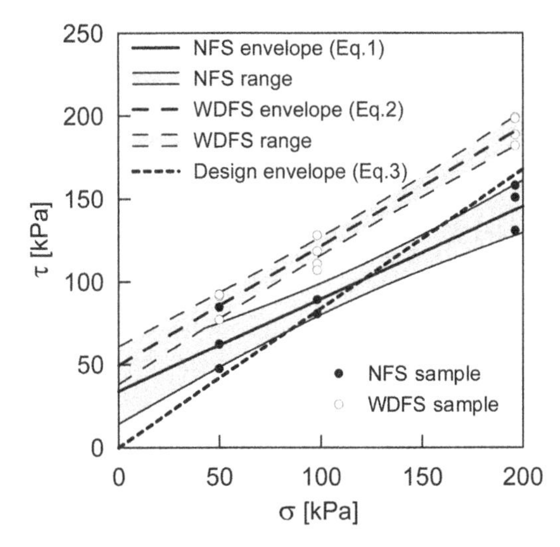

Figure 5. Failure envelope for drained direct shear box.

Wetting/Drying sequence seems to provide to the stabilized soils a general increase of both strength and stiffness for any of the investigated stress levels. However, this effect appears to be more evident at $\sigma'_v = 200$ kPa where, all the stress – displacement curves lie well above the envelope of NF samples (shaded area in Fig. 4c).

Figure 5 shows the interpretation of direct shear tests for both NFS and WDFS in terms of classical Coulomb shear strength criterion.

The linear fitting on the strength data provided the following equations for NF and WDF samples respectively:

$$\tau = 34 + \sigma \tan 29° \text{ (kPa) } (r^2=0.865) \quad (1)$$

$$\tau = 50 + \sigma \tan 36° \text{ (kPa) } (r^2=0.969) \quad (2)$$

The higher correlation coefficients r^2 obtained for the WD samples confirms, again, that the multiple wetting/drying sequences reduce the range of variability of the stabilized-soil strength.

More particularly, from Figure 5 it can be also better noticed the effect of multiple WD on the overall strength of treated materials, which may be characterized by a friction angle significantly higher with respect to that measured on natural soil (see Tab. 1).

Figure 5 sketches also the strength criterion (eq. 3) customarily required by the Italian Railway Authority for the design of embankments using stabilized- fine grained soil (i.e. under site condition, characterized by alternating wet and dry periods).

The experiments carried out so far demonstrate that eq. (3) is overconservative in the design of embankments stressing the foundation soil up to 120 kPa, that is, for embankment approximately 6.0–7.0 m high. In other words, the stabilized soil forming the bank is characterized by a effective higher strength with respect to that assumed commonly and precautionary in stability calculations.

4 CONCLUSIONS

From the experimental research carried out so far the following main conclusions can be drawn:

- The minimum quicklime content necessary to stabilize fine grained soils for the construction of Milan-Bologna high speed railway embankments turned out equal to 1.2%.
- On the basis of this result a percentage of 2.5% was used as better compromise between economical aspects and engineering performance of stabilized soil;
- Even if a large data scatter, probably due the natural soil variability and mechanical testing procedure

was measured, a very satisfactory increase of shear strength was obtained in the laboratory tests using lime treatment.

- To evaluate the conservation of such shear strength properties over long times, a new testing procedure was proposed by the Italian Railway Authority consisting of five cycles of wetting/drying, which should simulate the actual field climatic variation.
- Laboratory tests on wetted/dried samples showed a general increase of shear strength (i.e. cohesion rather than friction), probably due to the acceleration of the pozzolanic reaction of the lime with the natural clayey soil;
- Investigations on stabilized samples taken from the embankments of the Milan-Bologna line confirmed the effectiveness of quicklime stabilization even with a larger range of gained strength, probably due to seasonal factors, with respect to that measured under controlled laboratory conditions;
- The multiple wetting/drying sequence on field samples seems to minimize the seasonal effects during construction on both strength and stiffness, providing a relatively stable long-term resistance to the soil;
- Finally, the experimental investigation carried out so far demonstrates that stability calculation, customarily carried out without taking into account cohesion (see. Eq. (3) is probably much conservative in the design of embankments stressing the foundation soil up to 120 kPa, that is, for embankment approximately 6.0–7.0 m high.

To validate such results, not only in static but also in dynamic condition (typical situation for high speed railways) further researches are of course necessary, that should investigate also the effectiveness of the multiple wetting/drying technique to simulate field aging under climatic variations.

ACKNOWLEDGEMENTS

The author's wishes to extend special thanks to the Italian Railways Authority, to the Consortium of CEPAV Uno and to everybody of the technical staff involved in the construction of the new Italian high speed railways.

REFERENCES

Åhnberg, H. 2007. *On yield stresses and the influence of curing stresses on stress paths and strength measured in triaxial testing of stabilized soils*. Canadian Geotechnical Journal, 44, pp. 54–66.

AASTHO, 1986. *Standard Specification fro Transportations Materials and Methods of Sampling and Testing*, vol. I Specification, Washington.

Balasubramaniam, A.S., Buensuceso, B., Oh, E.Y.N., Bolton, M., Bergado, D.T. & Lorenzo, G. 2005. Strength degradation and critical state seeking behavior of lime treated soft clay. In Deep Mixing'05.*Proceedings of International Conference on Deep Mixing – Best Practice and Recent Advances, Stockholm, 23–25 May 2005*. Swedish Geotechnical Institute (SGI), Linköping, Sweden, pp. 35–40.

Consoli, N.C., Rotta, G.V. & Prietto, P.D.M. 2000. *Influence of curing under stress on the triaxial response of cemented soils*. Geotechnique, 50(1), pp. 99–105.

Eades, J.L. & Grim, R.E. 1966. A Quick Test to Determine Lime Requirements for Lime Stabilization. *Highway Research Record 139 Behavior Characteristics of Lime-Soil Mixture*, Highway Research Board, 1996.

Thompson, M.R. 1967. *Factors Influencing the Strength of Lime-Soil Mixtures*. Bulletin 492, Engineering Experiment Station, University of Illinois, Urbana, 1967.

Ingles, O.G. & Metcalf, J.B. 1972. *Soil stabilization, theory and practice*. Butterworth Pty. Limited, Sydney, Australia.

Khattab, S.A.A., Al-Mukhtar, M. & Fleureau, J.M. 2007. *Long-term Stability Characteristics of a Lime-Treated Plastic Soil*. Journal of materials in civil engineering, ASCE, April 2007.

Little, D.N. 1996. *Fundamentals of the stabilization of soil with lime*. National Lime Association Bulletin No. 332, Arlington, Va.

Liu, E.L. & Shen, Z.J. 2006. *Experimental Study on the Mechanical Behavior and Destructured Process of Artificially Structured Soils in Triaxial Compression*. In Ali Porbaha, Shui-Long Shen, Joseph Wartman, Jin-Chun Chai, Ground Modification and Seismic Mitigation; Geotechnical Special Publication no. 152, ASCE, 2006.

Rogers, C.D.F., Glendinning, S. & Dixon, N (1996). *Lime Stabilization*. Proceedings of the seminar held at Loughborough University Civil & Building Engineering Department on 25 September 1996, Thomas Telford 1996.

Sherwood, P.T. 1993. *Soil Stabilization with Cement and Lime. State-of-the-Art Review*, Transportation Research Laboratory, HMSO, London 1993.

Advances in Transportation Geotechnics – Ellis, Yu, McDowell, Dawson & Thom (eds)
 ISBN 978-0-415-47590-7

Stabilization of poor soil by paper sludge mixing

H. Shigematsu, Y. Demura & M. Otomo
Department of Civil Engineering, Ishikawa National College of Technology, Japan

Y. Fujiwara
NTT InfraNet, Japan

ABSTRACT: Paper sludge absorbs much water contained in poor soil and stabilizes the soil quickly. Also, even if paper sludge is mixed with soils, there is no problem with soil environment. In this study, in order to clarify the influence of the paper sludge on stabilization, the poor soils mixed with paper sludge are investigated by a series of laboratory tests. The following conclusions are obtained from the present study: (1) The tendency of strength reduction of the soil mixed with paper sludge due to adding water fluctuates with changes in the activity of the poor soil. (2) The tendency of strength recovery of the soil by paper sludge mixing is fluctuant with changes in the poor soil. (3) If paper sludge is mixed with different soil with different activity, the interaction between fibers and soil particles is different.

1 INTRODUCTION

A lot of paper sludge is generated during the process in which recycled papers are reproduced from used papers every year. Recently, paper sludge is often utilized effectively at construction works such as improvement of liquid mud (Kato et al., 2005), application to base course materials of road (Saito et al., 2002; Takeda et al., 2001) and treatment of discharged soils in shield tunneling constructions (Nakanan et al., 1994). Even if paper sludge is mixed with poor soil, it does not solidify chemically like cement or lime. However, it absorbs much water contained in the soil and stabilizes the poor soil ground quickly. In the case of earthworks at the poor ground (high water content state) such as diatomaceous earth, peat and volcanic cohesive soil etc., the trafficability goes down remarkably by machine-based construction, as shown in Fig 1. This paper examines the improvement effect of poor soil by paper sludge mixing.

In this study, in order to clarify the influence of the paper sludge on strength properties, poor soils mixed with paper sludge are investigated by different type of experiments. Details about the method of the experiments are explained at section 2.3. The tendency of strength reduction of the soils (low water content state) mixed with paper sludge in advance due to adding water is firstly discussed. Secondary, the tendency of strength recovery of the soils (high water content state) by paper sludge mixing is discussed. Finally, the interaction between soil particles and paper sludge is also investigated based on the scanning electron microscope (SEM) of the specimens.

Figure 1. Decline of trafficability in poor soil ground (Shigematsu et al., 2001).

2 MATERIALS AND LABORATORY TESTS

2.1 *Paper sludge*

The types of paper sludge used for experiments are PMF (Paper Micro Fiber) and PSA (Paper Sludge

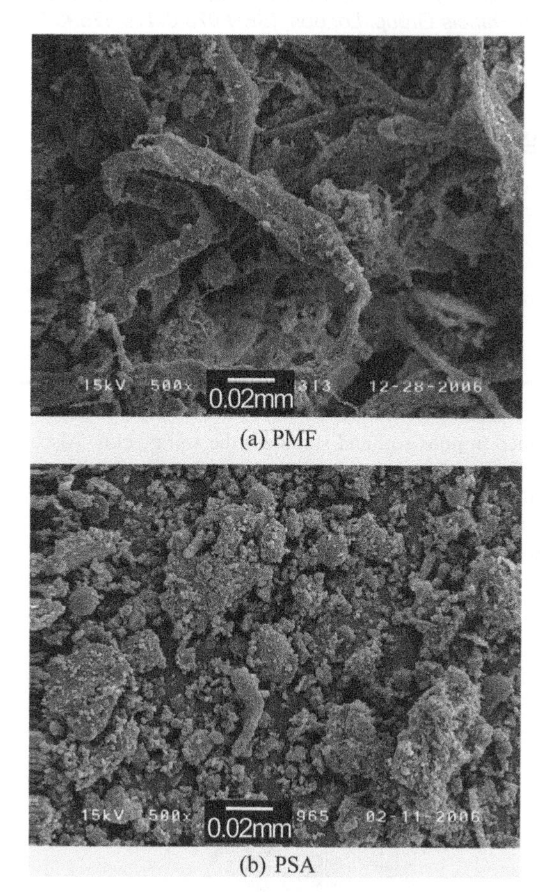

(a) PMF

(b) PSA

Figure 2. SEM micrographs of paper sludge.

Ash). PMF is the material in which some impurities in the paper sludge were removed, and passed through the 2 mm sieve after dried. PSA is the ash of the burnt paper sludge. The SEM micrographs of PMF and PSA are shown in Fig. 2. PMF contains many fibers damaged and PSA almost composes of alumina (Al_2O_3) and quartz (SiO_2). Fundamental characteristics of PMF and PAS are shown in Table 1. The absolute density of PMF and PSA were determined after the testing method for density of soil particles. The absolute density of PSA (=2.556 g/cm^3) is higher than that of PMF (=2.045 g/cm^3) and shows the almost same value as the density of the soil particles (=2.65 g/cm^3). The apparent density of PMF and PSA were determined by the testing method for minimum and maximum densities of sands. Both of them are much lower than density of water (=1.0 g/cm^3). From pH test results of paper sludge, PMF (pH = 8.2) showed low alkalinity and PSA (pH = 12.8) showed high alkalinity. Also, leaching test results of PMF and PSA show that, even if paper sludge is mixed with soils, there is no problem with soil environment.

Table 1. Fundamental characteristics of paper sludge.

	PMF	PSA
Absolute density (g/cm^3)	2.045	2.556
Apparent density (g/cm^3)	0.363	0.696
pH	8.2	12.8

Table 2. Physical and chemical properties of the soil samples.

	Unoke	Kitayokone	Kaolin
Density of soil particles (g/cm^3)	2.657	2.671	2.746
Liquid limit w_L (%)	63.8	65.5	67.5
Plastic limit w_P (%)	27.7	29.9	29.1
Plasticity index I_P	36.1	35.5	38.4
Sand fraction (%)	46.3	12.5	0*
Silt fraction (%)	38.5	49.4	1.4*
Clay fraction (%)	15.2	38.1	98.6*
Activity A	3.41	1.23	0.39
pH	5.7	6.1	5.2
Ignition loss L_i (%)	5.0	6.6	4.2

* Based on laser diffraction method.

2.2 *Poor soil samples*

The poor soil samples used for experiments are two kinds of natural sedimentary soils and powdered Kaolin. The sedimentary soils were obtained at Unoke area of Kahoku city in Ishikawa Prefecture, Japan and at Kitayokone area of Tsubata town in Ishikawa Prefecture, Japan. Based on the name of these sites, the soil samples are called Unoke soil and Kitayokone soil respectively from now on. The retrieved Unoke and Kitayokone soils were dried until water content becomes to be 15% or less, and then passed once through the 4.75 mm sieve. Physical and chemical properties of the soil samples are shown in Table 2. Although the plasticity indices, I_P of Unoke soil, Kitayokone soil and Kaolin show similar values, grain size distributions of those soils are different each other. Therefore, Unoke soil, Kitayokone soil and Kaolin can be classified into activated clay ($A > 1.25$), ordinary clay ($0.75 < A < 1.25$) and non-activated clay ($A < 0.75$) respectively by activity A, as shown in Fig. 3.

2.3 *Laboratory tests*

In order to understand the tendency of strength reduction of the soils (low water content state) mixed with paper sludge in advance due to adding water and the tendency of strength recovery of the soils (high water

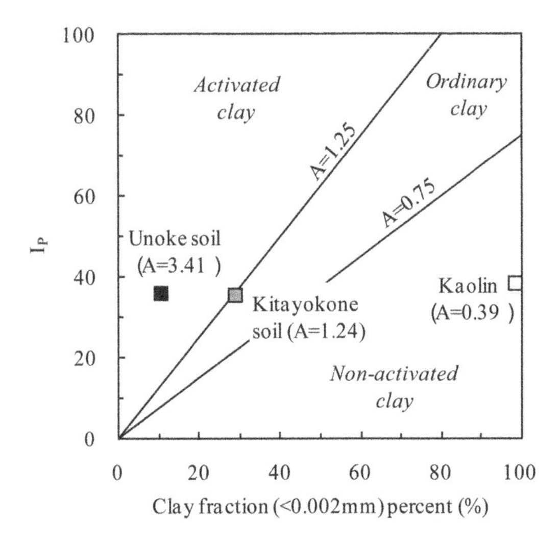

Figure 3. Relationships between plasticity index and clay fraction (<0.002 mm).

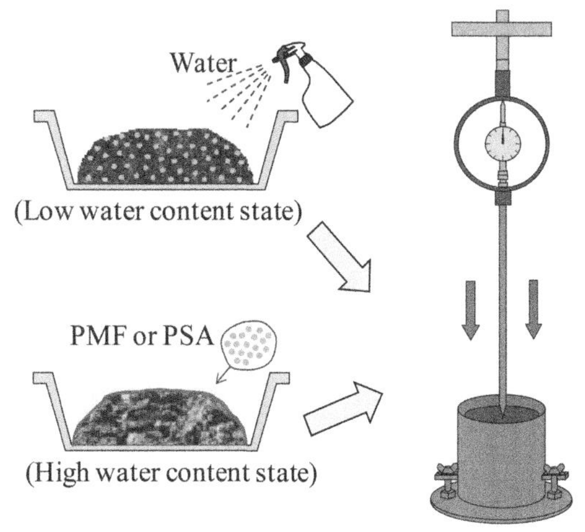

Figure 4. Schematic illustration of cone index tests.

Table 3. Mixture rate of Paper sludge.

	Dried mass per 100 g	
Mixture rate	Soil sample	Paper sludge
0%	100 g	0 g
20%	80 g	20 g
40%	60 g	40 g

content state) by paper sludge mixing, two types of experiments were carried out.

The specimens used for the first experiments were made by the testing method for soil compaction using a rammer after mixed with paper sludge into the soil samples at different mixture rate (20% and 40%). The mixture rate was decided based on the dried mass of the whole sample with mixed paper sludge and soil, as shown in Table 3. Then cone index tests on the specimens were conducted during the process in which water is added gradually, as shown in Fig. 4. Until the cone index, q_c becomes to be 0.1 MN/m^2 or less, the tests were continued repeatedly.

The specimens used for the second experiments were made by the same method after initial water content of the soil samples was controlled. The initial water content w_0 of the soil samples was decided based on the liquid limit w_L. The initial water content w_0 of the specimens was adjusted such as $w_0 = w_L$, $w_0 = 0.8\,w_L$ and $w_0 = 0.6\,w_L$. Then cone index tests on the specimens were conducted during the process in which paper sludge is mixed gradually, as shown in Fig. 4. Until q_c becomes to be 1.2 MN/m^2 or more, the tests were continued repeatedly.

3 EXPERIMENTAL RESULTS

3.1 *Tendency of strength reduction of poor soils mixed with paper sludge by adding water*

Cone index test results are shown in Figs. 5–7. These figures show the relationships between the cone index, q_c and the water content at different mixture rate of paper sludge. The values of 0% shown in those figures mean the experimental result of the soil sample without paper sludge. The q_c-water content curves can be approximated by the exponential curves. From q_c-water content relations of Unoke soil (activated clay) mixed with paper sludge shown in Fig. 5, the strength of the soil mixed with PMF can be maintained at higher water content state than the soil mixed with PSA. Moreover this tendency can be seen clearly at higher mixture rate of paper sludge. However, the tendency of strength reduction of other soil samples (Kitayokone soil and Kaolin) mixed with paper sludge is widely different. For Kitayokone soil (ordinary clay), if the mixture rate of paper sludge is 40%, the strength of the soil mixed with PMF can be maintained at higher water content state than the soil mixed with PSA, as shown in Fig. 6(b). On the other hand, if the mixture rate of paper sludge is 20%, the difference of the water content is not large as shown in Fig. 6(a). For Kaolin (non-activated clay), regardless of the mixture rate of paper sludge, the q_c-water content relation of the soil mixed with PMF is similar to the relation of the soil mixed with PSA, as shown in Fig. 7.

From the above findings, it is concluded that the tendency of strength reduction of the soil mixed with paper sludge due to adding water fluctuates with the difference of activity of the soils.

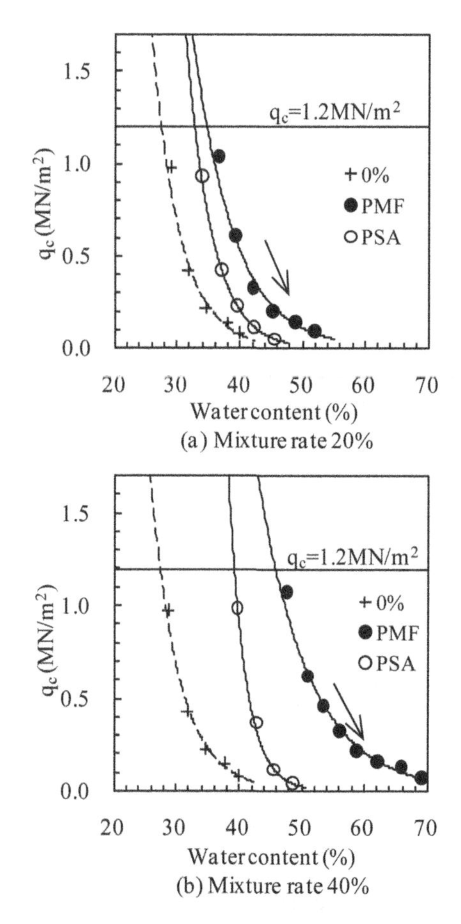

Figure 5. Cone index test results of Unoke soil.

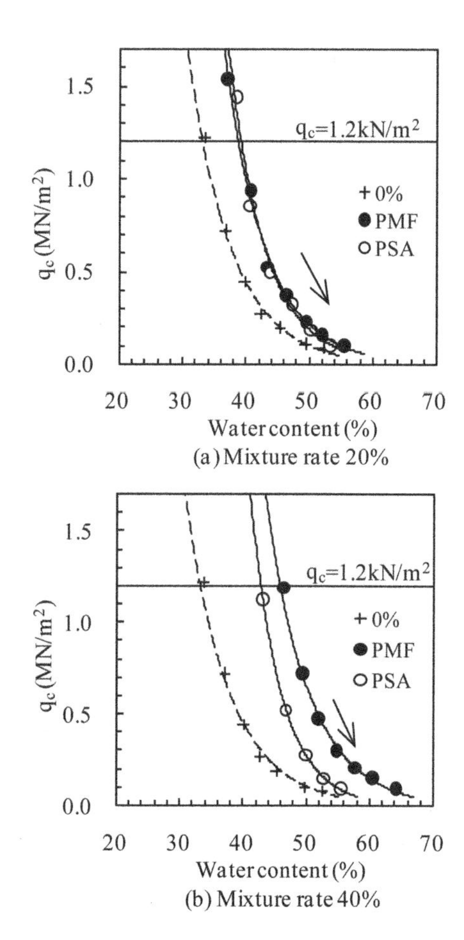

Figure 6. Cone index test results of Kitayokone soil.

3.2 *Tendency of strength recovery of poor soils by paper sludge mixing*

Cone index test results are shown in Figs. 8-10. These figures show the relationships between q_c and the mixture rate at different initial water content w_0. The q_c-mixture rate curves as well as Figs. 5-7 can be approximated by the exponential curves. From experimental results shown in Fig. 8, the strength of Unoke soil (activated clay) mixed with PMF can be recovered at lower mixture rate than the soil mixed with PSA. However, the tendency of the strength increase for the other soil samples (Kitayokone soil and Kaolin) by paper sludge mixing is different. For Kitayokone soil (ordinary clay), regardless of the initial water content w_0, the difference between q_c-mixture rate relation of the soil mixed with PMF and the relation of the soil mixed with PSA is not observed, as shown in Fig. 9. On the other hand, for Kaolin (non-activated clay), regardless of the initial water content w_0, the strength increase of the soil mixed with PSA can also be recovered at lower mixture rate than the soil mixed with PMF, as shown in Fig. 10. In other words, the results of kaolin mixed with paper sludge are completely different from the results of Unoke soil mixed with paper sludge.

From the above discussions, it was found that the tendency of strength recovery of the soil by paper sludge mixing is fluctuant largely with the activity of the soils.

3.3 *SEM micrographs of specimens*

In order to understand the interaction between soil particles and paper sludge, the specimens used for the experiments were observed by the scanning electron microscope (SEM). SEM micrographs of the specimens (Unoke soil and Kaolin) mixed with PMF (40%) are shown in Fig. 11. For Unoke soil mixed with PMF, some fibers twining around soil particles like a rhizome can be seen. For Kaolin mixed with PMF, many microscopic soil particles adhering to the surface of

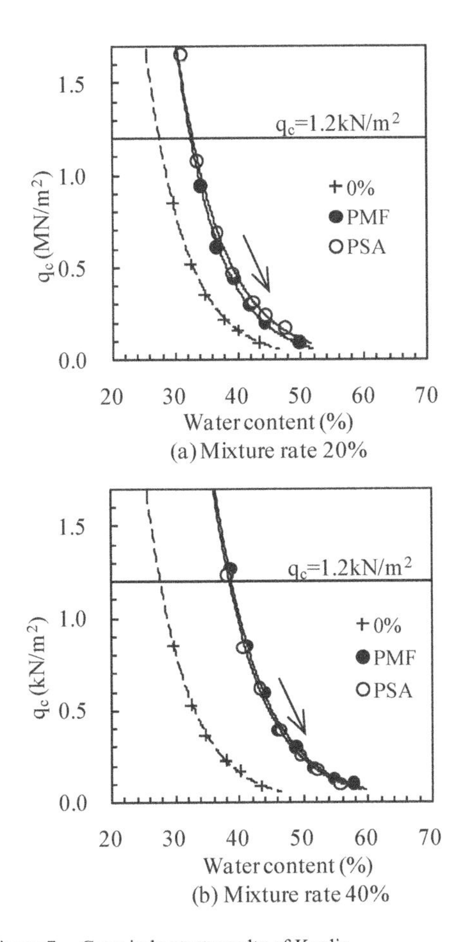

Figure 7. Cone index test results of Kaolin.

fibers can be considered. SEM micrographs of the specimens (Unoke soil and Kaolin) mixed with PSA (40%) are also shown in Fig. 12. Although peds of PSA and soil particles can be seen in Unoke soil, those cannot be observed in Kaolin.

From above the findings, if PMF or PSA is mixed into the soils with different activity, the interaction between fibers and soil particles is largely different.

4 CONCLUSIONS

The main conclusions obtained in this study are summarized as follows.

(1) The tendency of strength reduction of the soil mixed with paper sludge due to adding water fluctuates largely with the difference of activity of the soils. For Unoke soil classified into activated clay, the strength of the soil mixed with PMF can be maintained at higher water content state than the soil mixed with PSA. However, for kaolin classified into non-activated clay, regardless of the mixture rate of paper sludge, the q_c-water content relation of the soil mixed with PMF is similar to the relation of the soil mixed with PSA.

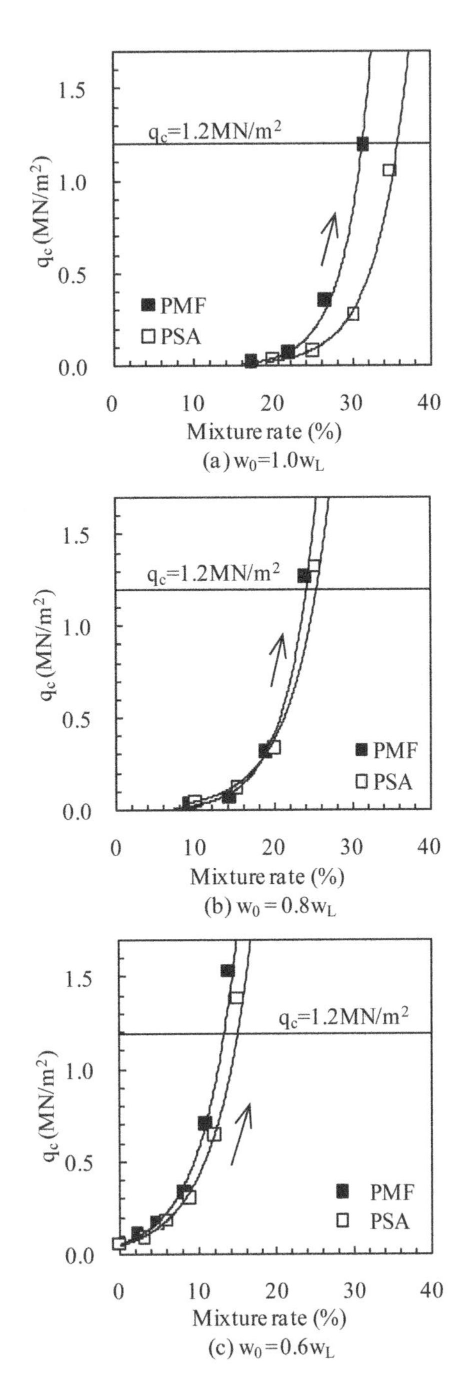

Figure 8. Cone index test results of Unoke soil.

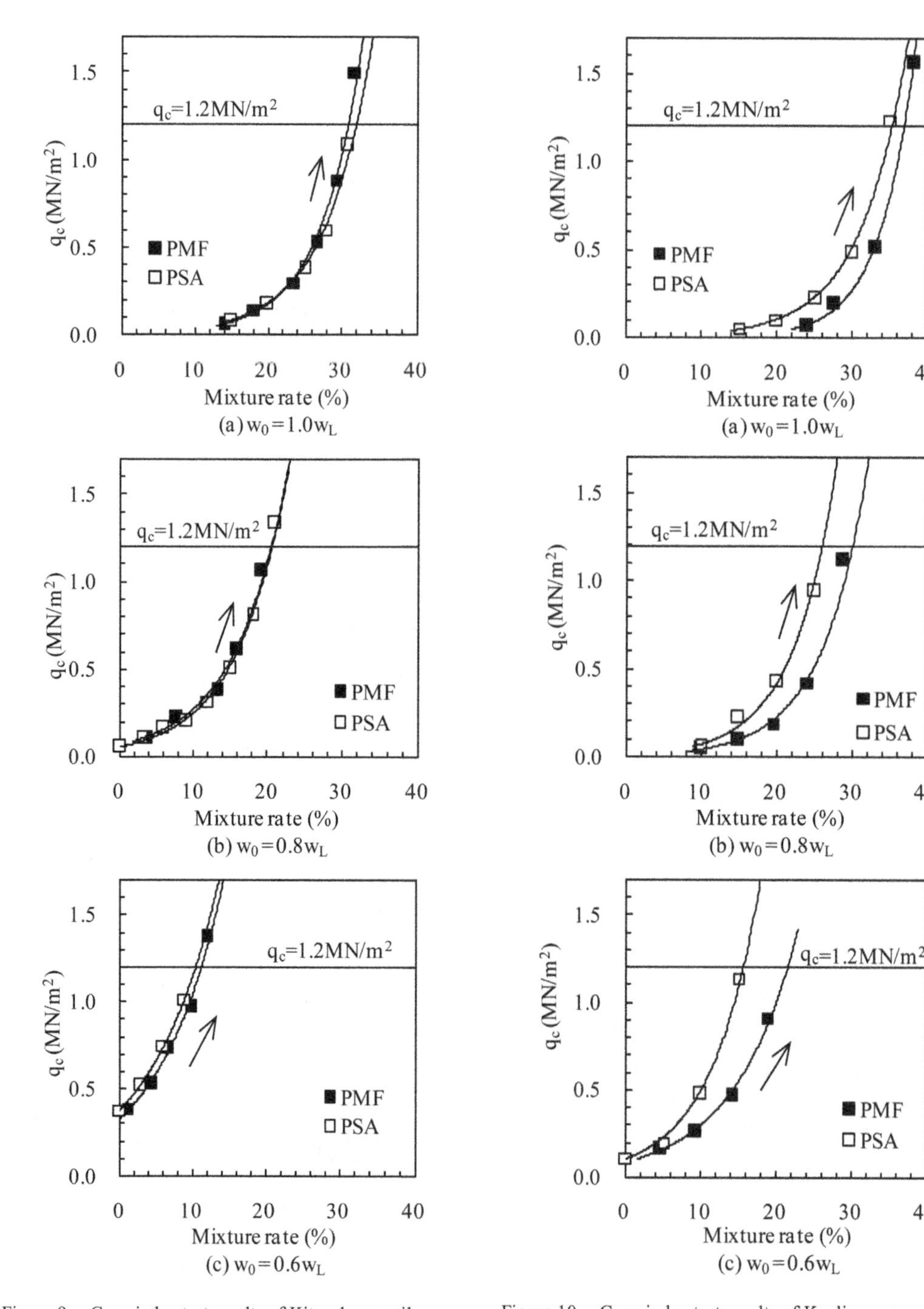

Figure 9. Cone index test results of Kitayokone soil.

Figure 10. Cone index test results of Kaolin.

(2) The tendency of strength recovery of the soil by paper sludge mixing is fluctuant with the difference of activity of the soils. The strength of Unoke soil (activated clay) mixed with PMF can be recovered at lower mixture rate than the soil mixed with PSA. Regardless of initial water content, the strength increase of Kaolin (non activated clay) mixed with PSA can also be recovered at lower mixture rate than the soil mixed with PMF.

(a) Unoke soil mixed with PMF (mixture rate 40%).

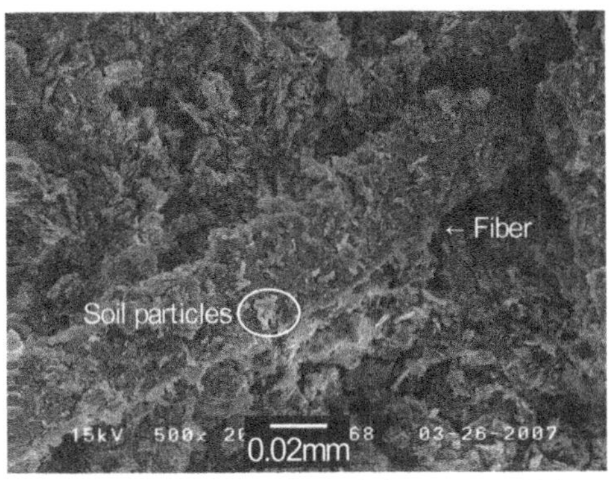

(b) Kaolin mixed with PMF (mixture rate 40%).

Figure 11. SEM micrographs of specimens.

(a) Unoke soil mixed with PSA (mixture rate 40%).

(b) Kaolin mixed with PSA (mixture rate 40%).

Figure 12. SEM micrographs of specimens.

(3) If PMF or PSA is mixed into the soils with different activity, the interaction between fibers and soil particles is different greatly.

REFERENCES

Kato, Y., Imai, G., Ohmukai, N., Mochizuki, Y., Saito, E. and Yoshino, H. 2005. Studies on improvement of liquid mud by use of paper sludge ash, *Proceedings of 40th Japan National Conference on Geotechnical Engineering*: 677–678 (in Japanese).

Mitchell, J.K, (1993): Fundamentals of Soil Behavior, Second edition, 185–186.

Nakanan, I., Otani, K., Ishimoto, K. and Shimizu, E. 1994. An experimental study on earth pressure balanced shield using the paper, *Proceedings of 49th annual conference of the Japan society of civil engineers*: 58–59 (in Japanese).

Saito, E., Mochizuki, Y., Yoshino, H., Tanaka, T., Tyazono, Y. and Watanabe, T. 2002. Effective utility for base course using paper sludge ash, *Proceedings of 57th annual conference of the Japan society of civil engineers*: 1587–1588 (in Japanese).

Shigematsu, H., Yashima, A., Nishio, M., Saka, Y. and Hatanaka, S. 2001. Geotechnical properties of diatomaceous earth in northern Gifu Prefecture and cut slope stability, *Journal of geotechnical engineering, JSCE*: 139–154 (in Japanese).

Takeda S., Mochizuki, Y., Saito, E. and Ogata, T. 2001. Study of Base course using Paper Sludge Ash, *Proceedings of 56th annual conference of the Japan society of civil engineers*: 568–569 (in Japanese).

10 Characterisation and recycling of geomaterials

Advances in Transportation Geotechnics – Ellis, Yu, McDowell, Dawson & Thom (eds)
© 2008 Taylor & Francis Group, London, ISBN 978-0-415-47590-7

Beneficial use of brick rubble as pavement sub-base material

T. Aatheesan, A. Arulrajah & J. Wilson
Swinburne University of Technology, Melbourne, Victoria, Australia

M.W. Bo
DST Consulting Engineers, Thunder Bay, Ontario, Canada

ABSTRACT: Brick rubble, crushed concrete and crushed rock are commonly obtained from construction and demolition activities. Construction and demolition materials occupy a major proportion in landfills as waste materials. Recycled brick rubble, crushed concrete and crushed rock can be used as viable substitutes for natural aggregates in civil construction applications such as pavement sub-base materials. This paper reports on the California Bearing Ratio (CBR) laboratory tests for various brick rubble blends with crushed concrete and crushed rock. The CBR values obtained were compared with existing local road authority specifications to investigate the potential use of brick rubble with crushed concrete and crushed rock as pavement sub-base material. The aggregates were collected from a construction and demolition material recycling site in Victoria, Australia.

1 INTRODUCTION

Recycling and reuse of waste materials is a topic of global concern and of great international interest. The urgent need for recycling is driven mainly by environmental considerations due to the increasing scarcity of natural resources as well as the increasing cost of land fill in most countries. Construction and demolition (C&D) materials are generated by regeneration of infrastructures and demolition activities and accounts for the major proportion of the waste materials present in landfills. Recycled brick, concrete and glass for instance are viable substitute materials for natural construction materials in engineering applications such as pavement sub-base material. Certain countries have been using recycled C&D materials in civil engineering applications but there is still scope for wider engineering applications of such recycled materials.

This paper primarily focuses on brick rubble blends with crushed concrete and crushed rock. This paper also investigates the application and usage of these blends as pavement sub-base material based on laboratory tests carried out in Victoria, Australia.

The engineering properties of brick rubble blended with crushed concrete and crushed rock was investigated in this study. A suite of laboratory tests were conducted on blend mixes of 10%, 15%, 20%, 25% and 30% of brick rubble with crushed concrete and crushed rock. The suite of laboratory tests undertaken included particle size distribution, modified compaction and California Bearing Ratio (CBR). The California Bearing Ratio (CBR) is a critical parameter in pavement design and construction. CBR values from the laboratory tests were compared with the existing local authority specifications for flexible pavement base and sub-base.

2 SAMPLING AND LABORATORY TESTING

2.1 *Site description*

Samples of crushed concrete class CC3, crushed rock class CR3 (CC3 and CR3 are specifications applicable for pavement sub-base applications in Victoria, Australia) and brick rubble were collected from the Alex Fraser Recycling site at Laverton North, Victoria which is located approximately 20 km to the west of Melbourne, Australia. The Alex Fraser Group produces recycled materials such as crushed concrete, crushed rock, pavement base, crushed glass in different classes.

2.2 *Laboratory testing*

The laboratory tests on blend mixtures were conducted at Swinburne University of Technology, Melbourne.

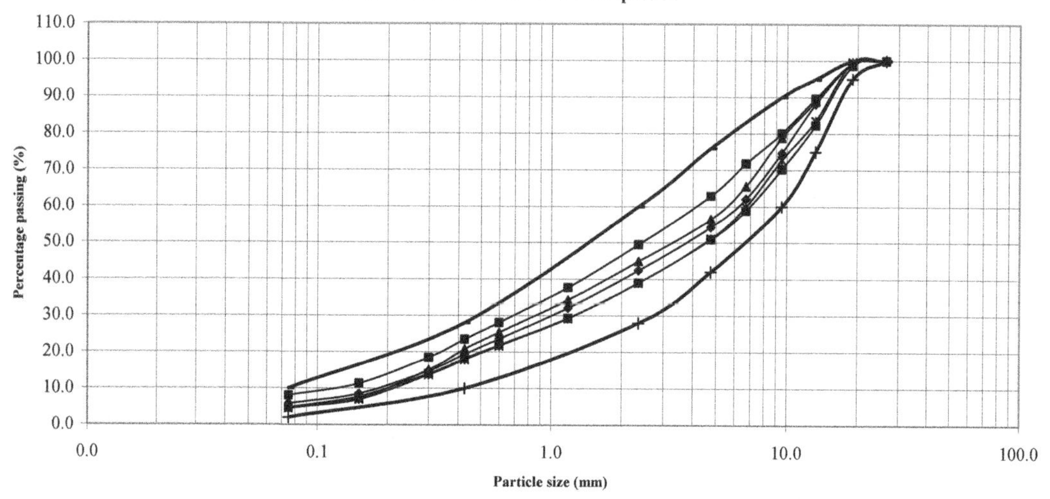

Figure 1. Particle size distribution for brick rubble blended with crushed concrete.

The blended mixtures comprised of brick rubble with crushed concrete (CC3) and brick rubble with crushed rock (CR3). A suite of blended mixtures of 10%, 15%, 20%, 25% and 30% content of brick rubble were tested for particle size distribution, modified compaction and CBR. The blend mixtures were prepared by hand mixing to required percentage by weight and all samples had a maximum aggregate size of 20 mm. Brick rubble from the recycling site typically comprises of 70% brick rubble and 30% other materials such as asphalt, concrete and rock.

3 BRICK RUBBLE BLENDS WITH CRUSHED CONCRETE

3.1 *Particle size distribution*

The particle size distribution of each blend mixture was obtained by wet sieving using 5 different blend mixtures prepared in accordance with Australian Standards (AS 1141, 1996).

The grading curves for each blend of brick rubble with crushed concrete are illustrated in Figure 1.

3.2 *Modified compaction*

The maximum dry density and optimum moisture content of blended samples were obtained from tests using modified compaction in accordance with Australian Standards (AS 1289, 2003).

Table 1. Summary of optimum moisture content and maximum dry density for brick rubble blended with crushed concrete.

Sample description	Brick content (%)	Optimum moisture content (%)	Maximum dry density Mg/m^3
70CC3*	30	12.5	1.95
75CC3	25	12.0	1.94
80CC3	20	11.7	1.95
85CC3	15	11.7	1.99
90CC3	10	12.0	1.95

* For example, 70CC3 refers to 30% brick rubble content blended with 70% crushed concrete by weight.

The test results of these parameters for brick rubble blended with crushed concrete are summarised in Table 1.

3.3 *California Bearing Ratio (CBR)*

California Bearing Ratio of each blends were performed under four day soaked conditions following compaction at their optimum moisture contents in accordance with Australian Standards (AS 1289, 1998).

The CBR values for brick rubble blended with crushed concrete are summarised in Figure 2.

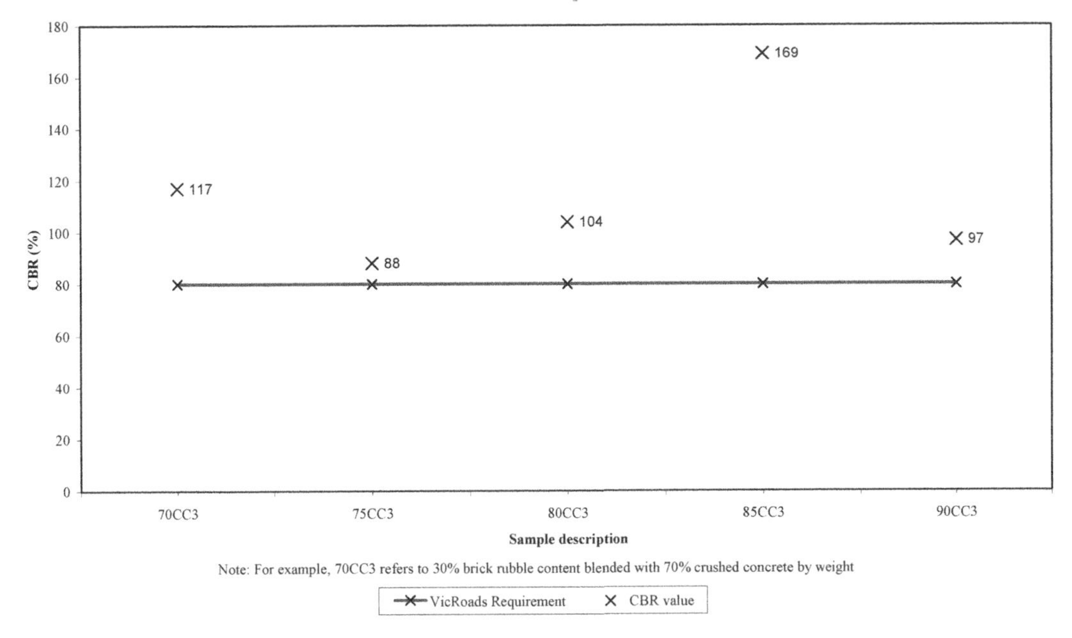

Figure 2. CBR values for brick rubble blended with crushed concrete.

4 BRICK RUBBLE BLENDS WITH CRUSHED ROCK

4.1 *Particle size distribution*

The particle size distribution of each blend mixture was similarly obtained by wet sieving using 5 different blend mixtures prepared in accordance with Australian Standards (AS 1141, 1996).

The grading curves for brick rubble blended with crushed rock are shown in Figure 3.

4.2 *Modified compaction*

The modified compaction parameters, maximum dry density and optimum moisture content are tabulated in Table 2.

Table 2. Summary of optimum moisture content and maximum dry density for brick rubble blended with crushed rock.

Sample description	Brick content (%)	Optimum moisture content (%)	Maximum dry density Mg/m^3
70CR3 *	30	9.0	2.12
75CR3	25	9.2	2.14
80CR3	20	9.0	2.15
85CR3	15	9.0	2.17
90CR3	10	9.0	2.22

* For example, 70CR3 refers to 30% brick rubble content blended with 70% crushed concrete by weight.

4.3 *California Bearing Ratio (CBR)*

The samples were prepared by modified compacting effort at their corresponding optimum moisture content and tested after four day soaked condition.

The CBR values obtained for brick rubble blended with crushed rock are summarised in Figure 4.

5 COMPARISON WITH LOCAL ROADWORK SPECIFICATION

VicRoads is the authorised governing body for local road work and bridge work in Victoria, Australia. VicRoads classifies recycled crushed concrete for pavement sub-base as Class CC3 (VicRoads, 2007). It further classifies crushed rock as CR3 for sub-base applications (VicRoads, 2006). These classifications are based on the physical and mechanical properties of crushed concrete and crushed rock. Minimum California Bearing Ratio (CBR) value specified by VicRoads for crushed concrete (CC3) and crushed rock (CR3) is 80%.

The CBR values of the tested blends were above 80% and found to satisfy the VicRoads requirement on CBR for crushed concrete CC3 and crushed rock CR3 materials. The test results indicate the potential

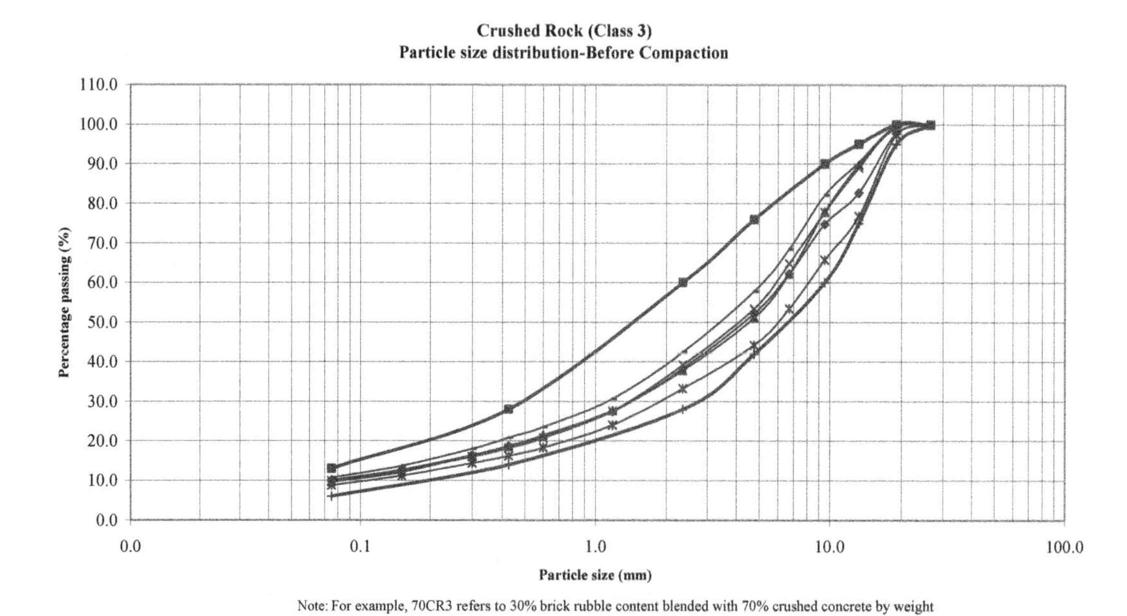

Figure 3. Particle size distribution for brick rubble blended with crushed rock.

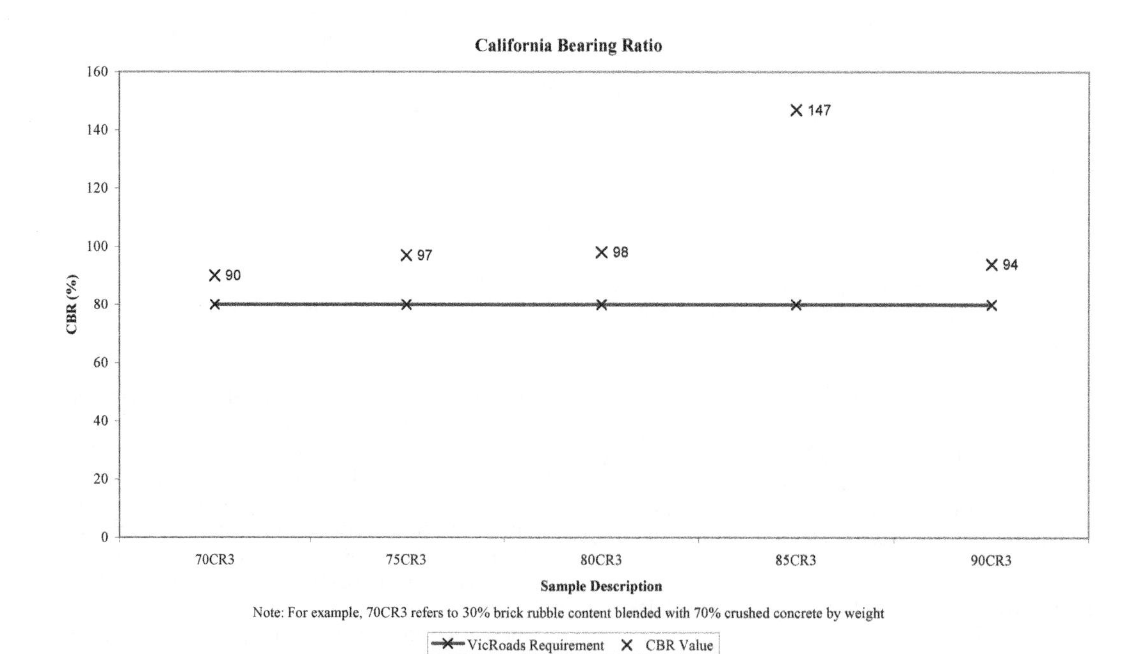

Figure 4. CBR values for brick rubble blended with crushed rock.

of using up to 30% of brick rubble content in blends with crushed concrete and crushed rock for pavement sub-base. The grading limits of all blends were within VicRoads upper and lower bounds for 20 mm Class CC3 crushed concrete and 20 mm Class CR3 crushed rock.

6 CONCLUSION

California Bearing Ratio tests were undertaken on various proportions of brick rubble blended with crushed concrete and crushed rock obtained from a recycling site in Victoria, Australia. The various blends of brick rubble with crushed concrete and crushed rock were found to be within the requirements of the local road authority.

The CBR values indicate that the replacement levels of crushed concrete and crushed rock with brick rubble can be potentially increased to 30%. The CBR test results indicate the potential of brick rubble to be used in blends with recycled crushed concrete and crushed rock as a pavement sub-base material.

Further laboratory tests such as repeat load cyclic triaxial tests, wet and dry strength, permeability and Los Angeles abrasion loss will be undertaken in the next phase of this research to validate these findings.

ACKNOWLEDGEMENT

The authors would like to acknowledge Sustainability Victoria for funding this research project (Contract No: 3887).

REFERENCES

Australian Standards (AS 1141, 1996). AS 1141.11, Method for sampling and testing aggregates, Standards Australia, Standards Association of Australia.

Australian Standards (AS 1289, 1998). AS 1289.5.2.1, Method of testing soils for engineering purposes, Standards Australia, Standards Association of Australia.

Australian Standards (AS 1289, 2003). AS 1289.6.1.1, Method of testing soils for engineering purposes, Standards Australia, Standards Association of Australia.

VicRoads (2006). Standard Specifications for Road works and Bridge works, *Section 812, Crushed rock for base and subbase pavement,* July.

VicRoads (2007). Standard Specifications for Road works and bridge works, *Section 820, Recycled crushed concrete for pavement subbase and light duty base*, January.

Advances in Transportation Geotechnics – Ellis, Yu, McDowell, Dawson & Thom (eds)
© 2008 Taylor & Francis Group, London, ISBN 978-0-415-47590-7

Some observations on determining CBR and the use of stiffness as an alternative

J. Booth, G.P. Keeton & R.C. Gosling
Soil Mechanics, Doncaster, UK

ABSTRACT: This paper presents some observations from direct and indirect testing to obtain California Bearing Ratio. It raises questions about reliability of published correlations between the direct and indirect methods based on concerns raised about their origins and on data from recent case histories. It also describes use of the Light Weight Deflectometer to obtain Stiffness Modulus and the need for a standardised method for use in the field.

1 PAVEMENT THICKNESS DESIGN

1.1 *Current basis*

The traditional approach to pavement thickness design in the UK has been based on the California Bearing Ratio (CBR). For roads the design is likely to have been carried out following HD 25/94 which until recently formed Volume 7 Section 2 Part 2 of the Design Manual for Roads and Bridges (DMRB).

During 2007 this portion of the DMRB, as published on a Highways Agency (HA) website, was replaced with an Interim Advice Note (IAN 73/06 Draft HD 25) which continues to use CBR but also offers Stiffness Modulus as an alternative.

As noted by Powell et al (1984) CBR is a generally accepted practical measure that can be used to give an estimate of both stiffness and strength.

Direct determinations of CBR can be made in the laboratory or in situ following BS 1377 (1990). However, the BS applies constraints such that it may not be possible to obtain CBR values from the direct test method.

Over recent years there appears to have been an increasing trend towards obtaining CBR values using alternative types of test. The available indirect test methods include static plate tests and the dynamic cone penetrometer (DCP).

The correlations between results from these two types of test and CBR are reproduced in various publications and are therefore viewed as having legitimacy. However, the origins of the correlations are obscure and it is difficult to obtain their provenance.

Recent experience in carrying out plate and DCP tests has cast doubt on the reliability or universal applicability of the correlations.

Although not stated explicitly in the various references, it has been assumed that the correlations relate to unsoaked rather than soaked CBR.

1.2 *Alternative basis*

It is arguable that CBR is not a very satisfactory basis for pavement design and reservations have been expressed by many authors. An alternative and perhaps more rational basis would be the use of Stiff-ness Modulus (E).

The various design charts in IAN 73/06 give layer thickness against the dual scale of CBR and E. The IAN reproduces a correlation between CBR and E with a caveat acknowledging a degree of uncertainty.

Methods of obtaining E include the Falling Weight Deflectometer (FWD) or Light Weight Deflectometer (LWD). The main drawback of this approach is the lack of a laboratory scale means of measuring E; a number of approaches have been considered but none has yet been adopted for routine laboratory use.

2 DIRECT CBR TESTS

2.1 *Testing in the laboratory*

CBR tests in the laboratory can be performed on recompacted specimens to establish design parameters or on undisturbed specimens recovered from the finished works as proof testing. Recompacted samples can be used as an alternative to the latter provided that the in situ conditions are correctly replicated and it is recognized that laboratory test values may overestimate the in situ CBR value due to the idealized compaction process and confinement in the steel mould.

Particle sizes of greater than 20 mm (coarse gravel and larger) are removed from the material prior to test. In some soils this means that the material tested will not have the same grading as that forming the subgrade in the construction works. In the extreme case this means that laboratory CBR tests are not possible in coarse granular soils. Removal of the coarser fraction might be expected to reduce the CBR value determined.

CBR values are normally calculated from the forces required to push the 50 mm dia. plunger 2.5 mm and 5.0 mm respectively into the soil. Laboratory set ups are capable of measuring CBR values of 100% or more but in these circumstances require loads in excess of 20 kN (approximately 2 tonnes).

2.2 *Testing in situ*

In situ CBR tests are, in principal, exactly the same as laboratory tests with the exception that there is no confinement other than by the surrounding and underlying materials. It might be expected that the absence of a rigid confining mould would lead to lower CBR values being obtained in situ compared with those in the laboratory.

It is essential that the tests are performed on well prepared ground and at depths appropriate to the proposed works. In particular the plunger must be located to avoid material greater than 20 mm wherever possible.

It is important that the rigid kentledge is adequate (say at least 1.5 times the load to be applied in testing). In practice however a typical set up using a loaded four wheel drive vehicle as kentledge will only be capable of measuring CBR values of up to about 30%.

2.3 *Constraints on testing*

It is evident from the discussion in 2.1 and 2.2 that one major constraint is the removal/avoidance of particles larger than 20 mm. These are large relative to the 50 mm diameter plunger and would significantly influence the result thereby rendering it unrepresentative. Thus determining CBR values in soils where coarse gravel and larger particles are prevalent can be a problem. It could be postulated that this has been one of the reasons for indirect methods being developed.

3 INDIRECT METHODS OF OBTAINING CBR

3.1 *Static plate tests*

This approach is using a test which measures stiffness rather than strength.

Generally the diameter of the plate is much larger than even the coarsest soil particles. Thus this method would appear to have attractions in coarse granular soils where direct testing is restricted but this begs the question of how the correlation with CBR could have been developed in the first place.

3.1.1 *Published correlations*

Obtaining CBR value from plate tests involves a correlation with modulus of subgrade reaction (k). In addition the k value requires conversion to that for a 762 mm (30 inch) diameter plate if a different size was used for the test.

3.1.2 *CBR from k*

An equation to provide CBR value from k_{762} was given in HD 25/94, Chapter 4. The same equation is reproduced in IAN 73/06 although without the example calculation of the earlier HD 25/94 which contained an arithmetic error (result should be CBR = 2.4% not 24%). Both of these Department for Transport (DfT) documents state the equation is an approximate empirical relationship. However, neither gives a formal reference or provides any other indication of where the correlation originates.

It seems reasonable to suppose that the equation was obtained from earlier work by DfT in house and/or Transport and Road Research Laboratory (TRRL) on their behalf to which Croney (1977) would have had access. However, in his book Croney restricts the presentation of the correlation to a linear relation shown on a log/log plot without showing the raw data points. Nevertheless the IAN/HD equation seems to be a tolerable approximation to the trend line as far as can be judged by measuring from the small scale plot Croney presents.

The DfT approach comprises a single equation irrespective of material type or grading. In contrast the relationships given by MoD (2006) show a distinct grading influence. Admittedly the data was collated to allow k to be estimated from CBR rather than vice versa. Nevertheless at the point where the curves for two material types overlap, ie a CBR value of 10%, the CBR/k factor for a cohesive subgrade or sand with a significant secondary fine constituent is 50 compared with factor of only about 35 for a coarse grained non-plastic subgrade.

The authors' impression from some isolated experiences (not presented here) is that material type and/or grading do have an impact.

3.1.3 *Conversion of k for plate size*

The conversion to k_{762} from other plate diameters is presented graphically in IAN 73/06 and HD 25/94. Both publications also give an equation; they are of the same form as each other but with slightly different constants. Both equations are tolerable approximations to the curve presented in Croney who states that that it is based on American experimental evidence.

Croney quoted 760 mm as the reference plate diameter (presumably a "soft" metrication of 30 inch).

This is however a trivial difference from 762 mm and would not account for the greater variance between the Croney plot and either of the two equations than the difference between the lines defined by the equations themselves. Thus quoting revised factors in IAN 73/06 compared with HD 25/94 appears to be of no consequence and probably introduces an unnecessary refinement.

3.1.4 *Plate test procedure*

Both IAN 73/06 and HD 25/94 refer to BS 1377 as giving details of the test and to the 600 Series in Volume 1 of the Specification for Highway Works (MCHW) published by HA for its use.

The reference to BS 1377 is however somewhat disingenuous in as much as the BS is very generalized for this test and offers two alternative procedures; namely the constant rate of penetration and the incremental loading test. The unwary might assume that the former should be selected given that it would be a quasi-replication of the direct CBR test. However, reference to other documents, see below, indicates that the latter is appropriate.

The reference to MCHW is also potentially misleading because the description of the plate test in this document is for use for the determination of constrained modulus in association with earthworks for buried steel structures. While MCHW does indicate a maintained, incremental, load test it talks in terms of high imposed pressures which, as discussed below, would not be rational for the determination of CBR.

Croney describes the test for the purpose of determining k which is calculated at a mean plate deflection of 1.25 mm, the difference between this and 1.27 mm (a "hard" metrication of 0.05 inch) is inconsequential. However, it is important to arrange the loadings to give three or four increments leading up to and one or two just beyond the target settlement to allow sensible interpolation of its value. The example test result in Croney and in the earlier book by TRRL (1952), which is in Imperial units, demonstrate the matter of the increments being selected on the basis of the strain induced, rather than stress imposed. The point is made explicitly in Reference Specification Appendix J produced by British Airports Authority (BAA).

In any event working at relatively small strains exacerbates the potential for bedding errors. Furthermore any non linearity would introduce uncertainty and again the BAA specification gives clear guidance on how to evaluate the test curve in these circumstances which is absent from the TRRL and DfT publications.

3.2 *Dynamic cone penetrometer*

This approach is using a test which essentially measures strength rather than stiffness, given that like all penetrometers they can be deemed to shear the soil through which they are advanced.

The DCP is reportedly robust enough to sustain hard driving. Thus it should push coarser particles to one side. Once again however it is difficult to see how correlations could have been established in coarse granular soils.

3.2.1 *Background documents*

The DCP is not included in Part 9 (in situ testing) of BS 1377. There is a minimal description of it in HD 25/94. This description is reproduced in IAN 73/06 although with enhanced detail being given in Annex A (sic Appendix A,) and the Draft Clauses for MCHW. Neither document gives a formal reference.

The DCP equipment most commonly used in the UK is manufactured by Farnell who designate it as their Model A2465. In the absence of a formal published standard for the test, it is necessary to fall back on the manufacturers operating instructions.

3.2.2 *History of development*

The penetrometer was written up in Australia by Scala (1956) who cites a 1951 publication as the basis for its original design. Further development took place in southern Africa with several papers being published in the early 1980s. Since then it has also been used elsewhere in the world as evinced for example by publications in Canada (Standard Test Procedures Manual, Saskatchewan Highways and Transportation, 1992) and Sri Lanka (Karunaprema and Edirisinghe, Electronic Journal of Geotechnical Engineering, 2002).

According to Farnell, their model is similar to the South African equipment. The Farnell hammer is an 8 kg weight dropping 575 mm which differs slightly from the Australian equipment (values quoted in Imperial units). The Farnell model is supplied with a 20 mm diameter 60° apex angle cone whereas earlier publications describe both a 60° and a 30° cone as having been used. The Farnell model driving energy and cone correspond to the equipment specification in IAN 73/06.

3.2.3 *Published correlations*

Farnell's operating instructions quote four equations for correlating DCP mm/blow to CBR value. All are in the same form although the constants differ.

Two of these equations (Smith & Pratt and Van Vuuren) were established for 30° cones. The other two equations (Kleyn & Van Heerden and TRRL) were established for a 60° cone, ie the same as the Farnell equipment. However, attempts by the authors to obtain confirmation of these equations independent of Farnell's operating instructions have to date been frustrated.

The Kleyn & Van Heerden (1983) paper referenced by Farnell does not actually quote the equation. It does however present a graph to a log/log scale of CBR versus a mm/blow trend line. The trend is linear (on the log/log plot) over the range 100 to 2 mm/blow. Given the accuracy of scaling from the graph, the constants assigned by Farnell seem to sensibly agree with the linear portion of the trend in the 1983 paper.

The 1983 paper references several earlier publications. Two of these, ie Kleyn, Maree & Savage (1982) and Kleyn and Savage (1982), have very similar titles to each other which appears to have resulted in a jumbled reference in the Farnell operating instructions. In any event neither provides any useful addition to the 1983 paper as far as the provenance of the equation is concerned. A third reference, ie Kleyn (1975), apparently presents the derivation of the graphical correlation on the basis of some 2000 data points but is not readily available in the UK even from the British Library.

The TRRL equation quoted by Farnell is reproduced from Overseas Road Note 8 (1990). However, the road note is actually a user manual for a computer program to analyse DCP data. While this document does give the TRRL equation explicitly (and also reproduces the Kleyn equation) it gives no indication of or reference to the source data for its derivation. Enquiries by the authors to the Overseas Unit of TRRL have failed to elicit a substantive response.

The differences between CBR values determined by the Kleyn & Van Heerden or the TRRL equations for the same mm/blow can be perceived by some as being significant. However, IAN 73/06 gives only the TRRL equation; it does not quote a formal reference or provide any other substantiation of its reliability.

4 COMPARISONS OF DIRECT AND INDIRECT TESTING

4.1 *Case history 1*

At this site earthworks fill had been placed with controlled compaction for an access road forming part of a housing development. The fill material was described as borderline gravelly clay / clayey gravel; water was ponding at fill surface confirming a relatively high fines content.

CBR values were required to be determined by plate tests and due to the granular content of the material a 450 mm diameter plate was selected. The plate tests were undertaken on the surface of the fill, a bed of sand being used to level the plate. They were performed as incremental tests, increasing the loadings by 20 kPa in five stages and allowing settlement to stabilize at each. The CBR values were calculated via k in accordance with HD25/94 and were in the range of 1 to 2%, with one exceptional value of about 4%.

These values seemed extraordinarily low for the material and inconsistent with the perceived condition and trafficability of the site. Consequently a series of in situ CBR tests were carried out subsequently, these being at depths between 0.2 and 0.7 m below the surface. At several of the direct test locations the kentledge provided by the loaded four wheel drive vehicle was inadequate, see 2.2 above. The remainder of the tests gave CBR values in the range of about 25 to 30%.

The in situ CBR tests were not located adjacent to the plate tests, precluding comparison of individual results, but overall they gave values greater than those from the plate tests by a factor of about 20. CBR values from the plate tests are plotted against those from the nearest direct tests on Figure 1.

There were many factors affecting the test results from this site. In retrospect the most important may have been a stress rather than strain controlled definition of each stage, ie not following the BAA philosophy, see 3.1.4 above. Nevertheless this experience raised doubts about the propriety of the interpretation of CBR values from plate tests for low to medium strength cohesive materials.

4.2 *Case history 2 and 3*

Following Case history 1 opportunities arose at two other sites under much more controlled conditions which allowed the comparison of plate test derived CBR values with those from direct determinations. In particular, the materials were much less granular (up to 30% gravel only) than at Site 1 and were compacted such as to allow a wide range of CBR values to be investigated. The total thickness of the earthworks fill tested was known to be in excess of 2 m.

At each location on the two sites a 450 mm diameter plate test and an adjacent in situ CBR test was performed. For the plate tests only sufficient sand was used such as to ensure full contact of the plate with the underlying soil. The tests were carried out in incremental stages but by measuring the force required to produce stabilized settlements of 0.25, 0.5, 0.75, 1, 1.25, 1.5 and 2 mm (where reaction load allowed). The CBR values were calculated via k in accordance with HD25/94. In total six tests were performed at Site 2 and eight tests at Site 3. A comparison of the CBR values from the two types of test plotted against each other is shown on Figure 1.

It can be seen that these controlled tests provide a reasonable trend over a wide range of CBR value. However, the log scale does minimize the visual impact of the discrepancies. These results indicate that the CBR value calculated from the plate tests is generally less than that from the direct tests by a factor of between about 0.8 and 0.5.

Arguably there is a weak trend for the discrepancy to increase as the direct CBR value reduces. This may be due to the lower values being associated with more cohesive materials for which the conventional formulae may not remain valid. The difference between the two tests is far smaller at high CBR values and this is consistent with plate tests on free draining granular

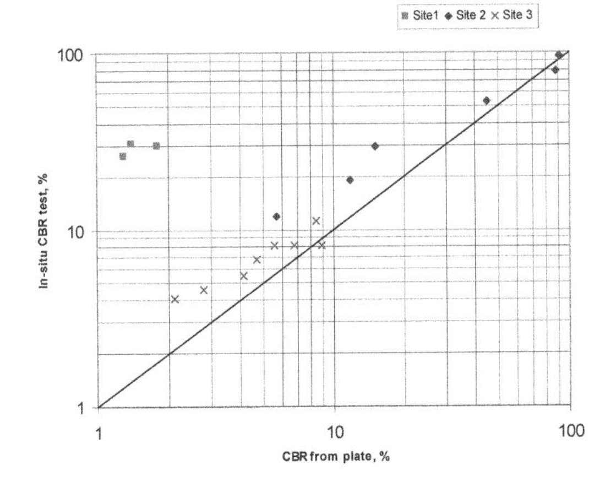

Figure 1. Comparison of CBR values from direct in situ tests and derived from plate tests.

Table 1. Comparison of CBR values by direct and indirect methods.

Material type	CBR lab (%)	Factors		
		Plate/lab	DCP/lab	DCP/plate
Silt	2 to 6	1.1 to 1.3	1.2 to 3.1	1.6 to 4.7
Sand	10 to 43	0.3 to 0.5	0.4*	1.6 to 8.2

*Ignoring one result.

materials (compacted limestone) from another site which have readily produced an equivalent CBR value in excess of 100%.

4.3 *Case history 4*

Ground conditions at the site comprised glacial till which could be divided into two material types; namely sandy slightly gravelly silt and silty very gravelly sand. The fines (combined silt and clay) content was about 30% and 15% in the silt and sand facies respectively.

Direct CBR tests were carried out in the laboratory on samples recompacted with a 4.5 kg rammer. These were actually relationship suites and the CBR values interpolated at natural moisture content have been selected for the comparison. Indirect tests to obtain CBR comprised DCP and plate bearing tests using a 750 mm diameter plate. Both the laboratory and the plate tests give unique values whereas some judgment has to be exercised over which depth range from the DCP record should be selected.

There were a total of nine plate tests and ten DCPs but only five laboratory tests. A comparison between the results is summarized in Table 1.

There is an overall trend in all three of the test methods for the CBR values, as might be expected, to be higher in the sand than in the silt. However, it can be seen that when CBR values at individual locations by the three methods are compared with each other there is considerable disparity.

The tabulated data for CBR values derived from both the plate test and DCP when compared with those from laboratory tests appear to support the contention given earlier that material type and grading does influence the correlation. The plate test to DCP correlation is however more disconcerting.

5 FALLING WEIGHT DEFLECTOMETER

5.1 *Published correlation CBR against E*

The correlation between CBR and E now embodied in DfT road design guidance first appears in Powell, et al (1984). It was developed by collating data from a number of sources, but mainly from measurements of wave propagation in a variety of materials. The relationship is:

$$E = 17.6(CBR)^{0.64} \quad MPa \qquad (1)$$

and is described by Powell et al. as "a lower bound relationship, valued between 2 and 12%".

As stated in IAN 73/06, there is no definitive relationship between CBR and E. The equation may be used to give an estimate of E acknowledging a degree of uncertainty.

5.2 *Current DfT road design methodology*

Following Powell et al. road design, including road foundation design, is based on multi layer linear elastic modeling. The parameters input to the model are stiffness moduli; the CBR – E correlation being developed to allow the continued use of CBR. The guidance in IAN73/06 provides design charts scaled for CBR (up to about 30%) and E (up to 150 MPa).

5.3 *Use of Lightweight deflectometer*

In theory at least the use of the LWD sidesteps the issue of CBR-E correlation by providing direct reading results in terms of Stiffness Modulus. The device measures stiffness using a dynamic method with a pulse time of around 15 mS. It is capable of measuring the strength of a wide variety of materials with stiffnesses up to about 400 MPa.

Current LWD devices available in the UK are proprietary designs and there are no established test methods beyond the manufacturers instructions.

The main theoretical drawback in using LWD to produce a directly measured E is that the model inputs are layer stiffness moduli whereas the device only measures surface moduli. Given sufficient thickness of stratum, which can be assumed in many subgrades, the main difference between LWD and FWD is that there is no surcharge pressure applied when using the former. The FWD, with its larger zone of influence and

Table 2. Comparison of E values by static and LWD methods.

Material type	Static test method	Mean static result[1] MPa	Mean LWD result MPa
Saturated chalk	600 mm plate	26	32
Gravelly sand[2]	in situ CBR	86	43
Gravelly Clay[2]	600 mm plate	75	102

[1] Calculated from CBR result using Equation (1).
[2] recompacted fill materials.

more sophisticated geophone array, is able to measure the layer stiffnesses of entire pavement construction. The authors are not aware of any formal studies of the differences in stiffness measurements made by the two devices.

The LWD has been used recently by the authors in situations where direct CBR or plate tests would have been impractical. It had also been used in some instances where the traditional approach was viable.

It has become clear from these experiences that a standard test method for the LWD is needed. This subject was raised during the "Lightweight deflectometer" conference at Loughborough in April 2007. The main points that need to be covered are:

a) the LWD needs to be properly seated; this still requires some work, particularly when using the device in the base of holes/small pits cored/excavated through existing road pavements.
b) the response curve needs to be examined on each drop to avoid anomalous readings; this is essential, the worst anomalies observed were when testing railway ballast and the central geophone was inadvertently place over a void resulting in completely erroneous readings.
c) avoiding permanent deformation when testing weaker materials; this arises because of the need to test the soil in elastic conditions, the procedure adopted by the authors is to start with a high load, and to keep reducing drop heights until no change in the measured modulus observed, this is the correct modulus to report.

5.4 *Case histories*

Opportunities to compare the dynamic and static methods have been limited to date and have involved only a few data points, see Table 2.

Direct comparisons have been hampered by differing plate sizes which mean that like for like comparisons are not possible. However, it can be seen that the largest discrepancy occurs where the loaded area is smallest, ie with the CBR.

6 CONCLUDING REMARKS

This paper summarizes some observations from experience of testing by methods used in practice for pavement thickness design. It is not a detailed study of the theory and takes as read the complications arising from the dimensions of the testing equipment, depth of the stressed zone, magnitude of strain etc. However it is hoped that the questions raised may act as a catalyst for future work in this vexatious area.

The authors wish to thank Soil Mechanics for the encouragement to prepare this paper and for the provision of the resources.

REFERENCES

BS 1377. 1990. *Methods of test for soils for civil engineering purposes.* British Standards Institution

Croney, D. 1977. *The design and performance of road pavements.* London: Transport and Road Research Laboratory

Kleyn, E.G. 1975. *The use of the dynamic cone penetrometer (DCP).* Transvaal Roads Department Report L2/74

Kleyn, E.G. Maree, J.H. & Savage P.F. 1982. *Application of a portable pavement dynamic cone penetrometer to determine in situ bearing capacities of road pavement layers in South Africa.* Proceedings of the Second European Symposium on Penetrometer Testing (ESOPT II), Amsterdam

Kleyn, E.G. & Savage, P.F. 1982. *The application of the pavement DCP to determine the bearing properties and performance of road pavements.* International Symposium on Bearing capacity of Roads and Airfields, Trondheim Norway

Kleyn, E.G. & Van Heerden. 1983. *Using DCP to optimise pavement rehabilitation.* Transvaal Roads Department Report LS/83 and Annual Transportation Convention, Johannesburg

MoD. 2006. *A guide to airport pavement design and evaluation.* Design and Maintenance Guide XX, 2nd Edition, Construction Support Team, Defence Estates, Ministry of Defence

Powell, W.D. Potter, J.F. Mayhew, H.C, & Nunn, M.E. 1984. *The structural design of bituminous roads.* Report LR1132. Transport and Road Research Laboratory

Scala, A.J. 1956. *Simple methods of flexible pavement design using cone penetrometers.* Proceedings of 2nd Australian-New Zealand Conference on Soil mechanics and Foundation Engineering

TRRL. 1952. *Soil mechanics for road engineers.* London: Her Majesty's Stationary Office

TRRL. 1990. *Overseas Road Note 8 – A users manual for a program to analyse dynamic cone penetrometer data.* Transport and Road Research Laboratory Overseas Unit

Advances in Transportation Geotechnics – Ellis, Yu, McDowell, Dawson & Thom (eds)
 ISBN 978-0-415-47590-7

Experimental observations and theoretical predictions of shakedown in soils under wheel loading

S.F. Brown
University of Nottingham, UK

S. Juspi
Ove Arup and Partners, UK

H.S. Yu
University of Nottingham, UK

ABSTRACT: Shakedown is the term given to the state of a structural system subjected to cyclic loading when its response has become resilient in nature and no further accumulation of plastic strains occurs. An experimental study was conducted with a view to observing the shakedown phenomenon of various soils and granular materials under moving wheel loading. The experiments were conducted using two laboratory wheel tracking devices with test specimens compacted into moulds. The results are presented along with the theoretical predictions derived from parallel work based on use of a lower bound theory. Monotonic triaxial tests were conducted to determine the Mohr-Coulomb failure parameters, as these formed part of the input for the theoretical predictions. The permanent deformation, which developed under the wheel as loading progressed, was monitored. Three types of response were identified, from which, a range of values for measured shakedown were identified. These were found to compare reasonably well with the theoretical predictions.

1 INTRODUCTION

One of the principal deterioration mechanisms for flexible pavements and ballasted railway track is the development of permanent surface deformation under repeated wheel loading. This can cause ruts in roads and differential settlement in railway track, both of which are major considerations in design and are caused by the accumulation of plastic strain under repeated loading. A review of these issues by Brown (1996 and 2004) described how observations from repeated load triaxial testing had demonstrated the existence of a threshold deviator stress level, below which the magnitude of accumulated plastic shear strain is quite small. The same phenomenon is apparent in the theoretical concept of shakedown, derived from structural engineering and first applied to pavements by Sharp and Booker (1984). The concept is that a material or structure will 'shakedown', after a certain number of load application cycles, into a state that responds in an entirely resilient, albeit non-linear, mode with no further accumulation of plastic strain. If this threshold or shakedown stress level can be reliably predicted or easily determined experimentally, then it provides a very useful component for design of pavement and railway track foundations.

The results presented in this paper summarize some preliminary, simple experiments designed to identify 'shakedown' under moving wheel loading applied directly to a compacted layer of soil and to compare the results with theoretical predictions based on use of the lower bound theory developed by Yu (2005). This represents the first stage of a project that went on to study more realistic layered systems of soils and granular materials.

2 SOIL CHARACTERISTICS

Four materials were used; three soils and a crushed rock, descriptions of which are in Table 1.

The materials were selected on the basis of ready availability and to provide a range of characteristics.

Standard vibrating hammer compaction tests to BS 1377-2 (1990) were conducted to determine optimum water contents for use in the wheel tracking experiments. The results are summarized in Table 2 along with the specific gravities of the soil solids (G_s).

Triaxial tests were carried out on the three soils using 38 mm diameter × 76 mm high specimens and for the crushed rock using 150 mm × 300 mm specimens. All were compacted as close as possible to

Table 1. Descriptions of materials.

Type	Description
Silty-clay	$w_L = 30\%$, $w_P = 16\%$. Mercia Mudstone
Sandy-silt	71% sand, 16% silt, 13% clay. Gravel washings
Sand	Poorly-graded building sand, max. particle size = 6 mm
Crushed rock	Granite to type 1 sub-base specification, 37 mm max particle size

w_L = liquid limit, w_P = plastic limit.

Table 2. Compaction characteristics of soils.

Soil type	G_s	Optimum water content (%)	Max. dry density (kg/m^3)
Silty-clay	2.70	15.5	1,882
Sandy-silt	2.62	15.1	1,723
Sand	2.66	4.2	1,813
Crushed rock	2.77	4.8	2,193

Table 3. Triaxial test results.

Soil	w (%)	ρ (kg/m^3)	S_r (%)	c_u (kPa)	c (kPa)	φ (°)
Silty-clay	15.2	2,162	94	55	–	–
Sandy-silt	15.5	1,694	52	–	14	38
Sand	4.2	1,860	23	–	9	36
Crushed rock	4.0	2,141	32	–	13	49

w = mean water content, ρ = mean bulk density, S_r = mean degree of saturation, c_u = undrained cohesion, c = apparent cohesion, φ = angle of shearing resistance.

optimum water content. For the silty-clay, the specimens were prepared by trimming with a wire saw from a slab compacted in three layers within a mould. The sand and silt specimens were prepared by compaction in a 38 mm diameter cylindrical mould using five layers with a small tamping rod having a 30 mm diameter foot. The crushed rock was compacted with a vibrating hammer in a 150 mm diameter split mould using six layers.

Standard monotonic load triaxial tests were carried out to determine the Mohr-Coulomb failure parameters for each material using confining stresses in the range 10 to 100 kPa. Drainage conditions were chosen to approximately match those likely to occur under a moving wheel load, so undrained conditions were used for the silty-clay, whereas drained tests were adopted for the other materials. Strain rates of 2%/min and 10%/hr were used for the undrained and drained tests respectively. The results are summarized in Table 3.

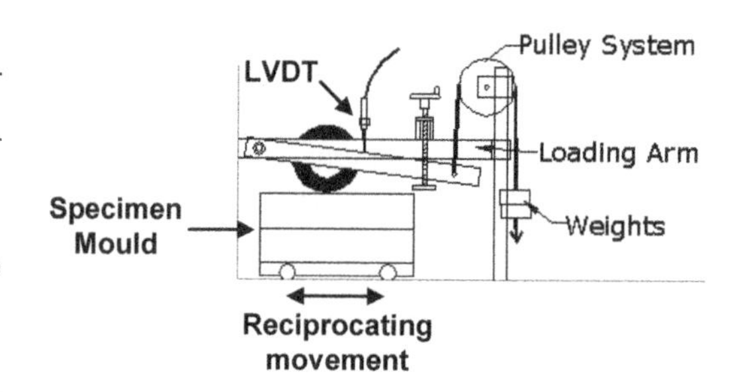

Figure 1. The wheel tracking apparatus.

3 THE WHEEL TRACKING EQUIPMENT

The three soils were tested in a small wheel tracking apparatus (SWT) developed originally for testing asphalt slabs (Jacobs, 1977). Each test specimen was contained in a mould 400 mm long × 280 mm wide and most specimens were compacted to a depth of 125 mm with one set of sand specimens compacted to 250 mm. Wheel loading was applied by a 200 mm diameter steel wheel with a solid rubber tyre of 50 mm width. The slab, in its mould, was reciprocated relative to the wheel at a rate of 40 passes per minute using a mechanism driven by an electric motor. The load was applied to the wheel using a lever arm and weight hanger and the surface permanent deformation was measured using an LVDT. The details can be seen in Figure 1. Each test specimen was compacted in three layers within the mould at optimum water content using a vibrating hammer with a 100 mm diameter foot.

A larger apparatus, known as the Slab Test Facility (STF), was used to test the specimens of crushed rock which were 1 m long × 0.6 m wide × 0.18 m deep. They were also contained in a steel mould into which they were compacted using a vibrating hammer in three layers. The apparatus is shown in Figure 2 and was originally developed by Brown et al (1985) to test reinforced asphalt slabs. The mechanism involves load application by a hydraulic actuator operating at one end of a reaction beam. The wheel carriage is oscillated using a hydraulic motor and the applied load is monitored by a load cell which forms part of a servo-loop that controls the actuator load to maintain a constant value at the wheel. The wheel speed was 1.4 km/hr and the pneumatic tyre inflation pressure was 276 kPa. The permanent deformation caused in the wheel track was measured periodically during each test with the loading interrupted so that a straight edge and rule could be used.

Although it has been shown (Brown and Chan, 1996) that unidirectional wheel loading, which is the situation in a road pavement, causes smaller rut depths than bidirectional, the latter was adopted for these experiments. This was a limitation of the SWT and was

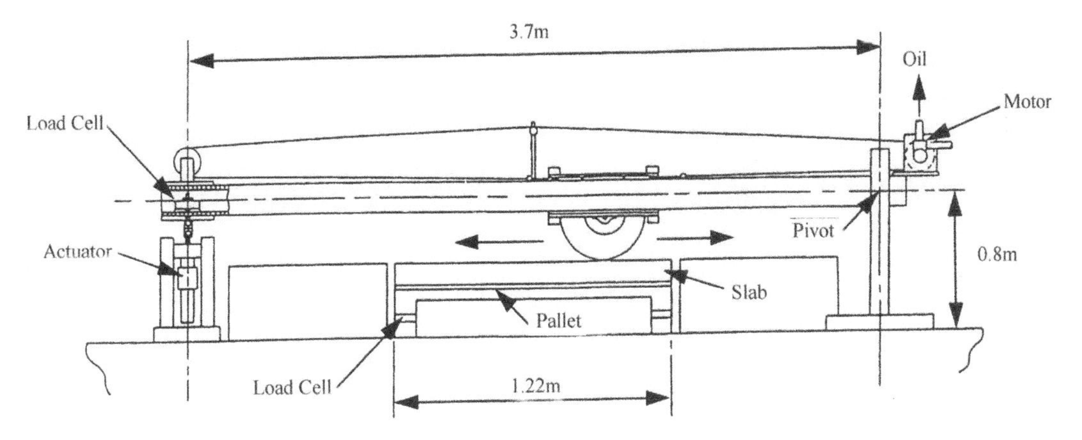

Figure 2. The slab test facility.

adopted for both facilities to speed up the experiments in order to minimize drying of the test specimens, which were unsealed.

The contact areas between the wheels and the soil surfaces were determined by measuring the footprint generated on a piece of graph paper using ink applied to the tyre with the paper placed on the soil surface. The size of the area for each apparatus was a function of the applied load and the soil type because of soil distortion, so values were recorded over a wide range of conditions.

Part of the input for the theoretical prediction of shakedown is the ratio of normal to shear load applied at the soil surface. For the SWT, the normal load was determined using a load cell beneath the wheel to calibrate the lever arm loading system. In order to determine the surface shear load, direct measurements were made of the small force require to move the mould on its bearings relative to the wheel. Full details are described by Juspi (2007). For the STF, the tangential force required to rotate the wheel was measured. This included the force needed to pull the carriage along on its bearings. A ratio of shear to normal load of 0.08 was determined for the SWT and 0.12 for the STF.

4 EXPERIMENTAL RESULTS

4.1 *Test conditions and measurements*

Since access to the surface of the SWT test specimens during experiments was restricted, the permanent vertical surface deformation that developed in the wheel track was measured indirectly using an LVDT, via the loading lever arm that moved downwards with the wheel. The measured deformation was the difference between the original surface level of the soil and the bottom of the rut, so any shoulders that developed on either side were ignored. Measurements were taken at intervals, as loading proceeded, up to 16,000 load passes (10,000 for the STF), unless failure, in terms of very large deformation, developed earlier. A few tests were continued for 50,000 cycles. When shakedown was exceeded, it was apparent quite early in the tests.

A summary of the test conditions is given in Table 4. It will be seen that a range of applied wheel contact pressures was applied. These were selected to embrace the shakedown value so that its magnitude could be deduced from the results. Two sets of tests were carried out on the sand; PS1 used a specimen thicknesses of

Table 4. Test conditions.

Soil type	ρ (kg/m^3)	w (%)	S_r (%)	Contact pressure (kPa)
Sand (PS1)	1,889	4.2	24	100
	1,889	4.0	23	111
	1,890	4.4	25	119
	1,880	4.2	24	127
	1,883	3.9	22	154
Sand (PS2)	1,889	4.2	24	100
	1,889	4.2	24	111
	1,889	4.1	23	119
	1,885	4.0	23	127
	1,890	4.2	24	154
Silty-clay	2,169	15.2	95	225
	2,167	15.2	94	237
	2,162	15.0	93	269
	2,164	15.1	93	301
Silt	1,731	15.4	54	193
	1,734	15.2	54	229
	1,736	15.0	53	244
	1,736	14.9	53	251
	1,732	15.3	54	257
	1,736	15.2	54	261
Crushed rock	2,172	4.0	34	289
	2,192	3.9	35	355
	2,200	4.2	37	372
	2,234	4.0	38	384

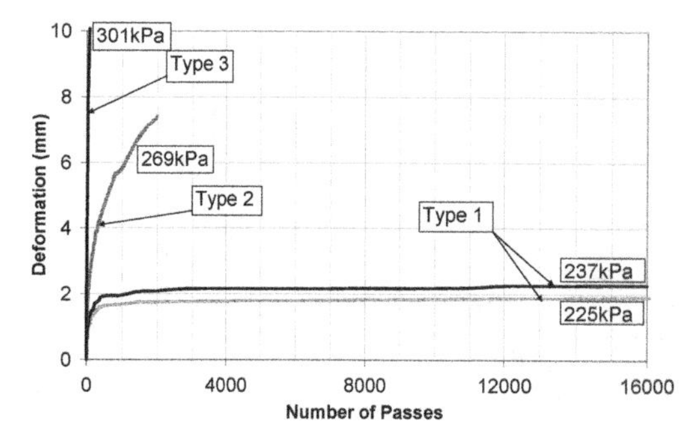

Figure 3. Relationships between permanent surface deformation and wheel passes for SWT tests on the silty-clay.

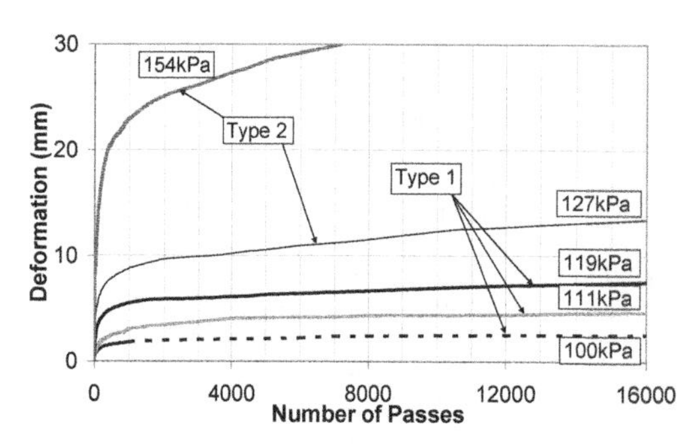

Figure 4. Relationships between permanent surface deformation and wheel passes for SWT tests on the sand (PS2).

250 mm, whereas PS2 and all tests on the silty-clay and the silt used 125 mm.

4.2 *Results*

Various patterns of deformation development were observed, so three categories were defined to assist with the interpretation of results. Figure 3 shows SWT data for the silty-clay which exhibited each category. When deformation developed very quickly, exceeding a rate of 0.018 mm/pass after 500 passes, it was classified as Type 3 and was clearly in excess of shakedown. When the rate approached a minimal value of 0.001 mm/pass after 1,000 passes it was Type 1, regarded as below shakedown. Intermediate cases were classified as Type 2 which will have included the shakedown situation. When stabilization to an approximately constant deformation rate occurred for Types 1 and 2, it did so after between 200 and 500 wheel passes. This procedure was similar to that taken by Werkmeister et el (2001) for analyzing the results of repeated load triaxial tests.

Measured data from the other SWT tests using 125 mm thick specimens is shown in Figures 4 and 5. The stress levels shown against each test result are the applied contact pressures beneath the wheel. The

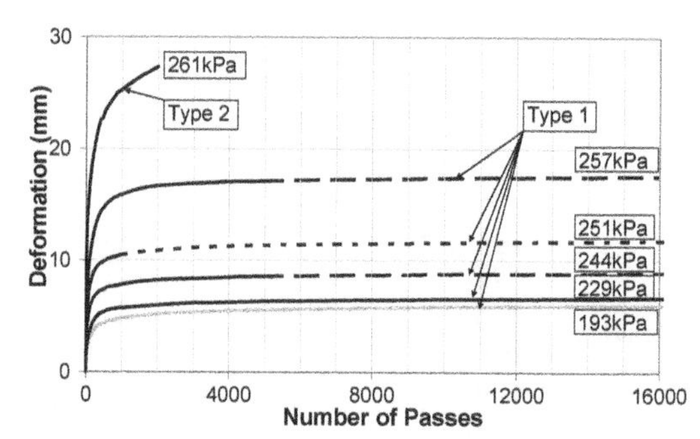

Figure 5. Relationships between permanent surface deformation and wheel passes for SWT tests on the silt.

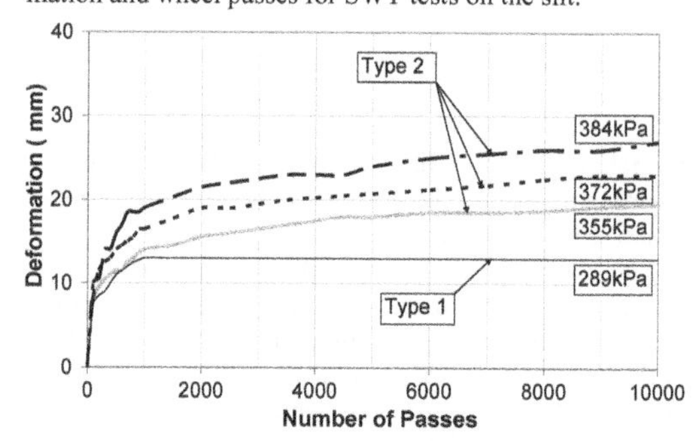

Figure 6. Relationships between permanent surface deformation and wheel passes for STF tests on the crushed rock.

crushed rock results from the STF are shown in Figure 6.

Figures 7 and 8 show photographs of specimens towards the end of testing to indicate the appearances for each of the defined conditions, Types 1 to 3. Because of the large rut shoulders for the sand specimen shown in Figure 7(a), the photograph gives the impression of high deformation although the depression from the original surface was less than 4 mm. This specimen was a PS1 (Table 4). Clearly this is not a realistic condition for a pavement foundation being directly trafficked, since the rut shoulders would be important. Equally, however, it would not be normal to design for traffic operating directly on the soil. The results were considered a preliminary exercise to check validity of the theory for the simplest case prior to proceeding with more realistic layered systems.

5 THEORETICAL PREDICTIONS

The theoretical values for shakedown loads were computed using the solution developed by Yu (2005) based on the principle set out by Melan (1938). He stated that for a load to be below the shakedown limit, its magnitude had to be such that at no location could

(a) Sand after 8,000 passes at 100kPa (Type 1, 3.4 mm deformation)

(b) Sandy-silt after 16,000 passes at 229kPa (Type 1, 6.7 mm deformation)

Figure 7. Examples of Type 1 results.

(a) Crushed rock after 10,000 passes at 355kPa (Type 2, 19.5 mm deformation)

(b) Silty-clay after 650 passes at 301kPa (Type 3, 37.1 mm deformation)

Figure 8. Examples of Types 2 and 3 results.

the yield condition be exceeded when considering a combination of residual stresses in the soil and load-induced values. Yu used the Hertz loading condition shown in Figure 9 with applied circular normal and shear stress distributions and assumed an elastic half-space. He demonstrated that the critical stress location for defining shakedown was located on the x-z plane at $y = 0$, ie, on the axis of symmetry in the direction of the wheel load travel. He used the Mohr-Coulomb yield criterion and his solutions allowed the shakedown condition to be determined as a function of the applied shear to normal load ratio (Q/P in Figure 5), the radius of the loaded area and the Mohr-Coulomb parameters c and φ.

Values of the normal load at shakedown were computed for each of the four materials under the loading conditions in the wheel tracking tests. This was done by using the measured values of c and φ from the triaxial tests (Table 2), the measured contact areas, assuming they were circular and the measured ratio of 0.08 (for the SWT) or 0.12 (STF) between the shear and normal loads. The results are compared with the experimental data in Table 5.

6 DISCUSSION

It is apparent from the experimental data that the value of normal wheel load contact pressure at shakedown is better defined for some materials than others. This is an observation previously made by Cheung (1994) in relation to the threshold stress for various soils. For those tested here, the shakedown value for the sandy-silt was most closely defined. Figure 10 shows the results in graphical form plotted against the theoretical predictions. This indicates that the predicted values do indeed provide a lower bound and are within about 20% of those measured. Overall, the results provided sufficient confidence for the investigation to proceed with the study of layered systems using various combinations of soils and granular materials more closely

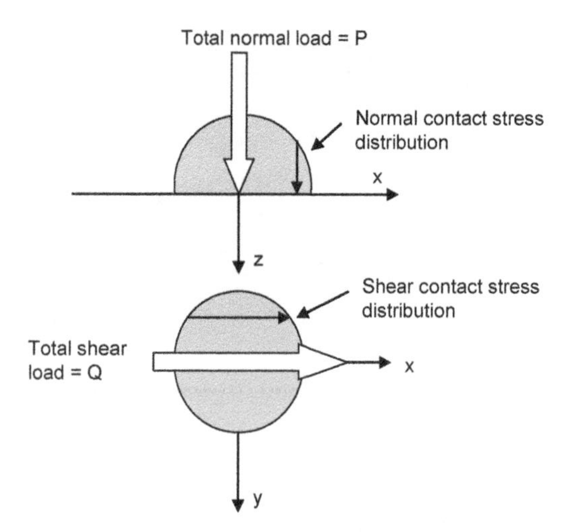

Figure 9. Theoretical wheel loading assumptions.

Table 5. Comparison of theoretical shakedown predictions and measured values.

Soil type	p_{max} (kPa)	p_{min} (kPa)	p at shakedown (kPa)
Silty-clay	237	269	233
Sandy-silt	257	261	217
Sand	119	127	122
Crushed rock	289	355	290

p_{max} = maximum contact stress for Type 1 response, p_{min} = minimum contact stress for Type 2 response, p = contact pressure.

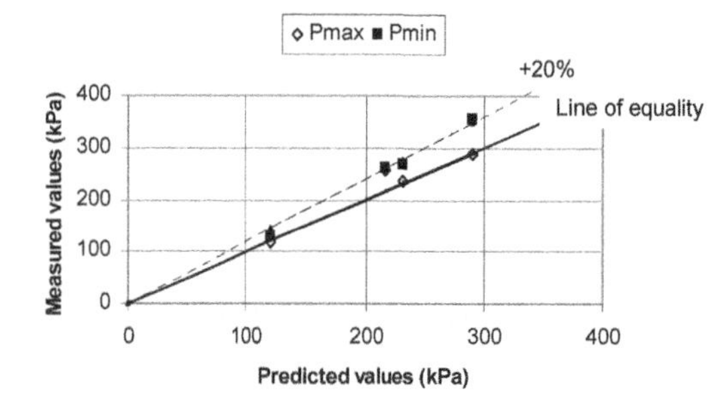

Figure 10. Comparison of predicted and measured values of wheel contact pressures at shakedown.

resembling road or railway structures. These results will be presented elsewhere.

7 CONCLUSIONS

The simple experiments described herein demonstrated that the lower bound shakedown theory developed by Yu (2005) for single layer systems under moving wheel loading gives realistic predictions.

ACKNOWLEDGEMENTS

The research described in this paper was supported by a research grant from the UK Engineering and Physical Sciences Research Council. The Authors are grateful to Mr B V Brodrick for his assistance with the experimental facilities.

REFERENCES

British Standards Institute. 1990. Methods of test for soils for civil engineering purposes. Classification tests. BS1377:2.

Brown, S.F. 1996. 36th Rankine Lecture: Soil Mechanics in Pavement Engineering, *Géotechnique* 46(3): 383–426.

Brown, S.F. 2004. Design considerations for pavement and rail track foundations. *Proc. Inst. Seminar on Geotechnics in Pavement and Railway Design and Construction*, Athens: 61–72.

Brown, S.F, Brunton, J.M, Hughes, D.A.B & Brodrick, B. V. (1985). Polymer grid reinforcement of asphalt, *Proc. Association of Asphalt Paving Technologists* 54: 18–41.

Brown, S.F & Chan, F.W.K. 1996. Reduced rutting in unbound granular pavement layers through improved grading design. *Proc. Inst. of Civil Engineers Transport* 117(): 40–49.

Cheung, L.W. 1994. Laboratory assessment of pavement foundation materials. *PhD thesis, University of Nottingham.*

Jacobs, F. A. (1977). Properties of rolled asphalt and asphaltic concrete at different states of compaction. *TRRL Report SR 288.*

Juspi, S. 2007. Experimental validation of the shakedown concept for pavement analysis and design. *PhD thesis, University of Nottingham.*

Melan, E. 1938. Der spannungsgudstand eines Henky-Mises schen Kontinuums bei Verlandicher Belastung. *Sitzungberichte der Ak Wissenschaften Wie (Ser.2A)* 147(73).

Sharp, R.W. & Booker, J.R. 1984. Shakedown of pavements under moving surface loads. *Journal of Transportation Engineering, ASCE* 110(1): 1–13.

Werkmeister, S. Dawson, A.R. & Weliner, F. 2001. *Permanent deformation behaviour of granular materials and the shakedown theory.* Transp. Research Record 1757. Transportation Research Board: 75–81.

Yu, H.S. 2005. Three dimensional analytical solutions for shakedown of cohesive-frictional materials under moving surface loads. *Proceedings of the Royal Society Series A,* 461 (2059): 1951–1964.

Advances in Transportation Geotechnics – Ellis, Yu, McDowell, Dawson & Thom (eds)
© 2008 Taylor & Francis Group, London, ISBN 978-0-415-47590-7

The response of ground penetrating radar (GPR) to changes in temperature and moisture condition of pavement materials

R.D. Evans
Jacobs, Derby, UK/Loughborough University, UK

M.W. Frost & N. Dixon
Loughborough University, UK

M. Stonecliffe-Jones
Jacobs, Derby, UK

ABSTRACT: The use of geophysical techniques to assess geotechnical and pavement structures can provide much useful information to the engineer. The development of ground penetrating radar (GPR) in recent years has led to its increasing use for pavement and geotechnical investigations, and the technique involves recording the amplitude and travel time of electromagnetic GPR signals reflected from features within the ground or structure of interest. Depths can be determined, and features of interest such as different layers, excess moisture, voids and changes in materials can be identified. The interpretation of GPR data depends largely on the 'dielectric constant' of the material(s), which governs the passage of GPR signals through a material and the amount of signal energy reflected from features within a structure.

This paper reports an investigation of pavement material samples, conducted under controlled conditions, using GPR. The effect of changes in material moisture and temperature on the dielectric constant, and hence the passage of GPR signals, was investigated. Core samples of bituminous material obtained from highway pavement sites were used to conduct a series of laboratory tests, in which the temperature of the material was controlled in the range from −5 to +45 degrees C, and the dielectric constant and GPR signal velocity were determined. Also, the materials dielectric constant and signal velocity were determined under dry and soaked moisture conditions.

The test programme allowed an assessment of the effect of changes in materials temperature and moisture condition to the response of data obtained during GPR investigations. The results of the testing showed that both moisture and temperature can have a significant effect on the data obtained from GPR surveys of pavement structures.

1 INTRODUCTION

1.1 *Ground penetrating radar (GPR) and dielectric permittivity*

One of the most useful techniques used for pavement investigation is ground penetrating radar (GPR), which involves recording reflections of electromagnetic waves transmitted into the pavement structure. GPR is completely non-destructive, and relatively quick to conduct compared to many other investigation techniques. The uses of GPR in pavement investigation include determination of layer thicknesses, location of construction changes, areas of high moisture, voids, reinforcement and other discrete objects.

The ability to use GPR data to assess pavement properties relies on the response of materials to the passage of electromagnetic waves transmitted from the GPR antenna. There are a number of important processes that can affect the propagation of GPR signals, and Olhoeft (1998) describes the electrical, magnetic and geometrical properties that are of importance in determining the performance of GPR. Important factors in the response of a material are its electromagnetic properties, namely dielectric permittivity (ε), magnetic permeability (μ) and electrical conductivity (σ).

The dielectric properties of pavement materials are of great importance when conducting GPR investigations, and Daniels (2004) provides a good overview. Dielectric substances are those which are poor conductors of electricity, but support electrostatic fields well, and the dielectric permittivity of a substance refers to its ability to store (i.e. 'permit') an electric field which has been applied to it.

Each material has a 'dielectric constant', which is a measure of its relative (to a vacuum) dielectric permittivity, and it is a critical parameter in the practical application of GPR data. The dielectric constant affects the velocity at which GPR signals will travel through the material, affects the 'reflection coefficient' (governing how much energy is reflected when there is a change in material), and also affects the resolution of data that is obtained. Therefore, understanding what factors affect the dielectric constant, and the degree to which they affect it, can assist in conducting GPR surveys and interpreting and understanding the data.

The measurement of the dielectric properties of asphalt materials can also be useful to assess density and provide information on compaction quality control. The presence of relatively low dielectricity air (in air voids) within materials will affect the overall dielectric properties of an asphalt mix, so it is possible to assess the compaction (i.e. density) of pavement material by measuring dielectric properties of the asphalt (as described by Saarenketo, 1997).

In GPR investigations, the dielectric constant can be determined by calculating the GPR signal velocity within the material. The velocity is related to the dielectric constant by the relationship shown in Equation 1, below:

$$v = \frac{c}{\left(\sqrt{\mu_r \varepsilon_r}\right)} \tag{1}$$

Where v = velocity of electromagnetic (i.e. GPR) wave through the material; c = velocity of light in free space (vacuum) = approximately 300,000 kms^{-1}; μ_r = relative magnetic permittivity (=1 for non-magnetic materials); and ε_r = dielectric constant (relative permittivity).

Reflections of GPR signals occur when the materials in the pavement have contrasting dielectric properties. In this scenario some of the radar energy passing from one material to the other is reflected back from the material boundary to the antenna. The amount of radar energy reflected is indicated by the reflection coefficient (which depends on the contrast in dielectric properties of the materials) and is given by:

$$\rho = \frac{\left(\sqrt{\varepsilon_1}\right) - \left(\sqrt{\varepsilon_2}\right)}{\left(\sqrt{\varepsilon_1}\right) + \left(\sqrt{\varepsilon_2}\right)} \tag{2}$$

Where ρ = reflection coefficient; ε_1 = dielectric constant of the upper material; and ε_2 = dielectric constant of the lower material.

1.2 *Temperature and moisture effects on dielectric properties*

Measurements of the dielectric properties of various types of materials can prove useful. The dielectric properties of wood can be used to determine density and moisture content non-destructively, and Kabir et al (2001) showed that the dielectric constant of wood increases with increased temperature. Previous work has also shown that, at GPR frequencies, both temperature and moisture affects the dielectric properties of pavement materials. Jaselskis et al (2003) investigated the dielectric properties of a number of asphalt samples in the frequency range from 100 Hz to 12 GHz, whilst researching the use of a microwave pavement density sensor. It was observed that the dielectric constant of asphalt samples slightly increased with temperature, and also increased with moisture (greatly at low frequencies and slightly at higher GPR frequencies).

An overview of the work conducted on the effect of moisture content on the dielectric constant of materials at GPR frequencies is given by Daniels (2004). Water has a dielectric constant of approximately 80, so a relatively small increase in moisture content can cause the bulk dielectric constant of asphalt materials (with dielectric constant of approximately 2–12) to be greatly altered. Methods such as time domain reflectometry (TDR) rely on this moisture-dielectric relationship to assess soil moisture content from dielectric constant measurements.

Shang et al. (1999) conducted a series of tests using an electromagnetic wave apparatus to assess the dielectric constant of a number of laboratory prepared asphalt samples in dry and 'soaked' (up to 1.25% moisture content) conditions. It was found that moisture content was a dominant factor, with the dielectric constant of samples increasing linearly by 0.62 for each 1% increase in moisture content. Further work by Shang & Umana (1999) indicated that beyond a moisture content of 1.2%, the effect on dielectric constant was greater and non-linear.

1.3 *Investigation synopsis*

A series of laboratory tests were conducted, so that the dielectric constant could be calculated for bitumen bound pavement material at a range of temperatures (with constant moisture content), and also for material in soaked and dry conditions (at constant temperature). The test methodology employed was to use material taken from in-service pavements and use GPR equipment and analysis procedures in a similar manner to that employed during in situ pavement investigations, so that it would be possible to best relate the findings of the study in a practical engineering context.

For testing at various temperatures, it was important to use material samples that were dry, as any removal of moisture from the material (as temperatures were increased) would affect results. Also, for the testing at different moisture conditions it was important to test dry and soaked material at the same temperature to avoid any influence temperature may have on the results.

2 METHODOLOGY

2.1 *Materials*

One of the underlying themes of the work was to be able to apply the results as well as possible to in-service materials and conditions. Therefore, the pavement material used in the study was obtained from cores taken from in-service trunk road pavements in the UK. A total of 20 core samples (each of 150 mm diameter) were obtained from bituminous bound pavements, consisting of various thicknesses of hot rolled asphalt (HRA) and dense bitumen macadam (DBM) layers. The core samples ranged from approximately 220 mm to 420 mm in depth, and were typical of trunk road bituminous pavement constructions existing in the UK.

2.2 *Laboratory preparation of samples*

2.2.1 *Temperature*

10 core samples were selected for testing at different temperatures. The dielectric constant was determined for each core sample at seven discrete temperatures, ranging from −5 to +45 degrees centigrade, chosen to give a typical range of potential temperatures a pavement may be subjected to in the UK. The presence of water in the samples would have an affect on the test results, so it was essential that the material was dry during testing at different temperatures. Initially, the cores were dried for 48 hours in a climate chamber, to ensure that any free water in the material had been removed. After the drying period, each core was conditioned at the desired temperature (starting with the highest) for 48 hours in the climate chamber, before being removed from the chamber and immediately tested using GPR. Once the testing at the desired temperature had been completed, the core was placed back into the chamber and conditioning at the next temperature was undertaken.

2.2.2 *Moisture content*

Prior to the temperature testing, all 20 of the core samples were tested under differing moisture conditions. When the cores had initially been extracted from the in-service pavements, they had been stored (at room temperature) for a number of weeks. This led to the core materials being in a 'dry' condition, although no attempt had been made to remove all free moisture from the material by deliberate drying. The cores were initially tested in this 'dry' condition, at room temperature. Following 'dry' testing, the cores were submerged in a water filled tank for 48 hours, and re-tested in this 'soaked' condition (at room temperature).

2.3 *GPR test procedures*

A GPR system operating a dipole antenna at a centre frequency of 1.5 GHz was used to collect test data. The GPR recorded the travel time of signals transmitted into the top of the core samples and reflected back from their base.

Following conditioning at the required temperature or moisture condition, the core samples were placed upright, with a metal plate placed at the base. The metal plate provided a perfect reflector for GPR signals to ensure easy identification of the base of the core sample. GPR pulses were emitted from the antenna transmitter downwards into each sample, travelling along its full depth before being reflected back to the antenna receiver from the metal plate at the base of the core. For each GPR pulse the travel time of the reflected signal were recorded to within 0.03 ns (nanoseconds).

The average velocity of the GPR signal within each of the core samples was the distance travelled divided by the time taken. The length of each core was measured, and the travel time of the signal was determined from the data recorded by the GPR system. Hence, the dielectric constant of the material can be determined by substituting into Equation 1, giving:

$$d = \frac{ct}{\left(\sqrt{\varepsilon_r}\right)} \quad (3)$$

Where d = depth (i.e. length) of core sample; t = one-way travel time of reflected signal. μ_r (in Equation 1) can be taken as being = 1for bituminous pavement materials. For each core (at each temperature or moisture condition) the travel time of reflected signals was recorded and the dielectric constant calculated.

3 RESULTS

3.1 *Dielectric constant variation with temperature*

10 individual core samples of bituminous pavement material (referenced Core 1 to Core 10) were scheduled for testing at −5, zero, 5, 15, 25, 35 and 45 degrees centigrade. During the conditioning of the samples at the highest temperature the bituminous binder of the material of Cores 5 and 7 softened to the point where the material suffered partial collapse. Hence, Cores 5 and 7 were not used during the test programme.

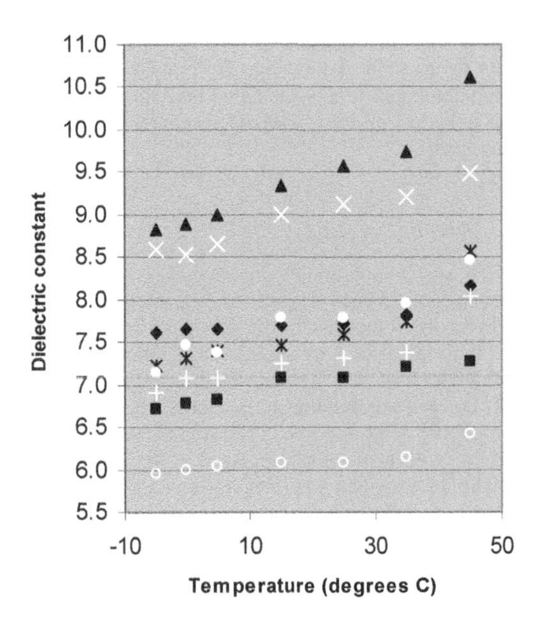

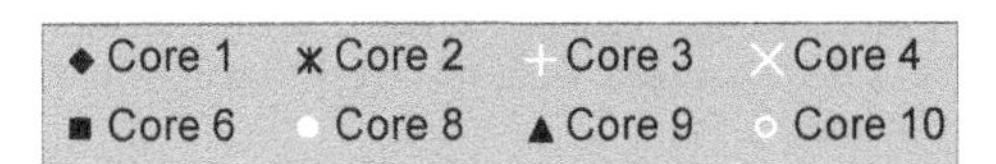

Figure 1. Dielectric constant of bituminous core samples determined at 1.5 GHz in temperature range −5°C to 45°C. (Core samples 5 and 7 damaged during testing).

Figure 1 shows the results of the testing, and it can be seen that for each individual core material sample there was an overall increase in dielectric constant as the temperature of the material increased.

The rate of increase in dielectric constant varied between specific material samples but most showed a similar trend. The size of the increases in the calculated dielectric constant values over the temperature range from −5 to +45 degrees centigrade were between 7.3% and 20.2%. The average rates of increase in dielectric constant (for the whole temperature range) were between 0.14% and 0.40% per degree centigrade increase in temperature. However, some of the results indicate that there may be a non-linear trend, as several (although not all) of the samples showed larger than average rates of increase for the dielectric constant between 35 and 45 degrees centigrade.

From the collected data it can also be observed that for the 8 samples tested there was a large range of values of dielectric constant determined at each temperature. At 15 degrees centigrade, for example, the dielectric constant of the core samples ranged between 6.1 and 9.3, showing that generically similar materials at the same moisture and temperature condition could posses quite different dielectric properties.

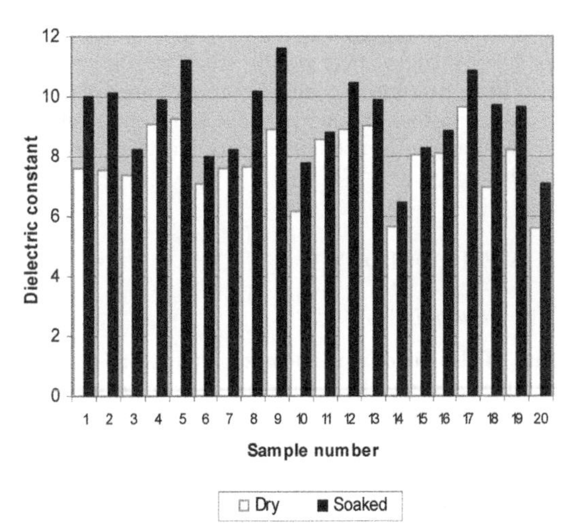

Figure 2. Dielectric constant of bituminous material samples, in 'dry' and 'soaked' condition, determined at 1.5 GHz.

3.2 *Dielectric constant variation with moisture*

The results of the tests conducted on the core samples at different moisture contents are shown in Figure 2. The dielectric constant was determined for 20 individual samples, in both the 'dry' and 'soaked' conditions (see Section 2.2.2) at room temperature.

The data shows that for each individual sample, the dielectric constant was greatest when the material was in the 'soaked' condition. However, the magnitude of the increase in dielectric constant varied greatly between samples, from the smallest increase of 3% (Sample 11) to the greatest increase of 39% (Sample 18).

The average dielectric constant of the material samples when dry was 8.0, and when soaked was 9.3 (a difference of 16%) which corresponds to a decrease in signal velocity from approximately 0.106 m/ns to 0.098 m/ns (see Equation 1).

4 *DISCUSSION*

4.1 *Temperature*

The data from the temperature testing shows that there is a relationship between the dielectric constant of bitumen bound pavement materials and temperature, under the conditions tested. The data collected, however, is limited in its scope and so further investigation is also needed to more comprehensively asses this relationship.

The results plotted in Figure 1 indicate that there is a general trend, with an average increase in dielectric constant of 0.27% per degree centigrade increase in temperature, but the specific trend for each individual material was over a range of values and requires further investigation.

The mechanism for the increase in dielectric constant with temperature has been previously investigated in relation to non-pavement materials. Studies such as Hrubesh & Buckley (1997) and Satish et al (2002) have investigated the effect of temperature on the dielectric properties of silica aerogels (for use as a low dielectric material in electronics) and ceramic-polymer composites. In these studies the dielectric constant was observed to increase with temperature, and it was concluded that the cause was the greater mobility of molecules within the material (caused by the elevated temperatures) allowing dipoles to re-orient more readily, and thus causing an increase in the ability of the material to support electromagnetic fields. In practical terms, this means the increasing temperatures facilitated a mechanism that increases the dielectric constant. It is thought that this process is also occurring within the asphalt material investigated in this study.

The data shows that, for the material samples at the same condition (temperature) there is a range of dielectric constant values. This agrees with much previous work, including that of Evans et al (2007), which shows that generically similar asphalt materials, in similar conditions, can have different dielectric constant values.

4.2 *Moisture*

The entire range of dielectric constants during the testing process was 5.6 to 11.6, which gives an indication of the potential range of variation in dielectric properties that may be present in typical bituminous pavement materials, depending on their temperature and moisture condition.

Greater dielectric constants (i.e. lower signal propagation velocities) were recorded for each material when the material was soaked. Low dielectricity air ($\varepsilon_r \sim 1$) in voids present in materials are replaced within the material matrix by relatively high dielectricity water ($\varepsilon_r \sim 80$), causing the overall bulk dielectric constant of the material to increase.

The individual nature of the core sample materials will have had a great effect on the change in dielectric constant. The amount of moisture present in the 'soaked' samples compared to the 'dry' samples will affect the magnitude of the difference in dielectric constant between these two states. The amount of interconnected air voiding within the materials will govern the amount of moisture that the samples could absorb during the soaking process, so it is likely that the % increase in dielectric constant between dry and soaked states is an indicator of the amount of voiding present in the material.

Moisture testing showed large range (3% to 39%) of dielectric constant increase when material was soaked. From a qualitative view, the core logs for the materials which tended to have lower increases in dielectric constant appeared to be slightly more voided, but as no quantitative assessment of voiding has yet been undertaken, no firm conclusions can be drawn. Core logs showed varying degrees of voiding within materials, but further testing (which is planned) will establish the actual amount of voiding present within the material samples.

4.3 *General*

The dielectric constant of material can vary depending on signal frequency (although only by a small amount within the GPR signal frequency range), but it should be noted that the results in this study are related to a specific signal frequency (1.5 GHz).

The nature of the materials used in this study means that it is not possible to provide specific conclusions to some aspects of the work, because the precise mixes and materials used in each core sample were different. The material samples obtained were nominally similar (i.e. HRA surface course with DBM binder course and base), but each individual sample was taken from a different site on the UK trunk road network. Although every sample used can be described as a bitumen bound asphalt material, the individual nature of the aggregates and bitumen binder used has meant that there is a range of dielectric properties present in the materials tested.

Although the use of samples taken from different sites allows less control over the specific nature of the material used in the study, the use of such material, however, means that the results obtained can be taken to be more representative of the degree of variability likely for in-situ bituminous materials, than might be otherwise obtained from laboratory prepared samples.

Within the results obtained, there will be certain uncertainties and potential errors. Experimental work and analysis was conducted to minimise these as much as possible. This included repeat testing of several of the material samples (which re-produced test results to an acceptably high degree), but factors remain which, given the methodology employed, can not be influenced. For example, the GPR equipment used is capable of measuring signal travel times to within 0.03 ns. This precision could lead to an error in individual dielectric constant calculation of approximately 0.07. Whilst it is important to note such potential for error, the level of uncertainty in the results of this study is not considered to have significantly affected the data or the conclusions drawn.

The use of unmodified GPR equipment, and the use of material samples obtained from in-service pavements, allows the study data to realistically represent the degree of accuracy that may be obtained from in-situ GPR pavement investigations, and shows that practical application of the results is possible.

Whilst the effect of moisture on dielectric constant has been widely investigated (and is used by some methods to assess material moisture content) the effect of temperature variation on asphalt materials is less researched. It is hoped to build further on the initial investigations described in this paper to more fully investigate the changes in dielectric properties with both temperature and moisture.

5 CONCLUSIONS

Within the range of conditions investigated, the results from the study show that:

- The dielectric constant of asphalt materials increases with temperature.
- The mechanism for the increase in dielectric constant with temperature is likely to be the increased re-orientation ability of dipoles, resulting from the increase in thermal energy.
- The dielectric constant of asphalt materials increases with moisture.
- The response of an asphalt material to wetting, and the resulting effect on its dielectric constant, is likely to be governed by the amount of air voids initially present in the material matrix.
- The dielectric properties are material specific, and generically similar bituminous materials do not necessarily have the same dielectric constant.
- Bearing the previous point in mind, the calibration of GPR data to the correct signal velocity (which is governed by the dielectric constant) is required on a site specific basis, to ensure the accuracy of data analysis and interpretation of in-situ GPR pavement investigation data.
- The work conducted for this paper is limited in some aspects, and further work is required and planned to address these issues.

The use of GPR for investigation of bituminous pavement material relies on the dielectric properties of the material. The specific type of material and the condition it is in have an effect on the dielectric properties. A wider understanding of the dielectric response of bituminous materials, under conditions which might be expected in-situ, allows a more comprehensive understanding of the significance of data obtained by GPR.

REFERENCES

Daniels DJ. 2004. *Ground penetrating radar*, 2nd edition. The Institution of Electrical Engineers, London. ISBN 0 86341 360 9.

Evans RD, Frost MW, Stonecliffe-Jones M & Dixon N. 2007. Assessment of the in-situ dielectric constant of pavement materials. *Transportation Research Board 86th Annual Meeting: compendium of papers*, Paper 07-0095. Washington DC, USA.

Jaselskis EJ, Grigas J & Brilingas A. 2003. Dielectric properties of asphalt pavement. *Journal of Materials in Civil Engineering*. Volume 15, No 5, ASCE, pp. 427–434.

Hrubesh LH & Buckley SR. 1997. Temperature and moisture dependence of dielectric constant for silica aerogels. Materials Research Society Meeting, March 31–April 4, 1997, San Francisco, USA.

Kabir MF, Daud WM, Khalid KB & Sidek HAA. 2003. Temperature dependence of the dielectric properties of rubber wood. *Wood and Fiber Science*, 33 (2), pp. 233–238.

Olhoeft GR. 1998. Electrical, magnetic and geometric properties that determine ground penetrating radar performance. In *Proc. of GPR '98, Seventh International Conference on Ground Penetrating Radar*, May 27–30, 1998, The University of Kansas, Lawrence, USA, pp. 177–182.

Saarenketo T. 1997. Using ground penetrating radar and dielectric probe measurements in pavement density quality control. *Transportation Research Record: Journal of the Transportation Research Board* No. 1575, TRB, National Research Council, Washington D.C. pp. 34–41.

Satish B, Sridevi K & Vijaya MS, 2002. Study of piezoelectric and dielectric properties of ferroelectric PZT–polymer composites prepared by hot-press technique. Journal of Physics D: Applied Physics, Volume 35, Number 16, pp. 2048–2050.

Shang JQ, Umana JA, Bartlett FM & Rossiter JR. 1999. Measurement of complex permittivity of asphalt pavement materials. *Journal of Transportation Engineering*, Volume 125, Issue 4, pp. 347–356.

Shang JQ & Umana JA. 1999. Dielectric constant and relaxation time of asphalt pavement materials *Journal of Infrastructure Systems*, Volume 5, No. 4, pp. 135–142.

Advances in Transportation Geotechnics – Ellis, Yu, McDowell, Dawson & Thom (eds)

Impact of changes in gradation of bases on performance due to pulverization

A. Geiger
Soils and Aggregates Branch, Materials and Pavements Section, Texas Department of Transportation, Austin, TX

J. Garibay, S. Nazarian, I. Abdallah & D. Yuan
Center for Transportation Infrastructure Systems, The University of Texas at El Paso, El Paso, TX, USA

ABSTRACT: Pulverization of pavement base materials is routinely carried out for rehabilitation of roads through full-depth reclamation (FDR). The pulverized materials are then, either as-is or mixed with a stabilizer, re-compacted in place. The optimum stabilizer type and content are currently determined either based on experience or through a series of laboratory tests that evaluates the strength, stiffness and durability of the mix. For lab testing, base materials are retrieved from the site way before pulverization. An extensive laboratory study was performed to determine the impact of changes in gradation on the performance of bases in order to achieve the desirable performance. It was found that the change in gradation indeed impacts the properties of the mix and the amount of stabilizer needed, and should be considered in the design stages of FDR.

1 INTRODUCTION

Many highway agencies use full-depth reclamation (FDR) along with soil stabilization to rehabilitate their roads more economically (Mallick et al. 2002). A major concern during the construction of base layers is the degradation and crushing of aggregates due to handling, pulverization and placement. Changes in gradation may adversely affect the strength and stiffness of the final product with or without stabilizing agents. The selection of the type and determination of the percentage of additive depend on the gradation of the material.

An aggregate base must meet its purpose as the main structural layer of a pavement by performing the following three functions: subgrade protection, support for surfacing, and as a construction platform. Changes in the strength/stiffness characteristic of the base layer may result in premature failure of the pavement (Dawson 2003).

2 LABORATORY STUDY

In order to evaluate the impact of changes in gradation on base performance several strength and stiffness tests were performed. The results from a dolomite base in El Paso, TX stabilized with 2% to 6% either cement, fly ash or lime are shown in detail in this paper. The aggregates were characterized in a number of ways.

Table 1. Results from ACV and AIV Tests.

		Constituents		
Test	Value	Gravel Ret. #4 Sieve %	Sand %	Fines Pass. #200 Sieve %
AIV	22	78	19	3
ACV	19	66	31	3

One promising method for characterizing the crushing potential of the aggregates during pulverization was the Aggregate Crushing Value (ACV) and Aggregate Impact Value (AIV) as per British Standard BS-812 (Garibay et al. 2007). After each test, the materials were also subjected to a sieve analysis to determine the constituents of the mix. The results are shown in Table 1. During these tests, most of the crushed aggregates are converted to sand. This observation concurred with careful field observation as reported by Garibay et al. (2007).

2.1 *Development of gradation curves*

In the development of the gradation curves, the Texas Department of Transportation (TxDOT) Item 247 for flexible bases was assumed as the control gradation. As discussed in Geiger et al. (2007), three additional gradation curves were developed that contained excessive

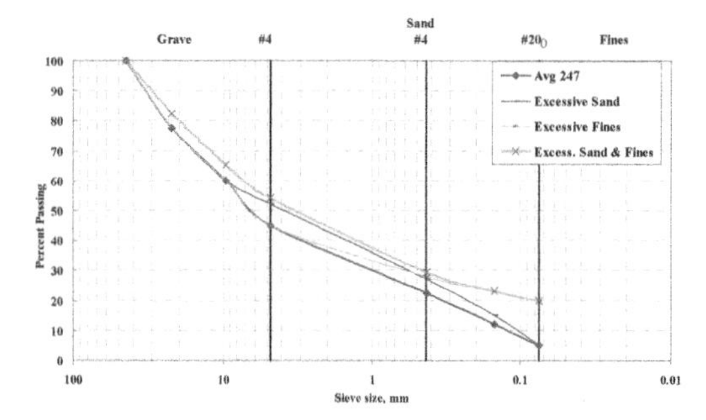

Figure 1. Gradation Curves for Blends Used in This Study.

Table 2. Variation in Optimum Moisture Content with Aggregate Blend and Stabilizer Type.

Gradation	Avg. 247	Excess Sand	Excess Fines	Excess Sand and Fines
0% Cement	7.2	7.5	7.7	7.1
2% Cement	8.4	7.5	7.9	7.6
4% Cement	8.6	7.7	7.4	7.4
6% Cement	7.5	8.1	8.1	7.9
0% Lime	7.2	7.5	7.7	7.1
2% Lime	7.8	8.0	8.1	7.1
4% Lime	7.5	8.6	8.3	7.1
6% Lime	8.9	8.7	8.4	7.6
0% Fly Ash	7.2	7.5	7.7	7.1
2% Fly Ash	7.8	8.0	8.1	7.1
4% Fly Ash	7.5	8.6	8.3	7.1
6% Fly Ash	8.9	8.7	8.4	7.6

sand (ES), excessive fines (EF), and excessive sand and fines (ESF) to simulate the possible scenarios that occur during pulverization. These three and the control (called Avg 247) gradations are shown in Figure 1.

2.2 *Moisture-density relationships*

For each gradation and stabilizer content, the moisture-density (MD) relationships were determined.

The optimum moisture contents (OMC) from all gradations and the three stabilizers are shown in Table 2. The OMC's for the four blends with no additives or with cement were fairly similar with an average value of about 7.4%. The OMC increased as the stabilizer increased for the materials stabilized with lime and fly ash. However, the fly ash specimens with excess sand and fines demonstrated a reduction in the OMC as the fly ash content increased.

The maximum dry densities (MDD) did not vary much with the aggregate blend and stabilizer type and concentration. The average MDD was 2,195 Kg/m^3 with a coefficient of variation (COV) of 2.4%.

To maximize the use of the specimens compacted to establish the MD curve, specimens were wrapped

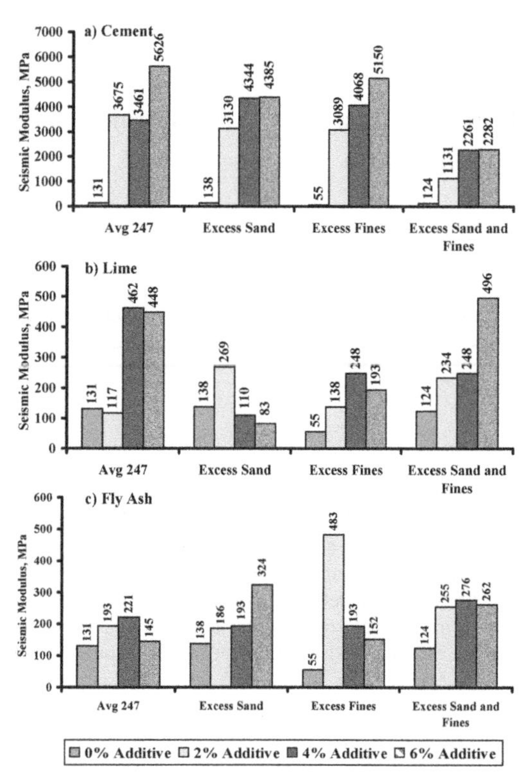

Figure 2. Variation in Seismic Modulus at OMC with Aggregate Blend and Stabilizer Type.

in a membrane for 24 hours so their modulus could be determined using the free-free resonant column (FFRC) test (Nazarian et al., 2003). The FFRC device provides the seismic modulus of the specimen rapidly and nondestructively. Hilbrich and Scullion (2007) have shown that this method is quite reliable for measuring the moduli of stabilized materials. Figure 2 presents the variations in seismic modulus (SM) at OMC for different blends at varying stabilizer contents. These values corresponded best with those measured in the field during quality management.

As shown in Figure 2a, the addition of cement seems to increase the seismic modulus significantly. The lime and fly ash are not as effective in increasing the seismic modulus of this base, as shown in Figures 2b and 2c.

2.3 *Triaxial strength*

The triaxial strength parameters of the four blends were first determined. The angles of internal friction were 59° (Avg 247), 55° (ES), 50° (EF) and 53° (ESF). The cohesions were approximately 55 kPa, 50 kPa, 70 kPa and 60 kPa, respectively. This indicates that the three alternative blends are less strong than the control blend, with the mix with EF being the weakest.

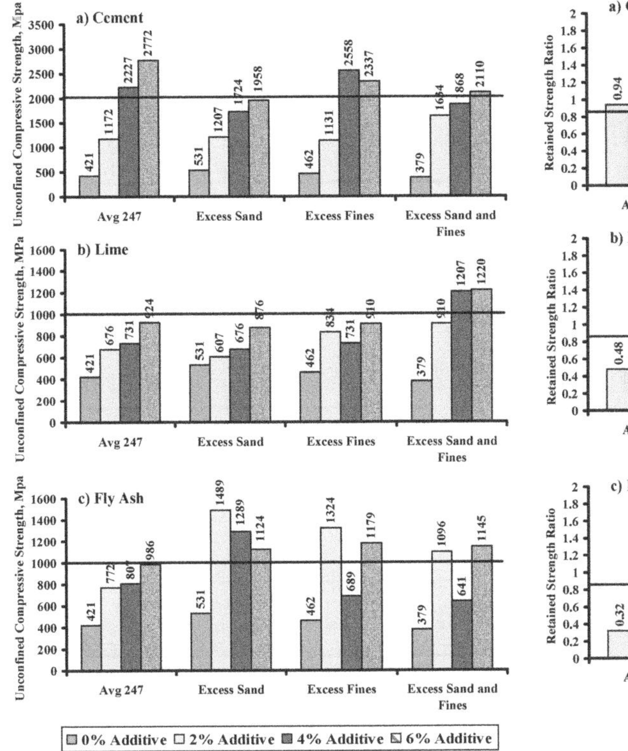

Figure 3. Unconfined Compressive Strength of Soil-Lime and Soil-Fly Ash Specimens.

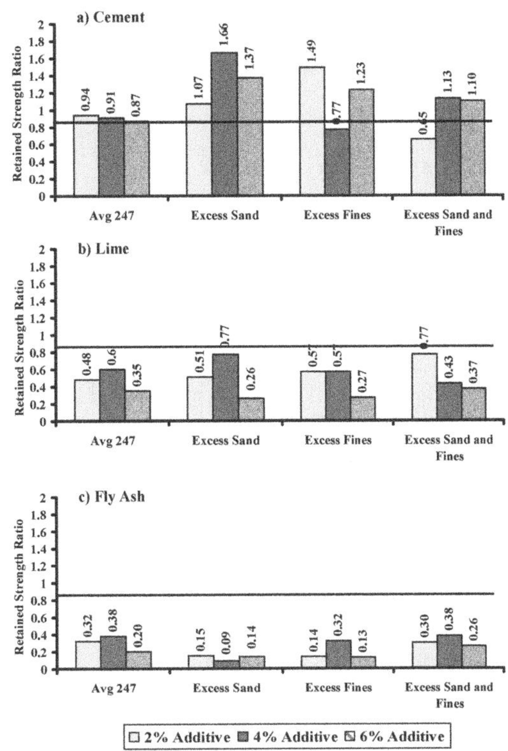

Figure 4. Retained Strength Ratios for Capillary Moisture Conditioning.

The unconfined compression strength (UCS) tests were performed on specimens prepared with different blends and stabilizer contents. Each specimen was prepared at the corresponding optimum moisture content for the given blend and stabilizer content and cured for seven days before the UCS tests. The UCS results for cement stabilized mixes are summarized in Figure 3a. According to Scullion et al. (2003), a UCS of close to 2000 MPa for cement and 1000 MPa for lime and fly ash mixes are desirable. The amount of cement needed to reach 2000 MPa depended on the gradation. The Avg. 247 and Excess Fines blends required about 4% cement, while the blend with Excess Sand and Fines required about 6% cement. The blend with excess sand required more than 6% cement.

None of the specimens prepared with lime achieved the desired 1000 MPa strength as shown in Figure 3b, except for the blend with the Excess Sand and Fines. Several of the fly ash mixes achieved the target strength.

2.4 *Retained strength*

The retained strength concept corresponds to the strength of the mixtures after being subjected to moisture conditioning. Duplicate sets of specimens used to estimate the strengths were subjected to ten days of capillary saturation after curing for seven days. The retained strength ratio (RSR) was determined using:

$$RSR = \frac{\text{Compressive Strength after Moisture Conditioning}}{\text{Compressive Strength without Moisture Conditioning}} \quad (1)$$

A retained strength ratio of above 85% is usually considered desirable for stabilized mixtures.

The retained strength ratios are shown in Figure 4. For almost all cases where cement was used, the RSRs are greater than 85%. The EF blend with 4% cement only achieved an RSR of 77%. Aside from experimental errors, the reason for this matter is unknown. The RSR for the ESF blend with 2% cement is 65%, perhaps because of the low amount of stabilizer and high amount of fines.

In Figure 4b, the RSRs for the lime-treated specimens fail to reach the 85%. The maximum RSR was about 77%. The fly ash specimens (Figure 4c) yielded substantially smaller RSRs with a maximum RSR of 38%.

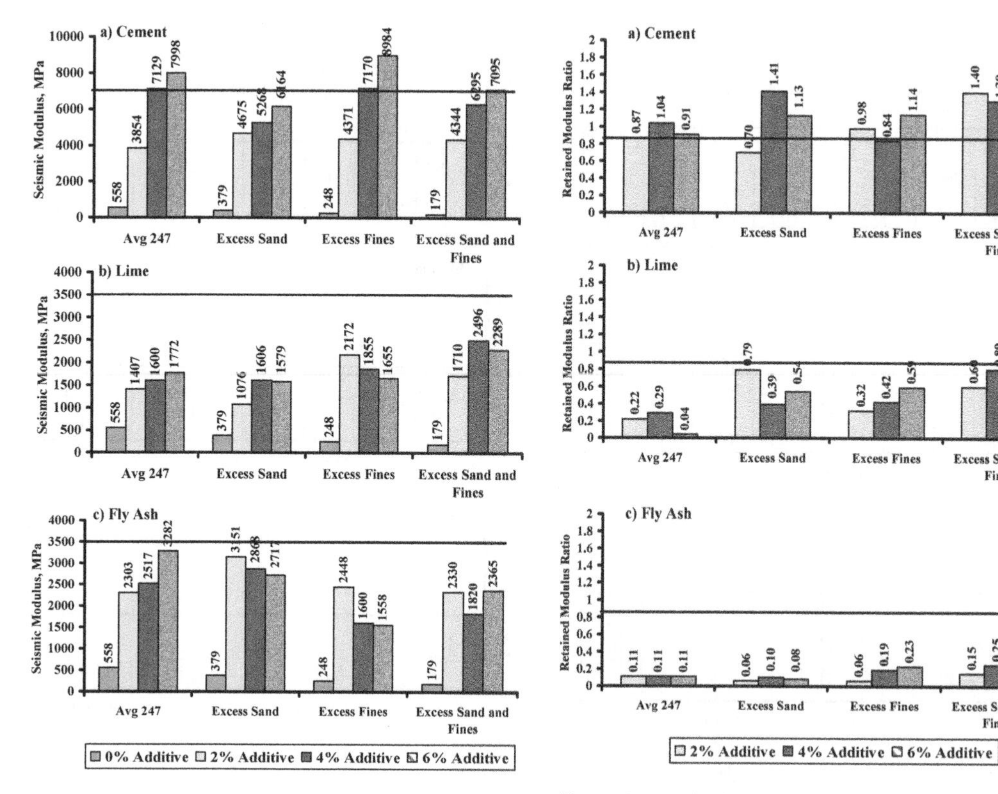

Figure 5. Seismic Modulus vs. Stabilizer Content.

Figure 6. Retained Modulus Ratios for Capillary Moisture Conditioning.

2.5 *Modulus*

Aside from strength, the modulus of each mix should be ideally determined. Modulus is one of the most important parameters considered in the structural design of flexible pavements. The traditional method of determining modulus is to perform resilient modulus or repeated-load triaxial tests. Based on the recommendation of Hilbrich and Scullion (2007), only the seismic moduli of the specimens were measured. The benefit of this type of measurement is that the same specimens prepared for the UCS tests can be tested, therefore minimizing the number of the specimens that have to be prepared.

A minimum seismic modulus of 7 GPa is perceived reasonable for soil-cement mixes. For lime or fly ash treated soils, a minimum modulus of 3500 MPa is selected. The variations in seismic modulus with the gradation and cement content are shown in Figure 5a. The results are quite consistent with those from the UCS tests shown in Figure 3. The recommended cement contents based on the UCS also hold for the modulus criterion of 7 GPa. As shown in Figures 5b and 5c, none of the lime or fly ash gradation-stabilizer combinations met the desired modulus of 3500 MPa.

2.6 *Retained modulus*

The retained modulus idea follows the same requirements set by the retained strength ratio. The retained modulus ratio (RMR) is determined by:

$$RMR = \frac{\text{Seismic Modulus after Moisture Conditioning}}{\text{Seismic Modulus without Moisture Conditioning}} \quad (2)$$

The RMR for the cement stabilized specimens are presented in Figure 6a. The RMRs of almost all the blends with cement exceed the 85% value with one exception. The ES blend with 2% cement achieved RMRs of 70%.

Figure 6 also presents the RMRs for the lime and fly ash mixes. For the lime specimens, none of the mixes achieved an RMR of 85% (Figure 6b). The maximum RMR is about 80% in three cases. For the fly ash-stabilized blends, the RMRs are significantly less than 85% as well. The highest RMR in this case is about 25%. The RMR and RSR values demonstrate the same trends. To minimize the number of specimens necessary to perform appropriate mix design, the RMR based on the nondestructive FFRC tests can be used to minimize the number of specimens.

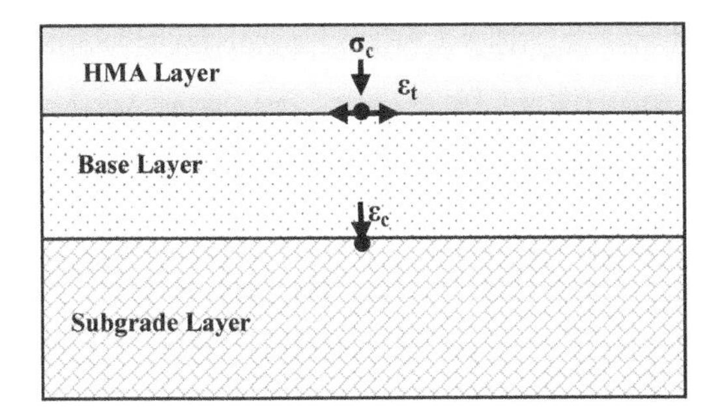

Figure 7. Critical Stresses and Strains in a Three Layer Flexible Pavement System.

This study demonstrates that the changes in gradation due to pulverization do impact the performance of the mixes.

2.7 *Implication on structural capacity*

The implications of changes in the modulus of the stabilized base due to change in gradation were analyzed to estimate the required pavement thickness in order to obtain similar performance in subgrade rutting, hot mix asphalt (HMA) rutting and fatigue cracking from all blends. Figure 7 summarizes the critical stresses and strains that were used to quantify the fatigue cracking and rutting.

Two different three-layer pavement sections with typical cross-sections encountered in Texas were considered. A thin (50 mm) and a thick (125 mm) HMA layers were considered. A modulus of 3.5 GPa and a Poisson's ratio of 0.33 were assigned to the HMA. The thickness of the pulverized base was assumed to be 300 mm with a Poison's ratio of 0.35. The modulus of this layer varied based on the laboratory results obtained for the four blends with 4% cement. The seismic modulus was multiplied by 0.7 to obtain the equivalent resilient modulus as recommended by Hilbrich and Scullion (2007). The subgrade was assumed to have a modulus of 70 MPa with a Poisson's ratio of 0.40.

The base modulus from the Avg. 247 blend with 4% cement was used as the control value. The moduli from the other three blends (ES, EF and ESF) with 4% cement were then used to determine the equivalent layer thicknesses for similar performance to the control base.

In order to determine the equivalent pavement thicknesses for the bases, the critical stresses and strains were determined for the control blend first. Using the resilient moduli of the alternative blends, the thickness of either the HMA or the base was then changed until the critical stresses and strains were equal or less than those from the control section. The equivalent thicknesses are reported in Table 3. The

Table 3. Layer Thicknesses Required for Equivalent Performance.

	HMA Thickness when Base Thickness Maintained Constant		Base Thickness when HMA Thickness Maintained Constant	
	50-mm HMA	125-mm HMA	50-mm HMA	125-mm HMA
Blend	mm	mm	mm	mm
ES	115	190	355	380
EF	50	125	300	300
ESF	90	165	330	330

EF blend seems to be performing as well as the control blend. However, the ESF and ES blends require thicker base or HMA to provide the same performance. Unfortunately, Garibay et al. (2007) has shown that for a number of field cases, the pulverization activity turn the gravel to sand. As such, the change in gradation should be considered in the mix design to ensure intended performance from the pulverized materials.

3 CONCLUSIONS

Based on the knowledge gained from this study, the following observations can be made:

- The optimum moisture content and the maximum dry unit weight for the stabilized materials may differ by as much as 2% and 160 Kg/m^3 as the gradation and the stabilizer content changes.
- As the fine sand content of the mix increases, the strength and stiffness of the stabilized mix decreases. As such, more additives are required if the pulverization turns gravel to sand. When the fines content increases, the strength and stiffness of the mix is slightly compromised.
- The UCS for cement stabilized material consistently increased as the cement content increased. Yet, for lime and fly ash stabilized specimens, the UCS decreased as the stabilizer content increased after the specimen was subjected to moisture conditioning. When specimens were tested prior to moisture conditioning, the lime specimens showed an increase in UCS with an increase in lime content.
- The retained strength ratio (RSR) of 85% for cement stabilized soil was readily achieved regardless of the blend or cement content. The lime and fly ash specimens did not achieve the RSR of 85%.
- Changes in gradation during pulverization impacts the performance of the pavement. The change in gradation should be evaluated and considered during mix design.

ACKNOWLEDGEMENTS

The authors wish to acknowledge the support of the Texas Department of Transportation.

REFERENCES

Dawson, A.R. (2003), "The Unbound Aggregate Pavement Base" <http://www.nottingham.ac.uk/~evzard/carpap2a.pdf> (October 3, 2005).

Geiger, A. Yuan, D. Nazarian, S. and Abdallah, I. (2007), "Effects of Pulverization on Properties of Stabilized Bases", Research Report 0-5223-1, The University of Texas at El Paso, El Paso, Texas.

Garibay J.L., Yuan D., Nazarian S., and Abdallah I. (2007) "Guidelines for Pulverization of Stabilized Bases," Research Report 0-5223-2, The University of Texas at El Paso, El Paso, Texas.

Hilbrich and Scullion (2007) "A Rapid Alternative for Lab Determination of Resilient Modulus Input Values for the AASHTO M-E Design Guide," Transportation Research Board Annual Meeting CD, Washington, DC.

Mallick, R.B., Bonner, D.S., Bradbury, R.L., Andrews, J.O., Kandhal, P.S., Kearney, E.J., (2002), "Evaluation of Performance of Full Depth Reclamation (FDR) Mixes", Transportation Research Board, National Research Council, Washington, D.C. <http://cee.wpi.edu/ce_faculty/faculty_pages/ FDR_field3.pdf> (May 18, 2005).

Nazarian, S., Williams, R., and Yuan, D. (2003), "A Simple Method for Determining Modulus of Base and Subgrade Materials," ASTM STP 1437, ASTM, West Conshohocken, PA, pp. 152–164.

Scullion T., Guthrie W.S., Sebesta S.D. (2003), "Field Performance and Design Recommendations for Full Depth Recycling in Texas," Research Report 0-4182-1, Texas Transportation Institute, College Station, TX.

Advances in Transportation Geotechnics – Ellis, Yu, McDowell, Dawson & Thom (eds)
© 2008 Taylor & Francis Group, London, ISBN 978-0-415-47590-7

Evaluation of a tropical soil's performance in laboratory

T.B. Pedrazzi & C.E. Paiva
Doctor degree UNICAMP-BR & doctor UNICAMP-BR

ABSTRACT: This work has the objective to develop a study of the mechanical characteristics of one tropical soil typical of the agricultural region metropolitan city of Campinas, in São Paulo, under the condition of traffic of vehicles of average load and with speed very low due to the constant holes and the lacks of any type of maintenance.

1 INTRODUCTION

In Brazil, most (90%) (DNIT, 2007) of highway network is not paved and is under municipal jurisdiction. This majority, the absolute predominance is of soil road, mainly of tropical soils in its surface and with thicknesses upper the 1,0 meter, and without any type of covering. This preference is, mainly, due to the generalized procedure of maintenance adopted for the municipal organisms in which, when defects in the road surface occur it is carried out one cut and removal of all the superficial layer until the finish of the eventual holes. The new surface is displayed to the traffic until presenting problems again and then, being regularized in an inferior quota. This procedure continues until the depth of 1,0 meter, when points of superficial deformation occur and the draining is engaged, compelling the execution of located interventions and the construction of a new road platform with imported material and in superior quota, for retaking the practical common of maintenance adopted. This work has the objective to develop a study of the mechanical characteristics of one tropical soil typical of the agricultural region metropolitan city of Campinas, in São Paulo, under the condition of traffic of vehicles of average load and with speed very low due to the constant holes and the lack of any type of maintenance.

2 CHARACTERISTICS ESSENTIAL TECHNIQUES OF A LAND ROAD

Good capacity of support and good conditions of rolling and tack are basic characteristics that a land road must present to guarantee satisfactory conditions of traffic. Being that, the capacity of support this related to the property of the materials that constitute the surface of rolling in remaining cohesive front to the requests of the traffic. The conditions of rolling and tack are related with the regularity of the track and the attrition it enters the tires of the vehicles and the track.

The deficiencies techniques located in the underlayer and/or the layer of reinforcement are responsible for the problem of the lack of legal capacity of support. The adequate or material ground use to granulate and an efficient compacting are essential factors to the execution of a slope road with adjusted support capacity.

The very fine ground use (without treatments), substances to free granulate on the track and deficiencies in the maintenance processes they implicate harms conditions of traffic to the vehicles and drastically diminish the comfort and security of the users, mainly in rain stations.

3 GEOTECHNICAL CHARACTERISTICS OF THE GROUND OF THE REGION METROPOLITAN OF CAMPINAS – SP

The subsoil of the regional metropolitan of the city of Campinas, in São Paulo, is formed, in its majority, for one ground tropical whose grain size (NBR 7181, 1984) is defined as an clayed fine average sand; occurring variations in the percentage of the fraction clay.

In accordance with methodology Tropical Compact Miniature – MCT (Nogami, 1995), this ground typical of the region is classified with a lateritic clayed sand (LA').

According to Nogami and Villibor (1995), this ground in its natural conditions possesss low specific mass drought, low support capacity and can be collated for immersion in water, but, when duly compact it presents high capacity of support, raised resilience module, low permeability, small contraction for loss of humidity, reasonable cohesion and small

Table 1. The geotechnical characterization of the ground typical of the region metropolitan of Campinas.

Property	Description
Grain Size (NBR 1781, 1984)	Clayed fine average sand
Unified Soil Classification System	Clayed sand (SC)
Highway Research Board	Clayed sand (A-6)
MCT Classification	LA'
Density of the soil In Situ (NBR 9813, 1987)	$\gamma = 1,74\,g/cm^3$
Humidity (In Situ) (DNER-ME 213, 1994)	w = 13,09%
Limit of Liquidity (NBR 6459, 1984)	LL = 29,75%
Limit of Plasticity (NBR 7180, 1984)	LP = 15,51 %
Index of Plasticity	IP = 14,24%
Normal Energy Proctor (NBR 7182, 1986)	$\gamma_{dmax} = 1{,}793\,g/cm^3$ $w_{ot} = 12{,}6\%$

expansibilities for immersion in water. The properties these that make possible its use in bases and sub-bases of pavements and in compacted earthwork.

(a)

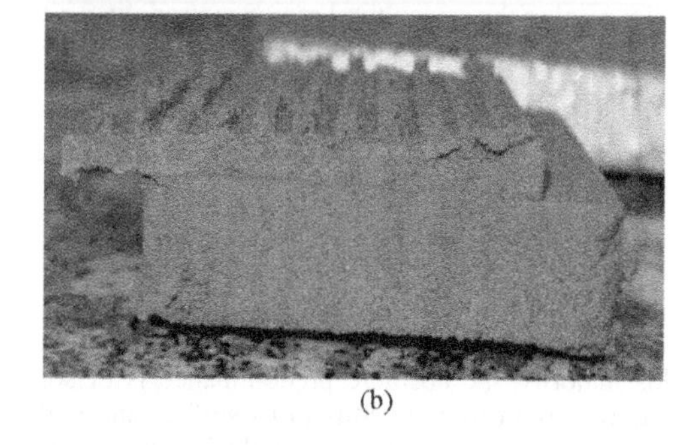

(b)

Photo 1. First serie with the bodies of test not saturated: (a) Compact in the normal energy; (b) Conditions of field.

4 DIRECT SHEAR STRENGTH OF THE GROUND OF THE REGION METROPOLITAN OF CAMPINAS – SP

Although to constitute one ground of excellent quality to constitute compacted slope, practical common of construction and the maintenance used for the locals agencies: to cut virgin land does not apply a process from compacting and directly does not receive no type of covering in its surface of rolling, displayed to the traffic.

With the objective to evaluate the resistance of this ground to the ticket of the traffic, assays of direct shear in fastest way with the soil in the conditions of field and the conditions of compact had been executed in the normal energy (NBR 7182, 1986), a first serie with the bodies of test not saturated (Photo 1) and another one with the saturated (Photo 2) bodies of test. Table 2 presents the results of shear strength direct gotten.

In order to verify the equivalence of conditions of field with the conditions of compacting in the normal energy (NBR 7182, 1986), the ground was compact in the cylinder of Proctor, 3 layers, with the small socket with different energies below of the normal (26 blows). It was obtained that the field conditions are equivalent, approximately, to a compacting with 2 blows/layer and optimal values had been $\gamma_{dmax} = 1,49\,g/cm^3$ and $w_{ot} = 18{,}39\%$.

(a)

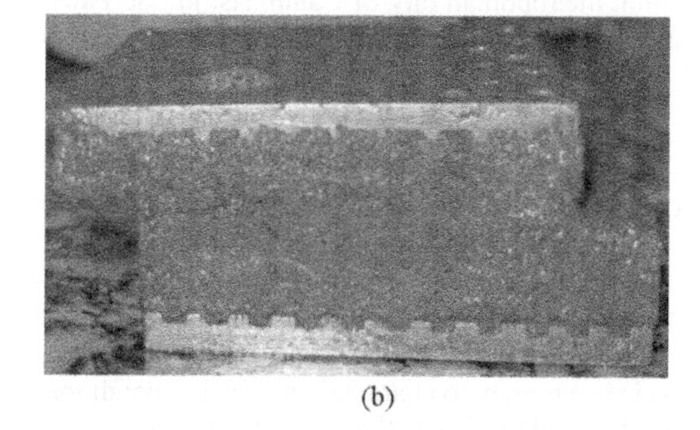

(b)

Photo 2. Second serie with the bodies of test saturated: (a) Compact in the normal energy; (b) Conditions of field.

Table 2. Results of the assay of Direct Shear (cohesion × angle of internal friction).

		Not satured	Satured
Condition In Situ	Cohesion (kg/cm^2)	0,9	0,55
	Angle of Internal Friction (°)	37,35	34,08
Compacted in Normal Energy	Cohesion (kg/cm^2)	0,34	0,32
	Angle of Internal Friction (°)	36,41	34,47

Table 3. Results of lost percent of direct shear strength.

		Not satured	Satured
Condition In Situ	Cohesion (%)	−62,22	−41,82
	Angle of Internal Friction (%)	−2,51	−1,74

5 CONCLUSION

The rupture of ground is almost a phenomenon of direct shear. The shear strength of ground essentially must to the attrition between particles. However, in tropical ground the cohesion parcel provokes resistance of significant value (Pinto, 2002).

Of table 2, shear strength observes itself that the ground loses cohesion and angle of internal friction when not compact if compared with the ground compact, as much in the state saturated as not saturated, as it shows table 3.

How much to the analysis of the rupture of this ground they are necessary studies you add that they will be made and published future. The results presented in this article technician are part of a thesis of doctor degree in progress in the UNICAMP-BR.

REFERENCES

Departamento Nacional de Infra-Estrutura de Transporte – DNIT. Disponível em: <www.dnit.org.br>

Associação Brasileira de Normas Técnicas – ABNT. *NBR 7181 Solos: Análise Granulométrica.* Rio de Janeiro, 1984.

Nogami, J. S; Villibor, D. F. *Pavimentação de Baixo Custo com Solos Lateríticos.* São Paulo, 1995.

Associação Brasileira de Normas Técnicas – ABNT. *NBR 9813 Solo: Densidade Aparente In Situ.* Rio de Janeiro, 1987.

Departamento Nacional de Estradas de Rodagem – DNER. *DNER-ME 213 Solo: Determinação do Teor de Umidade.* Rio de Janeiro, 1994.

Associação Brasileira de Normas Técnicas – ABNT. *NBR 6459 Solo: Limite de Liquidez.* Rio de Janeiro, 1984.

Associação Brasileira de Normas Técnicas – ABNT. *NBR 7180 Solo: Limite de Plasticidade.* Rio de Janeiro, 1984.

Associação Brasileira de Normas Técnicas – ABNT. *NBR 7182 Solo: Ensaio de Proctor.* Rio de Janeiro, 1986.

Pinto, C. S. *Curso Básico de Mecânica dos Solos em 16 Aulas.* São Paulo, 2002.

Advances in Transportation Geotechnics – Ellis, Yu, McDowell, Dawson & Thom (eds)
© 2008 Taylor & Francis Group, London, ISBN 978-0-415-47590-7

Qualifying soils for low volume roads based on soil and pavement mechanics

V. Peraça, W.P. Núñez, L.A. Bressani, J.A. Ceratti
Federal University of Rio Grande do Sul, Porto Alegre, Brazil

R.J.B. Pinheiro
Federal University of Santa Maria, Santa Maria, Brazil

ABSTRACT: Due to their paramount importance in unsurfaced or thinly surfaced roads, it seems worthwhile to qualify soils based on both Pavement and Soil Mechanics principles. Such an approach is used in this article to discuss the mechanical behavior of a residual soil found in Southern Brazil. The experimental program included tests traditionally used in Pavement Engineering, such as grain size distribution, Atterberg Limits, compaction, CBR, resilient modulus and permanent deformation under repeated loading. The influence of water content and compaction degree on soil strength and stiffness was quantified. Besides, direct shear tests were carried out to define strength parameters, used to verify the layer safety against shear failure. Water retention curves relating soil suction to saturation degree were used to analyze permanent deformation evolution under repeated loads. The interpretation of test results confirmed the suitability of the approach here suggested with the purpose of qualifying soils for roads construction.

1 INSTRUCTION

As in most developing countries, Brazilian economy strongly relies on the agricultural production. More than 90% of the country's road network (nearly 1,560,000 km) is unpaved, and local materials (mostly soils, but also weathered rocks, laterites, industrial wastes, etc.) are used to improve roads rideability.

Due to their paramount importance in unpaved or thinly paved roads, it might be convenient to qualify soils based not only on Pavement Mechanics concepts, but also on Soil Mechanics well known to geotechnical engineers.

In this context, this article reports the results of a research carried out with the objectives of:

a) developing an approach to qualify soils and unbound materials for unpaved or thinly paved low volume roads; and
b) validating that approach by means of laboratory tests carried out in samples and compacted specimens of an acid volcanic rock residual soil, commonly found in the Central Region of Rio Grande do Sul state (Southern Brazil).

2 BRIEF OVERVIEW

As stated by Bulman (1980), in designing and constructing quite simple roads the pavement engineer has to take into account the strength and moisture characteristics of soils. Therefore, it is hard to understand that only a small number of publications have discussed the importance and application of Soil Mechanics concepts to Pavement Engineering.

As Brown (1996) pointed out, Pavement Mechanics has developed, to some extent, in isolation from mainstream Geotechnics. Although much of the research on repeated loading of soil and granular materials has been quite sophisticated and comparable in quality with that developed in other fields of Soil Mechanics, procedures used in current practice remain empirical and backward. Brown's wide-ranging review outlined the techniques that are available to study the problem using theory, laboratory testing and field experiments. The well-known scholar concluded that the background knowledge accumulated from research presents an opportunity for improving current practice.

Nowadays, several researchers and practitioners share the idea that there is a need to refine the design procedures of flexible pavements using the Soil Mechanics of Unsaturated Soils, due to the reason that roads are constructed using compacted soils with degree of saturation between 75 to 90%.

Based on the results of comprehensive laboratory investigations, Côté & Konrad (2003) proposed a practical methodology for pavement engineers to assess the saturated and unsaturated hydraulic characteristics of base-course materials.

Several studies, like those reported by Gupta & Ranaivoson (2007), Yang et al. (2005) and Ceratti et al. (2004), have discussed variation of resilient modulus and shear strength with soil suction for compacted subgrades.

A TRB research project carried out from 2001 to 2004 had as one of its objective to extend the results of this kind of studies, such that modifications can be proposed to the available design charts for determining the thickness of various layers of unsurfaced and surfaced roads using the Soil Mechanics for Unsaturated Soils.

Recently, the US Federal Highway Administration released a publication (FHWA, 2006) discussing aspects in relation to construction, construction specifications, monitoring and performance measurements. The manual includes details on geotechnical exploration and characterization of in place and constructed subgrades as well as unbound base/subbase materials. The influence and sensitivity of geotechnical inputs are reviewed with respect to the requirements on past and current AASHTO design guidelines and the new mechanistic-empirical approach.

3 THE PROPOSED APPROACH

The approach suggested in this paper includes the following tests traditionally used by pavement engineers:

a) Grain size distribution and Atterberg Limits, used for soil classification purpose;
b) X-ray diffraction, in order to determine the soil (aggregate) mineralogy, more specifically to detect the presence of expansive clay minerals;
c) The determination of the soil compaction parameters: optimum moisture content (OMC) e maximum dry unit weight (γ_d), for laboratory specimens molding;
d) California Bearing Ratio computation, since this is an index well known to pavement engineers;
e) Soil resilient Modulus modeling as a function of stress state, with pavement design purpose; and
f) Analyzing water content effect on the soil permanent deformation characteristics (an initial permanent strain, ε_{pi} and permanent deformation ratio, PDR).

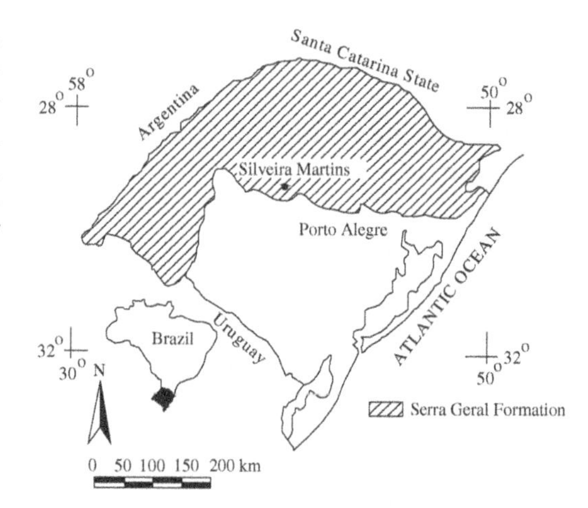

Figure 1. Rio Grande do Sul state Central Region.

Due to the crucial importance of soils in unpaved or thinly paved roads, it is suggested the accomplishment of:

a) direct shear tests on soaked specimens of compacted materials to define effective shear strength parameters (cohesive interception, c′ and internal friction angle ϕ'), that will be used in the verification of layer safety against shear failure;
b) water retention curves using the filter paper method (Chandler & Gutierrez, 1986), to relate saturation degree (or water content) to soil suction, a variable which strongly affects the strength and stress-strain behavior of unsaturated soils.

It is highly advisable to perform resilient modulus and permanent deformation tests on specimens molded at various water contents (e.g. OMC and OMC ± 2%) and compaction degrees, since it is known that slight variations of these parameters may remarkably reduce soils strength and stiffness.

It is also recommended to carry out direct shear tests on specimens with high and low initial void indexes, because in low volume roads sometimes compaction is neglected, and it is necessary to quantify the effects of lower compaction degrees on strength parameters.

In order to extrapolate results it is highly advisable the insertion of the sampling site in Geology and Pedology maps.

4 APPYING THE APPROACH TO A RESIDUAL SOIL USED IM WEARING COURSE OF UNPAVED ROADS

The approach proposed in the previous chapter was used to qualify soils for unpaved or thinly surfaced roads of the Central Region of Rio Grande do Sul state, in Southern Brazil, shown in Figure 1.

Table 1. Characterization tests results.

Gravel (%)	Sand (%)	Silte (%)	Clay (%)	LL	PI	γ_s (kN/m^3)
21	64	15	0	46	13	27.83

Table 2. Compaction parameters and bearing capacity.

OMC (%)	γ_d (kN/m^3)	CBR$_{max}$ (%)
20.1	16.05	21

4.1 *Geological and pedological information*

The Central Region of Rio Grande do Sul state is part of a geological formation called Serra Geral, basically composed by a series of lava spills. Residual soils resulting from chemical weathering of acid volcanic rocks are commonly used in wearing courses of unpaved roads.

One of such soils was sampled in a site placed on acid rocks, including rhyolites, rhyodacites and dacites, in Silveira Martins county.

In that soil profile, surface horizons are rather thin and, because of that, just the thickest horizon was sampled. That is a genuine residual soil keeping the relict structure of the rock.

4.2 *Soil characterization and classification*

Table 1 presents the results of grain size distribution (sieving and sedimentation) and Atterberg Limits of the studied soil.

The soil is a granular material with only 15% of fines, classified as silty sand with gravel. According to ASSHTO classification system it is a A-2-7(0).

It is surprising for a soil with 0% clay content to present a Plasticity Index of 13%. Moreover, X-ray diffraction tests carried out by Peraça (2007) revealed the presence of quite active clay minerals (smectite family). According to Cozzolino & Nogami (1993), the clay minerals agglomerated in crumbs (with silt size) are not identified in grain size distribution, and are responsible for saprolitic soils plasticity.

4.3 *Compaction and bearing characteristics*

Table 2 presents compaction parameters obtained in Proctor's Normal Energy and maximum CBR value.

In spite of the grain size distribution results, the soil compaction parameters (OMC = 20.1% and γ_{dmax} = 16.05 kN/m^3) suggest the presence a clay fraction. The CBR maximum value of 21%, rather low for well-graded granular soils, corroborates this finding.

Table 3. Compaction degree and moisture content of specimens compacted for resilient modulus testing.

CD	100 and 90%
w	OMC and OMC + 2%

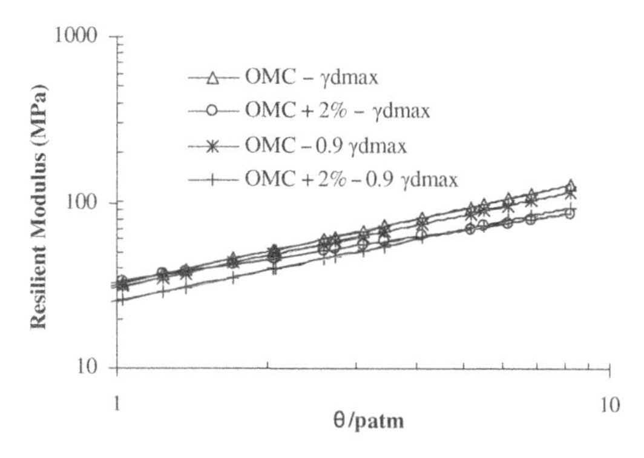

Figure 2. Soil resilient modulus as a function of stress state.

4.4 *Resilient modulus test*

Resilient Modulus tests were carried out on following AASHTO TP46-94 standard, with loading frequency of 1 Hz and loading time of 0.1 s.

The specimens, 10.0 cm diameter and 20.0 cm high, were dynamically compacted in five layers. As shown in Table 3, two compaction degrees (CD) and two moisture contents (w) were targeted.

Soil resilient modulus was modeled as a function of stress state, shown below:

$$RM = k_1\left(\frac{\theta}{p_{atm}}\right)^{k_2} \tag{1}$$

where RM is the soil resilient modulus; θ is the sum of principal stresses and p_{atm} is the atmosphere pressure; and k_1 and k_2 are the model parameters.

Figure 2 and Table 4 present tests results. High determination coefficient values (R^2) demonstrate that the model fits tests results rather well.

Figure 2 suggests that the compaction degree does not significantly affect the soil elastic behavior. However, it must be kept in mind that the conditioning stage at the beginning of the test (when high stress levels lead the soil specimen to a yield surface) imposes additional compaction, thus reducing the differences among the initial dry unit weights shown in Table 4.

Conversely, increasing moisture content reduces the soil resilient modulus, especially at lower stress states. For higher values of θ, the modulus curves tend to converge. In fact, as shown in Table 4, the stress state dependency in that compaction condition dry unit

Table 4. Resilient modulus model parameters for various compaction conditions.

Specimen compacted		Model Parameter		
w (%)	CD (%)	K_1 (MPa)	K_2	R^2
OMC	100	32	0.66	0.98
OMC +2%	100	32	0.46	0.88
OMC	90	31	0.64	0.98
OMC +2%	90	25	0.63	0.98

weight close to maximum and moisture content 2% above OMC) is lower ($k^2 = 0.46$).

All in all, test results show that the soil elastic behavior, compared to other Brazilian residual soils, is rather poor, with modulus lower than 100 MPa for any state stress. This fact somehow limits its use in pavements with thick asphalt layers, where fatigue is the main distress mechanism. However, as previously sated, this soil was evaluated for unsurfaced or thinly surfaced low volume roads, where rutting and not fatigue is the most common distress.

4.5 *Permanent deformation under repeated loading*

Due o the lack of Brazilian standards, the repeated loading permanent deformation tests were carried out following procedures adopted in studies previously performed in the Pavement Laboratory of the Federal University of Rio Grande do Sul (Núñez et al., 2004).

Considering the high stress state acting in the upper layers of unsurfaced or thinly surfaced pavements, the confining stress (σ_3) and the deviation stress (σ_d) applied to the specimens were equal to 105 kPa and 315 kPa, respectively.

The specimens, 10.0 cm diameter and 20.0 cm high, were compacted at two moisture contents (OMC and OMC+2%), in order to achieve two different dry unit weights (γ_{dmax} and 0.9 γ_{dmax}). Approximately 80,000 loading cycles were applied to each specimen, at a frequency of 1 Hz.

Figure 3 shows the effects of specimen's compaction degree and moisture content axial permanent strain (ε_p).

The permanent axial strains evolution in specimens compacted at OMC and γ_{dmax} and at OMC+2% and 0.9 γ_{dmax} follows a pattern known as plastic shakedown. After a quite high initial buildup, deformation increases at a quite lower constant rate.

The effect of water content on permanent strain is clearly understood comparing the curves corresponding to samples compacted at γ_{dmax} and water contents equal to OMC and OMC+2%. While at the end of

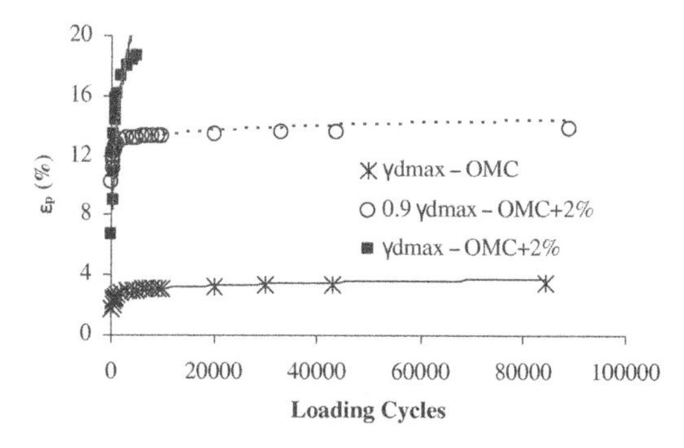

Figure 3. Permanent deformation evolution in specimens with different compaction degrees and moisture contents.

the test, the permanent axial strain of the specimen compacted at OMC was close to 4%, in the specimen compacted at OMC+2% the permanent strain continuously increased up to failure; a pattern known as incremental collapse.

Since the specimen compacted at the same water content (OMC+2%) and lower dry unit weight (0.9 γ_{dmax}) did not fail (even though the initial deformation was extremely high), it seems that the specimen's saturation degree (which combines water content and void ratio) should be considered instead.

In fact, the saturation degrees of the specimens compacted at OMC+2% and γ_{dmax} and OMC+2% and 0.9 γ_{dmax} were equal, respectively, to 83% and 66%. Test results showed the development of excess pore-pressure in well-graded granular materials with saturation degrees close to 85%.

Further discussion of these results, based in water retention curves, is presented in section 4.7.

The sensitivity of permanent deformation behaviour to slight variations of water contents was also noticed by Núñez (2005) when testing a granite well-graded granular residual soil. A specimen compacted just 1% above OMC, collapsed under triaxial repeated loading.

Those findings are extremely important since Brazilian standards for soils and unbound aggregates bases and sub-bases establish that the materials must be compacted at γ_{dmax} and water content equal to OMC ±2%. Consequently, bases or sub-bases compacted in strict observance of the specification may present quite bad performance, even failing.

4.6 *Direct shear tests*

Direct shear tests were carried out following procedures established I BS 1377-90 and ASTM D3080–90 and also recommendations proposed by Head (1982).

Specimens, 6.0 cm diameter and 2.0 high, were compacted to achieve "dense" (1.04 γ_{dmax}) and "loose" (0.84 γ_{dmax}) conditions. In view of the high stresses acting in granular layers of unsurfaced or thinly

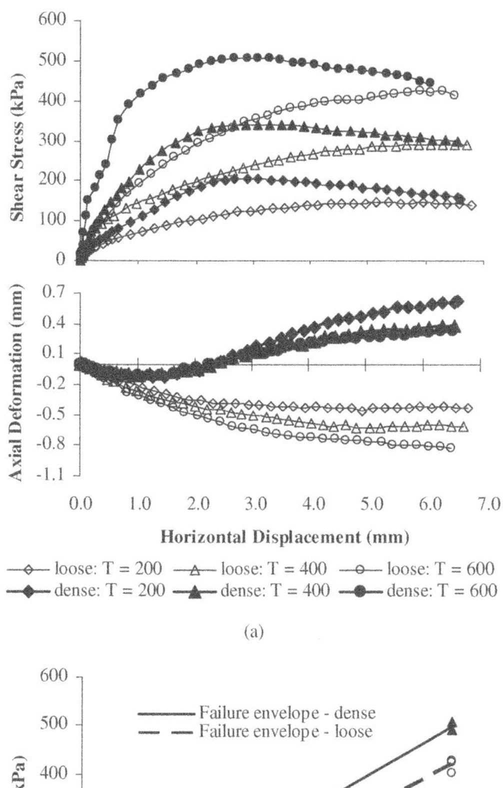

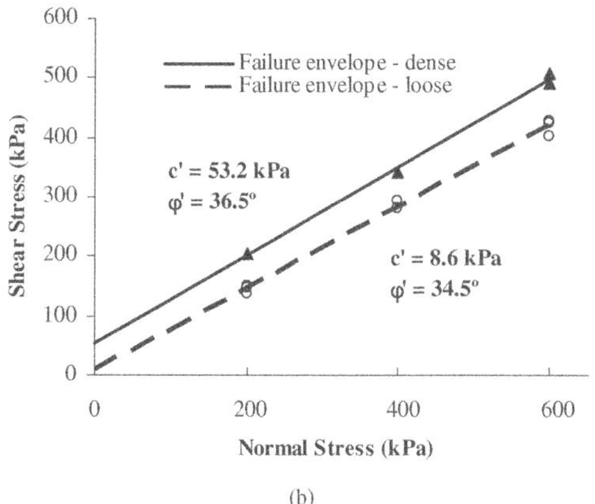

Figure 4. Direct shear test results: a) curves of shear strength and vertical deformation as a function of horizontal displacement; b) shear failure envelopes.

surfaced roads, the specimens were sheared under normal stresses of 200; 400 and 600 kPa.

Figure 4 shows the results of the direct shear tests carried out on soaked specimens at a shear rate of 0.048 mm/min.

Figure 4(a) shows that, as expected, for both "dense" and "loose" specimens shear strength increased along with normal stress. Even when the "dense" specimen is tested under high normal stress (600 kPa), shear strength does not remarkably decrease after peak. While the volume of the "loose" specimen simply decreased, the "dense" specimen, after suffering an

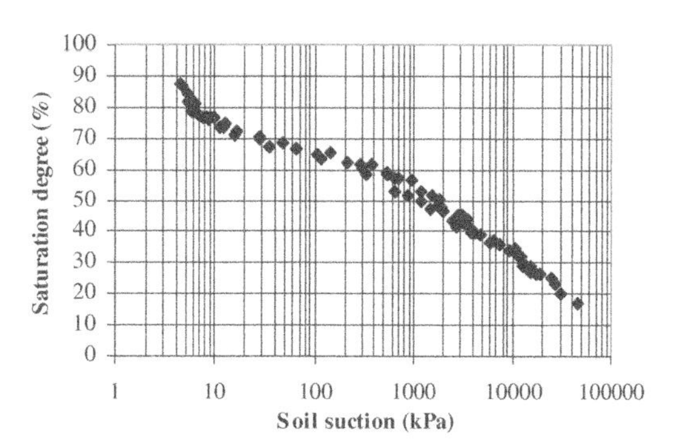

Figure 5. Soil water retention curve (specimen compacted at γ_{dmax}).

initial volume decrease, expanded. Such behaviours are typical of granular materials.

The effects of specimens' compaction degree on shear strength and strength parameters (cohesive interception, c' and internal friction angle, ϕ') may be observed in Figure 4(b). In the "loose" specimen (with higher void ratio) the failure envelope presents $c' = 8.6$ kPa and $\phi' = 34.5°$. Increasing the specimens' compaction degree (thus reducing void ratio) causes a sharp increase of cohesive interception ($c' = 53.2$), but only a minor growth of internal friction angle ($\phi' = 36.5°$).

All in all, directs shear tests results confirms the importance of properly compacting soils in order to provide high strength bases and sub-bases. Those results are used in section 5 to check soil layers safety against shear failure.

4.7 *Water retention curve*

Previous studies carried out in the Federal University of Rio Grande do Sul made clear the need of considering soil suction when analyzing the strength, hydraulic conductivity and stress-strain be behavior of unsaturated soils (Ceratti et al., 2004; Feuerharmel, 2007).

In order to understand why slight variations in water content remarkably affect the soil permanent deformation behavior, a water retention curve, shown in Figure 5, was plotted using the filter paper technique (Chandler et al., 1992).

Four specimens, 5.0 cm diameter and 1.9 cm high, were statically compacted at OMC to achieve γ_{dmax}. Two specimens followed drying paths (were dried) and two others wetting paths (were moistened).

It may be seen in Figure 5 that when the soil achieved a saturation degree of 76% (specimen compacted at γ_{dmax} and OMC) soil suction of 10 kPa was measured. But, for a specimen's moisture content of OMC+2% (saturation degree of 83%) soil suction is reduced in 50% (5.0 kPa). As shown in Figure 3, such a reduction in soil suction dramatically modified the soil permanent deformation behavior. While the specimen compacted at OMC and γ_{dmax} accumulated some 4% axial permanent strain, in the specimen compacted at moisture content 2% higher the permanent strain continuously grew until failure experience.

Although a reduction of 5.0 kPa in soil suction may seem rather insignificant, it must be considered that effective geostatic stresses in pavement granular layers and subgrades are quite low (for instance, the geostatic effective vertical stress 40.0 cm under compacted layers of the studied soil will not exceed 6.0 kPa). Therefore, a difference of 5.0 kPa in soil suction is not negligible, and, although further research is needed, it seems reasonable to explain the soil permanent deformation behavior based on soil suction.

5 APPLYING TEST RESULTS IN VERYING LAYERS SAFETY AGAINST SHEAR FAILURE

Theyse at al. (1996) proposed to safeguard unbound bases or sub-bases against shear failure or excessive gradual shear deformation by limiting the shear stress. The allowable shear stress in the layer can be calculated from the maximum single load shear strength, expressed in terms of the Möhr-Coulomb strength parameters c′ and ϕ' and a selected "factor of safety". These parameters can be determined from laboratory triaxial or direct shear tests.

The Factor of Safety at any point in the layer can be defined as:

$$F = \frac{\text{maximum shear strength}}{\text{working shear stress}}$$

It has been shown that:

$$F = \frac{\sigma_3\left\{K\left[\tan^2\left(45+\frac{\phi'}{2}\right)-1\right]\right\}+2Kc'\tan\left(45+\frac{\phi'}{2}\right)}{(\sigma_1-\sigma_3)} \quad (2)$$

where F is factor of safety; σ_1 and σ_3 are calculated major and minor principal stress acting at a select position in the layer (compressive stresses are positive and tensile stresses negative); c′ and ϕ' are cohesion and angle of internal friction; K is a constant: 0,65 for saturated and 0,95 for normal conditions.

The allowable factor of safety varies according to the road category and the design traffic. For lightly trafficked rural roads the equation (3) may be used:

$$N = 10^{\left(\frac{F+1.472}{0.371}\right)} \quad (3)$$

where N is the recommended number of load applications to safeguard against shear failure for Road and F is the Factor of safety.

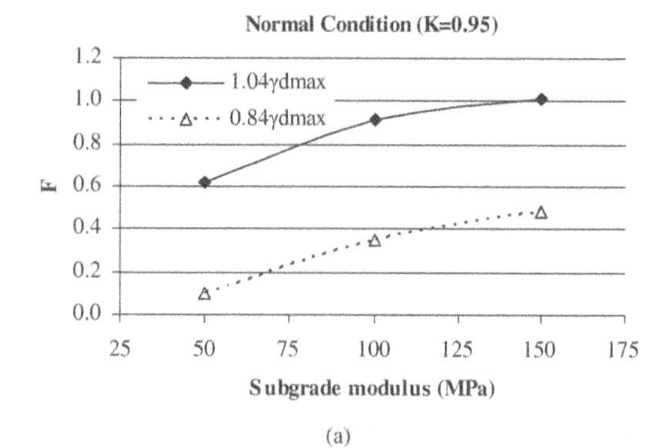

(a)

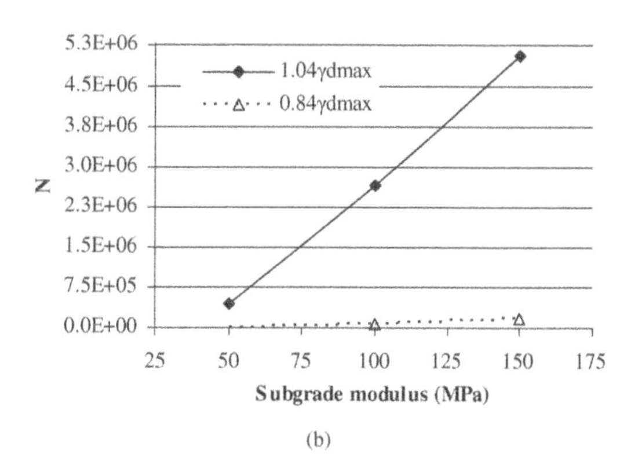

(b)

Figure 6. (a) Factors of Safety against shear failure and (b) allowable numbers of loads as functions of subgrade resilient modulus under normal condition (K = 0.95).

In order to illustrate that approach, a very lightly trafficked road is here considered. A 15.0 cm-thick compacted layer of the studied soil is considered as wearing course on subgrades with resilient modulus of 50.0; 100.0 and 150.0 MPa.

EVERSTRESS 5.0, a well-known software, and the soil resilient modulus models presented in section 4.4, were used to compute critical values of principal stresses σ_1 and σ_3, under 82 kN axle load.

The major and minor principal stresses and the effective shear strength parameters corresponding to "dense" (and "loose" (0.84 γ_{dmax}) conditions, shown in Figure 4(b), were used to compute F values in both saturated (K = 0.65) and normal (K = 0.95) conditions, and the corresponding allowable numbers of equivalent single axle loads.

Figures 6 and 7 show Safety Factors as functions of subgrade resilient modulus for saturated and normal conditions. The effect of specimens' saturation and compaction state ("dense" and "loose") is clearly seen. Higher F values imply higher allowable number of load applications, as also shown in Figures 6 and 7.

Figure 6(b) shows that, if compacted at OMC and 1.04 γ_{dmax} over a subgrade with resilient modulus of

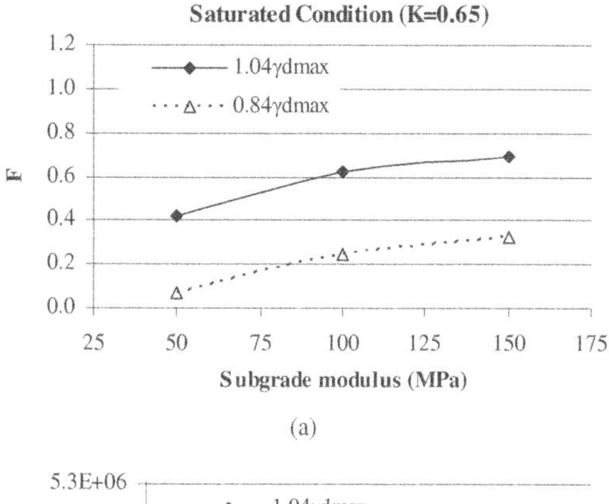

(a)

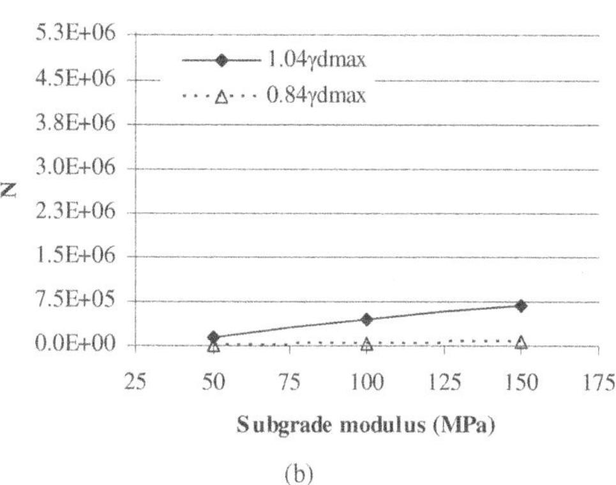

(b)

Figure 7. (a) Factors of Safety against shear failure and (b) allowable numbers of loads as functions of subgrade resilient modulus under normal condition (K = 0.65).

150 MPa, the residual soil 15.0 cm-thick layer will resist nearly 5,000,000 equivalent axle loads before shear failure. This allowable number will be dramatically reduced if drainage is neglected and the layer is saturated (K = 0.65). Figure 7(b) shows that, even if properly compacted (1.04 γ_{dmax}) on a rather stiff subgrade (150 MPa), the layer will barely withstand 750,000 axle loads.

The effect of compaction state is still more striking. If compaction is neglected (0.84 γ_{dmax}), in spite of a stiff subgrade (150 MPa) and adequate drainage condition (K = 0.95), the residual soil layer will fail after some 200,000 axle loads.

If both compaction and drainage fail (0.84 γ_{dmax} and K = 0.65), the layer performance will be awful even if built on a stiff subgrade. The acceptable number of axle loads before failure (N ≅ 80,000) would be less than 2% of the allowable traffic corresponding to "dense" (1.04 γ_{dmax}) and well-drained (K = 0.95) conditions.

This analysis, based on direct shear test results, makes clear the essential need of properly compacting and draining soils layers in unsurfaced or thinly surfaced low-volume roads.

6 CONCLUSIONS

This paper proposed an approach to qualify soils and unbound materials for unsurfaced or thinly surfaced low volume roads. The approach attempts to establish a bridge between Soil and Pavement Mechanics and was validated by means of laboratory tests carried out in samples and compacted specimens of an acid volcanic rock residual soil, commonly found in the Central Region of Rio Grande do Sul state (Southern Brazil).

It was shown that grain size distribution not always permits to identify soil fractions and that X-ray diffraction is essential to detect the presence of expansive clays in saprolitic soils.

The paramount importance of compaction degree and water content in the shear strength and stiffness of that residual soil was demonstrated. The approach made possible to quantify the effects of compaction state and drainage efficiency on the allowable traffic on unsurfaced or thinly surfaced roads.

The importance of considering soil suction in the analysis of unsaturated soils used in low volume roads was made clear. A water retention curve was used to explain the dramatic variation of soil shear strength and permanent deformation with slight variations of water content. It was shown that the saturation degree plays a more important role than water content and that paving standards should not specify water content ranges for compaction.

All in all, it seems that the proposed approach, based on Soil and Pavement Mechanics tests, shows high potential for qualifying soils for unsurfaced and lightly surfaced low volume roads.

REFERENCES

ASTM D3080–90. 1990. *Standard Test Method for Direct Shear Test of Soils Under Consolidated Drained Conditions*. Presents. American Society for Testing and Materials.

BS 1377–90. (1990). *British Standard Methods of Test for Soils for Civil Engineering Purposes, Part 7 (total stress): Shear strength test*. Presents British Standards Institution.

Brown, S.F. 1996 Soil mechanics in pavement engineering. *Géotechnique:* 46 (3): 383–426.

Bulman, J.N. 1980. Aspects of soil mechanics of particular importance to highway engineering in Africa. *7th Regional Conference for Africa on Soil Mechanics and Foundation Engineering*. Ghana: Accra.

Ceratti, J. A.; Wail, Y.Y.G. & Núñez, W. P. 2004. Seasonal variations of a subgrade soil resilient modulus em southerm Brazil. In: Geology and Properties of Earth Materials (ed). *Journal of the Transportation Research Board* (1874): 165–173. Washington, D.C.

Chandler, R.J; Crilly, M.S. & Montgomery-Smity, G. 1992. A low-cost method of assessing clay desiccation for low-rise buildings. In: Proc. Of the Institute of Civil Engineering, (ed). 92 (2): 82–89.

Chandler, R.J. & Gutierrez, C.I. 1986. The filter-paper method of suction measurement. *Technical Note, Géotechnique*: 36 (2): 265–268.

Cote, J. & Konrad, J.M. 2003. Assessment of the hydraulic characteristic of unsaturated base-course materials: a pratical method for pavement engineers. *Canadian Geotechnical Journal* (40): 121–136. Canada.

Cozzolino, V.M.N. & Nogami, J.S. 1993. MCT geotechnical classification for tropicals soils. *Soils and Rocks*. 16 (4): 77–91.

Christopher, B.R; Schwartz, C. & Bourdreau, R. 2006. *Geotechnical Aspects of Pavement.* U.S. Department of Transportation, Federal Highway Administration – FHWA. Washington D.C.

Feuerharmel, C. 2007. *A study on the shear strength and hydraulic conductivity of unsaturated soils of the Serra Geral Formation.* PhD Thesis, Federal University of Rio Grande do Sul, Porto Alegre.

Gupta, S. & Ranaivoson, A. 2007. *Pavement Design Using Unsaturated Soil Technology.* Minesota Department of Transportation. Minesota.

Head, K.H. 1982. *Manual of Soil Laboratory Testing:* (2). London. Pentech Press.

Núñez, W.P. 2005. *Reason for the preamture failure of a urban pavement.* Confidential report. Porto Alegre.

Núñez, W.P.; Malysz, R.; Ceratti, J.A.; Gehling, W.Y.Y. 2004. Shear strength and permanent deformation of unbound aggregates used in Brazilian pavements. In: *6th International Symposium on unbound aggregates in roads, 2004, Nottingham, U. K.. Pavement Unbound. Leiden*: 23–31. Neetherlands: Balkema.

Peraça, V. 2007. *Qualifying soils for Road Wearing Courses: An Approach Based on Mechanics of Pavements and Soil Mechanics.* Master Dissertation, Federal University of the Rio Grande do Sul. Porto Alegre.

Theyse, H.L. 2006. The suction pressure, yield strength and effective stress of partially saturated unbound granular pavement layers. *CSIR Built Environment, Pretoria, Gauting.* South Africa.

Theyse, H.L.; De Beer, M. & Rust, F.C. 1996. Overview of South African mechanistic pavement design method. *Transportation Research Record (1539):* 6–17.

Vanapalli, S. 2004. Design of flexible pavements using the soil mechanics for unsaturated soils. *Paper submitted for publication to the Transportation Research Board.*

Yang, S.H.; Huang, W.H. & Tai, Y.T. 2005. Variation of resilient modulus with soil suction for compacted subgrade soils. *Transportation Research Board.* (1913): 99–106.

Advances in Transportation Geotechnics – Ellis, Yu, McDowell, Dawson & Thom (eds)

In-ground stress-strain condition beneath center and edge of vibratory roller compactor

R.V. Rinehart, M.A. Mooney & J.R. Berger
Colorado School of Mines, Golden, Colorado, USA

ABSTRACT: Roller-integrated measurement of soil stiffness is now commonly used during continuous compaction control (CCC). Soil stiffness is estimated from measured drum acceleration and eccentric position. To better understand the relationship between roller measured stiffness and soil behavior, a study was conducted to capture in-ground stress strain behavior during roller passes. Experimental results show that vertical stresses beneath the drum center are greater than under the drum edge. Combined with supporting theoretical analysis, the results illustrate that plane strain conditions exist under the drum center to a depth equal to 0.25 of the drum length. Beyond that depth and at the drum edge, plane strain conditions do not exist.

1 INTRODUCTION

Continuous compaction control (CCC) involves the measurement of soil properties based on drum vibration characteristics together with the documentation of roller position (initially via wheel encoders and now via GPS). First generation measurement indices such as the compaction meter value (Forssblad 1980) are phenomenological and empirically related to increases in soil density and stiffness. More recently, measurement indices provide an assessment of soil stiffness, particularly on fully compacted soils in a proof roll – type capacity (e.g., Kröber et al. 2001, Anderegg & Kaufmann 2004, Adam & Kopf 2004, Mooney & Rinehart 2007). There is potential for these stiffness measures to be related to design parameters in mechanistic-empirical (M-E) pavement design, e.g., layered moduli, resilient modulus. There would be a considerable benefit if roller measured stiffness parameters were strongly tied to soil properties used in M-E design and performance prediction. To date, however, the meaning behind soil stiffness values measured by CCC rollers is not thoroughly understood.

To develop an understanding of soil properties reflected through roller-based stiffness values, a thorough assessment of in-situ stress-strain response imparted by CCC rollers must be performed. One important aspect, and the subject of this paper, is the determining whether to treat the problem in a two-dimensional (2D) or three-dimensional (3D) sense. The variation of the stress-strain response across the roughly 2 m length of the drum (y-direction in Fig. 1) will reveal which treatment is necessary. We will show here that the loading condition beneath the drum in combination with stress-dependent soil modulus leads to stiffer soil response beneath the drum center than under the drum edge. The implications on field use of CCC and on modeling are described.

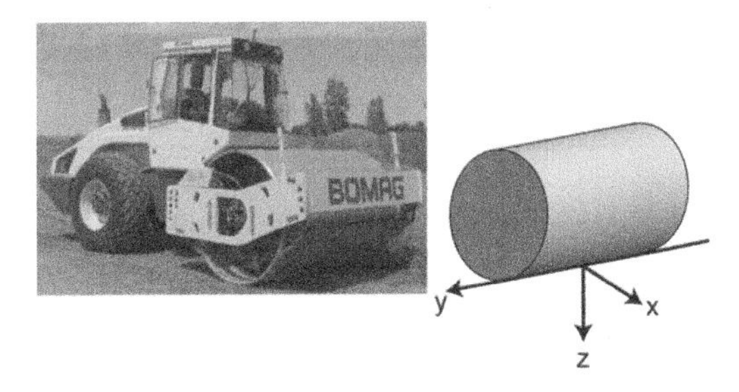

Figure 1. Bomag vibratory roller compactor and coordinate system used here.

2 TESTING PROGRAM

A *crushed rock over silt* test bed was constructed within sandy silt embankment fill material (ML per Unified Soil Classification System). A 3.5 m wide by 35 m long trench was excavated to a depth of approximately 1.5 m in the ML soil. A 30 cm layer of ML soil was placed and compacted. In-ground stress and strain sensors were installed in this layer of material (see Fig. 2) to measure total normal vertical stress (σ_z) and strain (ε_z) and total normal longitudinal stress (σ_x) and strain (ε_x). Subsequently, a 30 cm thick layer of screened crushed rock from a local quarry (SM per USCS) was placed, moisture conditioned to optimum and compacted. Stress

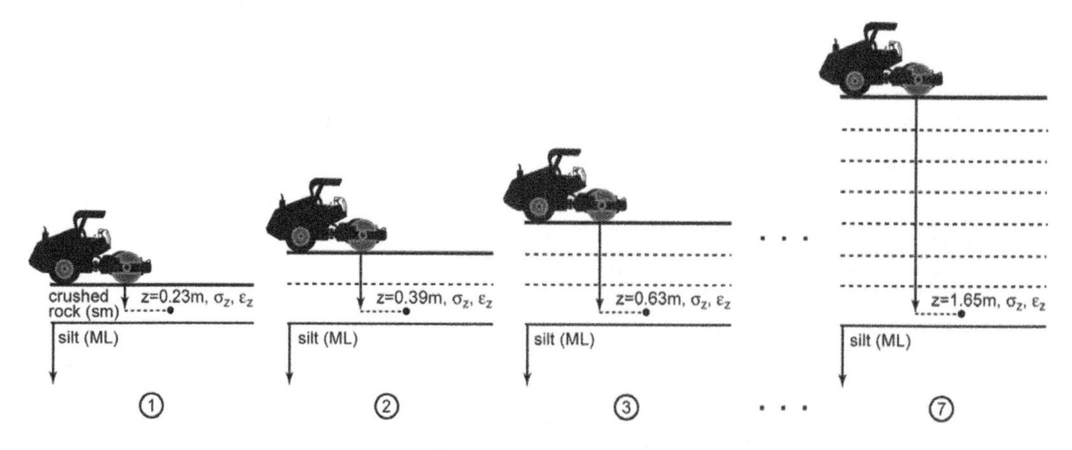

Figure 2. Layout of crushed rock over silt test bed. Sensors were placed in one crushed rock layer and one silt layer. Superposition of test results on subsequently placed layers of crushed rock produces stress - strain profiles with depth.

Table 1. Summary of Bomag Vibratory Roller Parameters.

Model	BW213 DH-4 BVC
Static Mass (kg)	14,900
Static Drum Linear Load (kN/m)	42.4
Operating Frequency (Hz)	28
Eccentric Force Range (kN)	0–365
Eccentric Force used (kN)	102

and strain sensors were placed in this layer to measure σ_x, σ_z, ε_x and ε_z (see Fig. 2). Six additional 30 cm thick (approximate) layers of the crushed rock were placed and compacted. Instrumentation was not installed in these additional layers.

Geokon earth pressure cells (EPCs) were used to measure stress and LVDTs were utilized to measure strain. Both types of sensors are capable of capturing the static (due to roller weight) and cyclic (due to drum vibration) response of the soil. The construction of the test beds, installation of sensors, and verification of sensor efficacy are described in detail elsewhere (Rinehart & Mooney 2008, Miller et al. 2007). In this paper, z is considered positive downward from the soil surface, x is positive in the direction of roller travel and y is positive to the roller operator's right.

A Bomag roller compactor, capable of static rolling and vibration at many different eccentric force settings was used for this study (see Table 1).

3 RESULTS

3.1 *Stress and strain response*

Roller passes were performed at typical operating speeds (0.5–1.0 m/s) over each layer of the test bed. It is worth noting that the objective of this study was to characterize in-situ stress strain behavior during roller passes over fully compacted soil, e.g., similar to a proof roll application. The ultimate goal is to use roller-based assessment of the properties of the constructed earth structure to predict performance. Therefore, the focus of the aforementioned results is to characterize stress-strain fields induced by static and vibratory rollers on fully compacted soil.

Figure 3 presents total vertical stress σ_z and strain ε_z data measured from the same sensor set as the layers were built up. The number of 30 cm layers above the sensors is indicated to the right of each figure (and referenced to Figure 2). Compressive stresses and strains are positive here and elsewhere in this paper. Data is provided for both vibratory passes (eccentric force is 102 kN, operating frequency is 28 Hz) and static passes (no vibration). The forward speed was about 1.0 m/s in all passes. Reported σ_z and ε_z are measured over a 100 mm EPC dimension or strain sensor gage length, and therefore reflect average values. The difference between these average values and values over infinitesimal dimensions has been shown to be minimal (Mooney & Miller 2008) and does not affect the interpretation of the results.

It should also be noted that we are measuring total stresses. These soils are partially saturated and suction induced pore pressures do exist. The measurement of pore water and air pressures under static and dynamic loading is extremely challenging. The accurate characterization of these pore pressures and effective stress conditions is one ultimate goal and may be achievable within a decade or so, for now, there is considerable value in characterizing total stresses.

The abscissa in Figure 3 reflects the relative position of the drum to the sensor location. The roller is traveling left to right; thus, negative x indicates that the drum is approaching the sensors. The drum is directly above the sensors at $x = 0$ and is moving away from the sensors for $x > 0$. The σ_z and ε_z measurements reflect

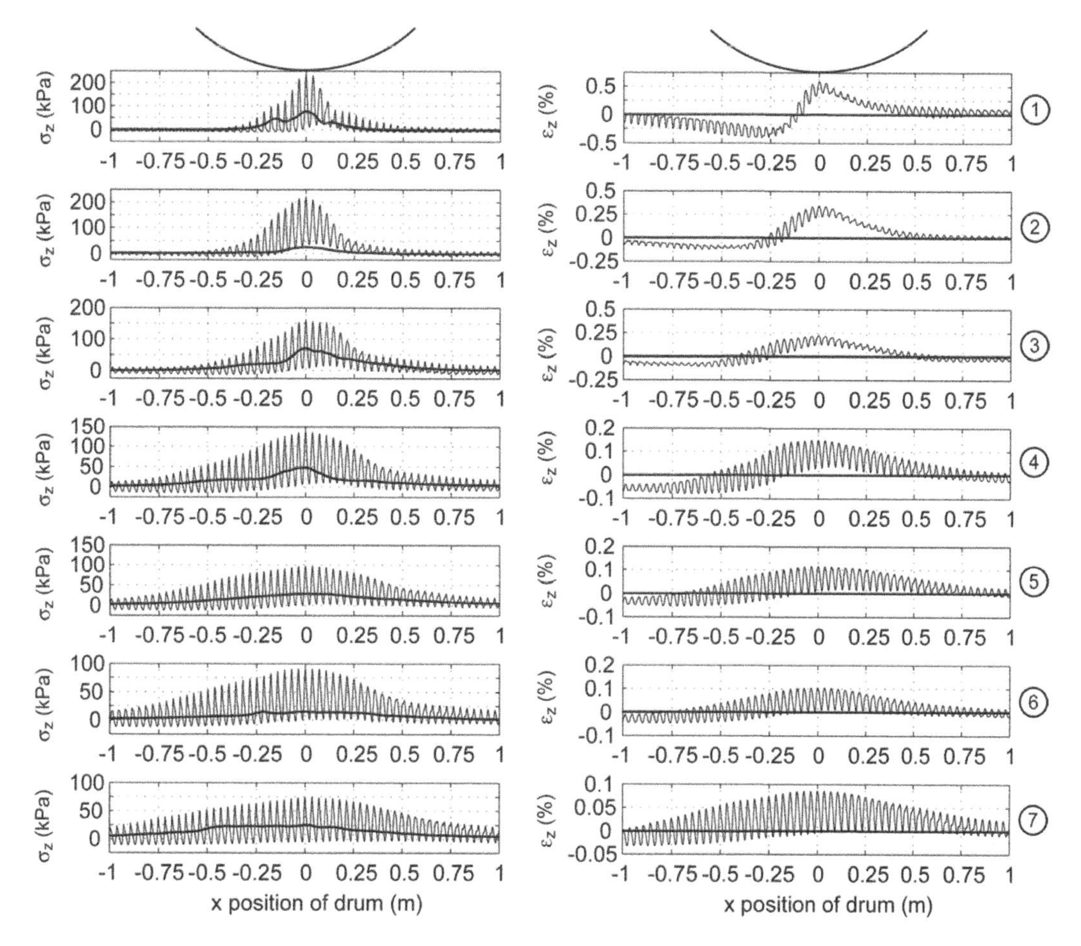

Figure 3. Vertical stress and strain histories during vibratory and static roller passes at multiple depths (see Figure 2).

both the static weight of the roller and the dynamic forces generated by the vibrating drum (Rinehart & Mooney 2008). The effects of overburden have been removed so the stress and strain measurements presented here are due to the roller alone.

The σ_z and ε_z response due to vibration oscillates about the response due to static loading as was generally anticipated. The soil immediately in front of the drum experiences vertical (z) extension and longitudinal (x) compression (not shown). This bow wave phenomenon is due to the traveling nature of the roller. The vertical extension is noticeable in the top layer and diminishes with depth. The zone of soil influenced by the drum widens with depth. With respect to the superimposed drum shape, the width of the influenced zone is on the order of centimeters near the surface and grows to meters at depth.

The relatively large cyclic stresses and small cyclic strains observed are also noteworthy and can be explained by the curved drum. The curved surface of the drum results in the loading area increasing with increasing penetration into the soil. This is manifested as a hardening response where stress increases more rapidly than strain (Rinehart & Mooney 2008).

3.2 *Center vs. edge stress – strain response*

The response in Figure 3 was recorded beneath the center of the roller drum. The peak σ_z and ε_z values at $x = 0$ from Figure 3 are presented vs. depth in Figure 4. Both static and vibratory values are plotted. The lines in Figure 4 are best fit approximations. Also shown are the peak σ_z and ε_z values measured beneath the edge of the drum. Values of σ_z are greater under the drum center than under the drum edge. Values of ε_z are less under the drum center than under the drum edge for depths < 1 m. With a drum length of 2.1 m, it is reasonable that center and edge strains are equal for depths greater than 1 m.

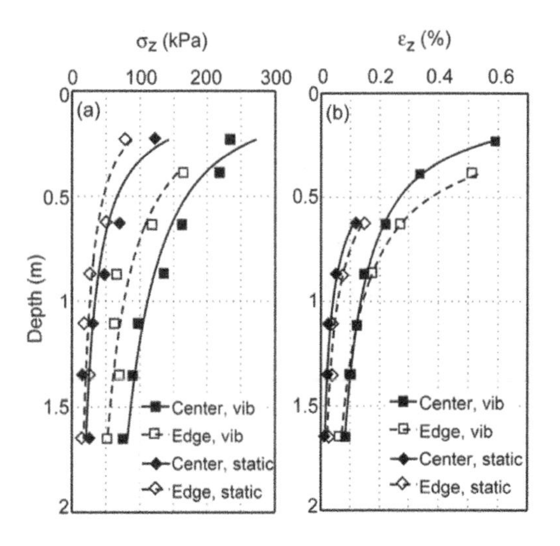

Figure 4. Peak vertical stress and strain measured beneath the center and edge during static and vibratory roller passes. Lines reflect best fit. Data from one sensor set as test bed was constructed in layers (recall Figure 2).

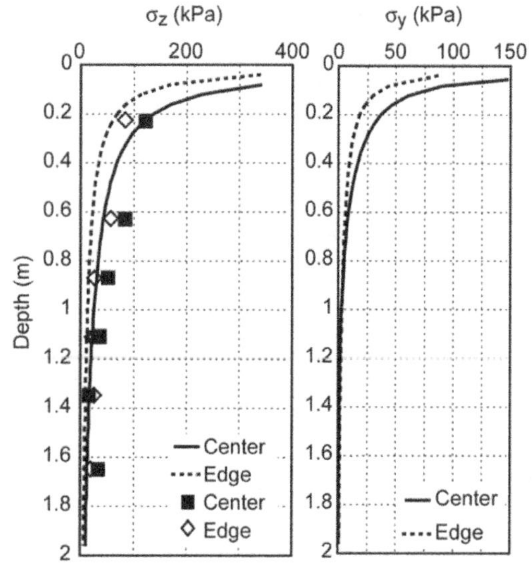

Figure 5. Theoretical vertical stress distributions based on a finite line load.

3.3 *Comparison with elastic theory*

The observed difference in σ_z under the drum center and drum edge can be interpreted using elastic solutions for half-space problems. Here, we investigate the stress state due to a finite-length line load with the well-known Boussinesq solution and the stress state due to an infinite length line-load (two-dimensional plane strain) with the Flamant solution (e.g., Johnson 1987). Granted, the test bed is layered and the contact loading is not strictly a line load; however, this analysis is sufficient to explain the observed behavior. To implement the Boussinesq solution for a finite line-load, the stress solution was numerically integrated over the 2.1 m length of the line load, and stresses σ_z, σ_y, and σ_x were computed as functions of depth z. Similarly, the stresses σ_z and σ_x were computed using the Flamant solution. The force used in both cases is the static linear load under the roller drum, 42.4 kN/m (Table 1).

Figure 5 shows the σ_z and σ_y distributions with depth under the center and edge of the drum resulting from the finite length line load. The measured σ_z data (Figure 4) is shown for comparison. Figure 5 shows that elastic half-space theory and line load representation generally agrees with the measurements and corroborates the finding that the stresses are higher under the center of the drum. The difference between center and edge σ_z decreases with depth and become reasonably similar below 1.5 m. Note that the force transmitted to the soil during vibratory operation is unknown and therefore comparisons between the data measured during vibratory passes and the elastic half-space theory are not possible. It is also important to note that representing the roller loading with a line load is a significant simplification. Currently, the exact contact mechanics between the drum and soil are unknown and the topic of ongoing research.

It follows from the general agreement between measured and theoretical σ_z that the theoretical distribution of σ_y with depth might shed further light on center vs. edge conditions. The right plot of Figure 5 illustrates that σ_y is also greater under the drum center than under the edge to a depth of approximately 1.0 m. Here, experimental σ_y data is not available for comparison.

A comparison of the stresses determined via the integrated Boussinesq solution under the drum center with the plane strain, infinite line load Flamant solution can shed light on whether plane strain conditions exist. Other analysis (Rinehart & Mooney 2008) show that plane strain conditions may exist at the center of the drum. Figure 6 presents theoretical σ_z and σ_y profiles for the 3D integrated Boussinesq solution under the drum center to the 2D Flamant solution. Of particular interest here is the depth at which loss of plane strain occurs. Under plane strain conditions, $\sigma_y = \nu\ (\sigma_x + \sigma_z)$, where ν is Poisson's ratio. Therefore, we may determine when plane strain conditions are lost as a function of depth by comparing the values of σ_y determined using $\nu(\sigma_x + \sigma_z)$ from the Flamant solution (plane strain) with the value of σ_y from the integrated Boussinesq solution. These values are shown in Figure 6, where we can estimate that plane strain conditions exist to a depth of approximately 0.5 m, 25% of the roller width (based on a threshold of 10% deviation between the 3D and 2D cases).

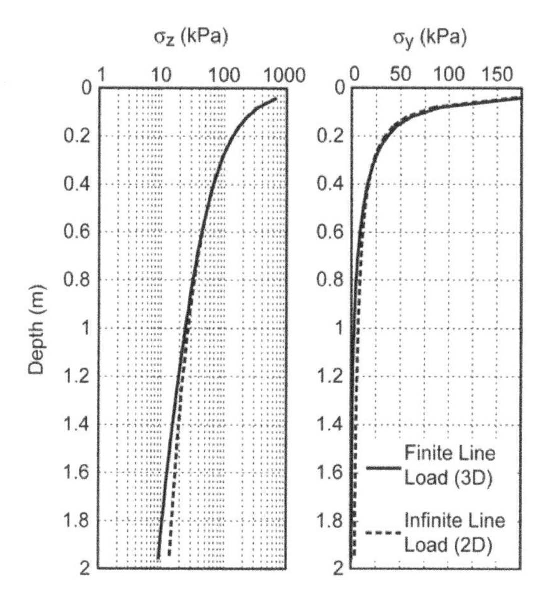

Figure 6. Comparison of the integrated Boussinesq solution to a finite line load (3D) under the drum center and the plane strain Flamant solution to an infinite line load (2D).

3.4 *Implications*

Roller-integrated soil stiffness is based on the measured vertical vibration of the drum and the position of the eccentric mass (see Mooney & Adam 2007). The resulting stiffness is a composite reflection of the vertical stress – strain response in the soil. Here, we have shown that, within a 1 m depth of the roller, σ_z values are greater beneath the drum center than the drum edge. Elasticity theory explains these findings are a consequence of loading/boundary conditions. Theory also shows that σ_y is greater at the drum center than beneath the drum edge. While not shown here, the measurement of σ_y during similar test bed studies reveal a similar finding (Rinehart & Mooney 2008). As a result of these stress conditions and the stress-dependent nature of soil (e.g., Brown 1996), measured ε_z values were found to greater beneath the drum edge than the center. Combining these findings reveals that the soil beneath the drum center responds to roller loading with greater stiffness than does the soil beneath the drum edge.

These findings have two key implications. First, in current practice roller-measured soil stiffness is often correlated with spot test results, e.g., static or dynamic plate load test modulus. Since the geometry of the roller influences the roller-measured soil stiffness, care should be taken when comparing to spot test results. Second, the findings show that the stress and strain conditions vary across the width of the drum and therefore roller drum/soil interaction should be appropriately modeled in three dimensions.

4 CONCLUSIONS

A crushed rock over silt test bed was constructed with in-ground sensors to capture soil stress-strain behavior during vibratory and static roller passes. Measurements made during static (no vibration) and vibratory passes reveal that vertical stresses are greater under the center of the drum than under the edge of the drum. The corresponding measured vertical strains were greater under the drum edge than under the drum center. The analysis of a homogeneous, isotropic, linear-elastic half-space subjected to line loading supports the measured vertical stress results and leads to the conclusion that plane strain conditions exist to a depth of 0.5 m under the center of the drum but not under the drum edges. This suggests the drum/soil interaction should be modeled as a 3D problem. The center vs. edge results shown here also indicates that the soil under the center of the drum will exhibit a stiffer response than under the drum edge.

REFERENCES

Adam, D. & Kopf, F. 2004. Operational Devices for Compaction Optimization and Quality Control. *Proc. Int. Seminar on Geotechnic Pavement and Railway Design and Construction*, December 2004, Athens, Greece, 97–106.

Anderegg, R. & Kaufmann, K. 2004. Intelligent Compaction with Vibratory Rollers: Feedback Control Systems in Automatic Compaction and Compaction Control.*Transportation Research Record* No. 1868, Transportation Research Board, Washington, D.C., 124–134.

Brown, S.F. 1996. Soil mechanics in pavement engineering, *Geotechnique,* Vol. 46, No. 3, pp. 383–426.

Forssblad, L. 1980. Compaction meter on vibrating rollers for improved compaction control. *Proc. Intl. Conf. on Compaction*, Vol. II, Paris.

Johnson, K.L. 1987. *Contact Mechanics*. Cambridge: Cambridge University Press.

Kröber, W., Floss, R. & Wallrath, W. 2001. Dynamic soil stiffness as quality criterion for soil compaction. *Geotechnics for roads, rail tracks, and earth structures*, Balkema, Lisse.

Miller, P.K., Rinehart, R.V. & Mooney, M.A. 2007. Measurement of Soil Stress and Strain using In-ground Instrumentation, *Proc. of GeoDenver*, Feb. 18–21, Denver, CO.

Mooney, M.A. & Adam, D. 2007. Vibratory Roller Integrated Measurement of Earthwork Compaction: An Overview, *Proc. 7th Intl. Symp. Field Measurements in Geomechanics*, ASCE, Boston, MA, Sept. 24–27, 2007.

Mooney, M.A. & Rinehart, R.V. 2007. Field Monitoring of Roller Vibration during Compaction of Subgrade Soil, *J. Geotechnical and Geoenvironmental Engineering*, ASCE, 133(3), 257–265.

Mooney, M.A. & Miller, P.K. 2008. Analysis of Light Falling Weight Deflectometer Test based on In-situ Stress and Strain Response, *J. Geotech. & Geoenv. Eng.*, ASCE, in press.

Rinehart, R.V. & Mooney, M.A. 2008. Measurement of roller compactor induced soil stresses and strains, *Geotechnical Testing Journal*, ASTM (in review).

Advances in Transportation Geotechnics – Ellis, Yu, McDowell, Dawson & Thom (eds)
© 2008 Taylor & Francis Group, London, ISBN 978-0-415-47590-7

Mechanical properties of paving material of pedestrian friendly pavements using coal ash

K. Sato, T. Fujikawa & K. Kawahara
Fukuoka University, Fukuoka, Japan

M. Ishida
NIPPO Corporation Kyushu Branch, Fukuoka, Japan

ABSTRACT: This study aims to develop of materials for eco-friendly pavement. According to increase of coal ash (FA: Fly Ash, PFBC: Pressurized Fluidized Bed combustion ash), we tried to develop effective utilization of coal ash as a stabilizer because disposal of coal ash has been getting more difficult. Furthermore, in order to prevent from a pavement crack, we tried to use materials mixing bamboo chips crushed to fiber. Therefore, we carried out the tests to suggest design work of material for pedestrian-friendly pavement with evaluation of mechanical properties and toughness gained from the result of unconfined compression test. The result is following (1) Material with PFBC had higher solidification than material with FA, (2) Using FA and cement, strength more than same level of PFBC was gained, (3) By mixing 10% bamboo chips, improvement of brittle failure was observed.

1 INTRODUCTION

These days, pedestrian-friendly soil pavement design has been drawing people's attention in terms of environmental protection, global warming and alleviation of heat-island effect. The current pedestrian-friendly pavement design decides the composition of the pavement according to soil selected on past experience. Therefore, in the current situation, there are no clear criteria for a required performance of pavement. The main fields of pedestrian-friendly pavement are limited to the park, garden, path, and river bed in terms of strength of the pavement materials and being comfortable for pedestrians. However, if durability of pavement is improved, its utilization of pavement material would be expanded also coal ash from facilities including the fire power stations in 2005 reached about 11.15 million tons, and increase shows about 300,000 tons, compared with the previous year. In addition, the demand for coal power generation in the electric industry has been increasing due to the influence of rising price of crude oil in recent years. Considering these matters, coal ash is expected to continue increasing. Effective use of coal ash accounts for about 96% of total. However, the rest of ash has been reclaimed in the disposal site. In future, because it has become difficult to secure a large scale of disposal sites, expanding effective utilization becomes crucial issue. It is known that coal ash becomes a solidification material similar to cement by combustion technology. In this study, we focused on the fiber as materials for remedy of brittle failure.

On the other hand, regarding bamboo chips shoot, price of domestic bamboo shoots has been in slump and production of bamboo shoots has been decreasing because import increased and alternate materials were developed. Therefore, a lot of bamboo groves have been ignored, felling bamboo is done to protect bamboo groves, and the way of disposal has become problematic. However, as the bamboo has properties of corrosion-inhibiting and being fibered, it is difficult to be disposed and effectively used. Then, in this study the new effective utilization method was examined by using coal ash as a solidification material for the pedestrian-friendly pavement. In this case, the behavior of brittle failure was expected to be shown as well as a general solidification treatment soil. Therefore, after paving, it was assumed that a crack by material hardening and partial destruction by walking load could happen. Then, it is necessary to think about adding materials with toughness as a remedy of brittle failure. This study aims not only the remedy of brittle failure but also expansion of effective use of bamboo by adding bamboo material (bamboo chips) crushed to the fiber state. Considering these thoughts, we focus on examining the application method of coal ash as pavement materials, to focus on a remedy of brittle failure by adding bamboo and on the expansion effective use of bamboo. This report was made by the result of unconfined compression tests for mechanical properties of the pedestrian-friendly pavement mixing decomposed granite soil with two kinds of coal ash with a different combustion system. Additionally, it reports the result of examining the application of the bamboo chips and a toughness assessment of mixing materials with bamboo chips.

2 SAMPLE AND TESTING PROCEDURE

2.1 *Sample*

The soil used for experiments are decomposed granite soil of Dazaifu Fukuoka. And, the coal ash used as solidification materials are FA generated by pulverized coal firing system and PFBC generated by Pressurized Fluidized Bed Combustion system. Also, with expectation of strength growth by adding cement as sub-materials, blast-furnace slag cement B was used.

Figure 1 shows the particle-size curves of decomposed granite soil, FA and PFBC. Table 1 shows the physical properties of soil, coal ash and bamboo chips.

Photo 1 and Photo 2 shows SEM of FA and PFBC.

The decomposed granite soil has wide particle size distribution. Coal ash contains more fine-grained fractions than decomposed granite soil. The particle-size of FA is smaller than PFBC. Furthermore, Photo 1 and Photo 2 also show that FA has smaller and more spherical. Table 2 shows the chemical components of two kinds of coal ash.

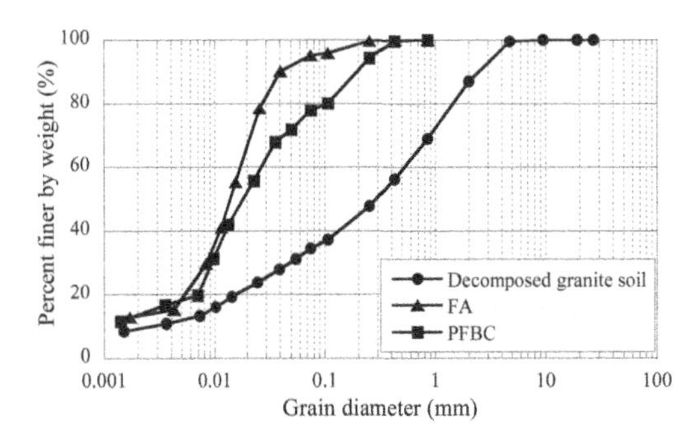

Figure 1. Particle-size curves for sample.

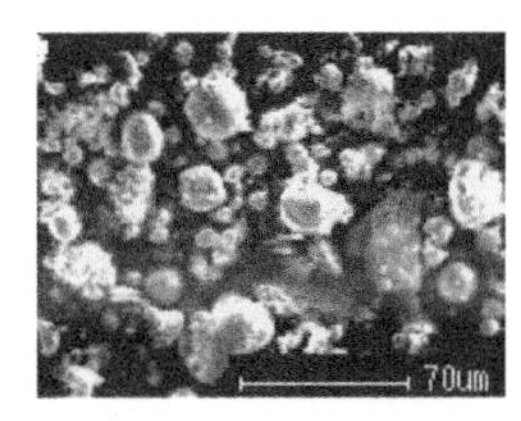

Photo 1. SEM of FA. Photo 2. SEM of PFBC.

Table 2. Chemical component of coal ash (%).

	SiO_2	Al_2O_3	CaO	Fe_2O_3	MgO	Na_2O	K_2O	Ig-Loss
FA	54.7	27.6	7.6	6.4	0.9	0.1	0.1	2.3
PFBC	25.2	19.7	40.5	4.0	1.5	0.4	0.2	7.2

Photo 3. Bamboo chips.

The components of SiO_2 and Al_2O_3 account for more than 80% of FA. On the other hand, it contains more CaO by 40% than FA. Moreover, it is major difference that PFBC contains SiO_2 which is less than half compared with FA and contains little Al_2O_3. Photo 3 shows the bamboo chips. The bamboo chips were used in the state of being fibered and moisturize.

2.2 *Testing procedure*

Table 3 shows mixing conditions of samples. Samples were composed in volume ratio to each 1 m^3. Case1 shows no additional solidification material in

Table 1. Physical properties of sample.

Sample name	ρ_s (g/cm^3)	w (%)	Dmax (mm)	Fines (%)	Sand frection (%)	Gravel faction (%)	Uc (%)	Uc′ (%)
Decomposed granite soil	2.612	2.5	9.5	34.5	52.5	13	208	1.78
FA	2.297	0.4	0.85	95.1	4.9	0	–	–
PFBC's	2.919	0.04	0.85	77.9	22.1	0	–	–
Bamboo chips	1.514	21.7	–	–	–	–	–	–

Table 3. Mixing condition (%).

Case No.	Coal ash	Coal ash	Cement	Bamboo chips
Case1	FA	0	0	0
Case2		10	0	0
Case3		30	0	0
Case4		10	10	0
Case5		30	10	0
Case6	PFBC	10	0	0
Case7		30	0	0
Case8		10	0	5
Case9		30	0	5
Case10		10	0	10
Case11		30	0	10
Case12		0	0	5
Case13		0	0	10

order to comprehend the strength characteristic of soil. Case2 to Case7 was done in order to comprehend solidification with curing by changing additive ratio of solidification material. In Case8 to Case13 shows toughness assessment by changing additive ratio of bamboo chips from 5% to 10% to sample added PFBC. Case12 and Case13 were performed to understand toughness of soil without addition of solidification material (PFBC) as Case1.The experiment examined a compaction test (JIS A 1210, A-c) to decide density of the specimen and obtained optimum water content and maximum dry density. Here, abscissa axis of compaction curve is not general water content but rate of addition water when mixing was done. In compaction test, as coal ash is a porous material with tendencies of particle breakage. Non cyclic method was selected in terms of causing less damage from particle breakage. The specimen was sealed up by the lap and cured in a thermostatic room ($20 \pm 3°$) Curing period was 0, 7, 28 days.

After curing period, unconfined compression test was carried out to examine effect not only of mixing ratio of solidification material and but also strength at the different curing period.

3 RESULTS AND DISCUSSION

3.1 *Compaction characteristic*

Figure 2 shows the compaction test result of Case2 to Case7 added only solidification materials. Moreover, Table 4 shows the optimum water content and maximum dry density gained from compaction test result.

Case1 shows compaction samples to understand influence by addition of solidification materials. The compaction characteristic of the materials mixed coal

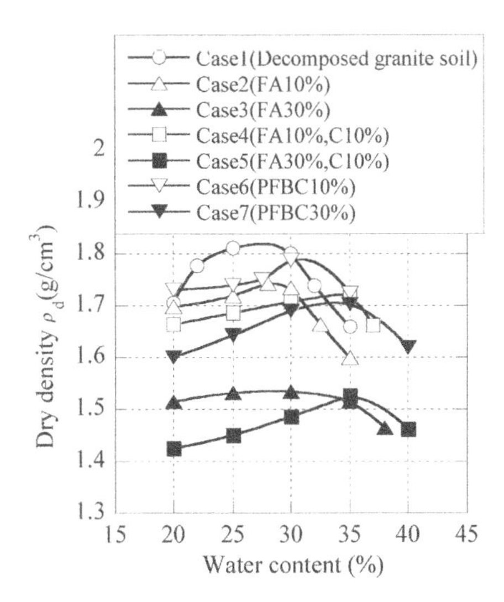

Figure 2. Result of compaction test (Case2 to Case7).

Table 4. Result of compaction test (Case2 to Case7).

Case No.	Mixing condition	ρ_{dmax}(g/cm^3)	w_{opt}(%)
Case1	Decomposed granite soil	1.786	27.3
Case2	FA10%	1.742	28.6
Case3	FA30%	1.535	28.5
Case4	FA10%,Cement10%	1.720	34.1
Case5	FA30%,Cement10%	1.523	35.2
Case6	PFBC10%	1.789	30.9
Case7	PFBC30%	1.706	33.6

ash shows that by increasing additive rate of solidification materials, maximum dry density was decreased and optimum water content was increased by mixing FA and PFBC. It is presumed that adding solidification materials causes increase of fine fraction of sample. Regarding compaction characteristics added the sample with FA and with PFBC, it was shown that maximum dry density of the sample added PFBC increased and compaction characteristics were good under the same condition. It is presumed that PFBC attribute good compaction characteristics to large particle-size than FA and porous particle configuration as Figure 1 shows particle-size curves and Photo 1 and Photo 2 by electron microscope.

Figure 3 shows the compaction test result of Case8 to Case13 added solidification materials and bamboo chips. Table 5 shows the optimum water content and maximum dry density gained from result of compaction test.

Figure 3 and Table 5 show that dry density decreased rapidly by the addition of the bamboo chips. It is presumed that it was caused by decreasing soil ratio

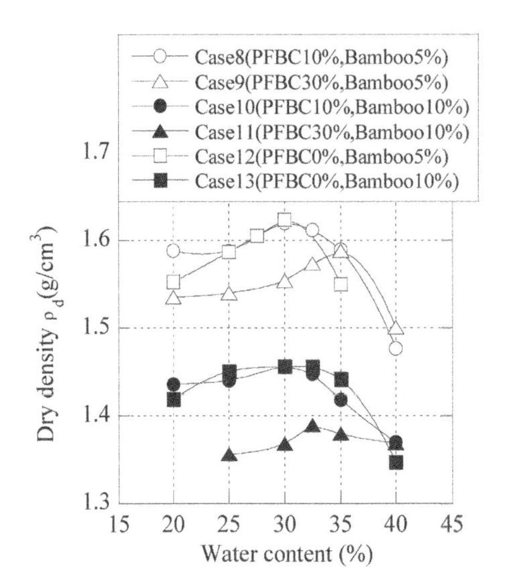

Figure 3. Compaction test result from adding bamboo chips.

Table 5. Compaction test result from bamboo chips. (Case8 to Case13)

Case No.	Mixing condition	ρ_{dmax} (g/cm³)	w_{opt} (%)
Case8	PFBC10%,Bamboo chips5%	1.618	30.6
Case9	PFBC30%,Bamboo chips5%	1.585	35.0
Case10	PFBC10%,Bamboo chips10%	1.456	30.3
Case11	PFBC30%,Bamboo chips10%	1.387	32.6
Case12	PFBC0%,Bamboo chips5%	1.620	30.3
Case13	PFBC0%,Bamboo chips10%	1.456	30.3

while increasing additive rate of bamboo chips with low density. Compaction characteristics of materials added bamboo chips depend on the additional amount of bamboo chips regardless of the additional amount of coal ash.

3.2 *Unconfined compression characteristic by mixing solidification materials*

Figure 4(a), (b) and (c) shows unconfined compression test results of each case of 28 days curing.

Materials mixed soil with FA and cement and materials mixed soil with PFBC causes a remarkable appearance of strength by curing. On the contrary, it was found that materials only mixed FA can hardly be expected any appearance of strength. Especially, Case5 (FA30% and cement10%) indicate the biggest strength. Materials mixed FA with cement is found to be very effective in appearance of strength of materials, compared with material mixed only FA.

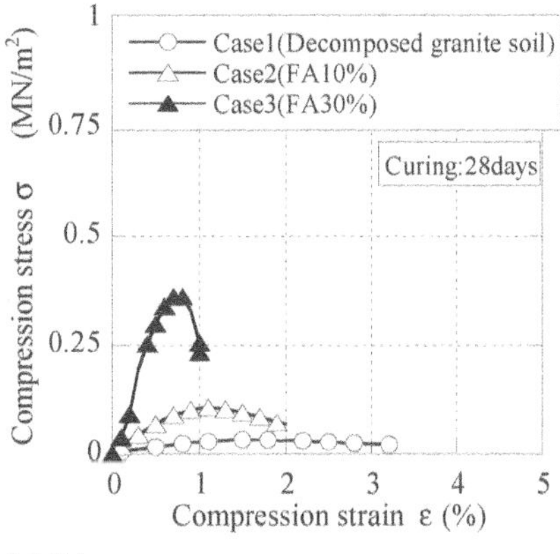

(a) FA

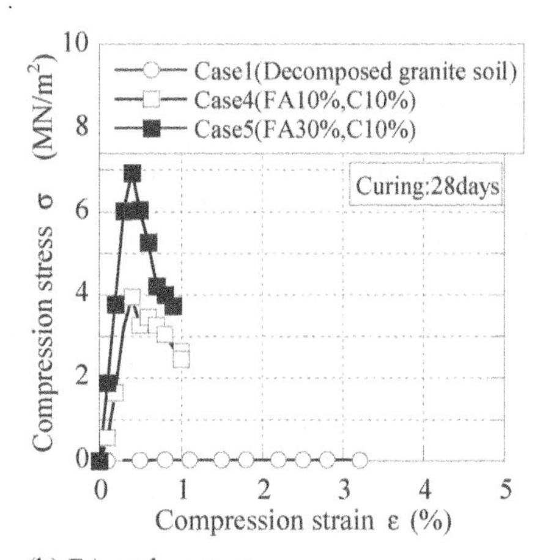

(b) FA and cement

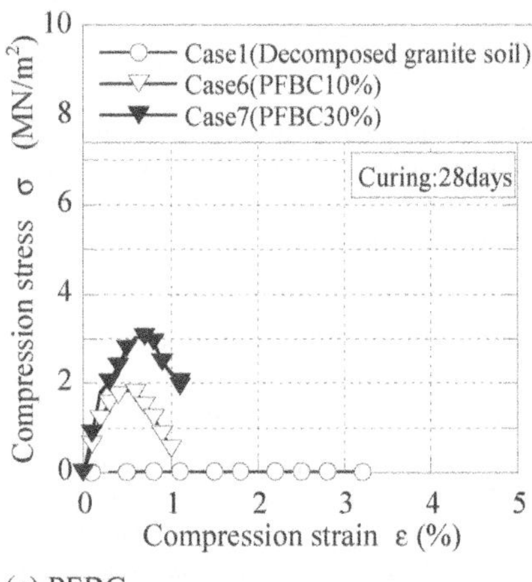

(c) PFBC

Figure 4. Unconfined compression test results (Mixing solidification materials).

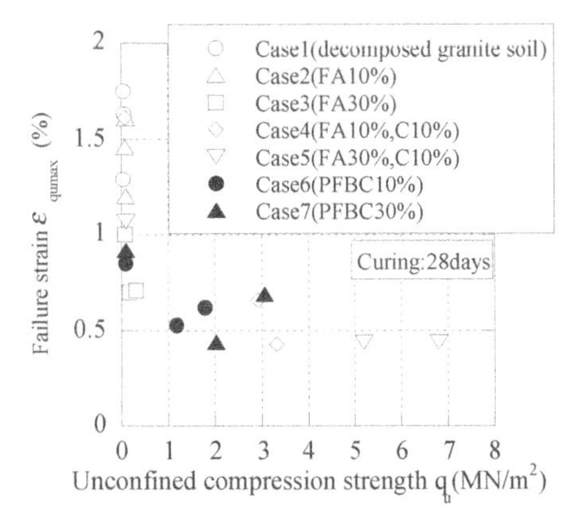

Figure 5. Failure strain of pavement materials on unconfined compression test.

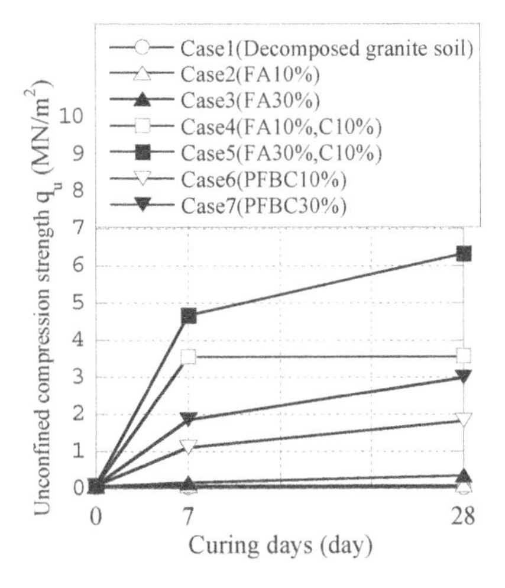

Figure 6. Relation between curing days and q_u.

Figure 5 shows relations between unconfined compression strength and failure strain with each case of 28 days curing. It is shown that once unconfined compression strength increased, the failure strain reached failure area in the vicinity 0.5%. It is shown that as there is appearance of strength by mixing FA with cement and PFBC and it becomes highly rigid, the failure strain becomes small. And, materials with remarkable appearance of strength were understood that failure strain showed less than 1% and materials showed brittleness failure because of rigidity. Those properties indicate that walking load could cause a crack and pavement would be uncomfortable for pedestrians.

Figure 6 shows relation between curing days and unconfined compression strength. Materials mixed FA with cement and materials mixed PFBC are shown that there is highly increased strength in each case of 7day curing. The sample mixed FA30% with cement 10%

Table 6. Define index toughness assessment[1), 2)].

Parameter	Define index of toughness assessment	Appellation
Stress	q_{ures}/q_{umax}	Toughness degree
Performance	$\frac{W}{(q_{u\max} \times (\varepsilon_{qures} - \varepsilon_{qu\max}))}$ [*Correspond of* $P_1/(P_1+P_2)$]	Toughness potential

$$W = \int_{\varepsilon_{qu\max}}^{\varepsilon_{qures}} (q_f \sim \varepsilon)\, d\varepsilon$$

(Case5) and with PFBC gradually increased strength even after 7 days had passed. On the contrary, the sample mixed FA 10% with cement 10% (Case4) stopped increasing movement of strength in about one week. This showed that it is necessary to consider appearance of strength of 7 day curing in case of paving.

3.3 *Toughness assessment by mixing bamboo chips*

a) Evaluation method of toughness

As materials in this test were added solidification material, low index of failure strain was shown as well as general solidification treatment soil. Furthermore, the result of examination showed brittleness failure caused by rapid decreasing stress after failure. Therefore, if it is used as pavement material, there is a possibility that a crack due to hardening and partial destruction would occur. Then, this study examined the remedy of brittleness failure by mixing bamboo chips.

Here, toughness means tenacity of materials which can not be expressed only by strength. Then, in order to evaluate these properties quantitatively, toughness assessment was done using stress-strain curve gained from unconfined compression test. We defined index of toughness assessment as Table 6[1),2)].

Failure point shows unconfined compression strength as q_{umax} and the strain at this moment was described as ε_{qumax}. In residual state, under unconfined compression strength when the strain was 10%, residual strength was q_{ures} and strain in residual state was ε_{qures}. And, Table 6 shows stress and performance as parameters. Stress parameters is q_{umax} and q_{ures} for this evaluation.

Performance can be shown by the area enclosed by the stress-strain curve. As samples with different curing days and different density are necessary

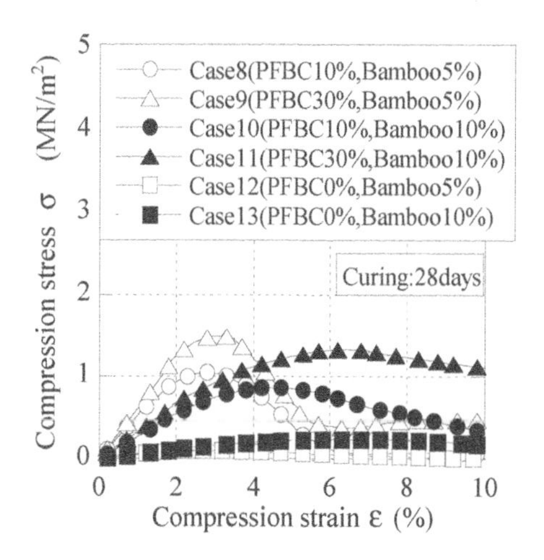

Figure 7. Unconfined compression test results mixing bamboo chips.

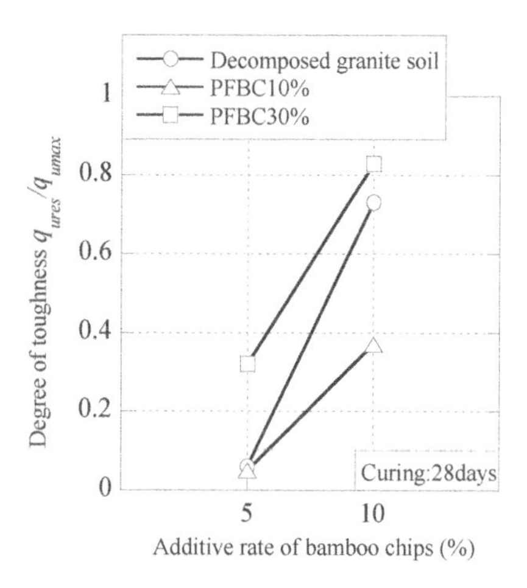

Figure 8. Relation between additive rate of bamboo chips and degree of toughness.

to be compared, we decided to think of toughness, considering stress-strain curve regulated by unconfined compression strength. Also, toughness degree showed that the level of stress decreased after material failure. On the other hand, toughness potential showed the level of bearing force after material failure. Both showed that the nearer value approached 1, the more toughness same sample has.

b) Understanding of fracture morphology of sample mixed bamboo chips

Figure 7 shows unconfined compression test result of materials added bamboo ships and cured for 28 days. The failure strain is remarkably grown by the addition of bamboo chips. It is understood that the fracture morphology transforms into ductile failure, compared with Figure 4(c). Furthermore, it is understood that failure strain become higher, compared with two types of materials mixed with 5% and 10% bamboo chips.

c) Toughness assessment by mixing bamboo chips

Figure 8 shows relation between additive rate of bamboo chips and toughness degree. According to Figure 8, each sample shows that stress decrease is low after failure in residual state because toughness degree increases by raising additive rate of bamboo chips.

Figure 9 shows relation between additive rate of bamboo chips and toughness potential. This figure shows that bearing force of materials increases because the value of toughness potential is grown by increasing additive rate of bamboo chips. It shows that an increase of about 0.3 to 0.4 (the value of toughness potential) is observed when the additive rate of bamboo chips is increased. It was shown that adding bamboo chips was effective in enhancing toughness of materials.

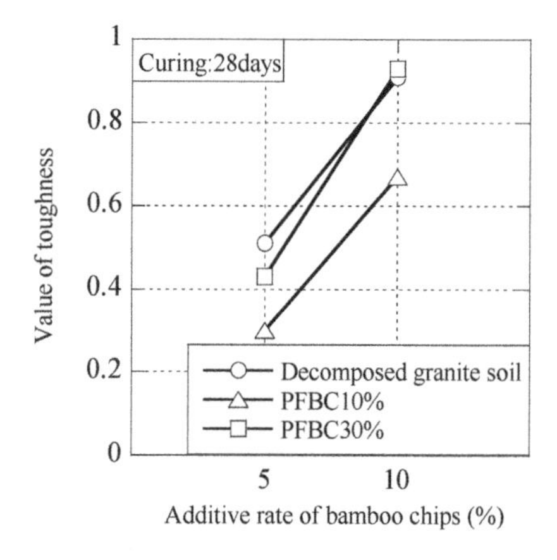

Figure 9. Relation between additive rate of bamboo chips and toughness potential.

4 CONCLUSIONS

This study aims to develop pedestrian-friendly pavement materials by mixing decomposed granite soil with two kinds of coal ash from different furnaces. According to these thoughts, we tried to comprehend solidification of materials and mechanical properties by the atmospheric curing test.

Moreover, toughness assessment from the result of unconfined compression tests was carried out in order to prevent pavement materials from a crack caused by dry and shrinkage of pavement materials.

Summary of this study is following.

(1) After studying solidification of material, we found that remarkable appearance of strength was

caused by mixing FA and cement with decomposed granite soil.

(2) Regarding sample mixed PFBC with decomposed granite soil, it showed higher solidification than sample mixed only FA with soil.

(3) When thinking of strength of pavement, it was clarified to have appearance of strength in 7day curing.

(4) The result of toughness assessment mixing bamboo chips showed that brittle failure was improved by an addition of 10% bamboo chips.

REFERENCES

1) Tomohiro Morisawa, Yoshio Mitarai and Kazuya Yasuhara: Effect of toughness improvement of cement treated soil with tire-chips, *6th Environmental Geotechnical Engineering Symposium* (in Japanese), pp.359–364, 2005.

2) Eisaku Takayama and Hideo Ochiai: Improvement on Brittle Behavior of Foamed Light-weight soil by Mixing Various Types of Short Fibers, *Symposium on Utilization technology of artificial ground material* (in Japanese), pp. 55-58, 2005.

Advances in Transportation Geotechnics – Ellis, Yu, McDowell, Dawson & Thom (eds)
© 2008 Taylor & Francis Group, London, ISBN 978-0-415-47590-7

Study of suction in unsaturated soils applied to pavement mechanics

B.A. Silva
Federal University of Rio de Janeiro – COPPE / Military Institute of Engineering, Rio de Janeiro, Brazil

L.M.G. Motta
Federal University of Rio de Janeiro – COPPE, Rio de Janeiro, Brazil

A. Vieira
Military Institute of Engineering, Rio de Janeiro, Brazil

ABSTRACT: High groundwater table exerts detrimental effects on the roadway pavement. This paper presents a real application of Soils Mechanics and Geotechnical Engineering in tropical climates, showing that the performance of a pavement is so influenced by its pore pressure and resilient modulus values. The measurement of resilient modulus has become increasingly accepted for characterizing engineering properties of pavement materials. This paper aims to correlate the phenomena of suction (matric, osmotic or both) with parameters used by Pavements Mechanics, studying the soils behavior when submitted to road loads, refining the mechanical (or rational) design of pavements inside of the climatic and regional conditions found in real situations of work. It presents an Experimental Program Methodology using a test-pit facility to study a Brazilian typical pavement design behavior. Paper filter is the technique which will be employed to obtain suction values due to its great range. Time Domain Reflectometry (TDR) will be also used to measure soil moisture along the test-pit profile.

1 INTRODUCTION

The moisture existing in a road pavement layer defines its mechanical behavior. Thus, depending on the subgrade material, as well as the local environmental characteristics and the road design, it is shown a picture regarding resilient modulus and suction alterations, what in the latest analyses regulate the mechanical behavior and fatigue life of the road pavement.

In this paper a typical Brasilian road subgrade was submitted to moisture changes in a concrete test-pit sufficiently monitored, and which will be later analyzed all the predicted performance of a full-scale pavement submitted to a groundwater table (GWT) oscillation and capillary rise.

2 THEORETICAL PRINCIPLES

Ping et al (2002) tested three types of granular subgrades, submitting them to moisture oscillations and observing the damage it causes to mechanical performance of its base. They observed that an A-3 soil type, when used as subgrade, supplied bigger protection than the other two tested soils (A-3 and A-2-4). In addition, no matter the bigger suction values in A-2-4 soil, it was more sensible to a moisture variation than the A-3 one. The authors also affirm that the percentile of fine existing in the subgrade soil governs the moisture influence in resilient modulus.

Edil & Motan (1979) affirm that it has been growth the idea of correlating soil resilient behavior with its suction value, because it is more appropriated than the use of moisture or saturation degree as indicators of the resilient behavior of subgrades.

Medina & Motta (2005) mention the mechanisms by which the moisture can interferes in pavement, showing that the groundwater table oscillation, due to the rain water access, can cause moisture variations in subgrade in case the groundwater table is in a small depth (about 1,0 meter or less).

Medina & Motta (2005) also affirm that the migration of salt existing in soil or in aggregates of pavement layers can occur by capillarity of water (with dissolved salts) to upper layers; when It reaches the surface, the water evaporates depositing the salts. This phenomenon, besides generating stains in asphalt concrete's cracks, confirms the capillarity ascension in pavements layers.

Perera et al (2003) collected foundation pavement soils of hole USA and studied the moisture equilibrium of foundation road soils. They concluded that moisture directly influences the stress and strain in pavement layer systems. The authors also concluded that the matric suction measured in situ show a good correlation with Thornthwaite Index and with soil type, a very similar result obtained previously by Russam & Coleman (1961).

Ping & Ling (2000) show the problems faced by constructs in Florida. They complain that the typical constituents of pavement foundation (A-1, A-3 e A-2-4) concentrate excessive amount of water and thus turn hard to dry and compact. The experimental results shows that the soil suction and its relative humidity have direct effect in the soil drainage tax. The suction values vary significantly with the amount of water when the percentage passing in #200 sieve is greater than 20%. They have also concluded that the drainage tax decreases with the increase of soil suction for each type of soil and also decrease with the increase of fine percentile. The authors suggest that the influence of moisture in the drainage tax of soil is more significantly than that of temperature.

Silva (2003) says that dry and wet processes of soils give variability in suction values which changes the resilient characteristics of pavement layers constituent materials. These changes are so significant that drainage process of a material until a moisture correspondent to a 2% under the optimum moisture can duplicate its resilient modulus. On the other hand the opposite is also valid, i.e., a moisture gain, especially in fine soils (more sensible to moisture changes), can take it to lose a great part of its resilient potential.

Sadasivam & Morian (2006) studied the effect of the groundwater table in expected pavement performance (cracking from top to bottom, bottom to top and sinking of wheel trail) using simulations in four different levels of groundwater table (1,52 m, 3,05 m, 6,10 m e 9,15 m), six different subgrade soils and two typical structures of pavement. They pointed out that the groundwater table level is an entrance parameter in AASHTO Design Guide 2002 and in EICM software. The subgrade permanent deformations forecast are compared with Ayres Model for permanent deformations in no cohesive materials. They yet concluded that:

- With the growth of groundwater table (GWT), the cracking from top to bottom in pavement surface decreases, while the cracking by fatigue in the base of the bituminous layer increases;
- The groundwater table does not influence significantly the sinking of wheel trail of the bituminous layer;
- the subgrade permanent deformation is much more affected by changes in groundwater table, depending on the subgrade soil type (silt or clay).

Liang & Rabab'ah (2007) say that the majority of fine soils show a decreasing resilient modulus path with the increase of its moisture, and that the matric suction has been a good state variable to foresee the dependence of resilient modulus with moisture for cohesive soils.

3 MATERIALS AND METHODS EMPLOYED

3.1 *Test-pit*

The test-pit used for this paper has internal dimensions of 2.0 m × 2.0 m, It was done with concrete walls of 20.0 cm thickness. Two of these walls (opposed) are doubled, with water entrance between them and with a hole (15.0 cm of height) in its bottom to allow water flow from bottom to top.

A load application structure was built with metallic beams in order to support until 200.0 ton in the middle of the 2.0 m space. In one of the walls It was done holes to each 20.0 cm to allow passage of instruments threads which will be accommodated into the pavement structure.

It also has an external piezometer to verify the exact level of water into the test-pit, and a compressed air installation for applying dynamic load.

Two faucets and two drains were put in order to control entrance and exit of water in order to verify the histeresis of the phenomena (resilient modulus and suction) studied. A reinforced blanket and a 10.0 cm stone layer put upon it work as a filter which avoid the blockage of the drain by thin particles of soil.

For this paper, It only studied the subgrade layer. But the main interest is to study all the pavement profile submitted to an oscillation of groundwater table (GWT) level and to dynamic and static load applications.

Below there is a schematic design of test-pit. It shows the TDR's installation and the load application for the hole pavement profile. As said previously, in this paper only the subgrade was studied.

3.2 *Materials employed*

The subgrade material employed is a residual soil coming from Rio de Janeiro, Brazil.

Its physical and chemical characteristics are s-specified in tables below.

3.3 *Method employed*

3.3.1 *Initial preparation for soil*

The soil, as received from its origin was stored outdoors, covered with a plastic blanket to protect it from moisture oscillation (wetting or drying). Thus, it remains until its compaction into test-pit.

A little amount of soil (about 5.0 kg) was picked out to be used for Physical and Chemical Characterizations.

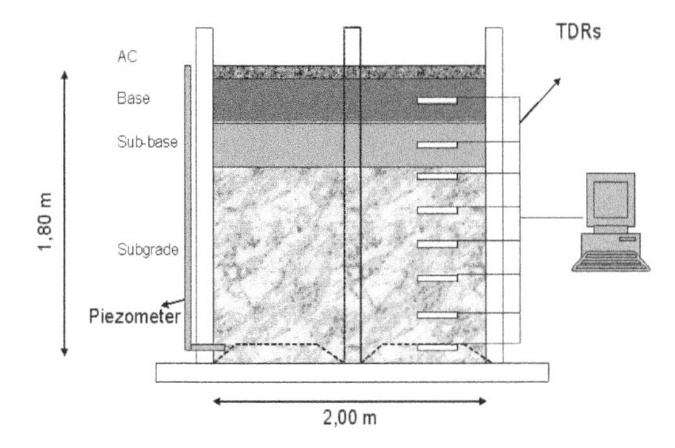

Figure 1. Lateral view of test-pit with the installations of TDRs.

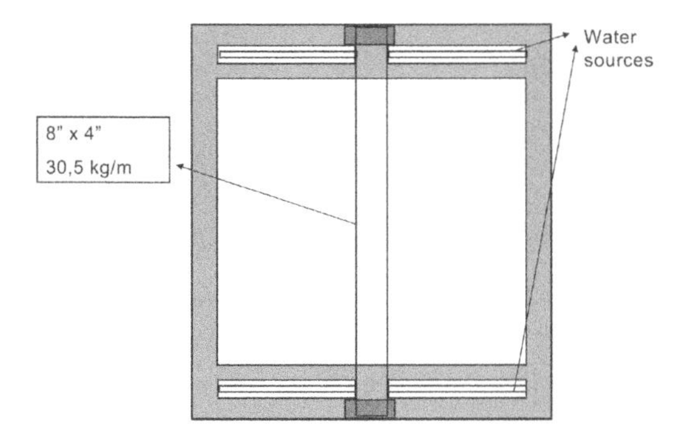

Figure 2. Upper view of test-pit with the metallic beam to support the load (dynamic and static).

Table 1. Granulometric Distribution of Subgrade.

Granulometric Distribution (%) (ABNT Scale)					
		Sand			
Clay	Silt	Fine	Med	Great	Grave
5	8	12	19	19	37

Table 2. Subgrade physical characteristics.

Max. Dry Dens. (g/cm^3)	Opt. Moist. Cont. (%)
1970,0	12,0

3.3.2 *TDR's calibration*

Using the compacting soil curve and the Optimum Moisture Content it was compacted 05 (five) samples of subgrade soil. The dimensions of these cylindrical samples are 10.0 × 20.0 cm (diameter and height).

These samples were treated as follows:

- 1 (one) sample was submitted to a dry cycle until the moisture of OMC−2%;
- 1 (one) sample stayed in OMC;
- 1 (one) sample was submitted to a wet cycle by capillarity until a equivalent moisture of OMC+2%;
- 1 (one) sample was submitted to a wet cycle by capillarity for 3 days;
- 1 (one) sample was submitted to a wet cycle by capillarity for 5 days.

Thus, all the moisture ranges which the subgrade will work with was reached. This procedure is fundamental to obtain a calibration curve of TDR's probes.

Observe that in the TDR's calibration it was employed a kind of method for wetting soil which differs from compacting them in bigger moistures. This is because it had to be also obeyed to work density material.

The TDRs probe used are German products of IMKO Co. It is also necessary the use of a signal conditioning (TRIME – IMKO) and a 9–24 V electric source. Figure 3.

After molding the samples in needed moistures, they were prepared to receive the TDRs probes. This preparation consists of making 2 holes in the sample, using a supplied (by IMKO) form. Figure 4.

In order to have a good repeatability of the moisture measures, it was done others two holes, orthogonally to the first ones and repeated the last procedure (Figure 5). This procedure confirms the former obtained values and the repeatability suggested by IMKO (0,2%).

It was also done a test to verify the influence of the metallic mold used to compact the samples (Figure 6). It was observed that it does not have considerable mistakes when using the TDRs probes with that metallic mold. The electromagnetic generated field done by TDR's probes concentrates in the middle (between) its legs.

With the values obtained in probes' measures and the real values obtained by drying the samples in an oven (60°C) for 24 hours, it obtained the calibration. IMKO supplied a software (Wincal) specific to calibrate these probes. So the values (real and measured) are inserted into the software and it saves the calibration curves in probes' chips.

3.3.3 *Analyses of obtained moistures*

For the first situation (tested subgrade immediately before compaction), without any capillarity ascension (with GWT equal to zero) the moisture stayed constant, as can be seen in Figure 7 below.

For the case of GWT situated in the middle of subgrade height, the moistures distribution is as shown in Figure 8.

Table 3. Subgrade chemical characteristics.

PH			Sulfuric Attack							
H_2O	KCl 1M	ΔP (%)	SiO_2 %	Al_2O_3 %	Fe_2O_3 %	K_2O %	Res. %	Ki	Kr	O.M. %
5,85	3,57	2,57	13,6	8,0	5,2	1,3	68,3	2,89	2,04	0,15

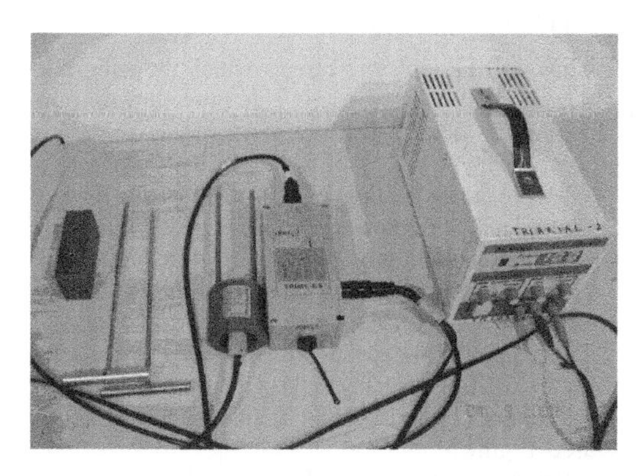

Figure 3. IMKO equipment employed to calibrate TDRs probes.

Figure 4. IMKO supplied form to make the holes.

The procedure used to defines the equilibrium of the humidity capillarity path, which define the plate test begin, was to maintain the same moisture profile static for two days without any alteration. To reach the equilibrium of this subgrade soil It took 12 days.

3.3.4 *Soil compacting in test-pit*

With the compacting test results of soil, where It was determined the subgrade moisture content, It was proceeded its homogenization aided by a big cement mixer with 520 liters of capacity.

The soil thus homogenized was put into the test-pit in layers of 30.0 cm of thickness and compacted with a manual compactor (Figure 9). The compaction control

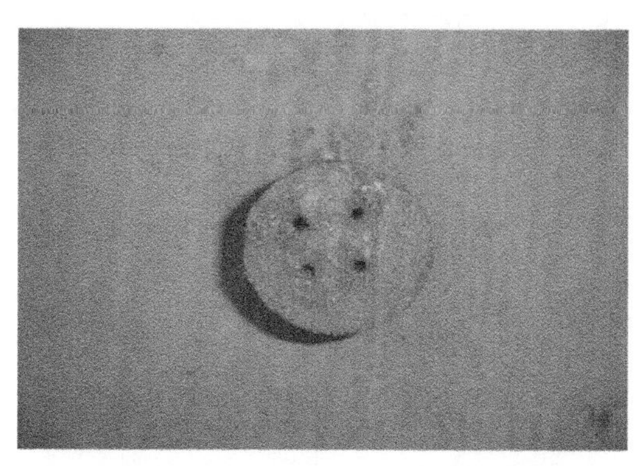

Figure 5. Other two orthogonal holes to confirm the repeatability of measures.

Figure 6. Verification of measures done with and without the metallic model.

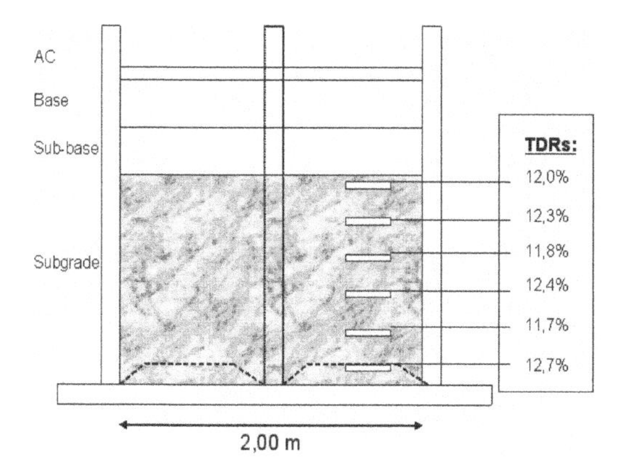

Figure 7. Moistures variation through subgrade layer without GWT.

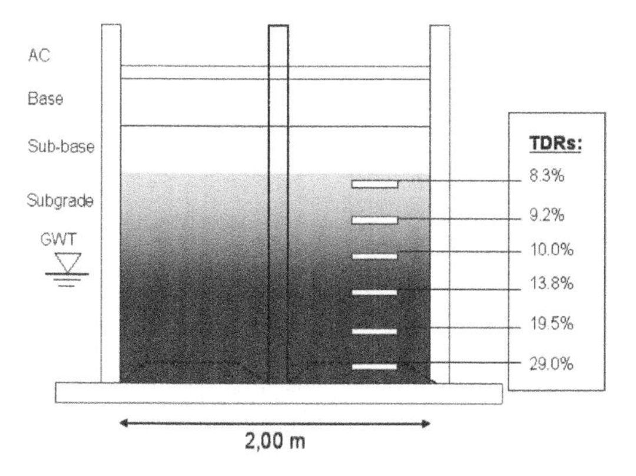

Figure 8. GWT in the middle of subgrade.

Figure 9. Manual compactor and GeoGauge equipment.

was done by a sand flask and with an equipment called GeoGauge with nuclear principle of work (Figure 9).

When the position of TDRs was reached, it stopped the compaction and installed them conveniently.

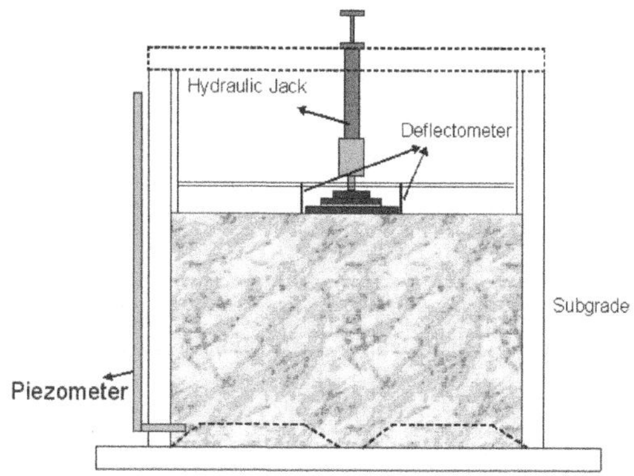

Figure 10. Plate Load Test apparatus.

3.3.5 *Plate load test*

The plate load test (ASTM D 1196/93) is a stress controlled test, where load stages are successively applied after the stabilization of its correspondent deformation.

Silva (2001) describes the basic equipments and the execution of the test.

In this paper it was used a pneumatic cylinder for load application, which was applied on a hydraulic jack with 1000.0 KN of capacity and a manometer with 20.0 KN of precision.

The plates are circular with decrease diameters of 80.0 cm, 45.0 cm e 30.0 cm, with 2.54 cm of thickness. They are put in a pyramidal position one above the other, with the 80.0 cm plate in contact with soil (Figure 10).

Graduated deflectometers (0.01 mm) were installed above the last plate surface.

For the test is necessary that a thin layer of fine sand, which purpose is to level the contact surface with soil.

In the beginning of the test it is applied a first load to accommodate the plate above soil. This load is that which produces a medium deformation from 0.25 mm until 50.50 mm, and than unloaded. After unloaded and the stabilization of deflectometer's points, it applies a load half of the first one to obtain the beginning references mark of deflectometers.

It begins the load with increments from 0.015 MPa until 0.020 MPa. The load should be maintained until the registered deflections do not show oscillations bigger than 0.02 mm during 2 consecutives minutes. Register the readings.

It continues in successive loads until it reaches 0.20 MPa or 0.25 MPa,

Finally, slowly unload the jack, doing two or three intermediate return readings and a final one, with a load equal of that used in the beginning of the readings, in the beginning of the load.

The stress-strain curve allows knowing the relations between total strain and correspondent stress, as

Table 4. K values for two GWT conditions.

GWT	K (MPa/m)
Without GWT	90,6
In the middle of subgrade layer	72,4

anelastic and elastic strains to maximum load applied too. The K value (Reaction Coefficient of Soil) is determined, in this curve, for the deformation of 0.127 cm.

For this subgrade soil the results obtained for K, with the soil only compacted in the optimum moisture content, correspond to Normal Proctor energy, and for both used conditions of GWT are shown in Table 4.

Thus, it can be observed that there is a reduction of 20% in K value due to the saturation of part (half) of the subgrade layer.

3.3.6 *Total suction measurement*

Suction has two portions, one matric (or capillary) and other osmotic. The matric suction is the suction influenced by the soil matrix, by intergranular spaces forms and dimensions. The osmotic portion is that coming from variation in dissolved salt concentration in soil water. It is so studied in case of soil contamination. But in the majority of road engineering problems the osmotic portion is so small that we should ignore it, and the total suction can be considered equal to matric suction.

3.3.7 *The paper filter method*

The paper filter method for soil suction measure was initially used by Agronomy. Recent experiences were conduced in subgrade of airport pavements and expansive potential profile by Mckeen (1985).

The paper filter method is said to be an indirect method of suction measurement. It is based in equilibrium that a default paper establishes with the soil moisture with a determined suction.

In this paper it was used the E-832 specification from ASTM, with Whatman No. 42 paper. The calibration curve showed below came from Fredlund & Rahardjo (1993) studies.

For paper filter moistures (w_f) under 45%:

$$\text{Log Suction (kPa)} = 5{,}327 - 0{,}0779\ wf \quad (1)$$

For paper filter moistures (w_f) above 45%:

$$\text{Log Suction (kPa)} = 2{,}412 - 0{,}0135\ wf \quad (2)$$

The apparatus used for suction measuring in subgrade is shown in Figure 11. The complete method to obtain the suction values of soil using paper filter method can be consulted in Silva (2003).

Figure 11. Equipment used for suction measuring in pavement layers.

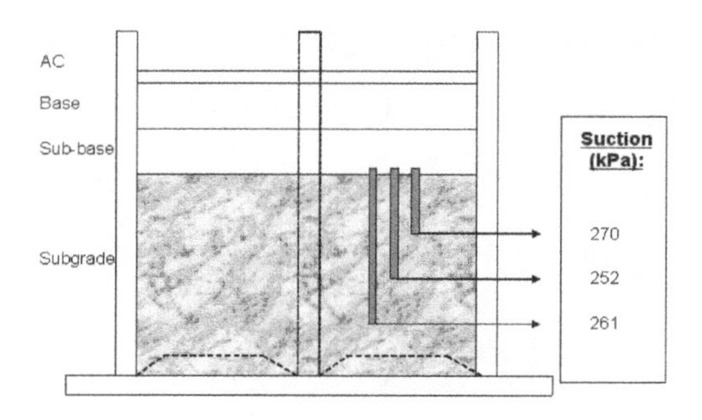

Figure 12. Suction values through subgrade layer without GWT.

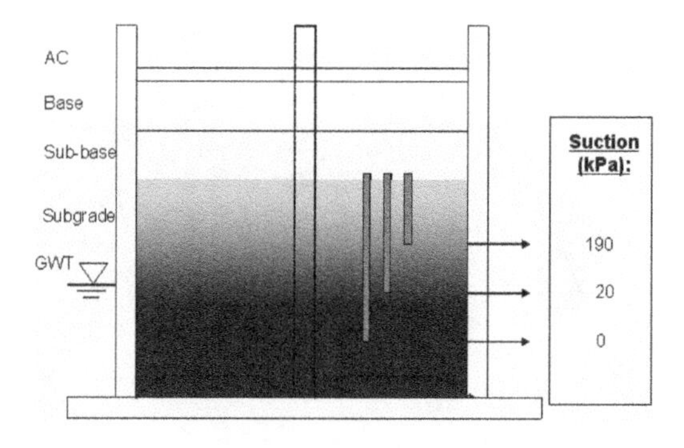

Figure 13. Suction values through subgrade layer with GWT in the middle of subgrade.

In this work, for both equilibrium situations studied for GWT, it obtained the suction values shown in Figure 12 and Figure 13 below.

4 CONCLUSIONS

Thus, for the same type of soil, in this case a real future pavement foundation, the moisture variation due to the ascension of GWT (capillarity), diminishes in 20% the K value.

Considering the coarsed characteristic of the studied soil (% silt and clay), it verifies that capillarity rise is not high, retaining the saturation zone next to the GWT level. This fact reinforces the literature when it says that the finer the soil, the bigger is its capillarity rise, and the coarser is the soil, the smaller will be its capillarity rise.

These obtained results for subgrade can, without any doubt, be considered for other pavement layers performance (sub-base and base), i.e., depending on the magnitude of GWT oscillation, or moisture variation due to other factors (infiltration through surface cracks, poor drainage, etc.) the support capacity of pavement layers is affected, with a consequent decrease of its resilient modulus and a reduction in AC fatigue life.

The suction reached values also share with those expressed in literature, i.e., the reached values and the gradient obtained by capillarity phenomenon justify the support lost, occurred by the foundation material of pavement, which takes so many damage consequences to its useful life.

The experience shows that, depending on the material employed and of the regional climatic conditions the capillarity grows up from GWT until 2 or 3 meters of height. Classical Geotechnic studies had yet discovered the existence of capillarity elevations at about 15,0 meters height. Facts like these should be considered by design engineers, who sometimes concentrate their researches in traffic counting and pavement materials, despising an ambiental analyze, due to humidity interferences in road pavement.

REFERENCES

Edil, T.B., Motan, S.E., *Soil Water Potential and Resilient Behaviour of Subgrade Soils*, TRR No 705, TRB, National Research Council, Washington D.C., 1979, pp. 54–63.

Fredlund, D.G., Rahardjo, H., *Soil Mechanics for unsaturated soils*, John Wiley & Sons, Inc. Canada, 1993.

Liang, R.Y., Rabab'ah, S., *Predicting Moisture-dependent Resilient Modulus of Cohesive Soils Using Soil Suction Concept*, XIII CPMSIG, Isla Margarita, Venezuela, 2007.

Mckeen, R.G., *Validation of procedures for Paviment Design on Expansive Soils*, Final report, U.S. Dep. of Transportation, Washington, DC, 1985.

Medina, J., Motta, L.M.G., *Mecânica dos Pavimentos*, 2ª Ed., Rio de Janeiro, 2005.

Ping, W.V., Ling, C., *Influence of Soil Suction and Environmental Factors on Drying Characteristics of Granular Subgrades Soils*, 79th TRB, Washington, D.C., 2002.

Ping, W. V., Liu, H., Zhang, C., Yang, Z., *Effects of High Groundwater Table and Capillary Rise on Pavement Base Clearance of Granular Subgrades*, 81st TRB, Washington, D.C., 2002.

Russam, K., Coleman, J.D., *The Effect of Climatic Factors on Subgrade Moisture Conditions*, Geotechnique, Volume XI, No. 1, 1961.

Sadasivam, S., Morian, D., *Effects of Groundwater Table Depths on Predicted Performance os Pavements*, 85th TRB, Washington, D.C., 2006.

Silva, B.A., *Aplicação do método mecanístico de dimensionamento de pavimentos a solos finos do norte do Mato Grosso*, Master degree dissertation, Instituto Militar de Engenharia, Rio de Janeiro, 2003.

Silva, P.D.E.A., *Estudo do Reforço de Concreto de Cimento Portland (Whitetopping) na Pista Circular Experimental do Instituto de Pesquisas Rodoviárias*, Doctor Thesis, UFRJ-COPPE, Rio de Janeiro, 2001.

Advances in Transportation Geotechnics – Ellis, Yu, McDowell, Dawson & Thom (eds)
© 2008 Taylor & Francis Group, London, ISBN 978-0-415-47590-7

Large-scale direct shear testing of geocell reinforced soil

Y.M. Wang & Y.K. Chen
School of Civil and Transportation Engineering, South China University of Technology, Guangzhou, China

C.S. Wang & Z.X. Hou
Guangdong Meihe Expressway Ltd, Guangdong, China

ABSTRACT: The shear strength test results of geocell reinforced soil that are carried out with large-scale direct shear tests are reported in this paper. Three types of specimens which are unreinforced silty gravel soil, geocell reinforced silty gravel soil, and geocell reinforced cement stabilizing silty gravel soil were investigated. The comparison of large-scale shear test results and triaxial compression test results involving unreinforced silty gravel soil was conducted to evaluate the influence of testing methods. The test results showed that the unreinforced soil and geocell reinforced soil gave similar nonlinear features on the behavior of shear stress and displacement, while there was a quasi-elastic characteristic for the geocell reinforced cement stabilizing soil in the case of higher normal stress. With the reinforcement of geocell, the cohesive strength of silty gravel soil was found considerably increased, but the friction angle of the geocell reinforced soil did not change obviously.

1 INTRODUCTION

Geocell reinforced soil has gained a considerable popularity in highway subgrade engineering in the recent past. It has been found to be useful on reinforcement of embankment, steep slopes, retaining walls and abutment backfills because of improving the load-bearing capacity, increasing the strength and stiffness, reducing settlement, and saving cost and time in construction (Bathurst & Jarrett 1988, Cancelli et al. 1993, Sekine et al. 1994, Bathurst & Crowe 1994, Mhaiskar & Mandal 1996, Rajagopal et al. 1999, Latha et al. 2006, Mengelt & Edil 2006).

Since the first use of cellular confinement systems to improve road bases over weak subgrades (Bathurst & Jarrett 1988), the stability of geocell reinforced soil has required considerations on the interface shear strength between soil and geocell. A test method for determining the interface shear capacity of geosynthetic reinforced soil was first introduced by ASTM D 5321-92 (1992), *Standard Test Method for Determining the Coefficient of Soil and Geosynthetic or Geosynthetic and Geosynthetic Friction by the Direct Shear Method* and currently is the revised edition ASTM D 5321-02 (2002).

The methods of test introduced above are now used to provide the shear parameters of a geosynthetic against soil, or a geosynthetic against another geosynthetic, under a constant rate of deformation. It is applicable for all kinds of geosynthetics. However, limited investigations were reported in literature on the shear behavior of geocell reinforced soil. (Bathurst & Karpurapu 1993, Rajagopal et al 1999, Cheng & Jonathan 1993, Yan, et al. 2006, Jing, et al. 2006).

This paper reports the results from serials of large-scale direct shear tests carried out on silty gravel soil confined with or without multiple geocells. The comparisons of large-scale shear test with triaxial compression test for the same type of soil are conducted to evaluate the influences of testing method on the shear strength indices as well. Meanwhile, the testing results of geocell reinforced cement stabilizing silty gravel soil are studied and analyzed as well.

2 EXPERIMENT METHOD

2.1 *Test instrumentations*

The general test arrangement of large-scale direct shear test is shown in Figure 1. The test system was built in the structure engineering laboratory at South China University of China. The shear box includes two stacked parts. The inner size of each part has a dimension of 500 mm × 500 mm × 200 mm (length × width × height). Each part can be disassembled easily for installing and unloading soil specimen. During the directly shearing test, the bottom half was fixed on the ground, and the top half was driven by a 30-tonne capacity horizontal loading system with force transducers collecting data constantly by computer software.

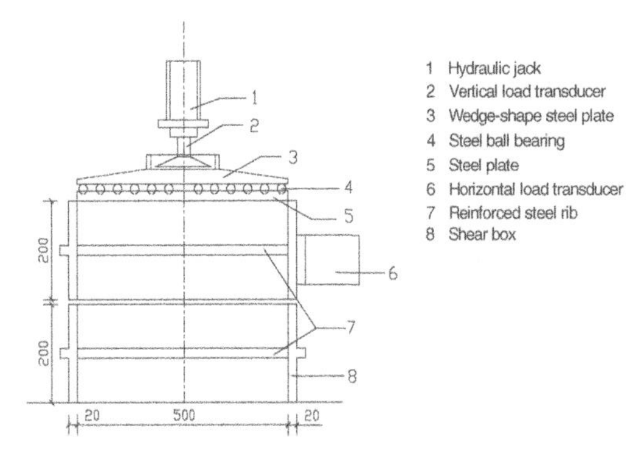

Figure 1. General arrangement of large-scale direct shear test.

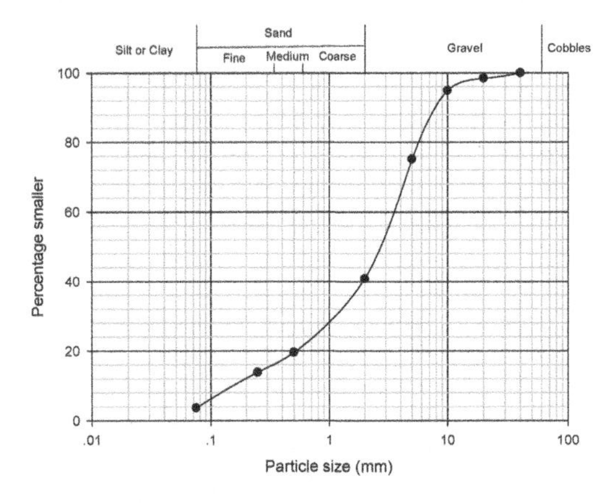

Figure 2. Particle size distribution of backfill soil.

The vertical loading system was driven by a 100-tonne capacity MTS hydraulic actuator. There were two steel plates placed on the top of the soil specimen to bear the vertical load. Two lines of steel linear bearings were seated between these two steel plates in order to reduce friction of plates caused by the horizontal movement of shear box. The normal load was measured by an electronic load transducer.

Two 0.01 mm-sensitivity dial indicators were located on the left side and right side of top-half shear box to measure the horizontal displacement manually.

2.2 *Test materials*

2.2.1 *Soil*

The soil used in the tests was well graded with a maximum particle size of 40 mm, coefficient of curvature of 2.98, coefficient of uniformity of 20.6 and fines content less than 4% by weight. The particle size distribution is shown in Figure 2. The maximum dry density of the soil is 2.049 g/cm^3. The optimum water content is 9.057%. Soil properties are summarized in Table 1. This soil can be classified as GM according to the Unified Soil Classification System.

2.2.2 *Geocell*

The geocell reinforcement material used in the test was a commercially available industrial product manufactured by Netlon China Limited. It was polypropylene stripes welded together to give an open-cell construction that had a cell area of 625 cm^2 and depth of 10 cm. The cellular materials came in panels expanded to cover an area of 4 m $\times$ 12.5 m. The properties of geocell are summarized in Table 2.

2.2.3 *Cement*

Portland cement has been used as stabilizer to increase the strength of coarse soil for a long time. In this study, the cement stabilized soil was used as backfill materials together with geocell reinforcement in the abutment. Table 3 shows properties of the cement used in the study.

2.3 *Test procedures*

2.3.1 *Geocell reinforced soil*

The test method for geocell reinforced soil in this paper was large-scale direct shear test. The procedures of large-scale direct shear test were described as following:

First of all, the large-scale direct shear device must be calibrated to measure the internal resistance to shear inherent to the device. Then, the geocell reinforced soil specimens were prepared. The geocell reinforced soil specimens were molded within the shear box which had a volume of 0.1 m^3 (500 mm long $\times$ 500 mm wide $\times$ 400 mm high). Generally, three or four specimens were required in one direct shear test. After finishing the specimen molding, a confining stress was applied vertically to the specimen, and the upper box was pushed horizontally at a rate of 1 mm/min. until the sample failed, or got to a general strain of 5%. Three specimens were tested at varying confining stresses to determine the shear strength parameters including the soil cohesion (c) and the friction angle (φ).

2.3.2 *Unreinforced soil*

Large-scale direct shear tests and triaxial compression tests were carried out to investigate the shear feature of unreinforced soil and the influence of different testing methods.

The procedure of unreinforced soil large-scale direct shear test was similar to the geocell reinforced soil large-scale direct shear test except without paving the geocell material.

The triaxial compression tests were carried on through a LoadTrac II triaxial compression test apparatus made by Geocomp Company in the U. S. The size of each specimen was 61.8 mm $\times$ 132.7 mm

Table 1. Soil properties.

Liquid limit	Plastic Limit	Plasticity index	Optimum water content %	Maximum dry density g. cm^{-3}	CBR %	Unconfined compressive strength MPa
38.8	26.2	12.6	9.057	2.049	19.8	0.487

Table 2. Geocell properties.

Product type	Length mm	Width mm	Height mm	Thickness mm	Unit elongation %	Breaking elongation %	Stripe tensile strength $kN.m^{-1}$	Welding tensile strength $kN.m^{-1}$
TGLG-PP-100-500	250	25	10	0.07	7.6	9.8	311	130

Table 3. Properties of Portland cement.

			Coagulating time			Compressive strength		Rupture strength	
Cement type	Fineness* %	Specific gravity $g.cm^{-3}$	Initial set min.	Final set min.	Invariability (Boiling)	R_3 MPa	R_{28} MPa	R_3 MPa	R_{28} MPa
Portland cement	3.8	3.1	212	280	Eligibility	31.3	59.2	5.6	8.3

* Large than 0.08 mm sieve

(diameter × height). Consolidated – Undrained test method was taken with the triaxial apparatus. Three specimens were tested for each type of soil with confining pressures of 50 kPa, 200 kPa, and 400 kPa respectively.

2.3.3 *Geocell reinforced cement stabilizing soil*

The test method for geocell reinforced cement stabilizing soil in this paper was large-scale direct shear test. A dosage of 5% cement (by dry soil weight) was added into the well soaked soil and mixed fully with the soil before the soil mixture was molded to the specimen of large-scale direct shear test. Other procedures were the same as described in section of *geocell reinforced soil.*

2.3.4 *Unreinforced cement stabilizing soil*

For the purpose of investigating the influence of geocell on cement stabilizing soil, several unreinforced cement stabilized soil triaxial compression tests were carried on. The test procedures are similar to above. A total of 15 tests which include 9 large-scale direct shear tests and 6 triaxial compression tests were carried out in this study. Considering silty gravel and cement stabilizing silty gravel which reinforced with geocell or not. If the soil is stabilized with cement, the dosage of cement is 5%.

3 TEST RESULTS AND COMPARISONS

3.1 *Large-scale direct shear tests*

The shear stress-displacement behaviors observed from large-scale direct shear tests on different reinforcement soil are shown in Figures 3a, b,c. The results of unreinforced soil show that the shear stress increases with increasing shear displacement when the normal stress is 200 kPa. As increasing the normal stress to 400 kPa and above, the shear stress-displacement response is softening pattern which shows an increase in shear stress with increasing shear displacement in the beginning and a decrease in shear stress with increasing shear displacement in the end. In the case of geocell reinforced soil (Fig. 3b), all the curves of shear stress-displacement with different normal stresses appear in softening pattern. There is a characteristic that the shear stress increases to a peak with increasing the shear displacement from zero to a certain magnitude and thereafter, it decreases gradually

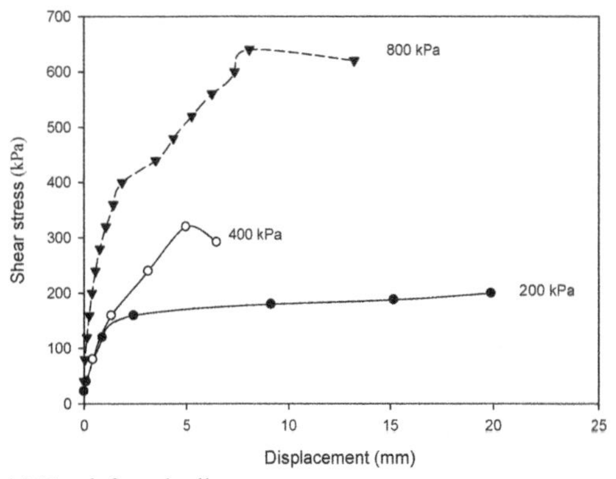

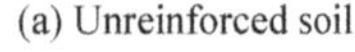
(a) Unreinforced soil

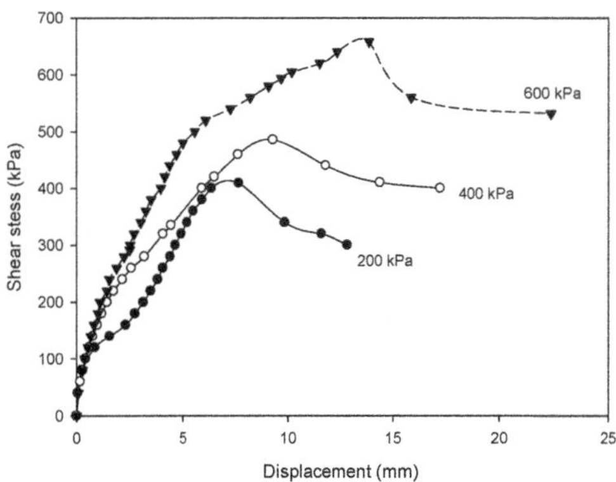

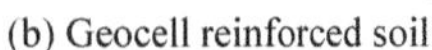
(b) Geocell reinforced soil

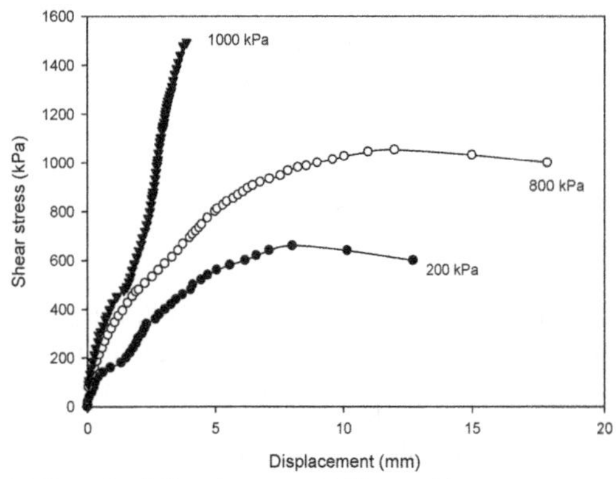

(c) Geocell reinforced cement stabilizing soil

Figure 3. Shear stress- displacement behavior of different reinforcement soil.

to the residual shear stress with the displacement on the increase. The reason for this correspondence is that there is a considerable degree of interlocking on dense soil and there is an additional friction on the interface between soil and geocell reinforcement. Therefore, before shear failure can take place this interlocking

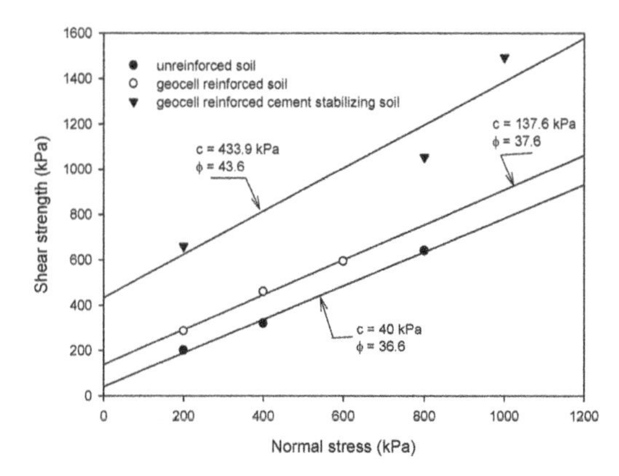

Figure 4. Shear capacity envelops.

and additional friction must be overcome in addition to the frictional resistance at the points contact. As the interlocking and additional friction is progressively overcome, the shear stress necessary for additional deformation decreases. However, the result of geocell reinforced cement stabilizing soil shows different feature when the normal stress comes up to 1000 kPa (Fig. 3c). The shear failure takes place with a relatively high shear stress and low deformation. It is clear from the data in this figure that there is a quasi-elastic characteristic on the behavior of shear stress and displacement. This is mainly due to the 5% cement stabilizing soil possessing a considerable degree of rigidity. Each plot in Figure 3 shows that there is a transformation in shear stiffness. It increases with increasing the normal stress for all tests.

The linear shear capacity envelops deduced from the peak or maximum recorded shear stress versus normal stress data for different reinforcement soils plot together as illustrated in Figure 4. It shows that there is a good correspondence for unreinforced soil and geocell reinforced soil. But in the case of geocell reinforced cement stabilizing soil, there is some scatter of the data points about the linear regression line.

This is due to the unavoidable small variations in the setup of the shear surface between the two parts of shear boxes exactly at the interface between geocell reinforcement soil and soil. The correlation coefficient of the linear regression on geocell reinforced cement stabilizing soil is 0.91, and on unreinforced soil and geocell reinforced soil both are 0.99. Based on the linear regression lines presented in Fig. 6 the parameters of shear strength which include cohesion "c" and friction angle "φ" can be derived. The results show that there is a considerable increase on cohesion from unreinforced soil to geocell reinforced soil. Comparing the cohesion of unreinforced soil with that of geocell reinforced soil, cohesion of geocell reinforced soil increases 244%, and the cohesion of geocell

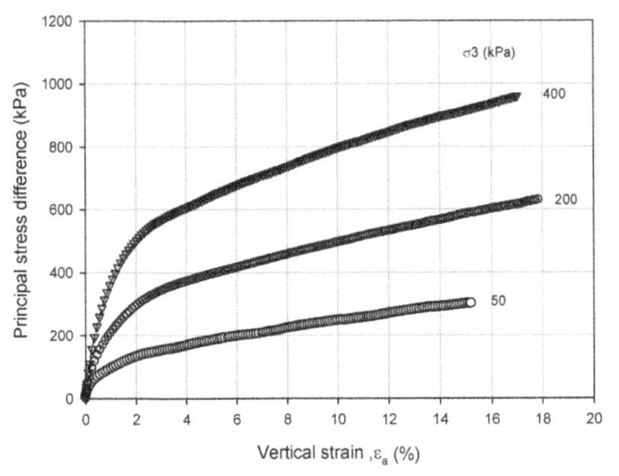

(a) Unreinforced soil

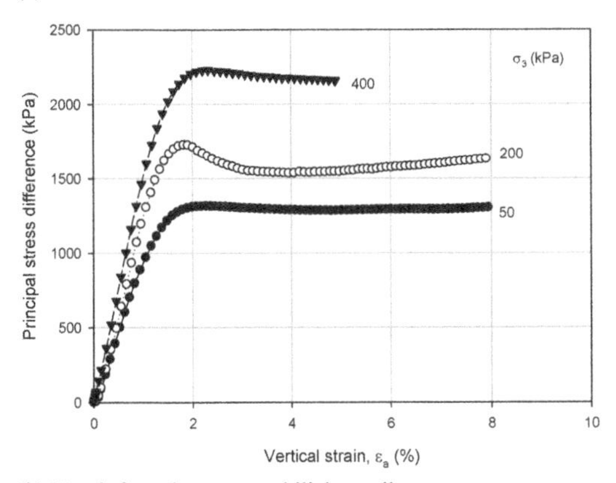

(b) Unreinforced cement stabilizing soil

Figure 5. Triaxial behavior of principle stress difference versus vertical strain.

reinforced cement stabilizing soil is 11 times as big as that of unreinforced soil. However, the friction angle does not change dramatically. Based on the data presented on Figure 4 it can be argued that the geocell reinforced soil develops a large amount of cohesion on the interface shear strength but the friction angle.

3.2 *Comparison of results with triaxial compression tests*

Figure 5 shows the behavior of principle stress difference versus vertical strain of consolidated-undrained triaxial compression tests for unreinforced soil and unreinforced cement stabilizing soil. Comparing the results of unreinforced soil with that of unreinforced cement stabilizing soil, it seems that these two types of soil have different features of stress- strain. With the same vertical strain of 2%, the principal stress difference of unreinforced cement stabilizing soil reaches the peak point of the curve, but the principal stress difference of unreinforced soil increases

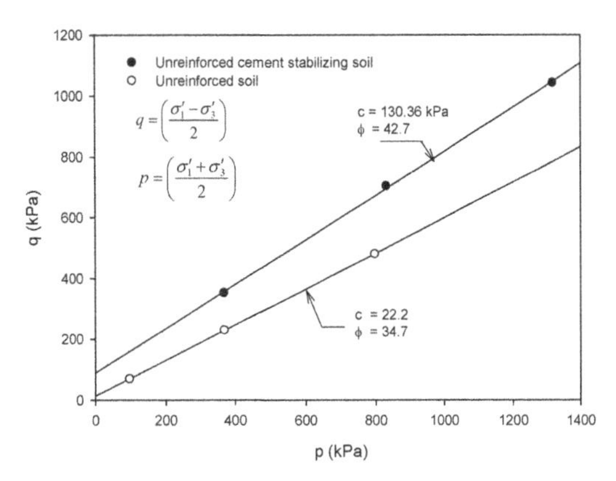

Figure 6. Failure strength envelopes.

with increasing vertical strain and after that, the stiffness turns to less. Nevertheless, the linear shear failure strength envelopes deduced from the maximum recorded effective principal stress ratio was plotted as Figure 6.

The p and q in this diagram are the mean effective normal stress and effective shear stress. It appears a very good correlation for both unreinforced soil and unreinforced cement stabilizing soil. Based on the regression linear lines and test data, the consolidated undrained shear strength parameters are interpreted and summarized on Table 4.

It is observed that the effective cohesion of unreinforced cement stabilizing soil is nearly 6 times as big as that of unreinforced soil while the frictional angle increases 19.2%.

Comparing the results of triaxial compressive shear test to that of large-scale direct shear test on unreinforced soil, it shows that there is a significant difference in the value of the cohesion. The triaxial compressive shear tests gave a cohesion of 22.2 kPa, and the large-scale direct shear tests exhibited a relatively higher result of 40 kPa. However, there is only a marginal decrease in the friction angle which is 34.7 and 36.6 respectively. This result can be attributed to the different specimen size that was used in the tests. With a dimension of 500 mm × 500 mm × 400 mm (length × width × height), the large-scale shear test specimen has much more area of soil confined by shear box. Therefore, the degree of interlocking in the soil can be developed to a higher lever which results in the increase of cohesion. Based on the test results in this study, it may be concluded that the cohesion of soil is mainly responsible for the increase of shear strength due to the influence of test methods. Comparing to the results of triaxial compression test, the large-scale direct shear test provides a higher cohesion and an approximate friction angle.

Table 4. Shear strength parameters with different test methods.

	large-scale direct shear test		triaxial compression test	
Type of reinforcement soil	c kPa	ϕ °	c kPa	ϕ °
Unreinforced soil	40	36.6	22.2	34.7
Geocell reinforced soil	137.6	37.6	–	–
Geocell reinforced cement stabilizing soil	433.9	43.6	–	–
Unreinforced cement stabilizing soil	–	–	130.36	42.7

4 CONCLUSIONS

This paper reports the shear strength test results of geocell reinforced soil carried out with large-scale direct shear tests. Unreinforced silty gravel soil, geocell reinforced silty gravel soil, and geocell reinforced cement stabilizing silty gravel soil were investigated. The comparisons of large-scale shear tests with triaxial compression tests for the same type of soil are conducted to evaluate the influences of testing methods on the shear strength indices as well. The following conclusions can be drawn from the results reported here:

(1) The use of large-scale direct shear test device provides a valuable method to gain the shear strength parameters for geocell reinforced soil.
(2) The behaviors of stress-strain on both unreinforced soil and reinforced soil with large-scale direct shear tests are nonlinear. When the normal stress comes up to 1000 kPa, there is a quasi- elastic characteristic on the behavior of shear stress and displacement for the geocell reinforced cement stabilizing soil. The shear stiffness increases with increasing the normal stress for all tests.
(3) With the reinforcement of geocell, the cohesive strength of silty gravel soil considerably increases and it increases much more highly for the cement stabilizing soil.
(4) The friction angle of the geocell reinforced soil does not change obviously. The failure envelops of both the unreinforced soil and the reinforced soil are nearly parallel to each other.

REFERENCES

Bathurst, R.J. & Jarrett, P. 1989. Large-scale model tests of geocomposite mattresses over peat subgrades. *ASTM Transportation Research Record 1188, Transportation Research Board,* Washington, D. C.: 28–36

Bathurst, R.J. & Kappurapu, R. 1993. Large-scale triaxial compression testing of geocell-reinforced granular soils. *Geotechnical Testing Journal,* 16: 296–303

Bathurst, R.J. & Crowe, R.E. 1994. Recent case histories of flexible geocell retaining walls in North America. *The International Symposium on Recent Case Histories of Permanent Geosynthetic-reinforced Soil Retaining Walls,* Tokyo, Japan

Cancelli, A. et al. 1993. Index and performance tests for geocells in different applications. *ASTM Special Technical Publication,* 1190: 64–75

Cheng, S.C. & Jonathan. 1993. Symposium on geosynthetic soil reinforcement testing procedures. *ASTM Special Technical Publication,* 1190: 243

Jing, H. & Yu, M. 2006. Application of geocell to strengthen expressway subgrade in desert. *Journal of Chang'an University (Natural Science Edition)*, 26: 15–18

Latha, G. et al. 2006. Experimental and theoretical investigations on geocell-supported embankments. *International Journal of Geomechanics*, 6: 30–35

Mengelt, M. et al. 2006. Resilient modulus and plastic deformation of soil confined in a geocel. *Geosynthetics International,* 13: 195–205

Mhaiskar, S.Y. & Mandal, J.N. 1996. Investigations on soft clay subgrade strengthening using geocells. *Construction and Building Materials,* 10: 281–286

Rajagopal, K. et al. 1999. Behavior of sand confined with single and multiple geocells. *Geotextiles and Geomembranes.* 17: 171–184

Sekine, E. et al. 1994. Study on properties of road bed reinforced with geocell. *Quarterly Report of RTRI (Railway Technical Research Institute) (Japan)*, 35: 23–31

Yan C. et al. 2006. Testing and application of geocell in loess slopes of highway. *Chinese Journal of Rock Mechanics and Engineering*, 25, n SUPPL. 1

Advances in Transportation Geotechnics – Ellis, Yu, McDowell, Dawson & Thom (eds)
© 2008 Taylor & Francis Group, London, ISBN 978-0-415-47590-7

Influence of hydraulicity on resilient modulus and Poisson's ratio of hydraulic, graded iron and steel slag base-course material

N. Yoshida
Research Center for Urban Safety and Security, Kobe University, Kobe, Japan

T. Sugita
Rock Fields, Co., Ltd., Kobe, Japan

T. Miyahara
Graduate School of Science and Technology, Kobe University, Kobe, Japan

E. Hirotsu
Nomura Jimusho, Inc., Tokyo, Japan

ABSTRACT: In this study, repeated loading triaxial compression tests were carried out on the samples compacted with an optimum water content and a maximum dry density. The samples were either cured for a year or not cured at all after aging treatment. It is shown that resilient modulus depends on both mean effective principal stress and deviator stress, just as previous studies, and that the magnitude and the stress dependence have increased after one-year curing. Resilient Poisson's ratio for the samples without curing is also stress-dependent but it is in an inverse manner with resilient modulus; moreover, after one-year curing, its stress-dependency appears to fade away. Both resilient modulus and Poisson's ratio can be expressed with a power function of mean effective principal stress and deviator stress.

1 INTRODUCTION

In Japan, the structural design method of asphalt pavement is moving toward a semi-empirical, mechanistic one based on a multi-layered elastic structural analysis (Japan Road Association, 2001). For this, resilient deformation characteristics, modulus and Poisson's ratio of pavement material are becoming more important input data than ever.

Iron and steel slag aggregate for use in road construction was standardized and added to Japanese Industrial Standards in 1979 (Japanese Standards Association, 1992). Since then, iron and steel slag aggregate has been used in base-course and hot asphalt mix, at least where it is available and economically transportable. For instance, according to Japan Slag Association (2006), in 2005, 24.8 million tons of blast-furnace slag were produced in Japan, and its 4.9 million tons were slowly air-cooled slag and the remaining was water-granulated slag. The yearly production of blast-furnace slag ranges from 23.1 million tons to 24.8 million tons for the last ten years. On the other hand, steel slag was produced by 13.4 million tons, of which 9.9 million tons were converter slag and the remaining was electric arc furnace slag. The yearly production ranges from 12.1 million tons to 14.1 million tons for the last ten years. About 66% of blast-furnace slag was utilized in cement production, 16% in road construction material, mainly as base-course and asphalt-mix aggregates, 12% in concrete aggregates, and so on. For converter slag, about 52% was used in harbor and public work construction, 22% as base-course and asphalt-mix aggregates and 17% was reused. About 38% of electric arc furnace slag was used as base-course and asphalt-mix aggregates for road construction, 34% as in harbor and public work construction, *etc.* Considering the fact that obtaining good quality natural aggregates are becoming difficult in Japan, more attention will be given to iron and steel slag as an alternative aggregate.

This paper first provides a basic information of iron and steel slag used for road construction and related to this study. Then, the resilient modulus and Poisson's ratio of hydraulic, graded iron and steel slag base-course material are discussed, based on repeated loading triaxial compression tests on the samples with and without curing after aging treatment.

2 HYDRAULIC, GRADED IRON AND STEEL SLAG BASE-COURSE MATERIAL

In Japan, the national industrial standard of iron and steel slag for use in road construction was established in 1979 mainly for blast-furnace slag, and steel slag was included in the standard in 1992 (Japan Standards Association, 1992).

Iron and steel slag possesses hydraulic nature to a greater or lesser extent. In case of granulated blast-furnace slag, for instance, hydraulicity results from hydrates formed with calcium oxide, silica dioxide and alumina oxide seeped out from glass portion in the slag upon contact with water and in an alkaline environment. Air-cooled blast-furnace slag and steel slag also possess grass portions but only a little; thus hydraulic nature is much smaller than the granulated blast-furnace slag. Average chemical components of each slag in Japan are given on Table 1 (Japan Slag Association, 2004).

Presently, five types of iron and steel slag are standardized: hydraulic, graded iron and steel slag (designated as HMS); graded iron and steel slag (MS); crusher-run iron and steel slag (CS); single-graded steel slag (SS); and crusher-run steel slag (CSS). Presently, HMS and MS are used in base-course and CS in subbase-course while SS is used in hot asphalt mixtures and CSS in asphalt stabilization (Japanese Standards Association, 1992).

Table 2 summarizes the specifications required for these slag. In the table, "aging" is a treatment required for sulfur in blast-furnace slag to be oxidized to overcome its yellow water problem and for free lime in steel slag to be hydrated to stabilize its expansion problem. It is specified in terms of time as shown in the table. A basic aging treatment is simply to store slag in open yard but in present practice, warm water or vapour are applied for accelerating the both reactions (Japan Slag Association, 2004). "Coloration" and "immersion expansion ratio" are for confirming that

Table 1. Average chemical compositions of iron and steel slag used in Japan (modified after Japanese Slag Association, 2004).

		SiO_2	CaO	Al_2O_3	T Fe*	MgO	S	MnO	TiO_2
Blast-furnace slag		33.8	42.0	14.4	0.3	6.7	0.84	0.3	1.0
Steel slag	Converter slag	13.8	44.3	1.5	17.5	6.4	0.07	5.3	1.5
	Electric arc furnace oxidizing slag	19.0	38.0	7.0	15.2	6.0	0.38	6.0	0.7
	Electric arc furnace reducing slag	27.0	51.0	9.0	1.5	7.0	0.50	1.0	0.7

* ferric oxide, *etc.* (in percent by mass)

Table 2. Iron and steel slag for road construction (modified after Japanese Standards Association, 1992).

Type	Hydraulic graded iron and steel slag	Graded iron and steel slag	Crusher-run iron and steel slag	Single-graded steel slag	Crusher-run steel slag	Notes
Designations	HMS-25	MS-25	CS-40, CS-30, CS-20	SS-20, SS-5	CSS-30, CSS-20	
Usage	Base-course	Base-course	Subbase-course	Hot asphalt mix	Hot asphalt-stabilization	
Coloration	No coloration	No coloration	No coloration	–	–	only for blast-furnace slag
Immersion expansion ratio (%)	1.5 or smaller	1.5 or smaller	1.5 or smaller	2.0 or smaller	2.0 or smaller	only for steel slag
Unit weight (kg/liter)	1.5 or larger	1.5 or larger	–	–	–	
Uniaxial compressive strength (MPa)	1.2 or larger (13-day cured)	–	–	–	–	
Modified CBR (%)	80 or larger	80 or larger	30 or larger	–	–	
Specific gravity in saturated surface-dry condition	–	–	–	2.45 or greater	–	
Water absorption percentage (%)	–	–	–	3.0 or smaller	–	
Abrasion (%)	–	–	–	30 or smaller	50 or smaller	
Aging	6 months or more	6 months or more	6 months or more	3 months or more	3 months or more	for steel slag
Plasticity index (%)	–	–	–	–	–	

sulfur and free lime in slag are sufficiently stabilized, respectively.

Hydraulic, graded iron and steel slag is composed of solely blast-furnace slag, a mixture of blast-furnace slag and steel slag, or a mixture of other combinations of slag with or without additives. Hydraulic, graded iron and steel slag tested in this study is composed of air-cooled blast-furnace slag alone without additives and meets the specifications listed on Table 2. Its specific gravity is 2.369, and its optimum water content and maximum dry density are 11.5% and 2.13 g/cm^3, respectively. Figure 1 shows the grading curve of the sample together with the range designated in the industrial standard. Not to mention but this material satisfies all the quality and environmental requirements designated in the industrial standard.

3 SAMPLE PREPARATION AND REPEATED LOADING TRIAXIAL COMPRESSION TEST

Samples were prepared in the following manner. The water content of material was first adjusted to its optimum water content. Then, a prescribed amount of the material was placed into a mould in 5 to 6 stages, in each stage compaction being performed, in such a way that the resulting dry density becomes its maximum dry density. Note that grains not passing a 19.1 mm sieve were excluded for the sample preparation considering the sample size with a diameter of 100 mm and a height of 200 mm. Each sample with a mould was double-wrapped up with polyethylene bags and a weight of 49 N was placed on its top until it was served to testing. In this study, two types of samples, in terms of curing time, were tested: one was cured in a dark room with a moisture of about 60% and a temperature of about 20°C for a year, and another was not cured at all. For reference, Figure 2 shows uniaxial compressive strength obtained from these samples, together with the past test results on a different hydraulic, graded iron and steel slag material compacted in the same manner as in this study. It seems that hydraulicity has developed sufficiently in one-year cured samples.

The repeated loading triaxial compression test system consists of axial loading, lateral loading, triaxial cell and control units (*e.g.*, Yoshida *et al.*, 2003). Compressed air is used for axial loading and is converted to water pressure for lateral loading. Repeated axial load is applied onto the specimen with a double-action Bellofram air cylinder by controlling the compressed air by an electro-pneumatic transducer through a servo-amplifier. Any load waveform can be set using a functional synthesizer. A load cell is installed between the loading piston and the cap inside the triaxial cell, and the axial displacement of sample is measured using non-contact displacement sensors and targets attached on the side of the sample, as shown in Figure 3. Total 8 sensors, 4 on the upper portion and 4 on the lower portion, are used. The range of measurement and resolution are 10 mm and 1 μ, respectively for the upper 4 sensors, and those for the lower 4 sensors are 6 mm and 0.5μ, respectively. The radial displacement of sample is measured at the mid-height of sample using a radial displacement ring manufactured with strain gauges, steel spring, *etc.* at a local machine shop (Sugita, 2006).

For repeated loading, the loading and pausing durations are 0.4 and 1.2 seconds, respectively, due to constraints from the use of compressed air, and the applied waveform is a harversine shape. The loading sequence basically follows AASHTO Designation T292-91 (AASHTO, 1998); but as shown in Figure 4, the magnitude of the applied deviator and confining stresses slightly differ: the deviator stress ranges from

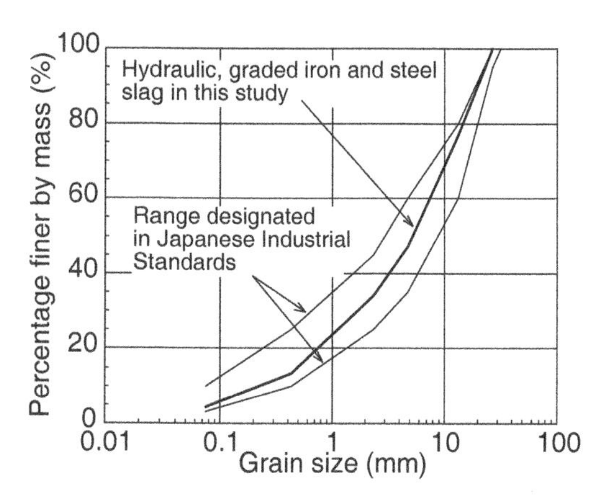

Figure 1. Grain size distribution of hydraulic, graded iron and steel slag tested in this study.

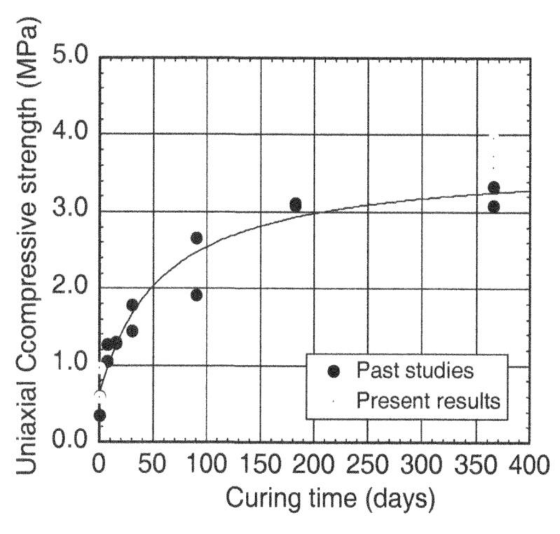

Figure 2. Relationships of uniaxial compressive strength with curing time.

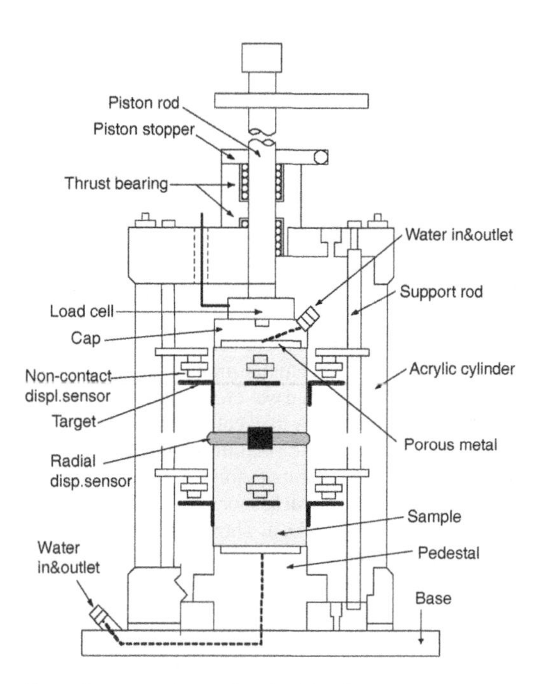

Figure 3. Arrangement for triaxial cell.

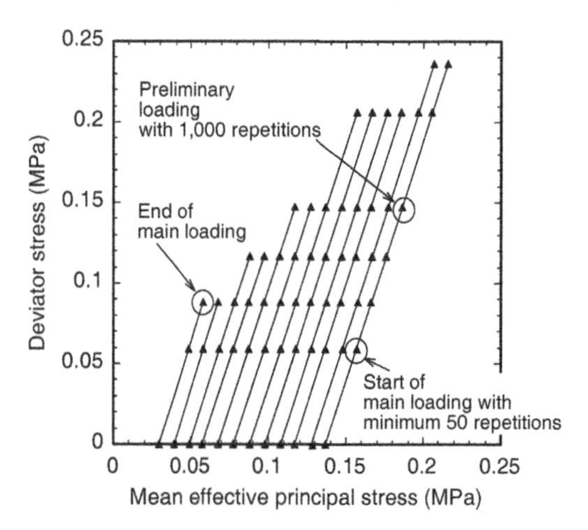

Figure 4. Applied stress conditions.

0.059 to 0.236 MPa and the mean principal effective stress from 0.049 to 0.216 MPa (Yoshida *et al.*, 2003).

Three samples were made and tested for each curing condition.

4 RESULTS AND DISCUSSIONS

The resilient modulus, M_r is computed as a ratio of repeated deviator stress, $q(=\sigma_1'-\sigma_3')$ to resilient axial strain, ε_a and the resilient Poisson's ratio, ν_r as a ratio of

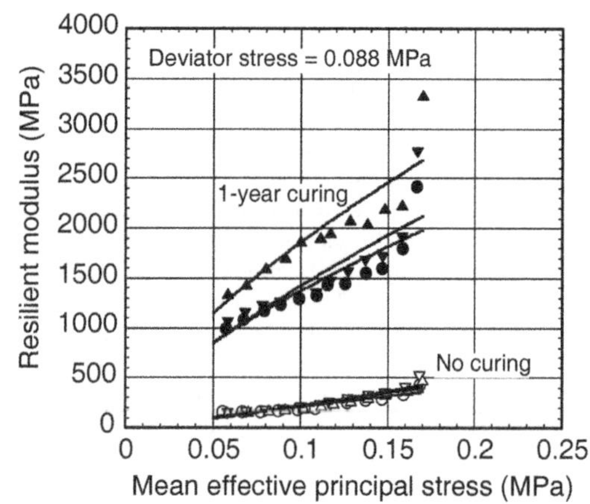

Figure 5. Relationships of resilient modulus with mean effective principal stress.

resilient horizontal strain, ε_r to resilient axial strain, ε_a. Here, in computing the resilient horizontal strain, the following converted horizontal displacement, $d_{cylinder}$ is used, instead of the measured horizontal displacement, d_{barrel}. Note that $d_{cylinder}$ and d_{barrel} are both in millimeters.

$$d_{cylinder} = 50 - \sqrt{\frac{8}{15}d_{barrel}^2 - \frac{200}{3}d_{barrel} + 2500} \tag{1}$$

Figure 5 shows relationships between resilient modulus and mean effective principal stress for a deviator stress nearly equal to 0.088 MPa. The results for one-year cured samples are moderately scattered but it can be said that the resilient modulus increases as the mean principal effective stress increases. Resilient modulus of the samples without curing is much smaller than that of one-year cured ones. It also depends on mean effective principal stress but its degree is smaller compared to the one-year cured samples.

Relationship of resilient modulus with deviator stress is shown in Figure 6 for a mean effective principal stress nearly equal to 0.157 MPa. The resilient modulus of one-year cured samples is seen to decrease with deviator stress but that of no cured samples seems to decreases slightly with deviator stress.

These observations are also made for other stress states adopted in this study. The lines drawn in the both figures are regression curves obtained by performing a multiple regression analysis for each sample using the following equation:

$$M_r = K\frac{p^M}{q^N} \tag{2}$$

where Mr is resilient modulus (MPa), p is mean effective principal stress (MPa), q is repeated deviator stress

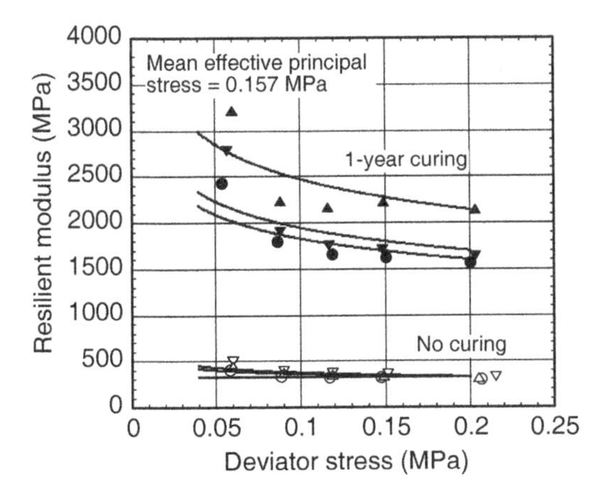

Figure 6. Relationships of resilient modulus with deviator stress.

Table 3. Regression constants, K, N and M for resilient modulus.

K	M	N	R^2
One-year cured samples			
4206	0.688	0.192	0.870
4956	0.749	0.197	0.819
5499	0.689	0.207	0.859
No cured samples			
2095	0.987	−0.005	0.905
2393	1.240	0.197	0.920
1894	1.094	0.162	0.912

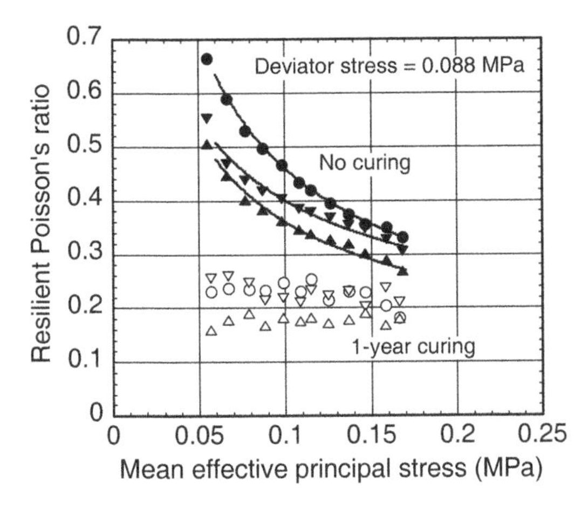

Figure 7. Relationships of resilient Poisson's ratio with mean effective principal stress.

(MPa), and K, M, N are regression constants, of which values are given on Table 3. From the coefficient of determination, R^2, in the table, the approximation can be said fairly good.

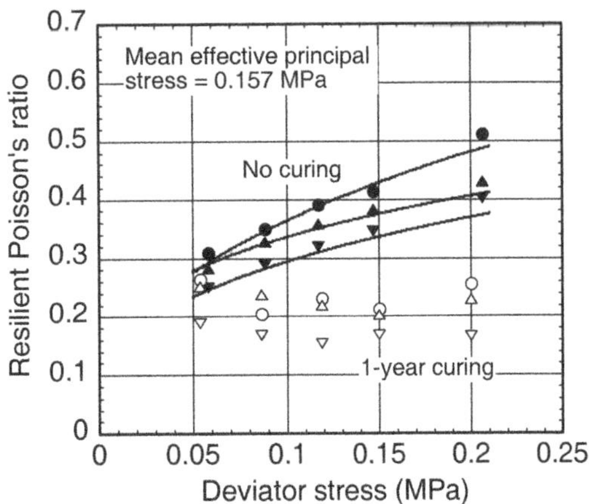

Figure 8. Relationships of resilient Poisson's ratio with deviator stress.

Table 4. Regression constants, K, N and M for resilient Poisson's ratio.

K	M	N	R^2
0.287	−0.629	−0.402	0.977
0.229	−0.545	−0.329	0.895
0.266	−0.465	−0.272	0.905

For resilient Poisson's ratio, its relationship with mean effective principal stress is shown in Figure 7 for a deviator stress of 0.088 MPa, and Figure 8 is the relationship with deviator stress for a mean effective principal stress of 0.157 MPa. It is seen from Figure 7 that no-cured samples exhibit larger Poisson's ratio than one-year cured ones and its dependency on mean effective principal stress is greater in no-cured ones. Regarding the dependency of resilient Poisson's ratio on mean effective principal stress, the resilient Poisson's ratio of one-year cured samples seems not to vary with mean effective principal stress and may be considered constant: the average value over three samples is 0.21. A similar observation can be made on relationship of resilient Poisson's ratio with deviator stress; that is, one-year cured samples seem not to exhibit its dependency on deviator stress and may be considered constant.

The relationships of resilient Poisson's ratio with mean effective principal stress and deviator stress are opposite to those of resilient modulus but the way resilient Poisson's ratio changes with the two stress variables appears similar to the one observed in resilient modulus. Thus, the same form of equation as for resilient modulus, Equ. (2), may be applied to the resilient Poisson's ratio data for no-cured samples. The results of a multiple regression analysis conducted on no-cured samples are drawn in Figures 7 and 8 and the regression constants are given on Table 4. High values of the coefficient of determination, R^2, seem to suggest that Equ. (2) can be used to express resilient Poisson's ratio.

5 CONCLUSIONS

This paper first described a background information related to hydraulic, graded iron and steel slag base-course material of interest in this study, then resilient modulus and Poisson's ratio were presented which were obtained from repeated loading triaxial compression tests performed on one-year cured and no-cured samples.

From the test results, the followings are pointed out.

- The resilient modulus depends on both mean effective principal stress and deviator stress regardless of curing, just as previous studies. The magnitude of resilient modulus and the degree of stress dependence are both larger in the one-year cured samples than no-cured ones.
- The resilient Poisson's ratio for the no-cured samples is also stress-dependent but it decreases with mean effective principal stress and increases with deviator stress. After one-year curing, resilient Poisson's ratio becomes smaller and its stress dependency appears to fade away.
- Both resilient modulus and Poisson's ratio can be expressed with a power function of mean effective stress and deviator stress.

REFERENCES

AASHTO, 1998. Standard Method of Test for Resilient Modulus of Subgrade Soils and Untreated Base/Subbase Materials, AASTHO Designation T 292-91. *Standard Specifications for Transportation Materials and Methods of Sampling and Testing*, 19th Edition, pp. 1057–1071.

Japan Slag Association, 2004. Application of Iron and Steel Slag to Road Construction, *Technical Brochure*, fs-89. (in Japanese)

Japan Slag Association, 2006. Iron and Steel Slag, *Annual Report*, Japan Slag Association. (in Japanese)

Japanese Standards Association, 1992. *Iron and steel slag for road construction*, JIS A 5015.

Japan Road Association, 2001. *Pavement Design and Construction Guide*. (in Japanese)

Yoshida, N., Sugisako, Y., Nakamura, H. and Hirotsu, E., 2003. Resilient modulus of hydraulic mechanically stabilized slag base-course material, *Deformation Characteristics of Geomaterials*, Di Benedetto et al. (eds), Swets & Zeitlinger, pp. 293–298.

Sugita, T. 2006. Resilient Characteristics of Hydraulic Base-Course Material. *M.Sc. Thesis*, Graduate School of Science and Technology, Kobe University. (unpublished)

 ISBN 978-0-415-47590-7

Author Index